Burger's Medicinal
Chemistry and Drug Discovery

Dr. Alfred Burger

BURGER'S MEDICINAL CHEMISTRY AND DRUG DISCOVERY

Fifth Edition
Volume 2: Therapeutic Agents

Edited by

Manfred E. Wolff

Technipharm Consultants
Laguna Beach, California

A WILEY-INTERSCIENCE PUBLICATION

JOHN WILEY & SONS, Inc., New York · Chichester · Brisbane · Toronto · Singapore

This text is printed on acid-free paper.

Library of Congress Cataloging in Publication Data:
Burger, Alfred, 1905–
    [Medicinal chemistry]
    Burger's medicinal chemistry and drug discovery. -- 5th ed. /
edited by Manfred E. Wolff.
        p.   cm.
    "A Wiley-Interscience publication."
    Contents: v. 2. Therapeutic agents
    Includes bibliographical references and index.
    ISBN 0-471-57557-7   1000784835
    1. Pharmaceutical chemistry.   I. Wolff, Manfred E.   II. Title.
III. Title: Medicinal chemistry and drug discovery.
RS403.B8   1994
615'. 19--dc20   94-12687

Printed in the United States of America

10 9 8 7 6 5 4 3 2 1

Preface

The final four volumes of this 5th edition of *Burger's* will contain discussions of the important individual drug classes in modern medicine. One of the major decisions considered in undertaking this was whether to confine the chapters to a particular disease category, such as antihypertensive agents, or to a particular therapeutic modality, such as beta blockers. A consistent choice between these two alternatives was sought and the opinions of numerous thought leaders were obtained. But in the end we "decided to be undecided," as Churchill once complained about a British government. It seemed best to structure each chapter in a manner that appeared most appropriate to the topic under consideration, based on the views of the authors. This approach has led to situations where occasionally the same topic is considered from different perspectives in multiple chapters. But we feel it is important to develop chapters that are able to stand alone in delivering their information.

A second question that had to be considered was whether to collect each subject area, such as cardiovascular agents, into a single volume. Although this might have been advantageous with regard to convenience of use, it would have required delaying the production of each volume until the very last chapter in a given area was completed. In the interests of publishing the chapters in a timely manner, a decision was made to divide each subject area into two parts, which could be published in an early and a later volume.

The revolution that has taken place in drug discovery and in medicinal chemistry is in clear evidence in this volume. In every chapter, one can read of the remarkably strengthened biological understanding of each disease category under discussion. Much of this new information has been obtained by the application of the techniques of molecular biology to unsolved questions in biochemistry and pharmacology. These studies have been extraordinarily fruitful for drug discovery since an understanding of biological events at the molecular level, in the end, is essential for the discovery and design of new drug molecules to influence those events.

In addition, the growing army of scientists working in all of the areas of drug discovery and its underlying basic sciences has produced a body of research notable not only for its sophistication, but also for its sheer volume. Thus, both the quantity and the quality of research related to drug discovery has served to enhance our understanding and our capability in this field greatly. Even so, much remains to be done. In the chapter, "Cardiac Drugs," one can read of the remarkable progress made in our knowledge of the action of the cardiac glycosides since the time of William Withering in 1785. At the same time, one can see the great gaps in our perception of this area that still remain. Another valuable lesson to be drawn from this chapter is the difficulty, and importance, of designing a definitive clinical trial in a multifactorial disease.

A notable point that should be made is the economic advantage provided to society by the new drugs whose discovery and development is chronicled in this series. In this volume alone, three enormously significant areas, cardiovascular disease, gastrointestinal disease, and tubercular disease are given consideration. In the past two centuries, tuberculosis is said to have killed one billion people. Today, disease is the most important killer in western-style countries. The costs of healthcare and hospitalization for such diseases is reduced tremendously by the availability of safe and effective medicines. Both for tuberculosis and duodenal ulcer, for example, new drugs virtually eliminated the need for surgical intervention, with its attendant high costs. In addition, the benefits of these pharmaceuticals can easily be brought to any individual on the globe. A crucial medicine discovered through the application of the high technology discussed in these chapters can readily be made available in a remote African village, unlike an MRI installation or other advanced instrument.

Generally we have still not achieved ideal pharmaceuticals, and in numerous areas we are faced with a moving target. Particularly in the field of chemotherapy, drug resistance by pathogenic organisms has seriously eroded the usefulness of existing medicines. Thus, the discovery and development of new agents are required. In doing this, it is important that the lessons of the past not be forgotten, even through they may have been learned in a time of empiricism. The results of the patient molecular manipulations made by medicinal chemists in former years outlined in the chapters on cholinergic and anticholinergic drugs, for example, can still be studied with profit by workers using the modern methods of drug discovery and design reviewed in Volume I.

Once again, it is my pleasant duty to acknowledge the efforts of those individuals who have made this volume of the series possible. Most of all I thank my friends, the dedicated authors, a number of whom had previously contributed to the 4th edition, who generously took time from their already overcrowded schedules to pass their expert knowledge on to others. I am grateful to Michalina Bickford, Managing Editor with John Wiley & Sons, Inc. for all of her work in connection with this series. As always I thank my wife, Gloria, for her steadfast support and encouragement in everything I do.

MANFRED E. WOLFF

Laguna Beach, California

Contents

Part I   Gastrointestinal Drugs, Pt 1, 1

25. CHOLINERGICS, 3

Joseph G. Cannon
The University of Iowa
Iowa City, Iowa, USA

26. ANTICHOLINERGIC DRUGS, 59

B. V. Rama Sastry
School of Medicine
Vanderbilt University
Nashville, Tennessee, USA

27. GASTRIC PROTON PUMP INHIBITORS, 119

Andreas W. Herling and Klaus Weidmann
Hoechst AG
Frankfurt/Main, Germany

Part IIA   Cardiovascular Drugs, Pt 1, 152

28. CARDIAC DRUGS, 153

Richard E. Thomas
University of Sydney
Sydney, Australia

Part IIB   Cardiovascular Drugs, Pt 1, 263

29. ANTIHYPERTENSIVE AGENTS, 265

Pieter B. M. W. M. Timmermans and Ronald D. Smith
DuPont-Merck Research Laboratories
Wilmington, Delaware, USA

30. PHENOXYACETIC ACID URICOSURIC DIURETICS, 323

Hiroshi Koga, Haruhiko Sato, and Takashi Dan
Fuji-Gotemba Laboratories
Chugai Pharmaceutical Company, Ltd.
Gotemba-shi, Shizuoka, Japan

31. DIURETIC AND URCOSURIC AGENTS, 363

Cynthia A. Fink and Lincoln Werner
Ciba Geigy Corp.
Summit, New Jersey, USA

viii

Contents

Part III   Chemotherapeutic Agents, Pt 1, 461

32. AMINOGLYCOSIDE, MACROLIDE, GLYCOPEPTIDE, AND MISCELLANEOUS ANTIBACTERIAL ANTIBIOTICS, 463

Herbert A. Kirst
Lilly Research Laboratories
Greenfield, Indiana, USA

33. SULFONAMIDES AND SULFONES, 527

Nitya Anand
Central Drug Research Institute
Chattar Manzil Palace
Lucknow, India

34. ANTIMYCOBACTERIAL AGENTS, 575

Giuliana Gialdroni Grassi
University of Pavia
Pavia, Italy

Piero Sensi
University of Milan
Milan, Italy

35. ANTIFUNGAL AGENTS, 637

Eugene D. Weinberg
Indiana University
Bloomington, Indiana, USA

INDEX, 653

Burger's Medicinal
Chemistry and Drug Discovery

PART I
GASTROINTESTINAL DRUGS

CHAPTER TWENTY-FIVE

Cholinergics

JOSEPH G. CANNON

The University of Iowa
Iowa City, Iowa

CONTENTS

1 Introduction,   4
2 Cholinergic (Acetylcholine) Receptors,   5
3 Acetylcholine and Analogs,   6
   3.1 Variations of the quaternary ammonium group,   7
   3.2 Variations of the acyl group,   9
   3.3 Variations of the ethylene bridge,   10
   3.4 Substitution of the ester group by other groups,   11
   3.5 Prodrug to acetylcholine,   13
4 Cholinergics Not Closely Related Structurally to Acetylcholine,   13
   4.1 Nicotine and analogs and congeners,   13
   4.2 Muscarine, muscarone, and related compounds,   15
   4.3 Pilocarpine and analogs and congeners,   20
   4.4 Arecoline and analogs and congeners,   20
   4.5 Oxotremorine and analogs and congeners,   28
   4.6 Miscellaneous, structurally unique muscarinic agonists,   36
5 Conformation-Activity Relationships of Cholinergic Agonists,   37
6 Anticholinesterases,   42
   6.1 Reversible inhibitors, 44
   6.2 Carbamate-derived inhibitors, 45
   6.3 Organophosphorus-derived inhibitors, 47
   6.4 Reversible, noncovalent inhibitors related to 1,2,3,4-tetrahydro-9-aminoacridine, 50
   6.5 Miscellaneous inhibitors, 51
7 Acetylcholine-release Modulators,   52

Burger's Medicinal Chemistry and Drug Discovery,
Fifth Edition, Volume 2: Therapeutic Agents,
Edited by Manfred E. Wolff.
ISBN 0-471-57557-7   © 1996 John Wiley & Sons, Inc.

1 INTRODUCTION

The transmission of impulses throughout the cholinergic nervous system is mediated by acetylcholine (**1**), and compounds that produce their pharmacologic effects by mimicking or substituting for acetylcholine are called cholinergics or parasympathomimetics.

$$CH_3-CO-O-CH_2-CH_2-\overset{+}{N}(CH_3)_3$$

(**1**)

Compounds that inhibit or inactivate the body's normal hydrolysis of acetylcholine by acetylcholinesterase in nervous tissue and/or by cholinesterase (pseudocholinesterase, butyrylcholinesterase) in the blood are called anticholinesterases. The gross observable pharmacological effects of both types of compounds are quite similar. More recently, compounds have been found that enhance the release of acetylcholine from cholinergic nerve terminals, thus (like the anticholinesterases) producing cholinergic effects by an indirect mechanism.

Choline is taken into the nerve terminal from the synaptic cleft by a sodium-dependent active transport process, which is the rate-limiting step in biosynthesis of acetylcholine in the nerve terminal (1). In the nerve terminal, choline reacts with acetyl coenzyme A in a process catalyzed by choline acetyltransferase. The acetylcholine thus synthesized is sequestered in the synaptic storage vesicles in the nerve terminal for future use as a neurotransmitter. The active transport of acetylcholine into the storage vesicles has been reviewed (2).

Therapeutic indications for cholinergics, anticholinesterases, and/or acetycholine-releasing agents in contemporary practice include the following:

1. Relief of postoperative atony of the gut and the urinary bladder. In such conditions, cholinergic stimulation may relieve the stasis by stimulating peristaltic movements of the intestine and ureters and by constriction of the bladder.

2. Reduction of intraocular pressure in some types of glaucoma by increasing the drainage of intraocular fluid through the canal of Schlemm.

3. Relief of muscular weakness in myasthenia gravis. This condition reflects a failure of an appropriate amount of acetylcholine to reach cholinergic receptors on the postmyoneural junctional membrane following rapidly repetitive nerve impulses. The reduced level of acetylcholine may result from excessive enzyme-catalyzed hydrolysis of it or from diminished production or release; the etiology of the disease remains obscure.

4. Relief of symptoms of Alzheimer's disease and some other types of senile dementia.

A deficiency of functional cholinergic neurons, particularly those extending from the lateral basalis, has been observed in patients with progressive dementia of the Alzheimer type (3). Cholinomimetic therapy has been directed at compensating for the inadequate cholinergic activity in these neurons. However, clinical results with cholinergics and anticholinesterases often have been disappointing or inconsistent (4) due, in some instances, to the inability of quaternary ammonium derivatives to penetrate the blood–brain barrier or to a lack of specificity or selectivity of the drug for the cholinergic receptor(s) involved in the pathological condition. There has been (and continues to be) great emphasis on the search for and study of nonquaternary ammonium molecules (having greater lipophilic character) that will penetrate the blood–brain barrier and interact with appropriate acetylcholine receptors in the brain. Thus older tertiary amine drugs such as pilocarpine and arecoline, which demonstrate only modest cholinergic activity and are classed as partial agonists, have been the subjects of intense structure–activity studies. It has been speculated (5) that partial agonists at M_1 receptors probably have less predisposition to cause receptor desensitization than full agonists, which may make partial agonists potentially more valuable from a therapeutic point of view. The utility of cholinergics in correction of other types of deficits in memory and learning has been investigated for many years (6), with largely inconclusive results. However, this remains a fascinating and a potentially significant area of research.

2 CHOLINERGIC (ACETYLCHOLINE) RECEPTORS

Acetylcholine receptors have been subdivided into two pharmacological types (muscarinic and nicotinic), based on their selective response to two alkaloids: muscarine (**2**) and nicotine (**3**).

(2)

(3)

Neither nicotine nor muscarine is a normal physiological component of the mammalian body; hence the muscarinic/nicotinic classification of acetylcholine receptors is artificial. While it seems well established that

muscarine is a true cholinergic agonist, it is widely accepted that nicotine has little agonist effect in some parts of the nervous system; its peripheral actions are largely indirect and probably involve presynaptic release of acetylcholine (7–11).

Muscarinic receptors occur peripherally, e.g., at parasympathetic postsynaptic sites on glands and smooth (nonstriated) muscles, and they are involved in gastrointestinal and ureteral peristalsis, pupillary constriction, peripheral vasodilatation, reduction of heart rate, and promotion of glandular secretion. Autonomic ganglia also contain muscarinic receptors. Peripheral nicotinic receptors are found postsynaptically on striated (voluntary) muscle fiber membranes and in all autonomic ganglia (sympathetic as well as parasympathetic). There are also nicotinic and muscarinic pathways in the central nervous system.

On the basis of pharmacological data, muscarinic receptors have been subcategorized as M_1, M_2, and M_3. By definition, M_1 receptors occur in the cerebral cortex; corpus striatum; hippocampus, where they may be involved in cognitive processes relevant to Alzheimer's disease, in particular short-term memory (12); and autonomic ganglia where they are involved in membrane depolarization, which is mediated by stimulation of phospholipase C and subsequent production of inositol-1,4,5-triphosphate and diacyl glycerol (13). M_2 receptors are present in the cerebellum, heart, and ileum; those muscarinic receptors in secretory glands and smooth muscle have been tentatively classed as M_3. The central nervous system contains all known subtypes of muscarinic receptors (13). Molecular cloning studies have revealed that there are at least five subtypes of muscarinic receptors, designated m_1–m_5 (14). Correlation between these receptor subtypes, cloned from different tissues and those identified using classical pharmacological methods (M_1–M_3) is not clear, although m_1–m_3 are generally accepted to have pharmacological characteristics identical to those of M_1–M_3, respectively (14).

Muscarinic receptors are glycoproteins with molecular weights of approximately 80,000. They are located on the outer surface of the cell membrane, and they are of the G-protein–linked type; their stimulation affects intracellular production of second messenger substance(s). It is widely believed that the pathophysiology of Alzheimer's disease involves M_1/m_1 receptors. A recent review (15) describes the molecular basis of muscarinic receptor function.

Nicotinic receptors are of the ion channel type. They are pentameric proteins that are composed of at least two distinct subunits, each of which contains multiple membrane-spanning regions, and the individual subunits surround an internal channel (16). Nicotinic receptors are subcategorized as N_M, found postsynaptically at the (striated) neuromuscular junction. Stimulation produces membrane depolarization and skeletal muscle contraction and N_N, found in autonomic ganglia. Stimulation of these receptors produces depolarization (a result of cation channel opening) and firing of the postganglionic neuron. The nicotinic receptor glycoprotein has been isolated and extensively studied (17–19). Reviews of the structure of nicotinic receptor(s) are available (20,21).

Although acetylcholine has no center of asymmetry and is optically inactive, its *in vivo* receptors exhibit discrimination between enantiomers of synthetic and naturally occurring cholinergic stimulants. Both central and peripheral muscarinic receptors are highly stereospecific; peripheral nicotinic receptors seem to be less so, although these usually show some preference for one or the other member of enantiomeric pairs. Central nicotinic receptors frequently demonstrate a higher degree of stereoselective binding character than is noted with the peripheral receptors. Understanding of nicotinic receptor stereoselectivity and specificity is complicated by the likelihood

that many nicotinic agents (in addition to nicotine) are not agonists but rather function indirectly by promoting presynaptic release of acetylcholine (7). However, Casy (7) has presented data suggesting that nicotine may also have a direct (agonist) component of action.

3 ACETYLCHOLINE AND ANALOGS

Although it admirably serves its physiological role in the body, acetylcholine is a poor therapeutic agent. Its rapid rate of hydrolysis in the gastrointestinal tract precludes oral administration, and a similarly rapid hydrolysis by the esterases in the blood and by acetylcholinesterase in the nervous tissue limits its usefulness by injection. Acetylcholine has virtually no clinical uses.

The need for therapeutically satisfactory cholinergic agents and the simple and easily synthesized structures necessary for cholinergic activity have stimulated preparation and biological study of a great number of derivatives, analogs, and congeners of acetylcholine. The following types of structural variations have been addressed:

1. Alteration of the quaternary ammonium head.
2. Replacement of the acetyl group by other acyl moieties.
3. Alteration of the ethylene bridge connecting the quaternary ammonium and the ester groups.
4. Substitution of another group for, or elimination of, the ester moiety.

The "five atom rule," first suggested by work of Alles and Knoefel (22) and stated more formally by Ing (23), proposes that, for maximum muscarinic activity, there should be attached to the quaternary nitrogen atom, in addition to three methyl groups, a fourth group with a chain of five atoms, as illustrated for acetylcholine: C-C-

O-C-C-N. This empirical observation has been found to be valid for a large number of molecules, regardless of the precise nature of the five atoms involved.

Synthesis of compounds and examination of their biological activities have supplied considerable information on structural requirements for cholinergic activity, but especially in the older literature, these data must be examined and interpreted with caution. They have been obtained using a variety of *in vivo* and *in vitro* testing procedures and biological preparations in a variety of animal species. Often, different biological properties associated with stimulation of the cholinergic nervous system were measured. Furthermore, the observed effectiveness of a cholinergic agent in producing a biological response depends, e.g., on its inherent potency and intrinsic activity as well as on the rate at which it is metabolically inactivated (in the case of esters, hydrolysis by acetylcholinesterase and/or by blood esterases). Frequently, especially in the older literature, these individual factors have not been separately and individually assessed. This problem has been cited (24) with respect to lack of consistency among laboratories in the methods used to determine cholinergic receptor subtype selectivity. Therefore, in the following discussion of the relationship of chemical structure to cholinergic activity, only generalized (and tentative) conclusions can be made, and these are frequently based on a composite of the cholinergic activities for which the compound was tested.

3.1 Variations of the Quaternary Ammonium Group

Two types of structural alterations of the quaternary head have been studied: replacement of the nitrogen atom by other atoms and replacement of the *N*-methyl groups by hydrogen, alkyl, nitrogen, or

oxygen. Acetylphosphonocholine (**4**) (23), acetylarsenocholine (**5**) (23), and acetylsulfonocholine (**6**) (25) exhibit muscarinic effects, but they are considerably less potent than acetylcholine.

$$CH_3-CO-O-CH_2-CH_2-R$$

(**4**) R = $^+P(CH_3)_3$ (**7**) R = $C(CH_3)_3$

(**5**) R = $^+As(CH_3)_3$ (**8**) R = $^+N(CH_3)_2NH_2$

(**6**) R = $^+S(CH_3)_2$ (**9**) R = $^+N(CD_3)_3$

Ing (23) noted that the potencies of acetylcholine analogs containing other charged atoms than nitrogen (phosphorus, arsenic, sulfur) are in inverse order to the volumes occupied by these atoms. The carbon isostere (**7**) of acetylcholine exhibits no cholinergic activity, but it is an excellent substrate for acetylcholinesterase (26). Studies of the role of nitrogen substituents in the acetylcholine molecule strongly indicate that the *N,N,N*-trimethyl quaternary ammonium pattern of acetylcholine itself is optimum for potency and activity. The acetate esters of *N,N*-dimethylethanolamine, *N*-methylethanolamine, and ethanolamine possess weak muscarinic activity, and they show no nicotinic activity (27). The tertiary amine congener of carbachol exhibits greatly diminished nicotinic and muscarinic effects compared with the *N,N,N*-trimethyl quaternary compound (28) (Number 19 Table 25.1). These conclusions seem valid for cholinergic agents having, like acetylcholine, a high degree of molecular flexibility. In contrast, in certain acetylcholine congeners in which the nitrogen is a part of a relatively rigid ring system (pyrrolidine, morpholine, piperidine, quinuclidine), tertiary amine salts are more potent muscarinics than their quaternary derivatives (41). The enhanced activity of the tertiary amines has been rationalized on conformational grounds. It must be as-

Table 25.1 Representative Esters of Choline

$$R\text{-}O\text{-}CH_2\text{-}CH_2\text{-}\overset{+}{N}(CH_3)_3$$

Number	R	References
1	HCO	29
2	$BrCH_2CO$	30
3	C_2H_5CO	29, 31
4	H_2NCH_2CO	32
5	$n\text{-}C_3H_7CO$	29, 31
6	$i\text{-}C_3H_7CO$	31
7	$n\text{-}C_4H_9CO$	29, 31
8	C_6H_5CO	31
9	$C_6H_5CH_2CO$	31
10	$C_6H_5CH{=}CHCO$	31
11	$(C_6H_5)_2C(OH)CO$	33
12	$CH_3(CH_2)_{10}CO$	34
13	$CH_3(CH_2)_{14}CO$	34
14	$HOCH_2CO$	29
15	$CH_2{=}CHCO$	35
16	CH_3COCO	29
17	$CH_3CHOHCO$	36
18	O_2N	37
19	H_2NCO	38, 39
20	$(CH_3O)_2PO$	40

sumed that the tertiary amines are protonated at their *in vivo* sites of action.

Replacement of one *N*-methyl group of acetylcholine by ethyl permits retention of most of the cholinergic activity, but as more *N*-methyl groups are replaced by ethyl, there is a progressive loss of cholinergic effect (42). When one *N*-methyl is replaced by *n*-propyl or *n*-butyl, there is almost complete loss of cholinergic activity (25). The hydrazinium congener (**8**), in which one *N*-methyl is replaced by NH_2, was less active than acetylcholine in all assays performed (43). The pyrrolidine congener (**10**) is 20–33% as potent as acetylcholine (44); this compound can be viewed as a cyclic congener of acetyl *N,N*-diethylcholine, and

$$CH_3-COO-CH_2-CH_2-\overset{+}{\underset{\underset{H_3C}{\diagup}}{N}}\diagup$$

(10)

it is decidedly more potent than the diethylcholine ester.

However, in general, incorporation of the choline moiety into a heterocyclic ring markedly lowers potency compared with acetylcholine (45,46). The *tris*-(trideuteromethyl) congener (**9**) showed similar potency to acetylcholine in a dog blood pressure assay (47).

Replacement of one *N*-methyl by methoxyl in acetylcholine and in three congeners (**11–14**) permits retention of some cholinergic effects, and in certain compounds, nicotinic or muscarinic activities are enhanced over the parent *N,N,N*-trimethyl system (48).

$$R-CO-O-CH-CH_2-\overset{\overset{\displaystyle CH_3}{|}}{\underset{\underset{CH_3}{|}}{N}}{}^{+}\!\!-OCH_3$$
$$\qquad\qquad\quad\underset{R'}{|}$$

(**11**) R = CH_3 ; R' = H

(**12**) R = R' = CH_3

(**13**) R = H_2N ; R' = H

(**14**) R = H_2N ; R' = CH_3

The reverse *N*-alkoxy systems (**15**) demonstrated only extremely weak muscarinic activity (49). Amine oxide analogs of cholinergic agonists (**16–19**) exhibit little or no cholinergic effect, and they are not substrates for cholinesterases (50).

$$R-CO-O-CH-CH-O-\overset{+}{N}(CH_3)_3$$
$$\qquad\qquad\underset{R'}{|}\;\;\underset{R''}{|}$$

(**15**) R = CH_3 or NH_2 ;
   R', R'' = combinations of H, CH_3

The observed biological effects of several variations of the quaternary head of acetylcholine and its congeners may be rationalized by invoking results of molecular orbital calculations (51), which indicate that in both muscarine and acetylcholine, the nitrogen atom is nearly neutral and a

$$R-CO-O-\underset{\underset{R'}{|}}{CH}-CH_2-\underset{\underset{CH_3}{|}}{\overset{\overset{CH_3}{|}}{N}}{\overset{+}{\quad}}-O^{-}$$

(16)　R = CH$_3$;　R' = H

(17)　R = R' = CH$_3$

(18)　R = H$_2$N ;　R' = H

(19)　R = H$_2$N ;　R' = CH$_3$

large part (70%) of the formal charge is distributed among the three attached methyl groups, which form a large ball of spreading positive charge. Furthermore, Kimura and co-workers (52) determined that chain extension of one alkyl group of tetramethylammonium produces a great decrease in the charge density on the nitrogen, and they proposed that cholinergic agonist activity for a quaternary ammonium compound requires a minimum level of charge density on the nitrogen.

3.2 Variations of the Acyl Group

Qualitatively, choline has the same pharmacological actions as acetylcholine, but it is far less active (53). Esterification of the alcohol function frequently increases the potency. However, formylcholine is less potent than acetylcholine, and homologation of the acetate methyl group of acetylcholine generally produces compounds that are less potent than acetylcholine (see Table 25.1). Polar groups such as OH (number 17 in Table 25.1) and NH$_2$ (number 4) markedly decrease muscarinic potency, but a C=O group (number 16) permits retention of considerable activity. Bromoacetylcholine (number 2) is a direct muscarinic and nicotinic agonist (54), and under reducing conditions, it covalently binds to nicotinic receptors but not to muscarinic receptors (30). Acrylylcholine (number 15), which has been isolated from

tissues of a marine gastropod (55), has relatively high cholinergic activity. Higher fatty acid esters (numbers 12 and 13) were prepared for testing as hemolytic agents, but they have apparently never been evaluated for cholinergic activity. A study of acetylcholine congeners derived from relatively high molecular weight acids (56), most of which contained a benzene ring, revealed that as the molecular weight of the acid increases parasympathetic stimulant activity diminishes, and there is a gradual change to atropinelike (muscarinic-blocking) activity.

In general, carbamic acid esters of choline and its congeners are more potent and more toxic than the corresponding acetates. Carbamoyl choline (number 19 in Table 25.1) is a potent muscarinic agent, and it demonstrates pronounced nicotinic stimulant effects at autonomic ganglia. It is likely that these ganglionic actions are due, at least in part, to release of endogenous acetylcholine from the terminals of cholinergic fibers (57). Carbamoyl choline is completely resistant to hydrolysis by acetylcholinesterase and by nonspecific cholinesterases (58). The nitrate ester (number 18) has marked nicotinic and muscarinic agonist effects and, in addition, an intense paralyzing nicotine action. The dimethylphosphate ester (number 20) displays powerful nicotinic action but little muscarinic effect.

The "reversed ester" congener (**20**) of acetylcholine exhibits weak muscarinic and no nicotinic effects (59).

In contrast, the carbonate congener (**21**) is a full agonist at muscarinic and nicotinic receptors, with an activity approximately

$$CH_3-O-CO-CH_2-CH_2-\underset{+}{N}(CH_3)_3$$

(**20**)

$$CH_3-O-CO-O-CH_2-CH_2-\underset{+}{N}(CH_3)_3$$

(**21**)

one order of magnitude less than that of acetylcholine (60).

Acetylthiocholine (**22**) and acetylseleno-choline (**23**) exert acetylcholine-like effects on the guinea pig ileum and on the frog rectus abdominis, but they are somewhat less potent than acetylcholine (61).

$$CH_3-CO-X-CH_2-CH_2-\overset{+}{N}(CH_3)_3$$

(**22**) X = S

(**23**) X = Se

(**24**) X = NH

Unesterified thiocholine and selenocholine display a relatively high degree of acetylcholine-like activity compared with their acetate esters, in contrast to the dramatic potency difference between choline and acetylcholine. The biological effects of these unesterified thiols and selenols have been suggested to be due to their oxidation to disulfide and diselenide derivatives (61). The amide congener (**24**) of acetylcholine has little or no cholinergic activity (38). Acetylthionocholine, in which the carbonyl oxygen of acetylcholine is replaced by sulfur, displays some acetyl-choline-like effects in an electroplax preparation (62).

3.3 Variations of the Ethylene Bridge

The distance between the ester moiety and the cationic head of acetylcholine seems to be critical. Acetoxytrimethylammonium (**25**), completely lacking the ethylene bridge of acetylcholine, showed a pharma-cological profile qualitatively quite similar to acetylcholine (63), but it was much less potent.

$$CH_3-CO-O-\overset{+}{N}(CH_3)_3$$

(**25**)

$$CH_3-CO-O-CH_2-\overset{+}{N}(CH_3)_3$$

(**26**)

Acetoxymethyltrimethylammonium (**26**) appeared to have little or no muscarinic effect on a guinea pig ileum preparation (63). The profound instability of this com-pound in solution precluded collection of quantitative data. An older report (37) indicated that structure **26** has "intense muscarine action" and "marked nicotine stimulant action." Acetyl γ-homocholine (**27**) is decidedly less potent and/or active than acetylcholine (56). 4-Acetoxybutyl-trimethylammonium (**28**) exhibits extreme-ly weak nicotinic and muscarinic effects (64).

$$CH_3-CO-O-(CH_2)_n-\overset{+}{N}(CH_3)_3$$

(**27**) n = 3

(**28**) n = 4

Replacement of one or more of the hydrogen atoms of the ethylene bridge with alkyl groups produces marked changes in potency and activity. Acetyl-β-methyl-choline (**29**) is a more potent muscarinic agonist than acetylcholine, but it has a much weaker nicotinic action (58).

$$CH_3-CO-O-\underset{\underset{R'}{|}}{CH}-\underset{\underset{R}{|}}{CH}-\overset{+}{N}(CH_3)_3$$

(**29**) R = H ; R' = CH_3

(**30**) R = CH_3 ; R' = H

(**31**) R = R' = CH_3

Acetyl-α-methylcholine (**30**) is a more po-tent nicotinic than a muscarinic, but both effects are decidedly less than for acetylcholine (65); it is hydrolyzed by acetylcholinesterase at a rate similar to that of acetylcholine (66). A factor in the ob-served potency of acetyl-β-methylcholine is its slower rate of hydrolysis by acetylcholinesterase, due to poor affinity of the compound for the enzyme's catalytic site (67) and to its extremely high resistance to hydrolysis by nonspecific serum cholinesterases. Compound (29) and the carbamate ester of ($\pm$)-β-methylcholine (32) (bethanechol) are useful therapeutic agents. The introduction of the C-methyl into the acetyl α- and β-methylcholine molecules creates a chiral center, and the enantiomers exhibit different properties (Table 25.2).

$$H_2N - CO - O - CH - CH_2 - \overset{+}{N}(CH_3)_3$$
$$\underset{CH_3}{|}$$

(32)

S-(+)-Acetyl-β-methylcholine (the eutomer) is hydrolyzed by acetylcholinesterase at about half the rate of acetylcholine; the R-($-$)-enantiomer is a weak inhibitor of the enzyme (65). Work of Ringdahl (68) suggests that the markedly lower activity of R-($-$)-acetyl-β-methylcholine is a consequence both of lower affinity for the muscarinic receptor(s) and lower intrinsic activity. The antipodes of carbamoyl-β-methylcholine (bethanechol) (32) displayed a eudismic ratio of 740 at rat jejunum sites (69) and of 915 using guinea pig intestinal muscle (70), with the S antipode as the eutomer.

($\pm$) - $Erythro$ - Acetyl - α,β-dimethylcholine (31) exhibits 14% of the muscarinic potency of acetylcholine, and it is almost completely resistant to acetylcholinesterase; the ($\pm$)-$threo$ isomer is inert as a cholinergic and is a poor substrate for acetylcholinesterase (71). These racemic mixtures have apparently never been resolved. gem-Dimethyl substitution of acetylcholine, either on the α,α or the β,β positions of the choline moiety, greatly reduces but does not abolish muscarinic activity (72). Both compounds are relatively poor substrates for bovine erythrocyte acetylcholinesterase. Replacement of the C-methyl groups in the ethylene bridge by longer chains causes an increase in toxicity and a reduction in muscarinic activity, e.g., the acetate esters of β-n-propyl and β-n-butylcholines (73,74).

3.4 Substitution of the Ester Group by Other Groups

The ester moiety of acetylcholine does not appear to be essential for cholinergic activity. In general, alkyl ethers of choline and of thiocholine are less potent and less active than acetylcholine (45); thio ethers are less potent than the corresponding oxygen compounds. Contrary to some earlier literature reports, the vinyl ether of choline (34) is

Table 25.2 Muscarinic Activities of Acetyl C-Methyl Cholines[a]

Substituent	Stereochemistry	Number Moles equivalent to 1 mole of AcCh as Agonist in Guinea Pig Ileum	Activity Ratio, $\dfrac{(+)}{(-)}$
α-CH$_3$	RS	49	
	S-($-$)-	232	8
	R-(+)-(eutomer)	28	
β-CH$_3$	RS	1.58	
	S-(+)-(eutomer)	1.01	240
	R-($-$)-	240	

[a]Adapted from ref. 66. Courtesy of Plenum Press.

not a more potent muscarinic agent than the ethyl ether (**33**) (both are weak muscarinics), but it is a better nicotinic agent, displaying higher potency than acetylcholine (75).

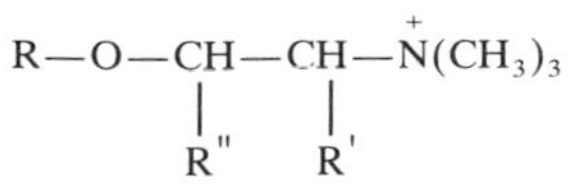

$$R-O-CH-CH-\overset{+}{N}(CH_3)_3$$
$$\qquad\quad | \qquad |$$
$$\qquad\quad R'' \qquad R'$$

(**33**) R = C_2H_5 ; R' = R'' = H

(**34**) R = CH_2=CH ; R' = R'' = H

(**35**) R = C_2H_5 ; R' = H ; R'' = CH_3

(**36**) R = C_2H_5 ; R' = CH_3 ; R'' = H

(**37**) R = CH_2=CH ; R' = H; R'' = CH_3

(**38**) R = CH_2=CH ; R' = CH_3; R'' = H

α-Methyl substitution of choline in its ethyl and vinyl ethers (compounds **36** and **38**) greatly diminishes muscarinic potencies but retains potent nicotinic effects. β-Methyl substitution (compounds **35** and **37**) permits retention of some degree of muscarinic effect, but nicotinic effects are completely abolished (75). A series of open-chain congeners of muscarine, typified by structure **39**, exhibited low muscarinic potency, which has been ascribed by Friedman (45) to the compounds' stereochemical heterogeneity.

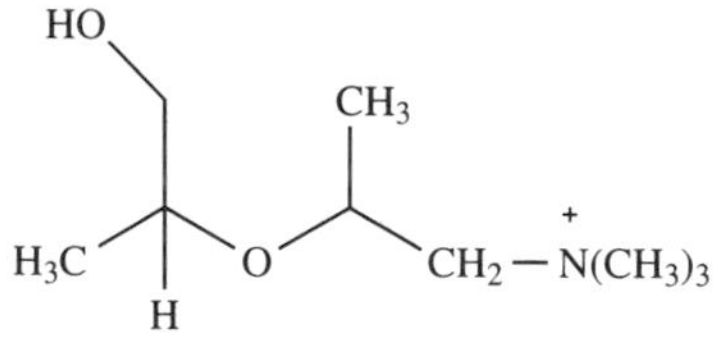

(**39**)

An open chain analog (**40**) of desmethylmuscarine lacking chiral centers exhibited extremely low muscarinic activity (76).

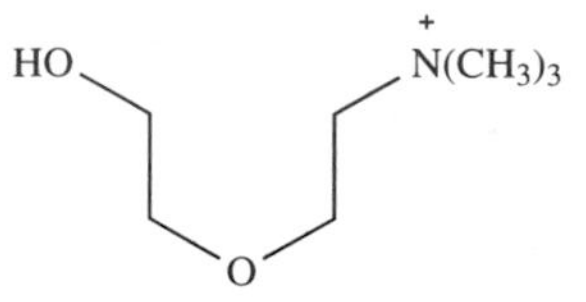

Some aromatic ethers of choline display marked nicotinic activity, but they are inactive at muscarinic sites (77). The *o*-tolyl ether of choline is a potent ganglionic

(**40**)

stimulant (78), but the 2,6-xylyl ether of choline is inert as a nicotinic (79). Additional ring-substituted phenyl ethers of choline were described by Hey (80). Clark and co-workers (81) studied conformationally restricted racemic bicyclic choline phenyl ethers (**41**) in which the choline moiety was a part of the ring system.

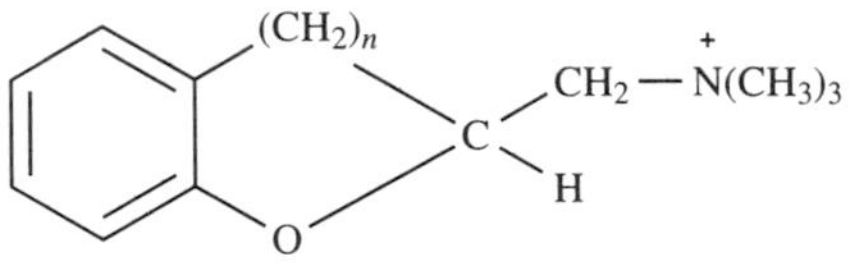

(**41**) *n* = 1-3

On the basis of biological data on these compounds, it was concluded that the nicotinic activity of choline phenyl ether and of choline *o*-tolyl ether is a reflection of the ability of the molecule to assume a "planar" conformation when interacting with the ganglionic nicotinic receptor. In contrast, the inactive 2,6-xylyl ether of choline cannot assume this planar disposition. An additional series of conformationally restricted aryl choline ethers (**42–45**) demonstrated that only the piperidine derivative (**45**) is a ganglionic stimulant (82).

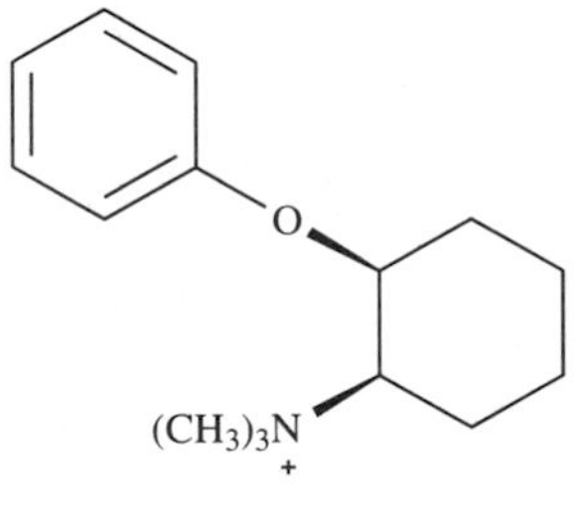

(**42**)

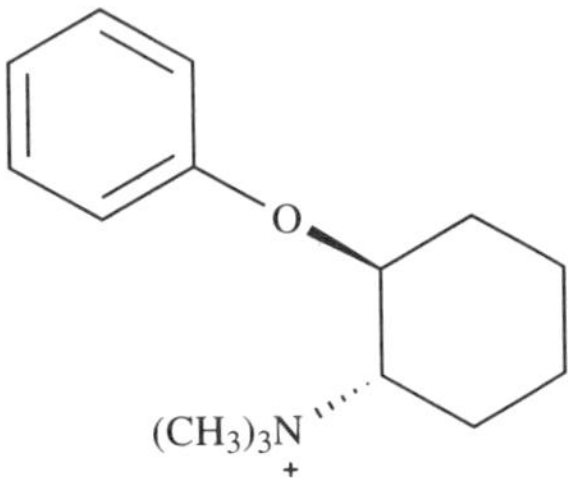

(43)

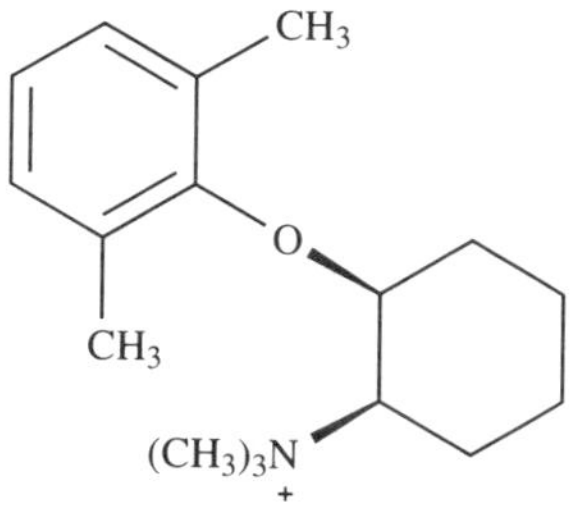

(44)

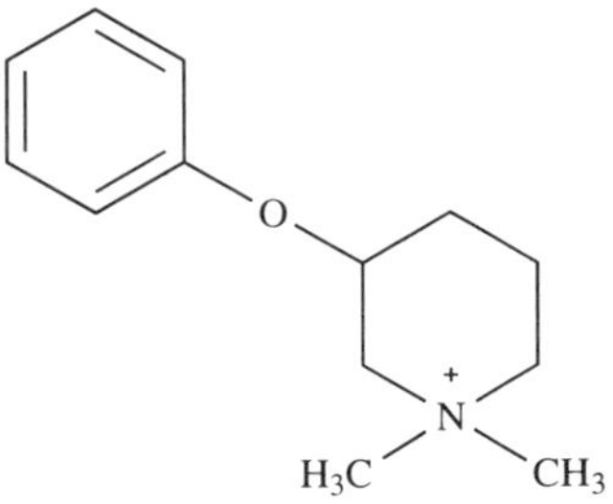

(45)

$$R - CO - alkylene - \overset{+}{N}(CH_3)_3$$

(46a) R = CH$_3$; alkylene = CH$_2$-CH$_2$

(46b) R = CH$_3$; alkylene = (CH$_2$)$_3$

(46c) R = C$_2$H$_5$; alkylene = CH-CH$_2$
                                              |
                                           CH$_3$

$$R - \overset{+}{N}(CH_3)_3$$

(47)

structure **47**, between methyl and *n*-amyltrimethylammonium a minimum activity occurs in most assays at the ethyl or *n*-propyl group, and maximum muscarinic potency is demonstrated at the *n*-amyl chain (45,46). Above heptyl, the compounds become antagonists to acetylcholine. Numerous examples of ketones and *N*-alkyl congeners have been tabulated (45,46).

3.5 Prodrug to Acetylcholine

N,*N*-Dimethylaminoethanol, HO–CH$_2$–CH$_2$–N(CH$_3$)$_2$ (Deanol) (**48**), was proposed to be a possible prodrug for acetylcholine in the central nervous system (CNS). However, it was found (83) that compound **48** does not produce an increase in acetylcholine levels in rat brain slices *in vitro*, nor does it increase acetylcholine levels in whole brains of intact rats. Clinical results with, this compound in treatment of CNS acetylcholine deficiency conditions have been unimpressive.

The inactivity of structures **42** and **44** suggested the need for a transoid arrangement of the O-C-C-N$^+$ system for nicotinic activity (discussed below), although the trans isomer (**43**) demonstrated no greater pharmacologic effect than the cis systems (**42**) and (**44**). Ketonic systems, typified by compounds **46a–46c**, carbon isosteres of the ester moiety, display weak activities, and they are predominantly more nicotinic than muscarinic (45).

The secondary alcohol analogs of these ketones are even weaker, and the thio ketones are also weak (46).

In congeners in which an alkyl chain replaces the acetate ester moiety, as in

4 CHOLINERGICS NOT CLOSELY RELATED STRUCTURALLY TO ACETYLCHOLINE

4.1 Nicotine and Analogs and Congeners

Naturally occurring levorotatory nicotine (**3**) has the *S*-absolute configuration. Its enantiomer, (*R*)-(+)-nicotine, was decidedly less potent than the naturally occurring material in assays for peripheral effects (84). The pyrrolidine methyl quaternary derivative of S-(−)-nicotine (**49**) shows peripheral activity comparable to that of nicotine itself.

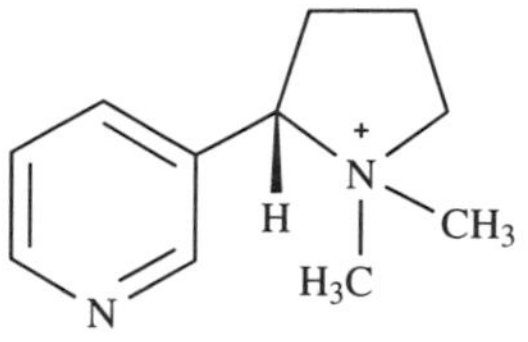

(**49**)

Nornicotine, in which the *N*-methyl is replaced by hydrogen, is somewhat less potent or active than nicotine in most assays (86). *R*- and *S*-nornicotine are equipotent with *R*-(+)-nicotine, the less active, unnatural enantiomer, in a rat brain membrane-binding assay (87). In that study, *S*-(−)-nicotine was 13 times more potent than its *R*-enantiomer. Replacement of the *N*-methyl of nicotine with ethyl or *n*-propyl causes an exponential loss of peripheral nicotinic effect (86). Synthetic compounds (**50**) representing structures in which each of the bonds of the pyrrolidine ring of nicotine is cleaved, one by one, produced a series of "seconicotines" (86).

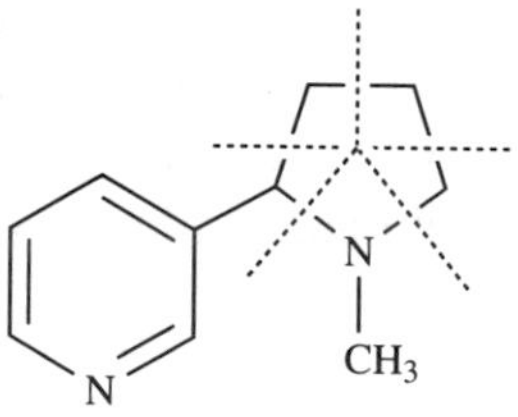

(**50**)

Only the open chain congeners (**51** and **52**) display nicotine-like activity. The potency of structure **51** is increased in its *N,N*-dimethyl congener; in contrast, the *N,N*-dimethyl derivative of structure of **52** is inert. A nicotine structural isomer (**53**) retains a

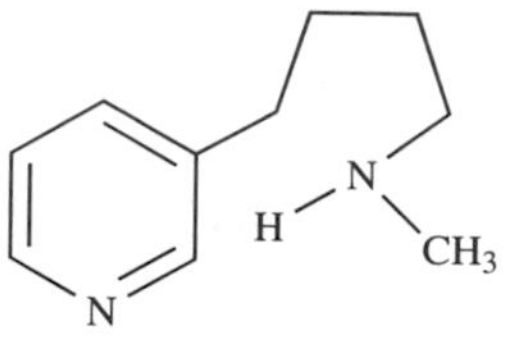

(**51**)

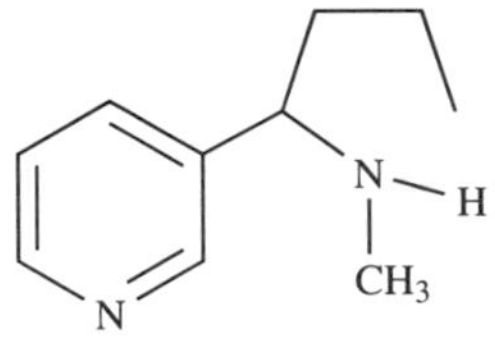

(**52**)

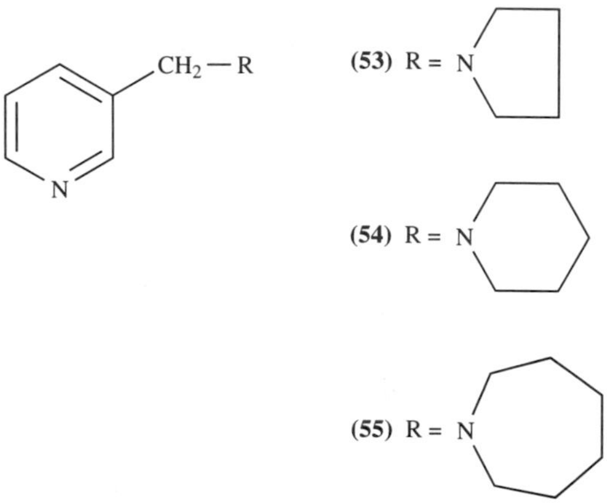

(**53**) R = pyrrolidine

(**54**) R = piperidine

(**55**) R = azepane

considerable degree of nicotine-like activity (86).

The piperidine congener (**54**) is somewhat less potent and active, and the perhydroazepine congener (**55**) is inert. The pyrrolidine ring and piperidine ring *N*-methyl quaternary derivatives of structures **53** and **54** are slightly less active than the corresponding tertiary bases (85). Further modifications of structure **53** are illustrated in structure **56**.

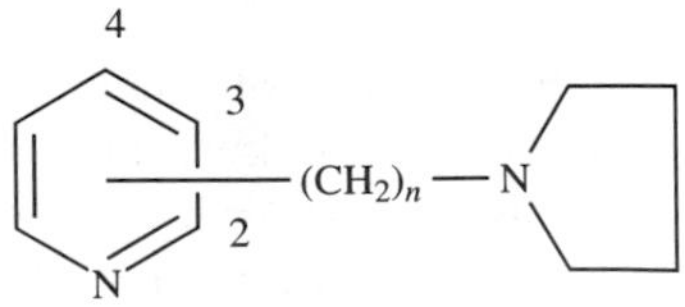

(**56**)

When $n = 2$, all positions of attachment to the pyridine ring (carbon 2, 3, or 4) result in extremely low potency and activity. When $n = 1$, attachment to positions 2 and 4 produces practically inert compounds (86). Replacement of the pyridine ring of structure **53** by bioisosteric benzene, 2-thienyl, 2-furanyl, and 2-pyrrolyl ring systems abolishes almost all nicotine-like activity.

The finding that (S)-nicotine enhances memory and learning performance in animals and improves performance of behavioral tasks in Alzheimer's disease patients has stimulated further interest in nicotine analogs and congeners (88). A series of 3'-, 4'-, and 5'-substituted nicotine analogs (**57**) was evaluated as ligands of the neuronal nicotinic (N_N) receptor from rat brain membranes (89).

(**57**)

Only a small substituent is tolerated at position 4'; the (2'S,4'R)-methyl congener is the most potent of the entire series, but it is somewhat less potent in the binding assay than S-nicotine itself. None of the 3'- or 5'-substituted analogs approach this binding affinity. In a series of substituted 2-arylpyrrolidines in which the substituted aryl group is a bioisosteric replacement for the pyridine ring of nicotine, it was found that the isoxazole derivative (**58**) is a potent cholinergic channel activator (90).

(**58**)

All substitutions on the pyrrolidine ring diminished the binding affinity compared with structure **58**. The primary metabolism of compound **58** involves oxidation at the 5' position. It was, therefore, unexpected that the 5'-methyl congener of structure **58** has *in vitro* half-lives equivalent to or shorter than structure **58**.

4.2 Muscarine, Muscarone, and Related Compounds

The structure and absolute configuration of naturally occurring (+)-(2S,4R,5S)-muscarine (**2**) has been confirmed by stereospecific synthesis (91). The molecule may be viewed as a cyclic analog of acetylcholine in which the carbonyl and β-carbons are linked by a bimethylene bridge (cf. structures **59** and **60**).

The natural (+) isomer of muscarine is one of eight stereoisomers of structure **59**. The enantiomer of (+)-(2S,3R,5S)-muscarine (**2**) is almost inert, as are both enantiomers of the other three diastereomers of structure **59** (epimuscarine, allomuscarine, and epiallomuscarine) (92). A point of interest is the absolute configurational identity of the C-2 position of muscarine (**2**) and of the S-(+)-eutomer of acetyl-β-methylcholine (**61**). The oxidation product of (+)-(2S,3R,5S)-muscarine, (−)-(2S,5S)-muscarone (**62**), shows even more structural analogy to the acetylcholine molecule; it is an active muscarinic agonist, and it also exhibits a nicotinic component of activity not possessed by muscarine.

For many years, the literature (93) consistently presented misleading information concerning the stereochemistry of the muscarone molecule, and there were accompanying hypotheses, based on this misinformation, concerning the relationship of the absolute configuration of muscarone enantiomers to their pharmacologic properties and to the pharmacologic properties

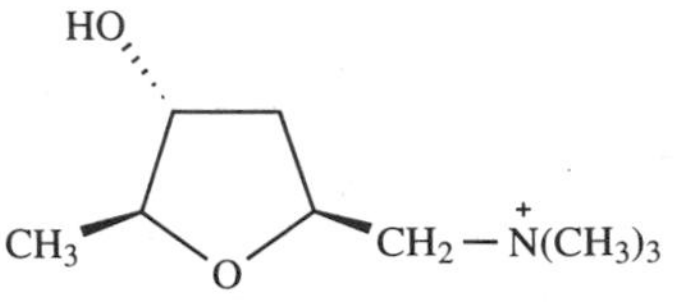

(2) 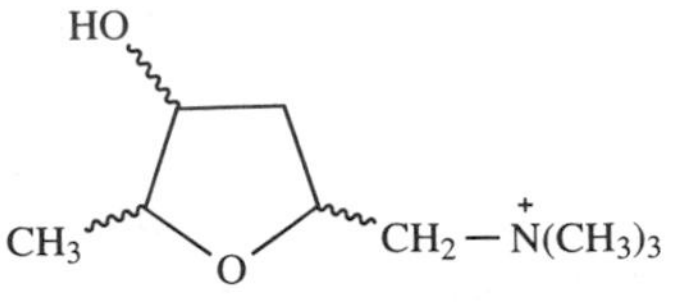   (59) 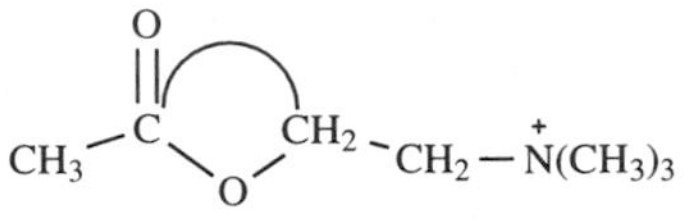   (60)

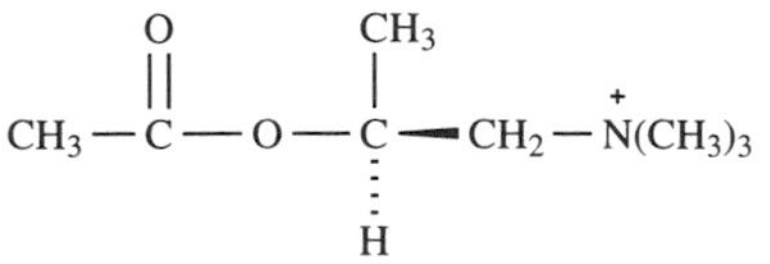

(61) 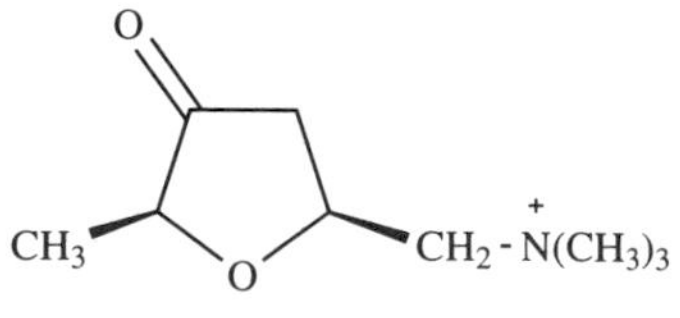   (62)

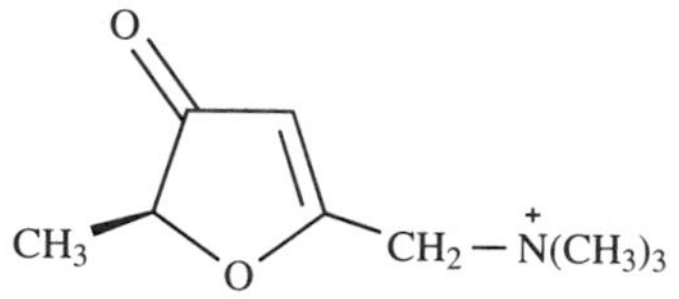

of the corresponding enantiomers of muscarine and of other cholinergic agonists. This confusion was in part alleviated by the demonstration (94) that the eutomer of muscarone has chirality at C-2 and at C-5 ($2S,5S$) identical with natural muscarine. The earlier literature (92,95) had reported small eudismic ratios (2.4–10.1) for the muscarone enantiomers, in contrast to the large values established for muscarine enantiomers. This pharmacological inconsistency has been explained by De Amici and co-workers (96) on the basis of optical heterogeneity of the muscarone enantiomers used in the earlier studies. These workers performed enantiospecific syntheses to obtain the two muscarone enantiomers in >98% enantiomeric excess. In both binding and functional tests, $(-)$-$(2S,5S)$-muscarone (**62**) was the eutomer, and the eudismic ratios of the muscarone enantiomers were in the range of 280–440, which is quantitatively similar to those for muscarine. Beckett and co-workers (97) speculated that the approximate equivalence of muscarinic action shown by enantiomers of 4,5-dehydromuscarone (**63**) is related to enolization phenomena. On the basis of the work described for muscarone enantiomers, the validity of this explanation might be questioned.

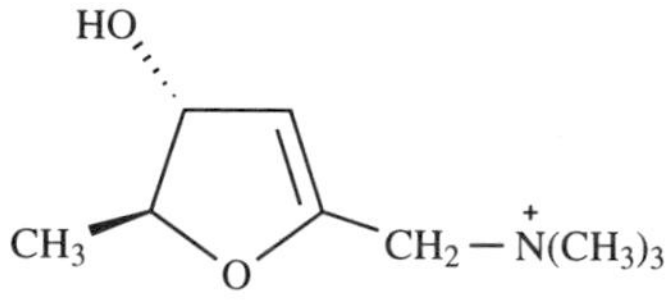

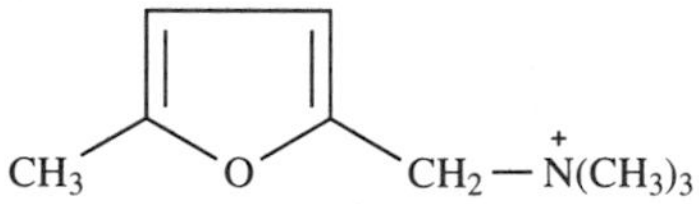

dl-Dehydromuscarine (**64**) retains considerable muscarinic agonist activity, but it

(63)

(64)

(65)

shows no effects at nicotinic receptors (92). Incorporation of elements of the muscarine structure into an aromatic ring has produced some systems, such as structure **65**, which approach acetylcholine in muscarinic potency (25). Activity is lowered by changing the C-2 methyl of structure **65** to ethyl (98) and by replacing the C-2 methyl with hydrogen (25). The data on these furan derivatives are consistent with the "rule of five."

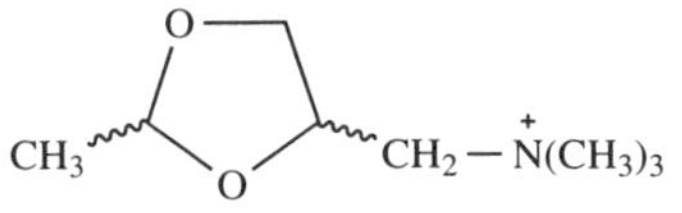

(66)

(68)

(69)

Muscarinic activity of 2-methyl-4-tri-methylammoniummethyl-1,3-dioxolane systems (66) resides in the cis isomer (99); stereospecific synthesis of the two enantiomers of this cis isomer revealed that the L(+) enantiomer (C-4 = R) is more than 100 times more potent than the D(−) isomer (C-4 = S), and is approximately six times more potent than acetylcholine in a guinea ileum assay. The more active L(+) compound is related configurationally to the most potent muscarine stereoisomer (2), although it should be noted that several authors in the older literature incorrectly assigned the (S) absolute configuration to position 4 of L(+)-cis- structure 66, apparently through misapplication of priority rules. 2,2-Dialkyl analogs of these dioxolanes are much weaker muscarinic agonists than the parent systems (66), and the difference in potency between the C-4 (R) and (S) enantiomers diminishes sharply with increasing size of substituents at C-2 (100). Both enantiomers of the cis- and trans-oxathiolane system (67), bioisosteres of the dioxolanes, were evaluated for nicotinic and muscarinic effects (101).

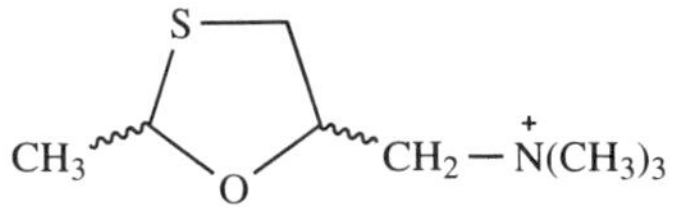

(67)

The (+)-cis isomer of structure 67 was the most potent muscarinic of all of the isomers. It demonstrated a high eudismic ratio, which was of the same order of magnitude as that for muscarine and the dioxolanes. This (+)-cis enantiomer has the same absolute configuration as the muscarinically most active L-(+)-muscarine (2)

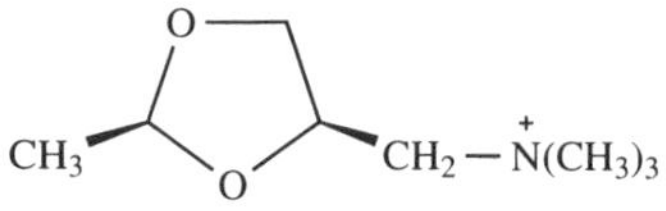

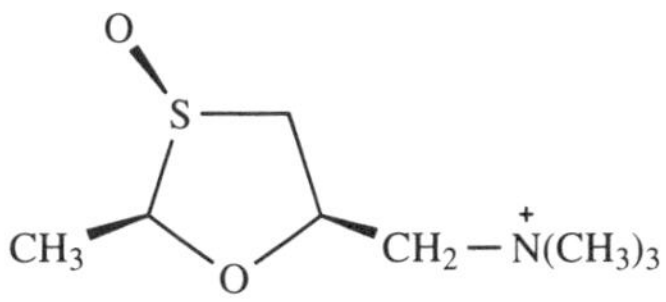

and the (+)-cis-dioxolane (68). The other isomers represented by structure 67, while much less potent than the (+)-cis isomer, also demonstrated a degree of muscarinic agonist effect. All four isomers of structure 67 showed similar nicotinic potency and activity, close to that of carbachol, and eudismic ratios were low. Studies on the diastereomeric cis-sulfoxides (69) (2R, 3S, 5R) and (70) (2R, 3R, 5R) indicated that structure 70, which has the same absolute configuration as (+)-(2S,3R,5S)-muscarine (2), is a potent and selective muscarinic agent with a high eudismic ratio (102). (Note that the presence of the sulfur atom reverses the R, S designations of the chiral centers, compared with muscarine.) Compound 69 is exponentially less potent. Both compounds demonstrate low nicotinic potency and activity. The sulfone congeners of the enantiomers (2R, 5R and 2S, 5S) of cis-structure 71 are weak muscarinics with a eudismic ratio of unity. Neither is an extremely potent nicotinic, although the 2R, 5R enantiomer is more potent than the 2S, 5S.

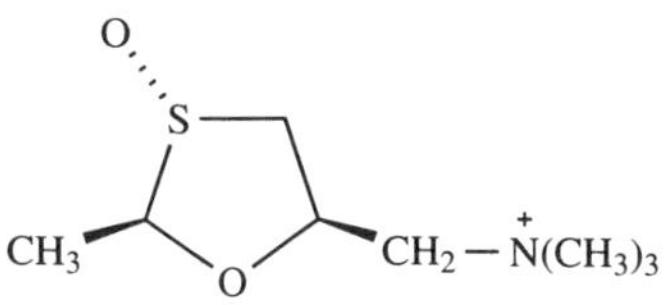

(70)

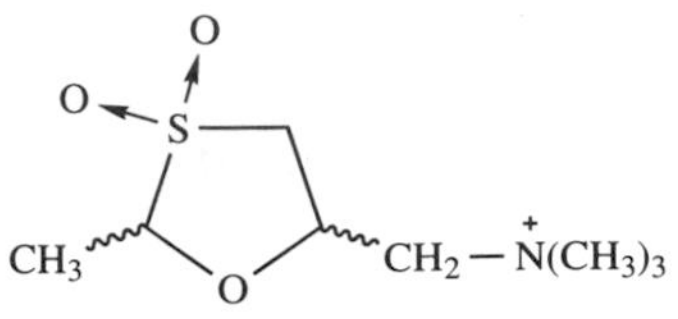

(71)

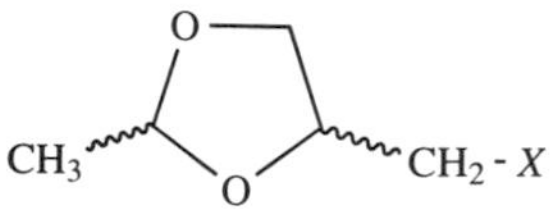

(72) $X = {}^+S(CH_3)_2$

(73) $X = {}^+P(CH_3)_3$

(74) $X = {}^+As(CH_3)_3$

Cis and trans isomeric mixtures of dioxolane congeners bearing sulfur, phosphorus, or arsenic cationic heads (**72–74**) display lower muscarinic effects than the corresponding nitrogen system (103).

Both of the racemic *cis*-1- and *cis*-3-desether dioxolane systems (**75** and **76**) demonstrate muscarinic activity not substantially lower than that of the "supermuscarinic" L-(+)-*cis*-dioxolane (**68**) (103). It was suggested (104) that occupation of only one of the two receptor sites proposed to be reacting with the ring oxygens of the *cis*-dioxolane (**68**) is sufficient to induce muscarinic activity.

The moderately high potency ($\frac{1}{10}$ acetylcholine) of the spirodioxolane (**77**) compared with the low muscarinic activity of a more flexible system (**78**), a mixture of isomers ($\frac{1}{300}$ acetylcholine), led Ridley and co-workers (105) to speculate that the rigid molecule of structure **77** may approximate

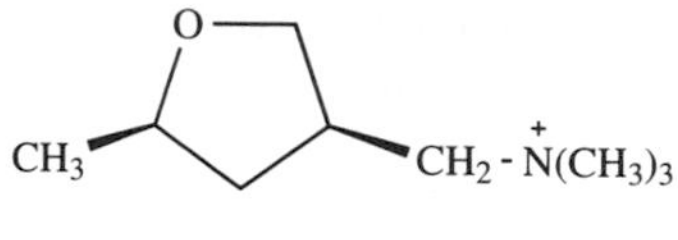

(75)

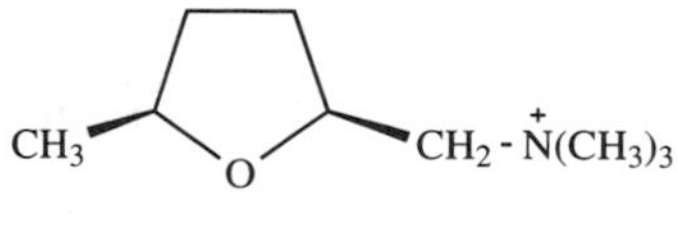

(76)

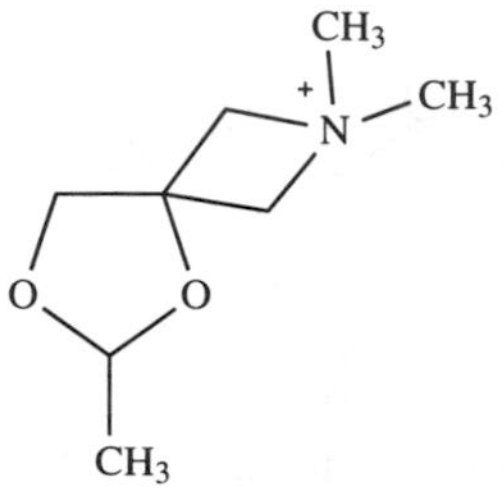

(77)

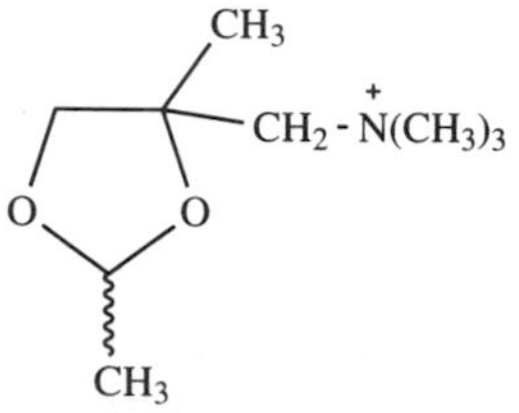

(78)

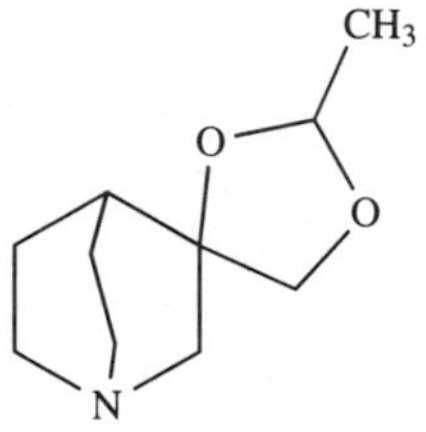

(79)

the conformation of the L-(+)-*cis*-dioxolane (**68**) when it binds to muscarinic receptor(s).

A more complex spirodioxolane molecule (**79**), bearing a tertiary amine rather than a quaternary moiety, was resolved into its four possible stereoisomers (106). The $3(R),2'(S)$ isomer (**80**) is the most potent in binding studies, but the $3(R),2'(R)$ isomer (**81**) displays the largest selectivity between M_1 receptors (ganglion) and M_2 receptors (heart). These compounds illustrate a previously cited phenomenon: a tertiary amine may be a more potent cholinergic agonist than the corresponding quaternary system, if the nitrogen head is a part of a rigid ring. The $3(R)$ stereochemistry of the most potent isomers (**80** and **81**) is consistent with that of other potent 1,3-dioxolane deriva-

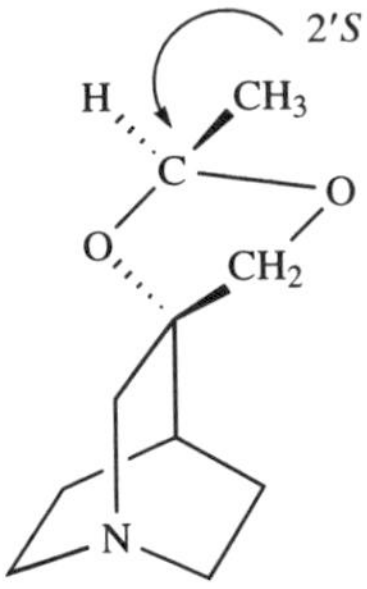

(80) 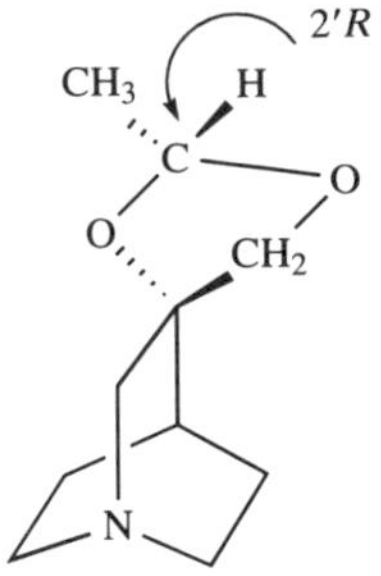　　　　　　(81)

(84)

(85)

(86)

tives. Extension of these studies (107) led, e.g., to a racemic 1,2,4-oxadiazole derivative (82) that has high affinity and efficacy at central muscarinic receptors.

Additional 1,2,4-oxadiazole derivatives containing 1-azanorbornane (83) and isoquinuclidine (84) rings were studied (108). These compounds can exist as geometric isomers, and the exo-1-azanorbornane isomer (83b) was described as one of the most potent and efficaceous muscarinic agonists known.

A carbocyclic muscarine analog, (±)-desethermuscarine (85), exhibits striking muscarinic effects (109), although the compound is considerably less potent than was originally reported (110).

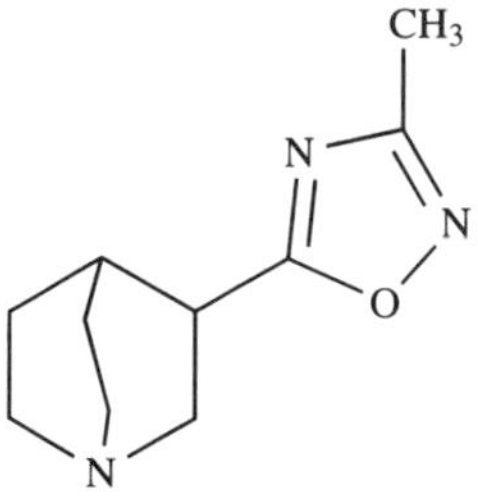

(82)

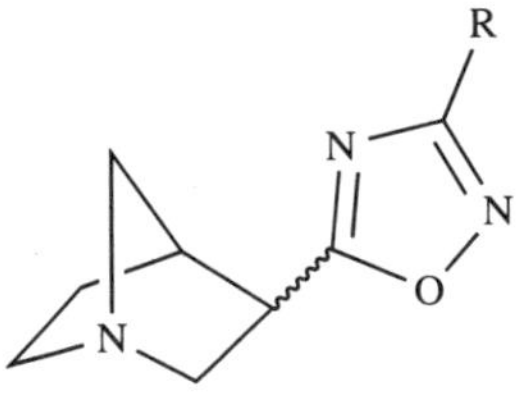

(83a) R = CH₃
(83b) R = NH₂

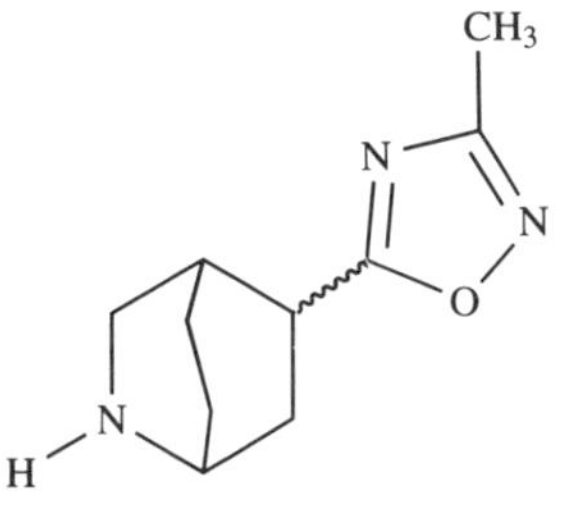

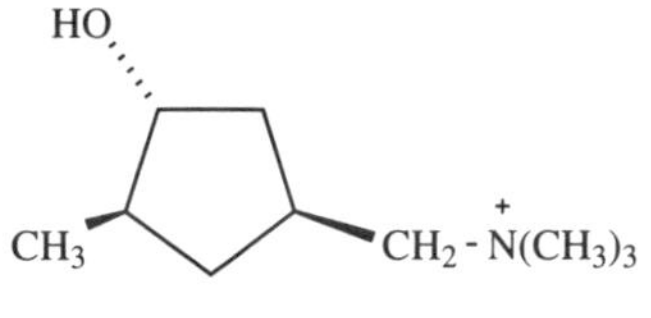

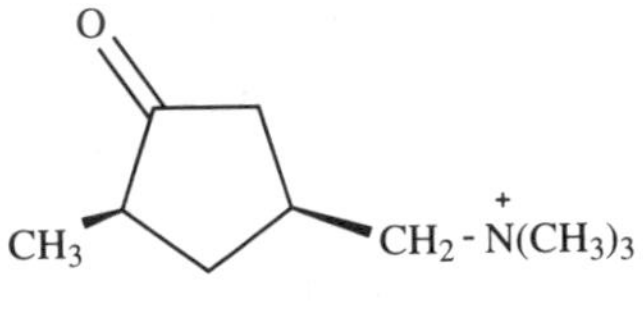

The other three geometric isomers of desethermuscarine (epi-, allo-, and epiallo-) are weaker cholinergic agents (111,112). Two attempts (109,113) to obtain (±)-desethermuscarone (86) resulted in inseparable mixtures of epimers, which were reported (109) to be equipotent to acetylcholine in assays for nicotinic and muscarinic effects. The high potencies of desethermuscarine (85) and of the epimeric mixture of desethermuscarone (86) suggest that the ring oxygens in muscarine and muscarone may not play a critical role in agaonist–muscarinic receptor interactions. Beckett and co-workers (97) had suggested prime importance of the keto group of muscarone (compared with the ring oxygen), and the importance of the ring C-methyl group has been cited previously.

Cyclohexane analogs of desethermuscarine and desethermuscarone show greatly diminished muscarinic activity (114). These compounds, however, lack the presumably important ring C-methyl group.

4.3 Pilocarpine and Analogs and Congeners

Pilocarpine (**87**), the chief alkaloid from the leaflets of shrubs of the genus *Pilocarpus*, has a dominant muscarinic action, but it causes anomalous cardiovascular responses, and the sweat glands are particularly sensitive to the drug (115).

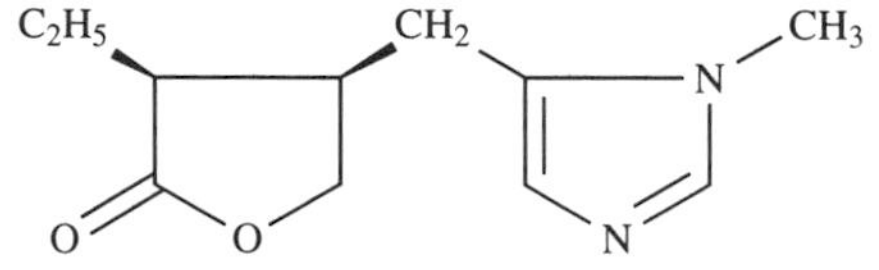

(**87**)

Its structure is distinguished by the lack of a quaternary ammonium head; however, it is presumed that a nitrogen-protonated cation is the biologically active species. Pilocarpine has been described as a muscarinic partial agonist (116). Structural and conformational analogies and interatomic distance similarities between pilocarpine and muscarinic agonists such as acetylcholine, acetyl β-methylcholine, muscarine, and muscarone have been invoked (117) to rationalize pilocarpine's pharmacologic properties. The potential utility of pilocarpine in treatment of glaucoma is coupled with its low ocular bioavailability. A double prodrug strategy (118,119) involves cleavage of the lactone ring of pilocarpine and esterification of the freed carboxyl and alcohol groups (**88**) to produce derivatives with a much greater lipophilic character. In the presence of human plasma or rabbit eye tissue homogenates, pilocarpine is formed from these derivatives in quantitative

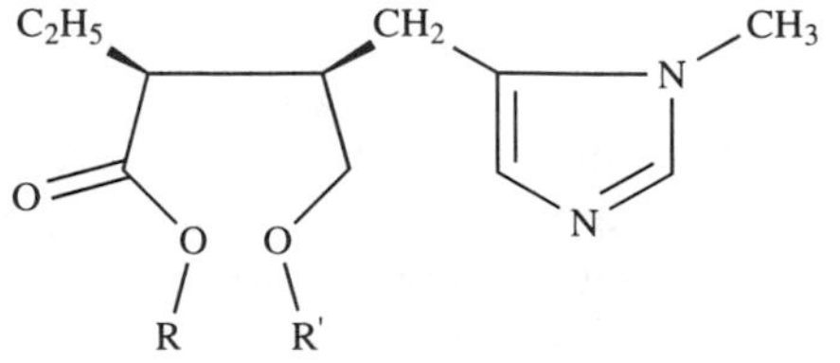

(**88**)

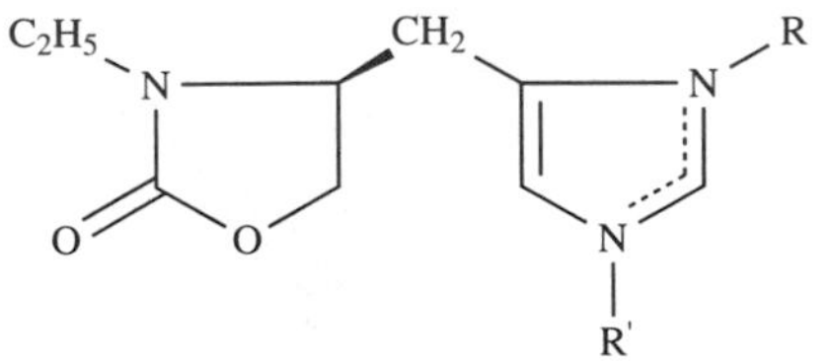

(**89**) R = CH$_3$; R' = H

(**90**) R = R' = H

(**91**) R = H ; R' = CH$_3$

amounts, due to the action of tissue esterases. Cyclic carbamate analogs (**89–91**) of pilocarpine were designed (120) to correct pilocarpine's short duration of action, due to its rapid metabolic inactivation by hydrolytic cleavage of the lactone ring. Analog (**89**), having the same substitution pattern as pilocarpine, was equipotent to pilocarpine in a guinea pig ileum assay. *In vitro*, base-catalyzed epimerization of pilocarpine at the *C*-ethyl group position forms the diastereometer isopilocarpine in which pharmacologic activity is lost (121).

4.4 Arecoline and Analogs and Congeners

Arecoline (**92a**), an alkaloidal constituent of the seeds of *Areca catechu*, is a cyclic "reverse ester" bioisostere of acetylcholine (cf. compound **20**).

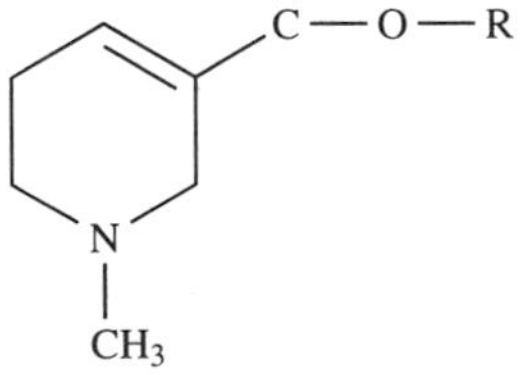

(**92a**) R = CH$_3$

(**92b**) R = C$_2$H$_5$

(**92c**) R = n–C$_3$H$_7$

(**92d**) R = CH$_2$–CH=CH$_2$

(**92e**) R = CH$_2$–C≡CH

(**92**)

In contrast to pilocarpine, arecoline acts at nicotinic receptors as well as at muscarinic sites; it has been described (122) as a partial agonist at M$_1$ and at M$_2$ receptors.

Arecoline is equipotent to its quaternary analog, *N*-methylarecoline, as a muscarinic agonist (122). The secondary amine, norarecoline, is a somewhat weaker muscarinic agonist than arecoline (123,124). The muscarinic activity of esters of arecaidine (the free carboxylic acid derivative of arecoline) varies with the nature of the alcohol: from the methyl ester (**92a**) to the ethyl (**92b**), the affinity for the muscarinic receptor(s) increases (125), but there is a sharp drop in affinity and intrinsic activity with the *n*-propyl ester (**92c**). However, the allyl ester (**92d**) is more potent and active than the *n*-propyl (although less potent than the standard compound, carbachol); the propargyl ester (**92e**) is more active than carbachol and indeed was described (125) as a more potent muscarinic agonist than acetylcholine. Data were presented suggesting that the triple bond of the propargyl ester contributes to receptor binding. Reduction of the ring double bond in the methyl (**92a**) (arecoline) and ethyl (**92b**) esters of arecaidine causes a 250- to 1000-fold reduction in muscarinic receptor affinity. The ester group positional isomer (**93**) is less potent and active than arecoline in the guinea pig ileum assay (126), and the *N,N*-dimethyl quaternary derivative of structure **93** is slightly more potent than the tertiary amine.

The five-membered ring congener (**94**) of arecoline is approximately 50% as muscarinically potent as arecoline in a guinea pig assay, and the five-membered ring ana-

(95)   (96)

log (**95**) of dihydroarecoline is approximately 1% as potent as arecoline in this assay (125). The sulfur bioisostere (**96**) of arecoline ($R = CH_3$) is more potent and active than its *N,N*-dimethyl quaternary ammonium congener, being approximately equipotent to arecoline itself (126). The ester group positional isomer of the sulfur bioisostere (**97**) ($R = CH_3$) retains muscarinic effects, but it is somewhat less potent and active than the 3-substituted compound (**96**). In both structure **96** and structure **97** the ethyl and *n*-propyl esters are inferior to the methyl.

Further appliction of the bioisosteric-replacement strategy substituted the tetrahydropyridine ring of arecoline with a tetrahydropyrimidine moiety (**98a–98f**) (127).

(97)

It was proposed that the amidine moiety of the tetrahydropyrimidine ring would be a suitable ammonium bioisostere, lacking the permanent cationic head present in a quaternary ammonium system. This might facilitate penetration of the blood–brain

(93)   (94)

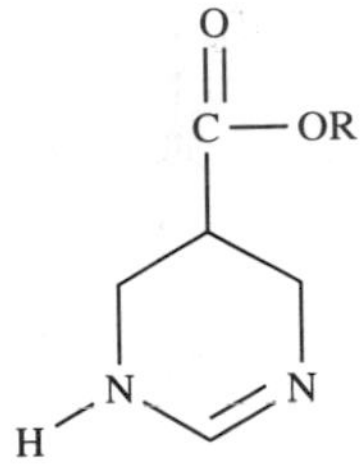

(98a) R = CH$_3$
(98b) R = C$_2$H$_5$
(98c) R = n-C$_3$H$_7$
(98d) R = 2-C$_3$H$_7$
(98e) R = CH$_2$-C≡CH
(98f) R = CH$_2$–C$_6$H$_5$

(99)   (100)

barrier. Of this series, the methyl ester (**98a**) shows high affinity for muscarinic receptors in rat brain, and it stimulates phosphoinositide metabolism in the rat hippocampus. It ameliorated memory deficits associated with lesions in the septohippocampal cholinergic system in rats.

The potential utility of arecoline in producing significant cognitive improvement in Alzheimer's patients (128) and in enhancing learning in normal young humans and in aged nonhuman primates (129,130) is largely negated by its short duration of action, which has been ascribed (131) to rapid *in vivo* hydrolysis of the ester group. This metabolic lability stimulated preparation of metabolically stable aldoxime derivatives (**99**) (132); compounds were prepared in which R and R' = alkyl, cycloalkyl, olefinic, or acetylenic groups. Derivatives of structure **99** in which R = CH$_3$ and R' = CH$_3$ or propargyl are muscarinic agonists both *in vitro* and *in vivo*. These compounds are two to three orders of magnitude more potent than arecoline, and they have a longer duration of action. They are orally effective. A large number of carbamate derivatives (**100**), where R = aryl or alkyl, were prepared as possible prodrugs of the aldoximes (**99**). The derivative where R = p-chlorophenyl had high potency and activity (133); it was more active in an assay for CNS effect than in an assay for peripheral cholinergic effect. Thus this compound was proposed to demonstrate a separation of CNS from peripheral effects. Structural variations based on structure **101** in which R' = CH$_3$ and the R group is such that the carbonyl

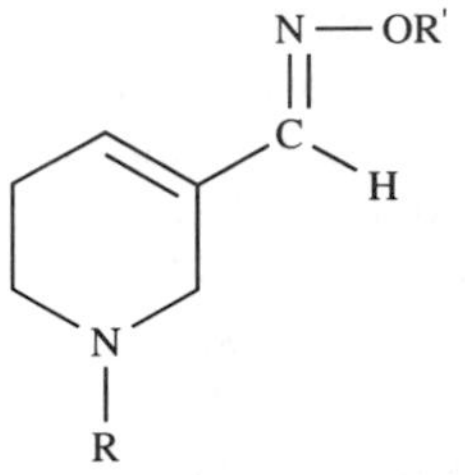

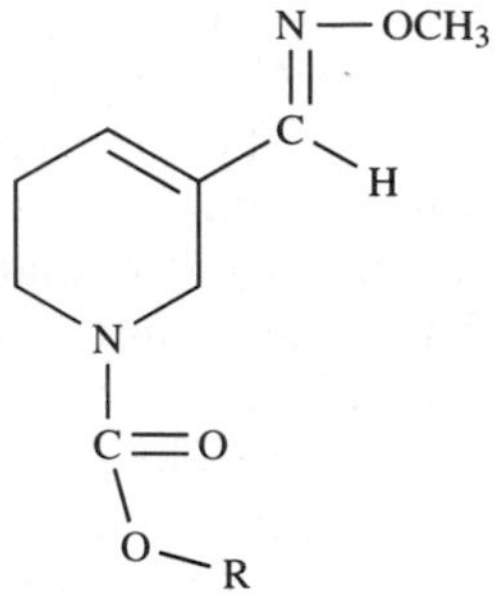

moiety is a part of a carbonate, carbamate, or carboxylate ester provided some compounds having muscarinic activity. Some of the carbamate derivatives also show *in vitro* cholinesterase inhibitory actions.

Both enantiomers of the reversed ester congener (**102**) of arecoline are (approximately equally) weak muscarinic agonists (125), as are their N-methyl quaternary derivatives.

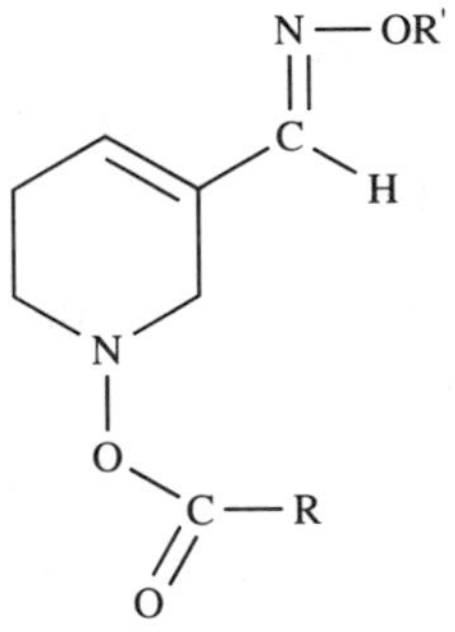

(101) 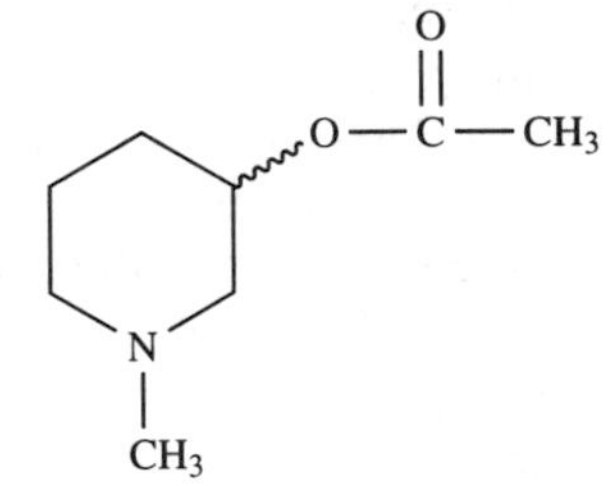   (102)

In contrast, both enantiomers of 3-acetoxyquinuclidine (**103**) are potent muscarinics, and the S-enantiomer is only approximately one order of magnitude less potent than acetylcholine. All four isomers of the thianium system (**104**) have been prepared and studied (125). The sulfur atom in sulfonium salts may form a chiral center, and steroisomers can be isolated, since the energy barrier to pyramidal inversion is substantially higher (ca. 100 kJ/mol) than it is in the case of the corresponding am-

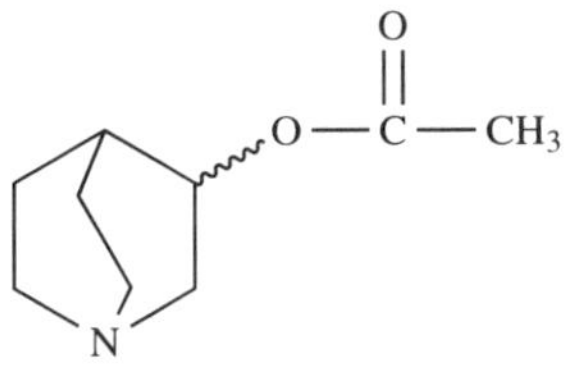

(103)

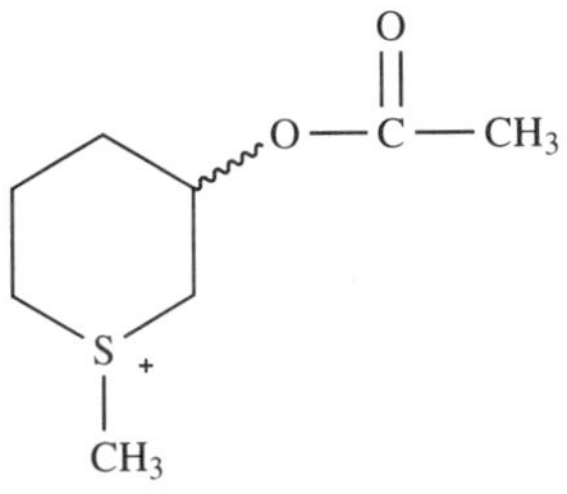

(104)

monium compounds (ca. 38 kJ/mol). The (+)-*trans*-thianium isomer (**104**) demonstrated high muscarinic potency, slightly greater than that of the *S*-quinuclidine (**103**), but the (−)-*trans*-enantiomer (**104**) and the (±)-*cis*-isomer (**104**) demonstrated low potency. These data on the piperidine, quinuclidine, and thianium derivatives were rationalized (125) on conformational bases.

In an alternate strategy to provide resistance to *in vivo* ester cleavage of arecoline (134), bioisosteric replacement of the methyl ester groups of arecoline and norarecoline by a 3-alkyl-1,2,4-oxadiazole ring (**105**) was investigated (135).

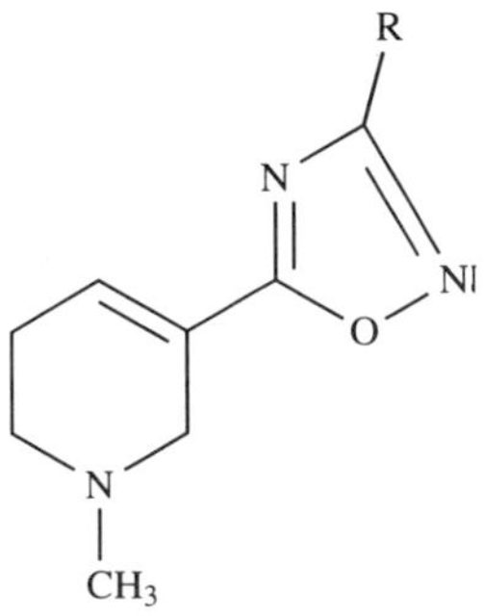

(105)

Analogs of structure **105**, where $R =$ unbranched C_{1-8} alkyl, are muscarinic agonists, and most show strong affinity in two binding assays in rat brain membranes. Derivatives of structure **105**, in which $R =$ a branched alkyl chain or a cyclic system, are muscarinic antagonists. Analogs in which the R group contains an ether moiety (e.g., CH_2-O-CH_3) are also muscarinic agonists, but they have lower receptor binding affinity than the alkyl derivatives. Congeners of structure **105**, in which the 1,2,5,6-tetrahydropyridine ring was replaced by quinuclidine or tropane, are potent antagonists with high affinity for central muscarinic receptors. Introduction of a methyl substituent at position 5 or 6 of the tetrahydropyridine ring of structure **105** ($R = n$-C_4H_9) destroyed agonist effect and produced a muscarinic antagonist in the single example reported. The *N*-desmethyl analog of structure **105**, where $R = n$-butyl, retains potent muscarinic agonism. Additional members of the series (**105**) have been reported (136), and molecular mechanics calculations indicated a preference for the *E* rotameric form (**106**).

In continuation of efforts to identify M_1 selective muscarinic agonists capable of crossing the blood–brain barrier, the 3-carbomethoxy group of arecoline was replaced by bioisoteric 1,2,5-oxadiazole (**108**) or by 1,2,5-thiadiazole rings, with oxygen ether substituents at position 3 (**109**) or with thioether substituents at position 3 (**110**) (137).

The ring-oxygen bioisosteres (**108**) ($R = n$-butyl or *n*-hexyl) show low affinity for central muscarinic receptors. However, all members of the thiadiazole oxygen ether series (**109**), where R varied from CH_3 through n-C_8H_{15} and also included some branched chain C_6 alkyl groups, demonstrate high potency in displacing tritiated oxotremorine-M (a nonselective muscarinic agonist) and tritiated pirenzepine (a selective M_1 antagonist) from rat brain membrane tissue. The *n*-butyloxy and *n*-

(106)   (107)

E Rotamer   *Z* Rotamer

pentyloxy substituents provide maximal pharmacological effects. The alkylthio analogs (**110**) demonstrate a similar structure–activity relationship to the alkoxy series (**109**); however, the thio ethers have higher receptor affinity and higher potency. Thus these systems (**109** and **110**) show a higher degree of selectivity for M_1 receptors than for M_2. The unsubstituted system (**111**) (R = H) is a potent but nonselective muscarinic agonist. Derivatives of structure **111**, where R = *n*-propyl, *n*-pentyl, *n*-heptyl, or *n*-octyl, have 10 to 100 times less affinity for central muscarinic receptors than the corresponding alkoxy and alkylthio derivatives.

Study of a series of arecoline congeners (**112**) in which the carbomethoxy group is replaced by a pyrazine moiety (R = CH_3 – *n*-C_7H_{15}) reveals that M_1 agonist activity is related to chain length, with *n*-hexyl providing maximum activity (138).

A comparison of M_1 agonist efficacy of these pyrazines and related 1,2,5-thiadiazoles (**109**) and 1,2,5-oxadiazoles (**108**) suggested that M_1 efficacy may be related to the magnitude of electrostatic potential located over the nitrogens of the respective heterocyles. The heteroatom directly attached to the 3 position of the pyrazine or the 1,2,5-thiadiazole markedly influences the M_1 efficacy of the compounds by determining the energetically favorable conformers for rotation about the bond connecting the tetrahydropyridyl ring and the heterocycle. A three-dimensional model for the M_1 agonist pharmacophore was proposed as a result of these studies.

Fusion of the tetrahydropyridine ring with a 3-alkoxyisoxazole moiety (**113a**–**113g**) was accomplished (Table 25.3) (124). Both **113a** and **113b** are muscarinic agonists in a guinea pig ileum assay, and as with arecoline and norarecoline, the *N*-methyl

(108)   (109)

(110)   **(111)**   **(112)**

tertiary amine congener **113a** is somewhat more potent than the nor derivative **113b**. Variation of the *O*-alkyl (R′) group in the nor series (**113b–113g**) produces pharmacological activities and potencies parallel to those described for the esters of arecaidine (**92a–92e**): the *O-n*-propyl and *n*-butyl homologs **113d** and **113e**, respectively, demonstrate only weak muscarinic agonist effect; however, the *O*-allyl and propargyl homologs display prominent muscarinic agonism, with the *O*-propargyl compound being one order of magnitude more potent than the *O*-allyl. However, as illustrated in

Table 25.3 *In vitro* **Muscarinic Effects of Isoxazole Bioisosteres of Arecoline and Norarecoline**[a]

(113)

Number	R	R′	Binding to Rat Brain Membranes *In Vitro* (IC$_{50}$, μM[b])	Muscarinic Agonism (EC$_{50}$, μM[c])
Carbachol (number 19 in Table 25.1)			9.0	0.07
Arecoline (**92a**)			5.6	0.1
Norarecoline			30	0.3
(**113a**)	CH$_3$	CH$_3$	27	0.8
(**113b**)	H	CH$_3$	45	1.8
(**113c**)	H	C$_2$H$_5$	6.0	3.4
(**113d**)	H	*n*-C$_3$H$_7$	10	>50
(**113e**)	H	*n*-C$_4$H$_9$	15	>50
(**113f**)	H	CH$_2$–CH=CH$_2$	9.6	13
(**113g**)	H	CH$_2$–C≡CH	5.8	1.2

[a]Adapted from ref. 124. Courtesy of the American Chemical Society.
[b]Inhibition of binding of [^{3}H]-*N*-propylbenziloylcholine mustard.
[c]Guinea pig ileum, *in vitro*.

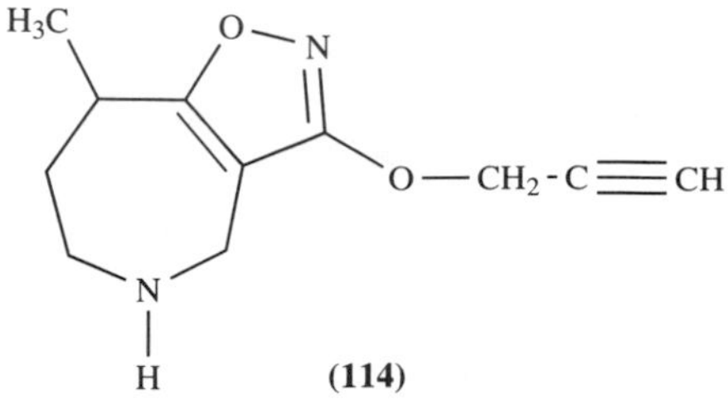

(114)

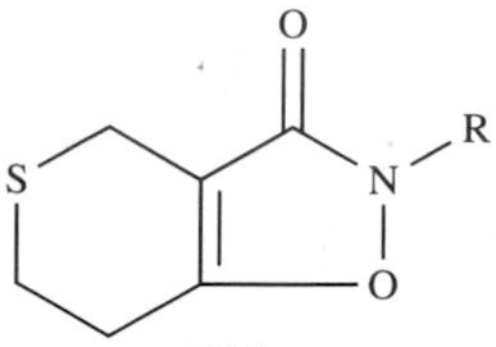

(116)

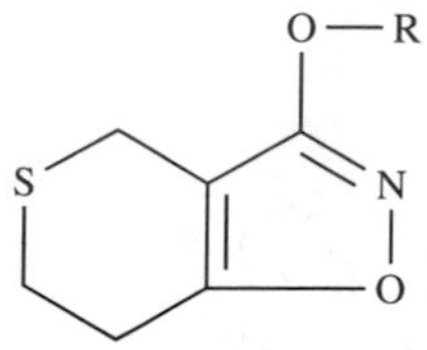

(117)

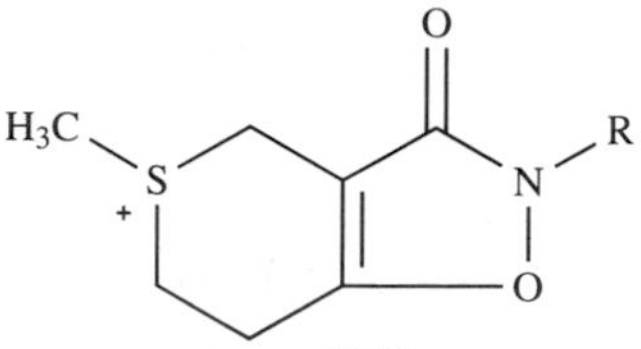

(118)

a R = CH₃   **b** R = C₂H₅
c R = CH(CH₃)₂   **d** R = CH₂-C≡CH

Table 25.3, there is no correlation between the effects of the compounds on central and peripheral cholinergic receptors. It was speculated (124) that the effects observed in the ileum preparation are mediated primarily by M₂ receptors, whereas the rat brain membrane-binding data may represent a nondescriminate binding to all types of muscarinic binding sites. The tetrahydroazepine congener (114) of the tetrahydropyridine isoxazole systems (113) is said to possess high affinity for the central M₁ receptor, coupled with only limited toxicity (139).

Another tetrahydroazepine congener (115) displays higher affinity for muscarinic receptors but somewhat lower efficacy than the analogous fused piperidine compounds (113) (5); it was described as a partial agonist. Studies (140) of sulfur analogs and congeners (116–119) of the isoxazolotetrahydropyridines (113) demonstrated that the thiopyran derivatives (116a and 117a) are inactive as muscarinic agonists.

However, the S-methyl sulfonium derivative (119a) binds to brain and heart muscarinic receptors, albeit not as strongly as arecoline. Compound 119a is also inferior in potency and activity to "sulfoarecoline" (96). However, the numerical-

ly large ratio of agonist activity at M₁ receptors to that at M₂ receptors for structure 119a is slightly greater than for arecoline or sulfoarecoline, which is a desirable parameter for therapy of Alzheimer's disease. The O-ethyl homolog (119b) is a muscarinic antagonist, in contrast to its tetrahydropyridine bioisostere (113c), which is described as a muscarinic partial agonist (140). The S-methylsulfonium homologs (119c and 119d) demonstrate pharamacological properties similar to those of the O-methyl homolog (119a). The single lactam derivative tested (118a) is a weak muscarinic agonist in all assays, but it demonstrates a decided preference for M₁ receptors over M₂ receptors.

Further modification of the arecoline structure involved replacement of the ester group of tetrahydropyrimidine derivatives (as in structure 98a–98f) with a 1,2,4-oxadiazole ring (structures 120a–120h) (141).

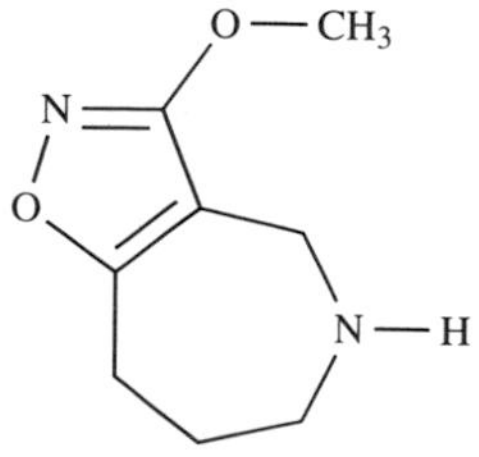

(115)

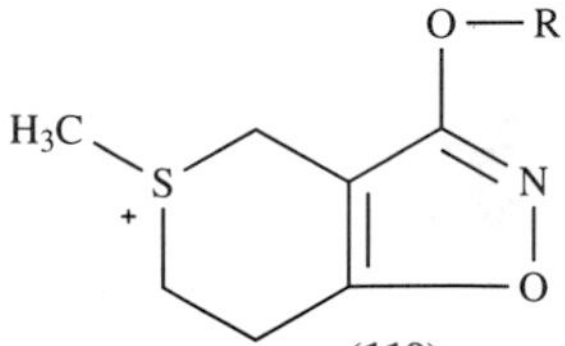

(119)

a R = CH₃   **b** R = C₂H₅
c R = CH(CH₃)₂   **d** R = CH₂-C≡CH

(120a) R = CH$_3$
(120b) R = C$_2$H$_5$
(120c) R = n - C$_3$H$_7$
(120d) R = n - C$_4$H$_9$
(120e) R = n - C$_5$H$_{11}$
(120f) R = n - C$_6$H$_{13}$
(120g) R = n - C$_7$H$_{15}$
(120h) R = n - C$_8$H$_{17}$

(122a) R = C$_2$H$_5$
(122b) R = n - C$_3$H$_7$
(122c) R = CH$_2$-C≡CH

(123a) R = CH$_3$
(123b) R = C$_2$H$_5$
(123c) R = n - C$_3$H$_7$
(123d) R = CH$_2$-C≡CH
(123e) R = 2-C$_3$H$_7$

Each of the test compounds **120a–120h** binds with high affinity to muscarinic receptors from rat brain. The 3-methyl homolog (**120a**) displays high efficacy at muscarinic receptors coupled to phosphoinositide metabolism in the rat cortex and hippocampus. Increasing the length of the alkyl substituent (**120b–120h**) increases affinity for muscarinic receptors, albeit not in a linear fashion, yet decreases activity in the phosphoinositide turnover assay. It was concluded that at low concentrations, compound **120a** selectively stimulates M$_1$ receptors.

Regioisomers of an exocyclic amidine system bioisosteric with arecoline (**121a–121d**) were tested as their racemic modifications (142).

Only the 5-carbomethoxy isomer (**121c**) displays high affinity and activity at muscarinic receptors coupled to phosphoinositide metabolism in rat cortex. Evaluation of other alkyl esters (**122a–122c**) of the 5-carboxylic acid revealed that only the propargyl derivative (**122c**) retains

substantial agonist activity. In a series of cyclic guanidines (2-aminotetrahydropyrimidines, **123a–123e**), all members show high binding affinities in a rat brain membrane assay (142). However, only the methyl and propargyl esters (**123a** and **123d**) show high muscarinic agonist activity in the phosphoinositide metabolism assay. Computational chemical studies revealed a common minimum energy conformation for all of the muscarinically active members of the series (**121**, **122**, **98**, and **123**), which suggests that all of the subject compounds in this study interact with muscarinic receptors in a similar fashion. The utility of amidine systems as suitable replacements for the quaternary ammonium group in acetylcholine in developing ligands for M$_1$ receptors is supported by these studies.

The spiropiperidine systems (**124**, diastereomers, stereochemistry unspecified, and **125**) are hybrids of the arecoline molecule and the spirodioxolanes (**80** and **81**). Compounds **124a** and **124b** (the spiro-3-piperidine series) show weak binding ability in rat cortex (143); compounds **125a** and **125b** (the spiro-4-piperidine derivatives) show marked muscarinic agonist effects in a phosphatidyl inositol turnover assay. Structures **125a** and **125b** also show moderate binding ability against [³H]-N-methylscopolamine and [³H]oxotremorine in rat cortex. Compound **125a** compares favorably

(121a) R = 3-COOCH$_3$
(121b) R = 4-COOCH$_3$
(121c) R = 5-COOCH$_3$
(121d) R = 6-COOCH$_3$

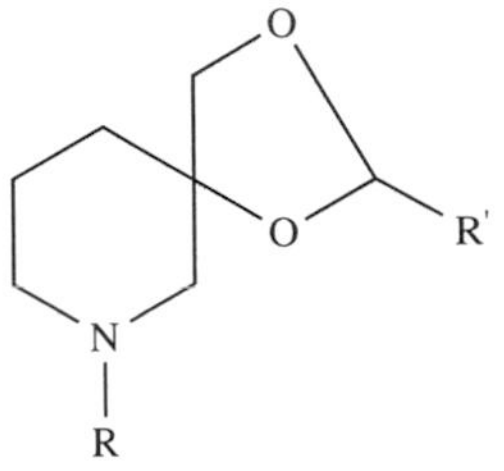

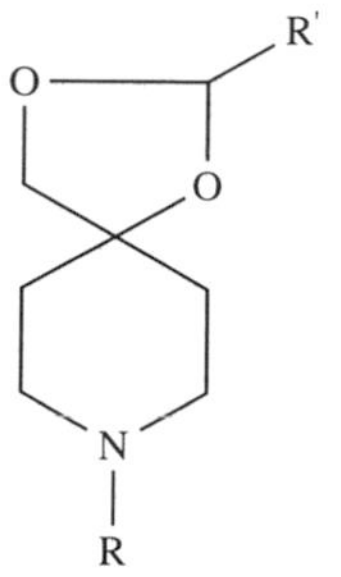

(124a) R = R' = CH₃
(124b) R = H ; R' = CH₃

(125a) R = R' = CH₃
(125b) R = H ; R' = CH₃
(125c) R = CH₃ ; R' = C₂H₅

(130)   (131)

(132) n = 1 or 2

with arecoline in terms of receptor efficacy, although it, like compound **125b**, demonstrates much lower receptor affinity. The 2-ethyl homolog (**125c**) demonstrates decidedly lower receptor efficacy than structures **124a** or **124b**, and it also demonstrates lower receptor affinity. Variations of structure **125** are represented by structures **126** and **127**, which are agonists at rat central M_1 receptors (144). Studies of extended series of other spiropiperidine derivatives demonstrated that structure **128** is a partial muscarinic agonist that reverses carbon dioxide–induced impairment in mice (145), and that structure **129** is a muscarinic agonist with affinity for cortical M_1 receptors (146). The (±)-spiroquinuclidine derivative (**130**) is a selective M_1 agonist (12).

Some quinuclidine and azanorbornane derivatives bearing oximino or heterocyclic

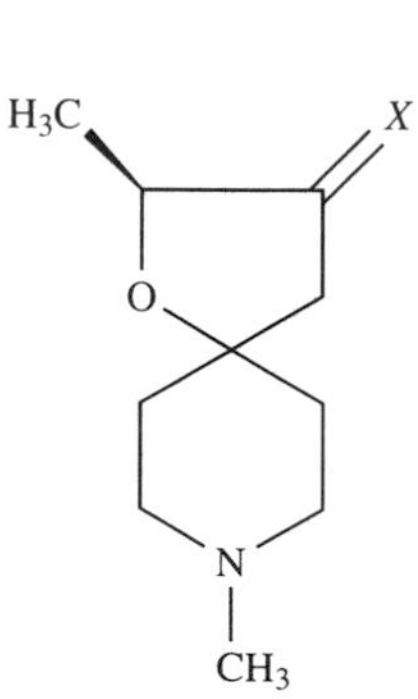

ring substituents at position 3 (e.g., **131** and **132**) demonstrate potent muscarinic agonism. Some compounds of these types are selective for M_2 and M_3 receptors (147–151).

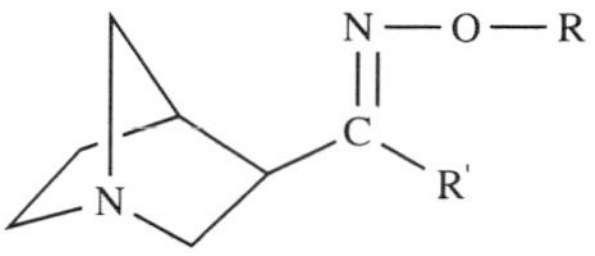

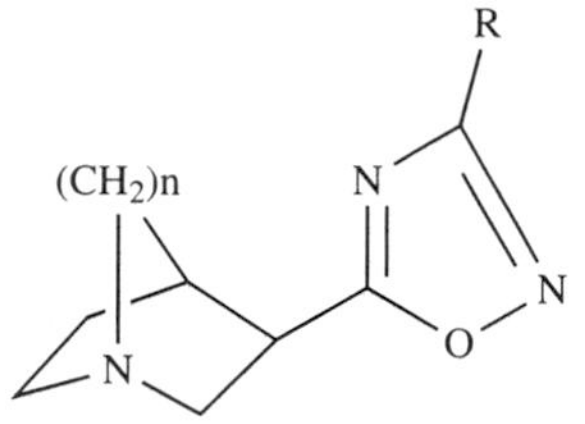

4.5 Oxotremorine and Analogs and Congeners

Tremorine (**133**), a synthetic compound with weak cholinergic activity (152,153) is metabolized to the lactam oxotremorine (**134**), which is approximately equipotent to acetylcholine as a muscarinic agent but lacks nicotinic effects (154).

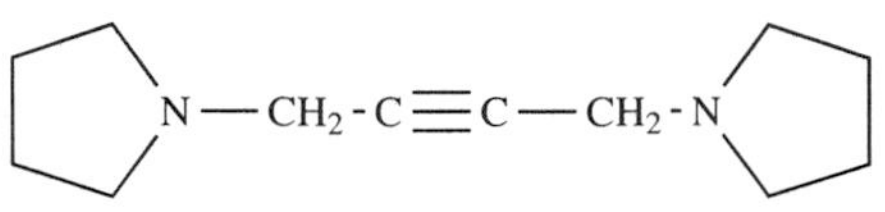

(126) X = O
(127) X = CH₂

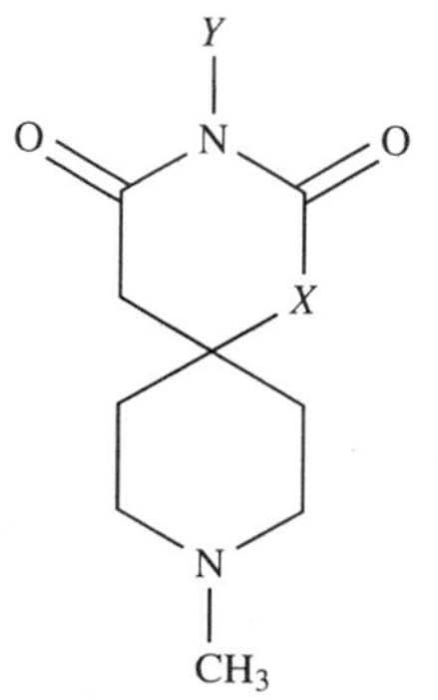

(128) X = CH₂ ; Y = OCH₃
(129) X = O ; Y = C₂H₅

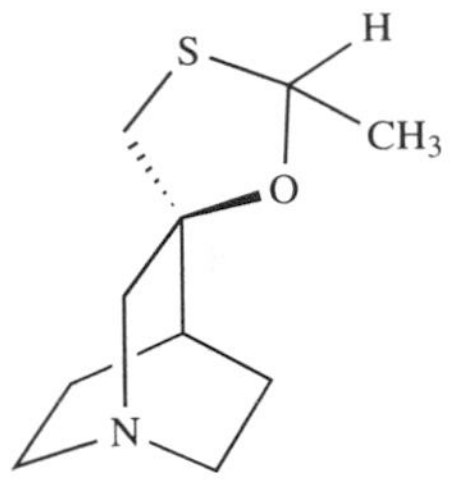

(133)

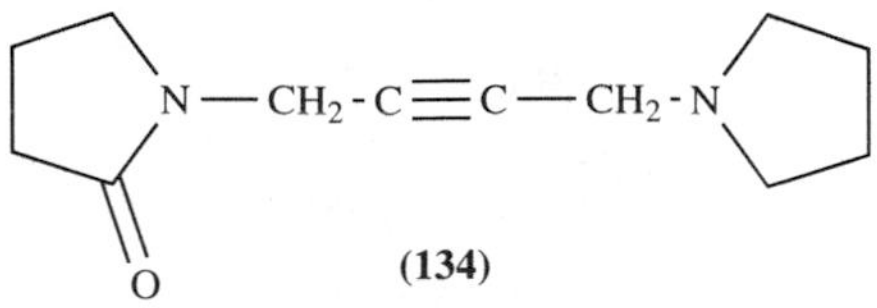

(134)

As with pilocarpine and arecoline, increased interest in pharmacotherapy of Alzheimer's disease and other deficits in memory has led to renewed and expanded studies of oxotremorine. This compound has little or no effect on serum or red cell cholinesterase. Oxotremorine has been described as a potent partial muscarinic agonist; its peripheral actions, including effects on cardiovascular mechanisms, have been ascribed (122) to preferential activation of M_2 receptors. Brimblecombe (152,153) reported pharmacological data on a large number of tremorine–oxotremorine derivatives, some of which are listed in Table 25.4, and from which the following conclusions were drawn.

1. The carbonyl group of oxotremorine is essential, as evidenced by comparison of oxotremorine (134) with tremorine (133).

2. The pyrrolidine nitrogen is a satisfactory replacement for a trimethylammonium quaternary group, resulting in only a small loss of activity (cf. structures 134 and 135).

3. The acetylenic bond is essential; partial or complete saturation results in complete loss of activity.

4. Replacement of the pyrrolidine ring with dimethylamino or diethylamino results in partial or complete loss of activity.

5. Increase in the size of the lactam ring results in a change from agonism to antagonism.

To assess the validity of item 2 above, Brimblecombe (152) compared the trimethylammonium quaternary moiety with the tertiary amino N-substituted pyrrolidine group (Table 25.5). Variations are apparent in the activities of the quaternary ammonium salts, but the variation is not nearly as great as that demonstrated by the tertiary amines (pyrrolidines). The Brimblecombe group (155) described an additional series

of some 23 tremorine–oxotremorine congeners, but none of these compounds shows significant muscarinic agonist effects in the guinea pig ileum or the cat blood pressure assay.

In a series of compounds (142–144) in which the pyrrolidine ring of oxotremorine is replaced by imidazole, the parent compound (142) resembles oxotremorine in its muscarinic efficacy (156); addition of a methyl group to the imidazole ring (143) greatly decreases muscarinic activity (157), while addition of methyl groups to both imidazole and pyrrolidinone rings (144) produces a potent muscarinic antagonist (156).

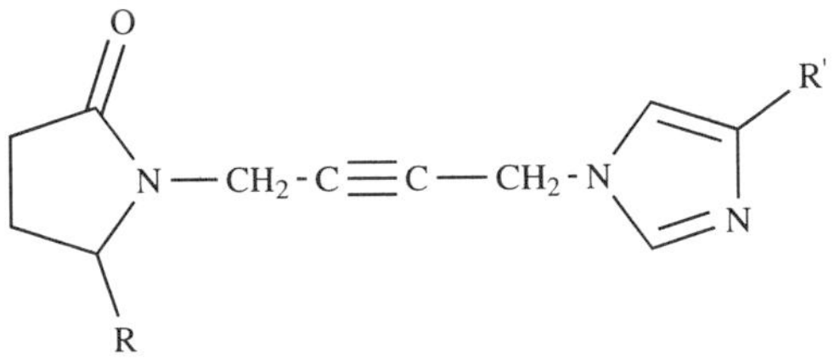

(142) R = R' = H
(143) R = H ; R' = CH$_3$
(144) R = R' = CH$_3$

A series of oxotremorine analogs (145a–145e), in which the pyrrolidone ring is contracted to a β-lactam moiety (158), reveals diminished potency.

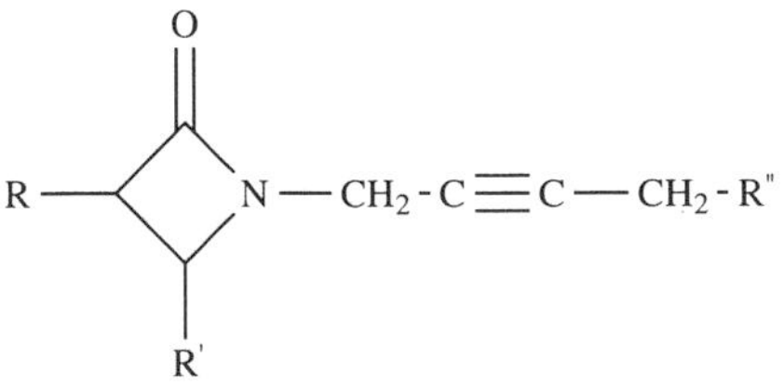

(145a) R = R' = H ; R" = NC$_4$H$_8$
(145b) R = CH$_3$; R' = H ; R" = NC$_4$H$_8$
(145c) R = H ; R' = CH$_3$; R" = NC$_4$H$_8$
(145d) R = H ; R' = CH$_3$; R" = N(CH$_3$)$_2$
(145e) R = H ; R' = CH$_3$; R" = $^+$N(CH$_3$)$_3$

Table 25.4 Muscarinic Actions of Oxotremorine-like Compounds[a]

Number	Structure	Muscarinic Activity (Isolated Guinea Pig Ileum, Acetylcholine = 1)
(134)	oxotremorine	1.48
(133)	tremorine	0.01
(135)("oxotremorine-M")	$N-CH_2-C\equiv C-CH_2-\overset{+}{N}(CH_3)_3$ (2-oxopyrrolidine)	1.74
(136)	$N-CH_2-C\equiv C-CH_2-N(CH_3)_2$ (2-oxopyrrolidine)	0.16
(137)	$N-CH_2-CH=CH-CH_2-N$ (2-oxopyrrolidine, pyrrolidine)	0.01
(138)	$N-CH_2-CH_2-CH_2-CH_2-N$ (2-oxopyrrolidine, pyrrolidine)	0.01
(139)	$N-CH_2-C\equiv C-CH_2-N(C_2H_5)_2$ (2-oxopyrrolidine)	0.01
(140)	$N-CH_2-C\equiv C-CH_2-N$ (2-oxopiperidine, pyrrolidine)	antagonist
(141)	$N-CH_2-C\equiv C-CH_2-N$ (succinimide, pyrrolidine)	0.06

[a]From ref. 152.

Table 25.5 **Muscarinic Activities of Oxotremorine Congeners and Analogs Containing Trimethylammonium or Pyrrolidino Groups**

$$R-CH_2-C{\equiv}C-CH_2-R'$$

Compound Number	R	Muscarinic Activity (Isolated Guinea Pig Ileum, Acetylcholine = 1)	
		$R' = \underset{+}{N}(CH_3)_3$	$R' = $ 1-pyrrolidino
1		1.74	1.48
2		0.11	0.03
3		0.06	0.001
4	$CH_3CON(CH_3)-$	0.87	1.15
5	$(CH_3)_2NCON(CH_3)-$	0.03	0.04
6	$(CH_3)_2NCOO-$	0.15	0.004
7	CH_3COCH_2-	2.10	0.005
8	CH_3COO-	0.47	0.005

Compound **145a** is the most potent muscarinic agonist of the series, being 6-fold less potent than its pyrrolidone congener (**134**). Compound **145b** is a weak partial muscarinic agonist; compound **145c** is a muscarinic antagonist; compounds **145d** and **145e** are muscarinic agonists, 220-fold less potent than compound **145a**. An *N*-acetylated piperidine derivative (**146**) is a potent muscarinic agonist (24). Conversion of the pyrrolidone ring of oxotremorine and some congeners into an imidazolidone ring (**147**) produces compounds with a variety of effects at muscarinic receptors, generally of low potency (159).

Studies (160,161) of methyl group substitution into the oxotremorine molecule, structures **148a–148g** (all of which were tested as the racemates), revealed that the 3′-methyl isomer (**148a**) has weak muscarinic stimulant activity in intact mice.

(**146**)

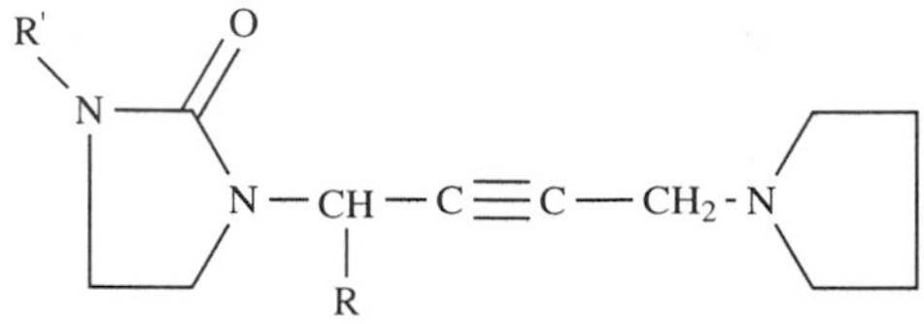

(147) R = H or CH_3 ; R' = H, CH_3, CH_3CO, or HCO

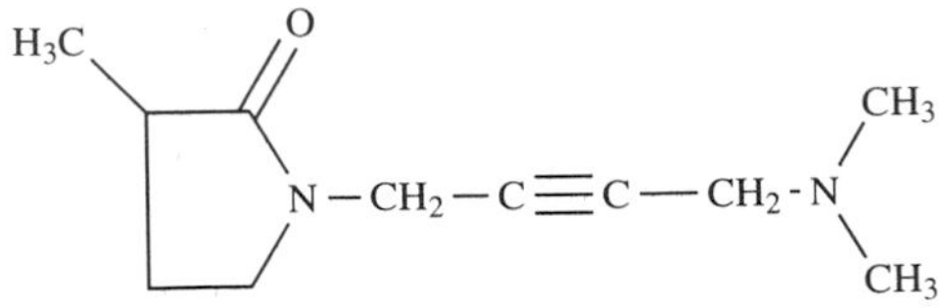

(149)

(150)

Similarly, the *N,N*-dimethyl congener (**149**) displays weak oxotremorine activity. The remaining isomers (**148b–148g**) have central atropine-like antimuscarinic effects but only weak peripheral parasympatholytic actions. Compounds **148d–148f**, as well as the C-1, C-2″ dimethyl derivative **150**, were resolved and the enantiomers were evaluated (162). Some compounds are agonists, some are partial agonists, and some are antagonists at muscarinic sites. In some instances, the eudismic ratio is large, and in some instances, the ratio is relatively small. It is difficult to draw structure–activity correlations from these data. Amstutz and co-workers (163) resolved the 5′-methyl pyrrolidine derivative (**148c**) as well as the *N,N*-dimethyl tertiary amine (**151a**) and its quaternary ammonium derivative (**151b**). The pyrrolidine derivative ((*R*)-**148c**) is a central and peripheral muscarinic antagonist, as had been reported earlier for the racemate. Compounds (*R*)-**151a** and -**151b**

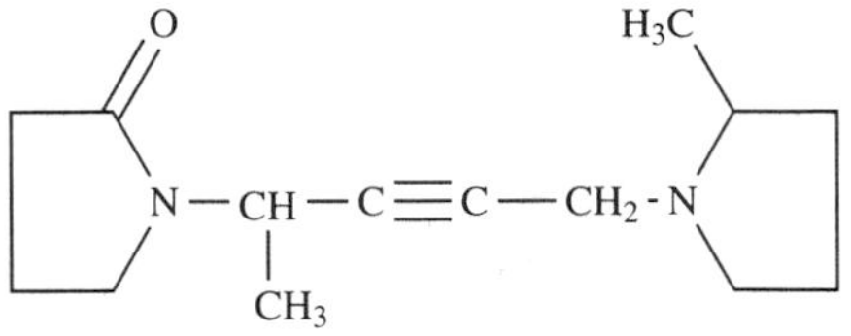

are potent muscarinic agonists in the guinea pig ileum. Compound (*R*)-**151a** (the tertiary amine) shows both central muscarinic (hypothermia) and central antimuscarinic activity (antagonism of oxotremorine-induced tremor) *in vivo*. These central agonist–antagonist properties may reflect interactions of the drug with different subpopulations of muscarinic receptors. For all three compounds in this study, the (*R*)-enantiomers are considerably more potent than the (*S*)-enantiomers, both *in vivo* and *in vitro*, irrespective of whether agonist or antagonist effects were measured. Substitution of a benzene ring into various positions of the oxotremorine molecule destroys muscarinic activity; these derivatives are competitive muscarinic antagonists (164). The *N*-methylacetamide congener (**152**), closely related to compound 4 in Table

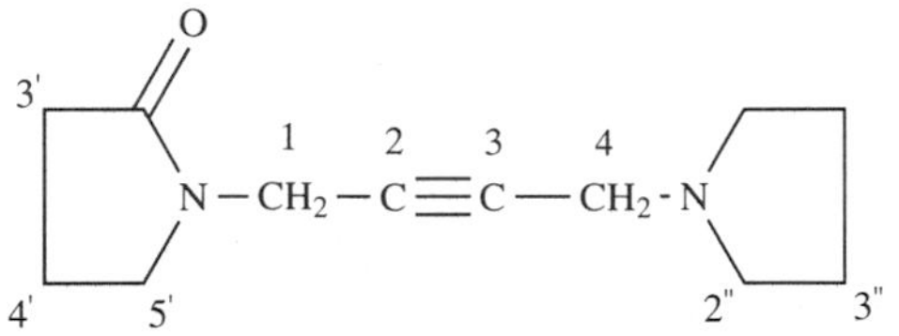

(**148a**) 3′ - CH_3
(**148b**) 4′ - CH_3
(**148c**) 5′ - CH_3
(**148d**) 3″ - CH_3
(**148e**) 2″ - CH_3
(**148f**) 1 - CH_3
(**148g**) 4 - CH_3

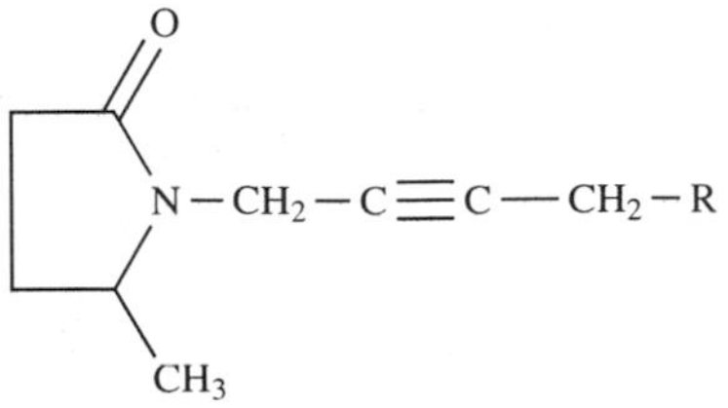

(**151a**) R = $N(CH_3)_2$
(**151b**) R = $^+N(CH_3)_3$

(152)

(154)

25.5, was reported (165,166) to be a pre-synaptic antagonist and a postsynaptic agonist at muscarinic receptors *in vitro* and *in vivo*.

Racemic **152** (167), as well as its pure enantiomers (168), blocks oxotremorine-induced tremors. In other regions of the brain, e.g., those involved in analgesia and hypothermia, compound **152** acts as a muscarinic agonist (169). A series of congeners (**153**) addressed replacement of the acetamide moiety of structure **152** by methanesulfonamido, trifluoroacetamido, methylsulfonimido, or acetimido; introduction of a methyl into the C-1 position of the 2-butyne chain; and variation of the C-4 tertiary amino or quaternary group (170). Replacement of the acetyl group or the *N*-methyl group in structure **152** and its analogs by a methanesulfonyl group abolishes efficacy and decreases affinity at guinea pig ileal receptors. The trifluoro-acetamide analogs of structure **152** also exhibit diminished affinity and efficacy. Substitution of an acetyl group for the *N*-methyl group of structure **152** decreases efficacy, but has little effect on affinity for the receptor(s). Most of the tertiary amines show central antimuscarinic effects. Bioisosteres of structure **152** bearing a urea

(153)

R = CH$_3$SO$_2$; CF$_3$CO; CH$_3$CO
R'= CH$_3$, CH$_3$SO$_2$, CH$_3$CO
R"= H, CH$_3$
R'"= 1 - pyrrolidyl, N(CH$_3$)$_2$, N(C$_2$H$_5$)$_2$, $^+$N(CH$_3$)$_3$

moiety (structure **154**) in which the R, R', and R" groups were combinations of CH$_3$ and H, display muscarinic agonism, partial agonism, or antagonism (159); a structure–activity relationship is not apparent in this series. Conformationally restricted analogs of compound **152** have been described (171) in which the acetyl or *N*-methyl substituent of the acetamide moiety is connected with the methyl substituent on the butynyl chain (structures **155a–155i**).

(155a) R = CH$_3$; X = O ; R' = NC$_4$H$_8$
(155b) R = CH$_3$; X = O ; R' = N(CH$_3$)$_2$
(155c) R = CH$_3$; X = O ; R' = $^+$N(CH$_3$)$_3$
(155d) R = CH$_3$CO ; X = O ; R' = NC$_4$H$_8$
(155e) R = CH$_3$CO ; X = O ; R' = N(CH$_3$)$_2$
(155f) R = CH$_3$CO ; X = O ; R' = NC$_4$H$_8$
(155g) R = CH$_3$CO ; X = H$_2$; R' = NC$_4$H$_8$
(155h) R = CH$_3$CO ; X = H$_2$; R' = N(CH$_3$)$_2$
(155i) R = CH$_3$CO ; X = H$_2$; R' = $^+$N(CH$_3$)$_3$

These structural modifications resulted in decreased affinity for rat cerebral cortex tissue and in most cases abolished efficacy at both central and peripheral muscarinic receptors. Other conformationally re-stricted analogs of compound **155** in which the amide moiety and the methyl group on the butynyl chain were joined to form a six- or seven-membered ring preserved affinity for muscarinic receptor(s), but abolished efficacy (172).

A nitrogen mustard congener (**156**) of oxotremorine is a potent and selective muscarinic agonist (173).

(156) X = Cl
(157) X = Br

(159)

When administered to an intact animal, the signs of muscarinic stimulation are followed by a phase of long-lasting antimuscarinic effects (174). Both the stimulatory and the blocking effects are elicited by the aziridinium ion (158) formed by *in vivo* cyclization of the parent 2-chloro-alkylamine. This aziridinium system is closely related structurally to "oxotre-morine-M" (compound 135), which is an extremely potent muscarinic agonist. The blocking activity is correlated with alkylation (covalent bond formation) of the muscarinic receptor(s) by the aziridinium ion. The bromine-derived mustard (157) shows threefold greater *in vitro* muscarinic stimulant activity than the chlorine compound (156), and this can be rationalized on the basis that bromine is a better leaving group, and aziridinium ion formation is more facile with structure 157 than with 156. Neither of these mustards displays a significant amount of effect at nicotinic receptors. These compounds may be useful in receptor inactivation studies. The slower rate of cyclization of the chloro compound (156) may permit its penetration of the blood–brain barrier before formation of the aziridinium ion. The racemic nitrogen mustard (159) (X = Cl or Br) is a derivative of structures 151a and 151b, for which potent muscarinic activity was described. Cl- and Br-structure 159 demonstrate similar phar-

macologic effects to the mustards 156 and 157 (175). Cl- and Br-159 display a higher potency than the *C*-desmethyl systems 156 and 157, and this was ascribed (175) to greater receptor affinity rather than to a greater rate constant for alkylation of the muscarinic receptors. Tertiary 3- and 4-haloalkylamine analogs of oxotremorine were investigated in mice as prodrugs of muscarinic agonists (176) (Table 25.6).

The azetidinium cyclization product of the 3-halopropylamine moiety (compounds 3–5 in Table 25.6) and the pyrrolidinium product of the 4-halobutylamine moiety of compound 6 in Table 25.6 should be much less susceptible to nucleophilic attack *in vivo*, and hence these quaternary systems, unlike the aziridinium moiety, should have little or no tendency to bond covalently with the muscarinic receptor(s) to produce blockade. Central muscarinic effects (tremors, analgesia) correlated well with the k_1 and $t_{1/2}$ data. The slow cyclization rate of compound 3 and the extremely rapid cyclization rate of compound 6 were reflected in weak or no CNS-related activities. As might be predicted, compound 3 showed relatively weak peripheral muscarinic activity (salivation) whereas compound 6 was potent. Compounds 4 and 5 (Table 25.6) were cited as meriting further study.

The older literature (177) described McN-A-343 (160) as a potent stimulant of muscarinic receptors in sympathetic ganglia but as only a weak stimulant of the heart and smooth muscles.

It was proposed that structure 160 is a selective M_1 agonist, but this explanation for the selectivity of action has been challenged (178). Of a series of congeners of

(158)

Table 25.6   Apparent First-order Rate Constants for Cyclization of ω-Haloalkyl Oxotremorine Congeners

Number	n	X	k_1, min^{-1}	$t_{1/2}$, min
1	2	Cl	0.019	36.5
2	2	Br	0.850 ± 0.035	0.8
3	3	Cl	0.0016 ± 0.0003	436
4	3	Br	0.0610 ± 0.0041	11.4
5	3	I	0.0491 ± 0.0055	14.1
6	4	Cl	>2	>0.4

compound **160**, structures ($\pm$)-**161** and ($\pm$)-**162** are muscarinic partial agonists, showing 5- and 16-fold higher potency than compound **160**, respectively (179). The S-enantiomers of structures **161** and **162** exhibit low eudismic ratios (1.5 and 4.9, respectively). These compounds retain selectivity for ganglionic muscarinic receptors (presumably M_1), and they show relatively weak nicotinic activity.

Dimethylsulfonium (**163a** and **163b**) and thiolanium (**163c** and **163d**) analogs of oxotremorine demonstrate higher affinities for peripheral muscarinic receptors than the corresponding trimethylammonium and N-methylpyrrolidinium compounds (180).

However, the sulfur compounds have lower intrinsic activities than their nitrogen analogs. The sulfur compounds also demonstrate potent affinity for rat cerebrocortical

(160)

(161) R = Cl ; R' = H
(162) R = H ; R' = Cl

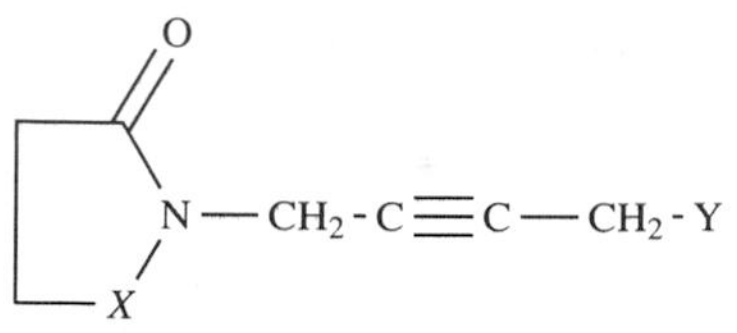

(163a) $X = CH_2$; $Y = {}^+S(CH_3)_2$

(163b) $X = CO$; $Y = {}^+S(CH_3)_2$

(163c) $X = CH_2$; $Y = {}^+SC_4H_8$

(163d) $X = CO$; $Y = {}^+SC_4H_8$

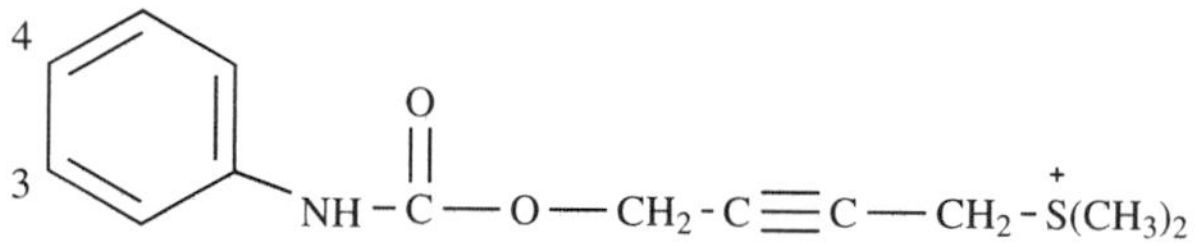

tissue in a $(-)$-$[{}^3H]$-*N*-methylscopolamine displacement assay. Sulfonium congeners (**164**) bearing chlorine or bromine at position 3 or 4 of the benzene ring retain selectivity for ganglionic muscarinic receptors, but they were concluded to be partial agonists when compared with the nitrogen system (**160**) (178). Nitrogen mustard congeners of structure **160** (structures **165a** and **165b**) demonstrate analogous effects to those described previously for other oxotremorine-based nitrogen mustards (181).

(164)

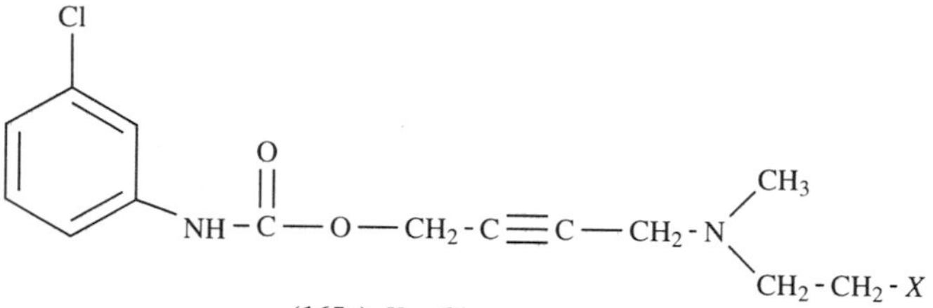

(165a) $X = Cl$
(165b) $X = Br$

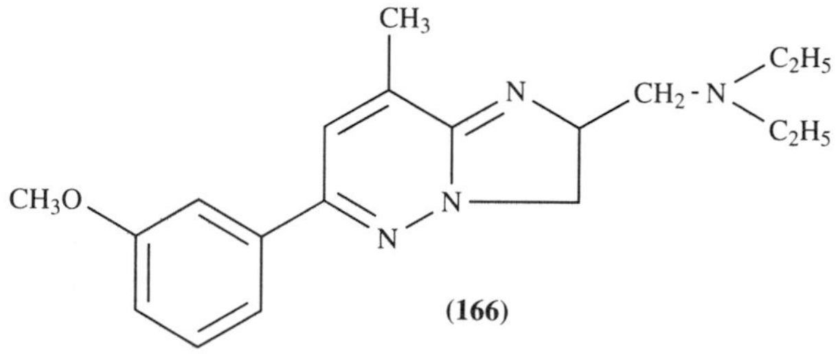

(166)

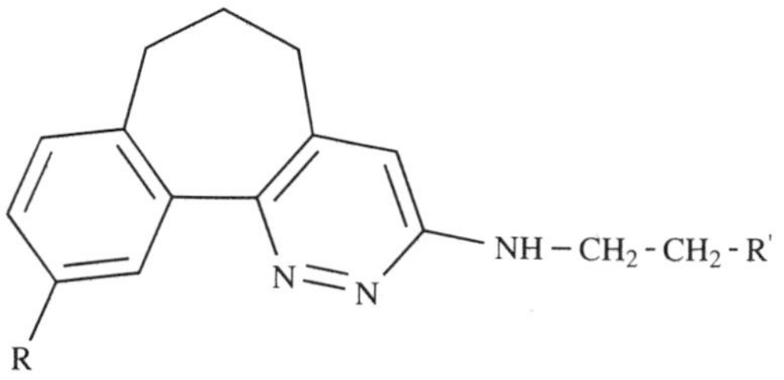

4.6 Miscellaneous, Structurally Unique Muscarinic Agonists

Several structurally-unique muscarinic agonists, typified by structures **166–168**, have been described.

Compound **166** is claimed to be highly selective for the M_1 receptor (139); structures **167** and **168** have been described as selective M_1 agonists devoid of classical muscarinic side effects (139,182).

(167) R = H ; R′ = morpholino
(168) R = NO_2 ; R′ = $N(C_2H_5)_2$

5 CONFORMATION-ACTIVITY RELATIONSHIPS OF CHOLINERGIC AGONISTS

Conclusions about pharmacologically significant conformations of cholinergic agonists have been largely based on x-ray diffraction studies and NMR (chiefly ^1H) studies. Casy (183) noted that in the cholinergic field, NMR evidence complements the results of x-ray studies. Casy (184) proposed four questions to which conformational studies of cholinergic agonists must be directed:

1. Does the "active" conformation of a cholinergic ligand correspond to its preferred stereochemistry or is an energetically less favored form bound to the receptor?

2. Is there a unique mode of ligand binding to cholinergic receptors or do multiple modes exist?

3. May the dual effects (nicotinic and muscarinic) of acetylcholine be explained in terms of conformational isomerism?

4. Do agonist and antagonist ligands occupy the same or different binding sites (with one or more features common to both)?

These challenging questions remain, some 20 years later, basically unanswered, despite a large body of chemical and biological work and a voluminous literature.

Indeed, the establishment of the existence of subpopulations of both muscarinic and nicotinic receptors renders question 3 even more formidable.

The three-dimensional steric disposition of acetylcholine and its congeners is defined on the basis of torsion angle τ. In a system X-C-C-Y, τ is defined as the angle between the plane containing the C-C and C-X bonds and the plane containing the C-C and C-Y bonds, and is illustrated with Newman projections (185) in Figure 25.1. The smaller rotation needed to make the front ligand eclipsed with the rear one is the torsion angle τ. If this rotation is clockwise, it is assigned a (+) sign, if counterclockwise, a (−) sign (185).

Structure **169** illustrates and defines relevant torsion angles in the acetylcholine molecule.

$$C^7 - C^6 - O^1 - C^5 - C^4 - \overset{\overset{\textstyle C^1}{|}}{\underset{\underset{\textstyle C^3}{|}}{N}} - C^2$$
$$\overset{\|}{O}$$

$$\tau_1 = C^5\text{-}C^4\text{-N-}C^3$$
$$\tau_2 = O^1\text{-}C^5\text{-}C^4\text{-N}$$
$$\tau_3 = C^6\text{-}O^1\text{-}C^5\text{-}C^4$$
$$\tau_4 = C^7\text{-}C^6\text{-}O^1\text{-}C^5$$

(169)

The torsion angles τ_1 and τ_4 usually fall close to 180°; the values of τ_2 and τ_3 are more useful in defining pharmacologically significant conformations for acetylcholine and related molecules. From x-ray studies (186), it was noted that in most cases, τ_3 values fall in the range $180 \pm 36°$ (antiplanar), placing the quaternary head and the

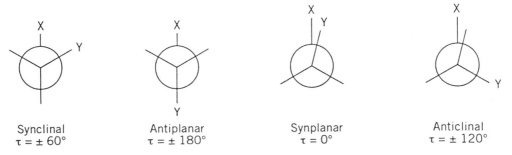

Synclinal	Antiplanar	Synplanar	Anticlinal
$\tau = \pm 60°$	$\tau = \pm 180°$	$\tau = 0°$	$\tau = \pm 120°$

Fig. 25.1 Definition of torsion angles (τ). From ref. 185.

acetyl group far apart, and the τ_2 angle commonly has a value of 73–94°, so that N and O functions are approximately synclinal (gauche). Many compounds with the O-C-C-N$^+$ moiety, where the oxygen function is hydroxy or acyloxy, prefer the τ_2 synclinal (gauche) N/O disposition in the solid state: L(+)-muscarine iodide (**2**), the (4R)-(+)-cis-dioxolane (**66**), and the furan derivative (**65**). NMR data suggested (187) that acetyl β-methylcholine (**29**) exists in solution in a τ_2 synclinal conformation. However, there are many exceptions: the potent muscarinic agonists carbamoylcholine (number 19 in Table 25.1) and (+)-trans-ACTM (**170**) (186) in the crystal state prefer the τ_2 anticlinal τ_3 antiplanar conformations, as do the weakly active thio and seleno analogs of acetylcholine (**22** and **23**). L-(+)-Muscarone (**62**) exhibits a τ_2 that is antiplanar (188).

(**170**)

The crystal structures of certain nicotinic agents, for example, acetyl α-methylcholine (**30**) and lactoylcholine (number 17 in Table 25.1), have torsion angle (τ_2 and τ_3) features similar to most muscarinic agents (189). In contrast, some cyclic analogs of aryl choline ethers exhibit maximum nicotinic effects when the τ_2 is antiplanar ("transoid") (82).

However, there is no assurance that any of the preferred conformations determined experimentally by x-ray or solution NMR methods, or by molecular orbital calculations (190) represent the geometry of the agonist at cholinergic receptors; barriers to rotation in molecules such as acetylcholine are low (191,192), and there is considerable rotational freedom in the muscarine and muscarone systems (184). It seems well established as a broadly applicable working hypothesis that an agonist molecule may interact with its receptor(s) in other than its (the agonist's) lowest energy conformation. The energy expended by the agonist's assuming a higher energy conformation is compensated for by the energy advantage of the agonist–receptor interaction itself. An alternate strategy is the study of conformationally restrained analogs in which the presumed pharmacologically significant portions of the molecule (e.g., the quaternary ammonium head and the ester function) are in relatively "frozen" positions, so that the three-dimensional geometry of the pharmacophoric moieties is known. This approach presents the disadvantage of requiring molecules that are larger and more structurally complex than the parent, with anticipated reduced receptor affinity and different solvent partition characteristics a likely consequence. Casy (193) has discussed this difficulty in more detail.

Schueler (194) suggested that the muscarinic and nicotinic effects of acetylcholine are mediated by different conformers of the flexible molecule, and he evaluated structures (±)-**171** ("transoid") and **172** ("cisoid") as examples of conformational extremes.

(**171**)   (**172**)

Both structure **171** and structure **172** exhibit only feeble cholinergic effects, and the difference in activity between the two is not great. The enantiomers of structure **171** are also far less potent than acetylcholine at muscarinic sites (195), but they have intrinsic activities (compared with acetylcholine) approaching unity. The *S*-enantiomer is somewhat more effective than the *R*-, demonstrating a *S*:*R* eudismic ratio of 13.4. However, because of the ability of the piperidine ring to undergo ring inversion (conformational flip) with concomitant change in the τ_2 torsion angle, the piperidine ring of structure **171** is not an ideal template for acetylcholine conformational studies. Casy (196) summarized literature conformational studies on structure **171**.

The *trans*-sulfur isostere (**173**) of the 3-acetoxypiperidinium system (**171**) is a potent muscarinic agonist (6–10% as potent as acetylcholine), but the cis isomer (**174**) is virtually inert (197).

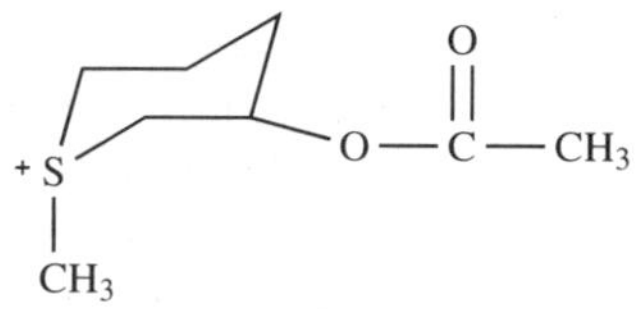

(173)

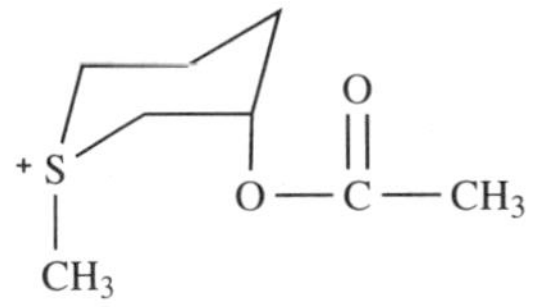

(174)

In contrast, both the cis and the trans isomers (**173** and **174**) display nicotinic action; the cis isomer is approximately 20% as potent as acetylcholine and is approximately seven times as potent as the trans. In the trans system (**173**), the τ_2 angle (S-C-C-O) is described as anticlinal–antiplanar on the basis that the chair conformer has an axial *S*-methyl and an equatorial acetoxy (197). However, the ability of the thianium ring to undergo ring inversion (125) presents the same difficulty that was cited for the piperidinium ring of structure **171**.

Quaternary tropyl acetates (**175** and **176**), in which a greater degree of molecu-

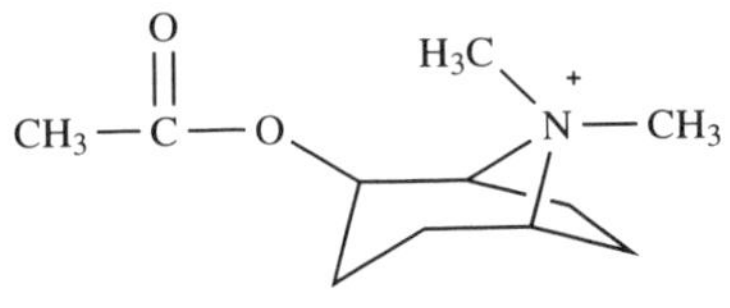

(175)

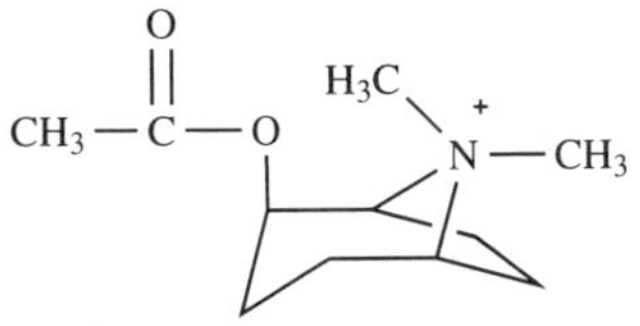

(176)

lar rigidity is imposed, are extremely weak muscarinics, but both exhibit potent nicotinic effects (198).

However, as demonstrated by Hardegger and Ott (199), the tropane ring is capable of assuming a conformation in which the piperidine ring portion is a boat. Therefore, the possibility of ring inversion in structures **175** and **176** cannot be precluded, and as with piperidine and thiane rings, conformational integrity is questionable. Stereoisomers of the *trans*-decahydroquinoline (**177**) (200) and the *trans*-decalin (**178**) (71) display extremely low orders of muscarinic effect, with the 2,3-*trans*-diaxial isomer of

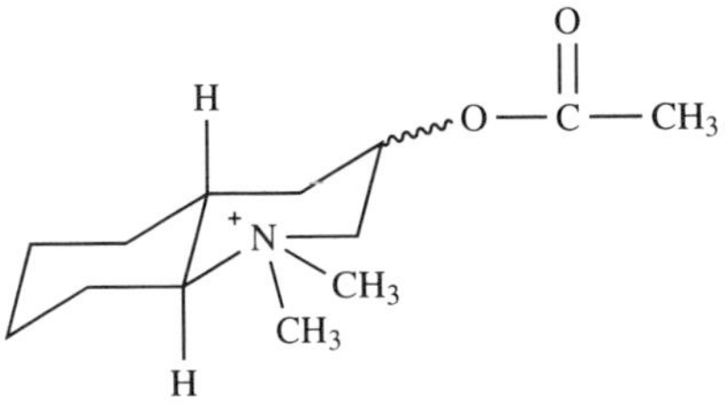

(177)

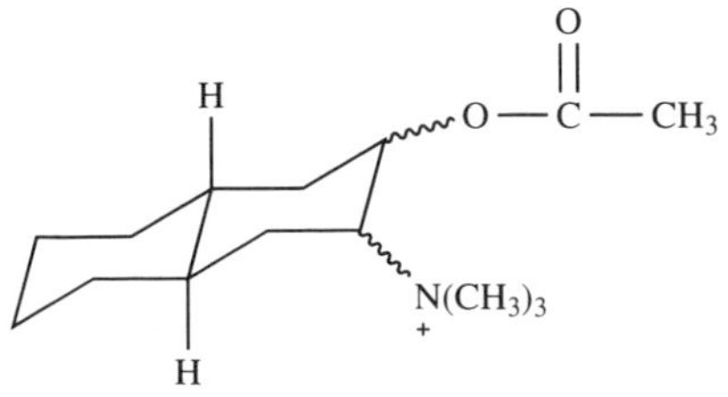

(178)

structure **178** being the most active of the four stereoisomers of this structure (0.06% the activity of acetylcholine). A possibly serious defect in the design of compounds such as the piperidinium (**171**), the morpholinium (**172**), the tropines (**175** and **176**), and the decahydroquinoliniums (**177**) is that these molecules do not bear a trimethylammonium cation characteristic of acetylcholine, but rather the quaternary head is a part of a ring system. It was indicated previously that incorporation of the nitrogen of acetylcholine itself into a ring is detrimental to cholinergic activity and potency. These compounds also illustrate a not infrequent observation that the imposition of conformational integrity by incorporation of a pharamcophore into a complicated molecule is frequently at the expense of pharmacologic activity and potency.

In a series of C-methylated *trans*-decalin congeners of structure **178**, the 2,3-dimethyl compound (**179**) is the most potent muscarinic, but it is only as 2% as active as acetylcholine (201).

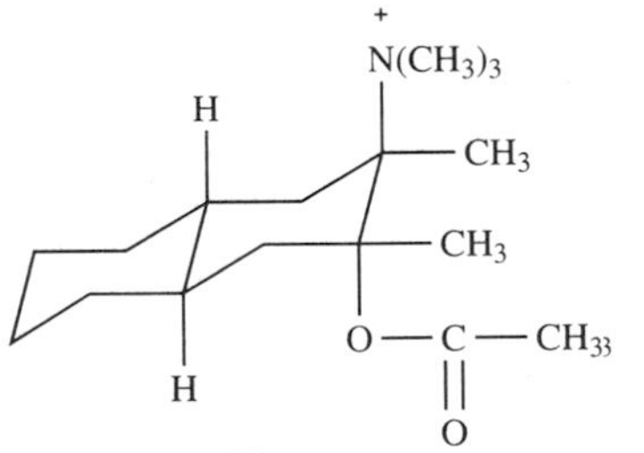

(179)

It is noted that all of these active decalin-derived molecules have antiplanar τ_2 stereochemistry. In the series of cyclohexane-derived compounds (**180**), the (1*R*, 2*R*)-(−)-trans system is an extremely weak muscarinic, and the (±)-cis isomer is completely inert (202). Casy (184) suggested that an unfavored *trans*-diaxial conformer (antiplanar τ_2) for structure **180** may be the pharmacologically active form of the molecule. However, inclusion of a *t*-butyl group

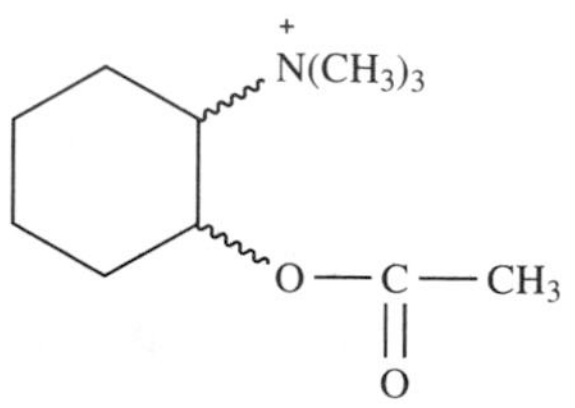

(180)

into the cyclohexane system to stabilize the 1,2-*trans*-diaxial geometry does not lead to greatly increased muscarinic effect (203). The *cis*- and *trans*-cyclopentane systems (**181**) have been described (184) as feeble spasmogenics with a τ_2 angle near anticlinal. Congeners of acetyl γ-homocholine (**27**) and of 4-acetoxybutyltrimethylammonium (**28**), in which the amino alcohol entity is a part of cyclopropane or cyclobutane ring systems, exhibit feeble muscarinic and appreciable nicotinic effects (204,205), and these compounds provide little insight into active conformations for acetylcholine. The low cholinergic potencies and activities of the compounds probably preclude their being used as a basis for acetylcholine conformation–activity hypotheses in cholinergic agonism, because "almost any compound with a quaternary nitrogen has some stimulant or inhibitory activity at cholinergic receptor sites" (206).

The cyclopropane ring has been exploited as the smallest system capable of conferring conformational rigidity on an acetylcholine analog (207,208); (+)-*trans*-ACTM (**170**) equals or surpasses acetylcholine's muscarinic potency in two test systems.

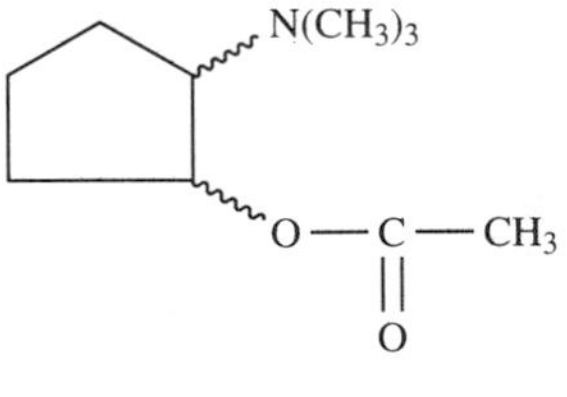

(181)

The (−)-trans enantiomer is several hundred times less potent, and the racemic cis system (182) is almost inert. All isomers are feeble nicotinics. X-ray analysis of (+)-*trans*-ACTM (170) (209) established the τ_2 angle as 137° (which is within the anticlinal range), and because of the rigidity of the cyclopropane ring, this value probably closely approaches the solution conformation. The (1*S*,2*S*) configuration of (+)-*trans*-ACTM superimposes on the equivalent centers in the potent muscarinic agonists (*S*)-(+)-acetyl-β-methylcholine (29) (see Table 25.2) and (2*S*,4*R*,5*S*)-(+)-muscarine (2) (210). A racemic cyclobutane analog (183) of *trans*-ACTM is much less potent than (±)-*trans*-ACTM (211). No conformational study on structure 183 has been reported, and inspection of molecular models does not reveal convincing structural differences between the cyclopropane and cyclobutane systems. The racemic *cis*-cyclobutane isomer (184) is equipotent to the racemic *trans* isomer (183) as a muscarinic receptor stimulant (212).

Chothia and Pauling (209) defined the following molecular parameters for muscarinic agonism in acetylcholine congeners, based on their conformational analysis of (+)-*trans*-ACTM (170): $\tau_1 = 180°$; $\tau_2 = +73°$ to $+137°$; $\tau_3 = 180 \pm 35°$; $\tau_4 = 180°$ or $-137°$. Interatomic distances were defined as N$^+$-O^1 = 360; N$^+$-C^6 = 450; N$^+$-C^7 = 540 pm. Low potency or inactivity of certain acetylcholine derivatives is attributed to deviations from one or more of these parameters. However, Casy (184) cited examples of deviations from these values in which potent agonist activity is manifested.

(182)

(183)   (184)

Schulman and co-workers (213) used quantum mechanical calculations to construct a theoretical model to deduce the pharmacologically active conformation of acetylcholine and other agonists interacting with the muscarinic receptor "of the parasympathetic and central nervous systems," based on studies on acetylcholine and a series of eight extremely potent muscarinic agonists, including some rigid and semirigid congeners. The usual dihedral angles, which were described above to define conformations of cholinergic drugs, were replaced in this study with two new geometric parameters that were proposed to be more suitable for describing the muscarinic pharmacophore (Fig. 25.2): a characteristic distance [PQ] and a dihedral angle PNOQ.

The left diagram of Figure 25.2 shows the quaternary head of acetylcholine interacting with the receptor's carboxylate oxygen and the carboxyl oxygen of acetylcholine interacting with an electrophilic group on the receptor, such as a hydrogen-bonding proton. In the right diagram of Figure 25.2 the carboxylate oxygen is represented symbolically by P, while the electrophilic site is located at the point of minimum electrostatic potential near the ester oxygen, designated by Q. The interaction dihedral angle [PNOQ] is indicated on the right side of the figure. The significance of these studies may be compromised by the fact that they are based on the premise that the so-called muscarinic receptor is a single entity rather than a heterogeneous group of muscarinic recep-

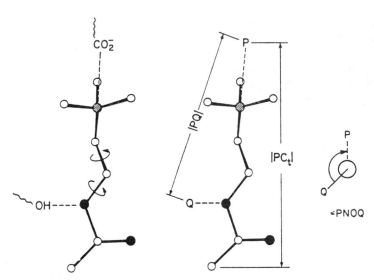

Fig. 25.2 Proposed muscarinically active conformation of acetylcholine. From ref. 213. *Left:* Acetylcholine interacting with the receptor's carboxylate oxygen and an electrophilic group such as a hydrogen-bonding proton. *Right:* Oxygen is indicated symbolically by P, while the electrophilic site is located at the point of minimum electrostatic potential near the ester oxygen Q. The interaction dihedral angle PNOQ is indicated at the right of the structures. The distances [PQ] and [PC$_t$] are also shown. Courtesy of the American Chemical Society.

tor subtypes and by the assumption that all of the agonists used in the study are interacting with the same receptor in the same fashion.

It has not yet been possible to demonstrate unequivocally that acetylcholine assumes different conformations for interaction at nicotinic and muscarinic receptors and/or at the subpopulations of each major receptor type; neither has this theory been disproved by the body of chemical and biological data. Chothia (214) proposed that there is only one conformation of acetylcholine relevant to both nicotinic and to muscarinic receptors, but that different sides of the molecule react with the nicotinic and muscarinic receptors: the "methyl side" activating muscarinic and the "carbonyl side" activating nicotinic receptors. This concept was invoked to explain specificity of action of several cholinergic agonists, but it has been criticized (184, 215) from experimental as well as from theoretical considerations. The absence of nicotinic properties in many of the con-

formationally restricted congeners of acetylcholine has impeded definition of specific geometry with respect to nicotinic effects. In addition, as noted previously, many nicotinic agents (including nicotine itself) are probably not agonists but have indirect modes of action involving such phenomena as triggering the release of endogenous acetylcholine from synaptic storage vesicles and from nerve terminals. It must be concluded that the relationship of acetylcholine's molecular geometry to its physiological roles is still not understood.

6 ANTICHOLINESTERASES

Symptoms resulting from an inadequate supply of acetylcholine may be relieved by blocking the body's acetylcholine-deactivating mechanism. Interest in this category of agents has increased greatly over the past few years, a result of recognition of their potential value in therapy of Alzheimer's disease as well as of other defects in memory and learning. Sussman and co-workers (216) have shown that the catalytic site of acetylcholinesterase is located at the bottom of a deep and narrow gorge surrounded by 14 aromatic amino acids. Figure 25.3 is a simplified representation of the catalytic region of acetylcholinesterase (217). Significant features are an anionic site that anchors the quaternary head of the substrate; a serine residue, the primary alcohol moiety of which participates in a transesterification reaction with acetylcholine, resulting in acetylation of the enzyme; and an imidazole ring (part of a histidine residue) that, as a neighboring group, participates in and facilitates the acetyl group transfer. The resulting acetylated serine moiety is extremely labile and rapidly undergoes spontaneous hydrolytic cleavage to liberate acetate anion and to regenerate the active catalytic surface.

Taylor (218) described three classes of acetylcholinesterase inhibitors, based on their mechanism of action.

1. Reversible inhibitors.
2. Agents having a carbamate ester linkage that is hydrolyzed by acetylcholinesterase but much more slowly than acetylcholine.
3. Organophosphorus inhibitors, which are true hemisubstrates for acetylcholinesterase.

Both the carbamates and the organophosphorus derivatives form a covalent (ester) bond with serine OH of the enzyme in essentially the same manner as does acetylcholine. Taylor (218) stated that the terms *reversible* and *irreversible* as applied to the carbamate and organophosphorus anticholinesterase agents, respectively, reflect only quantitative differences in rates

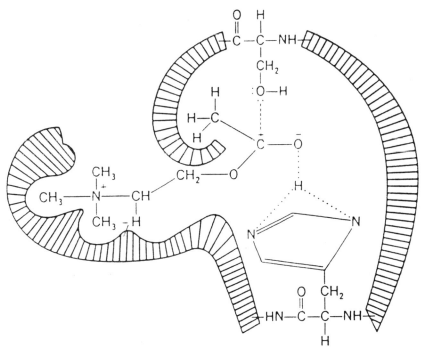

Fig. 25.3   Simplified diagram of the catalytic area of acetylcholinesterase. From ref. 217. Courtesy of Williams & Wilkins.

of cleavage of the esterified enzyme and not an actual difference in mechanism. An additional mechanism for acetylcholinesterase inhibition has been identified: it was concluded (219) that there exists on the acetylcholinesterase molecule, closely adjacent to the anionic subsite, a conformationally flexible, hydrophobic area that tends readily to assume a near planar form. The dimensions of this area are apparently unknown, but it is proposed to be of adequate size to accommodate large, nearly planar polycyclic ring systems such as 1,2,3,4-tetrahydro-9-aminoacridine (**197**) which is a reversible, noncovalently bonded inhibitor of the enzyme.

6.1 Reversible Inhibitors

Simple quaternary compounds such as tetramethylammonium cation combine with the anionic site of the catalytic surface of acetylcholinesterase and deny acetylcholine's access to this site. These compounds have a short duration of action due to the facile reversibility of their binding and rapid renal elimination (218), and thus they have minimal therapeutic utility. Cohen and Oosterbaan (220) tabulated a comprehensive list of tetraalkyl quaternary ammonium acetylcholinesterase inhibitors and described the kinetic studies on which the designation of "reversibility" for these compounds was based. Homologation of the methyl groups on the tetramethylammonium molecule tends to increase activity (Table 25.7) (221). Attempts to correlate biological test data with the calculated diameter of the unhydrated quaternary head led to inconclusive results.

Belleau (222) calculated entropies and enthalpies of binding for a homologous series of alkyltrimethylammonium compounds $RN^+(CH_3)_3$, where $R = CH_3$ through $n\text{-}C_{12}H_{15}$. The observed relative biological potencies in the series were rationalized on the basis of hydrophobic

Table 25.7 Inhibition of Acetylcholinesterase by Quaternary Ammonium Ions[a]

Ion	Relative Anticholinesterase Potency
$(CH_3)_4N^+$	1
$(C_2H_5)_4N^+$	5
$(n\text{-}C_3H_7)_4N^+$	100
$(n\text{-}C_4H_9)_4N^+$	50
$n\text{-}C_3H_7N^+(CH_3)_3$	5
$n\text{-}C_4H_9N^+(CH_3)_3$	7.5
$n\text{-}C_5H_{11}N^+(CH_3)_3$	10
$n\text{-}C_7H_{15}N^+(CH_3)_3$	25

[a] Adapted from ref. 221. Courtesy of Elsevier Publishers, B.V.

bonding phenomena coupled with the ability of the alkyl chains to displace ordered water from the acetylcholinesterase surface. Edrophonium (**185**) combines a quaternary head (for interaction with the anionic site of the enzyme) with a phenolic OH, which presumably hydrogen bonds to a portion of the esteratic area.

(**185**)

However, even this compound displays a rapidly reversible inhibition of the enzyme and its duration of action is short.

Holmsted (223) tabulated an extended series of *bis*-quaternary ammonium compounds that have been evaluated for anti-acetylcholinesterase activity; compound **186** is representative of this category, which includes some of the most effective enzyme inhibitors. Additional examples of anticho-

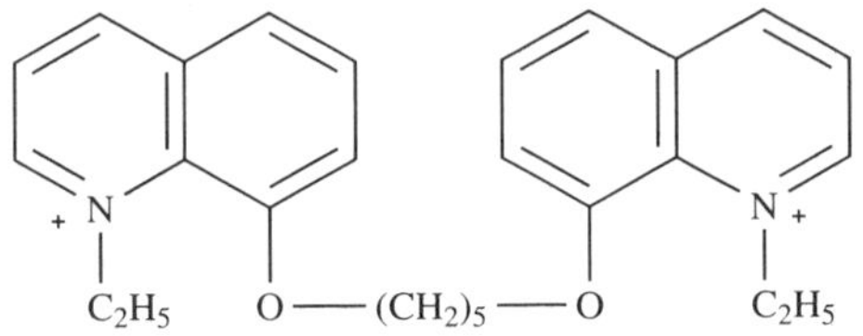

(186)

linesterase *bis*-quaternary ammonium compounds were reported by Fulton and Mogey (224) and by Cavallito and Sandy (225), who noted that there is a gradual increase in antiacetylcholinesterase activity as the chain joining the two quaternary heads increases. The optimum connecting chain length seems to be five to six carbons. In addition, enzyme inhibitory activity is maximal in those molecules in which the substituent(s) on the quaternary nitrogen are decidedly lipophilic. It has been suggested (226) that the *bis*-quaternary systems bind in chelate fashion at two anionic sites on the enzyme molecule.

6.2 Carbamate-derived Inhibitors

The prototype carbamate-type acetylcholinesterase inhibitor is physostigmine (**187**), isolated from the seeds of the Calabar bean, *Physostigma venenosum*.

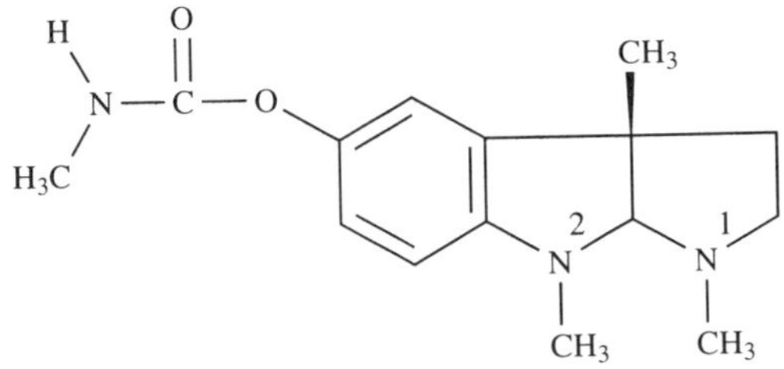

(187)

The absolute configuration of naturally occurring physostigmine has been established by spectral and chemical methods (227). At the pH of the body fluids, a significant proportion of physostigmine

molecules is protonated at N^1, and this cationic species forms a complex with the catalytic surface of acetylcholinesterase, following which transfer of the *N*-methylcarbamoyl moiety to the OH of the serine residue occurs, analogous to the process described for the acetyl moiety of acetylcholine. The carbamoylated enzyme is much more stable than the acetylated enzyme ($t_{1/2}$ for hydrolysis of the carbamoylated enzyme is 15–30 min compared to less than a millisecond for the acetylated serine moiety) (218). Sequestration of the enzyme in its carbamoylated form prevents enzyme-catalyzed hydrolysis of acetylcholine for prolonged periods of time. *In vivo*, the duration of enzyme inhibition by agents such as physostigmine is 3–4 h (218). Watts and Wilkinson (228) developed a kinetic scheme that was offered as a more adequate explanation for carbamate–acetylcholinesterase reactions and that explains the observed catalysis of ester cleavage of carbamoylated acetylcholinesterase by excess carbamate. Physostigmine has been reported (229) to have memory-enhancing effects in Alzheimer's disease patients. However, for clinical use it has a relatively short half-life, variable bioavailability, and a small therapeutic index.

(+)-Physostigmine, the enantiomer of the naturally occurring compound, has little effect on acetylcholinesterase *in vitro* (230, 231); it is a weak centrally acting cholinergic agonist. Rubreserine (**188**), an oxidation product of the ester-cleaved physostigmine molecule, is approximately 0.4% as active as physostigmine as an acetylcholinesterase inhibitor (232).

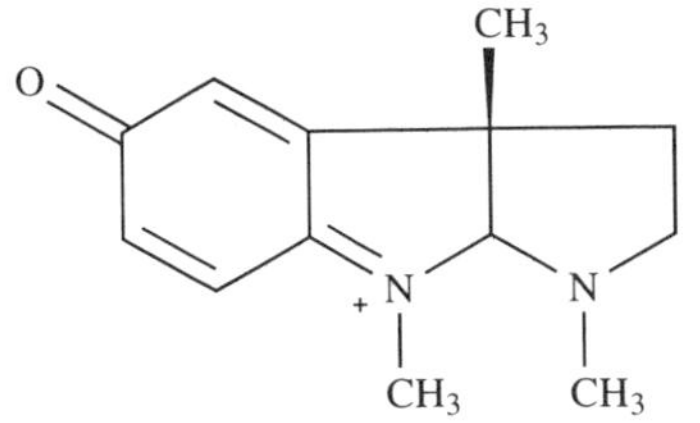

(188)

Brossi (233) prepared two short series of physostigmine (**187**) homologs, one in which the substituent(s) on the nitrogen of the carbamate moiety was (were) varied, and one in which the substituent on N^1 was varied. Both (+)- and (−)-enantiomers of the first series were prepared, and only the (−)- enantiomer of the second series was reported. Several of the (−)-enantiomers (same absolute configuration as physostigmine) show high potency in inhibition of acetylcholinesterase and of cholinesterase from a variety of sources.

Heptylphysostigmine (**189**) is a more lipophilic analog and is reported (234) to be less toxic than physostigmine, while retaining its *in vitro* acetylcholinesterase inhibiting potency. It was postulated (235) that replacing the N^2 of the physostigmine nucleus with a methylene group would increase the molecule's chemical and metabolic stability by modification of the less stable aminal group to a more stable amino group. A series of 8-carbaphysostigmine congeners (**190**) was studied, in which R′ was H or CH_3; R″ was CH_3, C_2H_5, n-C_3H_7, or benzyl; and R‴ was variety of C_1–C_7 alkyl chains, phenyl, or benzyl.

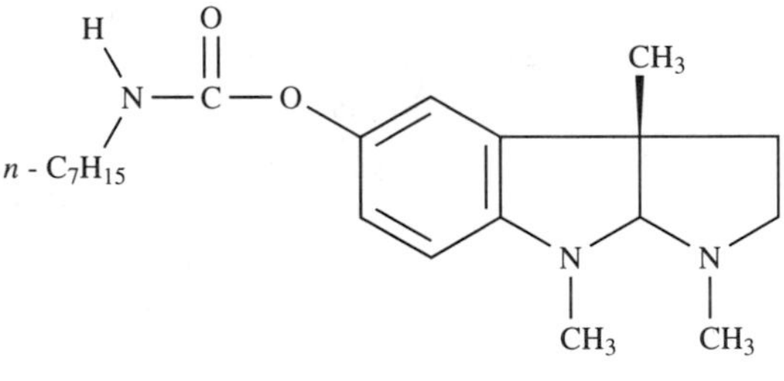

(189)

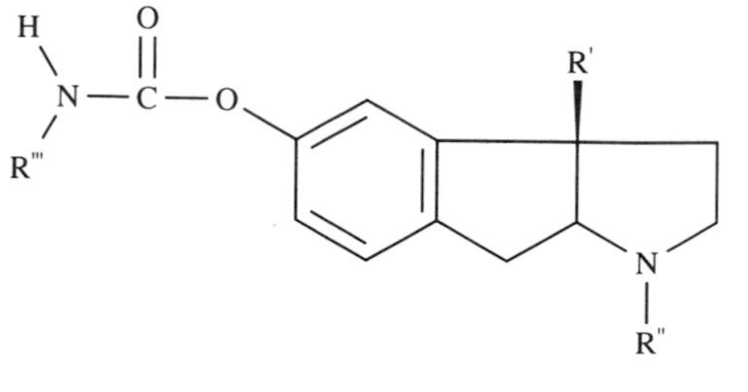

(190)

All compounds were tested as their racemic modifications; selected ones were resolved, and both enantiomers were studied. Two (−)-enantiomers, R′ = CH_3, R″ = C_2H_5, R‴ = n-C_7H_{15}; and R′ = CH_3, R″ = C_2H_5, R‴ = n-C_6H_{13}, were more potent than physostigmine or structure **189** in inhibition of acetylcholinesterase, and they were 6- to 12-fold more potent than their respective (+)-enantiomers. These more active enantiomers had the same absolute configuration as physostigmine itself.

Studies aimed at incorporation of the carbamate ester moiety and the cationic site of physostigmine into simpler organic molecules have led to quaternary ammonium compounds based on structure **191**.

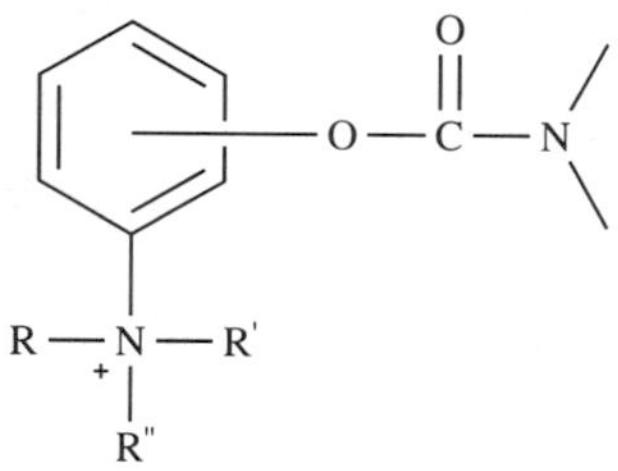

(191)

meta-Substituted systems frequently exhibit high miotic action (which is taken as a reflection of anticholinesterase activity), whereas *ortho*- and *para*-substituted molecules are inert (232). Foldes and co-workers (236) concluded that the optimum N^+– C=O interatomic distance for compounds of the type **191** is 4.7 Å. The *meta* isomers of these compounds meet this requirement, as does pyridostigmine (**192**), which is used clinically. Molecular models demonstrate that the active *m*-quaternary ammonium phenylcarbamate systems can assume reasonable conformations in which their cationic heads and C=O groups coincide with analogous groups in acetylcholine. Long (232) described the *ortho*-substituted system (**193**) as a potent acetylcholinesterase inhibitor, whereas the *meta*- and *para* isomers are much less active. Compound

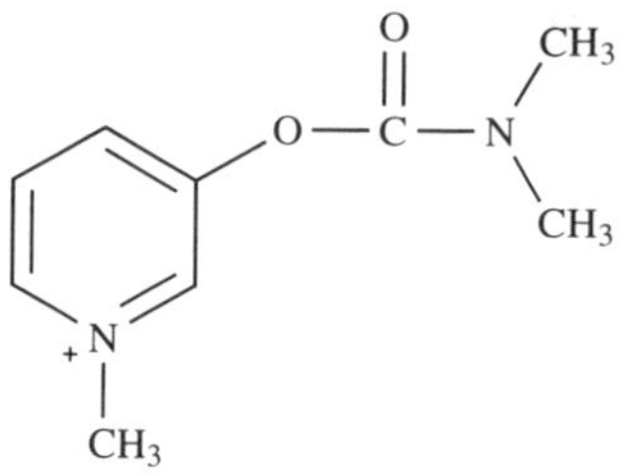

(192)

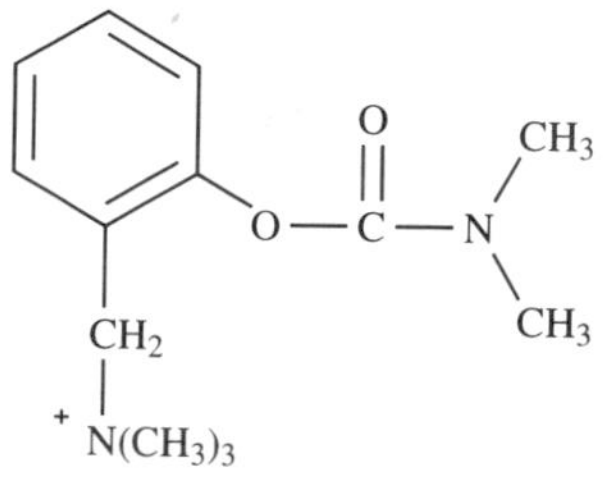

(193)

193 conforms to the 4.7 Å N^+ to C=O distance requirement, but its other two positional isomers do not. It is noteworthy that physostigmine (**187**) and its congeners, e.g., structures **189** and **190**, deviate considerably from the proposed 4.7 Å requirement; the N^+–C=O distance is considerably greater (on the order of 8–8.5 Å).

6.3 Organophosphorus-derived Inhibitors

Among the most powerful anticholinesterases (inactivating both acetylcholinesterase and plasma cholinesterase) are phosphorus-containing compounds, most commonly derivatives of orthophosphoric acid or of phosphonic acids. Certain of these compounds are extremely toxic, and much of the developmental work in the area was done with the object of preparing chemical warfare agents ("nerve gases"). Compounds in this category are potent and useful insecticides. The reaction between acetylcholinesterase and most organophosphorus inhibitors occurs only at the esteratic site of the enzyme, and the reaction

here is a transesterification, comparable to that involving the carbamate esters and acetylcholine itself. The reaction at the esteratic site of acetylcholinesterase is enhanced by the geometry of the tetrahedral phosphates, which resemble the transition state for acetate ester hydrolysis. The resulting serine-phosphorylated or -phosphonylated enzyme is extremely stable; if the R groups attached to the phosphorus (structure **194**) are methyl or ethyl, regeneration of the enzyme by hydrolytic cleavage requires several hours.

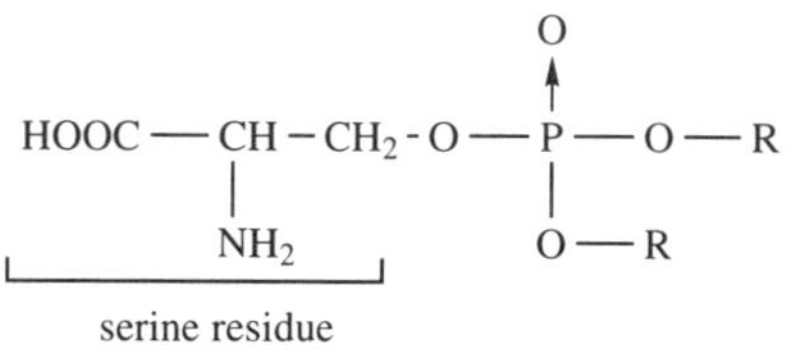

serine residue

(194)

If the R groups are isopropyl, essentially no hydrolysis occurs and reestablishment of enzyme activity can occur only after *de novo* synthesis of the enzyme. A characteristic structural feature of the anticholinesterase phosphorus compounds is the grouping P-Z, where Z is an electronegative moiety, a good leaving group, and the cleavage of the P-Z bond is accompanied by liberation of a large amount of energy. The P-Z bond is eminently susceptible to attack by nucleophiles, such as the serine OH of the esteratic site of acetylcholinesterase. In general, the enzyme inhibitory potency of the organophosphorus compounds parallels the ease of nucleophilic attack on the phosphorus atom. Compounds in this category include ester and amide derivatives of orthophosphoryl halides, pyrophosphate esters and amides, alkyl and aryl phosphonic acid derivatives, and thiophosphoric acid derivatives. Representative compounds are shown in Table 25.8.

Holmsted (223,237) and Hayes (238)

Table 25.8   Some Phosphorus-containing Acetylcholinesterase Inhibitors

Number	Structure	Chemical, Proprietary, or Generic Name(s)
1	$i\text{-}C_3H_7 - O$ ... P ... O, $i\text{-}C_3H_7 - O$ ... F	DFP, diisopropyl fluorophosphate, diisopropyl phosphofluoridate
2	CH_3, $CH_3 - N$ ... O, $C_2H_5 - O$ ... P ... CN	Tabun, ethyl N-dimethyl-phosphoramidocyanidate
3	$i\text{-}C_3H_7 - O$ ... P ... O, H_3C ... F	Sarin (GB), isopropyl methylphosphonofluoridate
4	CH_3, $(CH_3)_3C - CH - O$ ... P ... O, H_3C ... F	Soman, pinacolyl methylphosphonofluoridate
5	$C_2H_5 - O$ ... O ... $P - O - P$ ... O ... $O - C_2H_5$, $C_2H_5 - O$ ... $O - C_2H_5$	Tetraethyl pyrophosphate, TEPP
6	$(CH_3)_2N$ ... O ... $P - O - P$ ... O ... $N(CH_3)_2$, $(CH_3)_2N$ ... $N(CH_3)_2$	Octamethylpyrophosphoramide, OMPA
7	$C_2H_5 - O$ ... O, P, $C_2H_5 - O$ ... $S - CH_2\text{-}CH_2\text{-}\overset{+}{N}(CH_3)_3$	Echothiophate, diethoxyphorphinylthinocholine

Table 25.8 (*Continued*)

Number	Structure	Chemical, Proprietary, or Generic Name(s)
8	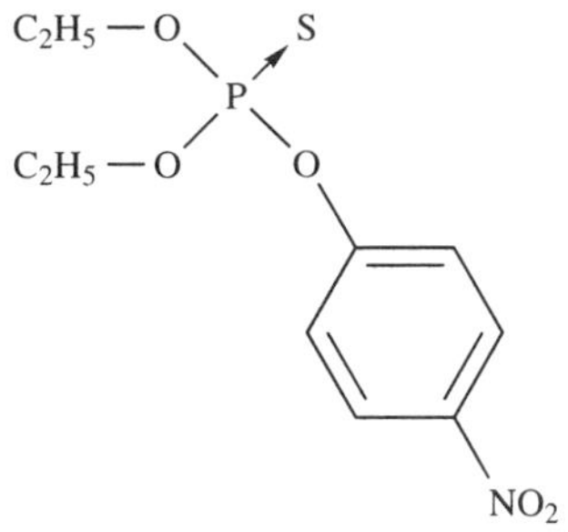	Parathion
9	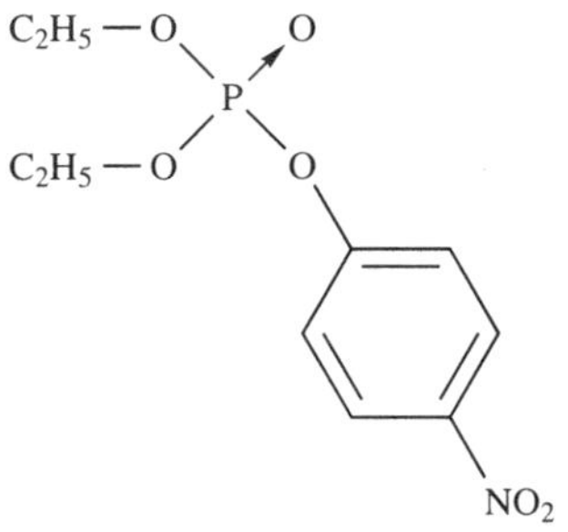	Paraoxon

have tabulated and discussed a large number of organophosphorus acetylcholinesterase inhibitors. Tabun, sarin, and soman (numbers 2, 3, and 4 in Table 25.8) are among the most toxic war "gases" known. OMPA (number 6) is inert as such, but it is metabolized to an *N*-oxide derivative that is the biologically active entity (239). Parathion (number 8) is inactive in inhibition of acetylcholinesterase *in vitro*; mixed function oxidases (in human liver) convert parathion into its oxygen bioisostere paraoxon (number 9), the pharmacologically active metabolite (218). Echothiophate (number 7) is representative of inhibitors that bind initially to the anionic site of acetylcholinesterase as well as to the esteratic area. This compound is used clinically.

The 1,3,2-dioxaphosphorinane (**195**) is representative of organophosphorus acetylcholinesterase inhibitors that were designed to have a short duration of pharmacological effect (240).

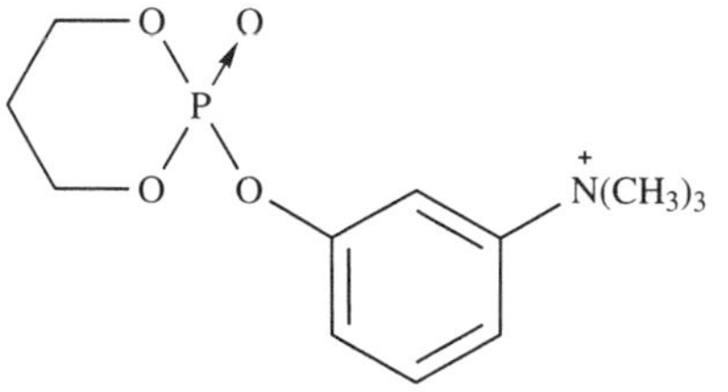

(**195**)

This compound inactivates acetylcholinesterase by formation of "an unstable covalent intermediate." The inhibited enzyme hydrolyzes spontaneously with $t_{1/2} \approx 10$ min. The covalent derivative of structure

195 with plasma cholinesterase is considerably more resistant to spontaneous hydrolysis. Compound **195** was proposed to be a useful adjunct prophylactic agent against the insecticide paraoxon and chemical warfare agents such as soman (240).

A remarkable neurotoxic natural product, anatoxin-a(s) (**196**), has been isolated from several biological sources, including a blue-green alga (241,242).

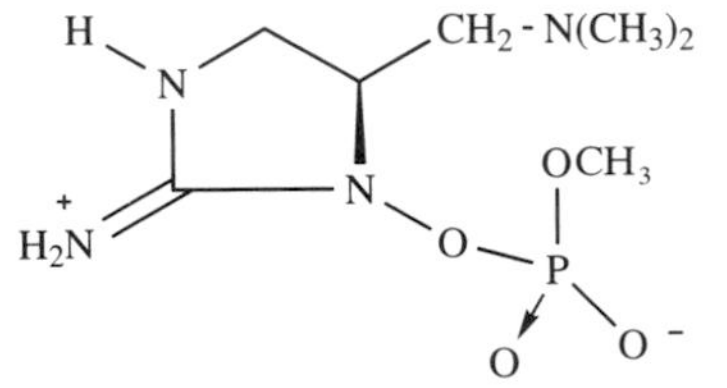

(196)

Its high toxicity ($LD_{50} = 20–40 \ \mu g/kg$ in mice) has been ascribed to anticholinesterase activity.

6.4 Reversible, Noncovalent Inhibitors Related to 1,2,3,4-Tetrahydro-9-aminoacridine

1,2,3,4-Tetrahydro-9-aminoacridine (**197**) (THA, tacrine) was described in 1961 (243) as a reversible inhibitor of acetylcholinesterase and an even more potent inhibitor of butyrylcholinesterase. On the basis of x-ray crystal studies it was reported (244) that a tryptophan residue (Trp 84) at the catalytic surface of acetylcholinesterase is the binding site for the aromatic ring of structure **197**.

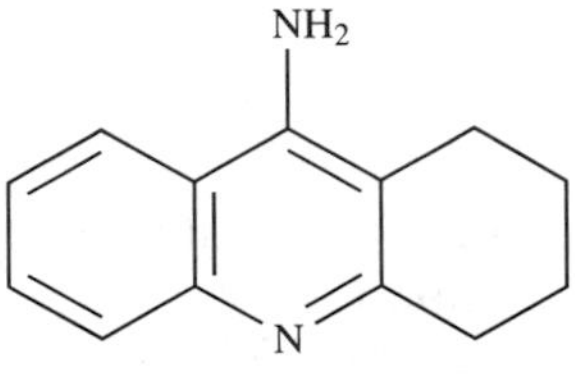

Clinical efficacy in relief of symptoms of Alzheimer's disease was claimed (245) for THA, but this positive finding is tempered by its tendency to produce hepatotoxicity (246). It was speculated (247) that the hepatotoxicity of structure **197** might be related to its lipophilic character, and a (±)-1-hydroxy derivative **198** was designed in the hope that the OH group would serve as a metabolic "handle" for glucuronidation and subsequent facilitated elimination.

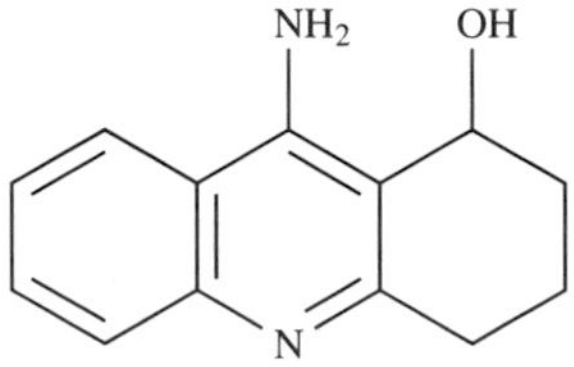

(198)

Compound **198** is somewhat less active *in vitro* against acetylcholinesterase than is THA (**197**). However, the two compounds are approximately equipotent in reversal of scopolamine-induced memory impairment in mice, a putative predictive model of activity in Alzheimer's disease. These data suggest that, in addition to acetylcholinesterase inhibition, there may be other biochemical components to the mechanism of action of compound **198**. Indeed, this speculation may be applicable to THA itself and to others of its active analogs and congeners.

An extensive structure–activity study (248) addressed modifications of the THA molecule, illustrated by structure **199**: $X =$

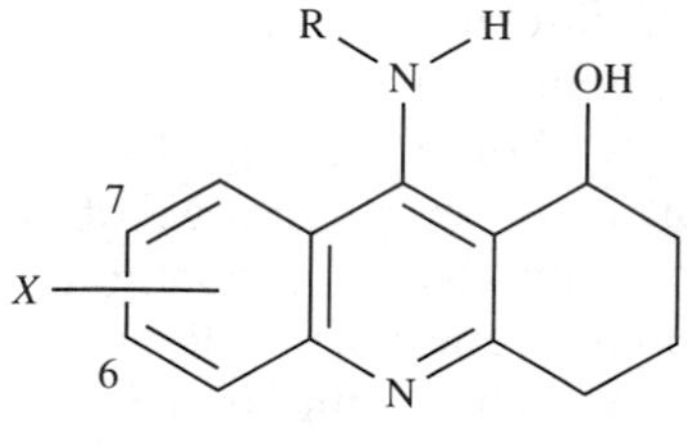

(197)

(199)

H, 6-Cl, 7-Cl, 6-F, or 6-CF$_3$; R = H, alkyl, benzyl, ring-substituted benzyl, or ω-phenoxyalkyl. Most of the compounds are inferior to compound **197** as inhibitors of acetylcholinesterase, but some are decidedly less toxic, and a few are equieffective or superior to compound **197** in an assay for their ability to reverse scopolamine-induced memory impairment.

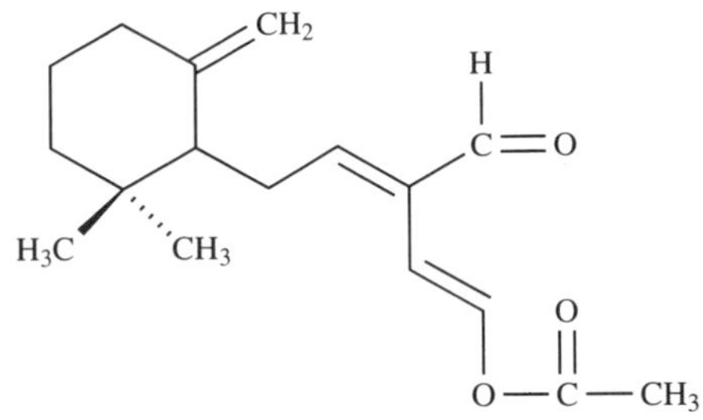

(202)

6.5 Miscellaneous Inhibitors

An α-chloro-β-phenethylamine (**200**) irreversibly inactivates acetylcholinesterase (249); the active pharmacophoric species is the aziridinium cation (**201**).

The quaternary ion nature of the aziridinium cation allows for reversible complex formation with the anionic site of the enzyme, which precedes slow alkylation of the nucleophilic (serine OH) site. Tetramethylammonium retards the irreversible inactivation of the enzyme by this compound.

Onchidal **202** is the principal constituent of a secretion of glands of a mollusk (250). This compound is an irreversible inhibitor of acetylcholinesterase. It is not a substrate for the enzyme, and its mechanism of action apparently does not involve acylation of the active site serine. It was speculated (250) that the α,β-unsaturated aldehyde moiety may be involved, through Michael addition, in covalent bond formation with the enzyme.

Based on molecular design derived from rationalizations concerning the topography of the catalytic area of acetylcholinesterase,

(203)

(204)

inhibitors based on structures **203** and **204** were found (251,252).

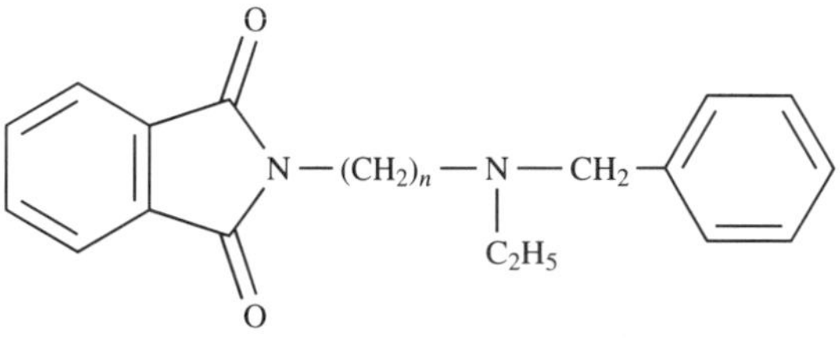

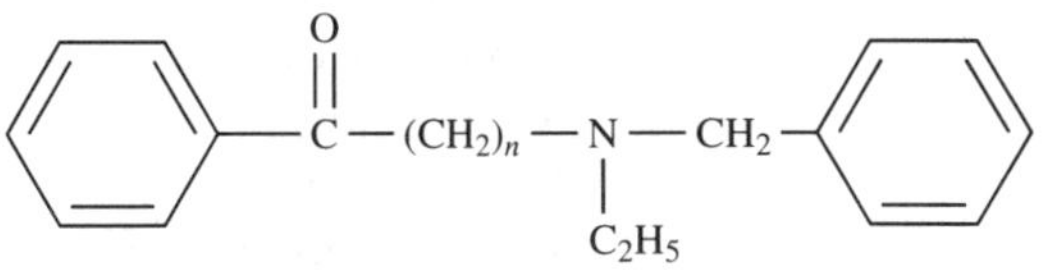

An extended series (**205**, congeners of **203** and **204**) revealed that one member (R = H, Y = H, $n = 3$, and attachment of the ketonic carbon at position 8 of the tetrahydrobenzazepine ring) demonstrates potent *in vitro* inhibition of acetylcholinesterase and is active in a series of *in vivo* assays for CNS cholinergic effects, but it produces no significant peripheral choliner-

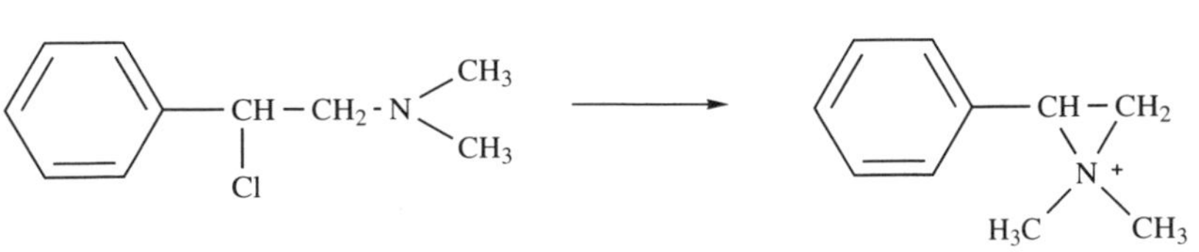

(200) (201)

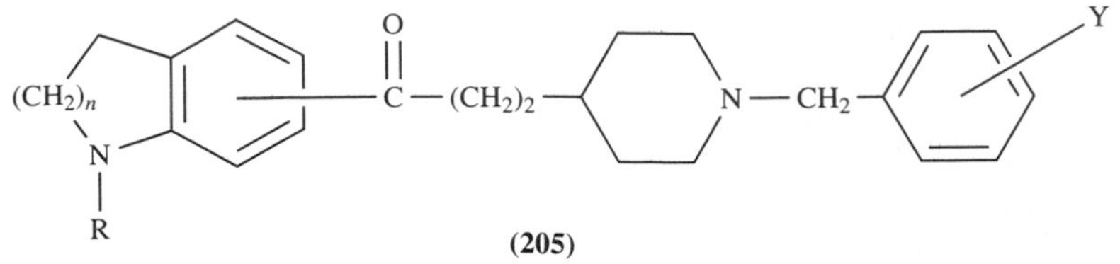

(205)

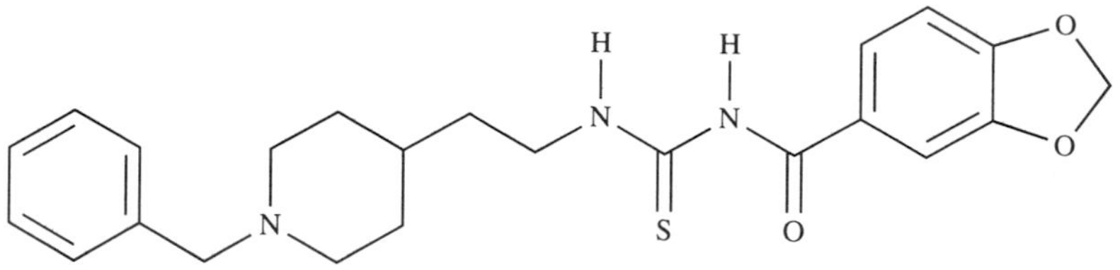

(206)

gic effects (253). Molecular modeling studies (docking analysis) of compounds in the series of compounds **203–205** indicated (254) that the *N*-benzyl group interacts with the same tryptophan residue (Trp 84) as the aromatic ring of tacrine (THA, **197**). The other aromatic ring in these inhibitors interacts with another trypothan residue (Trp 279) on the enzyme molecule, and hydrogen bonding interaction between the carbonyl group of the inhibitors **203–205** and a tyrosine (Tyr 121) hydroxyl on the enzyme seems to play an important role. These data should useful for the future design of more potent, more specific inhibitors.

Members of a series of 1-aroyl-3-[1-benzyl-4-piperidinyl)ethyl]thiourea derivatives (**206**) are potent (submicromolar range) acetylcholinesterase inhibitors (255). Comparable potency is retained by replacing the unsubstituted benzene ring with a bioisosteric 2-pyridyl group. The guanidine congener of structure **206** is almost inactive. In a passive avoidance test in rats, structure **206** had maximal antiamnesic activity at 0.03 mg/kg with a therapeutic ratio greater than 1000, and it displayed cholinergic side effects only at high doses. Its potential use as an antidementia agent was suggested (255).

7 ACETYLCHOLINE-RELEASE MODULATORS

Structure–activity studies (256) of 3,3-disubstituted oxindoles led to DuP-996 (**207**), which enhances potassium-evoked acetylcholine release in rat cortex, hippocampus, and caudate nucleus, *in vitro* (257,258).

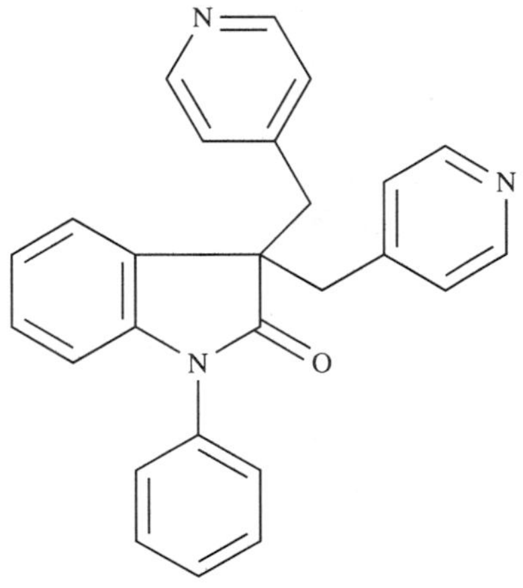

(207)

This enhancement of acetylcholine release from nerve terminals occurs only when the release has been triggered (259). Compound **207** is reported (259) to exert significant effects in the human central nervous system. Dopamine and serotonin release are also enhanced by this agent, but release of glutamate, GABA, and norepinephrine is unaffected. Further structure–activity studies of 3,3-disubstituted oxindoles have been reported (260), and the potential utility of this category of compounds in treatment of cognitive and neurological deficiencies was stressed. One of the possible advantages to a therapeutic strategy for Alzheimer's disease or other deficits in memory and learning using acetylcholine-releasing agents is that such a process would permit stimulation of both nicotinic and muscarinic receptors in the brain. There is evidence that stimulation of central nicotinic receptors is beneficial in Alzheimer's patients (261).

REFERENCES

1. L. A. Barker and T. W. Mittag, *J. Pharmacol. Exp. Ther.*, **192**, 86 (1975).
2. S. M. Parsons and G. A. Rogers, *Annu. Rep. Med. Chem.*, **28**, 247 (1993).
3. R. Katzman and L. J. Thal, in G.J. Siegel, B. Agranoff, R. W. Albers, and P. Molinoff, Eds., *Basic Neurochemistry: Molecular, Cellular, and Medical Aspects*, 4th Ed., Raven Press, New York, 1989, pp. 827–838.
4. P. Taylor, in A. G. Gilman, T. W. Rall, A. S. Nies, and P. Taylor, Eds., Goodman and Gilman's *The Pharmacological Basis of Therapeutics*, 8th Ed. New York, Pergamon Press, 1990, p. 147.
5. A. Lagersted, E. Falch, B. Ebert, and P. Krogsgaard-Larsen, *Drug Des. Discov.*, **9**, 237 (1993).
6. P. S. Anderson and D. Haubrich, *Annu. Repts. in Med. Chem.*, **16**, 51 (1981).
7. A. F. Casy, *The Steric Factor in Medicinal Chemistry. Dissymetric Probes of Pharmacological Receptors*, New York, Plenum Press, 1993, pp. 258, 327.
8. G. Lundgren and J. Malmberg, *Biochem. Pharmacol.*, **17**, 2051 (1968).
9. C. Y. Chiou and J. P. Long, *Arch. Int. Pharmacodyn. Ther.*, **182**, 269 (1969).
10. C. Y. Chou and J. P. Long, *Proc. Soc. Exp. Biol. Med.*, **132**, 732 (1969).
11. C. Y. Chiou, J. P. Long, R. Potrepka, and J. L. Spratt, *Arch. Int. Pharmacodyn. Ther.*, **187**, 88 (1970).
12. A. Fisher, Y. Karton, E. Heldman, D. Gurwitz, R. Haring, H. Meshulam, R. Brandeis, Z. Pittel, Y. Segall, D. Marciano, I. Markovitch, Z. Samocha, E. Shirinb, M. Sapir, B. Green, G. Shoham, and D. Barak, *Drug Des. Discov.*, **9**, 221 (1993).
13. R. J. Lefkowitz, B. B. Hoffman, and P. Taylor in ref. 4, p. 100.
14. P. Krogsgaard-Larsen, in P. Krogsgaard-Larsen and H. Bundgaard Eds. *A Textbook of Drug Design and Development*, Harwood Academic Publishers, Philadelphia, 1991, pp. 422–423.
15. J. Wess, *Trends Pharmacol. Sci.*, **14**, 308 (1993).
16. R. J. Lefkowitz, B. B. Hoffman, and P. Taylor in ref. 4, p. 99.
17. T. Nogrady, *Medicinal Chemistry*, 2nd ed., Oxford University Press, New York, 1988 pp. 141–150.
18. J.-P. Changeux, A. Devillers-Thiery, and P. Chemmuivilli, *Science*, **25**, 1335 (1984).
19. J. L. Popot and J.-P. Changeux, *Physiol. Rev.*, **64**, 1162 (1984).
20. J. Weiss, T. Buhl, G. Lambrecht, and E. Mutschler, in C. Hansch, P. G. Sammes, and J. B. Taylor, Eds., *Comprehensive Medicinal Chemistry*, Vol. 3. Oxford, Pergamon Press, 1990, pp 423–491.
21. A. Devillers-Thiery, J. L. Galzi, J. L. Eiselé, S. Bertrand, D. Bertrand, and J.-P. Changeux, *J. Membrane Biol.*, **136**, 97 (1993).
22. G. A. Alles and P. K. Knoefel, *Calif. Publ. Pharmacol.*, **1**, 187 (1939).
23. H. R. Ing, *Science*, **109**, 264 (1949).
24. J. C. Jaen and R. E. Davis, *Annu. Rep. Med. Chem.*, **29**, 23 (1994).
25. H. R. Ing, P. Kordik, and D. P. H. Tudor Williams, *Br. J. Pharmacol.*, **7**, 103 (1952).
26. J. Bannister and V. P. Whittaker, *Nature*, **167**, 605 (1951).
27. R. L. Stehle, K. I. Melville, and F. K. Oldham, *J. Pharmacol. Exp. Ther.*, **56**, 473 (1936).
28. J. Trzeciakowski and C. Y. Chiou, *J. Pharm. Sci.*, **67**, 531 (1978).
29. H. C. Chang and J. H. Gaddum, *J. Physiol. (Lond.)*, **79**, 255 (1933).
30. A. Karlin, *Cell Surf. Rev.*, **6**, 191 (1980).
31. R. Hunt and R. de M. Taveau, *Br. Med. J.*, 1788 (1906).
32. D. Bovet and F. Bovet-Nitti, *Médicaments du Système Nerveux Végétatif*, S. Karger, Basel, 1948.
33. H. R. Ing, G. S. Dawes, and L. Wajda, *J. Pharmacol. Exp. Ther.*, **85**, 85 (1945).

34. E. Fourneau and H. J. Pye, *Bull. Soc. Chim. Fr.*, **15**, 544 (1949).

35. V. P. Whittaker, *Biochem. Pharmacol.*, **1**, 342 (1959).

36. B. V. Rama Sastry, C. C. Pfeiffer, and A. Lasslo, *J. Pharmacol. Exp. Ther.*, **130**, 346 (1960).

37. R. Hunt and R. R. Renshaw, *J. Pharmacol. Exp. Ther.*, **25**, 315 (1925).

38. R. B. Barlow, J. B. Bremner, and K. S. Soh, *Br. J. Pharmacol.*, **62**, 39 (1978).

39. U.S. Pat. 2, 347, 367 (1945); R. T. Major and H. T. Bonnet. *Chem. Abstr.*, **39**, 4721 (1945).

40. R. R. Renshaw and C. Y. Hopkins, *J. Am. Chem. Soc.*, **51**, 953 (1929).

41. A. K. Cho, D. J. Jenden, and S. I. Lamb, *J. Med. Chem.*, **15**, 391 (1972).

42. P. Holton and H. R. Ing, *Br. J. Pharmacol.*, **4**, 190 (1949).

43. F. W. Schueler and C. Hanna, *Arch. Intern. Pharmacodyn. Ther.*, **88**, 351 (1951.

44. H. Kilbinger, A. Wagner, and R. Zerban, *Naunyn Schmiedebergs Arch. Pharmacol.*, **295**, 81 (1976).

45. H. L. Friedman, in A. Burger, Ed., *Medicinal Research Series*, Vol. 1, Dekker, New York, 1967, p. 79.

46. J. H. Welsh and R. Taub, *J. Pharmacol. Exp. Ther.*, **103**, 62 (1951).

47. B. Belleau, in L. Roth, Ed., *Isotopes in Pharmacology*, University of Chicago Press, Chicago, 1965, p. 469.

48. L. L. Darko, J. G. Cannon, J. P. Long, and T. F. Burks, *J. Med. Chem.*, **8**, 841 (1965).

49. G. Lambrecht, *Pharm. Zeit.*, **120**, 1411 (1975).

50. J. G. Cannon, R. V. Smith, G. A. Fisher, J. P. Long, and F. W. Benz, *J. Med. Chem.*, **14**, 66 (1971).

51. B. Pullman, Ph. Courrièrre, and J. L. Coubeils, *Mol. Pharmacol.*, **7**, 397 (1971).

52. I. Kimura, I. Morishima, T. Yonezawa, and M. Kimura, *Chem. Pharm. Bull.* (*Tokyo*), **22**, 429 (1974).

53. R. Marcus and A. M. Coulston in ref. 4, p. 1543.

54. C. Y. Chiou, *J. Pharm. Sci.*, **64**, 469 (1975).

55. V. P. Whitaker, *Ann. N. Y. Acad. Sci.*, **90**, 695 (1960).

56. A. Blankart, *Festschrift Emil C. Barell*, 284 (1936).

57. Ref. 4, p. 124.

58. Ref. 4, p. 123.

59. B. C. Barrass, R. W. Brimblecomb, D. C. Parkes, and P. Rich, *Br. J. Pharmacol.*, **34**, 345 (1968).

60. B. Jenssen and G. Lambrecht, *Naunyn Schmiedebergs Arch. Pharmacol.*, **307** (Suppl.), R54 (1979).

61. K. A. Scott and H. G. Mautner, *Biochem. Pharmacol.*, **13**, 907 (1964).

62. H. G. Mautner, *Annu. Rep. Med. Chem.*, 230 (1969).

63. W. B. Geiger and H. Alpers, *Arch. Int. Pharamcodyn. Ther.*, **148**, 352 (1964).

64. A. M. Lands and C. J. Cavallito, *J. Pharmacol. Exp. Ther.*, **110**, 369 (1954).

65. A. H. Beckett, *Ann. N. Y. Acad. Sci.*, **144**, 675 (1967).

66. Ref. 7, p. 232.

67. M. M.-L. Chan and J. B. Robinson, *J. Med. Chem.*, **17**, 1057 (1974).

68. B. Ringdahl, *Br. J. Pharmacol.*, **89**, 7 (1986).

69. C. De Micheli, M. De Amici, P. Pratesi, E. Grana, and M. G. Santagostino Barbone, *Farmaco*, **38**, 514 (1983).

70. H. Schwörer, G. Lambrecht, E. Mutschler, and H. Kilbinger, *Naunyn Schmiedebergs Arch. Pharmacol.*, **331**, 307 (1985).

71. E. E. Smissman, W. L. Nelson, J. LaPidus, and J. L. Day, *J. Med. Chem.*, **9**, 458 (1966).

72. G. H. Cocolas, E. C. Robinson, and W. L. Dewey, *J. Med. Chem.*, **13**, 299 (1970).

73. A. Simonart, *Arch. Int. Pharmacodyn. Ther.*, **48**, 328 (1934).

74. R. Hunt and R. de M. Taveau, *J. Pharmacol. Exp. Ther.*, **1**, 303 (1909).

75. J. G. Cannon, A. Gangjee, J. P. Long, and A. J. Allen, *J. Med. Chem.*, **19**, 934 (1976).

76. J. G. Cannon, P. J. Mulligan, J. P. Long, and S. Heintz, *J. Pharm. Sci.*, **62**, 830 (1973).

77. P. Hey, *Br. J. Pharmacol.*, **7**, 117 (1952).

78. R. Hunt and R. R. Renshaw, *J. Pharmacol. Exp. Ther.*, **58**, 140 (1936).

79. E. R. Clark and M. de L. S. A. Jana, *Br. J. Pharmacol. Chemother.*, **27**, 135 (1966).

80. P. Hey, *Br. J. Pharmacol. Chemother.*, **7**, 117 (1952).

81. E. R. Clark, P. M. Dawes, and S. G. Williams, *Br. J. Pharmacol. Chemother.*, **32**, 1123 (1968).

82. E. R. Clark, I. E. Hughes, and C. F. C. Smith, *J. Med. Chem.*, **19**, 692 (1976).

83. G. Pepeu, D. X. Freedman, and N. J. Giarman, *J. Pharmacol. Exp. Ther.*, **129**, 291 (1960).

84. M. D. Aceto, B. R. Martin, I. M. Uwaydah, E. L. May, L. S. Harris, C. Izazola-Conde, W. L. Dewey, T. J. Bradshaw, and W. C. Vincek, *J. Med. Chem.*, **22**, 174 (1979).

85. R. B. Barlow, in U. S. von Euler, Ed., *Tobacco Alkaloids and Related Compounds*, Macmillan, New York, 1965, p. 277.

86. F. Haglid, in ref. 85, p. 315.

87. C. Reavill, P. Jenner, R. Kumar, and I. P. Stolerman, *Neuropharmacology*, **27**, 235 (1988).

88. R. L. Elliott, K. B. Ryther, D. J. Anderson, J. L. Raszkiewicz, J. P. Sullivan, and D. S. Garvey, *Abstr. Am. Chem. Soc.*, (208 Mtg), Medi 198 (1994).

89. N.-H. Lin, G. M. Carrera Jr., and D. J. Anderson, *J. Med. Chem.*, **37**, 3542 (1994).

90. N.-H. Lin, Y. He, D. J. Anderson, J. T. Wasicak, R. Kasson, D. Sweeny, and J. P. Sullivan, *Abstr. Am. Chem. Soc.*, (208 Mtg), MEDI 199 (1994).

91. E. Hardegger and F. Lohse, *Helv. Chim. Acta*, **40**, 2383 (1957).

92. P. G. Waser, *Pharmacol. Rev.*, **13**, 465 (1961).

93. J. G. Cannon, in M. E. Wolff, Ed., *Burger's Medicinal Chemistry* Part III, 4th ed., Wiley-Interscience, New York, p. 346.

94. H. Bollinger and C. H. Eugster, *Helv. Chim. Acta*, **54**, 2704 (1971).

95. L. Gyermek and K. R. Unna, *J. Pharmacol. Exp. Ther.*, **128**, 30 (1960).

96. M. De Amici, C. Dallanoce, C. De Micheli, E. Grana, A. Barbieri, H. Ladinsky, G. B. Schiavi, and F. Zonta, *J. Med. Chem.*, **35**, 1915 (1992).

97. A. H. Beckett, B. H. Warrington, R. Griffiths, E. S. Pepper, and K. Bowden, *J. Pharm. Pharmacol.*, **28**, 728 (1976).

98. A. K. Armitage and H. R. Ing, *Br. J. Pharmacol.*, **9**, 376 (1954).

99. B. Belleau and J. Puranen, *J. Med. Chem.*, **6**, 325 (1963).

100. K. J. Chang, R. C. Deth, and D. J. Triggle, *J. Med. Chem.*, **15**, 243 (1972).

101. E. Teodori, F. Gualtieri, P. Angeli, L. Brasili, M. Gianella, and M. Pigini, *J. Med. Chem.*, **29**, 1610 (1986).

102. E. Teodori, F. Gualtieri, P. Angeli, L. Brasili, and M. Gianella, *J. Med. Chem.*, **30**, 1934 (1987).

103. J. G. R. Elferink and C. A. Salemink, *Arzneim.-Forsch.*, **25**, 1858 (1975).

104. C. Melchiorre, P. Angeli, M. Gianella, F. Gualtieri, M. Pigini, M. L. Cingolani, G. Gamba, L. Leone, P. Pigini, and L. Re, *Eur. J. Med. Chem.*, **13**, 357 (1978).

105. H. F. Ridley, S. S. Chatterjee, J. F. Moran, and D. J. Triggle, *J. Med. Chem.*, **12**, 931 (1969).

106. J. Saunders, G. A. Showell, R. Baker, S. B. Freedman, D. Hill, A. McKnight, N. Newberry, J. D. Salamone, J. Hirshfield, and J. P. Springer, *J. Med. Chem.*, **30**, 969 (1987).

107. J. Saunders, M. Cassidy, S. B. Freedman, E. A. Harley, L. L. Iversen, C. Kneen, A. M MacLeod, K. J. Merchant, R. J. Snow, and R. Baker, *J. Med. Chem.*, **33**, 1128 (1990).

108. L. J. Street, R. Baker, T. Book, C. O. Kneen, A. M. MacLeod, K. J. Merchant, G. A. Showell, J. Saunders, R. H. Herbert, S. B. Freedman, and E. A. Harley, *J. Med. Chem.*, **33**, 2690 (1990).

109. R. S. Givens and D. R. Rademacher, *J. Med. Chem.*, **17**, 457 (1974).

110. K. G. R. Sundelin, R. A. Wiley, R. S. Givens, and D. R. Rademacher, *J. Med. Chem.*, **16**, 235 (1973).

111. F. Gaultieri, M. Gianella, C. Melchiorre, M. Pigini, M. L. Cingolani, G. Gamba, P. Pigini, and L. Rossini, *Farm. Ed. Sc.*, **30**, 223 (1975).

112. C. Melchiorre, P. Angeli, M. Gianella, M. Pigini, M. L. Cingolani, G. Gamba, and P. Pigini, *Farm. Ed. Sc.*, **32**, 25 (1977).

113. F. Gualtieri, M. Gianella, C. Melchiorre, and M. Pigini, *J. Med. Chem.*, **17**, 455 (1974).

114. P. Angeli, C. Melchiorre, M. Gianella, and M. Pigini, *J. Med. Chem.*, **20**, 398 (1977).

115. P. Taylor in ref. 4, p. 127.

116. P. Krogsgaard-Larsen in ref. 14, p. 427.

117. W. H. Beers and E. Reich, *Nature* (*London*), **228**, 917 (1970).

118. H. Bundgaard in ref. 14, p. 180.

119. H. Bundgaard, E. Falch, C. Larsen, G. L. Mosher, and T. J. Mikkelson, *J. Med. Chem.*, **28**, 979 (1985).

120. P. Sauerberg, J. Chen, E. WoldeMussie, and H. Rapoport, *J. Med. Chem.*, **32**, 1322 (1989).

121. M. A. Nunes and E. J. Brockmann-Hanssen, *J. Pharm. Sci.*, **63**, 716 (1974).

122. P. Krogsgaard-Larsen in ref. 14, pp. 425–426.

123. D. Bieger, E. Krüger-Thiemer, H. Lüllmann, and A. Ziegler, *Eur. J. Pharmacol.*, **9**, 156 (1970).

124. P. Sauerberg, J.-J. Larsen, E. Falch, and P. Krogsgaard-Larsen, *J. Med. Chem.*, **29**, 1004 (1986).

125. G. Lambrecht and E. Mutschler, in F. G. De Las Heras and S. Vega, Eds., *Medical Chemistry Advances*, Pergamon Press, Oxford, U.K., 1981, pp. 119–125.

126. U. Moser, G. Lambrecht, and E. Mutschler, *Arch. Pharm.*, **316**, 670 (1983).

127. W. S. Messer, Jr., P. G. Dunbar, T. Rho, S. Periysamy, D. Ngur, B. R. Ellerbrock, M. Bohnett, K. Ryan, G. J. Durant, and W. Hoss, *Bioorg. Med. Chem. Lett.*, **2**, 781 (1992).

128. J. E. Christie, A. Shering, J. Ferguson, and A. I. M. Glen, *Br. J. Psychiatry*, **138**, 46 (1981).

129. R. T. Bartus, R. L. Dean, and B. Beer, *Neurobiol. Aging*, **1**, 145 (1980).

130. N. Sitaram, H. Weingartner, and J. C. Gillin, *Science* (*Washington, D. C.*), **201**, 274 (1978).

131. O. Nieschulz and P. Schmersahl, *Arzneim.-Forsch.*, **18**, 22, (1968).

132. E. Toja, C. Boretti, A. Butti, P. Hunt, M. Fortin, F. Barzaghi, M. L. Formento, A. Maggioni, A. Nencioni, and G. Galliani, *Eur. J. Med. Chem.*, **26**, 853 (1991).

133. *Ibid.*, **27**, 519 (1992).

134. J. Saunders, A. M. MacLeod, K. J. Merchant, G. A. Showell, L. J. Street, R. J. Snow, and R. Baker, *Abstr. Am. Chem. Soc.* (196 Mtg), MEDI 68 (1988).

135. P. Sauerberg, J. W. Kindtler, L. Nielsen, M. J. Sheardown, and T. Honoré, *J. Med. Chem.*, **34**, 687 (1991).

136. G. A. Showell, T. L. Gibbons, C. O. Kneen, A. M. McLeod, K. Merchant, J. Saunders, S. B. Freedman, S. Patel, and R. Baker, *J. Med. Chem.*, **34**, 1086 (1991).

137. P. Sauerberg, P. H. Olesen, S. Nielsen, S. Treppendahl, M. J. Sheardown, T. Honoré, C. H. Mitch, J. S. Ward, A. J. Pike, F. P. Bymaster, B. D. Sawyer, and H. E. Shannon, *J. Med. Chem.*, **35**, 2274 (1992).

138. J. S. Ward, L. Merritt, V. J. Klimkowski, M. L. Lamb, C. H. Mitch, F. P. Bymaster, B. Sawyer, H. E. Shannon, P. H. Oleson, T. Honoré, M. J. Sheardown, and P. Sauerberg, *J. Med. Chem.*, **35**, 4011 (1992).

139. Eur. Pat. 318, 166 (1989); P. Krogsgaard-Larsen, E. Falck, and H. Pedersen, cited by M. R. Pavia, R. E. Davis, and R. D. Schwarz, *Annu. Rep. Med. Chem.*, **25**, 21 (1990).

140. P. Sauerberg, E. Falch, E. Meier, H. L. Lembol, and P. Krogsgaard-Larsen, *J. Med. Chem.*, **31**, 1312 (1988).

141. P. G. Dunbar, G. J. Durant, Z. Fang, Y. F. Abuh, A. A. El-Assadi, D. O. Ngur, S. Periyasamy, W. P. Hoss, and W. S. Messer, Jr., *J. Med. Chem.*, **36**, 842 (1993).

142. P. G. Dunbar, G. J. Durant, T. Rho, B. Ojo, J. J. Huzl, III, D. A. Smith, A. A. El-Assadi, S. Sbeih, D. O. Ngur, S. Periyasamy, W. Hoss, and W. S. Messer, Jr., *J. Med. Chem.*, **37**, 2774 (1994).

143. J. Saunders, G. A. Showell, R. J. Snow, R. Baker, E. A. Harley, and S. B. Freedman, *J. Med. Chem.*, **31**, 486 (1988).

144. F. Wanibuchi, T. Konishi, M. Harada, M. Terai, K. Hidaka, T. Tamura, S. Tsukamoto, and S. Usuda, *Eur. J. Pharmacol.*, **187**, 479 (1990).

145. Y. Ishihara, H. Yukimasa, M. Miyamoto, and G. Goto, *Chem. Pharm. Bull.*, **40**, 1177 (1992).

146. S.-i. Tsukamoto, M. Ichihara, F. Wanibuchi, S. Usuda, K. Hidaka, M. Harada, and T. Tamura, *J. Med. Chem.*, **36**, 2292 (1993).

147. E. Pombo-Villar, K.-H. Wiederhold, G. Mengod, J. M. Palacios, P. Supavilai, and H. W. Boddeke, *Eur. J. Pharmacol.*, **226**, 317 (1992).

148. S. M. Bromidge, F. Brown, F. Cassidy, M. S. G. Clark, S. Dabbs, M. S. Hadley, J. M. Loudon, B. S. Orlek, and G. J. Riley, *Bioorg. Med. Chem. Lett.*, **2**, 787 (1992).

149. S. M. Bromidge, F. Brown, F. Cassidy, M. S. G. Clark, S. Dabbs, J. Hawkins, J. M. Loudon, B. S. Orlek, and G. J. Riley, *Biorg. Med. Chem. Lett.*, **2**, 791 (1992).

150. B. S. Orlek, F. E. Blaney, F. Brown, M. S. G. Clark, M. S. Hadley, J. Hatcher, G. J. Riley, H. E. Rosenberg, H. J. Wadsworth, and P. Wyman, *J. Med. Chem.*, **34**, 2726 (1991).

151. R. M. Eglen, G. C. Harris, A. P. Ford, E. H. Wong, J. R. Pfister, and R. L. Whiting, *Naunyn-Schmiedeberg's Arch. Pharmacol.*, **345**, 375 (1992).

152. R. W. Brimblecombe, *Drug Actions on Cholinergic Systems*, University Park Press, Baltimore, Md., 1974, pp. 24–30.

153. A. Bebbington, R. W. Brimblecombe, and D. Shakeshaft, *Br. J. Pharmacol. Chemother.*, **26**, 56 (1966).

154. A. K. Cho, W. L. Haslett, and D. J. Jenden, *J. Pharmacol. Exp. Ther.*, **138**, 249 (1962).

155. A. Bebbington, R. W. Brimblecombe, and D. G. Rowsell, *Br. J. Pharmacol. Chemother.*, **26**, 68 (1966).

156. M. W. Moon, C. G. Chidester, R. F. Heier, J. K. Morris, R. J. Collins, R. R. Russell, J. W. Francis, G. P. Sage, and V. H. Sethy, *J. Med. Chem.*, **34**, 2314 (1991).

157. V. Sethy, J. Francis, D. Hyslop, G. Sage, T. Olen, A. Meyer, R. Collins, R. Russell, R. Heier, W. Hoffmann, M. Piercey, N. Nichols, P. Schreur, and M. Moon, *Drug Dev. Res.*, **24**, 53 (1991).

158. B. M. Nilsson, B. Ringdahl, and U. Hacksell, *J. Med. Chem.*, **33**, 580 (1990).

159. B. M. Nilsson, H. M. Vargas, and U. Hacksell, *J. Med. Chem.*, **35**, 3270 (1992).

160. B. Ringdahl, Z. Muhi-Eldeen, C. Ljunggren, B. Karlén, B. Resul, R. Dahlbom, and D. J. Jenden, *Acta Pharm. Suec.*, **16**, 89 (1979).

161. B. Resul, B. Ringdahl, and R. Dahlbom, *Acta Pharm Suec.*, **16**, 161 (1979).

162. B. Ringdahl and D. J. Jenden, *Mol. Pharmacol.*, **23**, 17 (1983).

163. R. Amstutz, B. Ringdahl, B. Karlén, M. Roch, and D. J. Jenden, *J. Med. Chem.*, **28**, 1760 (1985).

164. B. M. Nilsson, H. M. Vargas, B. Ringdahl, and U. Hacksell, *J. Med. Chem.*, **35**, 285 (1992).

165. Ö. Norström, P. Alberts, A. Westlind, A. Unden, and T. Bartfai *Mol. Pharmacol.*, **24**, 1 (1983).

166. F. Casamenti, C. Cosi, and G. Pepeu, *Eur. J. Pharmacol.*, **122**, 288 (1986).

167. B. Resul, R. Dahlbom, B. Ringdahl, and D. J. Jenden, *Eur. J. Med. Chem. Chim. Ther.*, **17**, 317 (1982).

168. R. Dahlbom, D. J. Jenden, B. Resul, and B. Ringdahl, *Br. J. Pharmacol.*, **76**, 299 (1982).

169. B. Ringdahl, M. Roch, and D. J. Jenden, *J. Pharmacol. Exp. Ther.*, **242**, 464 (1987).

170. B. M. Nilsson, B. Ringdahl, and U. Hacksell, *J. Med. Chem.*, **31**, 577 (1988).

171. J. R. M. Lundkvist, B. Ringdahl, and U. Hacksell, *J. Med. Chem.*, **32**, 863 (1989).

172. J. R. M. Lundkvist, H. M. Vargas, P. Caldirola, B. Ringdahl, and U. Hacksell, *J. Med. Chem.*, **33**, 3182 (1990).

173. B. Ringdahl, B. Resul, F. J. Ehlert, D. J. Jenden, and R. Dahlbom, *Mol. Pharmacol.*, **26**, 170 (1984).

174. B. Ringdahl and D. J. Jenden, *J. Med. Chem.*, **30**, 852 (1987).

175. B. Ringdahl, M. Roch, E. D. Katz, and M. C. Frankland, *J. Med. Chem.*, **32**, 659 (1989).

176. B. Ringdahl, M. Roch, and D. J. Jenden, *J. Med. Chem.*, **31**, 160 (1988).

177. A. P. Roszkowski, *J. Pharmacol. Exp. Ther.*, **132**, 156 (1961).

178. C. Mellin, H. M. Vargas, and B. Ringdahl, *J. Med. Chem.*, **32**, 1590 (1989).

179. B. M. Nilsson, H. M. Vargas, and U. Hacksell, *J. Med. Chem.*, **35**, 2787 (1992).

180. B. Ringdahl, *J. Med. Chem.*, **31**, 164 (1988).

181. B. Ringdahl, C. Mellin, F. J. Ehlert, M. Roch, K. M. Rice, and D. J. Jenden, *J. Med. Chem.*, **33**, 281 (1990).

182. C. Schumacher, R. Steinberg, J. P. Kan, J. C. Michaud, J. J. Bourguignon, C. G. Wermuth, P. Feltz, P. Worms, and K. Biziere, *Eur. J. Pharmacol.*, **166**, 139 (1989).

183. Ref. 7, p. 252.

184. A. F. Casy, *Progr. Med. Chem.*, **11**, 1 (1975).

185. IUPAC Tentative Rules, *J. Org. Chem.*, **35**, 2849 (1970).

186. R. W. Baker, C. H. Chothia, P. Pauling, and T. J. Petcher, *Nature*, **230**, 439 (1971).

187. Ref. 7, pp. 255–257.

188. P. Pauling and T. J. Petcher, *Nature New Biol.*, **236**, 112 (1972).

189. C. Chothia and P. Pauling, *Proc. Natl. Acad. Sci. U. S. A.*, **65**, 477 (1970).

190. H.-D. Höltje, M. Hense, S. Marrer, and E. Maurhofer, *Prog. Drug Res.*, **34**, 9 (1990).

191. A. F. Casy, *Ann. Rep. Progr. Chem.* (Sect. B), 477 (1974).

192. D. Lichtenberg, P. A. Kroon, and S. I. Chan, *J. Am. Chem. Soc.*, **96**, 5934 (1974).

193. Ref. 7, pp. 258–259.

194. F. W. Schueler, *J. Am. Pharm. Assoc. Sci. Ed.*, **45**, 197 (1956).

195. G. Lambrecht, *Eur. J. Med. Chem.*, **11**, 461 (1976).

196. Ref. 6, pp. 259–260.

197. G. Lambrecht, *Eur. J. Med. Chem.*, **12**, 41 (1977).

198. S. Archer, A. M. Lands, and T. R. Lewis, *J. Med. Pharm. Chem.*, **5**, 423 (1962).

199. E. Hardegger and W. Ott, *Helv. Chim. Acta*, **36**, 1186 (1953).

200. E. E. Smissman and G. S. Chappell, *J. Med. Chem.*, **12**, 432 (1969).

201. E. E. Smissman and G. R. Parker, *J. Med. Chem.*, **16**, 23 (1973).

202. J. B. Kay, J. B. Robinson, B. Cox, and D. Polkonjak, *J. Pharm. Pharmacol.*, **22**, 214 (1970).

203. A. F. Casy, E. S. C. Wu, and B D. Whelton, *Can. J. Chem.*, **50**, 3998 (1972).

204. J. G. Cannon, A. B. Rege, T. L. Gruen, and J. P. Long, *J. Med. Chem.*, **15**, 71 (1972).

205. J. G. Cannon, Y. Lin, and J. P. Long, *J. Med. Chem.*, **16**, 27 (1973).

206. D. J. Triggle, *Chemical Aspects of the Autonomic Nervous System*, Academic Press, London, 1965, p. 83.

207. P. D. Armstrong, J. G. Cannon, and J. P. Long, *Nature*, **220**, 65 (1968).

208. C. Y. Chiou, J. P. Long, J. G. Cannon, and P. D. Armstrong, *J. Pharmacol. Exp. Ther.*, **166**, 243 (1969).

209. C. Chothia and P. Pauling, *Nature*, **226**, 541 (1970).

210. P. D. Armstrong and J. G. Cannon, *J. Med. Chem.*, **13**, 1037 (1970).

211. J. G. Cannon, T. Lee, V. Sankaran, and J. P. Long, *J. Med. Chem.*, **18**, 1027 (1975).

212. J. G. Cannon, D. M. Crockatt, J. P. Long, and W. Maixner, *J. Med. Chem.*, **25**, 1091 (1982).

213. J. M. Schulman, M. L. Sabio, and R. L. Disch, *J. Med. Chem.*, **26**, 817 (1983).

214. C. Chothia, *Nature*, **225**, 36 (1970).

215. E. Shefter and D. J. Triggle, *Nature*, **227**, 1354 (1970).

216. J. L. Sussman, M. Harel, F. Frolow, C. Oefner, A.

Goldman, L. Toker, and I. Silman, *Science*, **253**, 872 (1991).

217. J. E. Gearien, in W. O. Foye, Ed., *Principles of Medicinal Chemistry*, 3rd Ed., Lea & Febiger, Philadelphia, 1989, p. 337.

218. Ref. 4, pp. 132–135.

219. G. M. Steinberg, M. L. Mednick, J. Maddox, R. Rice, and J. Cramer, *J. Med. Chem.*, **18**, 1056 (1975).

220. J. A. Cohen and R. A. Oosterbaan, *Cholinesterases and Anticholinesterase Agents, Handbuch Exp. Pharmakol.*, Vol. 15, G. B. Koelle, Ed., Springer-Verlag, Berlin, 1963, p. 299.

221. F. Bergmann and A. Shimoni, *Biochim. Biophys. Acta*, **10**, 49 (1953).

222. B. Belleau, *Ann. N. Y. Acad. Sci.*, **144**, 705 (1967).

223. B. Holmsted, *Pharmacol. Rev.*, **11**, 567 (1959).

224. M. P. Fulton and G. A. Mogey, *Br. J. Pharmacol.*, **9**, 138 (1954).

225. C. J. Cavallito and P. Sandy, *Biochem. Pharmacol.*, **2**, 233 (1959).

226. M. Mooser, H. Schulman, and D. S. Sigman, *Biochemistry*, **11**, 1595 (1972).

227. Ref. 7, p. 279.

228. P. Watts and R. G. Wilkinson, *Biochem Pharmacol.*, **26**, 757 (1977).

229. B. W. Bolger, *Clin. Pharm.*, **10**, 447 (1991).

230. J. R. Atack, Q. S. Yu, T. T. Soncrant, A. Brossi, and S. I. Rapoport, *J. Pharmacol. Exp. Ther.*, **249**, 194 (1989).

231. F. J. Dale and B. Robinson, *J. Pharm. Pharmacol.*, **22**, 889 (1970).

232. J. P. Long, *Biochem. Biophys. Acta*, **10**, 374 (1953).

233. A. Brossi, *J. Med. Chem.*, **33**, 2311 (1990).

234. N. M. Rupniak, S. J. Tye, C. Brazell, A. Heald, S. D. Iverson, and P. G. Pagella, *J. Neurol. Sci.*, **107**, 246 (1992).

235. Y. L. Chen, J. Nielsen, K. Hedberg, A. Dunaiskis, S. Jones, L. Russo, J. Johnson, J. Ives, and D. Liston, *J. Med. Chem.*, **35**, 1429 (1992).

236. F. F. Foldes, E. G. Erdos, N. Baart, J. Zwart, and E. K. Zsigmond, *Arch. Int. Pharmacodyn. Ther.*, **120**, 286 (1959).

237. B. Holmsted in ref. 220, pp. 428–485.

238. W. J. Hayes, Jr., *Pesticide Studies in Man*, 2nd ed., Springer-Verlag, Berlin, 1982, pp. 284–435.

239. J. E. Casida, T. C. Allen, and M. A. Stahmann, *J. Biol. Chem.*, **210**, 607 (1954).

240. Y. Ashani, H. Leader, L. Raveh, R. Bruckstein, and M. Spiegelstein, *J. Med. Chem.*, **26**, 145 (1983).

241. S. Matsunaga, R. E. Moore, W. P. Niemczura, and W. W. Carmichael, *J. Am. Chem. Soc.*, **111**, 8021 (1989).

242. N. A. Mahmood and W. W. Carmichael, *Toxicon*, **24**, 425 (1986).

243. E. Heilbronn, *Acta Chem. Scand.*, **15**, 1386 (1961).

244. M. Harel, I. Schalk, L. Ehret-Sabatier, F. Bouet, M. Goeldner, C. Hirth, P. H. Axelsen, I. Silman, and J. L. Sussman, *Proc. Natl. Acad. Sci. U. S. A.*, **90**, 9031 (1993).

245. W. K. Summers, L. V. Majovski, G. M. Marsh, K. Tachiki, and A. Kling, *N. Engl. J. Med.*, **315**, 1241 (1986).

246. W. K. Summers, K. R. Kaufman, F. Altman, Jr., and J. M. Fischer, *Clin. Toxicol.*, **16**, 269 (1980).

247. G. M. Shutske, F. A. Pierrat, M. L. Cornfeldt, M. R. Szewczak, F. P. Huger, G. M. Bores, V. Haroutunian, and K. L. Davis, *J. Med. Chem.*, **31**, 1278 (1988).

248. G. M. Shutske, F. A. Pierrat, K. J. Kapples, M. L. Kornfeldt, M. R. Szewczak, F. P. Huger, G. M. Bores, V. Haroutunian, and K. L. Davis, *J. Med. Chem.*, **32**, 1805 (1989).

249. B. Belleau and H. Tani, *Mol. Pharmacol.*, **2**, 411 (1966).

250. S. N. Abramson, Z. Radic, D. Manker, D. J. Faulkner, and P. Taylor, *Mol. Pharmacol.*, **36**, 349 (1989).

251. Y. Ishihara, K. Kato, and G. Goto, *Chem. Pharm. Bull.*, **39**, 3225 (1991).

252. Y. Ishihara, M. Miyamoto, T. Nakayama, and G. Goto, *Chem. Pharm. Bull.*, **41**, 529 (1993).

253. Y. Ishihara, K. Hirai, M. Miyamoto, and G. Goto, *J. Med. Chem.*, **37**, 2292 (1994).

254. Y. Yamamoto, Y. Ishihara, and I. D. Kuntz, *J. Med. Chem.*, **37**, 3141 (1994).

255. J. L. Vidaluc, F. Calmel, D. Bigg, E. Carilla, A. Stenger, P. Chopin, and M. Briley, *Abstr. Am. Chem. Soc.* (208 Mtg), Medi 200 (1994).

256. R.A. Earl, M.J. Myers, C.Y. Cheng, V.R. Ganti, R.M. Scribner, V.J. Nickolson, S.W. Tam, and L. Cook, *Abstr. Am. Chem. Soc.* (196 Mtg), Medi 99 (1988).

257. V. J. Nickolson, S. W. Tam, M. J. Myers, and L. Cook, *Drug Dev. Res.*, **19**, 285 (1990).

258. W. S. Tam, D. Rominger, and V. J. Nickolson, *Mol. Pharmacol.*, **40**, 16 (1992).

259. B. Saletu, A. Darragh, P. Salmon, and R. Coen, *Br. J. Clin. Pharmacol.*, **28**, 1 (1989).

260. R. A. Earl, M. J. Myers, C. Y. Cheng, A. L. Johnson, R. M. Scribner, J. M. Smallheer, C. Amaral-Ly, V. J. Nickolson, S. W. Tam, M. A. Wuonola, G. A. Boswell, V. J. DeNoble, and L. Cook, *Abstr. Am. Chem. Soc.* (203 Mtg), Medi 104 (1992).

261. J. Varghese, I. Lieberburg, and E. D. Thorsett, *Annu. Rep. Med. Chem.*, **28**, 197 (1993).

CHAPTER TWENTY-SIX

Anticholinergic Drugs

B. V. RAMA SASTRY

School of Medicine
Vanderbilt University
Nashville, Tennessee

CONTENTS

1 Introduction, 60
　1.1 Types and selectivity of antispasmodics, 61
　1.2 Gastric secretion, peptic ulcer, and
　　　anticholinergics as antiulcer agents, 62
　1.3 Anticholinergics as mydriatics and
　　　cycloplegics, 66
　1.4 Anticholinergic drugs for premedication
　　　during anesthesia, 66
　1.5 Anticholinergic activity as a side effect of
　　　drugs and anticholinergic syndrome, 67
　1.6 Classification of anticholinergic agents based
　　　on subtypes of muscarinic receptors, 67
2 Biocomparative Assay of Anticholinergics, 67
　2.1 Antispasmodic activity, 68
　2.2 Antiulcer activity, 70
　2.3 Mydriatic and cycloplegic activities, 71
　2.4 Miscellaneous anticholinergic activities, 72
3 Solanaceous Alkaloids, 72
　3.1 History, 72
　3.2 Chemical structure, 73
　3.3 Preparative methods, 74
　3.4 Molecular factors in the absorption, fate, and
　　　excretion of atropine and related
　　　compounds, 75
　3.5 Semisynthetic derivatives of solanaceous
　　　alkaloids, 77
4 Synthetic Anticholinergics, 79
　4.1 Analogs of atropine, 79
　4.2 Receptor subtype selective
　　　anticholinergics, 82
　　　4.2.1 Tricyclic benzodiazipines, 82
　　　4.2.2 Benzothiazipines, 83
　　　4.2.3 Quinuclidines, 84
　　　4.2.4 Polymethylene tetramines, 84

Burger's Medicinal Chemistry and Drug Discovery,
Fifth Edition, Volume 2: Therapeutic Agents,
Edited by Manfred E. Wolff.
ISBN 0-471-57557-7 © 1996 John Wiley & Sons, Inc.

4.2.5 Indenes, 85
4.2.6 Sila-difenidols, 86
4.2.7 Diphenylacetoxy derivatives, 86
4.2.8 Himbacene alkaloids, 87
5 Structure–Activity Relationships, 87
5.1 Cationic head, 87
5.2 Cyclic moieties, 96
5.3 Length of the main chain connecting the cationic head and the cyclic groups, 98
5.4 Esteratic linkage, 99
5.5 Hydroxyl group, 100
5.6 Epoxy group, 101
5.7 Stereoisomerism and anticholinergic activity, 101
5.7.1 Optical isomerism, 101
5.7.2 Derivatives of tropine and pseudotropine, 102
5.7.3 Stereochemical configuration, 102
5.7.4 Dissociation constants of cholinergics and anticholinergics, 102
5.8. Compounds with dual action: cholinergic and anticholinergic activities, 104
6 Interaction of Anticholinergics at the Muscarinic Receptors, 104
6.1 Kinetic basis for the mechanism of action of anticholinergics, 105
6.2 Specificity of antagonism, 105
6.3 Molecular basis for the interaction of acetylcholine and anticholinergics at the muscarinic receptors, 105
7 Therapeutic Uses of Anticholinergics, 106
8 Molecular Basis for the Side Effects of Anticholinergics, 107
9 Profile of Anticholinergic Activities of Various Agents, 108
10 Nonanticholinergics as Antiulcer Agents, 111

1 INTRODUCTION

The role of acetylcholine as a parasympathetic neurotransmitter and its effects on smooth muscle and glands are reviewed in Chapter 25. Typical parasympathetic effects, in addition to cardiac inhibition and vasodilation in certain areas, are miosis and increased gastrointestinal motion and secretion. It is believed that acetylcholine is the common factor in many of these processes. Electrical stimulation of parasympathetic nerves causes the appearance of acetylcholine at the neuromuscular junction; presumably acetylcholine appears reg-

ularly during the spontaneous functioning of the postganglionic fibers of the parasympathetic nerves and is regularly kept from accumulating, owing to hydrolysis under the influence of acetylcholinesterase (Chapter 25). Spontaneous release of acetylcholine at parasympathetic nerve endings results in the involuntary contraction or spasm of the muscle. Therefore, the contractions of the stomach, intestinal tract, heart, certain blood vessels, and many other structures in various pathological situations are often attributed to the amounts of acetylcholine in excess of normal requirements. Gastric secretion salivation, micturition, lacrimation, sweating, and miosis are influenced by acetylcholine. The rates of these activities can be controlled by certain anticholinergic drugs.

Anticholinergic drugs interfere with physiological functions that depend on cholinergic nerve transmission. These drugs do not prevent acetylcholine from being released at nerve endings, but they may compete with the liberated neurohormone for cholinergic receptor sites. Acetylcholine is the chemical transmitter at postganglionic parasympathetic nerve endings as well as at autonomic ganglia and somatic neuromuscular junctions. Acetylcholine is also a chemical transmitter at certain synapses in the central nervous system and drugs acting at these central cholinergic sites are discussed elsewhere (Chapter 37). Different types of anticholinergic drugs antagonize the actions of acetylcholine at the three types of peripheral synapses mentioned above. Anticholinergic drugs that block somatic neuromuscular junction (curariform drugs) and autonomic ganglia (ganglionic blocking drugs) are described in Chapters 29 and 36. The pharmacological actions of anticholinergic drugs discussed in this chapter mimic the effects of cutting the parasympathetic nerve supply to various organs; therefore, they are designated as parasympatholytics. Muscarine mimics the actions of acetylcholine on the structures innervated by parasympathetic nerves; it is relatively inactive at autonomic ganglia and somatic neuromuscular junctions. Parasympatholytics that antagonize the actions of muscarine are also known as antimuscarinic agents.

The classic parasympatholytic agent is atropine; therefore, anticholinergic drugs used to be referred to as atropinic agents. Typical effects produced by atropine are mydriasis, tachycardia, decreased gastrointestinal peristalsis, and diminished secretions of gastric juice, saliva, and sweat. A large number of anticholinergic agents have been synthesized that have specific actions

and uses. Although all anticholinergics could be considered antispasmodics to different degrees, for convenience they are divided into three categories: (*1*) antispasmodics, which are specifically used to relieve spasms of the bowel (e.g., irritable colon and spastic colitis); (*2*) antiulcer agents, which reduce gastric secretion; and (*3*) mydriatics and cycloplegics, which relax the sphincter of the iris and the ciliary muscles.

1.1 Types and Selectivity of Antispasmodics

Substances patterned on atropine are widely used as antispasmodics of the gastrointestinal tract. Theoretically, any such substance that relaxes the acetylcholine-induced spasm of the smooth muscles in suitable doses can be termed an antispasmodic. In practice, not every anticholinergic agent can be used as an antispasmodic. The reason is that in addition to their spasmolytic action, anticholinergics influence the functions of other organs, including heart, sweat and salivary glands, and iritic and ciliary muscles, producing side effects. Moreover, a number of them in small doses cause undesirable disorders in the central nervous system. The same antispasmodic is not suitable for the spastic states of all organs. Furthermore, there are differences in the *in vitro* and *in vivo* efficacies of antispasmodics. Atropine abolishes the acetylcholine-induced spasm of guinea pig ileum completely; however, it is a familiar clinical experience that atropine does not completely antagonize the spasm caused by increased tone of the intestinal vagus nerve.

A number of agents cause spasm of the gastrointestinal tract. The spasm may be induced not only by acetylcholine but also by histamine, 5-hydroxytryptamine, or barium chloride. Atropine and other anticholinergics are effective mostly against acetylcholine-induced spasm and less so against the remaining three spasmogens. Against a spasm induced by acetylcholine, atropine is effective at the lowest concentrations, e.g., 10^{-9} g/mL. Higher concentrations are necessary to antagonize 5-hydroxytryptamine spasm (10^{-7}), histamine spasm (10^{-6}), and Ba^{2+} spasm (10^{-5}). Thus atropine is a highly specific anticholinergic neurotropic spasmolytic.

Barium ion acts on all smooth muscles regardless of innervation and is called a musculotropic spasmogen. Drugs that relieve the spasm produced by barium ions are called musculotropic spasmolytics. Papaverine and nitrites are typical members of this class. However, various drugs that resemble atropine manifest both kinds of spasmolytic action in widely varying situations.

The ideal atropine-like antispasmodic should be specific for the spasmogen, should have selectivity for smooth muscles, and should abolish completely the spasm induced by the stimulation of the parasympathetic nerve to the organ. Furthermore, the atropine-like antagonist should be specific for the subtype of muscarinic receptor localized in the organ. None of the available antispasmodics satisfies all these requirements. However, a great many compounds have been synthesized with the hope of developing drugs that will exhibit more selective antispasmodic action and have fewer side effects than atropine. Some of these antispasmodics show relative selectivity toward the subtype of muscarinic receptor localized in smooth muscle cells.

1.2 Gastric Secretion, Peptic Ulcer, and Anticholinergics as Antiulcer Agents

The pathophysiology of peptic ulcer is not fully known, and in the present state of knowledge, it is not possible to present the pertinent normal physiology briefly. For a detailed discussion on the physiology and chemistry of gastric secretion and the pathologic physiology of peptic ulcer, reference should be made to reviews on the subject (1–5). The following is a brief summary of the gastric secretion and its relationship to peptic ulcer, a knowledge of which is necessary to understand the problems of developing antiulcer agents.

Gastric juice contains a mixture of water, inorganic ions, hydrochloric acid, pepsinogens, mucus, various polypeptides, and the intrinsic factor. Pepsinogens are precursors of the proteolytic enzymes, pepsins. They are readily converted into the corresponding pepsins by either acid or pepsin itself. Conversion by acid is instantaneous at pH 2.0. In humans, gastric juice contains hydrochloric acid during the period of interdigestive secretion as well as during the period of digestive secretion. Although the mechanisms of interdigestive secretion are not known, they depend partly on the tonic activity of the vagus. The gastric secretory activity during the period of digestive secretion may be divided into three phases: cephalic, gastric, and intestinal. Each phase is named to denote the region in which the stimuli act to induce gastric secretion.

In the cephalic phase the stimuli are initiated in the central nervous system. The stimuli are the sight, smell, taste, and thought of food, which act through conditioned and unconditioned reflexes. The final efferent path is the vagus nerve. The impulses in the vagus nerve stimulate the secretory cells in the gastric glands. Acetylcholine, which is released from the postganglionic nerve endings, exerts a direct action on the secretory cells, Administration of atropine abolishes this phase. The secretion is high in acid and pepsinogens, and its concentration of mucus is lower than that of the basal secretion; mucus output rises 8–10 times as volume increases.

The gastric phase of secretion begins copiously as soon as the food enters the stomach, and it may continue 3–4 h with a total volume of 600 mL or more of strongly acid juice, containing a high concentration of pepsinogens. The gastric phase of secretion is caused by local and vagal responses to distension and by the hormone gastrin, which is released by the mucosa of the pyloric gland area. The local nerves of the pyloric area are confined to the mucosa and are cholinergic. Irrigation of the pyloric gland area with acetycholine releases gastrin, and this liberation of gastrin is abolished by atropinization. There is a synergism between gastrin and acetylcholine at the target cells; the effect of injected gastrin on both acid and pepsinogen secretion is increased twofold to eightfold by subthreshold parasympathomimetic stimuli, and it is strongly inhibited by atropinization.

The intestinal phase, which begins when chyme passes from the stomach to intestine, contributes about 10% of the total response to a test meal. Protein and its digestion products, milk, dilute alcohol, and acid itself are effective stimulants. Although there may be a nervous component, the intestinal phase includes humoral stimulation of secretion by unknown agents. Gastrin released from the small intestine may be involved. The response to whatever humoral agent comes from the intestine is greatly increased when subthreshold doses of cholinergic drugs are given.

A number of humoral inhibitors of gastric secretion arise in the small intestine. They are termed the enterogastrones. An enterogastrone is present in the jejunum and duodenal mucosa. It is released in the presence of fat and inhibits gastric secretion and motility. The hormone secretin, which stimulates pancreatic secretion, is an enterogastrone. It is produced in the proximal duodenum and inhibits gastric secretion in the presence of acids. Cholecystokinin, which is the same as pancreozymin, and gastrin share the same terminal tetrapeptide. Given alone, cholecystokinin is only a mild stimulant of gastric acid secretion. It is

a competitive inhibitor of the receptor for gastrin, which is a powerful stimulant of gastric acid secretion. Therefore, in the presence of gastrin, cholecystokinin decreases the total output of acid. Glucagon (and possibly enteroglucagon) reduces the gastrin-induced acid secretion by noncompetitive inhibition of the receptors to gastrin. A gastric inhibitory polypeptide (GIP) that is present in duodenal mucosa inhibits histamine- as well as gastrin-induced acid secretion. A vasoactive intestinal peptide (VIP), which has been isolated from small intestinal mucosa, inhibits histamine-induced acid secretion. GIP and VIP are two possible enterogastrones, whose significance has yet to be established.

Histamine, the exact role of which is not clearly understood, stimulates secretion of gastric juice that is rich in hydrochloric acid. Recently, histamine receptors have been divided into three types, H1, H2 and H3 (Chapter 27 and the chapter on Histamine H1 Antagonists). Stimulation of H2 receptors by histamine results in increased gastric acid secretion. H2 receptor antagonists (burimamide, metiamide, cimetidine) inhibit histamine-induced gastric acid secretion in humans and animals. In humans, H2 antagonists inhibit not only histamine but also pentagastrin (a synthetic analog of gastrin) stimulated gastric acid secretion. This suggests that, at least in humans, gastrin acts partially via histamine. Blockage of acetylcholine receptors by atropine and histamine receptors by H2 antagonists results in reduction of the effectiveness of gastrin to induce acid secretion. Therefore, there seems to be a complex interaction among the three receptors (acetylcholine receptors, H2 receptors, and gastrin receptors) involved in the acid secretion by parietal cells.

Among local hormones and messengers, prostaglandins (PGE_1, PGA_1, (see the chapter on Prostaglandins)) cyclic adenosine $3',5'$-monophosphate (cAMP) inhibit both pentagastrin- and histamine-stimulated gastric secretion. According to present evidence, all hormones that reduce gastric acid secretion increase both adenyl cyclase and intracellular cAMP activity. Conversely, all hormones that primarily stimulate gastric acid secretion reduce intracellular cAMP levels. Therefore, cAMP is involved in the final links of gastric acid secretion.

The interplay among various neuronal and hormonal factors in the gastric acid secretion by the parietal cell during cephalic, gastric, and intestinal phases are schematically shown in Figure 26.1. In addition to being inhibited by atropine-like agents and enterogastrones, the acid secretion is inhibited by gastrone in the mucus of human stomach, by urogastrone isolated from the urine of men and dogs, by strongly acid solutions in the duodenum, and by stimulation of the sympathetic nervous system.

Peptic ulcer occurs in the pyloric region of the stomach or the first few centimeters of the intestine. The gastroduodenal mucosa is exposed constantly to mechanical, physical, and chemical insults, some of which have already been described. A peptic ulcer does not develop without the presence of a pepsin-containing juice of such low pH that it can exert a peptic influence on the gastric wall itself. The extent of this insult is determined by the number of acid- and pepsinogen-producing cells, their irritability, and/or the magnitude of the stimuli that reach them. These stimuli are partly nervous (vagal) and partly hormonal (gastrin, corticosteroids).

The healthy stomach does not digest itself. Counteracting the aggression are defensive factors such as buffering and dilution by food, inhibition of the secretion of gastric juice, and drainage of gastric contents. In addition, however, the local condition of the mucosa (the mucosal resistance) is also of importance. Some of the determinants of mucosal resistance are the mucous barrier, the local circulation, and

the healing capacity of the mucosa. A peptic ulcer forms when the insult is more powerful than the defense. In the case of duodenal ulcers, the powerful irritation is often the important factor; in gastric ulcers it is the insufficient defense.

The ideal agent for the treatment of the peptic ulcer would be one that selectively inactivates pepsin or inhibits the output of hydrochloric acid to maintain the pH of the gastric contents at about 4.5 for long periods after its oral ingestion. It should produce no, or only minimal, side effects, induce no tolerance, and be inexpensive. It should be effective during all periods and phases of gastric secretion and prevent the formation of ulcers.

Atropine-like anticholinergics do not satisfy all requirements of an antiulcer agent. They block acetylcholine at the neuroeffector junction of the vagus. They give relief to patients with a peptic ulcer by their antisecretory and antispasmodic effects. They decrease the basal hydrochloric acid and pepsin secretion, thereby allowing the healing of ulcers. The antispasmodic effects of atropine-like agents are as consistent as their antisecretory effects. Motor activity is closely related to ulcer pain, and the pain-relieving action of anticholinergic agents seems to be related to their effect on depressing motor activity (antispasmodic effect).

An "effective" atropine-like anticholinergic drug is capable of favorably influencing the excessive gastric secretion under certain conditions. It exerts a significant effect on acid secretion during the basal and interdigestive night secretion to the point of abolishing it completely for many hours (6). Its effect on secretion during the feeding of milk and cream is significant (7,8). However, anticholinergic drugs do not effectively perform a "medical vagotomy," and they do not effectively reduce gastric acidity to the extent of achlorhydria when patients are fed (8,9). The effective anticholinergic agent as an antiulcer drug

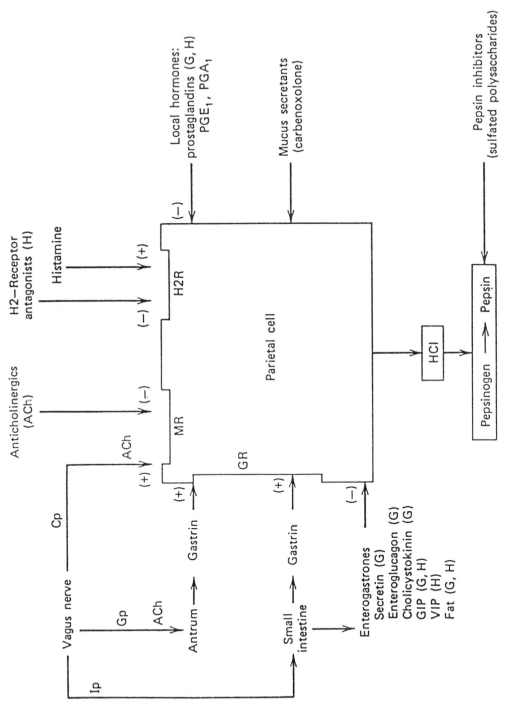

Fig. 26.1 Interactions among neuronal and hormonal factors and pharmacological agents during cephalic (Cp), gastric (Gp), and intestinal (Ip) phases of gastric acid secretion by parietal cell. ACh, acetylcholine; GIP, gastric inhibitory peptide; VIP, vasoactive intetinal peptide; MI, muscarinic receptor; H2, histamine H2 receptor; GR, gastrin receptor; (+), stimulation of acid secretion; (−), inhibition of acid secretion. In parentheses, next to the inhibitory agents, the blocked stimulant agent is indicated (ACh, acetylcholine; H, histamine; G, gastrin).

should be selective for the subtype of muscarinic receptors localized on the secretory cells of gastric glands as well as mucosa of pyloric gland area. Anticholinergics selective for muscarinic receptors of M_1 subtype are useful to decrease gastric acid secretion.

1.3 Anticholinergics as Mydriatics and Cycloplegics

The size of the pupil is determined by the balance of forces exerted by the dilator muscles fibers (sympathetically innervated and radially arranged) and the constrictor muscle fibers (parasympathetically innervated and circularly arranged) of the iris. Normally both sets of muscle fibers have a constant degree of tonus and act reciprocally to dilate or constrict the pupil. Any substance that paralyzes the constrictor muscle fibers (parasympatholytic) allows the unopposed tone of dilator muscle fibers to widen the pupil.

Acetylcholine is the transmitter between the constrictor muscle fibers and the parasympathetic nerve, which innervates them. Therefore, acetylcholine and its congeners stimulate the constrictor muscle fibers of the iris and constrict the pupil. Atropine and related compounds paralyze the constrictor muscle fibers and cause widening or dilatation of the pupil.

The ciliary muscle is innervated by the parasympathetic nerve, and acts to decrease the tone on the supporting muscle fibers of the lens and thus increases the accommodative power of the eye. Acetylcholine and its congeners constrict the ciliary muscle fibers, and atropine and related compounds paralyze the ciliary muscle.

Mydriatics are drugs that dilate the pupil but have minimal effect on the ciliary muscle and thus on accommodation. Cycloplegics are drugs that partially or completely paralyze accommodation. Most of the anticholinergics have both properties to varying degrees. For mydriatics other than anticholinergics and for drugs that constrict the pupil (miotics), see the chapter on Antiglaucoma and Anticataract Drugs.

Mydriatics and cycloplegics are special types of antispasmodics. In clinical practice mydriasis is produced by local instillation of the chosen drug into the conjunctival sac. This enables one to produce the desired effects on the eye with minimal systemic effects. However, such compounds should possess properties that allow them to penetrate the cornea in effective concentrations. There are no significant differences between the muscarinic receptors of the guinea pig ileum or the rabbit iris as judged by the binding characteristics of potent anticholinergic agents. Muscarinic receptors in both tissues are possibly of M_3 subtype. If anticholinergic drugs were available that are selective for muscarinic receptors on constrictor muscles and ciliary muscles, mydriatic and cycloplegic effects could be produced by different drugs.

1.4 Anticholinergic Drugs in Premedication During Anesthesia

Prevention of some undesirable side effects during anesthesia has been considered a function of premedication with anticholinergic drugs. For example, atropine is a popular anticholinergic agent, which has been used for its antisialagogic, antibradycardia, and antiemetic effects (10). The emphasis of using anticholinergic drugs during premedication has been changing over the years due to availability of better inhalational anesthetic agents than ether. An anticholinergic agent (e.g., atropine), although no longer regarded as an essential premedicant under all circumstances, does have specific applications for injured patients and children. Atropine (0.6 mg, i.v.) blocks the muscarinic actions of suxamethonium (succinylcholine), bradycardia and salivation, during crash induction of anesthesia in an injured patient (10). Administration of an anticholinergic drug to prevent bradycardia in children in response to suxamethonium or tracheal intubation is desirable.

1.5 Anticholinergic Activity as a Side Effect of Drugs and Anticholinergic Syndrome

Many of the drugs used in current medical practice, especially anesthetic drugs and other drugs used as adjuvants to anesthesia, have properties that disturb patient recovery due to their anticholinergic effects on the central nervous system (11). These effects are termed central anticholinergic syndrome (CAS) and are discussed in different chapters on centrally acting drugs (Volumes 3 and 5). These effects can be reversed with the centrally active cholinesterase inhibitor physostigmine, which has a sparing effect on acetylcholine molecules at muscarinic receptor sites. Increased number of acetylcholine molecules displace the molecules of anticholinergic drug from the muscarinic receptor sites.

1.6 Classification of Anticholinergic Agents Based on Subtypes of Muscarinic Receptors

Acetylcholine produces its parasympathomimetic effects by binding to cholinergic receptor sites termed muscarinic receptors. The classical anticholinergic agent, atropine, binds to the same muscarinic receptors and prevents the binding of acetylcholine to these receptors to elicit muscarinic responses. Based on modern developments in the design of relatively sensitive antagonists for muscarinic receptors in different tissues (12–16), muscarinic receptors have been subdivided into three (possibly five) subtypes: M_1 to M_5 (Table 26.1). All muscarinic receptors are glyco-

proteins of molecular weight of 80,000 and have seven membrane-spanning regions. All of the receptors have a slow response time (100–250 ms) and are coupled to G proteins (14,15). They act directly on ion channels or are linked to second messenger systems, attenuation cAMP formation (17,18) and formation of inositol triphosphate and diglyceride (17,19). The final effect of activation of these receptors can be to open or close K^+ channels, Ca^{2+} channels or Cl^- channels. These multiple channel activities lead to either depolarization or hyperpolarization of the cell membrane. The final responses are either excitatory or inhibitory. Atropine blocks all of these activities and does not distinguish subtypes. Selective muscarinic agonists and antagonists that can distinguish different subtypes are needed. Furthermore, it will be a major advance to obtain information to indicate that each subtype produces a specific function. Then it will be possible to develop specific anticholinergic drugs that are useful only as antispasmodics, antisecretory, or mydriatic agents. With certain anticholinergic agents, some degree of selectivity (not specificity) has been attained to produce antispasmodic, antisecretory, or mydriatic effects (Table 26.2). No antagonist has a potency on one receptor subtype that is more than 10 times higher than its potency on other subtypes. All receptor subtypes have K_d values for (−)-N-methyl scopolamine (NMS) and (−)-3-quinuclidinyl benzilate (QNB) of less than 1.0 nM (14). NMS and QNB are standard anticholinergic agents in addition to atropine to compare anticholinergic potencies at muscarinic receptors.

2 BIOCOMPARATIVE ASSAY OF ANTICHOLINERGICS

Many of the methods of obtaining experimental evidence for the antispasmodic, antiulcer, and mydriatic activities do not

Table 26.1. Provisional Division of Muscarinic Receptors (M), Their Agonists and Antagonists into Five Subtypes[a]

M Subtype	M_1	M_2	M_3	M_4	M_5
Previous name	$M_{1\alpha}$	$M_{2\alpha}$ Cardiac M_2	$M_{2\beta}$ Glandular M_2	$M_2(?)$	
Tissue location	Lower esophageal sphincter, gastric glands, CNS[b] ganglia	Heart	Glands smooth muscle, CNS		
Selective agonists[c]	McN-A-343				
Selective antagonists	Pirenzepine, (+)-telenzepine	Methoctramine, AF-DX 116, himbacine	HHSID[d] p-F-HHSID[e]	Himbacine (high affinity)	
Effector pathway	IP_3/DG[f]	cAMP↓ K^+ channel↑	IP_3/DG	cAMP↓	IP_3/DG
Gene	m_1	m_2	m_3	m_4	m_5
Amino acids (human)	460	466	590	479	532

[a]Summarized from refs. 12–19.
[b]Central nervous system sites.
[c]Selective agonists for receptors M_2–M_5 are not available.
[d]Hexahydrosiladifenidol.
[e]p-Fluoro-hexahydrosiladifenidol.
[f]Inositol-1,4,5-triphosphate/diglyceride.

measure precisely and selectively only one type of pharmacological activity. However, the techniques that are available (23,24) if used with an understanding of their scope and limitations, can provide useful information in the development of anticholinergic agents and their structure–activity relationships.

2.1 Antispasmodic Activity

In studying drugs more or less like atropine, it is customary to test their antispasmodic action on smooth muscles, such as the isolated guinea pig ileum, duodenum, or jejunum of rabbit or rat intestine. Acetylcholine or any one of the cholinergics may be used as a spasmogen, and the ability of the antispasmodic to inhibit or abolish the cholinergic induced spasm may be measured. Helical strips of blood vessels with intact endothelium, e.g., strips of rat aorta, can be used also to evaluate the antispasmodic activity of anticholinergic drugs (25). The antagonistic activities may be expressed as affinity constants or relative molar activities in relation to a standard antagonist. The selectivity of the antispasmodic activity may be determined by using different spasmogens, e.g., histamine, 5-hydroxytryptamine, nicotine, and barium chloride.

Thiry-Vella fistulas, prepared at various levels of the gastrointestinal tract, have been used in the conscious dog for determining motility by (1) placing an indigestible bolus in the oral end of the fistula and determining the traverse time before and after treatment with drugs; (2) placing a balloon containing water and attached to a kymographic recording system in the fistula and recording the pressure waves and their alterations by the action of drugs;

Table 26.2.   Derivatives of Solanaceous Alkaloids and Their Semisynthetic Substitutes[a]

Generic Name	Trade Name(s)	Dose or Preparation	Advantage of Molecular Modification	Therapeutic Use
Atropine sulfate		0.5 mg (oral i.v or s.c.); 0.5–1.0% ophthalmic solution		Mydriatic with long recovery period; preanesthetic medication to decrease secretions, treatment of Parkinsonism, and anti-ChE poisoning
Atropine tannate	Atratran	1–2 mg (tablet)	Slow absorption with sustained release of the alkaloid	Antispasmodic in ureteral and renal colic
Ipratropium[b] bromide	Atrovent	Inhaler	Low systemic absorption	Bronchodilator in asthma
Atropine N-oxide hydrochloride	X-tro, Genatropine	0.5–1.0 mg (capsule)	Slow release of the alkaloid	Same as atropine for oral use
Hyoscyamine hydrobromide		0.25–1.0 mg	Possibly fewer central effects than atropine, owing to small doses administered	Same as atropine for oral use
Methylatropine bromide	Mydriasine	0.5–2% solution	Mydriatic with short recovery period	Mydriatic
Methylatropine nitrate	Metropine	1–5% solution	Same as above	Mydriatic
Scopolamine hydrobromide		0.6 mg (oral, (s.c.); 0.2% solution	Central depressant (twilight sleep)	Sedative during preoperative or postoperative gynecologic care
Genescopolamine hydrobromide		1–2 mg	Gradual release of alkaloid	Same as above
Methscopolamine bromide	Pamine, Lescopine	2.5–5.0 mg (oral); 0.25–1.0 mg (s.c. or i.m.)	Parasympatholytic without central effects	Antisecretory and antispasmodic in peptic ulcer
Methscopolamine nitrate	Skopolate, Skopyl	2–4 mg (oral); 0.25–0.5 mg (s.c. or i.m.)	Same as above	Same as above
Homatropine hydrobromide		1–2% solution	Mydriatic with recovery period less than that of atropine	Mydriatic
Homatropine methyl bromide	Novatropine, Mesopin	5 mg (oral)	Parasympatholytic without central effects	Antisecretory and antisposmodic
Anisotropine methyl bromide[c]	Valpin, Endo	10 mg (oral)	Parasympatholytic without central effects	Antisecretory and antispasmodic

[a]For details of the preparations and their uses standard references in pharmacology should be consulted (20–22).

[b]8-Isopropylnoratropine methobromide.

[c]8-methyl-3-(2-propylpentonoyloxy) tropinium bromide octatropine bromide.

and (*3*) placing a French catheter in the aboral end of the fistula, connecting it to a suitable recording system, and thus making a record of normal pressures and those occurring after treatment (23). Other qualitative and quantitative methods to study the antispasmodics have been described (23). These include the fluoroscopic study of the gastrointestinal motility and the use of an ingestible pressure-sensitive radio-telemetering capsule (Transensor) for measuring the pressure in the gastrointestinal tract.

The subtype of muscarinic receptor in the smooth muscle has been characterized as M_3, using selective anticholinergics and different smooth muscle preparations from different species. These smooth muscle tissues include the trachea (26), ileum (27,28), uterine artery (29), and submucosal arterioles (30) of guinea pig; the aorta (31) and coronary artery (32) of rabbit; and the trachea (33) aorta (34) and iris (35) of the rat. Human uterine arteries (36) airways (37), and ciliary muscles (38) have also been shown to contain M_3-type muscarinic receptors.

2.2 Antiulcer Activity

The problems encountered in testing drugs for antiulcer activity result in part from a lack of complete understanding of the physiological and biochemical mechanisms involved in the formation of ulcers and in part from the testing of drugs for activity on normal or quasi-normal animal preparations, though they are ultimately applied to abnormal or pathological human states. There are various methods for producing ulcers in experimental animals (23).

A preparation developed by Shay and co-workers (39) has been used to test for antiulcer activity on an all-or-none basis. The ligation of the pylorus of rats, previously fasted for 48–72 h, leads to the accumulation of acid gastric contents and ulceration of the stomach 17–19 h after the operation. The antiulcer agents are given subcutaneously or intraduodenally at the time of ligation of the pylorus, or orally 1 h before. The animals are killed 17–19 h after pyloric ligation, and the stomach contents are collected for examination. The stomach is opened along the greater curvature, and the ulcers are examined and scored by a suitable scheme such as 0 = normal, 1 = scattered hemorrhagic spots, 2 = deeper hemorrhagic spots and some ulcers, 3 = hemorrhagic spots and ulcers, and 4 = perforation. Variable results have been reported by investigators using this technique.

Production of chronic experimental peptic ulcers in dogs (or rats) by the Mann-Williamson (40) procedure is one of the standard methods. The gastric juice is diverted into the intestine some distance from the pancreatic and biliary secretions. The objective is achieved by isolating the duodenum from the pylorus and the jejunum. The oral end of the duodenum is closed, and its distal end is anastomosed with a loop of ileum, to discharge the pancreatic and biliary secretions into the lower portion of the bowel. The cut end of the jejunum is then anastomosed to the pylorus. About 95% of dogs so prepared developed typical chronic peptic ulcers just distal to the gastric anastomosis with the jejunum. With similar operative procedures 85% of rats develop gastric, marginal, or jejunal ulcers.

The complete reversal of the duodenum in dogs produces chronic peptic ulcers in about 6 months (23). These animals maintain their weight until the development of ulcerations and is a useful preparation for detecting and comparing antiulcer activity.

Stress produces ulcers in the rats and could be used to test antiulcer activity of drugs (41). Rats fasted for 48 h and immobilized in a galvanized screen cage under light ether anesthesia develop ulcers in the glandular region of the stomach after 4 h of restraint. The estimate of severity can be all or none or may be coded in the same way as the Shay preparation.

One of the side effects of adrenocorticotropic hormone (ACTH) and corticoid therapy in humans is the development or reactivation of gastroduodenal ulcers. Daily subcutaneous administration of cortisol or Δ^1-cortisol to rats for 4 days results in the regular development of gastric ulcers (42). This procedure has been adapted to testing antiulcer activity (43). There are certain differences between steroid ulcers and "natural" ulcers in localization, rate of development, and severity (44).

The antisecretory activities of anticholinergics are as important as their antiulcer activities for their therapeutic usefulness. The Pavlov gastric pouch (45), with intact vagal and sympathetic nerve supply, and a modified Heidenhain pouch (46), which is essentially denervated, are prepared from dog stomach and have been used for determining the action of drugs on gastric secretion. Histamine or a test meal is usually used as a stimulus. Similar methods for the preparation and use of chronic total gastric fistulas and chronic denervated gastric pouches have been described for determining drug action on gastric secretion in rats (47–49).

There are a significant number of reports in which antisecretory and antimobility effects of anticholinergic drugs have been evaluated in ulcer patients (9,58). The antisecretory potency can be measured best in the duodenal ulcer patient in whom the acid output is already high. Ability of the drug to abolish or diminish acid output under histamine stimulation is a stringent test of activity, but the test has limited physiological relevance. The effect of the drug on the amount of acid secreted under ordinary clinical conditions is the most pertinent of all tests in relation to therapeutic application. In clinical comparison, not only the degree of effects but also the duration of the antisecretory effects should be compared.

Despite extensive research, certain aspects of ulcer disease are not clearly elucidated. Due to the multiple processes that control acid and pepsin secretion and defense and repair of gastroduodenal mucosa, it is more likely that causes of ulceration differ among individuals (3,5). Two other factors have been acknowledged as risk factors in pathophysiology of peptic ulcers: nonsteroidal antiinflammatory drugs (NSAID) and helicobacter pylori infection (4,51–53). NSAIDs induce a significant number of gastric and duodenal ulcers, possibly due to inhibition of prostaglandin synthesis with consequent loss of protective effects. H. pylori has been recognized as a risk factor in the ulcerative process, similar to acid and pepsin. Duodenal ulcer is typified by H. pylori infection and duodenitis and possibly impaired duodenal bicarbonate secretion in the face of moderate increases in acid and peptic activity. Increased peptic activity with decreased duodenal buffering capacity possibly leads to enhanced mucosal injury and finally results in gastric metaplasia. In the presence of antral H. pylori, the gastric metaplasia becomes colonized and inflamed. The inflammation and infection disrupts mucosal defense and regenerating mechanisms resulting in ulceration. The combination of inflammation, protective deficiencies, and moderate amounts of acid and pepsin may be enough to induce ulceration. Several groups of drugs, including anticholinergic agents, have been developed to antagonize risk factors causing ulcer disease. A good animal model that incorporates all variable causes of ulcer disease is yet to be developed.

The muscarinic receptors of the parietal cells are of M_1 subtype. The specific anticholinergic agents for M_1 receptors are considered to be effective for the treatment of ulcer disease (52). The muscarinic receptors on the duodenal smooth muscle are possibly of M_3 subtype. Anticholinergics at M_3 may partially decrease pain of duodenal ulcers by decreasing the motility of duodenum.

2.3 Mydriatic and Cycloplegic Activities

A simple and relatively accurate test for mydriatic activity has been described (55). The method requires mice and a binocular microscope magnifying about 10 times and provided with a scale in the eyepiece with which to examine and measure the diameter of the pupil of the mouse. A strong light shining into the eye of the mouse must be attached to the microscope. The diameter of the pupil is measured at the peak effect after administration of the anticholinergic agent by intraperitoneal injection. The duration of the effect is also important, since one of the most characteristic and valuable properties of atropine and analogous compounds is the prolonged effect that they produce in the eye.

Entopic pupillometry is an accurate and practical method for measuring the size of the pupil in human beings (54). With a Cogan entopic pupillometer, the normal size of the pupil and the near and far points before and after instillation of the drug in the conjunctival sac can be measured at different time intervals. The amount of light entering the eye is quite small, and the movements of the eye during the measurement do not interfere with the test.

2.4 Miscellaneous Anticholinergic Activities

A number of other methods are available for comparing the activities of anticholinergic agents, of which the antitremor and antisalivary effects are widely used. Arecoline or pilocarpine may be used to induce tremor or salivation in a suitable species that can be blocked by an anticholinergic agent. There seems to be good correlation between anticholinergic and antitremor effects (57). Recovery of the salivary gland from cholinergic block may conceivably precede that of the gastric glands and two effects may, therefore, not necessarily parallel each other in duration (9).

3 SOLANACEOUS ALKALOIDS

The older anticholinergic drugs are the various galenical preparations of belladonna, hyoscyamus, and stramonium, all of which are derived from plants of the potato family, the Solanaceae. The species used as drugs include *Atropa belladonna* (one of several plants known colloquially as deadly nightshade) *Hyoscymus niger* (black hen-

bane), and *Datura stramonium* (jimson-weed, jamestown weed, or thorn apple). The active principles in all these plants consist mostly of (−1)-hyoscyamine, with smaller, variable amounts of (−)-scopolamine (hyoscine). Atropine is (±)-hyoscyamine.

3.1 History

The poisonous nature of solanaceous alkaloids has been known for many centuries (58). The toxic properties of deadly nightshade are evident when children eat the black berries, which look attractive in a fall hedgerow in England. The children become delirious, and their eyes have widely dilated pupils. The deadly nightshade was used by the poisoners of the Middle Ages to induce obscure and often delayed poisoning. Therefore, Linné, in 1753, named the shrub *Atropa belladonna* after Atropos, the oldest of the Three Fates, who cuts the thread of life. *Belladonna* does not refer to Atropos, who is considered as a grim and awesome female, but to the Italian name ("handsome women") of the plant, which was used by Venetian ladies to give them dilated pupils ("sparkling eyes").

Datura has an ancient history, for it is said to have been used at the oracular shrine of Apollo in his temple at Delphi. Here the priestess of the god, Pythia, sat on a tripod uttering incoherent words in a divine ecstasy, in reply to the questions that were asked. Pythia was intoxicated by the fumes from burning datura leaves; her replies were interpreted by a priest in the form of a verse. The more common uses of datura were for robbery or conspiracy. Indian courtesans were known to place datura in their visitor's wine, so that they could be robbed without interference. As recently as 1908, there was a plan to poison the European garrison in Hanoi, Vietnam, using datura. Those in the conspiracy intended to stupefy the soldiers and then to kill them.

The pharmacological actions of atropine and related alkaloids are intimately connected with our knowledge of the organization and function of the autonomic nervous system. Schmiedeberg and Koppe (59) were the first in 1869 to focus attention on the similarity between a drug effect and electrical stimulation, when they pointed out that muscarine and vagus stimulation affected the heart in the same fashion and that the actions of both were antagonized by atropine. Furthermore, they recommended atropine as an antidote for mushroom poisoning. As early as 1887, Kobert and Sohrt (60) provided experimental proof for both similarities and dissimilarities between atropine and scopolamine.

Atropine was isolated by Mein (61) in 1831, and since then both atropine and scopolamine have been synthesized (62,63). A biogenetic scheme for the synthesis of atropine alkaloids in *Datura* species starting from ornithine has been described (64).

3.2 Chemical Structure

All the solanaceous alkaloids are esters of the dicyclic amino alcohol 3-tropanol (tropine, 1). Atropine is an ester of tropic acid and tropine. In scopolamine, the organic base is scopine. Scopine differs from tropine in having an oxygen bridge between C6 and C7. There are some other alkaloids that are members of the solanaceous alkaloids (e.g., apoatropine, noratropine, belladonnine), but they are not of sufficient therapeutic value to be discussed in this context.

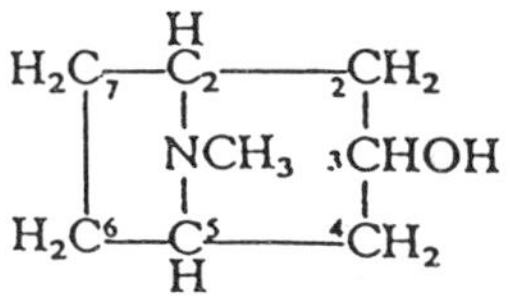

The carbon α to the carboxyl group of tropic acid is asymmetric and easily racemized during the isolation of the solanaceous alkaloids. Atropine and atroscine are racemic forms. The corresponding levo isomers, ($-$)-hyoscyamine and ($-$)-scopolamine (hyoscine) occur naturally in the solanaceous plants.

The absolute configuration of ($-$)-tropic acid has been established by its correlation with ($-$)-alanine (65). According to the Cahn-Ingold-Prelog (66) convention, natural ($-$)-tropic acid possesses the (S) configuration. Accordingly, ($-$)-hyoscyamine and ($-$)-hyoscine have an (S) configuration (67).

The piperidine ring system can exist in two principal conformations. Its chair form has the lowest energy requirement. However, the alternate boat form can also exist, because the energy barrier is not great. The formula of 3-hydroxytropine indicates that, even though there is no optical activity because of the plane of symmetry, two stereoisomeric forms, tropine (2) and pseudotropine (3), can exist because of the rigidity imparted to the molecule through the ethane chain across the 1,5 positions (68). In tropine, the axially oriented hydroxyl group, trans to the nitrogen bridge, is designated as α, or anti, and the alternate, equatorially oriented hydroxyl group

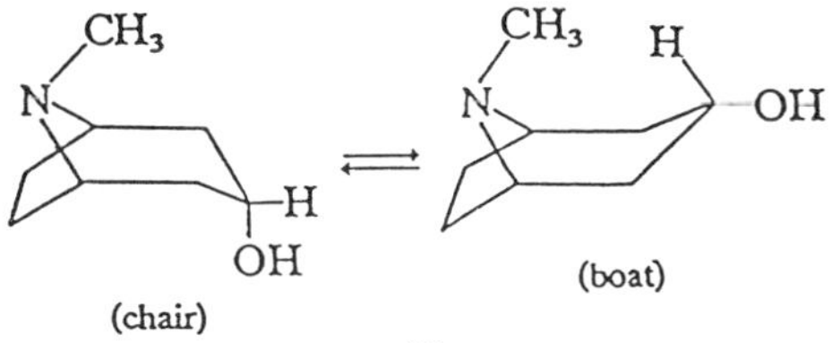

(2)

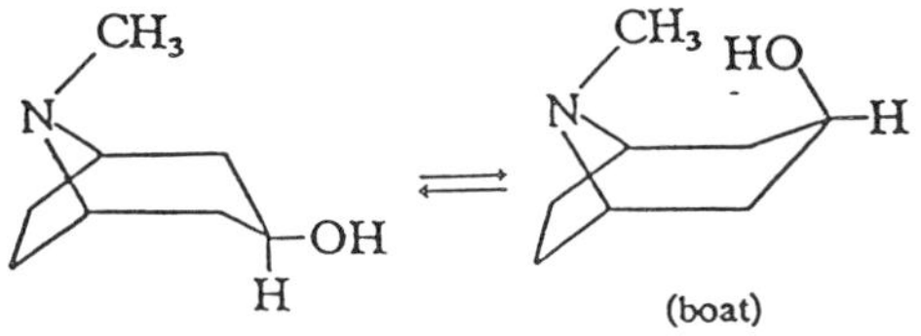

(1)

(3)

as β or syn. It is generally considered that cycloheptane is fixed through an $\_N(CH_3)\_$ bridge in the structures of tropine and pseudotropine. Therefore, a chair conformation is ascribed to the piperidine ring system in tropine and pseudotropine. However, there is only a seeming difference between the two conformations of tropane derivatives (69). The tropane system can be considered with equal justification as a piperidine twisted through the $\_CH_2CH_2\_$ bridge or as a cycloheptane fixed through an $\_N(CH_3)\_$ bridge. When the tropane system is structured by the chair form of piperidine, it represents also the boat form of cycloheptane. Similarly, the boat form of piperidine is at the same time a chair form of the cycloheptane ring. Therefore, it may be assumed that both forms are present in a state of equilibrium (68). Based on the conformations of the tropane system, the structure of atropine (**4**) can be represented by structures **5** and **6**, of which structure **5** is more generally accepted.

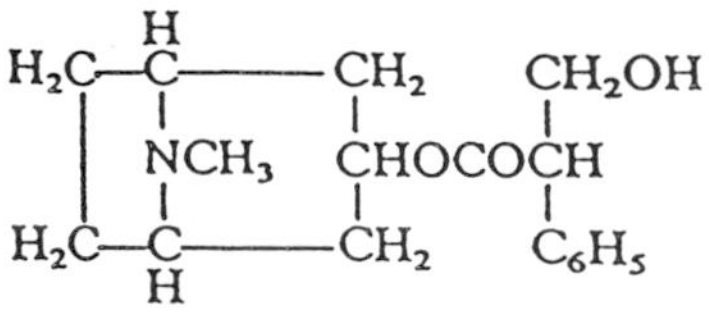

(4)

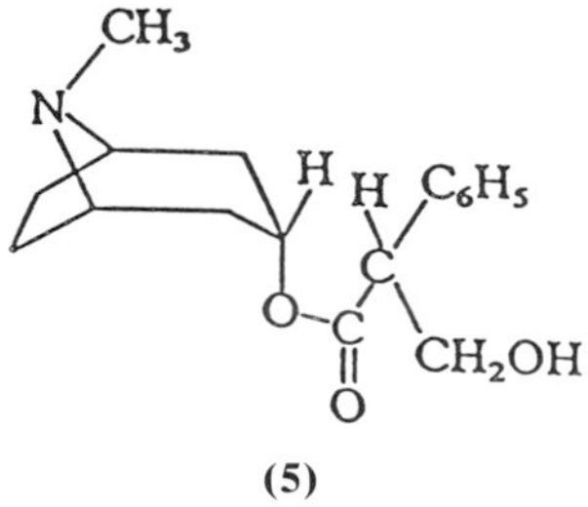

(5)

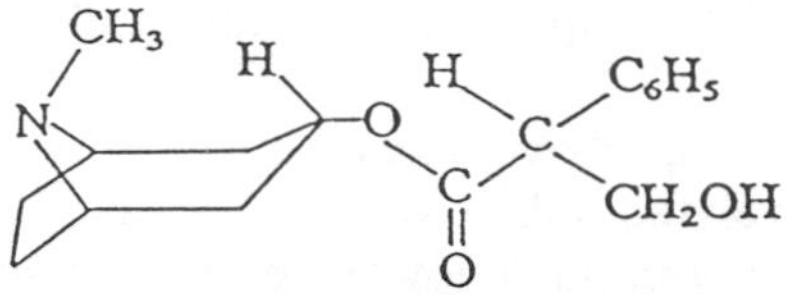

(7)

(6)

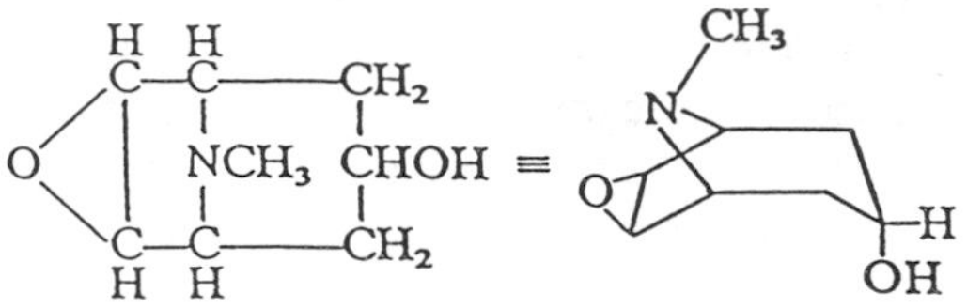

(8)

The amino alcohol derived from scopolamine (**7**), that is, scopine (**8**), has the axial orientation of the 3-OH group but, in addition, has a β-oriented epoxy group bridged across the 6,7 positions.

3.3 Preparative Methods

Conventional methods of alkaloid isolation are used to obtain a crude mixture of atropine and (−)-hyoscyamine from the plant products. This crude mixture of alkaloids is racemized to atropine by refluxing in chloroform or by treatment with cold dilute alkali (70).

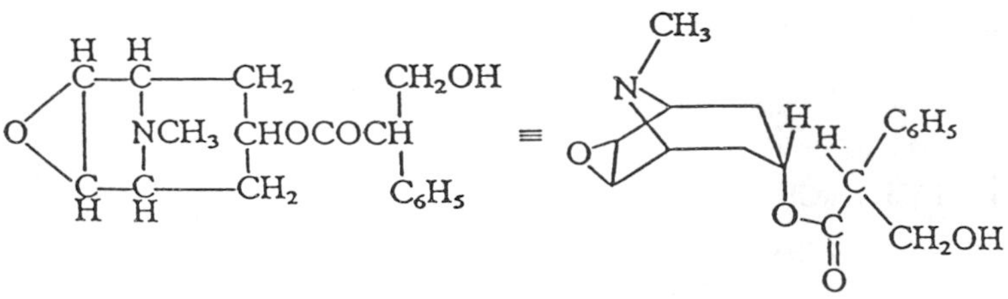

Atropine can be synthesized from troponin and tropic acid as starting materials. Troponin can be prepared by Robinson's (71) synthesis and reduced under proper conditions to tropine. (±)-Tropic acid can be prepared from ethyl phenylacetate (72,73) or acetophenone (74). The acetyl derivative of tropyl chloride is condensed with tropine hydrochloride to yield the *O*-acetyl derivative of atropine hydrochloride, from which the acetyl group is split spontaneously in aqueous solution (75).

One of the commercial sources for (−)-hyoscyamine is Egyptian henbane (*Hyoscyamus muticus*), in which it occurs to the extent of 0.5%. Another method for extraction of the alkaloid uses *Duboisia* species. It is prepared from the crude plant material in a manner similar to that used for atropine and is purified as the oxalate. (±)-Tropic acid can be resolved through its quinine salt and the separate enantiomorphs can be converted into (+) and (−)-hyoscyamines.

(−)-Scopolamine (hyoscine) is isolated from the mother liquor remaining after the isolation of hyoscyamine and is marketed as its hydrobromide. Scopolamine is readily racemized to atroscine, when subjected to treatment with dilute alkali. The synthesis of scopolamine differs from that of atropine in the synthesis of the amino alcohol, scopine portion of the molecule. Fodór and co-workers (63,76,77) have synthesized scopine starting from 6-β-hydroxy-3-tropanone. Esterification of scopine with acetyltropyl chloride and mild hydrolysis of the acetylscopolamine give scopolamine.

3.4 Molecular Factors in the Absorption, Fate, and Excretion of Atropine and Related Compounds

The belladonna alkaloids are absorbed rapidly after oral administration (78). They enter the circulation when applied locally to the mucosal surfaces of the body. Atropine absorbed from inhaled smoke of medicated cigarettes can abolish the effects of intravenous infusion of methacholine in humans. The trans conjunctival absorption of atropine is considerable. About 95% of radioactive atropine is absorbed and excreted following subconjunctival injection

in the rabbit. The total absorption of quaternary ammonium derivatives (Section 3.5) of the alkaloids after an oral dose is only about 25%. The liver, kidney, lung, and pancreas are the most important organs that take up the labeled atropine. The liver probably excretes metabolic products of atropine via the bile into the intestine (in mice and rats).

Because most synthetic antispasmodic and antiulcer agents are administered orally, their absorption from the gastrointestinal tract limits their therapeutic usefulness. There are striking differences in the absorption of tertiary amines and quaternary ammonium compounds (79–81). The tertiary amines (e.g., noroxyphenonium, mepiperphenidol) are absorbed completely from rat intestinal loops, whereas the maximal absorption of the corresponding quaternary ammonium compounds is about one-fifth of the total dose. The poor absorption of quaternary ammonium compounds may be partly due to the positive charge, which promotes the formation of a nonabsorbable complex with mucin. The ready absorption of tertiary amines may be explained partly by their permeability through lipid membranes (82).

Considerable species variations have been reported for the metabolic detoxification of atropine in mammals (83–94). These differences seem to be more quantitative than qualitative. At least four types of molecular modifications occur in the urinary excretion products of atropine (Fig. 26.2). Cleavage of the ester bond takes place in the rabbit and the guinea pig (87), whereas para and meta-hydroxylation of the benzene ring of tropic acid occur in the mouse and the rat (83,85). The tropine moiety of atropine is also chemically modified in urinary excretion products in humans and mice and further unidentified, "tropine-modified atropines" are excreted in humans and mice (86). Tropic acid itself does not undergo metabolic alteration for urinary excretion in all species mentioned

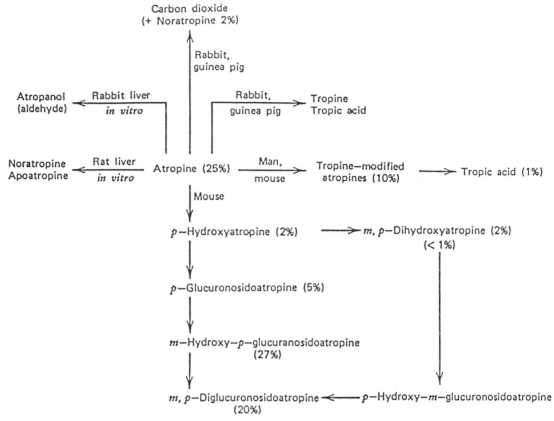

Fig. 26.2 Metabolism of atropine and its variations in different species.

above. The metabolic conversions of tropine itself are not fully investigated. However, demethylation of atropine-N-$^{14}CH_3$ (or tropine-N-$^{14}CH_3$) has been reported in a number of species with consequent exhalation of $^{14}CO_2$ (93). The possible metabolic changes of atropine are schematically represented in Figure 26.2.

After intravenous injection of atropine, approximately 25% of the dose is excreted in the mouse urine as atropine, more than 50% as conjugates with glucuronic acid, and the remaining 20–25% as intermediate oxidation products (probably p-hydroxy atropine and 3,4-dihydroxyatropine) and tropine-modified atropines. Rats are known to metabolize atropine in a manner similar to mice. In humans about 50% of the administered dose of atropine is excreted

unchanged in the urine and about 33% as unknown metabolites, which are esters of tropic acid. Neither hydroxylation of the tropic acid moiety nor glucuronide formation has been demonstrated in humans (86). Only less than 2% appears as tropic acid in urine.

It has been known for more than a century that rabbits can tolerate large quantities of atropine (87,88). The cause of this observation is the ability of the serum of some, but not all, rabbits to hydrolyze atropine into tropic acid and tropine. The hydrolysis is due to an enzyme, atropinesterase, that is found in most other tissues as well as the serum of these rabbits. Highest activities are found in the liver and intestinal mucosa; only the brain and aqueous humor of the eye contain no enzyme. The enzyme is also found in the liver of the guinea pig and accounts for the appearance of tropic acid in the urine of the rabbit and guinea pig, but not of other animals, following the administration of atropine.

The presence of atropinesterase in rabbits is inherited through an incompletely dominant gene (87). This gene is associated with another gene that influences the color of the fur, causing "extension of black pigment in the fur." Atropinesterase can also hydrolyze homatropine and scopolamine. This enzyme is stereospecific for (S)-$(-)$-hyoscyamine which is split; the more inert (R)-$(+)$ isomer is not readily hydrolyzed (87).

3.5 Semisynthetic Derivatives of Solanaceous Alkaloids

Early attempts to modify the atropine molecule (4) were aimed at converting solanaceous alkaloids containing a tertiary nitrogen into quaternary ammonium compounds and N-oxides. Later developments have been to retain the tropine (or scopine) portion of the molecule and substitute various acids for tropic acid. In this way, a series of tropeines have been synthesized, among which a number of active compounds have been found (89–95). Of the tropeines, mandelyl tropeine (homatropine, 9), has survived as a therapeutic agent to the present.

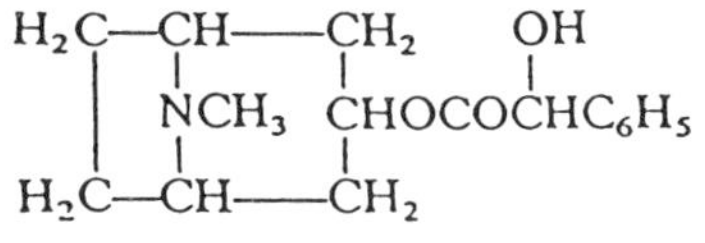

$$H_2C\text{—}CH\text{——}CH_2 \quad OH$$
$$| \quad NCH_3 \quad CHOCOCHC_6H_5$$
$$H_2C\text{—}CH\text{——}CH_2$$

(9)

Methylatropine nitrate (bromide, 10) is a synthetic quaternary derivative of atropine. Atropine oxide (atropine N-oxide) is known as a genatropine (11) and may be prepared by oxidation of the alkaloid with hydrogen peroxide.

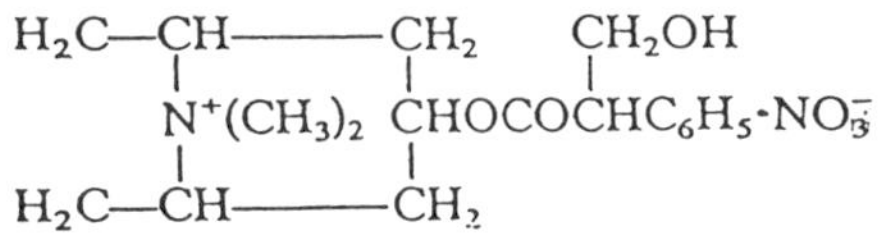

$$H_2C\text{—}CH\text{——}CH_2 \quad CH_2OH$$
$$N^+(CH_3)_2 \quad CHOCOCHC_6H_5 \cdot NO_3^-$$
$$H_2C\text{—}CH\text{——}CH_2$$

(10)

$$H_2C_7\text{——}CH\text{—}_2CH_2 \quad CH_2OH$$
$$O \leftarrow NCH_3 \quad CHOCOCHC_6H_5$$
$$H_2C^6\text{——}CH\text{——}CH_2$$

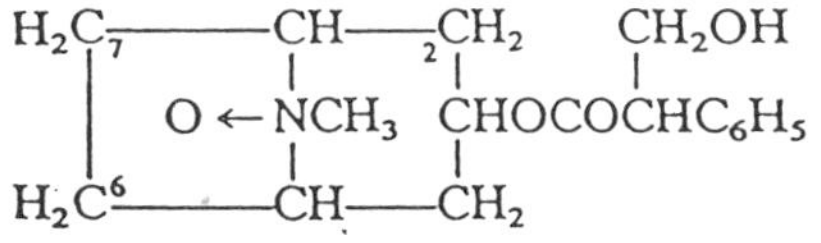

(11)

Derivatives of scopolamine (7) prepared by similar methods are available commercially. These include methscopolamine bromide (12), methscopolamine nitrate, and genoscopolamine (scopolamine N-oxide, 13).

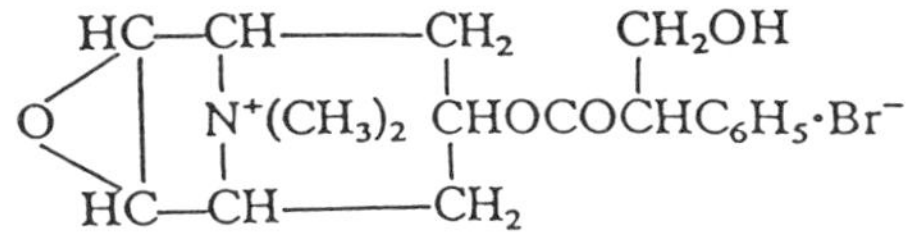

$$HC\text{—}CH\text{——}CH_2 \quad CH_2OH$$
$$O \quad N^+(CH_3)_2 \quad CHOCOCHC_6H_5 \cdot Br^-$$
$$HC\text{—}CH\text{——}CH_2$$

(12)

$$HC_7\text{——}CH\text{—}_2CH_2 \quad CH_2OH$$
$$O \quad O \leftarrow NCH_3 \quad CHOCOCHC_6H_5$$
$$HC^6\text{——}CH\text{——}CH_2$$

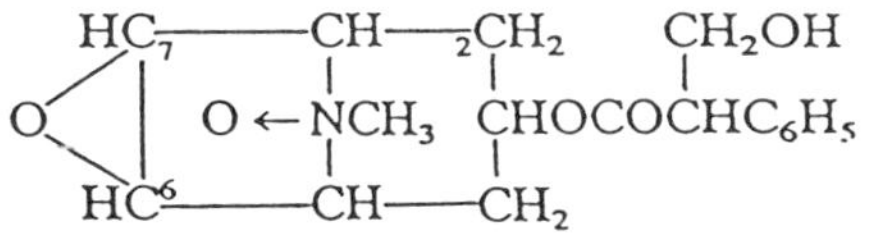

(13)

Homatropine (9) is prepared by evaporating tropine with mandelic and hydrochloric acids. Homatropine methyl bromide (14) may be prepared from homatropine by treatment with methyl bromide.

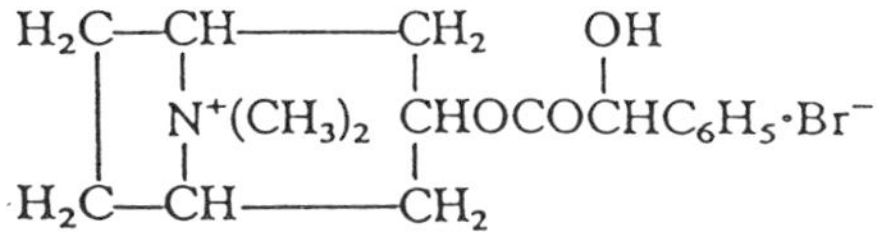

$$H_2C\text{—}CH\text{——}CH_2 \quad OH$$
$$N^+(CH_3)_2 \quad CHOCOCHC_6H_5 \cdot Br^-$$
$$H_2C\text{—}CH\text{——}CH_2$$

(14)

Quaternization may or may not increase all types of anticholinergic activity significantly. However, there are certain practical advantages to quaternization. The quaternary ammonium derivatives usually penetrate central nervous system less readily than the corresponding tertiary amine analogs. Therefore, quaternization serves as a useful technique to avoid or minimize the side effects caused by the stimulation of the central nervous system by the tertiary amines when the drugs are used for their peripheral actions. When a tertiary alkaloidal base is converted to the quaternary form, the latter is also less readily absorbed through the intestinal wall (96,97). This is not a disadvantage when the drug is used for its effects in the gastrointestinal tract. However, if the drug is used for its systematic or central actions the effect becomes erratic and unpredictable because the absorption is poor after oral administration. The tertiary amines are preferred for ophthalmic use, because they penetrate the cornea better than their quaternary ammonium derivatives. However, when a drug (e.g., atropine) has a long-lasting mydriatic effect, its recovery period can be shortened by quaternization. Therefore, the selection of the type of the derivative (Table 26.2) depends on the specific purpose for which it is used, the mode of its administration, and the duration of the desired effect.

Ipratropium bromide (15) is a useful quaternary derivative of atropine (98,99). It is marketed as a metered-dose inhaler for treatment of bronchospasm associated with chronic obstructive pulmonary disease. Because it is administered as an aerosol preparation, its droplets are deposited in upper airways and produce bronchodilation within minutes. Only less than 5% of the drug enters the systemic circulation. Ipratropium is an example of a useful drug developed for a specific purpose by minor molecular modification of atropine.

The *N*-oxides are converted to the corresponding tertiary bases *in vivo*. Atropine *N*-oxide and scopolamine *N*-oxide are slowly reduced to atropine and scopolamine in the animal body. Therefore, *N*-oxidation is a convenient technique for prolonging the duration of the action of the alkaloidal bases. The *N*-oxides are said to be less toxic.

Solanaceous alkaloids have a wide and variable spectrum of anticholinergic activities and are widely used in therapeutics. The various derivatives of atropine, scopolamine, and homatropine are listed in Table 26.2. Because of their chemical differences and the resulting biological interactions, different derivatives are preferred for antispasmodic, antisecretory, mydriatic, or central effects. Scopolamine produces the same type of depression of the parasympathetic nervous system as does atropine and homatropine, but it differs markedly from atropine in its action in the central nervous system. Whereas atropine stimulates the CNS, causing restlessness, scopolamine acts as a narcotic and sedative. It also causes temporary amnesia ("twilight sleep") when used along with morphine in obstetric and gynecologic procedures. For a detailed discussion of the anticholinergic activities of solanaceous alkaloids and their related compounds, the standard textbooks or reviews in pharmacology should be consulted (20–22,78).

(15)

4 SYNTHETIC ANTICHOLINERGICS

Although atropine and its related alkaloids are potent anticholinergics they have a wide spectrum of pharmacological activities. Therefore, therapeutic administration of these alkaloids to elicit a particular desired activity invariably results in some undesirable side effects. For this reason, the search for compounds possessing one or another of the specific desirable actions has been an active field of investigation in medicinal chemistry. The ideal specificity of action has not been attained in these attempts; perfect atropine substitutes with predominantly antispasmodic, antisecretory, or cycloplegic actions have yet to be synthesized. However, some progress has been made since the discovery of multiple subtypes of functional muscarinic receptors (M_1–M_5) and the cloned muscarinic receptors (m_1–m_5) have been identified (Table 26.1). Several antagonists that show selectivity to one subtype of muscarinic receptors over others have been introduced, and they have become useful in delineation of subtypes of muscarinic receptors in various tissues. Some of these agents may become useful as antiulcer agents, antispasmodics, or mydriatics.

4.1 Analogs of Atropine

The synthetic anticholinergic drugs can be considered analogs of atropine or antagonists of acetylcholine. Most of these compounds were designed using broad principles of molecular modification such as (*1*) scission of the atropine molecule into simpler molecules containing the essential pharmacodynamic groups, (*2*) molecular modification by introducing "blocking" moieties into cholinergics, and (*3*) changes in other anticholinergics using principles of bioisosterism.

The structure of atropine has been the basis for a large number of synthetic anticholinergic agents. However, no significant changes have been made to effect the "ester-complex" grouping because of atropine-like properties. Another probable consideration is that it is far simpler to synthesize a fairly complex molecule from two halves by esterification than by any method. Therefore, many esters of amino alcohols and carboxylic acids have been synthesized as atropine substitutes, in which the structures of either one or both halves have been changed. For example, in homatropine, the tropic acid moiety has been replaced by mandelic acid. The complex and massive amino alcohol moiety (tropine, **1**) of atropine has afforded unusually rich opportunities for the synthesis of anticholinergics (Fig. 26.3). Scission of its piperidine ring at point X gives the derivatives of hydroxyalkylpyrrolidines (**16**), and scission of its pyrrolidine ring at point Y makes it possible to proceed to derivatives of 4-hydroxy piperidine (**17**). The scission of both rings at Z leads to dialkylaminoalkanol derivatives (**18**). Furthermore, simplification and alteration of these three groups of amino alcohols has resulted in the synthesis of esters containing structural features more or less similar to atropine.

Antagonists of acetylcholine often have chemical structures resembling that of acetylcholine, although they differ from it by greater complexity of the molecule and higher molecular weight. Acetylcholine is a quaternary ammonium compound; atropine and tropine contain a tertiary nitrogen. Therefore, a number of atropine-like compounds having quaternary nitrogen atoms have been synthesized. In some of them, the acetyl group of acetylcholine has been replaced by acid moieties containing blocking groups (e.g., diphenylacetic acid).

The principles used in the design of antimetabolites have been applied to synthesize atropine-like compounds. The ester group in atropine-like compounds has been

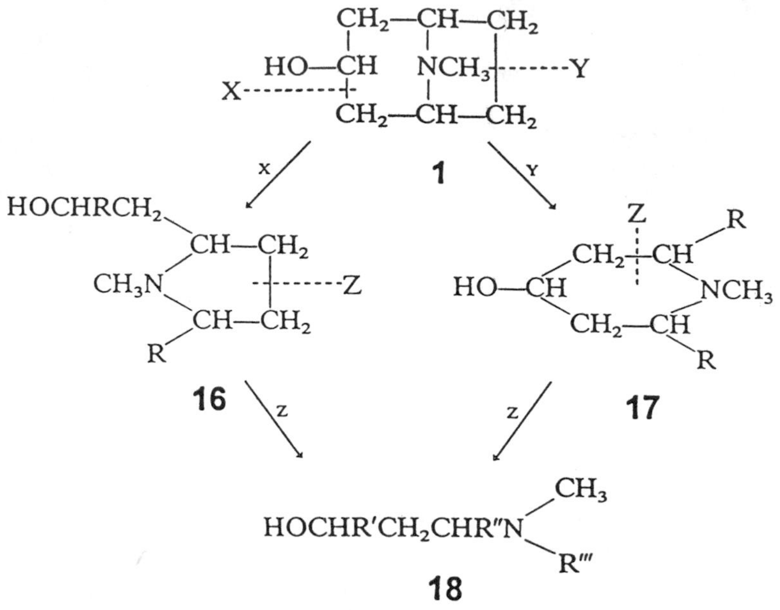

Fig. 26.3 Scissions of the tropane ring.

Table 26.3. Classification of Synthetic Anticholinergics

Group	Characteristic Group in the Main Chain	Atoms in the Chain	Additional Pharmacophoric Groups that may Be present
1	Ester	$\overset{\overset{\text{O}}{\|\|}}{\text{C—C—O—C}}$	—OH
2	Thioester	$\overset{\overset{\text{O}}{\|\|}}{\text{C—C—S—C}}$	—OH
3	Amide	$\overset{\overset{\text{O}}{\|\|}}{\text{—C—}}\overset{\overset{\text{H}}{\|}}{\text{N—C}}$	—OH
4	Carbamate	$\overset{\overset{\text{O}}{\|\|}}{\text{>N—C—O—C}}$	
5	Alkane amino alcohols amides	—C—C—C—	—OH —CONH$_2$
6	Alkene	—C=C—	

replaced by a thioester, an amide, an ether group, or a chain of methylene carbons (Table 26.3).

All synthetic anticholinergic agents have some structural features in common. In most, the molecule has bulky "blocking moieties," often cyclic radicals, linked by a chain of atoms of limited length, to a positively charged amine nitrogen (Fig. 26.4). The length and structure of the main chain have considerable influence on the anticholinergic activity of the substance. At the same time, the chemical nature of the main chain determines the class of organic substances to which a given substance belongs. Therefore, the classification of synthetic anticholinergics in Table 26.3 is based on the structure of the main chain of the molecule, taking into consideration, wherever necessary, the presence or absence of any additional pharmacophoric groups (OH, $CONH_2$). It is beyond the scope of this Chapter to consider all compounds that belong to each group. However, several examples from drugs used as therapeutic agents are discussed at appropriate places. These compounds may be classified differently, and the same compound may be placed in more than one group. Each one of them may be considered to be an agent with optimum anticholinergic activities among a series of structurally related compounds whose structure–activity relationships have been evaluated for different types of pharmacological effects. Compounds with the same or similar

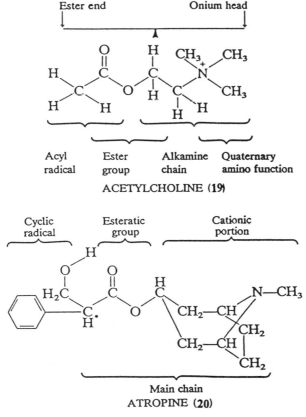

Fig. 26.4 Structural features of acetylcholine and atropine. The asymmetric carbon in atropine is marked with an asterisk (*).

structural features may exhibit other pharmacological effects as side effects. For example, a large number of compounds have been synthesized containing an ether link in the main chain. These compounds are useful as anti-Parkinsonian drugs (Chapter 78) and antihistaminic agents (Chapter 82).

4.2 Receptor Subtype Selective Anticholinergics

The muscarinic actions of acetylcholine can be stimulatory or inhibitory. Acetylcholine stimulates secretion and contraction of the gut, but it inhibits the contraction of the heart and relaxes the smooth muscle of blood vessels. Acetylcholine can inhibit adenylate cyclase and activate guanylate cyclase. In cortical neurones, muscarinic agents cause a slow depolarization mediated by closing potassium channels. Acetylcholine opens potassium channels in the heart, causing hyperpolarization and a reduced rate of firing of the nodal tissue. In many tissues, calcium channels are opened and probably intracellular calcium is mobilized. Like many other transmitters, acetylcholine increases the turnover of phosphoinositides. Therefore, there is much expectation that selective agents will be found among muscarinic antagonists that will be useful to block one particular physiological or biochemical response to acetylcholine. Thus several classes of agents have been synthesized that have structural features similar to those of atropine-like agents: cyclic blocking moieties linked by a chain of atoms of limited length to a positively charged nitrogen atom.

Based on the cyclic blocking moieties and other substituents, subtype selective muscarinic antagonists can be classified into eight groups: (*1*) tricyclic benzodiazipines, (*2*) benzothiazepines, (*3*) quinuclidines, (*4*) polymethylene tetramines, (*5*) indenes, (*6*) sila-difenidols, (*7*) diphenylacetyloxy derivatives, and (*8*) himbacine akaloids.

4.2.1   TRICYCLIC BENZODIAZIPINES.   Significant side effects of tricyclic antidepressants like imipramine are antimuscarinic effects. Benzodiazipines also cause some antimuscarinic effects like dry mouth at therapeutic concentrations. Some of the well-known anticholinergic agents, Banthine, Pro-Banthine, and Trest, have tricyclic bulky moieties at the ends of their molecules. Some molecular features of these three types of pharmacological agents are present in tricyclic pyridobenzodiazipines. In these compounds, portions of benzene and pyridine (or other rings) are fused to a seven-membered diazepine ring in the middle. The tricyclic bulky moiety containing benzene, diazepeine, and pyridine rings (or other rings) at the end of a molecule satisfies one of the requirements of an anticholinergic agent. Further substitutions on the iminonitrogen atom has resulted in selective M_1 receptor antagonists, which are useful as antiulcer agents.

Pirenzepine (**21**) was the first M_1 receptor antagonist that was shown to inhibit gastric secretion (100,101). This drug (100–150 mg/day) is used in several countries to decrease gastric secretion and achieve maximal rates of ulcer healing. At these doses, incidence of dry mouth and blurred vision are not significant. Because it has a low lipid solubility and limited permeability into the central nervous system, it does not cause any CNS side effects.

(21)   (22)

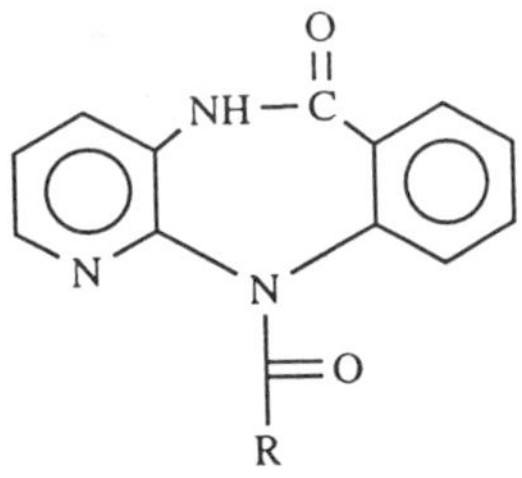

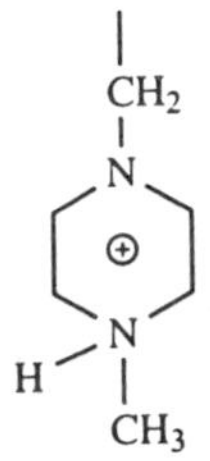

(21) 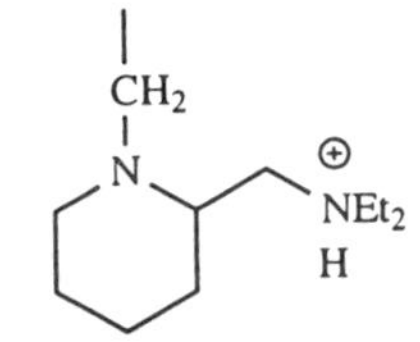   (23) 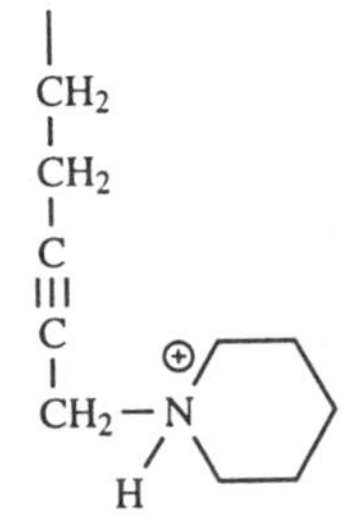   (24)

Telenzepine (**22**), an analog of pirenzepine, is 4–10 times more potent for
inhibition of gastric secretion. AF-DX-116
(**23**) is an analog of pirenzepine and differs
markedly in its profile of muscarinic activities (102,103). It has greatest affinity for
cardiac M_2 receptors. Its cardioselectivity is
also observed in humans, and it may become useful in sinus bradycardia and AV
block of vagal origin. AX-RA 513 (**24**) is
another analog of pirenzepine and exhibits
selectivity toward M_2 receptors (104). The
spatial orientation of the protonated side
chain nitrogen atom in relation to the
tricycle seems to be of major importance
for M_1–M_2 selectivity. AQ-RA 741 (**25**) is
an analog of pirenzepine that exhibits higher affinity to chimeric m_2 and M_4 receptors
than for m_5 receptors (105). VH-AH-37
(**26**), a pirenzepine derivative, exhibits
higher affinity to chimeric m_5 receptors
than for m_2 receptors.

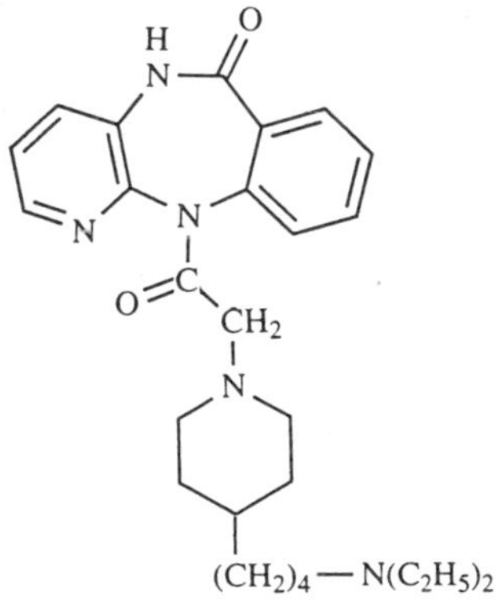

4.2.2 BENZOTHIAZIPINES. Benzothiazipines
are closely related compounds to benzo-
diazepines. The nitrogen atom in the
diazepine ring is replaced by a sulfur atom

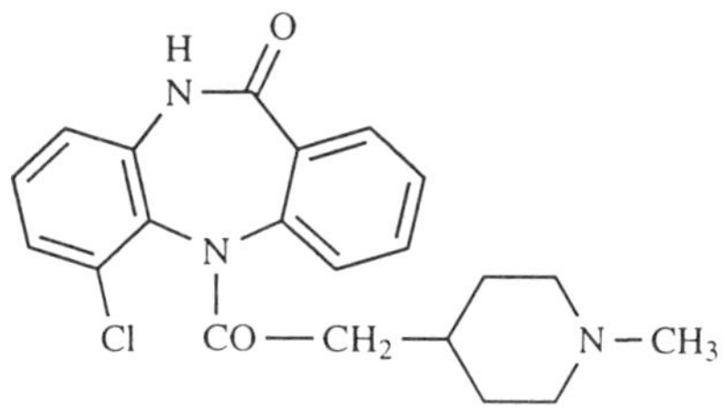

(25)

(26)

(27)

Table 26.4. K_i **Values of 2-R-Quinucliin-2-enes at Muscarinic Receptors**

R Group	K_i (Values)[a]		
	M_1	M_2	M_3
Benzofuranyl	9.6	31	59
Benzothienyl	81	270	420
Benzoxazolyl	100	400	720
Benzothiazoyl	170	600	1100

[a]Reciprocals of K_i values give relative affinities at the receptors. Summarized from ref. 107.

(27). Among these compounds BTM-1086 (*cis*(−)-2,3-dihydro-3-(4-methylpiperazinyl)-2-phenyl-1,5-benzothiazepin-4(5*H*)-ones monohydrochloride) was found to be a M_1 receptor antagonist (106). It inhibits acetylcholine release from parasympathetic nerves and also gastric secretion.

4.2.3 QUINUCLIDINES. A series of achiral 3-heteroaryl substituted quinuclidin-2-ene derivatives (**28–31**) was synthesized by Haskell and co-workers (107) who determined their dissociation constants (K_i) at different subtypes of muscarinic receptors. Among these compounds 2-benzoforanyl quinuclin-2-diene exhibits highest affinity ($K_i = 9.6$ n*M*) at M_1 receptors, and has lower affinity at M_2 ($K_i = 31$ n*M*) or M_3 ($K_i = 59$ n*M*) receptors. This antagonist is well accommodated within the defined model (108) of the m_1 receptor. The quinuclidin-2-ene ring will be located in an area of the receptor defined by val-102, ala-160 and val-385 where the quinuclidine ring of potent agonists binds (108). Substitution of the benzofuranyl group (**28**) by benzothienyl (**29**), benzoxazolyl (**30**), or benzothia-zoylyl (**31**) group decreased the affinity at the M_1 receptor (Table 26.4).

(30)   **(31)**

There is a good correlation between the magnitude of the electrostatic potential in the benzene nucleus and the M_1 receptor affinity. Furthermore, future work may yield more selective M_1 antagonists, which will be useful in the treatment of ulcers.

4.2.4 POLYMETHYLENE TETRAMINES. Several polymethylene tetramines were developed as selective inhibitors of M_2 receptor antagonists (109,110). Among these methoctramine is a prototype compound (**32**). The selectivity and affinity of methoctramine-like compounds at M_2 receptors depend on a tetramine backbone and the nature of substituents on the terminal nitrogens (**33, 34**). This selectivity is improved by introduction of *N*-methyl groups into the

(28)   **(29)**   **(32)**

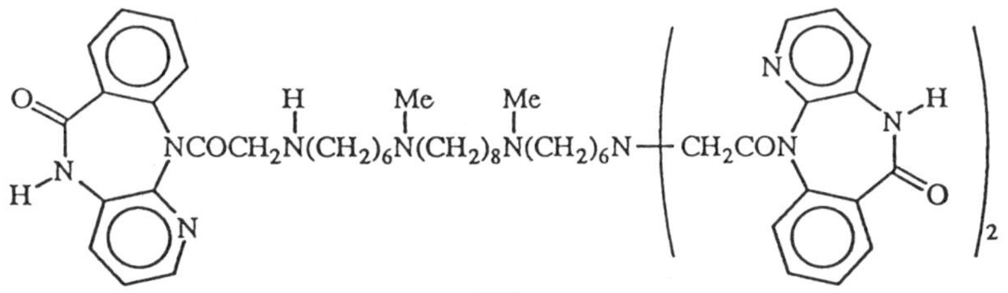

(33)

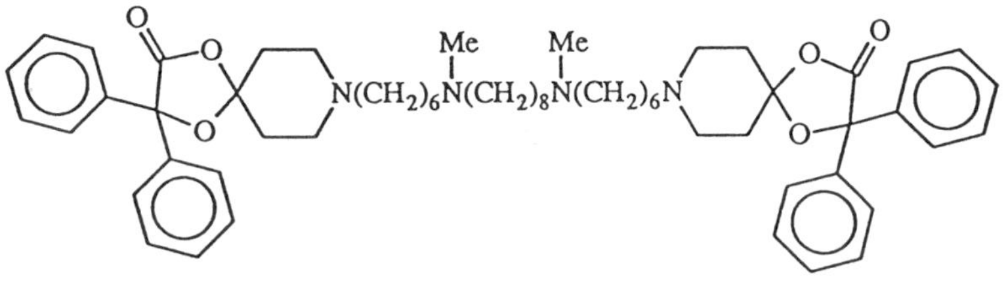

(34)

tetramine backbone and introduction of a tricyclic system on the terminal nitrogens of the tetramine backbone. Among a series of tetramines, tripitramine (33) is a potent and selective M_2 receptor antagonist (Table 26.5).

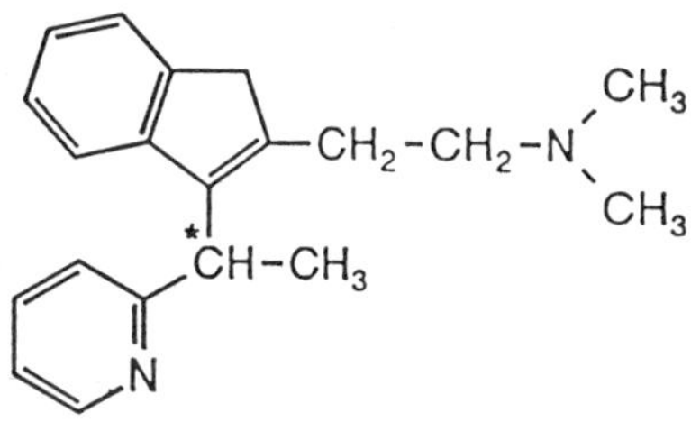

4.2.5 INDENES. Dimethylpyrindene (dimethindene, 35) was first introduced as an H_1 receptor antagonist (see the chapter on Histamine H_1 Antagonists). Subsequentially, it was found to be a M_2 receptor antagonist. Due to the presence of asymmetric carbon in the molecule, it occurs

(35)

in two optical forms. In general, (S)-dimethindene is more potent than the (R)-enantiomer at muscarinic receptor subtypes M_1, M_2, and M_3. However, the stereoselec-

Table 26.5. Methoctramine-related Tetramines and Their Selectivities at M_2 and M_3 Subtypes of Muscarinic Receptors[a]

| | pA$_2$ | | Selectivity Ratio[b] |
Antagonist	M_2	M_3	M_2:M_3
Methoctramine	7.78	6.28	32[c]
Tripitramine	9.69	6.50	1550[c]
4-DAMP	8.53	9.19	0.22[d]

[a]Test system for M_2 receptors = guinea pig left atria. Test system for M_3 receptors = guinea pig ileum.
[b]The selectivity ratio is the antilog of the difference between pA$_2$ values on two different systems.
[c]Data quoted from ref. 109.
[d]Data quoted from ref. 110.

Table 26.6   Activities of Enantiomers of Dimethindene at Subtypes of Muscarinic Receptors

Receptor Subtype[a]	Test System[b]	pA$_2$ Values of Isomers	
		(R)	(S)
M$_1$	Rabbit vas deferens[a]	5.81	6.83
M$_1$	Rat duodenum[a]	5.49	6.36
M$_2$	Guinea pig atria	6.25	7.86
M$_2$	Rabbit vas deferens	6.22	7.74
M$_3$	Guinea pig ileum	5.61	6.92
M$_3$	Guinea pig trachea	5.59	6.96

[a]Different selective agonists were used to stimulate the receptors.
[b]Some test systems contain more than one subtype of muscarinic receptor.

tivity (31- to 41-fold) is greatest at M$_2$ receptors (Table 26.6). (S)-Dimethindene is more specific for muscarinic receptors than for receptors of other biogenic amines (norepinephrine, dopamine, and 5-HT). It penetrates the blood–brain barrier in humans and, therefore, may become a valuable tool in Alzheimer's disease or evaluation of M$_2$ receptors of the central nervous system by PET studies.

4.2.6 SILA-DIFENIDOLS. Studies on anticholinergic agents of procyclidine (kemadrin, Table 26.3) and the defenidol type have shown that substitution of the central carbon atom (R_3-C-OH) by a silicon atom (R_3-Si-OH) leads to drugs with increased antimuscarinic potency and increased selectivity to M$_3$ receptors (36). Hexahydrosiladifenidol (HHSID) and its p-fluoro-derivative have been used to characterize M$_3$ receptors in smooth muscle (111,112).

HHSID shows a 15 to 30-fold higher antimuscarinic potency at M$_3$ receptors of guinea pig ileum and urinary bladder than at M$_2$ receptors of the rat heart and vascular endothelium.

4.2.7 DIPHENYLACETYLOXY DERIVATIVES. Barlow and co-workers (113) have synthesized several muscarinic antagonists and tested them at the muscarinic receptors in the heart and the smooth muscle. One of these compounds, 4-[(diphenylacetyl)oxy]-1,1-dimethyl-piperidium (4-DAMP, **37**), was about 10 times more potent on M$_3$ receptors of the smooth muscle than M$_2$ receptors of the heart (114). It has become very useful to identify M$_3$ in several tissues, especially smooth muscles of trachea, ileum, vascular tissue and cilia of different species.

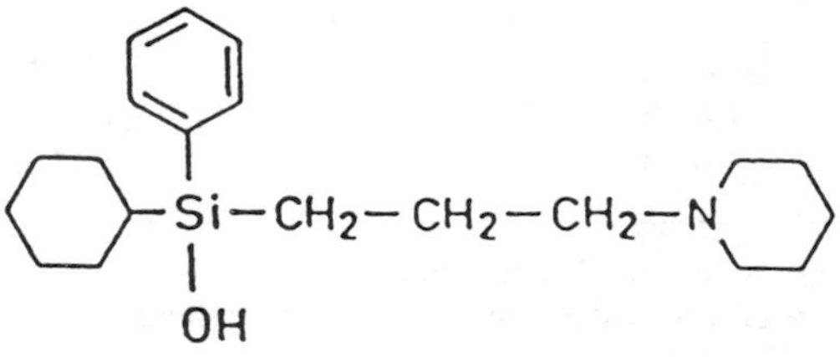

(36)

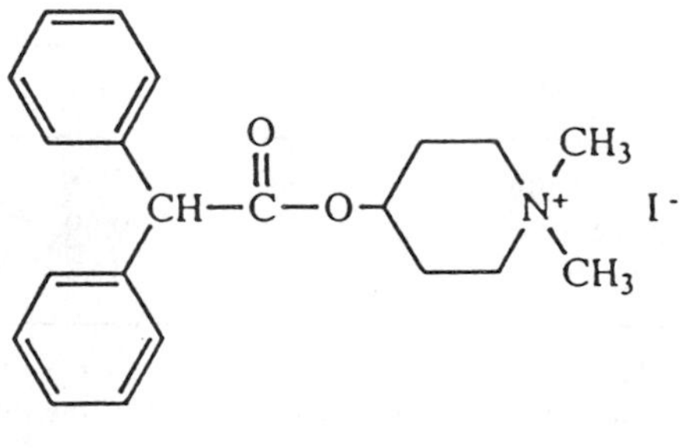

(37)

(38)

4.2.8 HIMBACENE ALKALOIDS. Himbacene akaloid has a tricyclic structure (38). It is considered to be selective for the cardiac M_2 receptor (115). However, it binds to all five cloned muscarinic receptor subtypes in the following order of potencies (116): $hM_2 = hM_4 > hM_3 > hM_1 > hM_5$. Its K_d values at these receptors are 4, 7, 59, 83, and 296 nM, respectively. It is a potent blocker of oxotremorine-induced and muscarinic receptor–mediated cyclic AMP inhibition in rat striatum ($K_d = 4.4$ nM) and N1E-115 neuroblastoma cells ($K_d = 10.6$ nM), responses that are considered to be mediated by M_4 receptors. It inhibits oxotremorine-induced acetylcholine release from rat hippocampal tissue ($K_d = 8.6$ nM). The subtype of muscarinic receptor involved in this response is possibly of the M_2 or M_4 type. At the postsynaptic putative M_1 and M_3 receptors involved in the phosphoinositide turnover in the rat cortex, himbacine has low activity and higher K_d (181 nM). It appears that himbacine is a more potent muscarinic antagonist at M_2 or M_4 receptors than at M_1 or M_3 receptors (115,116).

5 STRUCTURE–ACTIVITY RELATIONSHIPS

Although atropine-like agents are antagonists of acetylcholine at one type of cholinergic receptor (muscarinic receptor) that is specific for activation by L(+)-muscarine, they may demonstrate many other pharmacological properties (ganglionic blocking, neuromuscular blocking, musculotropic, central stimulant, or depressant activities). The following discussion of the relationships of structure to activity is limited to their inhibitory actions at the muscarinic receptors. Certain structural features are common in the many anticholinergic agents that have been synthesized and evaluated pharmacologically (Fig. 26.4). Some of these features also appear in cholinergics (19). A typical atropine-like anticholinergic agent (20) contains a cationic head and a heavy blocking moiety (cyclic groups) that are connected by a chain of atoms of definite length (117–123). Their molecules include essential constituent groups (cationic head, cyclic radicals) as well as nonessential but contributing anchoring groups (e.g., hydroxyl). The steric factors that are related to the essential groups influence the anticholinergic activity significantly. Several anticholinergic agents incorporating the above structural features are listed in Table 26.7.

5.1 Cationic Head

The cationic head is an essential group in a large number of anticholinergic and cholinergic compounds. Ordinarily, this is a substituted ammonium group or, less frequently, a sulfonium or a phosphonium group. The mechanism of the cholinergic or anticholinergic action of substances has long been linked to such cationic groups (117–123). It is reasonable to assume that the cationic head with its positive charge is attracted by the negatively charged field (anionic center) of the muscarinic receptor. Thus the cationic head seemingly starts the process of the adsorption of the substance at the receptor. Following the attraction of the oppositely charged groups, the weaker

Table 26.7. Synthetic Anticholinergics (Atropininic agents)

Nonproprietary Name	Selected Proprietary Name(s)	Chemical Name of Salt	Structure of Base	References for Synthetic Procedures
		Tertiary Amines—Characteristic Group in the Main Chain: Ester		
Adiphenine	Trasentine	2-Diethylaminoethyl diphenyl-acetate hydrochloride	$(C_6H_5)_2CHCO_2CH_2CH_2N(C_2H_5)_2$	124,125
Amprotropine	Syntropan	3-Diethylamino-2,2-dimethyl-propyl tropate phosphate	HOH_2C—$C(CH_3)(C_6H_5)$—$CHCO_2CH_2CH_2N(C_2H_5)_2$	126,127
Amino-carbofluorene		2-Diethylaminoethyl-9-fluorene-carboxylate hydrochloride	fluorene-$CHCO_2CH_2CH_2N(C_2H_5)_2$	128
Cyclopentolate	Cyclogyl	β-Dimethylaminoethyl (1-hydroxycyclopentyl)-phenylacetate hydrochloride	(1-hydroxycyclopentyl)(C_6H_5)$CHCO_2CH_2CH_2N(CH_3)_2$	129
Dicyclomine	Bentyl	2-Diethylaminoethyl bicyclo-hexyl-1-carboxylate hydrochloride	bicyclohexyl-$CO_2CH_2CH_2N(C_2H_5)_2$	130
Eucatropine	Euphthalmine	4-(1,2,2,6-Tetramethyl-piperidyl) mandelate hydrochloride	tetramethylpiperidyl-$C_6H_5CH(OH)CO_2$	131,132

Oxphencyclimine	Daricon, Vio-Thene	1-Methyl-1,4,5,6-tetrahydro-2-pyrimdylmethyl α-cyclo-hexyl-α-phenylglycolate hydrochloride		133
Piperidolate	Dactil	1-Ethyl-3-piperidyl diphenyl-acetate hydrochloride	$(C_6H_5)_2CHCO_3$	134
Pipethanate	Sycotrol	2-(1-Piperidino)ethyl benzilate hydrochloride	$(C_6H_5)_2C(OH)CO_2CH_2CH_2N$	135

Quaternary Ammonium Compounds–Characteristic Group in the Main Chain: Ester

Glycopyrrolate	Robinul	3-Hydroxy-1,1-dimethyl-pyrrolidinium bromide α-cyclopentylmandelate		136
Mepenzolate	Cantil	N-Methyl-3-piperidyl benzilate methylbromide	$(C_6H_5)_2C(OH)CO_2$	137,138
Methantheline	Banthine	β-Diethylaminoethyl 9-xanthenecarboxylate methobromide	$CHCO_2CH_2CH_2N^+(C_2H_5)_2CH_3$	128
Oxyphenonium	Antrenyl	Diethyl(2-hydroxyethyl)-methylammonium-α-phenyl-α-cyclohexylglycolate bromide	$HOCCO_2CH_2CH_2N^+(C_2H_5)_2CH_3$	139

(Continued on page 90)

Table 26.7. (*Continued*)

Nonproprietary Name	Selected Proprietary Name(s)	Chemical Name of Salt	Structure of Base	References for Synthetic Procedures
Penthienate	Monodral	2-Diethylaminoethyl α-cyclopentyl-2-thiophene-glycolate methobromide	$HOCCO_2CH_2CH_2N^+(C_2H_5)_2CH_3$ (cyclopentyl and 2-thienyl substituents)	140,141
Pipenzolate	Piptal	*N*-Ethyl-3-piperidyl benzilate methobromide	$(C_6H_5)_2C(OH)CO_2-$ piperidinium ring N^+—CH_3, C_2H_5	134
Poldine	Nacton	2-Hydroxymethyl-1,1-dimethylpyrrolidinium methyl-sulfate benzilate	$(C_6H_5)_2C(OH)CO_2CH_2-$ pyrrolidinium $N^+(CH_3)_2$	142
Propantheline	Pro-Banthine	β-Diisopropylmethylaminoethyl 9-xanthenecarboxylate bromide	xanthene $CHCO_2CH_2CH_2N^+CH_3$ with C_3H_7, C_3H_7	128
Valethamate	Murel	2-Diethylaminoethyl-3-methyl-2-phenylvalerate methobromide	$CH_3CHCHCO_2(CH_2)_2N^+(C_2H_5)_2CH_3$ with C_2H_5, C_6H_5	143,144

Characteristic Groups in the Main Chain: Thioester, Amide, or Carbamate

Thioester

Thiphenamil — Trocinate — β-Diethylaminoethyldiphenyl-thiolacetate hydrochloride — $(C_6H_5)_2CHCOSCH_2CH_2N(C_2H_5)_2$ — 145,146

Amide

Tropicamide — Mydriachyl — N-Ethyl-2-phenyl-N-(4-pyridylmethyl) hydracryl-amide — 147

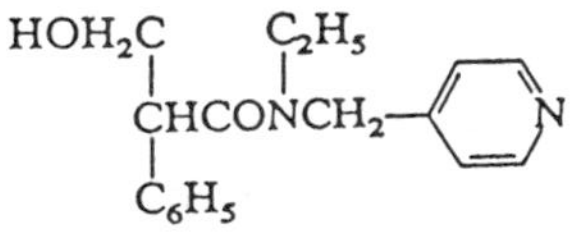

Carbamate

Dibutoline — Dibuline — Bis[dibutylcarbamate of ethyl-(2-hydroxyethyl)-dimethylammonium]sulfate — 148,149

C_4H_9 C_4H_9 NCO$_2$CH$_2$CH$_2$N$^+$(CH$_3$)$_2$C$_2$H$_5$

Characteristic Groups in the Main Chain: Alkane

Amino alcohol containing quaternary nitrogen

Hexocyclium — Tral — N-(β-Cyclohexyl-β-hydroxy-β-phenylethyl)-N'-methyl piperazine methylsulfate — 150

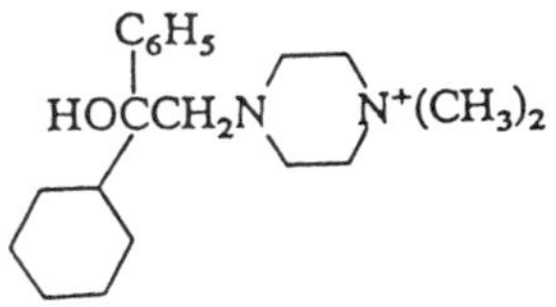

Mepiperphenidol — Darstine — 5-Methyl-4-phenyl-1-(1-methylpiperidinium)-3-hexanol bromide — 151

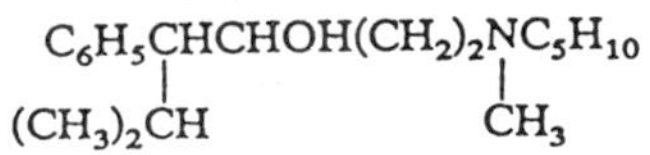

$C_6H_5CHCHOH(CH_2)_2NC_5H_{10}$ CH_3
$(CH_3)_2CH$

Tricyclamon — Elorine, tricoloid — 1-Cyclohexyl-1-phenyl-3-pyrrolidino-1-propanol methochloride — 152–154

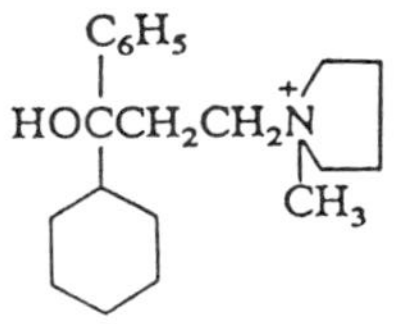

Tridihexethyl — Pathilon — 3-Diethylamino-1-phenyl-1-cyclohexyl-1-propanol ethiodide — 155,156

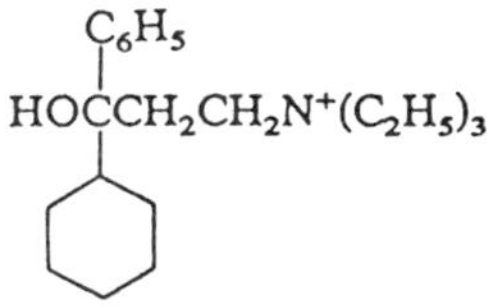

$HOCCH_2CH_2N^+(C_2H_5)_3$ C_6H_5

(Continued on page 92)

Table 26.7. (*Continued*)

Nonproprietary Name	Selected Proprietary Name(s)	Chemical Name of Salt	Structure of Base	References for Synthetic Procedures
Amino alcohol containing tertiary nitrogen Procyclidine	Kemadrin	1-Cyclohexyl-1-phenyl-3-pyrrolidino-1-propanol hydrochloride	C_6H_5, HOCCH$_2$CH$_2$N (cyclohexyl, pyrrolidine)	152–154
Amino amide containing quaternary nitrogen Isopropamide	Darbid	(3-Carbamoyl-3,3-diphenyl-propyl)diisopropylmethyl-ammonium iodide	$H_2NCOCCH_2CH_2\overset{+}{N}(C_3H_7)(CH_3)(C_3H_7)$, C_6H_5, C_6H_5	157
Amino amide containing tertiary nitrogen Aminopentamide	Centrine	α,α-Diphenyl-γ-dimethylaminovaleramide	C_6H_5, $H_2NCOCCH_2CH_2CH(CH_3)N(CH_3)_2$, C_6H_5	158–161
Miscellaneous Methixene	Trest	1-Methyl-3-(thioxanthen-9-ylmethyl)piperidine hydrochloride hydrate	(thioxanthene, CH$_2$, N–CH$_3$ piperidine)	162

Characteristic Group in the Main Chain: Alkene

Diphemanil	Prantal	4-Diphenylmethylene-1,1-di-methylpiperidinium methylsulfate	$(C_6H_5)_2C=$ (piperidinium) $N^+(CH_3)_2$	163

dipole–dipole, hydrophobic, and van der Waals forces go into action; if there are many of them, especially in the case of anticholinergics, they contribute to the stability of the drug–receptor complex. In such an interaction not only the charge of the cation head but also its size and shape are of vital importance.

The basicity of different amino derivatives, and consequently the degree of their ionization at physiological pH, varies over a broad range. The more ions of the anticholinergic ammonium compound or amine in solution, the greater the probability of their interaction with the anionic center of the muscarinic receptor to form the drug–receptor complex. In addition, the stability of the drug–receptor complex that has formed should depend on the basicity, inasmuch as rate of hydrolysis of salts is inversely proportional to the base strength.

Thus high basicity should favor the anticholinergic activity of a substance. Although the logic of this conclusion is simple, its proof involves great difficulties. In a series of anticholinergics, transition from one derivative to another is associated with stepwise changes in basicity as well as steric factors. In this respect the N-oxides, which

are obtained through the oxidation of the corresponding tertiary amines, have lower basicities and also lower anticholinergic activities (161–163). The N-oxides are closer to the corresponding quaternary ammonium compounds than to the tertiary ammonium ions in steric respect and partition between aqueous and organic phases. Alkylation converts the N-oxides into typical quaternary ammonium compounds. By this procedure, both the basicity and anticholinergic activity of the substance increase sharply (Table 26.8).

The influence of a steric factor is more evident among compounds in which the size of the substituents at the nitrogen atom is varied both in the series of anticholinergic and cholinergic compounds. Progressive replacement of the N-methyl groups of acetylcholine with ethyl groups leads to a stepwise reduction in muscarinic activity (164). Likewise, maximal anticholinergic or blocking activity (Table 26.9) is obtained by replacing the N-methyl groups of β-dimethylaminoethyl benzilate methylchloride with ethyl groups (118). Further increases in size to butyl or larger alkyl groups reduce or abolish the activity (118,166–171). Therefore, it seems that for

Table 26.8   Basicity and Anticholinergic Activity of Substituted Aminoethyl Esters of Benzilic Acid

$$(C_6H_5)_2C(OH)COOCH_2CH_2\overset{+}{N}(CH_3)_2R$$

R	Basicity, pK_a	Van der Waals Radius[a] of $N-R$ (Å)	Dose[b] Required to Eliminate the Effects of Arecoline in Mice, $\mu mol/kg$	
			Salivation	Tremor
H	8.08	2.25	5.0	6.8(6.7)
CH$_3$	10.87	3.09	0.48	
OH	4.68	3.01	94.8	284.5
OCH$_3$	10.18	4.37	4.5	

[a] An estimate of the steric volume and, therefore, steric hindrance for the interaction of the cationic head at the muscarinic receptor.

[b] Calculated from the values reported in ref. 68.

Table 26.9 Influence of the Number, Size, and Structure of Alkyl Groups in the Cationic Head on the Anticholinergic Activity

Compound Pair	Name or Structure of Compound[a]		Test system	Activity Ratio[b] B·A	Reference
	Series A	Series B			
1	$RN(CH_3)_2$	$RN(C_2H_5)_2$	Cat: salivation	1.63	118
			Cat: blood pressure	2.09	
			Mouse: mydriasis	0.45	
2	$R\overset{+}{N}(CH_3)_2nC_3H_7$	$R\overset{+}{N}(CH_3)_2isoC_3H_7$	Cat: salivation	2.00	118
			Cat: blood pressure	2.38	
			Mouse: mydriasis	4.09	
3	$R\overset{+}{N}(CH_3)_2nC_3H_7$	$R\overset{+}{N}(CH_3)_2nC_4H_9$	Cat: salivation	0.49	118
			Cat: blood pressure	0.52	
			Mouse: mydriasis	0.63	
4	$R\overset{+}{N}(CH_3)_2C_2H_5$	$R\overset{+}{N}CH_3(C_2H_5)_2$	Cat: salivation	1.06	
			Cat: blood pressure	1.31	
	(lachesine)		Mouse: mydriasis	0.60	
5	$R\overset{+}{N}CH_3(C_2H_5)_2$	$R\overset{+}{N}(C_2H_5)_3$	Cat: salivation	1.00	118
			Cat: blood pressure	0.79	
			Mouse: mydriasis	1.33	165
6	Atropine ($>NCH_3$)	N-Ethylnoratropine	Cat: blood pressure	0.04	165
7	Homatropine	N-Isopropylnor-homatropine	Cat: blood pressure	0.12	165
8	Atropine	N-Allylnora-tropine	Cat: blood pressure	0.04	166
			Rat: mydriasis	0.13	

[a] $R = (C_6H_5)_2C(OH)CO_2CH_2CH_2-$.

[b] In Tables 26.9–26.11, 26.13 and 26.15., the influence of the molecular modification on the pharmacological activity is expressed as activity ratios. An activity ratio represents the ratio of the relative molar activities of two substances, whose activities are compared with a standard substance. A ratio of 1.0 indicates that the molecular modification that converts the compound in series A to the corresponding compound in series B does not change the pharmacological activity. An activity ratio of greater than unity indicates that the molecular modification has increased the activity; when it is less than unity the molecular modification has decreased the activity.

stimulant activity the small cationic head must fit into a definite space and must aid the neutralization of the charge of the anionic site of the receptor. The inhibitory action is obtained when large enough groups are substituted on the cationic portion to prevent close contact with the receptor and hence the neutralization of the charge (172,173). Thus the cationic portion of the blocking agents provides the electrostatic forces necessary to orient the molecules toward the receptor and hold them in place.

The anticholinergic activity depends not only on the number and the molecular weight of the alkyl radicals, which are connected to the nitrogen atom, but also on their structure. In contrast to di-n-propylamino derivatives, di-isopropylamino derivatives have an anticholinergic activity close to or higher than the activity of diethylamino derivatives (135,174–177). The close correlation of the activities of the diethyl and diisopropyl derivatives could be related with the equal linear lengths (from the nitrogen atom) of these radicals.

In the case of cyclic amino alcohols where nitrogen enters into the composition of the cycle, the optimal anticholinergic effect is produced not by the N-ethyl, N-isopropyl, or N-allyl, but by N-methyl radical, as is apparent from a comparison of the esters of tropine (Table 26.4–26.6). It may be that the elements of the cyclic structure occupy a sufficiently large space besides the nitrogen atom. As a general rule, quaternization with a small alkyl group increases activity (Table 26.10), although a few exceptions have been reported (186,187).

Besides the charge on the cationic head of anticholinergics (and cholinergics), other factors seem to contribute to the interaction between the muscarinic receptor and the anticholinergics. The substituents at the nitrogen atom apparently participate actively in the process. This is evident from the anticholinergic action of the 3,3-dimethylbutylester of benzilic acid, $(C_6H_5)_2$ $C(OH)CO_2CH_2CH_2C(CH_3)_3$, which contains no nitrogen and consequently is not ionized but which has in the corresponding position a t-butyl radical which sterically imitates the trimethylammonium group (188). A similar replacement of a trimethyl-

Table 26.10 Differences Between the Anticholinergic Activities of Tertiary and Quaternary Ammonium Compounds and Atropine-like Agents

Compound Pair	Series A: Tertiary Ammonium Compounds	Series B: Quaternary Ammonium Compounds	Test System	Activity Ratio, B:A	References
1	Atropine	Methylatropine	Guinea pig: ileum	2.10	118
			Mouse: mydriasis	2.30	121
2	(−)-Hyoscyamine	(−)-Methylhyoscyamine	Mouse: mydriasis	2.70	122
3	(−)-Scopolamine	(−)-Methscopolamine	Guinea pig: ileum	7.60	178–180
			Mouse: mydriasis	1.00	
4[a]	Tertiary analog of methantheline $XN(C_2H_5)_2$	Methantheline $\overset{+}{X}N(C_2H_5)_2CH_3$	Rabbit: intestine	2.83	181
					182
5[b]	Tertiary analog of penthienate $XN(C_2H_5)_2$	Penthienate $\overset{+}{X}N(C_2H_5)_2CH_3$	Rabbit: ileum	1.24	121
			Rabbit: salivation	30.80	
6	(±)-Procyclidine	(±)-Tricyclamol	Guinea pig: ileum	18.3	183,184
			Mouse: mydriasis	13.5	
7	(±)-Benzhexol[c]	Methyl analog of (±)-benzhexol	Guinea pig: ileum	2.64	185
			Mouse: mydriasis	8.89	
8[d]	$RN(CH_3)_2$	$\overset{+}{R}N(CH_3)$	Cat: salivation	17.9	118
			Cat: blood pressure	10.3	
			Mouse: mydriasis	2.41	
9[d]	$RN(C_2H_5)_2$	$\overset{+}{R}N(CH_5)_2$	Cat: salivation	15.1	118
			Cat: blood pressure	9.06	
			Mouse mydriasis	14.2	

[a]For complete structures see Table 26.8.
[b]For complete structures see Table 26.8.
[c]1-Piperidino-3-phenyl-3-cyclohexyl-propan-1-ol.
[d]$R = (C_6H_5)_2C(OH)CO_2CH_2CH_2^-$.

ammonium group with a *t*-butyl radical in acetylcholine leads to its "carbon analog," $CH_3CO_2CH_2CH_2C(CH_3)_3$, which is similar to acetylcholine in its behavior toward cholinesterase (189).

5.2 Cyclic Moieties

The introduction of two phenyl groups into a molecule of acetylcholine or a cholinergic substance (i.e., $CH_3CO_2(CH_2)_2N(CH_3)_3$ or $CH_3(CH_2)_4N(CH_3)_3$) changes the compound to an anticholinergic agent [$(C_6H_5)_2CHCO_2(CH_2)N(CH_3)_3$ and, $(C_6H_5)_2CH(CH_2)_4N(CH_3)_3$, respectively]. Anticholinergics contain varied cyclic structures, the phenyl group being the most common (68). Often, one encounters cyclohexyl and cyclopentyl radicals and the corresponding unsaturated groups (cyclohexenyl, cyclopentenyl). Substances containing α-, or, less frequently, β-thienyl radicals may possess high anticholinergic activity. Often, unbranched (methyl, ethyl) or branched (isobutyl, isoamyl) groups are located at the same carbon atom together with one or two cyclic groups. The anticholinergic activities of substances that contain only aliphatic radicals are lower than those of the corresponding compounds with cyclic substituents.

The most common and, as a rule, the most active anticholinergics contain two cyclic substituents as blocking groups at the same carbon atom (Table 26.11), but a third cyclic substituent lowers the anticholinergic activity (190). When these cyclic groups are too large, such as biphenyl and naphthyl, the compounds have low anticholinergic activities. A sufficiently large number of anticholinergics that contain only one cyclic group on carbon are known; however, there is usually also an aliphatic radical or, even better, a hydroxyl group present in such a case. Examples of such compounds are the esters of tropic acid. The introduction of a second phenyl into the α-carbon of tropic acid lowers the anticholinergic activity of its aminoalkyl esters (68).

It is difficult to assess which cyclic substituents contribute the most for the anticholinergic activity. It could be that the effect of one or another moiety depends on the substituents already present and on other characteristics of the substance. An overwhelming majority of the therapeutically most active anticholinergics contain at least one phenyl group (Table 26.7). The

Table 26.11 The Influence of Cyclic Radicals on Anticholinergic Activity (Test System: Rabbit Intestine)

Compound Pair	Name or Structure of Compound[a]		Activity Ratio, B:A	References
	Series A	Series B		
1	$C_6H_5CH_2R$	$(C_6H_5)_2CHR$ (adiphenine)	6.7	190
2	$C_6H_5CH_2R$	$(C_6H_5)_3CR$	0.7	190
3	$CH_2(OH)R$	$C_6H_5CH(OH)R$	23.3	190
4	$C_6H_5CH(OH)R$	$(C_6H_5)_2C(OH)R$	114	190
5	Adiphenine (transentine)	$(C_6H_5)CH(C_6H_{11})R$ (transentine-H)	3.3	191,192
6	Adiphenine	Dicyclomine	10.0	193

[a]$R = CO_2CH_2CH_2N(C_2H_5)_2$.

second cyclic group, where there is one, need not be a phenyl. It is even better if, for example, it is cyclohexyl, cyclopentyl, or any other cyclic structure. Such un-symmetrical doubly substituted compounds have higher anticholinergic activities and lower toxicities (68,191–192). This is a situation similar to 5,5-disubstituted bar-bituric acid hypnotics and anticonvulsants.

A question might arise whether the aromatic (flat surface) nature of one of the cyclic radicals is essential for anticholiner-gic activity, because such a large number of anticholinergics contain a phenyl group. The sufficiently high activity of compounds in which both substituents are alicyclic (e.g., cyclohexyl or cyclopentyl) provides a basis for asserting that the aromatic nature of the substituents is not essential in anticholin-ergics (68).

Not only the number and the character of the cyclic group but also the mode of linking of the substituents is important for anticholinergic activity. Two phenyl nuclei are linked differently in 2-diethylamino-ethyl esters of diphenylacetic acid (**39**), fluorene-9-carboxylic acid (**40**), and *p*-bi-phenylacetic acid (**41**). Of these, the di-phenylacetic acid derivatives have the high-est anticholinergic activity (190).

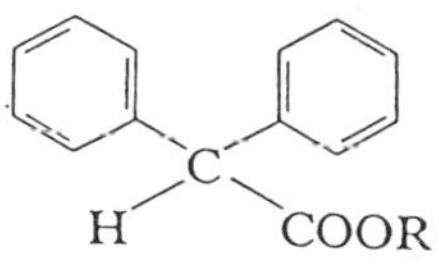

(**39**)

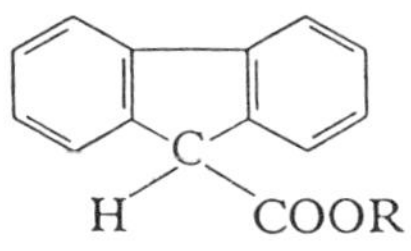

(**40**)

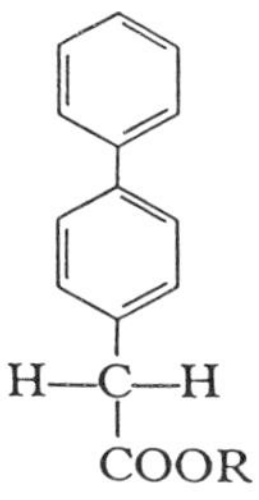

(**41**)

The importance of the cyclic nature of the substituent and not simply of its mass is evident from the comparison of the anticholinergic activities of 1-cyclohexyl-1-phenyl-3-piperidino-1-propanol (**42**) and 1-(*n*-hexyl)-1-phenyl-3-piperdino-1-propanol (**43**), of which structure **42** is an active anticholinergic whereas structure **43** is not effective (173).

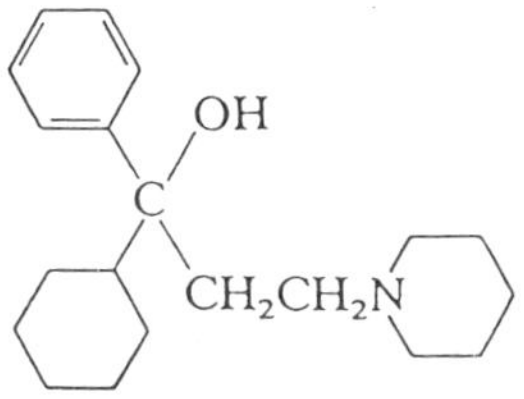

(**42**)

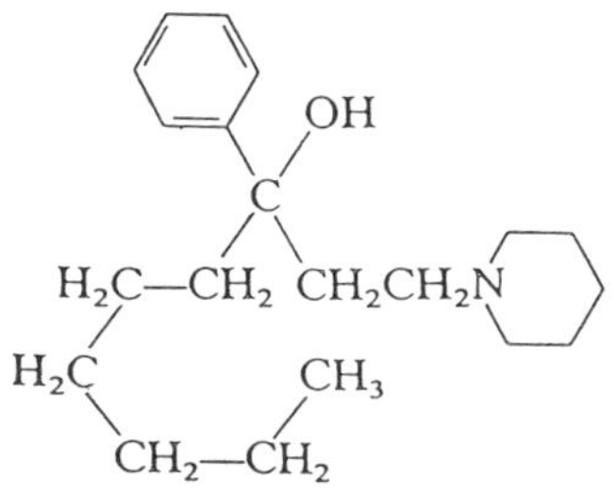

(**43**)

As far as the contribution of cyclic structures to anticholinergic activity is concerned, the introduction of cyclic groups into acetylcholine or a cholinergic com-pound leads to a change in the pharmaco-logical properties that, without lowering and possibly even strengthening its affinity for the muscarinic receptor, abolishes or blocks the action of the chemical transmit-ter. This phenomenon is similar to the transition from a metabolite to an an-timetabolite. It has been suggested that the cyclic groups of the anticholinergic agent form an additional contact with the mus-carinic receptor by hydrophobic or van der Waals forces; as a result, this contact is strengthened and the muscarinic receptors are protected from approaching molecules of acetylcholine (173,194). Cyclic groups of substantial size can create a kind of protective screen that sterically hinders the approach of molecules of acetylcholine not only to the given active site but also to the vicinity of the active sites of the receptor. Tricyclic anticholinergic agents may fall under this category.

5.3 Length of the Main Chain Connecting the Cationic Head and the Cyclic Groups

The presence of the cationic head and of cyclic groups is not sufficient for optimal anticholinergic activity of a compound. The activity depends on the mutual distribution of these groups. This establishes the basic requirements for a chain of atoms that connects the cationic head and the cyclic moieties; these apply to the length and form of the chain, lateral branching, and functional groups in the chain, if any.

A considerable number of the anticholinergics belong to the group of aminoalkyl esters of substituted acetic acids. In an overwhelming majority of cases, the substi-

tuted esters of β-aminoethanol are more active as anticholinergics than the corresponding derivatives of γ-aminopropanol (118,169,174,195). Further increase of the chain length of the amino alcohol leads to a decrease or disappearance of the anticholinergic activity. The aminoalkyl esters of diphenylacetic acid are more active anticholinergics than the corresponding aminoalkyl esters of β,β-diphenylpropionic acid (196). Therefore, in all these esters with high anticholinergic activity the main chain connecting the cyclic moieties and the cationic head contains five atoms (Table 26.12, series 1–3). In an homologous series of compounds in which the ester group is replaced by a chain of carbon atoms (Table 26.12, series 4–9), there are three atoms in the main chain in compounds with maximal anticholinergic activity. To explain the differences in the anticholinergic activities of different series of compounds, the ability of their structures to exist in different spatial conformations must be taken into account.

Acetylcholine and related esters can exist in two conformations (Fig. 26.5): skewed and extended (197) (e.g., **44** and **45**,

Table 26.12   The Chain Length between Cationic Head and Cyclic Radicals among Anticholinergics

Number	Series	Value of n for High Anticholinergic Activity	Total number of Atoms in the Chain	Test System	Reference
1	$(C_6H_5)_2C(OH)CO_2(CH_2)_nNC_5H_{10} \cdot HCl$	2^a	5	Rabbit: mydriasis	195
2	$(C_6H_5)_2C(OH)CO_2(CH_2)_n\overset{+}{N}C_5H_{10} \cdot CH_3Br$	2^a	5	Rabbit mydriasis	195
3	$(C_6H_5)_2C(OH)CO_2(CH_2)_n\overset{+}{N}(C_2H_5)(CH_3) \cdot$ CH$_3$Cl	2^a	5	Mouse: mydriasis	118
4	$(C_6H_5)(C_2H_5)C(OH)(CH_2)_nN(C_2H_5)_2 \cdot$ HCl	2	3	Rabbit: ileum	155
5	$(C_6H_5)(C_2H_5)C(OH)(CH_2)_nNC_5H_{10} \cdot$ HCl	2	3	Rabbit: ileum	155
6	$(C_6H_5)_2C(OH)(CH_2)_nNC_5H_{10} \cdot HCl$	2	3	Rabbit: ileum	155
7	$(C_6H_5)_2CH(CH_2)_nN(C_2H_5)_2 \cdot CH_3I$	2	3	Mouse: salivation	
8	$(C_6H_5)_2C(OH)(CH_2)_nN(C_2H_5)_2 \cdot HCl$	2	3	Mouse: salivation Mouse: tremor	
9	$(C_6H_5)_2C(OH)(CH_2)_nN(C_2H_5)_2 \cdot CH_3I$	2	3	Mouse: salivation	

^{a}No exact values are available for esters with $n-1$.

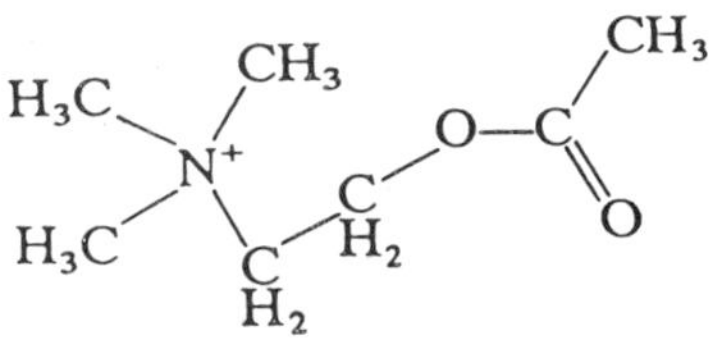

Fig. 26.5 Conformations of cholinergics and anticholinergics.

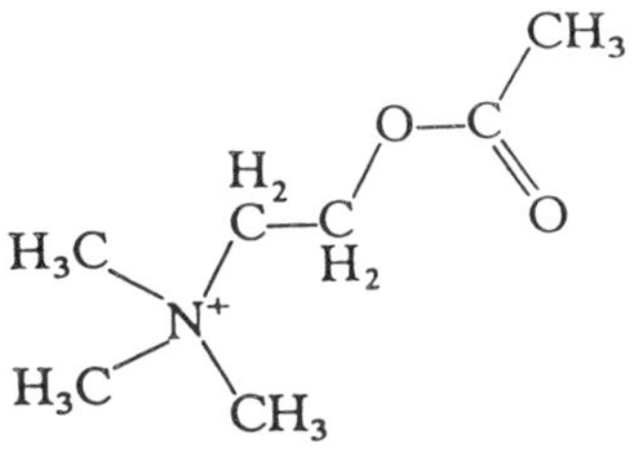

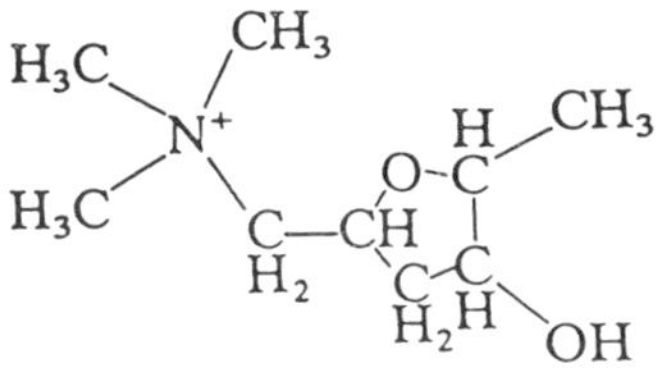

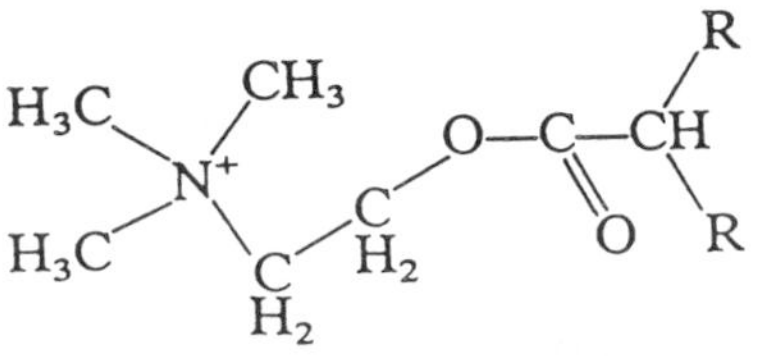

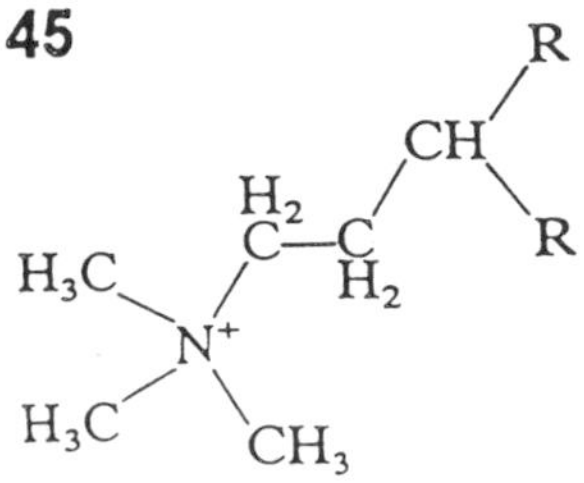

respectively, for acetylcholine). The skewed form (**44**) of acetylcholine is closely related to the structure of muscarine (**46**) (197). Similarly, the substituted aminoethyl esters, which are anticholinergics, may exist in two conformations. The skewed form of acetylcholine (**44**), muscarine (**46**), the skewed form of aminoethyl esters (**47**), and the extended form of aminopropane derivatives (**48**) all interact at the same muscarinic receptors. In the former two compounds, the interatomic distance between the quaternary nitrogen and the ether oxygen atom is nearly the same, and both of them are agonists. In structures **47** and **48** the interatomic distance between the quaternary nitrogen and the carbon atom to which the cyclic radicals are attached is the same, and both of them are antagonists (68). Thus anticholinergic activity depends not only on the length of the main chain of the molecule but also on its ability to adopt a certain conformation that is favorable for the interaction of the substance with the receptor.

There is some information about the influence of branching of the main chain on the anticholinergic activity. Esters with a methyl group α to the ester oxygen in the amino alcohol part are less active than compounds without the methyl group (174,195,198). Similarly, the derivatives of 1,3-aminopropanol, aminopropane, and γ-aminobutyronitrile (198,199) that contain a branch at the carbon atom β to the nitrogen are less active anticholinergics than the compounds without the branching. The negative influence of such a side chain has been explained by steric hindrance at the receptor (173).

The inclusion in the main chain of optimum length of other atoms such as oxygen, sulfur, nitrogen, and other functional groups changes any anticholinergic activity (68). However, such compounds are considerably potent.

5.4 Esteratic Linkage

The question of the importance of complex ester grouping in anticholinergics, and even more of its role, has not been cleared up sufficiently. A great importance was attached to the complex esters in the initial period of the search for atropine-like substances, when active compounds were known only among the esters of amino alcohols and carboxylic acids. However, the presence of this grouping is not necessary for the manifestation of anticholinergic activity. Presently, a large number of substances are known that belong to different chemical structures and that possess high anticholinergic activity (Table 26.7).

The influence of an ester link can be assessed by comparing similar compounds that do not contain pharmacophoric groups other than the anchoring groups (amino nitrogen, cyclic radicals). Comparative data on the anticholinergic activities of the 2-diethylaminoethyl ester of diphenylacetic acid $[(C_6H_5)_2CHCO_2CH_2CH_2CH_2N(C_2H_5)_2]$ and 1,1-diphenyl-5-diethylaminopentane $[(C_6H_5)_2CHCH_2CH_2CH_2CH_2N(C_2H_5)_2]$ indicate that they are equally active (68). Thus the complex ester group is not essential for anticholinergic activity; however, it may contribute to optimal activity when it is present in atropine-like compounds (122). It may influence the conformation of a molecule that in turn determines the effectiveness of the interaction of the essential anchoring groups, the cationic head, and the cyclic radicals (68) with the muscarinic receptor.

5.5 Hydroxyl Group

The anticholinergic compounds that contain a hydroxyl group in a certain position of a molecule possess considerably higher activity than similar compounds without the hydroxyl. That position is of great importance. For esters of amino alcohols and hydroxycarboxylic acids, maximum activity is achieved if the hydroxyl is β to carboxyl. Atropine is about 10 times more active than homatropine. However, esters with an α-hydroxyl also possess considerable anticholinergic activity. In anticholinergic amino alcohols the hydroxyl on the third

carbon atom from the nitrogen gives optimal activity (Table 26.7). The location of the hydroxyl group in relation to the cyclic radicals is also of vital importance. In the great majority of anticholinergics, they are located at the same carbon atom or at adjacent carbons.

The hydroxyl group in anticholinergics can be replaced by CN and $CONH_2$ groups while preserving some degree of activity. However, replacement of the hydroxyl by methoxy or an acetoxy group lowers the activity (200,201).

The hydroxyl group may interact by hydrogen bonding with a site on the muscarinic receptor, which is rich in electrons. In support of this statement (68), hydroxylated anticholinergics form complexes in solution with substances such as amines, which contain electron donor atoms. In a series of structurally related compounds the anticholinergic activity was proportional to their capacity for molecular association by way of the hydroxyl group. There is a direct relationship between the anticholinergic activity and the mobility of the hydrogen atom of the hydroxyl group as determined by the rate of acetylation. The contribution of the hydroxyl group to the free energy of adsorption is quite quite high, of the order of 2 kcal; it is apparently independent of the number of methyl groups attached to the cationic head (202).

Although a hydroxyl group increases the activity of an anticholinergic, it does not convert a cholinergic substance into an anticholinergic. Propionyl-, α-hydroxypropionyl-, and α,β-dihydroxypropionylcholines possess cholinergic properties (203). α-Hydroxy substitution decreases the original muscarinic activity to about one-third, whereas the introduction of both α- and β-OH functions decreases the muscarinic activity to about one-tenth.

The introduction of a hydroxyl group into 2-diethylaminoethyl phenylacetate approximately doubles its activity (Table 26.13), whereas the same structural change in the corresponding ester of diphenylacetic

Table 26.13   The Influence of the Hydroxyl and Epoxy Groups on Anticholinergic Activity

Group	Name or Structure of Compound		Test System	Activity Ratio B:A	Reference
	Series A	Series B			
Hydroxyl	$C_6H_5CH_2CO_2CH_2CH_2N(C_2H_5)_2$	$C_6H_5CH(OH)CO_2CH_2CH_2N(C_2H_5)_2$	Rabbit: intestine	2.3	190
	$(C_6H_5)_2CHCO_2CH_2CH_2N(C_2H_5)_2$	$(C_6H_5)_2C(OH)CO_2CH_2CH_2N(C_2H_5)_2$	Rabbit: intestine	143	190
Epoxy	(−)-Hyoscyamine	(−)-Scopolamine	Guine pig: ileum	0.24	204
			Mouse: eye	2.70	
			Cat: eye	5.80	
			Cat: blood pressure	0.64	
			Cat: salivation	0.77	
	(−)-Methylhyoscamine	(−)-Methylscopolamine	Mouse: eye	1.00	204
			Cat: eye	3.33	
			Cat: blood pressure	0.80	
			Cat: salivation	0.80	

acid increases its activity about 140 times. The positive influence of the hydroxyl group has been observed in a large number of anticholinergics (Table 26.7). The exceptions are those cases in which the cyclic groups are too large or are connected in such a way that they can sterically prevent the interaction of the hydroxyl with the surface of the muscarinic receptor.

5.6 Epoxy Group

The presence of an epoxy group seems to increase the mydriatic activity (Table 26.13). However, scopolamine (**7**) which contains an epoxy group, is a central depressant, as indicated by drowsiness, euphoria, amnesia, and dreamless sleep. Atropine (**4**), which does not contain an epoxy group, stimulates the medulla and higher cerebral centers. In clinical doses (0.5–1.0 mg), this effect is usually confined to mild vagal excitation. Toxic doses of atropine cause restlessness, disorientation, hallucinations, and delirium.

5.7 Stereoisomerism and Anticholinergic Activity

5.7.1.   OPTICAL ISOMERISM.   Atropine (**4–6**) is the racemic form of hyoscyamine, which is the (S)-(−)-tropyl ester of 3α-tropanol (**2**). The carbon α to the carbonyl group is asymmetric. (S)-(−)-Hyoscyamine is more active than (R)-(+)-hyoscyamine as an anticholinergic (Table 26.11). The alkaloid scopolamine (**7**) is the (S)-(−)-tropyl ester of scopine (**8**); again the (S)-(−) isomer is more active than (R)-(+) isomer in its anticholinergic activities.

A considerable number of synthetic anticholinergic agents patterned after the structure of atropine contain an asymmetric carbon atom corresponding to the position of the asymmetric carbon in atropine. In all compounds examined, the asymmetric carbon is located in the acyl moiety and is connected with the cyclic and the hydroxyl groups (directly or through a methylene group). The (−) isomers are often more active than (+) isomers (Table 26.14), indicating some apparent stereospecificity with respect to the carbon atom α to the carbonyl group of atropine.

The atropine-like activities of some compounds in which the asymmetric carbon atom is considerably closer to the amino group have been described. In 1,1-diphenyl-3-piperidino-1-butanol, the carbon α to the nitrogen is asymmetric. In this case the (+) isomer seems to be more active than the (−) isomer. In 3-quinuclidinyl diphenylacetate, the carbon atom β to the nitrogen is asymmetric; the (−)isomer has more atropine-like activity than the (+) isomer.

Table 26.14   Optical Isomerism and Anticholinergic Activity

Number	Compounds Whose (+) and (−) Isomers Are Tested	Test System	Position of the Asymmetric Carbon	Active Isomer	Isomeric Ratio[a]	Reference
1	Hyoscyamine	Dog: salivation		(−)	30	204
		Cat: salivation	α-Carbon to	(−)	20	
		Cat: blood pressure	the carbonyl	(−)	23	
		Guinea pig: ileum	group	(−)	32	
		Rabbit: ileum		(−)	110	
2	Scopolamine	Dog: salivation	α-Carbon to	(−)	17	204
		Rabbit: intestine	the carbonyl	(−)	15	
			group	(−)		
3	Tricyclamol	Guinea pig: ileum	Carbon with	(−)	160	205
			cyclic radical	(−)	62	
4	Benzhexol	Guinea pig: ileum	Carbon with	(−)	10	205
		Rabbit: intestine	cyclic radical	(−)	160	122
		Mice: mydriasis		(−)	5	205
5	Procyclidine	Guinea pig: ileum	Carbon with	(−)	49	205
		Mice: mydriasis	cyclic radical	(−)	18	205
6	1,1-Diphenyl-3-piperidino-1-butanol hydrochloride	Rabbit: intestine	α to N	(+)	84	206 207
7	Methiodide of number 6	Rabbit: intestine	α to N	(+)	3	206 207
8	Diphenylacetate of 3-quinuclidinol	Rabbit: intestine	β to N	(−)	24	208

[a]Activity ratio between the enantiomers.

5.7.2. DERIVATIVES OF TROPINE AND PSEUDOTROPINE. The configuration of the 3-OH group in the tropine part of the molecule has significant influence on the activity at the muscarinic receptor (Table 26.15). The derivatives of ψ-tropine (pseudotropine, **3**) are less active; the activity ratio for the ψ compound relative to the isomeric tropine varies from 2 to 13, but more information is needed on this point.

5.7.3. STEREOCHEMICAL CONFIGURATION. The acetylcholine-like cholinergics and atropine-like anticholinergics contain similar pharmacodynamic groups. Various hypotheses have assumed that both stimulant and blocking drugs interact with the muscarinic receptor through the essential pharmacodynamic groups. The tropic acid por-

tion of atropine contains an asymmetric carbon, and the muscarinic receptor is stereospecific for the carbon α to the carbonyl group in anticholinergics. Acetylcholine does not contain such an asymmetric carbon atom. Lactoylcholine is an agonist that contains an asymmetric carbon (203), and the muscarinic receptor is stereospecific for the carbon α to carbonyl group among lactoylcholine-like cholinergics (211). Owing to the structural similarities in tropic and lactic acids, it has even been suggested that a lactoylcholine-like parasympathetic neurohormone may occur in animal tissues (203,212), however, this has not been corroborated.

5.7.4. DISSOCIATION CONSTANTS OF CHOLINERGICS AND ANTICHOLINERGICS. The

Table 26.15 Relative Anticholinergic Activities of the Esters of Tropine and Pseudotropine

Isomer Pair	Pair of Structural Isomers		Test System	Activity Ratio, A:B	Reference
	A	B			
1	Atropine	Tropyl-ψ-tropine	Cat(?): blood pressure	2	209
2	Benzoyl-tropine HCl	Benzoyl-ψ-tropine	Rabbit or Guinea pig: intestine	3	210
3	CH$_3$I of number 2	CH$_3$I of number 2	Guinea pig: intestine	13	210
4	C$_2$H$_5$Br of number	C$_2$H$_5$Br of number 2	Guinea pig: intestine	4	210

absolute configuration ((R) and (S)) is self-consistent for a molecule in question and cannot be used to relate a series of compounds. The configuration in relation to a standard substance (D and L) is useful to compare a series of compounds. For example, the pharmacological parameters of all D compounds in a series can be compared with those of the L compounds, provided each one of the compounds contains a single asymmetric carbon (213).

In a number of studies on structure–activity relationships, the pharmacological activities are expressed in terms of potencies or relative molar activities, which are derivatives of their ED$_{50}$s and whose reciprocals do not give exact measures of affinities (214). Affinities are required to make valid conclusions on the stereoisomer–receptor interactions and the nature of receptor surfaces. The differences in the potencies of a pair of stereoisomers may be due to the differences in their affinities or intrinsic efficacies. For these reasons, the following information is necessary to make definite conclusions for delineating receptor surfaces using stereoisomer–receptor interactions (213): (*1*) the dissociation constant of agonists (K_A) and antagonists (K_B) act at the same receptors, (*2*) the absolute configuration of the compounds, and (*3*) the interrelationships between the configura-

tions of agonists and antagonists acting at the same receptors.

The dissociation constants of some cholinergic agonists and antagonists have been determined (Table 26.16). D(+)- and L(−)-Lactoylcholines are agonists at muscarinic receptors, and there is no significant difference in their efficacies. The K_A of D(+)-lactoylcholine is lower than the K_A of the L(−) isomer at the muscarinic receptor. Therefore, the D(+) isomer has a higher affinity to the muscarinic receptor than the L(−) isomer (215). Mandeloylcholines and tropinoylcholines are competitive antagonists of acetylcholine and lactoylcholine at the muscarinic receptors (217,219). Among these anticholinergics, the D isomer has a higher affinity ($1/K_B$) than the corresponding L isomer. The above anticholinergics did not exhibit significant intrinsic efficacies at the muscarinic receptors. The carbon α to the carbonyl carbon of the ester group is asymmetric in agonists (lactoylcholines) and their competitive antagonists (mandeloylcholines and tropinoylcholines). Therefore, the D isomers have the preferred relative configuration, which comes into definite spatial position with the muscarinic receptor. Similarly D(−)-hyoscyamine has higher affinity to the muscarinic receptor than the L(+) isomer and has the preferred configuration (220).

Table 26.16 Dissociation Constants and Intrinsic Efficacies of Cholinergic and Anticholinergic Agents

Cholinergic or Anticholinergic and Configuration	Activity			
	Type of Receptor[a] (Test System)	Dissociation Constant[b] (K_A or K_B)	Relative Intrinsic Efficacy	References
Acetylcholine	Muscarinic	1.08×10^{-6}	1.00	215
	Nicotinic	2.17×10^{-6}	1.00	215
(R)-D$(+)$-Lactoylcholine	Muscarinic	7.3×10^{-5}	0.52	215
	Nicotinic	1.85×10^{-5}	1.07	215
(S)-L-$(-)$-Lactoylcholine	Muscarinic	3.02×10^{-4}	0.30	215
	Nicotinic	8.08×10^{-5}	1.15	215
(R)-D-$(-)$-Acetyl-β-methylcholine	Muscarinic	Inactive		216
(S)-L-$(+)$-Acetyl-β-methylcholine	Muscarinic (active isomer)	1.24×10^{-6}		216
(R)-D-$(-)$-Mandeloylcholine	Muscarinic	3.00×10^{-6}	NS[c]	217,218
(S)-L-$(+)$-Mandeloylcholine	Muscarinic	5.22×10^{-6}	NS	217,218
(S)-D-$(-)$-Tropinoylcholine	Muscarinic	2.15×10^{-8}	NS	217,219
(R)-L-$(+)$-Tropinoylcholine	Muscarinic	3.26×10^{-7}	NS	217,219
(S)-D-$(-)$-Hyoscyamine	Muscarinic	4.47×10^{-10}	NS	220
(R)-L$(+)$-Hyoscyamine	Muscarinic	1.41×10^{-8}	NS	220

[a]Muscarinic activities are tested on the guinea pig longitudinal ileal muscle in all cases except acetyl-β-methylcholine, which is tested on the circular muscle from fundus of rabbit stomach. Nicotinic activities are tested on the frog rectus abdominis muscle.
[b]Moles/liter.
[c]Not significant.

5.8 Compounds with Dual Action: Cholinergic and Anticholinergic Activities

In several groups of atropine-like agents, derived from acetylcholine-like compounds, agonist activity is replaced by partial agonist activity and eventually antagonist activity with increasing substitution (216,221–223). For example, a transition between cholinergic and anticholinergic properties occurs when the acyl group of acetylcholine is progressively lengthened. Cholinergic activity decreases from formylcholine to butyrylcholine and the higher members of the series are anticholinergics (Table 26.17). It has been demonstrated that hyoscyamine and atropine at small dose levels exhibit cholinergic properties (224,225).

6 INTERACTION OF ANTICHOLINERGICS AT THE MUSCARINIC RECEPTORS

It is generally accepted that acetylcholine and atropine interact with the same postganglionic muscarinic receptors. Whereas acetylcholine stimulates these receptors, atropine blocks them. Although considerable progress has been made in understanding these interactions of stimulant and blocking drugs, some aspects of drug–receptor interaction are not clear. For detailed discussions of cholinergic and anticholinergic drugs at the muscarinic receptors see original papers and reviews on the subject (226–232).

Table 26.17 Cholinergic and Anticholinergic Activities of Choline Esters

$$RCO_2CH_2CH_2\overset{+}{N}(CH_3)_3 \cdot I^-$$

R	Intrinsic Activity, α	pD_2	pA_2	Test System	Reference
H	1	5.2		Rat: intestine	1
CH_3	1	7.0		Rat: intestine	
CH_3CH_2	1	5.3		Rat: intestine	
$CH_3(CH_2)_2$	0.5	5.1		Rat: intestine	
$CH_3(CH_2)_3$	0		4.7	Rat: intestine	
$CH_3(CH_2)_5$			4.7	Guinea pig: ileum	223
$CH_3(CH_2)_6$			4.7	Guinea pig: ileum	
$CH_3(CH_2)_7$			5.0	Guinea pig: ileum	
$CH_3(CH_2)_8$			5.5	Guinea pig: ileum	
$CH_3(CH_2)_9$			6.0	Guinea pig: ileum	
$CH_3(CH_2)_{10}$			6.5	Guinea pig: ileum	
	0		5.4	Rat: intestine	221

6.1 Kinetic Basis for the Mechanism of Action of Anticholinergics

The major action of a number of anticholinergics is a competitive antagonism to acetylcholine and other cholinergic agents. The antagonism can, therefore, be overcome by increasing the concentration of acetylcholine at receptor sites of the effector organs. Thus anticholinesterases partially reverse the antagonism of anticholinergics by sparing acetylcholine at the receptor sites. The anticholinergics can inhibit all muscarinic actions of acetylcholine and other choline esters. Responses to postganglionic cholinergic nerve stimulation may also be inhibited, but less readily than responses to administered choline esters. The differences in the ability of anticholinergics to block the effects of exogenous choline esters and the effects of endogenous acetylcholine liberated by the postganglionic parasympathetic nerves may be due to the release of the chemical transmitter by the nerve at the receptors in relatively inaccessible sites where diffusion limits the concentration of the antagonist.

6.2 Specificity of Antagonism

Atropine is a highly selective antagonist of acetylcholine, muscarine, and other cholinergic agents on the smooth and cardiac muscles and glands. This antagonism is so selective for cholinergic agents that atropine blockade of the actions of other types of drugs has been taken as evidence for their actions through cholinergic mechanisms. For example, the smooth muscle of guinea pig ileum is stimulated by muscarine, 5-hydroxytryptamine, histamine, and barium chloride. Atropine is more specific in blocking the stimulant effects of muscarine and acetylcholine at lower dose levels than those of the other three stimulant agents.

6.3 Molecular Basis for the Interaction of Acetylcholine and Anticholinergics at the Muscarinic Receptors

Structure–activity relationships among muscarinic agents (or cholinergics) indicate the existence on the receptor of two active sites separated by a distance $3.2 \pm 0.2 \, \text{Å}$ (215,232–235). One of them is an anionic site with which the quaternary ammonium group interacts to induce stimulant or blocking actions. The ether oxygen of muscarine and the ester oxygen of acetylcholine interact with the second site. There are some similarities between the active sites on acetylcholinesterase and the muscarinic receptor (Chapter 25). The amine portion of anticholinergics interacts at the same anionic site as the quaternary group of acetylcholine and atropine. Several facets of acetylcholine–atropine antagonism are well known:

1. One molecule of atropine blocks one molecule of acetylcholine. Atropine is a larger molecule than acetylcholine and either mechanically or electrostatically inactivates receptors engaged by it.

2. Atropine has greater affinity than acetylcholine for the receptor. Its intrinsic activity is not significant, whereas acetylcholine has high intrinsic activity. Substances with intermediate intrinsic activities behave either as cholinergics or as anticholinergics depending on the nature of their influence on the receptor. Among such substances are partial agonists with "dual action"; cholinergic activity precedes the anticholinergic activity. The partial agonists can be detected in a homologous series by gradually proceeding from agonists to antagonists with increasing molecular weight.

3. Besides the cationic head, bulky cyclic groups are essential constituents of compounds with anticholinergic activity. It seems clear that the van der Waals or hydrophobic binding of the planar cyclic groups together with the binding of the amine group produce a stable drug–receptor complex that effectively blocks the close approach of acetylcholine to the receptor.

4. Acetylcholine increases potassium efflux and causes depolarization of the membrane, both of which effects are blocked by atropine.

5. The receptor proteins on the membrane may undergo molecular disorientation during the interaction of acetylcholine with the cholinergic receptor, and this change in the receptor proteins may be prevented by a suitable blocking agent (236).

As the five subtypes of muscarinic receptors are isolated and their amino acid sequences are determined, future investigations may reveal the molecular nature of interactions of anticholinergics with muscarinic receptors (Chapter 25).

7 THERAPEUTIC USES OF ANTICHOLINERGICS

The chief use of most of the antispasmodic agents is as an adjunct in the management of the peptic ulcer; this group of drugs includes adiphenine, aminopentamide, amprotropine, dibutoline, diphemanil, glycopyrrolate, hexocyclium, homatropine methylbromide, methscopolamine bromide, methscopolamine nitrate, oxphencyclimine, oxphenonium, penthienate, pipenzolate, piperidolate, pipethonate, propanthelin, tricyclamol, and trihexethyl (231). Pirenzipine is a leading compound in this group of M_1 receptor antagonists that decrease acid secretion. The anticholinergic agents that are useful as adjuvants in the management of the functional disorders of the bowel (e.g., irritable colon, spastic colitis, ulcerative colitis, and diverticulitis) include dicyclomine, hexocyclium, mepenzolate, and valethamate.

The mydriatic and cycloplegic activities of anticholinergics in humans are listed in Table 26.18. Atropine is recommended in situations requiring complete and prolonged relaxation of the sphincter of iris and the ciliary muscle. Mydriatics like

Table 26.18   Mydriatic and Cycloplegic Activities of Anticholinergics in Humans[a]

Number	Drug[b]	Strength of Solution, %	Mydriasis		Cycloplegia	
			Maximal, min	Recovery, days	Maximal, h	Recovery, days
1	Atropine sulfate	1.0	30–40	7–10	1–3	8–12
2	Oxyphenonium bromide	1.0	30–40	7–10	1–3	8–12
3	Scopolamine hydrobromide	0.5	20–30	3–5	0.5–1	1–2
4	Atropine methyl nitrate	1.0–5	30	2	1	2
5	Homatropine bromide	1.0	10–30	0.25–4	0.5–1.5	0.5–2
6	Cyclopentolate hydrochloride	0.5–1.0	30–60	1	0.5–1	1
7	Dibutoline sulfate	5.0–7.5	60	0.25–0.5	1	0.25–0.5
8	Tropicamide[c]	1.0	20–35	0.25	0.5	0.08–0.25
9	Eucatropine hydrochloride	5–10	30	0.25–0.5	none	

[a]For details see refs. 237–239. The values should be considered approximate.
[b]One instillation of one drop, unless otherwise specified.
[c]Two drops at 5-min intervals.

cyclopentolate, eucatropine, and homatropine bromide, with a shorter duration of action, are usually preferred for measuring refractive errors because of the relative rapidity with which their cycloplegic effects are terminated.

Atropine and scopolamine are used for medication before the administration of some inhalation anesthetics to reduce excessive salivary and bronchial secretions. Atropine and related agents have been used in the treatment of renal colic and hyperhidrosis and to control sweating that may aggravate certain dermatologic disorders. Atropine also may be used to counteract the toxicity of certain cholinergic drugs and anticholinesterase agents.

Certain drugs with anticholinergic effects are used for the symptomatic treatment of Parkinson's disease (paralysis agitants) and related syndromes of the extrapyramidal tracts (see Chapter 37). (Of the presently available drugs, none is useful in all cases of Parkinsonism). Despite claims of superiority for newly introduced synthetic agents,

none possesses outstanding efficacy and freedom from adverse side effects when compared clinically with atropine and scopolamine (237).

8 MOLECULAR BASIS FOR THE SIDE EFFECTS OF ANTICHOLINERGICS

The most widely used mode of approach in the design of anticholinergics is based on the use of tropine alkaloids as models of prototypes, from which congeners or homologs or analogs have been designed. Tropine alkaloids have many pharmacological activities and interact at many cholinergic sites. In drug design, the main purpose is to increase one pharmacological action at one particular site of action while suppressing other pharmacological activities at other sites. It is not always possible to abolish all pharmacological effects other than the desired activity by molecular modification. Though the desired activity is useful in its therapeutic applications, other pharmacological activities manifest themselves as side effects. For example, atropine, scopolamine, and cocaine are structurally related, each having a tropine nucleus. They differ in some of their pharmacological activities. Atropine stimulates the central nervous system, scopolamine depresses the central nervous system, and cocaine is a local anesthetic and CNS stimulant.

By molecular modification, it has been possible to produce a series of anticholinergics having qualitative effects resembling those produced by parasympathectomy to a particular organ. Although these drugs exert specific therapeutic effects at one organ, they exert side effects at other organs. Recent developments on the design of anticholinergics selective for subtypes of muscarinic receptors and identification of subtypes of muscarinic receptors in a number of organs have partially provided a solution for this problem.

The untoward effects associated with the use of anticholinergics are manifestations of their pharmacological actions and usually occur on excessive dosage. The effects include dryness of mouth, blurred vision, difficulty in urination, increased intraocular tension, tachycardia, and constipation. Most of these side effects are lessened when the quaternary anticholinergics are administered orally in the treatment of peptic ulcer because of low absorption into the systematic circulation. In the case of tertiary amines, the central side effects of euphoria, dizziness, and delirium may be observed because the drugs can cross the blood–brain barrier.

Many synthetic quaternary ammonium compounds may block acetylcholine at ganglia at high doses. Ganglionic blocking agents cause impotence as a side effect. High doses of methantheline may also cause impotence, an effect rarely produced by pure antimuscarinic drugs and indicating ganglionic blockade. Toxic doses of quaternary ammonium compounds (e.g. menthantheline, propantheline, and oxyphenonium) block acetylcholine at the somatic neuromuscular junction and paralyze respiration.

Adiphenine and amprotropine have local anesthetic activities, and anesthesia of the oral mucosa results when tablets of these drugs are chewed. It should be remembered that local anesthetic esters and amides exert their action by anticholinergic mechanisms, probably essentially at the nodes of Ranvier.

The central side effects have appeared among children even when cyclopentolate, tropicamide, and other anticholinergics are used as mydriatics. All anticholinergics increase intraocular pressure in most patients with simple glaucoma.

Some of the cyclic groups in anticholinergics are pharmacophoric moities for other types of activities. For example, the compounds containing a phenothiazine nucleus exhibit central depressing and antihistaminic side effects. These side effects are of advantage in the treatment of Parkinson's disease. The side effects of certain drugs, which result from their anticholinergic activities, are prominent among some analgesics (e.g., meperidine), antihistamines (e.g., promethazine), psychosedatives (e.g. benactizine), and psychotomimetics (e.g., dexoxodrol).

9 PROFILE OF ANTICHOLINERGIC ACTIVITIES OF VARIOUS AGENTS

The relative anticholinergic activities of the well-known therapeutic compounds are listed in Table 26.19. Although it is difficult to justify collecting the results of a wide variety of experiments, it seems likely that Table 26.19 gives some idea of their relative antisecretory, antispasmodic, and mydriatic activities relative to atropine. Ratios less than unity indicate that the drugs are more active than atropine.

Table 26.19 Relative Activities of Anticholinergics[a]

		Equipotent Molar Ratios Relative to Atropine[c]			Total Dose per day in Humans[f]		
Number	Drug[b]	Antisecretory[d]	Antispasmodic[e]	Mydriatic	mg	μmol	References
		Solanaceous Alkaloids and Semisynthetic Substitutes					
1	Atropine sulfate	1.0	1.0	1.0	0.8–2.0	2.3–5.8	
2	Methylatropine	0.48 c 3.02 r,a	0.47 g	0.44 m			204
3	(−)-Hyoscyamine	0.56 c	0.31 g	0.54 m			
4	(+)-Hyoscyamine	11.0 c	10.0 g				
5	(−)-Methylhyoscamine	0.25 c		0.20 m			
6	(−)-Scopolamine	0.73 c	1.3 g	0.20 m			
7	(−)-Methscopolamine	0.29 c	0.17 g	0.21 m	5–10	13–25	
8	(±)-Homatropine	0.44 r,a 30 d	8.5 g	7.7 c			244
9	(±)-Methylhomatropine	1.8 rb	7. 2 rb	2.4 m			241
		Synthetic Anticholinergics: Esters, Quaternary					
10	Glycopyrrolate	0.9 r,a	1.0 g	1.0 m	2–6	5–15	242
11	Lachesine	0.39 c	0.96 rb	0.96 m			118
12	Mepenzolate				100	238	
13	Methantheline	0.37 c	0.48 g	3.0 m	400	952	243
14	Methyleucatropine	1.6 c		36 m			180
15	Oxyphenonium	1.0 rb 1.0 d,a	1.0 g	1.0 rb	40	93	244,245
16	Penthionate	0.26 rb	0.39 rb		20–40	48–9	121
17	Pipenzolate		1.0 g		20–25	46–58	246
18	Poldine methyl sulfate	1.0 c	1.0 g	1.0 m	8–48	18–106	247
19	Propantheline	0.26 c	0.40 g	1.6 m	75–240	167–536	243
20	Valethamate				30–80	78–207	
		Esters, Tertiary					
21	Adiphenine		42 rb		300–600	865–1729	248
22	Amprotropine phosphate	20 rb	55 rb	20	200–400	494–988	248
23	Benactyzine	5.6 c	3.5 rb	17 m	3–9	8.3–24.8	118
24	Carbofluorene	100 d	7.5 rb		500	1449	249
25	Cyclopentolate				see Table 26.14		
26	Dicyclomine	60 r,a	8.0 rb	8.8 r	60–80	173–231	193
27	Eucatropine	29 c		250 m	see Table 26.14		118,180
28	Oxyphenycyclimine				20–50	53–132	
29	Piperidolate				200	557	
30	Propivane	40 rb	29 rb	5000 c	400	1274	248 249
		Thioesters, Tertiary					
31	Triphenamil		6.0 rb		800	2204	250
		Carbamates, Quaternary					
32	Dibutoline sulfate	10.8 d,a 43 h,a		7.7 h	75–100	233–311	251

Table 26.19. (*Continued*)

Main Chain: Alkane, Quaternary							
33	Hexocyclium methyl sulfate				100	234	
34	Isopropamide iodide		0.85 rb		10	21	252
35	Mepiperphenidol	0.37 c	0.48 g		200–500	541–1351	252,254
36	Tricyclamol chloride		1.2 g	2.3 m	200–300	593–890	205
37	Trihexethyl chloride	1.0 r,a	2.17 rb		75–200	212–567	255
Main Chain: Alkane, Tertiary							
38	Aminopentamide		2.1 rb		2	6.8	256
39	Benzhexol		3.7 g	16 m			205
40	Procyclidine		22 g	31 m	20	62	
41	Methixene				3–6	8–16	
Main Chain: Alkene, Quaternary							
42	Diphemanil methyl	0.74	8.0 g		400–600	1026–1538	257

[a]No comparative studies of all anticholinergics in the same animal species or on the same test system are available. The above data were assembled or cross-calculated from information reported in a number of sources; therefore, the activities are relative and approximate. However, the information is useful for comparing the available anticholinergic agents.

[b]All quaternary salts are bromides, and all tertiary amines are listed as hydrochlorides, unless otherwise specified.

[c]The compounds were tested in different species. The following abbreviations are used to indicate the species: c, cat; d, dog; h, human; g, guinea pig; m, mouse; r, rat; rb, rabbit. Values less than unity indicate that these drugs are more active than atropine.

[d]Antisecretory activities are on salivation unless otherwise indicated. An a after the species indicates inhibition of acid secretion.

[e]All antispasmodic activities are inhibition of the contraction of intestine using cholinomimetic as spasmogen.

[f]The total dose includes the initial dose as well as maintenance dose used orally (except dibutoline, which is administered subcutaneously) in humans.

Atropine itself is an active substance in all three types of activities. (−)-Methscopolamine seems to be the most active of all compounds; it is about five times as active as atropine. None of the synthetic compounds are more active than (−)-methscopolamine, and few of them are more active than atropine. The compounds with high antisecretory activities also exhibit some degree of antispasmodic and mydriatic activities. Therefore, there is no complete dissociation between the three types of anticholinergic activities. Compounds with only one type of anticholinergic activity have yet to be synthesized. Only among weak compounds is there any dissociation between the antispasmodic and the mydriatic activities (e.g., propivane).

However, this difference may be related to the mode of administration. Mydriatic activities are measured after instillation into the eye, whereas antispasmodic and antisecretory activities are measured after parenteral administration to the animal or on *in vitro* preparations. To establish a claim that one compound has only one type of anticholinergic activity, two types of data should be available: (*1*) all types of activities should be measured in the same animal after the drug is administered by the same route and (*2*) the exact concentrations of the drugs at the sites of their action should be known. Such information is not available for most compounds in published literature.

For their antispasmodic and antisecretory activities in humans, the drugs are administered orally. A comparison of their oral doses (micromoles) indicates that atropine is the most active compound. In clinical experience, all three types of anticholinergic activities are exhibited by all compounds. The principal advantage of the available quaternary ammonium compounds lies in the fact that they are relatively free of any of the CNS effects that may be seen with atropine. This may permit the administration of sufficient quantities of the compounds to achieve a more fully effective peripheral anticholinergic action.

Studies on the effectiveness of anticholinergic agents in the ulcer disease complicated by Helicobacter pylori infection are hampered by the lack of suitable animal models (258). Marchetti and co-workers (259) developed a mouse model of H. pylori infection that mimics human disease. The pathogenesis of H. pylori infection *in vivo* was investigated using fresh clinical isolates of bacteria to colonize the stomachs of mice. The gastric pathology resembling human disease was observed with cytotoxin-producing strains but not with noncytotoxic strains. Oral immunization with purified H. pylori antigens protected mice from bacterial infection. This model will be useful for the development of therapeutic regimen involving vaccines against H. pylori and anticholinergic agents.

10 NONANTICHOLINERGICS AS ANTIULCER AGENTS

The interplay of various neuronal, hormonal, and other factors in gastric acid secretion are shown in Figure 26.1. Pharmacological agents can be used to decrease gastric acid secretion by their action at different sites. So far, the principal medications other than anticholinergics and antacids to treat peptic ulcers are limited. Experimental and clinical investigations are in progress on a number of agents that can

decrease the volume and acidity of gastric secretion through mechanisms other than blockade of the cholinergic nervous system (260–268). These include (*1*) histamine H2-receptor antagonists (262–267), (*2*) gastrin inhibitors (261), (*3*) pepsin inactivators (261), (*4*) mucus producers (72), (*5*) prostaglandin analogs (72), (*6*) enterogastrone and its analogs (261,268), (*17*) noncholinergic antispasmodics (72,260), and (*8*) gastric H^+/K^+ ATPase inhibitors (Chapter 27).

Histamine H2 receptor antagonists are popular for the treatment of peptic ulcer (262–267). A single dose of cimetidine, an H2 receptor antagonist, has a maximum effect on nocturnal acid output in humans. No further effect is obtained by adding poldine, an anticholinergic agent, to cimetidine (265). Cimetidine is also an effective drug in healing gastric and duodenal ulcers. Anticholinergic drugs and antacids help control symptoms, but they do not accelerate healing. Promethazine, an antihistaminic, inhibits the release of gastrin in the dog and human (261).

Pepsin inhibitors, sulfated amylopectin (Depepsin), and carrageenin decrease acid secretion in experimental animals and protect animals against histamine-induced ulcers (261). They must be studied further to ascertain adequately their therapeutic usefulness.

Carbenoxolone and cimetidine are complementary in their contribution to the healing of peptic ulcers, and the use of both may be better than either singly for some patients. Carbenoxolone accelerates healing by helping the defense mechanisms of the body. It stimulates extramucus secretion and prolongs cell life in the gastric epithelium. Cimetidine reduces gastric actid secretion. It would be interesting to know if anticholinergic drugs are more effective for the treatment of peptic ulcer in the presence of carbenoxolone.

Carbenoxolone (Biogastrone, Duogastrone) is the disodium salt of glycyrrhetinic acid hemisuccinate. It is prepared by hydrolysis of glycyrrhizic acid, a glycoside in licorice root. It increases the secretion of mucus and accelerates the healing of gastric ulcers. This drug is now under clinical investigation in the United States.

Pharmacological doses of several prostaglandins and their analogs inhibit gastric acid secretion. 15-(R)-Methyl-PGE$_2$ in small doses (100–200 μg, oral) reduces gastric acid secretion and output in humans and other animals, and it is currently being studied in the treatment of peptic ulcer.

Among the nonanticholinergic antispasmodics, alverine citrate (Spacolin) and isometheptene (octin) hydrochloride or mucate are available on the market. They relax smooth muscle by nonspecific actions. They exert little effect on gastric acid secretion. They are most useful in the symptomatic treatment of gastrointestinal disorders characterized by hypermotility and spasm.

ACKNOWLEDGEMENTS

The author is supported by USHHS-NIH-Research Grant DA-06207; the Council for Tobacco Research, U.S.A., Inc.; the Smokeless Tobacco Research Council, Inc.; and the Center for Anaesthesia Toxicology of Vanderbilt University.

REFERENCES

1. J. K. Siepler, K. Maha Kian, and W. T. Trudeau, *Clin. Pharm.*, **5**(2), 128 (1986).

2. A. H. Soll, *J. Clin. Gastroenterol.*, **11**(Suppl. 1), 51–55 (1989).

3. H. R. Mertz and J. H. Walsh, *Med. Clin. North Am.*, **75**(4), 799 (1991).

4. M. L. Partipilo and P. S. Woster, *Pharmacotherapy*, **13**(4), 330 (1993).

5. M. F. Dixon, *Scand. J. Gastroenterol.*, **201**(Suppl.), 7 (1994).

6. D. C. H. Sun, H. Shay, and J. L. Ciminera, *JAMA*, **158**, 713 (1955).

7. D. C. H. Sun and H. Shay, *Arch. Intern. Med.*, **97**, 442 (1956).

8. J. E. Lennard-Jones, *Br. Med. J.*, **5232**, 1071 (1961).

9. W. H. Bachrach, *Am. J. Dig. Dis.*, **3**, 743 (1958).

10. R. S. J. Clarke, J. F. Nunn, J. E. Utling, and B. R. Brown Jr., Eds., *General Anaesthesia*, Butterworth, London, 1989, pp. 412–118.

11. J. Rupreht and B. Dworacek, in ref. 10, pp. 1141–1159.

12. Nomenclature Committee, *Tips* (Dec. Suppl.), VII (1989).

13. Nomenclature Committee, *Tips* (Jan. Suppl.), 19 (1990).

14. J. R. Cooper, F. E. Bloom, and H. E. Roth, *The Biochemical Basis of Neuropharmacology*, Oxford University Press, New York, 1991, pp. 190–219.

15. B. V. R. Sastry, *Anaesthetic Pharmacol. Rev.*, **1**, 6 (1993).

16. A. Ashkenazi and E. G. Peralta, in S. J. Peroutka, Ed., *Handbook of Receptors and Channels, G Protein-Coupled Receptors*, CRC Press, Boca Raton, Fla., 1994, pp. 1–28.

17. J. Lechleiter, E. Peralta, and D. Clapham, *Tips* (Dec. Suppl.), 34 (1989).

18. C. C. Stephan and B. V. R. Sastry, *Cell. Mol. Biol.*, **38**, 601 (1992).

19. C. C. Stephan and B. V. R. Sastry, *Cell. Mol. Biol.*, **38**, 701 (1992).

20. American Medical Association, (AMA), *Drug Evaluations*, AMA, Chicago, 1992, pp. 355–357, 856–859.

21. U.S. Pharmacopeia, *Drug Information for Health Care Professional*, Rockville, Md., 1992, pp. 309–327.

22. R. A. Lehne, *Pharmacology for Nursing Care*, W. B. Saunders, Philadelphia, 1994, pp. 128–133.

23. D. D. Bonnycastle, D. R. Laurence and A. L. Bacharach, Eds., *Evaluation of Drug Activities: Pharmacometrics*, Vol. 2, Academic Press, Inc., New York, 1964, pp. 507–520.

24. E. G. Vernier in Ref. 23, Vol. 1, pp. 301–311.

25. M. A. Horst, B. V. R. Sastry, and E. J. Landon, *Arch. Int. Pharmacodyn. Ther.*, **288**(1), 87 (1987).

26. N. Watson and R. M. Eglen, *Br. J. Pharmacol.*, **112**(1), 179 (1994).

27. R. M. Eglew, N. Adham, and R. I. Whiting, *J. Autonomic Pharmacol.*, **12**(3), 137 (1992).

28. O. Soejima, T. Katsuragi, and T. Furukawa, *Eur. J. Pharmacol.*, **249**(1), 1 (1993).

29. A. Jovanovic, L. Grbovic, D. Drekic, and S. Novakovic, *Eur. J. Pharmacol.*, **258**(3), 185 (1994).

30. E. Bungardt, E. Vockert, F. R. Moser, R. Tacke,

E. Mutschler, G. Lambrecht, and A. Suprenant, *Eur. J. Pharmacol.*, **213**(1), 53 (1992).

31. N. Jaiswal, G. Lambrecht, E. Mutschler, R. Tacke, and K. V. Malik, *J. Pharmacol. Exp. Ther.*, **258**(3), 842 (1991).

32. N. Jaiswal and K. V. Malik, *Eur. J. Pharmacol.*, **192**(1), 63 (1991).

33. A. D. Fryer and E. E. El-Fakahamy, *Life Sci.*, **47**(7), 611 (1990).

34. C. M. Boulanger, K. J. Morrison, and P. M. Vanhoutle, *Br. J. Pharmacol.*, **112**(2), 519 (1994).

35. K. Shiraishi and I. Takayanagi, *Gen. Pharmacol.*, **24**(1), 139 (1993).

36. A. Jovanovic, L. Grbovie, and I. Tulic, *Eur. J. Pharmacol.*, **256**(2), 131 (1994).

37. A. F. Roffel, C. R. Elzinga, and J. Zaagsma, *Pulmon. Pharmacol.*, **3**(1) 47 (1990).

38. I. H. Pang, S. Matsumoto, E. Tamm, and L. DeSantis, *J. Ocular Pharmacol.*, **10**(1), 125 (1994).

39. H. Shay, S. A. Komarov, S. S. Fels, D. Meranze, M. Gruenstein, and H. Siplet, *Gastroenterology*, **5**, 43 (1945).

40. F. C. Mann and C. S. Williamson, *Ann. Surg.*, **77**, 409 (1923).

41. H. M. Hanson and D. A. Brodie, *J. Appl. Physiol.*, **15**, 291 (1960).

42. A. Robert and J. E. Nezamis, *Proc. Soc. Exp. Biol. Med.*, **99**, 443 (1958).

43. T. A. Lynch, W. L. Highley, and A. G. Worton, *J. Pharm. Sci.*, **51**, 529 (1962).

44. O. J. T. Thije, in L. Meyler and H. M. Peck, Eds., *Drug Induced Diseases*, Excerpta Medica Foundation, Leyden, 1965, pp. 30–34.

45. J. Markowitz, J. Archibald, and H. G. Downie, *Experimental Surgery*, Williams & Wilkins, Baltimore, Md., 1959.

46. R. A. Gregory, *J. Physiol.*, **144**, 123 (1958).

47. R. S. Alphin and T. M. Lin, *Am. J. Physiol.*, **197**, 257 (1959).

48. T. M. Lin, R. S. Alphin, and K. K. Chen, *J. Pharmacol. Exp. Ther.*, **125**, 66 (1959).

49. S. A. Komarov, S. P. Bralow, and E. Boyd, *Proc. Soc. Exp. Biol. Med.*, **122**, 451 (1963).

50. G. Dotevall, G. Schroder, and A. Walon, *Acta Med. Scand.*, **177**, 169 (1965).

51. K. J. Ivey, *Alimentary Pharmacol. Ther.*, **5**(Suppl. 1), 91 (1991).

52. H. R. Clearfield, *Alimentary Pharmacol. Ther.*, **5**(Suppl. 1), 1 (1991).

53. A. J. DeCross and B. J. Marshall, *Am. J. Med. Sci.*, **306**(6), 381 (1993).

54. A. Bettarello, *Dig. Dis. Sci.*, **30**(Suppl. 11), 365 (1985).

55. P. Pulewka, *Arch. Exp. Path. Pharmakol.*, **168**, 307 (1932).

56. D. G. Cogan, *Am. J. Opthalmol.*, **24**, 1431 (1941).

57. A. Ahmad and P. B. Marshall, *Br. J. Pharmacol.*, **18**, 247 (1962).

58. H. Burn, *Drugs, Medicines and Man*, Scribner, New York, 1962, pp. 225–232.

59. O. Schmiedeberg and R. Koppe, *Das Muscarin Das Giftige Alkaloid des Fliegenpilzes*, Vogel, Leipzig, 1869, pp. 27–29.

60. R. Kobert and A. Sohrt, *Arch. Exp. Path. Pharmakol.*, **22**, 396 (1887).

61. P. Mein, *Annalen*, **6**, 67 (1833).

62. K. W. Bentley, *The Alkaloids*, Vol. 1, Interscience Publishers, New York, 1957, pp. 10–24.

63. G. Fodór, J. Toth, I. Koczor, and I. Vincze, *Chem. Ind.*, 1260, (1955).

64. E. Leete, in P. Bernfeld, Ed., *Biogenesis of Natural Compounds*, Pergamon Press, New York, 1963, pp. 745–746.

65. G. Fodór and G. Csepreghy, *Tetrahedron Lett.*, **7**, 16 (1959).

66. R. S. Cahn, C. K. Ingold, and V. Prelog, *Experientia*, **12**, 81 (1956).

67. G. Fodór and G. Csepreghy, *J. Chem. Soc.*, 3222, (1961).

68. S. G. Kuznetsov and S. N. Golikov, *Synthetic Atropine-like Substances*, OTS 63-22078, U.S. Government Printing Office, Washington, D.C., 1965.

69. G. Fodór, *Experientia*, **11**, 129 (1955).

70. W. Will, *Chem. Ber.*, **21**, 1717 (1888).

71. R. Robinson, *J. Chem. Soc.*, **111**, 762 (1917).

72. E. Müller, *Chem. Ber.*, **51**, 252 (1918).

73. W. Wislicenus and E. A. Bilhüber, *Chem. Ber.*, **51**, 1237 (1918).

74. A. McKenzie and J. K. Wood, *J. Chem. Soc.*, **115**, 828 (1919).

75. R. Wolffenstein and L. Mamlock, *Chem. Ber.*, **41**, 723 (1908).

76. G. Fodór, J. Tóth, I. Koczor, and I. Vincze, *Chem. Ind.*, 764, (1956).

77. G. Fodór, J. Tóth, A. Romeike, I. Vincze, P. Dobó, and G. Janzsó, *Angew. Chem.*, **69**, 678 (1957).

78. J. H. Brown, A. G. Gilman, J. W. Rall, A. S. Nies, and P. Taylor, Eds., *The Pharmacological Basis of Therapeutics*, Pergamon, New York, 1990, pp. 150–165.

79. R. M. Levine, M. R. Blair, and B. B. Clark, *J. Pharmacol. Exp. Ther.*, **114**, 78 (1955).

80. R. M. Levine and B. B. Clark, *J. Pharmacol. Exp. Ther.*, **121**, 63 (1957).

81. R. M. Levine and E. M. Pelikan, *J. Pharmacol. Exp. Ther.*, **131**, 319 (1961).

82. T. H. Wilson, *Intestinal Absorption*, W. B. Saunders, Philadelphia, 1962, pp. 241–254.

83. R. E. Gosselin, J. D. Gabourel, S. C. Kalser, and J. H. Wills, *J. Pharmacol. Exp. Ther.*, **115**, 217 (1955).

84. S. C. Kalser, J. H. Wills, J. D. Gabourel, R. E. Gosselin, and C. F. Epes, *J. Pharmacol. Exp. Ther.*, **121**, 449 (1957).

85. J. D. Gabourel and R. E. Gosselin, *Arch. Int. Pharmacodyn.*, **115**, 416 (1958).

86. R. E. Gosselin, J. D. Gabourel, and J. H. Wills, *Clin. Pharmacol. Ther.*, **1**, 597 (1960).

87. W. Kalow, *Pharmacogenetics*, W. B. Saunders, Philadelphia, 1956, pp. 54–56.

88. G. Werner and R. Wurker, *Naturwissenschaften*, **22**, 627 (1959).

89. V. Evertsbusch and E. M. K. Geiling, *Arch. Int. Pharmacodyn.*, **105**, 175 (1956).

90. H. L. Schmidt and G. Werner, *Proc. Meet. Coll. Int. Neuro-Psychopharmacol, 3rd Munich*, 1962, pp. 427–432, (1964); *Chem. Abstr.*, **65**, 7818 (1966).

91. K. Matsuda, *Niigata Igakkai Zasshi*, **80**(2), 53 (1966).

92. G. Werner, P. C. Bosque, and J. C. Quevedo, *Abh. Deut. Akad. Wiss Berlin, K1 Chem. Geol. Biol.*, (3) 541–544, 629–636, (1966).

93. G. Werner, *Planta Med.*, **9**, 293 (1961).

94. R. Truhaut and J. Yonger, *C.R. Acad. Sci. Paris Ser. D*, **264**(21), 2526 (1967).

95. W. F. Von Oettingen, *The Therapeutic Agents of the Pyrrole and Pyridine Group*, Edwards, Ann Arbor, Mich. 1936, p. 130.

96. B. B. Brodie and C. A. M. Hogben, *J. Pharm. Pharmacol.*, **9**, 345 (1957).

97. L. S. Schanker, *J. Med. Pharm. Chem.*, **2**, 343 (1960).

98. W. Deckers, *Postgrad. Med. J.*, **51**(Suppl. 7), 76 (1975).

99. M. Abramowicz, *Med. Lett.*, **29**(745), 71 (1987).

100. A. A. Carmine and R. N. Brogden, *Drugs*, **30**, 85 (1985).

101. W. Londong, *Scand. J. Gastroenterol.*, **21**, 55 (1986).

102. E. Giraldo, R. Hammer, and H. Ladinsky, *Tips*, 80 (Feb. Suppl) (1986).

103. H. F. Pitschner, B. Schulte, M. Schlepper, D. Palm, and A. Wellstein, *Life Sci.*, **45**, 493 (1989).

104. E. Eberlein, W. Engel, G. Trummitz, G. Milm, N. Mayer, and K. Hasselbach. *Tips*, 76 (Feb. Suppl.) (1988).

105. J. Wess, D. Gdula, and M. R. Braun, *Mol. Pharmacol.*, **41**, 369 (1992).

106. I. Takayanagi and K. Koike, *Tips*, 88 (Feb. Suppl.) (1988).

107. U. Hacksell, B. M. Nilsson, G. Nordvall, G. Johansson, S. Sundquist, and L. Nilvebrant, *Life Sci.*, **56**(11–12), 831 (1995).

108. G. Nordvall and U. Hacksell, *J. Med. Chem.*, **36**, 967 (1993).

109. C. Melchiorre, A. Minarini, R. Budriesi, A. Chiarini, S. Spampinato, and V. Tumiatti, *Life Sci.*, **56** (11–12), 837 (1995).

110. V. Tumiatti, M. Recanatini, A. Minarini, C. Melchiorre, A. Chiarini, R. Budriesi, and M. L. Bolognesi, *Farnaco*, **47**, 1133 (1992).

111. G. Lambrecht, U. Moser, E. Mutschler, J. Wess, H. Linoh, M. Strecker, and R. Tacke, *Naunyn Schmeideberg Arch. Pharmacol.*, **325**(Suppl.), R62 (1984).

112. F. Lambrecht, *Tips* (Feb. Suppl.), 91 (1986).

113. R. B. Barlow, K. J. Berry, P. A. M. Glenton, N. M. Nikolau, and K. S. Soh, *Br. J. Pharmacol.*, **58**, 613 (1976).

114. N. J. Birdsall and E. C. Hulme, *Trends Autonomic Pharmacol.*, **3**, 17 (1985).

115. J. H. Miller, P. J. Aagaard, V. A. Gibson, and M. McKinney, *J. Pharmacol. Ex. Ther.*, **263**(2), 663 (1992).

116. S. Anwarul, H. Gilani, and L. B. Cobbinn, *Naunyn Schmiedeberg Arch. Pharmacol.*, **332**, 16 (1986).

117. C. C. Pfeiffer, *Science*, **107**, 94 (1948).

118. H. R. Ing, G. S. Dawes, and J. J. Wajda, *J. Pharmacol. Exp. Ther.*, **85**, 85 (1945).

119. H. R. Ing, *Science*, **109**, 264 (1949).

120. A. M. Lands, *J. Pharmacol. Exp. Ther.*, **102**, 219 (1951).

121. F. P. Luduena and A. M. Lands, *J. Pharmacol. Exp. Ther.*, **110**, 282 (1954).

122. J. P. Long, F. P. Luduena, B. F. Tuller, and A. M. Lands, *J. Pharmacol. Exp. Ther.*, **117**, 29 (1956).

123. F. W. Schueler, *Arch. Int. Pharmacodyn.*, **93**, 417 (1953).

124. Swiss Pat. 190,541 (1937); *Chem. Abstr.*, **32**, 589 (1938).

125. Ger. Pat. 653,778 (1937); *Chem. Abstr.*, **32**, 2956 (1938).

126. H. Horenstein and H. Pählicke, *Chem. Ber.*, **71**, 1644 (1938).

127. U.S. Pat. 1,987,546 (1935), A. Blankart (to Hoff-

mann-LaRoche Inc.); *Chem. Abstr.*, **29**, 1432 (1935).

128. R. R. Burtner and J. W. Cusic, *J. Am. Chem. Soc.*, **65**, 1582 (1943).

129. G. R. Treves and F. C. Testa, *J. Am. Chem. Soc.*, **74**, 46 (1952).

130. C. H. Tilford, M. G. Van Campen, Jr., and R. S. Shelton, *J. Am. Chem. Soc.*, **69**, 2902 (1947).

131. C. Harries, *Ann. Chem.*, **296**, 341 (1897).

132. C. Harries, *Chem. Ber.*, **31**, 665 (1898).

133. J. A. Faust, A. Mori, and M. Sahyun, *J. Am. Chem. Soc.*, **81**, 2214 (1959).

134. J. H. Biel, H. L. Friedman, H. A. Leiser, and E. P. Sprengeler, *J. Am. Chem. Soc.*, **74**, 1485 (1952).

135. A. H. Ford-Moore and H. R. Ing, *J. Chem. Soc.*, 55, (1947).

136. B. V. Franko and C. D. Lumford, *J. Med. Pharm. Chem.*, **2**, 523 (1960).

137. J. H. Biel, E. P. Sprengeler, H. A. Leiser, J. Horner, A. Drukker, and H. L. Friedman, *J. Am. Chem. Soc.*, **77**, 2250 (1955).

138. J. P. Long and H. K. Keasling, *J. Am. Pharm. Assoc. Sci. Ed.*, **43**, 616 (1954).

139. Swiss Pat. 259,958 (1949) (to Ciba); *Chem. Abstr.*, **44**, 5910 (1950).

140. F. F. Blick and M. U. Tsao, *J. Am. Chem. Soc.*, **66**, 1645 (1944).

141. U.S. Pat. 2,541,634 (1951), F. F. Blick (to Regents, Univ. of Michigan); *Chem. Abstr.*, **46**, 538 (1952).

142. F. P. Doyle, M. D. Mehta, G. S. Sach, and J. L. Pearson, *J. Chem. Soc.*, **1958**, 4458.

143. D. Krause and D. Schmidtke-Ruhnau, *Arzneim Forsch.*, **5**, 599 (1955).

144. D. Krause and S. Schmidtke-Ruhnau, *Arch. Exp. Path. Pharmakol.*, **229**, 258 (1956).

145. H. G. Kolloff, J. H. Hunter, E. H. Woodruff, and R. B. J. Moffett, *J. Am. Chem. Soc.*, **71**, 3988 (1949).

146. R. O. Clinton and V. J. Salvador, *J. Am. Chem. Soc.*, **68**, 2076 (1946).

147. Br. Pat. 728,579 (1955), (to Roche Products); *Chem. Abstr.*, **50**, 5773 (1956).

148. U.S. Pat. 2,408,893, K. C. Swan and N. G. White; *Chem. Abstr.*, **41**, 775 (1947).

149. U.S. Pat. 2,432,049 (1947), K. C. Swan and N. G. White; *Chem. Abstr.*, **42**, 1962 (1948).

150. U.S. Pat. 2,907,765 (1959), A. W. Weston (to Abbott Laboratories); *Chem. Abstr.*, **54**, 7746 (1960).

151. U.S. Pat. 2,665,278 (1954), E. M. Schultz (to Merck & Co.); *Chem. Abstr.*, **49**, 5525 (1955).

152. U.S. Pat. 2,891,890 (1959), D. W. Adamson (to Burroughs Wellcome Inc.). *Chem. Abstr.*, **54**, 1546 (1960).

153. U.S. Pat. 2,826,590 (1958), E. M. Bottorff (to Eli Lilly & Co.); *Chem. Abstr.*, **52**, 11124 (1958).

154. D. W. Adamson, P. A. Barrett, and S. Wilkinson, *J. Chem. Soc.*, **1951**, 52.

155. J. J. Denton and V. A. Lawson, *J. Am. Chem. Soc.*, **72**, 3279 (1950).

156. U.S. Pat. 2,698,325 (1954), D. W. Adamson (to Burroughs Wellcome Inc.); *Chem. Abstr.*, **50**, 1919 (1956).

157. P. Janssen, D. Zivkovic, P. Demoen, D. K. De-Jongh, and E. G. Van Proosdij-Hartzema, *Arch. Int. Pharmacodyn.*, **103**, 82 (1955).

158. L. C. Cheney, W. B. Wheatley, M. E. Speeter, W. M. Byrd, W. E. Fitzgibbon, W. F. Minor, and S. B. Binkley, *J. Org. Chem.*, **17**, 770 (1952).

159. U.S. Pat. 2,647,926 (1953), M. E. Speeter (to Bristol Laboratories); *Chem. Abstr.*, **48**, 9405 (1954).

160. W. B. Wheatley, W. F. Minor, W. M. Byrd, W. E. Fitzgibbon, Jr., M. E. Speeter, L. C. Cheney, and S. B. Brinkley, *J. Org. Chem.*, **19** 794 (1954).

161. R. B. Moffett and B. D. Aspergren, *J. Am. Chem. Soc.*, **79**, 4451 (1957).

162. Swiss Pat. 358,081 (1961), J. Schmutz (to Dr. A. Wander, A.G.); *Chem. Abstr.*, **57**, 13731 (1962).

163. N. Sperber, F. J. Villani, M. Sherlock, and D. Papa, *J. Am. Chem. Soc.*, **73**, 5010 (1951).

164. P. Holton and H. R. Ing., *Br. J. Pharmacol.*, **4**, 190 (1949).

165. L. Gyorgy, M. Doda, and K. Nador, *Acta Physiol. Hung.*, **17**, 473 (1960).

166. K. Nador, L. Gyorgy, and M. Doda, *J. Med. Pharm. Chem.*, **3**, 183 (1961).

167. R. Meier and K. Hoffman, *Helv. Med. Acta*, **7**(Suppl. 6), 106 (1941).

168. A. M. Lands, V. L. Nash, and K. Z. Hooper, *J. Pharmacol. Exp. Ther.*, **86**, 129 (1946).

169. R. R. Burtner and J. W. Cusic, *J. Am. Chem. Soc.*, **65**, 262 (1943).

170. D. K. de Jongh, E. G. van Proosdij-Hartzema, and P. Janssen, *Arch. Int. Pharmacodyn.*, **103**, 120 (1955).

171. M. H. Ehrenberg, J. A. Ramp, E. W. Blanchard, and G. R. Treves, *J. Pharmacol. Exp. Ther.*, **106**, 141 (1952).

172. A. M. Lands and C. J. Cavalitto, *J. Pharmacol. Exp. Ther.*, **110**, 369 (1954).

173. A. M. Lands and F. P. Luduena, *J. Pharmacol. Exp. Ther.*, **116**, 177 (1956).

174. J. G. Cannon and J. P. Long, in A. Burger, Ed.,

Drugs Affecting the Peripheral Nervous System, Marcel Dekker, New York, 1967, p. 133.

175. J. W. Cusic and R. A. Robinson, *J. Org. Chem.,* **16,** 1921 (1951).

176. R. F. Feldkamp and J. A. Faust, *J. Am. Chem. Soc.,* **71,** 4012 (1949).

177. J. Krapcho, C. F. Turk, and E. J. Pribil, *J. Am. Chem. Soc.,* **77,** 3632 (1955).

178. E. Nyman, *Acta Med. Scand.,* **118,** 466 (1944).

179. P. B. Marshall, *Br. J. Pharmacol.,* **10,** 354 (1955).

180. E. Büllbring and G. S. Dawes, *J. Pharmacol. Exp. Ther.,* **84,** 177 (1945).

181. G. Lehmann and P. K. Knoeffel, *J. Pharmacol. Exp. Ther.,* **80,** 335 (1944).

182. W. E. Hambouger, D. L. Cook, M. M. Winbury, and H. B. Freese, *J. Pharmacol. Exp. Ther.,* **99,** 245 (1950).

183. Montuschi, J. Phillips, F. Prescott, and A. F. Green, *Lancet,* **1,** 583 (1952).

184. H. M. Lee, W. Gibson, W. G. Dinwiddle, and J. Mills, *J. Am. Pharm. Assoc. Sci. Ed.,* **43,** 408 (1954).

185. R. W. Cunningham, B. K. Harned, M. C. Clark, R. R. Cosgrove, N. S. Daughterty, C. H. Hine, R. E. Vessey, and N. N. Yuda, *J. Pharmacol. Exp. Ther.,* **96,** 151 (1949).

186. R. Foster, P. J. Goodford, and H. R. Ing, *J. Chem. Soc. (London),* 3575, (1957).

187. L. H. Strenbach and F. Kaiser, *J. Am. Chem. Soc.,* **75,** 6068 (1953).

188. A. B. Funke and R. F. Rekker, *Arzneim. Forsch.,* **9,** 539 (1959).

189. L. A. Mounter and V. P. Whittaker, *Biochem. J.,* **47,** 525 (1950).

190. A. M. Lands, J. O. Hoppe, J. R. Lewis, and E. Ananenko, *J. Pharmacol. Exp. Ther.,* **100,** 19 (1950).

191. J. D. P. Graham and S. Lazarus, *J. Pharmacol. Exp. Ther.,* **69,** 331 (1940).

192. J. D. P. Graham and S. Lazarus, *J. Pharmacol. Exp. Ther.,* **70,** 165 (1940).

193. B. B. Brown, C. R. Thompson, G. R. Klahm, and H. W. Werner, *J. Am. Pharm. Assoc. Sci. Ed.,* **39,** 305 (1950).

194. A. M. Lands and F. P. Juduena, *J. Pharmacol. Exp. Ther.,* **117,** 331 (1956).

195. F. F. Blicke and C. E. Maxwell, *J. Am. Chem. Soc.,* **64,** 428 (1942).

196. A. A. Goldberg and A. H. Wragg, *Chem. Soc.,* 4823, (1957).

197. F. Jellinek, *Acta Crystallogr.,* **10,** 277 (1957).

198. H. G. Kolloff, J. H. Hunter, and R. B. Moffett, *J. Am. Chem. Soc.,* **72,** 1650 (1950).

199. J. J. Denton, V. A. Lawson, W. B. Neier, and R. J. Turner, *J. Am. Chem. Soc.,* **71,** 2050 (1949).

200. K. Fromherz, *Arch. Exp. Path. Pharmakol.,* **173,** 86 (1933).

201. F. F. Blicke and C. E. Maxwell, *J. Am. Chem. Soc.,* **64,** 431 (1942).

202. R. B. Barlow, K. A. Scott, and R. P. Stephenson, *Br. J. Pharmacol.,* **21,** 509 (1963).

203. B. V. R. Sastry, C. C. Pfeiffer, and A. Lasslo, *J. Pharmacol. Exp. Ther.,* **130,** 346 (1960).

204. R. B. Barlow, *Introduction to Chemical Pharmacology,* John Wiley & Sons, Inc., New York, 1964, p. 211.

205. W. M. Duffin and A. F. Green, *Br. J. Pharmacol.,* **10,** 383 (1955).

206. Y. Kasuya, *Chem. Pharm. Bull,* **6,** 147 (1958).

207. Y. Kasuya, *J. Pharm. Soc.,* **78,** 509 (1958).

208. L. O. Randall, W. M. Benson, and P. L. Stefko, *J. Pharmacol. Exp. Ther.,* **104,** 284 (1952).

209. C. Liebermann and L. Limpach, *Chem. Ber.,* **25,** 927 (1892).

210. L. Gyermek, *Nature,* **171,** 788 (1953).

211. B. V. R. Sastry and J. V. Auditore, *Proc. First Int. Pharmacol. Congr.,* **7,** 323 (1963).

212. C. C. Pfeiffler, *Int. Rev. Neurobiol.,* **1,** 195 (1959).

213. B. V. R. Sastry, *Am. Rev. Pharmacol.,* **13,** 253 (1973).

214. R. F. Furchgott, *Adv. Drug Res.,* **3,** 21 (1966).

215. B. V. R. Sastry and H. C. Cheng, *J. Pharmacol. Exp. Ther.,* **180,** 326 (1972).

216. R. F. Furchgott and P. Bursztyn, *Ann. N. Y. Acad. Sci.,* **144,** 882 (1967).

217. B. V. R. Sastry and H. C. Cheng, *Toxical. Appl. Pharmacol.,* **19,** 367 (1971).

218. H. C. Cheng and B. V. R. Sastry, *Arch. Int. Pharmacodyn. Ther.,* **223,** 246 (1976).

219. B. V. R. Sastry and H. C. Cheng, *J. Pharmacol. Exp. Ther.,* **201,** 105 (1977).

220. P. B. Marshall, *Br. J. Pharmacol.,* **10,** 270 (1955).

221. J. M. van Rossum and J. A. Th.M. Hurkmans, *Acta Physiol. Pharmacol. Neerl.,* **11,** 173 (1962).

222. E. J. Ariens and A. M. Simonis, *Acta Physiol. Pharmacol. Neerl.,* **11,** 151 (1962).

223. R. Schneider and A. R. Timms, *Br. J. Pharmacol.,* **12,** 30 (1957).

224. A. Teitel, *Nature,* **190,** 814 (1961).

225. A. Ashford, G. B. Penn, and J. W. Ross, *Nature,* **193,** 1082 (1962).

226. A. Bebbington and R. W. Brimblecombe, *Adv. Drug. Res.,* **2,** 143 (1965).

227. D. J. Triggle, *Chemical Aspects of Autonomic Nervous System,* Academic Press, Inc., New York, 1965, pp. 108–159.

228. E. J. Ariens, A. M. Simonis, and J. M. Van Rossum, in E. J. Ariens, Ed., *Molecular Pharmacology*, Vol. 1, Academic Press, Inc., New York, 1964, pp. 156–169.

229. W. D. M. Paton and H. P. Rang, *Proc. R. Soc.*, **163B**, 488 (1966).

230. H. P. Rang, *Ann. N. Y. Acad. Sci.*, **144**, 756 (1967).

231. C. D. Thron, *J. Pharmacol. Exp. Ther.*, **181**, 529 (1972).

232. R. W. Brimblecombe, *Drug Actions on Cholinergic Systems*, University Park Press, Baltimore, Md., 1974, pp. 19–42.

233. L. B. Kier, *Mol. Pharmacol.*, **3**, 487 (1967).

234. C. Y. Chiou and B. V. R. Sastry, *Arch. Int. Pharmacodyn. Ther.*, **181**, 94 (1969).

235. C. Y. Chiou and B. V. R. Sastry, *J. Pharmacol. Exp. Ther.*, **172**, 351 (1970).

236. B. Csillik, *Functional Structure of the Post-synaptic Membrane in the Myoneural Junction*, Publishing House of the Hungarian Academy of Sciences, Budapest, 1965, pp. 95–112.

237. R. G. Janes and J. F. Stiles, *Arch. Opthalmol.*, **62**, 69 (1959).

238. AMA Council on Drugs, *New Drugs*, American Medical Association, Chicago, Ill. 1967, p. 441.

239. W. H. Havener, *Ocular Pharmacology*, C. V. Mosby, St. Louis, Mo., 1966, pp. 177–267.

240. A. R. Cushny, *J. Pharmacol. Exp. Ther.*, **15**, 105 (1920).

241. R. L. Cahen and K. Tvede, *J. Exp. Pharmacol. Ther.*, **105**, 166 (1952).

242. B. V. Franko, R. S. Alphin, J. W. Ward, and C. D. Lunsford, *Ann. N. Y. Acad. Sci.*, **99**, 131 (1962).

243. E. A. Johnson and D. R. Wood, *Br. J. Pharmacol.*, **9**, 218 (1954).

244. A. J. Plummer, W. E. Barrett, R. Rutledge, and F. F. Yonkman, *J. Pharmacol. Exp. Ther.*, **108**, 292 (1953).

245. D. M. Brown and R. M. Quinton, *Br. J. Pharmacol.*, **12**, 53 (1957).

246. J. Y. P. Chen and H. Beckman, *J. Pharmacol. Exp. Ther.*, **104**, 269 (1952).

247. P. Acred, E. M. Atkins, J. G. Bainbridge, D. M. Brown, R. M. Quinton, and D. Turner, *Br. J. Pharmacol.*, **12**, 447 (1957).

248. G. Lehmann and P. K. Knoefel, *J. Pharmacol. Exp. Ther.*, **74**, 217, 274 (1942).

249. B. N. Halpern, *Arch. Int. Pharmacodyn.*, **59**, 149 (1938).

250. H. Ramsey and A. G. Richardson, *J. Pharmacol. Exp. Ther.*, **89**, 131 (1947).

251. K. C. Swan and N. G. White, *Arch. Opthalmol.*, **33**, 16 (1945).

252. A. Jageneau and P. Janssen, *Arch. Int. Pharmacodyn.*, **106**, 199 (1956).

253. S. C. McManus, J. M. Bochley, and K. H. Beyer, *J. Pharmacol. Exp. Ther.*, **108**, 364 (1953).

254. J. D. McCarthy, S. O. Evans, H. Ragins, and L. R. Dragstedt, *J. Pharmacol. Exp. Ther.*, **108**, 246 (1953).

255. A. C. Osterberg and W. D. Gray, *Arch. Int. Pharmacodyn.*, **137**, 250 (1962).

256. J. B. Hoekstra and H. L. Dickison, *J. Pharmacol. Exp. Ther.*, **98**, 14 (1950).

257. S. Margolin, M. Doyle, J. Giblin, A. Markovsky, M. T. Spoerlein, I. Stephens, H. Berchtold, G. Belloff, and R. Tislow, *Proc. Soc. Exp. Biol. Med.*, **78**, 576 (1951).

258. L. S. Tompkins and S. Falkow, *Science*, **267**, 1621 (1995).

259. M. Marchetti, B. Arico, D. Burroni, N. Figura, R. Rappuoli, and P. Ghiara, *Science*, **267**, 1655 (1995).

260. D. E. Butler, R. A. Purdon, and P. Bass, *Am. J. Dig. Dis.*, **15**, 157 (1970).

261. P. Bass, *Adv. Drug. Res.*, **8**, 205 (1974).

262. W. L. Burland and M. A. Simkins, Eds., *Cimetidine: Proceedings of the Second International Symposium on Histamine H2-Receptor Antagonists*, Excerpta Medica, Amsterdam, 1977.

263. G. H. Durant, J. C. Emmett, and C. R. Ganellin in ref. 262, pp. 1–12.

264. G. O. Barbezat and S. Bank in ref. 262, pp. 110–121.

265. R. E. Pounder, J. G. Williams, R. H. Hunt, S. H. Vincent, G. J. Milton-Thompson, and J. J. Misiewicz in ref. 262, pp. 189–204.

266. E. Aadland, A. Berstad, and L. S. Semb in ref. 262, pp. 87–97.

267. W. S. Blackwood and T. C. Northfield in ref. 262, pp. 124–130.

CHAPTER TWENTY-SEVEN

Gastric Proton Pump Inhibitors

ANDREAS W. HERLING
KLAUS WEIDMANN

Hoechst AG
Frankfurt/Main, Germany

CONTENTS

1 Introduction,   120
2 Gastric Acid Secretion and Its Inhibition,   120
  2.1 Medical need of gastric acid inhibitors, 120
  2.2 Mechanism of gastric acid secretion, 121
  2.3 Inhibition of gastric acid secretion, 123
3 Test Assays for Studying Gastric Acid
  Inhibitors,   124
  3.1 *In vivo*, 125
  3.2 *In vitro*, 125
4 Irreversible Gastric Proton Pump Inhibitors,   126
  4.1 Chemistry and biochemistry, 126
      4.1.1 Compounds, 126
      4.1.2 Mechanism of action, 127
      4.1.3 Structure-activity relationships, 134
  4.2 Pharmacology, 135
      4.2.1 Animal pharmacology, 135
      4.2.2 Human pharmacology, 138
  4.3 Pharmacokinetics, 139
  4.4 Metabolism, 139
  4.5 Toxicology, 140
      4.5.1 Enterochromaffin-like cell
            proliferation, 140
      4.5.2 Thyroid toxicity, 141

Burger's Medicinal Chemistry and Drug Discovery,
Fifth Edition, Volume 2: Therapeutic Agents,
Edited by Manfred E. Wolff.
ISBN 0-471-57557-7   © 1996 John Wiley & Sons, Inc.

5 Reversible Gastric Proton Pump Inhibitors, 142
   5.1 Chemistry and biochemistry, 142
      5.1.1 Compounds, 142
      5.1.2 Mechanism of action, 142
   5.2 Pharmacology, 144
6 Other Inhibitors, 144
7 Conclusions, 145

1 INTRODUCTION

The methods of treatment of peptic ulcer disease have changed dramatically during the last 25 years. Numerous compounds having different modes of action have been used to treat upper gastrointestinal ulcers. Healing of ulcers may be achieved by at least three different modes (1). The first involves the stimulation of regeneration of cells surrounding the ulcer base, as postulated for licorice extracts. Second, the ulcer can be protected from gastric acid and pepsin by coating agents like sucralfate or bismuth. Third, the acidity of the gastric juice can be reduced, as can be achieved by inhibitors of gastric acid secretion or by antacids, which neutralize gastric acidity, and by prostaglandins which inhibit not only gastric acid secretion but additionally stimulate endogeneous gastric bicarbonate secretion.

The most successful method to treat upper gastrointestinal ulcers has been the inhibition of gastric acid secretion. During the last 20 yr numerous clinical trials and clinical experience have demonstrated that inhibition of gastric acid secretion is superior in promoting ulcer healing and relieving ulcer symptoms to all the other possible methods mentioned above.

The era of gastric acid inhibitors started experimentally in the early 1970s with the histamine-H_2-receptor antagonists. This was followed by gastric proton pump inhibitors about five years later. The purpose of this chapter is to review the present knowledge about gastric proton pump inhibitors, which interfere with the gastric H^+/K^+-ATPase, the so-called gastric proton pump.

2 GASTRIC ACID SECRETION AND ITS INHIBITION

2.1 Medical Need of Gastric Acid Inhibitors

Acid secretion is a physiological process of the stomach. Gastric acid production is caused by a highly specialized cell type in the fundic mucosa, the gastric parietal cell. Gastric acid is of direct as well as of indirect physiological importance, via acid-induced pepsinogen activation, to initiate the digestive process, kill bacteria and other microbes, and to ensure a stable intragastric environment (2). However, under certain circumstances gastric acid and pepsinogen may injure the gastroduodenal mucosa. Until now, only little has been known about the pathogenesis of gastric or duodenal ulcer disease (3). In recent years, *Helicobacter pylori* has gained additional interest in research to elucidate the pathogenesis of gastric and duodenal ulcer diseases. However, it has been known for about 20 yr that inhibition of gastric acid secretion is a useful therapeutic principle for the treatment of ulcers of the upper gastrointestinal tract. Histamine-H_2-receptor antagonists, such as, cimetidine, ranitidine, famotidine, nizatidine (4,5), and roxatidine (6), are widely used in clinical settings to inhibit gastric acid secretion. The different methods of vagotomy proximal to the inervation of the parietal cells by a surgical operation (7) is still practiced in patients with a high rate of relapse of ulcers or a therapy-resistant ulcer history. From the pharmacological point of view, the administration of an antimuscarinic drug resembles the situation after vagotomy, but the use of antimuscarinic drugs for inhibition of gastric acid secretion in ulcer patients is less common. Until recently, gastrin receptor antagonists, e.g., proglumide, have not been potent enough to be of clinical usefulness.

Although histamine-H_2-receptor antagonists are quite effective in some peptic acid disorders, e.g., uncomplicated duodenal ulcers or gastric ulcers, their effectiveness in other disorders is less apparent. Therefore, the prolonged and potent reduction of acid secretion caused by gastric proton pump inhibitors is necessary to treat patients with Zollinger-Ellison syndrome and gastroesophageal reflux disease (4). This results in the superiority of omeprazole, the first gastric proton pump inhibitor on the market, over histamine-H_2-receptor antagonists (8,9) in these disorders. Similar superiority over histamine-H_2-receptor antagonists has been reported for lansoprazole (10–12) as well as for pantoprazole (13). Furthermore proton pump inhibitors have been used in surgery because of their effective and fast onset of gastric acid inhibition. In case of emergency surgery, a single dose of a proton pump inhibitor prior to operating, e.g., caesarean section, has been shown to prevent aspiration pneumonitis (14,15).

2.2 Mechanism of Gastric Acid Secretion

Gastric acid secretion appears in the fundic region of the stomach. The oxyntic gland represents the secretory unit of the fundic mucosa. The acid-secreting parietal cells are located in the wall of its midsection. In addition to parietal cells, these glands consist of pepsinogen-secreting chief cells, mucous-secreting superficial and neck cells, endocrine, and somatostatin cells (2).

Regulation of gastric acid secretion is complex and involves local, peripheral, and central influences (16). In addition to the stimulatory receptors, there are also different types of inhibitory receptors (e.g., prostaglandin and somatostatin) on the basolateral membrane of the parietal cell. There are three different types of stimulatory receptors on the basolateral membrane of the parietal cells (Fig. 27.1): muscarinic receptors, gastrin receptors, and histamine-H_2-receptors (2,17).

Acetylcholine originates in the vagus nerves and cholinergic neurons in the wall of the stomach; its action is neurocrine. The muscarinic receptor on the parietal cell does not belong to the M_1-receptor subclass. This finding is based on studies with selective muscarinic antagonists in isolated parietal cells. However, with selective M_1-receptor antagonists, acid secretion *in vivo* can be inhibited indicating that M_1-receptors are involved in acetylcholine-induced gastric acid secretion possibly within the postsynaptic neurons (18).

Gastrin is synthesized and stored in antral G-cells of the stomach, and its secretion into the blood is stimulated by amino acids in the gastric juice. Acid at the surface of the antral mucosa inhibits gastrin release and represents the natural brake for gastrin release (19). Neutral pH in the antrum caused by pharmacologically induced gastric acid inhibition or food intake does not actively stimulate gastrin release from the antral G-cells but represents the removal of the natural brake of gastrin release. Gastrin mediates its action on the parietal cell in an endocrine manner. The secretion of gastrin from G-cells is associated with suppression of somatostatin release from D-cells. Besides gastrin receptors on parietal cells

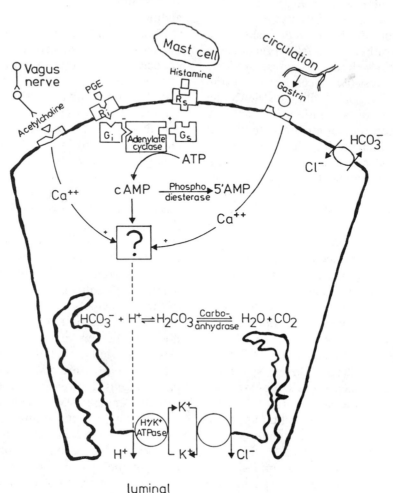

Fig. 27.1 Schematic representation of the gastric parietal cell, showing the pathways by which gastric hydrochloric acid is generated and secreted. Reproduced courtesy of Elsevier Science from ref. 34.

(20), additional locations for gastrin receptors have been discussed, e.g., histamine-releasing cells mediating acid secretion (21).

Histamine is released by mast cells or mast cell-like cells of the gastric mucosa, depending on the animal species, and reaches the parietal cell in a paracrine manner (19). The histamine receptor on the parietal cell belongs to the H_2-receptor subclass.

The binding of acetylcholine or gastrin to their specific receptors increases intracellular calcium. The binding of histamine to its H_2-receptor on the parietal cell activates the membrane-bound adenylate cyclase with a corresponding increase in intracellular cyclic AMP levels. Increased levels of both intracellular calcium (gastrin/acetylcholine) and cyclic AMP (histamine) finally cause acid secretion (2). Furthermore there is an interaction between these intracellular pathways (22). The protons for acid formation are generated intracellularly by the carbonic anhydrase reaction.

The final step of acid secretion is mediated by H^+/K^+-ATPase (E.C. 3.6.1.3.), the so-called gastric proton pump. This enzyme has a molecular weight of 114,012 and consists of 1,033 amino acids (23). The amino acid sequence of H^+/K^+-ATPase shows about 62% homology to that of the Na^+/K^+-ATPase. Both enzymes are phosphorylated by ATP, and potassium binding causes dephosphorylation and a conformational change in the protein. Na^+/K^+-ATPase is widely distributed in all mammalian cells. In contrast H^+/K^+-ATPase, the proton pump, is predominantly located at the apical membrane of the parietal cell, although a similar enzyme has been reported to be located in distal colonic epithelial cells of the rabbit (24,25) and in renal collecting duct segments of rats (26) and rabbits (27). Furthermore, the effect of some proton pump inhibitors on bone resorption by osteoclasts in bone slices has also been studied. From these results it has been concluded that osteoclasts contain an H^+/K^+-ATPase-like enzyme, which shows some differences from the gastric one (28).

The architecture of the apical membrane of the parietal cell is extremely variable. It forms, under nonsecretory conditions, tubulovesicular structures and, under secretory conditions, secretory cannaliculi. Its morphology depends on the secretory state of the parietal cell (29). The lumen of the cannaliculi belongs to the extracellular compartment under secretory conditions (Fig. 27.1) and contains hydrochloric acid with a pH of about 1. H^+/K^+-ATPase exchanges protons for potassium ions across the apical surface. Thereby, the gastric proton pump pumps out the protons against a proton gradient of 1:1,000,000.

A potassium chloride cotransporter must be closely related to the proton pump during proton secretion. Potassium and chloride ions move across the apical membrane together with secreted protons (Fig. 27.1) (30–32). Potassium is recycled while hydrochloric acid of the gastric juice is formed by chloride ions together with the secreted protons

Stimulation of gastric acid secretion across the apical membrane may predominantly reflect the activation or insertion of an active potassium chloride cotransporter rather than direct activation of the gastric proton pump (2).

2.3 Inhibition of Gastric Acid Secretion

Gastric acid secretion can be inhibited pharmacologically by specific antagonists of the stimulatory receptors (muscarine-M_1/M_2, gastrin, histamine-H_2), by agonists to inhibitory receptors (prostaglandin, somatostatin), by carbonic anhydrase inhibitors, and by proton pump inhibitors.

In a well-defined pharmacological model, the stomach-lumen perfused rat, acid secretion can be dose dependently induced by the infusion of carbachol, gastrin, or histamine (33). In this model atropine causes inhibition only during carbachol-induced gastric acid secretion without causing any inhibitory effects during histamine or gastrin stimulation. Proglumide, a weak but selective antagonist of gastrin and cholecystokinin receptors, only inhibits gastrin-stimulated gastric acid secretion. Cimetidine inhibits histamine- and gastrin-induced gastric acid secretion but does not inhibit carbachol stimulation (34) (Table 27.1).

Omeprazole, the first gastric proton pump inhibitor used clinically, causes a comparable inhibition of stimulated gastric acid secretion irrespective of the kind of stimulation (34). This inhibitory behaviour is in clear contrast to the inhibitory effect of the different types of receptor antagonists (Table 27.1). In pylorus-ligated rats, ID_{50} values for omeprazole during carba-

Table 27.1 The Inhibitory Profiles of Gastric Acid Inhibitors with Different Modes of Action[a]

| Inhibitor | Stimulant | | | |
| | Carbachol | Gastrin | Histamine | Forskolin/IBMX intracellular |
	receptor level			
Atropine	+	−	−	−
Proglumide	−	+	−	−
Cimetidine	−	+	+	−
Omeprazole	+	+	+	+

[a] + Inhibitory effect; − no inhibitory effect.

chol, gastrin, and histamine stimulation as well as during basal conditions are not significantly different (Table 27.2).

Acid secretion can be induced in the stomach-lumen perfused rat model by an initial injection of isobutyl methylxanthine (IBMX, a phosphodiesterase inhibitor) followed by an infusion of forskolin (direct stimulation of adenylate cyclase). This kind of stimulation represents an induction of gastric acid secretion on a subreceptor level *in vivo* similar to dibutyryladenosine 3,5-cyclic monophosphate (dbcAMP) stimulation of acid formation *in vitro* in isolated gastric glands or isolated parietal cells. Omeprazole also inhibits IBMX-forskolin induced gastric acid secretion (33) (Table 27.1). The inhibition of the gastric proton pump, representing the last step of acid formation within the parietal cell, is up to now, the most efficient method of blocking gastric acid secretion. Gastric proton pump

inhibitors can abolish acid secretion stimulated by any secretagogue (35,36).

3 TEST ASSAYS FOR STUDYING GASTRIC ACID INHIBITORS

The biomedical characterization of a compound for its inhibitory effectiveness on gastric acid secretion requires *in vivo* studies in laboratory animals as well as *in vitro* studies in isolated stomachs, fundic mucosa, gastric glands, cells, and subcellular fractions (enzymes, receptors). Predominantly *in vitro* studies are performed to elucidate the mechanism of action of any compound of interest, whereas animal studies are used to proof a compound's practical usefulness.

In general many prerequisites have to be fulfilled by a compound showing pharmacological activity. These cannot be studied in

Table 27.2 ID$_{50}$ Values for Omeprazole and Cimetidine Given Intraduodenally on Stimulated and Basal Gastric Acid Secretion in Pylorus-Ligated Rats[a]

| Compound | Stimulant | | | |
	Carbachol	Gastrin	Histamine	Basal
Omeprazole, mg/kg	1.1 ± 0.2	0.8 ± 0.1	0.7 ± 0.2	1.2 ± 0.4
Cimetidine, mg/kg	30.8 ± 4.7	13.2 ± 0.7	9.1 ± 1.5	not done

[a] Values represent mean $\pm$ SEM: $n = 8$.

their complexity by *in vitro* techniques. After oral administration to laboratory animals, the compound should be stable enough to pass through the aggressive environment of the gastric juice (chemical stability) or to get absorbed by the stomach wall in its active form. After having passed through the stomach, the compound should be absorbed from the gut so as to become systemically available (bioavailability). Then the compound should be distributed in the body to reach the target organ and should be present at the target organ for a sufficient length of time to exert the desired pharmacological effect. Afterwards it should be eliminated as completely as possible either in its original form or following metabolism.

Normally *in vitro* studies are used to elucidate the mechanism of action of a compound active *in vivo* or to screen for a defined target. In the latter case, a compound active *in vitro* with a defined mechanism of action must subsequently fulfill all of the above mentioned criteria for practical usefulness in animal studies. If not, a rational drug optimization has to be started to synthesize derivatives of the compound and to test them *in vitro* and *in vivo* in a rational test hierarchy.

3.1 *In Vivo*

Gastric acid secretion can be studied *in vivo* in rats and dogs. Conscious rat models are rats which have a chronic gastric fistula (37), are pylorus-ligated (38), or involve anesthetized rats with stomach-lumen perfusion (39). Gastric acid secretion can be studied under basal conditions as well as during stimulation of gastric acid secretion following an intravenous infusion or subcutaneous injection of a secretagogue, i.e., carbachol, gastrin, or histamine.

From the biomedical experience with different kinds of gastric acid inhibitors, the conscious dog seems to be the most relevant animal species for the prediction of the antisecretory potential of a test compound in humans. Gastric acid secretion can be studied in a dog that has been surgically prepared with a chronic gastric fistula or with a Heidenhain pouch (40). This is mostly done during stimulatory conditions, e.g., a continuous intravenous infusion with histamine. As the Heidenhain pouch is vagally denervated, stimulation of gastric acid secretion with gastrin needs a small amount of carbachol.

Originally cytoprotection was defined as the potential of a test compound to protect the gastric mucosa of rats against necrotizing agents, such as, absolute ethanol, 0.6 N hydrochloric acid, 0.2 N sodium hydroxide, 25% sodium chloride, or boiling water in nonantisecretory doses (41). Several prostaglandins caused cytoprotection, particularly in rats, over a dose range which had no antisecretory activity. However, clinical experience with prostaglandins has shown that ulcer healing is only achieved at antisecretory doses (42). Therefore, it seems very likely that the cytoprotective property of a compound in rats has very limited relevance to predicting its ulcer healing potential in humans if cytoprotection is really separated from the antisecretory potential.

3.2 *In Vitro*

The effect of gastric proton pump inhibitors on H^+/K^+-ATPase activity (ATP cleavage) can be studied *in vitro* with partially purified H^+/K^+-ATPase preparations (43). This assay has been used more effectively to study the mechanism of action of gastric proton pump inhibitors in detail than to study the structure activity relationship of such inhibitors (42). Since the enzyme assay should be performed at neutral pH values and since proton pump inhibitors of the omeprazole type need acid activation, a preincubation period at a pH no lower than about 6 was used to initiate the acidic conversion of the test compound into its active principle. This reflects the chemical instability of the test compound of neutral pH values more than its effect during conditions of much higher acidity within the secretory cannaliculus of the parietal cell during acid secretion. Many chemically labile inhibitors, therefore, are very active in this test system. However, they do not cause an inhibition in more complex test systems and, consequently, are without any practical usefulness (42).

More suitable for studying the mechanism of action, acid-conversion, and the structure activity relationship of proton pump inhibitors are proton-transport studies in intact gastric vesicles. Such vesicles form a pH gradient similar to those *in vivo* (44–46).

Acid formation *in vitro* has been studied very intensively in isolated parietal cells from guinea pigs (47), dogs (48), and rabbits, as well as in whole gastric glands from rabbits (48) and humans (49). Measurement of acid formation was achieved indirectly by means of the accumulation of the weak base, ^{14}C-aminopyrine (pK_a 5.0), within the secretory compartment of the parietal cell (50). Due to its nature as a weak base, ^{14}C-aminopyrine accumulates in acidic compartments. At pH 7.4 (test medium and cytosol of the parietal cells), it can pass freely through biological membranes in its unionized form but becomes trapped immediately within the secretory cannaliculi because of ionization. Furthermore, oxygen consumption correlates with acid formation (50) and has been very useful to identify artifacts of inhibition of ^{14}C-aminopyrine accumulation through neutralization of the acidic compartment by the basic nature of the test compound (45). During artifact conditions, ^{14}C-aminopyrine accumulation is reduced by neutralization of the acidic compartment by the test compound even when the proton pump is still active and oxygen consumption is un-

inhibited. The ^{14}C-aminopyrine accumulation technique has been widely used to study structure activity relationships of gastric acid inhibitors as well as to separate the inhibitory effect of proton pump inhibitors from that of receptor antagonists. For instance, histamine-H_2-receptor antagonists act only during histamine stimulation but cause no inhibitory effect during dbcAMP-stimulated ^{14}C-aminopyrine uptake. In contrast, proton pump inhibitors inhibit both kinds of stimulation (43,51).

4 IRREVERSIBLE GASTRIC PROTON PUMP INHIBITORS

Timoprazole (**1**), a substituted benzimidazole, is the prototype of this kind of gastric acid inhibitor (Fig. 27.2). Proton pump inhibitors structurally related to timoprazole have been defined as irreversible inhibitors that inhibit the gastric proton pump in a noncompetitive manner by covalent binding of their active intermediates to the enzyme protein.

4.1 Chemistry and Biochemistry

4.1.1 COMPOUNDS. In the 1970s, many pharmaceutical companies started to look for inhibitors of gastric acid secretion, due to the significant superiority of histamine-H_2-receptor antagonists in the therapy of gastric and duodenal ulcers over other established treatments, e.g., antacids, diet, and surgery. At AB Haessle, Sweden, systematic synthesis of inhibitors of gastric acid secretion started at that time with 2-pyridylthioacetamide (CMN 131) as a lead compound (42,52). It was soon recognized that this series of compounds had a different mechanism of action from receptor antagonism. In 1974 this group synthesized timoprazole (**1**), the first well-defined inhibitor of the gastric proton pump, followed by picoprazole (**2**) in 1976 and omeprazole (**3**)

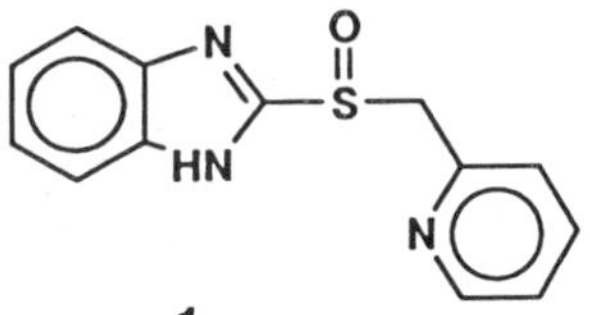

1

Timoprazole (Haessle)

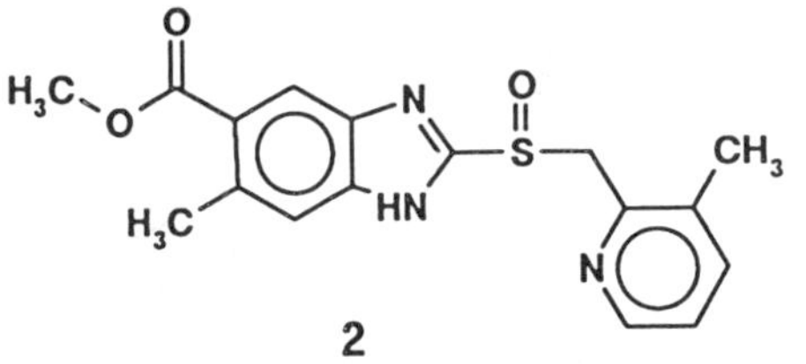

2

Picoprazole (Haessle)

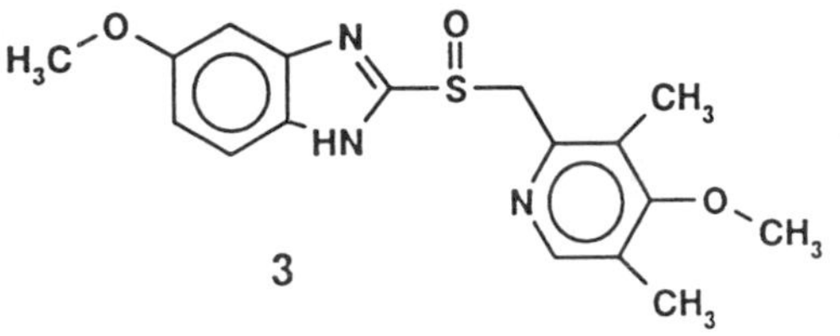

3

Omeprazole (Haessle)

Fig. 27.2 2-[(Pyridylmethyl)sulfinyl]-1*H*-benzimidazoles as irreversible gastric proton pump inhibitors.

in 1979 (Fig. 27.2) (43,52,53). Chemically, the basic structure consists of a substituted benzimidazole ring and a substituted pyridine ring connected to each other by a methylsulfinyl chain. In the meantime, other pharmaceutical companies have reported their gastric proton pump inhibitors. Most of the known omeprazole-like proton pump inhibitors are 2-[(2-pyridylmethyl)-sulfinyl)]-1*H*-benzimidazoles, showing that the possibilities for structural modifications are very restricted. Only in a few analogs have the heterocyclic rings been exchanged by other moieties, e.g., the benzimidazole by thienoimidazole or the 2-pyridylmethyl group by a 2-aminobenzyl group.

The following proton pump inhibitors are based on the framework of omeprazole: lansoprazole (**4**, AG-1749, Takeda) (54), pantoprazole (**5**, BY 1023, Byk Gulden) (55), rabeprazole sodium (**6**, E-3810, Eisai) (56), TY-11345 (**7**, Toa Eiyo) (57), disu-prazole (**8**, Upjohn) (58,59), Ro 18-5364 (**9**, Hoffmann-La Roche) (60), and SK&F 95601 (**10**, SK&F) (61) (Fig. 27.3). Savi-prazole (**11**, HOE 731, Hoechst) (62) and S 1924 (**12**, Hoechst) (63) contain a thieno

[3,4-d] imidazole group instead of the benz-imidazole moiety of the omeprazole framework (Fig. 27.4). In contrast isomers with a thieno[2,3-d]imidazole moiety (e.g., **13**), claimed in patent applications by Chemie Linz (64) and Pfizer (65) as irreversible proton pump inhibitors, showed only weak inhibitory effectiveness on gastric acid secretion (62) (Fig. 27.4). NC-1300 (**14**), NC-1300-B (**15**), leminoprozole (**16**, NC-1300-0-3, Nippon Chemiphar) (66,67), and S 3337 (**17**, Hoechst) (63) are substituted benzimidazole derivatives in which the 2-pyridylmethyl moiety has been replaced by a 2-aminobenzyl group (Fig. 27.5). BY 308 (Byk Gulden) (68) represents a sulfide basing on the omeprazole framework and it has been characterized as a prodrug, which requires oxidation *in vivo* to the corresponding sulfoxide to be active.

4.1.2 MECHANISM OF ACTION. At AB Haessle, Sweden, it was soon recognized that the gastric proton pump is the site of action for substituted benzimidazoles (43,53) and that enzyme inhibition parallels inhibition of gastric acid secretion in lab-

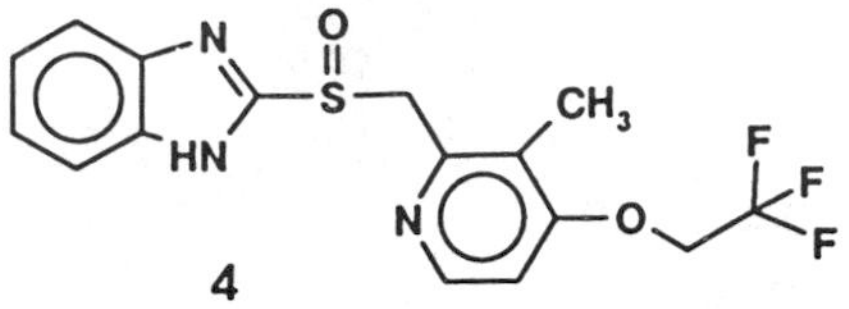

4

Lansoprazole (Takeda)

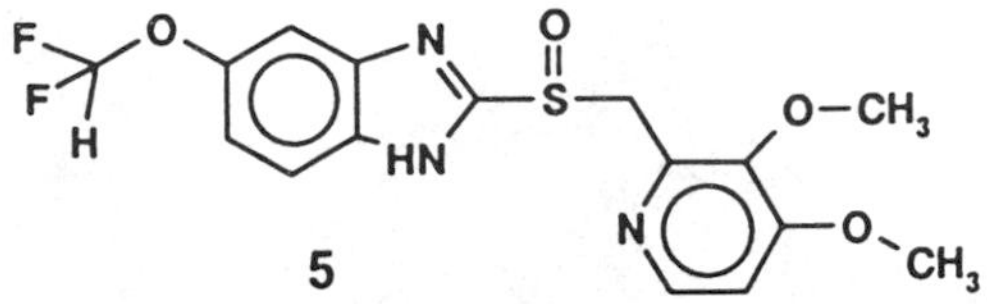

5

Pantoprazole (Byk Gulden)

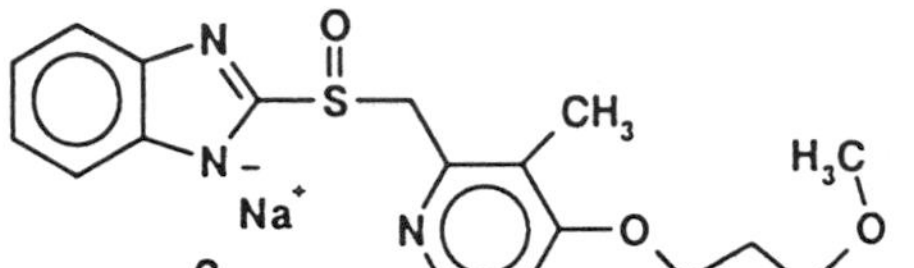

6

Rabeprazole sodium , E-3810 (Eisai)

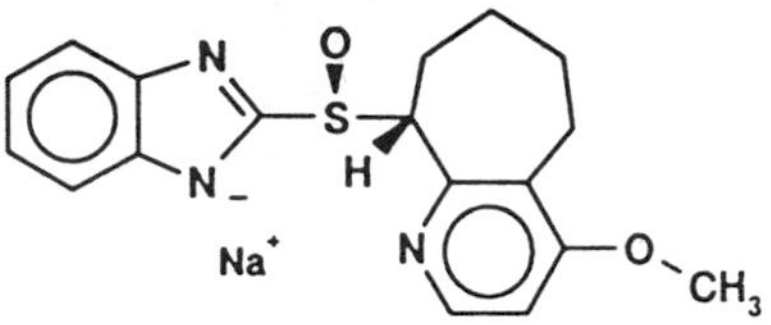

7

TY-11345 (Toa Eiyo)

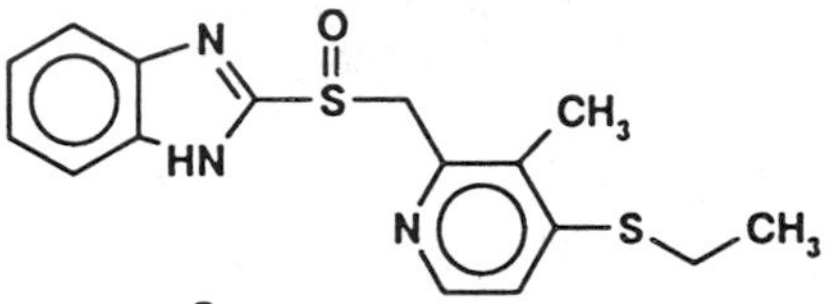

8

Disuprazole (Upjohn)

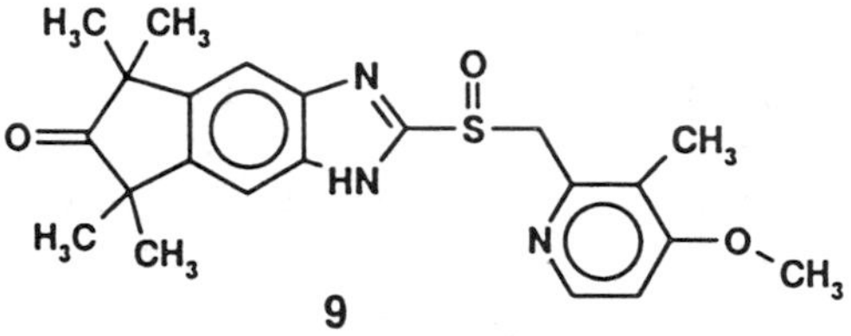

9

Ro 18-5364 (Hoffmann-La Roche)

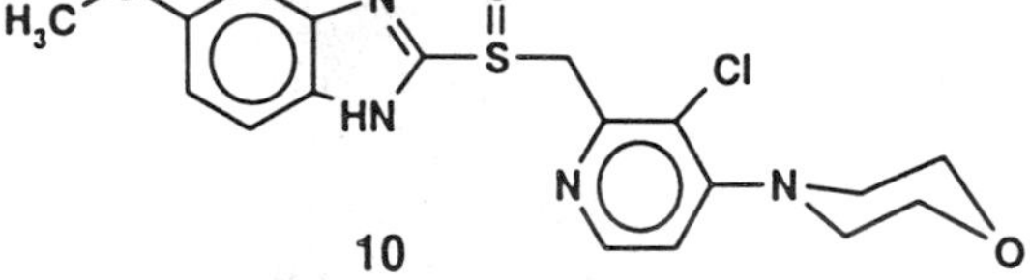

10

SK&F 95 601 (SK&F)

Fig. 27.3 Omeprazole analogues with different substitution patterns as irreversible gastric proton pump inhibitors currently or previously under development.

oratory animals (37). Physico-chemically, omeprazole represents a weak base with a pK_a of 4. It can pass through biological membranes as a result of its lipid permeable properties (52). At physiological pH it is predominantly unionized and this neutral form passes freely across biological membranes. However, in an acidic environment with a pH below 4, it is predominantly protonated. This results in a limited per-

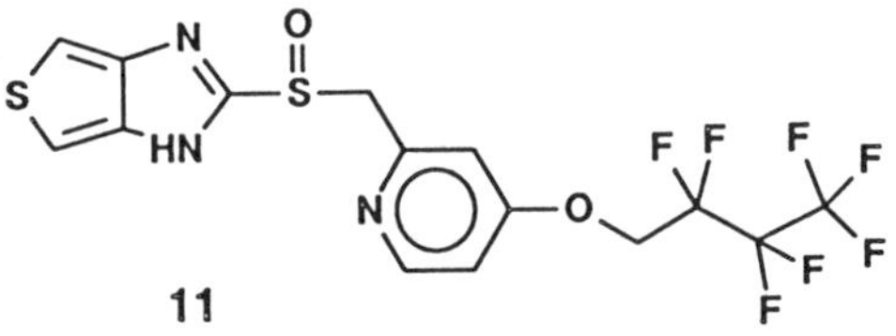

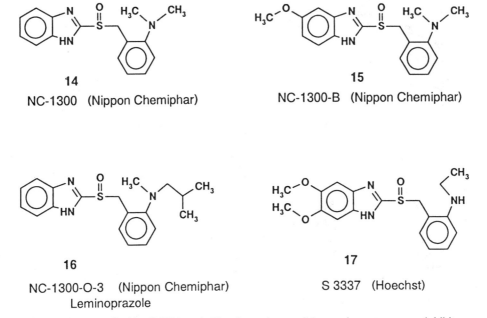

11

Saviprazole   (Hoechst)

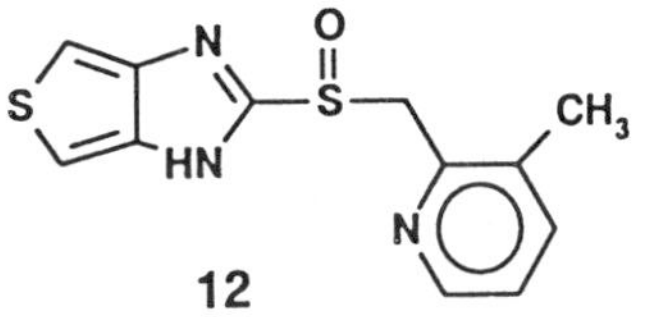

12

S 1924   (Hoechst)

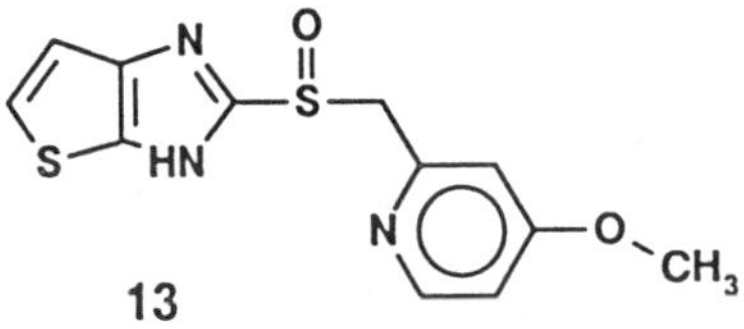

13

(Chemie Linz , Pfizer)

Fig. 27.4   2-[(Pyridylmethyl)sulfinyl]-1*H*-thienoimidazoles as irreversible gastric proton pump inhibitors currently or previously under development.

14

NC-1300   (Nippon Chemiphar)

15

NC-1300-B   (Nippon Chemiphar)

16

NC-1300-O-3   (Nippon Chemiphar)
Leminoprazole

17

S 3337   (Hoechst)

Fig. 27.5   2-(2-Aminobenzyl)sulfinyl]-1*H*-benzimidazoles as irreversible gastric proton pump inhibitors currently or previously under development.

meability of the drug (42). Due to the unique structure of the gastric parietal cell, with its acidic compartment within the secretory cannaliculi, omeprazole is trapped within these cannaliculi in its protonated form and undergoes a chemical transformation process (69,70) (Fig. 27.6). After initial protonation and nucleophilic attack of the pyridine-N atom on C-2 of the benzimidazole, the C-SO-bond in **18** is cleaved to yield the sulfenic acid (**19**), which cyclises to the pyridinium sulfenamide (**21**). The sulfenamide represents the active enzyme inhibitor (70–74) and binds covalently to suflhydryl groups of cysteines (**20**) of the proton pump. The formation of a disulfide linkage from the reaction of enzyme sulfhydryl groups with the unstable sulfenic acid (**19**), which could not be isolated experimentally, might also be involved (70,74). The covalent binding of the active inhibitor to the H^+/K^+-ATPase followed by structural alteration of the enzyme finally inactivates the catalytic function of the proton pump (42).

Two molecules of the active intermediate of omeprazole bind to one active site of the gastric proton pump (75,76). This binding is a disulfide linkage and can be reversed and prevented by the addition of mercaptane (77–79). Detailed investigations of three reactions of H^+/K^+-ATPase enzyme cycle have shown that the K^+-stimulated ATPase activity, formation of phosphoenzyme and *p*-nitrophenol-phosphatase (pNPPase) activity are also inhibited (75,80). Pantoprazole binds to the hog gastric proton pump with a stoichiometry of 3 nmol/mg protein causing 94% inhibition of ATPase activity (81).

In summary omeprazole represents a prodrug which is itself inactive. It needs the acidic conditions of the parietal cell to ensure its inhibitory effect on acid secretion. During multiple dose studies, omeprazole facilitates its own enteral absorption by its specific effect on gastric acid secretion. Under these conditions, the inhi-

bition of gastric acid secretion and resulting neutral pH values within the gastric lumen prevents acid-induced degradation of omeprazole. To prevent acid-induced transformation within the gastric lumen and binding to superficial sulfhydryl groups of the gastric mucosa, omeprazole must be administered orally in a galenic formulation that prevents acid-induced activation within the gastric juice during its passage through the stomach (82) (Fig. 27.7). In general, this mechanism of action for omeprazole is also valid for all other omeprazole-like proton pump inhibitors.

The structure-activity relationships of proton pump inhibitors of the omeprazole-type are based on the balance between chemical stability at neutral pH values and acid-induced conversion into the active sulfenamide. Chemically very stable derivatives do not show any inhibitory effect either *in vitro* or *in vivo*. Derivatives which are unstable at neutral pH, are very active in the test assay of partly purified H^+/K^+-ATPase. This assay is performed under physiological pH conditions at pH 7.4 after preincubation at pH 6 of the enzyme protein with the derivative to be tested. The high activity is therefore the result of the conversion of the derivative in solutions of neutral pH values (pH 6 to 7.4) to its active intermediate and this does not reflect the situation of high acidity (pH 1-2) within the secretory compartment of the parietal cell (42). Derivatives that are very unstable at neutral pH do not inhibit gastric acid secretion *in vivo* because their transformation had already taken place prior to the active principle reaching the target enzyme.

Due to the close chemical relationship between all proton pump inhibitors of the omeprazole-type (Figs. 27.2–27.5), it seems very likely that they share the same or a very similar mechanism of action to that of omeprazole (42) (Fig. 27.6). The sulfenamide has been identified to be the active principle for most of the irreversible proton pump inhibitors: lansoprazole

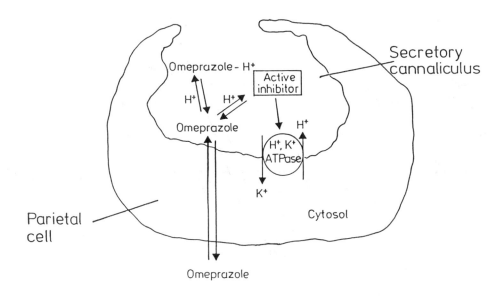

Fig. 27.6 Acid-catalyzed activation of omeprazole to the active pyridinium sulfenamide, which reacts with essential sulfhydryl groups of the gastric proton pump. Reproduced courtesy of Elsevier Science in a redrawn design from ref. 34. Asterisk indicates that only one isomer is shown (isomerization is caused by the 5-methoxy substitution in the benzimidazole ring).

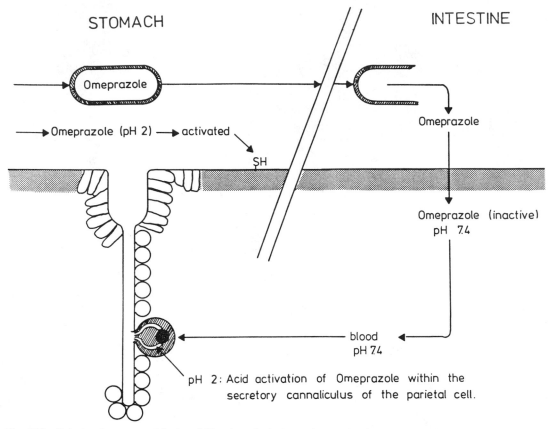

Fig. 27.7   Relation between acidic instability, intestinal absorption, and acid transformation of omeprazole-like gastric proton pump inhibitors. Reproduced courtesy of Elsevier Science from ref. 34.

(83,84), pantoprazole (55,81,85), saviprazole (34), and SK&F 95601 (61) as well as for rabeprazole sodium (86). Pantoprazole has a grater chemical stability at neutral pH values than omeprazole, whereas its corresponding inhibitory effect on gastric proton pump in different vesicle models was less pronounced (87). Similarly, greater chemical stability at physiological pH has also been claimed for SK&F 95601 but nevertheless acid-induced transformation to the sulfenamide as the reactive intermediate (61) is essential for its effectiveness. Acid-induced transformation has also been reported for the 2-aminobenzyl substituted benzimidazoles (88,89).

The acid-induced transformation process for saviprazole (**11**), a slightly less stable

proton pump inhibitor than omeprazole, has been elucidated (34) (Fig. 27.8). In the absence of mercaptanes, the activation induced by aqueous hydrochloric acid yields a complex mixture of products. However, with aqueous HBF_4 (50%) in methanol at $-50°C$, the sulfenamide pyridinium salt (**23**) could be isolated in high yield in analogy to omeprazole (70). This compound is highly reactive and unstable in solution. When sulfenamide (**23**), the active principle of saviprazole, is treated at $-30°C$ with L-cysteine (L-Cys), which has been chosen as a model for enzyme sulfhydryl-groups, the L-cysteine disulfide pyridinium salt (**22**) is formed. Treatment of saviprazole with L-cysteine in methanol ($-25°C$) in the presence of aqueous hydro-

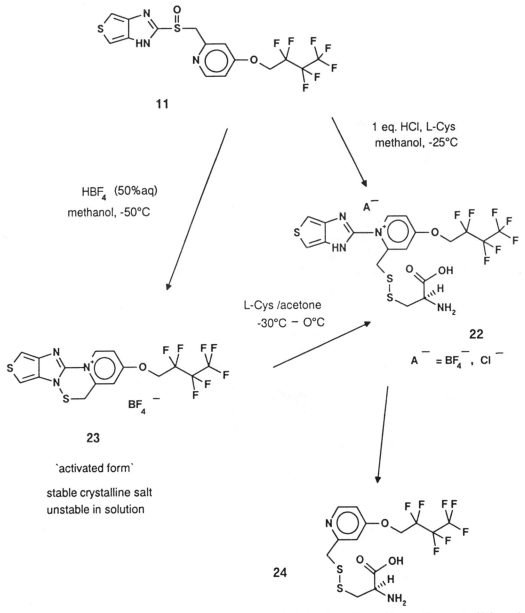

Fig. 27.8 Acid activation of saviprazole to its active principle, the sulfenamide pyridium salt (**23**), and its L-cysteine (L-Cys) addition product (**22**) as a model for the enzyme inhibitor complex.

chloric acid also results in the formation of the pyridinium disulfide (**22**). Similar experiments with omeprazole and beta-mercaptoethanol in acidic media have been reported (70,74). In contrast to omeprazole, the saviprazole-L-cysteine adduct (**22**) is sensitive to cleavage into the 2-pyridyl disulfide (**24**) and an unidentified thienoimidazole fragment (Fig. 27.8) The instability of the primary L-cysteine addition product (**22**), as an example of the enzyme inhibitor complex, might be correlated to the pharmacological profile of saviprazole, the noncomplete inhibition of gastric acid secretion (62,90).

With respect to the sulfinyl group, gastric proton pump inhibitors exist as a racemic mixture of both enantiomers. Although the chirality is lost in the corresponding pyridinium sulfenamide, there is still the question whether one enantiomer is more susceptible towards acid activation than the other. However, both enantiomers of lansoprazole inhibit dbcAMP-induced aminopyrine uptake in isolated canine parietal cells as well as H^+/K^+-ATPase activity in canine gastric microsomes with equal activity (91). Similar findings were observed for the enantiomers of Ro 18-5364 (92).

As a result of the inherent chemical instability of proton pump inhibitors of the omeprazole-type (a prerequisite for their acid-induced transformation process), there have been some attempts to stabilize the substituted benzimidazole molecule. The introduction of appropriate prodrug moieties on the benzimidazole nitrogen improve stability, solubility, and other physico-chemical properties. The omeprazole-type inhibitor is then generated from these prodrugs by enzymatic or acidic cleavage (59).

4.1.3 STRUCTURE–ACTIVITY RELATIONSHIPS. Because of omeprazole's highly specific mode of action, the possibilities for structural modifications are very restricted. It has been claimed by Brändström et al. (52) that the three structural elements of omeprazole—the substituted pyridine ring, the substituted benzimidazole moiety, and the methylsulfinyl chain connecting these two—are essential for the biological effect, and that only a few compounds with ring systems closely related to benzimidazole showed weak pharmacological activities.

The attempts of competitors to replace the benzimidazole ring in timoprazole/omeprazole by a heterocycle annellated imidazole moiety led to the thienoimidazole series. The thieno[3,4-d]-imidazoles (e.g., **12**) have been disclosed in patent applications by Hoechst (93) and Pfizer (65). Hoechst has focused its activity on the 4-(heptafluorobutyloxy) pyridine compound saviprazole (**11**). Surprisingly the thieno[2,3-d]imidazole isomers (e.g., **13**) disclosed in patent applications by Chemie Linz (64) and Pfizer (65), have only weak activities in all biological assays probably as a result of their pronounced chemical stability (62) (Fig. 27.4).

Omeprazole-type proton pump inhibitors have to be highly reactive at low pH, although the prerequisite for their selective biological activity is a relatively high chemical stability under physiological conditions around neutral pH (Fig. 27.7). Chemical handling also demands a certain degree of chemical stability. One of the major problems is that the biological activity of the proton pump inhibitors frequently correlates with their chemical lability. Many compounds showing excellent inhibitory data *in vitro* are often too unstable to be of further interest. Chemical stability is an important additional criterion for the evaluation of a potential gastric proton pump inhibitor.

In the 2-[(2-pyridylmethyl)sulfinyl]-1*H*-benzimidazole and the 2-[(2-pyridylmethyl)sulfinyl]-1*H*-thieno[3,4-d]imidazole series, the biological activity and the chemical stability of the compounds largely depends on their substitution pattern.

Pyridine Substitution. Compared to timoprazole (**1**), the prototype of irreversible gastric proton pump inhibitors, an additional 4-methoxy group in the pyridine ring increases the biological activity by enhancing the nucleophilicity of the pyridine nitrogen atom. Since, in the first event of the acid activation process, the pyridine N-atom binds to C-2 of the benzimidazole group, a more nucleophilic pyridine renders the acid-catalyzed activation process to the cyclic pyridinium sulfenamide to be more effective (Fig. 27.6).

A 4-fluoroalkoxy substitution, combining lipophilicity and electron demanding properties, results in compounds with strong inhibitory activity as illustrated by the 4-(2.2.2-trifluoroethyloxy) compound, lansoprazole (4), and the 4-(2.2.3.3.4.4.4-heptafluorobutyloxy) compound, saviprazole (11). The 3-methoxypropoxy substituent in rabeprazole sodium (6) has a stronger electron-donating character and is even more activating, but using rabeprazole as a sodium salt increases the chemical stability.

A further increase of the pyridine nucleophilicity by a 4-amino substituent leads to practically useful compounds only if the electron-donating effect on the pyridine is modified by an additional inductive electron demanding substituent (chloride or methoxy) in the 3- or 5-position, e.g., SK&F 95601 (10). This favorable effect of 3-substituents on the chemical stability could also be observed in pantoprazole (5) and the thieno[3,4-d]imidazole series (62).

Benzimidazole Substitution. Irreversible gastric proton pump inhibitors have an unsubstituted benzimidazole or thieno[3,4-d]imidazole moiety (timoprazole (1), lansoprazole (4), saviprazole (11)) or a benzimidazole ring substituted with an electron-donating 5-alkoxy group (omeprazole (3), SK&F 95601 (10)) (94). The 5-difluoromethoxy group in pantoprazole (5) is closely related. In contrast, if the benzimidazole moiety is substituted by an electron-acceptor substituent (5-nitro, 5-methylsulfinyl, or 5-trifluoromethyl), the resulting compounds have poor chemical stability and are not practically useful. Their chemical behavior is dominated by a nonprotic activation process at neutral pH (94). Thus the substitution pattern in each omeprazole-like proton pump inhibitor, both on the benzimidazole and the pyridine ring, is optimized towards chemical stability at neutral pH and towards biological activi-

ty, depending on the effective acid catalyzed activation process which yields the cyclic pyridinium sulfenamide.

4.2 Pharmacology

4.2.1 ANIMAL PHARMACOLOGY, INHIBITORY EFFECT ON GASTRIC ACID SECRETION. The inhibition of gastric acid secretion caused by omeprazole has been studied predominantly in rats and dogs. Omeprazole causes dose-dependent inhibition of basal and stimulated gastric acid secretion in these species. Because of its unique mechanism of action, the inhibition of the last enzymic step of acid formation, omeprazole causes an inhibition of gastric acid secretion which is independent of the kind of stimulation (37,95). ID_{50} values are not statistically significantly different when comparing dose–response curves in rats either during basal or during carbachol-, gastrin-, or histamine-stimulated gastric acid secretion (Table 27.2).

Under experimental pharmacological conditions, omeprazole was less effective when administered orally as either solution or suspension in comparison to intraduodenal or intravenous administration (37,96). This is in accordance with its pH-dependent conversion which occurs to a significant extent within the gastric juice under these experimental conditions after oral administration. Here, the active principle derived from acid-induced transformation cannot reach the target enzyme at the apical membrane of the parietal cell (Fig. 27.7). Similar findings have been reported for lansoprazole (97,98) and saviprazole (90).

Substituted thienoimidazoles, such as, S 1924 and saviprazole, show a more pronounced residual secretion rate than omeprazole after a single administration of doses producing more than 90% inhibition in rats and dogs (63,90). Saviprazole and omeprazole are equally effective in inhibiting gastric acid output in pylorus-ligated rats within three hours at a dose of 10 mg/kg intraduodenally. However, increasing the dose up to 30 mg/kg causes no further increase in the inhibitory level for saviprazole, whereas the inhibitory level for omeprazole increases up to 100%. In pylorus-ligated rats, gastric acid output over three hours starting 16 h after compound administration was less after saviprazole than omeprazole at both doses tested. This indicates a higher residual secretion rate and a shorter duration of action for saviprazole (Fig. 27.9). In dogs (Heidenhain pouch) saviprazole, like omeprazole, causes an immediate and complete inhibition of histamine-stimulated gastric acid secretion after an intravenous dose of 1 mg/kg. However, this complete inhibition caused by saviprazole only lasts for 30 minutes. Acid output then rises to approximately 10% of maximally induced acid secretion. In contrast, the inhibitory effect of omeprazole rises to approximately 5% of maximally

stimulated acid secretion over the study period of four hours after compound administration. The residual secretion rate after saviprazole was significantly higher than that after omeprazole (62,90).

Despite its significantly lower potency compared to omeprazole in Heidenhain-pouch dogs, SK&F 95601 has been selected for clinical trials because of its higher chemical stability at physiological pH (61). It has been shown that rabeprazole sodium is twice as potent in dogs than omeprazole (99). Compounds, in which the 2-pyridylmethyl group is replaced by a 2-aminobenzyl moiety, e.g., NC-1300, NC-1300-B (66,100), leminoprazole (101), and S 3337 (63), inhibit the gastric proton pump and gastric acid secretion in pylorus-ligated rats at the same concentration and dose range as omeprazole. Furthermore the duration of action of NC-1300-B is much longer than that of omeprazole. NC-1300-B inhibits gastric acid output in conscious pylorus ligated rats after a single pretreat-

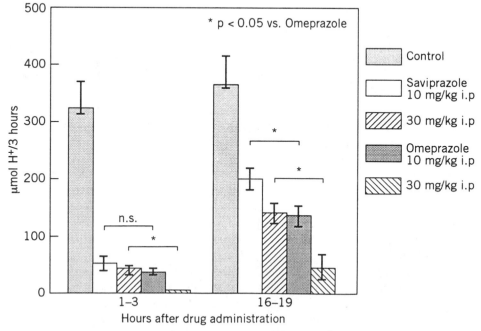

Fig. 27.9 Pharmacodynamic difference between saviprazole and omeprazole at high doses on basal gastric acid secretion in pylorus-ligated rats; $n = 8$ rats.

ment of 72 h (66). In contrast to the ome-prazole-like potency *in vitro* and in pylorus ligated rats, S 3337 (63), NC-1300 (102), and compounds with substitutions on the nitrogen atom of the 2-aminobenzyl moiety (89) show a significantly lower potency on gastric acid secretion in dogs compared to omeprazole. Recently, there have been attempts to replace the benzimidazole by an imidazole moiety in connection with an N-substituted 2-aminobenzyl instead of the pyridine ring, resulting in effective gastric proton pump inhibitors with good antisec-retory potencies even in dogs (102).

Gastric acid inhibition lasts 3–4 d after a single administration of omeprazole to dogs at a dose which causes maximal inhibition on the day of administration (37,96). This long-lasting inhibition is in contrast to clini-cally available histamine-H_2-receptor an-tagonists, which inhibit histamine-induced gastric acid secretion for only 4–18 h. This long duration of acid inhibition caused by omeprazole results from the irreversible covalent binding of the acid-induced sul-fenamide to sulfhydryl groups of the gastric proton pump. Gastric acid secretion reap-pears only when new enzyme is synthesized. The half-life of the gastric H^+/K^+-ATPase in rats is about 48 h (103). Therefore, acid secretion inhibited by omeprazole reap-pears within 24 h and returns to normal in about 3–4 d (19). Since there is a significant residual inhibition after 24 h the degree of inhibition increases over the first few days of repeated daily administration. Steady state (constant inhibitory level) is achieved after about four days of treatment (104).

In contrast to omeprazole, it has been claimed that for lansoprazole *de novo* syn-thesis of the gastric proton pump does not participate in the process of recovery from inhibition but that gluthathione is somehow involved in the reactivation of the enzyme (84). For saviprazole it has also been sug-gested that the fading inhibitory profile, which led to the more pronounced residual secretion rate (90,105), is caused by per-turbations of the cellular glutathione level (46). The duration of action of rabeprazole sodium is significantly shorter compared to omeprazole in dogs (99) or lansoprazole in rats (106). It is suggested that endogenous extracellular glutathione is involved in the reactivation process *in vivo* (107).

As gastric proton pump inhibitors of the omeprazole type need acid for their conver-sion into the corresponding active princi-ples, any condition in which gastric acid secretion is inhibited results in reduced drug-induced inhibition. For example, if gastric acid secretion is inhibited by a short-acting histamine-H_2-receptor antago-nist and omeprazole is administered during conditions of maximal inhibited acid forma-tion at a dose which causes a 3–4 d inhibi-tion, the omeprazole-induced inhibition will be significantly reduced. The inhibitory effect of omeprazole depends on the secre-tory state of the parietal cell (79,108). Simi-lar findings have been reported for panto-prazole (109).

Cytoprotection. In addition to ome-prazole's antisecretory effect, it has been reported that in rats omeprazole has cyto-protective properties which are not medi-ated by its inhibitory effect on gastric acid secretion (110,111). It has been suggested that the anti-ulcer property of pantoprazole is somehow related to its antisecretory effect (112). In addition to their antisecre-tory effects, a stimulatory effect of NC-1300 (113) and NC-1300-0-3 (114) on mucus glycoprotein secretion could be responsible for their ability to protect the stomach against necrotizing agents.

The anti-ulcer effect of lansoprazole in rats is 3 to 10 times more pronounced than that of omeprazole (97). However, from clinical experience, there is no doubt that omeprazole exerts its effect on ulcer heal-ing in humans predominantly via its an-tisecretory activity (36,115,116). This fact is confirmed by clinical experiences with omeprazole and lansoprazole: both compounds are equally effective in ulcer healing in humans at similar doses (36,115,117) to those which express their antisecretory effect, although lansoprazole is 3 to 10 times more potent in rat ulcer models. Therefore, it seems justified to conclude that the rat ulcer models have only a very minor relevance for ulcer healing in humans.

4.2.2 HUMAN PHARMACOLOGY. For these agents, human pharmacology parallels animal pharmacology, especially that of the dog, with respect to the effective dose-range and duration of action. Omeprazole causes dose-dependent inhibition of basal and stimulated gastric acid secretion and a reduction of intragastric acidity in healthy volunteers after a single administration (118–120) or repeated daily administrations (118). This is also seen in patients with peptic ulcer disease (121–124). The inhibitory effect after a single administration lasts 2–3 d (118). Gastric acid output can be reduced effectively without tachyphylaxis in Zollinger Ellison syndrome patients treated for nine years with omeprazole (125).

Lansoprazole causes dose-dependent inhibition (15, 30, 60 mg) of stimulated and basal gastric acid secretion in healthy volunteers. It increases intragastric pH and has a long duration of action (126,127). Morning administration for lansoprazole is recommended due to circadian differences in bioavailability (128). Lansoprazole and omeprazole at doses of 30 and 20 mg, respectively, seemed to be equally effective in increasing gastric pH, inhibiting gastric acid output, and increasing serum gastrin (129). Similar inhibitory activity has been reported for rabeprazole sodium (20 mg) in healthy volunteers (130). The administration of 40 or 60 mg pantoprazole for five days maintained gastric pH above 3 for 33% or 58% of time, respectively, compared with 15% of time after placebo (131). In another study, 40 mg pantoprazole for five days increased gastric pH to about 4

(132). The dose of 40 mg/d pantoprazole has been recommended for the treatment of acid-related diseases (133). A single intravenous dose of SK&F 95601 caused dose-dependent inhibition of pentagastrin-stimulated gastric acid output in healthy volunteers at doses between 16 and 96 mg (134).

In clinical studies, the effectiveness of proton pump inhibitors has been studied in various acid-related disorders of the upper gastrointestinal tract: duodenal ulcers, gastric ulcers, Zollinger Ellison syndrome, reflux esophagitis. Omeprazole fulfills the therapeutic requirements of rapid and reliable therapeutic effect, safety, and simple treatment regimen (135). However, the high recurrence rate of healed duodenal or gastric ulcers is still a significant clinical problem. If a healed duodenal ulcer is left untreated. The relapse rate is 50–80%/yr. The relapse rate seems to be independent of the compound used to cure ulcer disease initially (136,137). However, long-term treatment with low doses of omeprazole prevents duodenal ulcer relapse (138) or relapse in reflux esophagitis (139).

There is growing evidence that the relapse rate of peptic ulcers is associated with colonization of the gastric mucosa with *Helicobacter pylori*. Treatment with omeprazole alone is not sufficient to eradicate the bacteria (140–143), although it has been reported that omeprazole and lansoprazole both have inherent selective antibacterial activity against *Helicobacter pylori* (144) due to blockage of SH-groups of its urease (145). The combination of omeprazole plus the antibiotic amoxycillin for up to four weeks is effective in eradicating *Helicobacter pylori* (146–148) with subsequent reduced ulcer recurrence (149, 150). In many clinical studies, omeprazole was combined with various antibacterial agents (e.g., metronidazole) and successful eradication of *Helicobacter pylori* was demonstrated (116). Similar clinical results have been reported for lansoprazole (151).

4.3 Pharmacokinetics

After administration of omeprazole intraduodenally to dogs or orally as buffered suspensions to humans, omeprazole is rapidly absorbed from the gut reaching plasma peak levels within 30 min (37,118). Due to its inherent-chemical instability in acidic conditions, omeprazole-like proton pump inhibitors have to be administered orally as a galenic formulation which prevents acidic degradation during passage through the stomach (Fig. 27.7). After omeprazole administration in an enteric-coated formulation intestinal absorption of omeprazole is delayed and plasma concentration–time curves are flat and broad with low peak plasma concentrations (82). Subsequently, omeprazole is eliminated from the plasma with a half-life of 40–60 min depending on the species (37,118). There is no correlation between the duration of the inhibitory effect caused by omeprazole (up to four days) and its plasma concentration at a given time ($t_{1/2}$ less than 60 min). However, the area under the plasma-concentration curve (AUC) during 0–4 h correlates well with the inhibitory effect both in dogs (37) and humans (118). Therefore, the shape of the plasma concentration-time curve is of no importance, and different formulations resulting in equal AUCs cause equal inhibitory effects (82). This indicates that the amount of omeprazole being absorbed from the gut correlates with the amount of omeprazole being transported by the blood to the parietal cell and the amount being available for inhibition of the gastric proton pump (42).

In this context the results of autoradiographic studies in mice with ^{14}C-omeprazole are very impressive. After intravenous injection of ^{14}C-omeprazole, radioactivity was present 16 h only in the gastric wall (152). This indicates that the active enzyme inhibitor is present for many hours only at the site of its action. This correlates closely to its long-lasting duration of action while the parent-compound rapidly disappears from the blood. Similar results were obtained for saviprazole (34).

Plasma half-life of between one and two hours in healthy volunteers has been reported for lansoprazole (117), pantoprazole (153), and rabeprazole sodium (154).

4.4 Metabolism

Omeprazole is rapidly and completely metabolized after intestinal absorption. Oxidative processes of metabolism are predominant and three main metabolites of omeprazole have been identified: the sulfone, sulfide, and hydroxyomeprazole. Omeprazole sulfone is further metabolized or it is eliminated in the feces. Renal excretion is the predominant route of elimination of omeprazole metabolites (36,42,155). The metabolism of lansoprazole in humans is comparable to that of omeprazole, affording hydroxylansoprazole, lansoprazole-sulfone, lansoprazole-sulfide, and the hydroxylated sulfone (156).

Omeprazole inhibits cytochrome P450-mediated metabolic reactions *in vitro* in hepatic microsomes. This effect is comparable to the inhibition caused by cimetidine with respect to the extent of the inhibitory effect and the effective concentrations (155). In primary cultures of human hepatocytes, omeprazole and lansoprazole are mixed inducers of CYP1A and CYP3A (157). Only under well-defined clinical conditions, interactions in drug metabolism between omeprazole and diazepam, phenytoin, (158) or warfarin (159) have been detected. However, clinically significant drug interactions appear to be unlikely (160). It has been reported that the interaction of pantoprazole with cytochrome P450 *in vitro* (161) and *in vivo* (162) is less than that of omeprazole or lansoprazole, whereas the overall antisecretory potency and efficacy of pantoprazole is similar to that of omeprazole (163). In addition, pantoprazole does not influence the disposition kinetics of theophylline in humans (164). The data available so far indicate that omeprazole, lansoprazole, and particularly pantoprazole at therapeutic doses should be relatively free of clinically significant drug interactions (165,166).

4.5 Toxicology

Omeprazole was well tolerated in chronic toxicological studies in rats or dogs up to the highest oral doses tested (414 mg/kg or 138 mg/kg, respectively) (42). Even in a cancerogenicity study in mice (two years of treatment) no results of concern were observed.

4.5.1 ENTEROCHROMAFFIN-LIKE CELL PROLIFERATION. In contrast to the lack of findings in the cancerogenicity study in mice, in rats there was a dose-dependent increase in gastric carcinoids. This was more pronounced in female than in male rats (104,167). Carcinoids in rats consist of enterochromaffin-like cells (ECL-cells), which are predominantly formed by histamine-releasing cells in the rat gastric mucosa (168) and differ, therefore, from carcinoids in humans, which are normally formed by serotonin-producing cells (104). Similar results have been reported for long-acting histamine-H_2-receptor antagonists: BL-6341 (169), loxtidine (170,171), and SK&F 93479 (172). These findings are more likely caused by the common pharmacodynamic effect of these histamine-H_2-receptor antagonists and omeprazole causing a long-term inhibition of gastric acid secretion than an inherent carcinogenic effect of these different compounds (173).

Initially there were some doubts (174–176) concerning the attempts to explain the occurrence of carcinoids in rats by the drug-induced excessive hypergastrinaemia. These initial doubts have not been proven to be valid because the subsequent results support the gastrin hypothesis (173,177) convincingly. ECL-cells are functionally and trophically under the control of gastrin (168). The release of gastrin, synthesized in antral G cells, is mediated by the amino acid composition of the gastric juice as well as by its acidity. Acidic pH values inhibit antral gastrin release. Neutral pH values do not stimulate gastrin release but represent the removal of the natural brake for gastrin release (19). Any pharmacologically induced inhibition of gastric acid secretion, therefore, must finally lead to increased antral G-cell density with corresponding increased serum gastrin levels and ECL-cell proliferation (Fig. 27.10). This has been demonstrated for omeprazole (178–182), lansoprazole (183), saviprazole (34), ranitidine (179–182), and even antacids (184,185). Even different degrees of surgical removal of acid-secreting mucosa (75%, 90%, 100% fundectomy) in rats without any administration of gastric acid inhibitors finally resulted in a stepwise increase in serum gastrin, indicating that the antral acid load (or pH) appears to be the major factor for gastrin release (186). In addition antrectomy, the removal of the gastrin-releasing mucosa, prevented ECL-cell proliferation in omeprazole-treated rats (187), although it did not accelerate reversal of omeprazole-induced ECL-cell hyperplasia (188). The long-lasting hypergastrinaemia during life-long treatment with omeprazole in rats finally induced carcinoids. Therefore, the occurrence of carcinoids in the cancerogenicity study in rats was not caused by an inherent carcinogenic effect of omeprazole but was pharmacodynamically induced by its long-lasting effect of gastric acid inhibition with subsequent excessive hypergastrinaemia (173,177).

In humans, omeprazole treatment as well as treatment with histamine-H_2-receptor antagonists causes a moderate increase in serum gastrin, which reflects the degree of gastric acid inhibition (189,190). The

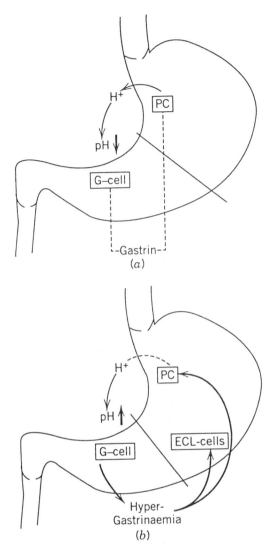

Fig. 27.10 (a) Relation between intragastric pH, gastrin release, and gastrin stimulation on gastric acid secretion. (b) The hypergastrinaemia hypothesis to explain the effect on ECL cells during inhibitory condition of gastric acid secretion.

coadministration of enprostil (191) or pirenzepine (192) can reduce the serum gastrin levels in ulcer patients during treatment with omeprazole. There were no detectable omeprazole-induced pathological changes in the gastric mucosa in patients with peptic ulcer or reflux esophagitis in whom therapy was extended to 1–4 yr (193,194). From the clinical experience of

chronic gastritis progressing over years, long-lasting achlorhydria is accompanied by a slight hyperplasia but no dysplasia of the ECL-cells in the gastric mucosa. This observation parallels the findings when omeprazole is given for up to five years. Initially it was recommended to monitor plasma gastrin levels in patients receiving long-term treatment (195) and to reduce the dose of omeprazole if plasma gastrin values were to increase above five times the upper limit of normal (196). However, with increasing clinical experience with proton pump inhibitors, it has been concluded that monitoring of plasma gastrin levels and fundic ECL cells are of no clinical relevance even during long-term therapy (166). Clinical experience of long-term treatment with potent acid-suppressing drugs has demonstrated that possible consequences of long-lasting acid suppression in the stomach, e.g., gastric bacterial overgrowth or hypergastrinaemia, are without clinical relevance (197).

4.5.2 THYROID TOXICITY. It has been reported that high doses (138 and 430 mg/kg) of omeprazole to rats interfere with the peripheral conversion of thyroxine (T4) to triiodothyronine (T3), resulting in the decrease of serum T3, unchanged serum T4, and no change in the morphology of the thyroid gland (167). The effects of omeprazole, lansoprazole, saviprazole, SK&F 65601, and rabeprazole sodium on thyroid parameters (total serum T4 and T3, free serum T4 and T3, thyroid weights) have been studied after oral treatment for 14 d at doses of 50, 150, and 450 mg/kg (198). The selected doses were 10 to 100 times higher than the intraduodenal dose effective in inhibiting gastric acid secretion in pylorus-ligated rats.

Omeprazole and SK&F 65601 caused no changes in thyroid weights. Slightly elevated serum T4 and T3 values were not dose-dependent (34,198). The treatment period of 14 d was too short to effect a decrease in total serum T3 for omeprazole, although this was obvious at the end of the three-month toxicity study (167). Rabeprazole sodium caused significantly elevated serum T4 and T3 values with an inverse dose-dependency without changes of the thyroid weights. Saviprazole and lansoprazole decreased total serum T4 in a dose-dependent manner without affecting serum T3 levels. Although both compounds showed the same degree of lowering of serum T4, there was a significant increase in the thyroid gland weight only after the highest dose of saviprazole. The highest dose of lansoprazole showed a comparable tendency to increase thyroid weight after the treatment period was prolonged up to six weeks (34). In clinical studies at therapeutic doses up to 60 mg lansoprazole, no influence on serum hormone levels could be observed (199).

Further studies with saviprazole to elucidate the mechanism of the decrease in serum T4 revealed that the compound does not interfere with hormone synthesis, storage, or release. The decrease in peripheral T4 was caused by an increased hepatobiliary elimination of T4 resulting from induction of hepatic T4-metabolizing enzymes. The increase in thyroid weight induced by 450 mg/kg saviprazole in rats could be prevented by the exogenous substitution of T4 (34) in a dose-dependent manner.

5 REVERSIBLE GASTRIC PROTON PUMP INHIBITORS

Reversible inhibitors of the gastric proton pump differ from omeprazole-like compounds in their chemical structure as well as in their mode of interference with the enzyme. No acid-induced transformation is necessary, and their enzyme kinetics show that they bind competitively in a noncovalent manner to specific sites of the gastric proton pump.

5.1 Chemistry and Biochemistry

5.1.1 COMPOUNDS. SCH 28080 (**25**, Shering-Plough) (200) is the prototype of a reversible proton pump inhibitor (Fig. 27.11). It was soon recognized that SCH 28080 has antisecretory properties in various *in vitro* and *in vivo* assays (201) and cytoprotective properties in different rat ulcer models (202). As early as 1983 it was suggested that the antisecretory effect might involve the gastric proton pump (201), although during subsequent years the predominant scientific interest concentrated on its cytoprotective potential. The clinical development of SCH 28080 was discontinued because of liver toxicity in animals and elevated liver enzyme activity in the serum of human volunteers (203). In rats the cytoprotective effect of the follow-up compound, SCH 32651 (**26**, Shering-Plough (204) (Fig. 27.11), was four times more potent (205) than the antisecretory potential (206). From results obtained in guinea pig isolated fundic mucosa it has been suggested that the antisecretory effect of SCH 32651 is caused by a direct influence on the parietal cell at, or near, the site of the gastric proton pump (207).

Systematic synthesis was done by Smith Kline & French to identify freely reversible, non covalent inhibitors of the gastric proton pump with a mode of action comparable to SCH 28080. Influenced at that time by the appearance of carcinoids in long-term studies in rats treated with omeprazole, it was expected that gastric proton pump inhibitors with a shorter duration of action could be of therapeutic interest (208,209). This research resulted in choosing compound SK&F 96067 (**27**, SK&F) (210) and SK&F 97574 (**28**, SK&F) (211) for further development (Fig. 27.11).

5.1.2 MECHANISM OF ACTION. The assumption that SCH 28080 and SCH 32651 cause their antisecretory effect by direct interference of the gastric proton pump was confirmed in 1985 (212). SCH 28080 inhibits

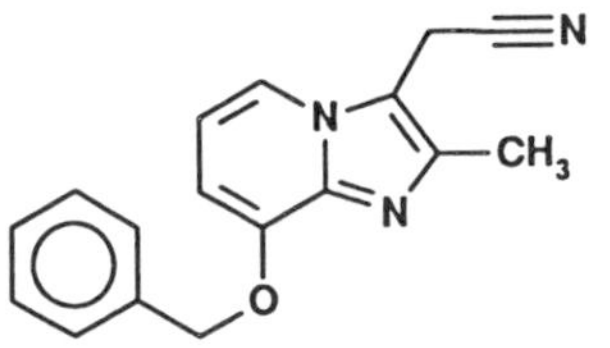

25

SCH 28080 (Schering-Plough)

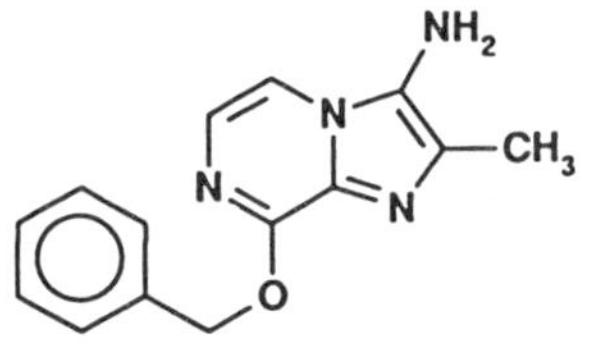

26

SCH 32651 (Schering-Plough)

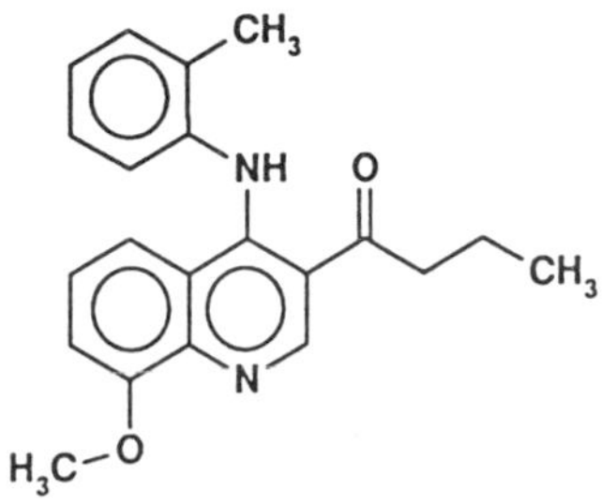

27

SK&F 96067 (SK&F)

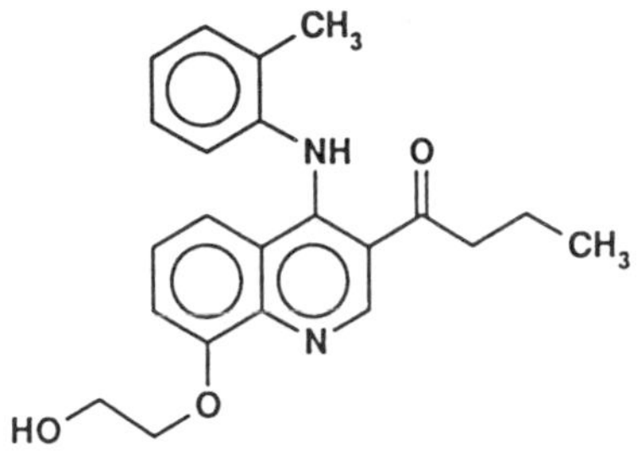

28

SK&F 97574 (SK&F)

Fig. 27.11 Reversible gastric proton pump inhibitors currently or previously under development.

the enzyme activity in a concentration-dependent manner with an IC_{50} in the micromolar range comparable to that of omeprazole. The effect of SCH 32651 was obviously less pronounced (212,213).

The difference between omeprazole and SCH 28080 in their ability to inhibit gastric proton pump depends on their inhibition kinetics. In contrast to omeprazole, SCH 28080 is a competitive inhibitor of the high affinity K^+ site of the gastric proton pump. Although it was not very specific for the gastric proton pump, its effect on Na^+/K^+-ATPase activity is much less pronounced in comparison to its effect on gastric H^+/K^+-ATPase activity (213,214). SCH 28080 is a protonatable weak base ($pK_a = 5.6$). Therefore, like omeprazole and related compounds, SCH 28080 accumulates in acidic compartments of the parietal cell in its protonated form (215). However, SCH 28080 is chemically stable and, after proto-

nation, is itself active (216) and does not need an acid-induced transformation such as that required by omeprazole-like irreversible inhibitors. Therefore, in proton transport studies, SCH 28080 inhibits the initial rate of H^+/K^+-ATPase mediated H^+ accumulation and the steady-state proton concentration. This is in contrast to omeprazole, which first needs accumulation of acid within gastric vesicles to generate an interior of low pH to facilitate the acid-induced transformation prior to being able to inhibit the gastric proton pump (217). SCH 28080 binds to the lumenal side of the proton pump (215,218). Its inhibitory effect cannot be prevented by the addition of sulfhydryl reducing agents but can be reversed by dilution or washing (219). SK&F 96067 is also active in its protonated form and binds competitively to the lumenal K^+ binding site of the gastric proton pump (220).

5.2 Pharmacology

As a consequence of its mechanism of action, the inhibition of the last enzymic step of acid formation, SCH 28080 inhibits gastric acid secretion under basal conditions as well as during all kinds of stimulation. In isolated guinea pig fundic mucosa, SCH 28080 inhibits methacholine-, histamine-, or dbcAMP-stimulated acid secretion in a concentration-dependent manner with IC_{50} values in the micromolar range (207). Similar results have been obtained *in vivo*. Pentagastrin-, histamine-, and dimaprit-induced gastric acid secretion in dogs is inhibited by similar doses (202). The antisecretory effect of SCH 28080 in human volunteers was first demonstrated in 1982 (221) at a time when its unique mechanism of action had not been elucidated.

In rats and dogs SK&F 96067 causes a dose-dependent inhibition of gastric acid secretion. Its duration of action is longer than that of cimetidine but much shorter than that of omeprazole (210,222). In human kinetic and pharmacological studies, SK&F 96067 was well tolerated without serious adverse events (223). A dose of 300 mg twice daily led to an increase in gastric pH which was more effective than 150 mg of ranitidine twice daily (211). At present, the profile of acid inhibition by reversible proton pump inhibitors resembles that of histamine-H_2-receptor antagonists (12).

6 OTHER INHIBITORS

The inhibitory effects of verapamil, a calcium channel antagonist, on gastric acid secretion *in vitro* and *in vivo* are controversial. There are reports that verapamil causes a significant inhibition of acid secretion in humans (224), and in rats (225,226), but others claim that it has no effect at all (227,228).

It is generally accepted that the interference with acid formation in gastric parietal cells by verapamil is not caused by its calcium channel antagonism (229). As early as 1983, it was reported that verapamil and gallopamil inhibit gastric acid formation in isolated parietal cells. The inhibitory effect *in vitro* was independent of the kind of stimulation (histamine, dbcAMP, and K^+). From these results, it was concluded that verapamil and gallopamil inhibit gastric acid secretion by interfering with the proton pump of the parietal cell (230). Even early kinetic studies indicated that verapamil inhibits the gastric proton pump in a competitive manner with respect to K^+ (231). Similar findings were obtained with trifluoperazine, an antipsychotic drug, and TMB-8, an antiarrhytmic agent (231). However, in isolated rabbit gastric glands, verapamil has nonspecific inhibitory effects on acid formation with various IC_{50} values depending on the agonist used (232). These nonspecific inhibitory effects have been characterized as being of a protonophoric (233) and detergent (234) nature. The last authors did not find any inhibitory effect of verapamil on gastric acid secretion in stomach-lumen perfused rats or in dogs. Only in the *in vivo* model of pylorus-ligated rats could an inhibition of gastric acid secretion be demonstrated. However, this effect was more due to its cardiovascular effect followed by compensatory processes of the autonomic nervous system than to a direct inhibitory effect on the gastric parietal cell (234).

The interference of some neuroleptics and antidepressants, such as, trifluoperazine, doxepin, and trimipramine, with the gastric proton pump is caused by an allosteric mechanism at the K^+ site of the enzyme (217,235).

It has been suggested that stimulation of gastric acid secretion across the apical membrane of the gastric parietal cell predominantly reflects the insertion of an active potassium and chloride transporter/ channel rather than direct activation of the gastric proton pump (2). However, the mechanisms involved in K^+ and Cl^- transport into the gastric parietal cell are still controversial. The presence of a Cl^- channel in the apical membrane has been confirmed by using established Cl^- channel blockers. 9-Anthracene carboxylate *in vitro* in isolated rabbit gastric parietel cells and the more potent diphenylamine-2-carboxylate inhibit gastric acid formation in a concentration-dependent manner irrespective of the kind of stimulation (236). The authors' experience with Cl^- channel blockers is the absence of any detectable effect in *in vivo* models. Thus Cl^- channel blockers can be defined *in vitro* as indirect gastric proton pump inhibitors.

7 CONCLUSION

In many clinical studies, predominantly in patients with Zollinger-Ellison syndrome, reflux esophagitis, or complicated gastric and duodenal ulcers, the superiority of omeprazole-like irreversible gastric proton pump inhibitors above histamine-H_2-receptor antagonists has clearly been demonstrated. Their clinically relevant advantages are related to their different mode and longer duration of action. The interference with the last enzymic step of acid formation is the most efficient method to block gastric acid secretion because the inhibitory effect is independent of the kind of acid stimulation. The longer duration of action ensures a more comfortable treatment scheme for patients given one tablet daily.

Reversible gastric proton pump inhibitors share the secretagogue independent mode of action with the irreversible inhibitors but show a duration of action that is comparable to that of histamine-H_2-receptor antagonists. Their clinically relevant advantage over irreversible inhibitors has still to be proven. Other gastric proton pump inhibitors, such as, Cl^- channel blockers, are at present of experimental interest only.

At present the possible use of proton pump inhibitors in indications other than the treatment of acid-related disorders of the upper gastrointestinal tract is only of experimental status. One possible indication could be the specific inhibition of the H^+/K^+-ATPase-like proton pump in osteoclasts in the treatment of osteoporosis (28).

ACKNOWLEDGEMENT

The authors wish to thank Dr. P. Godden for revising the manuscript.

REFERENCES

1. W. L. Peterson and C. T. Richardson in M. H. Sleisenger and J. S. Fordtran, eds., *Gastrointestinal Disease*, Saunders, Philadelphia, Vol. 1, 1983, pp. 708–724.
2. M. M. Wolfe and A H. Soll, *N. Engl. J. Med.*, **319**, 1707–1715 (1988).
3. A. H. Soll and J. I. Isenberg in M. H. Sleisenger and J. S. Fordtran, eds., *Gastrointestinal Disease*, Saunders, Philadelphia, Vol. 1, 1983, pp. 625–672.
4. M. Feldman and M. E. Burton, *N. Engl. J. Med.*, **323**, 1672–1680 (1990).
5. M. Feldman and M. E. Burton, *N. Engl. J. Med.*, **323**, 1749–1755 (1990).
6. D. Murdoch and D. McTavish, *Drugs*, **42**, 240–260 (1991).
7. P. H. Jordan in M. H. Sleisenger and J. S. Fordtran, eds., *Gastrointestinal Disease*, Saunders, Philadelphas, Vol. 1, 1983, pp. 739–749.
8. S. Holt, *South. Med. J.*, **84**, 1078–1087 (1991).
9. C. B. H. W. Lamers, T. Lind, S. Moberg, J. B. M. J. Jansen, and L. Olbe, *N. Engl, J. Med.*, **310**, 758–761 (1984).
10. C. J. Hawkey, R. G. Long, K. D. Bardhan, K. G. Wormsley, K. M. Cochran, J. Christian, and I. K. Moules, *Gut*, **34**, 1458–1462 (1993).
11. K. D. Bardhan, J. Ahlberg, W. S. Hislop, C. Lindholmer, R. G. Long, A. G. Morgan, S. Sjostedt, P. M. Smith, R. Stig. K. G. Wormsley, *Aliment. Pharmacol. Ther.*, **8**, 215–220 (1994).

12. J. W. Freston, *Aliment. Pharmacol. Ther.*, **7**(Suppl. 1), 68–75 (1993).

13. J. P. Balder and J. C. Delchier, *Aliment. Pharmacol. Ther.*, **8**, 47–52 (1994).

14. L. Ng-Wingtin, D. Glomaus, F. Hardy, S. Phil, *Anaesthesia*, **45**, 436–438 (1990).

15. T. Gin, M. C. Ewart, G. Yau, and T. E. Oh, *Br. J. Anaesth.*, **65**, 616–119 (1990).

16. M. L. Schubert, *Current Opinion in Gastroenterol.*, **7**, 849–855 (1991).

17. A. H. Soll, *J. Clin. Invest.*, **61**, 381–389 (1978).

18. R. A. North, B. E. Slack, and A. Surprenant, *J. Physiol. (London)*, **368**, 435–452 (1985).

19. G. Sachs, *Current Opinion in Gastroenterol.*, **6**, 859–866 (1990).

20. A. H. Soll, D. A. Amirian, L. P. Thomas, T. J. Reedy, and J. D. Elashoff, *J. Clin. Invest.*, **73**, 1434–1447 (1984).

21. C. S. Chew and S. J. Hersey, *Am. J. Physiol.*, **242**, G504–G512 (1982).

22. C. S. Chew, *Am. J. Physiol.* **250**, G814–G823 (1986).

23. G. E. Shull and J. B. Lingrel, *J. Biol. Chem.*, **261**, 16788–16791 (1986).

24. J. D. Kannitz and G. Sachs, *J. Biol. Chem.*, **261**, 14005–14010 (1986).

25. M. Takeguchi, S. Asano, Y. Tabuchi, and N. Takeguchi, *Gastroenterology*, **99**, 1339–1346 (1990).

26. J. D. Gifford, L. Rome, and J. H. Galla, *Am. J. Physiol. Renal. Fluid Electrolyte Physiol.*, **262**, F692–F695 (1992).

27. L. C. Gard and N. Narang, *J. Clin. Invest.*, **81**, 1204–1208 (1988).

28. R. Sarges, A. Gallagher, T. J. Chambers, and L. A. Yeh, *J. Med. Chem.*, **36**, 2828–2830 (1993).

29. D. R. DiBona, S. Ito, Berglindh, and G. Sachs, *Proc. Natl. Acad. Sci.* **76**, 6689–6693 (1979).

30. D. H. Malinowska, J. Cuppoletti, and G. Sachs, *Am. J. Physiol.*, **245**, G573–G581 (1983).

31. J. Cuppoletti and G. Sachs, *J. Biol. Chem.*, **259**, 14952–14959 (1984).

32. S. Asano, M. Inoi, and N. Takeguchi, *J. Biol. Chem.*, **262**, 13263–13268 (1987).

33. A. W. Herling and M. Bickel, *Eur. J. Pharmacol.*, **125**, 233–239 (1986).

34. A. W. Herling and K. Weidmann in G. P. Ellis and D. K. Luscombe, eds., *Progress in Medicinal Chemistry*, Elsevier Science B.V., Amsterdam, Vol. 31, 1994, pp. 233–264.

35. E. Fellenius, T. Berglindh, G. Sachs, L. Olbe, B. Lander, S.-E. Sjöstrand, and B. Wallmark, *Nature*, **290**, 159–161 (1981).

36. S. P. Clissold and D. M. Campoli-Richards, *Drugs*, **32**, 15–47 (1986).

37. H. Larsson, E. Carlsson, U. Junggren, L. Olbe, S. E. Sjöstrand, I. Skandberg, and G. Sundell, *Gastroenterology*, **85**, 900–907 (1983).

38. H. Shay, D. C. H. Sun, and M. Gruenstein, *Gastroenterology*, **26**, 906–913 (1954).

39. A. M. Barrett, *J. Pharm. Pharmacol.*, **18**, 633–639 (1966).

40. R. V. DeVito and H. N. Harkins, *J. Appl. Physiol.*, **14**, 138–139 (1959).

41. A. Robert, J. E. Nezamis, C. Lancaster, and A. J. Hanchar, *Gastroenterology*, **77**, 433–443 (1979).

42. P. Lindberg, A. Brändström, B. Wallmark, H. Mattson, L. Rikner, and K.-J. Hoffmann, *Med. Res. Reviews*, **10**, 1–54 (1990).

43. B. Wallmark, B.-M. Jaresten, H. Larsson, B. Ryberg, A. Brändström, and E. Fellenius, *Am. J. Physiol.*, **245**, G64–G71 (1983).

44. H. C. Lee and J. G. Forte, *Biochim. Biophys. Acta*, **508**, 339–356 (1978).

45. J. Fryklund and B. Wallmark, *J. Pharmacol. Exp. Ther.*, **236**, 248–253 (1986).

46. W. Beil, U. Staar, and K.-Fr. Sewing, *Eur. J. Pharmacol.*, **187**, 455–467 (1990).

47. K.-Fr. Sewing, P. Harms, G. Schulz, and H. Hannemann, *Gut*, **24**, 557–560 (1983).

48. A. H. Soll, and T. Berglindh in L. R. Johnson, ed., *Physiology of the Gastrointestinal Tract*, Raven Press, New York, Vol. 1, 1987, pp. 883–909.

49. B. Elander, R. Fellenius, R. Leth, L. Olbe, and B. Wallmark, *Scand. J. Gastroenterol.*, **21**, 268–272 (1986).

50. T. Berglindh, H. Helander, and K.-J. Öbrink, *Acta Physiol. Scand.*, **97**, 401–414 (1976).

51. A. W. Herling, M. Becht, W. Kelker, M. Ljungström, and M. Bickel, *Agents Actions*, **20**, 35–39 (1987).

52. A. Brändström, P. Lindberg, and U. Junggren, *Scand. J Gastroenterol.*, **20**(Suppl. 108), 15–22 (1985).

53. E. Fellenius, B. Elander, B. Wallmark, H. F. Helander, and T. Berglindh, *Am. J. Physiol.*, **243**, G505–G510 (1982).

54. K. Kubo, K. Oda, T. Kaneko, H. Satho, and A. Nohara, *Chem. Pharm. Bull.*, **38**, 2853–2858 (1990).

55. B. Kohl, E. Sturm, J. Senn-Bilfinger, W. A. Simon, U. Krüger, H. Schaefer, G. Rainer, V. Figala, and K. Klemm, *J. Med. Chem.*, **35**, 1049–1057 (1992).

56. E-3180, *Drugs of the Future*, **16**, 19–22 (1991).

57. S. Yamada, T. Goto, E. Shimanuki, and S. Narita, *Chem. Pharm. Bull.*, **42**, 718–720 (1994).

58. W. T. Stolle, J. C. Sih, and R. S. P. Hsi, *J. Labelled Comp. Radiopharmaceuticals*, **25**, 891–900 (1988).

59. J. C. Sih, W. B. Im, A. Robert, D. R. Graber, and D. P. Blakeman, *J. Med. Chem.*, **34**, 1049–1062 (1991).

60. K. Sigrist-Nelson, R. K. M. Müller, and A. E. Fischl, *FEBS Letters*, **197**, 187–191 (1986).

61. R. J. Ife, C. A. Dyke, D. J. Keeling, E. Meenan, M. L. Meeson, M. E. Parsons, C. A. Price, C. J. Theobald, and A. H. Underwood, *J. Med. Chem.*, **32**, 1970–1977 (1989).

62. K. Weidmann, A. W. Herling, H.-J. Lang, K. H. Scheunemann, R. Rippel, H. Nimmesgern, T. Scholl, M. Bickel, and H. Metzger, *J. Med. Chem.*, **35**, 438–450 (1992).

63. A. W. Herling, M. Bickel, K. Weidmann, M. Rösner, H. Metzger, R. Rippel, H. Nimmesgern and K.-H. Scheunemann, *Pharmacology*, **36**, 289–297 (1988).

64. Eur. pat. Appl. EP-A-0 201 094 (Dec. 17, 1986), D. Binder and F. Rovensky (to Chemie Linz).

65. Eur. Pat. Appl. EP-A-0 237 248 (Sept. 16, 1987), J. L. Lamattina and P. A. McCarthy (to Pfizer, Inc.).

66. S. Okabe, Y. Akimoto, S. Yamasaki, and H. Nagai, *Dig. Dis. Sciences*, **33**, 1425–1435 (1988).

67. S. Okabe, K. Shimosako, and H. Harada, *Gastroenterology*, **104**(Suppl.), A163 (1993).

68. W. Bohnenkamp, M. Eltze, K. Heintze, W. Kromer, R. Riedel, and C. Schudt, *Pharmacology*, **34**, 269–278 (1987).

69. G. Rackur, M. Bickel, H. W. Fehlhaber, A. Herling, V. Hitzel, H. J. Lang, M. Rösner, and R. Weyer, *Biochem. Biophys. Res. Comm.*, **128**, 477–484 (1985).

70. P. Lindberg, T. Nordberg, A. Alminger, A. Brändström, and B. Wallmark, *J. Med. Chem.*, **29**, 1327–1329 (1986).

71. P. Lindberg, A Brändström, and B. Wallmark, *Trends Pharmacol. Sci.*, **8**, 399–402 (1987).

72. V. Figala, K. Klemm, B. Kohl, U. Krüger, G. Ranier, H. Schaefer, J. Senn-Bilfinger, and E. Sturm, *J. Chem. Soc., Chem. Comm.*, 125–129 (1986).

73. J. Senn-Bilfinger, U Krüger, E. Sturm, V. Figala, K. Klemm, B. Kohl, G. Rainer, H. Schaefer, T. Blake, D. W. Darkin, R. J. Ife, C. A. Leach, R. C. Mitchell, E. S. Pepper, C. J. Salter, N. J. Viney, G. Huttner, and L. Zsolnai, *J. Org. Chem.*, **52**, 4582–4592 (1987).

74. A. Brändström, P. Lindberg, N. A. Bergman, T. Alminger, K. Ankner, U. Jungern, B. Lamm, P. Nordberg, M. Erickson, I. Grundevik, I. Hagin, K. J. Hoffmann, S. Johansson, S. Larsson, I Löfberg, K. Ohlson, B. Persson, I. Skanberg, and L. Tekenbergs-Hjelte, *Acta Chem. Scand.*, **43**, 536–548 (1989).

75. P. Lorentzon, R. Jackson, B. Wallmark, and G. Sachs, *Biochim. Biophys. Acta*, **897**, 41–51 (1987).

76. D. J. Keeling, C. Fallowfield, and A. H. Underwood, *Biochem. Pharmacol.* **36**, 339–344 (1987).

77. B. Wallmark, A. Brändström, and H. Larsson, *Biochim. Biophys. Acta*, **778**, 549–558 (1984).

78. P. Lorentzon, B. Eklundh, A. Brändström, and B. Wallmark, *Biochim. Biophys. Acta*, **817**, 25–32 (1985).

79. W. B. Im, J. C. Sih, D. P. Blakeman, and J. P. McGrath, *J. Biol. Chem.*, **260**, 4591–4597 (1985).

80. B. Wallmark, P. Lorentzon, and H. Larsson, *Scand. J. Gastroenterol.*, **20**(Suppl. 108), 37–51 (1985).

81. J. M. Shin, M. Besancon, A. Simon, and G. Sachs, *Biochim. Biophys, Acta*, **1148**, 223–233 (1993).

82. A. Philbrant and C. Cederberg, *Scand. J. Gastroenterol.*, **20**(Suppl. 108), 113–120 (1985).

83. H. Nagaya, H. Satoh, K. Kubo, and Y. Maki, *J. Pharmacol. Exp. Ther.*, **248**, 799–805 (1989).

84. H. Nagaya, H. Satoh, and Y. Maki, *J. Pharmacol. Exp. Ther.*, **252**, 1289–1295 (1990).

85. W. A. Simon, D. J. Keeling, S. M. Laing, C. Fallowfield, and A. Tylor, *Biochem. Pharmacol.*, **39**, 1799–1806 (1990).

86. M. Morii, H. Takata, H. Fujisaki, and N. Takeguchi, *Biochem. Pharmacol.*, **39**, 661–667 (1990).

87. W. Beil, U. Staar, and K.-F. Sewing, *Eur J. Pharmacol.*, **218**, 265–271 (1992).

88. S. Okabe, Y. Akimoto, H. Miyake, and J. Imada, *Jpn. J. Pharmacol.*, **44**, 7–14 (1987).

89. G. W. Adelstein, C. H. Yen, R. A. Haack, S. Yu, G. Gullikson, D. V. Price, C. Anglin, D. L. Decktor, H. Tsai, and R. H. Keith, *J. Med. Chem.*, **31**, 1215–1220 (1988).

90. A. W. Herling, T. Scholl, M. Bickel, H.-J. Lang, K.-H. Scheunemann, K. Weidmann, and R. Rippel, *Pharmacology*, **43**, 293–303 (1991).

91. H. Nagaya, N. Inatomi, A. Nohara, and H. Satoh, *Biochem. Pharmacol.*, **42**, 1875–1878 (1991).

92. K. Sigrist-Nelson, A. Krasso, R. K. M. Müller, and A. E. Fischl, *Eur. J. Biochem.*, **166**, 453–459 (1987).

93. European Patent Application EP-A-0 234 485 (Sept. 2, 1987), H. J. Lang, A. W. Herling, and R. Rippel (to Hoechst).

94. A. Brändström, P. Lindberg, U. Junggren, and B. Wallmark, *Scand. J. Gastroenterol.*, **21**(Suppl. 118), 54–56 (1986).

95. S. J. Konturek, N. Cieszkowski, N. Kwiecien, J. Konturek, J. Tasler, and J. Bilski, *Gastroenterology*, **86,** 71–77 (1984).

96. H. Larsson, H. Mattsson, G. Sundell, and E. Carlsson, *Scand. J. Gastroenterol.*, **20**(Suppl. 108), 23–35 (1985).

97. H. Satoh, N. Inatomi, H. Nagaya, I. Inada, A. Nohara, N. Nakamura, and Y. Maki, *J. Pharmacol. Exp. Ther.*, **248,** 806–815 (1989).

98. H. Nagaya, N. Inatomi, and H. Satoh, *Jpn. J. Pharmacol.* **55,** 425–436 (1991).

99. H. Shibata, H. Fujisaki, K. Oketani, M. Fujimoto, T. Wakabayashi, and I. Yamatsu, *Jpn. J. Pharmacol.*, **49,** Abstr. 0–188 (1989).

100. S. Okabe, E. Higaki, T. Higuchi, M. Sato, and K. Hara, *Jpn. J. Pharmacol.*, **40,** 239–249 (1986).

101. M. Masuda, A. Uchida, H. Matsukura, and T. Kamishiro, *Folia. Pharmacol. Jpn.*, **104,** 325–335 (1994).

102. T. Yamakawa, H. Matsukura, Y. Nomura, M. Yoshioka, M. Masaki, H. Igata, and S. Okabe, *Chem. Pharm. Bull.*, **39,** 1746–1752 (1991).

103. W. B. Im, D. P. Blakeman, and J. P. Davis, *Biochem. Biophys. Res. Comm.*, **126,** 78–82 (1985

104. E. Carlsson, H. Larsson, H. Mattson, B. Ryberg, and G. Sundell, *Scand J. Gastroenterol.*, **21**(Suppl. 118), 31–38 (1986).

105. A. W. Herling, M. Becht, H.-J. Lang, K.-H. Scheunemann, K. Weidmann, T. Scholl, and R. Rippel, *Biochem. Pharmacol.*, **40,** 1809–1814 (1990).

106. Y. Tomiyama, M. Morii, and N. Takeguchi, *Biochem. Pharmacol.*, **48,** 2049–2055 (1994).

107. H. Fujisaki, H. Shibata, K. Oketani, M. Murakami, M. Fujimoto, T. Wakabayashi, I. Yamatsu, M. Yamaguchi, H. Sakai, and N. Takeguchi, *Biochem. Pharmacol.*, **42,** 321–328 (1991).

108. J. De Graef and M.-C Woussen-Colle, *Gastroenterology*, **91,** 333–337 (1986).

109. S. Postius., U. Bräuer, and W. Kromer, *Life Sciences*, **49,** 1047–1052 (1991).

110. S. J. Konturek, T. Brozowski, and T. Radecki, *Digestion*, **27,** 159–164 (1983).

111. H. Mattsson, K. Andersson, and H. Larsson, *Eur. J. Pharmacol.*, **91,** 111–114 (1983).

112. W. Kromer, S. Gönne, R. Riedel, and S. Postius, *Pharmacology*, **41,** 333–337 (1990).

113. S. Ohara, H. Ohkawa, K. Ishihara, K. Hotta, Y. Komuro, and H. Okiabe, *Scand. J. Gastroenterol*, **162**(Suppl.), 187–189 (1989).

114. T. Ischikawa, K. Ishihara, K. Saigenji, and K. Hotta, *Eur. J. Pharmacol.*, **251,** 107–111 (1994).

115. D. McTavish, M. M.-T. Buckley, and R. Heel, *Drugs*, **42,** 138–170 (1991).

116. M. I. Wilde and D. McTavish, *Drugs*, **48,** 91–132 (1994).

117. L. B. Barradell, D. Faulds, and D. McTavish, *Drugs*, **44,** 225–250 (1992).

118. T. Lind, C. Cederberg, G. Ekenved, U. Haglund, and L. Olbe, *Gut*, **24,** 270–276 (1983).

119. T. Lind, C. Cederberg, G. Ekenved, and L. Olbe, *Scand. J. Gastroenterol.*, **19,** 1004–1010 (1986).

120. W. Londong, V. Londong, C. Cederberg, and H. Steffen, *Gastroenterology*, **24,** 270–276 (1983).

121. S. J. Konturek, N. Kweicien, W. Obtutowicz, B. Kopp, and J. Olesky, *Gut*, **25,** 14–18 (1984).

122. J. Naesdal, G. Bodemar, and A. Walan, *Gastroenterology*, **19,** 916–922 (1984).

123. J. N. Thompson, J. A. Barr, N. Collier, J. Spencer, A. Bush, L. Cope, R. J. N. Gribble, and J. H. Baron, *Gut*, **26,** 1018–1024 (1985).

124. R. P. Walt, M. Gomes, E. C. Wood, L. H. Logan, and R. E. Pounder, *British Medical J.*, **287,** 12–14 (1983).

125. D. C. Metz, D. B. Strader, M. Orbuch, P. D. Koviack, K. M. Feigenbaum, and R. T. Jensen, *Aliment. Pharmacol. Ther.*, **7,** 597–610 (1993).

126. P. Müller, H. G. Dammann, U. Leucht, and B. Simon, *Aliment. Pharmacol. Therap.* **3,** 193–198 (1989).

127. M. Hongo, S. Ohara, Y. Hirasawa, S. Abe, S. Asaki, and T. Toyata, *Dig. Dis. Sci.*, **37,** 882–890 (1992).

128. S. W. Sanders, K. G. Tolman, P. A. Greski, D. E. Jennings, P. A. Hoyos, and J. G. Page, *Aliment. Pharmacol. Ther.*, **6,** 359–372 (1992).

129. S. Burley des Varannes, P. Levy, S. Lartigue, F. Dellatolas, and M. Lemaire, *Aliment. Pharmacol. Ther.*, **8,** 309–314 (1994).

130. M. Inoue, S. Kimura, Y. Horikawa, M. Ikeshoji, N. Matsumoto, W. Harada, C. Tao, M. Yameskaki, N. Katayama, H. Komatsu, T. Suenaga, T. Shirakawa, G. Kajiyama, T. Ishida, T. Shimatani, and R. Miyatake, *Jpn, Arch. Intern. Med.*, **41,** 143–150 (1994).

131. A. Hannan, J. Weil, C. Broom, and R. P. Walt, *Aliment. Pharmacol. Ther.*, **6,** 373–380 (1992).

132. K. E. L. McColl, A. M. E. Nujumi, C. A. Dorrian, and A. M. I. Macdonald, *Scand. J. Gastroenterol.*, **27,** 93–98 (1992).

133. W. Londong, *Aliment. Pharmacol. Ther.*, **8**(Suppl.), 39–46 (1994).

134. G. Acton, C. Broom, and K. Manchee, *Br. J. Clin. Pharmacol.*, **29,** 609P–610P (1990).

135. A. L. Blum, *Digestion*, **47**(Suppl. 1), 3–10 (1990).

136. S. C. Misra, S. Dasarathy, and M. P. Sharma, *Aliment. Pharmacol. Ther.*, **7,** 443–449 (1993).

137. D. Y. Graham, J. Colon-Pagan, R. S. Morse, T. L. Johnson, J. H. Walsh, A. J. McCullough, J. W.

Marks, M. Sklar, R. C. Stone, A. J. Cagliola, L. Walton-Bowen, T. J. Humphries, and the omeprazole duodenal ulcer study group, *Gastroenterology*, **102**, 1289–1294 (1992).

138. H. P. Festen, *Scand. J. Gastroenterol.*, **201**, 39–41 (1994).

139. J. Dent, N. D. Yeomans, M. Mackinnon, W. Reed, F. M. Narielvala, D. J. Hetzel, E. Solcia, and D. J. Shearman, *Gut*, **35**, 590–598 (1994).

140. E. J. Kuipers, E. C. Klingenberg-Knol, H. P. Festen, C. B. Lamers, J. B. Jansen, F. Nelis, and S. G. Meuwissen, *Scand. J. Gastroenterol.*, **28**, 978–980 (1993).

141. J. Weil, G. D. Bell, K. Powell, S. Morden, G. Harrison, P. W. Gant, P. H. Jones, and J. E. Trowell, *Aliment. Pharmacol. Ther.*, **5**, 309–313 (1991).

142. M. A. Daw, P. Deegan, E. Leen, and C. O-Morain, *Aliment. Pharmacol. Ther.*, **5**, 435–439 (1991).

143. P. Sherman, B. Shames, V. Loo, A. Matlow, B. Drumm, and J. Penner, *Scand. J. Gastroenterol.* **27**, 1018–1022 (1992).

144. T. Iwahi, H. Satoh, M. Nakao, T. Iwasaki, T. Yamazaki, K. Kubo, T. Tamura, and A. Imada, *Antimicrob. Agents Chemother.*, **35**, 490–496 (1991).

145. K. Nagata, H. Satoh, T. Iwahi, T. Shimoyama, and T. Tamura, *Antimicrob. Agents Chemother.*, **37**, 769–774 (1993).

146. A. T. Axon, *Scand. J. Gastroenterol.*, **201**(Suppl.), 16–23, (1994).

147. J. Labenz, E. Gyenes, G. H. Ruhl, and G. Borsch, *Am. J. Gastroenterol.*, **88**, 491–495 (1993).

148. J. Labenz, G. H. Ruhl, J. Bertrams, and G. Borsch, *Am. J. Gastroenterol.*, **89**, 726–730 (1994).

149. E. Bayersdorffer, G. A. Mannes, A. Somer, W. Hochter, J. Weingart, R. Hatz, N. Lehn, G. Ruckdeschel, P. Dirschedl, and M. Stolte, *Scand. J. Gastroenterol.*, **196**(Suppl.), 19–25 (1993).

150. C. J. McCarthy, R. Collins, S. Beattie, H. Hamilton, and C. O-Morain, *Aliment. Pharmacol. Ther.*, **7**, 463–466 (1993).

151. H. Lamouliatte, *Clin. Ther.*, **15**(Suppl.), 32–36 (1993).

152. H. F. Helander, C. H. Ramsey, and C. G. Regardh, *Scand. J. Gastroenterol.*, **20**(Suppl. 108), 95–104 (1985).

153. M. A. Pue, J. Laroche, I. Meineke, and C. de Mey, *Eur. J. Clin. Pharmacol.*, **44**, 575–578 (1993).

154. S. Yasuda, A. Ohnishi, T. Ogawa, Y. Tomono, J. Hasegawa, H. Nakai, Y. Shimamura, and N. Morishita, *Int. J. Clin. Pharmacol. Ther.*, **32**, 466–473 (1994).

155. C. G. Regardh, M. Gabrielsson, K. J. Hoffman, I. Löfberg, and I. Skanberg, *Scand. J. Gastroenterol.*, **20**(Suppl. 108), 79–94 (1985).

156. M. Tateno and N. Nakamura, *Rinsho Iyaku*, **7**, 51–62 (1991).

157. R. Curi-Pedrosa, M. Daujat, L. Pichard, J. C. Ourlin, P. Clair, L. Gervot, P. Lesca, J. Domergue, H. Joyeux, G. Fourtanier, and P. Maurel, *J. Pharmacol. Exp. Ther.*, **269**, 384–392 (1994).

158. R. Gugler and J. C. Jensen, *Gastroenterology*, **89**, 1235–1241 (1985).

159. T. Sutfin, K. Balmer, N. Bostrom, S. Eriksson, P. Hoglund, and O. Paulsen, *Ther. Drug. Monit.*, **11**, 176–184 (1989).

160. T. J. Humphries, *Dig. Dis. Sciences*, **36**, 1665–1669 (1991).

161. W. A. Simon, C. Büdingen, S. Fahr, B. Kinder, and M. Koske, *Biochem. Pharmacol.*, **42**, 347–355 (1991).

162. G. Hanauer, U. Graf, and T. Meissner, *Meth. Find. Exp. Clin. Pharmacol.*, **13**, 63–67 (1991).

163. W. Kromer, S. Postius, R. Riedel, W. A. Simon, G. Hanauer, U. Brand, S. Gönne, and M. E. Parsons, *J. Pharmacol. Exp. Ther.*, **254**, 129–135 (1990).

164. H. U. Schulz, M. Hartmann, V. W. Steinijans, R. Huber, B. Lührmann, H. Bliesath, and W. Wurst, *Int. J. Clin. Pharmacol. Ther. Toxicol.*, **29**, 369–375 (1991).

165. G. T. Tucker, *Aliment. Pharmacol. Ther.*, **8**, 33–38 (1994).

166. R. Arnold, *Aliment. Pharmacol. Ther.*, **8**(Suppl. 1), 65–70 (1994).

167. L. Ekman, E. Hansson, N. Havu, E. Carlsson, and C. Lundberg, *Scand. J. Gastroenterol.*, **20**(Suppl. 108), 53–69 (1985).

168. R. Hakanson, J. Oscarson, and F. Sundler, *Scand. J. Gastroenterol.*, **118**(Suppl.), 18–30 (1986).

169. R. S. Hirth, L. D. Evans, R. A. Buroker, and F. B. Oleson, *Toxicol. Pathol.*, **16**, 273–287 (1988).

170. R. T. Brittain, D. Jack, J. J. Reeves, and R. Stables, *Br. J. Pharmacol.*, **85**, 843–847 (1985).

171. D. Poynter, C. R. Pick, R. A. Harcourt, S. A. M. Selway, G. Ainge, I. W. Harman, N. W. Spurling, P. A. Fluck, and J. L. Cook, *Gut*, **26**, 1284–1295 (1985).

172. G. R. Betton, C. S. Dormer, T. Wells, P. Pert, C. A. Price, and P. Buckley, *Toxicol. Pathol.*, **16**, 288–298 (1988).

173. R. Hakanson and F. Sundler, *Eur. J. Clin. Invest.*, **20**, 65–71 (1990).

174. K. G. Wormsley, *Gut*, **25**, 1416–1423 (1984).

175. J. B. Belder, *Gut*, **26**, 1279–1283 (1985).

176. J. Penston and K. G. Wormsley, *Gut*, **28**, 488–505 (1987).

177. H. Larsson, R. Hakanson, H. Mattson, B. Ryberg, F. Sundler, and E. Carlsson, *Toxicol. Pathol.*, **16**, 267–272 (1988).

178. W. Creutzfeldt, F. Stöckmann, J. M. Conlon, U. R. Fölsch, G. Bonatz, and M. Wülfrath, *Digestion*, **35**(Suppl. 1), 84–97 (1986).

179. H. Larsson, E. Carlsson, H. Mattsson, L. Lundell, F. Sundler, G. Sundell, B. Wallmark, T. Watanabe, and R. Hakanson, *Gastroenterology*, **90**, 391–399 (1986).

180. F. Sundler, E. Carlsson, R. Hakanson, H. Larsson, and H. Mattsson, *Scand. J. Gastroenterol.*, **118**(Suppl.), 39–45 (1986).

181. F. Sundler, R. Hakanson, E. Carlsson, H. Larsson, and H. Mattsson, *Digestion*, **35**, 56–69 (1986).

182. B. Ryberg, H. Mattsson, H. Larsson, and E. Carlsson, *Scand. J. Gastroenterol.*, **24**, 287–292 (1989).

183. H. Lee, R. Hakanson, A. Karlsson, H. Mattson, and F. Sundler, *Digestion*, **51**, 125–132 (1992).

184. R. Arnold, H. Koop, H. Schwarting, K. Tuch, and B. Willemer, *Scand. J. Gastroenterol.* **21**(Suppl. 125), 14–19 (1986).

185. H. Koop, W. Spill, H. Schwarting, and R. Arnold, *Z. Gastroenterologie*, **26**, 643–647 (1988).

186. L. Lundell, A. E. Bishop, S. R. Bloom, K. Carlsson, H. Mattsson, J. M. Polak, and B. Ryberg, *Regulatory Peptides*, **23**, 77–87 (1988).

187. H. Larsson, E. Carlsson, R. Hakanson, H. Mattsson, G. Nilsson, R. Seensalu, B. Wallmark, and F. Sundler, *Gastroenterology*, **95**, 1477–1486 (1988).

188. J. Axelson, R. Hakanson, and F. Sundler, *Scand. J. Gastroenterol.*, **27**, 243–248 (1992).

189. S. Lanzon-Miller, R. E. Pounder, M. R. Hamilton, S. Ball, N. A. F. Chronos, R. Raymond, M. Olausson, and C. Cederberg, *Aliment. Pharmacol. Ther.*, **1**, 239–251 (1987).

190. H. Koop, C. Koch-Naumann, and R. Arnold, *Z. Gastroenterol.*, **28**, 603–605 (1990).

191. J. L. Meijer, J. B. Jansen, I. Biemond, I. J. Kuijpers, and C. B. Lamers, *Aliment. Pharmacol. Ther.*, **8**, 221–227 (1994).

192. A. Iwasaki, K. Miyazawa, Y. Kawamura, Y. Sakai, Y. Tashiro, K. Ito, Y. Arakawa, and Y. Matsu, *J. Gastroenterol.*, **29**, 398–402 (1994).

193. R. Lamberts, W. Creutzfeldt, F. Stöckmann, U. Jacubaschke, S. Maas, and G. Brunner, *Digestion*, **39**, 126–135 (1988).

194. G. Brunner, W. Creutzfeldt, U. Harke, and R. Lamberts, *Digestion*, **39**, 80–90 (1988).

195. W. Creutzfeldt, and R. Lamberts, *Scand. J. Gastroenterol.*, **26**(Suppl. 180), 179–191 (1991).

196. W. Creutzfeldt, *Digestion*, **39**, 61–79 (1988).

197. W. Creutzfeldt, *Drug. Saf.*, **10**, 66–82 (1994).

198. A. W. Herling, K. Weidmann, H. P Neubauer, and F. E. Beyhl, *Naunyn-Schmiedeberg–s Arch. Pharmacol.*, **347**, R99 (1993).

199. H. G. Dammann, A. von zur Mühlen, H. J. Balks, A. Damaschke, J. Steinhoff, U. Henning, J. A. Schwarz, and W. Fuchs, *Aliment. Pharmacol. Ther.*, **2**, 191–196 (1993).

200. J. J. Kaminski, J. A. Bristol, C. Puchalski, R. G. Lovey, A. J. Ellitott, H. Guzik, D. M. Solomon, D. J. Conn, M. S. Domalski, S. C. Wong, E. H. Göold, J. F. Long, P. J. S. Chiu, M. Steinberg, and A. T. McPhail, *J. Med. Chem.*, **28**, 876–892 (1985).

201. P. J. S. Chiu, C. Casciano, G. Tetzloff, J. F. Long, and A. Barnett, *J. Pharmacol. Exp. Ther.*, **226**, 121–125 (1983).

202. J. F. Long, P. J. S. Chiu, M. J. Derelanko, and M. J. Steinberg, *Pharmacol. Exp. Ther.*, **226**, 114–120 (1983).

203. J. J. Kaminski, J. M. Hilbert, B.N. Pramanik, D. M. Solomon, D. J. Conn, R. K. Rizvi, A. J. Elliott, H. Guzik, R. G. Lovey, M. S. Domalski, S. C. Wong, C. Puchalski, E. H. Gold, J. F. Long, P. J. S. Chiu, and A. T. McPhail, *J. Med. Chem.*, **30**, 2031–2046 (1987).

204. J. J. Kaminski, D. G. Perkins, J. D Frantz, D. M. Solomon, A. J. Elliott, P. J. S. Chiu, and J. F. Long, *J. Med. Chem.*, **30**, 2047–2051 (1987).

205. P. J. S Chiu, A. Barnett, C. Gerhard, M. Policelli, and J. Kaminski, *Arch. Int. Pharmacodyn.*, **270**, 128–140 (1984).

206. P. J. S. Chiu, A. Barnett, G. Tetzloff, and J. Kaminski, *Arch. Int. Pharmacodyn.*, 270, 116–127 (1984).

207. A. Barnett, P. J. S. Chiu, and G. Tetzloff, *Br. J. Pharmacol.*, **83**, 75–82 (1984).

208. T. H. Brown, R. J. Ife, D. J. Keeling, S. M. Laing, C. A. Leach, M. E. Parsons, C. A. Price, D. R. Reavill, and K. J. Wiggall, *J. Med. Chem.*, **33**, 527–533 (1990).

209. C. A. Leach, T. H. Brown, R. J. Ife, D. J. Keeling, S. M. Laing, M. E. Parsons, C. A. Price, and K. J. Wiggall, *J. Med. Chem.*, **35**, 1845–1852 (1992).

210. R. J. Ife, T. H. Brown, D. J. Keeling, C. A. Leach, M. L. Meeson, M. E. Parsons, D. R. Reavill, C. J. Theobald, and K. J. Wiggall, *J. Med. Chem.*, **35**, 3413–3422 (1992).

211. C. Broom, S. Eagle, S. Steel, M. Pue, and J.

Laroche, *Gastroenterology*, **104**(Suppl.), A46 (1993).

212. C. K. Scott, and E. Sundell, *Eur. J. Pharmacol.*, **112**, 268–270 (1985).

213. W. Beil, U. Staar, and K. F. Sewing, *Eur. J. Pharmacol.*, **139**, 349–352 (1987).

214. W. Beil, I. Hackbarth, and K. F. Sewing, *Br. J. Pharmacol.*, **88**, 19–23 (1986).

215. D. J. Keeling, S. M. Laing, and J. Senn-Bilfinger, *Biochem. Pharmacol.*, **37**, 2231–2236 (1988).

216. B. Wallmark, C. Briving, J. Fryklund, K. Munson, R. Jackson, J. Mendlein, E. Rabon, and G. Sachs, *J. Biol. Chem.*, **262**, 2077–2084 (1987).

217. W. Beil, U. Staar, P. Schünemann, and K. F. Sewing, *Biochem. Pharmacol.*, **37**, 4487–4493 (1988).

218. C. Briving, B.M. Andersson, P. Nordberg, and B. Wallmark, *Biochim. Biophys, Acta*, **946**, 185–192 (1988).

219. C. K. Scott, E. Sundell, and L. Castrovilly, *Biochem. Pharmacol.*, **36**, 97–104 (1987).

220. D. J. Keeling, R. C. Malcolm, S. M. Laing, R. J. Ife, and C. A. Leach, *Biochem. Pharmacol.*, **42**, 123–130 (1991).

221. M. D. Ene, T. Khan-Daneshmend, and C. J. C. Roberts, *Br. J. Pharmacol.*, **76**, 389–391 (1982).

222. M. E. Parsons, R. J. Ife, C. A. Leach, and C. Broom, *Gut*, **33**(Suppl. 2), 31–38 (1992).

223. S. Eagle, C. Gill, M. Chapelsky, J. Swagzdis, M. Pue, and C. Broom, *Br. J. Clin. Pharmacol.*, **36**, 175P–176P (1993).

224. R. E. Caldara, C. Masci, M. Barbieri, V. Sorghi, V. Piepoli, and A. Tittobello, *Eur. J. Clin. Pharmacol.*, **28**, 677–679 (1985).

225. M. Bouclier and M. Speeding, *Agents Actions*, **16**, 491–495 (1985).

226. C. W. Ogle, C. H. Cho, M. C. Tong, and M. W. L. Koo, *Eur. J. Pharmacol.*, **112**, 399–404 (1985).

227. E. Aadland and A. Berstad, *Scand. J. Gastroenterol.*, **18**, 969–971 (1983).

228. R. A. Levine, S. P. Petokas, A. Starr, and R. H. Eich, *Clin. Pharmacol. Ther.*, **34**, 399–402 (1983).

229. J. Nandi, R. L. King, D. S. Kaplan, and R. A. Levine, *J. Pharmacol. Exp. Ther.*, **252**, 1102–1107 (1990).

230. K. F. Sewing and H. Hannemann, *Pharmacology*, **27**, 9–14 (1983).

231. W. B. Im, D. Blakeman, J. Mendlein, and G. Sachs, *Biochim. Biophys. Acta*, **770**, 65–72 (1984).

232. C. S. Chew, *Biochim. Biophys. Acta*, **846**, 370–378 (1985).

233. W. Beil, R. J. Bersimbaev, H. Hannemann, and K. F. Sewing, *Pharmacology*, **40**, 8–20 (1990).

234. A. W. Herling and M. Ljungström, *Eur J. Pharmacol.*, **156**, 341–350 (1988).

235. W. Beil, P. Stünkel, A. Pieper, and K. F. Sewing, *Aliment. Pharmacol. Therap.*, **1**, 141–151 (1987).

236. D. H. Malinowska, *Am. J. Physiol.*, **259**, G536–G543 (1990).

PART IIA
CARDIOVASCULAR DRUGS

CHAPTER TWENTY-EIGHT

Cardiac Drugs

RICHARD E. THOMAS

Department of Pharmacy
University of Sydney,
Sydney, Australia

CONTENTS

1 Introduction,   155
2 Cardiac Physiology,   156
   2.1 Chemistry of myocardial contractility, 158
   2.2 Excitation and contraction coupling, 160
3 Transmembrane and Intracellular Signaling
   Systems,   165
   3.1 Voltage-dependent ion channels, 165
      3.1.1 Channel proteins, 166
      3.1.2 Channel gates, 167
      3.1.3 Inward voltage-dependent channels, 168
      3.1.4 Outward voltage-dependent
         channels, 170
   3.2 Ligand-medicated membrane
      transduction, 171
      3.2.1 Transduction systems, 171
      3.2.2 Second messenger systems, 173
   3.3 Intracellular Ca^{2+} signaling, 174
      3.3.1 Intracellular Ca^{2+} release, 174
      3.3.2 Propagation of intracellular Ca^{2+}
         signals, 174
      3.3.3 Replenishment of intracellular Ca^{2+}
         stores, 175
      3.3.4 Extrusion of intracellular Ca^{2+}, 175
   3.4 Na^+,K^+-ATPase, 177
      3.4.1 Enzymology, 177
      3.4.2 Sequencing, 180
      3.4.3 Na^+,K^+-ATPase isoforms, 185

Burger's Medicinal Chemistry and Drug Discovery,
Fifth Edition, Volume 2: Therapeutic Agents,
Edited by Manfred E. Wolff.
ISBN 0-471-57557-7   © 1996 John Wiley & Sons, Inc.

3.5 Autonomic regulation, 186
    3.5.1 Adrenergic control, 186
    3.5.2 Cholinergic control, 187
3.6 Endothelial-derived factors in the control of
    cardiovascular function, 188
    3.6.1 Leukocyte adhesion molecules, 188
    3.6.2 Endothelins, 188
    3.6.3 Endocardial endothelium factors, 189
4 Antiarrhythmic Agents, 189
  4.1 Mechanisms of cardiac arrhythmia, 189
    4.2.1 Disorders in the generation of the
      electrical signal, 189
    4.2.2 Disorders in the conduction of the
      electrical signal, 191
  4.2 Classification of antiarrhythmic drugs, 193
  4.3 Mechanism of action of antiarrhythmic
    drugs, 197
    4.3.1 Fast sodium channel blockers (class 1
      drugs), 197
    4.3.2 β-Adrenergic blocking agents (class II
      drugs), 200
    4.3.3 Agents that prolong the refractory
      period (class III drugs), 200
    4.3.4 Agents that inhibit Ca^{2+} channels (class
      IV drugs), 202
  4.4 Individual antiarrhythmic drugs, 202
    4.4.1 Lidocaine (lignocaine), 202
    4.4.2 Sotalol, 203
    4.4.3 Amiodarone, 203
5 Ischemic Heart Disease, 204
  5.1 Pathophysiology, 204
  5.2 Myocardial infarction, 205
  5.3 Secondary prevention of myocardial
    infarction, 206
  5.4 Angina pectoris, 207
6 Organic Nitrates, 208
  6.1 Transduction role of endogenous NO, 208
  6.2 Chemistry of organic nitrovasodilators, 210
  6.3 Bioactivation of nitrovasodilators, 212
  6.4 Nitrate tolerance, 213
  6.5 Pharmacokinetics of organic nitrates, 214
7 Calcium Channel Blockers, 214
  7.1 Pharmacology, 214
  7.2 Receptor sites for Ca^{2+} channel blockers, 217
8 β-Adrenoceptor Antagonists (β-blockers), 218
  8.1 Overview of effects of drugs on
    cardiovascular adrenergic receptors, 218
  8.2 Blockade of cardiac β receptors, 219
  8.3 Newer β-blockers, 221
  8.4 Use of β-blockers in heart failure
    (CHF), 222
9 Congestive Heart Failure, 223
  9.1 Pathophysiology, 223
  9.2 Symptoms and signs of heart failure, 225
  9.3 Treatment of heart failure, 226
10 Inotropes, 228

10.1 Catecholamines, 228
10.2 Dopaminergic agents, 229
10.3 Phosphodiesterase inhibitors, 229
11 Cardiotonic Steroids (Digitalis), 231
11.1 Source and use of cardiotonic steroids in medicine, 231
11.2 Chemistry, 232
11.3 Pharmacology, 235
11.4 Biochemical pharmacology, 237
11.5 Location of the digitalis binding site on Na$^+$,K$^+$-ATPase, 239
11.6 Structure–activity relationships (SAR), 240
11.6.1 Problems in the interpretation of SAR, 240
11.6.2 Guanylhydrazones, 242
11.6.3 Pregnane and related derivatives, 243
11.6.4 Erythrophleum alkaloids, 245
11.6.5 Forces that may bind cardiotonic steroids to the receptor site, 245
11.6.6 Role of the functional groups in the activity of cardiotonic steroids, 248
11.7 Endogenous digitalis-like factors (EDLF), 253

1 INTRODUCTION

Cardiovascular disease is responsible for about 50% of premature deaths in Western industrialized countries. The most common cardiovascular diseases are ischemic heart disease, heart failure, and hypertension; the first of these being responsible for almost 30% of premature deaths. These diseases are interrelated, and their treatment involves the use of many categories of therapeutic agent. This chapter deals in detail only with antiarrhythmic, antianginal drugs, inotropic drugs, and "cardiotonic" drugs, although for the sake of completion, brief reference is made to other cardiac drugs that are discussed in more detail elsewhere in this book.

Increasingly, medicinal chemists are using biotechnology to produce human enzymes and other proteins to act as screening agents to identify lead compounds and to maximize their efficacy and selectivity. However, of the three major groups of drugs discussed in this chapter, the properties of cardiotonic steroids (in the form of squill) were recognized by the ancient Egyptians; the antianginal nitrate of choice was introduced in the nineteenth century, and the different categories of antiarrhythmic drugs were discovered by accident, having been introduced for other purposes such as the treatment of malaria, epilepsy, local anaesthesia, or hypertension. To a considerable extent, medicinal chemistry remains "the science of serendipity", to quote the title of a recent review Stemp (1) who acknowledges the advances made in the use of more rational approaches to lead optimization but states; "However, we should not lose sight of the fact that it is in many cases an idea generated by a medicinal chemist, allied to serendipity, that ultimately leads to successful drug development."

Serendipity is most fruitful when aided by what Pasteur called the "prepared mind," which, in the case of new drug development, means a knowledge of how drugs act at the molecular level, coupled with careful observation of their pharmacological and clinical effects. This chapter provides the basis for preparing the mind for drug development in the categories mentioned above.

2 CARDIAC PHYSIOLOGY

For the heart to pump in an efficient manner there must be a high level of coordination of the contraction of the individual muscle fibers. Each muscle fiber consists of a single cell. Unlike skeletal muscle, which requires neuronal stimulation to contract, heart muscle contracts automatically. The heart *in situ* will continue to beat even though all nerves to the heart have been severed. A heart removed from the body and provided with nutrients will continue to beat, as will isolated sections of heart muscle. In the normal heart, the signal to contract develops in the sinoatrial (SA) node located at the top of the right atrium. This electrical impulse then passes from cell to cell, traveling along the cell membranes of the heart muscle fibers. Heart muscle (myocardium) consists of a variety of myocytes, of which there are three main types: nodal, conducting, and general (or nonspecialized).

Since the atria must contract before the ventricles, the impulse that travels over the atria is prevented from reaching the ventricles by a layer of nonconducting tissue that separates the atria from the ventricles, except at one part called the atrioventricular (AV) node. When the atrial impulse reaches this node, it is slowed so that when it emerges from the node into the ventricles the atria have finished contracting. If this did not happen, the atria and ventricles would push against each other. The conduction rate through the AV node is normally about 0.05 m/s, compared with 1 m/s in nonspecialized myocytes. Once the signal to contract has emerged from the AV node, it is picked up by conducting tissue known as the bundle of His (conduction rate about 1 m/s), which passes it to fast-conducting myocytes known as Purkinje fibers (conduc-

tion rate about 4 m/s), which in turn pass the signal to surrounding nonspecialized myocytes that conduct at a rate of about 1 m/s. It has been suggested that fast-conducting Purkinje-like fibers exist in the atria, but this is controversial. Conduction rates in the SA and AV nodes are increased by sympathetic stimulation and decreased by parasympathetic (vagal) stimulation (discussed in Section 3.5). Once a heart muscle fiber has contracted, it must enter a refractory period when it will not contract or accept a signal to contract. If this did not happen, the signal that originated in the SA node would not die away but just keep traveling around the heart, leading to completely disorganized contraction (fibrillation) in which there is no rest period to enable the heart to fill and hence no pumping action.

In summary, the pump action of the heart involves three principal electrical events: (*1*) the generation of a signal; (*2*) the conduction of the signal; (*3*) the dying away of the signal. Cardiac arrhythmia develop when any of the processes described above are disrupted or impaired.

The function of the heart is to maintain an adequate blood supply to all parts of the body. The blood, as it passes through the chambers of the heart, provides negligible nutrients to the tissues of the heart. The latter receive their blood supply from the coronary arteries that branch from the base of the aorta. The main coronary arteries circle the surface of the heart like a crown (hence the term *coronary*). Branches of the coronary arteries travel over other parts of the surface of the heart, sending further branches into the myocardium. Blockage of sections of the coronary arteries occurs in coronary artery disease (CAD), and this leads to myocardial ischaemia (lack of blood reaching sections of the myocardium). Ischaemic heart disease results in myocardial infarction (heart attack) and angina pectoris.

Muscles contract because of an interaction between the proteins actin and myosin. These proteins, in their various isoforms, are found in all living cells and are responsible for cell motility, intracellular translocations, and mitosis. In muscle cells, actin and myosin interact in a way that causes a shortening of the cell. There are three main types of muscle: skeletal, cardiac, and smooth. Both skeletal and cardiac muscle are striated (stiped). The striations (Fig. 28.1) arise because the regular alignment of actin and myosin filaments

gives rise to differences in refractive index along different parts of the filaments. In smooth muscle, actin and myosin are not arranged in regular arrays, so striations are not present.

Skeletal muscle normally will not contract until stimulated by a neurotransmitter, whereas cardiac and visceral smooth muscle can contract spontaneously. However, neuronal stimulation plays an important part in controlling the rate and intensity of contraction in these muscles. Cardiac and vis-

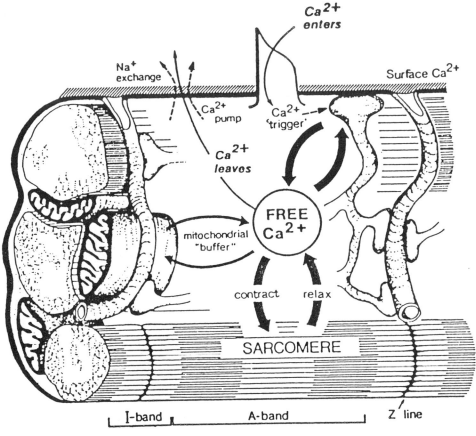

Fig. 28.1 Relationship between excitation and contraction in a heart muscle cell. The passage of the action potential along the cell membrane (sarcolemma) opens the L-type Ca^{2+} channel resulting in the passage of a small amount of extracellular Ca^{2+} into the cell. This triggers the release of a much larger amount of stored calcium located in the cysternae of the sarcoplasmic reticulum (SR). This may be augmented by calcium stores in the mitochondria. Free Ca^{2+} diffuses through the SR until it reaches the myofibrils, where it is transported into the sarcomeres (repeating units that make up the myofibril, see text and Fig. 28.2). Not shown in the diagram is the dense curtain of SR that encases each myofibril. Rising levels of Ca^{2+} within the myofibrils stimulates a pump that transports Ca^{2+} back into the SR, and then to the storage cysternae. Excess Ca^{2+} that entered the cell during membrane depolarization is removed from the cell by the Ca^{2+}/Na^+ exchanger and Ca^{2+} pump (depicted in Fig. 28.9) (2). Courtesy of Raven Press.

ceral smooth muscle also differ from skeletal and multiunit smooth muscle in that the cells are fused and function as a syncytium (a single mass of protoplasm containing multiple nuclei). Cardiac muscle consists of cylindrical cells (some which are branched) that abut very tightly end to end to form fused junctions known as intercalcated discs. This serves two functions: (*1*) when one muscle cell contracts it pulls on those attached to its ends and (*2*) when cardiac cells depolarize, the wave of depolarization travels along the cell membrane until it reaches the intercalated disc where it crosses to the next cell. Thus, when cells of the SA node (the normal pacemaker) depolarize, the wave of contraction spreads over the whole myocardium in such a way that the myocytes contract in a unified and coordinated fashion. Large channels known as gap junctions or connexons pass through the intercalated disc and connect adjacent myocytes. These channels play an important role in transmitting the action potential from one cell to another.

Figure 28.1 shows the principal features of a typical heart muscle cell. Each muscle fiber is a single cell and its cell membrane is called the sarcolemma. Within each cell are hundreds of cylindrical structures known as myofibrils. These consist of even smaller structures known as myofilaments, of which there are two types: thick and thin. The former is made of myosin, and the latter of actin and some other proteins. Individual myofilaments do not extend for the length of the myofibril but are arranged in compartments known as sarcomerers, which are separated from each other by dense material called Z lines. The sarcomere is thus the basic contractile unit (see Fig. 28.2) and comprises a Z line at each end to which are attached the thin filaments. The latter extend into a region known as the A band, which contains an array of thick filaments anchored along the M line. Each filament is surrounded by six filaments of the opposite type. When a muscle contracts, cross-

bridges form between the thick and thin filaments, pulling the latter toward the M line, thus shortening the sarcomere. When this happens extensively within the myocardium, the muscle contracts. Surrounding the sarcomere and permeating the body of the myocyte is the sarcotubular system, consisting of T tubules and sarcoplasmic reticulum (SR). These are shown in Figure 28.1, although they are not specifically labeled. The T tubules are transverse tubules that open onto the extracellular fluid (two of these are depicted in Figure 28.1). The SR tends to run longitudinally and comes in close contact with the cell membrane (sarcolemma), T tubules, and myofibrils. The regions of the SR that make contact with the sarcolemma and T tubules are enlarged to form terminal cisterae (Fig. 28.3), which are the main stores of bound intracellular Ca^{2+}. Depolarization of the sarcolemma of cardiac myocytes leads to small amounts of extracellular Ca^{2+} entering the cell (Fig. 28.1). This triggers the release of large stores of intracellular Ca^{2+}, mainly from the SR cysternae, and this triggers the interaction between actin and myosin. Calcium thus forms the link between membrane depolarization and contraction.

2.1 Chemistry of Myocardial Contractility

The chemistry of contraction in all muscle cells is similar: actin and myosin are normally primed to interact with each other but are prevented from doing so by a protein called tropomyosin (Fig. 28.2). A rise in the level of free Ca^{2+} in the vicinity of the contractile elements overcomes the inhibition of tropomyosin, allowing interaction between the actin and myosin. The reuptake of Ca^{2+} into storage vesicles causes free Ca^{2+} levels to fall, enabling tropomyosin to reexert is inhibitory action. In all muscle cells, depolarization of the cell membrane is the trigger that causes stored

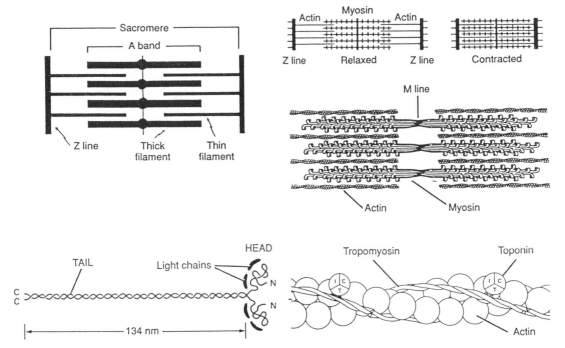

Fig. 28.2 *Top left:* The arrangement of thin (actin) and thick (myosin) filaments in skeletal muscle. *Top right:* The sliding of actin into the myosin matrix during contraction causes the Z lines to move closer together. *Middle right:* Detail of the arrangement of actin and myosin—note that myosin filaments reverse their polarity at the M line in the middle of the sarcomere. *Bottom left:* Detail of the myosin-II molecule, showing the two intertwined heavy chains and four light chains. *Bottom right:* Diagrammatic representation of the arrangement of actin, tropomyosin, and the three subunits of troponin (I, C and T) (3, p. 58). Courtesy of Appleton & Lange.

intracellular Ca^{2+} to be released. However, in cardiac and smooth muscle, an influx of extracellular Ca^{2+} is also required to initiate contraction. This explains why calcium channel blockers affect cardiac and smooth muscle but not skeletal.

As mentioned, actin and myosin are proteins found in all living cells. The myosin of muscle cells is myosin-II, which comprises two heavy chains and four light chains. The light chains and the *N*-terminal portions of the heavy chains combine to give a globular head (Fig. 28.2). Together they form the thick filaments. The thin filaments of skeletal and cardiac muscle comprise three types of protein; globular actin (G actin), tropomyosin, and troponin (which appears to be absent in smooth muscle). The approximate molecular weights of myosin, G-actin, and tropo-

myosin found in striated muscle are 460,000, 43,000, and 70,000, respectively, and for troponin, 18,000–35,000. The globular heads that protrude from the thick filaments contain an ATPase and an actin-binding site. Thin filaments are made of two helical strands of fibrous actin (F-actin), each comprising 300–400 subunits of G-actin. The sites on actin that bind to the myosin heads lie in the helical groove of the double-stranded F-actin but, in the relaxed state, are covered by tropomyosin molecules, 40–60 of which extend along the length of one thin filament (Fig. 28.2). In skeletal and cardiac muscle, a third protein, troponin, is also part of the thin filament. Troponin consists of three subunits, designated I, T, and C. Troponin T binds the trimer to tropomyosin, troponin I (through its influence on tropomyosin) inhibits the

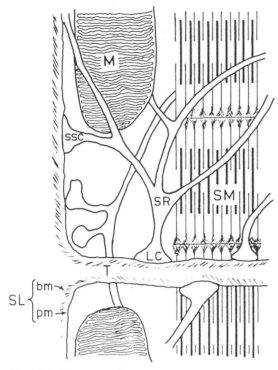

Fig. 28.3 Diagrammatic representation of portion of a myocardial cell (myocyte). M, mitochondrion (myocytes have large numbers of large mitochondria); SM, sarcomere; SR, sarcoplasmic reticulum; LC, longitudinal cisterna of the SR; SSC, subsarcolemmal cisterna of the SR; T, T Tubule; SL, sarcolemma, comprising a plasma membrane (p.m.) and a basement membrane (b.m.).

mere. Once the myosin head has swiveled, it dissociates from the binding site, returns to its former orientation, rebinds, swivels, and then binds again. Each cycle of binding and swiveling shortens the sarcomere by about 1%, and 30 or more cycles may occur before reuptake of Ca^{2+} causes tropomyosin to cover the binding sites, leading to relaxation of the myocyte.

The following experiment provides the model for how the contractile machinery works. If isolated molecules of myosin, or even cleaved myosin heads, are placed in a medium containing ATP and Mg^{2+}, ATP binds to the myosin heads, but little or no hydrolysis to ADP occurs. If actin is then added, hydrolysis proceeds in an uncontrolled fashion until all the ATP is consumed. If troponin and tropomyosin are also added, hydrolysis of ATP is inhibited. If Ca^{2+} is then added in increasing concentrations, the effects of troponin and tropomyosin are progressively overcome. This effect is negligible at $\sim 10^{-7} M\, Ca^{2+}$ and complete at $\sim 10^{-5} M\, Ca^{2+}$, these being the concentrations of free Ca^{2+} believed to be present in the sarcomere during muscle relaxation (diastole) and contraction (systole).

interaction between actin and myosin, and troponin C binds calcium. When the concentration of free Ca^{2+} in the vicinity of the sarcomere rises, Ca^{2+} binds to troponin C. This is believed to change the conformation of troponin T so that binding of the troponin trimer to tropomyosin is changed with the result that troponin I no longer exerts its inhibitory action on tropomyosin. When this occurs, tropomyosin moves laterally, uncovering the myosin-binding sites that lie in the helical groove of F-actin. Myosin heads then bind to these sites, triggering the hydrolysis of ATP, which in turn causes the heads to move, pulling the thin filaments into the matrix of thick filaments, thereby shortening the sarco-

2.2 Excitation and Contraction Coupling

The action potential that originates in the myocardial pacemaker cells (usually those of the SA node) travels along the cell membranes of the myocytes, passing from cell to cell via the intercalated discs and extending into the T tubules. Depolarization of T tubule membranes activates dihydropyridine receptors that open voltage-gated L-type Ca^{2+} channels in the T tubule membrane (the different types of Ca^{2+} channels are discussed in Section 3.1.3.2. In cardiac myocytes, this leads to a small influx of extracellular Ca^{2+} which triggers the release of a much larger amount of stored calcium from the SR cisternae that lie adjacent to the T tubules. Free Ca^{2+} diffuses through the SR system and crosses the SR membrane into the cytoplasm via channels activated by the ryanodine receptor. Since the SR forms a dense curtain around the myofibrils (not shown in Fig. 28.1), the myofibrils and hence the sarcomeres receive a sudden pulse of free Ca^{2+} that binds to troponin C and initiates contraction, as described earlier. The dihydropyridine and ryanodiine receptors are so-named because they are inhibited, respectively, by dihydropyridine and ryanodine. (see Section 3.1.3.2 for further details). Once the levels of free Ca^{2+} in the cytoplasm reach a critical concentration, it activates Ca^{2+},Mg^{2+}-ATPase, which pumps free cytoplasmic Ca^{2+} back into the SR, thus terminating muscle contraction.

The biophysical property that connects excitation and contraction is the electrical potential difference that exists across cell membranes: the intracellular face of the membrane is electrically negative with respect to the extracellular face. This situation is the result of several factors, principally:

1. The intracellular fluid (i), is rich in K^+ and poor in Na^+, and the reverse applies to the extracellular fluid (o). Respective concentrations in mol/L are approximately $[K^+]_i = 150 \times 10^{-3}$; $[K^+]_o = 4.0 \times 10^{-3}$; $[Na^+]_i = 6$–12×10^{-3}; $[Na]_o = 140 \times 10^{-3}$.

2. The membrane is more permeable to K^+ than it is to Na^+.

3. The anions of the intracellular fluid are largely organic and fixed and do not diffuse through the membrane.

4. A process of active transport exists that maintains the steady-state levels of Na^+ and K^+.

As a result of the first three factors, there occurs, at least initially, a net loss of intracellular K^+ that is not balanced electrically by either an efflux of anions or an influx of other cations. As predicted by the Nernst equation, the loss of intracellular K^+ is limited by the development of an inward-directed electrical gradient that balances the outward directed chemical gradient. However, there also exists a slow passive influx of Na^+ along its electrochemical gradient, and this in time would balance the efflux of K^+. This trend toward electrical neutrality is prevented by (1) the existence of a Na^+/K^+ pump (Na^+,K^+-ATPase) that actively transports Na^+ out of the cell in exchange for a smaller ratio of K^+ ions and (2) the presence of fixed anions within the cell. A steady-state situation is, therefore, reached in which the extracellular fluid has a slight excess of cations and the intracellular fluid a slight excess of anions, giving rise to the transmembrane potential difference. These potential differences can be measured by means of microelectrodes inserted on either side of the membrane. When this is done, it has been found that the potential difference across various cell membranes ranges from -20 to -100 mV (the intracellular side being negative). In most cardiac cells, the transmembrane potential difference is about -90 mV.

An appropriate stimulus, electrical or chemical, can depolarize the membrane, presumably by causing conformational changes that open certain selective ion channels in the membrane, enabling certain ions, particularly Na^+, to flow into the cell and reduce the negative charge on the inner surface. In the cells of most tissues, this effect if localized but may be of great importance in providing a stimulus to trigger events within the cell. In the cells of excitable tissue, depolarization can give rise to more than just a localized effect. If the stimulus reduces the transmembrane potential to a threshold value, it produces an action potential that is then transmitted in an all-or-none fashion along the entire membrane. An all-or-none effect means

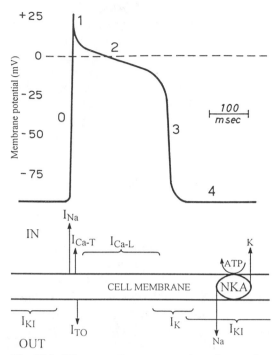

Fig. 28.4 Diagrammatic representation of an action potential of a nonautomatic ventricular cell, showing the principal ion fluxes involved in membrane depolarization and repolarization. The membrane potential in millivolts is given on the vertical axis. This denotes the electrical potential of the inner face of the membrane relative to the outer face. Phase 0 is due mainly to the opening of "fast" Na^+ channel giving rise to the I_{Na} current. As the membrane depolarizes (inner face becomes less negative) a transient (T-type) Ca^{2+} channel opens giving rise to the I_{Ca-T} current that makes a small contribution to phase 0. Usually, depolarization leads to an overshoot with the inner face of the membrane becoming positive. Depolarization triggers the opening of a K^+ channel that gives rise to the transient outward K^+ current (I_{TO}) that is responsible for the partial repolarization that constitutes phase 1 of the action potential (note that there are at least eight K^+ channels in the myocardium, see also Table 28.1). Depolarization also triggers the opening of the L-type Ca^{2+} channel (L for large or long-acting, also known as the slow Ca^{2+} channel), which gives rise to the I_{Ca-L} current that is principally responsible for the phase 2 plateau period of the action potential. During phase 2, another K^+ channel opens slowly, producing the outward (delayed)-rectifying K^+ current (I_K) that is principally responsible for the repolarization of the membrane (phase 3). Phase 4 represents the resting membrane potential (about -90 mV) and is maintained largely by the action of the inward rectifying potassium current (I_{KI}) and the action of the

that the action potential, once triggered, is transmitted in such a way that its size and rate of travel are independent of the initial stimulus (although it may be modified by other factors such as the effects of neurotransmitters). (see Section 3.5). As the action potential travels along the cell membrane, it induces a rise in the levels of free or "activator" Ca^{2+} within the cell. As discussed previously, this initiates the interaction between actin and myosin.

The action potential of a typical nonautomatic ventricular myocyte is shown in Fig. 28.4 and is divided into five phase numbered 0–4. The action potentials of myocytes are of much longer duration (300–500 ms) than those of other exitable tissues and are characterized by a plateau period (phase 2) that is most pronounced in nonspecialized ventricular myocytes (Fig. 28.5). The rapid depolarization and overshoot (phase 0) is due mainly to the opening of the fast Na^+ channel, augmented by Ca^{2+} entering via the transient Ca^{2+} channel (in some cases, it may be further augmented by an outward Cl^- current). The Na^+ channels are blocked by tetrodotoxin and kept open by aconitine and the veratridine alkaloids; this is the basis for the repetitive action of the last two mentioned substances. Following depolarization (which takes about 1 ms), there is a brief initial repolarization (phase 1) due to the closing of the Na^+ channel and a short-lived efflux of K^+ via the transient outward K^+ channel. In nerve cells, depolarization is followed by rapid and complete repolarization due to the opening of another K^+ channel. However, in the heart, repolariza-

Na^+/K^+ pump (Na^+,K^+-ATPase or NKA) that transports Na^+ out of the cell in exchange for an inward movement of K^+. A more detailed discussion of the ion transporting mechanism is given in the text). Under certain circumstances, various other channels make significant contributions to the action potential (see text).

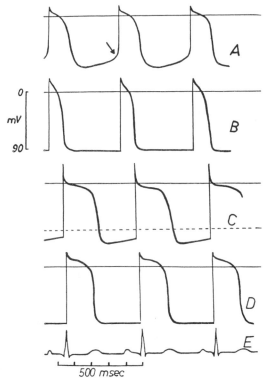

Fig. 28.5 Transmembrane action potentials recorded from cells of (*A*) the sinoatrial node, (*B*) atrial muscle, (*C*) Purkinje fibers, and (*D*) ventricular muscle. On the same time axis is the ECG (*E*). Note the differences in the diastolic (phase 4) depolarization and the level of threshold potential in rows A and C. The delay in AV conduction is indicated by the delay in depolarization (phase 0) between B and C (4). Courtesy of Williams & Wilkins.

tion is delayed by the influx of Ca^{2+} via an L-type Ca^{2+} channel. This gives rise to a plateau period (phase 2). Repolarization (phase 3) is due mainly to the opening of an outward-rectifying K^+ channel and the accompanying closure of the Ca^{2+} channel. At the end of phase 3, the transmembrane potential is restored to its resting value (phase 4), but the intracellular fluid has gained Na^+ and lost K^+. If this situation is not reversed, the prerequisites for excitation (an inward electrochemical gradient for Na^+ and an outward gradient for K^+) are lost. Steady-state levels of Na^+ and K^+ are restored by the expenditure of metabol-

ic energy that drives the Na^+K^+ pump (Fig. 28.4). This pump uses ATP to transport actively Na^+ out of the cell and K^+ into the cell. The pump requires the presence of Mg^{2+}, derives its energy from the hydrolysis of ATP, and is activated by rising concentrations of Na_i^+ and K_o^+. It is, therefore, called Na^+,K^+-ATPase and is specifically inhibited by cardiotonic steroids such as digoxin. The pump transports three Na^+ ions out of the cell in exchange for the inward transport of two K^+ ions. An inward-rectifying K^+ current also contributes to the maintenance of the phase 4 membrane potential. In addition, there are other channels that make significant contributions to the action potential under certain conditions (see Section 3 and Table 28.1).

Normally, the nonspecialized myocytes of the myocardium do not spontaneously depolarize (although they may do so under certain disease conditions). The cells of the nodal tissue and specialized conducting myocytes (e.g., Purkinje fibers) can spontaneously depolarize and generate an action potential that can spread to all cells of the myocardium. Such cells are said to show automaticity and to have pacemaker potential. In the nonspecialized myocytes, the phase 4 (resting potential) remains fairly constant at about -90 mV. This is due to the combined effects of a slow outward leak of K^+ that is greater than the slow inward leak of Na^+ and the action of Na^+,K^+-ATPase that results in a net cation loss from the cell. In cells that show automaticity, the outward leak of K^+ slows after repolarization, whereas Na^+ continues to leak into the cell along its electrochemical gradient. This results in a steady increase in intracellular cations, leading to a fall in negative membrane potential. Phase 4 of such cells is not flat as in Figure 28.4 but drifts upwards (becomes less negative) until it reaches a threshold value that triggers the opening of an L-type Ca^{2+} channel in nodal tissue or the Na^+ channel in conducting tissue (phase 0 in nodal tissue is thus due to

Table 28.1   Major K^+ Currents in the Heart (5)

Potassium Current	Functional Role
Inward (anomalous)-rectifying K^+ current (I_{K1})	Maintains phase 4 resting potential; closes with depolarization; prolongs phase 2
Outward (delayed)-rectifying K^+ current (I_K)	Opens at the end of phase 2 (plateau); initiates repolarization (phase 3)
Transient outward K^+ current (I_{to})	Opens briefly immediately after depolarization; contributes to early repolarization (phase I)
Ca^{2+}-activated K^+ current ($I_{K(Ca)}$)	Activated by high $[Ca^{2+}]_i$; accelerates repolarization in Ca^{2+}-overloaded heart
Na^+ activated K^+ current ($I_{K(Na)}$)	Activated by high $[Na^+]_i$; may promote repolarization in Na^+-overloaded heart
ATP-sensitive K^+ current ($I_{K(ATP)}$)	Normally inhibited by ATP; opens in energy-starved heart
Acetylcholine-activated K^+ current ($I_{K(ATP)}$)	Activated by G_i in response to vagal stimulation and adenosine; hyperpolarizes resting heart cells; slows SA pacemaker cells; shortens atrial action potential
Arachadonic acid-activated K^+ current	Activated by arachadonic and other fatty acids, especially at acid pH

the influx of Ca^{2+} and not Na^+ as in other myocytes). Figure 28.5 shows the action potentials for a selection of myocytes having spontaneous and nonspontaneous depolarizability. Figure 28.6 shows a more detailed depiction of the various action potentials of the myocardium, although the upward drift of phase 4 of automatic cells has not been accurately drawn. The electrical activity of individual myocytes combine to produce an electrical current that can be measured by electrodes placed on the skin. A recording of this activity is called an electrocardiogram (ECG) (Fig. 28.6), and its shape is influenced by where the electrodes are placed on the skin and by abnormalities in the generation and conduction of the various action potentials.

The time it takes for an automatic myocyte to depolarize spontaneously depends on the maximum value of the membrane potential and the slope of phase 4. Normally, the cells of the SA node depolarize before other potential pacemaker cells because the maximum value of the trans-

membrane potential is only about $-60\,mV$ and the upward slope of phase 4 is steep. The SA node is thus normally the pacemaker for the rest of the myocardium. If, for any reason, the signal from the SA node is slowed or blocked or if the process of repolarization is accelerated in other automatic cells, non-SA cells may initiate a wave of depolarization that may replace that from the SA node or act in conflict with it. Heartbeats that originate from non-SA pacemaker activity are called ectopic beats.

The means whereby the signal is propagated may be described in simple terms as follows. When any particular segment of membrane is depolarized, the inner face of that membrane becomes positive with respect to adjoining regions of the *inner* face. Negative charges then flow from the adjoining region into the depolarized region, causing the transmembrane potential of the adjoining region to fall to its threshold value, thereby triggering an action potential. Thus an action potential, once gener-

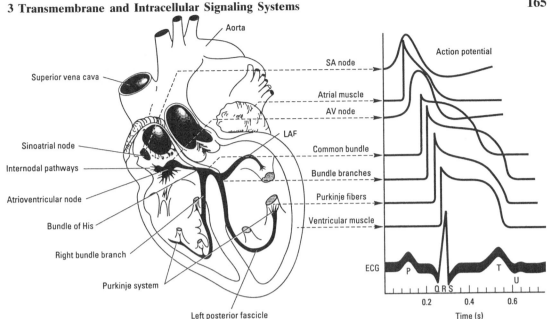

Fig. 28.6 Action potentials and the conducting system of the heart. Shown are typical transmembrane action potentials of the SA and AV nodes, specialized conducting myocytes, and nonspecialized myocytes. Also shown is the ECG, plotted on the same time scale (3, p. 499). Courtesy of Appleton & Lange.

ated, leads to depolarization of an adjoining segment of membrane, and this, in turn, leads to depolarization of a further membrane segment, and so on, in domino fashion, until the wave of depolarization has spread throughout the entire myocardium, passing from cell to cell via the nexus of the intercalated disc. The rate at which the action potential is propagated depends on many factors, such as the capacitance of the membrane (influenced greatly by the number of T tubules per unit of membrane surface), the speed and magnitude of the depolarization current (phase 0), and the diameter of the fiber.

The spontaneous discharge rate of automatic cells depends on (*1*) the slope of phase 4; (*2*) the magnitude of the maximum diastolic (resting) potential; and (*3*) the magnitude of the threshold potential. Changes in all these values can occur in disease states or as a consequence of drug action. Adrenergic drugs increase heart rate by increasing the slope of phase 4 of pacemaker cells. Cholinergic drugs slow the heart by decreasing the slope of phase 4. It follows that drugs such as atropine that block the parasympathetic nervous system increase heart rate, whereas drugs that block the sympathetic nervous system slow the heart.

3 TRANSMEMBRANE AND INTRACELLULAR SIGNALING SYSTEMS

3.1 Voltage-dependent Ion Channels

The molecular biology of transmembrane ion transport is a new and rapidly developing field, any description of which is in need of constant update. However, the following points seem to be well established (5–8).

1. Ion channels are constructed from a family of closely related glycoproteins that span the cell membrane and may have evolved from a common ancestor.
2. Ion channels are ion-selective, i.e., their

structure is such that the passage of only one type of ion is facilitated, although there are some channels where smaller proportions of other ions may co-travel with the principal ion.

3. In most cases, ions move passively through water-filled ion channels, diffusing from regions of high or low concentration. This passive movement is only possible if the channels are "open."

4. The opening of ion channels involves conformational changes in the channel proteins, a process known as gating. At least two possible gates are involved: designated h and m gates in the case of the Na^+ channel.

5. Gating mechanisms give rise to at least three possible states: (1) closed but able to open rapidly ("resting state"), (2) open; and (3) closed but not able to open (i.e., refractory).

6. Stimuli that cause channel opening include (1) changes in membrane potential (voltage gating); (2) direct or indirect chemical stimuli (ligand gating), and (3) mechanical deformation.

7. Ion channels are classified according to the (1) location of the channel (e.g., in sarcolemma, SR, mitochondria), (2) type and direction of ion transport, (3) type of gating stimuli (e.g., voltage-gated, ligand-gated), and (4) kinetics of channel opening and closure (e.g., fast, slow).

8. Channel closure may require a separate stimulus from that which opens the channel, or it may occur spontaneously, even when the stimulus to open is maintained.

9. Channel closure may be influenced by membrane potential (voltage dependence) or the time elapsed after a change in membrane potential (time dependence).

10. Ion channels contain receptors for regulatory substances that modulate ion channel activity.

11. Some channels conduct more effectively in one direction than the other, regardless of the size of the transmembrane potential. Such channels are said to be rectified.

3.1.1 CHANNEL PROTEINS. Ion channels are formed by an association of transmembrane proteins known as channel subunits, designated α, β, γ, etc. The α or major subunit of the Na^+, Ca^{2+}, and K^+ channels (Fig. 28.7) is a glycosylated protein that spans the cell membrane and appears to have evolved from a common ancestor (7,9). For all three channels, the subunit

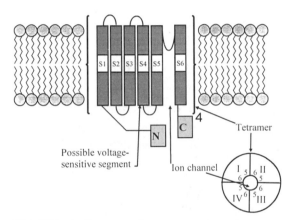

Fig. 28.7 A diagrammatic representation of the α subunit protein of Na^+, K^+, and Ca^{2+} channels. Although the α subunit contains the ion pore, other protein subunits are usually needed for activity (see text). The α subunit of all three channels is made of a tetramer, with each member of the tetramer consisting of six transmembrane segments (S1–S6). In the case of the Na^+ and Ca^{2+} channels, the four members of the tetramer are covalently linked, whereas the α subunit of the K^+ channel consists of an association of four non–covalently linked domains. The α subunit of the Ca^2 channel is called the α_1 subunit to distinguish it from a smaller α_2 subunit that also forms part of the Ca^{2+} channel protein complex, see Fig. 28.9). It is believed that the tetramer is arranged so that the S5 and S6 segments of each member of the tetramer make up the wall of the channel as indicated by the rosette-like structure in the lower portion of the figure. Each unit of the tetramer is designated as a "domain", and the four domains are identified by Roman numerals I–IV (see text for a definition of the term *domain*).

exists as a tetramer (indicated by the subscript 4 in the upper portion of Figure 28.7 and the rosette in the lower portion). In the case of the Na^+ and Ca^{2+} channels, each member of the tetramer is linked by covalent bonds, whereas the usual state for the K^+ channel is a non-covalently bound association of the four members of the tetramer. The protein structure shown enclosed in brackets in Figure 28.7 is a single domain that appears to consist of six transmembrane segments (designated S1 to S6) plus a seventh segment, shown as a loop within the ion channel. Intracellular and extracellular loops connect the transmembrane segments. The term *domain* is used to describe a string of amino acids that appears as a single compact portion of a protein. In the case above, it has been applied to the whole of one of the tetramer units, but it could be applied to portions of the unit such as the individual transmembrane segments. In this chapter, the four domains that make up the tetramer are referred to as domains I to IV. Some recently described K^+ channels have been reported to contain domains with only two transmembrane segments (10,11).

The major ion-channel protein (Fig. 28.7) consists of four domains (either separate associations as in the case of the K^+ channel or covalently bound units as in the case of the Na^+ and Ca^{2+} channels). It is believed that the four domains associate in such a way that the channel regions (between the S5 and S6 segments) of each of the four domains associate to line the actual channel (shown in the lower portion of Fig. 28.7). The S4 segment of each domain contains a sequence of positively charged arginine and lysine residues that are believed to respond to changes in membrane potential, causing a conformational change in the protein that contributes to the opening of the ion channel (12). As discussed below, the S4 unit probably constitutes the m gate. The h gate that inactivates Na^+ channels may consist of the

cytoplasmic polypeptide chain that links the S6 segment of domain III to the S1 segment of domain IV (13,14). Smaller protein subunits designated β, γ, and δ are also present in many ion channels. The role of these units is not understood: they are present in some channels but not others and may play a regulatory role in some subclasses of channel and/or contribute to the positioning and conformation of the α subunit within the membrane. Functional Na^+ channels have been identified that contain zero, one, or two β subunits (4,7). Recombinant DNA technology has been used to determine the amino acid sequence of a number of ion channel proteins (4). All α subunits so studied have been shown to display considerable homology (same amino acid sequence) especially in the intramembranous segments. Marked homology exists between channels for different ions and for the same ion channel in different tissues, which explains why many channel-affecting drugs show low selectivity (see Table 28.3). Determination of the significant differences in channel proteins, especially in the gating regions, could provide the basis for the design of drugs with increased selectivity.

3.1.2 CHANNEL GATES. The term *gating* refers to the process whereby external stimuli cause conformational changes in membrane proteins leading to the opening or closing of ion channels. The term does not necessarily imply the movement of a gatelike flap of polypeptide across the channel pathway, although something like this has been suggested for certain K^+ channels where the internal N-terminal sequence of the channel domain (N in Fig. 28.7) may occlude the inner entrance to the pore (15). Sodium channels have long been thought to contain at least two gates, known as m and h gates, both of which must be open for ions to pass through the channel (16). According to this model, channel gating cycles through three states:

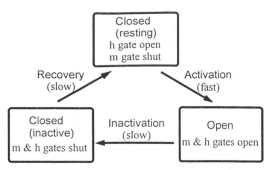

Fig. 28.8 Simplified representation of the gating mechanism in voltage-activated Na^+, K^+, and Ca^{2+} channels. The model posits three states (closed resting, open, and closed inactive) and two gates (h and m) both of which must be open for ions to pass through the channel. Recent work has shown that there are a number of substates, at least within the closed forms. The model depicted in the figure is thus a "lumped" representation with substates lumped within the states depicted. Nevertheless, it depicts what are generally considered to be the essential features of gating, namely a closed "resting" state that is capable of rapidly opening in response to changes in membrane potential followed by a refractory period in which the channel slowly returns to the resting state.

(1) closed resting; (2) open; and (3) closed inactive (Fig. 28.8). In the closed resting state, the h gate is open and the m gate shut. Membrane depolarization switches the m gate (believed to be the S4 segment, see Fig. 28.7) to the "open" position (channel activation). This is a rapid process and enables the fast passage of ions through the channel.

Depolarization also initiates the inactivation process so that the channel moves automatically from the open to the closed inactive state, although this proceeds more slowly than the activation process. In the closed inactive state, both m and h gates are shut, and the channel will not respond to a further repolarization until it has moved back to the closed resting state in which the h gate is again open but the m gate shut. This process is known as channel recovery. As detailed in Section 3.1.1, once both gates are open, passage of ions through the water-filled channel is a passive process with ions moving along their electrochemical gradients. Recent work has shown that the gating process is more complex than described here with various substates existing within each of the three main states (17,18). The number of substates may vary for different channels of the same type (17). Even so, the gating model shown in Figure 28.8 remains a useful basis for describing drug action.

3.1.3 INWARD VOLTAGE-DEPENDENT CHANNELS.

3.1.3.1 Sodium Channels. As stated earlier, ion channels consist of protein subunits designated α, β, γ, and δ. Usually, several subunits are present, but the α subunit is the main subunit and sometimes, it seems, the only unit. The α subunit of voltage-dependent Na^+ channels consists of about 2000 amino acids, subdivided into four covalently linked domains (Fig. 28.7). These subunits contain binding sites for tetrodotoxin, lidocaine, and other Class I antiarrhythmic drugs (discussed later). There is evidence that the α subunits of voltage-dependent Na^+ channels in the heart are associated with an additional subunit, although they can function alone (19). There is also evidence that the α subunits of cardiac Na^+ channels exist in at least two isoforms (8). Activation of the "fast" Na^+ channel in cardiac cells produces the rapid influx of Na^+ (the I_{Na} current) that depolarizes the membrane (phase 0) in all cardiac myocytes, except nodal tissue where voltage-activated Na^+ channels are absent or few in number (20). A voltage-independent Na^+ channel may exist in SA nodal cells, where it provides a slow leak of Na^+ into these cells during diastole. Initially, the current generated by this channel (I_{Na-B}) would offset by the outward leak of K^+, but since the latter declines with time, the inward diastolic passage of Na^+ would contribute to the upward shift of phase 4 of SA cells (Figs.

28.5 and 28.6) and thus to spontaneous depolarization.

Sodium ions may also enter myocytes by two nonselective cationic channels, about which little is known, including their structures. One of these channels generates the I_f current that occurs in nodal and His-Purkinje cells and is gated by high membrane potentials (21). The other is the I_{NS} current that is carried by Na^+ and gated by $[Ca^{2+}]_i$ during Ca_i^{2+} overload conditions.

3.1.3.2 Calcium Channels. There are at least four types of voltage-dependent Ca^{2+} channels designated L (large, or long, in conductance), T (transient in duration of opening), P (Purkinje), and N (neuronal) (22–24). T-type Ca^{2+} channels are activated at relatively negative membrane potentials (low voltage) and mediate a small transient Ca^{2+} current (I_{Ca-T}), which contributes to depolarization. N- and P-type channels are activated at higher voltages than T-type (about $-40\,mV$); have a conductance intermediate between T and L; and are found mainly in neurones, where they are involved in neurotransmitter release. L-type channels are activated above $-40\,mV$; have the largest conductance of all Ca^{2+} channels; and are the principal Ca^{2+} channel in skeletal, cardiac, and smooth muscle. They are activated slowly and are sometimes called slow calcium channels. They differ from the other Ca^{2+} channels in being blocked by calcium-channel blocking drugs such as verapamil and nifedipine (discussed later). L-type channels are activated by partial depolarization of the cell membrane and inactivated by full depolarization and by rising $[Ca^{2+}]_i$. Inactivation occurs relatively slowly so that a significant Ca^{2+} current (I_{Ca-L}) flows once the channel is activated. In conducting and nonspecialized myocardial cells, I_{Ca-L} contributes significantly to the plateau region (phase 2) of the action potential. In nodal tissues, where fast sodium channels are absent or sparse, I_{Ca-T} and I_{Ca-L} are responsible for depolar-

ization (phase 0); the former also contributes to the latter stages of phase 4. In nodal tissues, these two Ca^{2+} currents are activated in sequence either by spontaneous depolarization during phase 4 (the normal situation in cells of the SA node) or by the arrival of a conducted action potential (the normal situation in the AV node). Once the membrane has been depolarized and $[Ca^{2+}]_i$ rises, the channels are deactivated.

So far, most study of Ca^{2+} channels has involved the L-type channel, which is obtained readily from skeletal muscle. L-type Ca^{2+} channels are made of a complex of five protein subunits, comprising an α_1 subunit ($M_r \sim 212\,kDa$), an α_2 subunit ($M_r \sim 143\,kDa$), a β subunit ($M_r \sim 55\,kDa$), a γ subunit ($M_r \sim 30\,kDa$), and a δ subunit ($M_r \sim 27\,kDa$) (25). The arrangement of these subunits is depicted in Figure 28.9). The α_2 and γ subunits are joined by a disulfide bridge and are considered as one unit by some authors. The α_1 subunit is the principal functional component of the complex and contains the actual Ca^{2+} channel. Like the α subunit of Na^+ channels, it is a tetramer, each unit (or domain) comprizing six transmembrane sections as depicted in Figure 28.7. These four domains form a rosette with the channel in the center, the walls of the channel consisting of S5 and S6 sections. The α_1, γ, and δ subunits are hydrophobic and are believed to be membrane-spanning or embedded. The α_2 and β subunits are hydrophilic and may not be located within the membrane. In the depiction given in Figure 28.9, the α_2 unit is also shown located within the membrane. This could occur if its association with the α_1 and γ subunits stabilized such an arrangement. It has been suggested that the α_2, γ, and δ subunits are important in positioning the α_1 subunit in the membrane, and the β subunit, which is probably located on the cytoplasmic side of the membrane, may have a role in gating (27). Numerous isoforms of the L-type Ca^{2+} channel exist,

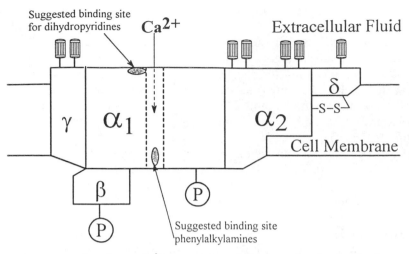

Fig. 28.9 Suggested structure of an L-type Ca^{2+} channel from skeletal muscle, showing the five protein subunits that comprise the channel. The pore is associated with the α_1 subunit, which has the general structure depicted in Figure 28.7. Phosphorylation sites are indicated by P. Also shown are suggested binding sites for the phenylalkylamine and dihydropyridine calcium channel blockers. The former are believed to bind to a protein loop within the pore near the cytoplasmic entrance. The latter probably bind between domains III and IV of the α_1 subunit (the arrangement of domains around the pore is shown in Fig. 28.7). This binding site is approached from the extracellular side of the channel (26). Courtesy of *Trend. Pharmacol. Sci.*

which may explain variations in the action potentials of various tissues.

The α and α_1 subunits of Na^+ and Ca^{2+} channels are phosphorylated by cyclic AMP–dependent protein kinases that mediate the effects of sympathetic stimulation on the channels. This adds an additional element of gating control.

3.1.4 OUTWARD VOLTAGE-DEPENDENT CHANNELS.

3.1.4.1 Potassium Channels. The major potassium channel subunit consists of four domains (Fig. 28.7), but unlike the α and α_1 subunits of Na^+ and Ca^{2+} channels, the domains are not covalently linked. At least 10 genes code for the K^+ channel domains, which means that hundreds of combinations of four-domain channels can be constructed. This is reflected by the multiplicity of K^+ channels found in cardiac and other tissues (28,29). A selection of such channels is shown in Table 28.1. Many K^+ channels are classified as rectifying, which means (1)

that they allow transport in one direction and (2) that their ability to pass current varies with membrane potential.

The inward-rectifying K^+ current (usually designated I_{K1}) allows K^+ to pass outward from the cell during phase 4 but is closed by depolarization, whereas the outward-rectifying K^+ current (usually designated I_K) is opened by depolarization. The former makes a major contribution to the value of the resting (phase 4) membrane potential as well as prolonging phase 2(30). The I_K current is activated slowly during the plateau period and is the major outward current contributing to repolarization (phase 3). Since it is activated slowly, it is often called the delayed rectifier (31). It turns off slowly once the membrane has repolarized and hence contributes to the initial stage of the phase 4 membrane potential. There may be a family of I_K currents with at least two members I_{Kr} and I_{Ks}; the former is small and rapid in activation (hence *r*) and the second is larger and slower in activation (hence *s*) (32). This distinction may be relevant to the action of certain type III antiarrhythmic drugs (see Section 4.3.3). The transient outward K^+ current (I_{to}) is activated by depolarization and contributes to the initial (phase 1) repolarization. The $I_{K(Ca)}$ current is activated by increased $[Ca^{2+}]_i$ and contributes to phase 3 repolarization; it is particularly important if there is $[Ca^{2+}]_i$ overload. In a similar fashion, the Na^+-activated K^+ channel contributes to repolarization, being especially important in $[Na^+]_i$ overload. The ATP-sensitive K^+ channel is normally closed but becomes active if the levels of ATP in the heart fall, as may occur in myocardial ischaemia (33). When activated, these channels give rise to the $I_{K(ATP)}$ current, which increases the rate of onset of repolarization, thus shortening the phase 2 plateau period and hence the inward flow of Ca^{2+}. This has the effect of reducing myocardial contracility and thus conserving energy for basic cell survival processes. This is protective in myocardial ischemia but may be a problem if there is significant pump failure, for example, following extensive infarction.

The parasympathetic nervous system, acting through the vagus nerves, has long been known to slow the heart due to its effects on nodal tissue. Vagal stimulation causes release of acetylcholine that binds to muscarinic (M_2) receptors and activates inhibitory G proteins (G_i) that stimulate the opening of the acetylcholine-activated K^+ channel, giving rise to the $I_{K(ACh)}$ current (34). This increases (makes more negative) the resting (phase 4) membrane potential in SA cells, which thus take longer to depolarize to depolarize and hence to generate a signal, thereby slowing the heart. Adenosine, binding to purine receptors also activates this channel. The arachadonic, or fatty acid–activated K^+ channel, opens when the concentration of fatty acids (including arachadonic acid) rise, as occurs in myocardial ischemia. The resulting fall in intracellular pH activates the channel,

which protects the heart in the manner described for the ATP-sensitive K^+ channel.

Blockers of voltage-gated K^+ channels include certain peptide toxins (charybdotoxin, dendrotoxin, and mast-cell deganulating peptide) and small charged organic molecules such as tetraethylammonium (TEA), 4-aminopyridine, and quinine. All of these compounds act extracellularly, but TEA may also act intracellularly (18). The binding of TEA has been extensively studied, and it would appear that the molecule binds to the peptide loop (known as the H5 region) that connects segments 5 and 6 in the channel protein (18).

Of considerably therapeutic interest are K^+ channel openers that act on the ATP-sensitive K^+ current ($I_{K(ATP)}$) (35). As discussed earlier, opening of these channels can be cardioprotective in myocardial ischaemia, and this has been observed in laboratory animal studies with a number of K^+ channel–opening drugs, including pinacidil (36), nicorandil (37), cromakalim (38), and aprikalim (39). The drugs appear to increase the sensitivity of thc channel to ATP (40). The antiischemic effects of nicorandil have also been observed clinically. Drugs that open these channels also relax smooth muscle and may have a use in increasing coronary blood flow in angina.

3.1.4.2 Chloride Channels.

A small outward Cl^- current (I_{Cl}) contributes to the action potential. Although not normally significant, the current can be increased by adrenergic stimulation (41). Under some circumstances, this channel can also generate an inward current.

3.2 Ligand-mediated Membrane Transduction

3.2.1 TRANSDUCTION SYSTEMS. As with voltage gating, most ligand-mediated membrane transductions involve a relatively small number of molecular mechanisms with a limited number of protein families that have evolved to transduce many extracellular signals. Two such mechanisms are particularly relevant to heart muscle contractility: ligand-gated channels and G-protein signaling systems.

Ligand-gated channels, like voltage-gated channels, comprise a number of protein subunits that traverse the cell membrane and form a channel that enables the passage of certain ions. For example, the nicotinic acetylcholine receptor is a pentamer comprising two α, one β, one γ, and one δ polypeptide subunit, each of which crosses the membrane four times. When the ligand binds to the α subunit, a water-filled channel opens and Na^+ enters the cell and initiates some process.

G proteins are a large family of GTP-binding proteins that transduce and amplify signals across cell membranes. Each G protein is a trimer consisting of α, β, and γ subunits. When activated by a ligand-receptor interaction, the β and γ subunits dissociate from the α subunit, which retains the bound GTP and is the active species. In the heart, G proteins are involved in dromotropy (conduction velocity), inotropy (force of muscle contraction), lusitropy (muscle relaxation), chrontropy (rate of contraction), and cardiac growth. Alterations in G proteins have also been implicated in heart failure. A comprehensive review of G proteins and the heart is available (42).

G-protein signaling systems comprise three distinct functional units located within or associated with the cell membrane: (1) a ligand-binding protein with a receptor facing the extracellular fluid; (2) a G protein located on the inner face of the cell membrane; and (3) an effector element. The binding of an appropriate ligand to the receptor induces a conformational change in the receptor protein that is passed to the G protein, which sheds GDP and binds

GTP. The activated G protein (G-GTP) changes the activity of the effector element. The latter can be an ion channel or an enzyme, and its activity can be stimulated or inhibited by G-GTP. Typically, enzymes when stimulated produce second messengers that relay the signal to some specialized mechanism within the cell. There are two main second messenger families: cyclic AMP (cAMP) and the Ca^{2+}–phosphoinositide system. A third second messenger system involving cGMP is important in some cells. The ability of this relatively limited signaling system to produce a vast array of responses throughout the body is explained by the fact that cells are usually programed to perform one specialized function (apart from their own general maintenance activities). Thus a cell when stimulated will, for example, contract or secrete a hormone or generate an electrical current. This specialized function is associated with specialized biochemistry. For example, activation of β adrenergic receptors of smooth muscle results in phosphorylation of a specific kinase (myosin light chain kinase) that relaxes smooth muscle.

An analogy would be a series of kitchen appliances, each programmed to do one specialized function, for example, boil water, mix food, or emit microwaves. The one external signal (or first messenger), namely the finger, when applied to the one type of receptor, namely the powerpoint switch, generates the same second messenger, namely the electric current running through the cord. However, the effector agent (kettle, microwave oven, etc) does different things when so stimulated, because it is programmed, or built, to do different things. In a similar way, a relatively few receptor types and even fewer signal transducing systems initiate the millions of specialized functions performed by different cells of the body.

Because of evolutionary changes, receptors and signal-transducing systems have developed into families that often show small but important differences (often seen between similar systems in different tissues), thus providing a further basis for ligand selectivity. Most receptors coupled to G proteins are structurally related and contain seven transmembrane segments (or domains). Because of this, they are said to belong to the serpentine family of receptors and are believed to have evolved from a common ancestor.

Once a ligand has bound to a receptor and induced the binding of GTP to the G protein, the effects of G-GTP persist for as long as this complex remains stable, usually for about 10 s. A ligand that remains in contact with the receptor for a few milliseconds can thus stimulate an effect that lasts for many seconds. The role of the G protein is thus to amplify enormously the consequences of the drug–receptor interaction. The receptor signal terminates when G-GTP is dephosphorylated to G-GDP.

3.2.2 SECOND MESSENGER SYSTEMS.

3.2.2.1 cAMP Systems. Cyclic adenosine monophosphate (cAMP) is produced when a G protein activates adenylyl cyclase (formerly known as adenylate cyclase). When activated, this enzyme converts AMP to cAMP, which usually acts by stimulating cAMP-dependent kinases. These enzymes phosphorylate other proteins, which may be located anywhere in the cell (e.g., myosin light chain kinase, in the example quoted earlier, acts on the light chains of the thick filaments, see Fig. 28.2). The specificity of the cell's response to cAMP production arises from its specialized function and associated specialized biochemistry. G proteins that stimulate cAMP production belong to the G_s family because they stimulate adenylyl cyclase. G proteins that inhibit adenylyl cyclase belong to the G_i family (*i* for inhibition). This is somewhat confusing since members of the G_i family stimulate production of the Ca^{2+}–phosphoinositide system in some cells. The effects initiated

by the production of cAMP are terminated by a complex system of enzymes that degrades cAMP and reverses the phosphorylations that it initiated.

3.2.2.2 Ca^{2+}–Phosphoinisitide Systems. The Ca^{2+}–phosphoinositide second messenger system is activated by the binding of a ligand to a receptor on the extracellular face of membrane protein, which is either linked to a G protein or functions as a tyrosine kinase. In either case, the result is activation of the membrane-bound enzyme, phospholipase C, which hydrolyzes another membrane component, phosphatidylinositol-4,5-biphosphate (PIP_2), to produce two second messengers: diacylglycerol (DAG) and inositol-1,4,5-triphosphate (IP_3). DAG acts within the membrane, where it activates protein kinase C. The water-soluble IP_3 diffuses through the cytoplasm, where it releases stored Ca^{2+} from internal vesicles. The released Ca^{2+} binds to the protein calmodulin, which then modulates other cell activities. This system is far more complex than the cAMP messenger system because of the great diversity of its components and functions in different cells and even within the same cell. Specialized Ca^{2+} and calmodulin kinases with limited substrate specificity exist in certain cells together with broad-specificity forms. There are also at least four distinct forms of protein kinase C. The Ca^{2+}–phosphoinositide system is thus receiving much attention as a basis for the design of selectively acting drugs. Multiple mechanisms exist to terminate and reverse the effects of the Ca^{2+}–phosphoinositide signaling system once the external signal is withdrawn.

3.2.2.3 cGMP Systems. The cGMP second messenger system is analogous to the cAMP system, except that binding of the ligand leads to the production of cGMP via stimulation of guanylyl cyclase. The system has been found in only a few tissue types, including vascular smooth muscle. It is important in cardiovascular therapy, since cGPM production relaxes vascular smooth muscle by activating a kinase that dephosphorylates myosin light chains. Ligands that stimulate this reaction in vascular smooth muscle include the atrial natriuretic factor (ANF) and a substance known as the endothelial relaxing factor (EDRF), which is probably nitric oxide. As discussed later, production of NO seems to be the basis for the antianginal action of the organic nitrates. Other ligands are acetylcholine and histamine. As with the other signaling systems, multiple mechanisms exist to terminate and reverse the effects of cGMP production once the receptors are vacated.

3.3 Intracellular Ca^{2+} Signaling

3.3.1 INTRACELLULAR CA^{2+} RELEASE. The release of stored intracellular Ca^{2+} plays a major role in intracellular signaling, especially in tissues such as the myocardium where Ca^{2+} is stored mainly in the longitudinal cysternae of the SR (Fig. 28.3). According to present knowledge, stored Ca^{2+} is released into the cytoplasm by passage through two structurally distinct but similarly functioning receptor-operated channels involving either the ryanodine receptor or the IP_3 receptor (43). Both channels are regulated by endogenous Ca^{2+} and both seem to require one or more cofactors. Activation of the IP_3 receptor-operated channel requires the presence of IP_3, Ca^{2+} and ATP but is inhibited by high Ca^{2+} concentrations (44). Heparin and other polysulfate compounds also inhibit the binding of IP_3 to the receptor and subsequent Ca^{2+} release. IP_3 receptors probably exist in all plant and animal cells. Although IP_3 is required for activation, it is probably Ca^{2+} that provides the signal for opening and closing the IP_3 channel (45).

Ryanodine receptors were thought to exist only in skeletal and cardiac muscle but have now been found in a wide variety of cell types. The receptor is activated by a number of physiological mechanisms, including rising concentrations of Ca^{2+}. Recent evidence indicates that some, if not all, ryanodine receptors require the presence of cADP as a cofactor (46). As discussed in Section 2.2, depolarization of the T tubules in cardiac muscle activates the dihydropyridine receptor in the T tubule membrane, which allows extracellular Ca^{2+} to enter the cell and trigger the release of stored Ca^{2+}. This then diffuses through the SR and activates the ryanodine receptor, which, in turn, opens the channel that allows Ca^{2+} to pass from the SR to the myofibrils. Caffeine and low concentrations of ryanodine also activate the ryanodine receptor (high concentrations of ryanodine inhibit). High concentrations of Ca^{2+} (of the order of $10^{-3} M$) will also inhibit the channel but, at physiological concentrations, the ryanodine receptor channel is Ca^{2+} activated. Caffeine and heparin have been used routinely to study the IP_3 and ryanodine receptors, but recent studies have shown that these agents are not as specific as previously thought and some revision of concepts about the pathways involving these receptors may be necessary (42).

3.3.2 PROPAGATION OF INTRACELLULAR CA^{2+} SIGNALS. Calcium signaling within cells is organized along spatial and temporal lines, but the basis for this is poorly understood. Once a signal (e.g., the binding of noradrenaline to a β receptor) initiates a Ca^{2+} pulse within a cell, it is propagated throughout the cell without diminution by a process of positive feedback. In the case of myocardial cells, the wave must be sustained along the complex system of myofibrils and even from cell to cell via the gap junctions of the intercalated discs. The basis for this undiminished propagation is the stimulatory effect that rising Ca^{2+} levels have on adjacent stores (referred to as Ca^{2+}-induced Ca^{2+} release). Once the stores have been depleted, there is a brief refractory period (at least for IP_3-dependent stores) to allow the store to refill, but how this happens in the presence of IP_3 is still not fully resolved but probably involves inhibition of the channel by the elevated levels of cytoplasmic Ca^{2+} as part of the process.

3.3.3 REPLENISHMENT OF INTRACELLULAR CA^{2+} STORES.

There seems little doubt that depletion of intracellar Ca^{2+} stores induces Ca^{2+} entry into the cell (47), but the nature of the signal between the depleted store and the cell membrane is not resolved (48). There is a growing evidence that depletion of intracellular Ca^{2+} stores directly stimulates an extracellular Ca^{2+} current into the cell that is activated simply by depletion of intracellular Ca^{2+} stores without the requirement of receptor occupancy (49). This current has been named the Ca^{2+}-release activated Ca^{2+} current (I_{CRAC}) (50) and appears to involve an ion channel rather than an ion carrier (51) and to be inhibited by rising levels of $[Ca^{2+}]_i$ (52). The physical connection between the Ca^{2+} stores and the CRAC channel has been the subject of much speculation. The most widely accepted hypothesis is that the depleted stores release a messenger substance that activates the membrane CRAC channel. Recently, such a factor (Ca^{2+}-influx factor, or CIF) has been isolated and shown to have the properties consistent with the putative messenger (53). However, there is still much clarification needed of the CRAC signaling systems, and alternative mechanisms and messenger substances have been proposed (49). It is also likely that multiple mechanisms exist in different tissues for replenishing intracellular Ca^{2+} stores. Calcium stores are also heterogeneous, so that different stores may employ different replenishment pathways.

3.3.4 EXTRUSION OF INTRACELLULAR CA^{2+}.

The rise in $[Ca^{2+}]_i$ that occurs in response to an appropriate signal results mainly from release of Ca^{2+} from intracellular stores

such as the cysternae of the SR, the mitochondria, and various fixed and soluble intracellular proteins. In some cells, this is augmented by passage of $[Ca^{2+}]_o$ via receptor-operated and voltage-operated channels (Fig. 28.10). The uptake of $[Ca^{2+}]_o$ is particularly important in myocardial and smooth muscle cells where the influx of $[Ca^{2+}]_o$ acts as the link between excitation and contraction. As described earlier, actin and myosin, in the presence of ATP and Mg^{2+}, would react in a spontaneous and uncontrolled fashion if this were not prevented, in the resting state, by tropomyosin. Rising levels of $[Ca^{2+}]_i$ overcome this block, thus acting like a switch causing the muscle to contract (Fig. 28.11).

Elevation of $[Ca^{2+}]_i$ normally triggers a counterreaction that rapidly returns $[Ca^{2+}]_i$ levels to resting values. The rapid decline is due mainly to reuptake by intracellular Ca^{2+} stores and binding by Ca^{2+}-buffering proteins. In those cells where $[Ca^{2+}]_o$ entered the cell during stimulation, the excess Ca^{2+} must be extruded.

About 50% of plasma Ca^{2+} is protein bound, giving a free extracellular concentration, $[Ca^{2+}]_o$, of about $1.25 \times 10^{-3} M$. Most intracellular calcium is bound. In myocardial cells, the concentration of free intracellular calcium $[Ca^{2+}]_i$, is about $0.7 \times 10^{-7} M$ (70 nM) during diastole, rising to about $10^{-5} M$ during systole. The ratio $[Ca^{2+}]_o$:$[Ca^{2+}]_i$ during diastole is thus about 18,000:1. The clearing of excess free Ca^{2+} from myocardial cells during diastole thus requires the movement of ions against an enormous chemical gradient.

There are two main routes whereby excess Ca^{2+} is transported out of most cells: (1) the Na^+/Ca^{2+} exchanger (antiport) and (2) the Ca^{2+} pump (Fig. 28.10). During systole, the former is the major mechanism. ATP is not required to energize the Na^+/Ca^{2+} exchanger but is required to phosphorylate the antiport proteins. In the phosphorylated state, the exchanger has high affinity for $[Na^+]_o$ and $[Ca^{2+}]_i$ and is

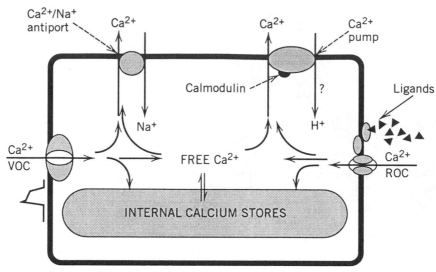

Fig. 28.10 Calcium transporting processes. Ca^{2+} enters cells along its electrochemical gradient through a variety of voltage-operated channels (VOC) and receptor-operated channels (ROC) and leaves cells against its electrochemical gradient by a variety of carrier systems, particularly the Ca^{2+}/Na^+ antiport (exchanger) and the Ca^{2+} pump (Ca^{2+}-ATPase). The Ca^{2+} pump can act at a basal calmodulin–independent level or in an activated calmodulin–dependent fashion. In the myocardium, the principal VOC is the L-type Ca^{2+} channel that generates the I_{Ca-L} current.

thus activated by the rise in $[Ca^{2+}]_i$ that follows depolarization. The stoichiometry of the antiport is disputed but is generally believed to be $3Na^+:1Ca^{2+}$.

The Ca^{2+} pump is the other main route for extruding Ca^{2+} during diastole. Compared with the exchanger, it has a higher affinity for Ca^{2+} but a lower capacity (53). The pump is activated by Ca^{2+} and Mg^{2+} and obtains energy from the hydrolysis of ATP. It is thus a Ca^{2+}-Mg^{2+}-ATPase (54). The Ca^{2+} pump is the main mechanism for extruding Ca^{2+} that leaks into the cell during the nonactivated state (55) (i.e., during diastole in the myocyte). The pump cycles through a series of conformational states associated with the binding of various ligands and the influence of other modulators. The pump, which, as indicated above, is an ATP hydrolyzing enzyme, is believed to span the full width of the plasma membrane. The model currently accepted is shown in Figure 28.12. According to this model, the enzyme (E) has a low

affinity Ca^{2+} binding state, represented by E_2, that exists in equilibrium with a high affinity Ca^{2+} binding state, E_1. In the absence of activators, more than 90% of the enzyme exists in the E_2 state. In the presence of Mg^{2+}, E_2 undergoes a conformation change to E_1, which is stabilized by a Ca^{2+}–calmodulin complex. Other activators include acidic phospholipids such as inositol; certain protein kinases, and some proteases, especially calpain. The E_1 state binds Ca^{2+} with high affinity to one or possibly two sites on its cytoplasmic surface. This promotes the binding of ATP, which, in the presence of Mg^{2+}, phosphorylates the enzyme to produce CaE_1P. In the presence of Mg^{2+}, CaE_1P undergoes a major conformational change to give CaE_2P in which the Ca^{2+} binding site has presumably been made accessible to the extracellular fluid. This site, which now has low Ca^{2+}-binding affinity, sheds Ca^{2+} to give E_2P, which undergoes ATP-stimulated dephosphorylation to give E_2 (57–59). As

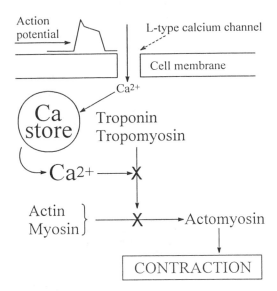

Fig. 28.11 Muscle contracts when two proteins, actin and myosin, combine to form actomyosin. In heart and skeletal muscle, this reaction would occur spontaneously but is prevented from doing so by the influence of two other proteins: troponin and tropomyosin. This block can be overcome by the binding of Ca^{2+} to one of the troponin subunits (troponin C). When the membrane is depolarized, a small amount of Ca^{2+} enters the cell mainly through the L-type Ca^{2+} channel. This acts as a trigger for the release of much larger amounts of free Ca^{2+} from the cell's internal stores. It is this Ca^{2+} that initiates contraction. Ca^{2+} thus provides the link between the action potential (excitation) and contraction.

shown in Figure 28.12, the pump is reversible and can carry Ca^{2+} either way across the cell membrane. Normally, its net effect is to extrude Ca^{2+} with a stoichiometry, according to the accepted model, of one Ca^{2+} ion transported per atom of ATP hydrolyzed. (Note: the ATP in step 5 of Figure 28.12 is not hydrolyzed). Since the pump does not function at its thermodynamic equilibrium, the full potential of the energy released by hydrolysis of ATP is not realized or, in some tissues, may be used in the countercurrent of H^+ (60). Although the stoichiometry of this exchange (if it occurs) is disputed, it is possible that the pump may function in an electrically neutral mode, exchanging one Ca^{2+} for two H^+

(61). In other tissues, including vascular smooth muscle, the pump appears to be electrogenic (55).

Most studies of the Ca^{2+} pump have been conducted using red blood cells, but similar properties to those described above have been reported for the cardiac sarcolemmal Ca^{2+} pump, including the role of calmodulin in activating the pump and the possible countercurrent of H^+ (62). It has been suggested that, for the pump to maintain $[Ca^{2+}]_i$ at about $70\,nM$ during diastole, at least one proton must by countertransported per Ca^{2+}, rising to at least three when the heart is ischemic (62).

Recombinant cDNA technology has been used to deduce the primary structure of the Ca^{2+},Mg^{2+}-ASTPase (i.e., the Ca^{2+} pump) (63–65), and this has revealed important features of this enzyme. It occurs as a number of isoforms and involves at least four genes located on several chromosomes. Additional variability arises from alternative splicing of mRNA transcripts. The enzyme contains 10 hydropathic domains leading to the suggestion that the enzyme makes 10 loops through the membrane (Fig. 28.13). The pump also shows sequence homology with other P-type ATPases, especially in those regions involving the phosphorylation and putative cation translocation sites, but differs in the structure of the regulatory sites. The bulk of the protein mass is located on the intracellular side of the plasma membrane. The mechanism whereby Ca^{2+} is transported by the enzyme across the plasma membrane has yet to be established but four of the transmembrane domains (possibly M_4, M_5, M_6 and M_8 in Fig. 28.13) may be involved in forming the transduction pathway. For further details see the legend to Figure 28.13.

3.4 Na^+,K^+-ATPase

3.4.1 ENZYMOLOGY. As shown in Figure 28.4, depolarization of myocardial cells,

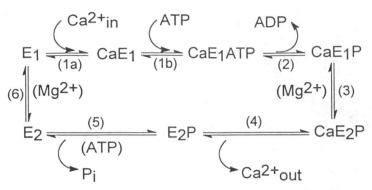

Fig. 28.12 Proposed reaction pathway for the plasma membrane Ca^{2+} pump (Ca^{2+}-ATPase). In the native state, the pump exists in a low Ca^{2+} affinity form (E_2) that can be converted to a high Ca^{2+} affinity state (E_1) by various modulators, including Mg^{2+}, Ca^{2+}, calmodulin, acidic phospholipids, and proteolysis (step 6). When $[Ca^{2+}]_i$ rises, intracellular Ca^{2+} binds to E_1 to give CaE_1 (step 1a). This stimulates the binding of ATP to a high affinity nucleotide binding site (step 1b). This leads to phosphorylation of the enzyme to give CaE_1P (step 2), which, in the presence of Mg^{2+}, undergoes a conformational change to CaE_2P (step 3). This lowers the affinity of the enzyme for Ca^{2+}, which is released into the extracellular fluid (step 4), leaving the enzyme in the E_2P state. In the presence of ATP in the range found in cells, E_2P dephosphorylates to E_2 (step 5), which then reenters the cycle (step 6) (55). Courtesy of *To-day's Life Science*.

other than those of the nodal tissues, results in an influx of Na^+ (phase 0) followed by an efflux of K^+ (phase 3). This, coupled with the diastolic movement of cations along their electrochemical gradients would soon result in the loss of membrane po-

tential and hence of excitability. Many other membrane activities essential for cell life would also be lost. This problem exists in all cells, and a mechanism is needed to restore the high $[K^+:Na^+]_i$ ratio that is essential to maintenance of the transmem-

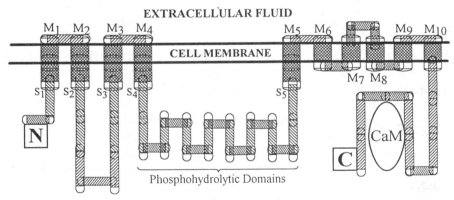

Fig. 28.13 Suggested model for the membrane Ca^{2+} pump (Ca^{2+}-ATPase). It is proposed that there are 10 transmembrane hydropathic domains designated M_1 to M_{10}. In five of these domains, the hydropathic domains extends as a "stalk" into the cytoplasm (regions designated S_1 to S_5). A large intracellular domain between M_4 and M_5 is the phosphohydrolytic domain that catalyzes the conversion of ATP to ADP (step 2 in Fig. 28.12). The similarity between the proposed structures of Ca^{2+}-ATPase and Na^+,K^+-ATPase (Fig. 28.18) is reflected by the considerable sequence homology between the two enzymes. A major difference between the two enzymes is the presence in Ca^{2+}-ATPase of the large C-terminal region that contains the calmodulin binding site that increases the affinity of the enzyme for Ca^{2+} (step 6 in Fig. 28.12).

brane potential difference. This is achieved by the action of the Na^+ pump otherwise known as Na^+,K^+-ATPase. The cells of all higher organisms, and possibly all eukaryocyte cells, depend on the continuous activity of this enzyme that transports $3Na^+$ out of the cell in return for the inward transport of $2K^+$. The importance of this enzyme is demonstrated by the proportion of total body ATP it consumes, estimated to be about 23% in humans. The enzyme has been extensively studied since the late 1950s and is still an active field of research. The enzyme has also generated much interest because it contains a receptor for cardiotonic steroids (digitalis) and may be the receptor for an endogenous digitalis-like substance that could be involved in the pathogenesis of essential hypertension. The milestones that led to our current knowledge of this enzyme have been reviewed extensively (66–69). Recent reviews covering specific aspects of Na^+,K^+-ATPase are available (70–77).

The precise mechanism whereby Na^+,K^+-ATPase actively exchanges $3Na_i^+$ for $2K_o^+$ is unknown. It is known that the enzyme is activated by rising $[Na^+]_i$ and $[K^+]_o$ and requires Mg^{2+} and ATP. The binding of ligands, including phosphorylation, drives the enzyme through a series of major conformational changes that are associated with the transport of Na^+ and K^+ (67). These changes are depicted in Figure 28.14.

Na^+,K^+-ATPase extends across the full width of the plasma membrane and consists of two polypeptide subunits, designated α and β. Both polypeptides and associated phospholipids must be present for phosphohydrolase and pump activity. The α polypeptide has a relative mass (M_r) of about 113 kDa and contains the ATP binding and phosphorylation sites and the binding sites for cations. The β subunit has a M_r of about 55 kDa and is a glycoprotein. Its role has not been established, but it is essential for activity. Its importance may be due to its effect on the conformation and positioning of the α (catalytic) subunit. The enzyme is unique in that it is specifically inhibited by cardiotonic steroids (digitalis) such as ouabain. According to the model shown in Figure 28.14 and originally proposed by Albers (78) and Post and co-workers (79) and elaborated on by others, the enzyme cycles through a series of conformations, of which there are at least four main states: E_1, E_1P, E_2P, and E_2. Recent studies have suggested that there may be at least three phosphorylated states and epicycles within the main cycle (80). The digitalis-binding site is associated with the E_2P state (the digitalis receptor is discussed later). In isolated vesicles, the pump functions as a dimer $[(\alpha, \beta)(\alpha, \beta)]$, although this may not be the case *in situ*. There has been much speculation as to how the dimer (if it exists as such in intact cells) might function. One possibility is a "flip-flop" model in which the two α subunits assume complementary conformations (81,82):

$$E_1\!\!-\!\!E_2 \rightleftharpoons E_2\!\!-\!\!E_1$$

where E is the conformation of the α subunit of Na^+,K^+-ATPase). A simplified representation of the above states is shown in Figure 28.15, which shows only the α subunits of the $[(\alpha, \beta)(\alpha, \beta)]$ dimer. Cooperativity between the members of the dimer has been reported (83), supporting the model shown in Figure 28.15, which posits that when one α subunit of the dimer is in the E_1 state, the other, of necessity, is in the E_2 form. Phosphorylation of the α subunit of Na^+,K^+-ATPase takes place through an asparate residue, as it does with other P-type ATPases. Electron microscopy of purified membrane-bound Na^+,K^+-ATPase indicates that major conformational differences exist between the E_1P and E_2P states (84). This work led to the three-dimensional model for the shape of the dimer depicted in Figure 28.16.

Because its activity involves a phos-

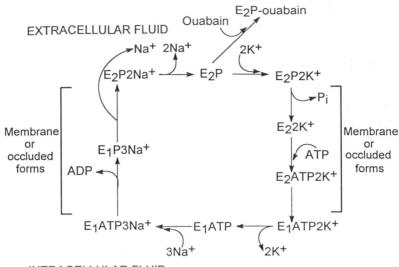

Fig. 28.14 Proposed reaction pathway for the plasma membrane Na^+/K^+ pump (Na^+,K^+-ATPase). The scheme refers to changes in the α subunit of the enzyme and is generally referred to as the Albers–Post model (75). The enzyme is believed to exist in three main states: (a) E_1 states in which the cation binding sites are accessible to the intracellular fluid, (b) E_2 states in which the cation binding sites are accessible to the extracellular fluid, and (c) a series of intermediate states in which the cation binding sites are occluded, either within the membrane or within enclosed globular regions of the protein. The sequence begins with the enzyme in the E_1 state, which in the presence of normal cytoplasmic levels of ATP exists as E_1ATP. This state has high affinity for Na^+ and as $[Na^+]_i$ rises, 3 Na^+ bind to give E_1ATP3Na^+. This promotes a conformational change that moves the bound Na^+ into an intramembrane or occluded state. This triggers phosphorylation of the enzyme, which leads to expulsion of one of the bound Na^+ ions into the extracellular fluid and a major conformational change in the enzyme, giving rise to the E_2P2Na^+ form in which the remaining Na^+ ions are moved out of the occluded state and presented on the extracellular face of the enzyme. These ions are then released, giving the E_2P form of the enzyme. It is this form that can bind digitalis (ouabain in the figure). If digitalis binds, the E_2P state is stabilized and the pump is inhibited (ceases to cycle). In the absence of digitalis, the E_2P state expresses its high affinity for K^+, which is taken from the extracellular fluid. The binding of 2 K^+ leads to dephosphorylation of the enzyme and occlusion of the bound K^+. This change triggers the binding of ATP and conversion of the enzyme to an E_1 state in which the cation binding site is again presented in a nonoccluded form to the intracellular fluid. K^+ is then released into the cytoplasm and the enzyme returns to the E_1ATP state and the beginning of the cycle.

phorylated intermediate, Na^+,K^+-ATPase is classified as a P-type ATPase. Other such ATPases include the Ca^{2+}-ATPases of the SR and plasma membrane and the H^+,K^+-ATPase of the stomach and colon. All P-type ATPases form an aspartyl phosphate during their cation transport cycle. They share other features that suggest that they evolved from a common ancestor. The α subunit of Na^+,K^+-ATPase shares a high degree of sequence homology with other P-type ATPases. Until recently, Na^+,K^+-ATPase was thought to be the only P-type

ATPase with a β subunit, but such units have now been found for H^+,K^+-ATPase and show high homology with the β subunit of Na^+,K^+-ATPase, once again suggesting a common ancestor.

3.4.2 SEQUENCING. Complementary DNA (cDNA) technology has been used to infer the amino acid sequences of the α and β subunits of Na^+,K^+-ATPase from a number of tissues from several species. Such studies involve the synthesis of one or more labeled probes that are then used to iden-

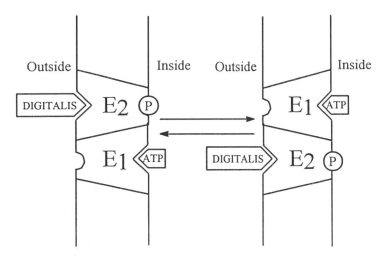

Fig. 28.15 It has been suggested that Na^+,K^+-ATPase functions as a dimer $(\alpha\beta)_2$ in which the α (catalytic) subunits operate in a coordinated cooperative fashion in such a way that when the conformation of one α subunit presents the cation binding sites to the intracellular fluid (E_1 states) the conformation of the other is such that the cation-binding sites are presented to the extracellular fluid (E_2 states). The former have bound ATP and the latter are phosphorylated. Since the ouabain (digitalis) binding site is associated with the E_2P state (Fig. 28.14), only one α subunit of the dimer can bind digitalis at any one time. Normally, a flip-flop oscillation takes place between these two states (see text), but when digitalis binds to the enzyme, this is arrested and the pumping action ceases (81).

tify appropriate colonies from a DNA library prepared using recombinant DNA technology. Once the appropriate DNA molecule has been identified, it is sequenced and the corresponding amino acid sequence deduced. The first step is to use classical biochemical techniques to identify a sequence of about eight amino acids from the protein under study. A DNA probe, labeled with ^{32}P, is then synthesized to correspond with the identified amino acid chain. Ideally, several such probes should be synthesized. A typical sequencing technique based on the work of Lingrel and co-workers (85,86) is shown in Figure 28.17. In this study, mRNA from sheep kidney (a rich source of Na^+,K^+-ATPase) was treated with reverse transcriptase to produce cDNA. This was then incorporated into plasmid DNA and transferred to *E. coli*. The bacteria were allowed to multiply on a plate to produce about 50,000 colonies, which were then screened with the probes to identify colonies containing DNA sequences corresponding to the probes.

These were isolated, and further screening techniques were used to identify a clone containing the gene for the α subunit. The clone was then cultivated, and the relevant DNA isolated and sequenced.

Lingrel and co-workers were the first to determine the amino acid sequences of an α subunit (85) and latter of a β subunit (86). They found that the α subunit from sheep kidney Na^+,K^+-ATPase contained 1016 amino acids ($M_r = 112,177$) and the β subunit from the same source contained 302 amino acids ($M_r = 34,937$). Lingrel's group and other groups have now sequenced the α and β subunits from many tissues and found that there are at least three α isoforms (designated $\alpha_1, \alpha_2, \alpha_3$) and two β isoforms (β_1, β_2) (87,88). The existence of isoforms of Na^+,K^+-ATPase was first detected by classical means by Sweadner (89) in 1979.

The topography of Na^+,K^+-ATPase has attracted much discussion. It is clear from classical studies that the bulk of the enzyme lies on the cytoplasmic side of the cell

Fig. 28.16 Three-dimensional model of the Na$^+$,K$^+$-ATpase dimer based on electron microscopy studies. The upper panel shows the model viewed from above (extracellular side) and depicts a deep cleft between the dimer units. The latter appear connected at the bottom of the cleft. The lower panel shows the model in a direction parallel to the membrane. The vertical line to the right indicates the probable level of the lipid bilayer (84). Courtesy of North Holland Pub.

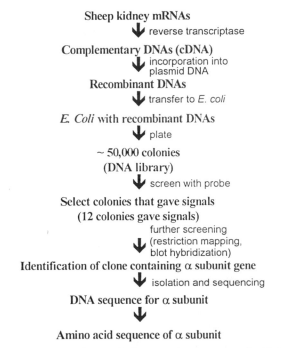

1. PREPARATION OF THE PROBE:

Obtain tryptic fragments of Na⁺,K⁺-ATPase α subunit

Identify amino acid sequence

Prepare DNA probe corresponding to above
(labelled with ^{32}P)

2. PREPARATION OF cDNA LIBRARY AND IDENTIFICATION
OF THE DNA SEQUENCE FOR THE NA⁺,K⁺-ATPASE GENE.

Sheep kidney mRNAs
reverse transcriptase

Complementary DNAs (cDNA)
incorporation into
plasmid DNA

Recombinant DNAs
transfer to E. coli

E. Coli with recombinant DNAs
plate

~ 50,000 colonies
(DNA library)
screen with probe

Select colonies that gave signals
(12 colonies gave signals)
further screening
(restriction mapping,
blot hybridization)

Identification of clone containing α subunit gene
isolation and sequencing

DNA sequence for α subunit

Amino acid sequence of α subunit

Fig. 28.17 Flow chart showing the steps used in a typical sequencing procedure for Na⁺,K⁺-ATPase (68).

membrane. Sequencing studies support this. Hydropathy analyses were originally interpreted as indicating that the α subunit passed through the membrane eight times with little protein protruding on the extracellular surface of the membrane. The β subunit appears to pass through the membrane once, and its bulk lies on the extracellular surface. More recently, doubt has been expressed about the number of transmembrane passes made by the α subunit. The problem is to identify hydropathic (water-heating) sequences of amino acids of sufficient length to traverse the membrane (about 4 nm) and to reconcile the suggested arrangement with other established properties of the enzyme. So far, no model that fits all criteria has gained general acceptance. Sweadner and Arystarkhova (72) suggested six plausible folding models, with the number of transmembrane traverses ranging from 7 to 14. Most of these models have some degree of experimental support. Lingrel and Kuntzweiler (76) have recently proposed a "working model" with 10 transmembrane

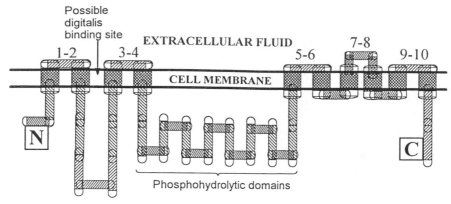

Fig. 28.18 Suggested model for the α subunit of the membrane Na^+/K^+ pump (Na^+,K^+-ATPase). The number of hydropathic transmembrane domains is disputed, but 10 seems a likely figure. These domains are designated 1 to 10, commencing from the C-terminal end. Alternatively, they are designated H_1 to H_{10} (H = hydropathic). The extracellular domains are designated 1–2, 3–4, etc. (or H_1–H_2, H_3–H_4, etc). The location of the digitalis binding site has not been unequivocally established, but there is strong evidence implicating a hydrophobic pocket that may lie between domains 2 and 3. The extracellular domains 1–2 and 3–4 have been strongly implicated as influencing the binding of cardiotonic steroids, but it has not been resolved whether this involvement is direct or indirect. Recent evidence suggests an indirect involvement (see text). The binding of ATP has been associated with a number of amino acids at various locations in the large cytoplasmic globular domain that exists between domains 4 and 5. Phosphorylation involves an aspartate residue located near the C-terminal end of the globular (phosphohydrolytic) domain. The depiction of the α subunit shown in the figure is highly diagrammatic and in reality the various domains are probably clumped together, as depicted in Figure 28.16.

zones (Fig. 28.18). A similar model has been proposed for Ca^{2+}-ATPase (Fig. 28.13).

There is general agreement that the four strongly hydrophobic α-helical sequences in the N-terminal third of the α subunit constitute transmembrane domains, these have been found to be highly conserved in all the ion-transporting ATPases so far examined (76). The middle third of the α unit contains no hydrophobic stretches long enough to form transmembrane domains (at least in the form of α helixes), and it has been assumed that this portion of the molecule forms a highly folded globular cytoplasmic domain. The arrangement shown in Figure 28.18 assumes that there are 10 transmembrane domains, numbered 1–10 from the N-terminal end of the protein. In considering linear depictions of transmembrane proteins such as those given in Figures 28.13 (calcium pump) and 28.18, it should be borne in mind that these are highly

diagrammatic and that the protein loops will aggregate to give a compact structure such as depicted in Figure 28.16.

A representation of the Na^+,K^+-ATPase reaction cycle is shown in Figure 28.14. As indicated, there are two main conformational states: E_1 and E_2. The cycle can be considered as commencing with the E_1ATP state. This state binds $3Na^+$ on the cytoplasmic surface of the α subunit, and this triggers phosphorylation of the enzyme, which then undergoes a conformational change that occludes the sodium, either within a pocket in the cytoplasmic portion of the protein or into an intramembranous site. The enzyme then undergoes a major conformational change that converts it to the E_2 state. There is evidence that this change is triggered by the release of one of the bound sodium ions into the extracellular fluid and that this leads to subsequent shedding of the remaining sodium ions and their replacement by K^+ (90). In between these two steps is a Na^+/K^+ free state (E_2P) that binds cardiotonic steroids such as ouabain, which then block the binding of K^+ and inhibit further cation pump activity.

The binding of K^+ triggers enzyme dephosphorylation, leading to a conformational change that occludes the bound K^+. A molecule of ATP then binds to the enzyme, and this induces a major conformational change that converts the enzyme to the E_1 state, whereby the bound K^+ is carried to the cytoplasmic side of the membrane and released to give the E_1ATP state. The energy needed to drive the cycle is provided by the hydrolysis of ATP that phosphorylates the enzyme following the binding of Na^+.

The sequencing of the α subunit has led to much speculation about the various binding sites depicted in Figure 28.14. The aspartate residue at position 369 in the sheep kidney α subunit (Asp-369) is the known site of phosphorylation. This is located near the beginning of the globular central domain that lies on the cytoplasmic side of the membrane. Before phosphorylation, a molecule of ATP is bound to the enzyme. The amino acids involved in the binding site have been the subject of much investigation. Studies using ATP analogues and fluorescent probes have implicated Cys-367, Cys-466, Lys-480, Lys-501, Gly-502, Asp-710, Asp-714, Lys-719, and Lys-767 (70,76). The implication of some of these amino acids may be artifactual; other amino acids may be important because of their secondary effects on protein conformation and hence on the topography of the actual binding site. Site-directed mutagenesis studies have indicated that Lys-501 and Lys-480 are not crucial for the binding of ATP, whereas Asp-710 and Asp-714 appear to be strongly implicated (91,92). It would appear then, that the central intracellular globular domain that runs from position 342 to position 779 in the sheep α_1 subunit chain contains a binding site for ATP that involves amino acids principally

in the 700 plus region, which, subsequent to the binding of $3Na^+$, phosphorylates the aspartate residue at position 369. The binding sites for Na^+ and K^+ have also been the subject of much study, and like the ATP binding site, the interpretation of results is plagued with the ambiguity of interpreting changes in cation binding affinity that could be due to direct and indirect modifications of the binding site as well as to changes that affect the amino acids involved in occluding and moving the cations. At the time of writing, there is little that can be said with confidence about the location of these sites. The third binding site of interest is that which binds the cardiotonic steroids such as ouabain. This will be discussed in Section 11.5.

3.4.3 NA$^+$,K$^+$-ATPASE ISOFORMS. The discovery by Sweadner (89) in 1979 of isoforms of Na^+,K^+-ATPase in rat brain was later confirmed by other workers for other tissues and species, and has since been the subject of sequencing studies. An excellent review by Levenson (75) in 1994 summarized this work with emphasis on the biological significance of multiple isoforms of Na^+,K^+-ATPase. The following statements are based on the Levenson article, although the author is responsible for any misinterpretations.

The isoforms of Na^+,K^+,-ATPase are the products of a multigene family that shows considerable chromosomal dispersion, suggesting that the isoforms are not redundant but have particular properties that fit particular physiological needs, even though the various isoforms seem to perform the same function in the cell, namely the maintenance of the transmembrane electrochemical gradient. Sequencing studies have shown a high degree of homology between subunit isoforms (for example, 85% for rat α subunits).

Each of the three α and two β subunit genes is expressed at different rates and in different degrees, depending on the type of tissue and the stage of its development. Current evidence indicates that any α subunit can associate with any β subunit, giving rise to six isoforms (and more if additional, unidentified, subunit isoforms exist). cDNA studies have identified genes with nucleotide sequences that could code for α subunit isoforms that do not correspond to any known α subunits. These genes (or pseudogenes) have been found in humans and other species. They may represent a backup system or they may have specialized roles, for example, performing cation transport (not necessarily of Na^+ and K^+) in particular tissues, tissue subtypes, or even in cellular organelles. The study of the functional roles of Na^+,K^+-ATPase isoforms has just begun and no definite conclusions can be drawn at this stage.

Because the enzyme is so important to the maintenance of cell function (consuming almost one-quarter of the energy used by a human at rest), the evolution of closely related isoforms of the enzyme can be regarded as a means of ensuring that mutational inactivation of one of the encoding genes does not result in cell death (i.e., provides a backup system). Later development may have used the availability of different isoforms to meet specific needs while maintaining the backup provided by the existence in each cell of the potential to express up to six different isoforms should the need arise.

As will be discussed in Section 3.4.3, different isoforms of Na^+,K^+-ATPase have different affinities for cardiotonic steroids. Also, the methods used for preparing the enzyme for structure–activity relationship (SAR) studies of cardiotonic steroids involves the loss of up to 99% of Na^+,K^+-ATPase activity during the purification process. This makes the interpretation of such SAR studies extremely difficult, since purification may selectively remove particular isoforms.

3.5 Autonomic Regulation

Although the denervated or isolated heart can beat spontaneously, the autonomic nervous system plays an important role in regulating the rate and force of contraction. Adrenergic, muscarinic, and purinergic receptors modulate the ion channels and carriers discussed in previous sections. All of the receptors for these systems involve G proteins that directly or indirectly affect the ion transporting system. The effects can involve all myocytes or may show selectivity for certain cell types, for example, nodal cells.

3.5.1 ADRENERGIC CONTROL. Adrenergic receptors in human myocardium include α_1, β_1, and β_2 subtypes, although the role of the last mentioned has not been clarified (93). Both β receptors are usually coupled via G proteins to adenylyl cyclase, but direct coupling to effector units may be involved in some cases. Stimulation of myocardial β_1 receptors affects ion currents as follows:

1. *L-type Ca^{2+} Channels.* $\uparrow I_{Ca-L}$ resulting from c-AMP-dependent protein kinase phosphorylation of channel protein leads to $\uparrow [Ca^{2+}]_i$ resulting in $\uparrow$ myocardial contractility (positive inotropic effect) and may trigger early afterdopolarizations and accelerate AV conduction (94).

2. *Phase 4 Inward Na^+ Current in Pacemaker Cells.* $\uparrow I_f$ leading to $\uparrow$ slope of phase 4 resulting in $\uparrow$ heart rate (positive chronotropic effect); may stimulate impulse formation in latent pacemaker cells in atrial and ventricular tissue giving rise to ectopic beats (95).

3. *Na^+,K^+-ATPase.* $\uparrow I_{Na-K\ pump}$ causing $\uparrow$ cation loss from cytoplasm leading to hyperpolarization of cell membrane.

4. *Outward Rectifying K^+ Channel.* $\uparrow I_K$ leading to $\uparrow$ rate of repolarization and

↓ refractory period; may trigger or accentuate reentry phenomena (discussed in Section 4.1); may shorten AV conduction time (96).

5. *Other K^+ Channels.* I_{to} is stimulated, but I_{KI} appears unaffected.

6. *Fast Sodium Channel.* ↑I_{Na} may occur under some circumstances, leading to increases in the rate of impulse conduction.

Arrhythmias caused or exacerbated by any of the above effects may be treated with β-adrenergic blocking drugs (β-blockers). The main effects of these drugs arise from their action on Ca^{2+} and K^+ channels. They thus (*1*) decrease heart rate by slowing impulse formation and conduction in the nodal tissue, (*2*) decrease force of contraction, and (*3*) prolong the refractory period. The effects on Ca^{2+} and K^+ channels are often complementary in achieving these effects. The β-blockers are thus of value in treating certain arrhythmias, preventing anginal episodes, and reducing the risk of sudden death in patients with coronary artery disease.

Several subsets of α_1 adrenergic receptors are present in the myocardium and are directly coupled through G proteins to Na^+,K^+-ATPase (97), some K^+ channels (98), and phospholipase C (99). Alpha-1 stimulation of Na^+,K^+-ATPase increases (hyperpolarizes or makes more negative) the phase 4 membrane potential, thus decreasing automaticity of non-SA automatic fibers. Alpha-1 agonists decrease I_{KI} and I_{TO}, leading to prolongation of repolarization. The arrhythmogenic potential of α-adrenergic stimulation in human myocardium has not been studied in much detail, but there is evidence that some arrhythmias may be induced by this mechanism, which would explain the efficacy of α-blockers such as phentolamine in treating certain postinfarction cardiac arrhythmias.

3.5.2 CHOLINERGIC CONTROL. Parasympathetic control of the heart is mediated by the vagus nerve and involves mainly M_2 muscarinic receptors, the density of which is two to five times greater in the atria than in the ventricles. Stimulation of M_2 receptors has the following effects:

1. *Channels Carrying the I_{Ca-L}, I_f, and I_K Currents.* ↑G_i activity resulting in inhibition of adenylyl cyclase, thus offsetting the effects of adrenergic stimulation on these currents in myocytes having high densities of M_2 receptors (100). Conduction through the AV node, in particular, is depressed by inhibition of I_{Ca-L} and I_K currents, whereas inhibition of the I_f current depresses the rate of spontaneous depolarization in the sinus node. The combined effects on all these currents is to slow the heart (negative chronotropic effect).

2. *Channels Carrying the $I_{K(ACh)}$ Current.* ↑K outflow during phase 4 resulting from stimulation of the channel by M_2 agonists. This increases the size of the phase 4 membrane potential (hyperpolarization) in SA and AV nodes, thereby adding to the negative chronotropic effects described under item 1 as well as shortening the action potential in atrial tissues (101).

Stimulation of M_2 receptors either by vagal activity or drug treatment protects the heart from arrhythmias involving the AV node and possibly from ventricular arrhythmia. Decreased vagal tone has been associated with increased mortality in postinfarction patients, as has the use of muscarinic blocking agents such as atropine. Atropine has a role in the infarcted patient if the heart rate is too slow (bradycardia) but has obvious risks. The antiarrhythmic effects of muscarinic stimulation are due partly to reduction in heart rate and partly to offsetting the broader range of effects due to adrenergic stimulation.

3.6 Endothelial-derived Factors in the Control of Cardiovascular Function

It is now well established that the vascular and myocardial endothelium play a major role in the control of cardiovascular function. Endothelial cells release a variety of factors that meet various normal physiological needs, including vasodilators, vasoconstrictors, and factors that inhibit or enhance coagulation and initiate tissue repair. If a blood vessel is cut, the release of vasoconstrictors and clotting factors assist in limiting blood loss, and the release of various cytokines encourages inflammatory cells, especially monocytes, to enter the tissue and initiate removal of damaged material. The release of growth factors such as the platelet-derived growth factor then stimulate tissue repair. Vasodilating factors, including the endothelium-derived relaxing factor (discussed in Section 6.1), are released in response to physiological demand, for example, increased shear forces arising from increased blood flow. In atherosclerosis, these processes become disrupted, leading to lesion production and other counterproductive actions that perpetuate the disease process. The nature of this pathology is discussed in more detail in Section 5.1. Some of the mechanisms that control vascular activity will now be discussed briefly.

3.6.1   LEUKOCYTE ADHESION MOLECULES. When a vessel is damaged, monocytes leave the blood and enter the tissues, where they convert to macrophages. This process is initiated by the expression on the epithelial surface of leukocyte adhesion molecules, of which three main classes have been identified: (*1*) endothelial-leukocyte adhesion molecule-1 (ELAM-1), which binds polymorphonuclear leukocytes; (*2*) vascular cell adhesion molecule-1 (VCAM-1), which

binds monocytes and lymphocytes; and (*3*) intercellular adhesion molecules (ICAM-1), which can interact with all leukocytes. Animal studies have shown that elevated serum cholesterol levels (induced by dietary saturated fats) stimulate the expression of leukocyte adhesion molecules that are analogous to VCAM-1, in that they stimulate the uptake of monocytes (which convert to macrophages) by vascular endothelium (102).

Abnormal vasoconstriction of coronary arteries is a major feature of most forms of ischaemic heart disease and is frequently superimposed on the atherosclerotic lesion or on areas of the vessel adjacent to the lesion. This explains why the clinical course of atherosclerotic disease does not correlate well with the degree of atherosclerosis. Normally, coronary vessels dilate in response to common stimuli such as exercise, mental stress, chilling, etc; but in atherosclerosed vessels, where the endothelium has been damaged, these stimuli can trigger vasoconstriction. Presumably, the production of vasodilating and antithrombotic substances such as EDRF and prostacyclin have been impaired in the damaged endothelium, whereas the opposing actions, many of them platelet derived, tip the balance in favor of vasoconstriction and thrombosis.

3.6.2   ENDOTHELINS. The endothelins are 21-residue peptides, of which three have so far been identified: endothelin-1, -2, and -3. The endothelins are found in many tissues such as brain, kidney, lungs, and intestine; but endothelin-1 is the only endothelin found in vascular endothelial cells. Endothelin-1 is the most potent vasoconstrictor peptide yet identified and has an extremely long activity for such a molecule, its pressor effects last for several hours following i.v. injection. At least two subtypes of endothelin receptor have been identified: ET_A and ET_B. The ET_A receptor is found on vascular smooth muscle and appears to mediate vasoconstriction, whereas the ET_B receptor is found on vascular endothelial cells and may mediate endothelial production of nitric oxide (see Section 6.1) and prostacyclin, both of which have vasodilator properties. Prostacyclin also inhibits platelet aggregation (103–105). The injection of endothelin-1 produces an initial vasodilation, presumably due to interaction with ET_B receptors, followed by a prolonged pressor effect, presumed to be due to interaction with ET_A receptors; the relative proportion of the two effects depends on the dose of endothelin and the vascular bed. The latter differ in their relative abundance of the two receptor types. The signal transduction following occupation of endothelin receptors is currently under investigation. In coronary artery preparations both phospholipase C activation and opening of the voltage-dependent L-type Ca^{2+} channel have been reported (106,107).

The role of endothelins in the pathogenesis of cardiovascular disease is under active investigation, and there is much that is controversial (108). Elevated levels of endothelin-1 have been associated with hypertension and have been suggested as a contributing factor in ischemic heart disease. In the latter, it is now recognized that coronary vasospasm is an important contributory factor to infarct-related ischaemia.

The discovery of endothelins and their apparent significance in cardiovascular disease has led to the rapid development of endothelin antagonists by a number of pharmaceutical companies, with at least eight antagonists under active development (109).

3.6.3 ENDOCARDIAL ENDOTHELIUM FACTORS. The endocardial endothelium lines the chambers of the heart and functions similarly to vascular endothelium in exerting active control over the underlying muscle. For example, it releases relaxing factors identical to NO. Several groups of investigators have shown that the endocardial endothelium releases a substance, tentatively named *endocardin*, that exerts a positive inotopic effect (110). This effect differs from that of other inotropic agents in that it does not affect the rate of development of force (dF/dt) but increases the maximum force developed and the duration of contraction (the latter being its main effect). The endocardial endothelium also *decreases* the duration of the inotropic effects of some other agents (e.g., 5-HT, endothelin, aggregating platelets). It thus has two opposing actions: a contraction-prolonging effect under some circumstances and an contraction abbreviating effect under other conditions. The suggestion has been made that the contraction-prolonging factor may be endothelin-1 (111), but this has been rejected by Smith and co-workers (110) who believe that a distinct substance, "endocardin," is involved. No mechanism has been found for this substance, but it may affect K^+ channels.

4 ANTIARRHYTHMIC AGENTS

4.1 Mechanisms of Cardiac Arrhythmia

As discussed earlier, the pumping action of the heart involves three main electrical events: (*1*) the generation of an electrical signal; (*2*) the conduction of the signal, and (*3*) the dying away of this signal. Cardiac arrhythmia arise when onc or morc of these processes are disrupted. The causes and mechanisms of such dysfunction are poorly understood, at least in the clinical setting. The following is a brief overview of the main mechanisms of cardiac arrhythmia as they are presently understood.

4.2.1 DISORDERS IN THE GENERATION OF THE ELECTRICAL SIGNAL. The heart has the capacity to beat spontaneously, although the rate of and force of contraction are influenced by neuronal activity. The spontaneous generation of an electrical signal in the heart is called impulse initiation and can be due to two causes: automaticity and triggered activity. Automaticity is a property possessed by those cardiac myocytes that will, if given time, spontaneously depolarize and generate an action potential. Under normal conditions, automaticity is limited to those cardiac myocytes where the resting (diastolic) transmembrane potential decreases over time (i.e., has an upward sloping phase 4; Fig. 28.5). These myocytes include those of the SA and AV nodes and the specialized conducting cells such as those of the Purkinje system. Given time, any of these cells will spontaneously depolarize and give rise to an action potential that would sweep over the whole myocardium. The time taken to depolarize spontaneously depends on (1) the magnitude of the maximum diastolic potential (i.e., its maximum negative value), (2) the rate of loss of negative potential during diastole (i.e., the slope of phase 4), and (3) the threshold potential that will initiate the depolarizing current (I_{Ca-L} in the case of nodal tissue and I_{Na} in the case of other myocytes). Conditions in the SA node are such that its cells normally depolarize first, and thus the SA node becomes the pacemaker, initiating signals (and hence heartbeats), normally at the rate of 60–100/min. If the rate of signal generation by the atrial node was depressed, other automatic cells (known as potential or latent pacemakers) would be the first to depolarize and initiate a beat. The normal depolarization rates at the junction of the AV node and bundle of His is 40–60/min, and in the His bundle branches it is about 20–40 impulses/min. Thus if a person had complete AV block, their ventricles would still beat at the slower rate of 20–40 beats/min. An electronic pacemaker may be attached to the ventricles of such a person to restore the ventricular beat to a normal value.

Disorders in impulse generation due to automaticity include the following.

1. *Automatic Arrhythmias of the SA Node.* If signals are generated more frequently than 100/min, the rest of the myocardium will follow and the condition is called a sinus tachycardia. If signals are generated less frequently than 60/min but not slow enough for latent pacemakers to be activated, the condition is called sinus bradycardia.

2. *Ecotopic Beats.* Beats due to signal generation other than in the SA node are called ectopic beats (the word *ectopic* means "away from the place," i.e., away from the place where it ought to be). Ectopic beats may develop because of (1) decreased automaticity of the SA node; (2) increased automaticity of latent pacemaker cells; (3) a conduction block that prevents the SA-generated impulse from reaching part of the myocardium, thus allowing automatic myocytes in that tissue to become pacemakers (this usually follows AV block); and (4) automaticity in nonspecialized myocytes. Nonspecialized myocytes normally have stable diastolic membrane potentials (i.e., phase 4 is flat, as shown in Fig. 28.4) and will not spontaneously depolarize. However, in certain disease states, especially in ischaemia where ATP levels may fall, the ability of these cells to maintain the resting potential via the Na^+ pump (Na^+,K^+-ATPase) and other mechanisms is impaired, and the cells may spontaneously depolarize and initiate an action potential.

Whereas disorders of impulse generation due to automaticity occur spontaneously, triggered disorders, as the name suggests, requires a triggering event to initiate a contraction signal. It is thought that, under some circumstances, an action potential can give rise to aberrant channel-opening responses such that an additional depolarization event occurs during or at the end of the primary action potential. In such cases, the action potential does not show the normal profiles given in Figs. 28.4–6 but has, superimposed on it or immediately following it, additional upsweeps known, respectively, as early after-depolarizations (EADs) and delayed after-depolarizations (DADs). These have been well characterized in electrophysiological studies of isolated cells, but their clinical significance has not been defined. It is possible that certain antiarrhythmic drugs, namely those belonging to classes Ia and III may give rise to EADs leading to a condition known as *torsade de pointes* "twisting of points," French). DADs have been associated with conditions in which there is an abnormally high $[Ca^{2+}]_i$, for example, due to digitalis toxicity or excessive sympathetic activity.

4.2.2 DISORDERS IN THE CONDUCTION OF THE ELECTRICAL SIGNAL. Disorders of conduction can lead to conduction block and reentry phenomena. Conduction block can be complete (no impulses pass through the block), partial (some impulses pass, e.g., every second impulse may pass); and bidirectional or unidirectional. A bidirectional block is one in which the impulse is blocked regardless of the direction from which it entered the cell, whereas a unidirectional block is one in which impulses from one direction are completely blocked but those from the opposite direction are propagated, although they are usually slower than normal.

4.2.2.1 Heart Block. The term *heart block* is used to describe AV block and is classified as first-, second-, and third-degree heart block, depending on whether the AV node, respectively, (*1*) conducts all impulses but at a slower rate than normal (PR interval is prolonged); (*2*) does not conduct all impulses from the atria to the ventricles (not all P waves are followed by QRS complexes); and (*3*) does not conduct any

impulses. In the last-mentioned case, the subject relies on automatic cells in the ventricles for beat initiation. Such people may be given artificial pacemakers, since the natural ventricular pacemakers depolarize too slowly.

4.2.2.2 Reentry Phenomena. By far the most important cause of life-threatening cardiac arrhythmias is a condition known as reentry. As mentioned earlier, the electrical signal generated by the sinus node is passed from cell to cell across the entire myocardium, triggering as it goes, the process of contraction. Once this has happened, the muscle must relax so that the heart can dilate and fill with blood ready for the next pumping action. Since the impulse passes from cell to cell, it could circle back and reactive cells that had earlier been activated. Normally, this does not happen because the cells become refractory (unable to accept a signal) for a period long enough for the original signal to die away. Under normal conditions, the cells will not contract again until reactivated by a new signal emerging from the sinus node. There are certain conditions in which this does not happen, and the impulse continues to circulate, either in a localized region or generally. The impulse is said to reenter myocardium that it had previously activated. There are a variety of circumstances that lead to this and a variety of consequences, hence the term *reentry phenomena*. However, the essential condition for reentry is a refractory period that is shorter than the conduction velocity or, to say the same thing, a conduction velocity that is longer than the refractory period. Thus any circumstance that shortens the refractory period or lengthens the conduction time could lead to reentry. Almost all tachyarrhythmias including fibrillation are due to reentry. The length of the refractory period depends mainly on the rate of activation of the I_K current and the rate of conduction depends on the rate of activation of the I_{Ca-L} current in nodal tissue and the I_{Na} current in other myocytes. The channels controlling these currents are thus the focus for developing drugs to suppress reentry but, since these currents are influenced by other currents, these too are receiving increasing attention.

Although many circumstances can lead to reentry, the most common is that depicted in Figure 28.19. The conditions needed for this type of reentry are as follows.

First, the existence of an obstacle around which the wave front can propagate is needed. The obstacle is any entity, for example, infarcted or scarred tissue, that physically or functionally cannot conduct the impulse.

The second condition is the existence around one side of the obstacle of a pathway that allows conduction at the normal rate and around the other side a pathway that has impaired conductivity. The impairment may be such that it allows conduction in only one direction (unidirectional block) or it may allow conduction to proceed at a greatly reduced rate such that when the impulse emerges from the impaired tissue, the normal tissue is no longer refractory. These pathways are usually localized, for example, within the AV node or the end-branches of part of the Purkinje system, but they can be more extensive and can give rise to daughter impulses capable of spreading to the rest of the myocardium.

Unidirectional block occurs in tissue that has been impaired, such that its ability to

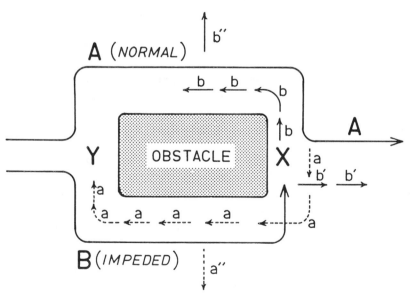

Fig. 28.19 Model for reentrant activity. A depolarization impulse approaches an "obstacle" (nonconducting region of the myocardium) and splits into two pathways (A and B) to circumvent the obstacle. If pathway B has impeded ability to conduct the action potential the following may occur. (*1*) If impulse B is slowed and arrives at cross junction X after the absolute refractory period of cells depolarized by A, the impulse may continue around the obstacle as shown by path b and/or follow A along path b'. In both cases, the impulse is said to be reflected. (*2*) If pathway B shows unidirectional block, impulse A may continue around the obstacle as shown by path a. If the obstacle is large enough, so that cells in cross region Y are repolarized before the return of a or b, then a circus movement may be established. Both (*1*) and (*2*) may propagate daughter impulses (a" and b") to other parts of the myocardium. These effects can give rise to coupled beats and, in the extreme situation where the phenomenon is widespread and random, fibrillation. The obstacle can be any anatomical or pathological feature, e.g., an infarcted area with an associated depressed myocardium; the infarct is the obstacle and the depressed myocardium is pathway B.

conduct is completely blocked in one direction but only slowed in the other. Because of unidirectional block, the impulse cannot proceed forward along path B (Fig. 28.19), the cells of which thus remain in a polarized state. However, when the impulse traveling along path A reaches a suitable cross-junction (point X on the figure), the impulse proceeds back along path a, although at a slower rate than normal (indicated by the dotted pathline). When this impulse reaches another cross-junction (Y), it is picked up by path A and conducted around the circle. If the obstacle is large enough, the cells in path a will have repolarized and the path will again be followed, giving rise to a continuous circular movement. The impaired tissue could be on the periphery of an infarct and the reason for the selective block a relative difference in capacity to conduct impulses coming from different directions (anisotropy). The essential condition for circular reentry is a wave train propagating along circular pathways where the conduction velocity is slower than the refractory period so that the wave front continuously encounters nonrefractory cells. When reentry occurs randomly in the myocardium, it gives rise to a completely disorganized situation known as fibrillation in which there is no coordination in the contraction of individual fibers and hence no pump action. This leads to cardiac arrest, and death will follow within minutes if rhythmic contraction is not restored. Cardiac arrest may result from fibrillation (shown by a chaotic trace on the ECG) or by cessation of all electrical activity (asystole) (shown by a straight line on the ECG).

4.2 Classification of Antiarrhythmic Drugs

Numerous attempts have been made to classify antiarrhythmic drugs in a way that incorporates their known mode of action and clinical applicability. None of these systems has gained universal endorsement. The most enduring is that proposed by Singh and Vaughan Williams (112,113) in the early 1970s and further elaborated over the next decade (114) and by Harrison and Bottorff (115) in 1992. Singh and Vaughan Williams based their classification on the results of microelectrode studies that suggested the existence of four main classes of antiarrhythmic activity: I, blocking of the fast Na^+ channel (I_{Na} current); II, blocking of β-adrenergic receptors; III, increasing duration of the action potential (prolonging the refractory period); and IV, blocking Ca^{2+} channels. Later, class I drugs were divided into three subgroups, depending on whether repolarization was (1) delayed, (2) accelerated, and (3) not significantly affected. Table 28.2 summarizes the main features of the Singh–Vaughan Williams classification and gives examples of the main drugs in each of these classes. The structural formulas of these drugs are given in Figure 28.20 While this classification of *action* has been well supported by experimental evidence, confusion has arisen because many drugs have more than one type of action. Amiodarone, for example, has actions that could place it in all four groups.

The Singh–Vaughan Williams classification has been strongly criticized by the Task Force of the Working Group on Arrhythmias of the European Society of Cardiology. Its report, published simultaneously in *Circulation* (116) and in the *European Heart Journal* (8), made the following criticisms of the Singh–Vaughan Williams classification:

1. The classification was a hybrid: classes I and IV refer to channel blocking, class II to receptor blocking, and class III to an effect on the action potential.

2. A particular class effect can be due to one or more mechanisms, for example, a class III effect can result from blocking any one of a number of different K^+

Table 28.2   The Singh–Vaughan Williams Classification of Antiarrhytmic Drugs

Class	Main Action	Examples[a]
I (also known as membrane-stabilizing drugs)	Block the fast Na^+ channel (I_{Na} current), leading to a decrease in slope of phase 0 (i.e., $\downarrow V_{max}$); divided into three classes	
Ia	moderate[b] $\downarrow V_{max}$; prolong refractory period	quinidine (**1**); disopyramide (**2**); procainamide (**3**)
Ib	weak[a] $\downarrow V_{max}$; accelerate repolarization; little effect on refractory period	lidocaine (**4**); mexiletine (**5**); tocainide (**6**); phenytoin (**7**); moricizine (**8**)
Ic	marked[a] $\downarrow V_{max}$; minimal effects on repolarization	flecainide (**9**); encainide (**10**); indecainide (**11**); propafenone (**12**)
II (β-adrenergic blocking drugs)	indirect reduction of $I_{Ca\text{-}L}$ current in SA and AV nodes	propranolol (**13**); atenolol; metopolol; esmolol; timolol
III	increase duration of action potential by blocking K^+ channels; little effect on I_{Na} current (V_{max})	sotalol (**14**); bretylium (**15**); amiodarone (**16**)
IV (calcium chanel blockers)	block $I_{Ca\text{-}L}$ current, thus slowing conduction in SA and AV nodes and depressing contractility in all heart myocytes	verapamil (**17**); diltiazem (**27**)

[a]See Figure 28.20 for structures.
[b]The effects on V_{max} refer to normal cells; the degree of effect can change in pathological states (see text).

channels or to affects on Na^+ and Ca^{2+} channels.

3. Some drugs have several classes of action, for example, amiodarone.
4. The classification is incomplete; it does not include cardiotonic steroids (digitalis), α-adrenergic blockers, cholinergic agonists; adenosine; and drugs that act by opening ion channels.
5. The classification is based on electrophysiological studies of isolated normal myocardial cells and thus may not represent cell behaviors in the diseased states where channels, receptors, and drug pharmacodynamics may be modified.
6. The classification does not provide a basis for selecting different strategies in treating arrhythmias, for example, prophylaxis, improving tolerance to certain arrhythmias (e.g., slowing tachycardias), and terminating other more malignant arrhythmias.
7. The classification can lead to inappriate use of antiarrhythmic drugs,

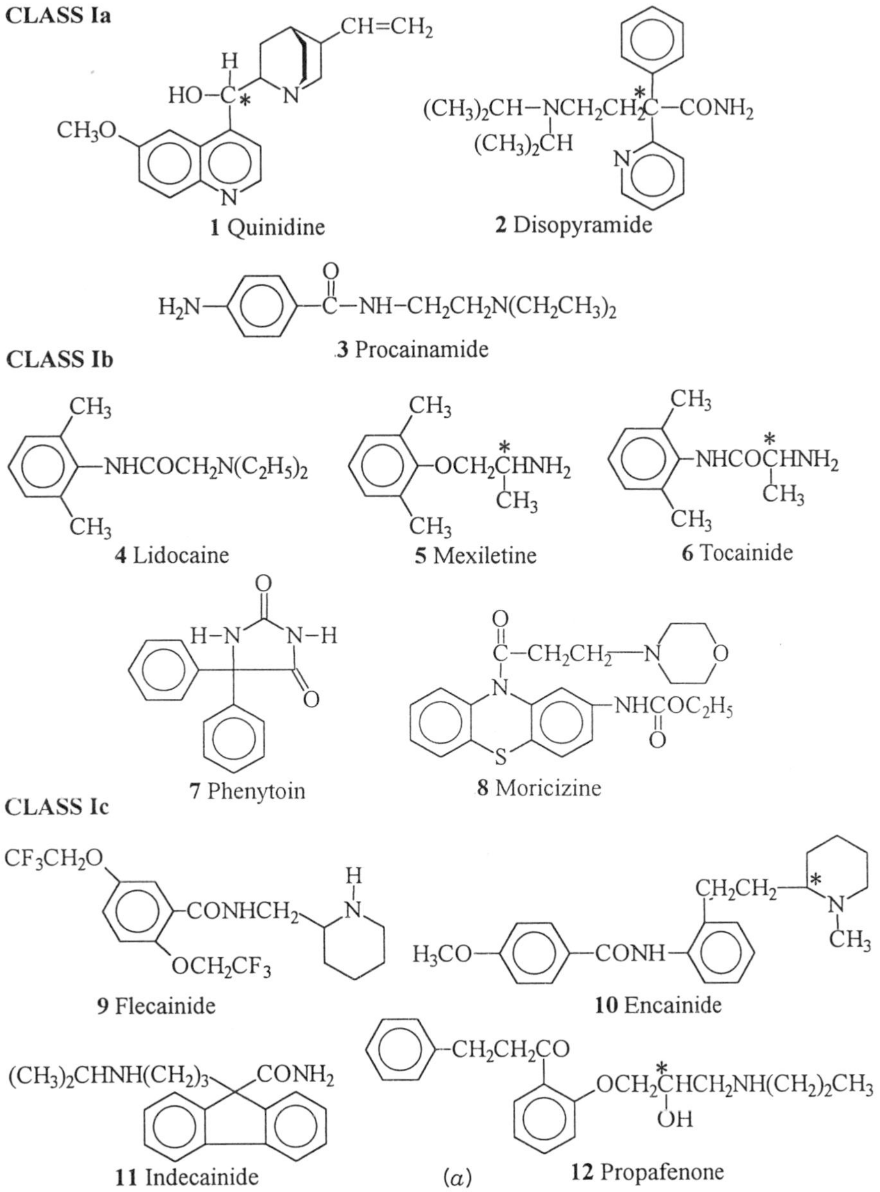

Fig. 28.20 (a) Examples of structural formulas of class I antiarrhythmic drugs. Chiral centers are indicated by an asterisk.

because it implies that drugs belonging to the same class have similar favorable and unfavorable characteristics.

The task force has proposed an alternative classification to that of Singh–Vaughan Williams, based on what it called the "vulnerable parameters" in cardiac arrhythmias. One of the reasons for this revision was the disastrous outcome of the CAST trial and others that dealt with the use of class Ic drugs for the prophylaxis of

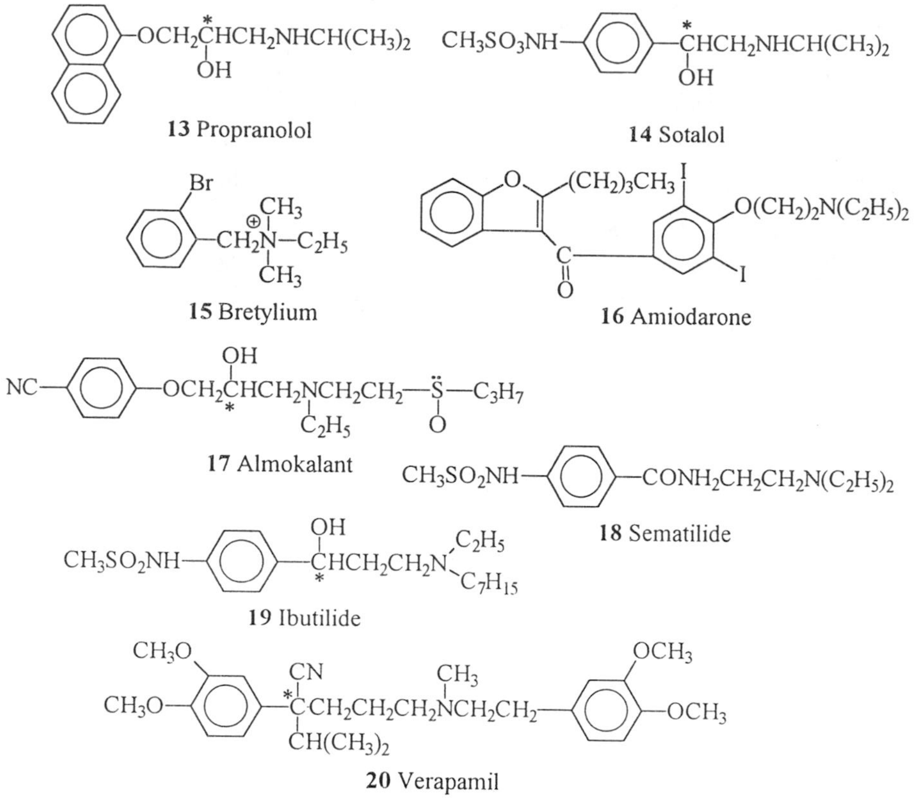

Fig. 28.20 (*b*)   Examples of structural formulas of antiarrhythmic drugs belonging to Classes II, III, and IV. Chiral centers are indicated by an asterisk.

cardiac arrhythmia (mentioned later). Vaughan Williams (117) has responded to the task force criticisms in a recent review. He refutes claims that the classification failed to predict adverse outcomes of recent antiarrhythmic trials and states that the task force classification, when "stripped of speculative elements, is shown to be similar to the original classification." Nattel (114) supported the Singh–Vaughan Williams classification by urging a return to its original purpose, namely to classify actions not drugs (a particular drug can then be said to have actions belonging to certain specified classes). In a 1993 article Nattel and Arenal (118) state that of the 32 drugs listed by the task force, 26 belong in one of the four

Singh-Vaughan Williams classes and most of the others are indirect acting.

The basic problem in classifying antiarrhythmic drugs is that of the interpretation and reconciliation of effects seen in normal isolated cells and those that occur in the diseased myocardium, functioning as a syncitium with its array of different cell types and complex interrelated activities. Workers in this field are still grappling with that basic problem in science of how to integrate reductionist and holistic observations. This is not meant as a criticism: it is a problem that all sciences face and one that can only be overcome by further and deeper study. In the present instance, this is no mere academic matter: ischemic heart disease is the biggest cause of premature death in Western industrial countries, and the main cause of such premature death is cardiac arrhythmia.

4.3 Mechanism of Action of Antiarrhythmic Drugs

Cardiac arrhythmias arise from disorders in the generation, speed of conduction, and termination of the electrical impulse that triggers contraction. Impulse generation depends on (*1*) the size of the resting (phase 4) membrane potential, (*2*) the rate at which this potential decreases during the resting state (relevant to automatic cells and to nonspecialized myocytes in certain disease states), and (*3*) the threshold potential at which the depolarizing channel opens (Ca^{2+} channel in nodal myocytes and Na^+ channel in other myocardial cells). The speed of conduction depends on the slope of phase 0 of the action potential, which depends on (*1*) the type of depolarizing current (slow I_{Ca-L} in nodal tissue and fast I_{Na} in other myocytes), (*2*) the size of the current (depends on the number of channels opening), and (*3*) the resistance capacity of the individual cells (size of the cell, number of T tubes per membrane surface area, rate of conduction across intercalated discs). The length of the refractory period depends on the shape of the action potential (duration of phase 2, rate of activation of the repolarizing currents).

Reentry is the cause of most malignant arrhythmias and occurs when the refractory period is shorter than the rate of impulse propagation in circumstances that give rise to circus and reflected impulses (Fig. 28.19). Traditionally, reentry arrhythmias have been treated with inhibitors of depolarization channels (type I or type IV antiarrhythmics) or inhibitors of repolarization channels (type III drugs). The rationale for this use will now be discussed.

4.3.1 FAST SODIUM CHANNEL BLOCKERS (CLASS I DRUGS). The mode of action of drugs that block the fast Na^+ channel (I_{Na} current) is not known. Originally, it was thought the drugs accumulated in membranes and disrupted the function of the channel proteins, a process known as membrane stabilization. This conclusion was based on the strong association between lipid solubility and potency and the fact that the earlier drugs, particularly quinidine, affected more than one ion channel (Table 28.3 shows the effects of class I drugs on K^+ channels). Of those class I drugs listed in the table, only lidocaine had no effect on the K^+ channels examined. However, now that the similarity between channel proteins has been firmly established (Fig. 28.7), it is not surprising that absolute specificity is not shown. If there is any "nonspecific" activity, it probably involves direct association with the channel proteins and not simply disruption of membrane fluidity. Recent studies support the idea that the Na^+ channel blockers react with specific receptors. It has been known for some time that the ability of the drugs to block the Na^+ channel depends on the frequency of depolarization (i.e., the drugs show phasic block). At normal heart rates, the drugs produce little or no block, depending on the particular drug and its concentration. As the heart rate increases, the degree of block increases. This type of phasic block is called use-dependent block and arises because of changes in receptor affinity or in access to the receptor or to both these reasons (119–124).

As illustrated in Figure 28.8, Na^+ channels cycle through a gating system. The figure illustrates a "lumped" gating system, since it is now believed that there are multiple closed states and two or more activated states. Which of these many states presents the drug with the most favorable binding conditions has not been established. Most studies have been conducted using lidocaine (**4**), and it has been sug-

Table 28.3   Specificity of Therapeutic Concentrations of Antiarrhythmic Drugs in Blocking K$^+$ Channels^a

Class	Agent	I_K	I_{KI}	I_{to}
		\multicolumn{3}{c}{K$^+$ Channel Blocked^b}		
Ia	Quinidine	+	+	+
Ia	Disopyramide	+	+	+
Ib	Lidocaine	−	−	−
Ic	Flecainide	+	−	−
Ic	Encainide	+	−	−
III	Sotalol	+	+	+
III	Amiodarone^c	+	+	−
III	Clofilium^c	+	+	−
III	Risotilide	+	−	−

^{a}Adapted from ref. 33.

^{b}+ = yes; − = no.

^{c}Blocks both I_{Kr} and I_{Ks}, whereas other class III drugs seem selective for I_{Kr} (see text for explanation of these symbols).

gested that this drug blocks one of the states that preceeds channel opening (125). However, there is much conflicting evidence, and all that can be concluded at this stage is that lidocaine decreases the probability that a Na$^+$ channel will open. With less channels opening, the slope of phase 0 (V_{max}) and hence the rate of conduction will be reduced. In nonpathological states this effect is moderate for class Ia drugs, weak for class Ib drugs and strong with class Ic drugs. As stated earlier, inhibitors of the Na$^+$ channel show use dependence and the degree of effect will increase with faster heart rates. In the resting state, most Na$^+$ blockers show low efficacy. Each time the channel passes through a depolarization cycle, conditions become optimum for binding, but once the channel enters closed (inactive) states, drug dissociates. Thus a steady state is set up that is determined by (*1*) the rate of channel activation and (*2*) the length of the closed inactive state; the former favors drug binding and the latter drug dissociation. In tachycardias, where the rate of depolarization is fast and the resting periods short, drug binding is favored, As this occurs, the proportion of blocked channels increases and the heart rate begins to slow.

The degree of drug binding also depends on the phase 4 (resting) membrane potential; the smaller this is, the greater will be the number of Na$^+$ channels that open when the cell is depolarized (126). Thus class I drugs will have greater effects on V_{max} and hence on conduction velocity in partially depolarized cells, such as those experiencing ischaemia. This is particularly marked with lidocaine, which is probably the reason for its effectiveness in treating ischemia-related reentry arrhythmias. In this setting, arrhythmias arise because of unidirectional block (Fig. 28.19). Lidocaine (also known as lignocaine) can depress conduction in ischemically impeded tissue to the extent that a unidirectional block is converted into a bidirectional block (thus removing the condition needed for the circus reentry arrhythmia) but has little effect on the rate of conduction in nonischemic tissue. This property makes it especially useful in treating reentry arrhythmia arising from infarcted tissue: conduction is depressed in the ischemic area (thus protecting against a reentry induced fibrillation), but the remaining myocardium is relatively unaffected.

Measurements of the diastolic dissociation time constants (τ) for dissociation of antiarrhythmic drugs from binding sites on or near the Na^+ channel have shown the following values for the Singh–Vaughan Williams class I subgroups (127):

Ia $1 s < \tau > 10 s$ (moderate affinity)
Ib $\tau < 1 s$ (weak affinity)
Ic $\tau > 10 s$ (strong affinity)

These values explain why the effects of drugs in the three subclasses on V_{max} in normal heart tissue are, respectively; moderate, weak, and strong (Table 28.2). The normal heart rate is 60–100 bpm, thus little of bound class Ic drugs would dissociate during the resting (diastolic) period. Factors that accelerate dissociation of drug include low molecular weight, high lipid solubility, and elevated external pH. These factors facilitate movement through cell membranes, which could indicate that such drugs have a tendency to leave binding sites on the channel protein and enter the cell membrane, subsequently diffusing from the membrane into the extracellular fluid. However, it has also been shown that hyperpolarization following binding slows dissociation (128). Hyperpolarization reduces the probability that the channel will open and this could mean either that the drugs' main route of dissociation is through the open channel or that affinity for the binding site is reduced when the protein is in the "open" conformation.

The beneficial therapeutic value of class I drugs in depressing the rate of impulse conduction has been discussed with respect to certain reentry arrhythmias. However, since the fundamental prerequisite for reentry is a conduction rate that is slower than the refractory period, class I drugs can, by slowing conduction, cause reentry (i.e., can be proarrhythmic). This was confirmed by the results of the Cardiac Ar-

rhythmia Suppression Trials (CAST 1 and 2), which found that the treatment of asymptomatic patients with the class Ic drugs flecainide (**9**) and encainide (**10**) caused a threefold increase in sudden death compared with controls, even though they were effective in suppressing ectopic beats (129,130). In a recently published meta-analysis of 138 trials involving 98,000 patients with myocardial infarction given prophylactic therapy with antiarrhythmic agents, the author (131) concluded that (*1*) the routine use of class I drugs is associated with increased mortality; (*2*) class II drugs (β-blockers) have been conclusively demonstrated to reduce mortality; (*3*) limited data on amiodarone appear promising; and (*4*) data on class IV drugs (calcium channel blockers) remain unpromising.

It seems that neither classical biochemical nor molecular biology techniques have unambiguously confirmed the existence of specific antiarrhythmic drug receptors on Na^+ channel proteins. It is known that neurotoxins such as tetrodotoxin bind to highly specific binding sites on Na^+ channel proteins. The use of radioactively labeled neurotoxins has identified at least five such sites (132). Class I antiarrhythmics have been shown to interfere with the binding of ^3H-batrachotoxinin A 20-α-benzoate ([^3H]BTXB) by what appears to be an allosteric mechanism (133).

In a recent study (134), the ability of antiarrhythmic drugs to inhibit binding of [^3H]BTXB was examined. Only class I drugs were found to inhibit binding, and this was thought to be due to noncompetitive allosteric binding. Of the class I drugs that show chirality, one enantiomer was usually more effective than the other in inhibiting binding. In the case of tocainide (**6**), it was found that the *R*(-) isomer was significantly more effective in inhibiting [^3H]BTXB binding than the *S*(+) isomer and that the IC_{50}s corresponded to the relative potencies of the tocainide isomers in prolonging conduction time in myocardial cells. Of the class I antiarrhythmics shown in Figure 28.20, seven have chiral centers. Of these, the two asymmetric class Ia drugs (**1** and **2**) have the chiral center attached directly to an aromatic ring, whereas those in classes Ib and Ic have a two-atom link between the aromatic ring and the chiral center. Since isomers involving both centers showed differences in binding affinity, it was concluded that the receptor showed stereochemical selectivity corresponding to both chiral locations.

4.3.2 β-ADRENERGIC BLOCKING AGENTS (CLASS II DRUGS).

As discussed earlier, activation of β adrenergic receptors can contribute to the generation of arrhythmia due to effects on a number of ion channels. These effects can lead to early afterdopolarizations (EADs), accelerated AV conduction; increased heart rate; increased automaticity (ectopic beats); shortened refractory period; hyperpolarization; and triggering or accentuation of reentry phenomena. β-Adrenergic blocking drugs thus have an obvious role in prevention and reversion (suppression) of arrhythmia due to excessive adrenergic activity or when normal levels of sympathetic activity are exacerbating arrhythmias due to other causes. β-blockers have additional properties that may contribute to their antiarrhythmic activity, including "membrane-stabilizing activity" (often referred to as quinidine-like activity), varying degrees of cardioselectivity and intrinsic sympathomimetric activity. To what extent these properties contribute to antiarrhythmic activity is not known, but they are probably not significant. Propranolol (**13**) is the prototype of this class of drug, but many β-blockers are effective in preventing and suppressing arrhythmias due to enhanced sympathetic tone. This has been shown conclusively in the use of these drugs for the secondary prevention of reinfarction and sudden death (131). The efficacy of β-blockers in this setting has made them one of the most effective groups of antiarrhythmic agents in reducing mortality associated with coronary artery disease. The β-blocker sotalol (**14**) (now usually classified as a class III drug) is considered by many people to be the drug of choice in the acute treatment of malignant or potentially malignant arrhythmias, replacing lidocaine in this respect. The D isomer of sotalol has no β-blocking activity but retains its class III activity

4.3.3 AGENTS THAT PROLONG THE REFRACTORY PERIOD (CLASS III DRUGS).

As stated earlier, the prime aim in treating most arrhythmias is to ensure that the myocytes remain refractory for a period longer than the time taken to propagate the depolarization wave. With class I drugs, the emphasis is on slowing the rate of conduction, although many class I drugs also lengthen the refractory period (Table 28.2). The aim with class III drugs is to prolong the refractory period. The increased rate of death associated with the prophylactic use of class Ic drugs, as confirmed by the CAST studies, has focused the attention of the pharmaceutical industry on the development of antiarrhythmic drugs that prolong the refractory period. Clinical, electrophysiological, and laboratory animal studies support the claim that class III action is more effective than class I action in suppressing cardiac arrhythmia and has less proarrhythmic potential (135). However, the tendency of class III drugs to cause the arrhythmia known as *torsades de pointes* is causing concern (136).

Structures **14–19** in Figure 28.20 are representative examples of class III drugs in clinical use or undergoing trial (as of 1994). There are some structural resemblances between these drugs, namely, like so many drugs, they have an aromatic ring separated from a nitrogen atom by linking groups. The search for SAR at this stage is not justified until it has been firmly established which channels are being modulated by the drugs (correlating structural features of drugs that act by different mechanisms is counterproductive). As shown in Table 28.3, several K^+ channels may be affected by a particular class III drug. A number of class III drugs show chirality but initial studies have shown no significant differences between enantiomers in their ability to prolong the action potential (135) (chiral centers are indicated by an asterisk in the structures in Fig. 28.20).

In general, class III drugs act by inhibiting the repolarizing K^+ currents (137), although one class III drug, ibutilide (**19**), appears to act by activating a slow inward Na^+ current (138). The fact that ibutilide contains the methanesulfonanilide group that is present in K^+ channel blockers such as sotalol (**14**), sematilide (**18**), and some other K^+ channel blockers shows that this group is not peculiar to K^+ channel blockade. In fact, the methansulfonanilide group is not an essential moiety for K^+ channel blockade; the class III drug bretylium (**15**) is a quaternary ammonium compound and probably acts like tetraethylammonium. The principal current blocked by type III drugs appears to be the outward (delayed) rectifying current (I_K) (see Table 28.1), but this may represent a subfamily of currents with at least two components: a small, rapidly activated current ("little K") (I_{Kr}) and a larger more slowly activated current ("big K") (I_{Ks}) (33). Amiodarone (**16**) and clofilium appear to inhibit both big K and little K whereas big K does not seem to be inhibited by other class III drugs (Table 28.3). The inward rectifying K^+ current (I_{KI}) and the transient outward K^+ current (I_{to}) are inhibited by some class III drugs (Table 28.3). Until more is known about the degree of channel selectivity and about the channels themselves, there is little point in speculating about SAR. The view has been expressed that it may not matter which K^+ channel is inhibited as long as there is a net decrease in the outward K^+ current during the phase 2 plateau period

(126). The pharmaceutical industry, on the other hand, is examining selectivity among K^+ channel blockers as a means of increasing therapeutic index.

One of the concerns about the use of drugs that prolong the refractory period is the development of early afterdepolarizations giving rise to a condition known as *torsade de points*. The points in question are the QRS vector of the ECG (shown in Fig. 28.6). The condition is usually caused by antiarrhythmic drugs and can lead to death. Drugs known to cause this problem are those that prolong the refractory period and include drugs in classes Ia and III. The mechanism of this arrhythmia is under active investigation but may be due to excessive Ca^{2+} influx during phase 2 (139). Drugs that block Ca^{2+} channels may, therefore, have reduced tendency to cause this problem, which may explain the low incidence of *torsade de points* associated with amiodarone use (this drug has classes I to IV activity). Likewise, drugs with β-blocking activity might be expected to have less tendency to cause this arrhythmia, since β-adrenergic activity stimulates the $I_{Ca\text{-}L}$ current (see Section 3.5.1). However, racemic sotalol causes *torsade de points* even though the L isomer has class II activity.

There is evidence that class III drugs show reverse use dependence, that is their ability to block K^+ channels decreases with heart rate (i.e., with the rate of the depolarization–repolarization cycle). It has been suggested that these properties would significantly reduce the effectiveness of class III drugs in terminating tachycardias and may precipitate arrhythmias by exacerbating bradycardias (125). However, other workers have disputed this conclusion (33).

In 1993, the *American Journal of Cardiology* published the results of a major symposium on antiarrhythmics that prolong the refractory period with particular emphasis on amiodarone and sotalol. Papers included a review of recent concepts about reentry phenomena and its confirmation as the most important mechanism giving rise to life-threatening arrhythmia (140). The opinion was expressed that it is unlikely that Na$^+$ channel blockers will continue to play a major role in the treatment of ventricular arrhythmias except in "structurally normal" hearts, and that emphasis will shift to agents that prolong the refractory period, with the major issue being whether "pure" type III compounds are preferable to complex (multiaction) molecules such as amiodarone and sotalol. Although this issue was not resolved, present evidence was claimed to favor multiaction compounds, especially those with β-blocking properties (141,142). Nevertheless, caution was expressed about the use of class III antiarrhythmics since there is still insufficient evidence about their efficacy and adverse effects, and also because much of the benefit of the drugs studied in trial may have been due to their class II activity (143). The limitations of electrophysiological and whole-animal laboratory studies for predicting clinical outcomes of antiarrhythmic therapy was reviewed and new animal models suggested (144).

4.3.4 AGENTS THAT INHIBIT CA^{2+} CHANNELS (CLASS IV DRUGS). Class IV drugs are generally known as calcium channel blockers. They block the "slow" calcium current (I_{Ca-L}) in a variety of tissues. On the cardiovascular system, calcium channel blockers have three principal effects: (*1*) vasodiation (reduction of smooth muscle contractility); (*2*) reduction of the rate of conduction in the SA and AV nodes (where I_{Ca-L} is responsible for phase 0); and (*3*) reduction in myocardial contractility due to reduction in the strength of the I_{Ca-L} current during phase 2. In therapeutic doses, the only calcium channel blocker that has significant effects on (2) and (3) above is verapamil (**20**). Diltiazem has moderate effects on these two features, whereas drugs such as nifedipine have no significant ef-

fects in therapeutic doses. Verapamil blocks both activated and inactivated calcium channels. Its effects are most marked on those myocytes that are depolarized by the Ca^{2+} channel, namely cells of the SA and AV nodes.

4.4 Individual Antiarrhythmic Drugs

The reader is referred to textbooks of pharmacology and to the general reviews quoted earlier for a comprehensive discussion of the properties of antiarrhythmic drugs, including pharmacokinetics and drug interactions. Some recent reports about three of these drugs will now be discussed briefly.

4.4.1 LIDOCAINE (LIGNOCAINE). Ventricular fibrillation (VF) following acute myocardial infarction (AMI) is the most common cause of premature death in Western industrialized nations. Once fibrillation develops, DC countershock is the only effective treatment. Unfortunately, death from AMI usually occurs before coronary care is available. The probability that VF will develop following admission to the coronary care unit (CCU) may be reduced by giving prophylactic antiarrhythmic therapy. Intravenous lidocaine has long been the drug of choice for this purpose, but its value has been challenged in view of the results of recent trials and meta analyses. Neurological symptoms are among the most common of the drug's noncardiac adverse effects. Its virtues include its rapid onset of action, relative efficacy, and apparently selective effects for ischaemic tissue. In a 1993 review, Nattel and Arenal (118) raised the question of whether lidocaine should still be considered useful in the prophylaxis of VF following AMI. The authors concluded that giving lidocaine as a prophylactic measure will not reduce mortality if prompt and effective cardioversion is available. However, in patients under 70 years, they believed that lidocaine prophylaxis was justified, provided appropriate precautions were taken to minimize adverse effects. Many cardiologists still use lidocaine as the first drug of choice for prophylaxis of VF, with procainamide and amiodarone as second-line treatments. However, sotalol is becoming increasingly preferred as the drug of first choice.

4.4.2 SOTALOL. Sotalol (**14**) is a nonselective β-adrenergic blocker that lacks intrinsic sympathomimetic activity and "membrane-stabilizing" properties. Unlike other β-blockers it prolongs the myocardial refractory period but has no effect on the Na^+ channel (V_{max}). Its effect on the action potential duration does not appear to be due to β-blockade, since d and l sotalol are equieffective in prolonging the refractory period but the l isomer is 50 times more active than the d in blocking β receptors *in vitro*. The mode of action, pharmacokinetics and therapeutic uses of sotalol were reviewed extensively in 1993 (145). Also published in 1993 was the report of a symposium on sotalol. Among the opinions presented was the view that antiarrhythmic drugs should only be used in life-threatening situations (146). A report was given on the safety of sotalol as evaluated by trial in more than 3000 patients. The overall mortality in patients treated with sotalol was 4.3%, but "only" in 0.8% was death attributed to the drug; this was regarded as favorable relative to other antiarrhythmics (147). In terms of efficacy, trials have indicated that sotalol was at least as effective or more effective in the management of life-threatening VA than other currently available drugs (148). Electrophysiological studies confirmed that both β-blockade and lengthening of the refractory period are important to the drug's beneficial effects (149). Sotalol has both high bioavailability after oral administration, a half-life that ranges from 10 to 18 h at steady state, and no active metabolites. These properties

make it useful for long-term prophylactic use (150).

4.4.3 AMIODARONE. Amiodarone (**16**) shows properties of all four Singh–Vaughan Williams classes. Initially, it was classified as a type III antiarrhythmic drug because of its ability to lengthen the refractory period. Subsequently, it was shown to inhibit inactivated Na^+ and Ca^{2+} channels and to noncompetitively block α and β-adrenergic receptors. In ischemic conditions, its class I effects ($\downarrow V_{max}$) predominate, but long-term use shows both increased refractory period plus $\downarrow V_{max}$. Amiodarone is effective against a wide range of arrhythmias but does have limitations, namely an extremely long elimination half-life (14 to more than 100 days) and adverse effects that include bradycardia, heart block, and a wide variety of extracardiac effects. The drug yields an active metabolite, desethylamiodarone, that has a half-life longer than the parent molecule. Optimal antiarrhythmic effects usually take several weeks to develop following oral administration. Amiodarone was the subject of an extensive review by Gill and co-workers (151) in 1992. The authors compared the European practice (where low dose amiodarone is widely used) with the North American practice (where the drug is used in high doses as a treatment of 'last resort' for life-threatening arrhythmias). However, amiodarone therapy is undergoing revaluation. There is no doubt about its efficacy in treating many arrhythmias, and its lack of significant adverse hemodynamic effects makes it valuable when there is significant pump failure. At the 1993 symposium on drugs that prolong the refractory period, considerable emphasis was given to amiodarone. The clinical settings for the appropriate use of amiodarone and sotalol were reviewed. The empirical use of both drugs for preventing ventricular arrhythmia and sudden death was considered justified in patients with coronary artery disease, but use in heart failure had yet to be defined; direct comparisons of amiodarone with sotalol were limited and not sufficient to establish which drug was superior overall (152). A report was given of the CASCADE trial which found that amiodarone was superior to other antiarrhythmics in preventing recurrence of VF in patients who had survived an episode of out-of-hospital VF (153). The symposium also discussed numerous other trials of amiodarone that are now under way in a number of countries but no final outcomes were presented.

5 ISCHEMIC HEART DISEASE

5.1 Pathophysiology

Ischemia means "deficiency in the supply of blood to a particular part of the body." Ischemic heart disease (IHD) arises because of narrowing of the coronary arteries, a condition known as coronary artery disease (CAD) or coronary heart disease (CHD). It is usually a consequence of the development of atherosclerotic lesions in the coronary arteries. The major complications of CAD are angina pectoris and myocardial infarction (MI). Athersclerosis is arteriosclerosis due to the formation of atheromatous plaque (*sclerosis* means "hardening"). *Atheroma* is the term given to the characteristic change that takes place in a blood vessel wall that begins with a fatty streak or deposition of lipid in the smooth muscle wall and develops into a fibrous plaque that becomes laden with lipids (particularly cholesterol and triglycerides) and cell debris and eventually with calcium. This plaque narrows the artery and also acts as a focus on which the thrombosis can develop.

Cholesterol is a lipid needed by all cells for the synthesis of cell membranes and, in some cells, for the synthesis of other steroids. Lipids are a diverse group of biochemical substances that have in common low solubility in water and high solubility in organic solvents. Some cholesterol is obtained from the diet but most is synthesized by the liver. A diet rich in saturated fats stimulates the liver to produce cholesterol, although the propensity for this is genetically determined.

Because of their low water-solubility, lipids are transported in the blood as protein complexes known as lipoproteins. Although the transport pathway is quite complicated, it may be simplified by stating that low density lipoproteins (LDLs) transport cholesterol from liver to other tissues, including a developing atheroma, whereas high density lipoproteins (HDLs) transport cholesterol from tissues and possibly from atheromas back to the liver to be metabolized. Triglycerides are transported to the tissues mainly as very low density lipoproteins (VLDLs). There is a strong correlation between high LDL levels and ischemic heart disease, and a negative correlation between HDL levels and this disease.

The following steps are believed to occur in the pathogenesis of atherosclerosis.

1. Damage occurs to endothelial lining of artery, which is a normal part of the body's wear and tear, particularly where arteries bifurcate giving rise to turbulence.

2. Monocytes enter the damaged endothelium and become macrophages. Under nonpathological conditions the endothelium will be repaired and no significant lesion will develop.

3. If LDL levels are high, cell receptors that normally facilitate uptake of the lipid will down regulate and the LDLs will remain in the blood and become oxidized, thus making them targets for phagocytosis.

4. When condition 3 applies, macrophages and smooth muscle cells of damaged vessels take up cholesterol from LDLs and triglycerides from VLDLs to form an atheroma.

5. The atheroma disrupts the repair process, leading to fibrosis, deposition of calcium, and hence to plaque formation and hypertrophy of arterial muscle. This leads to narrowing of the lumen of the vessel and to reduced blood flow.

6. Because of plaque formation, the production of prostacyclin and EDRF by vessel endothelium is reduced allowing the thromboxane A_2, endothelin, leukocyte adhesion molecules, etc. to become dominant, leading to further plaque development and platelet aggregation (endothelial-derived factors were discussed earlier).

7. Thrombosis and vasospasm may be triggered by the above events, leading to complete blockage of the artery. The tissue supplied by that artery then begins to die giving rise to an infarct.

Risk factors for the development of IHD include genetic predisposition, age, male sex, and a series of "reversible" risk factors such as smoking and lack of exercise. However, none of these factors is significant if serum cholesterol levels are below about 4.7 mmol/L. Once the levels rise above this value, the risk of IHD increases and the other risk factors then compound to increase the risk further. Contrary to previous reports, monounsaturated fats such as olive oil are as least as effective and probably more effective than polyunsaturated dietary fats in protecting the body from ischemic heart disease. Increased cholesterol levels are the main reversible determinant of risk of ischemic heart disease.

5.2 Myocardial Infarction

Myocardial infarction (MI) is necrosis (death) of a wedge-shaped portion of myocardium due to blockage of a coronary artery. The condition is also referred to as acute myocardial infarction (AMI) and is commonly called a heart attack. AMI is predominantly a disease of the left ventricle but may extend to the right ventricle. Death occurs as a result of one or more of the following: (1) cardiac arrest, usually due to ventricular fibrillation (VF); (2) acute left ventricular failure leading to acute pulmonary edema; (3) cardiogenic shock (massive heart failure due to extensive damage to myocardium); and (4) rupture of heart muscle (rarely).

Drug therapy for the treatment of AMI needs to be individualized according to presenting symptoms and signs. Drugs used in the treatment of AMI include oxygen; morphine; vasodilators; aspirin; heparin; thrombolytic enzymes; antiarrhythmic agents; and a variety of other adjuncts such as atropine, inotropic agents such as dopamine and dobutamine, laxatives, and sedatives. If there are no significant hemodynamic or rhythm disturbances, the patient is mobilized after 2–3 days and discharged after about 1 week. Most of the drugs used in the treatment of AMI are discussed in other chapters of this book. The following résumé is given for reasons of cross-reference.

Morphine is given routinely to relieve pain but also has other useful properties (reduces hyperventilation and is a mild vasodilator). However it can cause vomiting (an antiemetic such as i.v. metoclopramide may be needed), and it can cause hypotension and bradycardia (overcome by elevating extremities; if extreme atropine sulfate is given). Oxygen-enriched air is given to assist breathing; it may also limit infarct size (controversial). Vasodilators are valuable if there is significant pump failure, especially where this leads to acute pulmonary edema. Low dose aspirin (100–300 mg day) is given early if there are no contraindications. Heparin is used to prevent deep venous thrombosis (DVT) and pulmonary embolism, especially in patients requiring prolonged bedrest (effects on coronary thrombosis are controversial).

Thrombolytic enzymes, if given early, will greatly reduce mortality; streptokinase is usually the agent used, because it is effective and, at the moment, much cheaper than recombinant human tissue plasminogen activator (rTPA). Antibodies to streptokinase develop within 5 days of administration and last for about 1 year and will inactivate streptokinase if readministered during that period. Thus, if the patient has received streptokinase in the previous 12 months, rTPA will be used if there is a need for further treatment. Antiarrhythmic drugs are used when appropriate.

Most patients who survive long enough to be treated show some signs of cardiac arrhythmia. These are monitored on the electrocardiograph, and if they appear to be life-threatening or could develop into life-threatening conditions, appropriate antiarrhythmic drugs are given. The great concern is cardiac arrest due to fibrillation or asystole. If these conditions are not reversed within minutes, death will occur. Patients showing significant tachycardia, or salvos of ectopic beats, are likely to progress to fibrillation. In such cases, antiarrhythmic drugs may be commenced. Acute left ventricular failure will be discussed under Heart Failure (Section 9.3). Shock is a sudden and precipitous fall in blood pressure; when caused by pump failure, it is called cardiogenic shock. Mortality is high (>80%). Treatment involves use of inotropes and balloon counterpulsation.

5.3 Secondary Prevention of Myocardial Infarction

Patients who survive an MI are likely to suffer subsequent attacks. Strategies to prevent this are called secondary prevention of MI. The aim of treatment is to prevent life-threatening arrhythmia; prevent or limit coronary thrombosis; and limit continued progression of atherosclerosis. Generally,

the use of class I antiarrhythmic drugs in the secondary prevention setting has been unsatisfactory, although clear benefits have been shown for the use of class II drugs (β-blockers). Trials now under way of class III drugs may justify long-term prophylactic use of these drugs. Antiplatelet drugs, namely low dose aspirin, have been shown to be of value and should be given unless contraindicated. They should be withdrawn 2 weeks before surgery. The oral anticoagulants (vitamin K antagonists) are used when there is risk of thrombosis, otherwise their adverse effects seem to outweigh possible benefits. Hypolipidemic (lipid-lowering) drugs are receiving much attention. The older drugs are of questionable value but the HMG Co-A reductase inhibitors (simvastatin, fluvastatin, pravastatin, lovastatin, etc.) have been shown to be effective. These drugs inhibit 3-hydroxy-3-methylglutaryl-coenzyme A (HMG Co-A) reductase, an hepatic enzyme involved in the formation of mevalonic acid, a precursor for cholesterol and the rate-determining step in its synthesis. As a result, cholesterol synthesis falls and LDL receptors upregulate (increase in number). These drugs reduce LDL by 35–45%, reduce TG by 15–20%, and increase HDL by 5–10%. ADRs are usually not a problem and compliance is excellent. As with the treatment of AMI, the above résumé was given for purposes of cross-reference. The reader should consult other sections of this book for a detailed discussion of the drugs mentioned.

Epidemiological evidence suggests that the low incidence of coronary heart disease among populations who have a high intake of certain fish is due to the unique types of polyunsaturated fats found in these fish, namely omega-3 polyunsaturated fatty acids such as eicosapentaenoic acid, which seem to retard the development of IHD. The demographic evidence that fish oils protect against IHD is quite strong, although there are no definitive results from rigid controlled, randomized prospective studies. Almost certainly, they reduce plasma triglycerides and possibly cholesterol levels. They are also claimed to reduce platelet aggregation, vessel inflammation, and possibly blood pressure.

People with no previous history of MI but with a high level of risk factors are also given prophylactic treatment aimed at reducing cholesterol (diet and, when justified, HMG-CoA reductase inhibitors). Hypertension should be treated and lifestyle modification implemented (cease smoking, change diet, exercise and weight loss, if appropriate). Other disease states that predispose to IHD, e.g., diabetes, should be controlled.

5.4 Angina Pectoris

Angina is a tight strangling pain. Angina pectoris is angina of the chest (pectoral region). It is a form of temporary cardiac ischemia and arises when the coronary vessels do not supply sufficient blood (i.e., oxygen) to the myocardium. It is typically precipitated by exertion and relieved by rest or sublingual glyceryltrinitrate (GTN). Angina occurs because the blood supply to the myocardium via the coronary vessels is insufficient to meet the metabolic needs of the heart muscle for oxygen. The result is acute reversible left ventricular failure. There are two principal types of angina: exercise-induced and nonexercise induced (latter also known as variant or Prinzmetal's angina), the former is the most common.

Exercise stimulates the heart to beat more rapidly and more forcefully. To meet this increased work load, the myocardium needs more oxygen (i.e., blood). Normally this is achieved by dilation of coronary blood vessels. However this is inhibited in an antherosclerosed vessel. As a result, the myocardium becomes ischemic and begins to fail.

A vicious circle is set up in which the falling cardiac output causes the sympathetic nervous system to stimulate heart rate and contractility further, causing even greater ischemia and greater failure. The attack is usually self-terminating, but can proceed to a fatal outcome. The vicious circle can be broken by vasodilator drugs (notably, glyceryltrinitrate).

Nonexercise induced (variant) angina is thought to be due to vessel spasm. The mechanism is the same as described for exercise-induced angina, except that the initial ischemic event is induced by spasm of the vessel not demands of exercise and can occur when the person is at rest. It is less common than exercise-induced angina.

The treatment of angina includes (*1*) avoidance of precipitating factors, (*2*) treatment of the acute attack with GTN, (*3*) the use of drug therapy to prevent an attack (prophylaxis), (*3*) exercise within tolerance, and (*4*) coronary artery bypass graft (CABG). The nitrates are used for both the acute attack and for prophylaxis. Calcium channel blockers and β-blockers are used for prophylaxis.

An important strategy in the treatment of many cardiac conditions such as angina and heart failure is reduction of the work load of the heart. This may be defined in terms of the number of beats per minute and the work required per beat. Two terms are used to describe the cardiac workload associated with each beat: preload and afterload. These terms may be defined in many ways, but the most useful, in the context of discussing drugs, is to define *preload* as the volume of blood that fills the heart before contraction. Contraction of the great veins will increase preload, dilation of veins reduces preload. *Afterload* may be defined as the force that the heart must generate to eject blood from the ventricles. It is determined largely by the resistance in the arterial blood vessels. Contraction of these vessels increases afterload, dilation reduces afterload. There are more complete physiological definitions of preload and afterload than those given here but, for the purposes of this chapter, these definitions highlight the features essential for understanding drug action.

6 ORGANIC NITRATES

Nitrates are general vasodilators. Their main effect is venodilation, which reduces venous return (reduces preload), thereby reducing the volume of blood that the heart must pump per beat. This reduces the stress on ventricular walls and thus the myocardial demand for oxygen. At higher doses, the nitrates cause arteriodilation, which reduces the pressure against which the heart must eject its output (reduces afterload). There is also a small but useful dilation of coronary arteries, especially if the heart has developed collateral arteries to bypass the stenosed (narrowed) arteries. The overall effect is thus to reduce the workload of the heart and overcome its oxygen deficit, thus breaking the vicious circle referred to earlier. The nitrates also reduce coronary vessel spasm and decrease platelet aggregation. GTN is the nitrate of choice for treating an acute anginal attack. It is administered by the buccal route, either as tablets (placed under the tongue or in the pouch of the mouth) or by spray. This gives rapid absorption and avoids the extensive first-pass metabolism that occurs when GTN is ingested. GTN is also taken bucally as a prophylactic measure immediately before activities known to precipitate an attack. Sustained prophylaxis can be achieved by the use of transdermal GTN patches or oral longer-acting nitrates. Tolerance develops in sustained prophylactic therapy, and a nitrate-free period of about 8 h daily is needed to avoid this. The basis of nitrate tolerance is discussed later. Comprehensive reviews of the clinical pharmacology and therapeutic use of organic nitrates are available (154,155). The proceedings of a symposium on GTN therapy was published in 1992 in *The American Journal of Cardiology* (156).

The effectiveness of nitrates in the treatment of angina was first described in the nineteenth century, and until recently, the drugs have been regarded as a useful but rather uninteresting group of pharmacological substances. This has now changed with the discovery that the organic nitrates act by releasing nitric oxide (NO) and that NO is produced naturally by the endothelial cells of blood vessels and acts as a transducing agent that carries a signal from vascular endothelium to vascular smooth muscle, causing the latter to relax. Of further interest is the role of NO as a negative inotropic agent on the myocardium. Before considering the organic nitrates used therapeutically, the role of endogenous production of NO will be discussed.

6.1 Transduction Role of Endogenous NO

In 1980, Furchgott and Zawadzki (157) reported the existence of an endothelium-derived relaxing factor (EDRF). The nature of this relaxing factor remained unknown until 1987, when it was found that EDRF was the lipophilic gas NO that was produced in epithelial cells by the action of molecular oxygen on L-arginine. NO, because of its small size and lipophilic nature, diffuses rapidly from endothelial cells to the underlying vascular smooth muscle, where it binds to the haem of guanylyl cyclase and activates the synthesis of cGMP, which in turn reduces the levels of free intracellular Ca^{2+}, leading to muscle relaxation (158). Other biochemical actions have been attributed to NO, and its effects are not limited to vascular smooth muscle but are important in central and peripheral nervous systems, the immune system, and the myocardium (158). Among the actions described for NO are nitrosylation of thiol groups capable of modifying the action of enzymes and receptors (159). The synthesis of NO requires the presence of a series of cofactors, including calmodulin. At least three distinct NO synthase (NOS) enzymes have been described, two of which are $[Ca^{2+}]_i$-dependent and one is not. A $[Ca^{2+}]_i$-activated membrane-associated 135 kDa NOS has been cloned from endothelial cells (160,161); a $[Ca^{2+}]_i$-dependent 268 kDa neuronal NOS has been cloned and found in certain neuronal tissues and in skeletal muscle (162); and a $[Ca^{2+}]_i$-independent 130 kDa NOS produces NO in macrophages in response to stimulus by bacterial endotoxins or inflammatory cytokines (163). Similar $[Ca^{2+}]_i$-independent NOSs have been found in a wide variety of tissues, including cardiac myocytes.

The factors that stimulate production of endogenous NO are now under active investigation (164). It is likely that any mechanism, including voltage and ligand-stimulated processes, which increases $[Ca^{2+}]_i$, will activate NOS in cells where it is located. There is evidence that shear stress on vascular endothelium caused by increased blood velocity and viscosity stimulates NOS, causing the production of NO, which relaxes vascular smooth muscle to compensate for the stress (Fig. 28.21). It also seems likely that NO is not the only factor released by endothelium to relax smooth muscle, an endothelium-derived hyperpolarizing factor (EDHF) (Fig. 28.21) has been postulated as well as factors generated by NO that could be more powerful than NO in relaxing smooth muscle. Hyperpolarization raises the level of the stimulus needed to open Na^+ and Ca^{2+} channels

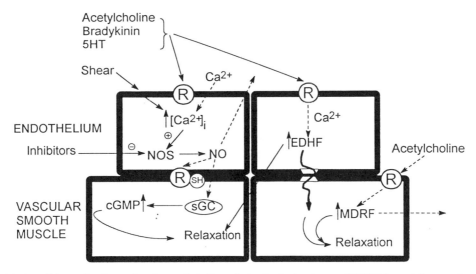

Fig. 28.21 Possible mechanisms whereby endothelium-derived relaxing factors (EDRFs) result in vascular smooth muscle relaxation. It is suggested that certain ligands (acetylcholine, bradykinin, 5HT) as well as shear forces caused by blood flow act on receptors in endothelium membranes to increase the levels of free intracellar $Ca^{2+}[Ca^{2+}]_i$. This stimulates Ca^{2+}-dependent NO synthase (NOS) to produce nitric oxide (NO), which then rapidly diffuses across cell membranes into smooth muscle and into the plasma. In smooth muscle, it stimulates soluble guanylyl cyclase to produce cGMP, which activates protein kinases that induce muscle relaxation. NO may also directly modify smooth muscle receptors by combining with thiol groups, thereby modulating the effects of other signal transducers. There are probably several other EDRFs: one suggested candidate is an endothelial-derived hyperpolarizing agent (EDHF) that may hyperpolarize the endothelial cell and transmit this effect through low-resistance pathways to smooth muscle cells. A muscle-derived relaxing factor (MDRF) has also been suggested, the synthesis of which is stimulated by direct action of a acetylcholine on smooth muscle membrane receptors. This may also diffuse to other smooth muscle cells (164). Courtesy of *Trend. Pharmacol. Sci.*

and initiate muscle contraction. It is conceivable that hyperpolarization initiated in an endothelial cell could spread to a myocyte by a low resistance pathway (electronic coupling) between cells (Fig. 28.21). Muscle-derived relaxing factors have also been suggested. There is also evidence for synergy between NO and other regulators of vascular tone.

There is now evidence that toxic shock (massive fall in blood pressure) consequent to bacterial infection may be due to excessive NO release (158). Lipopolysaccharides released from bacterial cell walls can stimulate the release of inflammatory cytokines, such as interleukin 1β, tumor necrosis factor, and γ interferon, which stimulate $[Ca^{2+}]_i$-independent NOS in blood vessel endothelium. The search for selective inhibitors of factors involved in delayed septic hypotension is now being undertaken by the pharmaceutical industry and includes inhibitors of $[Ca^{2+}]_i$-independent NOS and of selective cytokines such as the tumor necrosis factor. NO has been shown to combine with superoxide anion (O_2^-) to form peroxynitrite ($ONOO^-$) which may be involved in free-radical tissue injury, for example, following AMI (164). This observation could lead to the search for agents to protect against this reaction in ischemic disorders.

Another interesting line of investigation is the possible relationship between NO and the antihypertensive action of the ACE inhibitors. Bradykinin has long been known to lower blood pressure by reacting with receptors on endothelial cells, which then relay the response to the underlying smooth muscle. It has now been suggested that NO may be the transduction agent in this mechanism. Since the angiotensin-converting enzyme is known to inactivate bradykinin, it would follow, if the previous hypothesis is correct, that part, at least, of the antihypertensive effects of ACE inhibitors arises from protection of bradykinin from breakdown by ACE. In some people, this action is accompanied by the annoying side effect of bradykinin-induced cough. It is also possible that ACE inhibitors may stimulate basal production of NO (165). As discussed earlier, the development of an atheroma involves hypertrophy of vascular smooth muscle as well as lipid deposition. There is now evidence that such proliferation may be prevented by NO and that this may explain the effectiveness of ACE inhibitors in reversing vascular hypertrophy associated with hypertension and atherosclerosis (166).

NO reduces the force of myocardial contractility (acts as a negative inotrope), and there is evidence that this may be due to NO-stimulated formation of cGMP–protein kinase that inhibits the I_{Ca-L} current (167). In the heart, NO is synthesized not only be the endothelium of coronary vessels but by myocytes and by endocardial endothelial cells. The rate of synthesis of NO by the heart is high, and the rate of degradation is extremely high (about 100 times the rate in isolated vasculature) (168). It has been suggested that this high rate of breakdown is needed to protect the heart from the negative inotropic effects of NO (164). Reduction in cardiac function in systemic infection may be due to stimulation of $[Ca^{2+}]_i$-independent NOS by inflammatory cytokines.

6.2 Chemistry of Organic Nitrovasodilators

A number of organic nitrites (RONO), nitrates (RONO$_2$), and nitroso compounds (RNO$_2$) are, or have been, used in the treatment of angina. A selection of structures is given in Figure 28.22. Nitrogen has five electrons in its outer shell and thus can have oxidation states ("numbers") ranging from -3 to $+5$. The term *oxidation state* has no quantitative meaning. In covalent compounds, if an atom forms a bond with a more electronegative atom, the shared elec-

NITRITES

NaONO

21 Sodium nitrite

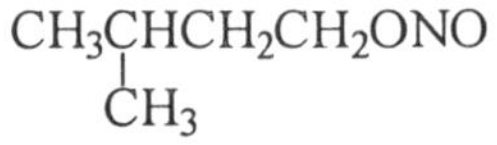

$$CH_3CHCH_2CH_2ONO$$
$$CH_3$$

22 Amyl nitrite

NITRATES

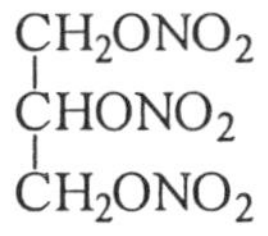

$$CH_2ONO_2$$
$$CHONO_2$$
$$CH_2ONO_2$$

23 Glyceryl trinitrate (GTN)
(nitroglycerin)

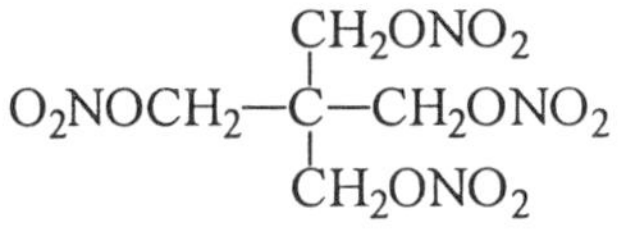

$$CH_2ONO_2$$
$$O_2NOCH_2-C-CH_2ONO_2$$
$$CH_2ONO_2$$

24 Pentaerythritol tetranitrate

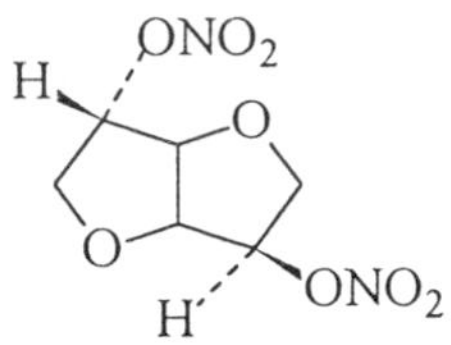

25 Isosorbide dinitrate

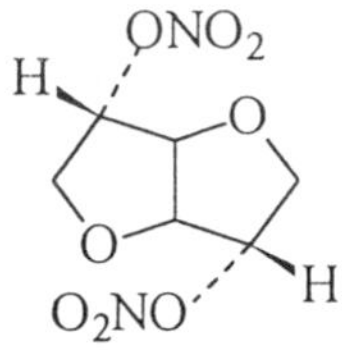

26 Isoidide dinitrate

Fig. 28.22 Organic nitrites and nitrates used in the treatment of angina pectoris.

trons will have a greater association with the latter. The less electronegative atom is then said to have undergone an increase of 1 in its oxidation state. Nitrogen is more electronegative than H or C but less than O. Thus, in most organic compounds, N shows a decrease in oxidation state (i.e., is reduced). The nitrogen in ammonia and amines has an oxidation number of -3. The oxidation states of N in NO is $+2$; in organic and inorganic nitrites, it is $+3$, and in organic and inorganic nitrates, it is $+5$. These oxidation states confer on N properties that are not associated with the reduced states. The only known reaction in mammalian biology where N changes its oxidation state is the conversion of arginine to NO.

The oxidized organic nitrogen compounds used as antianginal drugs have now been shown to be converted to NO in the body, but the processes do not involve the same enzymes as those involved in the endogenous synthesis of NO and seem to vary for different nitrovasodilators. Nitrites and nitroso compounds require a one-electron reduction, and this has been reported to occur nonenzymatically; whereas the nitrates require a three-electron reduction, and this, almost certainly, requires the involvement of enzymes. The nitroso compound nitroprusside ($[CN]_5Fe\text{-}N=O]^{2-}$) (oxidation state 3) yields NO *in vitro* in the presence of biological reducing agents such as glutathione without involvement of enzymes (169). The relevance of this to *in vivo* release of NO from nitrites has not been established, whereas enzymatic generation of NO from nitrites has been shown. Exposure of organic nitrates such as glyceryl trinitrate to high concentrations of certain thiols, notably cysteine, will also lead to spontaneous generation of NO and dinitroglycerin, but this has now been shown not to be relevant *in vivo* (170).

6.3 Bioactivation of Nitrovasodilators

There is now extensive evidence indicating that organic nitrates undergo enzymatic conversion to NO in the vasculature (171–174). The current situation is summarized in Figure 28.23 (175). Organic nitrates, once absorbed, may undergo some conversion to active substances in the blood, but the main site of biotransformation is in the vascular wall. There is general agreement that the process is enzymatic and that a variety of cofactors are involved. There is no agreement about the nature of the converting enzyme, except that it is not NO synthase (the enzyme involved in the endogenous generation of NO). Glutathione-*S*-transferases (176) and cytochrome P450 (177) have been reported to catalyze NO production from organic nitrates, but doubt has been expressed about the interpretation

of these results (178). It would appear that all nitrates are converted by a common, yet-to-be-identified enzyme and that this enzyme is not involved in the generation of NO from nitrites and nitroso compounds for which other enzymes have been implicated (175). Enzymatic liberation of NO from nitrites and nitroso compounds involves cytosolic enzymes, whereas membrane-bound enzymes are principally involved in the case of nitrates. These differences in enzyme type and location probably explain why the nitrates, in therapeutic doses, are primarily venodilators, whereas nitroprusside is mainly an arteriodilator, even though all the nitrovasodilators exert their activity through the release of NO.

It must be emphasized that the evidence that nitrates act by liberating NO, which then reacts with guanylyl cyclase to produce cGMP as depicted in Figure 28.23, is entirely circumstantial, even though compelling. The major problem is that it has not been possible, technically, to measure NO levels in smooth muscle at concentrations that would develop with therapeu-

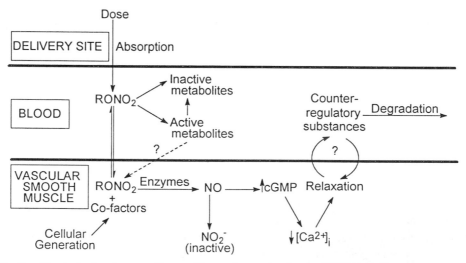

Fig. 28.23 Possible basis for the vasodilator effects of organic nitrates ($RONO_2$). The drugs may undergo metabolism in blood to produce inactive metabolites, principally nitrites (NO_2^-) and possibly active metabolites. Once the drug and possible active metabolites enter the vascular endothelium they are converted enzymatically to NO in the presence of a variety of cofactors. As detailed in Figure 28.21, production of NO leads to production of cGMP and hence to muscle relaxation. NO stands for one of the active species of this molecule ($NO\bullet$, NO^+, NO^- or a higher oxide; the actual active species has not been identified) (175). Courtesy of *Am J. Cardiol.*

tic levels of nitrate (178). Likewise, it is not known which species of NO is active. There are also major differences in the biochemical transformation of GTN when studied in whole tissues and homogenates of these tissues: major differences in stoichiometry, regioselectivity, and efficacy have been described (178). Furthermore, the enantiomers **25** and **26** are equipotent in homogenates but not in intact blood vessels (179). It is quite clear that cell homogenization alters the biotransformation of nitrates, and this is part of the reason for the conflicting reports about the biochemical mechanisms involved. Under biological conditions, NO can exist in a number of redox forms, including the free radical (NO$\bullet$), the nitrosonium ion (NO$^+$), and the nitroxyl anion (NO$^-$). Also under biological conditions, NO can give rise to higher oxides such as the nitrogen dioxide free radical ($\bullet$NO$_2$) and the peroxynitrite anion (ONOO$^-$). It is not known in which form NO is active biologically. The formation of NO from GTN also produces glyceryl dinitrate (GDN) as the other product. Two such compounds are possible: 1,2-GDN and 1,3-GDN. Preference for one or other of the reaction products is called regioselectivity. At GTN concentrations that correspond to therapeutic levels, the production of NO proceeds by pathways that are regioselective for 1,2-GDN, whereas 1,3-GDN levels increase as the concentration of GTN rises. This has been reviewed by Bennet and co-workers (178) who believe that the process that yields 1,2-GDN is the pathway that leads to cGMP production and muscle relaxation. It is quite possible that the therapeutic effects of GTN involves multiple mechanisms, even in the same tissue.

Although the major useful effect of organic nitrates in treating and preventing anginal attacks is venodilation (reduction in preload), useful dilation of coronary and other arteries contributes to efficacy. Atherosclerosis in coronary vessels usually involves certain vulnerable sections of the major coronary arteries, particularly where the arteries bifurcate, giving rise to turbulence and damage to the arterial wall. If plaque development is advanced, these sections of the arteries are not capable of dilation and a potential problem would arise if an agent dilated the minor coronary arteries. Such dilation would increase blood flow through areas where there was no obstruction, thus reducing the pressure needed to force blood through the restricted vessel into the microcirculation of the ischemic area. Such coronary "steal" does occur with some vasodilators and can precipitate or intensify an anginal attack. Fortunately, the nitrates act only on the large coronary vessels, and this has been found to be due to lack of ability of the minor vessels to convert nitrate to NO (170). NO has also been shown to reduce platelet aggregation.

6.4 Nitrate Tolerance

As suggested in Figure 28.23, the enzymatic conversion of nitrates appears to involve the participation of cofactors, and it has been suggested that these may be sulfydryl-donating compounds (170). This has led to the suggestion that depletion of intracellular thiols may be responsible for the development of tolerance to nitrovasodilators. This question has been considered in a recent review (180). While not ruling out the sulfhydryl hypothesis completely, these authors felt that sulfhydryl depletion did not explain other effects associated with the long-term use and sudden withdrawal of nitrovasodilator agents, particularly certain rebound phenomena. The authors suggest that counterregulatory vasoconstrictive processes may be involved in the development of tolerance (Fig. 28.23), for example, activation of the renin–angiotensin–aldosterone system. The evidence, at this stage, is conflicting, but overall, it supports counterregulatory mechanisms as being the major factor in the development of nitrate tolerance. Of particular significance is evidence that vascular sulfhydryl-generating systems are not reduced during the development of nitrate tolerance (181).

6.5 Pharmacokinetics of Organic Nitrates

The major feature dominating the pharmacokinetics of GTN and, to a lesser extent, the longer-acting nitrates such as isosorbide dinitrate (25) and pentaerythritol tetranitrate (24) is the existence of a high capacity hepatic organic nitrate reductase that removes the nitrate groups in stepwise fashion. The bioavailability of all orally administered nitrates is thus low and in the case of GTN renders the drug unavailable by this route. GTN and isosorbide dinitrate are both effectively absorbed by the buccal route and are used in this way for the treatment of an acute attack or for prophylaxis in situations known to precipitate an attack. Even when absorbed by nonenteral routes, the clearance of GTN is rapid (~50 L/min) and shows great variation between subjects, and even within the same subject when stressed by temperature, exercise, or changes in cardiac output. The duration of action of GTN taken bucally is 15–30 min. Isosorbide dinitrate and mononitrate have higher bioavailabilities and smaller systemic clearances, which account for their longer action. Isosorbide 5-mononitrate is a metabolite of the dinitrate and is now in clinical use in some countries because of its bioavailability by the oral route approaches 100%. Slow-release transdermal patches of GTN are now the most favored form of achieving prolonged nitrate levels for prophylaxis. Because of the development of tolerance (see above), a nitrate-free period of approximately 8 h/day is needed. This applies to both the transdermal administration and oral administration of longer-acting nitrates. The length of the nitrate-free period is thus independent of both the molecular form and route of administration and reflects the fact that the time required relates to the time taken to reverse the compensatory mechanisms or biochemical deficit that forms the basis of tolerance. Reviews of the pharmacokinetics of organic nitrates are available (154,175,182).

7 CALCIUM CHANNEL BLOCKERS

7.1 Pharmacology

Calcium channel blockers block the L-type Ca^{2+} channel (sometimes called the "slow" calcium channel) (see Section 3.1.3.2 and Figs. 28.9, 28.24). Calcium channel blockers are divided into three main chemical groups: phenylalkylamines, benzothiazepines, and dihydropyridines (Fig. 28.24). Moving across this list, there is decreasing cardiac effects and increasing vascular effects. The drugs have a wide use in cardiovascular medicine where they have three potential effects: (1) vasodilation (reduction in preload and afterload), (2) slowing of impulse generation and conduction in nodal tissue, and (3) reduction of myocardial contractility (negative inotropic effect). These effects are expected outcomes of the inhibition of the I_{Ca-L} current. However, the drugs vary in relative effects, and in clinical doses, the dihydropyridines (e.g., nifedipine) have no significant direct effects on the heart (they may cause reflex tachycardia), the benzothiazepines (e.g., diltiazem) have intermediate effects, and the phenylalkylamines (e.g., verapamil) have the most pronounced direct cardiac effects. The effects of verapamil on nodal impulse generation and conduction are useful in treating certain arrhythmia (see Section 4.3.4) and its effect on myocardial contractility can be a problem in patients with heart failure. Skeletal muscle does not require the entry of extracellular Ca^{2+} to support excitation contraction coupling and thus is not affected by calcium channel blockers in therapeutic doses.

PHENYLALKYLAMINES

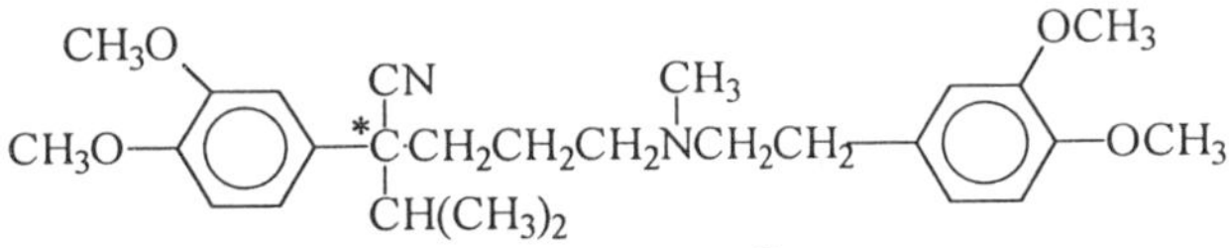

20 Verapamil

BENZOTHIAZEPINES

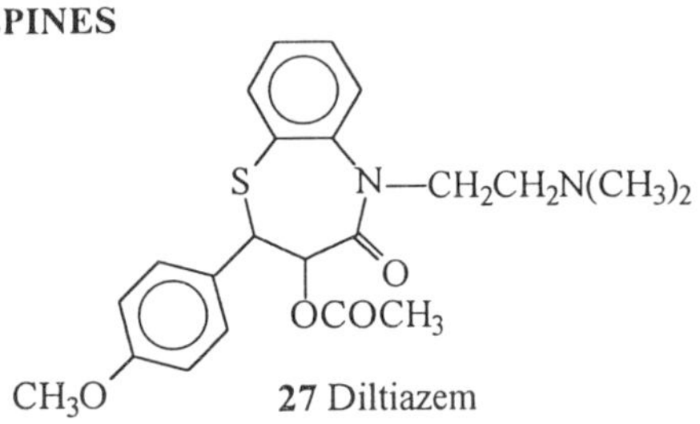

27 Diltiazem

DIHYDROPYRIDINES

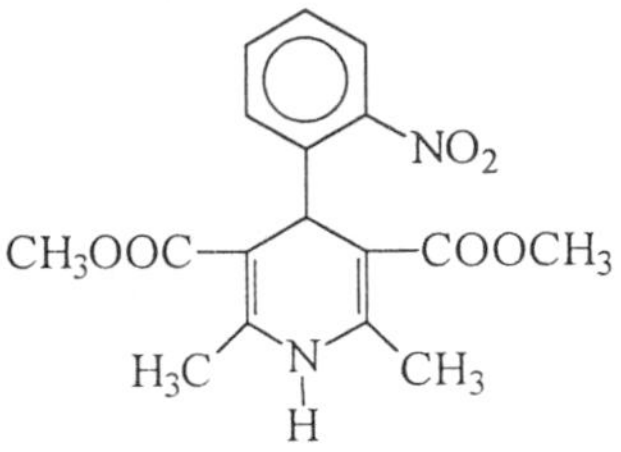

28 Nifedipine

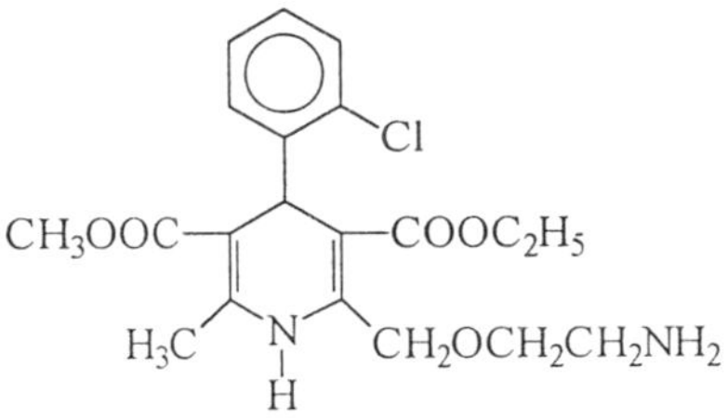

29 Amlodipine

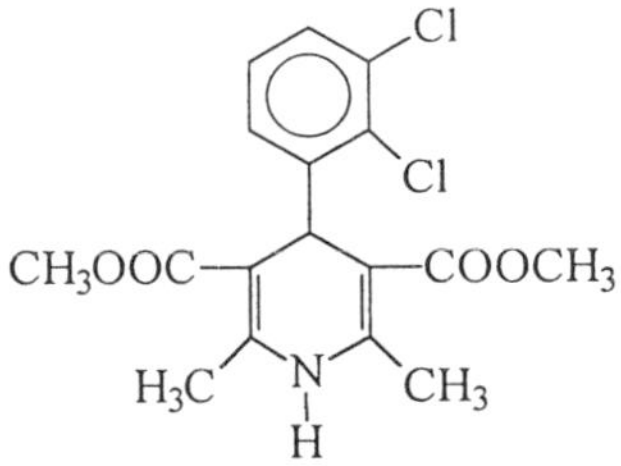

30 Felodipine

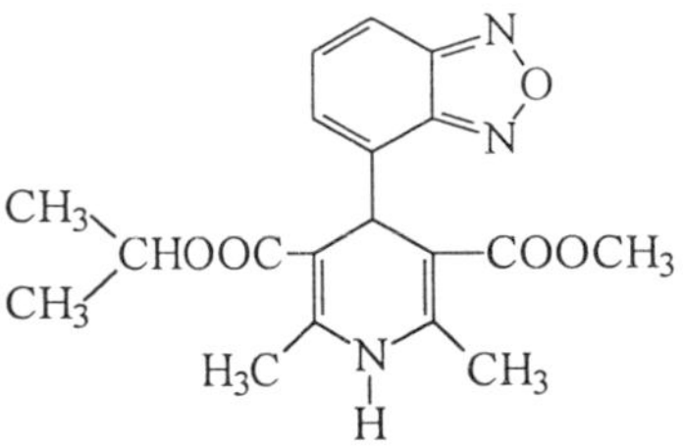

31 Isradipine

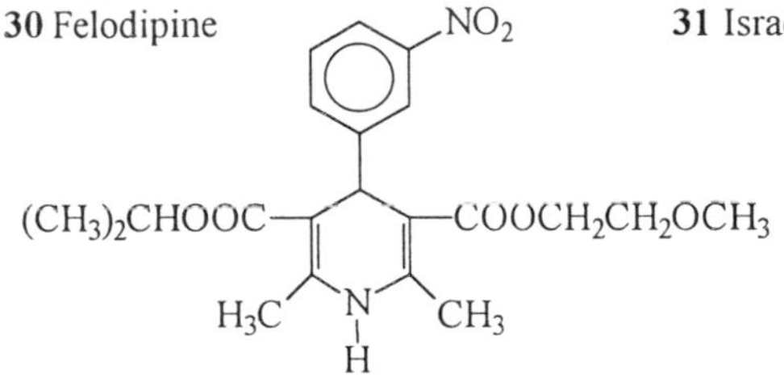

32 Nimodipine

Fig. 28.24 Structural formulae of a selection of calcium channel blocking drugs. These drugs block L-type Ca^{2+} channels (refer to Fig. 28.9).

Verapamil is a useful drug in the treatment and prevention of supraventricular tachyarrhythmias and in hypertensive patients not affected by its cardiodepressant effects; diltiazem is used mainly for the prophylaxis of angina but is also effective in treating supraventricular arrhythmias, coronary vessel spasm, and hypertension. Nifedipine is used for hypertension and angina.

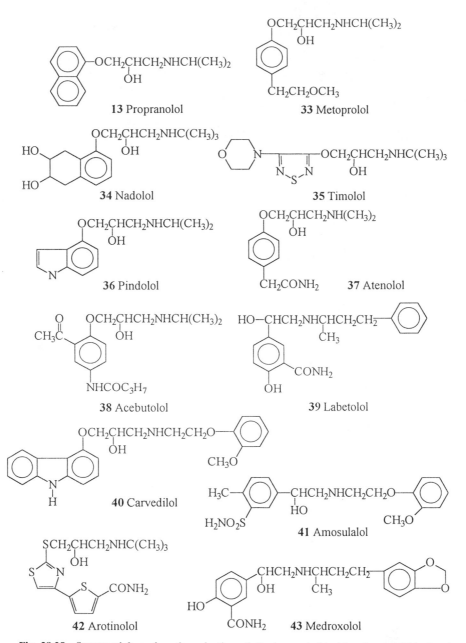

Fig. 28.25 Structural formulas of a selection of β-adrenergic blocking drugs (β-blockers).

Recent developments in the use of Ca^{2+} channel blockers have included the use of sustained-release formulations to achieve once or twice daily dosage and the development of new chemical entities, particularly dihydropyridines. Some properties of first- and second-generation dihydropyridines are given in Table 28.4. Most of the new drugs have longer elimination half-lives but, like nifedipine, have high rates of hepatic clearance and hence low bioavailabilities (183). An exception is amlodipine (**29**),

Table 28.4 Properties of Dihydropyridine Calcium Channel Blockers

Drug	Oral Bioavailability (%)	Elimination Half-Life (h)	Pharmacology (All Are Vasodilators)
Nifedipine (**28**)	45	3–4	Not selective for particular vascular beds; has slight negative inotropic effect; slow-release formulation available
Amlodipine (**29**)	64	34–58	Not selective for particular arterial beds; suitable for once daily dosing; no reflex tachycardia; little trough to peak variation
Felodipine (**29**)	16	10–12	Not selective for particular vascular beds; occasionally causes reflex tachycardia and angina; slow-release formulation available
Isradipine (**31**)	19	8	Not selective for particular arterial beds; shows potent antiatherogenic effects in animal studies, has little reflex tachycadia.
Nimodipine (**32**)	12	1	Selective for cerebral vessels; requires frequent dosing or i.v. infusion
Nitrendipine	16	8–10	Not selective for particular vascular beds; similar to nifedipine but longer acting
Nicardipine	15–40	11–12	Less potent than nifedipine but longer acting; has selectivity for coronary and cerebral arteries
Nisoldipine	4	15–16	Long duration of action; selectivity for coronary vessels; oral bioavailability poor

which has a much higher bioavailability (65%) and a long elimination half life (184).

The chronic use of Ca^{2+} channel blockers from all classes has been associated with a number of beneficial effects, including reduction in hypertension-induced left ventricular hypertrophy and antiatherogenic effects. Prophylactic treatment with diltiazem has been reported to have cardioprotective effects against AMI, but the dihydropyridines have variable and possibly harmful effects in this setting. Nimodipine (**32**) has relatively selective effects on cerebral vasculature and, with careful dosage, can achieve cerebral vessel dilation without reducing systemic blood pressure. This is useful in achieving improved cerebral perfusion in certain cerebral ischemic conditions. Overall, the properties of Ca^{2+} channel blockers are receiving renewed attention, and as the benefits of these drugs are confirmed and defined more precisely, future research will be directed toward exploiting and refining these properties. The use of second-generation Ca^{2+} channel blockers was the subject of a supplement to *The American Journal of Cardiology* in 1994 (185).

7.2 Receptor Sites for Ca^{2+} Channel Blockers

All calcium channel blockers appear to act by binding to the α_1 subunit of the L-type Ca^{2+} channel, but the receptors for the three classes of drugs are differently located. The phenylalkylamines, for example, verapamil (**20**), bind to a receptor that appears to be located at the mouth of the cytoplasmic end of the Ca^{2+} channel and may be located on the S6 transmembrane segment and adjacent intracellular amino acid residues of domain IV of the α_1 subunit (186). There is also evidence that phenylalkylamines only bind to the receptor when the channel is open (187), indicating that the receptor may lie within the mouth of the pore. As discussed in Section 3.1.4.1, certain K^+ channel blockers are thought to act by binding to the peptide loop (known as the H5 region) that connects segments S5 and S6 in the channel protein. Because of the similarity of the basic tetramer unit of K^+ and Ca^{2+} channels (Fig. 28.7) Catterall and Striessnig (26) suggested that phenylalkyamines may bind to the ends of these loops at a position near the cytoplasmic entrance to the ion pore (Fig. 28.9).

The dihydropyridine receptor site appears to lie in a hydrophobic environment (188) and is approached by the drugs from the extracellular side of the Ca^{2+} channel (189). Certain dihydropyridines, namely nifedipine (**28**), isradipine (**31**), and nitrendipine, are photoreactive and will covalently label the receptor when irradiated. These compounds label the α_1 subunit of the Ca^{2+} channel, but the location of the binding site, as indicated by traditional protein sequencing methods, is not compatible with other evidence (26). Studies using site-directed antipeptide antibodies to locate the binding site of covalently attached photoreactive dihydropyridines indicates that it is located in a cleft between the extracellular regions of domains III and IV of the α_1 subunit tetramer (190,191) (see Fig. 28.7 for a depiction of the rosettelike arrangement of tetramer units around the channel pore). It has been suggested that the binding of dihydropyridines to this site affects domain–domain interactions and thus influences the gating mechanism by an allosteric mechanism, resulting in failure of the gate to open in response to membrane depolarization (26). The benzothiazepine receptor site has not been extensively studied but

appears to be located on the α_1 subunit (192).

8 β-ADRENOCEPTOR ANTAGONISTS (β-BLOCKERS)

8.1 Overview of Effects of Drugs on Cardiovascular Adrenergic Receptors

Adrenergic receptors were initially divided into α and β. α-Receptors were further divided into postsynaptic (α_1) and presynaptic (α_2), and β receptors into β_1 (found mainly in the heart) and β_2 (found in vascular, bronchial, and uterine smooth muscle). These receptors are stimulated by endogenous catecholamines, namely noradrenaline (norepinephrine) and adrenaline (epinephrine), and are also responsible for some of the effects of dopamine. Stimulation of α_2 receptors results in inhibition of cell response. Their presynaptic location provides a negative feedback mechanism: noradrenaline released from the nerve ending combines with postsynaptic α_1 receptors, causing some stimulatory action (e.g., contraction of smooth muscle) but further release of noradrenaline is inhibited when the released transmitter combines with the presynaptic α_2 receptors. In some tissues, α_2 receptors are found postsynaptically, where they also exert an inhibitory action.

All adrenergic receptors are important sites of drug action in the cardiovascular system. Stimulation of α_1 receptors (e.g., by phenylephrine or methoxamine) causes contraction of vascular smooth muscle, which increases peripheral resistance (increases afterload) and reduces venous capacitance (increases preload). Together, these effects lead to an increase in blood pressure. Blockade of α_1 receptors (e.g., by prazosin or indoramin) causes vasodilation and lowers blood pressure, mainly by effects on afterload. Stimulation of α_2 receptors (e.g., by clonidine) leads to inhibition of transmitter release, and inhibition of

α_2 receptors (e.g., by phentolamine, which is also an α_1 blocker) prolongs the action of released transmitter. Activation of β_2 receptors (e.g., by salbutamol) causes relaxation of vascular smooth muscle and thus decreases both preload and afterload. Stimulation of β_1 receptors (e.g., by dobutamine) increases the rate and force of contraction of the heart, and stimulation of β_2 receptors (e.g., by isoproterenol (isoprenaline)) causes a fall in blood pressure. Most of the drugs mentioned above affect several classes of receptor. Thus isoproterenol affects both β_1 and β_2 receptors but has little effect on α receptors. Blockade of β receptors is discussed in the next section.

Blood vessels vary with respect to the density of α_1 and β_2 receptors on their smooth muscle cells. Blood vessels in the skin, for example, have mainly α_1 receptors (constrict when the sympathetic nervous system is activated), whereas blood vessels supplying the major tissues needed to deal with emergency situations have mainly β_2 receptors and dilate in response to sympathetic stimulation. Adrenergic effects on the heart are mediated mainly through β_1 receptors and result in an increased rate and force of contraction and thus in increased cardiac output. In summary, activation of the sympathetic nervous system increases blood pressure by effects on vascular α_1 and cardiac β_1 receptors and decreases blood pressure by effects on vascular β_2 receptors.

8.2 Blockade of Cardiac β Receptors

According to the above summary, β-blockers would be expected to decrease heart rate (which should be beneficial in angina pectoris), but their effects on blood pressure should be ambiguous since blockade of β_1 receptors should lower blood pressure

by reducing cardiac output. Blockade of vascular β_2 receptors, however, should increase blood pressure by increasing peripheral resistance. In theory, they should also be contraindicated in heart failure because they reduce the rate and force of myocardial contraction. Experience has shown that β-blockers are effective in (1) the treatment of hypertension (discussed elsewhere in this book), (2) the prevention of anginal attacks; (3) the suppression of certain cardiac arrhythmias; (4) the secondary prevention of myocardial infarction, and (5) possibly, amelioration of congestive heart failure.

β-Blockers were introduced initially as treatments for angina pectoris on the basis that they would prevent the sympathetic stimulation that precipitates and intensifies an anginal attack, as described in Section 5.4. Propranolol (13) was the first β-blocker to enter extensive use. Its effectiveness in the treatment of angina, hypertension, and secondary prevention of myocardial infarction were all reported in the mid-1960s. These discoveries led to the synthesis and testing of a huge number of β-blockers, many of which entered clinical use. The range of indications for β-blockers has also expanded and the drugs are used for cardiac arrhythmias, migraine, tremor, anxiety, schizophrenia, glaucoma, and hyperthyroidism. Figure 28.25 gives the structural formulas of some of the widely used older β-blockers and some of the newer members that have been developed as a result of renewed interest in these drugs.

β-Blockers are usually classified according to their cardioselectivity. Nonselective β-blockers block both β_1 and β_2 receptors and include propranolol (13), sotalol (14), pindolol (36), and oxprenolol. Cardioselective β-blockers have selectivity, but not specificity, for cardiac or β_1 receptors; examples include metoprolol (33), atenolol (37), and acebutolol (38). Because the latter group is not specific for β_1 receptors, they will cause some β_2-blockade and thus are contraindicated in asthmatics. Some β-blockers also block α receptors (e.g., labetolol, **39**).

β-Blockers may also be classified according to the presence or absence of other properties. Some have "intrinsic sympathomimetic activity," i.e., they are partial agonists (e.g., oxprenolol, pindolol, acebutolol); some have a membrane stabilizing or local anaesthetic activity (e.g., propranolol, oxprenolol, acebutolol); some have high lipid solubility (e.g., propranolol, oxprenolol, metoprolol, labetolol); others are water soluble (nadolol); and some are partially lipid soluble.

The differences between β-blockers are usually of little significance, but there are some guidelines. Nonselective β-blockers are more effective than cardioselective in treating tremor and anxiety, since these involve mainly β_2 receptors. Drugs without local anaesthetic activity (e.g., timolol) are best for treating glaucoma, and drugs lacking partial agonist activity seem best for treating migraine.

All β-blockers have a chiral carbon, and marked pharmacodynamic and pharmacokinetic differences exist between R and S enantiomers, depending on the activity examined. For example, in the case of propranolol, β-blockade is found only with the S isomer, whereas only the R isomer is active in inhibiting the conversion of thyroxine to triiodothyronine, and both isomers are equally active as class I antiarrhythmic agents. In addition, the plasma concentrations of R and S propranolol differ significantly after oral administration of the racemate. These differences are believed to apply to all β-blockers (only a few have been examined extensively) (193). The R and S enantiomers of β-blockers should thus be regarded as separate drugs and the administration of the racemate is equivalent to giving a mixture of different drugs, not just an active plus an inactive drug. These findings have great significance

for SAR studies, most of which have ignored the pharmacokinetic and pharmacodynamic differences between isomers.

The antihypertensive effects of β-blockers have never been satisfactorily explained. Suggested explanations include (*1*) reduction in cardiac output; (*2*) reduction in renin release, (*3*) reduction in sympathetic outflow from the CNS, and (*4*) presynaptic inhibition of neurotransmitter release (α_2 blockade). None of these explanations is capable of explaining the antihypertensive effects of all β-blockers. It is possible, that all contribute to a greater or lesser extent, depending on the particular β-blocker. This also confounds the interpretation of SAR.

The significance of the above statements for the medicinal chemist is that there exists no satisfactory pharmacological explanation of the action of β-blockers and no suitable animal models apart from general screening procedures. The only reliable guide is a well-conducted clinical trial. Even here, the drug designer should be aware of surrogate endpoints. For example, the lowering of blood pressure *per se* does not account for the exceptional ability of β-blockers in protecting against reinfarction in the secondary prevention of AMI, since many other effective classes of antihypertensives do not have this property. Similarly, there is little point in discussing structure–activity relationships when (*1*) the drugs have complex multiple effects that vary between drugs; (*2*) many of the effects seen in patients in the initial stages of drug treatment do not persist in chronic therapy; (*3*) the relevant activity to be tested is not clear; (*4*) the basis for this activity is not known; (*5*) marked differences exist between enantiomers with respect to pharmacokinetics, potency, and spectrum of activity; and (*6*) experimental laboratory models do not exist for most of the relevant biological activities, especially those relating to cardioprotective effects.

8.3 Newer β-Blockers

The so-called first-generation β-blockers developed in the 1960s and 1970s, all had significant adverse effects, including circulatory problems (Raynaud's phenomenon) and worsening of claudication (lameness, named after the Roman emperor Claudius), headache, unpleasant dreams, mental and physical lethargy, masking of hypoglycaemia (a problem in patients using hypoglycaemic drugs), and male impotence. New β-blockers have been developed to overcome these problems and to exploit further the spectrum of activity shown by these drugs. In 1994, almost 30 such drugs were in clinical trial and some have entered general use (194).

Of particular interest are β-blockers with vasodilating activity (195,196). This action is due to multiple effects, including β_2 agonism, α-receptor blockade, and possibly a vasodilator action that is independent of α and β receptors (Table 28.5). This demonstrates, once again, the complex action of this class of compounds. Some β-blockers have been developed to fill certain niches; for example, esmolol, which has a short elimination half-life, was developed for short-term intravenous use. The partial agonist xamoterol was developed with intended use in heart failure.

Other potentially useful properties have been found fortuitously, for example, carvedilol, apart from being a β-blocker, is an antioxidant and free-radical scavenger and has been shown to inhibit proliferation of vascular smooth muscle and to reduce infarct size in experimental animals (197).

Whether any of these new properties will prove to be of clinical benefit awaits long-term trials in large numbers of patients, but if the older β-blockers are any guide, the new features will prove disappointing. So far, the newer β-blockers have been tried in only small groups of patients, and the results have not been impressive. In a recent trial, carvediol, in spite of its vasodilating properties, was found not to be superior to atenolol in the treatment of mild to moderate hypertension (198). In another study, carvedilol and dilevalol, in spite of their potential vasodilating properties (Table 28.5), did not significantly lower peripheral resistance (measured by invasive techniques), and their antihypertensive effect seemed to be mediated mainly by reduction in heart rate (199). Celiprolol, which is a β_1-blocker and β_2 agonist, is more effective than atenolol in reducing peripheral resistance in hypertensive patients, but its overall antihypertensive effects do not seem superior to those of the older β-blockers, and its adverse effects are

Table 28.5   Vasodilating β-Blockers[a]

Drug	Suggested Mechanism of Vasodilation[b]
Labetotol (**39**)	α-Receptor blockade
Medroxolol (**43**)	α-Receptor blockade
Carvedilol (**40**)	α-Receptor blockade; direct vasodilation
Celiprolol	Partial β_2-receptor agonism; possibly α-receptor blockade; direct vasodilation
Bucindolol	α-Receptor blockade; partial β_2-receptor agonism
Dilevalol	β_2-Receptor agonism
Nebivolol	Direct vasodilation

[a]Adapted from ref. 194.

[b]In addition to the suggested mechanisms of vasodilation shown here, all the drugs listed were either selective or nonselective β-blockers with the effects associated with such activity.

similar (200). Nebivolol, which has direct vasodilator properties in addition to being a cardioselective β-blocker, has similar antianginal activity as atenolol (201). Larger trials may provide more evidence of some superior feature among the newer drugs, but in 13 major trials conducted between 1967 and 1992 of the older β-blockers involving more than 80,000 patients, nothing has emerged to identify any one β-blocker as being superior to any other in the treatment of hypertension (194).

The development of drugs with combined α- and β-blocking properties is receiving considerable attention. The combination is logical since α activity is associated with increased peripheral resistance and β_1 activity with increased cardiac output. Blocking both these activities should decrease blood pressure, whereas blocking β_2 activity should increase blood pressure. So far only two such drugs have received extensive study, the older labetolol and the newer carvedilol. Other drugs in this class include amosulalol (**41**), arotinolol (**42**), and medroxalol (**43**).

Labetolol (**39**) has two chiral centers, giving rise to four stereoisomers. The mixture of isomers (the form in which the drug is used) selectively blocks α_1 receptors but not α_2 and is a nonselective β blocker. Most of the α_1-blocking activity is associated with the *SR* isomer and most of the β_1-blocking activity with the *RR* enantiomer. This isomer is also known as dilevalol and has partial β_2 agonist activity. The *SS* and *RS* isomers have no adrenergic blocking activity. Labetolol is thus not a true multifunctional drug but a mixture of two active drugs with different activities plus two other compounds that seem devoid of activity. When given acutely, the labetolol mixture expresses both α and β-blocking activity, but only the latter seems to persist with chronic use (102).

Carvedilol (**40**) blocks α_1 receptors and is a nonselective β-blocker. Radioligand binding studies indicate that the drug has

some selectivity for blocking β_1 receptors, but this is not significant clinically. The drug has one chiral center, giving rise to two stereoisomers. It is used as the racemate. The isomers are equally effective in blocking α_1 receptors but β-blockade is due mainly to the *S* isomer. The ratio of α to β-blocking activity has been estimated as being between 1:10 and 1:100. Even so, the α-blocking activity is clinically significant and seems to be better maintained than with labetolol, although this has not been tested in chronic dosage (203).

In spite of the enormous disappointments that have been associated with the development of new β-blocker drugs, there does seem to be a basis for persisting with this type of research, especially since some of the properties of the newer drugs may attain greater significance when the drugs are tested over long periods. The use of β-blockers in hypertension is discussed elsewhere in this book.

8.4 Use of β-Blockers in Heart Failure (CHF)

Reports of the beneficial effects of β-blockers in heart failure date from 1975 (204) but it is only recently that the use of these drugs for CHF has attracted much attention. Before that, the general belief was that β-blockers were contraindicated in heart failure. Most experience has been with metoprolol, a β_1 selective blocker. A series of small studies has demonstrated its effectiveness (205), and withdrawal led to recurrence of symptoms (206). The possible benefits from the use of β-blocking drugs in CHF patients include reduction in sympathetic activity; restoration of β_1 receptors (these become down-regulated in CHF patients); and enhanced myocardial relaxation, which, together with reduced heart rate, leads to improved cardiac filling and improved efficiency. As with the ACE inhibitors, medicinal chemists can expect increased interest in using molecular modification to improve selectivity and enhance the properties that are emerging as useful in slowing the progression of CHF. Among the newer β-blockers undergoing trial in CHF is carvedilol (**40**), a drug that has vasodilatory properties in addition to blocking β receptors. The vasodilatory properties seem to be due to a combination of blocking α receptors plus a mechanism that does not seem to involve adrenergic receptors. So far, there have been no large-scale trials of the effects of β-blockers on survival of CHF patients, although their value in the postinfarcted patient has been established beyond doubt. Short-term hemodynamic studies are relatively easy to conduct, but experience has shown that these are surrogate endpoints, since the aim in treating heart failure is to improve the patient's quality of life and, it is hoped, delay progression of the disease. There is growing evidence that these outcomes can be achieved with ACE inhibitors (discussed later), which means that β-blockers will have to be shown to be equally effective or able to add benefit to concomitant ACE inhibitor therapy. Otherwise, β-blockers will serve only as a second-line drugs for patients who cannot tolerate ACE inhibitors, if they are found to have any value at all.

No animal model exists that will indicate whether β-blocker therapy has a role in treating CHF. Only long-term trials with the endpoints of increased exercise tolerance and increased survival will indicate whether β-blockers will prove useful in CHF, and to date, no such trial has been conducted on sufficient numbers of patients to reach a definite conclusion, although evidence of value was obtained in smaller trials. In 1994, Hampton (194) reviewed 14 trials of β-blockers in CHF. For 12 of these trials, the average number of patients per trial was 28, and for the 2 larger trials, the numbers were 380 and 516. Hampton estimated that 2000–3000 patients were needed

for any conclusion to be drawn with confidence about valid endpoints. Metaanalysis of the trials already conducted would not be valid because of (*1*) variations between the activities of the drugs studied; (*2*) variations in the etiology and severity heart failure in the different trials, and (*3*) variations in the endpoints and durations of the trials.

In summary, there are sufficient theoretical reasons and preliminary trial results to justify continued evaluation of β-blockers in the treatment of CHF. As will be discussed later, CHF is now regarded as ventricular disease involving a maladaptive hypertrophy that is triggered by some initial damage to the myocardium. Since there is now evidence that cyclic AMP may act as a stimulus for maladaptive hypertrophy, blocking its production by the use of β-blockers, especially in the early stages of heart failure, could be of great benefit. So far, nothing has emerged from the trial results to indicate what type of β-blocker is preferred, and hence no guidelines exist to aid the medicinal chemist in the design of new drugs for this purpose. Indeed, the situation may well be the same as in hypertension, namely that there is no basis for preferring one β-blocker to another, apart from certain limited niche situations.

9 CONGESTIVE HEART FAILURE

9.1 Pathophysiology

Congestive heart failure (CHF) is a condition in which the heart fails to pump sufficient blood to meet the body's needs. As a result, there is diminished blood flow through the arteries and congestion of blood in the veins. When the heart fails, a bank-up of blood occurs behind the affected ventricle leading to congestion in the circulatory system that supplies that ventricle. The symptoms and signs of CHF are largely associated with circulatory congestion and with attempts of the body to compensate for heart failure. Heart failure is due to ventricular disease, resulting from either a primary disease of the myocardium (i.e., a cardiomyopathy) or secondary to an excessive load placed on the heart. Heart failure secondary to excessive load can occur because of a volume load (excessive preload) or a pressure load (excessive afterload). If the load is placed on the left side of the heart, the patient develops left-sided heart failure (LSHF); if on the right side; right-sided heart failure (RSHF). However, failure on one side of the heart eventually causes failure on the other side (i.e., the patient develops biventricular failure). Some causes of heart failure are hypertension, heart valve stenosis (narrowing), cardiomyopathies, and ischemic heart disease.

When the heart fails, the body responds with a series of cardiac and circulatory adjustments designed to restore the fall in cardiac output. In chronic heart failure, these compensatory mechanisms initially maintain cardiac output, but as the disease progresses, they are overwhelmed by the progressive fall in contractility and, indeed, contribute to the worsening condition of the heart. In fact, the symptoms with which the patient presents are mainly the result of excessive compensation. The circulatory system attempts to compensate for a falling cardiac output by dilation of the heart; hypertrophy of the heart (increase in heart muscle mass); and activation of the sympathetic nervous system, leading to increased heart rate, increased contractility, salt and water retention, and raised venous pressure.

Serious manifestations of heart failure often appear suddenly as a result of some precipitating factor that places an additional load on the myocardium. Such factors include infection (particularly respiratory infection), pulmonary embolism, hyperthyroidism, anaemia, overexertion, excessive intake of sodium, excessive heat and humidity, emotional excesses, rapid hyper-

tension, and the development of cardiac arrhythmias.

CHF is a disease primarily of the elderly, and its prevalence will increase as the population ages unless preventative measures are taken. In the United States 3–4 million people suffer from CHF, and the rate of hospital discharge for this condition increased by 132% between 1970 and 1980 (207). This increase reflects mainly the fact that people with the end stages of the disease are being kept alive longer, rather than the aging of the population. However, the latter will have an impact with time, since the number of people in the 65+ age group will double over the next 40–50 years. Already this group accounts for about 40% of total drug costs. The development of means to prevent and treat CHF is thus an important challenge to the health care industry.

There is growing evidence that the progressive pathology of CHF arises from a neurohormonal impairment (208). CHF is associated with increased sympathetic activity and reduced parasympathetic activity. This could be secondary to CHF, or it could be a primary contributor, at least to the progression of the disease. Both experimentally and clinically, it appears that the neurohumoral imbalance associated with CHF begins with elevation of sympathomimetic activity and endothelin levels (209), followed some time later by activation of the renin–angiotensin–aldosterone system (210). Normally, endothelin is antagonized by the endothelium-derived relaxing factor (EDRF), but EDRF-dependent vasodilation is attenuated in CHF (211). There is also evidence for a specific cardiac renin–angiotensin system that is not affected by ACE inhibitors (212). Also relevant to the pathophysiology of heart failure is the possible presence in the body of an endogenous digitalis-like factor that appears elevated in CHF (213). The significance of this has yet to be determined. The endogenous factor may also play a role in the pathogenesis of hypertension (this is controversial). Impairment of the baroreceptor reflex seems to be a major contributor to the neurohormonal imbalance associated with CHF (214).

As the heart becomes damaged (usually due to ischemia or excessive preload and afterload), the cardiovascular system adapts in many ways, including myocardial hypertrophy (myocyte enlargement due to production of additional sarcomeres). Unlike normal cardiac hypertrophy (as occurs in young athletes), myocardial remodelling in damaged hearts is a maladaptive process that provides some initial benefit but accelerates the disfunction of the myocardium. This has long been known. There is now growing evidence that maladaptive myocardial remodelling begins in the early stages of the disease and is a major contributor to the process whereby heart failure begets heart failure. Attention of researchers is now being directed toward finding treatments that can be used in the early stages of heart failure to block or slow the development of myocardial hypertrophy. The hypertophied myocardium in CHF is associated with decreased capillary density, reduced number of mitochondria, increased amounts of connective tissue, and disruption of the connections (intercalated discs) between cardiac myocytes (important in signal conduction). Reversion to the production of fetal isoforms of some myocyte proteins is another feature of the maladaptive process. This probably enables larger amounts of protein to be produced but may shorten the life of the cell.

A number of major trials (e.g., SOLVD) indicate that ACE inhibitors added to established therapy with digoxin and diuretics can prolong life by an average of nine months. The possibility that earlier intervention could lead to greater prolongation is under investigation (215). One of the major areas of interest for the development of new drugs and new treatments for CHF will be that of preventing or slowing the processes that lead to maladaptive myocardial hypertrophy. The effectiveness of the ACE inhibitors in this regard has been mentioned, and there is increasing focus on the use of β-adrenergic blocking drugs to inhibit the excess sympathetic activity that is strongly associated with the initiation of the maladaptive process.

9.2 Symptoms and Signs of Heart Failure

The major symptoms and signs of heart failure arise as a consequence of venous congestion and cardiac compensation. It is useful to classify the symptoms of congestion in terms of whether the congestion is due to LSHF or RSHF. However, failure of one side of the heart usually causes eventual failure of the other side, so that the full spectrum of congestive symptoms may develop. Pulmonary venous congestion (consequent to LSHF) leads to dyspnea (difficulty in breathing), dyspnea on exertion (DOE), orthopnea (difficulty in breathing when lying down), paroxysmal nocturnal dyspnea (PND) (also known as cardiac asthma), cough, and chest rattles (rales). Systemic venous congestion (consequent to RSHF) leads to peripheral edema (excess fluid in tissues other than lung), distension of jugular veins, hepatomegaly (enlarged liver), jaundice, cirrhosis, ascites (fluid in peritoneal cavity), and proteinurea (protein in urine due to renal congestion). There are also a number of cardiac features, including enlarged heart (cardiomegaly), rapid heart rate (tachycardia), and changes in pulse and heart sounds as well as a variety of nonspecific features such as cold and clammy skin, cyanosis (pale skin), fatigue and cold intolerance, anorexia, cachexia (severe body wasting), confusion, and dulling of consciousness.

The term *congestion* means an abnormal accumulation of blood and, in the case of CHF, is due to the banking up of blood behind the failing ventricle. Congestion leads to edema (the passage of fluid from the vasculature into the surrounding tissue), and edema leads to effusions (the passage of fluid from tissues into body cavities such as the airways and peritoneal cavity).

9.3 Treatment of Heart Failure

Although heart failure has many causes, it is possible to enumerate common principles for its treatment. The management of heart failure involves (1) treatment of precipitating factors; (2) correction where possible of the cause (e.g., hypertension, defective valves); (2) control of complicating factors; and (4) improvement of myocardial function by improvement of myocardial contractility with positive inotropes; use of vasodilators to reduce preload and afterload; control of fluid retention with diuretics, and rest (physical and mental). In the application of these principles, the four major forms in which heart failure presents should be recognized. These are acute left ventricular failure, chronic CHF, resistant CHF, and pulmonary CHF (cor pulmonale).

Acute left ventricular failure giving rise to acute pulmonary edema is an emergency situation and requires vigorous and prompt treatment. It may occur as a consequence of myocardial infarction or it may represent sudden exacerbation of chronic congestive heart failure. The choice of drugs depends on the patient's hemodynamic profile and the presence of complicating factors. Drugs include morphine; furosemide (given i.v. to reduce blood volume), bronchodilators (e.g., i.v. aminophylline to assist breathing); vasodilators (e.g., i.v. nitroprusside) to shift the congested blood from the lungs to the general circulation, and other measures, depending on circumstances.

Chronic CHF is a steadily deteriorating disease with many disabling consequences. However, the skilful use of modern treatments can greatly improve the quality of life and may prolong life. Drugs used include: digitalis ("cardiotonic" agents), furosemide or other diuretics (to reduce fluid load), vasodilators (to reduce preload and/or afterload; ACE inhibitors are the drugs of choice); inotropic agents (used in special circumstances), spironolactone (if significant hyperaldosteronism is present); β-blockers, and K^+ supplements (which may be needed to compensate for diuretic-induced K^+ depletion but should not be used with ACE inhibitors unless blood K^+ levels low). In treating any form of heart failure, it is important to individualize the therapy.

The reader is referred to other chapters for a detailed discussion of diuretics, vasodilators, and β-adrenergic blocking drugs. Digitalis and inotropic agents will be discussed later in this chapter, but a few comments will now be made about the other drugs with respect to their use in CHF.

Diuretics are useful in mobilizing fluid, that is of overcoming the effects of congestion, edema, and/or effusions that are associated with untreated or severe CHF. They have contributed greatly to the comfort of patients and are lifesaving in acute episodes. However, in severe heart failure the response to diuretics declines and continuous infusion may be necessary. If overused, they can cause excessive fluid depletion, leading to a lowering of cardiac output and hepatic and renal failure. The alsosterone antagonist spironolactone (classified as a K^+-sparing diuretic) has attracted renewed interest because of experimental work that indicates that it may inhibit cardiac fibrosis (216). This finding, if confirmed clinically, will trigger further drug development in this area.

Vasoconstrictors can be beneficial in CHF, but their value depends on the class of vasoconstrictor with some having little or no benefit and some having detrimental effects (Table 28.6). The ACE inhibitors, if tolerated, are the vasodilators of choice in

Table 28.6 Effects of Various Classes of Vasodilator on Prognosis of Patients with Chronic CHF[a]

Vasodilator Class	Effect on Prognosis
ACE inhibitors	Marked improvement
Nitrates plus hydralazine	Some improvement
α-Adrenergic blockers	No improvement
Calcium channel blockers	Probably detrimental
Phospho-diesterase inhibitors	Detrimental

the treatment of CHF. These drugs inhibit the angiotension-converting enzyme (ACE), which converts angiotension I to angiotensin II. Angiotensin II causes vasoconstriction (i.e., increases afterload) and causes fluid retention (i.e, increases preload). ACE inhibitors, by blocking this, greatly reduce the load on the heart (Fig. 28.26). Angiotensin II is also a growth promoter, which stimulates cells via diacylglycerol and inositol phosphates. Thus the blocking of the production of angiotensin II or its receptors may have a direct effect in limiting myocardial hypertrophy. It is expected that much research and drug development in this area will occur over the

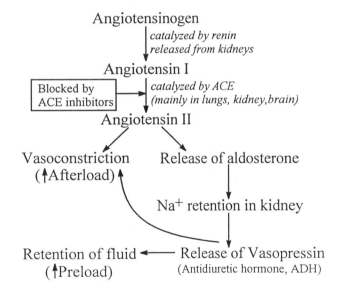

Fig. 28.26 The renin–angiotensin–aldosterone system. This system is part of the physiological mechanisms used to control blood pressure. If blood pressure falls (for example, because of blood loss) the volume of blood passing through the kidney diminishes, and this triggers the release of renin, which converts angiotensinogen to angiotensin I. This is further converted to angiotensin II by the angiotensin-converting enzyme (ACE). Angiotensin II has a direct vasoconstrictor effect (increases afterload) and also triggers the release of aldosterone, which increases salt retention and thus triggers the release of vasopressin (also known as the antidiuretic hormone, ADH). This causes fluid retention and thus increases preload and has direct vasoconstrictor effects. In disease states such as hypertension where preload and/or afterload are higher than is appropriate, inhibition of the conversion of angiotensin I to II by ACE inhibitors is an appropriate therapy. Similarly, in heart failure, where reducing the load on the heart will produce relief of symptoms, ACE inhibitors can useful.

next decade. The first major study to demonstrate conclusively the beneficial effects of ACE inhibitors was the CONSENSUS study of 1987, which showed a 31% reduction in mortality in patients with class IV heart failure treated with enalapril (218). Since then, a series of other studies have confirmed the effectiveness of ACE inhibitors and their superiority over other vasodilators in the treatment of CHF and following infarction. It seems clear that the beneficial effects of ACE inhibitors are not due simply to vasodilation but to other consequences of the lowering of angiotensin II levels. The cardioprotective effects of ACE inhibitors and their use in heart failure and ischemic heart disease were extensively reviewed (219,220).

The use of β-blockers in CHF is discussed in Section 8.4. Future directions in the development of drug therapy for CHF were recently reviewed (217). Because of the complex pathophysiology of CHF, it is unlikely that one drug will ever be adequate to provide optimum therapy.

10 INOTROPES

An inotropic drug is any agent that affects muscle contraction, regardless of the muscle concerned. Drugs that reduce muscle contraction, for example, β-blockers are called negative inotropes; drugs that increase muscle contraction (e.g., β-agonists) are called positive inotropes. When the term *inotropic* is used without qualification it is assumed to mean a positive inotropic agent.

The use of inotropic drugs in heart failure has always been controversial, since it is not clear whether the damaged heart needs rest or stimulation. If the decreased cardiac output of heart failure is an appropriate compensatory response to ventricular disease, stimulation may have adverse effects; but if depression of contractility is the prime pathophysiological event that

triggers the range of responses that leads to worsening of heart failure, then positive inotropes could alleviate symptoms as well as slowing the disease.

The usual physiological mechanism for increasing cardiac output is to reduce vagal stimulation (parasympathetic withdrawal) and, when this is not sufficient, to increase catecholamine release via sympathetic stimulation. Catecholamines stimulate β_1 receptors leading to increased production of cyclic AMP (cAMP), which in turn stimulates the voltage-gated L-type Ca^{2+} channel. This leads to an increased Ca^{2+} flux and hence to increased contractility as depicted in Figure 28.11. Cyclic AMP levels can be increased by direct stimulation of β_1 receptors or by slowing its metabolic breakdown by the use of phosphodiesterase inhibitors.

Digitalis is usually classified as a positive inotropic agent, but a growing number of authors do not believe that the concentrations of digitalis achieved by therapeutic doses results in a direct inotropic effect on the heart but, rather, that digitalis achieves all its beneficial hemodynamic effects as a result of its neurohumoral activity. This statement is controversial. Even so, digitalis is classed as a cardiotonic not an inotropic agent in this chapter, and it will be discussed later in Section 11.

10.1 Catecholamines

Catecholamines act by stimulating adrenergic receptors and, depending on selectivity, can produce some or all of the effects described in Section 8.1. Various orally active catecholamines such as prenalterol, salbutamol and L-dopa produce some short-term benefit as inotropic agents in heart failure, but they are generally unsatisfactory because of loss of efficacy and adverse effects, including tachycardia and cardiac arrhythmia. Dobutamine and dopamine can be valuable in intensive cardiac care but are inappropriate for chronic therapy.

In the doses used, dobutamine acts mainly as a positive inotropic agent, whereas dopamine has both vasopressor and positive inotropic activity. Recent work has greatly improved the knowledge of the clinical pharmacology of these drugs, namely the clinical settings in which the drugs are likely to be of most value and the dosage regimens appropriate to these settings. The reader is referred to a recent review for further details (221). This knowledge is leading to the design of new catecholamines, specifically to deal with certain acute cardiac situations. One such drug is dopexamine, which was developed to increase cardiac output by increasing stroke volume and myocardial contractility as well as reducing afterload (systemic vascular resistance). The drug has also been found to augment renal and visceral blood flow. Patients with advanced heart failure develop renal failure and anorexia because of reduced blood flow to these organs. Dopexamine acts mainly as a β_2 agonist augmented by milder stimulation of β_1 and DA1 and DA2 dopaminergic receptors (222). The value of dopexamine compared with dobutamine and dopamine has yet to be determined. It has both advantages and disadvantages compared with these drugs.

10.2 Dopaminergic Agents

Dopaminergic drugs may increase cardiac output and are thus often classified as inotropic drugs, but this effect is due mainly to reduction in preload and afterload (i.e., vasodilation) consequent to stimulation of presynaptic and postsynaptic dopaminergic receptors (DA2 and DA1 receptors). Drugs in this class include the prodrug levodopa (yields circulating dopamine) and ibopamine. Orally administered levodopa has been reported useful in the management of chronic CHF, but the trials were not sufficiently large or rigorous to enable any conclusions to be drawn (221). The benefit, if it exists, probably arises from stimulation of myocardial β_1 and vascular dopaminergic receptors. Ibopamine is another prodrug and is converted, after absorption, to the dopamine analogue N-methyldopamine (epinine). It has been evaluated in CHF patients and found to produce small improvements in hemodynamic function (slight increase in cardiac output, slight fall in systemic and pulmonary resistance, and no change or slight increase in heart rate). However, in the first hour after administration, there are unfavorable hemodynamic effects, including an increase in vascular pressure (221). It also has other undesirable effects, including a short duration of action, tachyphylaxis, and impairment of ventricular function in some patients. The drug does have one favorable property that could be the basis for further drug design, namely it improves renal function and may reduce plasma renin and noradrenaline levels, possibly due to activation of presynaptic DA2 receptors (223).

10.3 Phosphodiesterase Inhibitors

Cyclic AMP is degraded in the cytoplasm by phosphodiesterase (PDE), and hence inhibition of this enzyme is another method of elevating cAMP levels and stimulating the L-type Ca^{2+} channel. This possibility excited much interest in the 1970s and 1980s and resulted in a massive effort by a number of pharmaceutical companies to develop an inotropic agent that could replace digitalis and yet have a much wider margin of safety. However, the drugs proved a great disappointment in spite of some favorable early results. In general, they showed increased mortality rates, increased rates of cardiac arrhythmia, tachycardia, and attenuation of inotropic activity

Table 28.7   Subgroups of Phosphodiesterase III Inhibitors

Chemical Class	Examples
Bipyridines	Amrinone, milrinone
Imidazoles	Enoximone, piroximone
Benzimidazoles	Sulmazole, pimobendan, adibendan

on chronic use. The chemical classes studied are summarized in Table 28.7.

Amrinone and milrinone are examples of the bipyridine class of PDE inhibitors. Unlike theophylline, these drugs specifically inhibit only one of the three cardiac PDE isoenzymes (PDE III) (224). Both drugs are powerful inotropic agents but suffer from an unacceptable level of adverse effects (225,226). Amrinone has a short duration of action, and its oral form was withdrawn from further development, although an intravenous form was approved for clinical use. It is used in severe heart failure and must be given with a high initial bolus or infusion rate (227). Milrinone was developed in the same series as amrinone and initially appeared to have overcome some of the unfavorable effects of amrinone, including blood dyscrasias and short elimination half-life. Milrinone inhibits PDE III but probably has other activities that are potentially favorable in CHF. Small clinical trials seemed to confirm these advantages, but its performance in large trials was a disaster. In one major trial, milrinone alone or in combination with digoxin was not superior to digoxin alone in improving hemodynamics and showed more adverse effects than digoxin, especially ventricular arrhythmia (228). In another multicenter trial, in this case with more than 1000 patients, milrinone showed no clinical benefit compared with placebo but was associated with increased mortality (229). In the words of Armstrong and Moe (230), these

studies "helped close the door on this approach to the long-term therapy of heart failure." The tendency of these drugs to cause severe and often fatal arrhythmia is almost certainly due to the increase in cAMP, and given the known arrhythmogenic effects of such a maneuver and the association of excess sympathetic activity with the progression of CHF, it is not surprising that these problems occurred. However, because of the selectivity of the drugs for one isoform of PDE, it was possible that a useful inotropic effect could have been achieved without the well-known problems encountered with catecholamine therapy in the chronic treatment of heart failure.

Enoximone belongs to the imidazoline class of PDE III inhibitors and was developed for both acute and chronic treatment of CHF. In initial testing, it appeared to offer considerable hemodynamic benefit in CHF, but when subjected to rigorous trial it was not found superior to placebo (231) and probably has adverse effects on survival. Similar results have been obtained with the other imidazole, piroximone.

The benzimidazoles (sulmazole, pimobendan, and adibendan) also inhibit PDE III but may have additional properties that could be of value in CHF. For example, they appear to sensitize the sarcomeres to Ca^{2+}, but the clinical potential of this has yet to be established.

Vesarinone is a weak PDE III inhibitor with the added property of stimulating the Na^+ channel (69). The former activity is associated with higher doses of the drug, and in a trial using a 60- and 120-mg dose, the lower dose proved beneficial but the 120-mg arm of the trail had to be terminated prematurely because of a twofold increase in mortality.

In summary, the enormous effort to perfect new inotropic agents has yielded nothing of great promise and does not seem to be worth further effort, since its basic premise is probably flawed.

11 CARDIOTONIC STEROIDS (DIGITALIS)

11.1 Source and Use of Cardiotonic Steroids in Medicine

Cardiotonic steroids occur in a wide variety of plants and are secreted by the skin of certain toads. In plants, they usually occur as glycosides (known as cardiac glycosides) with one or more sugar groups attached at position 3 of the steroid. Removal of the sugar gives an aglycone or genin, which is usually much less potent than the parent glycoside. The best known cardiac glycosides are those obtained from *Digitalis* species (foxglove). The term *digitalis* is often used as a generic term for all cardiotonic steroids and authors usually refer to the cardiotonic steroid receptor on Na^+,K^+-ATPase as the "digitalis receptor," even though most of laboratory studies are conducted using the water-soluble glycoside, ouabain (**48**), a glycoside obtained from the seeds of *Strophanthus gratus*. Many hundreds of cardiotonic steroids have been obtained from natural sources or were prepared by synthetic or semisynthetic means. This subject was reviewed in 1990 (68). Some idea of the extensive chemistry of the cardiotonic steroids is given by the fact that the review (68) gave the structures of 249 cardiotonic steroids. A more concise review was published in 1992 (69).

Cardiotonic steroids are one of the oldest groups of drugs used in medicine. The glycosides from the bulb of the Mediterranean squill, *Drimia maritima*, were used by the ancient Egyptians and Italians who recognized the diuretic, emetic, and cardiotonic properties of the drug. Today, the best-known cardiotonic steroids are those from *Digitalis*, a genus of Scrophulariaceae found in Europe and Asia and known mainly by the species *Digitalis purpurea* (the purple foxglove) and *Digitalis lanata* (the hairy foxglove). The first recorded reference to the medicinal properties of the foxglove is believed to have been in A.D. 1250, although the drug was used long before that, particularly by the Welsh. Because of its high toxicity and variable content of active principals, the foxglove was rarely popular with physicians. The most notable exception was Dr. William Withering (232), who, in 1785 published a remarkable book titled *An Account of the Foxglove and Some of its Medicinal Uses* &c. Withering gave careful instructions on the use of the drug, but these were ignored and the drug fell into disuse. Withering believed that the main therapeutic benefits of digitalis arose from its diuretic effects (leading to reduced congestion and mobilization of fluids from edematous areas).

Initially, digitalis was used in the form of powdered plant, but with the isolation of its active principles, the pure steroids were used. Although many such compounds have been isolated and tested, only digoxin (**47**) remains in common use. Digoxin is used for two main indications: treatment of CHF and treatment of atrial tachyarrhythmias. In the case of the latter, the drug suppresses conduction through the AV node, thus protecting the ventricles from atrial arrhythmia. This effect is probably due to reflexogenic effects on afferent neurones, rather than a direct action on the nodal tissue.

In the period following World War I pharmacologists found that digitalis applied to isolated myocardium increased the force of contraction. It was, therefore, assumed that this was the basis for its useful effects in treating CHF, in spite of the fact that the concentrations needed greatly exceeded those used clinically, even when studied in species with sensitivities to digitalis comparable to humans. Although this issue is far from resolved, there is a growing body of evidence that indicates that therapeutic doses of digitalis are not significantly inotropic but improve hemodynamics mainly by a neurohumoral action. As discussed earlier, neurohormonal impairment is probably the basis for the progressive pathology of CHF. There is also growing acceptance of the Gillis-Quest hypothesis that the useful effects of digitalis arise, at least in part (if not entirely) as a result of reversing these neuronal effects. Malfunction of negative feedback mechanisms has been suggested as a cause of the loss of reflex control in CHF, and there is now clinical evidence that digoxin and certain other therapeutic measures act by improving reflex function (233,234). The data indicate that therapeutic doses of digoxin potentiate baroreceptor-mediated afferent regulation in CHF patients. Administration of digoxin in chronic CHF patients has also been shown to suppress the renin–angiotensin system (235) and to have sympatho-inhibitory effects (236). As discussed earlier, increased sympathetic activity is one of the earliest signs of neurohormonal disfunction in CHF.

The efficacy of digitalis in the treatment of CHF has been challenged by a series of studies that showed that withdrawal of digitalis did not significantly worsen cardiac function. Most of these studies had design faults, and more recent studies have restored the reputation of digitalis as a useful drug for treating CHF patients in normal sinus rhythm as well as in atrial tachycardia. Publications dealing with this controversy were reviewed in 1990 (68). In spite of its long use, it is only recently that digitalis has been tested using adequate trial conditions. These studies provided conclusive evidence for the efficacy of digoxin in the treatment of CHF (228,237–239). Of considerable significance was a 1993 trial that examined the effects of digoxin withdrawal on 178 patients with class II or class III CHF who were clinically stable on a combined therapy of diuretics, ACE inhibitors, and digoxin. Digoxin withdrawal from the triple therapy lead to a greater than fivefold increase in recurrent symptoms and morbidity (240). In 1992, *The American Journal of Cardiology* published the proceedings of a symposium dealing with the role of digoxin in the treatment of heart failure (241).

11.2 Chemistry

If *digitalis-like activity* is defined as the ability to inhibit Na^+,K^+-ATPase plus the ability to produce a positive inotropic effect, such activity is shown by a diverse group of natural, semisynthetic, and purely synthetic compounds. Naturally occurring digitalis-like substances belong to three groups: butenolides, pentadienolides, and erythrophleum alkaloids, representative examples of which are given in Figure 28.27a. The first two groups comprise what is usually meant by the term *cardiotonic steroid* or *digitalis*. They will be referred to as the classical cardiotonic steroids. Their unique features include a C/D cis ring junction, a 17β lactone, a 14β-OH, and usually an A/B cis ring junction. The cis–trans–cis steroid configuration found in most classical cardiotonic steroids is depicted in Figure 28.28 and explained in the legend. The numbering of the steroid ring system is shown in structure **44**. Naturally occurring cardiotonic steroids usually occur as glycosides with the sugar attached to carbon-3 (C3) of the steroid (a few sugars are bilinked to the steroid through C2 and C3). Steroids lacking sugar moieties are called genins or aglycones.

The sugars found in naturally occurring cardiac glycosides include common sugars such as glucose and rare sugars such as thevetose (**69**). Glucose (**78–81**) is a fully oxygenated hexose, but many of the sugars found in cardiac glycosides lack one or more oxygen atoms (deoxysugars). Usually, glycosides containing deoxysugars are more potent than those with fully oxygenated sugars. The glycoside sugars may occur in a variety of isomeric forms (Fig. 28.29) and may be present as monosides (e.g., ouabain,

BUTENOLIDES

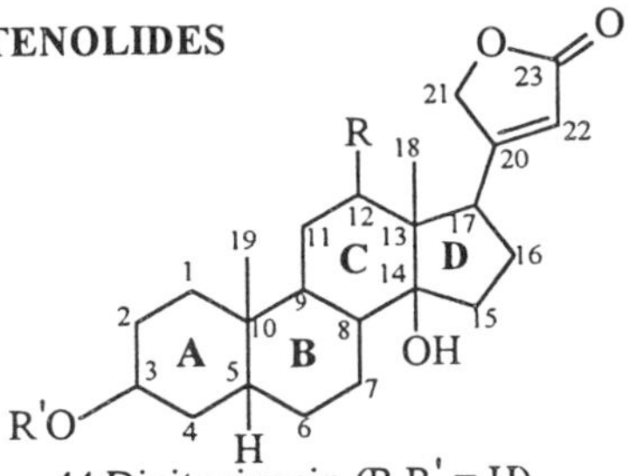

44 Digitoxigenin (R,R' = H)
45 Digitoxin (R = H; R' = digitoxose3)
46 Digoxigenin (R = OH; R' = H)
47 Digoxin (R = OH; R' = digitoxose3)

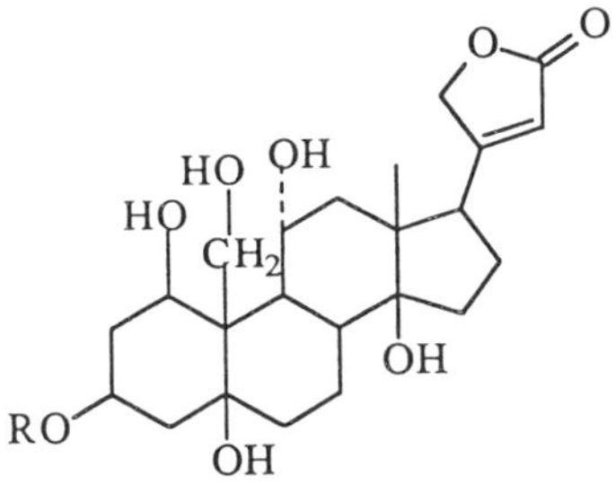

48 Ouabain (R = rhamnose)

PENTADIENOLIDES

49 Hellebrin

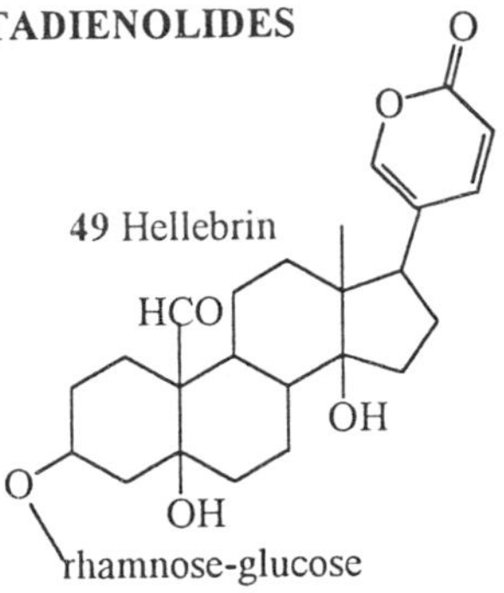

50 Bufalin

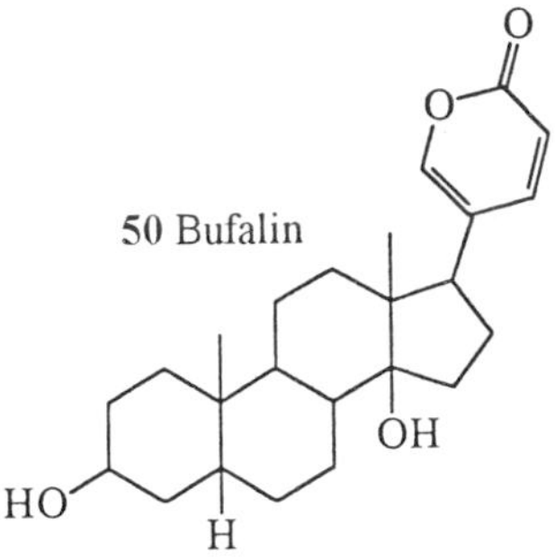

rhamnose-glucose

ERYTHTROPHLEUM ALKALOIDS

51 Cassaine

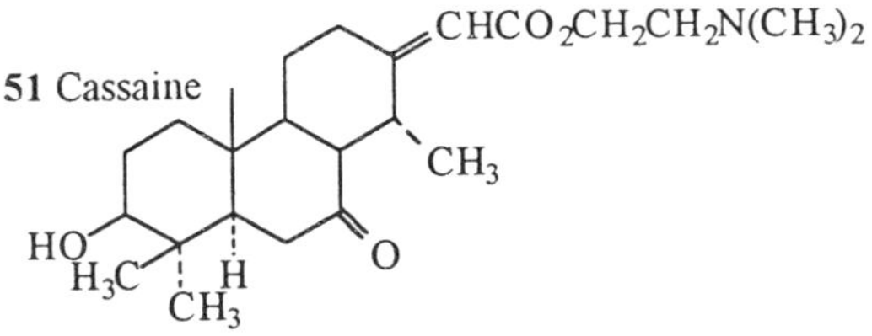

CHCO₂CH₂CH₂N(CH₃)₂

C17β–MODIFIED DERIVATIVES OF DIGITOXIGEN

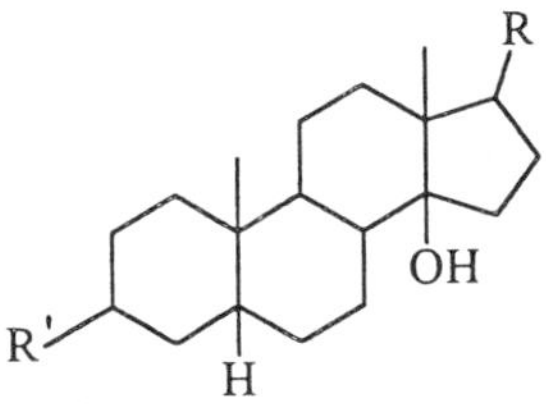

52 R: —CH=CH—CN
53 R: —CH=CH—COOCH₃
54 R: —CH=CH—COOCH₂CH₃
55 R: —CH=CH—COOCH₂CH₂CH₃
56 R: —CH=CH—COCH₃
57 R: —CH=CH—COOH
58 R: —CH=CH—CONH₂
59 R: —CH=N—NH—C=NH₂ (NH₂, ⊕)
60 R: —CH=N—NH—C—NH₂ (O)

R' = H: Genin
R' = glucose: Glucoside

(a)

Fig. 28.27 (a) Structural formulas of naturally occurring and semisynthetic compounds with digitalis-like activity.

BISGUANYLHYDRAZONES

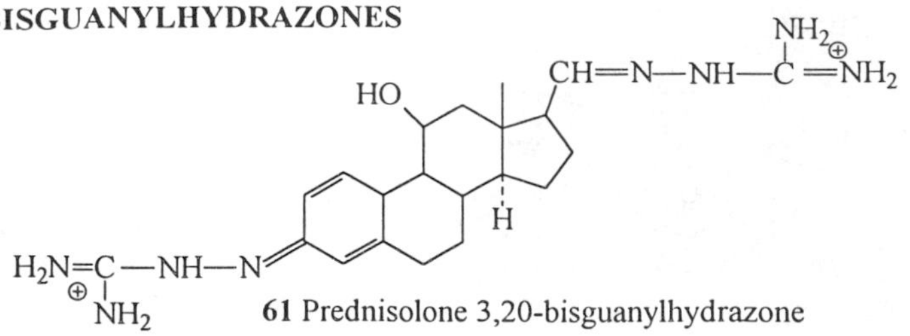

61 Prednisolone 3,20-bisguanylhydrazone

PREGNANE DERIVATIVES

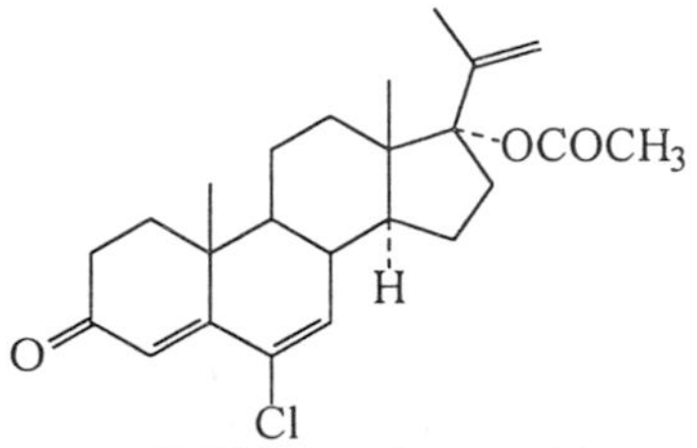

62 Chlormadione acetate

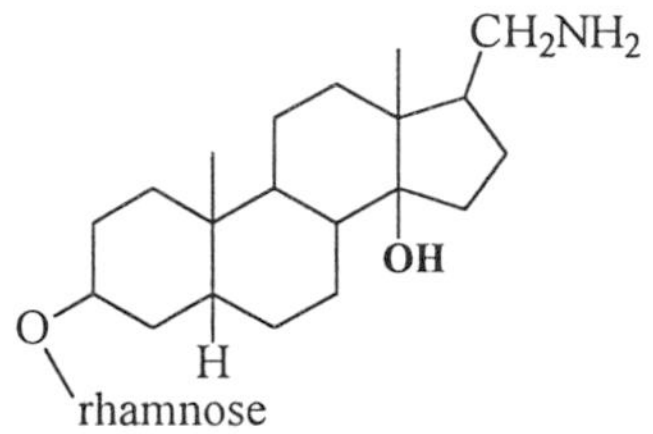

63 JT 246

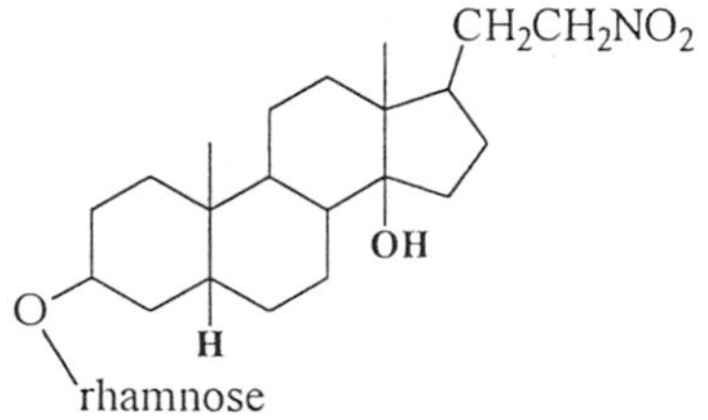

64 JT 253

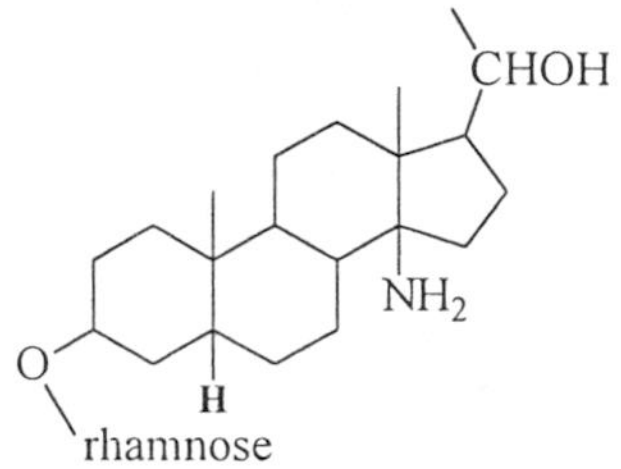

65 LND 623

N-CONTAINING 17β HETEROCYCLICS

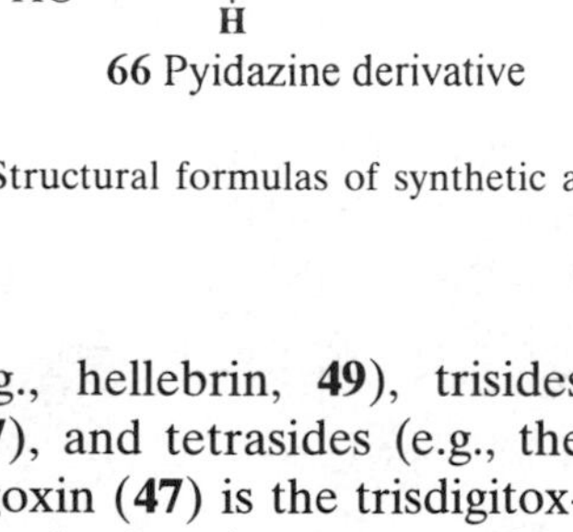

66 Pyidazine derivative

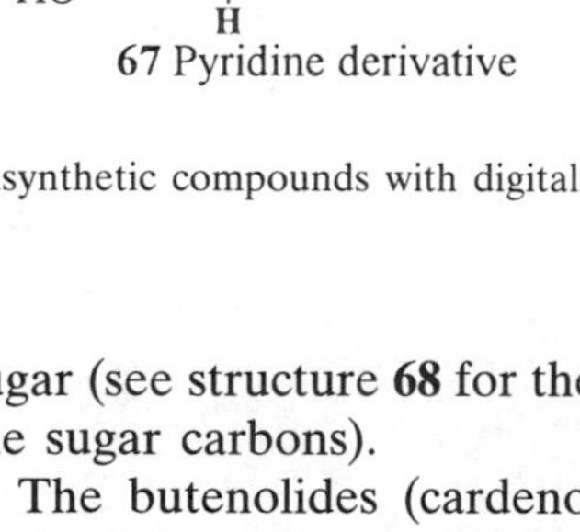

67 Pyridine derivative

(b)

Fig. 28.27 (b)   Structural formulas of synthetic and semisynthetic compounds with digitalis-like activity.

48), bisides (e.g., hellebrin, **49**), trisides (e.g., digoxin, **47**), and tetrasides (e.g., the lanatosides). Digoxin (**47**) is the trisdigitoxoside of digoxigenin, i.e., its sugar moiety consists of three digitoxose units (trisdigitoxose, **68**). Digitoxose is a 2,6-deoxy-

sugar (see structure **68** for the numbering of the sugar carbons).

The butenolides (cardenoides) (e.g., digoxin, **47**) contain an α,β-unsaturated lactone (derived from 4-hydroxybutenoic acid) attached to the 17β position of a 14β-

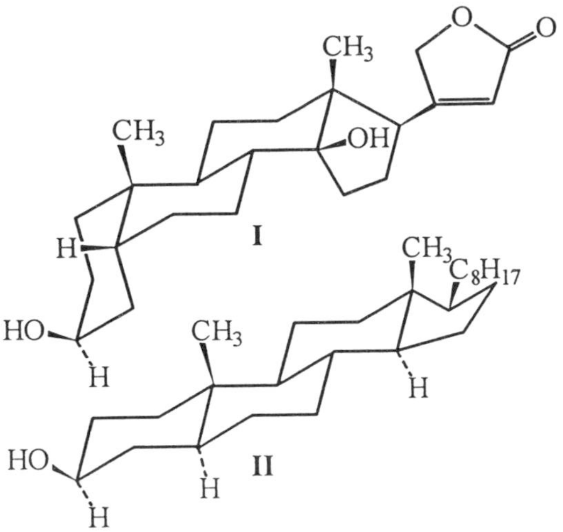

Fig. 28.28 Stereochemistry of (I) digitoxigenin and (II) cholestanol. The ring junctions in I are cis–trans–cis, and those in II are all trans. By convention, the configuration of steroidal substituent groups are designated as β if the group projects "upward" from the ring system (shown by the black wedges) and α when the group points "downward" (as shown by the dotted lines). Cis ring junctions are those where the nonring bonds of carbon atoms at the ring junction point in the same direction as in the A/B and C/D ring junctions of (I), whereas the reverse applies in trans ring junctions.

hydroxysteroid. Butenolides are the most common form of cardiotonic steroid found in plants and occur usually as a glycoside with one or more sugar residues linked through position C3 of the steroid. The pentadienolides (or bufadicnolides) contain a diunsaturated lactone (derived from 5-hydroxypentadienoic acid) attached to the 17β position of a 14β-hydroxysteroid. Pentadienolides are found in some plants and in toad venom. They include hellebrin (**49**), isolated from the roots of the Christmas rose, and bufalin (**50**), which occurs as the free genin and as the suberylarginine conjugate in the venom of the common toad (*Bufo vulgaris*). The naturally occurring cardiotonic steroids usually occur as complex mixtures comprising a group of related genins and a series of glycosides

based on these genins. The glycosides from foxglove include five genins and a group of sugar derivatives for each of the genin series. Although the sugars have no cardiotonic activity, their removal from the steroid usually results in a massive loss of potency.

Digitalis-like activity is also shown by a number of steroids and steroidlike substances that differ significantly in structure from the classical cardiotonic steroids. Alkaloids from *Erythrophleum* species, such as cassaine (**51**), are nitrogenous terpenes, and although they have a superficial resemblance to the butenolides and pentadienolides, they lack many of the structural features that were once thought essential for digitalis-like activity. A series of other synthetic and semisynthetic digitalis analogs also show significant structural differences from the classical cardenolides and pentadienolides but are positive inotropes, and most appear to be acting by the same mechanism (Fig. 28.27, structures **52-67**).

11.3 Pharmacology

The cardiac glycosides act on the heart by an array of direct and indirect effects to increase the force of contraction (positive inotropic effect), slow the rate of AV conduction, improve hemodynamic and renal function, and in toxic doses, produce a wide variety of cardiac arrhythmias. The basis of all these effects seems to be inhibition of Na^+,K^+-ATPase, but whether the useful therapeutic effects arise largely from inhibiting cardiac or neuronal Na^+,K^+-ATPase has not been resolved. Possibly, both contribute, but this has yet to be adequately investigated.

The glycosides seem to achieve their effects without a significant increase in oxygen consumption and, so far, are the only cardiac stimulants suitable for long-term use in CHF. However, they have a low therapeutic ratio, and much research

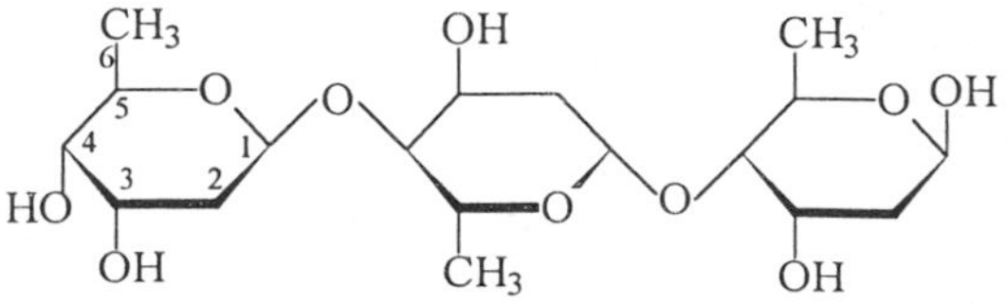

68 Tris-digitoxose

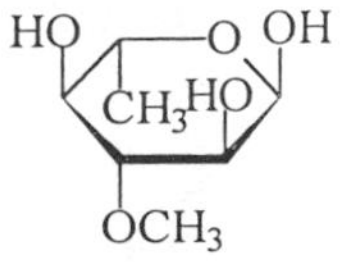

69 α-L-Thevetose

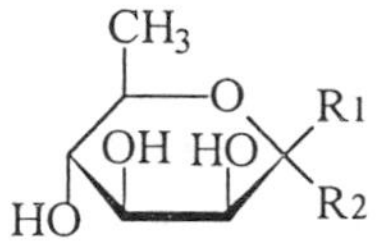

70 β-D-Rhamnose (R_1 = OH, R_2 = H)
71 α-D-Rhamnose (R_1 = H, R_2 = OH)

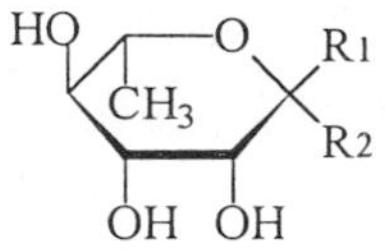

72 α-L-Rhamnose (R_1 = OH, R_2 = H)
73 β-L-Rhamnose (R_1 = H, R_2 = OH)

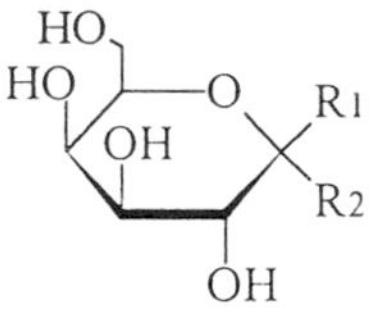

74 β-D-Galactose (R_1 = OH, R_2 = H)
75 α-D-Galactose (R_1 = H, R_2 = OH)

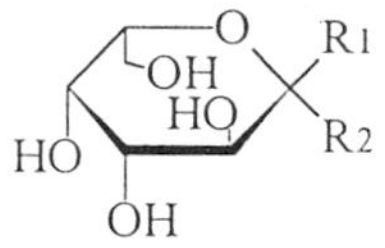

76 α-L-Galactose (R_1 = OH, R_2 = H)
77 β-L-Galactose (R_1 = H, R_2 = OH)

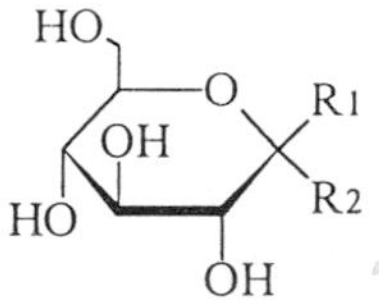

78 β-D-Glucose (R_1 = OH, R_2 = H)
79 α-D-Glucose (R_1 = H, R_2 = OH)

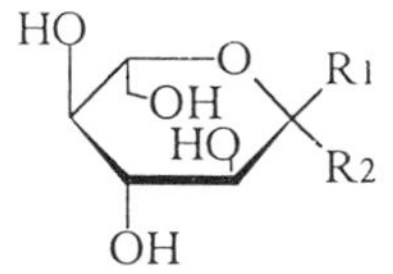

80 α-L-Glucose (R_1 = OH, R_2 = H)
81 β-L-Glucose (R_1 = H, R_2 = OH)

Fig. 28.29 Structural formulas of a selection of sugars found in naturally occurring cardiac glycosides. Not all of the possible isomers occur naturally. Although the sugars have no cardiotonic activity *per se*, their attachment to C3 of the steroid usually results in a manyfold increase in potency. Sugars lacking an OH group at C6 of the sugar (6-deoxy sugars) are particularly effective in increasing the potency of cardiotonic steroids.

has been devoted to finding a digitalis-like substance with a greater margin of safety. Two experimental defects have plagued the study of digitalis pharmacology and its mechanism of action:

1. Almost all laboratory studies have used concentrations that, on equilibrium, would produce toxic effects in the intact animal.
2. Almost all studies of new digitalis analogs have used isolated heart muscle preparations or worse still, isolated preparations of Na^+,K^+-ATPase.

Not only do such studies fail to detect the complex holistic effects of inhibiting Na^+,K^+-ATPase present in all cells of the body but the discovery that there are isoforms of Na^+,K^+-ATPase with varying sensitivity to cardiotonic steroids present (to different degrees) in different tissues in different species means that screening digitalis analogs using say, sheep kidney Na^+,K^+-ATPase that has been "purified" to the extent that 95% or more of the Na^+,K^+ATPase activity has been discarded, means that most of the SAR studies of digitalis analogs may have no validity as far as the therapeutic value of the drugs is concerned. In 1988, Dzimiri and Fricke (242) wrote,

Despite extensive investigations during the last two decades, the relationship between the inhibitory activities of cardiotonic steroids on myocardial Na^+K^+-ATPase (EC 3.6.1.3) and their positive inotropic actions still remains a subject of controversy. Opinions of different investigators seem to be polarizing rather than merging towards a common consensus.

Apart from their therapeutic value, the cardiotonic steroids are unique inhibitors of Na^+,K^+-ATPase, and since this enzyme plays such an important role in all eukaryotic cells (consuming about 25% of all energy used in a resting human), the manner in which digitalis inhibits the enzyme will continue to be of interest as a study in its own right.

11.4 Biochemical Pharmacology

The biochemical pharmacology of cardiotonic steroids was extensively reviewed in 1990 (68) and the reader is referred to that source for a comprehensive discussion and references. Briefly, there are two major explanations for improved hemodynamic effects seen when digitalis is used clinically.

The first explanation states that the effects are due to a direct action on the heart involving inhibition of myocardial Na^+,K^+-ATPase. Such an inhibition would lead to an increase in $[Na^+]_i$, and this would then inhibit the Na^+/Ca^{2+} exchanger (or antiport), thus leading to an increase in $[Ca^{2+}]_i$ and hence to stimulation of contraction as shown in Figure 28.11. This is known as the Na^+ pump lag hypothesis.

The second explanation states that the effects are due largely to reflexogenic effects of cardiotonic steroids on the autonomic nervous system. The reflexogenic sites considered important are the baroreceptors of the carotid sinus and aortic arch and the mechanoreceptors of the ventricle. Activation of these receptors, accord-

ing to the theory, produces an afferent response that leads to activation of efferent parasympathetic fibers and inhibition of efferent sympathetic fibers. These reflex effects are said to result in the following useful therapeutic actions: decrease in sinus rate (slowing of the heart); reduction in arterial and venous tone (due to inhibition of the sympathetic nervous system); diuresis (due to inhibition of the sympathetic nervous system and other effects); suppression of ventricular arrhythmia; and suppression of atrial arrhythmia by slowing of conduction in the AV node, decreased atrial automaticity, and decreased atrial refractory period.

Which of these mechanisms is responsible for the therapeutic effects of digitalis on the heart has yet to be resolved. There is no doubt that if cardiotonic steroids are applied to isolated heart muscle, a positive inotropic effect will be produced, but the drug concentrations required to produce this are at least an order of magnitude greater than what is required in the clinical setting. Nevertheless, there is an extensive body of direct and indirect evidence supporting the Na^+ pump lag hypothesis. Major reviews covering the three decades of research in this area are available (68,69,243–255).

Figure 28.30 summarizes the essential features of the Na^+ pump lag hypothesis, which attributes the inotropic and arrhythmogenic effects of cardiotonic steroids to a single mechanism, namely inhibition of myocardial Na^+,K^+-ATPase. As discussed earlier, depolarization of the sarcolemma leads to an influx of Na^+, which is followed by an efflux of K^+ during repolarization. The net effect is an increase in $[Na^+]_i$ and a decrease in $[K^+]_i$. If this were not reversed, the transmembrane potential would be lost and cell function would cease. The restoration of resting levels of $[Na^+]_i$ and $[K^+]_i$ is achieved mainly by the action of the Na^+ pump (Na^+,K^+-ATPase). According to the Na^+ pump lag hypothesis, therapeutic con-

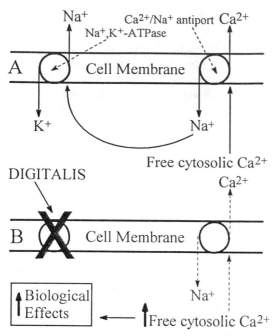

Fig. 28.30 Depiction of the Na$^+$ pump lag hypothesis. The relationship between the Na$^+$/K$^+$ pump and the Na$^+$/Ca^{2+} antiport are depicted in part *A* of the diagram. The transport of Na$^+$ out of the cell by Na$^+$,K$^+$-ATPase lowers [Na$^+$]$_i$ which promotes exchange of intracellular Ca^{2+} for extracellular Na$^+$. This lowers [Ca^{2+}]$_i$ and reduces the force of contraction (see Fig. 28.11). Part *B* of the figure shows the effect of inhibition of Na$^+$,K$^+$-ATPase by digitalis. Failure of the pump to remove cytoplasmic Na$^+$ leads to accumulation of this ion, and this slows the Na$^+$/Ca^{2+} antiport, leading to an increase in [Ca^{2+}]$_i$. According to the Na$^+$ pump lag hypothesis, therapeutic concentrations of cardiotonic steroids (digitalis) inhibit a moderate proportion of the cell's Na$^+$,K$^+$-ATPase units. Although the noninhibited units are stimulated by rising [Na$^+$]$_i$, they will still take longer to restore resting (diastolic) levels of cytoplasmic Na$^+$. This means that [Ca^{2+}]$_i$ remains elevated for longer periods than usual and hence enhances the contractile process. Nevertheless, in the presence of therapeutic concentrations of digitalis, cytoplasmic levels of Na$^+$ and Ca^{2+} are restored before the next impulse to contract reaches the cell, and there is no steady accumulation of Ca^{2+}. Toxic levels of digitalis are thought to produce extensive inhibition of Na$^+$,K$^+$-ATPase, leaving insufficient operating units to clear the excess Na$^+$ and Ca^{2+} from the cell before the next contraction event. This leads to a sustained increase in [Na$^+$]$_i$ and hence of [Ca^{2+}]$_i$ leading to toxic effects. Although there is little doubt that the mechanism described here accounts for the effects seen when digitalis is applied to isolated myocardium in con-

centrations of cardiotonic steroids inhibit a small proportion of Na$^+$,K$^+$-ATPase units (say, 25%). This means that it will take longer to clear the excess Na$^+$ that entered the cell during phase 0. Elevated levels of [Na$^+$]$_i$ inhibit the Na$^+$/Ca^{2+} antiport (see Fig. 28.10), thus prolonging the time it takes to clear excess Ca^{2+} that entered the cell during phase 2. Since the amount of stored intracellular Ca^{2+} released in response to this pulse depends on the size and duration of the pulse, inhibition of Na$^+$,K$^+$-ATPase should lead to higher levels of [Ca^{2+}]$_i$ and hence to an increase in force of contraction (positive inotropic effect). Essential to therapeutic benefit is that excess [Na$^+$]$_i$ be cleared before the next wave of depolarization. If this did not happen, Ca^{2+} would accumulate in the cell and eventually disrupt its function, particularly the production of ATP by the mitochondria. Energy needed to maintain resting membrane potentials would become inadequate and arrhythmia would develop. This is what is thought to happen when the cell is exposed to high (toxic) concentrations of cardiac glycosides. If there is extensive inhibition of Na$^+$,K$^+$-ATPase (for example, >50% of enzyme units inhibited at any time), the remaining uninhibited units cannot clear the excess Na$^+$ between depolarization cycles, and as a consequence, the Ca^{2+}/Na$^+$ antiport cannot clear excess Ca^{2+}, which would rise steadily to toxic levels. Thus the Na$^+$ pump lag hypothesis accounts for both the therapeutic and toxic effects of digitalis by a single mechanism, and if this is the only factor contributing to the action of cardiac glycosides in the

centrations of $10^{-8} M$ or higher, doubt has been expressed as to whether this mechanism explains the effects seen clinically where therapeutic effects are associated with digoxin concentrations in the range of $1 \times 10^{-9} M$ and toxic effects with $3 \times 10^{-9} M$ (see text).

clinical setting, then it is impossible to improve therapeutic ratios by chemical modification.

Evidence for and against the Na^+ pump lag hypothesis has been summarized (68). Many workers in the field believe that the sodium pump lag hypothesis has been adequately substantiated by experimental studies, but others reject it on the basis that it does not reflect effects seen at clinical concentrations of digitalis. The therapeutic effects of digoxin in humans are associated with steady-state plasma concentrations of $\sim 10^{-9} M$ and the toxic effects with concentrations of $\sim 3 \times 10^{-9} M$. Many laboratory animals are less sensitive to cardiotonic steroids than humans, but some have similar sensitivity. Regardless, the concentrations used in most experimental studies (usually of the order of $10^{-7} M$ or higher) are far too high. Those who reject or doubt the validity of the Na^+ pump hypothesis believe that the therapeutic and toxic effects of digitalis involve at least some contribution from extracardiac effects arising, in particular, from inhibition of neuronal Na^+,K^+-ATPase. References and a discussion of alternative theories are available (68). The belief is growing that digoxin in concentrations of $10^{-9} M$ has no direct effects on the heart and that both therapeutic and toxic effects arise from indirect effects on the heart mediated via neuronal activity.

11.5 Location of the Digitalis Binding Site on Na^+,K^+-ATPase

Regardless of whether cardiotonic steroids exert their effects mainly by direct action on the heart or by indirect neuronal effects (or a combination of both), there is general agreement that the receptor that mediates all of these effects is located on Na^+,K^+-ATPase (either myocardial, neuronal, or both). There has, therefore, been much effort devoted to characterizing this receptor. Photoaffinity labeling and other forms of chemical modification have been used extensively in these studies (70). From this and other work it would appear that the α subunit of Na^+,K^+-ATPase contains the binding site for cardiotonic steroids. The conformation of the α subunit has not been determined. The representation given in Figure 28.18 is only one of many possible arrangements (discussed in Section 3.4.2) and postulates 10 transmembrane domains. There is general agreement that there are four hydrophobic transmembrane domains between the N-terminus and the large globular domain that has been unambiguously associated with the hydrolysis of ATP. The influence of site-directed mutagenesis and chimeric constructs on the sensitivity of the α subunit to inhibition by cardiotonic steroids indicates that the digitalis binding site is located somewhere between the N-terminus and the phosphohydrolase site, most probably is in one or more of the four N-terminal hydropathic domains (domains 1–4) (256–258).

Initially, it was believed that the receptor may involve direct contact of the glycoside with the extracellular loops between domains 3 and 4 (3–4 on Fig. 28.18), with additional involvement of the extracellular loop between domains 1 and 2 (1–2 on Fig. 28.18). However, the amino acid strings of 3–4 were identical in some digitalis-sensitive and -insensitive species, and this led to the suggestion that the cardiotonic steroid biding site may lie between 1–2 and 3–4, as indicated in Figure 28.18 (68,69). Recent evidence has added indirect support for an intramembranous site. Monoclonal antibodies directed against the extracellular domains 1–2 and 3–4 stimulated rather than blocked the binding of [^{2}H]ouabain, indicating that cardiotonic steroids do not bind to the surface of these areas, and yet these domains appeared influential on binding (259,260). In a recent review, Sweadner

(73) suggested that a digitalis-binding site deep within a membrane cleft would explain many features of the binding of cardiotonic steroids, namely its slow onset, high affinity, and sensitivity to conformational changes occurring anywhere in the enzyme. Since the various transmembrane domains of the α subunit would not be strung out, as shown in Figure 28.18, but bunched together, as depicted in Figure 28.16, Sweadner suggested that cardiotonic steroids may penetrate into a pocket at the center of the complex thereby disrupting the alignment of proteins loops and blocking the conformational changes needed to translocate Na^+ and K^+. Studies using chimeric molecules formed from the α subunit of Na^+,K^+-ATPase and other P-ATPases, have shown that both the N-terminal and C-terminal halves of the molecule contribute to binding (261,262). It has yet to be resolved whether this indicates that domains in the C-terminal half come into direct contact with cardiotonic steroids or whether they influence binding by an allosteric mechanism. Sweadner's general statement about the binding of cardiac glycosides is not correct: cardiotonic steroids and their analogs show a wide spectrum of shapes, solubilities (ouabain is water soluble) and rates of binding and dissociation. Some potent cardiotonic steroids bind slowly, others, equally potent, show rapid onset of action.

Alice: "Dear, dear! How queer everything is today! And yesterday things went on just as usual. I wonder if I have been changed overnight? Let me think: was I the same when I got up this morning? I almost think I can remember feeling a little different."

"We're all mad here. I'm mad. You're mad." "How do you know I'm mad?" said Alice. "You must be," said the Cat, "or you wouldn't have come here."

—Lewis Carrol, *Alice in Wonderland*

11.6 Structure–Activity Relationships (SAR)

11.6.1 PROBLEMS IN THE INTERPRETATION OF SAR. The literature on the chemistry and SAR relationships of cardiotonic steroids is vast and cannot be covered in this chapter in any detail. Instead, some of the main issues will be identified and recent work discussed. The interested reader is referred to a 1990 review (68).

The naturally occurring ("classical") cardiac glycosides (Fig. 28.28) have a *cis*-C/D ring junction, a 14β-hydroxyl group, and a 17β-unsaturated lactone. The A/B ring junction is usually cis, although there are some active glycosides that have *trans*-A/B ring junctions. The steroids have 3β-hydroxyl groups to which sugars are attached when in the form of a glycoside. In addition to the naturally occurring cardiac glycosides, a range of other compounds, including the erythorphleum alkaloids (e.g., **51**) and a variety of synthetic and semisynthetic compounds (e.g., **52–67**) inhibit Na^+,K^+-ATPase, compete with classical cardiotonic steroids for binding to the receptor and elicit a positive inotropic effect on the myocardium. One of the challenges of SAR studies has been to determine whether these diverse structures can be reconciled in terms of a single model for the digitalis receptor.

The state of knowledge about the SAR of cardiotonic steroids can be summarized by Figure 28.31. This figure adequately describes the overall knowledge of the binding features of the classical cardiac glycoside structure:

1. The three main chemical moieties of the molecule (sugar, steroid, and 17β-substituent) all contribute to binding, in other words, binding appears to involve the whole molecule.

2. The biological activity of the molecule is extremely sensitive to minor changes in

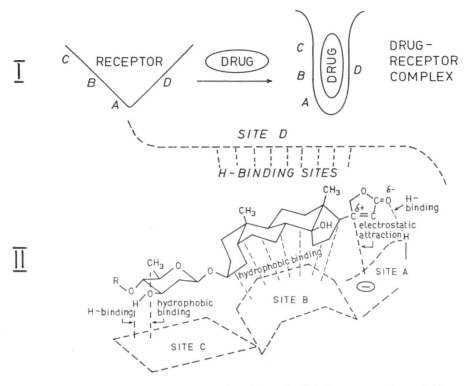

Fig. 28.31 Model depicting the binding of a cardiac glycoside to the digitalis receptor. The model incorporates a number of suggestions that have yet to be confirmed. SAR studies indicate that the molecule can be divided into three important binding regions: the lactone ring, the steroid, and the sugar moiety. All three regions seem to contribute to binding. In addition, there appears to be binding sites or constraints above and below the molecule, suggesting that the molecule becomes embedded in a deep cleft in which hydrophobic and electrostatic forces are involved.

molecular configuration, indicating that the drug–receptor interaction is highly stereoselective.

3. Functional groups both "above" and "below" the molecule appear to be involved in binding (or capable of disrupting binding), indicating that the drug is enclosed by a receptor cleft.

4. The 17β-side chain binds to regions at the base of the cleft, whereas the sugars bind toward the open end of the cleft.

In spite of the above highly specific requirements, a variety of structural types (Fig. 28.27) express digitalis-like activity, which could be explained in several ways: (*1*) in spite of major structural differences, the molecules present some, at least, of the key functional groups to appropriate receptor binding sites; (*2*) the drugs may bind in the same vicinity but at different locations; and (*3*) the receptor may be capable of making a variety of conformational changes to accommodate the different structures. The literature indicates that the last two explanations are applicable. To avoid confusion, the digitalis receptor is defined here as being that portion of the α subunit of Na^+,K^+-ATPase that binds the "classical cardiotonic" steroids such as digoxin and ouabain, encompassing the possibility that the digitalis receptor is capable of making conformational adjustments to maximize binding with variants of the classical structure.

A major problem in interpreting SAR results from different laboratories is the diversity of preparations and species used to determine potency. This coupled with the problem of variations in Na^+, K^+-ATPase isoform ratios in the tissues and enzyme preparations studied might suggest that SAR correlations among cardiotonic steroids is a fruitless task. However, there are reasons for believing that the binding site for cardiotonic steroids may be conserved across species and enzyme isoforms, at least as far as the cardinal binding site is concerned. The K_D values for the binding of a diverse group of cardiotonic steroids to beef heart Na^+,K^+-ATPase have been correlated with inotropic potency when tested on guinea pig atria (Table 28.8) (64). Although absolute potencies using the two preparations differed greatly, relative potencies remained much the same. When the two sets of data were compared in terms of relative potency, a correlation coefficient of 0.95 was obtained for the comparison of 22 5β-cardiotonic steroids and 0.96 for the comparison of five 5α-cardiotonic steroids (5α and 5β-steroids trended toward different regression lines, although the difference was not statistically significant) (263). The authors also reported a strong correlation ($r = 0.98$) for ability to bind to guinea pig Na^+,K^+-ATPase and positive inotropy in isolated guinea pig atria. Other studies (264,265) also concluded that relative potencies of cardiotonic steroids remained fairly constant across a wide range of species, in spite of great differences in species sensitivity. These workers concluded that the genin-binding site was probably homogeneous across species, but the sugar-binding site was less so, although still capable of recognizing the major sugar-binding groups.

Before considering SAR in more detail, it is appropriate to review the properties of those substances with digitalis-like activity whose structures differ significantly from those of the classical cardiotonic steroids.

11.6.2 GUANYLHYDRAZONES. In the 1960s the bisguanylhydrazones (e.g., prednisolone bisguanyhydrazone (**61**), PBG attracted much interest as possible therapeutic replacements for digitalis. The compounds produce positive inotropic effects and inhibit Na^+,K^+-ATPase, and yet they show none of the structural criteria that seem so critical for classical cardiotonic steroids. In a recent study, it was found that PBG differed from ouabain in that it did not show high and low affinities for Na^+,K^+-ATPase α subunit isoforms, although it did show marked differences in potency between species and it did inhibit [^{3}H]ouabain binding (266). This could mean that PBG binds to the digitalis receptor but involves binding moieties that are not sensitive to isoform differences or that it binds to a site other than digitalis receptor and induces conformational changes that inhibit ouabain binding. The guanylhydrazone group is a strong base, comparable to a quaternary ammonium compound. The bisguanylhydrazones may thus act like bisonium compounds, with the steroid playing a passive bridging role. This seems likely since their activity is not significantly affected by changes to the steroid ring system (243). Sequencing has revealed that the [1–2] and [3–4] extracellular domains of Na^+,K^+-ATPase (Fig. 28.18) each contain several amino acids with free carboxylic groups (76,85,86,256). These could form strong ionic interactions with each end of the bisguanylhydrazone molecule.

However, replacement of the lactone of digitoxigenin with a guanylhydrazone moiety gave a compound structure (**59**) with digitalis-like activity that, unlike the bisguanylhydrazones, did show high sensitivity to changes in the steroid ring system, indicating that it may be binding to the digitalis receptor as earlier defined (267). If

Table 28.8 Biological Activity of some Natural and Semisynthetic Cardiotonic Steroids[a,b]

| | Biological Activity | | | | |
| | Guinea Pig Myocardium (Relative Potencies) | | Binding to Beef Heart Na^+,K^+-ATPase | | |
Compound	Inotropy (Atria)	Na^+,K^+-ATPase	K^D ($\times 10^{-9} M$)	$\Delta G°$ (kJ/mol)	Relative Potency
	Natural				
Digitoxigenin (**44**)	100	100	10.27	−47.3	100
Digoxin (**47**)	182	324	3.57	−49.8	288
Digitoxin (**45**)	750	1446	0.81	−53.6	1268
Ouabain (**48**)	413	232	2.99	−50.3	343
	Semisynthetic				
Genins (R′ = H)					
Nitrile (**52**)	69				
Methyl ester (**53**)	29				
Methyl ketone (**56**)	20				
Acid (**57**)	inactive				
Amide (**58**)	slight activity				
Glucosides (R′ = glucose)					
Digitoxigenin	300	255	2.53	−51.1	406
Nitrile (**52**)	240	324	7.42	−48.2	138
Methyl ester (**53**)	170	249	4.63	−49.4	222
Ethyl ester (**54**)	21	35	12.69	−46.9	81
n-Propyl ester (**55**)	9	14	19.14	−45.6	54
i-Propyl ester[c]	22	47	16.32	−46.1	63
n-Butyl ester[c]	2	4	79.02	−42.0	13
i-Butyl ester[c]	2	3	88.90	−41.7	12
Methyl ketone (**56**)	80	139	13.95	−46.5	74

[a]Adapted from ref. 68.
[b]In calculating potencies, digitoxigenin was set at 100 for each data set.
[c]These esters are the higher analogs of structure **55**.

this were the case, it might indicate that the digitalis receptor contained a group capable of forming a strong electrostatic interaction with the guanylhydrazone cation. It is relevant that the closely related but nonionic semicarbazone (**60**) was devoid of activity. Monoguanylhydrazones based on the pregnane ring system have also been studied (268). These compounds inhibited Na^+,K^+-ATPase but produced only a transient increase in myocardial contractility, which was then followed by a profound negative inotropy. Preliminary studies (68) indicated the negative inotropy was caused by mitochondrial inhibition, but this needs to be confirmed.

11.6.3 PREGNANE AND RELATED DERIVATIVES. Pregnane derivatives with digitalis-like activity are of interest because they may provide leads to the putative endogenous digitalis-like factor that could be involved in the etiology of essential hypertension (discussed in Section 11.7). There is also evidence that some of these compounds may have better safety margins than the cardiotonic steroids in clinical use. Labella and co-workers (269,270) used the ability to compete with [³H]ouabain for binding to Na⁺,K⁺-ATPase as a screening test for endogenous substances that could be, or could resemble, the putative endogenous factor. In the course of these studies, they identified pregnane derivatives as being possible candidates, and this led to a systematic study of synthetic pregnane analogs. Chlormadinone acetate (CMA) (**62**) was found to be equipotent with digitoxigenin in the screening test(269,270). The authors suggested that the C6 substituent (Cl in the case of CMA) was distorting ring A toward a digitalis-like structure (i.e., like ring A of I in Fig. 28.28); see structure **44** for ring designations A to D.

Fullerton and co-workers (271) used X-ray crystallography, molecular graphics, and molecular mechanics computer programs to compare the structures of CMA with digitoxigenin. They found that when rings B and C were superimposed, the 20 carbonyl of CMA approximates closely the position of the lactone carbonyl of digitoxigenin. These studies provide strong evidence that the pregnane series are binding to the digitalis receptor on Na⁺,K⁺-ATPase and yet when tested for contractility, CMA was found to produce a negative inotropic effect, possibly because of additional intracellular effects. This was overcome by glycosidation, which may have reduced the ability of the compound to enter the cell and produce its negative inotropic effects (272).

Two interesting compounds in this series are the C20-amino and C21-nitro derivatives of 14β-hydroxy-5βH-pregnane-3β-rhamnoside (**63** and **64**). The former had about 8% the potency of digitoxin in inhib-

iting [³H]ouabain binding but appeared more than twice as potent when tested for inotropic activity in the anaesthetized dog. The nitro derivative had about 18% the potency of digitoxin in inhibiting ouabain binding and 70% the potency in the anaesthetized dog. Both compounds appeared to have greatly improved therapeutic to toxic ratios compared with digitoxin (273). These results, and others to be mentioned later, imply that the therapeutic ratios of cardiotonic steroids are not fixed (as would be predicted from the Na⁺ pump lag hypothesis). However, the results will need to be confirmed by chronic steady-state studies in conscious, instrumented animals before definite conclusions can be drawn.

Improved therapeutic ratios were also reported for the 14β-amino,20-hydroxy 5βH-pregnane 3β-rhamnoside (LND 623, **65**) and related compounds (274). It has been suggested that the toxic effects of the LND 623 and related steroids are due to inhibition of Na⁺,K⁺-ATPase and their inotropic effects to other mechanisms, including stimulation of L-type Ca²⁺ channels and possibly release of endogenous catecholamines (275).

One of the great problems with pharmacological studies of cardiotonic steroids is that there can be great variation between drugs in the rate and degree of tissue distribution and in the rate of drug–receptor binding and dissociation. Acute studies in anaesthetized animals are often conducted under nonequilibrium conditions, so that comparisons between drugs can be quite misleading and may reflect differences in tissue distribution and not efficacy. This is particularly so if the comparison is between fast onset–fast offset steroids and those that have slow onset and offset. Such studies can lead to quite misleading conclusions about potencies and therapeutic ratios. Under nonequilibrium conditions (for example, when the drug is infused continuously into an anaesthetized dog), cardiotonic steroids showing fast onset inotropy appear to have improved therapeutic to toxic ratios, but this may be an artifact of differing onset rates for inotropy and toxicity for different drugs. If there is a real improvement in therapeutic ratio superimposed on the apparent improvement seen in the above type of experiment, it must be studied under equilibrium conditions, preferably in human subjects with heart failure. In the laboratory, conscious animals with induced heart failure and instrumented for measuring increased generation of myocardial contractility is a suitable preclinical model.

11.6.4 ERYTHROPHLEUM ALKALOIDS. The erythrophleum alkaloids such as cassaine (**51**) are diterpene derivatives and lack the key structural features associated with the classical cardiotonic steroids. However, they are potent inotropic agents and inhibitors of Na^+,K^+-ATPase. These properties may be due to the presence of the α,β-unsaturated side-chain which has some resemblance to the α,β-unsaturated carbonyl system of the cardenolide lactone. SAR of erythrophleum alkaloids were reviewed in 1974 (243) and led to the suggestion that these compounds were binding to the same receptor as the classical cardiotonic steroids, but this has been disputed (276).

diterpene ring system

$$CH_2CH_2N(CH_3)_2$$

11.6.5 FORCES THAT MAY BIND CARDIOTONIC STEROIDS TO THE RECEPTOR SITE

11.6.5.1 Possible Binding Forces. **Hydrophobic bonding.** Water molecules tend to associate with each other by a random process whereby hydrogen bonds are continually being formed and broken. A hydrophobic solute disrupts this process, thereby decreasing its entropy. Because of this, hydrophobic molecules tend to leave an aqueous environment and aggregate with molecules of similar properties. The energy released when this happens is called hydrophobic bonding, although it is not bonding in a chemical sense. The body has abundant regions capable of accepting hydrophobic molecules, and the intriguing question is why drugs bind so selectively to specific regions known as receptors.

Electrostatic interactions. The forces that bind drugs reversibly to receptors are electrostatic and include ion–ion, ion–dipole, dipole–dipole, hydrogen bonds, and London dispersion forces (better known as van der Waals forces). The strength of electrostatic forces is inversely proportional to some power of the interatomic distance and the dielectric constant of the medium. Since the latter at the receptor surface is ambiguous and impossible to measure, estimates of the likely contribution of electrostatic effects to the strength of the drug–receptor interaction can only be approximations. Usually, electrostatic forces between drugs and receptors are significant only when they take place in some water-free cleft in the receptor. Ion–ion interactions are the strongest electrostatic forces and can contribute up to $20 \, kJ \, mol^{-1}$ of binding energy. Ion–dipole and dipole–dipole interactions are weaker and their power falls of, respectively, with the second and third power of the interatomic distance, whereas that of an ion–ion interaction falls with the first power. Ion–dipole and dipole–dipole interactions are thus far more specific than ion–ion interactions, since they depend on a much closer fit of the binding surfaces. This is relevant to understanding the SAR of the bisguanylhydrazones, which probably bind by ion–ion interactions. Hydrogen bonds are a special type of dipole–dipole interaction in which a hydrogen atom (covalently bound to a small electronegative atom such as O, N or

F) bonds electrostatically to another electronegative atom such as O or N. The bonds are short range and can be considered negligible if the ˙O·····O distance is much greater than about 0.3 nm. Hydrogen bonds are also highly directional, and their strength is greatly reduced if the hydrogen atom points more than 30° away from the acceptor atom. Under optimal conditions, a hydrogen bond can contribute 8–20 kJ mol^{-1} to the binding energy.

Charge-transfer bonds occur when a good electron donor comes in close contact with a good electron accepter. Among organic molecules, such complexes form most readily between planar conjugated systems. The energy of such bonds depends on the ionization potentials and geometry of the interacting substances. Significant binding occurs only if the receptive molecular orbitals overlap. They are thus very specific. Van der Waals forces can take place between any two atoms and are due to the fact that all atoms are continually producing transient dipoles. As two atoms approach, their respective dipoles become synchronized so that the negative pole of one tends to cancel the positive pole of the other. The maximum binding strength between any two atoms due to these "dispersion" forces is about 4 kJ mol^{-1} and is inversely proportional to the seventh power of the interatomic distance. Such bonds are thus highly specific, depending on a geometry that allows for close association of particular atoms. For large molecules, fitting closely together, the total van der Waals binding energy can be considerable. For example, the van de Waals binding energy between an antigenic determinant (epitope) and the binding site of an antibody consisting of a few amino acids can be as high as 120 kJ mol^{-1}.

11.6.5.2 Thermodynamics of the Drug–Receptor Interaction. The overall energy changes that drive chemical reactions, including drug–receptor interactions, is given by:

$$\Delta G = \Delta H - T \, \Delta S$$

where ΔG is the change in Gibbs free energy, ΔH is the enthalpy or change in internal energy, ΔS the change in entropy, and T is the absolute temperature. Reactions proceed spontaneously when ΔG decreases and can be enthalpy driven, entropy driven, or driven by both these changes. As mentioned later, it has been claimed that the interaction of cardiotonic steroids with their receptor is entropy driven. Bond energies are expressed in terms of the energy required to break the bond and thus equal $-\Delta G$ for the formation of the bond. This is related to the dissociation constant (K_D) as follows:

$$\Delta G = RT \ln K_D$$

where R is the universal gas constant and T is the absolute temperature.

Cardiotonic steroids bind to the digitalis receptor by reversible bonds. The most potent glycosides are considered to be relatively strongly bound, at least compared with most other drugs. However, as shown in Table 28.8, the total binding energy, even for potent glyclosides binding to sensitive enzymes, is only of the order of 50 kJ mol^{-1}. By comparison, the bond strength of a typical covalent bond is about 400 kJ mol^{-1}. Furthermore, the difference in binding energies between the most potent glycoside shown in Table 28.8 and the least potent is a mere 11.9 kJ mol^{-1} (roughly equivalent to one weak hydrogen bond or three van der Waals bonds), and yet they differ in potency by a factor of about 100.

These small energy differences do not seem to explain the selectivity of cardiotonic steroids for their receptor. Although the binding strength of one hydrogen bond can increase the affinity (i.e., equilibrium constant K) by a factor of up to 4000, this does not explain selectivity. In the biophase, there exists a wide range of binding opportunities, especially for hydrogen bonding, which has been suggested as being one of the key binding interactions between digitalis and its receptor. This apparent anomaly arises because the "binding energy" $(-\Delta G)$ is not a measure of the actual forces that bind the drug to the receptor but of the total energy changes in the system. When a drug interacts with a receptor protein, energy is usually required to bring about certain conformational changes in the protein as well as to make other changes such as the shedding of structural water. In addition, there is the contribution of the hydrophobic forces mentioned earlier. Thus ΔG is the final energy balance after all the energy changes have been summed. In many cases, it would seem that part of the energy changes that take place when a drug binds to a receptor are used to "pay" for unfavorable energy changes associated with changes in the receptor molecule. This concept was expressed (277) as follows:

$$\Delta G_{obs} = \Delta G_{int} + \Delta G_{D} - T\,\Delta S_{int}$$

where ΔG_{obs} = observed free energy change, ΔG_{int} = the intrinsic energy of binding (i.e., the summation of all electrostatic attractions that take place between the drug and the receptor surface), ΔG_{D} = "destabilization" energy changes (i.e., the changes needed to bring the receptor from its ground state to that which binds the drug), and $T\,\Delta S_{int}$ = the intrinsic entropy changes associated with the interaction of the drug with receptor surface. In many cases, ΔG_{D} is positive, i.e., it inhibits the formation of the drug–receptor complex, since the more negative the value of ΔG_{obs} the stronger is the binding of the ligand.

The equation above expresses the possibility that portion of the energy liberated by the formation of electrostatic bonds between the drug and receptor (ΔG_{int}) can be used to pay for the unfavorable allo-

steric changes that occur in the receptor protein as a consequence of the drug–receptor interaction. This concept helps to explain the specificity of drug action: only certain drug molecules are capable of binding to a receptor in such a way that the energy released can be used to bring about the conformational changes in the receptor protein that are needed for the expression (or inhibition) of the protein's biological activity. The concept of intrinsic energy of binding also emphasizes the fact that ΔG_{obs} (and hence K_D or any other measure of potency) cannot be used to deduce the strength of the electrostatic interactions between the drug and receptor. The $-\Delta G$ values given in Table 28.8 are thus measures of the overall balance of the binding event (i.e., of potency) but say little about the nature of the electrostatic interactions between drug and receptor.

The thermodynamics of the binding of cardiotonic steroids to the digitalis receptor have been extensively studied by Repke and co-workers. This group analyzed the enthalpy, entropy, and activation energy changes associated with the inhibition of Na^+,K^+-ATPase by a series of cardiotonic steroids (278). The authors found that the drug–receptor interaction was an endothermic process that was entropy driven. Activation energies ranged from 45 to 59 kJ mol^{-1} for the forward reaction and 84 to 99 kJ mol^{-1} for the dissociation reaction. The endothermic nature of the reaction would inhibit the reaction process, but this is more than offset by the large increase in entropy that arises, not only from hydrophobic bonding but mainly, according to the authors, from a relaxation in the conformational energy of Na^+,K^+-ATPase.

11.6.5.3 Relationship of Physical Properties to Binding. Several investigators have attempted to explain the interaction of cardiotonic steroids with the digitalis receptor in terms of certain physical properties. Bohl used a technique called the minimal topological difference (MTD) to determine the shape of the digitalis binding site. The method interprets SAR in terms of ability to fit the receptor cavity (279). Molecular modifications that lead to increased activity are considered to improve the fit for a particular part of the cavity; modifications that lead to falls in activity are considered to produce unfavorable steric interactions with the receptor wall, and modifications that have no effect on activity are considered to be sterically irrelevant. The technique is similar to a Free–Wilson analysis with the effect of each structural modification being regarded as independent of the influence of other structural modifications. The analysis indicated that the steroid fits into a tight cavity with strong steric repulsions occurring in regions of the receptor corresponding to the 11β, 12β, and 15β positions of the steroid. Receptor cavities were identified that corresponded to the 17β side chain, the steroid ring system, and the sugar moiety. A point of strong repulsion was identified in the sugar-binding site. Overall, the study emphasized that the whole molecule fits into a tight cavity. This could mean that van der Waals forces play a major part in binding the molecule.

Dzimiri and Fricke (242) tried to correlate lipophilicity with inotropic activity as determined in isolated guinea pig heart muscle preparations. The correlation was not good when glycosides of the digoxin series were included ($r = 0.20$) but was significantly improved when these and the cardanolides (saturated lactone derivatives) were omitted ($r = 0.54$). The authors concluded that lipophilicity made a major contribution to the tendency of cardiotonic steroids to bind to the digitalis receptor. They considered that the reduction in activity associated with the 12β-OH (digoxin) and 16β-OH (gitoxin) series was due to reduction in hydrophobicity and not steric factors. This is supported by the fact that derivatization of the 16β-OH increases

activity but is not consistent with the fact that activity falls when the 12β-OH is derivatized.

Kamernitzky and co-workers (280) determined the heteropolarity or biphilicity moments of a series of cardiotonic steroids and related these to activity. They concluded that these values depended on the location of substituent groups but were independent of the hydrophilicity of the molecule as a whole.

11.6.6 ROLE OF THE FUNCTIONAL GROUPS IN THE ACTIVITY OF CARDIOTONIC STEROIDS. The contribution of the functional groups of cardiotonic steroids to biological activity was extensively reviewed in 1990 (68) and 1992 (69), and by Repke and co-workers in 1993 (281). It is not possible, in the scope of this chapter, to deal with the huge volume of literature in a comprehensive fashion. Furthermore, great doubt now exists about many of the methods used for determining the biological activity of cardiotonic steroids. Since it now seems certain that digitalis, as it is used clinically, has a complex action of which inotropy may be only a small part, there is little point in dealing with SAR in any detail, other than to point out that all evidence indicates that the whole molecule is involved in binding and that this apparently takes place in a deep hydrophobic cleft in Na^+,K^+-ATPase.

That is not to say that studies of the effects on Na^+,K^+-ATPase are of no general value. The neurohumoral effects of digitalis, which are probably the main basis for its value in heart failure, are probably mediated by an effect on Na^+,K^+-ATPase, which could be stimulation or inhibition. It is not known where digitalis exerts its neuronal effects, but there could be a combination of central and peripheral effects. It is also not known whether digitalis is affecting afferent or efferent limbs of the neuronal system, and until its sites of action have been determined and the relevant isoforms of Na^+,K^+-ATPase identified,

SAR studies using surrogate end points such as inhibition of nonneuronal Na^+,K^+ATPase or inotropic effects using high doses in experimental animals can only give general indications of relevant SAR relationships.

For the purpose of SAR studies, cardiac glycosides are considered to have three distinct structural regions: the 17β side chain, the steroid ring system, and the 3β side chain. There is abundant evidence that all three regions are involved in binding the receptor.

11.6.6.1 17β Side Chain. One of the characteristic features of the naturally occurring cardiac glycosides is the unsaturated lactone attached to the 17β position of the steroid (Fig. 28.28). It was long thought that the unsaturated lactone was an essential feature for cardiotonic activity, although it was also known that the lactone by itself, or attached to simple carbocylic systems, was inactive (282). Thomas and co-workers (68) were the first to show that the lactone could be replaced by a series of open-chain groups (compounds **52–58** are representative examples). This study indicated that there were considerable structural restraints on the side chain that may influence both the fit of the side chain into the receptor cleft and/or its ability to form appropriate electrostatic interactions. It was suggested that the side chain was binding to the receptor by a hydrogen bond that was reinforced by an additional charge-transfer complex involving the α,β-unsaturated system.

Fullerton and co-workers (283) used a combination of x-ray crystallography, computer graphics, and molecular mechanics to determine the spatial orientation of the 17β side chains of cardiotonic steroids. From these studies they were able to determine the preferred orientation of a the 17β side chains and calculate the position of the potential hydrogen bonding group (usually a carbonyl group) in reference to that of digitoxigenin. They found an excellent correlation between the log of the I_{50} concentration needed for inhibition of Na^+,K^+-ATPase and the distance of the potential hydrogen-bonding group from that of the reference compound (283). From this, they concluded that the only contribution made by the 17β side chain was to correctly align a particular functional group, such as the carbonyl group, for a binding interaction, which they suggested was probably hydrogen bonding. However, there is a serious flaw in this work, namely it invokes the need for hydrogen bonds of up to 1 nm in length, which is far too great a distance (68,69). Of course, the receptor may make an accommodating conformational change to move the donor hydrogen atom closer to the side chain acceptor, but this would then confound the correlation. Alternatively, the carbonyl group on the 17β side chain may be acting simply as a surrogate marker for the general orientation of the side chain.

Bohl and Süssmilch (284) conducted a similar study using a different set of digitalis analogs. These workers concluded that maximum biological activity could be associated with different geometrical arrangements of steroid and receptor, and hence that receptor models based on rigid geometries would have limited applicability. This work has been reviewed (69).

There are also a number of compounds whose activity cannot be explained in terms of the 17β side chain acting solely as an entity for enabling hydrogen bond formation between the receptor and the steroid. The lack of activity of the amide (**58**) and the acid (**57**), together with fact that the guanylhydrazone (**59**) is active but the semicarbazone (**60**) is not, cannot be explained in terms of the hydrogen-bonding hypothesis. Thus, although hydrogen bonding, when possible, is probably the major binding interaction between the 17β side chain and the receptor, other binding interactions are also seen to contribute.

11.6.6.2 Steroid Ring System. There is good evidence for believing that not all compounds with digitalis-like activity bind to the receptor in the same way. The guanylhydrazones, for example, appear to bind to the digitalis receptor in a way that is quite different from the classical cardiotonic steroids. This discussion will, therefore, be limited to those cardiotonic steroids that appear to bind to the receptor in a way that resembles the latter group.

The essential feature of the steroid ring system is a carbocyclic system that is capable of fitting a constrained receptor cavity and providing sufficient binding energy to bring about the conformational changes needed for biological activity. Schönfeld and co-workers (276) concluded that the minimum requirement for the steroid was 5β, 14β-androstane-3β, 14β-diol (i.e., digitoxigenin with the lactone replaced by H), which had 0.03% the activity of the parent genin. Although the C/D cis ring junction seems essential, other studies indicate that the A/B cis (5βH) ring junction is not a requirement for potent cardiotonic activity. It was found that 5αH-digitoxigenin (uzarigenin) was roughly equipotent with digitoxigenin when tested for inotropy using guinea pig atria (285) and in binding studies using beef heart Na^+,K^+-ATPase (263). (Note: the terminology 5αH denotes a trans A/B ring junction such as in structure **II** in Fig. 28.28). It was also found that gomphoside, a glycoside with a trans A/B ring junction, had 23 times the potency of digitoxigenin, or almost five times the potency of ouabain (286). Although digoxigenin and its 5αH analog (uzarigenin) are approximately equipotent, attachment of sugars to the 3β-OH of both compounds produced dissimilar results. For example, glucosidation of digitoxigenin increased activity by almost 300% but reduced the activity of uzarigenin by 60% for the monoglucoside and by 80 to 90% for the diglucosides and triglucosides (284). On the other hand, it was found that the 6-deoxy-alloside and rhamnoside of uzarigenin were, respectively, 2.2 and 7.8 times more potent than the genin, clearly showing that the A/B cis ring junction is not an essential requirement for potent cardiotonic activity. These results led the authors to conclude that ring A was probably not important for binding the steroid but influenced activity by its effect on the orientation of the sugar moiety. This will be discussed further in the next section.

It was also found that the onset and washout times for the effects of uzarigenin on guinea pig atria were significantly slower than for digitoxigenin: 19 and 32 min compared with 4 and 5 min, respectively. Usually genins have fast onset and washout times, and the much slower times for uzarigenin suggested that the nature of the drug–receptor interaction was different for the two types of genin; possibly some form of induced fit was involved that was harder to achieve with the 5αH steroid but, once formed, was more stable than for the 5βH genin. Another possible explanation is that the approach to and from the binding site was less facile for the 5αH steroid. Comparisons of relative potencies for effects on guinea pig atrial contractility and binding to beef heart Na^+,K^+-ATPase for a series of cardiotonic steroids gave different regression lines for the 5αH and 5βH series (263), which further supports the idea that the two series interact with the receptor in different ways.

Rings A, B, and C of the steroid system all appear to adopt the chair conformation, but there is some question about ring D, which has a greater degree of flexibility than the other rings. X-ray crystallographic studies indicate that ring D can adopt envelope and half-chair conformations (283,287). However, it is possible that this reflects different crystal packing constraints and that one form predominates in solution.

Apart from the 17β side chain, the other principal substituents associated with high cardiotonic potency are the 14β-OH and the 3β-sugar moiety. The latter will be discussed in the next section. Replacement of the 14β-OH with 14β-H reduces activity by about 90% (288). The group is thus not "essential," but it makes an important contribution to activity. In fact, it is more important for potency than the sugar moieties. No known evidence has been obtained identifying the role this group plays in potentiating the activity of the steroid. The group projects from the β face of the steroid (Fig. 28.28) and may bind to the same receptor group that binds the 17β side chain, or it may exert a favorable influence on the conformation of the 17β side chain. The 3β-hydroxyl group does not seem to influence activity but serves as an attachment point for the sugar moieties. Further hydroxylation of the steroid reduces potency. For example, it was found that the K_D values for the binding of digitoxin and its 12β-OH and 16β-OH analogs (i.e., digoxin and gitoxin) to beef heart Na$^+$,K$^+$-ATPase were 1.1, 3.6, and $15 \times 10^{-9} M$, respectively. De Pover and Godfraind (289) found that formylation of the 16β-OH of gitoxin increased activity by a factor of 41. This modification does not just offset the hydrophilic effect of the 16β-OH but increases activity to several times that of digitoxin. It was suggested that the formyl group contributes to binding, possibly by interacting with the same site as the lactone. Some form of dipole interaction (which could include hydrogen bonding) was suggested. Fullerton and co-workers (283) found that the 16β-formyl group altered the conformation of the lactone, bringing the carbonyl into a more favorable position for binding, but that this effect was not sufficient to explain the effect on potency.

Recently, Haustein and Bauer (290) tested 16β-gitoxin in healthy human volunteers and found that the drug had an extremely rapid onset of action and no manifestations of toxicity. Inotropic effects

were inferred from the systolic time interval (duration of ventricular contraction) read from the ECG. The effects were only slight (<10% change in parameters) and required large doses (relative to digoxin), but there were no significant toxic effects. Studies in heart failure patients are needed before the value of this compound can be assessed. The drug could prove of value in cases where rapid improvement in cardiac function is required.

11.6.6.2 Sugar Moieties. The sugar moieties, as such, have no digitalis-like activity but, when attached to the 3-OH of the steroid, can modify greatly the activity of the parent genin. This is illustrated by the data shown in Table 28.9. Brown and Thomas found that certain sugars, namely thevetose and rhamnose, can increase potency by factors of 20 to 30, although some sugars, for example, mannose, had no

Table 28.9 Influence on Potency of Sugar Moieties Attached to the 3β-Position of Digitoxigenin[a]

Sugar Substituent[b]	Potency Enhancement Factor Relative to Digitoxigenin = 1[c] (Guinea Pig Atria)
Thevetose	27
Rhamnose	22
Monodigitoxose	15
Bisdigitoxose	10
Trisdigitoxose	9
Trisdigitoxose+acetyl +glucose[d]	3
Glucose	2
Mannose	1[e]
2'-Acetyl rhamnose	0.4
Triacetyl rhamnose	0.2

[a]Data from ref. 291; table adapted from ref. 68.
[b]See Figure 28.29 for structures.
[c]Relative potencies from log concentration response curves.
[d]Trivial name: lanatoside C.
[e]Factor of 1 menas no enhancement of activity.

effect on potency and certain substituted sugars decreased activity. Acetylation of the 2'-OH of digitoxigenin rhamnoside reduced activity by 98%, giving a compound with less than half the activity of the parent genin. The fact that such a relatively minor modification was able to offset much of the binding capacity of the rest of the molecule implies that the molecule is bound by forces that require a close drug–receptor interaction. Acetylation of all three rhamnosyl hydroxyls reduced activity even further, giving a compound with only one-fifth the activity of the parent genin (291).

Yoda and co-workers conducted an extensive study of the influence of sugar moieties on the binding of cardiac glycosides to Na^+,K^+-ATPase (68). They found that glycosides with 6-deoxy sugars (i.e., those with 5' methyl groups; see Fig. 28.29) were the most potent of all cardiotonic steroids and concluded that the methyl group played a key role in binding the sugar to the receptor. They found that genins bind and dissociate rapidly from the receptor and concluded that the binding of glycosides takes place in two steps: a rapid step involving binding of the genin portion followed by a slow binding of the sugar residue. The presence of an appropriate sugar was also thought to account for the low dissociation rate constants of potent glycosides.

Brown and Thomas (285,291) confirmed the importance of the 5'-methyl group and identified the 4'-hydroxyl and the α-L-glycosidic link as being associated with high potency in monosides. They concluded that the 2'-OH did not seem to be involved in essential binding, although, if acetylated, activity was lost, probably due to a steric factor. They concluded that either the 3'- or 4'-OH can contribute to binding, although neither is essential (263,286,291). As a general rule, the addition of an extra sugar unit to a monoside leads to a fall in activity. This is shown for the digitoxosides of digitoxigenin in Table 28.9. They also observed a similar fall in activity for the glucosides of uzarigenin. Dzimiri and Fricke (242) found somewhat similar results for the digitoxosides of digoxigenin: the mono and bis were roughly equipotent and the tris was less potent.

Fallon and Thomas attached a series of nonsugar open-chain esters to the 3β position of digitoxigenin (68). (See Fig. 28.32). These side chains also increased the potency of the genin, showing that enhancement of the potency of the steroid was not unique to sugar moieties. For examples, the methylglutarate methyl ester (**82**) had 3.5 times the potency of the parent digitoxigenin. It was thus more effective than glucose and mannose but less effective than digitoxose in potentiating the action of digitoxigenin. Since the side chain molecule has no protons capable of hydrogen bonding, such groups are clearly not essential for potentiation of activity. The fact that the side chain does not potentiate activity to the same extent as the 5'-methyl sugars could be due to its nonrigid structure, which would decrease the statistical probability of the groups being in the correct orientation for binding, or it could be that it lacks the equivalent of the sugar 3'- or 4'-OH groups, which, although not essential, could reinforce binding.

The amides of the series were also quite active. For example, structure **83** had 3 times the potency of the parent genin. Although the amide N-H can hydrogen bond, the corresponding methyl ester (**82**)

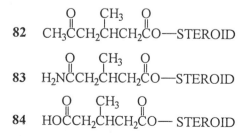

Fig. 28.32 Side chains attached to the 3β-OH of digitoxigenin as replacement for sugar moieties.

was equipotent. Hydrogen bonding through the amide does not, therefore, seem to be involved. One of the most interesting findings was that the free acids of the series (e.g., structure **84**) were all inactive. This was taken as evidence that an anionic group was present in the receptor near where the sugar binds. In 1974, Thomas and co-workers (243) drew a similar conclusion to explain the activity of the 3,20-bisguanylhydrazones, basing their argument on a report by Kyte (292) that an anionic site appeared to exist in the region of the receptor that binds the sugar moiety. If such an anionic group is located in the sugar-binding site, it may bind to the hydrogens of the 3′- or 4′-OH of appropriate sugars. This would enhance potency but is not essential, as the work of Fallon and Thomas has shown. Sequencing studies of Na⁺,K⁺-ATPase have shown that the extracellular domains 1–2 and 3–4 (Fig. 28.18) contain many free carboxylic acid moieties (a depiction of the amino acid sequences of these domains is available (68)).

It would seem, from what has been said, that the region of the digitalis receptor that binds the sugar residue closest to the steroid is sterically limited and largely hydrophobic, except for the possible presence of a carboxylic acid group. Van de Waals forces could be the basis of the binding interaction but seem to be enhanced by a strategically placed group capable of hydrogen bonding with either the 3′- or 4′-OH groups of sugar moieties.

One of the problems in studying the SAR of the sugar side chain is its flexibility around the glycosidic link. Although constraints to free rotation apply, these vary with different sugars, and in most cases, a reasonable degree of nonrestricted movement is possible. In 1986, Brown and Thomas (285) reported the activity of certain 5αH (i.e., A/B trans) glycosides from Asclepiadaceae species that had sugars that are bilinked to the steroid through posi-

tions 2 and 3 and thus are unable to rotate. Brown and co-workers (286) regarded these as useful reference compounds, especially since the most active glycoside, gomphoside was 23 times more potent than digitoxigenin. An interesting comparison is that of gomphoside with digitoxigenin rhamnoside and digitoxitenin monodigitoxoside. The authors found the potencies of these compounds, relative to digitoxigenin, to be 23, 22, and 15, respectively.

Chiu and Watson (293) showed that both the flexible glycosides had preferred conformations that placed the 3′-OH and 5′-methyl groups in locations identical to those of the rigid gomphoside but that the rotational freedom of the rhamnoside was more restricted, which meant there was a greater statistical probability that it would have the preferred conformation of gomphoside. This would account for the fact that the potency of the rhamnoside was higher than that of the digitoxoside and confirms the role of the 3′-OH and 5′-methyl groups in contributing to the binding of the sugar moieties. The study also supports previous conclusions that the configuration about the A/B ring junction was important mainly with respect to its influence on the orientation of the sugar side-chain.

11.7 Endogenous Digitalis-like Factors (EDLF)

The digitalis binding site on Na⁺,K⁺-ATPase is highly conserved across a wide spectrum of organisms, ranging from brine shrimp to mammals. This has led to speculation that the binding site is a receptor for an endogenous ligand, much in the same way as the opiate receptor led to the suggestion and subsequent discovery of endogenous opiate-like substances (endorphins and enkephalins). The synthesis of cardiotonic steroids, for example bufalin (**50**), by toads shows that vertebrates have the capability of synthesizing steroids with the unusual cis–trans–cis steroid ring junctions depicted in Figure 28.28, and bearing a 17β-lactone and a 4β-OH. Research in the 1950s and 1960s led to the conclusion that there existed an unidentified ("third") factor in addition to aldosterone and vasopressin (antidiuretic hormone) that was involved in regulating volume load. Finally, there was growing evidence that high salt intake may be a major triggering factor in the development of hypertension (68).

Hypertension is a modern disease (at least in terms of its high prevalence), and yet it has a strong genetic basis. The idea, therefore, developed that certain people were genetically capable of dealing with high salt intake without adverse effects (for example, by switching blood flow to shallow nephrons where salt reabsorption is low), whereas others were genetically less capable of making this adjustment and relied on hormonal action to deal with the volume overload induced by excess salt or other causes. An endogenous digitalis-like factor (EDLF) could provide such hormonal activity. Salt reabsorbed from renal tubules by tubular epithelial cells must be cleared from these cells, which would otherwise become engorged with salt and cease to function. Na^+,K^+-ATPase would serve such a role, since it would transport Na^+ from the epithelial cell into the interstitial fluid and hence into the general circulation. If an EDLF was released in response to the rising plasma salt and fluid levels, it would inhibit the passage of salt across the tubular epithelial cell and thus, presumably, reduce the overall reabsorption of salt from the tubule, reducing salt and volume overload in the body. Although the release of this factor may assist in clearing excess salt, it could be present in sufficient amounts to inhibit Na^+,K^+-ATpase in vascular smooth muscle, leading to a rise in free intracellular Ca^{2+} as depicted in Figure 28.30. This would cause vasoconstriction, leading to increased cardiac output and vascular resistance and thus to hypertension.

Speculation of the type described above led to the search for an EDLF and generated an extensive literature. The reader is referred to recent reviews by Kelly and Smith (294) and Schoner (295) for discussion of this work. These reviews examine the numerous reports of the presence of an EDLF, the levels of which were claimed to correlate with the presence of hypertension in human and laboratory animal models. Since Na^+,K^+-ATPase is found in all cells, the release of an EDLF could be regarded as being too indiscriminate in its action to function as an effective homeostatic agent in controlling renal function. This argument can be countered by invoking either a localized release or selective affinity for a particular isoform of Na^+,K^+,ATPase, at least under nonpathological conditions.

Many putative EDLFs have been described, all fitting at least some of the criteria for such a substance. The best substantiated candidate is ouabain (**48**) or an isomer of ouabain. This conclusion arises from the work of Hamlyn and co-workers (296–299) who isolated 53 nmol (approximately 32 μg) of "ouabain" by extracting 300 L of human plasma. The substance was purified by HPLC and affinity adsorption on insolated Na^+,K^+-ATPase. It had identical retention times on HPLC as ouabain, and the molecular mass as determined by fast atom bombardment was the same as oubain (m/z = 585.2). The compound also cross-reacted with antibodies against ouabain. However, the authors did not rule out the possibility that the compound may be an isomer of ouabain.

Kelly and Smith (294) remain sceptical of the whole concept of the EDLF factor and its role in hypertension. They reject the idea that the high conservation of the digitalis-binding site on Na^+,K^+-ATPase across a wide spectrum of phyla implies an endogenous ligand, pointing out that many of the drugs known to block ion channels bind to receptors for which there are no known ligands. The retention of the digitalis "receptor" could be just an expression of the retention of a key region of the enzyme needed for its action as a cation pump. These reviewers also express concern as to whether any of the putative EDLFs have been adequately characterized. They point to the possibility of contamination of samples used for extracting EDLFs and mention that trace amounts could have been obtained from dietary sources. Contamination in the laboratory is also a possibility, especially in laboratories that use reference samples of authentic ouabain. Kelly and Smith (294) also expressed doubt as to whether any of the evidence presented up to the time of their review (1994) established "unequivocally that an endogenous cardiac glycoside-like compound exists as a physiological regulator of Na^+,K^+-ATPase activity." Finally, these authors also point to the fact that cardiac glycosides in therapeutic doses are not normally hypertensive.

This point is also made by Schoner (295). Schoner also refers to other aspects of the supposed physiological and pathophysiological effects of the EDLF, particularly those affecting the heart and renal system, that are not adequately explained by the concentrations of EDLFs reported by various workers.

Nevertheless, there is much indirect evidence supporting the role of an EDLF in the pathogenesis of, at least, some forms of hypertension, but as Kelly and Smith (294) conclude, "The concept that either a circulating or locally acting cardiac glycoside-like hormone or autacoid will be identified and proven to have a physiologically relevant role requires substantial additional experimental support before it can be accepted as fact."

ACKNOWLEDGMENTS

Sections 11.6.5 and 11.6.6 of this chapter were based largely on the text of the author's in M. Bohl and W. L. Duax, Eds., *Molecular Structure and Biological Activity of Steroids*, CRC Press, Boca Raton, Fla., 1992. The author wishes to acknowledge the kind permission of CRC Press to use this material.

REFERENCES

1. G. Stemp, *Trend. Pharmacol. Sci.*, **15**, 350–351 (1994).

2. L. H. Opie, Ed., *Cardiovascular Disease*, Vol. 9, Raven Press, New York, 1984.

3. W. G. Gonong, *Review of Medical Physiology*, 17th ed., Appleton, Langer, East Norwalk, Conn., 1993, pp. 58, 499.

4. W. Trautwein, *Pharmacol. Rev.*, **15**, 279 (1963).

5. A. M. Katz, *N. Engl. J. Med.*, **328**, 1244–1251 (1993).

6. W. A. Catterall, *Science*, **242**, 50–61 (1988).

7. B. Hille, *Ionic Channels of Exitable Membranes*, 2nd ed., Sinauer, Sunderland, Mass., 1992.

8. Task force for the Working Group on Arrhythmias of the European Society of Cardiology, *Eur. Heart J.*, **12**, 1112–1131 (1991).

9. M. Strong, K. G. Chandy, and G. A. Gutman, *Mol. Biol. Evol.*, **10**, 31–38 (1993).

10. K. Ho and co-workers, *Nature*, **362**, 127–133 (1993).

11. Y. Kubo, T. J. Baldwin, Y. N. Jan, and L. Y. Jan, *Nature*, **362**, 127–133 (1993).

12. M. Noda and co-workers, *Nature*, **320**, 188–189 (1986).

13. W. Stühmer and co-workers, *Nature*, **339**, 597–603 (1989).

14. R. W. Aldrich, *Nature*, **339**, 578–579 (1989).

15. T. Hoshi, W. N. Zogotta, and R. Aldrich, *Science*, **250**, 533–538 (1990).

16. A. L. Hokin and A. F. Huxley, *J. Physiol. (Lond.)*, **117**, 500–544 (1952).

17. B. Bean, *Nature*, **348**, 192–193 (1990).

18. O. Pongs, *Trend. Pharmacol. Sci.*, **13**, 359–365 (1992).

19. L. L. Cribbs, J. Stain, H. A. Fozzard, and R. B. Rogart, *FEBS Lett.*, **275**, 195–200 (1990).

20. H. F. Brown, *Physiol. Rev.*, **62**, 505–530 (1982).

21. D. DiFrancesco, A. Ferroni, M. Mazzanti, and C. Tromba, *J. Physiol. (Lond.)*, **377**, 61–88 (1986).

22. D. Pelzer, S. Pelzer, and T. F. MacDonald, *Rev. Physiol. Biochem. Pharmacol.*, **114**, 107–207 (1990).

23. D. F. Slish, D. Schulz, and A. Schwartz, *Hypertension*, **19**, 19–24 (1992).

24. O. Krizanova, R. Diebold, P. Lory, and A. Schwartz, *Circulation*, **87**, VII44–VII48 (1993).

25. W. A. Catterall, *Science*, **253**, 1499–1500 (1991).

26. W. A. Catterall and J. Striessnig, *Trend. Pharmacol. Sci.*, **13**, 256–262 (1992).

27. A. Schwartz, *Am J. Cardiol.*, **73**, 12B–14B (1994).

28. K. G. Chandy, *Nature*, **352**, 26 (1991).

29. D. Escande and N. Standen, Eds., *K^+ Channels in Cardiovascular Medicine*, Springer-Verlag, Berlin, 1993.

30. W. R. Giles and Y. Imaizumi, *J. Physiol. (Lond.)*, **405**, 123–145 (1988).

31. M. C. Sanguinetti and N. K. Jurkiewicz, *J. Gen. Physiol.*, **96**, 195–215 (1990).

32. T. J. Colatsky, C. H. Follmer, and C. F. Starmer, *Circulation*, **82**, 2236–2242 (1990).

33. E. Carmeliet, L. Storms, and J. Vereecke, in D. P. Zipes and H. Jaife, Eds., *Cardiac Electrophysiology from Cell to Bedside*, W. B. Saunders, Philadelphia, 1990, pp. 103–108.

34. B. Sakmann, A. Norma, and W. Trautwein, *Nature*, **303**, 250–253 (1983).

35. D. Escande and I. Cavero, *Trend. Pharmacol. Sci.*, **13**, 269–272 (1992).

36. W. C. Cole, C. D. Mcpherson, and D. Sontag, *Circ. Res.*, **69**, 571–581 (1991).

37. G. J. Gross and co-workers, *Am. J. Cardiol.*, **63**, 11J–17J (1989).

38. G. J. Grover and co-workers, *J. Pharmacol. Exp. Ther.*, **257**, 156–162 (1991).

39. J. A. Auchampach, M. Maruyama, I. Cavero, and G. J Gross, *J. Pharmacol. Exp. Ther.*, **259**, 961–967 (1991).

40. D. Thuringer and D. Escande, *Mol. Pharmacol.*, **36**, 897–902 (1989).

41. R. D. Harvey, C. D. Clark, and J. R. Hume, *J. Gen. Physiol.*, **95**, 1077–1102 (1990).

42. T. Eschenhagen, *Cell Biol. Int.*, **17**, 723–748 (1993).

43. B. E. Ehrlich, E. Kaftan, S. Bezprozvannaya, and I. Bezprozvanny, *Trend. Pharmacol. Sci.*, **15**, 145–148 (1994).

44. S. Supattapone, P. F. Worley, J. M. Baraban, and S. H. Snyder, *J. Biol. Chem.*, **263**, 1530–1534 (1988).

45. C. W. Taylor and C. B. Marshall, *Trend. Biochem. Sci.*, **17**, 403–407 (1992).

46. A. Galioni, *Science*, **259**, 325–326 (1993).

47. J. W. Putney Jr., *Cell Calcium*, **7**, 1–12 (1986).

48. C. W. Taylor, *Trend. Pharmacol. Sci.*, **15**, 271–274 (1994).

49. C. Fasolato, B. Innocenti, and T. Pozzan, *Trend. Pharmacol. Sci.*, **15**, 77–83 (1994).

50. M. Hoth and R. Penner, *Nature*, **355**, 353–356 (1992).

51. M. Hoth and R. Penner, *J. Physiol.*, **465**, 359–386 (1993).

52. C. Randriamampita and R. Y. Tsien, *Nature*, **364**, 809–814 (1993).

53. P. Caroni and E. Carafoli, *Eur. J. Biochem.*, **132**, 451–460 (1983).

54. M. P. Blaustein, *J. Cardiovasc. Pharmacol.*, **12**, S56–S69 (1988).

55. K-I. Furakawa, Y. Tawada-Iwata, and M. Shigekawa, *J. Biochem.*, **106**, 1068–1073 (1989).

56. Roufogalis and Wang, *Today's Life Sci.*, 28–34 (Feb. 1993).

57. V. L. Lew, R. Y. Tsien, C. Miner, and R. M. Brookchin, *Nature*, **298**, 478–481 (1982).

58. H. P. Adamo, A. F. Rega, and P. J. Garrahan, *J. Biol. Chem.*, **265**, 3789–3792 (1990).

59. Y-H. Xu and B. D. Roufogalis, *Sodium Transport Inhibitors*, in G. S. Stokes and J. F. Marwood, Eds., *Progess in Biochemical Pharmacology*, Vol. 23, Karger, Basel, 1988, pp. 107–118.

60. J. Smallwood, D. M. Waisman, D. Lafreniere, and H. Rasmussen, *J. Biol. Chem.*, **258**, 11,092–11,097 (1983).

61. Y-H. Xu and B. D. Roufogalis, *J Membr. Biol.*, **105**, 155–164 (1988).

62. D. A. Dixon and D. H. Haynes, *J. Membr. Biol.*, **112**, 169–183 (1989).

63. G. E. Shull and J. Greeb, *J Biol. Chem.*, **263**, 8646–8657 (1988).

64. A. K. Verma and co-workers, *J. Biol. Chem.*, **263**, 14, 152–14, 159–(1988).

65. S. De Jaegere and co-workers, *Biochem. J.*, **271**, 655–660 (1990).

66. P. L. Jørgensen and J. P. Andersen, *J. Membr. Biol.*, **103**, 95–120 (1988).

67. J. M. Glynn and S. J. D. Karlish *Annu. Rev. Biochem.*, **59**, 171–205 (1990).

68. R. Thomas, P. Gray, and J. Andrews, in B. Testa, Ed., *Advances in Drug Research*, Vol. 19, Academic Press, Inc., London, 1990, pp. 313–562.

69. R. E. Thomas, in M. Bohl and W. L. Daux, Eds., *Molecular Structure and Biological Activity Of Steroids*, CRC Press, Boca Raton, Fla., 1992, pp. 399–464.

70. C. H. Pedemonte and J. H. Kaplan, *Am. J. Physiol.*, **258**, C1–C23 (1990).

71. N. S. Modyanov, E. Lutsenko, E. Chertova, and R. Efremov, *Soc. Gen. Physiol. Ser.*, **46**, 99–115 (1991).

72. K. J. Sweadner and E. Arystarkhova, *Ann. N. Y. Acad. Sci.*, **671,** 217–227 (1992).

73. K. J. Sweadner, *Trend. Cardiovasc. Med.*, **3,** 2–6 (1993).

74. O. I. Shamraj and co-workers, *Cardiovasc. Res.*, **27,** 2229–2237 (1993).

75. R. Levenson, *Rev. Physiol. Biochem. Pharmacol.*, **123,** 1–45 (1994).

76. J. B. Lingrel and T. Kuntzweiler, *J. Biol. Chem.*, **269,** 19,659–19,662 (1994).

77. J. B. Lingrel and co-workers, *Kidney Int.*, **45,** S32–S39 (1994).

78. R. W. Albers, *Annu. Rev. Biochem.*, **36,** 727–756 (1967).

79. R. L. Post and co-workers, *J. Gen. Physiol.*, **54,** 306s–326s (1969).

80. S. Yoda and A. Yoda, *J. Biol. Chem.*, **261,** 1147–1152 (1986).

81. G. Scheiner-Bobis, K. Fahlbusch, and W. Schoner, *Eur. J. Biochem.*, **168,** 123–131 (1987).

82. W. Schoner, H. Pauls, and R. Patzelt-Wenczler, G. Riecker, A. Weber, and J. Goodwin, Eds., *Myocardial Failure*, Springer-Verlag, Berlin, 1977, p. 115.

83. G. Scheiner-Bobis, E. Buxbaum, and W. Schoner, in J. C. Skou, J. G. Nørby, A. B. Maunsbach, and M. Esman, Eds., *The Na⁺, K⁺-Pump, Part A: Molecular Aspects*, Alan R. Liss, New York, 1988, p. 219.

84. H. Herbert, E. Skriver, and A. B. Maunsbach, *FEBS Lett.*, **187,** 182–186 (1985).

85. G. E. Shull, A. Schwartz, and J. B. Lingrel, *Nature*, **316,** 691–695 (1985).

86. G. E. Shull, L. K. Lane, and J. B. Lingrel, *Nature*, **321,** 429–431 (1986).

87. J. B. Lingrel, M. Orlowski, M. M. Shull, and E. M. Price, *Prog. Nucleic Acid Res.*, **38,** 37–89 (1990).

88. K. J. Sweadner, *Biochim. Biophys. Acta*, **988,** 185–220 (1989).

89. K. J. Sweadner, *J. Biol. Chem.*, **254,** 6060–6067 (1979).

90. D. W. Hilgerman, *Science*, **263,** 1429–1432 (1994).

91. K. Wang and R. A. Farley, *J. Biol. Chem.*, **267,** 3477–3480 (1992).

92. L. K. Lane, J. M. Feldmann, C. E. Flarsheim, and C. L. Rybcynski, *J. Biol. Chem.*, **268,** 17,930–17,934 (1993).

93. C. R. Jones, P. Molenaar, and R. J. Summers, *J. Mol. Cell. Cardiol.*, **21,** 519–535 (1989).

94. J. Hescheler and W. Trautwein, in H. M. Piper and G. Isenberg, Eds., *Isolated Adult Cardiomyocytes*, CRC Press, Inc, Boca Raton, Fla., 1989, pp. 128–154.

95. D. DiFrancesco, *Prog. Biophys. Mol. Biol.*, **46,** 163–183 (1985).

96. K. Yazawa and M. Kameyama, *Am. J. Physiol.*, **428,** 135–150 (1990).

97. A. Shah, I. S. Cohen, and M. Rosen, *Biophys, J.*, **54,** 219–225 (1988).

98. M. Apkon and J. M. Nerbonne, *Proc. Natl. Acad. Sci. U. S. A.*, **85,** 8756–8760 (1988).

99. U. Del Balzo and co-workers, *Circ. Res.*, **67,** 1535–1551 (1990).

100. A. Yatani, J. Codina, A. M. Brown, and L. Birnbaumer, *Science*, **235,** 207–211 (1987).

101. D. DiFrancesco, P. Ducouret, and R. B. Robinson, *Science*, **243,** 669–671 (1989).

102. M. I. Cybulski and M. A. Gimbrone Jr., *Science*, **251,** 788–791 (1991).

103. M. Nakamuta and co-workers, *Biochem. Biophys. Res. Comun.*, **177,** 34–39 (1991).

104. Y. Ogawa and co-workers, *Biochem. Biophys. Res. Comun.*, **178,** 656–663 (1991).

105. A. Sakamoto and co-workers, *Biochem. Biophys. Res. Comun.*, **177,** 34–39 (1991).

106. K. Goto and co-workers, *Proc. Natl. Acad. Sci. U. S. A.*, **86,** 3915–3918 (1989).

107. Y. Kasuya and co-workers, *Br. J. Pharmacol.*, **107,** 456–462 (1992).

108. T. Sakurai and K. Goto, *Drugs*, **46,** 795–804 (1993).

109. S. A. Douglas, T. D. Meek, and E. H. Ohlstein, *Trend. Pharmacol. Sci.*, **15,** 313–316 (1994).

110. J. A. Amith, A. M. Shah, S. Fort, and M. J. Lewis, *Trend. Pharmacol. Sci.*, **13,** 113–116 (1992).

111. J. Wang, G. Paik, and J. P. Morgan, *Circ. Res.*, **69,** 582–589 (1991).

112. B. N. Singh and E. M. Vaughan Williams, *Br. J. Pharmacol.*, **39,** 675–687 (1970).

113. B. N. Singh and E. M. Vaughan Williams, *Cardiovasc. Res.*, **6,** 109–119 (1972).

114. S. Nattel, *Drugs*, **41,** 672–701 (1991).

115. D. C. Harrison and M. B. Bottorff, in J. T. August, M. W. Anders, and F. Murad, Eds., *Advances in Pharmacology*, Vol. 23, Academic Press, Inc., San Diego, Calif., 1992, pp. 179–225.

116. Task Force of the Working Group on Arrhythmisas of the European Society of Cardiology, *Circulation*, **84,** 1831–1851 (1991).

117. E. M. Vaughan Williams, *J. Clin. Pharmacol.*, **32,** 964–977 (1922).

118. S. Nattel and A. Arenae, *Drugs*, **45,** 9–14 (1993).

119. B. Hille, *J. Gen. Physiol.*, **69,** 497–515 (1977).

120. L. M. Hondeghem and B. G. Katzund, *Biochim. Biophys. Acta*, **472,** 373–398 (1977).

121. A. O. Gtant, C. F. Starmer, and H. C. Strauss, *Circ. Res.*, **55**, 427–439 (1984).

122. L. M. Hondeghem and B. G. Katzung, *Ann. Rev. Pharmacol. Toxicol.*, **24**, 387–356 (1984).

123. C. F. Starmer and A. O. Grant, *Mol. Pharmacol.*, **28**, 348–356 (1985).

124. A. O. Grant and D. J. Wendt, *Trend. Pharmacol. Sci.*, **13**, 352–358 (1992).

125. F. R. Gilliam III, C. F. Starmer, and A. O. Grant, *Circ. Res.*, **65**, 723–739 (1989).

126. L. M. Hondeghem and D. J. Snyders, *Circulation*, **81**, 686–690 (1990).

127. T. J. Campbell, in E. M. Vaughan Williams, Ed., *Antiarrhythmic Drugs, Handbook of Experimental Pharmacology*, Vol. 89, Springer-Verlag, Berlin, 1989, pp. 135–155.

128. E. Cameliet, *Circ. Res.*, **63**, 50–60 (1988).

129. Cardiac Arrhythmia Suppression Trial (CAST), *N. Engl. J. Med.*, **321**, 406–412 (1989).

130. M. A. Butler, M. Iwasaki, F. P. Guengerich, and F. F. Kadlubar, *Proc. Natl. Acad. Sci. U. S. A.*, **86**, 7696–7700 (1989).

131. K. K. Teo, S. Yusuf, and C. D. Furberg, *JAMA*, **270**, 1589–1595 (1993).

132. W. A. Catterall, *Ann. Rev. Biochem.*, **55**, 953–985 (1986).

133. S. W. Postma and W. A. Catterall, *Mol. Pharmacol.*, **25**, 219–227 (1984)

134. R. S. Sheldon, H. J. Duff, and R. J. Hill, *Clin. Invest. Med.*, **14**, 458–465 (1991).

135. D. M. Roden, *Am. J. Cardiol.*, **72**, 44B–49B (1993).

136. J. Ben-David and D. P. Zipes, *Lancet*, **341**, 1578–1582 (1993).

137. T. J. Colatsky and C. H. Follmer, *Cardiovasc, Drug. Rev.*, **7**, 199–209 (1989).

138. K. S. Lee, *J. Pharmacol. Exp. Ther.*, **262**, 99–108 (1992).

139. C. T. January and J. M. Riddle, *Circ. Res.*, **64**, 977–990 (1989).

140. A. L. Wit and J. Coromilas, *Am. J. Cardiol.*, **72**, 3F–12F (1993).

141. B. N. Singh, *Am. J. Cardiol.*, **72**, 114F–124F (1993).

142. B. N. Singh, *Am. J. Cardiol.*, **72**, 18F–24F (1993).

143. J. W. Mason, *Am. J. Cardiol.*, **72**, 59F–61F (1993).

144. B. R. Lucchesi and co-workers, *Am. J. Cardiol.*, **72**, 25F–44F (1993).

145. A. Fitton and E. M. Sorkin, *Drugs*, **46**, 678–719 (1993).

146. J. Morganroth, *Am. J. Cardiol.*, **72**, 3A–7A (1993).

147. D. J. MacNeil, R. O. Davies, and D. Deitchman, *Am. J. Cardiol.*, **72**, 44A–50A (1993).

148. D. M. Roden, *Am. J. Cardiol.*, **72**, 51A–55A (1993).

149. B. M. Singh, *Am. J. Cardiol.*, **72**, 8A–18A (1993).

150. J. J. Hanyok, *Am. J. Cardiol.*, **72**, 19A–26A (1993).

151. J. Gill, R. C. Heel, and A. Fitton, *Drugs*, **43**, 69–110 (1992).

152. M. Nora and D. P. Zipes, *Am. J. Cardiol.*, **72**, 62F–69F (1993).

153. H. L. Greene, *Am. J. Cardiol.*, **72**, 70F–74F (1993).

154. J. Ahlner, R. G. Andersson, K. Torfgard, and K. L. Axelsson, *Pharm. Rev.*, **43**, 351–423 (1991).

155. H-L. Fung, *Br. J. Clin. Pharmacol.*, **34**, 5S–9S (1992).

156. *Am. J. Cardiol.*, [Special issue], **70**, 1B–98B (1992).

157. R. F. Furchgott and J. V. Zawadzaki, *Nature*, **288**, 373–376 (1980).

158. S. Moncada, R. M. J. Palmer, and E. A Higgs, *Pharmacol. Rev.*, **43**, 109–142 (1991).

159. S. A Lipton and co-workers, *Nature*, **364**, 626–632 (1993).

160. W. C. Sessa, C. M. Barber, and K. R. Lynch, *Circ. Res.*, **72**, 921–924 (1993).

161. S. Lamas and co-workers, *Proc. Natl, Acad. Sci. U. S. A.*, **89**, 6348–6352 (1992).

162. M. Nakane and co-workers, *FEBS Lett.*, **316**, 175–180 (1993).

163. C. R. Lyons, G. J. Orloff, and J. M. Cunningham, *J. Biol. Chem.*, **267**, 6360–6374 (1992).

164. R. Schulz and C. R. Triggle, *Trend. Pharmacol. Sci.*, **15**, 255–259 (1994).

165. G. Weimer, B. A. Schölkens, R. H. A. Becker, and R. Busser, *Hypertension*, **18**, 558–563 (1991).

166. R. D. Farhy, O. A. Carretero, K-H. Ho, and A. G. Scicli, *Circ. Res.*, **72**, 1202–1210 (1993).

167. P-F. Méry, S. M. Lohmann, U. Walter, and R. Fischmeister, *Proc. Natl. Acad. Sci. U. S. A.*, **88**, 1197–1201 (1991).

168. M. Kelm and J. Schrader, *Circ. Res.*, **66**, 1561–1575 (1990).

169. J. N. Bates, M. T. Baker, R. Guerra, Jr., and D. G. Harrison, *Biochem. Pharmacol.*, **42**, 5157–5165 (1991).

170. D. G. Harrison and J. N. Bates, *Circulation*, **87**, 1461–1467 (1993).

171. H-L. Fung, S. C. Sutton, and A. Kamiya, *J. Pharmacol. Exp. Ther.*, **228**, 334–341 (1984).

172. J. F. Brien and co-workers, *J. Pharmacol. Exp. Ther.*, **237**, 608–614 (1986).

173. S-J. Chung and H-L. Fung, *J. Pharmacol. Exp. Ther.*, **237**, 608–614 (1986).

174. S-J. Chung and H-L. Fung, *Biochem. Pharmacol.*, **45**, 157–163 (1993).

175. H-L. Fung, *Am. J. Cardiol.*, **72**, 9C–15C (1993).

176. S. Tsuchida, T. Maki, and T. Sata, *J. Biol. Chem.*, **265**, 7150–7157 (1990).

177. H. Schröder, *J. Pharmacol. Exp. Ther.*, **262**, 298–302 (1992).

178. B. M. Bennet, B. J. McDonald, R. Nigam, and W. C. Simon, *Trend. Pharmacol. Sci.*, **15**, 245–249 (1994).

179. B. M. Bennett and co-workers, *Circ. Res.*, **63**, 693–701 (1988).

180. H.-L Fung and co-workers, *Am. J. Cardiol.*, **70**, 4B–10B (1992).

181. M. Sakanashi, T. Matsuzaki, and Y. Aniya, *Br. J. Pharmacol.*, **103**, 1905–1908 (1991).

182. E. Kowaluk and H-L. Fung, in J. Abrams, C. J. Pepine, and U. Thadani, Eds., *Medical Therapy of Ischemic Heart Disease*, Little, Brown & Co., Boston, 1992, pp. 151–175.

183. J. G. Kelly and K. O'Malley, *Clin. Pharmacokinet.*, **22**, 416 (1992).

184. D. Murdoch and R. C. Heel, *Drugs*, **41**, 478 (1991).

185. W. H. Frishman, Ed., *Am. J. Cardiol.*, [Suppl. A] 73, (1994).

186. J. Striessnig, H. Glossman, and W. A Catterall, *Proc. Natl. Acad. Sci. U. S. A.*, **87**, 9108–9112 (1990).

187. R. W. Tsien, P. T. Ellinor, and W. A. Horne, *Trend. Pharmacol. Sci.*, **12**, 349–354 (1991).

188. L. G. Herbette, Y. M. H. Vant Erve, and D. G. Rhodes, *J. Mol. Cell. Cardiol.*, **21**, 187–201 (1989).

189. R. Kass, J. P. Arena, and S. Chin, *J. Gen. Physiol.*, **98**, 63–75 (1991).

190. H. Nakayama and co-workers *Proc. Natl. Acad. Sci. U. S. A.*, **88**, 9203–9207 (1991).

191. J. Striessnig, B. J. Murphy, and W. A. Catterall, *Proc. Natl. Acad. Sci. U. S. A.*, **88**, 10769–10773 (1991).

192. J. Striessnig and co-workers, *J. Biol. Chem.*, **265**, 363–370 (1990).

193. K. Stoschitzky, W. Lindner, and W. Klein, *Trend. Pharmacol. Sci.*, **15**, 102 (1994).

194. J. R. Hampton, *Drugs*, **48**, 549–568 (1994).

195. D. McAreavery, R. Vermeulen, and J. I. S. Robertson, *Cardiovasc. Drugs Ther.*, **5**, 577–588 (1991).

196. B. N. C. Pritchard, *J. Cardiovasc. Pharmacol.*, **19**, S1–S4 (1992).

197. D. McTavish, D. Campoli-Richards, and E. M. Sorkin, *Drugs*, **45**, 232–258 (1993).

198. INT-CAR-07 (UK) Study Group, *J. Cardiovasc. Pharmacol.*, **19**, S82-S85 (1992).

199. P. Omvik and P. Lund-Johansen, *Cardiovasc. Drugs Ther.*, **7**, 193–206 (1993).

200. K. D. Lamon, *Am. Heart J.*, **121**, 683–687 (1991).

201. G. Ruf and co-workers, *Int. J. Cardiol.*, **43**, 279–285 (1994).

202. A. Semplicini and co-workers, *Clin. Pharmacol. Ther.*, **33**, 278–283 (1983).

203. C. Giannattasio and co-workers, *J. Cardiovasc. Pharmacol.*, **19** (Suppl. 1), 18–22 (1992).

204. F. Waagstein, A. Hjalmarson, E. Varnauskas, and I. Wallentin, *Br. Heart J.*, **37**, 1022–1036 (1975).

205. R. S. Englemeier and co-workers, *Circulation*, **72**, 536–546 (1985).

206. K. Swedbert, *Am. J. Cardiol.*, **71**, 30C–38C (1993).

207. J. L. Flegg and E. G. Lakatta, *Int. J. Cardiol.*, **6**, 295–305 (1984).

208. M. Packer, *J. Am. Coll. Cardiol.*, **20**, 248–254 (1992).

209. R. J. Cody, *Heart Failure Index Rev.*, **5**, 15 (1993).

210. G. S. Francis and co-workers, *Circulation*, **82**, 1724–1729 (1990).

211. H. Drexler and co-workers, *Am. J. Cardiol.*, **69**, 1596–1601 (1992).

212. H. Urata and co-workers, *Circ. Res.*, **66**, 883–890 (1990).

213. S. S. Gottlieb and co-workers, *Circulation*, **86**, 420–425 (1992).

214. M. A. Greager, *Am. J. Cardiol.*, **69**, 10G–15G (1992).

215. G. S. Francis and K. M. McDonald, *Am. J. Cardiol.*, **69**, 3G–7G (1992).

216. C. G. Brilla, L. S. Matsubara, and K. T. Weber, *Am. J. Cardiol.*, **71**, 12A–16A (1993).

217. A. M. Katz, *Am. J. Cardiol.*, **70**, 126C–131C (1992).

218. CONSENSUS Trial Study Group, *N. Engl. J. Med.*, **316**, 1429–1435 (1987).

219. In ref. 155, pp. 1C–151C.

220. In ref. 184, pp. 1C–44C.

221. C. V. Leier, *Am. J. Cardiol.*, **69**, 120G–129G (1992).

222. C. V. Leier and co-workers, *Am. J. Cardiol.*, **62**, 94–99 (1988).

223. M. Wehling, J. Zimmerman, and K. Theisen, *Cardiology*, **77** (Suppl. 5), 81–88 (1990).

224. P. Honerjäger, *Am. Heart J.*, **121**, 1939–1944 (1991).

225. R. DiBianco and co-workers, *J. Am. Coll. Cardiol.*, **4**, 855–866 (1984).

226. B. Massie and co-workers, *Circulation*, **71**, 963–971 (1985).

227. P. T. Wilmshurst and co-workers, *Br. Heart J.*, **49**, 77–82 (1983).

228. R. DiBianco and co-workers, *N. Engl. J. Med.*, **320**, 677–683 (1989).

229. M. Packer and co-worker, *N. Engl. J. Med.*, **325**, 1468–1475 (1991).

230. P. W. Armstrong and G. W. Moe, *Circulation*, **88**, 2941–2952 (1993).

231. B. F. Uretsky and co-workers, *Circulation*, **82**, 774–780 (1990).

232. W. Withering, "An account of the foxglove and some of its medical uses, &c", M. Swinney, London pp i–xiii + 1–207, (MDCCLXXXV).

233. G. Mancia and co-workers, *Am. J. Cardiol.*, **69**, 17G–22G (1992).

234. D. W. Ferguson, *Am. J. Cardiol.*, **69**, 24G–232G (1992).

235. A. B. Covit and co-workers, *Am. J. Med.*, **75**, 445–447 (1983).

236. D. W. Ferguson and co-workers, *Circulation*, **80**, 65–77 (1989).

227. The Captopril-Digoxin Multocentre Research Group, *JAMA*, **259**, 539–544 (1988).

238. G. H. Guyatt and co-workers, *Am. J. Cardiol.*, **61**, 371–375 (1988).

239. R. Jaeschke, A. D. Oxman, and G. H. Guyatt, *Am. J. Med.*, **88**, 279–286 (1990).

240. M. Packer and co-workers, *N. Engl. J. Med.*, **329**, 1–7 (1993).

241. M. Gheorghiade, Ed., *Am. J. Cardiol.* [Special issue], **69**, 1G–150G (1992).

242. N. Dzimiri and U. Fricke, *Br. J. Pharmacol.*, **93**, 281–288 (1988).

243. R. E. Thomas, J. Boutagy, and A. Gelbert, *J. Pharm. Sci.*, **63**, 1649–1683 (1974).

244. T. Akera and T. M. Brody, *Pharmacol. Rev.*, **29**, 187–220 (1978).

245. A. Schwartz, G. E. Lindenmayer, and J. C. Allen, *Pharmacol. Rev.*, **27**, 1–134 (1975).

246. K. Greef, Ed., *Cardiac Glycosides, Part 1: Experimental Pharmacology*, Springer-Verlag, Berlin, 1981.

247. K. Repke and W. Schönfeld, *Trend. Pharmacol. Sci.*, **5**, 393–397 (1984).

248. T. W. Smith and co-workers, *Prog. Cardiovasc. Dis.*, **26**, 413–458 (1984).

249. T. W. Smith and co-workers, *Prog. Cardiovasc. Dis.*, **26**, 495–540

250. T. W. Smith and co-workers, *Prog. Cardiovasc. Dis.*, **27**, 21–56 (1984).

251. O. Hansen, *Pharmacol. Rev.*, **36**, 143–163 (1984).

252. B. M. Anner, *Biochem. J.*, **227**, 1–11 (1985).

253. E. Erdmann, K. Greef, and J. C. Skou, Eds., *Cardiac Glycosides 1785–1985*, Springer-Verlag, New York, 1986.

254. T. W. Smith, *N. Engl. J. Med.*, **318**, 358–365 (1985).

255. T. Akera and Y. C. Ng, *Life Sci.*, **48**, 97–106 (1991).

256. E. M. Price and J. B. Lingrel, *Biochemistry*, **27**, 8400–8408 (1988).

257. S. Noguchi and co-workers, *Biochem. Biophys. Res. Commun.*, **155**, 1237–1243 (1988).

258. J. R. Emanuel, S. Graw, D. Housman, and R. Levenson, *Mol. Cell Biol.*, **9**, 3744–3749 (1989).

259. I. Kano and co-workers, *Biochem. Cell Biol.*, **68**, 1262–1267 (1990).

260. E. A. Arystarkhova, M. Gasparian, N. N. Modyanov, and K. J. Sweadner, *J. Biol. Chem.*, **267**, 13,694–13,701 (1992).

261. R. Blostein, R. Zhang, C. J. Gottardi, and M. J. Caplan, *J. Biol. Chem.*, **268**, 10,645–10,658 (1993).

262. T. Ishii and K. Takeyasu, *Proc. Natl. Acad. Sci. U. S. A.*, **90**, 8881–8885 (1993)

263. L. Brown, E. Erdmann, and R. E. Thomas, *Biochem. Pharmacol.*, **32**, 2767–2774 (1983).

264. K. Ahmed and co-workers, *J. Biol. Chem.*, **258**, 8092–8097 (1993).

265. A. H. L. From, D. S. Fullerton, and K. Ahmed, *Mol Cell Biochem.*, **94**, 157 (1990).

266. Y-C. Ng, W-Y. Leung, and T. Akera, *Eur. J. Pharmacol.*, **155**, 93–99 (1988).

267. R. E. Thomas, J. Boutagy, and A. Gelbart, *J. Pharmacol. Exp. Ther.*, **191**, 219–231 (1974).

268. A. Gelbart and R. Thomas, *J. Med. Chem.*, **21**, 284–288 (1978).

269. R-S. Kim and co-workers, *Mol. Pharmacol.*, **18**, 402–405 (1980).

270. F. S. Labella and co-workers, *Fed. Proc.*, **44**, 2806–2811 (1985).

271. D. S. Fullerton and co-workers, *Curr. Top. Membr. Transp.*, **19**, 257–264 (1983).

272. J. Weiland and co-workers, *J. Enzyme Inhib.*, **2**, 31–36 (1987).

273. J. Templeton, personal communication, 1995.

274. J. M. Mauxent and co-workers, *Arneim.-Forsch.*, **42**, 1301–1305 (1992).

275. M. M. Adamanitidis, E. R. Honoré, and B. A. Dupuis, *Br. J. Pharmacol.*, **95**, 1063 (1988).

276. W. Schönfeld and co-workers, *Arch. Pharmacol.*, **329**, 414–426 (1985).

277. T. J. Franklin, in J. W. Lamble and G. A. Robinson, Eds., *Towards Understanding Receptors*, Elsevier/North Holland Biomedical Press, Amsterdam, 1981, pp 8–15.

278. J. Beer and co-workers, *Biochim. Biophys. Acta*, **937**, 335–346 (1988).

279. M. Bohl, *Z. Naturforsch.*, **40c**, 858–862 (1985).

280. A. V. Kamernitzky and co-workers, *J. Steroid Biochem.*, **32**, 857 (1989).

281. K. R. H. Repke, J. Weiland, R. Megges and R. Schön, "Approach to the Chemotopography of the Digitalis Recognition Matrix in Na^+/K^+-transporting ATPase as a Step in the Rational Design of New Inotropic Steroids", in G. P. Ellis and D. K. Luscombe, Eds., *Progress in Medicinal Chemistry* Vol. 30, Elsevier Science Publishers, Amsterdam, 1993, pp 135–202.

282. R. L. Vick, *J. Pharmacol. Exp. Ther.*, **125**, 40 (1959).

283. D. S. Fullerton and co-workers, in A. S. V. Burgen, G. C. K. Roberts and M. S. Tute, Eds., in *Molecular Graphics and Drug Design*, Elsevier, Amsterdam, 1986, pp. 257–284.

284. M. Bohl and R. Sümilch, *Eur, J. Med. Chem.*, **21**, 193–198 (1986).

285. L. Brown and R. E. Thomas, *Arneim. Forsch.*, **34**, 572–574 (1984).

286. L. Brown, R. E. Thomas, and T. Watson, *Arch. Pharmacol.*, **332**, 98–102 (1986).

287. K. Go and K. K. Bhandary, *Acta Cryst.*, **B45**, 306 (1989).

288. T. Shigei, H. Tsuru, Y. Saito, and M. Okada, *Experientia*, **29**, 449–450 (1973).

289. A. DePover and T. Godfraind, *Arch. Pharmacol.*, **321**, 135–139 (1982).

290. K.-O Haustein and R. Bauer, *Int. J. Clin. Pharmacol. Ther.*, **32**, 299–304 (1994).

291. L. Brown and R. E. Thomas, *Arneim-Forsch.*, **33**, 814–817 (1983).

292. J. Kyte, *J. Biol. Chem.*, **247**, 7634–7641 (1972).

293. C. K. Chiu and T. R. Watson, *J. Med. Chem.*, **28**, 509–515 (1985).

294. R. A. Kelly and T. W. Smith, *Adv. Pharmacol.*, **25**, 263–288 (1994).

295. W. Schoner, *Progress in Drug Research*, **41**, 249-291 (1993).

296. D. W. Harris and co-workers, *Hypertension*, **17**, 936–943 (1991).

297. J. H. Ludens and co-workers, *Hypertension*, **17**, 923–929 (1991).

298. W. R. Mathews and co-workers, *Hypertension*, **17**, 930–935 (1991).

299. S. Bova and co-workers, *Hypertension*, **17**, 944–950 (1991).

PART IIB
CARDIOVASCULAR DRUGS

CHAPTER TWENTY-NINE

Antihypertensive Agents

PIETER B. M. W. M. TIMMERMANS
RONALD D. SMITH

DuPont Merck Research Laboratories
Wilmington, Delaware, USA

Burger's Medicinal Chemistry and Drug Discovery,
Fifth Edition, Volume 2: Therapeutic Agents,
Edited by Manfred E. Wolff.
ISBN 0-471-57557-7   © 1996 John Wiley & Sons, Inc.

CONTENTS

1 Introduction,   266
2 The Physiology of Blood Pressure Control as the
   Basis of Antihypertensive Drug Action,   267
3 Definition and Epidemiology of
   Hypertension,   270
4 Rationale for the Drug Therapy of
   Hypertension,   272
5 Clinical Use of Antihypertensive Drugs,   275
   5.1 Treatment guidelines, 275
   5.2 Defining the "ideal" antihypertensive/blood
      pressure lowering drug, 276
6 Established Antihypertensive Drugs,   277
   6.1 History of antihypertensive drug
      discovery, 277
   6.2 Major classes of currently established
      antihypertensive drugs, 278
      6.2.1 Diuretics, 280
      6.2.2 Beta Blockers, 285
      6.2.3 Angiotensin Converting Enzyme (ACE)
         Inhibitors, 289
      6.2.4 Calcium Channel Blockers, 292
      6.2.5 Alpha-1 (α_1) Adrenergic Receptor
         Antagonists, 295
      6.2.6 Alpha-2 (α_2)/Imidazoline-1 Receptor
         Agonists, 297
      6.2.7 Innovative antihypertensive drugs, 298
   6.3 Innovative antihypertensive drugs, 300
      6.3.1 Angiotensin II receptor antagonists, 300
      6.3.2 Potassium channel openers, 304
      6.3.3 Renin inhibitors, 305
      6.3.4 Serotonin receptor ($5HT_2$)
         antagonists, 306

6.3.5   Neutral endopeptidase (NEP)
        inhibitors, 306
6.3.6   Adenosine receptor (A2)
        antagonists, 306
6.3.7   Vasopressin receptor (V_1)
        antagonists, 307
6.3.8   Endothelin receptor antagonists, 307
6.3.9   Neuropeptide Y receptor
        antagonist, 308
6.3.10 Protein kinase C inhibitors, 309
7 Novel Approaches to the Future the
  Pharmacological Treatment of Hypertension,   309
  7.1 Novel "mechanistic" approaches, 309
  7.2 Antisense molecules, 311
  7.3 Experimental models of genetic
      hypertension, 312
8 Perspectives for the Future of Medicinal Chemical
  Approaches to the Treatment of Hypertension or
  "Do we have all the antihypertensive drugs we
  need and, if not, can we afford the costs of the
  new science?",   313

1 INTRODUCTION

Antihypertensive drugs are playing an increasingly important role in the management of hypertensive patients (1,2). Controlled clinical trials have now clearly demonstrated that current antihypertensive drugs can decrease both the morbidity and mortality of cardiovascular disease. In less than 20 years, the number of antihypertensive agents available to the practicing physician has mushroomed. More importantly, however, the newer agents are more effective in lowering blood pressure and are much better tolerated than the earlier agents. The availability of multiple antihypertensive agents, however, has not meant that the problem of hypertension is solved. Although great strides have been made in educating the public to the risks associated with high blood pressure, in the understanding of the physiology of blood pressure control, in improving and standardizing the measurements of blood pressure, and in diagnostic procedures, we still do not know the etiology of essential hypertension. Further, the current antihypertensive drug therapy treats the symptom of

hypertension but does not treat the cause of hypertension. Thus, hypertensive patients must be treated for the rest of their lives. The "ideal" antihypertensive treatment has not been discovered. The challenge is to identify the factor(s) involved in the elevated pressure in each patient and to design drugs which correct or reverse the effect of that abnormal function. For the near term, it is likely that this will be done with small organic molecules targeted at the factors which control blood pressure.

The goal of the cardiovascular medicinal chemist and discovery pharmacologist is to design drugs which keep the heart and blood vessels alive longer than the cells/tissues/organs that they support. The design/development of drugs which lower blood pressure have proven an important first step. In the future, molecular biology techniques now being developed may permit the design of molecules that alter the "message" (the mRNA for a particular protein), the transcriptional event, or even the genome itself to prevent the onset or to alter the course of hypertensive disease (3,4). This article will highlight the wealth of knowledge about the wide array of chemical molecules which have been shown to provide beneficial blood pressure lowering effects in humans by many different mechanisms of action. It is hoped that this information will provide an understanding of how these drugs were discovered and how they act to modify the physiologic or pathophysiologic control of blood pressure and provide a basis for antihypertensive new drug discovery programs of the future.

2 THE PHYSIOLOGY OF BLOOD PRESSURE CONTROL AS THE BASIS OF ANTIHYPERTENSIVE DRUG ACTION

Hypertension is a pathologic state of elevated blood pressure which is a pathologic state of normal blood pressure control (Fig. 29.1). Hypertension has been described as a "quantitative disorder of blood pressure regulation" (5). Hypertension has also been described as an abnormality of Starling's concept of circulatory homeostasis involving changes in the volume of the

circulating blood affecting the renal output of extracellular fluid through changes in effective cardiac performance (6). The interrelation of a number of regulatory factors to control blood pressure and tissue perfusion was first described by Page in 1949 (7). According to this concept, tissue perfusion/pressure/resistance are interdependent upon "factors" designated chemical, reactivity, volume, vascular caliber, viscosity, cardiac output, elasticity, and neural, and "hypertension" is a multifactorial derangement of the normal equilibrium (see review in Ref. 8).

Blood pressure in its simplest terms is the force of the heart pumping action working against the resistance provided by the blood vessels. The purpose of the "blood pressure system" is to maintain blood flow to all of the tissues of the body at rest or during movements. To compensate for gravity, changing levels of activity, and other perturbations to the system, several interactive subsystems work to maintain tissue perfusion each with its own response time (9). The baroreceptors,

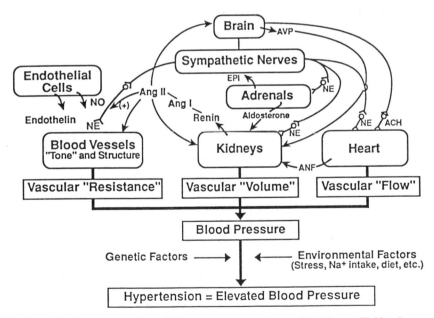

Fig. 29.1 The interplay between cardiovascular organ systems to regulate "normal" blood pressure and the genetic and environmental factors that lead to hypertension.

mainly in the walls of the aorta and the internal carotid arteries, act to rapidly adjust to changes in pressure (stretch) (response time in seconds). This is accomplished by activation of afferent nerves from the baroreceptors to the brain stem centers and modulation of efferent sympathetic nerve activity to peripheral blood vessels (norepinephrine release), to kidneys (renin release), or to the heart (norepinephrine release), and efferent parasympathetic nerve activity to the heart (acetylcholine release). Located adjacent to the baroreceptors are chemoreceptors which respond to changes in the amount of oxygen in the blood and activate central sympathetic outflow. The central ischemic receptors located in the cardiovascular centers of the brain stem respond to hypoperfusion of the brain by strongly activating the sympathetic nervous system to achieve maximum heart rate and force of contraction and vasoconstriction and may involve the adrenal medullary release of epinephrine.

The intermediate controllers of blood pressure (response time in minutes to hours) involve (1) the renin angiotensin system (RAS), namely the acute vasoconstrictor effects of Ang II and the Ang II-induced release of aldosterone to cause renal tubular sodium reabsorption and volume expansion, (2) a slow stretch relaxation of the blood vessel wall subjected to the higher pressure, and (3) a so-called capillary fluid shift mechanism whereby excess capillary pressure can cause fluid to traverse through capillary pores (9,10). The long-term (hours to days) controller of blood pressure according to Guyton (9) is the so-called "kidney-fluid system" for blood pressure control which has infinite gain to adjust blood pressure via renal fluid excretion. It is now clear that the integrating type of kidney-fluid system involves the RAS and sympathetic nervous system control mechanisms and others including atrial natriuretic factor (ANF), vasopressin,

endothelin, and nitric oxide (NO) (Fig. 29.1) (11,12). The "control of blood pressure" is not, however, limited to the kidney. The microcirculation of various tissues (13) and the activities (e.g., contraction, hormone release, transport) of individual cells within those tissues respond to the blood pressure-relations in each tissue (Table 29.1). Understanding of how blood pressure is controlled at both the systemic (Fig. 29.1) and at the cellular level is key to understanding the etiology of hypertension and how current antihypertensive drugs exert their effects. Further, many of the differences in the actions of the various classes of antihypertensive agents, e.g., on renal function, cardiovascular hypertrophy, or sympathetic activation, can be explained by differential effects at the cellular level of blood pressure control. In the future, we will need to know more about the next level of blood pressure control, that is, the intranuclear or genetic control of blood pressure. It can be anticipated that antihypertensive drugs of the future may seek to alter hypertension at the genetic level.

Essential hypertension is clearly multifactorial in origin (14) and the RAS, although important, would appear to be only one of the several factors involved. Extensive clinical experience with hypertensive patients profiled according to sodium output and plasma renin activity under carefully controlled conditions, however, has suggested that blood pressure represents the balance between the vasoconstrictor effects of Ang II and the volume (sodium excretion) controlling effects of Ang II via adrenal release of aldosterone (15,16). This so-called "vasoconstriction-volume" hypothesis proposes that most antihypertensive drugs intervene at one of the sites in the RAS, e.g., beta blockers, inhibiting renin release; ACE inhibitors, inhibiting Ang II synthesis; spironolactone, inhibiting aldosterone; diuretic inhibiting renal sodium reabsorption; and vasodilators and sympatholytics, acting as anticonstrictors. Fur-

Table 29.1 Cellular Sites of Blood Pressure Control

Organ/Tissue	Cell Type	Cellular Response	"Blood Pressure" Effect[a]
Blood vessels (including renal, coronary, mesenteric, and cerebral)	Smooth muscle cells	Contract	↑Resistance
	Endothelial cells	Release endothelin	Hypertrophy (↑resistance)
	Interstitial cells (fibroblasts)	Release NO	↓Resistance
		Collagen synthesis	↑Resistance
	Circulating blood cells (platelets)	Aggregate	↓↑Resistance
		Release cytokines	↓↑Capillary permeability
Kidneys	Afferent/arterioles	Contract	↑↓GFR
	Mesangial cells	Contract	↑↓GFR
		Release growth factors	Hypertrophy (↑↓GFR)
	Juxta glomerular cells	Renin release	↑Ang II synthesis (↑resistance)
	Tubular cells	Na^+, H_2O transport	↑↓Plasma volume
Adrenals	Zona glomerulosa cells	Release aldosterone	↑Na^+ reabsorption (↑volume)
	Medullary chromophin cells	Catecholamine release	↑Resistance, ↑CO
Heart	Myocytes	Contract	↑Flow
		Hypertrophy	↑↓Resistance
	Non-myocytes (fibroblasts)	Collagen synthesis	↑Resistance
Sympathetic nerves	Post-ganglionic cells	Norepinephrine release	↑Resistance, ↑CO
Brain	Neurohypophysis	AVP release	↑Na^+ reabsorption (↑volume)
	Neurones (medulla, hypothalamus. cerebral cortex)	↓↑Firing rate, transmitter release	↑↓Sympathetic outflow

[a]GFR = glomerular filtration rate
CO = cardiac output.

ther, it has been suggested that the selection of antihypertensive drug treatment for an individual patient can be made on the basis of renin profiling, e.g., low renin patients are most suited to diuretic treatment and high renin patients are most suited to RAS inhibitors, e.g., beta blockers, ACE inhibitors, or Ang II receptor antagonists (15). From this "hypothesis" has come an increasing awareness of the involvement of renin (actually Ang II) as a risk factor in cardiovascular disease (17), the need to subclassify hypertensive patients (16,18), and a further understanding of the mechanisms involved in the control of blood pressure. Although the clinical use of renin profiling to individualize treatment, e.g., beta blockers for high renin patients, has had mixed reviews (19), the concept of the vasoconstrictor-volume theory continues to provide a useful framework to challenge our understanding of blood pressure control and the profiling of hypertensive patients to optimize therapy (18).

3 THE DEFINITION AND EPIDEMIOLOGY OF HYPERTENSION

It is estimated that 15–25% of the adult population of most countries have elevated blood pressure (20,21). Essential hypertension, that is, hypertension of unknown origin, makes up the majority of these individuals. Although approximately two-thirds of these "hypertensives" have only mild hypertension, they have increased risk of cardiovascular disease (22,23). Cardiovascular disease is the main cause of death in most industrial countries (24) and hypertension is the most commonly reported and one of the most important risk factors (22,25). The consequences of untreated hypertension are expressed as an increased incidence of tissue and organ pathology involving the brain, heart, kidneys, and blood vessels (Table 29.2) (22,26). High blood pressure is a potential risk factor for cardiovascular disease independent of the presence or absence of other risk factors, e.g., smoking, diabetes, and/or hypercholesterolemia. The relative risk of hypertension, however, is dependent on other factors such as genetics, age, sex, race, diet, and environmental factors, e.g., stress and physical activity (27). Important-

ly, in a summary of nine major prospective observational trials totaling 420,000 individuals, no "threshold" for lower risk was identified (the lower the pressure, the lower the risk) (23).

Hypertension is arbitrarily divided into primary or so-called "essential" hypertension where the etiology is unknown and "secondary" hypertension where the etiology can be identified. It is likely that, as we understand more about the characteristics of these essential hypertensive patients and are able to identify their genomic profiles, a new classification may be required. The majority of antihypertensive drug trials have involved the evaluation of essential hypertensive patients. However, virtually all of the currently available antihypertensive drugs also lower blood pressure in other so-called secondary hypertensive states. Examples of secondary hypertension include the elevated blood pressure associated with renal disease (occlusive vascular and parenchymal), adrenal disease (phenochromocytoma, primary aldosteronism, Cushings disease), toxemia of pregnancy, centrally mediated neorogenic causes (encephalitis, increased intracranial pressure, aortic coarctation, thyrotoxcosis, and oral contraceptives.

Hypertension is also arbitrarily defined

Table 29.2 Pathological Risks of Uncontrolled Hypertension

Target Organ	Clinical Manifestation
Brain	Strokes (cerebral vascular accidents)
	Transient cerebral ischemia
Heart	Acute myocardial infarction
	Sudden coronary death
	Accelerated ischemic heart disease (angina, arrhythmias)
	Congestive heart failure
Kidneys	Renal failure
	Renal functional impairment
Blood vessels	Aortic aneurysm (fusiform, sacular, dissecting)
	Atherothrombotic obstruction and stenosis
	Ocular fundi damage (spasm to papilledema)
	Peripheral vascular disease and claudication

by the level of diastolic or systolic blood pressure at which treatment (pharmacologic or nonpharmacologic) is indicated (5,28). Blood pressure is commonly expressed as both systolic pressure (peak pressure at time of cardiac contraction and diastolic pressure (minimum pressure as time of cardiac relaxation). In simplest terms, the diastolic pressure reflects the tone (contractile state) of the blood vessels and the systolic pressure reflects the contractile state of the heart. Diastolic blood pressure is most often considered the risk factor of cardiovascular disease, but isolated systolic hypertension may represent an independent risk factor (29). Blood pressure, however, is not a single value for any animal or person, but rather a range of values through the 24-hour day. Blood pressure displays both circadian and ultradian rhythms, that is, blood pressure displays time-based and nontime-based related fluctuations (30,31). These fluctuations have been studied to determine if they have prognostic value or can be used to optimize drug treatment. Further, it has been hypothesized that there is a relationship between blood pressure variability and end-organ damage (32).

Blood pressure is one of the most commonly recorded clinical signs and the international efforts of the medical community to educate the public about the risks of hypertension have increased the routine monitoring of blood pressure. The majority of blood pressure determinations are performed by indirect measurement with mercury column sphygmomanometers. Comparison of casual (clinic or office) blood pressure determinations with home blood-pressure measurements with automatic ambulatory monitoring suggested that causal pressures alone may be inappropriate for antihypertensive drug trials and may explain some of the mixed results that have been reported (33). The use of ambulatory methods is further supported by a recent study of 1187 subjects followed for 7.5 years. In this study, ambulatory blood pressures identified high risk individuals independent of traditional factors. For example, women who had ambulatory hypertension and little or no fall in blood pressure at night were at very high risk of cardiovascular morbidity (34). Thus drug trials of the future will likely use ambulatory pressure measurement for patient selection and to monitor 24-blood pressure control.

The clinical assessment of casual blood pressure using indirect methods will remain the mainstay for diagnosing hypertension. The definition of "normotension" and of various stages of "hypertension" has been defined by clinical experts in various countries. The Joint National Committee and the World Health Organization/International Hypertension Society Subcommittee have published their own definitions of classifications of blood pressure (Tables 29.3 and 29.4). According to both of these

Table 29.3 Definition (Classification) of Blood Pressure by the Joint National Committee on Detection, Evaluation, and Treatment of High Blood Pressure[a],[b]

Category	Systolic (mm Hg)	Diastolic (mm Hg)
Normal[c]	<130	<85
High normal	130–139	85–89
Hypertension[d]		
Stage 1 (mild)	140–159	90–99
Stage 2 (moderate)	160–179	100–109
Stage 3 (severe)	180–209	110–119
Stage 4 (very severe)	≥210	≥120

[a]Ref. 35.

[b]Classification of blood pressure in adults (≥18 yrs of age).

[c]Optimal blood pressure with respect to cardiovascular risk is <120 mm Hg systolic and <80 mm Hg diastolic. However, unusually low readings should be evaluated for clinical significance.

[d]Based on the average of 2 or more readings taken at each of 2 or more visits after an initial screening.

Table 29.4   Definition (Classification) of Blood Pressure by the World Health Organization/International Society of Hypertension[a]

	SBP (mm Hg)		DBP (mm Hg)
Normotension	<140	and	<90
Mild hypertension	140–180	and/or	90–105
Subgroup: Borderline	140–160	and/or	90–95
Moderate and severe hypertension[b]	≥180	and/or	≥105
Isolated systolic hypertension	≥140	and	<90
Subgroup: Borderline	140–160	and	<90

[a]Ref. 36.
[b]Risk to be indicated by reporting the actual values of systolic (SBP) and diastolic blood pressure (DBP).

groups, the definition of the lowest level of "hypertension" is a systolic pressure of 140 mm Hg and a diastolic of 90 mm Hg. Both groups have established guidelines for the monitoring of blood pressure and the pharmacologic and nonpharmacologic management of these patients. As discussed below, these guidelines have importance in evaluating the need for new antihypertensive drugs.

4 RATIONALE FOR THE DRUG THERAPY OF HYPERTENSION

The use of drugs in the overall management of patients with a wide range of severity of hypertension has evolved from the major international efforts to detect high blood pressure, the availability of more effective and tolerated drugs and the increasing understanding of the risks associated with "unmanaged" hypertension (22,37).

The beneficial effects of drugs which lower blood pressure to decrease cardiovascular morbidity and mortality has been evaluated in numerous clinical trials over the last 30 years (1,18). The clinical endpoints of these trials have become increasingly more complex moving from simple blood pressure lowering in severely hypertensive patients to an evaluation of mortality and the quality of life in mild or borderline hypertensive patients. Hamilton et al., reported on one of the first clinical trials showing the benefit of antihypertensive drug treatment (38). In this study, antihypertensive drugs lowered blood pressure and prevented the complications of hypertension. Since that time, many trials have documented the efficacy of antihypertensive drugs in decreasing cardiovascular morbidity and mortality. The most convincing evidence for the use of drugs to treat hypertension comes from the results of a relatively few large controlled trials (Table 29.5). The results of these trials show that (1) antihypertensive drug treatment of patients with malignant hypertension dramatically improves survival, (2) the benefit of the treatment of patients with less severe hypertension is much less dramatic and appears to relate to the coexistence of other risk factors, (3) treatment of elderly hypertensives significantly decreases cardiovascular morbidity and mortality, and (4) current drugs are not effective in all hypertensive patients.

The five-year survival in untreated patients with malignant hypertension is very low, e.g., 0% in a study of 407 patients

Table 29.5 Major Double Blind-Controlled Clinical Trials with Antihypertensive Drugs

Study Name (No. of Patients)	Year	Entry DBP[a] (mm Hg)	Drug Treatment vs. Placebo	Mean Duration (years)	Principal Findings	Reference
Veterans Cooperative (380 patients)	1970	90–114	Reserpine, diuretic, and hydralazine	7–10	↓Incidence of cerebrovascular events ↓Progression to accelerated hypertension ↓Decreased left ventricular failure	(39)
Australian National Blood Pressure (3,427 patients)	1980	95–110	Thiazide, antiadrenergic (methyldopa, propranol, or pindolol) plus hydralazine (step care)	4.1	↓Cardiovascular morbidity ↓Cardiovascular mortality ↓Total mortality	(40)
Medical Research Council *Single blind (17,354 patients)	1981	90–109	Bendrofluazide or propranolol	5	↓Strokes ↓Total cardiovascular events ↓Progression to severe hypertension	(41)
IPPPSH[b] (6,357 patients)	1985	100–125	Half received diuretic plus other drugs (no beta blockers) Half received oxprenolol plus other drugs including diuretics	3–5	↓Cardiovascular mortality equivalent in each group	(42)
Systolic Hypertension in the Elderly Program (SHEP) (4,736 patients)	1991	>90 (SBP 160–219)	Chlorthalidone and atenolol (step care)	4.5	↓Strokes (fatal and nonfatal) ↓Major cardiovascular events	(29)
Treatment of mild hypertension (TOMHS, 1993) (902 patients)	1993	<100 mg Hg	Chlorthalidone, acebutolol, doxazosin, amlodipine, or enalapril (nutritional-hygienic treatment to all)	4.4	↓Blood pressure in all drug treatment groups ↓Major nonfatal cardiovascular events ↓Incidence of testing ECG abnormalities ↑Quality of life Drug treatment group differences minimal	(43)

[a] Entry diastolic blood pressure.
[b] IPPPSH = International Prospective Primary Prevention Study in Hypertension.

(44%). Antihypertensive (blood pressure lowering) drugs have been shown to increase five-year survival from 15–50% using early antihypertensive drugs including ganglionic blocking agents. Survival was 75% in one study with more modern drugs and the use of hemodialysis or renal transplant (see review of studies in Ref. 45). One important indicator of the effectiveness of modern antihypertensive drugs is the decreased incidence of malignant hypertension (46).

Data to support the use of drugs to treat patients with mild-to-moderate hypertension are less dramatic and may relate to a lower incidence of moribid events in this patient population. Several large and well-controlled trials have shown that a number of classes of antihypertensive agents can significantly decrease cardiovascular morbidity and mortality (Table 29.5). Fatal and nonfatal strokes are clearly reduced by antihypertensive drug treatment (18). Further, it is likely that the wide spread of antihypertensive drugs has contributed to the continuing decline in cerebral vascular accidents (47). Antihypertensive drug treatment has not consistently reduced the morbidity and mortality from coronary heart disease (18). No significant differences were seen in the Veterans Cooperative Trial, the Australian Trial, the MRC Trial, or the IPPPSH Trial. In the SHEP and TOMHS trials, the risk of coronary heart disease was significantly reduced by antihypertensive drug treatment (29,43). Surprisingly, several trials have shown increased coronary mortality with drug treatment (45,48). Most of these reports are in patients with pre-existing coronary heart disease and/or with abnormal electrocardiograms. In these trials, antihypertensive drug treatment did not reduce cardiovascular risk down to that of comparable normotensive subjects (45). The reasons for the failure of such treatments to reduce coronary heart disease may relate to inadequate dosing, drug-induced effects, e.g.,

elevated lipids or failure to meet target pressure (45,48). These negative data, however, have not changed the use of antihypertensive drugs for mild-to-moderate hypertension (1). Because of the clear definition of the cardiovascular risk associated with each incremental increase in blood pressure (37), patients with mild-to-moderate hypertension and other risk factors and those not controlled by nonpharmacological means continue to be candidates for antihypertensive drug treatment.

Three large multicenter trials (SHEP, MRC, and STOP trials) have clearly demonstrated that antihypertensive drugs decrease morbidity and mortality in the elderly (29,41,49). In the SHEP trial, for example, patients were 60 years of age and older and displayed SBP between 160–219 mm Hg in the absence of cardiac or cerebrovascular signs. In a stepped treatment scheme, approximately 48% of the patients were reduced by chlorthalidone alone, another 23% also received atenolol, and the remainder received other drugs. The cumulative cerebrovascular accident totals after five years were reduced from 9.2% in the placebo group to 5.5% in the active treatment group. Fatal and nonfatal coronary events were reduced from 184 in the placebo group to 140 in the active treatment group and all cardiovascular events were reduced form 414 to 289 (49). Similar decreases in stokes, fatal coronary events, and nonfatal coronary events were observed in each of the three trials (50).

The beneficial effects of current drugs cited above does not mean hypertension has been eliminated. It should be remembered that current drugs are not curative and do not control blood pressure in all patients. Therefore, there is a need for new antihypertensive drugs. This is the conclusion of several reviews of the clinical trials of antihypertensive drugs above (29,50,51). In the SHEP trial of highly selected hypertensive patients (1 in 10), for example, "treatment" controlled hypertension in only 50% of the patients screened. The more recent TOMHS trial (43) shows the problem of assessing the comparative antihypertensive effects of new drugs. In the TOMHS trial, five antihypertensive drugs (chlorthalidone, acebutolol, doxazosin, amlodipine, enalapril) were compared for the treatment of mild hypertension versus nutritional-hygiene treatment alone. Drug treatment significantly reduced cardiovascular and other events but no overall differences between the treatments could be demonstrated (43). Although all five drugs were extremely well tolerated, the "quality of life" scores improved more for those taking chlorthalidone and acebutolol than the other agents and the authors concluded that these findings supported the recommendation of the Fifth Joint National Committee report to initiate treatment with diuretics or beta blockers (see below (43).

5 CLINICAL USE OF ANTIHYPERTENSIVE DRUGS

5.1 Treatment Guidelines

National and international expert committees influence the practice of hypertension management by establishing guidelines for the definition of hypertension (Tables 29.3 and 29.4) and the selection of initial drug treatment. There are some differences in these guidelines from country to country (Table 29.6). The United States (35) recommends treatment at the lowest pressure and New Zealand's most conservative guideline recommends treatment for patients with pressures of ≥170 SBP/100 DBP. In both countries, however, diuretics and beta blockers are the recommended initial drug treatment. The basis of this recommendation is the long-term experience with these classes of agents and their demonstrated effects to reduce morbidity and mortality (Table 29.5). It has been argued that the alternative drugs, i.e., calcium antagonists, angiotensin converting enzyme inhibitors, alpha blockers, and the alpha beta blockers, are equally effective in reducing blood pressure but have not been tested in long-term, controlled studies (52). Cost is another factor influencing national guidelines. The older drugs are cheaper than the newer drugs. Thus, health economic factors are also being studied to evaluate the cost effectiveness of antihypertensive drug therapy (53,54).

The design of new antihypertensive drugs does not depend directly on treatment guidelines but the "unmet medical need" does influence the decision to invest in drug discovery research. The multi-bill-

Table 29.6 International Treatment Guidelines[a]

Country	When to Treat Systolic/Diastolic (mm Hg)	Initial Treatment
U.S.	≥140/90	Diuretic or beta blocker "preferred"
Canada	≥/100	Diuretic or beta blocker
Britain	≥160/100	Diuretic; beta blocker, all classes
Australia	≥160/100	All classes
New Zealand	≥170/100	Diuretic or beta blocker
WHO/ISH[b]	≥160/95	All classes

[a]Adapted from Swales (28).
[b]WHO = World Health Organisation/International Society of Hypertension.

ion dollar international market has supported the research and development to bring literally hundreds of new molecules to the marketplace. The current clinical practice guidelines suggest however, that there are multiple cheap and effective alternative therapies for hypertension and question the need for new drugs. Obviously, as will be discussed below, all classes of antihypertensive drugs are not the same. They all lower blood pressure but can have very different effects on different organ systems (18), e.g., cardiac hypertrophy (55) or renal function (56). The challenge of future new drug discovery is to design agents that act in patients unresponsive to current therapy. For example, in the placebo arm of the Helsinki Heart Study of gemfibrozil carried out in dyslipidemic men, the 8.5 year cardiovascular mortality rate was 10.5/1000 for patients with uncontrolled blood pressure (mean 154 SBP/101 DBP) versus only 4.7/1000 in patients whose pressures were controlled (mean 137/91) (57). It has been asked whether clinical practice should (can) face the challenge of identifying the small minority of patients that require drugs costing up to 100-fold more than thiazides (58). The answer is important because it also answers the question whether there is a need to develop new drugs.

It has been estimated that thiazide diuretics alone provide useful monotherapy in approximately 50–55% of patients and, in combination with other common antihypertensive agents, 85–95% of the patients are controlled (59). Such numbers may be somewhat optimistic in view of the 28–36% of those who did not reach target pressure in the SHEP, IPPSH, or Australian Trials (50). In the U.S., it is estimated that blood pressure is "uncontrolled" in 27% of patients on antihypertension medication. This would mean that >500,000 Americans age 15 or older who are taking current drugs still have blood pressures ≥140 SBP/≥90 DBP and are at increased risk of cardiovascular morbidity and mor-

tality [based on 1988–1991 estimates (21)]. Patient compliance, choice of drug treatment, drug dose, and additional risk factors can all influence the rate of control, but it is clear that current drugs are not effective in all patients with elevated blood pressure. Thus, the refractory hypertensive patient appears to represent an unmet medical need and an opportunity for new antihypertensive drug discovery.

5.2 Defining the "Ideal" Antihypertensive/Blood Pressure Lowering Drug

The evaluation of current antihypertensive agents and the goal of future new antihypertensive drug discovery is the definition of an "ideal" blood pressure lowering agent. The critical characteristics of such an "ideal" agent have increased in number with the availability of more effective and better tolerated antihypertensive agents (Table 29.7). In 1984, it was suggested that the antihypertensive drugs available before ACE inhibitors and calcium antagonists were "very good" but not "ideal" (60). Since that time, the ACE inhibitors and the calcium antagonists have proven to be even more useful antihypertensive agents and Ang II receptor antagonists are now on the horizon. With the introduction of each new class of drugs, the definition of "ideal" increases in complexity. Many factors must now enter into the decision to invest in new drug discovery or in the selection of a candidate for development.

The "mechanism of action" of future antihypertensive drugs will likely follow a breakthrough in the scientific understanding of blood pressure control at the cellular and genetic level. It is unlikely that the random screen of chemical libraries for blood pressure lowering effects in animals will yield drugs that will compete with those already on the market. Rather it is likely that the new and existing compounds

Table 29.7 Defining the "Ideal" Antihypertensive/Blood Pressure Lowering Drug

Mechanism of action	Specific site of drug action supporting expectation of efficacy and tolerability
Efficacy	Control blood pressure as oral monotherapy in the majority of essential hypertensive patients (mild to severe) Act in combination with other common antihypertensive agents No need to adjust dose in renally or hepatically impaired Decreased moribidity of cardiovascular disease (cardiac, renal, cerebral) Decreased mortality from cardiovascular disease
Tolerability	Better tolerated than existing drugs Improved quality of life No significant changes in heart rate or orthostatic hypotension No significant side effects (less than current drugs) No first dose effect or rebound hypertension No effect on sexual function Metabolically neutral No adverse interaction with common drugs (e.g., NSAIDS, cimetidine)
Physical form	Crystaline bulk drug stable in ambient light and heat Water soluble salt Nonracemic Small tablet size Neutral taste
Cost	Inexpensive to manufacture

will be screened against cellular or intracellular mechanistic targets identified as important to the etiology of hypertension. Whether the "targets" will be defined by genetic linkage experiments, by activity in genetically altered animal models, by antisense or gene therapy results, or by all other possible strategies remains to be determined. When we are better able to phenotype the individual hypertensive patient, it may be possible to plan new drug design and treatment strategy. We presently have a good idea of the hemodynamic and neurohumoral characteristics of hypertensive patients and our definition of the "ideal" antihypertensive fits our current level of understanding. As we explore the functions of the genes associated with hypertension, we may have to re-define this concept. Lowering blood pressure is only one goal in treating the hypertensive patient with multiple organ pathology. New drugs may have equivalent blood pressure lowering effect as current drugs but will likely have significantly greater tissue or cellular protective effects.

6 ESTABLISHED ANTIHYPERTENSIVE DRUGS

6.1 History of Antihypertensive Drug Discovery

"Lack of information regarding the etiology of most cases of hypertension has made the search for effective antihypertensive agents empirical" (61) appropriately states the early years of antihypertensive drug discovery. Although an association between renal disease and hypertension, for example, had been postulated in 1836, it was not until 1956 that angiotensin II ("hypertension") was identified as the vasoconstrictor factor in Goldblatt hypertension and not until 1988 that an orally active angiotensin II receptor antagonist, losartan, was identified [see review Timmermans (62)].

The "discovery" of most of our current antihypertensive drugs did not involve the targeted design of molecules to modify a blood pressure control system. Putting the "chemical" and "biological" ideas for antihypertensive drugs has been done in many different ways (Table 29.8). Clinical observations of the blood pressure lowering effects of reserpine and beta blockers led to their use as antihypertensives. Mechanistic approaches of ganglionic blockade and inhibition of DOPA decarboxylase to block the sympathetic nervous system thought to be involved in the etiology of hypertension did yield effective blood pressure lowering agents. Many of the early drugs such as the ganglionic blocking agents such as hexamethonium, however, were severely limited by mechanism-related side effects such as orthostatic hypotension. Interestingly, the first use of "drugs" to lower blood pressure was reported to be carried out with sodium thiocynate (CNNaS) in 1900, but its effectiveness was not demonstrated by clinical trial (63). The first drug treatment to lower blood pressure and reverse some of the symptoms of the malignant hypertension (neuroretinitis, relief of headache) was reported with the use of the antimalarial pentaquine (1) in 1947 (63,64).

Pentaquine (1)

The design of antihypertensive drugs was hampered by a lack of animal models of hypertension before the availability of the spontaneously hypertensive rat (SHR) (81). Before that time, the primary models involved modifying renal function (renal clamping, diet, nephritis). The compounds identified were primarily direct vasodilators which lowered blood pressure in both normotensive and hypertensive animals. As will be discussed in more detail below, retrospective analysis of these blood pressure lowering substances has led to the current concepts of "calcium channel antagonists" (82) and "potassium channel openers" (83). Many of these vasodilator drugs have proven useful therapeutic agents and tools for physiologists to further define the multiple mechanisms by which the tone of vascular smooth muscle is controlled. Circulating hormones, locally released paracine agents, ion channels, G-proteins, second messengers, and intracellular Ca^{2+} binding have all been evaluated as targets for new drug intervention and many different molecules have been shown to lower blood pressure (84,85). In the first 30 years of antihypertensive new drug discovery, "blood pressure lowering" was the therapeutic target. For future discovery programs, the effect on a new molecule at a novel mechanistic target will likely be required before the effect on blood pressure is considered.

6.2 Major Classes of Currently Established Antihypertensive Drugs

"Antihypertensive drugs" include several of the most widely prescribed drugs in clinical medicine and includes diuretics, beta blockers, ACE inhibitors, and calcium channel blockers (Table 29.9). These drugs are also used for other indications including heart failure, ischemic heart disease, and renal disease. Over 100 chemical entities are marketed internationally for the treatment of hypertension. For each class of

Table 29.8 Historical Highlights of Antihypertensive Drug Discovery

"Compound"	Year	Chemical Ideas	Biological Idea	Reference
Hexamethonium	1950	Tetraethyl ammonium	Antagonize ganglion stimulation in dogs and cats	Green, 1967 (65)
Hydralazine	1951	Pthalazine	Blood pressure lowering in conscious dogs	Moyer, 1953 (66)
Reserpine	1956	Plant extract	Clinical observation	Bein, 1956 (67)
Chlorothiazide	1957	Sulfonamide	*In vitro* carbonic anhydride inhibition; diuretic efficacy in rats	Beyer and Baer, 1961 (68)
Spironolactone	1957	Adrenocorticosteroids	Antagonism of aldosterone	Horisberger and Grebisch, 1987 (69)
Alpha methyldopa	1960	Dopa	*In vitro* dopa decarboxylane inhibitor	Stone and Porter, 1967 (70)
Verapamil	1962	Random screen	Coronary vasodilator in dogs	Haas and Busch, 1967 (71)
Diltiazem	1964	1,5-Benzothiazepine (thiazesum-antidepressant)	Coronary vasodilator in dogs	Nagao, 1982 (72)
Propranolol	1964	Dichloroisoproterenol	Isoproterenol antagonism in anesthetized cats	Shanks, 1966 (73)
Minoxidil	1965	Random screen	Blood pressure lowering in dogs	Campese, 1981 (74)
Clonidine	1966	Imidazoline	Clinical observation (screened as decongestant in anesthetized dogs)	Stahle, 1982 (75)
Practolol	1968	Propranolol	Selectively inhibition beta-1 (cardiac) receptors	Dunlop and Shanks, 1968 (76)
Prazosin	1968	2-Aminoquinoline	Blood pressure lowering in conscious hypertensive dogs	Scrabine et al., 1968 (77)
Nifedipine	1971	Hantzsch product	Coronary vasodilator in anesthetized dogs	Bossert and Vater, 1971 (78)
Captopril	1977	Benzyl succinic acid teprotide	*In vitro* ACE inhibition and Ang I antagonism	Ondetti, 1977 (79)
Losartan	1988	1-Benzylimidazole-5-acetic acids	Inhibition of Ang II binding and functional responses	Duncia, 1992 (80)

Table 29.9 Classes of Currently Established Drugs for the Treatment of Hypertension

"Class"		Example Compounds
Diuretics	Thiazide-type	Hydrochlorothiazide, chlorthalidone, bendroflumethiazide, trichlormethiazide
	Potassium sparing	Spironolactone, amiloride, triamterene
	"Loop"	Furosemide, ethacrynic acid, bumetanide, torasemide
Beta blockers	Nonselective (beta-1/beta-2)	Propranolol, timolol, nadolol, pindolol, cartetolol
	Selective beta-1	Atenolol, betaxolol, metoprolol, acebutolol
ACE inhibitors		Captopril, enalapril, lisinopril, fosinopril, ramipril
Calcium channel blockers	Dihydropyridine	Nifedipine, nicardipine, felodipine, amlodipine
	Phenylalkylamines	Verapamil, gallopamil
	Benzothiazepines	Diltiazem
Alpha-1 adrenergic receptor antagonists		Prazosin, doxazosin, terazosin, labetalol (also blocks beta-1)
Alpha-2 adrenergic receptor agonists		Clonidine, guanfacine, guanabenz
Miscellaneous		Alpha-methyldopa, neuronal blockers (bretylium, guanethidine), rauwolfia and derivatives (reserpine, deserpidine, rauwolfia whole root), ganglionic blockers (guanadrel, mecamylamine), nonspecific vasodilators (hydralazine, nitroprusside, diazoxide, minoxidil)

agents, there are virtually hundreds of patents and thousands of individual compounds identified. For example, between October of 1992 and March of 1993, nearly 100 patents and applications appeared for angiotensin II receptor antagonists alone (86). The most widely used antihypertensive drugs fall into six mechanistic "classes" (Table 29.9). Because of the importance of each of these classes, the discovery strategy, basic pharmacology, and chemistry for each class will be described. The remainder of currently available agents will only be mentioned briefly to highlight interesting biology or chemistry associated with an individual molecule.

6.2.1 DIURETICS

6.2.1.1 Discovery. The evolution of modern diuretics from the fortuitous discovery of the diuretic properties of the organomercurials to the more recent discoveries of the so-called "loop" or high ceiling diuretics has been the subject of several excellent reviews (87–90). In the 1950s, significant antihypertensive effect of intravenous organomercurial compounds such as meralluride was reported. The primary use of the organomercurials was in the treatment of congestive heart failure (88). The first orally active diuretics were "sulfonamide" containing molecules that inhibited carbonic anhydrase. The prototype of this class of agents was acetazolamide and, although it demonstrated weak diuretic effects in treatment of edema, had no significant antihypertensive effect. The discovery of acetazolamide, however, represents a classic putting together of a chemical idea (sulfonilamide has metabolic acidosis as a side effect) and the biologic idea (the enzyme carbonic anhydrase found in renal cortex) (90).

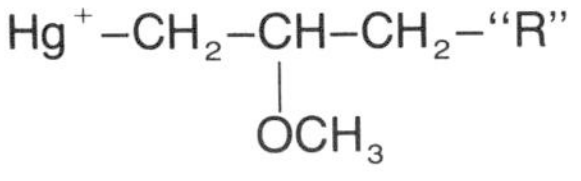

$$Hg^+-CH_2-CH-CH_2-\text{"R"}$$
$$|$$
$$OCH_3$$

R = Urea, aromatic or heterocyclic ring

(2)

Organomercurial general formula (91)

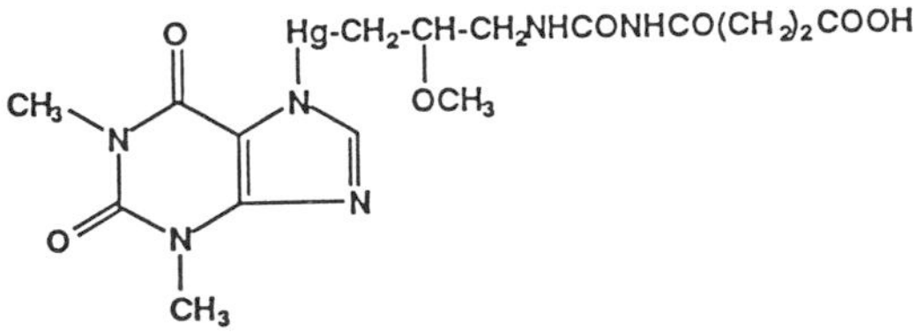

Meralluride (3)

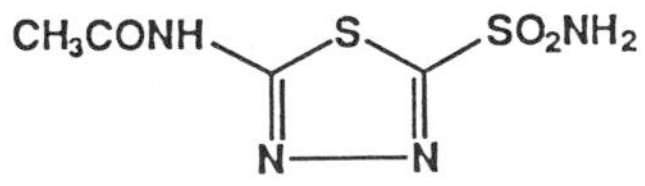

Acetazolamide (4)

The modern era of diuretic antihypertensives began with the sulfonamide derivative chlorothiazide (5). The strategy was to test renal electrolytic clearance of newly synthesized sulfonamide derivatives in conscious dogs and find a compound that was both natriuretic and chloruretic (e.g., induced saluresis). From this effort came a series of compounds with the well known benzothiadiazine heterocyclic nucleus including chlorothiazide and many other related compounds. Interestingly, this compound was purposely not tested in a hypertensive animal before it was given to

hypertensive patients. In the first clinical trials in hypertensive patients, chlorothiazide was added to existing therapy with excellent results and the modern era of antihypertensive drugs was begun (89).

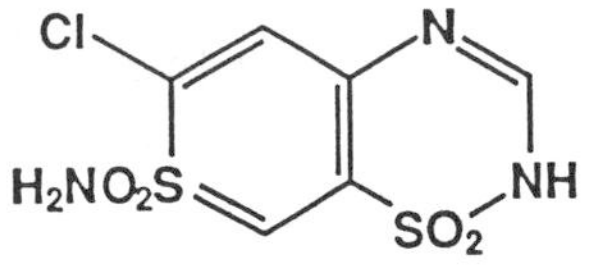

Chlorothiazide (5)

The dramatic clinical and commercial success of chlorothiazide led to literally thousands of new chlorothiazide-like compounds. These compounds, like hydrochlorothiazide, bendroflumethiazide, and trichlormethiazide, were more potent or longer acting but produced the same maximum sodium excretion and have been called "low ceiling" diuretics.

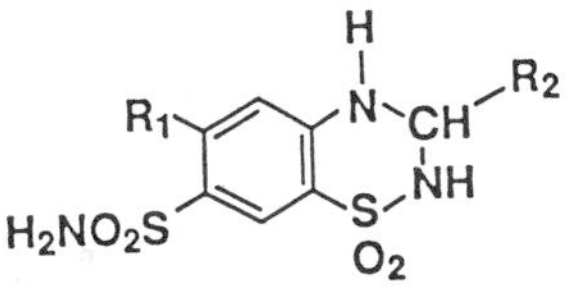

(6)

hydrochlorothiazide (R_1=Cl;R_2=H)

bendroflumethiazide (R_1=CF$_3$;R_2=CH$_2$⟨⟩)

trichlormethiazide (R_1=Cl;R_2=CHCl$_2$)

The next milestone in the development of diuretics came with the discovery of the sulfonamide (furosemide) (7) and the non-sulfonamide (ethacrynic acid) (8) which produced a much greater saluretic effect. These so-called "high ceiling" diuretics were shown to act on the ascending loop of Henle (Fig. 29.2) and were also designated "loop" diuretics. Loop diuretics are used by intravenous or oral administration but are usually limited to refractory hypertension or in those patients with co-existing heart failure.

The discovery and biological properties of the potassium sparing diuretics including

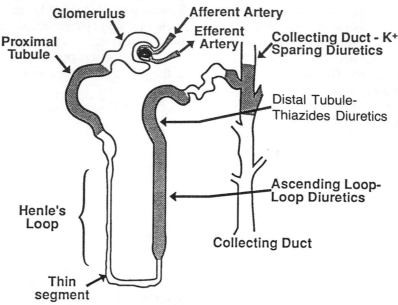

Fig. 29.2 A renal cortical nephron showing sites of action of the diuretics used in the treatment of hypertension.

spironolactone, amiloride, and triamterene has been extensively reviewed (69,90). After an observation that progesterone inhibited the sodium retaining effects of aldosterone, a series of steroidal derivatives, the spirolactones, were shown a similar aldosterone inhibitory effect (92). Spironolactone was subsequently shown to act competitively at the intracellular mineralocorticoid receptor site. The overall action of aldosterone is to increase the reabsorption of sodium and secretion of K^+. Thus, unlike thiazide and loop diuretics, which increase the output of both Na^+ and K^+, spironolactone was "potassium sparing" increasing sodium output more than potassium. Although the natriuretic response of spironolactone is limited by the basal level of aldosterone, it has antihyper-

tensive potency comparable to thiazides. From the medicinal chemistry search for nonsteroid potassium sparing diuretics came triamterene (**9**) and amiloride (**11**) which are not competitive aldosterone antagonists but which appear to act at the same distal tubular site (93). Because of a propensity to induce hyperkalemia, the use of these drugs in hypertension primarily is in combination with thiazide diuretics. The combination of thiazide and potassium sparing diuretics retain low ceiling diuretic and antihypertensive effects while not affecting serum potassium levels.

6.2.1.2 Pharmacology. The pharmacology profile of "diuretics" is limited primarily to their renal site of action. The thiazide, "loop," and potassium sparing diuretics

Furosemide (**7**)

Ethacrynic acid (**8**)

Triamterene (9)   Spironolactone (10)

Amiloride (11)

each act at distinct sites of the renal neph-rone (Fig. 29.2). The urinary volume composition and pH is different for each of these sites of action. A summary of the urinary output of these three classes of diuretics (90) suggests that thiazides increase urinary volume output (3X) and composition (mM): Na^+ (3X), K^+ (1.5X), Cl^- (2.5X), and HCO_{3^-} (25X) (pH 7.4). Loop diuretics increase urinary volume output (8X) and composition (mM): Na^+ (2.8X), K^+ (0.7X), Cl^- (2.5X), and no changes in HCO_{3^-} (pH 6.0). Potassium sparing diuretics increase urinary volume out (12X) and composition (mM): Na^+ (2.6X), K^+ (0.3X), Cl^- (1.8X), and HCO_{3^-} (15X) (pH 7.2).

Two effects of thiazide diuretics are potential limitations to their usefulness as antihypertensive drugs. These are (1) diuretic-induced hypokalemia and (2) diuretic-induced increase in serum choles-terol. The relation between thiazide-induced hypokalemia and sudden death was raided by the Multiple Risk Factor Inter-vention Trial (94) but not confirmed by the Hypertension Detection and Follow Up Program (95). The effects on serum choles-terol appear to be a short-term phenom-enon and should not limit the use of thiazide diuretics as antihypertensive agents (96). A recent review of several large trials supports this conclusion. In that analysis,

the total risk of strokes was reduced 25–42% in each trial and the total risk of coronary heart disease was reduced 16–32% although in two trials this was not statistically significant (97).

6.2.1.3 Chemistry. The medicinal chemi-cal discovery of chlorothiazide resulted from the synthesis of sulfonamides targeted at inhibition of carbonic anhydrase. A sim-ple *p*-carboxybenzone-sulfonamide was shown to be natriuretic and chloruretic. Two sulfamoyl groups mated to each other were better than one and the closure of the adjacent sulfamoyl and amino groups on the benzene ring gave the benzothiadiazine heterocyclic nucleus (1,2)4-benzothia-diazine1,1-chloride) abbreviated "thiazide" (89). Hydrochlorothiazide (chlorothiazide with unsaturated bond between C-3 and N-4) is 20 times more potent i.v. in the dog than orally in humans but displayed similar duration of action. The fluorosubstitution of R-6 (bendroflumethiazide) further in-creases potency. Dichloromethyl substitu-tion of the R-3 position (trichlor-methiazide) increased potency and doubled the duration of action. None of these chlo-rothiazide derivatives were particularly ac-tive against carbonic anhydrase (IC_{50} 3 × 10^{-4}–6 × 10^{-5} M). It has been suggested that the increase in potency (natriuretic activity) of the thiazides is directly related to their lipid solubility and sequestration into cells (89). The low ceiling agents such as chlorothalidone, metolazone, and in-dapamide are not benzothiadiazines but exhibit similar effects on the composition of urinary ion output suggestive of a similar site of action.

Chlorothalidone (12)

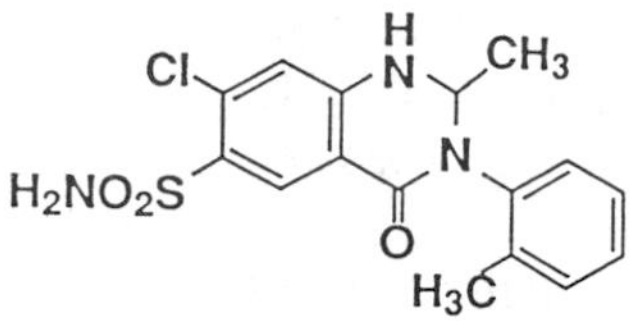

Metolazone (**13**)

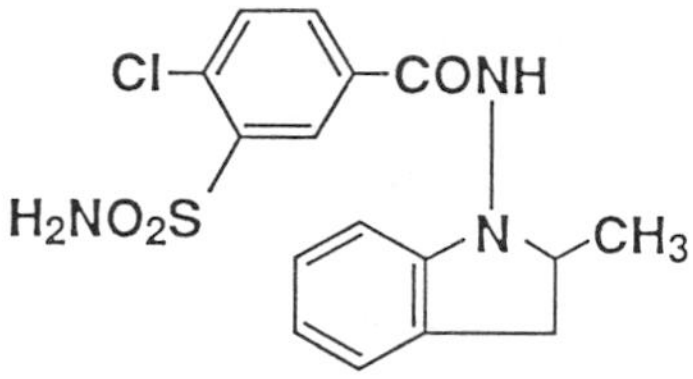

Indapamide (**14**)

The nonsulfonamide molecule diazoxide, related to thiazide molecule, has proven to have direct vasodilator effects and produces a rapid antihypertensive effect when given i.v. Diazoxide (**15**) has opposite effects than chlorothiazide on renal function and causes a decrease in urinary output. Reflex tachycardia and sodium and water retention have limited the usefulness of this compound (98,99).

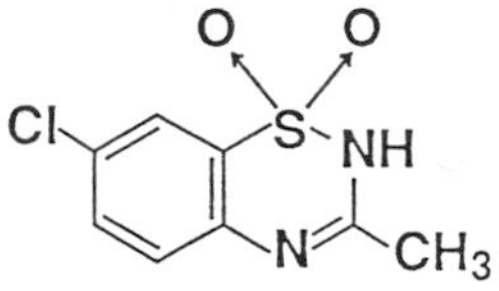

Diazoxide (**15**)

The loop diuretics appear to share a common site of action but do not appear to share common structural features. Furosemide, an anthranilic acid derivative and bumetanide (**16**), a 3-aminobenzoic derivative, are both sulfonamides and evolved from searches for a more potent thiazide. Ethacrynic acid was synthesized as the result of a search for nonmercurials that inhibited sulfhydryl catalyzed systems. The alpha/beta-unsaturated ketone moiety of ethacrynic acid confers a high degree of reactivity toward the sulfhydryl groups. It is unclear if this property is required for the

loop diuretic effects of ethacrynic acid and/or furosemide (89,90). New loop diuretics like torasemide (**17**) continue to be developed which may have less potassium wasting effects than furosemide (100,101).

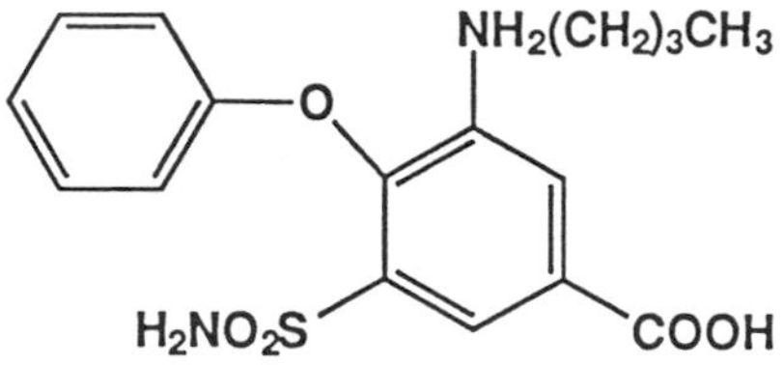

Bumetanide (**16**)

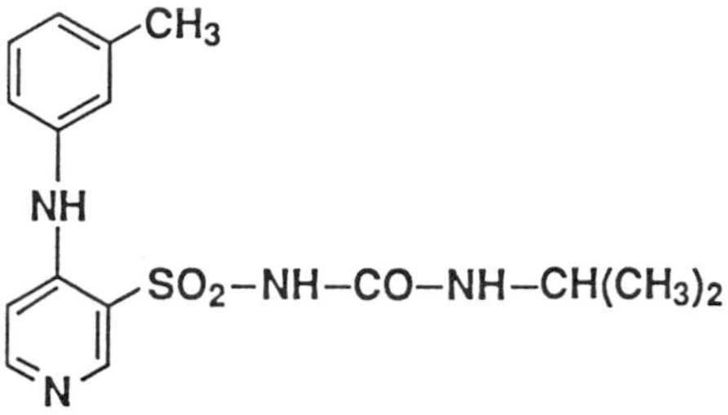

Torasemide (**17**)

Spironolactone (**10**) evolved from a search for steroids that mimic the aldosterone inhibitory effects of progesterone. Spironolactone is one of a series of 17-spironolactones that demonstrated this activity (92). An active metabolite of spironolactone, canrenone and its prodrug canrenoate have also been used clinically (90). Amiloride resulted from the random screening of 25,000 compounds using an acute fluid loaded rat model (102). Triamterene was "discovered" as the most active of a series of pteridine molecules identified from diuretic screening tests in rats (103). Triamterene has aldosterone antagonist properties but, unlike spironolactone, it retains its natriuretic and kaluretic activity in adrenalectomized animals (69).

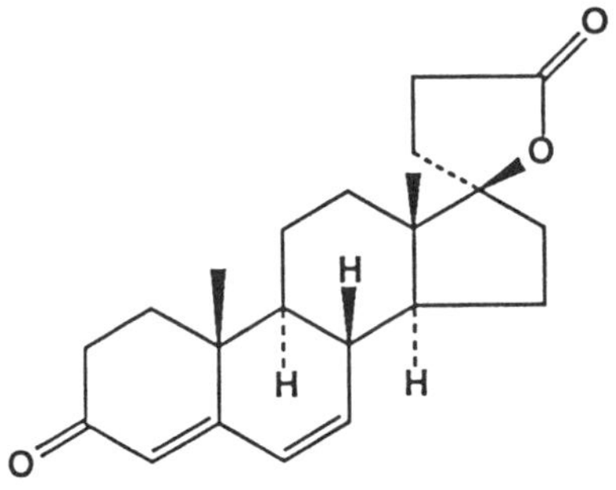

Canrenone (18)

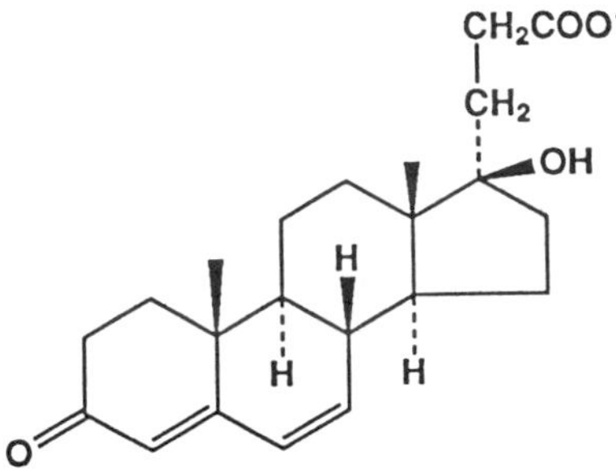

CH$_2$COO$^-$

CH$_2$

OH

Canrenoate (19)

6.2.2 BETA BLOCKERS

6.2.2.1 Discovery. Beta blockers were "discovered" from an effort to find inhibitors of the effects of the sympathetic nerve stimulation and the transmitters involved, e.g., norepinephrine (**20**). The first beta blocker was dichloroisoproterenol (**21**) (104) but this compound retained intrinsic agonist properties (so-called partial agonist or sympathomimetic activity) and was not developed.

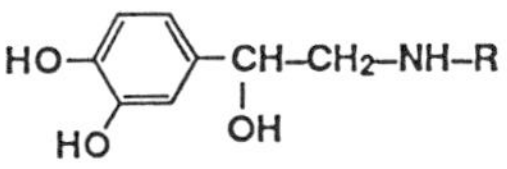

(20)

R
—H = norepinephrine
—CH$_3$ = epinephrine
—CH(CH$_3$)$_2$ = isoproterenol

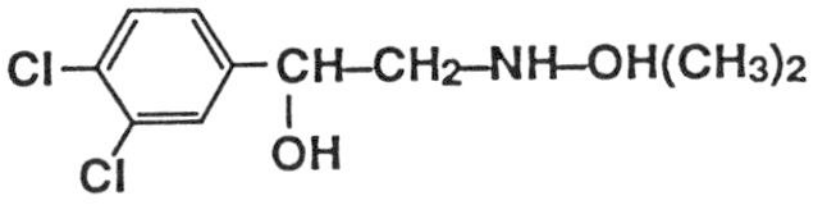

Dichloroisoproterenol (21)

The first beta blocker of therapeutic importance was propranolol (**23**) (73) that replaced the initial lead compound pronethalol (**22**) which produced thymic tremors in rats. Propranolol is the prototype of this important class of agents and is still widely used throughout the world. It was identified for its blockade of the cardiac effects of norepinephrine and was targeted for the treatment of angina.

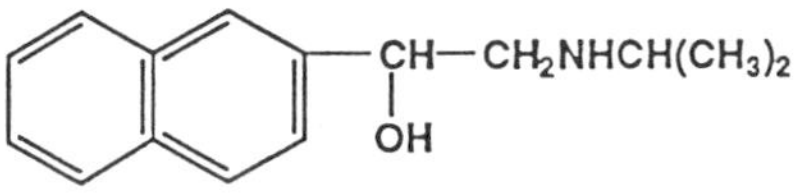

Pronethalol (22)

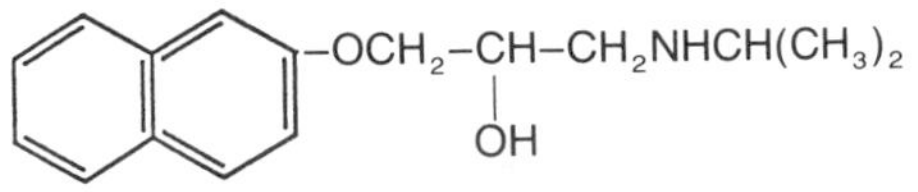

Propranolol (23)

The antihypertensive effects of propranolol were not widely accepted for sometime after its marketing for other indications (105). The mechanism of action of beta blockers to lower blood pressure has been the subject of much research and likely involves renal, cardiac, and central sites of action (Fig. 29.3).

The availability of agents which selectively blocked the cardiostimulatory and vasodilatory effects of the natural hormones epinephrine and norepinephrine solidifed the concept of beta receptors proposed by Ahlquist (106). Subsequently, studies by Lands et al. (107), characterized beta receptor subtypes, e.g., beta-1 for the heart and beta-2 for bronchial tissue and peripheral blood vessels. Propranolol and timolol (**24**) inhibited both subtype receptors and was designated nonspecific inhibitors. The identification of cardioselective (beta-1) selective antagonists was sought to reduce the peripheral vascular and bronchiolar side effects of the propranolol-like nonselective agents. The prototypes of this class of agents are metoprolol (**31**), atenolol (**33**), and

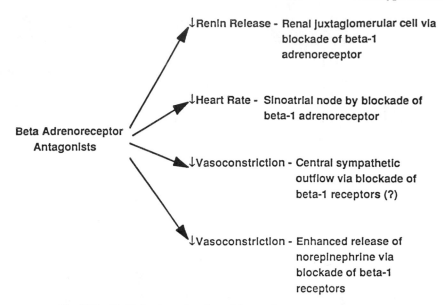

↓Renin Release - Renal juxtaglomerular cell via blockade of beta-1 adrenoreceptor

↓Heart Rate - Sinoatrial node by blockade of beta-1 adrenoreceptor

Beta Adrenoreceptor Antagonists

↓Vasoconstriction - Central sympathetic outflow via blockade of beta-1 receptors (?)

↓Vasoconstriction - Enhanced release of norepinephrine via blockade of beta-1 receptors

Fig. 29.3 Multiple sites of action of beta adrenoceptor antagonists.

acebutolol (**32**). All three of these compounds have proven to be effective and well-tolerated antihypertensive agents. To differentiate from existing compounds and to increase clinical efficacy, many molecules have been developed that combine beta blockade with other pharmacological properties. Labetalol (**26**) (see below) has affinity for both beta-1 and alpha-1 adrenoceptors. Carvedilol (**27**) is an example of molecules which have been developed that have both beta blocking and direct vasodilating properties (108). The rationale for this combination is that the beta blockade blocks the reflex cardiac stimulation normally seen with vasodilators. Other dipharmacophores have been discovered including molecules with alpha-1 adrenergic antagonist (peripheral) and serotonin (5HT$_{1A}$) agonist (central effects) such as urapidil (109).

Carteolol (**25**)

Labetalol (**26**)

Timolol (**24**)

Carvedilol (**27**)

Urapidil (28)

6.2.2.2 Pharmacology. The pharmacology of beta blockade relates to (*1*) specific beta receptor blockade, (*2*) nonspecific effects, e.g., membrane stabilizing effects, and (*3*) partial agonist activity at the beta receptor (isoproterenol-like cardiac stimulatory effects). Nonspecific antagonists like propranolol block sympathetically medicated cardiac stimulation, renin release, and lipolysis (beta-1 mediated) and bronchiolardilation, vasodilation, and glycogenolysis (beta-2 mediated). Metoprolol or atenolol being beta-1 selective can be expected to have less bronchiolar and vascular effects. Propranolol also has significant local anesthetic effects (seen with both *d-* and *l-*isomers, the *d-*isomer is 100-fold less potent than the *l-*isomer as a beta antagonist) and part of its antiarrhythmic effects may relate to this nonspecific effect. Propranolol is also lipid soluble and crosses the blood brain barrier. It is not clear whether the beneficial (anti-anxiety) or undesirable effects (vivid dreams) of propranolol are avoided with compounds which are more water soluble and, therefore, less CNS penetrating. Partial agonists effects are characteristic of the nonselective antagonist pindolol and the beta-1 selective antagonist celiprolol (**30**) (110). Theoretically, less depression of heart rate, peripheral vasoconstriction, and bronchiolar constriction are seen with these compounds. Although some evidence for such differences has been reported, the majority do not show compelling differences between the various "classes" of beta antagonists (111,112). "Neither potency nor efficacy or tolerability of more recent representatives are superior to those of beta blockers that have been available for a long time" was the conclusion of the late F. Gross (113) on the occasion of the introduction of another new beta antagonist betaxolol (114).

6.2.2.3 Chemistry. The chemical development of beta antagonists began with the modification of isoproterenol which is selective for the cardiac and vascular inhibitory (beta) effects of epinephrine. The dichloro-isoproterenol was the first beta adrenergic antagonist but it retained significant agonist effects. The dibenzyl compound, pronetholol, was a more potent antagonist but also had significant agonist activity (115). The diphenoxy 3-isopropylamino-2-propanol compound, propranolol, became the prototype for this class of therapeutic agents and the reference for most of the synthetic work that has followed. The aliphatic hydroxyl appears to be essential for activity and confers optical activity. Cardiac selectivity was accomplished by substitution of the para-position of the phenoxy 3-isopropyl amino-2-propanol, e.g., practolol (**34**) (-NHCOCH$_3$), atenolol (-CH$_2$CONH$_2$), or betaxolol (**35**) [cyclopropyl(methoxyl)ethyl]. Betaxolol was selected over the more potent corresponding cyclobutyl compound because of the lesser local anesthetic activity (116).

Pindolol (29)

Celiprolol (30)

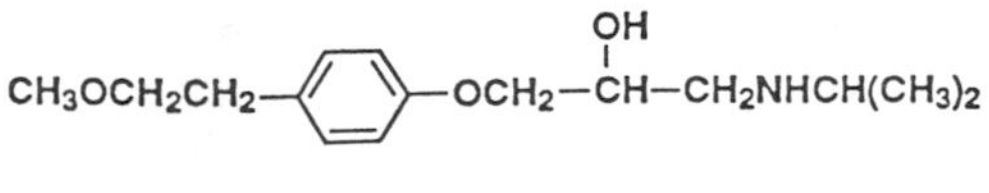

$CH_3OCH_2CH_2$—⟨benzene⟩—OCH_2—$\overset{OH}{\underset{|}{CH}}$—$CH_2NHCH(CH_3)_2$

Metoprolol (**31**)

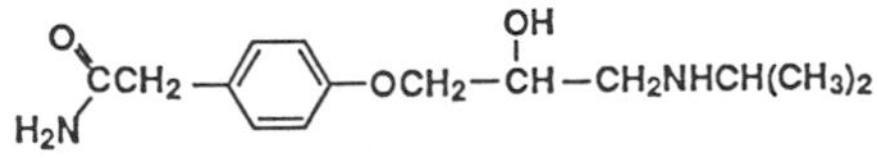

$\overset{O}{\underset{H_2N}{\overset{\|}{C}}}CH_2$—⟨benzene⟩—$OCH_2$—$\overset{OH}{\underset{|}{CH}}$—$CH_2NHCH(CH_3)_2$

Atenolol (**33**)

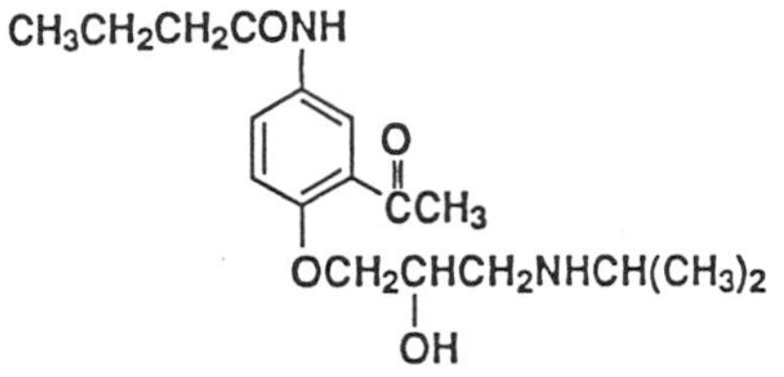

$CH_3CH_2CH_2CONH$—⟨benzene with $\overset{O}{\overset{\|}{C}}CH_3$ and $OCH_2\overset{}{C}HCH_2NHCH(CH_3)_2$ with OH⟩

Acebutolol (**32**)

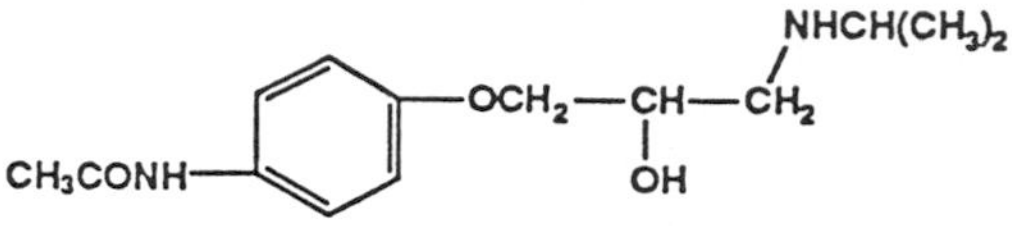

CH_3CONH—⟨benzene⟩—OCH_2—$\underset{OH}{CH}$—CH_2—$NHCH(CH_3)_2$

Practolol (**34**)

⟨cyclopropyl⟩—CH_2-OCH_2CH_2—⟨benzene⟩—OCH_2—$\overset{OH}{\underset{|}{CH}}$—$CH_2$$NHCH(CH_3)_2$

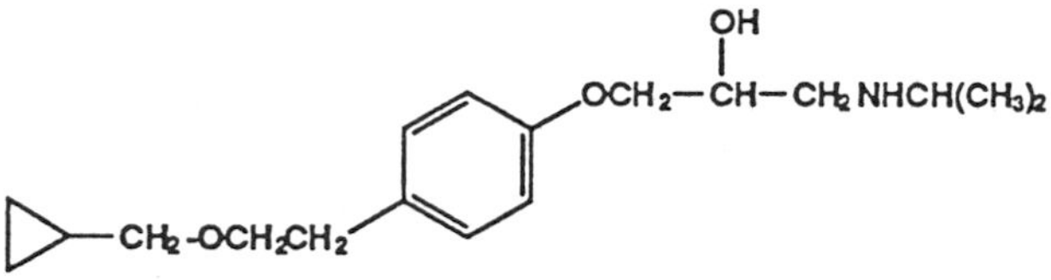

Betaxolol (**35**)

Most of the new beta blockers have sought to increase duration of action over that of propranolol which undergoes extensive first-pass metabolism. An alternative approach has been the development of esmolol, an ultrashort acting antagonist designed to provide rapid onset and offset of action with i.v. infusion in the critical care environment (117). The parent molecule esmolol is rapidly converted to the active moiety ASL-8123 (**37**) by blood hydrolysis of the labile aliphatic carboxyester group (118).

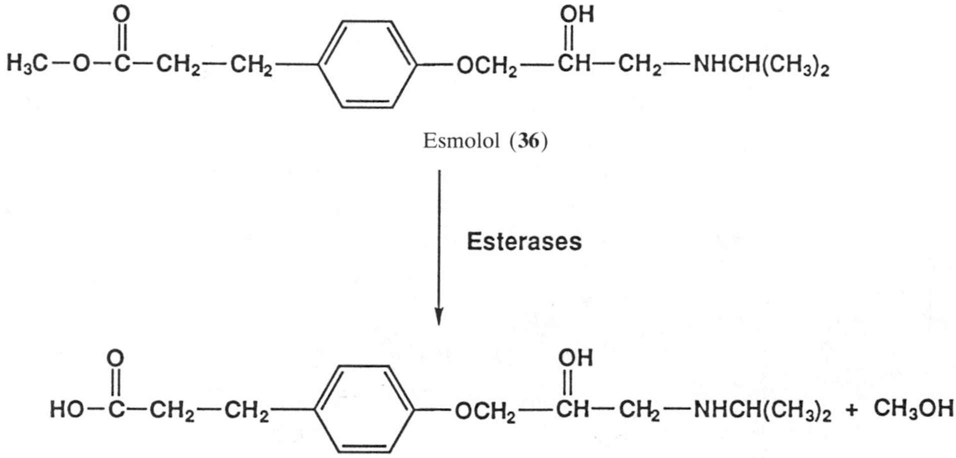

H_3C—O—$\overset{O}{\overset{\|}{C}}$—$CH_2$—$CH_2$—⟨benzene⟩—$OCH_2$—$\overset{OH}{\overset{\|}{C}H}$—$CH_2$—$NHCH(CH_3)_2$

Esmolol (**36**)

Esterases

HO—$\overset{O}{\overset{\|}{C}}$—$CH_2$—$CH_2$—⟨benzene⟩—$OCH_2$—$\overset{OH}{\overset{\|}{C}H}$—$CH_2$—$NHCH(CH_3)_2$ + CH_3OH

ASL-8123 (**37**)   Methanol

6.2.3.1 Discovery. The first ACE inhibitors were peptides isolated from the venom of *Bothrops jararaca* which had earlier been shown to contain bradykinin potentiating peptides (119). Peptides purified by two independent groups displayed both bradykinin potentiating and angiotensin I inhibitory actions. It was later shown that the same enzyme activity was responsible for both actions (ACE = kininase II). The nonapeptide, teprotide (Glu-Trp-Pro-Arg- Pro-Gln-Ile-Pro-Pro; SQ20881), was shown to have significant antihypertensive activity when administered i.v. to hypertensive patients (120). Teprotide, as a peptide, lacked oral bioavailability and displayed a very limited duration of action. The search was continued for nonpeptide ACE inhibitors suitable for the chronic oral treatment of hypertension. A novel *in vitro* assay was developed by which ACE activity was assessed using hippuryl-L-histidyl-L-leucine and measuring the stable product hippuric acid spectrophotometrically (121). The specificity of active inhibitors were then evaluated against the myotropic action of Ang I and II, bradykinin, acetylcholine, histamine, and PGE$_2$ (122). The breakthrough came when Byers and Wolfenden (123) showed that alpha-2 benzyl-succinic acid inhibited the ACE-related Zn$^+$ containing enzyme carboxy peptidase A. Benzyl-succinic acid "fit" the evolving model of the active site of ACE and succinyl-L-proline was found to be a weak ACE inhibitor but specific for Ang I in the guinea pig ileum. From this lead, the first orally active angiotensin converting enzyme inhibitor, captopril (SQ14225), quickly followed (121). The discovery of captopril was followed by enalapril, a compound which lacked the sulfhydryl group and the potential related side effects. Enalapril (MK 421), unlike captopril, is a prodrug and

must be de-esterified *in vivo* to the parent diacid moiety (124). The tremendous therapeutic and commercial success of captopril and enalapril has fueled extensive synthetic and patent activity that continues through today.

Captopril (**38**)

Enalapril (**39**)

6.2.3.2 Pharmacology. The pharmacological profile of ACE inhibitors is based on (*1*) its specificity for ACE to decrease Ang II formation, the presence or absence of the -SH-moiety, pharmacokinetic properties (oral bioavailability, duration of action, tissue penetration), and (*2*) its actions to inhibit kininase II to potentiate bradykinin by blocking its metabolism.

ACE inhibitors like captopril (**38**) and enalapril (**39**) (active moiety is enalaprilat) are very specific for ACE and inhibit carboxy-peptidase A or B, trypsin, or chymotrypsin only at >1000 times the concentrations needed to inhibit ACE. ACE inhibitors block the effects of exogenously administered Ang I but have no effect on Ang II. ACE inhibitors block the renin angiotensin (RAS) system by reducing the availability of Ang II to all Ang II receptors regardless of subtype and are therefore nonselective RAS antagonists. By blocking the availability of Ang II (the primary mediator of the RAS), ACE inhibitors reduce the Ang II-induced vasoconstriction and aldosterone release. Both of these effects exert a potentially powerful effect on blood pressure control (Fig. 29.1) and explain their beneficial effects in the treatment of hypertension.

ACE inhibitors also potentiate bradykinin which can interact with two bradykinin receptor subtypes. Bradykinin-1 (B_1) receptors mediate the vasoconstrictor and B_2 receptors mediate the vasodilator effects of bradykinin (Fig. 29.4). The widespread use of ACE inhibitors has shown them to be generally well tolerated. Cough and angioedema have been attributed to bradykinin and Ang II receptor antagonists may be even better tolerated. Although it remains controversial whether, or how much, bradykinin is involved in the antihypertensive effects of ACE inhibitor, the maximum blood pressure lowering effects, at least in experimental animals, are not different (126,127). A cardioprotective effect of bradykinin has been reported but the data remains controversial (128–130).

The presence of the sulfhydryl group of captopril may explain the penicillamine-like side effects (rash) as well as the beneficial effects on ischemic heart disease (131,132). The now well-established similarities between the actions of captopril (SH-containing) and enalapril (nonsulfhydryl), how-

ever, question the importance of this in the clinical use of these agents (133,134).

The possibility of independently-regulated local renin angiotensin systems has focused attention on possible differences between individual agents. Since ACE is localized on vascular endothelial cells or adjacent interstitial cells, tissue penetration is not needed to explain the actions of ACE inhibitors in peripheral tissues (135,136). Crossing the blood brain barrier appears to be related to the lipid solubility (137). Aside from differences in duration of action, e.g., captopril twice daily dosing and enalapril and lisinopril (**40**) once daily dosing, the actions of all members of this class are quite similar (134).

Lisinopril (**40**)

6.2.3.3 Chemistry. The ACE "active site" model consisting of Zn^{2+}, hydrogen bonding and positive charged ionic binding domains (138) "fits" all of the known ACE inhibitors (Table 29.10 (139,140). The zinc

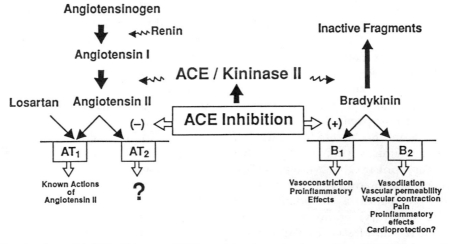

Fig. 29.4 Dual actions of ACE inhibitors to inhibit angiotensin converting enzyme (ACE/Kininase II) activity and the resultant decrease in levels of angiotensin II and increases in levels of bradykinin.

Table 29.10   ACE "Active Site" Model[a]

	Zn$^+$ Binding	Hydrogen Bonding	Positive Charge	ACE Inhibition IC$_{50}$ (μg)
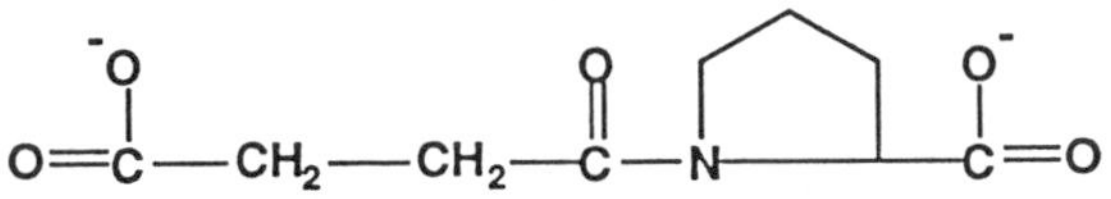 Succinyl-L-proline				330
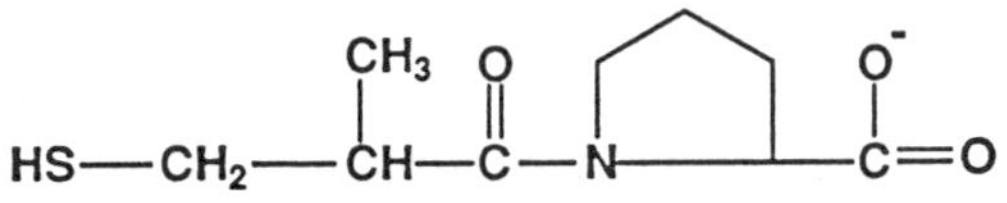 Captopril				0.023
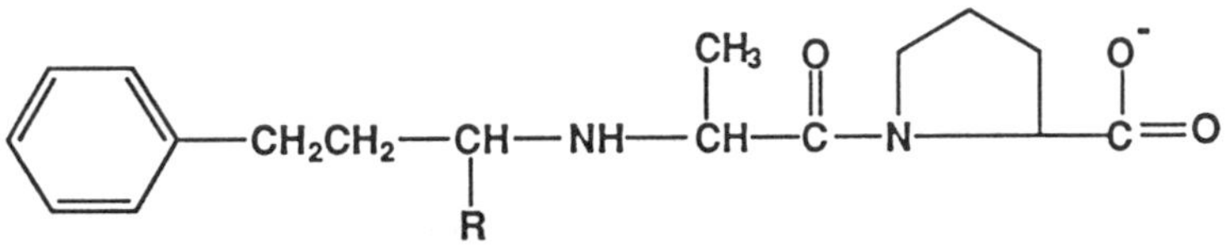 Enalapril R = $-CO_2C_2H_5$				1.2
Enalaprilat R = COOH				0.0012

[a]After refs. 121, 124.

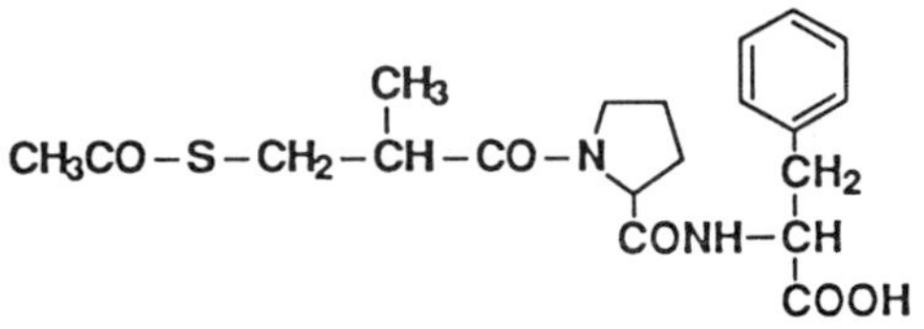

Alacepril (41)

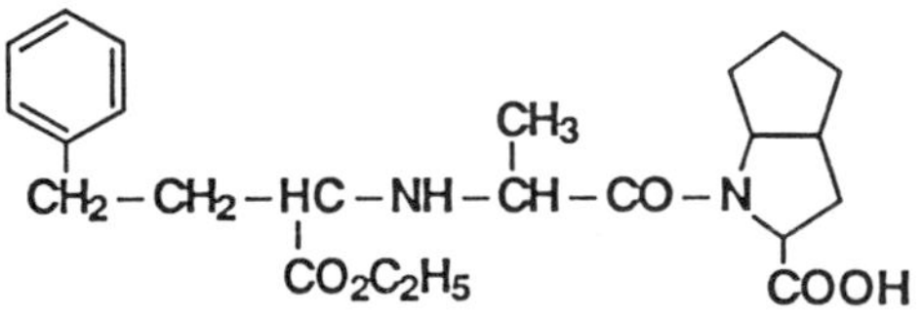

Ramipril (44)

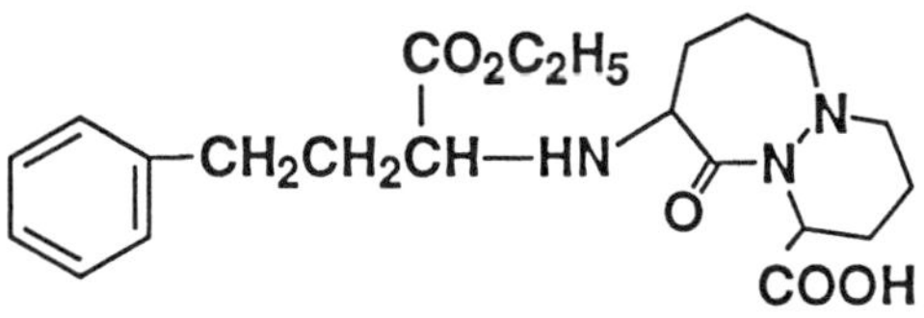

Cilazapril (42)

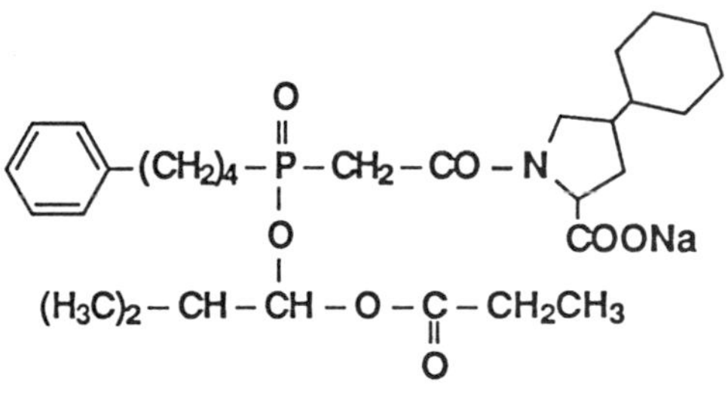

Fosenopril (45)

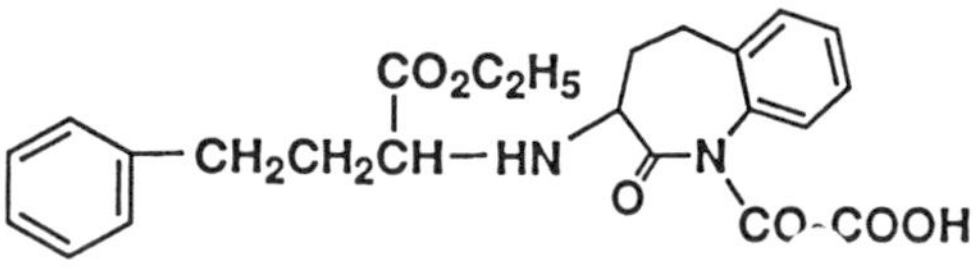

Benazepril (43)

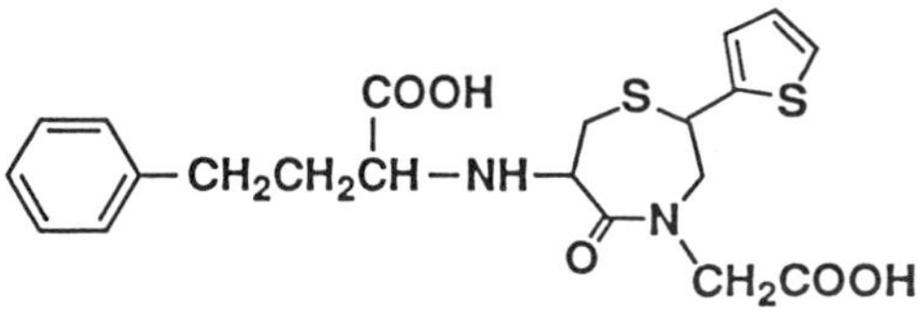

Temocapril (46)

ligand was "sulphydryl" for captopril and alacepril (DU-1219), "carboxy" for enalapril, lisinopril, cilazapril (Ro 31-2848), and ramipril (HOE-498), and "phosphorous" for fosinopril (SQ-28,555). The free terminal carboxylic acid is characteristic of all ACE inhibitors but the hydrogen bonding group has varied from the simple proline of captopril to more complex heterocycles, e.g., benazepril and temocapril. Lisinopril (**40**) contains a lysine residue in the 2-position conferring oral bioavailability to the methyl analog enalaprilat.

6.2.4 CALCIUM CHANNEL BLOCKERS

6.2.4.1 Discovery. Fleckenstein (82) in a series of classical studies in isolated cardiac tissue showed that certain coronary vasodilators like verapamil (**47**) and later, nifedipine, mimicked the effects of calcium withdrawal. These compounds were desig-

nated "calcium antagonists" (82). Calcium antagonists were initially found to have very dramatic effects in coronary heart disease but their potential as novel antihypertensive agents was soon realized (see review in Ref. 141). There are a large number of chemically diverse "small molecules" that inhibit the influx of extracellular calcium and a classification has been proposed (142). Accordingly, Group A antagonists and the three chemical types of calcium antagonists of most value as antihypertensives including verapamil (**47**), diltiazem (**48**), and nifedipine (Fig. 29.5) (and other dihydropyridines). Group B and C include less specific compounds such as bepridil, cinnarizine, or chlorpromazine that are not useful antihypertensives.

6.2.4.2 Pharmacology. Although verapamil, diltiazem, and nifedipine were all first identified *in vivo* coronary dilator screens, many of the compounds that have

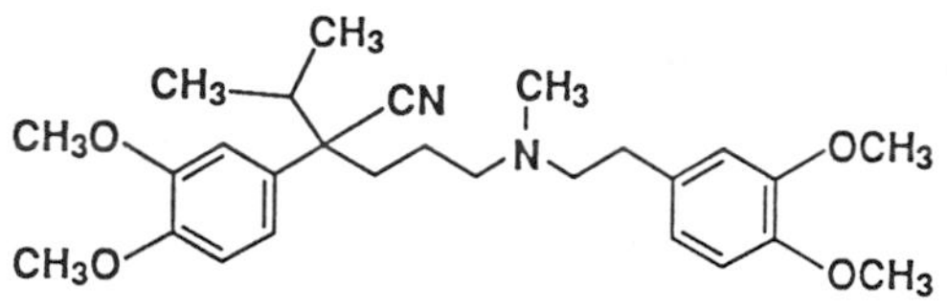

Verapamil (**47**)

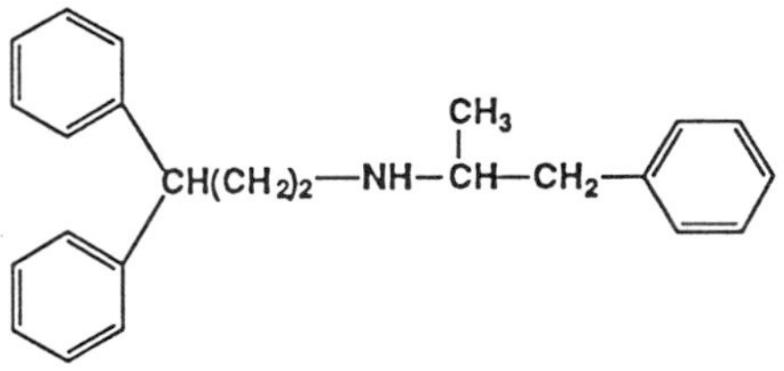

Prenylamine (**49**)

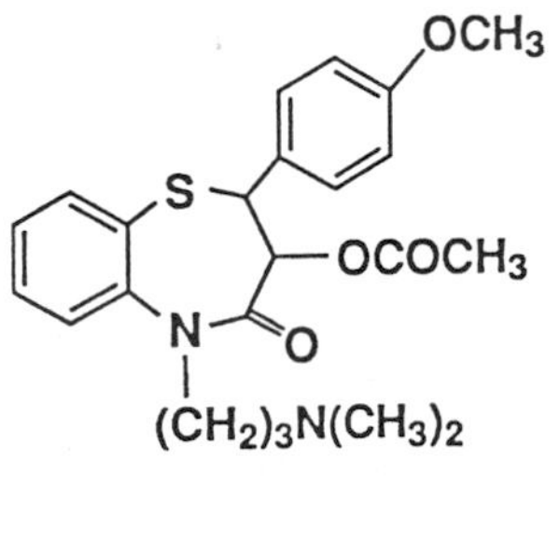

Diltiazem (**48**)

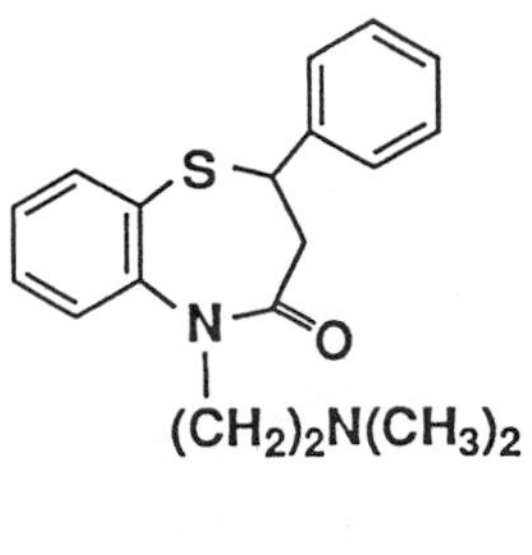

Thiazesim (**50**)

followed have been evaluated for their specific calcium antagonistic activity (141). The definitional characteristics of these agents include (*1*) inhibition of excitation-contraction coupling in isolated vascular smooth muscle, (*2*) inhibition of $^{45}Ca^{2+}$ uptake into microsomal preparation of vascular smooth muscle (*3*) inhibition of radiolabeled dihydropyridine (usually nitrendipine) binding, (*4*) inhibition of contractile force in isolated cardiac tissue, (*5*) electrophysiologic characterization of inhibition of the slow inward calcium current in isolated cardiac tissue, and (*6*) increased coronary blood flow *in vivo*. Calcium antagonists act by binding to specific allosterically linked receptor sites associated with the voltage-dependent calcium channels (143). The Group A-type compounds inhibit the so-called "L" (long-lasting; large current) subtype of voltage-dependent calcium channel and are inactive against "T"

(transient, tiny current) or N (neurally located) calcium channels (144). The compounds also inhibitor receptor operated calcium channels activated by vasoactive hormones like norepinephrine (Table 29.11).

The "discovery of calcium antagonists received worldwide acclaim because these agents are effective in patients with angina and hypertension and in many experimental models of tissue injury. It was clear that calcium antagonists were "atypical vasodilators" and, instead of decreasing renal function, they maintain or increase function (145). In experimental animals, calcium antagonists lower blood pressure in virtually every model of hypertension (SHR, renal, DOCA salt, Dahl-S, or SHR-stroke prone) (146). All known calcium antagonists are also potent coronary and cerebral vasodilators. Although vascular bed selectivity has been suggested for some dihydro-

Table 29.11   Calcium Antagonist Sites of Action

Tissue/Cell Type	Calcium Channel[a]	Ca^{2+} Dependent Response
Cardiac		
Sinoatrial node	VDC	Pacemaker activity
Atrioventricular node	VDC	AV conduction
Myocyte	VDC + ROC	Contractility/growth
Vascular smooth muscle (arteries and veins)	VDC + ROC	Contraction
Nonvascular smooth muscle (e.g., bronchiolar, gastronintestinal)	VDC + ROC	Contraction
Noncontractile cells		
Renal juxtaglomerular cells	VDC + ROC	Renin secretion
Adrenal medullar	VDC + ROC	Catecholamine secretion
Salivary gland	VDC	Salivation
Lacrimal gland	VDC	Tear formation

[a]VDC = Voltage-dependent channel.
ROC = Receptor-operated channel.

pyridines, at antihypertensive doses such selectivity is not likely to be important. In addition to vasodilation, however, calcium antagonists may also exert "vascular protective" effects by preventing excessive calcium uptake by vascular smooth muscle cells (147).

An important focus in the evaluation (and use) of calcium antagonists has been myocardial protection. By increasing coronary flow, decreasing peripheral resistance, preventing calcium overload, and exerting secondary cellular effects (inhibiting catecholamine release, platelet aggregation, leukocyte activation), calcium antagonists may provide myocardial and other tissue protective effects which support their use in hypertensive patients especially those with ischemic heart disease (148).

The dihydropyridine-type calcium antagonists can be differentiated from diltiazem and verapamil primarily on the basis of their effects on heart rate. Both diltiazem and verapamil possess direct membrane (quinidine-like) effects, tend to decrease heart rate and display antiarrhythmic effect on experimental animals. Nifedipine and related compounds are devoid of direct membrane effect, are most likely to cause a reflex increase in heart rate, and have much less antiarrhythmic effects in experimental animals (146,149).

6.2.4.3 Chemistry. The three classes of calcium antagonists represent distinct chemical types. They each share the common property of interfering with the influx of extracellular calcium via the calcium L channel (150). The core structure of the nifedipine-type structure is the aryl-1,4-dihydropyridine ring with an unsubstituted nitrogen (Fig. 29.5). Lower alkyl substitution on the 2- and 6-position were preferred. However, amlodipine has a large ester functionality which resulted in markedly enhanced duration (151). Asymmetrical esters at the 3- and 5-positions yield more potent compounds but create chiral centers. The (-) isomer of nitrendipine, for example, is a significantly more potent

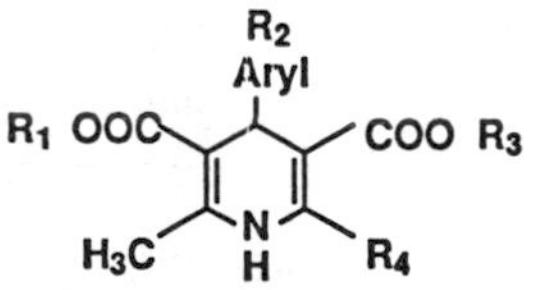

Compound	R_1	R_2	R_3	R_4
Nifedipine	$-CO_2CH_3$	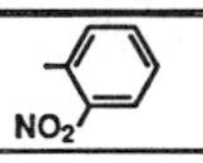 (aryl, NO_2)	$-CO_2CH_3$	CH_3
Nitrendipine	$-CO_2CH_3$	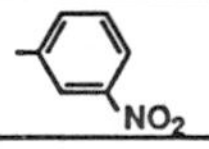 (aryl, NO_2)	$-CO_2C_2H_5$	CH_3
Nicardipine	$-CO_2CH_3$	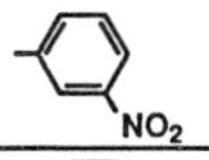 (aryl, NO_2)	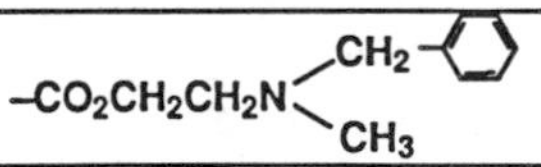 $-CO_2CH_2CH_2N(CH_2-, CH_3)$	CH_3
Felodipine	$-CO_2C_2H_5$	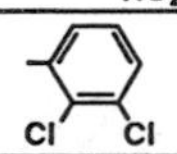 (aryl, Cl, Cl)	$-CO_2CH_3$	CH_3
Amlodipine	$-CO_2CH_3$	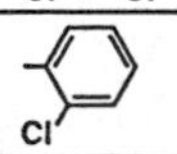 (aryl, Cl)	$-CO_2C_2H_5$	$CH_2OCH_2CH_2NH_2$
Bay k 8644	$-CO_2CH_3$	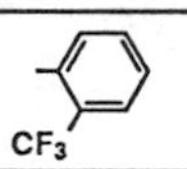 (aryl, CF_3)	$-NO_2$	$-CH_3$

Fig. 29.5 Core structure of the nifedipine-type.

inhibitor of calcium channel binding than the (+) isomer [IC_{50} (nM) 1.1 vs 30] (152). The substituted phenyl group in the 4-position of the 1,4-dihydropyridine is optimum with ortho- or meta-substituents (153). Interestingly, replacement of the 4-alkyl with an NO_2 yields a compound with marked Ca^{2+} channel activating properties. Bay K8644 was the first of a number of such compounds shown to increase the influx of extracellular calcium and to produce cardiac stimulation and vasoconstriction *in vivo* (154).

Verapamil and its methyl analog gallopamil (D-600) (**51**) were chosen from a series of prenylamine analogs (71). Verapamil, like the beta blocker propranolol, blocked the cardiac stimulatory effects of isoproterenol but had a greater effect on cardiac sympathetic nerve stimulation. It is of interest that Fleckenstein was asked to study verapamil and prenylamine to explain their cardiac depressant effects and from these studies came the concept of "calcium antagonism" (82). The (−) isomer of verapamil acts as specific calcium antagonists whereas the (+) isomer has little calcium but marked fast sodium channel effects (155).

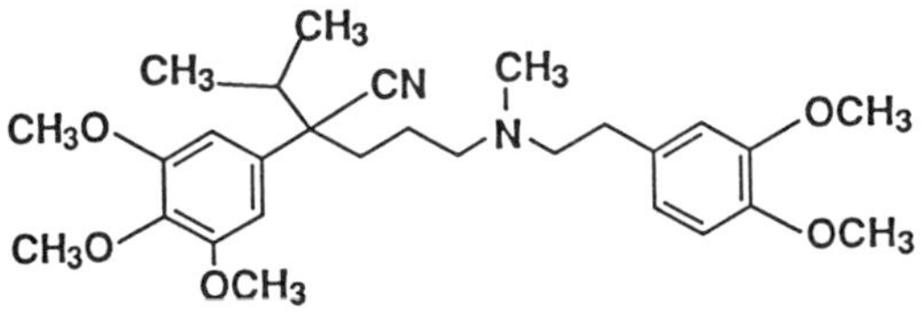

Gallopamil (**51**)

Diltiazem was identified from a random screen of novel 1,5-benzothiazepines related to the known compound reported to have antidepressant effects in animals, thiazesim. Diltiazem is the 2,3-cis-isomer which was more potent than the trans-isomer. Further, the *d*-cis-isomer is more potent and longer acting than the *l*-cis-isomer. At toxic doses, the *dl*-cis and *l*-cis isomers induced convulsions in mice whereas the

d-cis did not. All three isomers displaced similar local anesthetic effects (72).

6.2.5 ALPHA-1 (α_1) ADRENERGIC RECEPTOR ANTAGONISTS

6.2.5.1 Discovery. The "discovery" of prazosin followed the investigation of the blood pressure lowering effects of 2-amino-4(3H)-quinazolines. The candidate compound is the 4-aminoquinazoline, prazosin which lowered blood pressure involving "a component of sympathetic inhibition at peripheral site or central sites" (77). Unlike previous alpha blockers such as pentolamine, prazosin has significantly less cardiostimulatory effects (156). Such differences can now be explained on the basis of alpha adrenergic receptor subtypes.

Prazosin (**52**) is the prototype of an α_1 receptor antagonist and idazoxan is the prototype of an α_2 receptor antagonist (157). Prazosin selectively blocks postsynaptic α_1 receptors while having no effect on presynaptic receptors α_2 responsible for the inhibition of norepinephrine release from sympathetic nerve terminals (Fig. 29.6). Pentolamine, by contrast, nonselectively inhibits both α_1 and α_2 receptors resulting in a greater activation of sympathetic nerves (158).

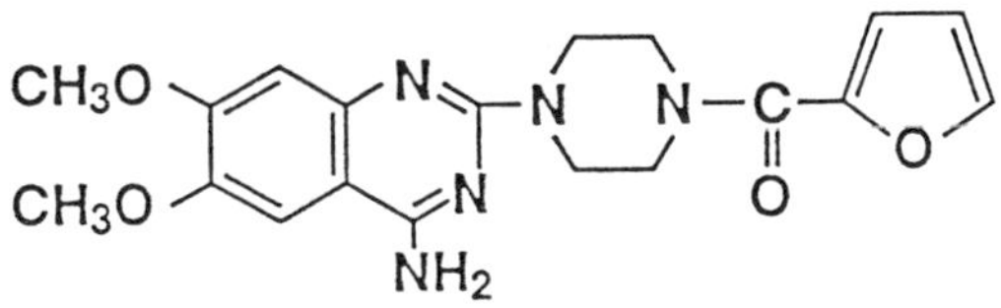

Prazosin (**52**)

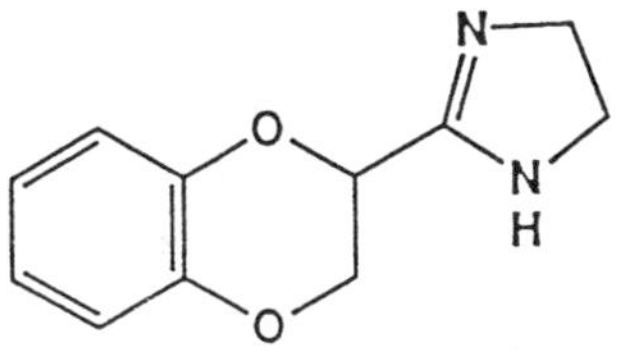

Idazoxan (**53**)

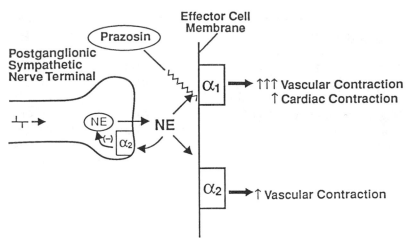

Fig. 29.6 The post-junctional site of action of the α_1 adrenergic receptor-specific antagonists such as prazosin in cardiac and vascular tissue.

6.2.5.2 Pharmacology. The pharmacology profile of prazosin and related compound relates to its block of sympathetic nerve function, so-called "sympatholytic effects." Prazosin has antihypertensive effects in both renal hypertensive and spontaneously hypertensive animals. The first dose effect (development of acute tolerance) observed with prazosin in man is also observed in SHR and appears to be due to activation of the RAS (159). The hallmark of the antihypertensive effect of prazosin is the lack of compensatory reflex tachycardia attributed to a balanced effect on both arterial and venous vascular beds. The most significant limitation in the clinical use of prazosin and related compounds is acute posterial hypotension related to blockade of the sympathetic nervous system (158,160).

6.2.5.3 Chemistry. The first series of adrenergic blocking agents that acted at what were later designated "alpha receptors" were the halolkylamine represented by phenoxybenzamine. These compounds were related to the nitrogen mustards and cyclized to form reactive ethylenimonium intermediates. In addition to alpha adrenoceptor blockade, these agents also inhibited

responses to serotonin, histamine, and acetylcholine (161). Phenoxybenzamine was used clinically by oral and i.v. administration but was limited by marked hypotension and reflex tachycardia. The next important advance came from a series of 2-substituted imidazolines observed to have histamine-like depressor activity. From these studies came phentolamine (**55**) which was shown to reverse the pressor response to epinephrine. Phentolamine is still being used in the management of

Phenoxybenzamine (**54**)

Phentolamine (**55**)

hypertensive emergencies but its use is limited by orthostatic hypotension and reflex cardiac stimulation (162). Although many other series of molecules have subsequently shown to have "sympatholytic" or "adrenolytic" activity, prazosin was the first alpha blocker suitable for the chronic oral treatment of essential hypertension. Prazosin was chosen from a series of dimethoxy-quinazolines for its potency and duration. The 2-furoyl piperazine was preferred over simpler esters or the simple dimethyl amine (77). The elimination of the half life of prazosin is approximately 3 hours. Simple saturation of the furan ring yields terazosin (163) with increased water solubility and with a half life of approximately 12 hours. Further increases in half life (reduced liver metabolism) were accomplished with doxazosin which contains a bulky replacement of the furan ring (164).

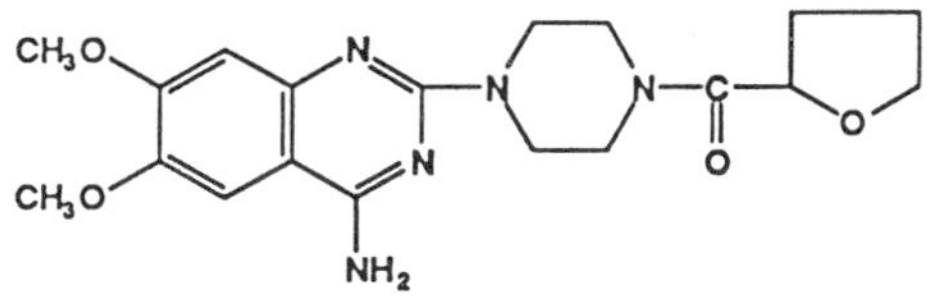

Terazosin (**56**)

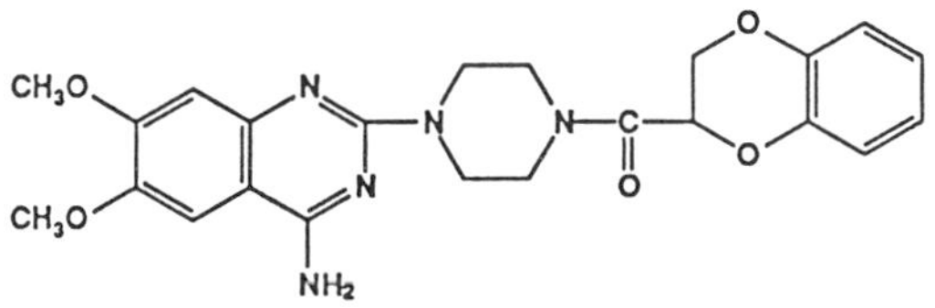

Doxazosin (**57**)

6.2.6 ALPHA-2 (α_2)/IMIDAZOLINE-1 RECEPTOR AGONISTS

6.2.6.1 Discovery. Clonidine (**58**) was originally synthesized for evaluation as a nasal decongestant and a few drops of an 0.3% solution was unexpectedly found to produce sedation, bradycardia, and long-lasting hypotension (165).

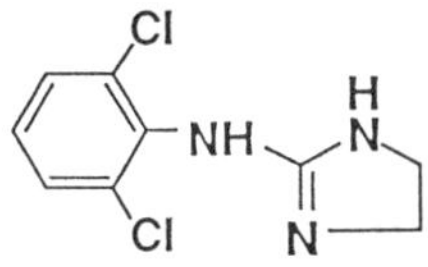

Clonidine (**58**)

Many different kinds of experiments were performed to characterize its complex actions *in vivo*. By i.v. bolus, clonidine produces an initial transient hypertensive effect followed by a prolonged hypotensive effect. In "decentralized" or ganglionic blocked dogs, the hypertensive effect was enhanced and the hypotensive effect was abolished. Intracisternal injection of 1 μg and i.v. administration of 30 μg of clonidine produced comparable reductions in blood pressure suggestive of a central site of action (165). Clonidine was then shown to act at pre- and post-ganglionic sites which were blocked by α_2 antagonists such as yohimbine (157). The antihypertensive actions of clonidine and structurally different compounds such as guanfacine (**59**), guanabenz (**60**), and methyldopa (**61**) have been attributed to activation of central α_2 receptors (166).

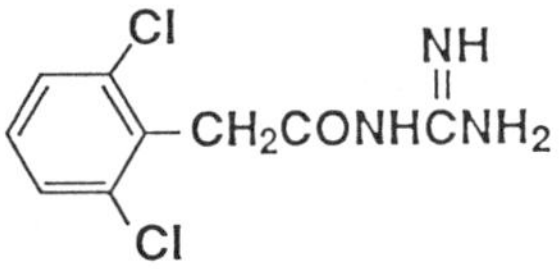

Guanfacine (**59**)

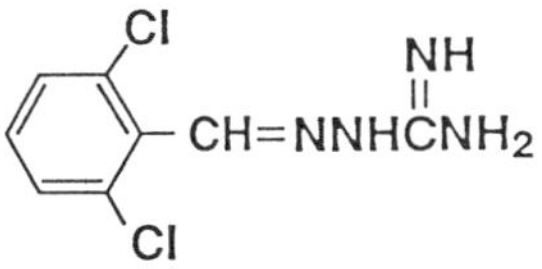

Guanabenz (**60**)

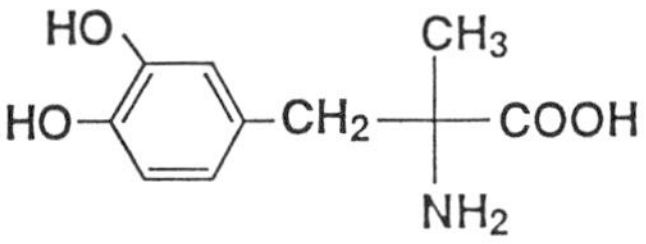

Methyldopa (**61**)

Recently a new concept has emerged whereby clonidine is thought to act at least in part by interaction with nonadrenergic so-called imidazoline receptors (167). The key observation was that the blood pressure lowering effect of i.v. clonidine was blocked by the "imidazoline" antagonist idazoxan but not by the α_2 blocker yohimbine (168). The clonidine derivatives, moxonidine and rilmenidine, which appear to have less CNS-depressant effects than the parent compound (169,170) display higher affinity for the nonadrenergic clonidine site than α_2 adrenoceptors in the kidney or brain but not in rabbit pulmonary artery and aorta (171). The non-adrenergic clonidine site has been designated a putative I_1-imidazoline receptor (172). I_1 receptors in the rostral ventrolateral medulla (RVLM) have been detected in rats, bovine, and humans and appear to have a role in blood pressure regulation (167).

6.2.6.2 Pharmacology. The pharmacology of clonidine as described above involves both central and peripheral effects and appears to activate both α_2 receptors (α_{2a} and α_{2b}) subtypes (173). Clonidine decreases salivation and gastric secretion in many species including humans, contracts isolated ileal smooth muscle, raises blood glucose levels, and exerts a saluretic effect in rats but not in dogs (165). It has recently been reported that control of insulin secretion by α_2 antagonists may reflect binding to a novel imidazoline receptor site in pancreatic islet cells (174). However, the effects of moxonidine injected i.c.v. to lower intraocular pressure in rabbits was blocked by the selected α_2 adrenoceptor

Rilmenidine (**63**)

antagonist, L-159,066 (175). Moxonidine (**62**) and rilmenidine (**63**) are being used to explore the special properties of the I_1 receptors (170,176). In rats, the natriuretic response to moxonidine involves both central and peripheral I_1 activation (177).

6.2.6.3 Chemistry. A great number of clonidine "analogs" have been described (165,176,178). Most of the early compounds were 2,6-dichloro-substituted congeners but the guanidine derivatives guanfacine (**59**) and guanabenz (**60**) displayed similar centrally mediated antihypertensive effects and somewhat more specificity for α_2 receptors and less CNS depressant effects. Although not structurally related to either series of compounds, methyldopa (**61**) is also thought to exert its antihypertensive effect at least in part via central α_2 receptors (179).

6.2.7 OTHER "MARKETED" ANTIHYPERTENSIVE DRUGS. The individual compounds that are marketed varies significantly from country to country based on many factors including the country of origin, the local regulatory climate, and clinical practice. The trends of drug use have changed as better drugs (more efficacy, better tolerated) become available (180). The number available is impressive, e.g., 75 single drug entities in the United States (1994 *Physician's Desk Reference*). It is beyond the scope to review the discovery, pharmacology, and chemistry of all of these compounds but these compounds are discussed in a number of reviews (84,86,181,182). Several compounds are available worldwide but their use has declined with the introduction of the newer agents (Table 29.12).

Moxonidine (**62**)

Table 29.12 Miscellaneous Marketed Antihypertensive Drugs

Compound	Mechanism of Action	Indication(s)	Reference
Alpha methyldopa	Multiple sites including α_2 agonism	Moderate-to-severe hypertension, eclampsia	(70,183)
Guanethidine	Sympathetic neuronal blocker	Refractory hypertension	(184)
Guanadrel	Sympathetic neuronal blocker	Refractory hypertension	(182)
Reserpine	Catecholamine release "smpatholytic"	Moderate-to-severe hypertension	(182)
Hydralazine	Nonselective vasodilation	Refractory hypertension	(66)
Nitroprusside	Nonselective vasodilation	Hypertensive emergencies	(185)
Diazoxide	Nonselective vasodilation	Hypertensive emergencies	(98)
Minoxidil	Nonselective vasodilation	Severe refractory hypertension	(74)

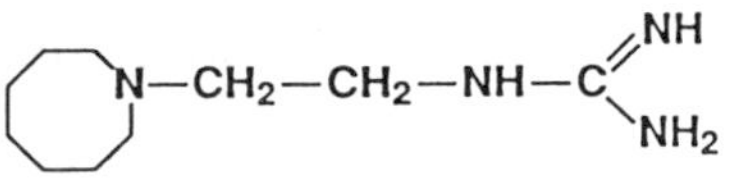

Guanethidine (64)

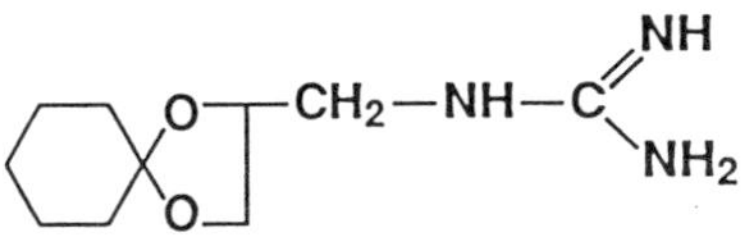

Guanadrel (67)

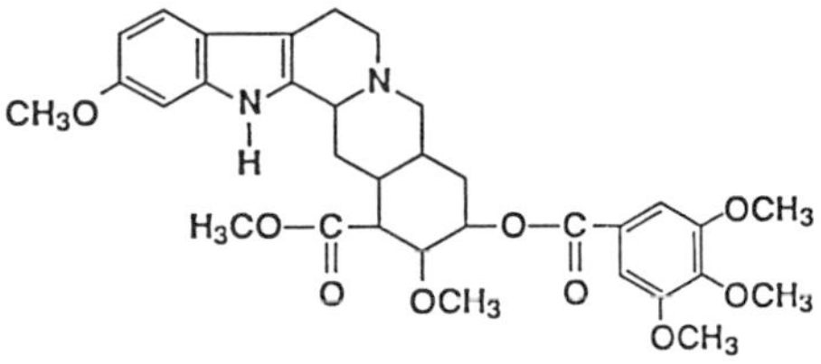

Reserpine (65)

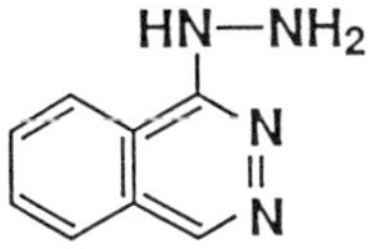

Hydralazine (68)

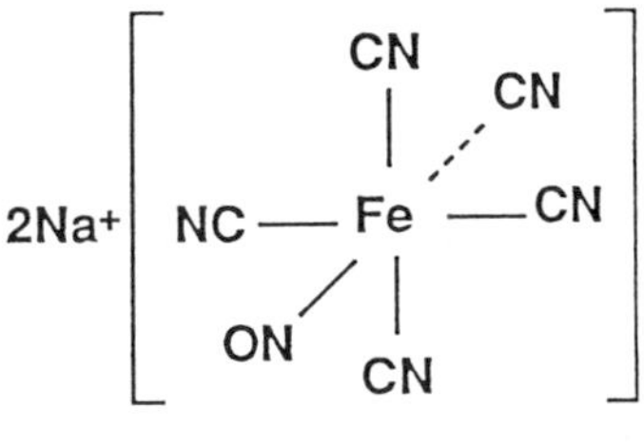

Sodium nitroprusside (66)

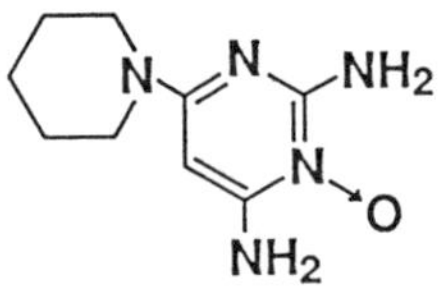

Minoxidil (69)

Many of these drugs are being used only for refractory or severe hypertension (99,162).

6.3 Innovative Antihypertensive Drugs

Several innovative targets for antihypertensive new drug discovery have been explored and novel molecules identified (Table 29.13). The compounds are in various stages of development and commercialization. The therapeutic importance of the lead compound or the viability of the "target" has not been determined for many of these compounds. The AT_1-selective nonpeptide Ang II receptor antagonists exemplified by losartan appears to have the greatest potential for the chronic oral treatment of essential hypertension.

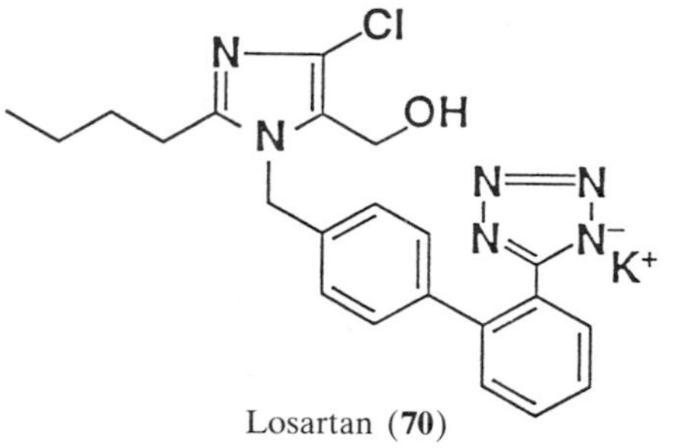

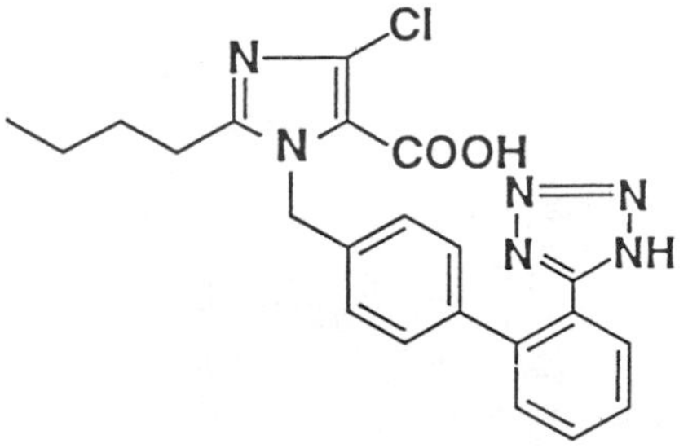

EXP3174 (71)

6.3.1 ANGIOTENSIN II RECEPTOR ANTAGONISTS. Losartan (70) has been identified as the first of a new class of nonpeptide angiotensin II receptor antagonists (80). These compounds block the effects of angiotensin II (Ang II) at its specific receptor sites (Fig. 29.7). The first inhibitors of the renin angiotensin system (RAS) were peptide analogs of Ang II. As peptides, these compounds were not orally active, very short acting, and possessed partial agonist activity. One such peptide, saralasin (Sar^1-Ala^8-Ang II), was extensively studied in hypertensive patients and was shown to lower blood pressure but its potential was limited by its peptide nature and partial agonist activity (86). Importantly, these studies proved the therapeutic effectiveness of Ang II receptor blockade. It was not until the

Losartan (70)

Table 29.13 Innovative Antihypertensive Drugs

"Mechanism of Action"	Prototype Compound(s)	Status
Ang II receptor (AT_1) antagonist	Losartan	NDA review, marketed in Europe
Potassium channel agonist	Pinacidil	Marketed worldwide
Serotonin receptor (5-HT_2) antagonist	Ketanserin	Phase III U.S., marketed in Europe
Renin inhibitor	Ro-42-5892	Phase III
Vasopressin receptor (VP1) antagonist	OPC-21268	Phase II
Endothelin (ET-1) receptor antagonist	BQ-123	Preclinical
Adenosine receptor (A_2) antagonist	CGS21680C	Preclinical
Neutral endopeptidase (NEP) inhibitor	SCH34826	Preclinical
Neuropeptide Y receptor antagonist	Atrinositol	Preclinical
Protein kinase C inhibitor	Ro-31-7549	Preclinical

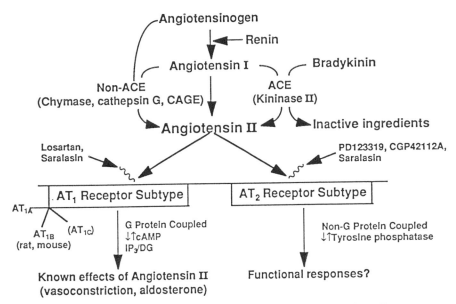

Fig. 29.7 Angiotensin II synthesized by ACE or non-ACE pathways interacts with AT_1 receptors to elicit the characteristic effects of vasoconstruction and aldosterone release.

orally active and long-acting nonpeptide receptor antagonist losartan became available, however, that the full potential of Ang II receptor blockade was realized (62). Losartan as the first of the new class of nonpeptide Ang II receptor antagonists has been an important experimental tool by which to validate many earlier findings with peptide molecules or with ACE inhibitors concerning the role of Ang II in cardiovascular pathology. Losartan serves as the prototype of an AT_1-selective antagonist. Losartan has been shown to have high affinity and selectivity for Ang II binding sites in vascular smooth muscle, in the kidney and adrenal gland. Unlike previously described "peptide" receptor antagonists like saralasin, losartan has no intrinsic (Ang II-lik) actions *in vivo*. Importantly, unlike ACE inhibitors, losartan does not inhibit ACE or potentiate bradykinin. Losartan is a competitive antagonist of the vasoconstrictor effects of exogenously administered Ang II both *in vitro* and *in vivo*. Further, losartan blocks the effects of endogenously-released Ang II *in vivo*, e.g., in renovascular hypertension. Losartan is metabolized

to E3174 (**71**) which is also a potent and specific Ang II receptor antagonist. In rats and humans, this conversion is sufficient for E3174 to contribute to the long-lasting inhibition of Ang II observed with losartan.

The discovery of losartan was paralleled by the independent discovery of a series of nonpeptide antagonists (prototype PD123319) and peptide agonists (prototype CGP42112A) which inhibit distinct Ang II binding sites. With these new tools, the earlier suggestions of Ang II receptor heterogeneity were confirmed. According to the current nomenclature, Ang II binding sites inhibited by losartan are designated AT_1 and those inhibited by PD123319 or CGP42112A are designated AT_2 (187,188). The peptide antagonist saralasin is nonselective and interferes with Ang II binding at both AT_1 and AT_2 sites. Although other receptor subtypes have been described in lower animals or cultured cells, in humans, only AT_1 and AT_2 subtypes are present. The distribution of Ang II receptor subtypes varies among species and among tissues within species. AT_1 receptors predominate in most tissues except in fetal tissue and in discrete areas of certain organs such as adrenals, kidneys, and brain. Virtually all of the well-known effects of Ang II including those on vascular tone, aldosterone release, renin release, norepinephrine release, and cellular growth are blocked by losartan (PD123319 (**72**) or CGP42112A (**73**) has no effect) and are by definition AT$_1$-mediated. The role of the AT$_2$ binding site remains controversial. Some evidence suggests a role for the AT$_2$ receptor in vascular or endothelial growth, in modulation of neuronal potassium channels, and cerebral autoregulation (189–191).

Many new nonpeptide AT$_1$-selective Ang II receptor antagonists have ben synthesized and several have been advanced to clinical trial (192–194). Most of the compounds which have been advanced for clinical trials contain a losartan-like bi-

phenyl tetrazole moiety. These include SR47436 (**74**), BIBR277 (**77**), and TCV-116 (**75**). Two exceptions are SKF108566 (**76**) which has a 5-acrylic acid substituted imida-

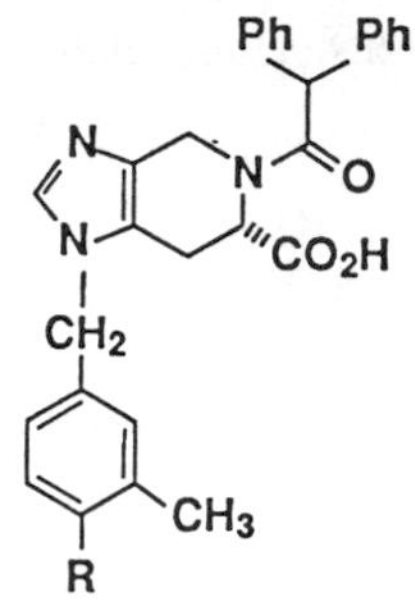

PD123319 (**72**)

R = –OCH$_3$	PD - 121981
R = –NH$_2$	PD - 123177
R = –N(CH$_3$)$_2$	PD - 123319
R = –H	PD - 124125

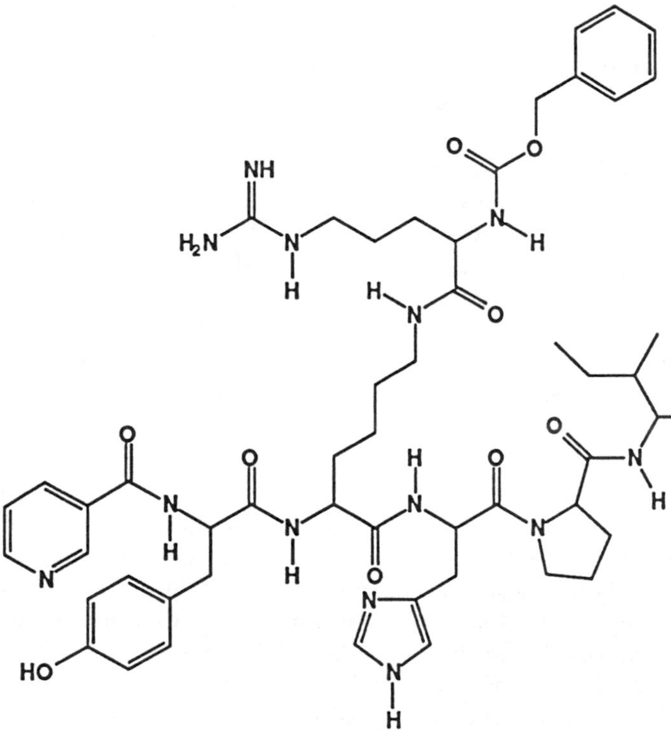

GGP42112 (**73**)

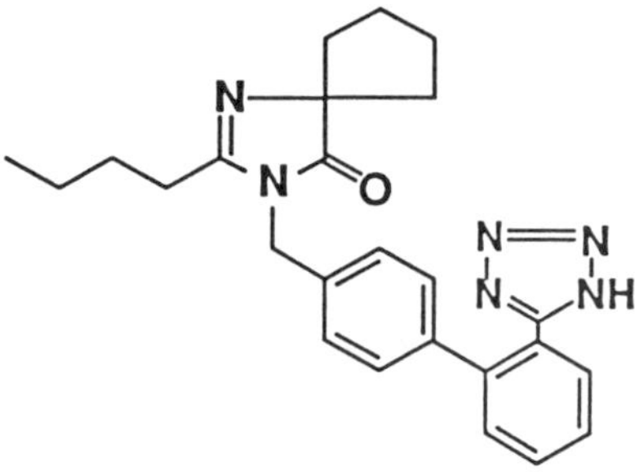

SR47436 (**74**)

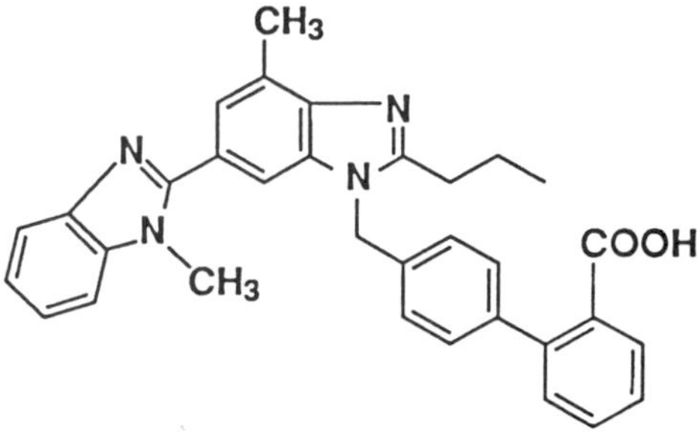

BIBR277 (**77**)

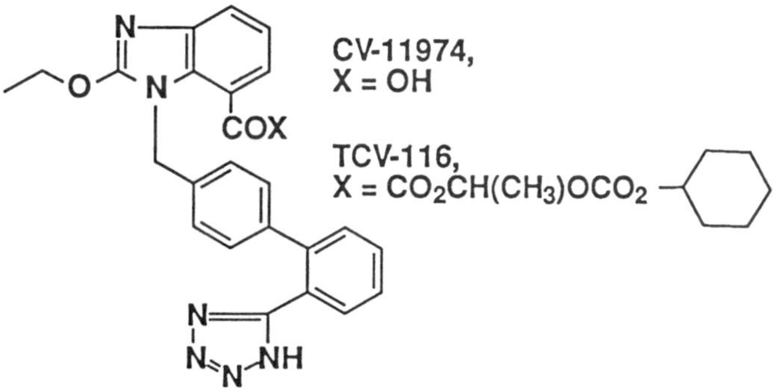

CV-11974,
X = OH

TCV-116,
X = CO$_2$CH(CH$_3$)OCO$_2$—

TCV-116, CV-11974 (**75**)

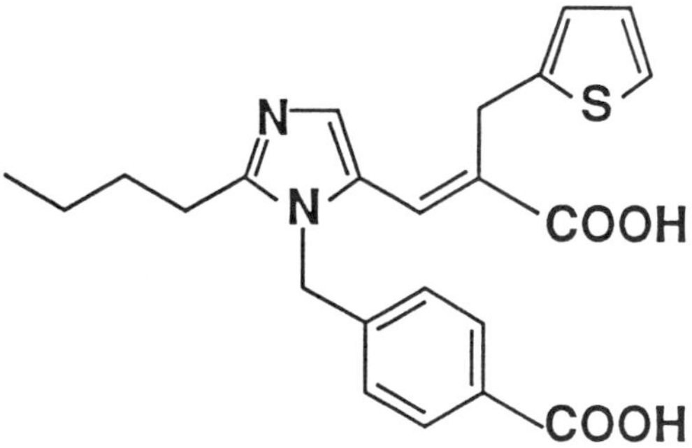

SKF108566 (**76**)

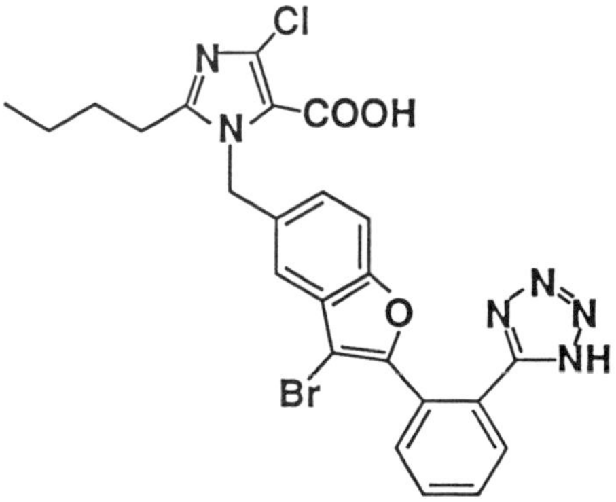

GR117289 (**78**)

zole (195) and GR117289 (**78**) with a much bulkier heterocycle replacing the biphenyl tetrazole. Losartan is active and has an active metabolite, whereas TCV-116 is a true prodrug itself and must be converted to the free acid to form CV-11974 (196).

Since AT$_1$-selective antagonists raise Ang II levels and leave AT$_2$ receptors unblocked, the search for an inhibitor of both receptors has been undertaken. BIBS39 and BIBS22 are compounds which have weak but significant affinity for the AT$_2$ site in addition to their affinity for AT$_1$ sites (197). XM953 (**80**) represents a new series of molecules with much increased affinity for the AT$_2$ site in which

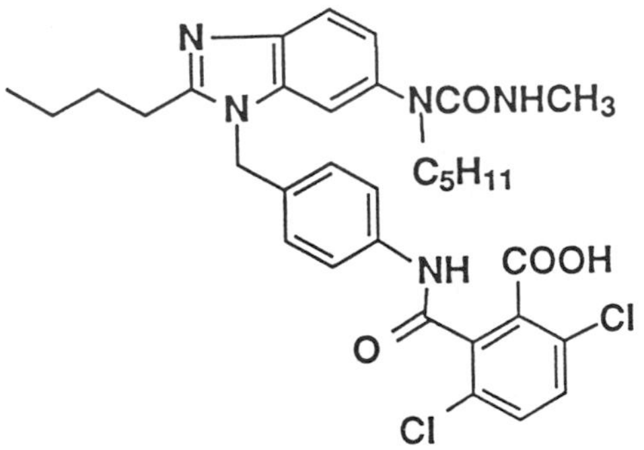

BIBS22 (**79**)

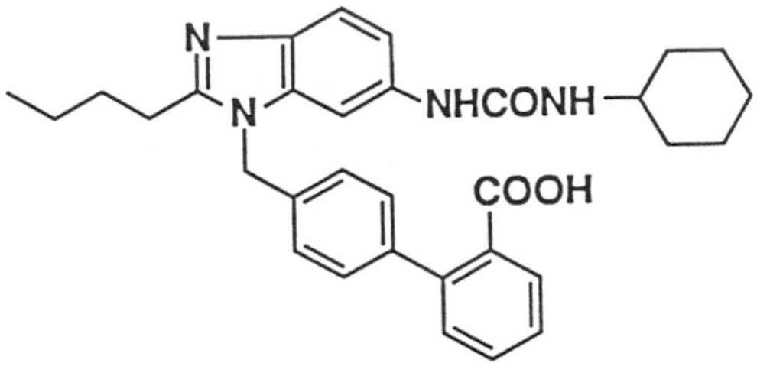

BIBS39 (**81**)

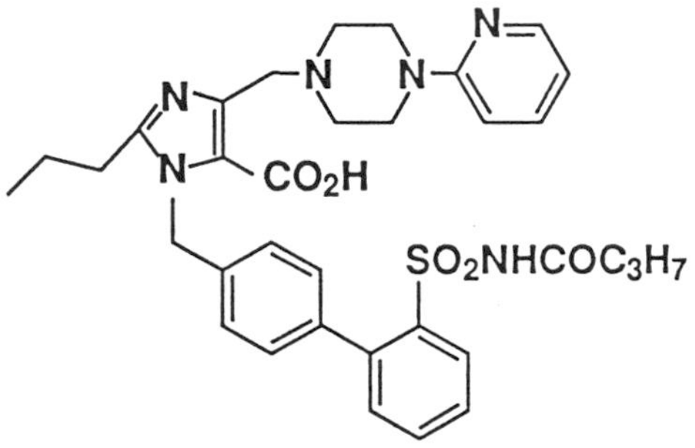

XM953 (**80**)

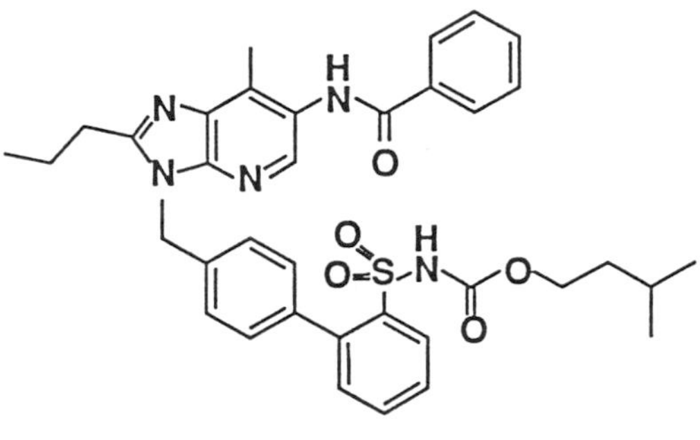

L-163,017 (**82**)

the tetrazole is replaced by a hexanoyl sulfonamide. The concentration of XM953 to inhibit AT_2 binding is 20 nM and the AT_1/AT_2 IC_{50} ratio is 3 (198). A "balanced affinity" for the two subtypes has been reported for L-163,017 (199). The blood pressure lowering effects of L-163,017 (**82**) in renal hypertensive rats and in sodium-deplete rhesus monkeys is similar to that seen with AT_1-selective agents (200).

6.3.2 POTASSIUM CHANNEL OPENERS. Potassium channel openers (KCOs) are a new class of therapeutic agents that have displayed therapeutic potential as antihypertensive, anti-ischemics, and anti-asthma agents (83,201). Advances in electrophysiology and molecular biology has facilitated the characterization of novel K^+ channels and defined the mechanism of

action of a number of "known" vasodilators such as diazoxide and minoxidil and of new vasodilators pinacidil and cromakalim.

Pinacidil was chosen from a series of N-alkyl-N''-pyridyl guanidines for its direct relaxant action on vascular smooth muscle (202). The KCO activity of pinacidil resides in the $(-, R)$ enantiomer. Non-KCO relaxant effects are also seen at higher concentrations which are not stereoselective (203). Cromakalim (BRL34915) was chosen from a novel series of 4-(cyclic amide)-2H-1-benzopyran-3-ols represents the focus of most synthetic activity in this area (201). The activity of cromakalim resides mainly in the $(-)$-enantiomer. As expected, cromakalim has demonstrated potent blood pressure lowering effects in experimental animals and in humans (204). As with other potent vasodilators, reflex tachycardia, increases in norepinephrine and renin, and fluid retention are common with the acute use of cromakalim. Controversy remains about the individual K^+ channels being affected by the KCOs including cromakalin (**84**) and the advantages of KCO, if any, compared to existing selective or nonselective vasodilators (83,201).

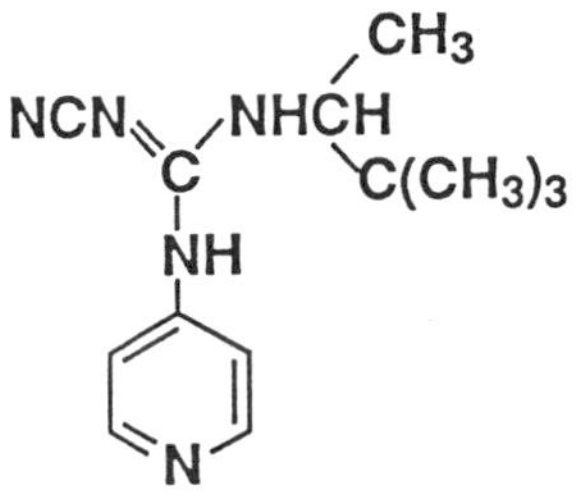

Pinacidil (**83**)

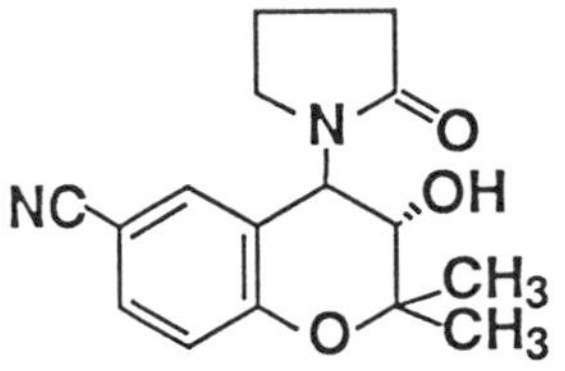

Cromakalim (**84**)

6.3.3 RENIN INHIBITORS. Renin inhibitors block the renin angiotensin systems (Fig. 29.7) by blocking the synthesis of angiotensin I from angiotensinogen. This rate limiting step in the synthesis of angiotensin II has been the focus of new drug discovery for 30 years. Extensive use of molecular modeling and active site analysis has been applied to design small molecular weight inhibitors (205). An international effort has yielded a very potent and specific inhibitors (206,207). Enalkiren (**85**) and Ro 42-5892 (**87**) have been most widely studied in humans (208,209). Enalkiren administered by i.v. infusion appears to lower blood pressure equivalent to ACE inhibitors but poor oral bioavailability limits its clinical usefulness. Although initial studies with i.v. Ro 42-5892 showed the expected reductions in blood pressure, subsequent studies with oral and i.v. doses up to 600 mg have failed to demonstrate significant reductions in blood pressure (208,210). The explanation of this poor efficacy is unexplained as animal studies predict antihypertensive ef-

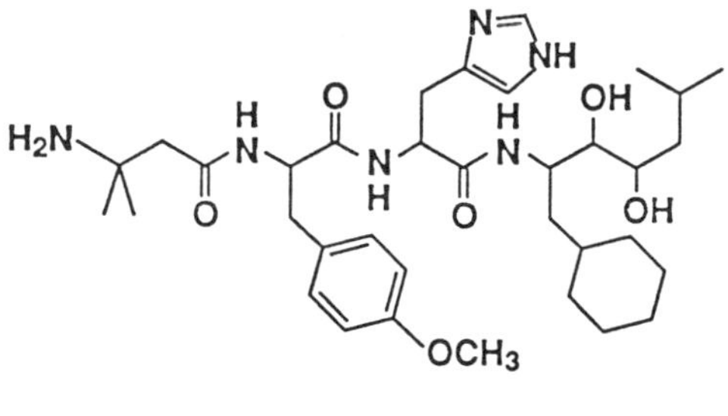

Enalkiren (**85**)

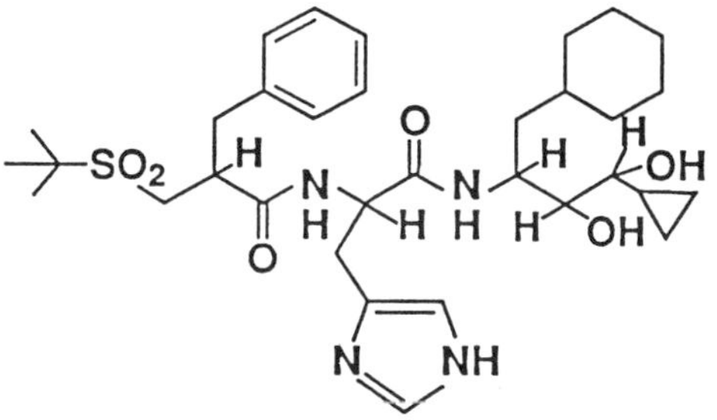

Ro 42-5892 (**87**)

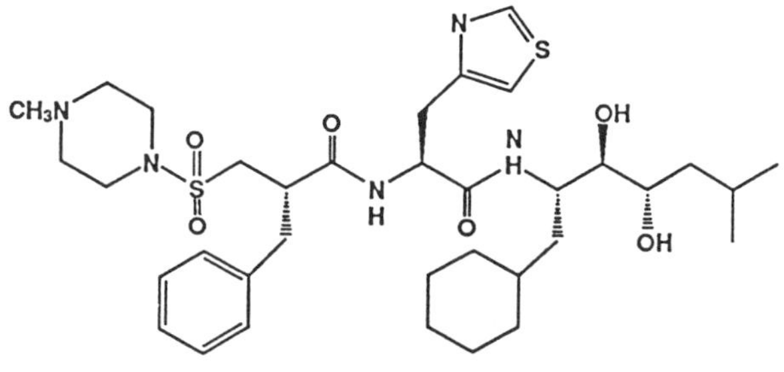

Zankiren (A-72517) (**86**)

fects comparable to other RAS inhibitors. Zankiren (A-72517) is a new compound which appears to have both oral bioavailability and blood pressure lowering effects at least in sodium deplete hypertensive patients (211). This compound also does not produce the cough associated with ACE inhibition (212).

6.3.4 SEROTININ RECEPTOR (5HT$_2$) ANTAGONISTS. Ketanserin is the prototype of series of quinazolinediones shown to have specific 5-HT$_2$-serotonergic blocking activity (213). At higher concentrations, ketanserin also inhibits α_1 adrenergic receptors and histamine receptors. Ketanserin is an effective antihypertensive in experimental animals and in humans. The mechanism of this hypotensive action cannot be explained by 5-HT$_2$-serotonergic or blockade of α_1 receptors alone (213). Interestingly, ketanserin appears to be more effective and well tolerated in elderly patients (214).

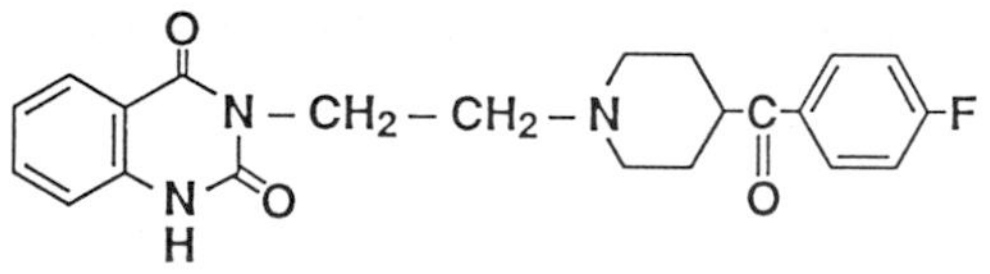

Ketanserin (**88**)

6.3.5 NEUTRAL ENDOPEPTIDASE (NEP) INHIBITORS. Atrial natriuretic factors (ANF) are endogenous peptides which have potent vasodilator and diuretic effects. The role of ANF in the control of fluid volume and electrolyte homeostasis has been extensively studied (215,216). ANF has been shown to have potential beneficial hemodynamic and renal actions in hypertension, congestive heart failure, and renal failure but its usefulness is limited by its peptide nature requiring constant i.v. infusion. An alternative approach is to inhibit the degradation of the endogenous ANF. UK69,578 (**89**) and Sch 34826 (**90**) represents a prototype of series of nonpeptide orally active compounds which inhibit NEP,

potentiate the effects of ANF, and demonstrate antihypertensive effects in experimental animals (217–219).

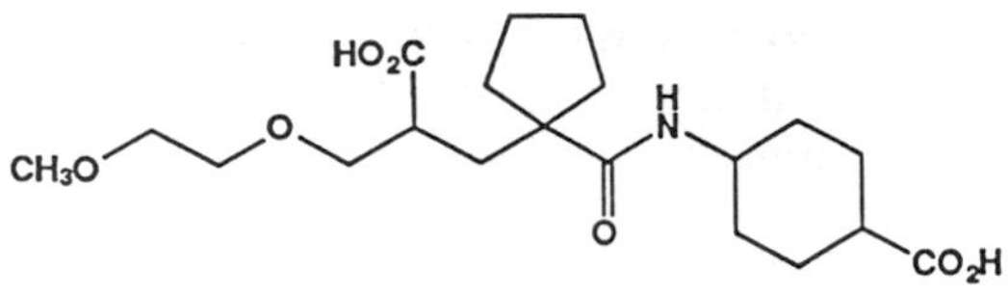

UK69,578 (**89**)

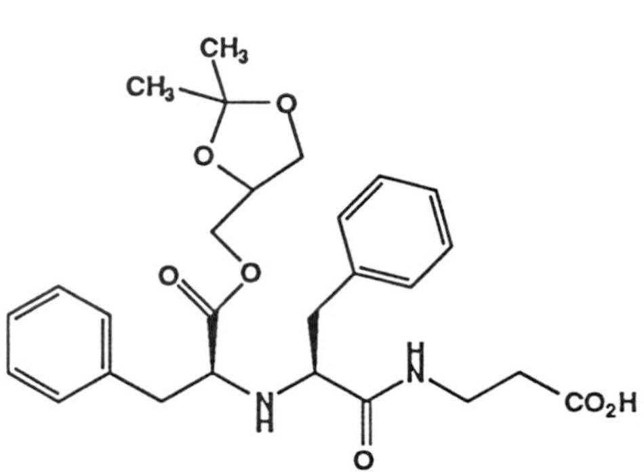

Sch 34826 (**90**)

Recently, "dipharmacophores" have been synthesized which inhibit both ACE and NEP (220). The lead compound, mixanpril, is a lypophillic prodrug with oral antihypertensive activity in conscious SHR.

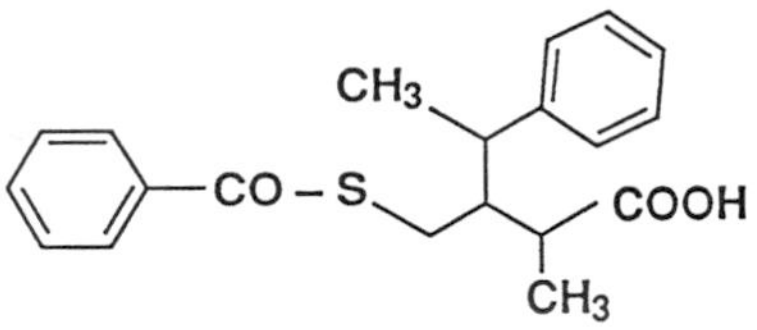

Mixanpril (**91**)

6.3.6 ADENOSINE RECEPTOR (A2) ANTAGONISTS. Adenosine can exert marked systemic cardiovascular effects by actions on A1 receptors in the cardiac sinoatrial or atrioventricular nodes, sympathetic nerves, or kidney and/or A2 receptors in vasculature or brain stem. The net effect of exogenously administered adenosine is bradycardia and inhibition of norepinephrine and renin release (221,222). A_1-selective agonists lower blood pressure and markedly reduce heart rate in normotensive and hypertensive rats (223). A2-selective antagonists such as CGS-21680 have been shown to lower blood pressure accompanied by reflex increases in heart rate and plasma renin activity (222). CGS21680 (**92**) is an adenosine agonist with high affinity for brain striatal adenosine (A2) receptors ($IC_{50} = 22 \, n$M). CGS21680 lowers blood pressure in SHR following oral administration. The prolonged reduction in blood pressure (up to 24 hours) is accompanied by a transient (less than 60 min) increase in heart rate (224). Chronic infusion of this compound (0.25 or 0.5 μg/kg/min i.v. for 2 weeks) led to the development of tolerance in SHR and, after 2 weeks, no significant blood pressure lowering effects were observed. Down regulation of A2 receptor or compensatory activation of the RAS are possible explanations (225). Adenosine receptors have been therapeutic targets not only in hypertension but in cardiac ischemia and cognition disorder (222). A2 agonists acting like nonselective vasodilators have the greatest blood pressure lowering effects. Whether the reflex increase in heart rate and PRA and/or development of tolerance are characteristic of CGS21680 or of the class of A2 agonists remains to be determined. Several chemical approaches have tried to mimic adenosine (226–228) but their viability as antihypertensive agents has not been established.

CGS-21680 (**92**)

6.3.7 VASOPRESSIN RECEPTOR (V_1) ANTAGONISTS. Although the role of vasopressin in the etiology of hypertension has not been established (229), the search for specific antagonists continues. OPC-21268 is a nonpeptide molecule that has been shown to competitively inhibit V_1 binding and vasopressin stimulated Ca^{2+} efflux (230) *in vitro* and block arginine vasopressin-induced vasoconstriction *in vivo* in rats (231). OPC-21268 at 0.3 to 3.0 mg/kg i.v. lowered blood pressure in SHR 25 weeks of age but not in SHR 15 weeks of age. In the malignant state of hypertension in SHR-stroke prone, the plasma vasopressin concentrations are elevated. In these animals, OPC-21268 also significantly reduced pressure (32 ± 8 mm at 3 mg/kg i.v.) (232). It remains to be established if this compound will lower blood pressure in essential hypertension in humans.

OPC21268 (**93**)

6.3.8 ENDOTHELIN RECEPTOR ANTAGONISTS. The "endothelin" family of peptides exert complex actions on vascular smooth muscle but their role in the pathophysiology of hypertension remains controversial (233). ET-1 is a more potent vasoconstrictor than angiotensin II and has high affinity for the ET_A receptor subtype which mediates the vasoconstriction characteristic of the endothelins. The ET_B receptor mediate vasodilation and antiplatelet effects (234).

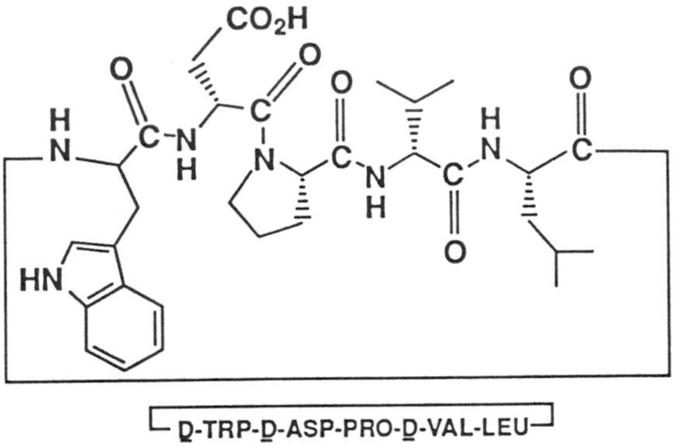

BQ123 (**94**)

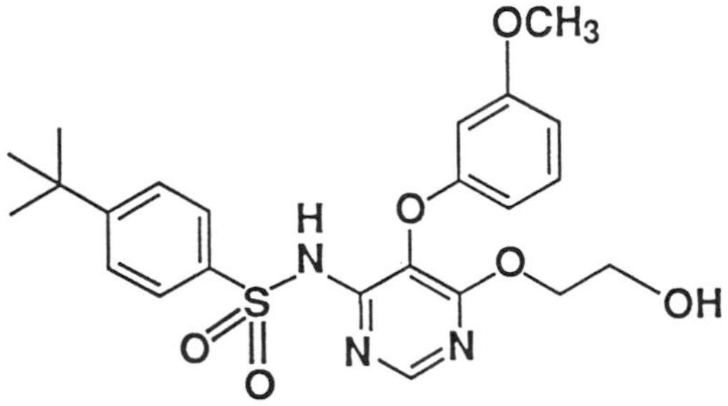

Ro 46-2005 (**95**)

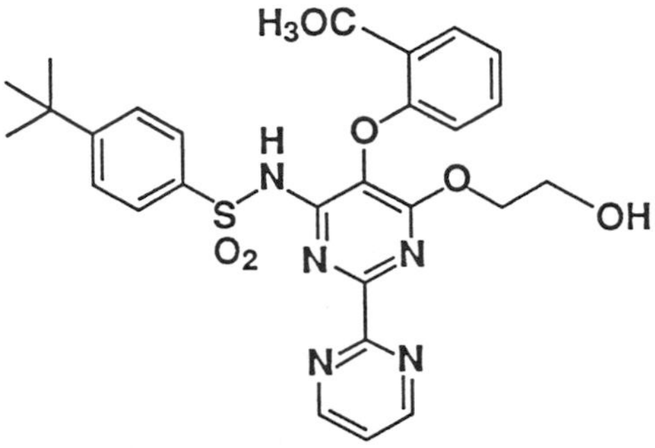

Ro 47-0203 (Bosentan) (**96**)

Chemical strategies involving linear peptide analogs, cyclic peptides (e.g., BQ-123), diphenyl ethers (discovered from screening fungal broths), and a number of nonpeptide series have been reported (see review in Ref. 235). Ro 46-2005 (**95**), a N-4-pyrimidinyl-benzene sulfonamide was introduced as the first nonpeptide ET_1 antagonists (nonselective for ET_A and ET_B) (**96**) (236). A related compound, Ro 47-0203, has recently been shown to have oral antihypertensive and antihypertrophic effects in deoxycorticosterone acetate-salt hypertensive rats (237).

6.3.9 NEUROPEPTIDE Y RECEPTOR ANTAGONIST. Atrinositol [PP56; D-myo-inositol 1,2,6-tris(dihydrogen phosphate)] represents one of the first molecules with selectivity for neuropeptide Y (NPY) receptors

(238). NPY is a 36-amino acid peptide and a potent vasoconstrictor hormone which may modulate sympathetic function. Elevated levels of NPY in hypertensive animals suggest that NPY antagonists may be efficacious as antihypertensives (see review in Ref. 238). Atrinositol has been shown to selectively inhibit the vasoconstrictor produced by NPY in peripheral arteries of rabbit and guinea pig without affecting the vasoconstrictor response to norepinephrine (239). However, more recent data suggest that atrinositol may act distal to the NPY receptor (240). Further, atrinosotol lowers basal blood pressure only in a dose range above that for specific NPY blockade (241). In a pilot study of hypertensives and healthy volunteers, atrinositol was infused at rest, during and after a maximal exercise test in a double-blind manner. No significant effects on pressure were noted. There was a trend toward a reduction in the maximum systolic pressure following exercise in the hypertensive patients (242). Current data with atrinositol do not appear to justify further development as an antihypertensive agent. Until a more potent and selective compound is available, however, the possibility of NPY antagonists as antihypertensive agents cannot be ruled out.

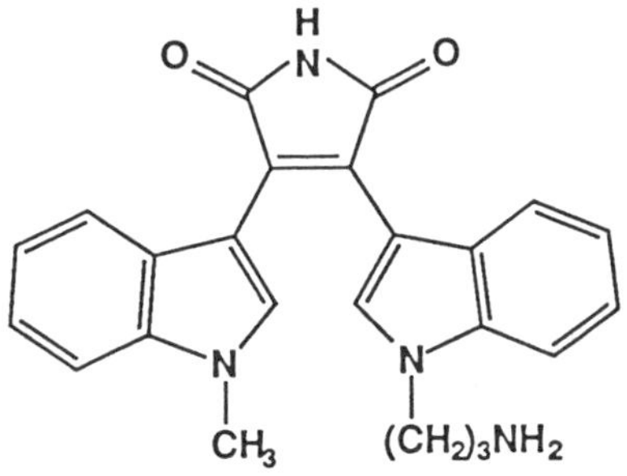

Ro 31-7549 (**98**)

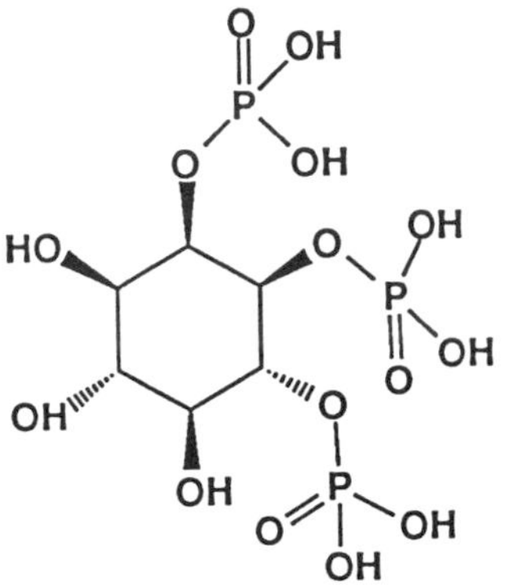

Atrinositol (PP56) (**97**)

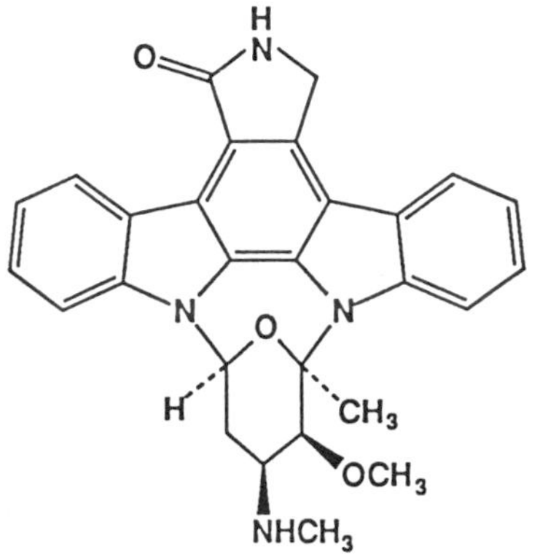

Staurosporine (**99**)

6.3.10 PROTEIN KINASE C INHIBITORS. Ro 31-7549 (**98**) is one of a series of a protein kinase C (PKC) inhibitors which may have therapeutic potential in hypertension, cancer, asthma, and inflammatory diseases (243,244). Protein kinase C consists of a family of closely rated enzymes (245) which are activated by both IgE-dependent and IgE-independent means (246). PKCs, by modulating the release of vasoactive substances such as leukotriene, ATO, and histamine could play a role in the etiology or expression of hypertension. Although the antiinflammatory potential of Ro 31-7549 has been shown, the antihypertensive effects of this compound have not yet been demonstrated. Staurosporine (**99**) has been widely used as an experimental tool to inhibit PKC but has not demonstrated therapeutic usefulness.

7 NOVEL APPROACHES TO THE FUTURE PHARMACOLOGICAL TREATMENT OF HYPERTENSION

7.1 Novel "Mechanistic" Approches

Understanding the etiology of hypertension is the first step of rational drug design. Current antihypertensive drugs lower blood pressure by altering the basic control mechanisms at the organ level (Fig. 29.1) or by altering the renal tubule or vascular smooth muscle function at the "cellular" level. The innovative approaches (Table 29.12) will likely provide additional probes into the basic mechanism(s) by which the genetic phenotype is expressed. Is it unlikely, however, that these known mechanisms will provide the "cure" of hypertension. If that is the case, toward what should the medicinal chemistry of the future be directed?

Total scans of the human genome are on the horizon and the association of hypertension with specific genes will become a reality. The proteins encoded by these genes will be interesting targets for the search for a cure for hypertension (247,248). What those targets are likely to be is conjecture at this point in time. Will we want to alter the genome itself or only the factor(s) that regulator transcription? The challenge will be to get drugs across several membranes into the nucleus compartment (Fig. 29.8). The glucocorticoids (steroid receptor) provides the model for the expectation that drugs can act at intranuclear sites (249).

Novel approches to antihypertensive drugs may also evolve from ongoing research in (1) calcium regulating hormones (250), (2) nitric oxide (234,251), (3) growth factors (252,253), (4) eicosanoid biosynthesis (254–256), and (5) vascular smooth muscle ion channels (257). It has been proposed that all forms of hypertension are associated with cytosolic-free calcium excess, intracellular magnesium depletion, or both. Parathyroid hormone, calcitonin, and vitamin D metabolism may all represent "mechanistic" targets (250).

The endothelium as a target and mediator of cardiovascular diseases including hypertension has been the focus of much research and debate (251,258). "Endothelin" was discussed above. The endothelium-derived relaxing factor nitric oxide is a potent vasodilator and is involved in modulator of vascular tone. The limitation of current nitrovasodilators, the other potentially cytotoxic effects of nitric oxide, and the limited potential of inhibitors of nitric oxide synthesis (which elevate blood pressure) have led to the conclusion that altering nitric oxide homeostasis is unlikely to provide therapeutic benefit (259). It is likely, however, that additional local factors will be discovered to keep the endothelium an important source for new potential mechanistic targets for antihypertensive drugs.

Growth factors as mechanistic targets have evolved from an increased knowledge of how these factors are regulated and the tissue protective effect of certain classes of antihypertensive drugs (252). Growth fac-

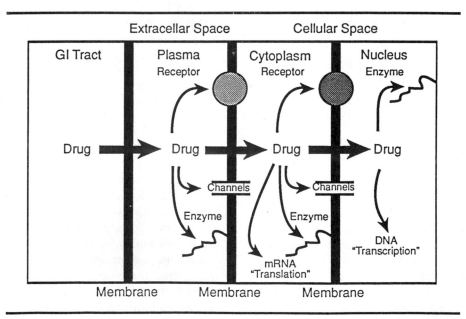

Fig. 29.8 Intranuclear new drug discovery targets require the passage of drugs across multiple cellular membranes.

tors such as platelet derived (PDGF), basic fibroblast (bFGF), insulin (IGF-1), and tumor necrosis (TNF-α) growth factors act on and are released by arterial smooth muscle cells, endothelial cells, macrophages, and T-lymphocytes. It is clear that lowering pressure is tissue protective, but it may be that bockade of growth factor release may provide enhanced tissue protection (252). A hypothesis has been advanced to suggest many new targets by which Ang II and other vasoactive substances not only play a role in blood pressure regulation but also participate in the remodeling of the arterial wall in hypertension by inducing changes in both smooth muscle cell phenotype and growth (253).

Eicosanoids are a family of natural hormones locally released from arachidonic acid (255). These hormones include prostacyclin, PGE_2 and thromboxane A_2, which can exert powerful effects directly on vascular and nonvascular smooth muscle and likely have modulatory roles on local blood flow/pressure relationships. Current research is exploring cytochrome P-450 dependent metabolites of archidonic acid and their relation to hypertension (254) and may provide new mechanistic targets.

Intracellular calcium regulation is accomplished by plasma membrane, mitochondria, and sarcoplasmic reticulum membrane sodium-calcium exchange or calcium transporter systems (260). As the studies of these transporters becomes known and the "hypertensive phenotype" identified, it may be possible to design specific drugs to reverse or modify the abnormal function. The obvious challenge will be to differentiate such compounds from the well-known calcium antagonist which act at the calcium voltage-dependent channel (see above).

7.2 Antisense Molecules

Many of the receptors, enzymes, and circulating hormones involved in blood pressure regulation and presumably in the expres-

sion (and possibly the cause) of hypertension are peptides. Each protein is coded by a sequence of pyrimide bases on the DNA structure which are first transcribed to complementary RNA and then translated into protein. The concept of "antisense" is to inhibit gene expression by introducing small, synthetic oligonucleotides which resemble single-strand DNA. In principle, the synthetic oligonucleotides bind to the DNA or RNA and render DNA incompetent to transcribe, or the mRNA to translate, the message and gene expression is inhibited. Although tremendous progress has been made in the application of antisense oligonucleotides to inhibit, biological system "chemical problems" remain such as stability and affinity of the oligonucleotide. In addition, problems remain concerning how to get the molecules into the nucleus (261). Most of the work to date has been modification of the phosphodiester backbone of DNA (**100**) including peptide nucleic acids which replace the entire ribophosphodiester backbone (262). Other modifications, e.g., replacing the natural D-2'-deoxyribose with the (L-) sugar to enhance resistance to nuclease degradation.

X = O, phosphodiester
X = S, phosphorothioate
B = Base (adenosine, thymidine, guanosine, cytidine)

Phosphodiester, phosphorothioate (**100**)

Peptide nucleic acid dimer (**101**)

Modification of the heterocyclic bases, e.g., substitution of the 5-methyl-2′-deoxycytidine (5-Me dc) and 5-bromo-2′-deoxyeytidine for 2′-deoxycytidine enhanced affinity to the target RNA (263).

Antisense to the angiotensin II-AT$_1$ receptor mRNA has been shown to block the drinking resonse to Ang II and down regulated Ang II receptor. The oligonucleotides administered directly into the brain were found to be rapidly taken up near the site of injection and to be inactive when administered i.v. (264,265).

The use of antisense oligonucleotides to inhibit proteins critical to the expression of hypertension is in its infancy but holds great promise in further understanding the etiology of hypertension. The design of antisense molecules with 3-D structural requirements has required rapid advances in chemical synthesis and the biophysical characterization of oligonucleotides (261). Whether these oligonucleotides will ultimately be replaced by small molecular mimics of the critical binding sites remains to be determined.

7.3 Experimental Models of Genetic Hypertension

Animal models of hypertension have played an important role over the years in selection of clinical candidates. Most early studies utilized anesthetized normotensive or renal hypertensive animals (rats or dogs) (181). The introduction of the spontaneously hypertensive rat (81) greatly facilitated the empirical screening of large libraries of fine chemicals. The advantage of the SHR was that it more closely mimicked human essential hypertension. The SHR has been followed by the development of Dahl, Lyon, and Milan genetically hypertensive strains. A review of the cellular membrane characteristics, renal, and neuronal mechanism, and the humoral backgrounds of these different strains concluded that although similarities exist none of these strains closely mimics man as we now understand the human essential hypertensive phenotype (266). It should be noted, however, that, although the SHR and other genetically hypertensive rats are now rarely used as a primary screen, they are used to verify that a new molecule active in an *in vitro* mechanistic screen does in fact lower blood pressure.

Linkage studies in segregating populations of offspring from the cross of hypertensive (SHR) and normotensive have been utilized to define the chromosome regions containing the genes responsible for hypertension (267,268). Once these genes are identified for man, it may be possible to create a genetic strains of rats that have hypertension because of an overexpression of the causative protein. Antihypertensive drug discovery targeted at those proteins might yield a new classes of therapeutic agent. The transgenic transfer of specific genes to alter the phenotype of the animal has already been accomplished. The (mREN2)27 transgenic rat, for example, containing the mouse REN2 gene has a renin-dependent hypertension. Unlike essential hypertension in man or SHR, however, this transgenic rat has enhanced production of adrenal renin production. Even though such problems with transgenics are common, animal models with specifically designed nucleotide alterations may provide a means by which to evaluate both the mechanistic target and potential new classes of antihypertensive drugs (267). With such targets, the chemistry will likely be quite different from that currently applied for current antihypertensive agents.

8 PERSPECTIVES FOR THE FUTURE OF MEDICINAL CHEMICAL APPROACHES TO THE TREATMENT OF HYPERTENSION OR "DO WE HAVE ALL THE ANTIHYPERTENSIVE DRUGS WE NEED AND, IF NOT, CAN WE AFFORD THE COSTS OF THE NEW SCIENCE?"

Medicinal chemists have provided the world with an impressive array of blood-pressure lowering substances. These molecules are widely used to treat cardiovascular disease and now generate several billions of dollars per year in sales. It seems clear that such an investment is appropriate in view of the lives saved. Even in the new health economic/cost containment climate, it is likely that more people will be diagnosed and treated for hypertension and that in the future more people will be started earlier and maintained on antihypertensive drug treatment for a longer time. If the medical need and the market are large and growing, it should follow that the search for even more effective, better tolerated drugs should also continue. The pharmaceutical industrial trend, however, is to reduce the investment in cardiovascular new drug discovery especially in antihypertensive drugs.

There are many reasons why investment in antihypertensive new drug discovery is being dramatically reduced. The availability of many effective drugs which, for the most part, are well tolerated, the availability of inexpensive drugs such as the diuretics and beta blockers, and finally, the most important consideration, the paucity of viable

new mechanistic targets. We now know that blood pressure is only one parameter of the hypertensive syndrome. Renal, cardiac, adrenal, cerebral, and vascular changes may both contribute to or result from the hypertension itself or from the associated circulating hormones, e.g., Ang II. We know that antihypertensives, e.g., vasodilators, for example, are not alike and that ACE inhibitors (and presumably Ang II receptor blockers) have quite different effects on cardiac hypertrophy or on renal function than nonspecific vasodilators. We know that we can block anywhere in the catecholamine synthesis-release-receptor cascade or the angiotensin II synthesis-release-receptor cascade and effectively lower blood pressure. We know that we can block any number of membrane, cellular, or intracellular events to relax vascular smooth muscle and effectively lower blood pressure. We can conservatively control 50–70% of hypertension with existing drugs or combinations of drugs. What's left do do?

The "new" science of atrial natriuretic factor (215), endothelin receptors (174,259), like neuropeptide Y (269), nitric oxide homeostasis (234), urodilan (270), or circulating ouabain-like factors (271) has not yet yielded useful new antihypertensive drugs. It is likely that there are many other "factors" or mechanisms of blood pressure that we do not yet know. Until these factors are clearly identified, however, new drug discovery will be slowed. We may have reached the end of the "physiological era" of our understanding of blood pressure control. For this level of understanding, we have the "nearly ideal" drugs (see Section 5.2). The future requires that we move to the molecular biology or "genome era" which will define new ways to look at how the hypertension syndrome is controlled, not just how high blood pressure is controlled. The early results with antisense, e.g., to the Ang II receptor, are not impressive but indicate what may be possible. Whether it is antisense or gene transfer, it is clear that the paradigm shift of antihypertensive therapy will ultimately focus on the human genome. The chemistry required will likely be a hybrid between classical organic medicinal chemistry and nucleotide chemistry. The breakthroughs of that future will come when small molecule non-nucleotides (analogous to current non-peptides) are discovered that accomplish the desired genome or "message" modulatory effects. Many of the drug discovery effects of the last 30 years have struggled with the bioavailability of drugs to the appropriate target cells. The drug discovery programs of the future will likely struggle with the bioavailability of drugs to the target cells of the nucleolar genomic library (Fig. 29.8).

Can society afford the costs of the new science necessary to move from the inexpensive drugs of the physiologic era to those of the genomic era? The answer is probably yes, if the free market is allowed to work. We know that when the pharmaceutical industry withdrew from anticonvulsant new drug discovery in the 1960's, no new anticonvulsant drugs emerged until the pharmaceutical industry got back into research. The same thing would likely happen with antihypertensive new drug discovery. When there are plausible mechanistic approaches to treating the now broader definition of hypertension, then medicinal chemistry will provide the appropriate molecules. The costs are only an issue during a period, like the present, when the science and technology are emerging and no clear cut "plausible" mechanistic approach can be identified. To prepare for this future for the genome era, the chemistry of peptide and nucleotide mimicry and of amide bond and phosphodiester backbone replacements will have to reach new heights.

References

1. D. Lyons, J. C. Petrie and J. L. Reid, *Brit. Med. Bull.*, **50**(2), 472 (1994).

2. S. MacMahon and A. Rodgers, *Cerebrovasc. Dis.*, **4**, 11 (1994).

3. V. J. Dzau, M. Mukoyama and R. E. Pratt, *J. Hypertens.*, **12**(Suppl 2), S1 (1994).

4. J. S. Kiely, "Recent advances in antisense technology", in Venuti, Ed., *Annuals Reports in Medicinal Chemistry*, 29th ed., Academic Press, Inc., 1994, pp. 297–308.

5. T. G. Pickering, *Am. J. Cardiol.*, **58**, 12D (1986).

6. J. G. G. Borst, *Lancet*, 677 (**March** 1963).

7. I. H. Page, *J. Am. Med. Ass.*, **140**, 451 (1949).

8. M. C. Khosla, I. H. Page and F. M. Bumpus, *Biochem, Pharmacol.*, **28**, 2867 (1979).

9. A. C. Guyton, *Science*, **252**, 1813 (1991).

10. J. E. Hall, A. C. Guyton and M. W. Brands, "Control of Sodium Excretion and Arterial Pressure by Intrarenal Mechanisms and the Renin Angiotensin System", in 2nd Ed., J. H. Laragh and B. M. Brenner, Eds., *Hypertension: Pathophysiology, Diagnosis, and Management*, Raven Press, New York, 1995, pp. 1451–1475.

11. J. E. Hall, *Clin. Cardiol.*, **14**(Suppl IV), 6 (1991).

12. K. D. Burns, T. Homma and R. C. Harris, *Seminars Nephrol.*, **13**(1), 13 (1993).

13. A. C. Shore and J. E. Tooke, *J. Hypertens.*, **12**, 717 (1994).

14. R. Ward, "Familial Aggregation and Genetic Epidemiology of Blood Pressure", in J. H. Laragh and B. M. Brenner, Eds., *Hypertension: Pathophysiology, Diagnosis and Management*, Raven Press, New York, 1990, pp. 81–100.

15. J. H. Laragh, *Am. J. Med.*, **55**, 261 (1973).

16. J. H. Laragh, *Arzneim. -Forsch. -Drug Res.*, **43**(1), 247 (1993).

17. M. H. Alderman, S. Madhovan, W. L. Ooi, H. Cohen, J. E. Sealey and J. H. Laragh, *N. Engl. J. Med.*, **324**, 1098 (1991).

18. M. C. Houston, *Am. Heart J.*, **123**, 1337 (1992).

19. J. D. Swales, "The Renin Angiotensin System in Essential Hypertension", in J. I. S. Robertson and M. G. Nicholls, Eds., *The Renin Angiotensin System*, Gower Medical Publishing, London, 1993, pp. 62.1–62.12.

20. Guidelines Subcommittee of the WHO/ISH Mild Hypertension Liaison Committee, *Hypertension*, **22**(3), 392 (1993).

21. American Heart Association, *Heart and Stroke Facts*, **Statistical Suppl**, 12 (1995).

22. W. B. Kannel, *J. Cardiovasc. Pharmacol.*, **13**(Suppl 1), S4 (1989).

23. S. MacMahon, R. Peto, J. Culter, R. Collins, P. Sorlie, J. Neaton, R. Abbott, J. Godwin, A. Dyer and J. Stamler, *Lancet*, **335**, 765 (1990).

24. T. J. Thom, F. H. Epstein, J. J. Feldmar, P. E. Leaverton and M. Wolz, *NIH*, (1992).

25. R. Collins, R. Peto, S. MacMahon, P. Hebert, N. H. Fiebach, K. A. Eberlein, J. Godwin, N. Qizilbash, J. O. Taylor and C. H. Hennekens, *Lancet*, **335**, 827 (1990).

26. W. C. Roberts, *Am. J. Cardiol.*, **60**, 1E (1987).

27. P. K. Whelton, *Lancet*, **344**, 101 (1994).

28. J. D. Swales, *J. Hypertens.*, **11**, 899 (1993).

29. J. Menard, M. Day, G. Chatellier and J. H. Laragh, *Am. J. Hypertens.*, **5**, 325 (1992).

30. M. W. Millar-Craig, C. N. Bishop and E. B. Raftery, *Lancet*, 795 (**April 15**, 1978).

31. A. Coca, *J. Hypertens.*, **12**(Suppl 5), S13 (1994).

32. G. Parati, A. Ravogli, A. Frattola, A. Gropelli, L. Ulian, C. Santucciu and G. Mancia, *J. Hypertens.*, **12**(Suppl 5), S35 (1994).

33. W. G. White, *Clin. Pharmacol. Ther.*, **45**(6), 581 (1989).

34. P. Verdecchia, C. Porcellati, G. Schillaci, C. Borgioni, A. Ciucci, M. Battistelli, M. Guerrieri, C. Gatteschi, I. Zampi, A. Santucci, C. Santucci and G. Reboldi, *Hypertension*, **24** 793 (1994).

35. Evaluation and Treatment of High Blood Pressure Joint National Committee on Detection, *Arch. Intern. Med.*, **153**, 154 (1993).

36. Guidelines Subcommittee, *J. Hypertens.*, **11**, 905 (1993).

37. E. A. Lew, *Am. J. Med.*, **55**, 281 (1973).

38. M. Hamilton, E. N. Thompson and T. K. M. Wisniewski, *Lancet*, **1**, 235 (1964).

39. Veterans Administration Cooperative Study Group on Antihypertensive Agents, *JAMA*, **213**, 1143 (1970).

40. Report by the Management Committee, *Lancet*, **1**, 1261 (1980).

41. Medical Research Council Working Party, *Brit. Med. J.*, **291**, 97 (1985).

42. The IPPPSH Collaborative Group, *J. Hypertens.*, **3**, 379 (1985).

43. J. D. Neaton, R. H. Grimm, R. J. Prineas, J. Stamler, G. A. Grandits, P. J. Elmer, J. A. Cutler, J. M. Flack, J. A. Schoenberger, R. McDonald, C. E. Lewis and P. R. Liebson, *JAMA*, **270**, 713 (1993).

44. G. W. Pickering, *High Blood Pressure*, Churchill, London, 1955, pp. 1–246.

45. L. Hansson, *Am. J. Hypertens.*, **1**, 414 (1988).

46. A. L. Dannenberg, T. Drizd, M. J. Horan, S. G. Haynes and P. E. Leaverton, *Hypertension*, **10**, 226 (1987).

47. M. J. Klag, P. K. Whelton and A. J. Seidler, *Stroke*, **20**, 14(1989).

48. N. M. Kaplan, M. H. Alderman, W. Flamenbaum, D. A. McCarron, H. M. Perry, E. Saunders and J. A. Schoenberger, *Am. J. Hypertens.*, **2**, 75 (1989).

49. SHEP Cooperative Research Group, *JAMA*, **255**, 3255 (1991).

50. J. Menard, *Am. J. Hypertens.*, **5**, 252S (1992).

51. R. C. Tarazi and F. M. Fouard-Tarazi, *Am. J. Cardiol.*, **58**, 3D (1986).

52. M. A. Weber, *Am. J. Cardiol.*, **72**, 3H (1993).

53. J. M. McKenney, *Drug Intel. Clin. Pharm.*, **19**, 629 (1985).

54. S. D. Paul, K. M. Kuntz, K. A. Eagle and M. C. Weinstein, *Arch.. Intern. Med.*, **154**, 1143 (1994).

55. B. Dahlof, *Cardiology*, **81**, 307 (1992).

56. N. K. Hollenberg, *Am. Heart J.*, **125**(2 Pt 2), 604 (1993).

57. M. Manttari, L. Tenkanen, V. Manninen, T. Alikoski and M. H. Frick, *Hypertension*, **25**, 47 (1995).

58. M. J. Brown, *Lancet*, **342**, 1374 (1993).

59. A. V. Chobanian, *Am. J. Cardiol.*, **59**(13), 1F (1987).

60. F. Gross, *Triangle*, **23**(1), 25 (1984).

61. M. Nickerson and B. Collier, "Drugs Inhibiting Adrenergic Nerves and Structures innervated by them", in 5th ed., L.S. Goodman and A. G. Gilman, Eds., *The Pharmacological Basis of Therapeutics*, MacMillan Publishing Co., Inc., New York, 1975, pp. 533–564.

62. P. B. M. W. M. Timmermans, P. C. Wong, A. T. Chiu, W. F. Herblin, P. Benfield, D. J. Carini, R. J. Lee, R. R. Wexler, J. M. Saye and R. D. Smith, *Pharmacological Review*, **45**(2), 205 (1993).

63. E. D. Freis, "Historical Development of Antihypertensive Treatment", in 2nd Ed., J. H. Laragh and B. M. Brenner, Eds., *Hypertension: Pathopysiology, Diagnosis, and Management*, Raven Press, New York, 1995, pp. 2741–2751.

64. E. D. Freis and R. W. Wilkins, *Proc. Soc. Exp. Biol. Med.*, **64**, 731 (1947).

65. A. F. Green, "Antihypertensive Drugs", in S. Garattini and P. A. Shore, Eds., *Advances in Pharmacology*, Academic Press, New York, 1962, pp. 162–225.

66. J. H. Moyer, *Arch. Intern. Med.*, **91**, 419 (1953).

67. H. J. Bein, *Pharmacol. Rev.*, **8**, 435 (1956).

68. K. H. Beyer and J. E. Baer, *Pharmacol. Rev.*, **13**, 518 (1961).

69. J. D. Horisberger and G. Giebisch, *Renal Physiol.*, **10**, 198 (1987).

70. C. A. Stone and C. C. Porter, "Biochemistry and Pharmacology of Methyldopa and some Related Structures", in N. J. Harper and A. B. Simmonds, Eds., in *Advances in Drug Research*, Academic Press, London, 1967, pp. 71–93.

71. H. Haas and E. Busch, *Arzneim. -Forsch. -Drug Res.*, **17**, 257 (1967).

72. T. Nagao, H. Narita, M. Sata, H. Nakajima and A. Kiyomoto, *Clin. Exp. Hyper. Theory and Practice,* **A4**(1–2), 285 (1982).

73. R. G. Shanks, *Cardiologia,* **49**(Suppl 11), 1 (1966).

74. V. M. Campese, *Drugs,* **22**, 257 (1981).

75. H. Stahle, "Clonidine", in J. S. Bindra and D. Lednicer, Eds., *Chronicles of Drug Discovery,* Wiley, New York, 1982, pp. 87–111.

76. D. Dunlop and R. G. Shanks, *Brit. J. Pharmacol.,* **32**, 201 (1968).

77. A. Scriabine, J. W. Constantine, H. J. Hess and W. K. McShane, *Experientia,* **24**, 1150 (1968).

78. F. Bossert and W. Vater, *Naturiv.,* **58** 578 (1971).

79. M. A. Ondetti, B. Rubin and D. W. Cushman, *Science,* **196**, 441 (1977).

80. J. V. Duncia, D. J. Carini, A.T. Chiu, A. L. Johnson, W. A. Price, P. C. Wong, R. R. Wexler and P. B. W. M. Timmermans, *Med. Res. Rev.,* **12**(2), 149 (1992).

81. K. Okamoto and K. Aoki, *Jpn. Circ. J.,* **27**, 282 (1963).

82. A. Fleckenstein, *Circ. Res.,* **52**(Suppl 1), 3 (1983).

83. U. Quast, *Fundam. Clin. Pharmacol.,* **6**, 279 (1992).

84. R. A. Buchholz, B. A. Lefker and M. A. R. Kiron, *Ann. Reports in Med. Chem.,* **28**, 69 (1993).

85. R. D. Smith and J.R. Regan, "Antihypertensive Agents", in 21st ed., J. A. Bristol, Ed. *Annual Reports in Medicinal Chemistry,* Academic Press, Inc., San Diego, 1986, pp. 63–72.

86. J. J. Baldwin, *Curr. Opin. Therap. Pat.,* **4**(5), 505 (1994).

87. A. Lant, *Drugs,* **31**(Suppl 4), 40 (1986).

88. P. J. Cannon, "Clinical Use of Diuretics in Hypertension", in G. Onesti, K. E. Kim and J. H. Moyer, Eds., *Hypertension: Mechanisms and Management,* Grune & Stratton, New York, 1973, pp. 261–272.

89. K. H. Beyer, *Hypertension,* **22**(3), 388 (1993).

90. I. M. Weiner, "Diuretics and Other Agents Employed in the Mobilization of Edema Fluid", in 8th Ed., A. G. Gilman, T. W. Rall, A. S. Nies and P. Taylor, Eds., *The Pharmacological Basis of Therapeutics,* Pergamon Press, New York, 1990, pp. 713–731.

91. G. H. Mudge, "Diuretics and Other Agents Employed in the Mobilization of Edema Fluid", in 5th Ed., L. S. Goodman and A. G. Gilman, Eds., *The Pharmacological Basis of Therapeutics,* MacMillan Publishing Co., Inc., New York, 1975, pp. 817–847.

92. J. A. Cella and C. M. Kagawa, *J. Am. Chem. Soc.,* **79**, 4808 (1957).

93. J. B. Puschett, *Cardiology,* **84**(Suppl 2), 4 (1994).

94. Multiple Risk Factor Intervention Trial, *JAMA,* **248** 1465 (1982).

95. Hypertension Detection and Follow-Up Program Cooperative Group, *JAMA,* **242**, 2562 (1979).

96. E. D. Freis and V. Papademetriou, *Drugs,* **30**, 469 (1985).

97. M. Moser, *Cardiology,* **84**(Suppl 2), 27 (1994).

98. F. A. Finnerty, Jr., "The Use of Diazoxide in Hypertension", in G. Onesti, K. E. Kim and J. H. Moyer, Eds., *Hypertension: Mechanisms and Management,* Grune & Stratton, New York, 1973, pp. 445–450.

99. J. Y. Garcia, Jr. and D. G. Vidt, *Drugs,* **34** 263 (1987).

100. T. Uchida, K. Yamanaga, H. Kido, Y. Ohtaki and M. Watanabe, *Cardiology,* **84**(Suppl 2), 14 (1994).

101. F. Fiaccadori, G. C. Pasetti, G. Pedretti, P. Pizzaferri and G. F. Elia, Cardiology, **84**(Suppl 2), 80 (1994).

102. E. J. Cragoe, Jr. "Pyrazine Diuretics", in Cragoe, Ed., *Diuretics: Chemistry, Pharmacology and Medicine,* Wiley & Sons, New York, 1983, pp. 303–341.

103. V. D. Wiebelhaus, J. Weinstock, A. R. Maass, F. T. Brennan, G. Sosnowski and T. Larsen, *J. Pharmacol. Exp. Ther.,* **149**, 397 (1965).

104. C. E. Powell and I. H. Slater, *J. Pharmacol. Exp. Ther.,* 122, 480 (1958).

105. F. R. Buhler, "Position Paper: Antihypertensive Actions of Beta-Blockers", in J. H. Laragh, F. R. Buhler and D. W. Seldin, Eds., *Frontiers in Hypertension Research,* Springer-Verlag, New York, 1981, pp. 423–435.

106. R. P. Ahlquist, *Am. J. Physiol.,* **153**, 586 (1948).

107. A. M. Lands, A. Arnold, J. P. McAuliff, F. P. Luduena and T. G. Brown, *Nature,* **214**, 597 (1967).

108. E. VonMollendorff, G. Sponer, K. Strein, W. Bartsch, B. Muller-Bechmann, G. Neugebauer, H. Czerwek, G. Bode and E. Schnurr, "Carvedilol", in A. Scriabine, Ed., *New Cardiovascular Drugs 1987,* Raven Press, New York, 1987, pp. 135–153.

109. P. A. VanZwieten, *J. Hypertens.,* **8**, 687 (1990).

110. B. N. C. Prichard and J. M. Cruickshank, "Beta Blockade in J. H. Laragh and B. M. Brenner, Eds., *Hypertension: Pathophysiology, Diagnosis and Management,* Raven Press, New York, 1995, pp. 2827–2859 2nd ed.

111. D. G. Shand, *Drugs,* **25**(Suppl 2), 92 (1983).

112. D. G. McDevitt, *J. Cardiovasc. Pharmacol.,* **8**(Suppl 6), S5 (1986).

113. F. Gross, "Growing Points of Blocking Agents",

in P. L. Morselli, J. R. Kilborn, I. Cavero, D. C. Harrison and S. Z. Langer, Eds., Betaxolol and Other 1 Adrenoceptor Antagonists, Raven Press, New York, 1983, pp. 1–12.

114. I. Cavero, F. Lefevre-Borg, P. Manoury and A. G. Roach, "*In Vitro* and *in vivo* Pharmacological Evaluation of Betaxolol, a New, Potent, and Selective 1 Adrenocoptor Antagonist", in P. L. Morselli, J. R. Kilborn, I. Cavero, D. C. Harrison and S. Z. Langer, Eds., *Betaxolol and Other 1 Adrenoceptor Antagonists*, Raven Press, New York, 1983, pp. 31–42.

115. J. W. Black and J. S. Stephenson, *Lancet*, **2**, 311 (1962).

116. P. Manoury, "Betaxolol: Chemistry and Biological Profile in Relation to its Physicochemical Properties", in P. L. Morselli, J. R. Kilborn, I. Cavero, D. C. Harrison and S. Z. Langer, Eds., *Betaxolol and Other 1 Adrenoceptor Antagonists*, Raven Press, New York, 1983, pp. 13–19.

117. P. Turlapaty, A. Laddu, V. Murthy, B. Singh and R. Lee, *Am. Heart J.*, **114**(4), 866 (1987).

118. C. Quon, K. Mai, G. Patil and H. Stampfli, *Drug Metabolism and Disposition*, **16**(3), 425 (1988).

119. M. A. Ondetti, N. J. Williams, E. F. Sabo, J. Pluscec, E.R. Weaver and O. Kocy, *Biochem.*, **10**(22), 4033 (1971).

120. H. Gavras, H. R. Brunner, J. H. Laragh, J. E. Sealey, I. Gavras and R. A. Vukovich, *N. Engl. J. Med.*, **291**, 817 (1974).

121. D. W. Cushman, H. S. Cheung, E. F. Sabo, B. Rubin and M. A. Ondetti, *Fed. Proc.*, **38**, 2778 (1979).

122. D. W. Cushman and M. A. Ondetti, *Biochem. Pharmacol.*, **29**, 1981 (1980).

123. L. D. Byers and R. Wolfenden, *J. Biol. Chem.*, **247**, 606 (1972).

124. C. S. Sweet, *Fed. Proc.*, **42**, 167 (1983).

125. J. M. Hall, *Pharmac. Ther.*, **56**, 131 (1993).

126. P. C. Wong, "Angiotensin Antagonist in Models of Hypertension", in J. M. Saavedra and P. B. M. W. M. Timmermans, Eds., *Angiotensin Receptors*, Plenum Press, New York, 1994, pp. 319–336.

127. V. Cachofeiro, T. Sakakibara and A. Nasjletti, *Hypertension*, **19**(2), 138 (1992).

128. W. Sunman and P. S. Sever, *Clin. Sci.*, **85**, 661 (1993).

129. A. D. Struthers, *Pharmac. Ther.*, **53**, 187 (1992).

130. M. R. Ujhelyi, R. K. Ferguson and P. H. Vlasses, *Pharmacotherapy*, **9**(6), 351 (1989).

131. J. McMurray and M. Chopra, *Brit. J. Clin. Pharmacol.*, **31**, 373 (1991).

132. G. J. Grover, P. G. Sleph, S. Dzwonczyk, P. Wang, W. Fung, D. Tobias and D. W. Cushman, *J. Pharmacol. Exp. Ther.*, **257**(3), 919 (1991).

133. K. R. Lees, I. B. Squire and J. L. Reid, *Clin. Exper. Pharmacol. Physiol.*, **19**(Suppl 19), 49 (1992).

134. J. J. Raia, Jr., J. A. Barone, W. G. Byerly and C. R. Lacy, *DICP Ann. Pharmacother.*, **24**, 506 (1990).

135. T. Unger, P. Gohlke, M. Paul, R. Rettig and R. Rainer, *J. Cardiovasc. Pharmacol*, **18**(Suppl 2), 520 (1991).

136. C. I. Johnston, *Hypertension*, **23**(2), 258 (1994).

137. R. E. Weishaar, R. L. Panek, T. C. Major, G. H. Lu, J. C. Hodges and D. T. Dudley, *Am. J. Hypertens.*, **3**(5, Pt. 2), 98A (1990) (Abstract).

138. C. R. Lines, G. C. Preston, C. E. Dawson, C. Brazell and M. Traub, *J. Psychopharmacology*, **5**(3), 228 (1991).

139. J. B. Kostis, *Am. J. Hypertens.*, **2**, 57 (1989).

140. T. Unger and P. Gohlke, *Cardio. Res.*, **28**(2), 146 (1994).

141. R. D. Smith, P. S. Wolf, J. R. Regan and S. R. Jolly, *Progress in Clinical Biochemistry and Medicine: The Emergence of Drugs which Block Calcium Entry*, 6th ed., Springer-Verlag, Berlin, 1988.

142. L. H. Opie, F. R. Buhler, A. Fleckenstein, L. Hansson, D. C. Harrison, P. A. Poole-Wilson, A. Schwartz and P. M. Vanhoutte, *Am. J. Cardiol.*, **60**, 630 (1987).

143. D. J. Triggle and R. A. Janis, *Annu. Rev. Pharmacol. Toxicol.*, **27**, 347 (1987).

144. E. W. McClesky, A. P. Fox, D. Feldman and R. W. Tsien, *J. Exp. Biol.*, **124**, 177 (1986).

145. G. P. Reams, "Renal Effects of Antihypertensive drugs", in W. M. Bennett, D. A. McCarron and J. H. Stein, Eds., *Contempory Issues in Nephrology. Pharmacology and Management of Hypertension*, 28th ed., Churchill Livingstone, New York, 1994, pp. 29–53.

146. R. D. Smith, *Fed. Proc.*, **42**(2), 201 (1983).

147. A. Fleckenstein, "Discovery and Mechanism of Action of Specific Calcium Antagonistic Inhibitors of Excitation Contraction Coupling in the Mammalian Myocardium", in *Calcium Antagonism in Heart and Smooth Muscle*, John Wiley & Sons, New York, 1983, pp. 34–108.

148. T. D. Giles, *Cardio. Drug Rev.*, **8**(2), 138 (1990).

149. A. M. Katz, W. D. Hager, F. C. Messineo and A. J. Pappano, *Am. J. Med.*, **79**(Suppl 4A), 2(1985).

150. A. Schwartz, *J. Mol. Cell. Cardiol.*, **19**(Suppl II), 49 (1987).

151. R. A. Burges and M. G. Dodd, *Cardio. Drug Rev.*, **8**(1), 25 (1990).

152. H. Meyer, F. Bossert, E. Wehinger, R. Towart and P. Bellemann, *Hypertension*, **5**(Suppl II), II–2 (1983).

153. D. Rampe, C. M. Su, F. Yousif and D. J. Triggle, *Brit. J. Clin. Pharmacol.*, **20**, 247S (1985).

154. M. Schramm, G. Thomas, R. Towart and G. Franckowiak, *Nature,* **303**, 535 (1983).

155. T. Ehara and R. Kaufmann, *J. Pharmacol. Exp. Ther.*, **207**, 49 (1978).

156. J. W. Constantine, W. K. McShane, A. Scriabine and H. J. Hess, "Analysis of the Hypotensive Action of Prazosin", in G. Onesti, K. K. Kim and J. H. Moyer, Eds., *Hypertension: Mechanisms and Management*, Grune & Stratton, New York, 1973, pp. 429–444.

157. S. Z. Langer, I. Cavero and R. Massingham, *Hypertension*, **2**, 372 (1980).

158. G. S. Stokes and J. F. Marwood, *Meth. and Find. Exptl. Clin. Pharmacol.*, **6**(4), 197 (1984).

159. R. D. Smith, D. K. Tessman and H. R. Kaplan, *J. Pharmacol. Exp. Ther.*, **217**(2), 397 (1981).

160. P. A. VanZwieten, "α-Adrenoceptor Blocking Agents in the Treatment of Hypertension", in J. H. Laragh and B. M. Brenner, Eds., *Hypertension: Pathophysiology, Diagnosis and Management*, 2nd ed., Raven Press, New York, 1995, pp. 2917–2935.

161. M. Nickerson, *Pharm. Rev.*, **1**, 27 (1949).

162. A. S. Hanson and S. L. Linas, "Refractory and Malignant Hypertension", in W. M. Bennett, D. A. McCarron and J. H. Stein, Eds., *Contemporary Issues in Nephrology. Pharmacology and Management of Hypertension*, 28th ed., Churchill Livingstone, New York, 1994, pp. 143–180.

163. J. J. Kyncl, R. C. Sonders, W. D. Sperzel, M. Winn and J. H. Seely, "Terazosin", in A. Scriabine, Ed., *New Cardiovascular Drugs 1986*, Raven Press, New York, 1986, pp. 1–18.

164. B. B. Hoffman and R. J. Lefkowitz, "Adrenergic Receptor Antagonists", in A. G. Gilman, T. W. Rall, A. S. Nies and P. Taylor, Eds., *The Pharmacological Basis of Therapeutics*, 8th ed., Pergamon Press, New York, 1990, pp. 221–243.

165. W. Kobinger, "Pharmacologic Basis of the Cardiovascular Actions of Clonidine", in G. Onesti, K. E. Kim and J. H. Moyer, Eds., *Hypertension: Mechanisms and Management*, Grune & Stratton, New York, 1973, pp. 369–380.

166. P. B. M. W. M. Timmermans, A. M. C. Schoop and P. A. VanZwieten, *Eur. J. Pharmacol.*, **70**, 7 (1981).

167. P. Dominiak, *Cardiovasc. Drug Therapy*, **8**(Suppl 1), 21 (1994).

168. B. I. Armah, E. Hofferber and W. Stenzel, *Arzneim. -Forsch. -Drug Res.*, 38, 1426 (1988).

169. B. N. Pritchard, *Cardiovasc. Drug Therapy*, **8**(Suppl 1), 49 (1994).

170. M. E. Safar, *Am. J. Med.*, **87**(Suppl 3C), 24S (1989).

171. M. Gothert and G. J. Molderings, *J. Cardiovasc. Pharmacol.*, **20**(Suppl 4), S16 (1992).

172. P. R. Ernsberger, K. L. Westbrooks, M. O. Christen and S. G. Schafer, *J. Cardiovasc. Pharmacol.*, **20**(Suppl 4), S1 (1992).

173. A. C. MacKinnon, M. Spedding and C. M. Brown, *Trends Pharmacol. Sci.*, **15**, 119 (1994).

174. N. G. Morgan, *Expert Opin. Invest. Drugs*, **3**(6), 561 (1994).

175. W. R. Campbell and D. E. Potter, *Prog. Neuro-Psych. Biol. Psych.*, **18**(6), P1051 (1994).

176. M. C. Michel and R. Schafers, *J. Cardiovasc. Pharmacol.*, **29**(Suppl 4), S24 (1992).

177. S. B. Penner and D. D. Smyth, *Cardiovasc. Drug Therapy*, **8**(Suppl 1), 43 (1994).

178. V. J. Gorbea-Oppliger and G. D. Fink, *Hypertension*, **23**(6 Pt 2), 844 (1994).

179. C. S. Tung, M. R. Goldberg, A. S. Hollister, B. J. Sweetman and D. Robertson, *Life Sci.*, **42**, 2365 (1988).

180. T. P. Gross, R. P. Wise and D. E. Knapp, *Hypertension*, **13**(Suppl 1), I113 (1989).

181. W. T. Comer, W. L. Matier and M. S. Amer, "Antihypertensive Agents", in M. E. Wolff, Ed. *Burger's Medicinal Chemistry*, 4th ed., John Wiley & Sons, New York, 1981, pp. 285–337.

182. J. G. Gerber and A. S. Nies, "Antihypertensive Agents and the Drug Therapy of Hypertension", in A. G. Gilman, T. W. Rall, A. S. Nies and P. Taylor, Eds., *The Pharmacological Basis of Therapeutics*, 8th ed., Pergamon Press, New York, 1990, pp. 784–813.

183. T. E. Gaffney, P. J. Privitera and S. Mohammed, "The Multiple Sites of Action of Methyldopa", in G. Onesti, K. E. Kim and J. H. Moyer, Eds., *Hypertension: Mechanisms and Management*, Grune & Stratton, New York, 1973, pp. 289–297.

184. R. L. Woosley and A. S. Nies, *N. Engl. J. Med.*, **295**, 1053 (1976).

185. J. N. Cohn and L. P. Burke, *Ann. Intern. Med.*, **91**, 752 (1979).

186. G. H. Anderson, Jr., D. H. P. Streeten and T. G. Dalakas, *Circ. Res.*, **40**(3), 243 (1977).

187. F. M. Bumpus, K. J. Catt, A. T. Chiu, M. DeGasparo, T. Goodfriend, A. Husain, M. J. Peach, D. G. Taylor, Jr. and P. B. M. W. M. Timmermans, *Hypertension*, **17**(5), 720 (1991).

188. M. DeGasparo, A. Husain, W. Alexander, K. J. Catt, A. T. Chiu, M. Drew, T. Goodfriend, J. W. Harding, T. Inagami and P. B. M. W. M. Timmermans, *A Proposed Update of the Nomenclature of Angiotensin Receptors, 1994*, (Unpub).

189. R. D. Smith and P. B. M. W. M. Timmermans, *Curr. Opin. Nephrol. Hypertens.*, **3**, 112 (1994).

190. M. DeGasparo, N. R. Levens, B. Kamber, P. Furet, S. Whitebread, V. Brechler and S. P. Bottari, "The Angiotensin II AT2 Receptor Subtype", in J. M. Saavedra and P. B. M. W. M. Timmermans, Eds., *Angiotensin Receptors*, Plenum Press, New York, 1994, pp. 95–118.

191. J. A. Keiser and R. L. Panek, "Pharmacology of AT2 Receptors" in J. M. Saavedra and P. B. M. W. M. Timmermans, Eds., Angiotensin Receptors, Plenum Press, New York, 1994, pp. 135–150.

192. D.T. Dudley and J. M. Hamby, *Curr. Opin. Therap. Pat.*, **3**(5), 581 (1993).

193. M. I. Steinberg, S. A. Wiest and A. D. Palkowitz, *Cardio. Drug Rev.*, **11**(3), 312 (1993).

194. D. Middlemiss and B. C. Ross, "Angiotensin II Receptor Antagonists: Non-biphenyl tetrazoles", in P. B. M. W. M. Timmermans and R. R. Wexler, Eds., *Medicinal Chemistry of The Renin Angiotensin System*, 21st ed., Elsevier, Lausanne, 1994, pp. 241–267.

195. R. M. Keenan, J. Weinstock, J. C. Hempel, J. M. Samanen, D. T. Hill, N. Aiyar, D. P. Brooks, E. H. Ohlstein and R. M. Edwards, "Nonpeptide Angiotensin II Receptor Antagonists: Design and SAR of Imidazole-5-Acrylic Acids:, in P. B. M. W. M. Timmermans and R. R. Wexler, Eds., *Medicinal Chemistry of the Renin Angiotensin System*, 21st ed., Elsevier, Lausanne, 1994, pp. 175–201.

196. K. Kubo, Y. Inada, Y. Kohara, Y. Sugiura, M. Ojima, K. Itoh, Y. Furukawa, K. Nishikawa and T. Naka, *J. Med. Chem.*, **36**, 1772 (1993).

197. J. Zhang, M. Entzeroth, W. Wienen and J. C. A. VanMeel, *Eur. J. Pharmacol.*, **218**, 35 (1992).

198. R. E. Olson, J. Liu, G. K. Lalka, M. K. VanAtten, R. R. Wexler, A. T. Chiu, T.T. Nguyen, D. E. McCall, P. C. Wong and P. B. M. W. M. Timmermans, *Bioorg, Med. Chem. Letter*, **4**(18), 2229 (1994).

199. R. S. L. Chang, T. B. Chen, S. A. O–Malley, R. J. Bendesky, P. J. Kling, V. J. Lotti, S. D. Kivlighn, P. K. S. Siegl, D. Ondeyka, N. B. Mantlo and W. J. Greenlee, *Can. J. Physiol. Pharmacol.*, **72**(Suppl 1), 132 (1994) (Abstract).

200. S. D. Kivlighn, G. J. Zingaro, R. A. Gabel, T. P. Broten, R. S. L. Chang, D. Ondeyka, N. B. Mantlo, R. E. Gibson, W. J. Greenlee and P. K. S. Siegl, *In Vivo Pharmacology of L-163,017: A Nonpeptide Angiotensin Receptor Antagonist with Balanced Affinity for AT1 and AT2 Receptors*, 1994 (Unpub).

201. J. M. Evans and S. D. Longman, "Potassium Channel Activators", in J. A. Bristol and D. W. Robertson, Eds., Annual Reports in *Medicinal*

Chemistry, 26th ed., Academic Press, Inc., San Diego, 1991, pp. 73–82.

202. E. Arrigoni-Martelli and J. Finucane, "Pinacidil", in A. Scriabine, Ed., *New Cardiovascular Drugs 1985*, Raven Press, New York, 1985, pp. 133–151.

203. N. S. Cook, *Trends Pharmacol. Sci.*, **9**, 21 (1988).

204. K. E. Andersson, *Pharmacol. Toxicol.*, **70**, 244 (1992).

205. C. Hutchins and J. Greer, *Crit. Rev. Biochem. Mol. Biol.*, **26**(1), 77 (1991).

206. S. Thaisrivongs, "Orally Active Renin Inhibitors", in *Current Drugs*, Current Patents Ltd., 1992, pp. B35–B49.

207. W. J. Greenlee and P. K. S. Siegl, "Angiotensin/ Renin Modulators", in J. A. Bristol and D. W. Robertson, Eds., *Annual Reports in Medicinal Chemistry*, Academic Press, Inc., San Diego, 1991, pp. 63–72.

208. E. F. Foote and C. E. Halstenson, *Ann. Pharmacother*, **27**(12), 1495 (1993).

209. R. J. Cody, *Drugs*, **47**(4), 586 (1994).

210. G. A. Rongen, J. W. M. Lenders, C. H. Kleinbloesem, C. Weber, H. Welker, E. Fahrner, H. Pozenel, A. J. J. Woittiez, G. Haug, M. S. Buchmann, C. E. K. Hoglund and T. Thien, *Clin. Pharmacol. Ther.*, **54** 567 (1993).

211. R. S., Boger, H. N. Glassman, R. Thys, S. K. Gupta, R. L. Hippensteel and H. D. Kleinert, *Am. J. Hypertens.*, **6**, 103A (1993).

212. D. E. Nelson, D. M. Moyse, J. M. O–Neil, R. S. Boger, H. N. Glassman and H. D. Kleinert, *Circulation*, **88**, 361 (1993).

213. J. M. VanNueten, P. A. J. Janssen, J. Symoens, W. J. Janssens, J. Heykants, F. DeClerck, J. E. Leysen, H. Vancauteren and P. M. Vanhoutte, "Ketanserin", in A. Scriabine, Ed., *New Cardiovascular Drugs 1987*, Raven Press, New York, 1987, pp. 1–56.

214. S. Oparil and D. A. Calhoun, " 1 and 2 Adrenergic Receptors", in W. M. Bennett, D. A. McCarron and J. H. Stein, Eds, *Contemporary issues in Nephrology: Pharmacology and Management of Hypertension*, 28th ed., Churchill Livingstone, New York, 1994, pp. 247–265.

215. P. Needleman, E. H. Blaine, J. E. Greenwald, M. L. Michener, C. B. Saper, P. T. Stockmann and H. E. Tolunay, *Annu. Rev. Pharmacol. Toxicol.*, **29**, 23 (1989).

216. T. Maack, "Receptors of Natriuretic Peptides: Structure, Function, and Regulation", in J. H. Laragh and B. M. Brenner, Eds., *Hypertension: Pathophysiology, Diagnosis and Management*, 2nd ed., Raven Press, New York, 1995, pp. 1001–1019.

217. P. J. S. Chiu, *Current Drugs*, B55 (**Dec**. 1991).

218. E. J. Sybertz, *Cardio. Drug Rev.*, **8**(1), 71 (1990).

219. E. J. Sybertz, P. J. S. Chiu, S. Vemulapalli, R. W. Watkins and M. F. Haslanger, *Hypertension*, (1995), in press.

220. M. C. Fournie-Zaluski, P. Coric, S. Turcaud, N. Rousselet, W. Gonzalez, B. Barbe, I. Pham, N. Jullian, J. B. Nichel and B. P. Roques, *J. Med. Chem.*, **37**(8), 1070 (1994).

221. I. Biaggioni, B. Bennett, B. Patel, E. K. Jackson and D. Robertson, *Clin. Res.*, **35**, 373A (1987).

222. I. Biaggioni and R. Mosqueda-Garcia, "Adenosine in Cardiovascular Homeostasis and the Pharmacologic control of its activity", in J. H. Laragh and B. M. Brenner, Eds., *Hypertension: Pathophysiology, Diagnosis and Management*, 2nd ed., Raven Press, New York, 1995, pp. 1125–1140.

223. R. Z. Gerencer, B. A. Finegan and A. S. Clanachan, *Brit. J. Pharmacol*, **107**, 1048 (1992).

224. A. J. Hutshison, R. L. Webb, H. H. Oei, G. R. Ghai, M. B. Zimmerman and M. Williams, *J. Pharmacol. Exp. Ther.*, **251**(1), 47 (1989).

225. R. L. Webb, M. A. Sills, J. P. Chovan, J. V. Peppard and J. E. Francis, *J. Pharmacol. Exp. Ther.*, **267**(1), 287 (1993).

226. A. Matsuda, M. Shinozaki, T. Yamaguchi, H. Homma, R. Nomoto, T. Miyasaka, Y. Watanabe and T. Abiru, *J. Med. Chem.*, **35**, 241 (1992).

227. N. P. Peet, N. L. Lentz, S. Sunder, M. W. Dudley and A. M. L. Ogden, *J. Med. Chem.*, **35**, 3263 (1992).

228. F. L. Belloni, C. I. Thompson, K. Fratta, J. L. S. Lewy, C. A. Schroeder and H. Machen, *Drug Dev. Res.*, **31**(4), 249 (1994).

229. I. Gavras and H. Gavras, "Role of Vasopressin in Hypertensive Disorders", in J. H. Laragh and B. M. Brenner, Eds., *Hypertension: Pathophysiology, Diagnosis and Management*, 2nd ed., Raven Press, New York, 1995, pp. 789–800.

230. X. Li, A. Kribben, E. D. Wieder, P. Tsai, R. A. Nemenoff and R. W. Schrier, *Hypertension*, **23**, 217 (1994).

231. Y. Yamamura, H. Ogawa, T. Chihara, K. Kondo, T. Onogawa, S. Nakamura, T. Mori, M. Tominaga and Y. Yabuuchi, *Science*, **252**, 571 (1991).

232. Y. Yamada, Y. Yamamura, T. Chihara, T. Onogawa, S. Nakamura, T. Yamashita, T. Mori, M. Tominaga and Y. Yabuuchi, *Hypertension*, **23**, 200 (1994).

233. P. Vanhoutte, *Hypertension*, **21**, 747 (1993).

234. Y. Takuwa, *Endocr. J.*, **40**(5), 489 (1993).

235. A. M. Doherty, "The Development of Endothelin Antagonists to Moderate Cardiovascular function" in J. H. Laragh and B. M. Brenner, Eds., *Hypertension: Pathophysiology, Diagnosis, and Management*, 2nd ed., Raven Press, New York, 1995, pp. 3115–3136.

236. M. Clozel, V. Breu, G. A. Gray and B. M. Loffler, *J. Cardiovasc. Pharmacol.*, **22**(Suppl 8), S377 (1993).

237. J. S. Li, R. Lariviere and E. L. Schiffrin, *Hypertension*, **24**, 183 (1994).

238. M. C. Michel and A. Buschauer, *Drugs of the Future*, **17**(1), 39 (1992).

239. M. Adamsson, B. Fallgren and L. Edvinsson, *Brit. J. Pharmacol.*, **105**(1), 93 (1992).

240. F. Feth, W. Erdbrugger, W. Rascher and M. C. Michel, *Life Sci.*, **52** (23), 1835 (1993).

241. X. Sun, L. Edvinsson and T. Hedner, *J. Hypertens*, **11**(9), 935 (1993).

242. H. Lind, L. Brudin, J. Castenfors, T. Hedner, L. Lindholm and L. Edvinsson, *Blood Press.*, **3**(4), 242 (1994).

243. P. D. Davis, L. H. Elliott, W. Harris, C. H. Hill, S. A. Hurst and E. Keech, *J. Med. Chem.* **35**(6), 994 (1992).

244. A. Gescher, *Brit. J. Cancer*, **66**(1), 10 (1992).

245. N. A. Turner, M. G. Rubsby, J. H. Walker, F. A. McMorris, S. G. Ball and P. F. Vaughan, *Biochem. J.*, **297**(Pt 2), 407 (1994).

246. U. Amon, E. VonStebut and H. H. Wolff, *Agents Actions*, **39**(1–2), 13 (1993).

247. R. P. Lifton and X. Jeunemaitre, *J. Hypertens.*, **11**, 231 (1993).

248. E. G. Nabel and G. J. Nabel, "Prospects for Gene Therapy in Cardiovascular Diseases", in J. H. Laragh and B. M. Brenner Eds., *Hypertension: Pathophysiology, Diagnosis and Management*, 2nd ed., Raven Press, New York, 1995, pp. 3137–3149.

249. M. A. Carson-Jurica, W. T. Schrader and B. W. O–Malley, *Endocrine Rev.*, **11**, 201 (1990).

250. L. M. Resnick, "Calcium-Regulating Hormones and Human Hypertension", in M. F. III Crass and L. V. Avioli, Eds., *Calcium Regulating Hormones and Cardiovascular Function*, CRC Press, Boca Raton, 1995, pp. 295–320.

251. J. B. Warren, F. Pons and A. J. B. Brady, *Cardio, Res.*, **28**, 25 (1994).

252. A. V. Chobanian, "Hypertension, Growth factors and their Relevance to Atherosclerotic Vascular Disease", in J. H. Laragh and B. M. Brenner, Eds., *Hypertension: Pathology, Diagnosis and Management*, 2nd ed., Raven Press, New York, 1995, pp. 515–521.

253. P. Pauletto, R. Sarzani, A. Rappelli, A. C. Pessina and S. Sartore, "Vascular Smooth Muscle Cell

differentiation and growth response in hypertension", in J. H. Laragh and B. M. Brenner, Eds., *Hypertension: Pathophysiology, Diagnosis and Management*, 2nd ed., Raven Press, New York, 1995, pp. 697–709.

254. J. Quilley, C. P. Bell-Quilley and J. C. McGiff, "Eicosanoids and Hypertension", in J. H. Laragh and B. M. Brenner, Eds., *Hypertension: Pathophysiology, Diagnosis and Management*, 2nd ed., Raven Press, New York, 1995, pp. 963–982.

255. K. H. Thierauch, H. Dinter and G. Stock, *J. Hypertens*, **11**, 1315 (1993).

256. W. L. Smith, *Am. J. Physiol.*, **263**(32), F181 (1992).

257. K. Hermsmeyer, P. Erne, "Vascular Muscle Ion Channels and Cellular Calcium Regulation in Hypertension", in J. H. Laragh and Brenner, Eds., *Hypertension: Pathophysiology, Diagnosis and Management*, 2nd ed., Raven Press, New York, 1995, pp. 673–683.

258. T. F. Luscher, *Eur. J. Clin. Invest.*, **23**, 670 (1993).

259. D. C. Lefroy, *Cardio. Res.*, **27**, 2105 (1993).

260. E. Carafoli, M. Chiesi and P. Gazzotti, "Membrane Carriers Related to Intracellular Calcium Regulation", in J. H. Laragh and B. M. Brenner, Eds., *Hypertension: Pathophysiology, Diagnosis and Management*, 2nd ed., Raven Press, New York, 1995, pp. 1245–1259.

261. J. F. Milligan, M. D. Matteucci and J. C. Martin, *J. Med. Chem.*, **36**(14), 1923 (1993).

262. J. Zhou, P. Ernsberger and J. G. Douglas, *FASEB J.*, **5**, No. 4(Pt 1), A870 (1991). (Abstract)

263. G. A. Knock, M. H. F. Sullivan, A. McCarthy, M. G. Elder, J. M. Polak and J. Wharton, "Angiotensin II (AT1) Vascular Binding Sites in Human Placentae from Normal Term, Preeclamptic and Growth Retarded Pregnancies," 1993, (Unpub)

264. D. D. Holsworth, J. S. Kiely, R. S. Root-Bernstein and R. W. Overhiser, *Peptide Research*, **7**(4), 185 (1994).

265. M. I. Phillips, D. Wielbo and R. Gyurko, *Am. J. Hypertens.*, **7**(4), 1419 (1994)). (Abstract)

266. P. Ferrari and G. Bianchi, "Lessons from Experimental Genetic Hypertension", in J. H. Laragh and B. M. Brenner, Eds., *Hypertension: Pathophysiology, Diagnosis and Management*, 2nd ed., Raven Press, New York, 1995, pp. 1261–1279.

267. E. M. St. Lezin, M. Pravenec and T. W. Kurtz, *Trends Cardiovasc. Med.*, **3**(4), 119 (1993).

268. F. Soubrier and F. Cambien, *Trends Endocrinol. Metab.*, **3**, 250 (1993).

269. V. Ralevic, G. Burnstock, "Neuropeptides in blood pressure control", in J. H. Laragh and B. M. Brenner, Eds., *Hypertension: Pathophysiology, Diagnosis and Management*, 2nd ed., Raven Press, New York, 1995, pp. 801–831.

270. M. Gunning and B. M. Brenner, "Urodilatin", in J. H. Laragh and B. M. Brenner, Eds., *Hypertension: Pathophysiology, Diagnosis and Management*, 2nd ed., Raven Press, New York, 1995, pp. 1021–1027.

271. F. J. Haddy, Jr. and V. M. Buckalew, "Endogenous Digitalis Like Factors in Hypertension", in J. H. Laragh and B. M. Brenner, Eds., *Hypertension: Pathophysiology, Diagnosis and Management*, 2nd ed., Raven Press, New York, 1995, pp. 1055–1067.

CHAPTER THIRTY

Phenoxyacetic Acid Uricosuric Diuretics

HIROSHI KOGA,
HARUHIKO SATO and
TAKASHI DAN

Fuji-Gotemba Research Laboratories
Chugai Pharmaceutical Company, Ltd.
Gotemba-shi, Shizuoka, Japan

CONTENTS

1 Introduction, 324
2 Phenoxyacetic Acid Family, 326
3 Xanthonyloxyacetic Acids, 328
4 Dihydrofuroxanthone-2-carboxylic Acids, 331
5 5,6-Dihydrofuro[3,2-*f*]-1,2-benzisoxazole-6-carboxylic Acids, 331
6 7,8-Dihydrofuro[2,3-*g*]-1,2-benzisoxazole-7-carboxylic Acids, 331
7 7,8-Dihydrofuro[2,3-*g*]benzoxazole-7-carboxylic Acids, 332
8 1,3-Dioxolo[4,5-*f*]-1,2-benzisoxazole-6-carboxylic Acids, 333
9 1,3-Dioxolo[4,5-*g*]-1,2-benzisoxazole-7-carboxylic Acids, 333
10 Sulfamoyl-2,3-dihydrobenzofuran-2-carboxylic Acids, 335
11 (4-Oxo-4*H*-1-benzopyran-7-yl)oxyacetic Acids, 338
12 Three-Dimensional Structure-Activity Relationships and Receptor mapping of Diuretic Activity of Phenoxyacetic Acids, 345
13 Pharmacology, 354
 13.1 A-56234, 354
 13.2 S-8666, 355
 13.3 AA-193, 356

Burger's Medicinal Chemistry and Drug Discovery,
Fifth Edition, Volume 2: Therapeutic Agents,
Edited by Manfred E. Wolff.
ISBN 0-471-57557-7 © 1996 John Wiley & Sons, Inc.

1 INTRODUCTION

By definition, diuretics are agents that increase urine formation. Initially, diuretics were used almost exclusively for the treatment of edema. However, now the most common application is for the treatment of hypertension. Consequently, these drugs must be safe and effective even when administered chronically. The side effects most frequently cited include hypokalemia, hyperglycemia, dyslipidemia, and hyperuricemia. However, in general, the frequency and severity of these problems are not great. Diuretics have been widely used as first-line antihypertensive therapy, because these drugs have been shown to reduce cardiovascular morbidity and mortality, especially fatal and nonfatal stroke (1,2). However, recent assessment of clinical trials have shown that while the risk of stroke is consistently lower with diuretic therapy, the same degree of success has not been demonstrated for coronary artery disease (CAD). Although there would be many explanations of why it has not been done as well in preventing CAD, one possibility is that the diuretic therapy has increased the patient's risk for CAD. There are many cardiovascular risk factors which include hypertension, hyperlipidemia, hyperglycemia (diabetes), and gout (3,4). Thus, the metabolic side effects, such as, hyperlipidemia, hyperglycemia, and hyperuricemia, attributable to diuretics may have increased the risk of CAD, counteracting the favorable antihypertensive action (3). Accordingly, ideal diuretics in the future would be devoid of their deficiencies and possess attractive new attributes. They would be isokalemic, isoglycemic, isolipidemic or hypolipidemic, and isouricemic or uricosuric.

The history of diuretics can be divided into six periods that, though overlapping in time, are characterized by major developments that occurred during each interval. The class of diuretics that distinguish each

period are mercurials, carbonic anhydrase inhibitors, thiazides, loop diuretics, antikaliuretics, and uricosuric diuretics (5,6).

Since the mercurial diuretic, merbaphen (Fig. 30.1) was incidentally discovered in 1919 (7), during the next 30 years, hundreds of mercurials were synthesized and evaluated as diuretics. They exhibited a proper urinary Na^+/Cl^- balance, high-ceiling saliuresis, and an acceptable K^+ excretion profile, together with uricosuric activity. However, due to their lack of oral activity, their toxicity, and their tendency toward tachyphylaxis, the mercurial diuretics have been all but universally replaced by other agents. By 1950, the carbonic anhydrase inhibitors, e.g., acetazolamide, were introduced. Although these agents were orally active, their tendency to produce metabolic acidosis led to their progressive ineffectiveness.

The discovery of the thiazide diuretic chlorothiazide in 1957 was the beginning of a new era in diuretics. The thiazides were orally active, mild diuretics with so few side effects that they were suitable for chronic use. The thiazide diuretics have been widely used for the treatment of hypertension and recommended as the drug of choice for initial therapy in all types of hypertension, because these agents deserve much of the credit for their success in reducing the incidence of stroke. However, the metabolic side effects attributable to the thiazides have been known since these agents came into wide usage. These side effects include disturbances in electrolytes, lipids, uric acid, glucose, and insulin, which may increase the risk of CAD (3).

In 1962, furosemide was discovered by continued exploration of the chemical lead afforded by the sulfonamide diuretics, such as, the thiazides. Simultaneously, an understanding of the mechanism of action of the mercurial diuretics provided the basis for the design of a novel type of diuretics, of which ethacrynic acid was the prototype. These drugs were categorized as high-ceil-

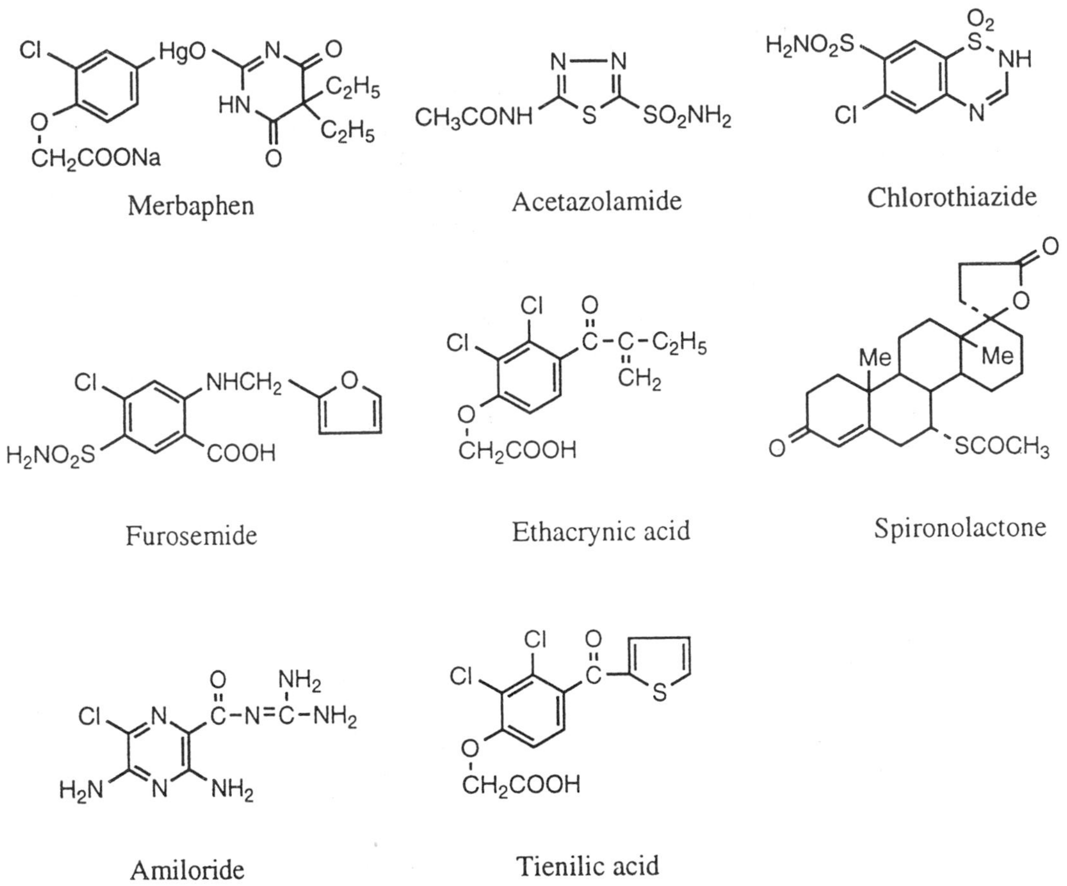

Merbaphen

Acetazolamide

Chlorothiazide

Furosemide

Ethacrynic acid

Spironolactone

Amiloride

Tienilic acid

Fig. 30.1 Representative types of modern diuretics.

ing loop diuretics, because they inhibited NaCl transport in the loop of Henle, inducing the increase in fractonal Na^+ excretion of over 20% of the filtered load. Furosemide and ethacrynic acid have replaced the mercurial diuretics, because they are equally potent, orally effective agents with less toxicity.

The hypokalemic propensity of the thiazides and the loop diuretics, which may cause dangerous arrhythmias (3), created the need for potassium-sparing diuretics. Spironolactone (1961), an aldosterone antagonist, showed natriuretic and antikaliuretic activities by competing with aldosterone for its receptor on the renal tubule. The action of aldosterone is to promote Na^+ reabsorption and K^+ and H^+ secretion. Amiloride (1964) is an example of another chemical class of diuretics, which displays antikaliuretic activity, but it is not a specific aldosterone antagonist. These drugs are weak diuretics and are more effectively used in combination with kaliuretic diuretics, such as, the thiazides and the loop diuretics, to avoid or reverse hypokalemia from kaliuretic diuretics.

Because most diuretics induce hyperuricemia, which occasionally increases the risk of gouty attacks, the need for agents with isouricemic or hypouricemic capability has been recognized for some time. Phenoxyacetic acid uricosuric diuretics exemplified by tienilic acid are one of the most recently developed diuretics and exhibit potent diuretic activity with uricosuric property.

2 PHENOXYACETIC ACID FAMILY

Ethacrynic acid (**2**) was the first drug of the phenoxyacetic acid family (Fig. 30.2). It was discovered during the course of searching for compounds that would mimic the chemical behavior of the mercury atom in the mercurial diuretics, such as, merbaphen (**1**), particularly in regard to its reaction with mercaptans, which was presumed to be responsible for the diuretic activity (8). It was speculated that the drug reacts with mercaptans in body fluids, especially cysteine and glutathione, to generate Michael adducts, and when these adducts reach the nephron of the kidney, a reaction occurs with the sulfhydryl group in the protein of the receptor, resulting in the diuresis. Ethacrynic acid (**2**) exhibited a ceiling saldiuresis of at least six times that of the thiazides. Although it mimics the mercurials in many respects, it differs from them in the other attributes, such as, oral activity, rapid onset and relatively short duration of action, pronounced saluresis, and induction of hyperuricemia.

To improve the drawbacks of ethacrynic acid (**2**), a number of structure-activity relationship (SAR) studies were undertaken. One of the most important results from these efforts was the discovery of tienilic acid (**3**). Tienilic acid (**3**) exhibits marked uricosuric activity, in contrast to ethacrynic acid (**2**), together with thiazide-like diuresis. This uricosuric effect is a rather unusual property for diuretics except for the mercurials. In fact, most diuretics in clinical use today are hyperuricemic. The importance of elevated serum urate levels is somewhat controversial. However, hyperuricemia is definitely associated with gout, which is also one of the cardiovascular risk factors (4). In addition, there is some evidence of an association between hyperuricemia and a variety of disorders including hypertension, hyperlipidemia, diabetes mellitus, obesity, renal lithiasis, renal failure, psoriasis, hypothyroidism, hyperparathyroidism, and alcoholism (8). Thus, some uricosuric acitivity would be a desirable attribute of a good diuretic agent. Although tienilic acid (**3**) was expected to be the first orally active uricosuric diuretic, a report of liver toxicity resulted in the suspension of its sale in most countries (8).

In the meantime, dihydroethacrynic acid (**4**) was found to possess diuretic activity with uricosuric activity, even though its diuretic activity was weaker than that of its unsaturated parent, ethacrynic acid (**2**). This indicated either that the reaction with sulfhydryl groups in the nephron is unnecessary for diuretic activity, or that more than one mechanism is involved. These results stimulated research on structurally related compounds. The systematic investigation of its annulated counterparts, indanyloxyacetic acids, led to the discovery of indacrinone (**5**). Many pharmacological, biochemical, and clinical studies with indacrinone (**5**) have been reported (8,9). It was shown to be an orally effective loop diuretic and to produce a sustained antihypertensive effect, more potent than hydrochlorothiazide or furosemide, in animals. In chimpanzees, it exhibited good uricosuric activity. Clinical studies revealed that indacrinone (**5**) was a well-tolerated loop diuretic, more active than furosemide, with little effect on serum urate levels in contrast to furosemide, which elevates urate levels. It produces a decrease in blood pressure equal to or greater than hydrochlorothiazide.

Since the discovery of tienilic acid (**3**) and indacrinone (**5**) as the prototype uricosuric diuretics, a number of their analogues have been investigated extensively. Among the earlier compounds **6**, **7**, HP 522 (**8**), and Abbott 49816 (**9**) were particularly interesting compounds, because of their good di-

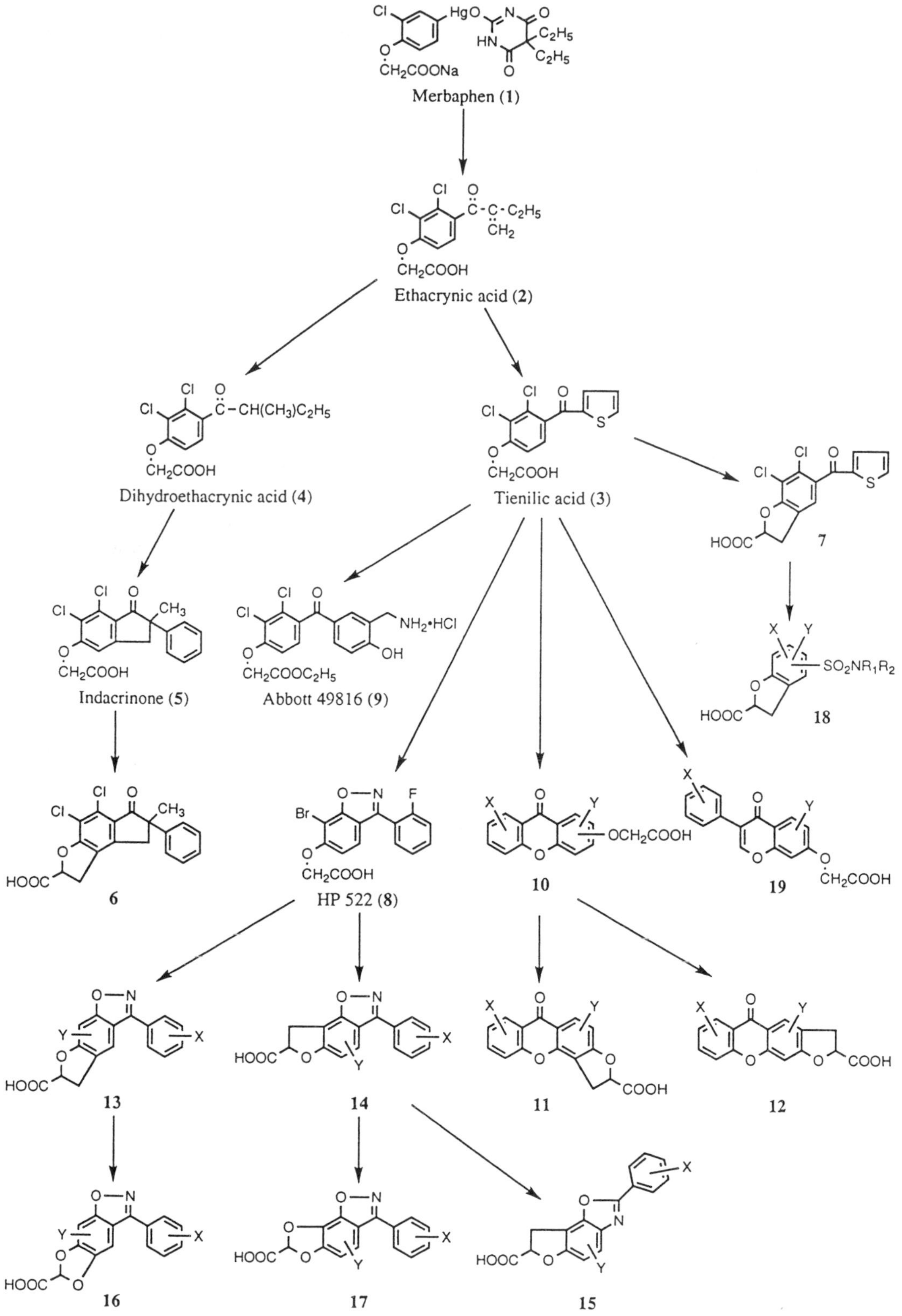

Fig. 30.2 Phenoxyacetic acid family.

uretic and uricosuric activities (8). Subsequently, the search for developing compounds with new scaffolds has been continued to find some novel types of phenoxyacetic acid series (**10–19**). Fig. 30.2 shows the various structural types that emanated from these studies. The arrows portray the chronology and flow of ideas for each member of the family, among which compounds **10–19** will be systematically discussed in this review, because the SAR and some pharmacology of the earlier compounds **1–9** have been already reviewed (8).

3 XANTHONYLOXYACETIC ACIDS

It was highly conceivable that the liver toxicity of tienilic acid (**3**) might be due to the metabolic change of the thienyl moiety in liver (10) and that the formation of metabolites might be reduced by changing

the thienyl group to a phenyl group and constructing more hydrophilic ring systems. Under these conceptions, xanthonyloxyacetic acids (**10a** and **10b**) have been designed as shown in Fig. 30.3. These compounds were derived by annulation to position 3 or 5 of 4-acyl group of tienilic acid (**3**), together with replacement of the thienyl group with a phenyl group (**11**). The substituents were generally limited to a halogen atom and methyl group from the reported results of SAR studies of the phenoxyacetic acid diuretics (8).

Diuretic and uricosuric activities in rats of the compounds **10a** and **10b** are shown in Table 30.1. Tienilic acid (**3**) and indacrinone (**5**) were used as the reference agents for the diuretic and uricosuric acivities. Tienilic acid (**3**) showed moderate diuretic and uricosuric activities, whereas indacrinone (**5**) showed potent diuretic and uricosuric activities. The diuretic activities of most of the compounds **10a** and **10b**

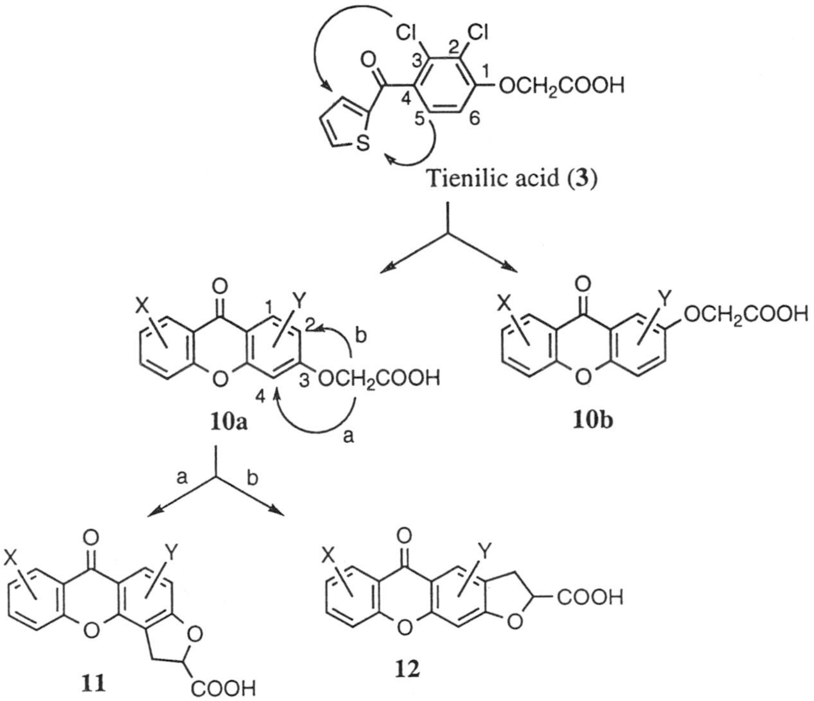

Fig. 30.3 Design of xanthonyloxyacetic acid and dihydrofuroxanthone-2-carboxylic acid derivatives.

Table 30.1 Diuretic and Uricosuric Activities of Xanthonyloxyacetic Acids[a]

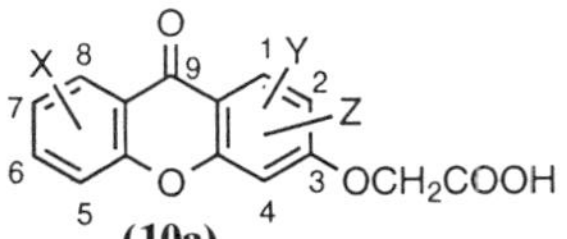

(10a)

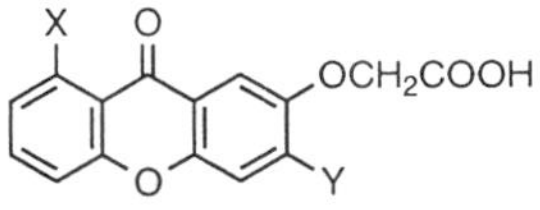

(10b)

Compound Number	X	Y	Z	Number of Animals	Diuretic[b] (0–6 h)	Uricosuric[b] (0–6 h)	Compound Number	X	Y	Z	Number of Animals	Diuretic[b] (0–6 h)	Uricosuric[b] (0–6 h)
(10aa)	H	H	H	5	153[c]	84	(10am)	H	2-Cl	H	30	200[c]	149[c]
(10ab)	8-F	H	H	5	127	NT[d]	(10an)	H	4-Cl	H	6	166	109
(10ac)	8-F	4-Cl	H	11	150[e]	128[e]	(10ao)	H	1-CH₃	H	5	203[f]	125[e]
(10ad)	8-F	2-Cl	H	6	166[e]	110	(10ap)	H	2-CH₃	H	10	207[c]	129[e]
(10ae)	8-F	4-CH₃	H	5	119	NT[d]	(10aq)	H	4-CH₃	H	5	134	NT[d]
(10af)	8-Cl	4-Cl	H	15	141[f]	126[e]	(10ar)	H	2-Br	H	5	188[c]	137[f]
(10ag)	7-Cl	4-Cl	H	5	215[f]	105	(10as)	H	1-CH₃	2-Cl	5	168	101
(10ah)	6-Cl	4-Cl	H	5	144[e]	105	(10ba)	F	Cl		5	123	129
(10ai)	5-Cl	4-Cl	H	5	146[e]	78	(10bb)	H	Cl		10	181[c]	102
(10aj)	H	1-Cl	2-Cl	9	180[c]	109	(10bc)	H	CH₃		5	123	99
(10ak)	H	2-Cl	4-Cl	10	178[c]	134[c]	Tienilic acid (3)				15	143[c]	118[e]
(10al)	H	1-Cl	H	5	263[c]	108	Indacrinone (5)				19	265[c]	156[e]

[a]From Ref. 11. Seven-week-old Wistar-Imamichi rats that had been fasted for 24 h were divided into groups of five heads so that the animals in each group would excrete almost the same amount of urine. After forced urination, the rats were orally administered the test compounds that were suspended in physiological saline containing 3% gum arabic in a dose volume of 25 mL per kg of body weight. The control rats were given only physiological saline containing 3% gum arabic. The animals were housed in separate metabolic cages and the urine excreted from each animal was collected over a period of 6 h following the administraton of the test compounds or physiological saline after complete starvation. The urine volume was cirectly read on a measuring cylinder after forced urination thereinto, and the amount of urine per kg of body weight was calculated. The amount of uric acid excreted in the urine was determined by the uricase-catalase method. Test compounds were administered at 100 mg/kg p.o. to Wistar-Imamichi rats and the activities are shown as relative activity (%) to the control (100%).

[b]Student's t-test: [c]$p < 0.001$ vs. control, [d]not tested, [e]$p < 0.05$, [f]$p < 0.01$; values without marks are not statistically significant.

329

Table 30.2 Diuretic and Uricosuric Activities of 1,2-Dihydrofuroxanthone-2-carboxylic Acids[a]

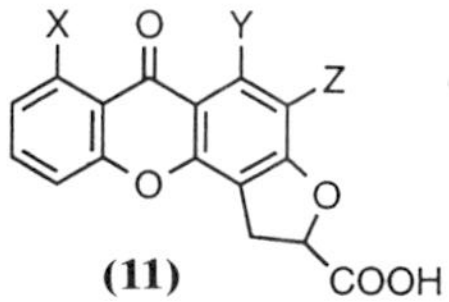

(11)

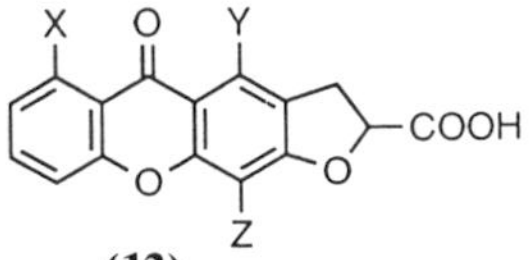

(12)

Compound Number	X	Y	Z	Number of Animals	Diuretic[b] (0–6 h)	Uricosuric[b] (0–6 h)
(11a)	H	H	H	5	203[c]	95
(11b)	F	H	H	5	146	129
(11c)	Cl	H	H	6	163[d]	95
(11d)	H	Cl	H	5	295[c]	100
(11e)	H	H	Cl	15	286[c]	130[c]
(11f)	H	CH$_3$	H	5	314[c]	112
(11g)	H	H	CH$_3$	10	282[c]	141[d]
(11h)	H	H	Br	5	203[d]	116
(11i)	H	CH$_3$	Cl	5	198[c]	97

Compound Number	X	Y	Z	Number of Animals	Diuretic[b] (0–6 h)	Uricosuric[b] (0–6 h)
(12a)	F	H	Cl	5	118	96
(12b)	F	H	CH$_3$	5	86	118
(12c)	Cl	H	Cl	5	100	95
(12d)	H	H	Cl	10	165[c]	126
(12e)	H	H	CH$_3$	15	164[c]	138[e]
Tienilic acid (3)				15	143[c]	118[e]
Indacrinone (5)				19	265[c]	156[c]

[a]From Ref. 11. Test compounds were administered at 100 mg/kg p.o. to Wistar-Imamichi rats and the activities are shown as relative activity (%) to the control (100%). Details of the test protocol are described in Table 30.1.

[b]Student's t-test: [c] p < 0.001 vs. control, [d] p < 0.01, [e] p < 0.05; values without marks are not statistically significant.

were comparable to or more potent than that of tienilic acid (**3**). Some of **10a**, having a substituent at the 1- or 2-position, showed potent diuretic activity almost comparable to that of indacrinone (**5**). Thus, introduction of a chlorine, bromine, or methyl group into the 1- or 2- position increased the diuretic activity, although the 1,2-disubstituted compounds (**10aj** and **10as**) showed only slight improvement in activity compared to the unsubstituted compound **10aa**. On the other hand, the introduction of these substituents into the 4-position gave little significant effect on the activity. Among the 5- to 8-substituted compounds, only the 7-chloro compound (**10ag**) showed potent diuretic activity. Although the uricosuric activity was not correlated with the diuretic activity and showed no apparent SAR, the 2-substituted compounds **10ak**, **10am**, and **10ar** possessed relative potent uricosuric activities.

As a result, compound **10am** was found to possess diuretic and uricosuric activities more potent than those of tienilic acid (**3**) and balanced diuretic and uricosuric activities better than those of indacrinone (**5**).

4 DIHYDROFUROXANTHONE-2-CARBOXYLIC ACIDS

Annulation to position 4 or 2 of the oxyacetic acid group of xanthon-3-yloxyacetic acids (**10a**) to 1,2-dihydrofuro[2,3-*c*]xanthone-2-carboxylic acids (**11**) or 2,3-dihydrofuro[3,2-*b*]xanthone - 2 - carboxylic acids (**12**) was performed in the hope of obtaining more potent uricosuric diuretics (8) (Fig. 30.3) (11). Table 30.2 shows the diuretic and uricosuric activities in rats of the compounds **11** and **12**. The SAR of compounds **11** was similar to that of **10a** and the diuretic activity was generally more potent than that of the corresponding compounds **10a**. The relative effectiveness of the substituents in producing diuretic activity was H > Cl > F (**11a** > **11c** > **11b**), CH$_3$ ≥

Cl > H (**11f** ≥ **11d** > **11a**), and Cl ≥ CH$_3$ > Br, H (**11e** ≥ **11g** > **11h**, **11a**) for X, Y, and Z, respectively. The disubstituted compound **11i** was less active than the monosubstituted compounds **11e** and **11f**. On the other hand, the diuretic activity of compounds **12**, except for **12e**, was comparable to or less than that of the corresponding compounds **10a**. These results show that annulation of phenoxyacetic acids does not necessarily lead to highly potent diuretics. Compounds **11e**, **11g**, and **12e** possessed uricosuric activity. Thus, among these compounds, only **11e**, **11g**, and **12e** had both diuretic and uricosuric activities.

5 5,6-DIHYDROFURO[3,2-*f*]-1,2-BENZISOXAZOLE-6-CARBOXYLIC ACIDS

5,6 - Dihydrofuro[3,2-*f*] - 1,2-benzisoxazole-6-carboxylic acids **13** were derived from annulation to position 5 of the oxyacetic acid group of HP 522 (**8**) (Fig. 30.4) (12–14). The diuretic activities of most of the compounds **13** were greater than that of tienilic acid (**3**) and were comparable to that of indacrinone (**5**) (Table 30.3). Compounds **13a**, **13b**, and **13c** possessed uricosuric activity. Thus, **13a**, **13b**, and **13c** had both diuretic and uricosuric activities, though the diuretic acivity was remarkably high, compared to the uricosuric activity.

Plattner et al. (14) also reported the preparation of compounds **13** and evaluation of their saluretic and uricosuric properties. They selected compound **13c** (A-56234) for further development as an uricosuric diuretic (14,15) (see Section 13).

6 7,8-DIHYDROFURO[2,3-*g*]-1,2-BENZISOXAZOLE-7-CARBOXYLIC ACIDS

7,8 - Dihydrofuro[2,3-*g*] - 1,2 - benzisoxazole - 7-carboxylic acids **14** were derived from

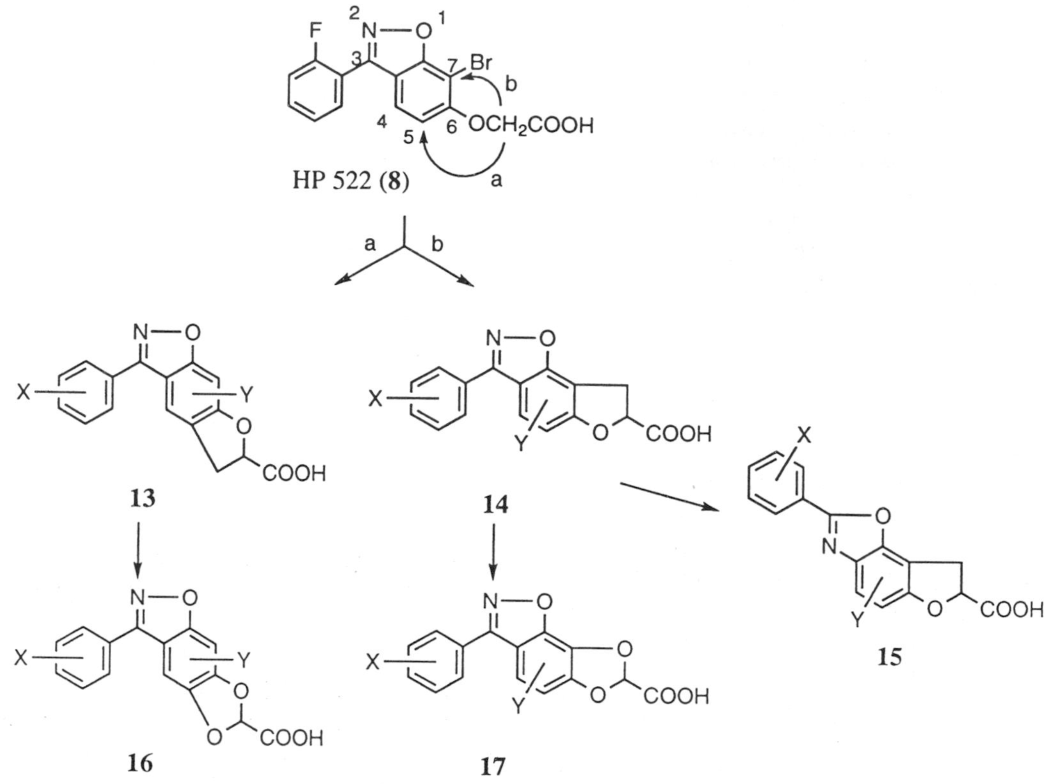

Fig. 30.4 Design of dihydrofuro-1,2-benzisoxazolecarboxylic acid, dihydrofurobenzoxazolecarboxylic acid, and dioxolo-1,2-benzisoxazolecarboxylic acid derivatives.

annulation to position 7 of the oxyacetic acid group of HP 522 (**8**) (Fig. 30.4) (12). The diuretic and uricosuric activities in rats of the compounds **14** are shown in Table 30.4 in comparison with tienilic acid (**3**) and indacrinone (**5**). In contrast to the isomers **13**, the diuretic activity of compounds **14** was very weak, and comparable to or less potent than that of tienilic acid (**3**). Some uricosuric activity was retained, again indicating that annulation does not necessarily lead to high-ceiling diuretics. Among these compounds, compounds **14c** and **14g** had uricosuric activity without diuretic activity. The uricosuric activity of **14c** was evaluated in detail in rats, mice, and Cebus monkeys in comparison to tienilic acid (**3**). Compound **14c** (AA-193) was selected for

further development as a new class of uricosuric agent from these preclinical results (see Section 13).

7 7,8-DIHYDROFURO[2,3-*g*]-BENZOXAZOLE-7-CARBOXYLIC ACIDS

7,8 - Dihydrofuro[2,3 - *g*]benzoxazole - 7 - carboxylic acids **15** are the isomers of **14** and formally derived by Beckmann rearrangement and annulation of the oxime of tienilic acid (**3**) (Figs. 30.2 and 30.4) (12). The diuretic and uricosuric activities in rats of the compounds **15** are shown in Table 30.5. The benzoxazoles **15** showed the highest level of diuretic activity without

Table 30.3 Diuretic and Uricosuric Activities of 5,6-Dihydrofuro[3,2-*f*]-1,2-benzisoxazole-6-carboxylic Acids[a]

(13)

Compound Number	X	Y	Number of Animals	Diuretic[b] (0–6 h)	Uricosuric[b] (0–6 h)
(**13a**)	H	H	50	285[c]	131[c]
(**13b**)	2-Cl	Na	5	300[c]	168[d]
(**13c**)	2-F	H	5	241[c]	134[d]
(**13d**)	3-F	H	5	215[c]	119
(**13e**)	4-F	H	5	334[c]	104
Tienilic acid (**3**)			15	143[c]	118[d]
Indacrinone (**5**)			19	265[c]	156[c]

[a]From Ref. 13. Test compounds were administered at 100 mg/kg p.o. to Wistar-Imamichi rats and the activities are shown as relative activity (%) to the control (100%). Details of the test protocol are described in Table 30.1.
[b]Student's *t*-test: [c]$p < 0.001$ vs. control, [d]$p < 0.05$; values without marks are not statistically significant.

significant uricosuric activity, except for **15d**, in contrast to the isomers **14**. Among these compounds, **15g** showed the most potent diuretic activity.

8 1,3-DIOXOLO[4,5-*f*]-1,2-BENZISOXAZOLE-6-CARBOXYLIC ACIDS

The replacement of the methylene group of the dihydrobenzofuran moiety of compounds **13** by an oxygen atom affords the more hydrophilic compounds, 1,3-dioxolo[4,5-*f*]-1,2-benzisoxazole-6-carboxylic acids (**16**) (Fig. 30.4) (16). From Table 30.6, it is likely that the dioxole compounds **16** possess higher diuretic activity than the corresponding dihydrofuran derivatives **13**, though the uricosuric activity was only marginal. Among these compounds, 8-chloro compounds **16c** and **16g** showed the highest level of diuretic activity.

9 1,3-DIOXOLO[4,5-*g*]-1,2-BENZISOXAZOLE-7-CARBOXYLIC ACIDS

The replacement of the dihydrofuran methylene group of compounds **14** by an oxygen atom also affords the more hydrophilic compounds, 1,3-dioxolo[4,5-*g*]-1,2-benzisoxazole-7-carboxylic acids (**17**) (Fig. 30.4) (16). The diuretic activity of the dioxole compounds **17** also seemed to be more potent than that of the corresponding dihydrofuran derivatives **14** (Table 30.7). Introduction of chlorine into the 4- or 5-position did not improve the diuretic activity. In the variation of the X substituent, the 4-fluoro analogue **17h** showed a high order of diuretic potency. Compounds **17a**, **17d**, and **17h** possessed uricosuric activity. Among these compounds, **17a** and **17h** had both diuretic and uricosuric activities.

As a result, compound **17a** was found to possess diuretic and uricosuric activities

Table 30.4 Diuretic and Uricosuric Activities of 7,8-Dihydrofuro[2,3-g]-1,2-benzisoxazole-7-carboxylic Acids[a]

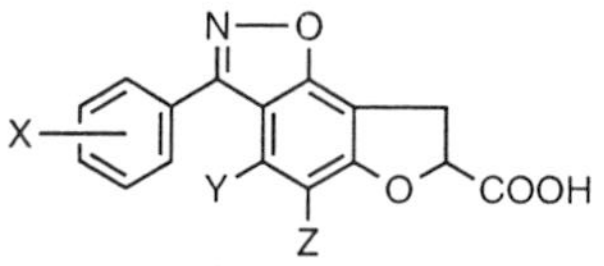

(14)

Compound Number	X	Y	Z	Number of Animals	Diuretic[b] (0–6 h)	Uricosuric[b] (0–6 h)	Compound Number	X	Y	Z	Number of Animals	Diuretic[b] (0–6 h)	Uricosuric[b] (0–6 h)
(14a)	H	H	H	5	102	122	(14j)	4-Cl	H	Cl	5	99	96
(14b)	H	H	Br	5	153	119[c]	(14k)	4-F	H	Cl	5	114	114
(14c)	H	H	Cl	49	107	138[d]	(14l)	4-OH	H	Cl	10	98	98
(14d)	H	H	CH$_3$	5	114	115	(14m)	2,6-F$_2$	H	Cl	5	125	115
(14e)	2-Cl	H	Cl	5	96	109	(14n)	2,4-F$_2$	H	Cl	5	123	131[e]
(14f)	2-F	H	Cl	5	88	113	(14o)	2-F	Cl	H	5	151[c]	104
(14g)	2-CH$_3$	H	Cl	5	96	142[c]							
(14h)	3-Cl	H	Cl	5	144[c]	102	Tienilic acid (3)				15	143[d]	118[c]
(14i)	3-F	H	Cl	5	149[e]	132[e]	Indacrinone (5)				19	265[d]	156[d]

[a]From Ref. 12. Test compounds were administered at 100 mg/kg p.o. to Wistar-Imamachi rats and the activities are shown as relative activity (%) to the control (100%). Details of the test protocol are described in Table 30.1.

[b]Student's t-test: [c]$p < 0.05$, [d]$p < 0.001$ vs. control, [e]$p < 0.01$; values without marks are not statistically significant.

334

Table 30.5   Diuretic and Uricosuric Activities of 7,8-Dihydrofuro[2,3-g]benzoxazole-7-carboxylic Acids^a

(15)

Compound Number	X	Y	Z	Number of Animals	Diuretic^b (0–6 h)	Uricosuric^b (0–6 h)
(15a)	H	H	Cl	5	248^c	101
(15b)	F	Cl	H	5	335^c	123
(15c)	F	Cl	Cl	5	262^c	102
(15d)	F	H	H	5	246^c	135^c
(15e)	F	H	Cl	5	242^c	102
(15f)	H	Cl	H	5	222^c	75^d
(15g)	H	Cl	Cl	5	419^c	97
Tienilic acid (3)				15	143^c	118^d
Indacrinone (5)				19	265^c	156^c

^{a}From Ref. 12. Test compounds were administered at 100 mg/kg p.o. to Wistar-Imamichi rats and the activities are shown as relative activity (%) to the control (100%). Details of the test protocol are described in Table 30.1.
^{b}Student's t-test: ^{c}p < 0.001 vs. control, ^{d}p < 0.05; values without marks are not statistically significant.

more potent than those of tienilic acid (3) and balanced diuretic and uricosuric activities.

10 SULFAMOYL-2,3-DIHYDROBENZOFURAN-2-CARBOXYLIC ACIDS

The discovery of 5-acyl-2,3-dihydrobenzofuran-2-carboxylic acids such as compound 7, with prominent diuretic and uricosuric activities (Fig. 30.2) (8), has stimulated the study to find more promising uricosuric diuretics by replacing the 5-acyl group with other functional groups, such as, the sulfamoyl group, which is commonly found in thiazide diuretics and probenecid (20), an uricosuric drug. Among the compounds synthesized, one of the most interesting is 5-sulfamoyl-2,3-dihydrobenzofuran-2-carboxylic acid 18a (Fig. 30.5) (17).

The diuretic activity of the compounds 18a was evaluated in rats and mice (Table

30.8). Tienilic acid (3) and indacrinone (5) were used as the reference agents for diuretic activity. Structural requirements for diuretic activity within these compounds were examined at three positions (5-, 6-, and 7-) of the 2,3-dihydrobenzofuran ring system. In variants of the X and Y groups at the 6- and 7-positions, the highest activity was found when both X and Y are Cl, and lower activities were observed for compounds having other substituent patterns, such as 18av–18az and 18aaa–18aad. Alkyl variants of the 5-sulfamoyl group (R$_1$ and R$_2$) showed the most potent activity in the case of the lower alkyl groups (18ab, 18ac, 18aaf, and 18aag), whereas the N-unsubstituted and monoalkyl-substituted sulfonamides (18aa and 18an–18au) were devoid of activity or exhibited only slight activity.

A method has been developed (20) to test for uricosuric activity using rats treated with potassium oxonate, which is known to be a uricase inhibitor. The test compounds were administered intraperitoneally and the uricosuric activity was evaluated in terms of the increase in fractional excretion of uric acid (FEua) and urine-excreted amounts of uric acid (UuaV) values. Uricosuric activity in oxonate-treated rats of the compounds, **18a**, are shown in Table 30.8. Tienilic acid (**3**) and probenecid (**20**), which were used as reference compounds, showed hyperuricosuric activities with increased FEua and UuaV values. Indacrinone (**5**), however, showed only an increase in UuaV. Furosemide showed a decrease of FEua, suggesting the possibility of hypouricosuric action. Among the compounds tested, marked increases of both FEua and UuaV were observed for **18aa–18ac**, **18af**, **18ak**, **18al**, **18an**, and **18ao** in which the *N*-substituents (R_1 and R_2) were hydrogen atom and/or lower alkyl groups. Indacrinone (**5**) has been reported to show urate-retaining activity as a result of potent and long-lasting diuresis in clinical trials (21). The

diuretic potency of **18ab** seems to be intermediate between those of indacrinone (**5**) and tienilic acid (**3**). Thus, it should show both uricosuric and moderate diuretic actions, which would allow its use in clinical antihypertensive therapy. Compound **18ab** (S-8666) was selected for clinical evaluation (see Section 13).

2,3-Dihydrobenzofuran-2-carboxylic acid derivatives **18b** with electron-withdrawing groups, such as, nitro, acyl, and sulfamoyl, at the 4- or 5- position were also synthesized (Fig. 30.5) (22). The diuretic activities in rats and mice of the compounds **18b** are shown in Table 30.9, in comparison to the reference compounds tienilic acid (**3**) and indacrinone (**5**). The diuretic activities of 5-nitro-substituted 2,3-dihydrobenzofuran-2-carboxylic acid derivatives (**18ba**, **18bf**, **18bg**, and **18bh**) were absent or weak, whereas those of the 4-nitro compounds (**18bb–18be**) were comparable to or more

Table 30.6   Diuretic and Uricosuric Activities of 1,3-Dioxolo[4,5-*f*]-1,2-benzisoxazole-6-carboxylic Acids[a]

(16)

Compound Number	X	Y	Z	R	Number of Animals	Diuretic[b] (0–6 h)	Uricosuric[b] (0–6 h)
(16a)	H	H	H	H	5	156[c]	102
(16b)	H	Cl	H	H	5	237[d]	98
(16c)	H	H	Cl	H	5	422[d]	77
(16d)	H	Cl	Cl	H	5	291[d]	91
(16e)	Cl	H	H	H	5	140[c]	72[c]
(16f)	F	H	H	H	5	193[e]	103
(16g)	F	H	Cl	H	5	328[d]	115
(16h)	CH_3	H	H	K	5	143	99
Tienilic acid (3)					15	143[d]	118[c]
Indacrinone (5)					19	265[d]	156[d]

[a]From Ref. 16. Test compounds were administered at 100 mg/kg p.o. to Wistar-Imamichi rats and the activities are shown as relative activity (%) to the control (100%). Details of the test protocol are described in Table 30.1.
[b]Student's *t*-test: [c]$p < 0.05$, [d]$p < 0.001$ vs. control, [e]$p < 0.01$; values without marks are not statistically significant.

Table 30.7 Diuretic and Uricosuric Activities of 1,3-Dioxolo[4,5-g]-1,2-benzisoxazole-7-carboxylic Acids[a]

(17)

Compound Number	X	Y	Z	R	Number of Animals	Diuretic[b] (0–6 h)	Uricosuric[b] (0–6 h)
(17a)	H	H	H	H	15	181[c]	124[d]
(17b)	H	Cl	H	H	10	157[c]	116
(17c)	H	H	Cl	H	5	155[e]	99
(17d)	2-Cl	H	H	K	5	122	126[e]
(17e)	4-Cl	H	H	H	5	151[d]	112
(17f)	2-F	H	H	H	5	174[e]	119
(17g)	3-F	H	H	H	5	193[c]	106
(17h)	4-F	H	H	H	10	343[c]	132[e]
(17i)	4-F	H	Cl	H	5	255[c]	110
(17j)	2-CH$_3$	H	H	K	5	93	101
Tienilic acid (3)					15	143[c]	118[e]
Indacrinone (5)					19	265[c]	156[c]

[a]From Ref. 16. Test compounds were administered at 100 mg/kg p.o. to Wistar-Imamichi rats and the activities are shown as relative activity (%) to the contol (100%). Details of the test protocol are described in Table 30.1.
[b]Student's t-test: [c]$p < 0.001$ vs. control, [d]$p < 0.01$, [e]$p < 0.05$; values without marks are not statistically significant.

potent than those of the reference compounds in rats, and were similar to that of tienilic acid (3) in mice. The 4-sulfamoyl compounds also showed significant diuretic activity. In the 4-sulfamoyl compounds, the activities of the *N*-methyl, **18bj**, and the *N,N*-dimethyl, **18bk**, were equivalent, differing somewhat from the results of the 5-sulfamoyl compounds (Table 30.8). Thus, the diuretic activity varied markedly according to the substituents and their positions. The potencies of diuretic activity relative to the position of substitution were 4-position ≫ 5-position and 4-position ≥ 5-position for the nitro and sulfamoyl groups, respectively.

Uricosuric activity in oxonate-treated rats of the compounds **18b** is also shown in Table 30.9. The 4-nitro compounds, **18bb**

and **18bc**, elicited a marked decrease in both FEua and UuaV, showing its hypo-uricosuric character. Compound **18be** was hyperuricosuric because both values increased, but its diuretic character was not potent enough to allow its use as a diuretic agent. Among the 4-sulfamoyl compounds, only **18bi** showed hyperuricosuric activity, and **18bk** and **18bl** increased the UuaV values but not the FEua values. The observed increase in FEua in **18bi** was only temporary (22).

As described above, some 5-sulfamoyl-2,3-dihydrobenzofuran derivatives **18a** showed interesting diuretic and unicosuric characteritics. In contrast, none of the 4-sulfamoyl and nitro derivatives used in this study showed a good balance of diuretic and uricosuric actions.

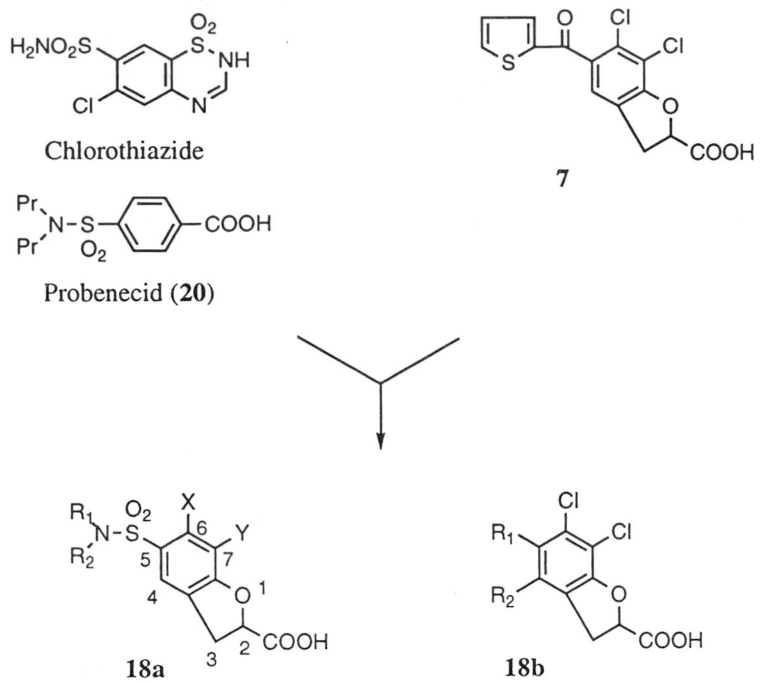

Chlorothiazide

7

Probenecid (20)

18a   18b

Fig. 30.5 Design of 5-sulfamoyl-2,3-dihydrobenzofuran-2-carboxylic acid and 6,7-dichloro-2,3-dihydrobenzo-furan-2-carboxylic acid derivatives.

11 (4-OXO-4H-1-BENZOPYRAN-7-YL)-OXYACETIC ACIDS

4-Oxo-4H-1-benzopyran derivative (**19a**), 2,3-dihydro-1-oxo-1H-isoindole derivative (**21**), 3,4-dihydro-4-oxo-3H-quinazoline derivative (**22**), benzofuran derivative (**23a**), and indene derivative (**23b**) were prepared based on the annulation concept (8) of the phenoxyacetic acid diuretics (Fig. 30.6) (23).

The diuretic and uricosuric activities in Wistar rats and antihypertensive activity in 11-deoxycorticosterone acetate (DOCA)-salt hypertensive rats of the compounds, **19a**, **21**, **22**, **23a**, and **23b**, are shown in Table 30.10. Tienilic acid (**3**) and inda-crinone (**5**) were used as reference com-pounds and **10al** (Table 30.1) (11) was used for comparison. Compound **19a** is the most potent diuretic; more potent than tienilic acid (**3**) and **10al**, and equipotent to inda-

crinone (**5**). The diuretic activity of **21** was lower than that of **19a**. Compounds **22**, **23a**, and **23b** caused no detectable change in the excretion of urine. Uricosuric activity was evaluated in rats using the renal clearance method (26). It was observed that urico-suric agents, such as, tienilic acid (**3**) and indacrinone (**5**), produced an increase in the fractional excretion rate of uric acid (FEua) without changing the concentration of uric acid in serum and urine. The uricosuric acivity (FEua) of **19a** was greater than that of tienilic acid (**3**), indacrinone (**5**), and **10al**. In contrast, **21** showed no uricosuric activity. In an antihypertensive study, the initial systolic blood pressure (SBP) in DOCA-salt hypertensive rats was approximately 110 mm Hg, rising to 191 ± 6 mm Hg ($n = 8$), three weeks later in the control group. The antihypertensive activity of **19a** was equivalent to that of inda-crinone (**5**) and **10al**, but **21** was less active

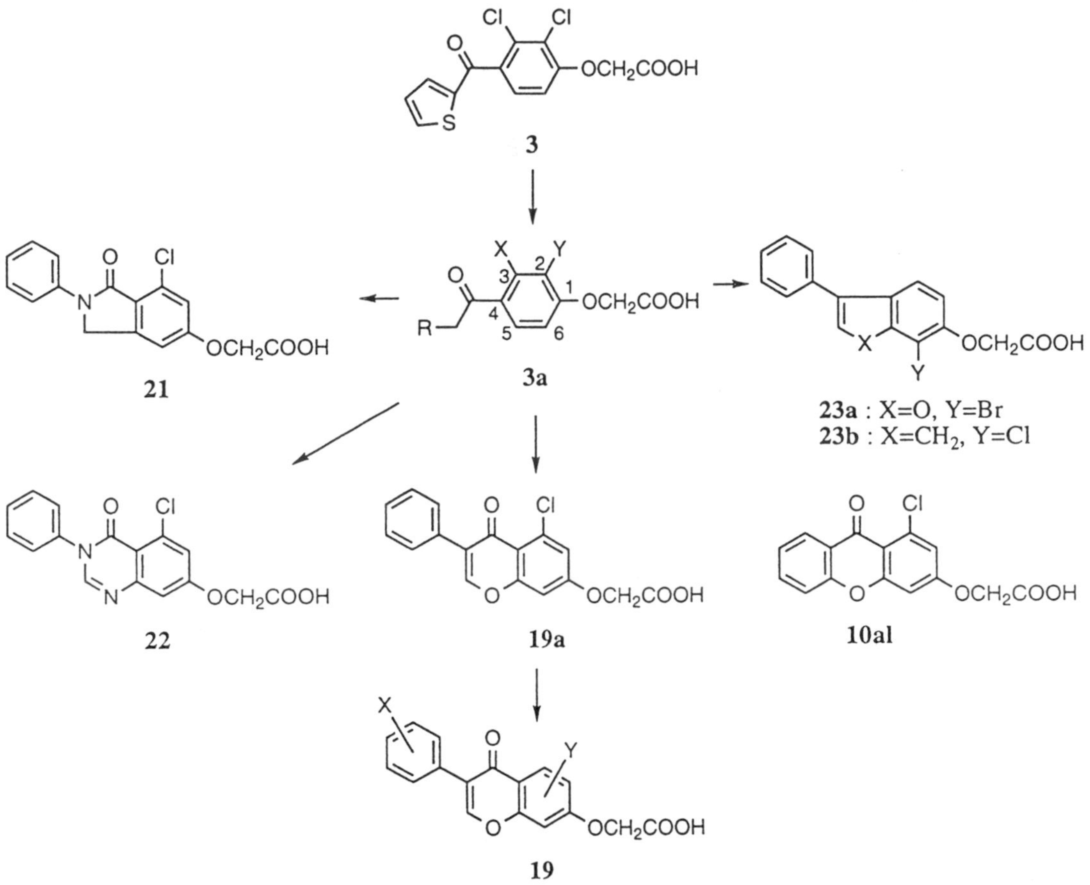

Fig. 30.6 Design of polycyclic phenoxyacetic acids.

than indacrinone (**5**), in accord with the observed diuretic activity. From these results, it is clear that among the new annulated compounds, **19a** has the most favorable profile as a uricosuric diuretic. Therefore, compound **19a** was selected as the lead compound.

The SAR data of compounds **19** is shown in Table 30.11 (27). Natriuretic activity in Wistar rats of the compounds **19** is shown and the results are represented as ratios to the controls. The excretion of Na$^+$ in the control groups was 0.33 meq/kg B.W. Tienilic acid (**3**), indacrinone (**5**), and furosemide were used as reference compounds. The relative effectiveness of the substituents at position 5 in producing natriuretic activity was Cl > CH$_3$ ≥ CF$_3$ ≫ F

(**19a** > **19l** ≥ **19o** ≫ **19p**). The 5,6-disubstituted compounds **19q** and **19r** lacked natriuretic activity. It was determined that the 5-Cl compounds showed the most potent activity in this series. Then variants at position 3 were evaluated. Concerning the substituent of a pendent phenyl group at the 3-position, only the 2-F group (**19f**) gave a good effect on the natriuretic response and the activity of **19f** was comparable with that of furosemide. Compounds **19c**, **19d**, **19e**, **19h**, and **19k** demonstrated almost equivalent activity to **19a**. On the other hand, introducing the 4-Cl (**19b**), 4-OH (**19g**), 2-OCH$_3$ (**19i**), or 2-OH (**19j**) group into the 3-phenyl ring markedly decreased the activity. In general, compounds **19** had almost equipotent uricosuric

Table 30.8 Diuretic and Uricosuric Activities of 5-Sulfamoyl-2,3-dihydrobenzofuran-2-carboxylic Acids[a]

(18a)

Compound Number	X	Y	R_1	R_2	R_3	Diuretic in Rats[b] Dose, mg/kg	Urine volume[e], mL/kg B. W.	Diuretic in Mice[c] Dose, mg/kg	Urine volume[e], mL/kg B. W.	Uricosuric in Oxonate-Treated Rats[d] Dose, mg/kg	Increase of UuaV[f], mg/kg min	Increase of FEua[f]
(18aa)	Cl	Cl	H	H	H	50	26 (N)	30	27 (N)	50	0.154	0.421
(18ab)	Cl	Cl	CH_3	CH_3	H	50	39 (1.4)	30	48 (1.9)	50	0.107	0.212
(18ac)	Cl	Cl	C_2H_5	C_2H_5	K	50	40 (1.6)	30	85 (2.9)	50	0.047	0.092
(18ad)	Cl	Cl	C_3H_7	C_3H_7	H	50	28 (1.3)	30	41 (2.0)	50	0.046	g
(18ae)	Cl	Cl	iso-C_3H_7	iso-C_3H_7	H	50	25 (N)	30	29 (N)			
(18af)	Cl	Cl	CH_3	C_4H_9	H	50	33 (1.4)	30	61 (2.3)	50	0.085	0.065
(18ag)	Cl	Cl	$PhCH_2$	$PhCH_2$	H	50	23 (N)	30	31 (N)			
(18ah)	Cl	Cl	CH_3	$PhCH_2$	H	50	36 (1.6)	30	64 (2.3)	50	0.032	g
(18ai)	Cl	Cl	CH_3	Ph	K	50	39 (1.6)	30	44 (1.7)			
(18aj)	Cl	Cl	CH_3	cyclo-C_6H_{11}	H	50	25 (N)	30	29 (N)			
(18ak)	Cl	Cl	$-(CH_2)_4-$		H	50	35 (1.6)	30	58 (2.1)	50	0.075	0.122
(18al)	Cl	Cl	$-(CH_2)_5-$		H	50	29 (N)	30	66 (2.3)	50	0.110	0.202
(18am)	Cl	Cl	$-(CH_2)_2O(CH_2)_2-$		H	50	25 (N)	30	39 (1.4)			
(18an)	Cl	Cl	H	CH_3	H	50	30 (1.2)	30	34 (1.3)	50	0.100	0.302
(18ao)	Cl	Cl	H	C_3H_7	H	50	28 (N)	30	47 (1.7)	50	0.090	0.339
(18ap)	Cl	Cl	H	iso-C_3H_7	H	50	26 (1.2)	30	38 (1.3)			
(18aq)	Cl	Cl	H	$PhCH_2$	H	50	24 (N)	30	26 (N)			
(18ar)	Cl	Cl	H	Ph	H	50	26 (N)	30	39 (N)			
(18as)	Cl	Cl	H	4-Cl-Ph	H	50	22 (N)	30	37 (1.3)			
(18at)	Cl	Cl	H	4-CH_3O-Ph	H	50	26 (N)	30	36 (1.3)			
(18au)[h]	Cl	Cl	H	$O\langle N(CH_2)_2\rangle_2$	H	50	23 (N)					

Compound											
(18av)	H	Cl	CH_3	H	50	27 (N)	30	24 (N)			
(18aw)	H	Cl	C_2H_5	H	50	29 (N)	30	31 (1.6)			
(18ax)	Cl	H	CH_3	H	50	30 (N)	30	24 (N)			
(18ay)	Cl	H	C_2H_5	H	50	29 (N)	30	26 (1.4)			
(18az)	H	Br	CH_3	H	50	29 (N)	30	29 (N)			
(18aaa)	H	Br	C_2H_5	H	50	27 (N)	30	28 (N)			
(18aab)	CH_3	CH_3	H	H	50	26 (N)	30	27 (N)			
(18aac)	CH_3	CH_3	CH_3	H	50	27 (N)	30	30 (N)			
(18aad)	CH_3	CH_3	CH_3	H	50	32 (N)	30	36 (1.3)			
(18aae)	Cl	Cl	$COOC_2H_5$	H	50	32 (1.4)	30	35 (N)			
(18aaf)	Cl	Cl	C_2H_5	Na	50	47 (1.6)	30	75 (3.8)			
(18aag)	Cl	Cl	C_3H_7	Na	50	44 (1.5)	30	75 (3.8)			
Tienilic acid (3)					100	39 (1.8)	100	36 (2.4)	100	0.123	0.055
Indacrinone (5)					50	34 (1.2)	50	72 (2.5)	50	0.063	g
Furosemide									50	0.028	−0.124
Probenecid (20)									50	0.124	0.070

[a] From Ref. 17.

[b] Male Sprague-Dawley rats, weighing about 250 g, at eight weeks of age, were used in this test. A few lumps of sugar in place of ordinary diet were given on the morning of the day before the test day and 20 mL/kg of 5% glucose solution was given orally at approximately 4 p.m. On the morning of the test day, a suspension or solution of the test compound in 2% gum arabic was orally administered to each rat at a dose of 20 mL/kg. The control group received only 2% gum arabic orally at 20 mL/kg. Immediately after the administration, the test animals were put in plastic cages for the metabolic tests and urine samples were collected for five hours. The cumulative urine volume, urinary sodium, and urinary potassium were measured.

[c] Female ddY mice, weighing about 20 g, were used for the test. The mice were fasted overnight but were allowed free access to water. On the morning of the test day, a suspension or solution of the test compound in 2% gum arabic was orally administered to each mouse at 30 mL/kg. The control mice received only the vehicle. Immediately after the administration, five mice of the treated group were put together in a plastic cage for the metabolic tests and urine was collected for four hours. The cumulative urine volume, urinary sodium, and urinary potassium were measured.

[d] Nine-week-old male rats were employed for the test. Potassium oxonate was intraperitoneally administered to the animals at a dose of 250 mg/kg to measure uric acid clearance and inulin clearance. Within two hours after the administration of the potassium oxonate, cannulae were placed in the right femoral artery, left femoral vein, and urinary bladder or each animal under pentobarbital anesthesia for blood collection, drug infusion, and urine collection, respectively. At two hours after the first administration, potassium oxonate was administered again at the same dosage and then 60% urethane (2 mL/kg) and 15% inulin (4 mL/kg) were subcutaneously injected. A mixture of 4% mannitol-1.5% inulin-0.9% saline was infused at the flow rate of 0.1 mL/min into each animal on a plate kept at 30°C. The animal was allowed 40 min to reach an equilibrium state, then arterial blood (0.2 mL each) samples were collected six times at 20-min intervals, and five 20-min urine samples were collected. Immediately after the collection of each blood sample, the serum was separated. The serum and urine samples were stored in a refrigerator. Immediately after collection of the first urine sample, a test compound suspended in 1% gum arabic was intraperitoneally administered at 2 mL/kg. Uric acid levels in the serum and the urine were measured by the method of Yonetani et al. (18). Inulin was measured essentially by the method of Vurek and Pegram (19). To analyze uric acid, 0.1 mL of a diluted solution of deproteinized serum or urine was admixed with a 1% dimedone-phosphoric acid solution and the resulting mixture was heated for five minutes. The mixture was then cooled in iced water and combined with 2.0 mL of acetic acid. The fluorescence was measured at 410 nm with excitation at 360 nm.

[e] Ratio to the control is shown in parenthesis; N indicates that the difference from the control was not statistically significant.

[f] Increases of urine-excreted amounts of uric acid (UuaV) and fractional excretion of uric acid (FEua) were calculated as the average value for 80 min after dosing.

[g] There was no difference compared with the control.

[h] HCl salt.

Table 30.9 Diuretic and Uricosuric Activities of 6,7-Dichloro-2,3-dihydrobenzofuran-2-carboxylic Acids[a]

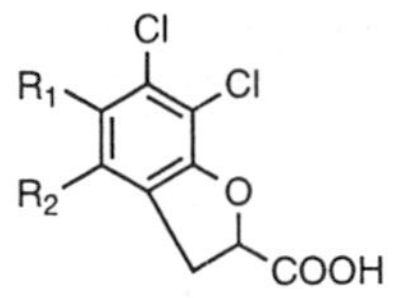

(18b)

Compound Number	R_1	R_2	Diuretic in Rats[b]		Diuretic in Mice[b]		Uricosuric in Oxonate-Treated Rats[b]		
			Dose, mg/kg	Urine volume[c], mL/kg B. W.	Dose, mg/kg	Urine volume[c], mL/kg B. W.	Dose, mg/kg	Increase of UuaV[d], mg/kg min	Increase of FEua[d]
(18ba)	NO_2	H	100	33 (N)	30	25 (N)			
(18bb)	$NHCOCH_3$	NO_2	100	40 (1.3)	30	45 (1.7)	50	−0.006	−0.188
(18bc)	$NHCOC_6H_4$-2-F	NO_2	100	48 (1.6)	30	42 (1.4)	50	e	−0.128
(18bd)	NH_2	NO_2	100	47 (1.5)	30	35 (1.3)			
(18be)	H	NO_2	50	38 (1.4)	30	36 (1.4)	50	0.126	0.259
(18bf)	NO_2	$NHCOCH_3$	50	33 (N)	30	30 (N)			
(18bg)	NO_2	$NHCOC_6H_4$-2-F	50	33 (N)	30	26 (N)			
(18bh)	NO_2	NH_2	50	30 (N)	30	37 (N)			
(18bi)	H	SO_2NH_2	50	35 (1.5)	30	44 (1.5)	50	0.029	0.097
(18bj)	H	SO_2NHCH_3	50	42 (1.8)	30	61 (2.2)			
(18bk)	H	$SO_2N(CH_3)_2$	50	48 (1.7)	30	60 (2.6)	50	0.048	−0.106
(18bl)	H	$SO_2N(CH_3)CH_2Ph$	50	34 (1.4)	30	39 (N)	50	0.032	e
(18bm)	H	$SO_2N(CH_3)Ph$	50	27 (1.1)	30	34 (1.4)			
Tienilic acid (3)			100	39 (1.8)	30	36 (2.4)	100	0.123	0.055
Indacrinone (5)			50	34 (1.2)	30	72 (2.5)	50	0.063	e
Furosemide							50	0.028	−0.124
Probenecid (20)							50	0.124	0.070

[a]From Ref. 22.

[b]The experimental details are described in Table 30.8.

[c]Ratio to the control is shown in parenthesis; N indicates that the difference from control was not statistically significant.

[d]Increases of UuaV and FEua were calculated as the average value for 80 min after dosing.

[e]There was no difference compared with the control.

342

Table 30.10   Biological Activities of Phenoxyacetic Acids[a]

Compound Number	Diuretic[b]	Uricosuric[c]	Antihypertensive[d]
(19a)	+ + +	+ + +	+ + +
(21)	+	−	+
(22)	−	NT	NT
(23a)	−	NT	NT
(23b)	−	NT	NT
Tienilic acid (3)	±	+	−
Indacrinone (5)	+ + +	+	+ + +
(10al)	+ +	+	+ + +

[a]From Ref. 23.

[b]Male Wistar rats weighing 120–180 g were starved for 18 h and deprived of drinking water for two hours before the test. The animals were orally loaded with 25 mL/kg of physiological saline, immediately followed by oral administration of the test drugs which were suspended in a 0.5% aqueous solution of carboxymethylcellulose sodium (CMC). The rats were housed singly in stainless steel metabolic cages with no access to food or water. Urine was collected during the five-hour period after dosing, and urine volume was measured. Urinary sodium and potassium contents were estimated by using an electrolyte analyzer with ion-selective electrodes (PVA-4, Photovolt, U.S.A.). Urine volume (mL/kg/5 h, 100 mg/kg, p.o.): $- \leq 0$, $0 < \pm \leq 10$, $10 < + \leq 20$, $20 < + + \leq 30$, $30 < + + +$.

[c]Male Wistar rats weighing 180–220 g were starved for 18 h and deprived of drinking water for two hours before the test. The animals were orally given both 25 mL/kg of saline and the test drugs. Sixty minutes after dosing of the test drugs, the animals were housed singly in the metabolic cages, and urine was collected for 30 min. Immediately after the 30-min collection of urine, blood was taken from the carotid artery under ether anesthesia. At the same time, the remaining urine in the bladder was directly collected by using a syringe, and total urine volume was measured. A blood sample was centrifuged within 30 min after collection, and resultant plasma was used for the measurement of uric acid and creatinine. Plasma and urinary uric acid were estimated by the uricase method (Uric acid β-test. Wako, Osaka, Japan), and creatinine content was determined by Jaffe's method (Creatinin Set, Wako, Osaka, Japan). FEua was calculated by using the following formula: FEua = Cua/Ccr, where Cua is uric acid clearance and Ccr is creatinine clearance. Excretion of uric acid (%, FEua, 200 mg/kg, p.o.): $- \leq 0$, $0 < + \leq 50$, $50 < + + \leq 150$, $150 < + + +$, NT: not tested.

[d]Four-week-old male rats (Sprague-Dawley) weighing 150–180 g were used. The left kidney of each rat was removed aseptically under ether anesthesia. From one week after the unilateral nephrectomy, the animals were treated with DOCA (15 mg/kg, subcutaneously, once a week) and received 1% sodium chloride as drinking water (24). Simultaneously, the test drugs were orally administered to the animals daily for three weeks. Systolic blood pressure (SBP) of the animals in a conscious state was measured weekly prior to the daily dosing, by a tail cuff method (25). SBP (Δ mmHg, 100 mg/kg/d, p.o.): $- \leq 10$, $10 < + \leq 20$, $20 < + + \leq 40$, $40 < + + +$, NT: not tested.

activity to tienilic acid (3) and indacrinone (5) (Table 30.11). It was difficult to relate the uricosuric activity to the structures and the uricosuric activities did not parallel the natriuretic activities.

In this series, [5-chloro-3-(2-methylphenyl)-4-oxo-4H-1-benzopyran-7-yl]oxyacetic acid (19e) showed a good balance of diuretic and uricosuric activities, and a good biological profile. Compound 19e (DR-3438) was selected for further investigation. The uricosuric and diuretic properties of the new agent DR-3438 (19e) were evaluated in several species (28). In the

conventional clearance studies in urate-loaded dogs, intravenous injection of DR-3438 (19e) (3–30 mg/kg) resulted in dose-related increases in fractional excretion of urate (FEua), urine flow, and sodium excretion. At doses causing similar natriuresis, tienilic acid (3) (50 mg/kg, i.v.) markedly increased the FEua value, whereas indacrinone (5) (1 mg/kg, i.v.) had no significant effect on it. Trichloromethiazide (1 mg/kg, i.v.) and furosemide (0.3 mg/kg, i.v.) tended to decrease the FEua. Thus, the uricosuric activity of DR-3438 (19e) (30 mg/kg) was 0.6-fold that of tienilic acid

Table 30.11 Natriuretic and Uricosuric Activities of (4-Oxo-4H-1-benzopyran-7-yl)oxyacetic Acids[a]

(19)

Compound Number	X	Y	Z	Natriuretic[b] (Rat)	Uricosuric[c]	Compound Number	X	Y	Z	Natriuretic[b] (Rat)	Uricosuric[c]
19a	H	Cl	H	3.37	++	**19l**	H	CH$_3$	H	2.76	++
19b	4-Cl	Cl	H	1.78	+[d]	**19m**	2-CH$_3$	CH$_3$	H	1.98	++[d]
19c	4-F	Cl	H	3.12	+	**19n**	2-Cl	CH$_3$	H	1.68	+[d]
19d	2-Cl	Cl	H	3.50	+	**19o**	H	CF$_3$	H	2.71	+
19e	2-CH$_3$	Cl	H	3.43	+	**19p**	H	F	H	1.27[e]	NT[f]
19f	2-F	Cl	H	4.15	+	**19q**	H	Cl	Cl	0.85[e]	NT[f]
19g	4-OH	Cl	H	0.80[e]	NT[f]	**19r**	H	Cl	CH$_3$	1.15[e]	NT[f]
19h	4-OCH$_3$	Cl	H	3.35	+	Tienilic acid (**3**)				1.80[g]	+
19i	2-OCH$_3$	Cl	H	0.97[e]	NT[f]	Indacrinone (**5**)				2.26	+
19j	2-OH	Cl	H	1.20[e]	NT[f]	Furosemide				4.60	NT[f]
19k	2-CH$_3$, 4-OCH$_3$	Cl	H	2.98	NT[f]						

[a]From Ref. 27.
[b]Ratio to control (treated/control value) is shown. Animal: SLC-Wistar, 100 mg/kg, p.o./5 h. Details of the test protocol are described in Table 30.10.
[c]Excretion of uric acid (%, FEua, 100 mg/kg, p.o. in rats): $-\leq 0$, $0 < + \leq 50$, $50 < ++ \leq 150$, $150 < +++$. Details of the test protocol are described in Table 30.10.
[d]200 mg/kg, p.o. in rats.
[e]The difference from the control is not statistically significant.
[f]Not tested.
[g]Dose: 300 mg/kg p.o.

(3) and 3.4-fold that of indacrinone (5). In contrast, in urate-loaded rabbits that exhibit net tubular secretion of urate, intravenous DR-3438 (19e) (30 mg/kg) produced a significant decrease in FEua. Stop-flow studies in dogs revealed that DR-3438 (19e) (30 mg/kg) blocks both urate reabsorption and p-aminohippurate secretion in the proximal segment of the nephron and strongly inhibits reabsorption of water, sodium, and potassium in the distal segments. These results suggest that DR-3438 (19e) exerts uricosuric activity through blocking urate transport in the proximal tubules, and diuretic and saluretic activities by inhibiting water and sodium reabsorption in the distal segment of the nephron.

12 THREE-DIMENSIONAL STRUCTURE-ACTIVITY RELATIONSHIPS AND RECEPTOR MAPPING OF DIURETIC ACTIVITY OF PHENOXYACETIC ACIDS

Although a number of suggestions regarding the biochemical basis for the ability of phenoxyacetic acid diuretis to modulate electrolyte transport have been made, little is known of their mechanism of action at the molecular level (8). Structure-activity relationship studies occasionally provide useful information about the molecular mechanism of drug action (29). A number of SAR studies have been made regarding the diuretic activity of phenoxyacetic acids qualitatively (8). Annulation to position 3 or 5 of 4-acyl groups of phenoxyacetic acids, such as, ethacrynic acid (2) and tienilic acid (3), afforded indacrinone (5) or HP 522 (8) (Figs. 30.2 and 30.7). On the other hand, compound 7 was derived by annulation to position 6 of the oxyacetic acid group of tienilic acid (3). These annulated compounds exhibited high-ceiling diuretic activity (8). From these findings, it has been suggested that annulation of phenoxyacetic acids leads to a high-ceiling diuretic (8,14). Also interesting is that indacrinone (5) has two enantiomers in which the diuretic activity of the R isomer is higher than that of the S isomer, and all

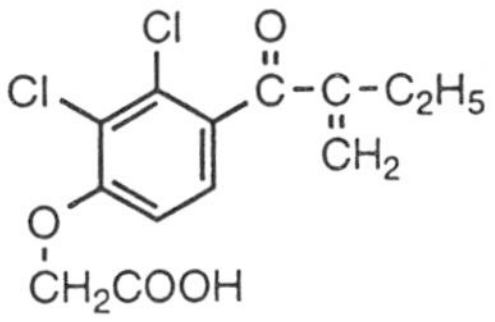

Ethacrynic acid (2)

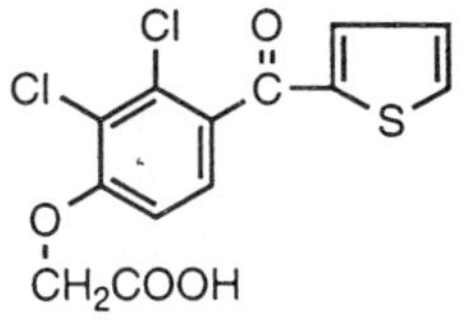

Tienilic acid (3)

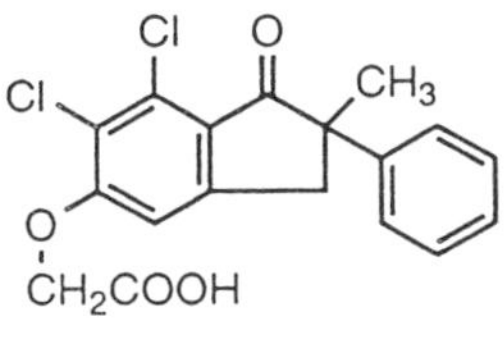

Indacrinone (5)

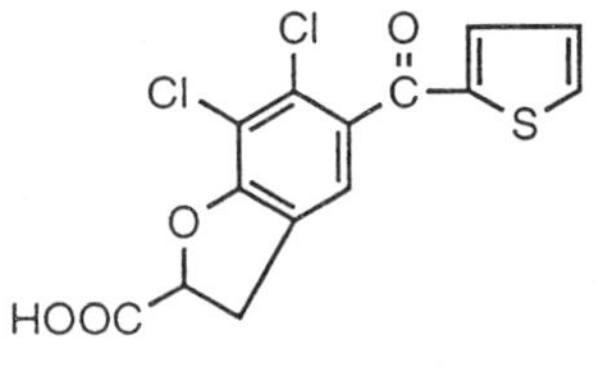

7

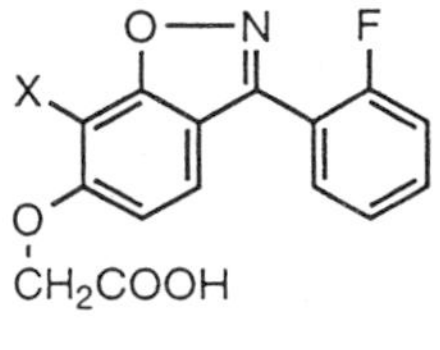

HP 522 (8): X = Br
8a: X = Cl

Fig. 30.7 Phenoxyacetic acid diuretics.

Table 30.12   Rat Diuretic Activity[a] for Phenoxyacetic Acid Diuretics[b]

Compound	Activity Score[c]	Compound	Activity Score[c]
Tienilic acid (**3**)	1	*S*-(+)-(**7**)	6
R-(−)-(**5**)	5	(±)-(**7**)	5
(±)-(**5**)	4	*R*-(−)-(**7**)	0
S-(+)-(**5**)	2	(**8a**)	3

[a]Test compounds were administered at 100 mg/kg p.o. to Wistar-Imamichi rats and the relative activity (%) to the control (100%) was calculated. See Table 30.1 and ref. 30 for details of test protocol.
[b]From Ref. 30.
[c]To simplify the data, the diuretic results were scored according to the following criteria. The scores 0, ±, 1, 2, 3, 4, 5, 6, and 7 represent the relative activity of ≤100%, 110–120%, 130–160%, 170–200%, 210–250%, 260–300%, 310–350%, 360–400%, and 410–450% to the control (100%), respectively.

the diuretic activity of compound **7** resides in the *S* isomer, whereas the *R* isomer is inactive (**8**). These results are summarized in Table 30.12 (30). Table 30.12 shows that the annulation of tienilic acid (**3**) leads to the more active compounds **5**, **7**, and **8a**, and that the steric effect of these annulated compounds is critical in explaining the diuretic activity. From these observations, it has been proposed that the indanone moiety of *R*-(−)-**5** and the dihydrobenzofuran-2-carboxylic acid moiety of *S*-(+)-**7** successfully mimic the receptor-bound conformations of the acyl and oxyacetate side chains, respectively, of **3** and the analogues. Accordingly the active model (receptor model) would be created with these moieties (Fig. 30.8) (30). In this model, the carboxylate group (anionic site) is thought

to interact with the cationic site of the receptor. The carbonyl oxygen seems to function as a hydrogen-bond acceptor and is located so that it is coplanar with the dichlorobenzene ring and on the same side as the chlorine atoms. This configuration seems to be important to form the hydrogen bond with the receptor favorably. The ether oxygen moiety seems to function as a hydrogen-bond acceptor and/or as a bridge, which is limited in size. Two chlorine atoms and the methyl and phenyl groups are thought to serve as hydrophobic groups, which are limited in size. The bridging methylenes of indanone and dihydrofuran moieties were removed in this model, because it seems to be necessary only to fix the conformations of these moieties favorably.

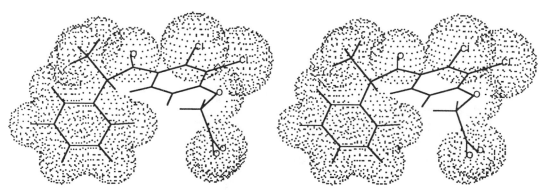

Fig. 30.8 Stereoview of the active model (receptor model) for diuretic activity of phenoxyacetic acids. Reproduced from ref. 30. Courtesy of the American Chemical Society.

To verify the validity of this active model, a three-dimensional SAR study of compounds **3**, **5**, **7**, and **8a** has been made. Thus, when a compound has high activity, it should fit this model well. The active conformation of compounds was defined so that among the lowest energy conformers, the conformer that resembles the active model was selected. The active conformations of **3** and enantiomers of **5** and **7** were superposed on the active model by matching the corresponding atoms in the benzene ring so that the carboxylate group occupies the same side of the benzene plane as that of the active model (Figs. 30.9 and 30.10). Compound **8a** was superposed on the model by matching the nitrogen atom of the isoxazole ring with the carbonyl oxygen of the model, in addition to the same procedure as described above (Fig. 30.10). Compounds R-($-$)-**5** and S-($+$)-**7** are very potent and have almost the same level of activity (Table 30.12), suggesting that the indanone moiety with the R configuration and the dihydrofuran-2-carboxylic acid moiety with the S configuration contribute to the great increase of the activity to almost the same extent (Figs. 30.9 and 30.10). The enantiomer S-($+$)-**5** has the extra region occupied by the more bulky phenyl group in the region of the methyl group of the active model (Fig. 30.9). This was thought to cause a reduction of activity for S-($+$)-**5** compared to that of R-($-$)-**5** (Table 30.12). On the other hand, tienilic acid (**3**) does not fit the active model well and occupies a different region in space than does the model (Fig. 30.9). This is thought to bring about a detrimental effect on the activity and to make the relevant binding with the receptor unfavorable. Enantiomer R-($-$)-**7** does not have the corresponding regions of hydrogen bonding and hydrophobicity of the receptor model, and the thienyl group penetrates deeply into the region where the chlorine atoms of the model just interact with the receptor (Fig. 30.10). Therefore, it is expected that it is inactive. In Fig. 30.10, it is shown that HP 522 derivative **8a** has a nitrogen that can accept a hydrogen bond in the region of hydrogen bonding of the active model, but it does not possess a hydrophobic group in the region occupied by the chlorine atom at position 3 of the model. Thus, it was possible to ratonalize the relation between the structure and the activity of the compounds **3**, **5**, **7**, and **8a** by using this active model qualitatively. These results show that the active model (receptor model) is very useful in explaining the three-dimensional SAR of phenoxyacetic acid diuretics, and should at least in part represent the delineation of the receptor.

In order to further evaluate the usefulness of the model, an attempt was made to explain the SAR of compounds **10–17** in Fig. 30.2 using the active model. The diuretic activity in rats of typical compounds is shown in Table 30.13 (11, 12, 16, 30). From these and related data (11, 12, 13, 16, 30), it has been suggested that the great difference in the activity between these compounds might be ascribed to the ring system rather than the substituent effect, and that the annulation hypothesis (8,14) described above does not necessarily apply to all of these compounds. The active conformers defined as mentioned previously were superposed on the model so that the carbon atom of the carboxylate group and the hydrogen-bonding and hydrophobic regions matched those of the model, without the aryl ring plane deviating from the dichlorobenzene ring plane of the model (Figs. 30.11–30.13).

Xanthones **10am** and **10bb** possess the same acivity (Table 30.13). The carbonyl oxygen of **10am** fits the carbonyl oxygen region of the active model well, whereas in the case of **10bb**, the oxygen at position 10 appears to function as a hydrogen bond acceptor (Fig. 30.11). Fig. 30.11 shows that both xanthones **10am** and **10bb** are predicted to possess activity similar to that of **8a**. Dihydrofuroxanthone **11e** (S isomer)

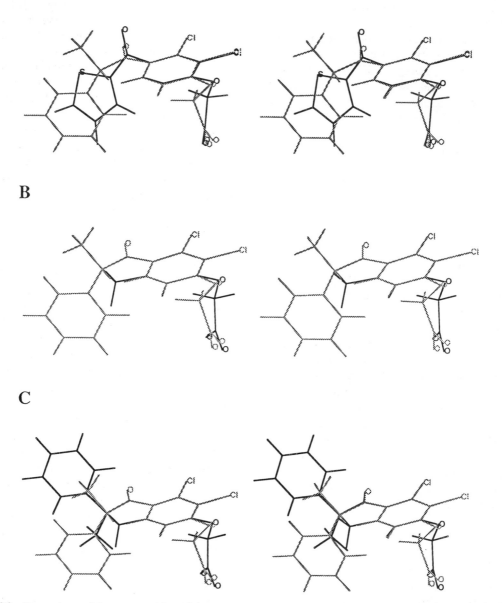

Fig. 30.9 Stereoviews of the superposition of the proposed active conformers of **3** (**A**, black), *R*-(−)-**5** (**B**, black), and *S*-(+)-**5** (**C**, black) and the active model (gray). Reproduced from ref. 30. Courtesy of the American Chemical Society.

fits the model better than xanthones **10am** and **10bb**. Consequently the activity of **11e** is expected to be superior to these xanthones **10am** and **10bb** (Fig. 30.11 and Table 30.13). In contrast, the isomer **12d** (*S* isomer) does not appreciably fit the model, as shown in Fig. 30.12, and the activity as the racemate is estimated to be less than that of **11e**. In the series of the benzisoxazoles, compound **13a** (*S* isomer) fits the

A

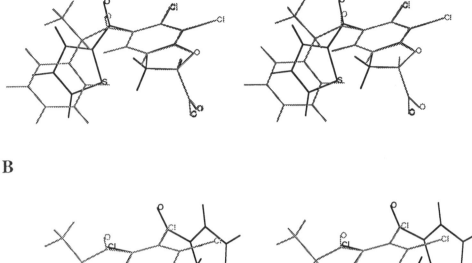

B

C

Fig. 30.10 Stereoviews of the superposition of the proposed active conformers of S-(+)-**7** (**A**, black), R-(−)-**7** (**B**, black), and **8a** (**C**, black) and the active model (gray). Reproduced from ref. 30. Courtesy of the American Chemical Society.

model well, whereas the isoxazole moiety of compound **14c** does not occupy the favorable regions for increasing the activity and the phenyl group stands out from the receptor model (Fig. 30.12). It seems that the difference in the activity between **13a** and **14c** shown in Table 30.13 is ascribable

to the degree of the fitting. Similarly, the difference in activity between compounds **16c** and **17c** seems to be attributable to the reason described above (Fig. 30.13). The dioxole derivatives **16c** and **17c** displayed diuretic activity of a level higher than that of the corresponding dihydrofuran deriva-

Table 30.13   Rat Diuretic Activity[a] for Phenoxyacetic Acids[b]

Number	Structure	Activity Score[c]	Number	Structure	Activity Score[c]
(10am)		2	(14c)		±
(10bb)		2	(16c)		7
(11e)		4	(17c)		1
(12d)		2	(15g)		7
(13a)		4			

[a]The conditions are the same as for Table 30.12, footnote [a].
[b]From Ref. 30.
[c]The criteria are the same as for Table 30.12, footnote [c].

tives **13a** and **14c**, respectively. As shown in Figs. 30.12 and 30.13, compounds **16c** and **17c** have an oxygen atom in the region of the methylene group of compounds **13a** and **14c**. This difference seems to make binding of **16c** and **17c** to the receptor or the transport into the active site more favorable compared to **13a** and **14c**. Benzoxazole derivative **15g** showed a high order of potency and the *S* isomer exhibited a high

degree of fitting to the active model, as illustrated in Fig. 30.13.

These analyses bring new and important insights into three-dimensional SAR for phenoxyacetic acid diuretics and define some structural requirements for increasing the diuretic activity. These include the regons occupied by the carboxylate group below the plane of the chlorobenzene ring, the (two) hydrogen-bonding group(s) in the

A

B

C

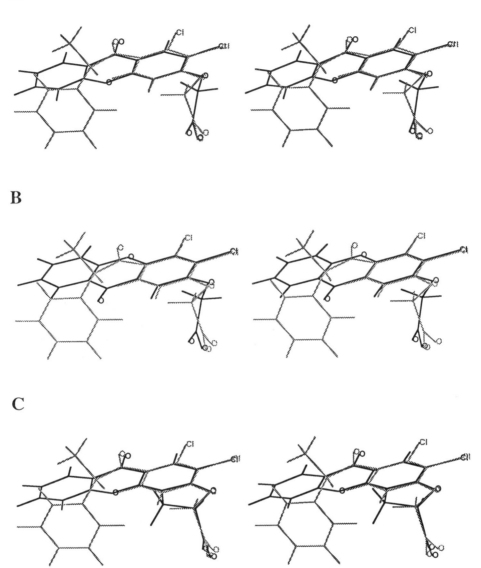

Fig. 30.11 Stereoviews of the superposition of the proposed active conformers of **10am** (**A**, black) **10bb** (**B**, black), and **11e** (**C**, black) and the active model (gray). Reproduced from ref. 30. Courtesy of the American Chemical Society.

same plane as that of the chlorobenzene ring, and the hydrophobic groups, which are limited in size. From these analyses, it is also apparent that the annulation hypothesis of phenoxyacetic acids can be applied only when the annulated compound satisfies the structural requirements for activity.

With these insights in mind, a modified receptor model was constructed, as shown in Fig. 30.14. The model was created with the active model shown in Fig. 30.8, the dioxole moiety in **16c**, and the 2-phenyloxazole moiety in **15g**. This model can be used for prediction of diuretic activity of

A

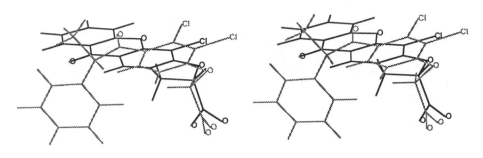

B

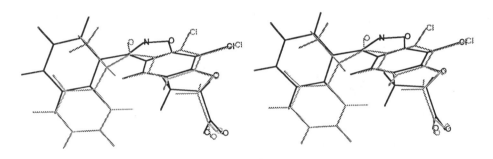

C

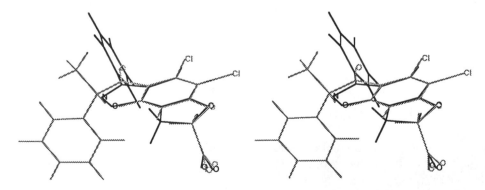

Fig. 30.12 Stereoviews of the superposition of the proposed active conformers of **12d** (**A**, black), **13a** (**B**, black), and **14c** (**C**, black) and the active model (gray). Reproduced from ref. 30. Courtesy of the American Chemical Society.

A

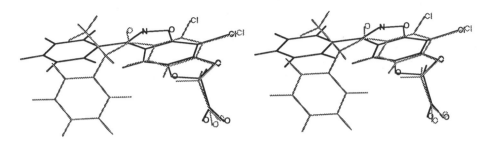

B

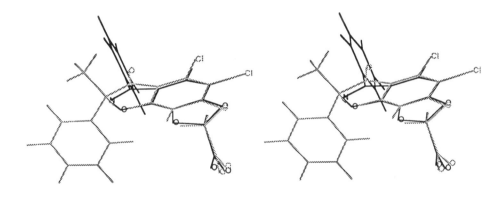

C

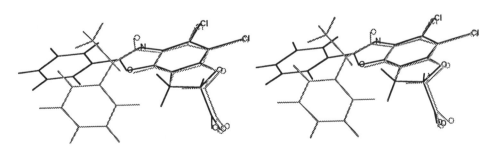

Fig. 30.13 Stereoviews of the superposition of the proposed active conformers of **16c** (**A**, black), **17c** (**B**, black), and **15g** (**C**, black) and the active model (gray). Reproduced from ref. 30. Courtesy of the American Chemical Society.

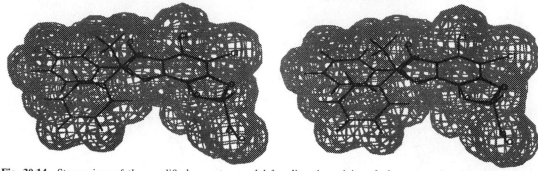

Fig. 30.14 Stereoview of the modified receptor model for diuretic activity of phenoxyacetic acids. Reproduced from ref. 30. Courtesy of the American Chemical Society.

phenoxyacetic acids and, furthermore, may permit one to design compounds with the desired degree of diuretic activity.

Plattner et al. (31) also reported the receptor mapping of phenoxyacetic acid diuretics Abbott 49816 (**9**) and the derivatives. However, because the model is thought to be a combined model for two separate classes of diuretics, the phenoxyacetic acids and the 2-(aminomethyl)-phenols (32), it will not be discussed further in this review.

13 PHARMACOLOGY

13.1 A-56234

13.1.1 DIURETIC AND URICOSURIC ACTIVITIES. It has been reported that Abbott has developed a new uricosuric diuretic, A-56234, for clinical trials (15). The diuretic, saluretic, and uricosuric activities of orally administered A-56234 (**13c**) were evaluated in several species in comparison to those of furosemide, indacrinone (**5**), hydrochlorothiazide, ethacrynic acid (**2**), and tienilic acid (**3**) (14,15). A-56234 was approximatley four and eight times less potent, respectively, than furosemide and indacrinone in the rat model. In saline-loaded male mice, the diuretic and natriuretic effects of A-56234 were equipotent to those of furosemide and indacrinone. A-56234 was

also equipotent to hydrochlorothiazide and significantly more potent than ethacrynic acid and tienilic acid, although the thiazide displayed a different diuretic and natriuretic profile, typical of a low-ceiling diuretic. In conscious female dogs, A-56234 was approximately six times more potent than furosemide and greater than 10 times more potent than indacrinone. Detailed studies showed no significant difference in the pattern of Cl^-, K^+, Ca^{2+}, or Mg^{2+} excretion. Urine volume was also increased in a dose-related manner. Onset of activity was dose-related and the duration of activity was four to six hours. Urate excretion was increased for A-56234 and tienilic acid but not for furosemide or indacrinone during the six-hour collection period. Tests on water-loaded conscious Cebus monkeys demonstrated significant uricosuric activity that was equal to that of indacrinone, but less than that of tienilic acid with single oral doses. Furosemide did not increase urate excretion in this assay. Intraduodenal administration of equivalent natriuretic doses of A-56234 and furosemide produced similar hemodynamic effects that were related to their potent diuretic activities. Both compounds lowered central venous pressure, maximum left ventricular dP/dt, and cardiac output. Pulmonary and systemic vascular resistances were increased by both compounds. In summary, studies to date indicate that A-56234 is a potent, high-ceiling diuretic with moderate uricosuric activity and has no adverse effects on the cardiovascular system. The resolved enantiomers S-(+)-A-56234 and R-(−)-A-56234 were tested in mice, dogs, and monkeys in comparison to the racemate A-56234. Diuretic activity in male mice indicated that the S-(+)-enantiomer possesses all the diuretic activity and is approximately twice as potent as the racemate. This enantiomeric separation of diuretic activity was also confirmed in the dog assay. In both dogs and conscious Cebus monkeys, both enantiomers exhibited equivalent uricosuric activity.

13.2 S-8666

13.2.1 URICOSURIC ACTIVITY. The uricosuric activity of S-8666 (**18ab**) was demonstrated by a clearance study using uricase-inhibited rats produced by the intraperitoneal (i.p.) administration of potassium oxonate under anesthesia. In the hyperuricemic rat model, S-8666 at a dose of 1 mg/kg, i.p., had no effect on the urinary uric acid excretion. However, this novel candidate uricosuric diuretic agent, at doses of more than 10 mg/kg, i.p., showed a uricosuric elevated fractional excretion of uric acid (FEua) which was obviously caused by the inhibition of net urate reabsorption in the renal tubules (33). The uricosuric activity of S-8666 in oxonate-treated rats was apparently more marked than those of known uricosuric agents, such as, probenecid, benzbromarone, tienilic acid, and indacrinone. S-8666 was also uricosuric in oxonate-untreated rats.

Extensive studies in various animals have shown species differences in the urate excretion mechanisms of mammalian kidneys and in the actions of uricosuric agents (34). Mechanisms of the renal excretion of urate have been studied extensively in the Cebus monkey, because of its similarity to

man and its sensitivity to drugs known to interfere with urate transport in man. The monkeys are, however, less sensitive to these drugs than man. On the other hand, chimpanzees are sensitive to all drugs known to be uricosuric in man and are generally more sensitive to the uricosuric agents than man. Therefore, the rank order of responsiveness in terms of uricosuric action is chimpanzees > man > Cebus monkeys. The uricosuric effect of S-8666 was evaluated by clearance-type studies using two chimpanzees. The drug at a dose of 10 mg/kg, p.o., showed obviously enhanced urate excretion without a marked change in plasma urate levels. The potency of uricosuric activity of S-8666 was similar to that of probenecid and less than that of indacrinone (33).

The effects of the enantiomers of S-8666 on urate excretion were studied using both a clearance technique in oxonate-treated rats under anesthesia and in vitro microperfusion technique of individual tubular segments isolated from the rabbit kidney. Yonetani et al. showed that the R-(+)-enantiomer of S-8666 exhibited elevation of FEua in oxonate-treated rats, whereas the S-(−)-enantiomer did not have uricosuric activity (33). In microperfusion studies, Shimizu et al. found that S-8666 in the lumen inhibited the lumen-to-bath urate flux in rabbit proximal tubules, whereas the drug in the bath reduced the bath-to-lumen urate flux (35). Moreover, they concluded that enantioselectivity was not found for these inhibitory effects of S-8666 on urate transport in isolated rabbit proximal tubules, although it was not determined whether S-8666 increased the net excretion of urate under their experimental conditions. The discrepancy between the effects of enantiomers of S-8666 on renal urate transport in rats and rabbits is not clear. However, the rabbit may not be a suitable species to investigate the effect of uricosurics on net urate reabsorption in renal tubules, because the animal shows the net urate secretion and the FEua exceeds the filtered load (34).

13.2.2 DIURETIC AND ANTIHYPERTENSIVE ACTIVITIES. In oxonate-treated rats under anesthesia, intraperitoneally administered S-8666 produced high-ceiling diuresis, and its efficacy at high doses was comparable with that of furosemide, which is a generally accepted loop diuretic (36). The natriuretic activity of S-8666 was determined using a commonly applied assay procedure in conscious laboratory animals that received the agent orally. Dose-dependent natriuresis was observed in male and female ddY mice (3–300 mg/kg), male SD rats (3–100 mg/kg), female beagle dogs (3–100 mg/kg), and female cynomolgus monkeys (2–50 mg/kg). The natriuretic activity of S-8666 was comparable to that of furosemide and indacrinone in mice and rats; however, extremely high activity was possessed by furosemide in dogs and by indacrinone in monkeys. The diuretic and natriuretic effects of S-8666 were also clearly less than those of furosemide and indacrinone in champanzees (33). Thus, S-8666 is a mild diuretic agent that shows little species difference in its natriuretic activity among laboratory animals. A comparative study on the duration of the natriuretic effect in rats was done between S-8666 and furosemide, which were intravenously administered at 1–10 mg/kg. The half-life was 16–20 min for S-8666, but only 8–9 min for furosemide. Thus, the natriuretic effect of S-8666 seems to last longer than that of furosemide. The natriuretic effect of both S-8666 and furosemide in rats was inhibited by combination with probenecid, and both agents also depressed free water reabsorption. The direct action of S-8666 on the cortical thick ascending limb of Henle's loop (CAL) was studied using an isolated single nephron from rabbits (37). The agent in the lumen reduced the lumen-positive voltage in CAL together with a reduction in ^{36}Cl flux from lumen to bath. However, the activity was much less potent than that of furosemide, whereas trichlormethiazide had no effect on CAL. These diuretic and natriuretic effects of S-8666 *in vivo* and *in vitro* resulted from the action of the S-(−)-enantiomer. The natriuretic activity of the S-(−)-enantiomer in conscious rats was two times higher than that of S-8666. This evidence led to the conclusion that S-8666 is a loop diuretic agent that has lower diuretic potency but a longer duration of action than furosemide.

S-8666 was diuretic and natriuretic in both normotensive and hypertensive animals, but its antihypertensive effect was not observed in normotensive animals, as it is with other diuretics. S-8666 produced an antihypertensive response in DOCA-salt hypertensive rats after consecutive treatment with 60–100 mg/kg/day p.o. for two weeks (38). The antihypertensive potency of S-8666 was three to four times that of tienilic acid and comparable with that of furosemide, but less than that of indacrinone. These experimental results led to the conclusion that S-8666 offers satisfactory antihypertensive effects in hypertensive rats compared to the available uricosuric diuretics. The prophylactic effects of S-8666 were also evaluated during the development of hypertension in DOCA-salt hypertensive rats, salt-loaded spontaneously hypertensive rats, and salt-loaded Dahl-S rats. The effective dose in these animals was 100 mg/kg/day p.o. (38). Thus, although S-8666 was not always potently antihypertensive, it possessed a favorable activity, comparable to or better than that of furosemide.

13.3 AA-193

13.3.1 URICOSURIC ACTIVITY. Although FEua and the effects of drugs influencing urate excretion are highly species-dependent in mammals, there are no major differences in the mechanisms of renal urate transport in different mammalian species that, at least, show the net urate reabsorption (39). The mechanisms of urate excretion in man have been historically assumed to result from the axial heterogeneity along nephron (40). The urate excretion mechanism in humans is considered to have four components: 1) uric acid filtrated in the glomeruli is 2) reabsorbed and 3) secreted, and 4) undergoes postsecretory reabsorption, and finally approximately 10% of the glomerular filtrate of urate is excreted in the urine (41). Although the 4-component model has not been certified directly, it appears to be the most reasonable description of the mechanisms of urate excretion and the different actions of uricosuric drugs. This model, which supposes the axial heterogeneity of urate transport along the nephron, is logically compatible with coextensive reabsorptive and secretory fluxes of urate in the proximal tubules (42).

AA-193 (**14c**) is chemically related to the uricosuric diuretics originating from tienilic acid (**3**). Despite this superficial relationship, AA-193 differs functionally from other phenoxyacetic acids in being a potent uricosuric agent without diuretic acivity in rats (43). Dan et al. characterized the uricosuric action of AA-193 by comparing it with the uricosuric properties of well-known drugs in rats, mice, and Cebus monkeys (42,44).

In conscious normal rats, probenecid and tienilic acid increased the urate excretion, but benzbromarone did not have the uricosuric acivity. Thus, the presecretory reabsorption of urate is probably dominant in rats. Dan et al. found that in rats AA-193 was the most potent uricosuric tested. In conscious DBA/2 mice, probenecid not only had so-called paradoxical actions but stimulated urinary urate wasting after administration of pyrazinamide. These data

suggest that the renal transport system of urate in the mouse is similar to that in man. AA-193, as well as benzbromarone, enhanced the urate excretion dose-dependently, but the effects were different in pyrazinamide suppression tests in mice. In conscious Cebus monkeys, the uricosuric and hypouricemic effects of AA-193 were more potent than those of probenecid and similar to those of tienilic acid, but less than those of benzbromarone. Benzbromarone had a considerable role in postsecretory reabsorption in the monkey, like in man. The uricosuric effect of AA-193 was entirely caused by the action of the S-(−)-enatiomer in rats (12,45). In rats, the S-(−)-enantiomer had uricosuric activity two times higher than that of AA-193, whereas R-(+)-enantiomer was not uricosuric. Similar stereospecificity of the uricosuric action of AA-193 was also confirmed in Cebus monkeys (46). These results described above suggest that AA-193 has a mode of action different from well-known uricosuric agents. AA-193 appears to be a new class of uricosuric agent that inhibits presecretory reabosrption in the proximal tubules.

It has been proposed that a urate-anion exchanger system in brush border membrane vesicles (BBMV), which mediates hydroxyl ion (OH⁻) gradient-dependent urate uptake, is the most likely route for the mediation of the urate transport in the first step of urate reabsorption in the proximal tubules. Dan et al. investigated the inhibitory effects of well-known uricosuric drugs on the OH⁻/urate exchange in BBMV (47). The rank order of potency was benzbromarone > tienilic acid > sulfinpyrazone > probenecid, which coincides well with clinical doses in man. AA-193 had the most potent inhibition on urate uptake ($Ki = 0.12 \mu M$). Dan et al. further evaluated the uricosuric action of AA-193 by comparing it with the effect of p-aminohippuric acid (PAH) secretion in rats, using *in vivo* and *in vitro* techniques. The intravenous administration of AA-193 elevated the FEua significaty in a dose-dependent manner at doses of 0.1–10 mg/kg (48). Only at the highest dose of 10 mg/kg, AA-193 caused a momentary decrease in FEpah. On the other hand, tienilic acid and probenecid reduced FEpah at uricosuric effective doses. To more directly compare the inhibitory effects of uricosurics on urate reabsorption and PAH secretion, Dan et al. investigated the effects of uricosurics on the OH^- gradient-dependent urate uptake in BBMV and the net PAH accumulation in cortical slices. The relation between the affinity of an uricosuric drug for urate and PAH transporters corresponds well with the difference between the effect on FEua and that on FEpah. The relative affinity of AA-193 for the urate uptake is 83-fold greater than that for the PAH accumulation. These results support the assumption that, in contrast with the other uricosurics, AA-193 has an extremely higher affinity for urate reabsorption than that for the common pathway of weak organic acids in rats.

To assess the hypouricemic effect of AA-193 derived from its uricosuric effect, Dan et al. used uricase-inhibited rats produced by oxonate feeding (49). In the hyperuricemic rat model, consecutive oral administrations of AA-193 at doses of 20, 50, and 100 mg/kg for seven days increased urate excretion and reduced the plasma urate level in a dose-dependent manner. In normal rats, consecutive administrations of AA-193 for seven days maintained the dose-dependent uricosuric activity without significant changes of the plasma urate level. Whereas plasma levels of AA-193 at two and eight hours after the first administration at a dose of 50 mg/kg p.o. were 0.18 and 0.07 mM, respectively, the level at 24 h was below the detectable limit. The time course of the plasma concentrations of AA-193 on day one was similar to that on day seven. In *in vitro* studes, 1 mM of AA-193 had no effect on liver uricase activity and

0.2 mM of AA-193 did not inhibit xanthine dehydrogenase activity. Therefore, it is unlikely that AA-193 at physiological doses has a significant effect on either the production or degradation of urate. Dan et al. concluded that AA-193 has a hypouricemic effect caused by increases in urate excretion in hyperuricemic rats.

The newly developed uricosuric agent, AA-193, was orally administered to 18 healthy adult male volunteers in a single administraton trial (10, 25, 100, 300, and 600 mg) (50). Although AA-193 at single-doses of 10–600 mg had no effect on urine volume and glomerular filtration rate, it increased urate excretion significantly in a dose-dependent manner at doses of 100–600 mg (Fig. 30.15). The elevaton of FEua suggests that the increase in urate excretion was obviously due to the inhibition of net urate reabsorption by AA-193. The correlation between the FEua and the plasma AA-193 level was observed, suggesting that the unchanged AA-193 itself may be an active form that inhibits the net reabsorption of urate in the kidney. The results of this study showed that there appears to be no particular problem regarding the further continuation of clinical trials of AA-193. Thus, AA-193 was orally administered to 32 healthy adult male volunteers in repeated administration trial (50, 100, 200, and 400 mg, once a day for seven days) (51). During the repeated administration days, the maximum plasma concentration (C_{max}) and the plasma half-life ($t_{1/2}$) of AA-193 did not show any large change. The area under the plasma concentration (AUC_{0-24}) and the amounts of daily urinary excretions of AA-193 after the seventh administration were about the same as those of after the first administration. Although AA-193 at multiple-doses of 50–400 mg/d for seven days had no effect on urine volume and glomerular filtration rate during the test periods, it increased urate excretion significantly in a dose-dependent manner at doses of 100–400 mg. As shown

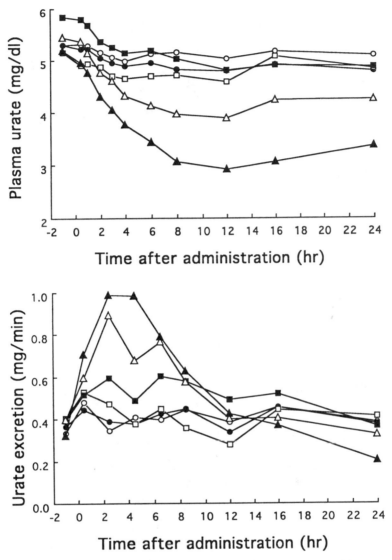

Fig. 30.15 Changes in plasma urate (upper) and urate excretion (lower) after oral administration of AA-193 (○: placebo; ●: 10 mg; □: 25 mg; ■: 100 mg; △: 300 mg; ▲: 600 mg) in healthy volunteers; (mean, $n = 4$–6).

in Fig. 30.16, the plasma urate concentration after the first administration gradually decreased and reached a minimum level at the third day, and thereafter the minum levels were maintained during the repeated administration days. At high doses (200 mg/d and 400 mg/d), the return to untreated plasma urate levels were observed 48 h after the final administration. In both single and repeated phase I studies, AA-193 had no abnormality in vital signs, including body temperature, blood pressure, pulse rate, respiration rate, and electrocardiogram. There was also neither an abnormal value nor a change attributable to the test drug in clinical laboratory tests. From the above results, it was considered that AA-193 should be highly useful as a

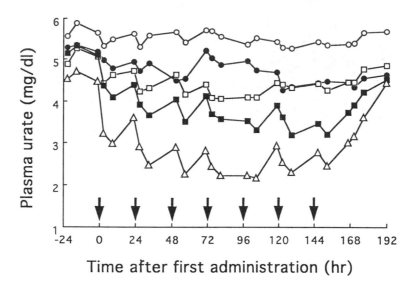

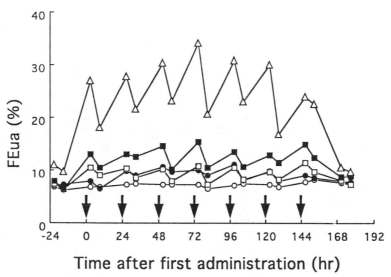

Fig. 30.16　Changes in plasma urate (upper) and FEua (lower) after repeated oral administration of AA-193 (○: placebo; ●: 50 mg; □: 100 mg; ■: 200 mg; △: 400 mg) for seven days in healthy volunteers; (mean, $n = 6$–8).

uricosuric agent, and there appears to be no particular problem regarding the further continuation of clinical trials.

REFERENCES

1. The Fifth Report of the Joint National Committee on Detection, Evaluation, and Treatment of High Blood Pressure (JNC V), *Arch. Intern. Med.*, **153**, 154 (1993).

2. 1993 Guidelines for the Management of Mild Hypertension: Memorandum from a WHO/ISH Meeting, *Hypertens. Res.*, **16**, 149 (1993).

3. H. R. Black, *Am. Heart J.*, **121**, 707 (1991).

4. R. D. Abbott, F. N. Brand, W. B. Kannel, and W. P. Castelli, *J. Clin. Epidemiol.*, **41**, 237 (1988).

5. E. J. Cragoe, Jr., Ed., *Diuretic Agents*, American Chemical Society, Washington, D.C., 1978.

6. E. J. Cragoe, Jr., Ed., *Diuretics. Chemistry, Pharmacology, and Medicine*, Wiley-Interscience, New York, 1983.

7. A. Vogl, *Am. Heart J.*, **39**, 881 (1950).

8. E. J. Cragoe, Jr. in Ref. 6, p. 201.

9. *Drugs Fut.*, **10**, 255 (1985).

10. S. D. Nelson, *J. Med. Chem.*, **25**, 753 (1982).

11. H. Sato, T. Dan, E. Onuma, H. Tanaka, and H. Koga, *Chem. Pharm. Bull.*, **38**, 1266 (1990).

12. H. Sato, T. Dan, E. Onuma, H. Tanaka, B. Aoki, and H. Koga, *Chem. Pharm. Bull.*, **39**, 1760 (1991).

13. T. Dan, E. Onuma, H. Tanaka, H. Sato, and H. Koga, personal communication (1984).

14. J. J. Plattner, A. K. L. Fung, J. A. Parks, R. J. Pariza, S. R. Crowley, A. G. Pernet, P. R. Bunnell, and P. W. Dodge, *J. Med. Chem.*, **27**, 1016 (1984).

15. *Drugs Fut.*, **14**, 9 (1989).

16. H. Sato, T. Dan, E. Onuma, H. Tanaka, B. Aoki, and H. Koga, *Chem. Pharm. Bull.*, **40**, 109 (1992).

17. H. Harada, Y. Matsushita, M. Yodo, M. Nakamura, and Y. Yonetani, *Chem. Pharm. Bull.*, **35**, 3195 (1987).

18. Y. Yonetani, M. Ishii, and K. Iwaki, *Jpn. J. Pharmacol.*, **30**, 829 (1980).

19. G. G. Vurek and S. E. Pegram, *Anal. Biochem.*, **16**, 409 (1966).

20. Y. Yonetani and K. Iwaki, *Jpn. J. Pharmacol.*, **33**, 947 (1983).

21. J. A. Tobert, V. J. Cirillo, G. Hitzenberger, I. James, J. Pryor, T. Cook, A. Buntinx, I. B. Holmes, and P. M. Lutterbeck, *Clin. Pharmacol. Ther.*, **29**, 344 (1981).

22. H. Harada, Y. Matsushita, M. Yodo, M. Nakamura, and Y. Yonetani, *Chem. Pharm. Bull.*, **35**, 3215 (1987).

23. M. Kitagawa, T. Mimura, and M. Tanaka, *Chem. Pharm. Bull.*, **39**, 2400 (1991).

24. II. Scylc, H. Stone, P. S. Timiras, and C. Schaffenburg, *Am. Heart J.*, **37**, 1009 (1949).

25. M. Gerold and H. Tschirky, *Arzneim.-Forsch.*, **18**, 1285 (1968).

26. A. R. Maass, I. B. Snow, M. Beg, and R. M. Stote, *Clin. Exp. Hypertension*, **A4**, 139 (1982).

27. M. Kitagawa, K. Yamamoto, S. Katakura, H. Kanno, K. Yamada, T. Nagahara, and M. Tanaka, *Chem. Pharm. Bull.*, **39**, 2681 (1991).

28. S. Tanaka, A. Kanda, and S. Ashida, *Jpn. J. Pharmacol.*, **54**, 307 (1990).

29. C. Hansch, P. G. Sammes, and J. B. Tailor, Eds., *Comprehensive Medicinal Chemistry*, Pergamon Press, Oxford, 1990.

30. H. Koga, H. Sato, T. Dan, and B. Aoki, *J. Med. Chem.*, **34**, 2702 (1991).

31. J. J. Plattner, Y. C. Martin, J. R. Smital, C-M. Lee, A. K. L. Fung, B. W. Horrom, S. R. Crowley, A. G. Pernet, P. R. Bunnell, and K. H. Kim, *J. Med. Chem.*, **28**, 79 (1985).

32. R. L. Smith in Ref. 6, p. 267.

33. Y. Yonetani, K. Iwaki, T. Shinosaki, A. Kawase-Hanafusa, H. Harada, and A. A. van ES, *Jpn. J. Pharmacol.*, **43**, 389 (1987).

34. F. Roch-Ramel and G. Peters in W. N. Kelley and I. M. Weiner, Eds., *Uric Acid: Handbook of Pharmacology*, Springer-Verlag, Berlin, Vol. 51, 1978, p. 211.

35. T. Shimizu, M. Nakamura, and M. Imai, *J. Pharmacol. Exp. Ther.*, **245**, 644 (1988).

36. H. Harada and Y. Yonetani, *Drugs Fut.*, **13**, 257 (1988).

37. T. Shimizu, M. Nakamura, and M. Imai, *J. Pharmacol. Exp. Ther.*, **245**, 651 (1988).

38. M. Matsuda, Y. Yonetani, M. Ishii, N. Kitajiri, M. Ueda, and H. Harada, *Jpn. J. Pharmacol.*, **43** (Suppl.), 193p (1987).

39. F. Roch-Ramel, *Clin. Nephrol.*, **12**, 1 (1979).

40. R. E. Rieselbach, *Adv. Exp. Biol. Med.*, **76B**, 1 (1977).

41. D. J. Levinson and L. B. Sorensen, *Ann. Rheum. Dis.*, **39**, 173 (1980).

42. T. Dan, H. Tanaka, and H. Koga, *J. Pharmacol. Exp. Ther.*, **253**, 437 (1990).

43. T. Dan, E. Onuma, H. Tanaka, H. Sato, B. Aoki, and H. Koga, *Jpn. J. Pharmacol.*, **46**(Suppl.), 224p (1988).

44. T. Dan, H. Koga, E. Onuma, H. Tanaka, H. Sato, and B. Aoki, *Adv. Exp. Med. Biol.*, **253A**, 301 (1989).

45. M. Onoma, T. Yoneya, E. Onuma, T. Dan, N. Imai, and K. Koizumi, *Jpn. J. Pharmacol.*, **61**(Suppl.), 306p (1994).

46. T. Dan, E. Onuma, H. Tanaka, H. Sato, and H. Koga, personal communication, 1989.

47. T. Dan and H. Koga, *Eur. J. Pharmacol.*, **187**, 303 (1990).

48. T. Dan, E. Onuma, H. Tanaka, and H. Koga, *Naunyn Schmiedebergs Arch. Pharmacol.*, **343**, 532 (1991).

49. T. Dan, T. Yoneya, M. Onoma, E. Onuma, and K. Ozawa, *Metabolism*, **43**, 123 (1994).

50. Y. Uji, S. Kashiwazaki, T. Dan, M. Sakai, T. Fukazawa, Y. Orikasa, H. Kamiyama, and A. Okazaki, *Rinshouiyaku*, **10**, 1037 (1994).

51. Y. Uji, S. Kashiwazaki, T. Dan, M. Sakai, T. Fukazawa, Y. Orikasa, H. Kamiyama, and A. Okazaki, *Rinshouiyaku*, **10**, 1057 (1994).

CHAPTER THIRTY-ONE

Diuretic and Uricosuric Agents

CYNTHIA A. FINK
LINCOLN H. WERNER

Ciba-Geigy Corporation
Summit, New Jersey

CONTENTS

1 Introduction, 364
2 Renal Physiology and Pharmacology, 365
  2.1 Renal physiology, 365
  2.2 Renal pharmacology, 372
    2.2.1 Classes of diuretic substances, 372
    2.2.2 Pharmacological evaluation of diuretics, 372
    2.2.3 Clinical aspects of diuretics, 374
3 Osmotic Diuretics, 375
4 Mercurial Diuretics, 376
  4.1 Historical, 376
  4.2 Structure–activity relationships, 376
  4.3 Pharmacology, 378
  4.4 Clinical application, 378
5 Carbonic Anhydrase Inhibitors, 378
  5.1 History, 379
  5.2 Structure–activity relationships, 380
  5.3 Clinical application, 382
6 Aromatic Sulfonamides, 383
  6.1 Introduction, 383
  6.2 Mefruside, 384
7 Thiazide and Related Compounds, 386
  7.1 Thiazides, 386
  7.2 Hydrothiazides, 388
    7.2.1 Structure activity relationships, 388
    7.2.2 Pharmacology and mechanism of action, 391
    7.2.3 Clinical application, 392
8 Other Sulfonamide Diuretics, 392
  8.1 Chlorthalidone, 394
  8.2 Hydrazides of *m*-sulfamoyl acids, 394
  8.3 1-Oxoisoindolines, 395

Burger's Medicinal Chemistry and Drug Discovery,
Fifth Edition, Volume 2: Therapeutic Agents,
Edited by Manfred E. Wolff.
ISBN 0-471-57557-7   © 1996 John Wiley & Sons, Inc.

8.4 Quinazolinone sulfonamides, 397
8.5 Mixed sulfamide diuretic-antihypertensive agents, 398
8.6 Tizolemide, 399
8.7 Bemitradine, 399
9 High-Ceiling Diuretics, 399
  9.1 Introduction, 399
    9.1.1 Pharmacology and mechanism of action, 400
  9.2 Ethacrynic acid, 400
  9.3 Indacrinone, 403
  9.4 Other aryloxy acetic acid high-ceiling diuretics, 404
  9.5 Furosemide, 405
  9.6 Bumetanide, 406
    9.6.1 Pharmacology, 411
  9.7 Piretanide, 412
  9.8 Azosemide, 412
  9.9 Xipamide, 412
  9.10 Triflocin, 413
  9.11 Torasemide, 415
  9.12 Muzolimine, 416
  9.13 MK 447, 417
  9.14 Etozolin, 417
  9.15 Ozolinone, 418
10 Steroidal Aldosterone Antagonis, 419
11 Aldosterone Biosynthesis Inhibitors, 426
12 Cyclic Polynitrogen Compounds, 426
  12.1 Xanthines, 426
  12.2 Aminouracils, 428
  12.3 Triazines, 428
  12.4 Potassium-sparing diuretics, 430
    12.4.1 Introduction, 430
    12.4.2 Triamterene, 431
    12.4.3 Other bicyclic polyaza diuretics, 434
    12.4.4 Amiloride, 437
    12.4.5 Azolimine and clazolimine, 440
13 Atrial Natriuretic Peptide, 440
  13.1 ANP clearance receptor blockers, 441
  13.2 Neutral endopeptidase inhibitors, 441
14 Uricosuric Agents, 444
  14.1 Introduction, 444
  14.2 Sodium salicylate, 445
  14.3 Probenecid, 446
  14.4 Sulfinpyrazone, 446
  14.5 Allopurinol, 447
  14.6 Benzbromarone, 448
15 Conclusion, 448

1 INTRODUCTION

Diuretics are among the most frequently prescribed therapeutic agents for the treat-ment of edema, hypertension, and conges-tive heart failure. These agents act primari-ly by inhibiting the reabsorption of sodium from the renal tubules.

Uricosuric agents increase the excretion of uric acid, one of the principal products of purine metabolism. These compounds are used in the treatment of gout, a condition where plasma levels of uric acid are elevated and, as a result, deposits of crystalline sodium urate form in connective tissues. Hyperuricemia is an adverse effect sometimes observed with diuretic treatment and arises from decreased extracellular volume and increased urate reabsorption. Compounds have been developed which have combined diuretic and uricosuric properties.

The medicinal use of diuretics dates back to 400 B.C. when Hippocrates administered metallic mercury to increase urine excretion. The modern era of diuretic therapy began in 1949 when sulfanilamide was discovered to posses diuretic and natriuretic properties (1). Since the 1950s, significant advances have been made in the discovery of new diuretic agents and their precise cellular mechanism of action. Today a heterogeneous array of diuretic compounds, possessing different structures and sites of actions, are available for safe and effective treatment of ˙edema and cardiovascular diseases.

2 RENAL PHYSIOLOGY AND PHARMACOLOGY

2.1 Renal Physiology

The kidneys are the principal organs of excretion and perform three major functions in maintaining homeostasis:

1. Remove water, electrolytes, products of metabolic waste, drugs, and other materials from blood;
2. Possess endocrine functions, i.e., secrete erythropoietin, renin, and 1,2,5-hydroxy-cholecalciferol; and
3. Selectively reabsorb water, electrolytes, and needed nutrients from the urine.

These reddish-brown organs are located on either side of the vertebral column just below the diaphragm; they are embedded in a protective mass of fat called the perirenal fat, and are covered by a fibrous capsule. Together the kidneys weigh about 300 g, which is about 0.4% of the total body weight. The kidneys can be divided into three major regions: the pelvis, the cortex, and the medulla (Fig. 31.1). The working unit of the kidney is the nephron. Each kidney contains about 1.2 million such structural units (2). The fundamental components that make up the nephron are the glomerulus and the tubules (Fig. 31.2). There are two types of nephrons: the cortical nephrons and the juxtamedullary nephrons. The cortical nephrons possess a short loop of Henle compared to juxtamedullary nephrons, and its parent glomerulus is positioned in the cortex. The juxtamedullary nephrons contain a longer loop of Henle, and its respective glomerulus is located near the corticomedullary boundary. About 80% of the nephrons found in the human kidney are cortical nephrons (3).

The glomerulus is comprised of a convoluted capillary network that is joined with connective tissue. The diameter of the afferent arterioles are larger than the efferent arterioles. As a result, the glomerular filtration pressure is estimated to be about 50 mm of mercury. This facilitates the rapid clearance of water and a variety of low to medium molecular weight solutes from the blood.

The Bowman's capsule surrounds the capillary network of the glomerulus, and its function is to collect the filtrate. It is composed of a single layer of squamous cells. The proximal tubule begins at Bowman's capsule. After several convolutions it descends towards the medulla to become the loop of Henle. For the juxtamedullary nephrons the transition from the proximal tube to the loop of Henle occurs at the boundary between the inner and outer

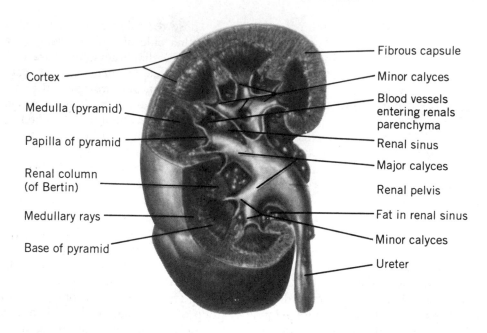

Cortex

Medulla (pyramid)

Papilla of pyramid

Renal column
(of Bertin)

Medullary rays

Base of pyramid

Fibrous capsule

Minor calyces

Blood vessels
entering renals
parenchyma

Renal sinus

Major calyces

Renal pelvis

Fat in renal sinus

Minor calyces

Ureter

Fig. 31.1   The kidney. Courtesy of CIBA Pharmaceutical Company, Division of CIBA-GEIGY Corporation.

stripe of the medulla. The tubule tapers at this point, descends into the inner zone of the medulla, makes a 'U' turn, and ascends again toward the cortex where it widens. The tubule becomes coiled as it returns towards the glomerular region and becomes the distal convoluted tubule. The arrangement of the cortical nephrons is slightly different. The descending loop of Henle tapers at the outer-inner stripe medulla boundary, widens out just before undergoing a hairpin turn, and reverts back towards the glomerulus.

The distal convoluted tubules straighten out and empty into the collecting ducts. These ducts travel back towards the medulla, merging with other ducts, and then enter the renal papillae.

An estimated 180 L of glomerular filtrate forms daily, which is about 60 times the total plasma content (4). Fortunately the reabsorption process begins immediately. Approximately 99% of the water and electroyltes are reabsorbed in the renal tubules. The glomerular filtrate is composed of

water, electrolytes (NH_4^+, Na^+, K^+, Ca^{2+}, Mg^{2+}, Cl^- and HPO_4^{2-}), glucose, amino acids, and nitrogenous wastes of metabolism. It actually has a similar profile to blood plasma, except it contains no blood cells and little or no plasma proteins. Reabsorption of the water and solutes occurs through the walls of the proximal and distal convoluted tubules, in the loop of Henle, and in the collecting tubules by active and passive transport systems (Fig. 31.3). From the use of micropuncture and isolated tubule techniques, much is known about the cellular and molecular mechanism of tubular reabsorption. Each particular segment of the nephron possesses its own characteristic ion transport systems. In general, these cells all contain a rate-limiting sodium entry transport system on the luminal membrane, which is coupled to a Na^+/K^+ ATPase on the basolateral membrane for sodium removal.

The reabsorption process begins in the proximal tubules. Approximately 50–55% of the filtered sodium and water along with

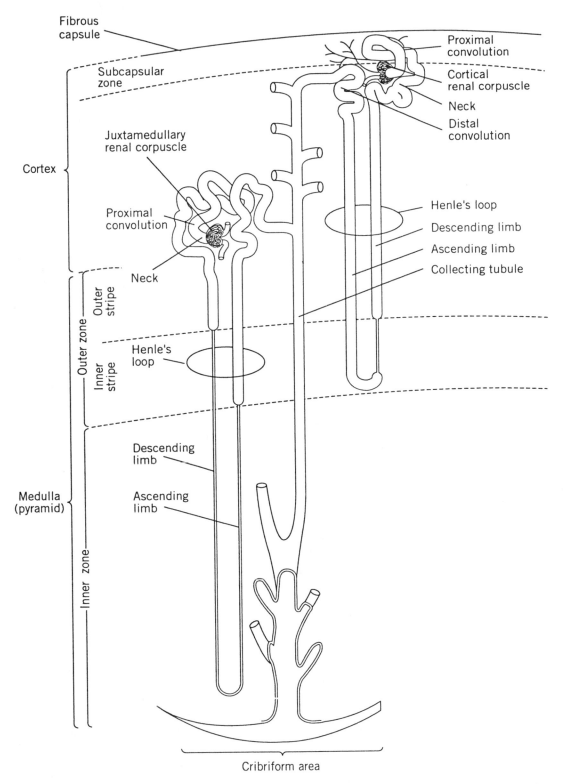

Fig. 31.2   The nephron. Courtesy of CIBA Pharmaceutical Company, Division of CIBA-GEIGY Corporation.

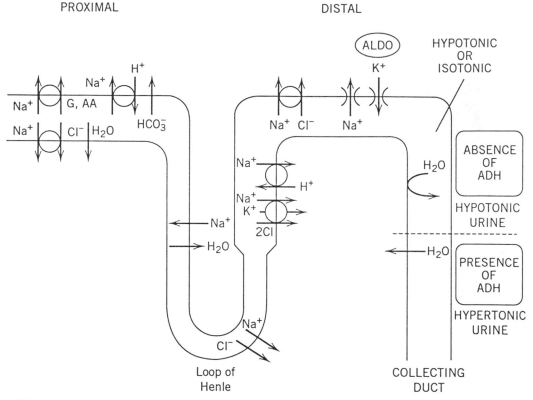

Fig. 31.3 Major transport mechanisms in the apical membrane of the tubule cells along the nephron (5). G, Glucose; AA Amino Acids; ADH, Antidiuretic hormone; ALDO, Aldosterone. Reproduced with permission from *Eur. J. Clin. Pharmacol.*

about 90% of the filtered amino acids, bicarbonate, glucose, and phosphate are reclaimed here (Fig. 31.4), (6). Glucose, phosphate, and the amino acids enter proximal tubule cells via electrogenic cotransport with sodium. The major route of sodium reentry into this tubule cell is the Na^+/H^+ exchanger. This transport system is also directly responsible for most of the proximal tubular reabsorption of bicarbonate and creates a favourable gradient that allows for both the active and passive transport of about 50% of the filtered chloride ion (7). The Na^+/H^+ exchanger has recently been cloned by a gene transfer approach (8). A Na^+/K^+ ATPase pump located on the basolateral side of the proximal cell removes the sodium to maintain a low intracellular sodium concentration (approximately one-tenth that of the luminal

fluid). Carbonic anhydrase in the cytoplasm indirectly catalyzes the intracellular formation of protons, which keeps the Na^+/H^+ exchanger active. These excreted protons also neutralize the bicarbonate in the tubule to form carbonic acid. Carbonic anhydrase located in the luminal brush border dehydrates the carbonic acid to form carbon dioxide and water.

As mentioned, chloride ion is removed from the proximal tubule by passive and active transport systems. As solutes are removed from the tubule, the osmotic gradient facilitates the reabsorption of water. This effectively increases the concentration of chloride ion above that found in the lateral intercellular space. This space is permeable to chloride, and it is passively absorbed across the junction (9). Chloride can also enter the cell through a chloride–

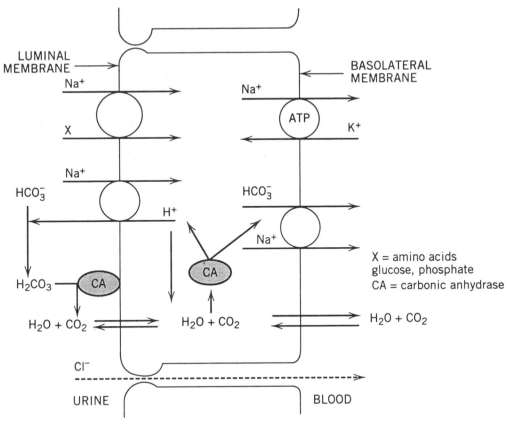

Fig. 31.4 Cell model of the proximal tubule.

formate exchanger (Fig. 31.5) (11,12). The basolateral membrane is also believed to contain a Na^+/HCO_3^- cotransporter (13).

About 35–40% of the filtered sodium is reabsorbed in the loop of Henle. The major luminal transporter of sodium in this region of the renal tubule is the $Na^+/K^+/2Cl^-$ electroneutral cotransporter located in the thick ascending region of the loop. The energy that drives the cotransporter arises

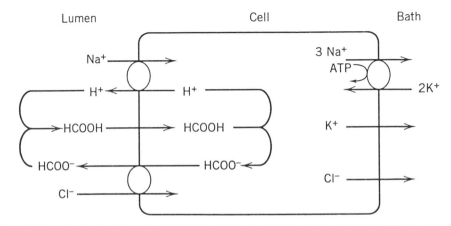

Fig. 31.5 Cell model of the chloride/formate exchange in the proximal tubule cell (10). Reproduced with permission from *J. Clin. Invest.*

from a concentration gradient generated from the Na^+/K^+ ATPase pump located on the basolateral membrane (Fig. 31.6). Chloride exits the basolateral side via a chloride channel and/or electroneutral KCl cotransporter (14). The potassium that enters the cell through the $Na^+/K^+/2Cl^-$ cotransporter can be recycled back through the lumen via a potassium channel to keep the tubule concentration of this ion high enough for the cotransporter to continue to function. The result of potassium leaving the luminal membrane and chloride at the basolateral side by conductive pathways generates a lumen-positive potential. This positive potential drives the flow of sodium ions out through a paracellular pathway. There is also a Na^+/H^+ exchanger on the apical membrane that plays a minor role in the reabsorption of sodium.

The distal convoluted tubule reabsorbs approximately 5–8% of the sodium contained in the glomerular filtrate. The major luminal transporter of sodium in this region is the neutral sodium chloride cotransporter (Fig. 31.7). A Na^+/K^+ ATPase pump is located on the basolateral membrane to remove sodium from the cell. Potassium can reenter the tubule through a barium-sensitive potassium channel. This region of the renal tubule is the site where calcium excretion and reabsorption are regulated. Parathyroid hormone and calcitriol are the main mediators of calcium reabsorption. They both increase distal calcium reabsorption; however, the exact mechanism is not well understood (15). Calcium is believed to exit the cell via a $Ca^{2+}/ATPase$ or a Na^+/Ca^{2+} exchanger on the basolateral membrane (16,17).

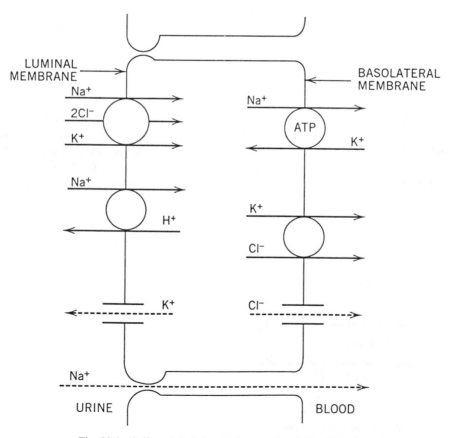

Fig. 31.6 Cell model of the thick ascending loop of Henle.

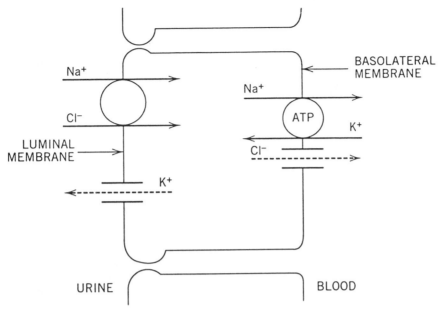

Fig. 31.7 Cell model of the distal convoluted tubule.

The collecting tubule is the last section of the renal tubule in which filtrate modification occurs. This region is responsible for 2–3% of sodium reabsorption. There are two major cell types in this region of the nephron: the principal cells and the intercalated cells. The principal cells are the predominant cell type, and they are responsible for sodium reabsorption and potassium secretion. Sodium enters the cell via a sodium channel and exits through the basolateral Na^+/K^+ ATPase pump (Fig. 31.8). Potassium can exit this cell on the luminal and basolateral sides through con-

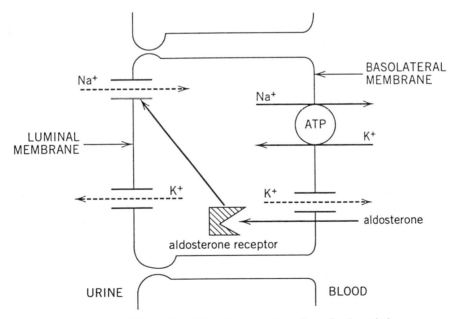

Fig. 31.8 Cell model of the principal cells in the collecting tubule.

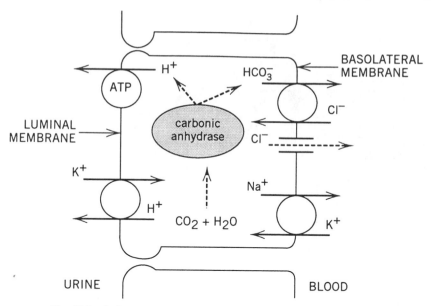

Fig. 31.9 Cell model of the intercalated cell of the collecting tubule.

ductance channels. The primary site of action of aldosterone is also in the principal cells. Aldosterone increases sodium reabsorption by opening sodium channels. The intercalated cells control hydrogen ion secretion and potassium reabsorption (Fig. 31.9). Protons are generated in the cell by the catalytic actions of carbonic anhydrase and are transported into the lumen through a H^+/translocating ATPase (18). An ATP-dependent K^+/H^+ exchanger may be responsible for potassium reabsorption in times of potassium depletion (19).

2.2 Renal Pharmacology

2.2.1 CLASSES OF DIURETIC SUBSTANCES. Diuretics compose a heterogeneous family of chemical compounds that all operate by inhibiting the reabsorption of sodium in the renal tubules. This diverse class of therapeutic agents includes organomercurials, polyols, sugars, thiazides, phenoxyacetic acids, aminomethylphenols, xanthines, aromatic sulfonamides, pteridines, pyrazines, and steroids. These agents have been classified in a variety of ways, including by

chemical structure, mechanism of action, tubular site of action, magnitude of natriuretic effect, and their effect on electrolyte depletion. Today no one classification system is commonly used. Frequently diuretics are grouped as loop, potassium-sparing, thiazide, and osmotic.

2.2.2 PHARMACOLOGICAL EVALUATION OF DIURETICS. The following three statements or generalizations are direct quotes from a paper by K.H. Beyer (20) and refer to the pharmacological evaluation of new drugs in general and to diuretics in particular.

The more closely one can approximate under controlled laboratory conditions the physiological correlates of clinically defined disease, the more likely one will be able to modulate it effectively.

In vitro experiments are apt to be inadequate and hence misleading when employed solely to anticipate the physiological correlates of complex clinical situations.

Often aberrations in function we call toxicity, relate more to changes a compound induced physiologically than to a direct, inherent, destructive effect of the agent, per se, on tissues.

Newer testing procedures have been designed with these generalizations in mind. Diuretics are generally evaluated in two species: the rat and the dog. The rat is used, as a rule, in initial screening for convenience and economy. Results obtained with diuretic agents in dogs, however, are generally more predictive of the response in humans than those obtained in the rat. However, many compounds exhibit diuretic activity in the rat, e.g., antihistamines, such as, tripelennamine, but are inactive in the dog and in humans. On the other hand, mercurial diuretics and ethacrynic acid are inactive in the rat; furosemide exhibits diuretic activity in the rat only at doses several times higher than those effective in the dog. Various modifications of the experimental procedure of Lipshitz et al. (21) are frequently used. Lipshitz measured urine volumes following administration of the drug to saline-loaded rats and compared the response to that obtained with a standard, i.e., urea. More recent modifications (22) measure urine volume and Na^+, K^+, and Cl^- excretion; e.g., male rats, fasted for 18 h, are given 5 mL of 0.2% NaCl solution/100 g body weight by stomach tube. The diuretic drugs are given by stomach tube at the time of fluid loading. The rats are placed in metabolism cages and urine volumes are measured at 30 min intervals over 3–5 h. period. Total amounts of Na^+ and K^+ excreted over the time period are determined by flame photometry; chloride can be determined in the Technicon Autoanalyzer using Skegg's modification of Zall's method (23).

A number of standard diuretics have also been tested in the mouse (25). In this animal, ethacrynic acid is a potent diuretic in contrast to the very low diuretic activity seen with this compound in the rat.

Diuretic activity in dogs has been evaluated according to one procedure as follows: at the start of each experiment, six mongrel dogs are given the drug by capsule, followed immediately by a subcutaneous injection of 100 mL of 0.9% saline. The urinary bladder is emptied by catheterization at the onset and two, four, and six hours after administration of the test compound. The dogs are kept in metabolism cages and urine volume measured; urine aliquots are analyzed for Na^+, K^+, and Cl^- content as described for the rat assay procedure. The colony of dogs is used repeatedly and only for diuretic tests. A control test with placebo is given prior to each test run.

The chimpanzee (26) has also been used to evaluate certain diuretics. In this animal, the effect of the compound on uric acid excretion can also be studied, because apes, like humans, are devoid of hepatic uricase and therefore maintain a relatively high level of circulating serum urate (27).

The evaluation of new diuretics requires more than an assessment of its ability to increase sodium and water excretion; its effect on potassium excretion, uric acid blood levels, and blood glucose levels must also be considered. The effect of a diuretic on Na^+, K^+, Cl^-, and water excretion can be studied readily in rats and dogs. The effect on uric acid blood levels cannot be determined in these animals because uric acid is rapidly metabolized by hepatic uricase in all mammals except the primates and humans. Chimpanzees have been used in the study of uricosuric agents (26), but tests with such large animals obviously presents numerous problems. Attempts have also been made to block the enzyme uricase in rats by administering potassium oxonate and thus to obtain higher serum uric acid levels (28).

Considerable evidence is available that demonstrates the tendency of certain benzothiadiazine diuretics to elevate blood glucose values in seemingly normal, as well as diabetic or prediabetic, individuals (29). A method using high doses of the compounds injected intraperitoneally into rats and determining blood glucose levels as compared to control values has been used to estimate a possible hyperglycermic effect of diuretics (30). Today the molecular and cellular mechanism of actions of diuretics are also investigated. Micropuncture and single nephron studies can provide insight into the exact site of action in the renal tubule and yield information regarding the specific transport mechanisms that are blocked by a drug.

2.2.3 CLINICAL ASPECTS OF DIURETICS. In a healthy human subject, changes in dietary intake or variations in the extrarenal loss of fluid and electrolytes are followed relatively rapidly by adjustments in the rate of renal excretion, thus maintaining the normal volume and composition of extracellular fluid in the body. Edema is an increase in extracellular fluid volume. In almost every case of edema encountered in clinical medicine, the underlying abnormality involves a decreased rate of renal excretion.

On of the factors influencing the normal relationship between the volume of interstitial fluid and the circulating plasma is the pressure within the small blood vessels. In diseases of hepatic origin, e.g., cirrhosis, the pressure relationships are disturbed primarily within the portal circulation and ascites results. In congestive heart failure, pressure-flow relationships may be disturbed more in the pulmonary or systemic circulation and edema may be localized accordingly.

There is overwhelming evidence to indicate that the primary disturbance of the kidney is in its ability to regulate sodium excretion, which underlies the pathogenesis of edema. Three approaches are available when edema fluid accumulates owing to excessive reabsorption of sodium and other electrolytes by the renal tubules. First, one can attempt to correct the primary disease if possible; second, one can reduce renal absorption of electrolytes by the use of drugs; and third, one can restrict sodium intake to a level that corresponds to the diminished renal capacity for sodium excretion. Cardiac decompensation is one of the most common causes of edema. Treatment

consists of full digitalization, which should be considered the primary therapeutic agent. Diuretic drugs have a secondary though very important role, for it has been shown that blocking excessive electrolyte reabsorption in the renal tubule alleviates the symptoms of cardiac failure and also improves cardiac function. Diuretics are also used in the treatment of hypertension.

Diuretic therapy may lead to a number of metabolic and electrolyte disorders. In general, these disturbances are mild but can be life-threatening in certain cases. Some of the common adverse effects observed with diuretic treatment are hypokalemia, hyperuricemia, and glucose intolerance. Diuretics that possess a site of action proximal to the collecting tubules, such as, the loop and thiazide diuretics, induce potassium loss; an average loss of 0.5 to 0.7 mEq/L generally results with long-term therapy (31). In most young hypertensive patients, this reduction does not present any problem; however in older patients or patients with preexisting heart disease, it may lead to the occurrence of ventricular arrhythmias. Combinations with potassium-sparing diuretics are frequently used to minimize the effect. Dietary potassium supplements may also be prescribed. In patients receiving long-term diuretic therapy, serum uric acid concentrations increase on average 1.3 mg/L due to a decrease in extracellular volume and increased urate reabsorption from the filtrate. Some patients may experience an attack of gout, or those with preexisting gout or excessive uric acid production may experience more frequent attacks. A number of studies have shown that some patients who have undergone long-term diuretic therapy have elevated blood levels of glucose and that their tolerance to glucose decreased (32). The mechanism of the diuretic-induced glucose intolerance is unknown. Some other side effects that are observed with diuretic treatment are increases in cholesterol levels in men and postmenopausal women (33), ototoxicity, and hypomagnesemia.

3 OSMOTIC DIURETICS

Osmotic diuretics all have several key features in common:

1. they are passively filtered at the glomerulus,
2. they undergo limited reabsorption in the renal tubules,
3. they are usually metabolically and pharmacologically inert, and
4. they have a high degree of water solubility.

These agents all function as diuretic agents by preventing the reabsorption of water and sodium from the renal tubules. The addition of a nonreabsorbable solute prevents water from being passively reabsorbed from the tubule, which in turn prevents a sodium gradient from forming thereby limiting sodium reabsorption. These actions hinder salt and water reabsorption from the proximal tubules; however, it has also been proposed that these agents have multiple sites and mechanisms of action (34,35).

Most of the osmotic diuretics are sugars and polyols (Table 31.1). Mannitol (Table 31.1, **5**) is the prototype of the osmotic diuretics and has been studied extensively.

Table 31.1   Osmotic Diuretics

No.	Generic Name	Trade Name	Structure
(1)	Ammonium chloride		NH_4Cl
(2)	Glycerin	Osmoglyn	
(3)	Glucose		
(4)	Isosorbide	Ismotic	
(5)	Mannitol	Osmitrol	
(6)	Sorbitol		
(7)	Sucrose		
(8)	Urea	Ureaphil	H_2NCONH_2

The compound is poorly absorbed following oral administration, and is therefore administered by intravenous infusion. It is freely filtered at the glomerulus and reabsorption is quite limited. The usual diuretic dose is 50–100 g given as a 25% solution.

These compounds are not prescribed as primary diuretic agents to edematous patients. One of the most important indications for the use of mannitol is the prophylaxis of acute renal failure. Following cardiovascular operations or severe traumatic injury, for instance, a precipitous fall in urine flow may be anticipated. Administration of mannitol, under such conditions, exerts an osmotic effect within the tubular fluid, inhibiting water reabsorption. A reasonable flow of urine can thus be maintained, and the kidneys can be protected from damage. Osmotic diuretics have also been prescribed in the relief of cerebral edema after neurosurgery, to lower intraoccular pressure in ophthalmologic procedures, and after a drug-overdose to maintain urine flow.

4 MERCURIAL DIURETICS

For approximately 30 years, mercurial diuretics were the most important diuretic agents. Since the introduction of orally active, potent, less toxic, nonmercuiral diuretics, beginning in 1950 with acetazolamide, their use has greatly declined. Today they represent only a small fraction of the injectable diuretics used, and injectable diuretics in turn are only a small portion of the total diuretic market.

4.1 Historical

The medicinal use of mercurial diuretics dates back to 400 B.C. when Hippocrates administered metallic mercury to increase urine excretion. Calomel (mercurial chloride) was used by Paracelsus (1493–1541

A.D.) as a diuretic. This information was lost until the nineteenth century when Jendrassik rediscovered the use of calomel as a diuretic agent (36). Calomel was an ingredient of the famous Guy's Hospital pill (calomel, squill, and digitalis). Calomel exerted a cathartic effect, and its absorption from the intestine was unpredictable. In 1919, Vogl (37) discovered the diuretic effect of merbaphen (9) following parenteral administration of this antisyphilitic agent.

(9)

General use of the drug as a diuretic was short-lived because of its toxicity. However it did lead to the synthesis of a large number of organomercurials between 1920 and 1950.

4.2 Structure–Activity Relationships

It is believed that the mechanism of action for these diuretic agents involves the *in vivo* release of mercuric ion in the renal tubules (38–40). This ion is thought to bind a sulfhydryl enzyme in the tubule membrane that is involved in sodium reabsorption. The mercuric ion reacts with a sulfhydryl group on the enzyme and another nucleophilic group in close proximity to form a bidentate complex that inactivates the enzyme and, as a result, the sodium reabsorption process (Fig. 31.10). Most mercurial diuretics have the general structure 10, in which Y is usually CH_3 and R is a complex organic moiety, usually incorporating an amide function or a urea group.

(10)

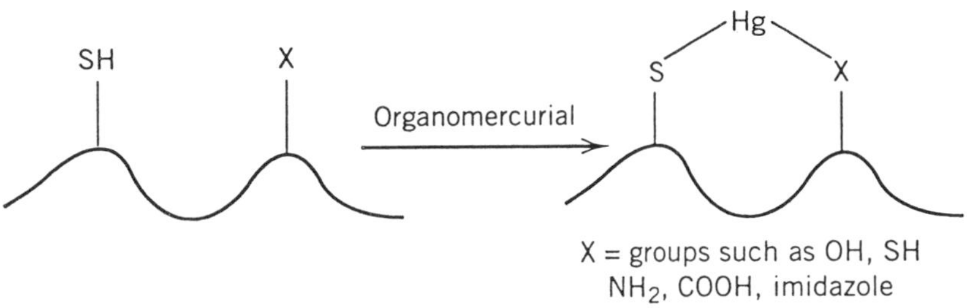

X = groups such as OH, SH
NH$_2$, COOH, imidazole

Fig. 31.10 Mercury enzyme complex.

R—CH$_2$CH$_2$Hg$^+$

(11)

All diuretic organic mercurials thus far examined are acid-labile *in vitro*. It is interesting to note that compounds of the related structure **11** with an unsubstituted β-carbon atom are acid-stable and do not exhibit duretic activity. Formula **12** shows the structural characteristics of mercurial diuretics. The most important mercurial diuretics are shown in Table 31.2. The R substituent largely determines the distribution and rate of excretion of the compound. The Y substituent, determined by the solvent in which the mercuration is carried out, generally has little effect on the prop-

erties of the compound (41,42). Among others, Y substituents, such as, H, CH$_3$, CH$_2$CH$_2$OH, and CH$_2$CH$_2$OCH$_3$, have been studied.

The nature of the X substituent affects the toxicity of the compound, irritation at the site of injection, and rate of absorption (41,42). Theophylline has been used commonly as an X substituent (43) or is commonly added by itself with the organomercurials. Because of its peripheral vasodilating effects, it can increase absorption of the mercurial diuretics at the site of injection. Theophylline is also weakly diuretic. When X is a thiol, such as, mercaptoacetic acid or thiosorbitol, cardiac toxicity and local irritation are reduced (44,45). Diglucomethox-

HOOC(H$_2$C)$_2$OCHNOCHN–CH$_2$CH(OCH$_3$)CH$_2$Hg-theophylline

R   Y   X

(12)

Table 31.2 Mercurial Diuretics

No.	Generic Name	Trade Name	Structure
(13)	Chloromerodrin	Neohydrin	H$_2$NOCHN–CH$_2$CH(OCH$_3$)CH$_2$HgCl
(14)	Meralluride USP	Mercuhydrin	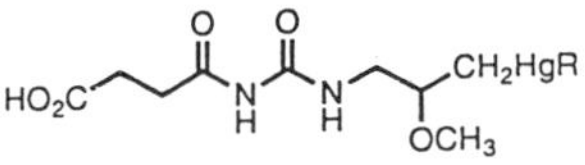 R = theophylline
(15)	Sodium mercaptomerin	Thiomerin	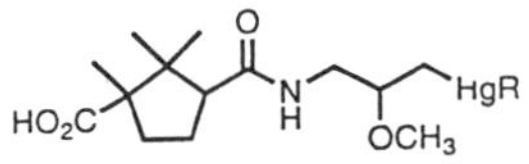 R = SCH$_2$CO$_2$Na
(16)	Mercurophylline NF XII	Mercupurin, Novurit	R = theophylline

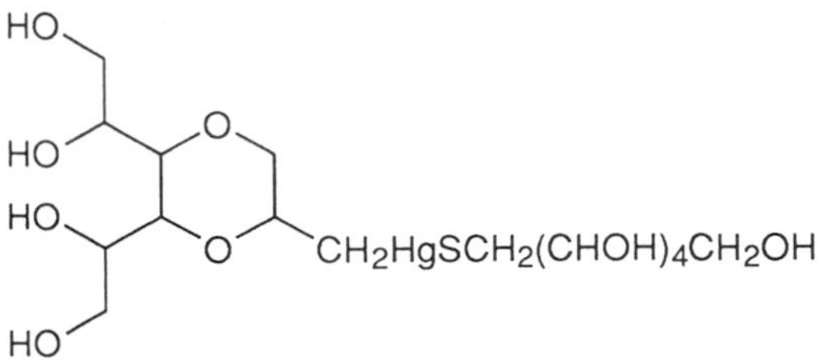

(17)

(18)

ane (**17**) (Mersoben®) (46) should be cited as a well-tolerated, potent mercurial diuretic that does not conform to the general structure **10**. Generally, mercurial diuretics are administered parenterally; chloromerodrin (Table 31.2, **13**), which lacks a carboxylic acid group, is orally effective but gastric irritation precludes its widespread use (47).

4.3 Pharmacology

Before the development of the loop diuretics, the organomercurials were the most potent diuretics available. Further studies have found that the major effect of organomercurials appears to be in the ascending limb of Henle (48–50). Organomercurials inhibit active chloride reabsorption in the thick ascending limb of Henle. During diuresis, the urine contains a high concentration of chloride ion matched by almost equivalent amounts of sodium (51). The effect of mercurial diuretics on potassium excretion is complex. They depress the tubular secretion of potassium and, for this reason, the diuresis is accompanied by significantly less potassium loss than occurs with other diuretics that do not inhibit the secretory mechanism. However, mercurials can have a paradoxical effect of increasing potassium excretion when initial excretory rates are low. Inhibition of organic acid secretion is seen in humans but not in the dog. In the chimpanzee and to a lesser extent in humans, mercurials, e.g., mersalyl (**18**), have an intense uricosuric action (52).

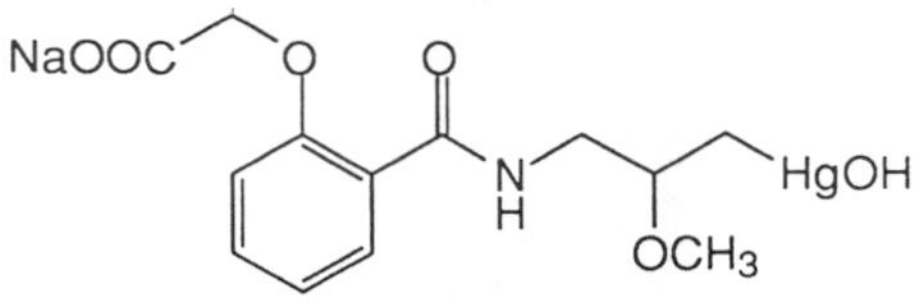

Side effects, such as, diarrhea, gingivitis, proteinuria, and stomatitis, can occur with organomercurial treatment. As with other diuretics, electrolyte imbalances are common after long-term use (hypochloremic alkalosis, hypokalemia, hyponatremia). In some cases following intravenous administration, severe hypersensitivity reactions may develop.

4.4 Clinical Application

Organomercurials are generally given intramuscularly. The usual dose is a 1 mL solution containing 40 mg Hg. In responsive, edematous patients, an increase in urine flow is evident in 1-2 h and reaches a maximum in 6–9 h. The effect is usually over within 24 h. A loss of about 2.5% of body weight represents an average response (51). Mercury is eliminated from the body in the urine, as a complex with cysteine. These diuretics therefore should not be prescribed to patients with renal insufficiency marked by adequate excretion of the mercury–cysteine complex.

5 CARBONIC ANHYDRASE INHIBITORS

Carbonic anhydrase in a zinc-containing enzyme that was first discovered in erythrocytes by Roughton in the early 1930s. This enzyme was subsequently found in many tissues, including the renal cortex, gastric mucosa, pancreas, eye, and central nervous system. Carbonic anhydrase catalyzes the reversible hydration of carbon dioxide and the dehydration of carbonic acid.

$$CO_2 + H_2O \rightleftharpoons H_2CO_3 \rightleftharpoons H^+ + HCO_3^-$$

These reactions can occur in the absence of the enzyme, however the rates are too slow for normal physiological function to occur. Normally the enzyme is present in the tissue in a high excess. Due to the levels of this enzyme in the kidney, approximately 99% of the enzyme's activity must be inhibited for physiological activity to be observed.

Hydrogen ion secretion takes place in the proximal tubule, the distal tubule, and the collecting duct. The driving force for H^+ secretion in the distal portions is the trans-tubular negative potential. In the proximal tubule, protons are actively excreted by the H^+/Na^+ exchanger. The source of cellular hydrogen ion is the hydration of carbon dioxide within the proximal tubular cells catalyzed by the action of cytosolic carbonic anhydrase to produce cellular H^+ and HCO_3^-. The hydrogen ion is secreted into the tubular lumen through the Na^+/H^+ exchanger, and then the Na^+ is reabsorbed and enters the peritubular fluid as $NaHCO_3$. In the proximal tubule, the secreted H^+ combines with HCO_3^- to form H_2CO_3, which is then dehydrated to CO_2 and H_2O. This reaction is also catalyzed by carbonic anhydrase located on the luminal border of the proximal tubular cells. The carbon dioxide diffuses back into the cell, where it is again hydrated and used as a source of hydrogen ion to drive the H^+/Na^+ exchanger (see Fig. 31.4).

Carbonic anhydrase inhibitors are among the best understood diuretics. Their major function is to inhibit the enzyme, carbonic anhydrase; although they are also believed to decrease cell membrane permeability for carbon dioxide and also to inhibit glucose 6-phosphate dehydrogenase (53). The administration of an inhibitor of carbonic anhydrase promptly leads to an increase in urine volume. The urinary concentrations of HCO_3^-, Na^+, and K^+ increase, whereas the normally acidic urine becomes alkaline and the concentration of chloride ion drops. In addition, there is a fall in titratable acid and ammonium ion excretion.

The net effect of the inhibitors in the proximal tubule is to prevent the reabsorption of bicarbonate. This can effect volume reabsorption in a number of ways:

1. Fewer protons are available for the H^+/Na^+ exchanger on the luminal membrane, thus sodium reabsorption is slowed.
2. The passive reabsorption process in the late proximal tubule is indirectly inhibited by the decreased chloride gradient.
3. Bicarbonate effectively becomes a non-reabsorbable anion and osmotically adds to the diuretic effect.

However, more than half of the bicarbonate that passes through the proximal tubule is reabsorbed in later segments of the renal tubule, thus attenuating the effectiveness of this class of diuretic agents. Carbonic anhydrase inhibitors also cause a significant kaliuresis, which can be attributed to the inhibition of distal proton secretion and high aldosterone levels, which result from volume depletion.

5.1 History

Building on earlier work by Strauss and Southworth in 1937 (54), Mann and Keilin (55), Davenport and Wilhelmi (56), and Pitts and Alexander in 1945 (57) proposed that the normal acidification of the urine results from the secretion of hydrogen ions by tubular cells. They confirmed that in dogs sulfanilamide (19) renders the urine

$$H_2N-\langle\rangle-SO_2NH_2$$

(19)

alkaline, perhaps because of the reduction of the availability of H^+ for secretion brought about by the inhibition of the enzyme carbonic anhydrase. The resulting increase in Na^+ and HCO_3^- excretion suggested to Schwartz (58) the diuretic potential of sulfanilamide. However this was of no practical significance because of the very high doses required to achieve diuresis.

5.2 Structure–Activity Relationships

Following the discovery of the carbonic anhydrase inhibitory activity of sulfanilamide, a variety of aromatic sulfonamides were found to exhibit the same type of activity (59). Aliphatic sulfonamides were much less active; substitution of the sulfonamide nitrogen in aromatic sulfonamides eliminated the activity. Roblin and co-workers (60,61), following the work of Schwartz (58), investigated a series of

heterocyclic sulfonamides. Compounds up to 800 times more active *in vitro* than sulfanilamide as carbonic anhydrase inhibitors were found (Table 31.3). An attempt to correlate pK_a values and *in vitro* carbonic anhydrase inhibitory activity in a series of closely related, 1,3,4-thiadiazole-2-sulfonamides (20) was not successful (66).

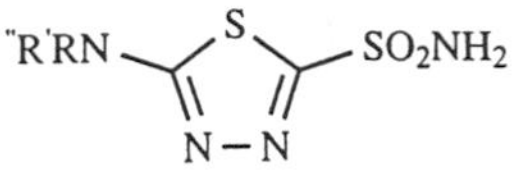

(20)

R' = lower alkyl, phenyl

R" = H, CH_3CO

The relationship between *in vitro* enzyme inhibition and *in vivo* diuretic potency was not very predictable owing to variation in drug distribution, binding, and metabolism (63), (Table 31.4), especially when different types of aromatic or heterocyclic sulfonamides were compared.

Certain derivatives of 1,3,4-thiadiazole sulfonamides were among the most active

Table 31.3 Carbonic Anhydrase Diuretics

No.	Generic Name	Trade Name	Structure	Ref.
(21)	Acetazolamide USP	Neohydrin	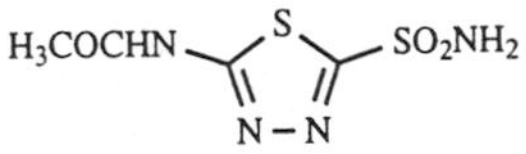	62
(22)	Dichlorophenamide USP	Daranide, Oratrol	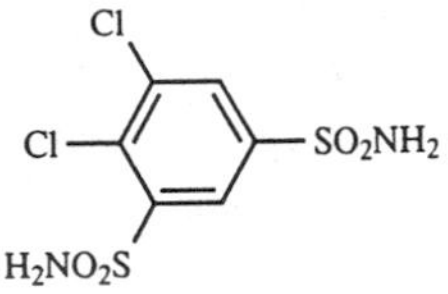	63
(23)	Ethoxzolamide USP	Cardrase, Ethamide	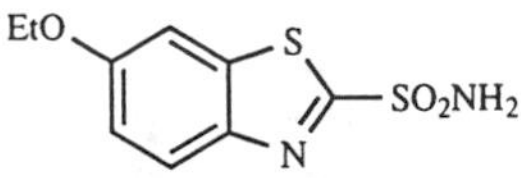	64
(24)	Methazolamide USP	Neptazane	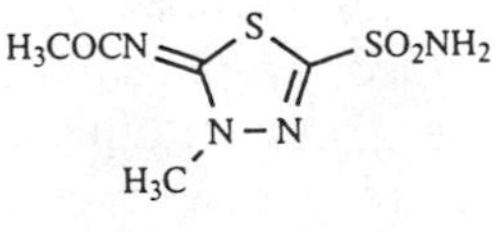	
(25)	Benzolamide		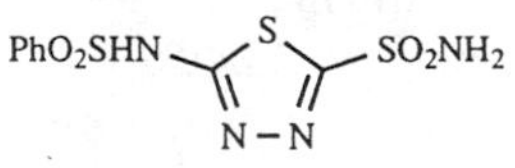	65

Table 31.4   Dissociation of Carbonic Anhydrase Inhibitory Activity *in Vitro* and Renal Electrolyte Effects in the Dog[a]

No.	Structure	Concentration Causing 50% Inhibition of Carbonic anhydrase, M	Dose (i.v.),[b] mg/kg	Urinary Excretion Rate μeq/min			
				Na	K	Cl	pH
(19)	H₂NO₂S—⟨⟩—NH₂	1.3×10^{-5}	C	14	14	3	6.8
			25	51	25	13	6.9
(27)	H₂NO₂S—⟨⟩—CO₂H	4.4×10^{-6}	C	26	53	7	6.6
			25	102	136	32	7.7
(28)	H₂NO₂S—⟨⟩—Cl, SO₂NH₂	1.4×10^{-7}	C	42	24	15	5.9
			2.5	468	101	303	7.4
(22)	H₂NO₂S—⟨⟩—Cl, Cl, SO₂NH₂	7.5×10^{-6}	C	23	37	21	5.9
			2.5	188	110	64	7.7
(21)	H₃COCHN—⟨S, N–N⟩—SO₂NH₂	7.2×10^{-8}	C	43	22	21	6.3
			2.5	186	128	53	8.0
(29)	Cl, NH₂, H₂NO₂S, SO₂NH₂	3.8×10^{-6}	C	5	20	19	5.5
			0.05	14	46	63	5.4
			0.25	54	57	96	5.5
(30)	Cl, N, NH, H₂NO₂S, S, O₂	1.7×10^{-6}	C	11	11	7	6.1
			0.05	20	24	7	6.6
			0.25	115	38	80	6.7
			1.25	308	65	236	6.9
(31)	Cl, H N, NH, H₂NO₂S, S, O₂	2.3×10^{-5}	C	62	34	43	6.5
			0.01	126	32	122	5.9
			0.05	265	33	291	5.5
			0.25	414	39	427	5.9

[a]From Ref. 63.
[b]C = Control phase (no drug), average of two or three 10-min clearances.

in vitro inhibitors of carbonic anhydrase with potencies several hundred times that of sulfanilamide (60,61). One of these, 5-acetylamino-1,3,4-thiadiazole-2-sulfonamide (acetazolamide, Tables 31.3 and 31.4, **21**) was studied in detail by Maren et al. (62) and became the first clinically effective diuretic of the carbonic anhydrase inhibitor class. A number of structural modifications of acetazolamide have been studied. An increase in the number of carbons in the acyl group is accompanied by retention of *in vitro* enzyme inhibitor activity and diuretic activity, but side effects become more pronounced. Removal of the acyl group leads to a markedly lower activity *in vitro* (67,68). Substitution on the sulfonamide nitrogen abolishes the enzyme inhibitory activity *in vitro*, but diuretic activity in animals is still present if the substituent is removable by metabolism (69). Two isomeric products **26** and **24** (methazolamide) are obtained on methylation of acetazolamide (66). Both these compounds are somewhat more active *in vitro* than acetazolamide but offer no advantages as diuretics over the parent compound.

(26)

(24)

A related sulfamoylthiadizolesulfonamide, benzolamide (**25**, Table 31.3), is about five times more active than acetazolamide. Clinical studies showed that 3 mg/kg p.o. **25** produces a full bicarbonate diuresis with increased excretion of sodium and potassium (65,70). Ethoxzolamide, a benzothiazole derivative, (Table 31.3, **23**) is a clinically effective diuretic carbonic anhydrase inhibitor (64). The compound lacking the ethoxy group is inactive as a diuretic when given orally to dogs, although it is a potent carbonic anhydrase inhibitor *in vitro* (71).

Dichlorophenamide (Tables 31.3 and 31.4, **22**), a benzenedisulfonamide derivative, is as active as acetazolamide *in vitro* as a carbonic anhydrase inhibitor and equally active as a diuretic (63). A large number of benzenedisulfonamide derivatives have been prepared and studied as diuretics. Some of these are very active as diuretics, although they are weak carbonic anhydrase inhibitors. In contrast to the compounds just discussed, Na^+ and HCO_3^- excretion is not increased; instead an approximately equal amount of chloride ion accompanies the sodium. These are described in the section on aromatic sulfonamides.

5.3 Clinical Application

Acetazolamide (**22**) is the prototypical carbonic anhydrase inhibitor. It is rapidly absorbed from the stomach, reaches a peak plasma level within 2 h and is eliminated unchanged in the urine within 8–12 h. The efficacious dose is 250 mg–1 g daily. During continuous administration of acetazolamide, the excretion of HCO_3^- leads to the development of metabolic acidosis. Under such acidic conditions, the diuretic effect of carbonic anhydrase inhibition is much reduced or completely absent, and therefore the effect of the drug is self-limiting (72). This is due to the fall in the level of plasma and filtered bicarbonate as the latter is lost in the urine. A state of equilibrium is reached when the small amount of hydrogen ion that is secreted in spite of carbonic anhydrase blockade is sufficient to reabsorb the reduced amount of filtered bicarbonate ion. Because of the development of kaliuresis, metabolic acidosis, and their self-limiting nature, these inhibitors are not generally prescribed for diuretic therapy. Their most common use today is to lower intraocular pressure in the treatment of glaucoma. There is a great deal of current research in this area (73–75). Carbonic anhydrase inhibitors also have some anticonvulsant activity in grand mal and especially petit mal epilepsy, but they have not gained wide acceptance for either of these indications (51). Acetazolamide may also be useful for treating acute mountain sickness (76).

6 AROMATIC SULFONAMIDES

6.1 Introduction

Beyer and Baer (77), in a paper published in 1975, discussed their early findings on the natriuretic and chloruretic activity of p-carboxybenzenesulfonamide (**27**, Table 31.4). Although the compound is considerably less active as a carbonic anhydrase inhibitor than acetazolamide, the way the kidney handled the carboxybenzenesulfonamide was considered to be more important and served to define the saluretic properties sought and ultimately found in chlorothiazide (**30**, Table 31.4).

A key discovery by Sprague (78) was that the introduction of a second sulfamoyl group *meta* to the first can markedly increase not only the natriuretic effect but also the chloruretic action of the compound. This is evident when the data for compounds **19** or **27** are compared with compound **28** (Table 31.4). Interestingly, the introduction of a second chlorine substituent as in compound **22** (Table 31.3), dichlorphenamide, produces a compound that is considerably less chloruretic with an excretion pattern that is typical of a carbonic anhydrase inhibitor.

The activity of 6-chlorobenzene-1,3-disulfonamide (**28**, Table 31.4) is further

enhanced by the introduction of an amino group *ortho* to the second sulfonamide group as in **29** (Table 31.4). Thus, 4-amino-6-chloro-1,3-benzenedisulfonamide (**29**) is an effective diuretic agent with a more favorable electrolyte excretion pattern than **28**. Chloride is the major anion excreted, and HCO_3^- excretion is low, since the urinary pH does not increase. The carbonic anhydrase inhibitory activity of **29** is only about three times that of sulfanilamide, **19**. The chlorine in the 6-position of **29** can be replaced by a bromine, trifluoromethyl, or nitro group without much change in activity; whereas a fluoro, amino, methyl, or methoxy group is less effective (79).

(**32**)

$R_1 = R_3 = H$, $R_4 = CH_3$ or CH_2CHCH_2
$R_1 = R_3 = H$, $R_4 = CO(CH_2)_{2-4}CH_3$
$R_1 = R_3 = R_4 = CH_3CO$
$R_1 = R_3 = CH_3$ or CH_2CH_3, $R_4 = H$

Substitution on the nitrogen atoms of 4-amino-6-chloro-1,3-benzenedisulfonamide gives compounds **32**. Methyl or allyl substitution of the aromatic amino group yields compounds with reduced oral activity. Acylation of the anilino group leads to an increase in activity when R_4 is a simple aliphatic acyl radical and reaches a maximum at four to six carbon atoms. Aromatic acyl derivatives are less active. Acylation with formic acid results in a cyclized product, 6-chloro-1,2,4-benzothiadiazine-7-sulfonamide-1,1-dioxide, chlorothiazide (**30**, Table 31.4) (80). The compound was the starting point for the development of the thiazide diuretics, one of the most important group of diuretics, which are discussed later. Complete acetylation ($R_1 = R_3 = R_4 = CH_3CO$) lowers the activity (79).

Methylation of both sulfonamide groups ($R_1 = R_3 = CH_3$ or CH_3CH_2; $R_4 = H$) gives a compound that has diuretic activity in the rat, but the observed activity is attributable to *in vivo* dealkylation of the sulfonamide function (81).

6.2 Mefruside

Horstmann and co-workers (82) at a later date realized and 6-chloro-1,3-benzene-disulfonamide (**28**, Table 31.4) had shown an excretion pattern combining Na^+, HCO_3^-, and a substantial amount of chloride ions even though it was an active carbonic

(**28**) R = NH$_2$
(**33**) R = NHCH$_3$
(**34**) R = N(CH$_3$)$_2$

anhydrase inhibitor. Investigation of derivatives substituted on the sulfonamide nitrogen para to the chloro substituent led to compounds with high diuretic activity. Introduction of a single methyl group (**33**) does not reduce the HCO_3^- excretion as compared to **28**, but disubstitution as in **34** leads to a substantially lower excretion of HCO_3^- and an improved Na^+/Cl^- ratio owing to less carbonic anhydrase activity. A large number of N-substituted benzenesulfonamide derivatives were prepared; the N-disubstituted compounds were of more interest because they exhibited primarily a saluretic effect. A selected number of these compounds are shown in Table 31.5. The threshold dose level and the increase in sodium excretion over control values in the rat are also given. These compounds are relatively weak carbonic anhydrase inhibitors with little effect on HCO_3^- excre-

tion, particularly those that are disubstituted on the sulfonamide nitrogen. A particularly favorable types of substituent is the tetrahydrofurfuryl group (**39–42**, Table 31.5). Activity is enhanced when the tetrahydrofuran ring bears a methyl substituent in the 2-position. A diuretic effect in rats is obtained with this compound (**41**, mefruside) at the low dose of 0.04 mg/kg. This compound has an asymmetric carbon atom; however, the difference in diuretic activity between the more active form and the racemate is not of practical significance. The action of mefruside is characterized by a prolonged increase in the rate of excretion of NaCl and water (83). The corresponding pyrrolidine derivatives are also active diuretics (**43** and **44**, Table 31.5).

Studies with ^{14}C-labeled mefruside have shown that the compound is almost completely metabolized *in vivo* to the lactone (**42**, Table 31.5) (84). This lactone has about the same diuretic activity as the parent compound and may be responsible for much of the observed activity of mefruside (82). Mefruside has been compared with chlorothiazide in the dog, and the findings suggest that mefruside has a mechanism of action similar to the thiazide diuretics without any unique features (85). Mefruside has also been studied in humans at doses of 25 mg and 100 mg. In volunteers undergoing water diuresis, the drug caused natriuresis and chloruresis extending for 20 h. Bicarbonate excretion was also increased whereas the acute excretion of potassium was slightly increased. *In vivo* carbonic anhydrase studies revealed 50% inhibition at $7.3 \times 10^{-7} M$ concentration (chlorothiazide; $1.7 \times 10^{-6} M$). The potency of mefruside and its effect on the renal concentrating the diluting mechanisms suggest that its action is similar to that of the thiazide diuretics (86,87). Studies in hypertensive patients also showed a close correlation with the thiazide diuretics in terms of both desirable and undesirable effects (88,89).

Table 31.5   Diuretic Activity in Rats of some 4-Chloro-3-sulfamoyl-*N*-substituted Benzenesulfonamides[a]

No.	R	Threshold dose, mg/kg p.o.	Increase in Na$^+$ Excretion at 80 mg/kg p.o., μeq/kg/6 h (Rat)
(35)	NHCH$_2$CH(CH$_3$)$_2$	2.4	4615
(36)		2.1	3000
(37)		6	3910
(38)		18	2900
(39)		0.9	3150
(40)		1.0	3000
(41)		0.04	4100
(42)		0.04 i.v.	3700 i.v.
		0.15 p.o.	3000 p.o.
(43)		0.15	3300
(44)		0.3	2515
(45)		0.04	3000

[a]Ref. 82.

7 THIAZIDE AND RELATED COMPOUNDS

7.1 Thiazides

Thiazides are a major class of diuretic agents that have been used for over 30 years. It was this group of therapeutic agents that first challenged and then replaced the mercurial diuretics that were used in the first half of the 20th century. Thiazide diuretics arose as an outgrowth of the carbonic anhydrase inhibitor area, in particular from the work of Novello and Sprague. Research in this area was based on the conviction of Beyer (90) that it should be possible to find a sulfonamide derivative that was saluretic and that increased Na^+ and Cl^- excretion in approximately equal quantities; in other words, a compound that did not act as a classical carbonic anhydrase inhibitor, which increased water, Na^+, and HCO_3^- elimination. A salurteic diuretic should permit a substantial reduction of edema without affecting the normal acid-base balance. Treatment of 4-amino-6-chlorobenzene-1,3-disulfonamide with formic acid resulted in the formation of the cyclic benzothiadiazine derivative, chlorothiazide (Table 31.4, **30**) (80). This compound was the first orally active, potent diuretic that could be used to the full extent of its functional capacity as a natriuretic agent without upsetting the normal acid-based balance. Chlorothiazide is saluretic with minimal side effects and fulfilled a clinical need.

Structure-activity relationships have been developed for the thiazides. The effect of varying the 3 and 6 substituents is shown in Table 31.6. Interestingly, compound **46**,

$R_6 = H$, has very little diuretic activity, whereas compounds where $R_6 = Cl$, Br, or CF_3 are highly active; and alkyl group in the 3-position decreases the activity slightly. The 3-oxo derivative of chlorothiazide also has weak diuretic activity (78). Interchanging the chlorine and sulfamoyl groups at positions 6 and 7 in chlorothiazide lowers the activity. Replacement of the 7-sulfamayl group by CH_3SO_2 or H gives compounds with little activity (79).

The degree of activity observed with compounds bearing an acyl or alkyl group on the 7-sulfamoyl group is in accord with the hypothesis that metabolic cleavage of the N-substituent occurs to yield the free sulfamoyl function (91,95). In the case of N_7-caproylchlorothiazide, urinary bioassay indicated that 50% of the excreted drug was present as chlorothiazide (63), whereas the N_7-acetyl derivative showed only weak saluretic activity and no detectable cleavage of the acetyl group (96). Substitution of the ring nitrogen atoms at position 2 or 4 with a methyl group reduces the activity and makes the heterocyclic ring more vulnerable to hydrolytic cleavage (79). The introduction of a more complex substituent in the 3 position, e.g., $R_3 = CH_2SCH_2C_6H_5$ (**57**, Table 31.6) led to a compound which was 8–10 times more potent on a weight basis than chlorothiazide (92,93). Similarly, the dichloromethyl and cyclopentylmethyl analogs (**58** and **59**, Table 31.6) were 10–20 times more potent, respectively, than chlorothiazide on a weight basis when tested in experimental animals (94). A number of "aza" analogs of chlorothiazide, derived from 2-aminopyridine-3,5-disulfonamide and 4-aminopyridine-3,5-disulfonamide, have been prepared (**60–67**, Table

(29)

(30)

Table 31.6   Comparative Effects of 3- and 6-Substituted Thiazides on Electrolyte Excretion in the Dog[a]

No.	R_6	R_3	Urinary Electrolyte Excretion[b]			Ref.
			Na^+	K^+	Cl^-	
(46)	H	H	+/−		+/−	
(30)	Cl	H	+ + + +	+	+ + + +	chlorothiazide
(47)	Br	H	+ + + +	+	+ + + +	
(48)	CH_3	H	+	+/−	+	
(49)	OCH_3	H	+ +	+/−	+ +	
(50)	NO_2	H	+ + +	+/−	+ + +	
(51)	NH_2	H	+/−		+/−	
(52)	Cl	CH_3	+ + +	+/−	+ + +	
(53)	Cl	$n\text{-}C_3H_7$	+ + +	+/−	+ + +	
(54)	Cl	$n\text{-}C_5H_{11}$	+ + +	+/−	+ + +	
(55)	Cl	C_6H_5	+/−	+/−	+/− +/−	
(56)	CF_3	H	+ + + +		+ + +	91
(57)	Cl	CH_2SBn	+ + + +	+	+ + + +	92,93
(58)	Cl	$CHCl_2$	+ + + +	+	+ + + +	94
(59)	Cl	$CH_2(C_5H_9)$	+ + + +	+	+ + + +	94

[a]From Ref. 78.
[b]Compounds administered p.o.

Table 31.7   Pyrido-1,2-4-thiadiazines 1,1-Dioxides[a]

No.	R
(60)	H
(61)	CH_3

No.	R
(62)	H
(63)	CH_3
(64)	NH_2
(65)	OH
(66)	Cl

No.
(67)

[a]From Ref. 97.

31.7). In general, the activity of each compound was comparable with, although somewhat less potent than that of its 1,2,4-benzothiadiadiazine analog (97).

In addition to diuretic activity, chlorothiazide and its congeners exert a mild blood pressure lowering effect in hypertensive patients and are frequently used clinically in the action. Initially, the antihypertensive effect was thought to be a consequence of the diuretic action. It was subsequently found, however, that removal of the 7-sulfonamide group from compounds of the chlorothiazide class eliminated the diuretic effect, but not the antihypertensive action (98,99). A compound of this type, diazoxide (**68**), is a much more effective antihypertensive agent. Surprisingly, salt and water retention has been observed with this compound (99).

(**68**)

7.2 Hydrothiazides

A new phase in the development of the thiazide diuretics was opened by the findings of deStevens et al. (100) that condensation of 4-amino-6-chloro-1,3-benzenedisulfonamide (**29**) with 1 mole of formaldehyde gives 6-chloro-3,4-dihydro-2H-1,2,4-benzothiadiazin-7-sulfonamide-1,1-dioxide (**31**), a stable crystalline compound. This compounds has been given the generic name hydrochlorothiazide. It was surprising that saturation of the 3,4-double bond in chlorothiazide leads to a compound that is 10 times more active in the dog (100,101) and humans (102–104) as a diuretic. Hydrochlorothiazide (**31**) has less than one-tenth the carbonic anhydrase inhibiting activity of chlorothiazide (**30**, Table 31.4). Like chlorothiazide, it also exerts a mild antihypertensive effect in hypertensive subjects (105).

7.2.1 STRUCTURE-ACTIVITY RELATIONSHIPS.
Structure-activity relationships have been extensively investigated by various groups (79,106–121). Substitution in the 6-position of hydrochlorothiazide follows the same rules as found for chlorothiazide; i.e., compounds of approximately equal activity result when the substituent in the 6-position is Cl, Br, or CF_3. Compounds where $R_6 =$ H or NH_2 are only weakly active. Substitution in the 3-position of hydrochlorothiazide has a pronounced effect on the diuretic potency, and compounds that are more than 100 times as active as hydrochlorothiazide on a weight basis have been obtained. It should be noted, however, that the maximal diuretic and saluretic effect that can be achieved with any of the thiazide diuretics is of the same magnitude, although the dose required may vary considerably.

Substituents in the 3-position of hydrochlorothiazide having the most favorable effect on activity were alkyl, cycloalkyl, haloalkyl, and arylalkyl, all of which maybe classified as hydrophobic in character. This is illustrated in Table 31.8 where the structures of derivatives of hydrochlorothiazide and the respective diuretic responses in the dog are shown. Table 31.9 lists the commercially available thiazide diuretics and the

(**29**)

(**31**)

Table 31.8 Hydrochlorothiazide Derivatives Structure Activity Relationships of Canine Studies

No.	Structure	Dose p.o., μg/kg	Average Excretion Per Dog Over 6 h			Natriurertic Activity (Approx.)	Partition Coefficient[a], Ether/Water
			Urine, mL	Na$^+$, meq	K$^+$, meq		
Control avg.		0	42.5–53.2	7.1–10.0	4.75		
(30)	*(benzothiadiazine: Cl, H$_2$NO$_2$S, N=, NH, SO$_2$)*	1250	102	18.5	6.5	1	0.08
(31)	R = H	20	69	12.1	4.35	10	0.37
		310	100	20.3	6.0		
		1250	125	26.1	8.6		
(69)	R = CH$_2$CH(CH$_3$)$_2$	1.3	59	13.8	4.3		
		20	100	20.7	6.8		
(70)	R = CH$_2$Cl	20	83	17.2	4.0		
		310	131	30.6	6.1		
(71)	R = CHCl$_2$	1.3	65	11.6	5.1	100	1.53
		20	113	22.5	6.5		
(72)	R = CH$_2$C$_6$H$_5$		74	18.6	4.0		
(73)	R = CH$_2$(C$_5$H$_9$)	1.3	95	22.7	4.9	1000	10.2

(Structure for the R-series: benzothiadiazine with Cl, H$_2$NO$_2$S, NH, N–H, R, SO$_2$)

[a]From Ref. 77.

Table 31.9 Benzothiadiazine Diuretics

No.	Generic Name	Trade Name	Clinical Dose, mg/kg p.o.	Structure	Ref.
(74)	Bendroflumethiazide NF	Bristuron, Naturetin	2–5		116
(57)	Benzthiazide NF	Aquatag, Exna	25–50		92,93
(69)	Buthiazide	Saltucin, Eunephran			106, 108
(30)	Chlorothiazide NF	Diuril	500–2000		79
(73)	Cyclopenthiazide	Navidrix	0.5–1.5		114,122
(75)	Cyclothiazide NF	Anhydron	1–6		114
(31)	Hydrochlorothiazide	Esidrix, HydroDiuril	25–100		99,106
(76)	Hydroflumethiazide	Saluron	25–50		116
(77)	Methyclothiazide	Enduron	5–10		113

Table 31.9 (*Continued*)

No.	Generic Name	Trade Name	Clinical Dose, mg/kg p.o.	Structure	Ref.
(**78**)	Polythiazide	Renese	4–8		18,123
(**79**)	Trichlorme thiazide	Methahydrin, Naqua	4–8		124,125

respective optimally effective doses per day in humans. Beyer and Baer (77) have studied four thiazides covering a thousand-fold increase in saluretic activity on a log-stepwise basis from chlorothiazide to hydrochlorothiazide, to trichlormethiazide, to cyclopenthiazide. This increase in activity appears to correlate with their lipid solubility (in terms of their/phosphate buffer partition coefficient), rather than their carbonic anhydrase inhibitory effect (Table 31.8).

7.2.2 PHARMACOLOGY AND MECHANISM OF ACTION. Chlorothiazide and hydrochlorothiazide are the prototypes of a group of related heterocyclic sulfonamides that differ among themselves mainly in regard to the dosage required for natriuretic activity. Examples of these compounds are shown in Table 31.8 and 31.9. The unique property of these drugs is their ability to produce a much larger chloruresis associated with a greater natriuretic potency than the carbonic anhydrase inhibitor, acetazolamide, and its congeners.

Following oral administration to normal subjects, hydrochlorothiazide is rapidly absorbed. Peak plasma levels are reached after $2.6 +/- 0.8$ h, and the drug is still detectable after 9 h. Approximately 70% of

a 65 mg dose was accounted for in urine after 48 h (126). Hydrochlorothiazide and other benzothiazine diuretics are excreted by the kidneys both through glomerular filtration and tubular secretion. The latter is shared with other organic acids and is specifically inhibited by probenecid. Concurrent administration of hydrochlorothiazide and probenecid did not modify the effects of hydrochlorothiazide on the urinary excretion of calcium, magnesium, and citrate. This combined therapy also prevented or abolished the increased serum uric acid levels associated with the use of thiazide diuretics (127). Thiazides are absorbed in the intestine in varying degrees following oral administration. Chlorothiazide is absorbed to the extent of only about 10%, however other members of this family have a much higher bioavailibility. Bile acid-binding resins, such as, cholestyramine, have been reported to bind to these drugs and therefore prevent their absorption (128). Generally these diuretics are highly plasma protein bound, primarily to albumin. Thiazides gain access to the renal tubule principally through proximal tubular secretion and to a small extent by glomerular filtration. Thiazide diuretics work by inhibiting the electroneutral Na^+/Cl^- co-transporter in the distal convoluted renal tubules. It is believed that these agents compete for the Cl⁻ binding site on the cotransporter (129).

As a class, the thiazides have an important action on potassium excretion. In most patients, a satisfactory chloruretic and natriuretic respond is accompanied by significant kaliuresis; this is also seen in dog diuretic studies (see Table 31.8) (130,131). At low doses, with some selected thiazides, a separation of natriuretic and kaliuretic effects has been observed, but at higher doses and repeated administration these differences disappear. The kaliuresis, although enhanced by the carbonic anhydrase activity of many of these compounds, is probably a consequence of increased delivery of sodium and fluid to the distal segment of the nephron. Elevated serum uric acid levels, which may be associated with gout resulting from decreased uric acid excretion during chronic thiazide administration, have been well documented (131). Calcium excretion is decreased, and the excretion of magnesium is enhanced by the administration of thiazide diuretics in normal subjects and in patients (127). Thiazide treatment has also been observed to increase plasma levels of cholesterol and triglycerides (132).

The metabolic fate of thiazides varies significantly. Chlorothiazide and hydrochlorothiazide undergo very little metabolism, whereas the more lipid-soluble drug, indapamide (**87**, Table 31.10), undergoes extensive degradation. Since 90% of the sodium is reabsorbed before it reaches the distal convoluted tubule, the effectiveness of this class of diuretics is limited.

7.2.3 CLINICAL APPLICATION. The thiazide diuretics have their greatest usefulness in the management of edema of chronic cardiac decompensation. In hypertensive disease, with or without overt edema, the thiazides have a mild antihypertensive effect. They are used with caution in patients with significantly impaired renal function. In some patients with nephrosis they have been effective, but their therapeutic usefulness in such cases has been unpredictable. The side effects observed with thiazide treatment can be divided into two types: hypersensitivity reactions and metabolic complications. Some of the common metabolic complications of these diuretics, as mentioned in section 7.2.2, are hypokalemia, magnesium depletion, hypercalcemia, hyperuricemia, and hyperlipidemia. Thiazides may also induce hyperglycemia and can aggravate a preexisting diabetic state. With respect to hypersensitivity, dermatitis, purpura, and necrotizing vasculitis have been observed. The thiazide diuretics are available as tablets; the wide range of dosages of the individual preparations are shown in Table 31.9. To minimize the possibility of potassium depletion, fixed combinations with potassium-sparing diuretics have been made available, e.g., hydrochlorothiazide-triamterene, and hydrochlorothiazide-amiloride.

Since 1957, the use of the thiazide diuretics was increased to a point where they represent one of the most commonly used diuretic agents. This is a reflection of their efficacy, safety, and ease of application.

8 OTHER SULFONAMIDE DIURETICS

This group of diuretics includes compounds that produce a pharmacological response similar to that seen with the thiazide diuretics, i.e., they are saluretics, and the maximally attainable level of urinary sodium excretion is in the same range as hydrochlorothiazide. The compounds in this group differ in chemical structure; however, most of them are derivatives of *m*-sulfamoylbenzoic acid.

Table 31.10 Other Sulfonamide Diuretics

No.	Generic Name	Trade Name	Clinical Dose, mg p.o.	Structure	Ref.
(80)	Alipamide		20–80		133
(81)	Chlorthalidone	Hygroton	50–200		134
(82)	Clopamide	Aquex, Brinaldix	10–40		135
(83)	Clorexolone	Flonatril, Nefrolan	25–100		136
(84)	Diapamide	Vectren	500		137
(41)	Mefruside	Baycaron	25–100		82
(85)	Metolazone	Zaroxolyn	2.5–20		138
(86)	Quinethazone	Hydromox	50–200		139
(87)	Indapamide	Ipamix, Natrilix	2.5–5		140

8.1 Chlorthalidone

An interesting class of compounds was developed from certain substituted benzophenones (134). Optimum diuretic properties were found in 3-(4-chloro-3-sulfamoylphenyl)-3-hydroxy-1-oxoisoindoline (**81**, chlorthalidone, Table 31.10), which is the isomeric form of an ortho-substituted benzophenone. The related compounds, **88** and **89**, are more potent carbonic anhydrase inhibitors than chlorthalidone but are less active as diuretics and have a shorter duration of action.

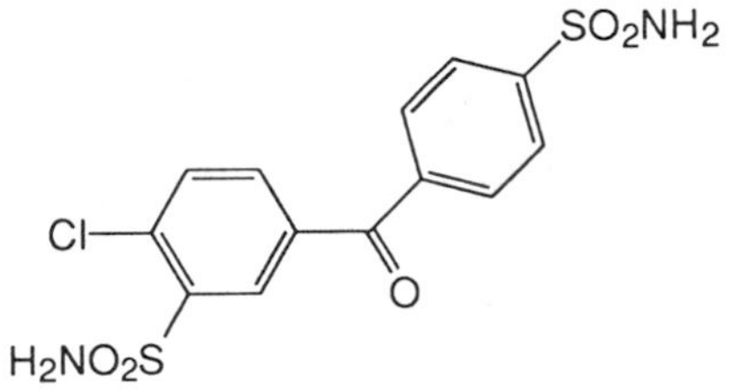

(88)

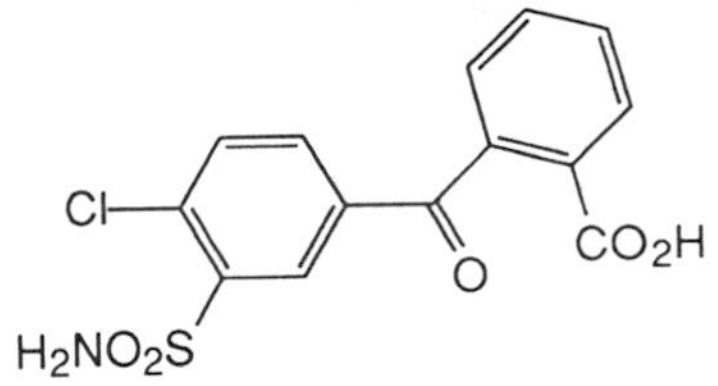

(89)

Chlorthalidone has shown good diuretic activity in dogs (141) and is characterized by an unusually long duration of action. It is about 70 times as active as hydrochlorothiazide as a carbonic anhydrase inhibitor *in vitro* and, although it induces primarily a saluresis, there is a increased output of K^+ and HCO_3^- at higher doses (63). Clinical studies have substantiated the pharmacological properties (142–145). As with the thiazide diuretics, a mild antihypertensive effect was seen (146). The recommended clinical dosage is 50–200 mg daily or every other day.

8.2 Hydrazides of *m*-Sulfamoylbenzoic Acids

The diuretic properties of a large series of compounds derived from 2-chlorobenzenesulfonamide, with a wide variety of functional groups in the 5-position (**90**), have been studied in the rat and in the dog (135). The R group includes such functional groups as substituted amines, hydrazines, pyrazoles, ketones, ester groups, substituted carboxamides, and hydrazides. The hydrazides are the most active group of compounds. One of these, *cis-N*-(2,6-dimethyl-1-piperidyl)-3-sulfamoyl-4-chlorobenzamide, clopamide (**82**, Table 31.10), has been studied in greater detail (135, 147, 148).

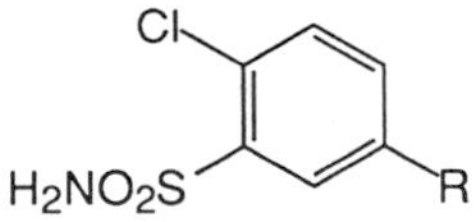

(90)

A dose-related diuretic response was seen in rats at doses of 0.01–1 mg/kg. In unanesthetized dogs, a diuretic response was seen with oral doses as low as 2 μg/kg. In anesthetized dogs, the natriuretic response observed was dose-related between 0.01 and 1 mg/kg administered i.v. The drug produced a prompt increase in urine flow and increased the excretion of sodium, potassium, and chloride. A small increase in bicarbonate excretion, which will not significantly alter plasma or urinary pH, was also noted. During maximal diuresis produced by hydrochlorothiazide, administration of clopamide had no effect on sodium excretion. Conversely, after a maximally effective dose of clopamide, hydrochlorothiazide was without effect, although in both cases an additional response to furosemide and spironolactone was observed. This suggests that although clopamide is not a thiazide diuretic, its natriuretic action closely resembles that of the thiazides. The recommended clinical dose is 10–40 mg/day.

A related hydrazide, alipamide (**80**, Table 31.10) is an effective diuretic agent in rats, dogs, monkeys, and humans (133). The suitable therapeutic dosage in humans is 20–80 mg/day. Alipamide exhibits primarily a saluretic action; carbonic anhydrase inhibition becomes an important factor only at high dose levels. A structure-activity study, in rats, showed that a hydroxamic acid moiety could replace a hydrazide group without loss of activity (149).

Similarly, diapamide (**84**, Table 31.10) is an effective saluretic agent in rats, dogs, and monkeys (137). In humans, the compound is comparable to the thiazides in terms of urine volume and electrolyte excretion (150). Elevated plasma urate and glucose levels accompany chronic administration. The clinically effective dose is 500 mg/day.

Indapamide (**87**, Table 31.10), a related sulphamoylbenzamide, possesses both saluretic and antihypertensive activity in rats, dogs, and humans after oral administration. In humans, 5 mg of indapamide administered daily produces a greater and more consistent lowering of blood pressure

than 500 mg of chlorothiazide given daily (140). Indapamide at a daily dose of 2.5 mg decreases serum potassium levels 0.5 mEq/L and increases uric acid levels about 1.0 mg/100 mL. Replacement of the indoline moiety of indapamide has also yielded compounds that possess potent diuretic activity. The 1-methylisoindoline analog, **91**, had a similar diuretic activity to indapamide; however it possessed an improved Na^+/K^+ excretion ratio (151). This compound was later found to cause blue pigmentation of the fur and some internal organs in rats and mice at 500 mg/kg p.o. (152). The *trans*-pyrrolidine derivative, **92**, has been found to be a more potent diuretic and natruretic agent than indapamide in the rat with an improved Na^+/K^+ ratio (153).

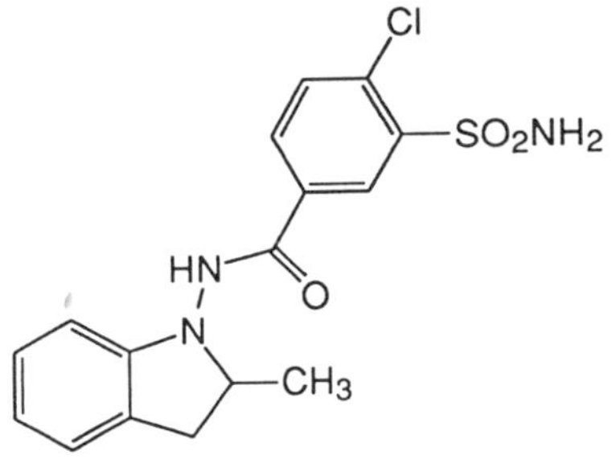

8.3 1-Oxoisoindolines

Interesting results have come from studies on a series of 4-chloro-5-sulfamoyl-N-substituted phthalimides (**93**). Maximum activity, about six times that of chlorothiazide on a weight basis, was seen when the N-substituent was a saturated ring containing six to eight carbons. Compounds in which R represented a smaller or larger ring were less active. When R was lower alkyl, decreased activity was also found. Reduction of one of the carbonyl groups to yield the corresponding 3-hydroxy-1-oxoisoindoline (**94**, R = cyclohexyl) resulted in a tenfold increase in potency. Complete reduction of the carbonyl function to a methylene yielded clorexolone (**83**, Table 31.10), which was

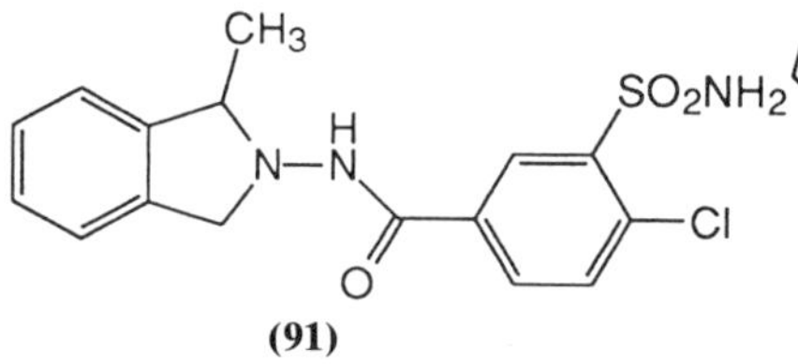

(87)

(91)

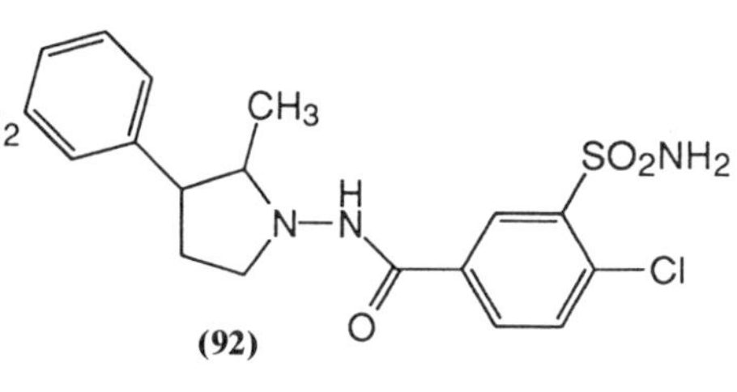

(92)

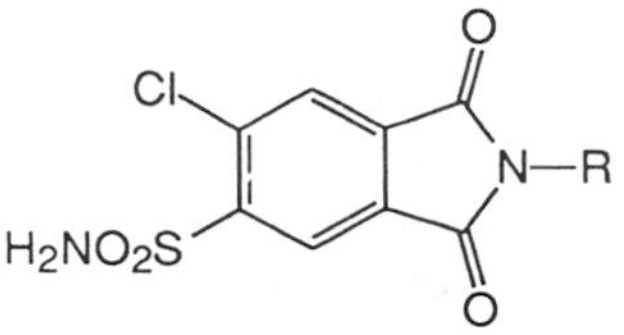

(93) 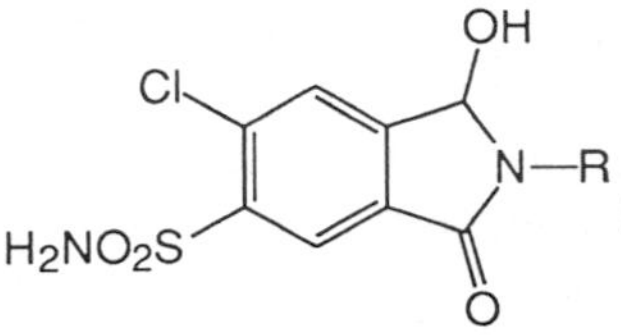   (94) 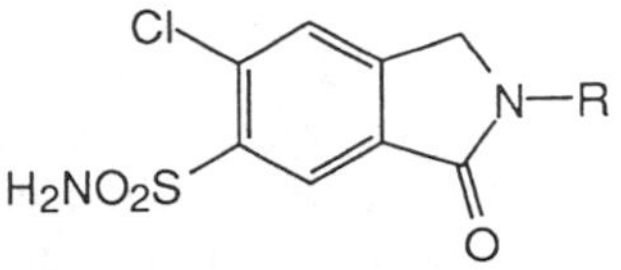   (83)

300 times more active than chlorothiazide on a weight basis when tested in the rat (136).

Interestingly, reduction of the other oxo group to yield the isomeric 6-chloro-5-sulfamoyl-1-oxoisoindoline, **95**, resulted in complete loss of activity. Structure–activity relationships for a number of 2-substituted 1-oxoisoindolines are shown in Table 31.11 (136).

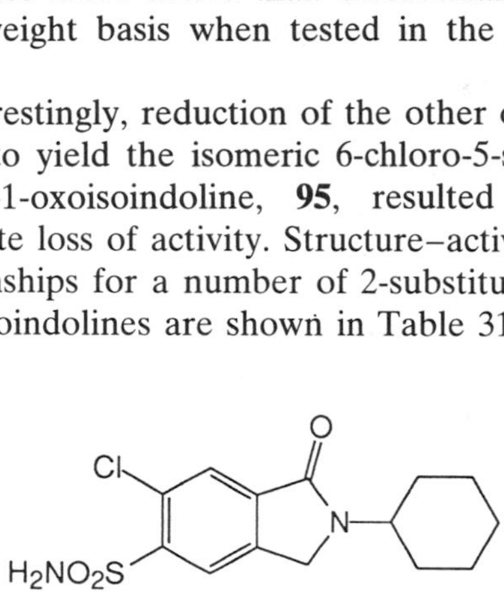

(95)

Methylation or acetylation of the sulfamoyl group of clorexolone decreases the activity by at least a factor of 10. In humans, clorexolone (**83**) is a potent diuretic. The clinical dose is 25–100 mg/day; the pattern of water and electrolyte excretion is similar to that caused by the benzothiadiazine diuretics. Urinary pH and HCO_3^- excretion remain unchanged after administration of clorexolone, indicating that there is no significant *in vivo* involvement of renal carbonic anhydrase inhibition. As is the case with the thiazide diuretics, elevated serum uric acid levels are seen, but there may be less propensity for hyperglycemia (154–156). In contrast to the thiazide di-

Table 31.11   Diuretic Activity in Rats of 2-Substituted 1-Oxoisoindolines[a]

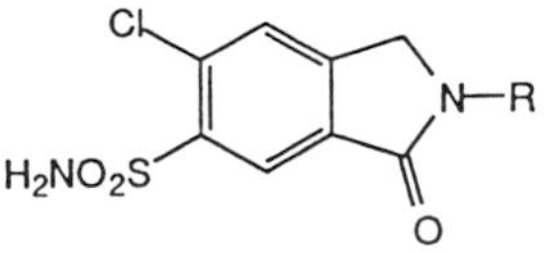

No.	R	Diuretic Activity (Chlorothiazide = 1)
(**96**)	Isobutyl	75–100
(**97**)	Cyclopentyl	100
(**83**)	Cyclohexyl	300
(**98**)	4-Methylcyclohexyl	100
(**99**)	3-Methylcyclohexyl	100
(**100**)	3,4-Dimethylcyclohexyl	200
(**101**)	Cycloheptyl	50–100
(**102**)	Cyclooctyl	100
(**103**)	Norborn-2-yl	100
(**104**)	Cyclohexylmethyl	200

[a]From Ref. 136.

uretics, insignificant amounts of the drug are excreted unchanged in humans and in dogs. The metabolites are compounds monohydroxylated on the cyclohexane ring. Neither the compound or its metabolites are stored in the body tissues (157).

8.4 Quinazolinone Sulfonamides

Replacement of the ring sulfone group in the thiazides by a carbonyl yields quinazolinones and dihydroquinazolinones (105, 106), respectively. These compounds produce nearly the same diuretic response as the parent thiazide derivatives; the dihydro-derivatives again are more active on a dose/kilogram basis. Substitution at R_3 by alkyl is disadvantageous in that it reverses the favorable Cl^- and K^+ excretion patterns seen in the few examples studied (139). The preferred member of the series was quinethazone (86, Table 31.10; 106, $R_2 = C_2H_5$, $R_3 = H$), which in humans has the same order of potency as hydrochlorothiazide with a high Na^+/K^+ excretion ratio. The duration of activity appears to be about 24 h (158). The recommended clinical dose is 50–200 mg/day.

(105)

(106)

A series of dihydroquinazolinones substituted in the 3-position (106, R_3 = aryl and arylalkyl) were studied and it was shown that some of these compounds are highly active diuretics. The more active derivatives have at least one hydrogen in the 2-position, a primary SO_2NH_2 group in the 6-position, and an *ortho* or *para* lower alkyl or CF_3-substituted aromatic ring in the 3-position of the quinazoline nucleus. The most interesting member of the series is metolazone (85, Table 31.10; 106, $R_2 = CH_3$, $R_3 = o$-CH_3-C_6H_4) (138). Studies in normal volunteers led to the conclusion that metolazone exerts its effect in the proximal tubule and in the cortical segment of the ascending limb of Henle of early distal convoluted tubule. The absence of significant bicarbonatriuria is evidence against carbonic anhydrase inhibition. Metolazone did not impair the ability to acidify the urine normally in response to an oral load of NH_4Cl, and it was concluded that metolazone has no effect on the distal H^+ secretory mechanism (159–161). In dogs, metolazone was found to be excreted by glomerular filtration and renal tubular secretion. The secretory mechanism was antagonized by probenecid; however, this did not affect the diuretic action of metolazone (162). A 10–15 mg dose of metolazone was approximately equivalent to 50 mg hydrochlorothiazide, and the time course of diuretic action was similar to hydrochlorothiazide. No acute elevation of urate or glucose or signs of toxicity were seen in a short-term study (163). The recommended clinical dose is 5–20 mg/day. In hypertensive patients a double-blind study compared a dose of 50 mg of hydrochlorothiazide with 2.5 mg and 5.0 mg of metolazone, and similar effects on blood pressure were observed. The effects on other parameters, e.g., body weight, electrolytes, serum uric acid, and blood sugar levels, were also comparable (164). In a study in patients with nonedematous, stable chronic renal failure, a high dosage of metolazone (20–150 mg) increased urine flow significantly. Its activity was greater than that of the thiazides, which are ineffec-

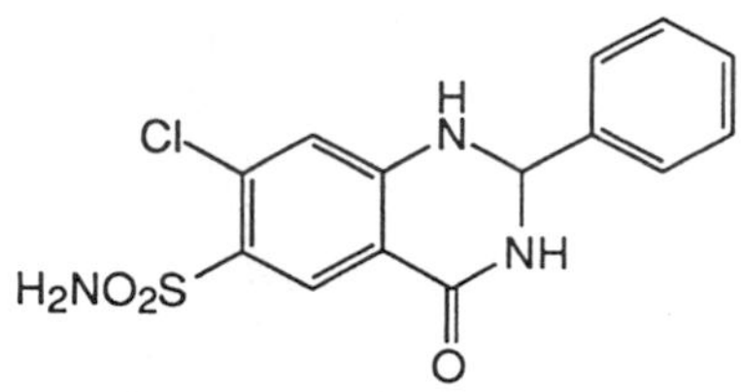

(107)

tive at glomerular filtration rates of less than 15–20 mL/min.

Fenquizone (**107**), which is structurally related to metolazone, has a thiazide-like diuretic profile, but it has less of an effect on carbohydrate and lipid metabolism (165) than the thiazides. In patients with mild essential hypertension, a 12-month study with fenquizone (10 mg/d) showed that the compound significantly lowered systolic and diastolic blood pressure. Serum levels of glucose, triglycerides, and cholesterol remained unchanged; however, uric acid levels increased slightly.

8.5 Mixed Sulfamide Diuretic-Antihypertensive Agents

Recently attempts have been made to combine an o-chlorobenzenesulfonamic diuretic

moiety with another molecule with known antihypertensive activity. Mencel and co-workers synthesized compounds where they covalently bonded enalapnilat, a known angiotensin converting enzyme (ACE) inhibitor, to several known thiazide diuretics and arylsulfonamides (166). Compound **108** was found to be a potent ACE inhibitor *in vitro*. In a sodium-depleted, spontaneously hypertensive rat (100 mg/kg i.p.), **108** reduced blood pressure by 41–42%. Compound **108** did elevate potassium and sodium excretion in the rat but it was found to be less potent than chlorothiazide in this model.

Fravolini and co-workers have covalently linked an o-chlorobenzesulfonamic diuretic to an propanolamine β-blocking pharmacophore (167). Compounds **109** and **110** posses both β-adrenergic antagonist and diuretic activity in the rat. At an equimolar dose, **109** produced a similar saluretic effect as hydrochlorothiazide, but as a β-blocker it was weaker than cartelol and propranolol.

BMY-15037-1 (**111**) is a chlorosulfamoylisoindolone derivative which has both diuretic and alpha-adrenoceptor antagonistic effects. In spontaneously hypertensive rats, oral administration of 0.3–

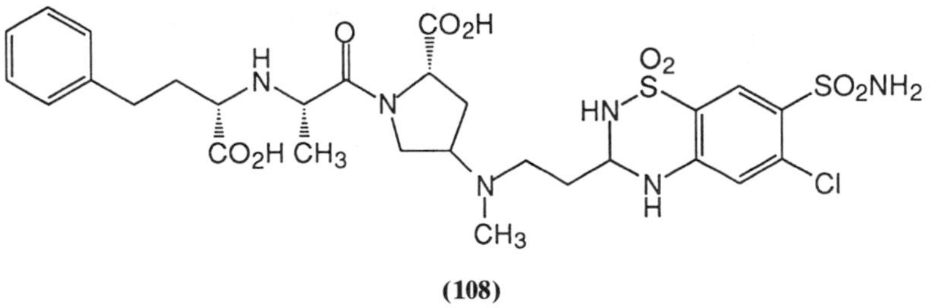

(108)

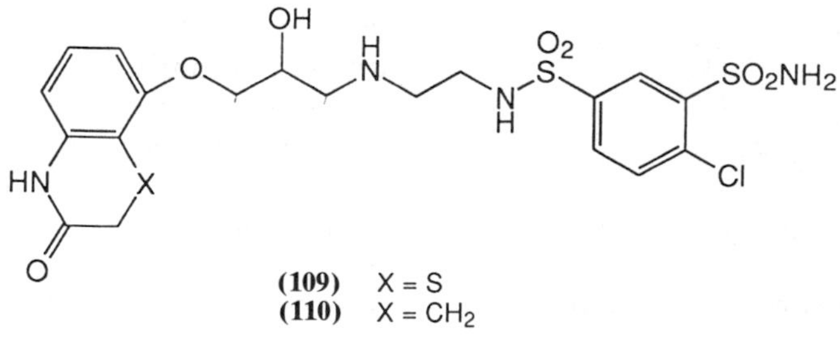

(109)   X = S
(110)   X = CH$_2$

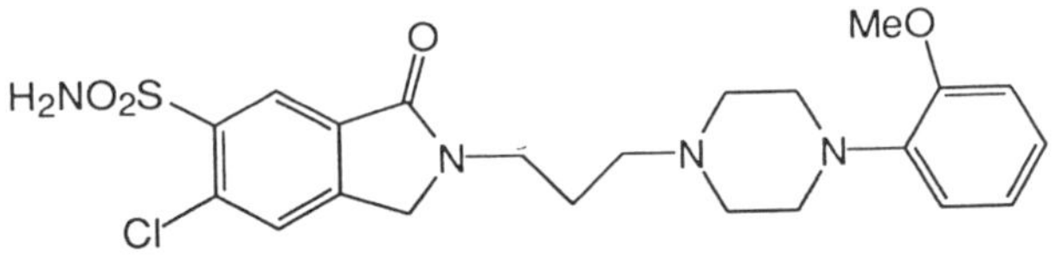

(111)

30 mg/kg of this compound decreased mean arterial pressure and induced saliuresis. The duration of action of BMY-15037-1 was similar to prazosine.

8.6 Tizolemide

A structurally novel type of sulfonamide diuretic was developed by Lang and co-workers at Hoechst (123). Tizolemide (112) was selected from a series of compounds for further investigation. Optimal activity was associated with an unsubstituted sulfamoyl group. In dogs the diuretic activity was similar to that of hydrochlorothiazide. Interestingly, tizolemide lowered serum uric acid levels in the cebus monkey, indicating a possible uricosuric effect.

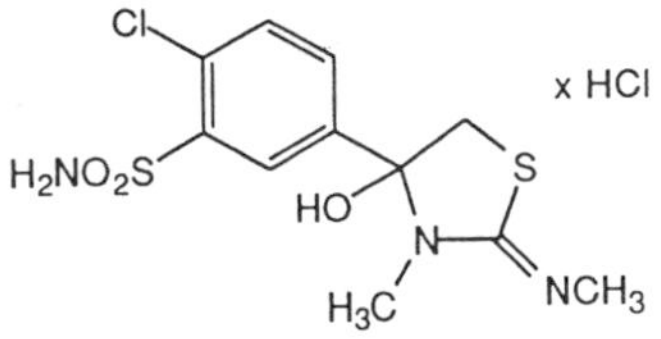

(112)

8.7 Bemitradine

Workers at Searle extended work on a series of previously described tetrazolopyrimidines that were shown to be antihypertensive agents in the rat and in humans (124,125). A series of triazolopyrimidines were prepared and SC-33643 was found to be the most potent. Bemitradine (SC-33643) has a thiazide-like profile of diuresis but is not a sulfonamide.

Bemitradine (113) was 5.5 times more potent than hydrochlorothiazide after oral administration in the unanesthetized dog (168) and significantly increased renal blood flow and glomerular filtration rates. Bemitradine is well absorbed following oral administration, however, the bioavailability is low due to hepatic first-pass metabolism.

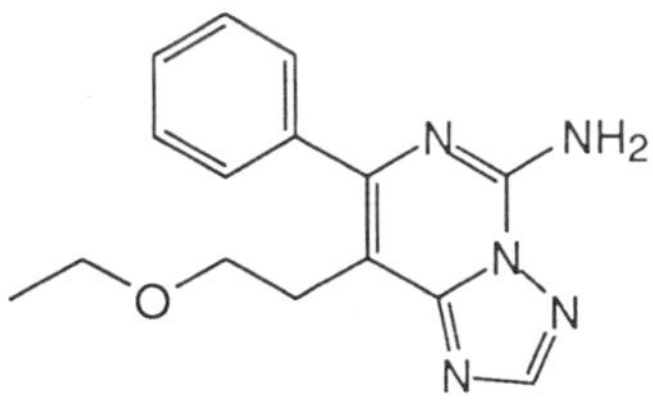

(113)

9 HIGH-CEILING DIURETICS

9.1 Introduction

The term, high-ceiling diuretics, has been used to denote a group of diuretics that have a distinctive action on renal tubular function. As suggested by their name, these drugs produced a peak diuresis far greater than that observed with other diuretics. These agents act primarily by inhibiting the reabsorption of sodium in the thick ascending loop of Henle and, thus, they are also commonly referred to as loop diuretics. This class of diuretic agents holds few structural features in common. Since they are most alike in potency and with respect to their renal site of action, they represent more a pharmacological rather than a chemical class of agents.

9.1.1 PHARMACOLOGY AND MECHANISM OF ACTION. It is currently believed that the renal site of action for these diuretics is the $Na^+/K^+/2Cl^-$ co-transporter located in the thick ascending loop of Henle. By binding to the chloride site of the cotransporter, these drugs inhibit the reabsorption of sodium, thus promoting their diuretic action (169). The carboxyl group common to many of these diuretics is essential for the binding activity, and it has only been successfully replaced by a sulfonamide.

Most of these drugs are readily absorbed in the gastrointestinal track. For example, furosemide and bumetanide have bioavailabilities of 65% and 100%, respectively. Generally these compounds are secreted from the blood to the urine via the organic acid transport system in the proximal tubule and travel through the renal tubule to their more distal site of action.

Increased potassium excretion and the elevation of plasma uric acid levels, as was observed with the thiazides, are also seen with the loop diuretics. These diuretics also increase calcium and magnesium excretion. The calciuric action of these agents has led to their use in symptomatic hypercalcemia (170).

Many of the hypersensitivity and metabolic disorders seen with the thiazides discussed in section 7.2.3 are also seen with loop diuretics. The development of transient or permanent deafness is a serious but rare complication observed with the class of agents. It is believed to arise from the changes in electrolyte composition of the endolymph (171). It usually occurs when blood levels of these drugs are very high.

The high-ceiling or loop diuretics now in use or currently being studied are shown in Table 31.12.

9.2 Ethacrynic Acid

Soon after the introduction of organomercurials as diuretics, the idea arose that their biological activity resulted from the blockade of essential sulfhydryl groups. From these early considerations, several series of highly active diuretics were developed that were thought to react selectively with functionally important sulfhydryl groups, or possibly other nucleophilic groups, that were essential for sodium transport in the nephron. These compounds generally contain an activated double bond attached to a moiety containing a carboxylic acid group of a type expected to assist transport into, or excretion by, the kidney. The general structure of these compounds is exemplified by formulas **114** and **115**. They are highly active in the dog when administered orally or parenterally, but are inactive in the rat.

A marked increase in diuretic activity was observed when chlorine was introduced *ortho* to the carbonyl group of the aryl side chain. Not only was the diuretic activity better, but the rate at which chemical addition of sulfhydryl compounds across the double bond in an *in vitro* system increased. The presence of two chlorines in positions 2 and 3 of the phenoxyacetic acid further increased the activity. A number of compounds corresponding to structures **114** and **115** and their diuretic activity in dogs are shown in Table 31.13 (174,181). Etha-

(114)   (115)

X, Y = H, Cl; R_1, R_2 = CH_3CO, NO_2, CN, alkyl

Table 31.12 High Ceiling Diuretics

No.	Generic Name	Trade Name	Clinical Dose, mg p.o.	Structure	Ref.
(140)	Bumetanide	Burinex, Lunetron	1–5		172,173
(120)	Ethacrynic acid	Edecrin	50–200		174
(132)	Furosemide	Lasix	20–80		175
(166)	Piretanide				176
(168)	Xipamide	Aquaphor	40–80		177
(167)	Azosemide	Luret			178
(180)	Torasemide	Unat, Toradiur	2.5–20		179

Table 31.13 Structure-Activity Relationships of Ethacrynic Acid Analogs

No.	Structure	Diuretic Activity (Dog i.v.)	$t_{1/2}^a$, min	Ref.
(116)	$H_2C=\overset{\underset{\mid}{CH_3}}{C}-CO-C_6H_4-OCH_2CO_2H$	+/−	90	181
(117)	$H_2C=\overset{\underset{\mid}{CH_3}}{C}-CO-C_6H_3(Cl)-OCH_2CO_2H$	+2	11	181
(118)	$H_2C=\overset{\underset{\mid}{CH_3}}{C}-CO-C_6H_3(Cl)-OCH_2CO_2H$	+1	27	181
(119)	$H_2C=\overset{\underset{\mid}{CH_3}}{C}-CO-C_6H_2(Cl)(Cl)-OCH_2CO_2H$	+3	1	181
(120)	$H_2C=\overset{\underset{\mid}{CH_2CH_3}}{C}-CO-C_6H_2(Cl)(Cl)-OCH_2CO_2H$	+6	<1	181
(121)	$H_3CHC=\overset{\underset{\mid}{CH_2CH_3}}{C}-CO-C_6H_2(Cl)(Cl)-OCH_2CO_2H$	+5	210	181
(122)	$\underset{H_3COC}{\overset{H_3COC}{>}}C=\overset{\underset{\mid}{H}}{C}-CO-C_6H_2(Cl)(Cl)-OCH_2CO_2H$	(+7)	2	181,182
(123)	$O_2NC=\overset{\underset{\mid}{H}}{\underset{R}{C}}-CO-C_6H_2(Cl)(Cl)-OCH_2CO_2H$ $R = CH_3 . CH_2CH_3$	+6	2	181
(124)	$H_3CCOC=\overset{\underset{\mid}{H}}{\underset{CH_3}{C}}-CO-C_6H_2(Cl)(Cl)-OCH_2CO_2H$			183

$^a t_{1/2}$ = Time in minutes required for one-half of a standard amount of test compound to react with excess mercaptoacetic acid at pH 7.4 and 25°C in DMF-phosphate buffer analogous to the procedure of Duggan and Noll (180).

crynic acid (**120**), Table 31.12 and 31.13, is the most interesting compound of this series and has been studied extensively. A report on (diacylvinylaryloxy)acetic acids has shown that compound **122**, Table 31.13, is approximately three times as active as ethacrynic acid (182). The corresponding (acylvinylaryloxy)acetic acids are less active, e.g., **124** (Table 31.13) (183). The 4-(2-nitropropenyl)phenoxyacetic acid derivatives (**123**, Table 31.13) are also three to five times as active as ethacrynic acid (184).

Ethacrynic acid is characterized by excellent oral absorption and rapid onset of action when administered orally or intravenously (185). Ethacrynic acid is extensively metabolized to its cysteine adduct after oral administration. It is believed that it is this metabolite that represents the pharmacologically active form of the drug since it has a much greater activity than the parent compound (186). About two-thirds of the compound is excreted by the kidney in the cysteine-adduct form along with parent compound and some unidentified metabolites. The remaining one-third undergoes biliary elimination, again as its cysteine-adduct.

Ethacrynic acid does not inhibit carbonic anhydrase *in vitro*. It has a steeper dose-response curve than hydrochlorothiazide, and the magnitude of its maximum saluretic effect is several times that of hydrochlorothiazide (185). The renal corticomedulary electrolyte gradient, after administration of ethacrynic acid and other high-ceiling diuretics, is virtually eliminated as a result of nearly total inhibition of Na^+ transport in the ascending limb of Henle (186). The clinical dose lies between 50 and 200 mg/day. In long-term studies, the antihypertensive effects of 100 mg ethacrynic acid were similar to 50 mg hydrochlorothiazide in patients with mild hypertension (187). Ethacrynic acid continues to be an effective diuretic even at very low glomerular filtration rates, and therefore is useful in the treatment of patients with chronic renal failure (188). Ototoxicity has been reported; this manifests itself as transient deafness (188). Permanent deafness has also been observed after treatment with high doses of ethacrynic acid in renal failure (189).

9.3 Indacrinone

Indacrinone (MK 196), (**125**) is an indanyl derivative of ethacrynic acid. Workers at Merck discovered that annulation of the unsaturated ketone side chain on the aromatic ring yielded compounds that retained their diuretic activity but which also possessed uricosuric properties. Both enantiomers of indacrinone possess uricosuric activity but the (−)-enantiomer is the more potent diuretic (190). Clearance studies in the rat indicated that MK 196 is a potent diuretic acting in the ascending limb of Henle, which results in significant increases in urinary excretion of sodium, calcium, magnesium, water (191), and uric acid (192).

(125)

The physiological disposition of ^{14}C-labeled MK 196 was studied in the rat, dog, monkey, and chimpanzee. The drug was well absorbed and showed minimal metabolism in the rat, dog, and monkey. Triphasic rates of elimination of drug and radioactivity were observed in these three species. In the dog, the terminal half-life was estimated to be about 68 h; in the monkey, there was a longer terminal half-life of approximately 105 h. The long terminal half-life of this compound may result in part from binding to plasma proteins. The major route of radioactivity elimination is via the feces for the rat (approximately 80%). In contrast, the monkey and the chimpanzee eliminate the majority of the dose via the urine. Minimal metabolism of MK 196 was observed in the rat, dog, and monkey; however, in humans and in the chimpanzee, there was extensive biotransformation. The major metabolite resulted from para hydroxylation of the phenyl group to yield [6,7-dichloro-2-(4-hydroxyphenyl)-2-methyl-1-oxo-5-indanyloxy]-acetic acid (**126**). This metabolite accounted for more than 40% of the 0–48 h urinary radioactivity; about 20% of the radioactivity was accounted for as unchanged drug (193,194).

(126)

In clinical studies in healthy subjects consuming a standard diet, 10 mg of MK 196 produced a slightly smaller diuresis than 40 mg of furosemide, and MK 196 did not influence uric acid excretion or 24 h urate clearance. A single dose of 40 mg of furosemide caused uric acid retention with a significant decrease of 24 h urate clearance; prolonged administration caused a statistically significant increase in plasma uric acid levels. Prolonged administration of MK 196 did not increase plasma uric acid levels, and the ratio of urate-creatinine clearance was indistinguishable from the values found in the placebo group. MK 196 thus appears to be a diuretic without uric acid-retaining properties (195). In a comparison of the diuretic effects of MK 196 and furosemide in normal volunteers receiving the drug every day for 14 days, an oral dose of 10 mg of MK 196 caused a gradual diuretic and saluretic response resulting in a maximal plateau during the period 4–7 h after drug administration, fol-

lowed by a slow return to baseline during the next 16–18 h. Although at the doses studied (10 mg and 20 mg, MK 196), the maximal response to furosemide (40 mg) was always higher than the maximal response to MK 196, the total 24 h saluresis following 10 mg of MK 196 was equivalent to that produced by furosemide. After 20 mg of MK 196 the 24 h response was greater than that with furosemide (196). A double-blind pilot study was conducted to compare the antihypertensive efficacy of two doses of MK 196 (10 mg and 15 mg) with 50 mg hydrochlorothiazide in patients with mild to moderate hypertension. Both doses of MK 196 lowered blood pressure as much as or more than 50 mg hydrochlorothiazide during the 24 h period following drug administration (197).

Clinical results indicated that MK 196 is a highly active diuretic with a gradual onset of action which reaches a plateau that persists for 4–7 h, then gradually returns to baseline values over the next 16–18 h. This is probably due to the long half-life of this compound as observed in animals. The antihypertensive effect is comparable to that observed with hydrochlorothiazide.

The cyclopentyl analog, MK-473 (**127**), has similar diuretic and uricosuric activity to indacrinone (**125**), but it also possesses substantial antihypertensive activity (198). In humans, this compound is well absorbed, but it undergoes extensive metabolism.

(127)

9.4 Other Aryloxy Acetic Acid High-Ceiling Diuretics

Since the discovery of ethacrynic acid and indacrinone many new members of the phenoxyacetic acid family of diuretics have been reported.

A series of [(3-aryl-1,2-benzisoxazol-6-yloxy] acetic acids were described by Shutske and co-workers at Hoechst (199). Of this group, HP 522 (**128**) was found to be a potent diuretic in mice and dogs with moderate uricouric properties in chimpanzees.

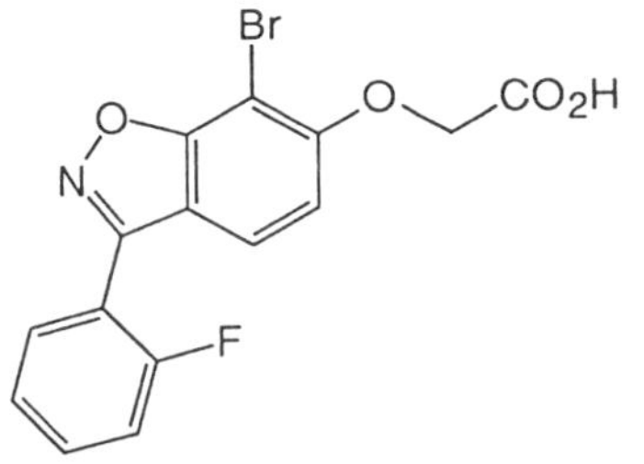

(128)

Workers at Merck reported on a series of 2,3-dihydro-5-acylbenzofurancarboxylic acids (200). The 5(2-thienylcarbonyl)-2-benzofurancarboxylic acid derivative (**129**) was found to have a higher natriuretic ceiling than hydrochlorothiazide and furosemide in the rat. Resolution of the enantiomers and testing in the chimpanzee revealed that the S-enantiomer is responsible for the compounds diuretic and saluretic activity.

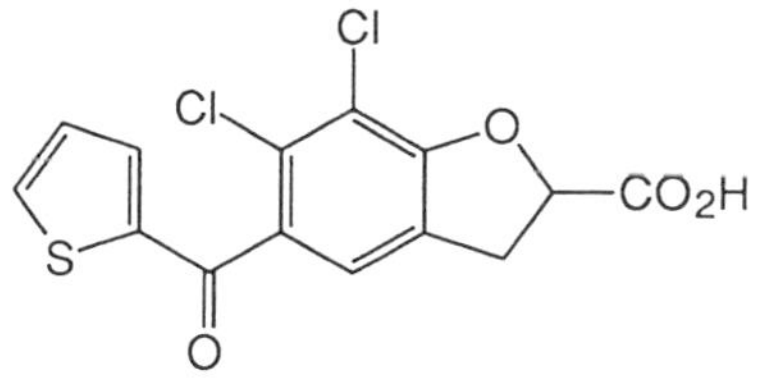

(129)

Plattner and co-workers at Abbott have disclosed a series of 5,6-dihydrofuro[3,2-f]-1,2-benzisoxazole-6-carboxylic acid high-ceiling diuretics (201). Abbott 53385 (**130**) had similar diuretic effects to furosemide in the saline-loaded mouse. In the conscious dog, it was about six times more potent than furosemide. Resolution of the enantio-

(130)

mers and pharmacological evaluation showed that only the S-isomer displays the diuretic and saluretic activity.

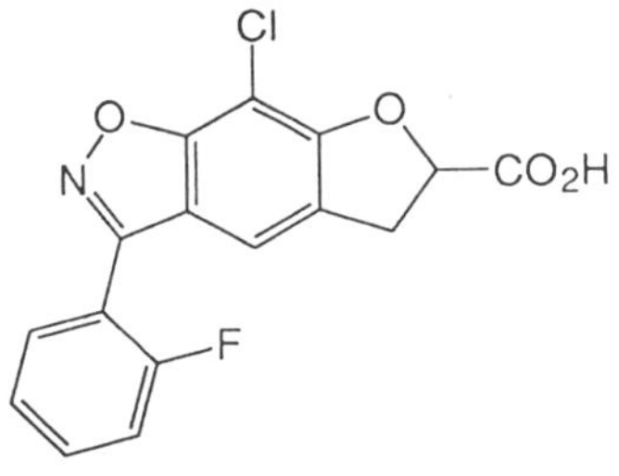

9.5 Furosemide

At the time work on ethacrynic acid was proceeding at Merck, Sharp, and Dohme, furosemide was being developed in the Hoechst laboratories in Germany. Investigation of a series of 5-sulfamoylanthranilic acids, **131**, substituted on the aromatic amino group showed that these compounds were effective diuretics. The isomeric series (**131a**) did not show saluretic properties (183,202).

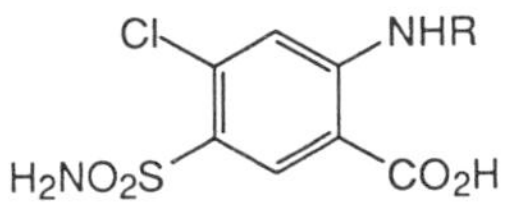

| (**131**) | (**131a**) |

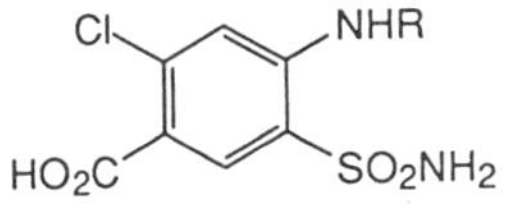

R =-CH$_2$C$_6$H$_5$, - CH$_2$(2-furyl), -CH$_2$(2-thienyl)

More than 100 variously substituted derivatives were studied pharmacologically, but only those that corresponded to the general structure **131** exhibited outstanding saluretic activity. The most active was furosemide **132** (R = furfuryl) (Table 31.12). In contrast to the dihydrobenzo-thiadiazine diuretics, where the substitutent in the 3-position of the heterocyclic ring can be varied to a considerable degree, the requirements for high activity in the 5-sulfamoylanthranilic acid series are much more stringent. On parenteral and oral administration to different species and to man, the degree of diuretic effect elicited, as measured by urine flow and Na^+ and Cl^- excretion, was several times that obtainable with the thiazide diuretics (203,204).

In a study undertaken to explore the effect of furosemide on water excretion during hydration and hydropenia in dogs, it was found that as much as 38% of filtered sodium was excreted during furosemide diuresis and both free water clearance (C_{H_2O}) and solute-free water reabsorption (TC_{H_2O}) were inhibited, indicating a marked effect in the ascending loop of Henle. Furosemide is largely excreted unchanged in the urine, but a metabolite, 4-chloro-5-sulfamoylanthranilic acid, has been identified (205,206). Studies by Hook et al. (207) and Ludens et al. (208) indicated that furosemide reduces renal vascular resistance and, thereby, enhances total renal blood flow in dogs. Clinical studies in normal subjects and in patients with edema of various etiologies have clearly shown that furosemide is an extremely potent saluretic drug (209,210).

Furosemide is rapidly absorbed following oral administration in man. It is highly bound to plasma protein, primarily to albumin, with bound fractions averaging about 97%. The bioavailability of furosemide is 50–70% in normal subjects. It is eliminated by renal, biliary and intestinal pathways. Furosemide is excreted as its glucuronide and is unchanged by the renal route.

The antihypertensive effects of furosemide were shown to be qualitatively and quantitatively similar to chlorothiazide in nonedematous patients with essential hypertension (211) Furosemide has been reported to produce moderate diuretic response in patients with renal disease and resistant edematous states when other diuretics, e.g., thiazides, mercurials, triamterene, and spironolactone, have failed. Doses up to 1.4 g/day may be required

(212,213). Ototoxicity has also been reported following large doses of furosemide (214).

9.6 Bumetanide

In a series of papers starting in 1970, Feit and co-workers in Denmark investigated the diuretic activity of derivatives of 3-amino-5-sulfamoylbenzoic acid **133**. In the initial series, R_2 was chlorine and R_1 was varied widely from alkyl to substituted benzyl. One of the most interesting compounds in this series was **134**, 3-butylamino-4-chloro-5-sulfamoylbenzoic acid, which approached the activity of furosemide when given i.v. (10 mg/kg in NaOH solution) to dogs. Interestingly, whereas in the anthranilic acid-furosemide series the N-furfuryl substituent afforded outstanding activity, this was not the case in this series (215).

(133)

(134)

Further investigation showed that compounds, in which $R_2 = -OC_6H_5$, $-NHC_6H_5$, or $-SC_6H_5$ (**133**), were very active diuretics; the structures and saluretic activity in dogs of the more active derivatives are shown in Table 31.14.

Compound **140** (Table 31.14), bumetanide, was the most interesting drug and has been studied extensively (172).

Table 31.14 Compounds Related to Bumetanide[a]

No.	Structure	Dose, mg/kg p.o.	Volume, mL/kg Urine	Urinary Excretion per 6 h (dog), meq/kg		
				Na$^+$	K$^+$	Cl$^-$
(135)	NHCH$_2$C$_6$H$_5$ / C$_6$H$_5$HN / H$_2$NO$_2$S / CO$_2$H	0.1	39	3.7	0.6	4.8
(136)	NH(CH$_2$)$_3$CH$_3$ / C$_6$H$_5$HN / H$_2$NO$_2$S / CO$_2$H	0.1	31	3.2	0.9	4.5
(137)	NHCH$_2$C$_6$H$_5$ / C$_6$H$_5$S / H$_2$NO$_2$S / CO$_2$H	0.1	38	3.1	0.9	5.0
(138)	NH(CH$_2$)$_3$CH$_3$ / C$_6$H$_5$S / H$_2$NO$_2$S / CO$_2$H	0.25	40	4.0	1.1	5.7
(139)	NHCH$_2$C$_6$H$_5$ / C$_6$H$_5$O / H$_2$NO$_2$S / CO$_2$H	0.1	36	4.0	2.1	5.9
(140)	NH(CH$_2$)$_3$CH$_3$ / C$_6$H$_5$O / H$_2$NO$_2$S / CO$_2$H	0.25	51	4.8	1.0	7.0
		0.1	31	3.3	0.49	4.5
		0.01	13	1.2	0.3	1.4
		0251i.v.	39	4.0	0.84	5.7
(141)	HN–furfuryl / C$_6$H$_5$O / H$_2$NO$_2$S / CO$_2$H	0.05	27	3.2	0.6	3.9
(132)	Cl / NH–furfuryl / H$_2$NO$_2$S / CO$_2$H	0.5	8	1.5	0.2	1.8

[a]From Ref. 172.

(159)   (143)

Further structure–activity studies uncovered related compounds with equally high diuretic potency; compounds **142–158** (Table 31.15) are representative of the series studied. It was found that a phenoxy group in the 4-position enhances activity in the anthranilic acid series as well as in the 3-amino-5-sulfamoylbenzoic acid series, e.g., compound **142** (216). The phenoxy group could be replaced by C_6H_5CO, $C_6H_5CH_2$ (217), and even a directly bonded C_6H_5 group (220), e.g., compounds **143**, **144**, and **150** (Table 31.15).

Interestingly, an equilibrium appears to exist between **143** and the corresponding benzoyl derivative **159**.

It is postulated, however, that at physiological pH the compound is present as its ring-opened benzoyl derivative **159**. Compounds **153** and **154** (Table 31.15), which do not have an amino substituent attached to the benzene ring of the sulfamoylbenzoic acid, were more stable and only cyclodehydrated on heating (220).

In the 3-amino-5-sulfamoylbenzoic acid series, the 3-amino substituent can be replaced by an OR or SR group (**147, 151, 154**, Table 31.5) (219,220); however, oxidation of the SR group to SO_2R (**148**, Table 31.15) eliminates the diuretic activity (218). Compound **154** (Table 31.15) is one of the most potent benzoic acid diuretics ever reported. It shows significant diuretic activity in dogs at 1 μg/kg, which represents a potency approximately five times as high as bumetanide (220). In the anthranilic acid series, the structural requirements are more exacting and the thiosalicyclic acid analog (**149**, Table 31.15) is only weakly active.

A series of compounds in which the sulfamoyl group was replaced by a methylsulfonyl group was also investigated (221). Many of the 5-methylsulfonylbenzoic acid derivatives showed considerable diuretic activity e.g., **155** and **157** (Table 31.15). The diuretic patterns of these compounds resemble those of previously discussed sulfamoylbenzoic acids. However, substitution of the sulfamoyl group by the spatially and sterically similar methylsulfonyl group generally led to decreased potency. Substitution of methylthio or methylsulfinyl for the methylsulfonyl group reduced the potency considerably, e.g, **156** (Table 31.15). The anthranilic acid analog (**158**, Table 31.15) of the highly active 3,4-substituted methylsulfonylbenzoic acid derivative, **155** (Table 31.15), was inactive at the dose tested, again confirming that the structural requirements in the anthranilic acid series are more demanding.

Interestingly, replacement of the chloro group in hydrochlorothiazide, by a C_6H_5S group (**160**) eliminated the diuretic activity (222). Similar results were found in the case of quinethazone and clopamide (**86, 82**, Table 31.10) (222).

(160)

Structural modifications of bumetanide (Table 31.14, **140**) were further explored by Nielsen and Feit (223). It was found that the carboxyl group in the 1-position of bumetanide could be replaced by a sulfinic

Table 31.15   Compounds Related to Bumetanide

No.	Structure	Dose, mg/kg i.v.	Volume, mL/kg Urine	Urinary Excretion per 3 h meq/kg			Ref.
				Na$^+$	K$^+$	Cl$^-$	
(142)		1.0 0.1	43 25	5.0 2.9	0.8 0.45	6.4 3.8	216
(143)		1.0 0.1	44 26	4.8 2.7	0.7 0.8	5.9 3.3	217
(144)		0.25	19	2.1	0.4	2.7	217
(145)		0.25	18	2.4	0.4	2.9	218
(146)		0.25 0.25^a	28 29	3.1 3.3	0.7 0.6	4.2 4.1	218
(147)		0.1 0.1^a	25.7 29.8	2.5 3.3	0.57 0.69	3.5 3.8	219
(148)		1.0		b	b	b	219
(149)		1.0	8	0.8	0.3	0.8	219
(150)		1.0	23	3.3	0.81	3.4	220

(Continued on page 410)

Table 31.5   (*Continued*)

No.	Structure	Dose, mg/kg i.v.	Volume, mL/kg Urine	Urinary Excretion per 3 h meq/kg			Ref.
				Na^+	K^+	Cl^-	
(151)		1.0	28	3.2	0.64	4.1	220
(152)		0.1	26	3.0	0.59	4.1	220
(153)		0.25	33	3.6	0.89	4.4	220
		0.1	12	1.0	0.42	2.0	
(154)		0.1	38	4.4	0.97	4.9	220
		0.01	23	2.5	0.53	3.5	
(155)		1.0	31	4.8	0.98	5.1	221
(156)		1.0	7	0.9	0.17	1.1	221
(157)		1.0[a]	33	3.9	0.64	4.7	221
(158)		10		[b]	[b]	[b]	221

[a]p.o.
[b]Same as control.

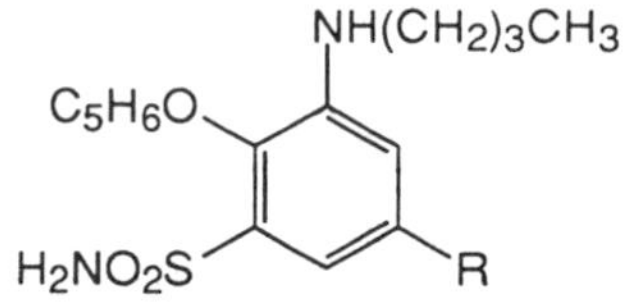

NH(CH₂)₃CH₃

C₅H₆O

H₂NO₂S — R

(161)　R = SO₂H or SO₃H
(162)　R = NHCH₂C₆H₅

or sulfonic acid group or converted to an aminobenzyl group (224) (**161** to **162**) with retention of diuretic activity.

In a further modification, the sulfamoyl group in the 5-position was replaced by a formamido group (225) (**163**); this compound has approximately one-tenth the activity of bumetanide. The electrolyte excretion pattern is still similar to that of bumetanide.

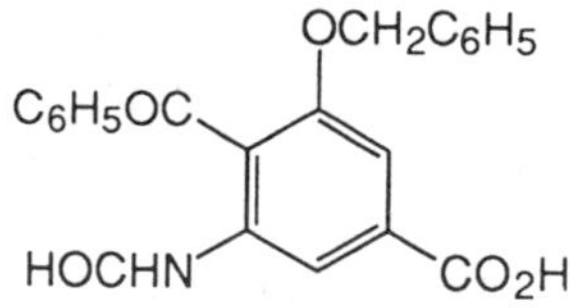

OCH₂C₆H₅

C₆H₅OC

HOCHN — CO₂H

(163)

9.6.1 PHARMACOLOGY. Bumetanide is a potent diuretic in dogs after both oral and intravenous administration. It is comparable to furosemide in its type of action and its maximum effect, but when given orally bumetanide is approximately 100 times more active. In dogs the drug is excreted rapidly by glomerular filtration and tubular secretion. No metabolites were detected in dogs (226). A parallel between bumetanide excretion and saluretic action in humans over the total period of response has been shown (27). The drug is a highly potent diuretic in patients with congestive heart failure (228,229) and in subjects with liver cirrhosis (230). Studies in humans indicate a major site of action in the ascending limb of Henle; a significant phosphaturia in-

duced during the period of maximum diuresis also suggests additional action in the proximal tubule (231,232). Bumetanide produces a rapid diuretic response with a pattern of salt and water excretion resembling that of furosemide. At the time of maximal diuresis, 13–23% of the filtered load of sodium is excreted; urinary calcium and magnesium also increase. As with other sulfonamide diuretics, hyperuricemia occurs following prolonged therapy.

The bioavailability of bumetanide after oral administration is 72–96% in normal subjects and diuresis usually begins within 30–60 min. Bumetanide is highly bound to plasma protein, (94–97%) and, therefore, probably gains access to its site of action by secretion into the proximal tubule. Probenecid, however, fails to alter bumetanide-induced diuresis in cats and in humans (233,234). Bumetanide is quickly eliminated by metabolism and urinary excretion from the body and has a plasma half-life of 1.5 h.

Several metabolites have been identified after oral administration of ^{14}C-labeled bumetanide in humans (235). All of the metabolites isolated, with exception of the conjugates, involved oxidation of the *n*-butyl side chain. About 80% of bumetanide is excreted from the body via the renal route. About 50% of the drug excreted by this route is unchanged. The major metabolite detected is the 3′-alcohol **164**. The remaining 20% of the drug is excreted by intestinal elimination. The 2′-alcohol (**165**) is the major metabolite found in the bile and feces.

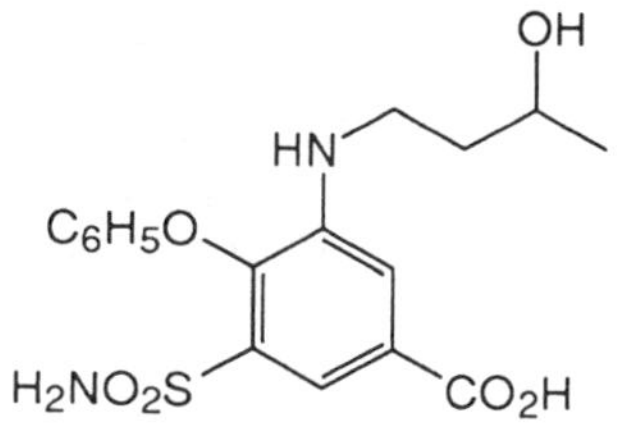

OH

HN

C₆H₅O

H₂NO₂S — CO₂H

(164)

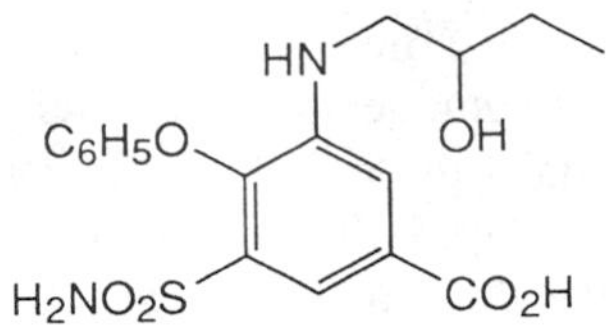

(165)

9.7 Piretanide

Workers at Hoechst synthesized a number of 3,4-disubstituted 5-sulfamoyl benzoic acids similar to bumetanide but utilizing a new synthetic method to incorporate the amine group in the presence of the other functional groups (236). Piretanide (**166**) (HOE 118) is the most interesting compound in the series. Micropuncture and clearance studies indicate that the primary site of action is in the thick ascending loop of Henle (237). Studies have shown that a 6 mg dose of piretanide provides equipotent diuresis as 40 mg of furosemide of 1 mg of bumetanide. However, compared to furosemide and bumetanide, piretanide causes a lower level of potassium excretion. Like bumetanide and furosemide, piretanide causes an increase in serum uric acid levels. Piretanide is well absorbed in the gastrointestinal track after oral administration, and its bioavailability is greater than 95% in normal subjects and in patients with renal failure (238).

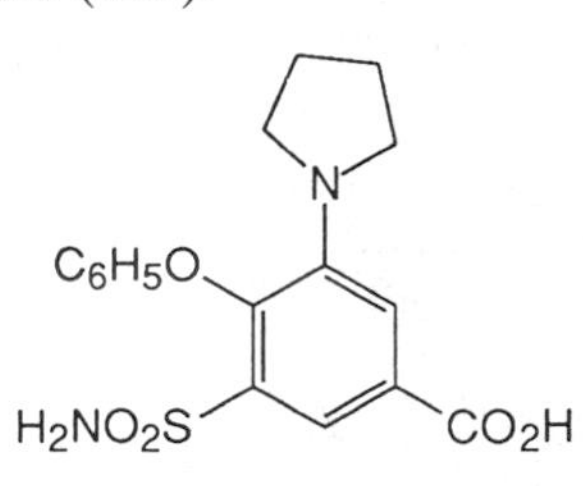

(166)

In patients with hypertension, long-term treatment with piretanide was shown to significantly lower blood pressure (239). From animal studies it appears that piretanide has a direct effect on the vascular tissue, but the mechanism is unknown (240).

9.8 Azosemide

Azosemide (**167**) is as sulfamoyl diuretic that was developed at Boehringer Mannheim (178). It is about five times more potent than furosemide after i.v. dosing, however it is equipotent after oral administration. Azosemide is poorly absorbed in

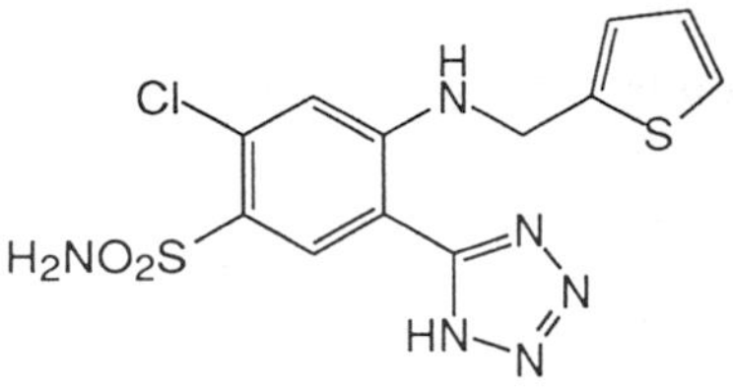

(167)

the gastrointestinal track and has a bioavailability of about 10% (238). Clearance studies have shown that the main site of action of azosemide is in the loop of Henle and to a lesser extent in the proximal tubule (241). In normal subjects, azosemide has a slower onset of action than furosemide, but at 4 h, 8 h, and 12 h, volume and sodium excretion levels were similar. The compound is extensively metabolized and only about 2% of the drug is excreted unchanged after oral dosing (242). The recommended dose to treat fluid retention is 80–160 mg orally or 15 mg i.v. The compound's poor bioavailability and slower onset of action may limit its use when rapid diuresis is required.

9.9 Xipamide

Xipamide (**168**, Table 31.12) is a derivative of 5-sulfamoyl salicyclic acid (177); 4-chlo-

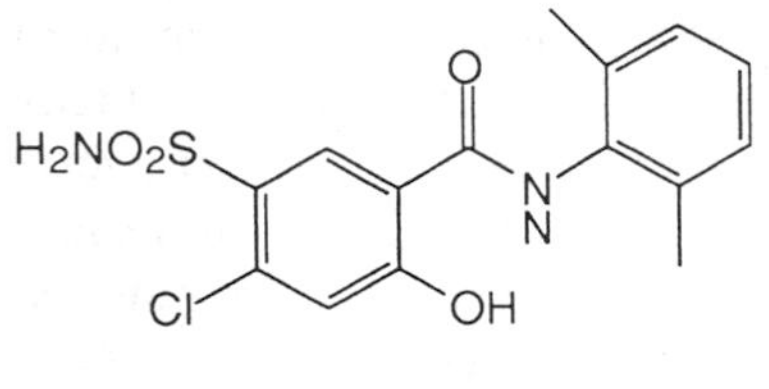

(168)

ro-5-sulfamoylsalicyclic acid itself had previously been reported by Feit and co-workers (222) to be a high ceiling diuretic. Investigation of esters; aliphatic, cycloaliphatic, aromatic and heterocyclic amides; ureides; and hydrazides of 4-chloro-5-sulfamoylsalicyclic acid showed that 4-chloro-2'-6'-dimethyl-5-sulfamoylsalicylanilide is the most active derivative. Replacement of the Cl group by Br, F, or CF_3 led to compounds with lower activity. The effects of modification of the anilide group are shown in Table 31.16 (177). Compounds methylated on oxygen and/or nitrogen are less active.

The hydrazide (177, Table 31.16) corresponding to xipamide is also only weakly active. Interestingly, the 2,6-dimethyl-piperidino derivative (178, Table 31.16) related to clopamide (82) is inactive (178).

Xipamide is active in rats and dogs following oral or intravenous administration. Application of 0.2 mg/kg i.v. in the dog, accompanied by a continuous infusion of 5% mannitol solution, led to a diuretic effect starting 40 min post-injection. After two hours, a diuretic effect could still be detected. An antihypertensive effect in spontaneous hypertensive rats was observed following 1 mg/kg p.o. (243). Xipamide is a weak carbonic anhydrase inhibitor $(ED_{50} = 1.1 \times 10^{-5} M)$ and is comparable to sulfanilamide $(ED_{50} = 1.3 \times 10^{-5} M)$ or hydrochlorothiazide $(ED_{50} = 2.3 \times 10^{-5} M)$ (Table 31.4) (244).

Xipamide is absorbed quickly from the gastrointestinal track and has a bioavailability of about 73%. At its therapeutic plasma level, it is 99% bound to plasma proteins (245). Twenty-four hours after oral administration, about 50% of the drug is excreted unchanged and about 25% as its conjugate.

Its diuretic profile is mixed; it behaves both as a loop diuretic and as a thiazide diuretic. In normal volunteers, doses of 0.5 mg/kg p.o. of xipamide is more effective than 0.5 mg/kg of chlorothalidone. It also produces a maximal natriuresis and kaliuresis similar to that of furosemide. The diuretic effects of xipamide lasts for about 24 h, with the maximum effect occurring during the first 12 h (246). In an evaluation in patients with edema of cardiac origin, xipamide was found to be an effective diuretic at 40 mg/day p.o.. Serum potassium levels were slightly lowered (247). In rats, serum potassium depletion could be avoided, and an equilibrated potassium balance was achieved in a 13 day study by combining xipamide with triamterene (248).

Although ototoxicity has been seen with salicyclic acid and with furosemide, studies with xipamide in guinea pigs did not reveal any ototoxicological properties (249).

9.10 Triflocin

The discovery of triflocin (179) resulted from a study of derivatives of flufenamic acid for possible antiinflammatory activity. Compounds incorporating a nicotinic acid moiety unexpectedly exhibited diuretic activity. Triflocin is structurally a novel and highly efficacious diuretic agent capable of promoting the excretion of as much as 30% of the sodium chloride filtered at the glomerulus. It was effective in the rat, rabbit, guinea pig, dog, and monkey. Triflocin is characterized by excellent oral absorption, rapid onset of action, and short duration of effect. The magnitude of diuresis produced by the compound is simi-

(179)

Table 31.16 Analogs of Xipamide[a]

No.	R	Urine Volume in rats, mL/kg	
		Dose, 1 mg/kg	Dose, 100 mg/kg
Controls		4–5	4–5
(169)		22.0	41.6
(170)		8.1	16.4
(171)		3.2	11.1
(172)		>10[b]	11.4
(173)		0.56[b]	13.8
(174)		0.60[b]	31.2
(175)		0.74[b]	17.8

Table 31.16 *(Continued)*

No.	R	Urine Volume in rats, mL/kg	
		Dose, 1 mg/kg	Dose, 100 mg/kg
(**176**)		ca. 5[b]	19.6
(**177**)		1.1	11.5
(**178**)			0

[a]From Ref. 177.
[b]Dose in mg/kg which increases 5 h urine volume by 50%.

lar to that seen with furosemide and ethacrynic acid. The renal sites of action are interpreted to be the proximal tubule and the ascending limb of Henle (250,251). Interestingly, a study of rats and dogs indicated that triflocin has no propensity for evoking hyperglycemia (252). The drug was studied in normal volunteers and found to be a markedly potent natriuretic agent; free water clearance (C_{H_2O}) was inhibited during water diuresis and solute-free water reabsorption (TC_{H_2O}) reduced during hydropenia, indicating a major site of action in the ascending limb in Henle. In addition, a fall in glomerular filtration rate of 10–15% was found at doses of 1 g given orally (253). Long-term toxicity studies revealed adverse effects; clinical studies were therefore discontinued (253).

9.11 Torasemide

Torasemide (**180**) is a pyridylsulfonylurea, high-ceiling diuretic that is structurally similar to triflocine (179). Its site of action is the $Na^+/2Cl^-/K^+$ cotransporter located in the loop of Henle. It also blocks chloride channels located on the basolateral side of the thick ascending limb (255). Torasemide is about 8–10 times more potent in dogs and in humans, has a longer duration of

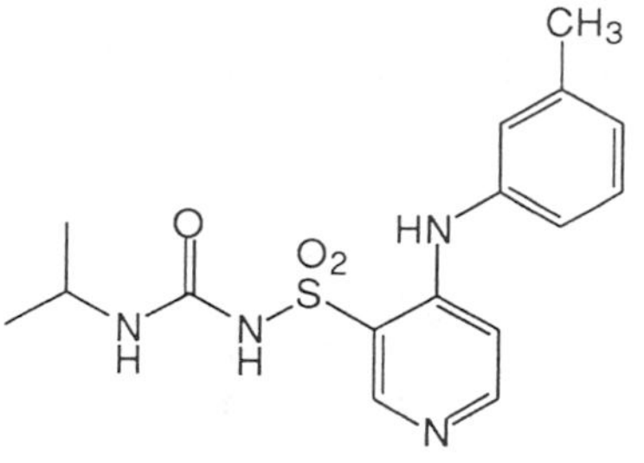

(180)

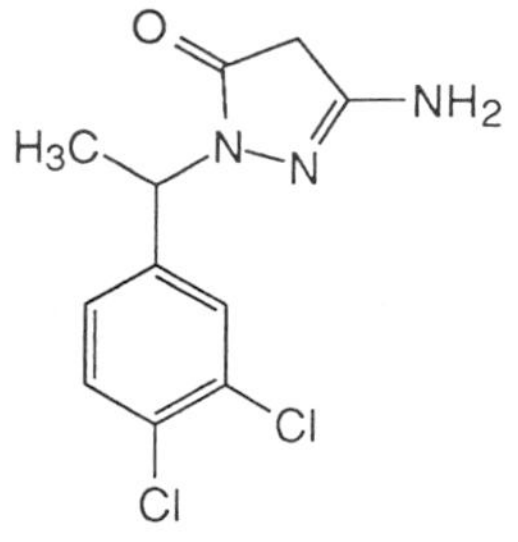

(181)

action, and causes less potassium excretion than furosemide.

Torasemide is quickly absorbed from the gastrointestinal track following oral administration and has a bioavailability of about 90%. It is highly bound to plasma protein (>95%) and has a half-life of about 2 h after intravenous dosing and 3 h after oral dosing in humans (256). The compound undergoes extensive metabolism in several species. In rats, less than 1% of the drug is excreted unchanged: most is excreted as a variety of hydroxylated metabolities (257). In humans, only 25% of the unchanged drug is excreted after 36 h. In hypertensive patients, a 2.5–20 mg dose of torasemide is effective. Torasemide was first introduced into the market in 1993 in Germany and Italy by Boehringer Mannheim. It is available in 2.5, 5, 10, and 20 mg tablets and 10 mg/2 mL ampules for injection.

9.12 Muzolimine

Workers at Bayer synthesized a series of 1-substituted pyrazol-5-ones. Some of the compounds prepared in this series were disclosed to be highly active diuretics. Muzolimine, Bay g2821 (181) was selected for further study (258).

The structure of this compound differs considerably from that of other high-ceiling diuretics, since it contains neither a sulfonamide nor a carboxyl group. It has a pk_a

of 9.2 and is very lipophilic. Clearance studies in dogs indicate that muzolimine does not increase the glomerular filtration rate, but has a saluretic effect similar to furosemide, induced by inhibition of tubular reabsorption in the ascending limb of Henle's loop (259).

Micropuncture studies in rat kidneys showed that muzolimine was effective only when given as a peritubular perfusion and not when administered intraluminally, in contrast to furosemide and bumetanide, which were effective when applied either peritubularly or intraluminally (259). Renal Na^+/K^+-ATPase activity in vitro is inhibited only at high concentrations; Mg^{2+}-ATPase activity was not affected (260).

Muzolimine is rapidly absorbed after oral administration and is estimated to have a bioavailability or greater than 90%. The plasma protein binding is 65%. This is lower than many of the other high-ceiling diuretics, and this may be the reason why the drug is effective in patients with advanced renal failure (261). Muzolimine also has a half-life of 10–20 h. It undergoes extensive metabolism in the liver; its major route of excretion is through the bile (262). Only about 10% of the drug is excreted unchanged.

Preliminary studies with muzolimine in patients showed that the drug is a high-ceiling diuretic with an onset of action and a peak diuresis similar to that of furosemide. The duration of action was 6–8 h as compared to 3–5 h after furosemide; 40 mg of muzolimine was more potent than

40 mg furosemide in all parameters investigated (263). In normal volunteers, the threshold dose was 10 mg and the dose-response curve for sodium was practically linear for doses up to 80 mg (264). Acute water diuresis and hydropenic studies carried out in seven normal volunteers suggested that muzolimine acts in the proximal tubule and in the medullary portion of the ascending limb of the loop of Henle (265). In July 1987, muzolimine was withdrawn from the market for toxicological reasons (polyneuropathy).

9.13 MK 447

Through screening procedures, workers at Merck found that 2-aminomethyl-3,4,6-trichlorophenol (182) (266) displayed significant saluretic-diuretic properties. Exploration of structure activity relationships showed that alkyl substituted, preferably alpha-branched, in position-4 and halo substitution in position-6 resulted in greatly enhanced activity. Substitution of the nitrogen and oxygen with groups resistant to hydrolysis greatly reduced the saluretic effects of these compounds (267). Also reorientation of the 2-(aminomethyl) group from the position ortho to the phenolic hydroxyl group to the meta and para positions results in loss of activity (268). Optimal activity was displayed by 2-aminomethyl-4-(1,1-dimethylethyl)-6-iodophenol, MK 447 (269), (183). The 5-aza analog displayed similarly activity to MK 447 in the rat, but was less active in the dog (270).

The saluretic effects of MK 447 in rats and dogs are generally superior, both qualitatively and quantitatively, to those of earlier high-ceiling loop diuretics. MK 447 was more effective than furosemide at 0.1–10 mg/kg p.o. in rats and dogs (271). A study in normal volunteers confirmed the high potency of the compound. Despite copious diuresis and natriuresis, no significant change in the elimination rate of potassium was observed (272). In humans, a 100 mg dose is equipotent to 80 mg furosemide, however its duration of action is longer.

Following oral administration, MK 447 is rapidly absorbed from the gastrointestinal track. It has a plasma half-life in rats and dogs of 1 h and 7.5 h, respectively. In humans, the half-life is 4–8 h (273). The compound undergoes extensive metabolism in rats, dogs, and humans. Interestingly, MK 447 does not have an effect *in vitro* when applied to the thick ascending loop of Henle segments of rabbit kidney. This suggests that a metabolite or metabolites may be responsible for the compounds diuretic activity (274).

MK 447 also possesses antiinflammatory activity. It is believed that the drug's antiinflammatory activity is due to its ability to inhibit the endoperoxide PGG_2 (275).

9.14 Etozolin

During the investigation of a series of 4-thiazolidones, some of which have chloleretic properties (276), a number of compounds with high-ceiling diuretic activity were found (277). Compound 184, piprozoline, is a choleretic compound without diuretic activity; compound 185, etozolin, is a highly active diuretic with weak choleretic properties. Minor deviations from structure 185 leads to a loss of diuretic activity. The different pharmacodynamic properties of 184 and 185 could not be traced to thermodynamic factors but rather must be re-

(182)   (183)

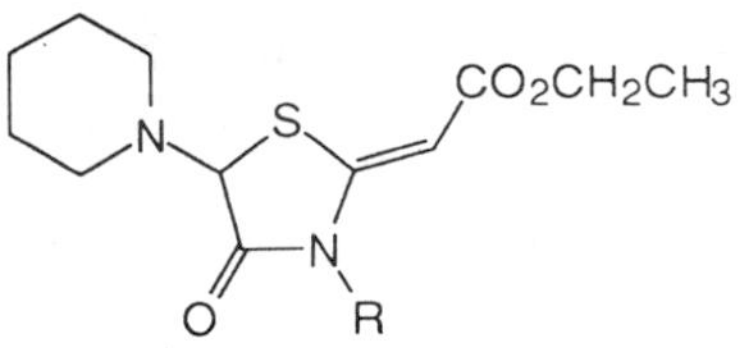

(184) R = CH_2CH_3
(185) R = CH_3

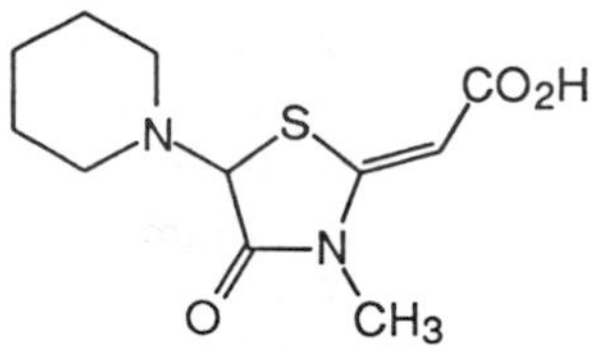

(186)

lated to closely defined receptor interactions (276).

Long-term toxicity studies in rats and dogs have shown that etozolin is well tolerated, and that it has a wide margin of safety (278). Studies in rats and dogs indicate that the compound is a potent saluretic agent with a relatively slow onset of action and prolonged activity. The maximal diuretic effect of etozolin lies between that of the thiazides and furosemide. Antihypertensive effects occur in the spontaneous hypertensive rat, DOCA, and Goldblatt rats. Etozolin does not appear to influence glucose tolerance in rats and dogs; these results are of particular interest because the tests were carried out in animals that had been treated with high doses of drug for 18 months and 12 months, respectively (279).

Clearance and micropuncture studies have shown that an initial dose of 50 mg/kg i.v. followed by 50 mg/kg/h i.v. results in a markedly increased urinary flow and sodium excretion, combined with a decreased glomerular filtration rate. Reabsorption in the proximal tubule is not affected significantly; however, fluid and electrolyte reabsorption in the loop of Henle is definitely decreased. Although etozolin differs chemically from furosemide and ethacrynic acid, it appears to share the same site of action in the nephron (280).

Absorption and metabolic studies with ^{14}C etozolin in the rat, dog and human indicate that at least 90% is absorbed following oral administration in humans. Etozolin is approximately 45% bound to plasma proteins. In the rat, blood levels could be described with a two-compartment body model, the absorption half-life was 0.6 h and the elimination half-life was determined to be approximately 6 h. In man the elimination half-life was 8.5 h; the blood levels followed with high probability a one-compartment body model (281).

The main metabolite of etozolin is the free acid, ozolinone (**186**) and its glucuronide. Other metabolites have also been detected (282,283).

In rats, etozolin is a more potent diuretic than hydrochlorothiazide, however in dogs it is less potent. In subjects with normal renal function, a 800 mg dose of the drug increased the excretion of water, chlorine, magnesium, potassium, and sodium without altering creatine clearance (284). In normal volunteers, a dose of 400 mg of etozolin was equipotent to 75 mg of a thiazide diuretic; 1200 mg was 2.8 times more effective than the 75 mg of the thiazide. Diuresis starts within 1–2 h after dosing, reaches a peak after 2–4 h, and then gradually decreases over the next 6 h (285). Owing to the long-lasting effect of etozolin, the compound would seem indicated for the treatment of cardiac and renal edema as well as for the treatment of hypertension (286). In hypertensive patients after a period of two weeks, treatment with 400 mg of etozolin daily significantly reduces systolic and diastolic blood pressure. The drug was introduced in 1977 as Elkpain.

9.15 Ozolinone

Ozolinone (**186**) is the major metabolite of etozoline, which is formed by enzymatic cleavage of the ethyl ester. It is reported to be more potent and less toxic than etozoline (287). Resolution of the enantiomers and examination of their biological activity revealed that the (−)-enantiomer possesses the diuretic activity (288). The (+)-enantiomer possess little diuretic activity and actually antagonizes furosemide activity (289). Ozolinone has a half-life of 6–10 h in humans (290) and has a plasma protein binding of 35%.

10 STEROIDAL ALDOSTERONE ANTAGONISTS

The adrenal cortex is responsible for synthesizing a number of biologically important steroids from cholesterol (Fig. 31.11). The mineralocorticoids, principally aldosterone (**187**), are important regulators of electrolyte and water balance in the body.

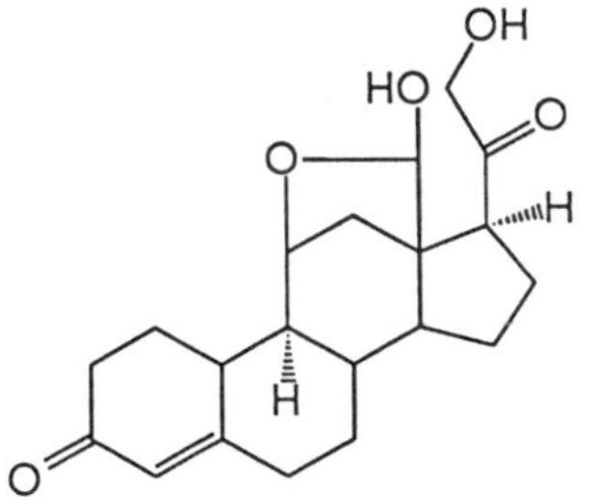

(187)

Aldosterone binds to a cytoplasmic receptor from the basolateral side located in the principal cells of the collecting tubule.

Translocation of the receptor complex to the cell's nucleus occurs, where a cascade of events leads to the generation of specific transport proteins. These proteins indirectly or directly increase the reabsorption of sodium and potassium excretion (291).

The renin–angiotensin–aldosterone system is an important mechanism for regulation of arterial blood pressure, fluid, and electrolyte homeostasis. When renal blood flow decreases, renin is excreted from the juxtaglomerular cells of the kidney into circulation. This enzyme is responsible for converting angiotensinogen to angiotensin I. Angiotensin-converting enzyme converts angiotensin I to angiotensin II. Angiotensin II is a potent vasoconstrictor and stimulates the secretion of aldosterone, which causes the retention of sodium by the kidney.

In some edematous conditions, there is increased tubular reabsorption of sodium, brought about by an excessive secretion of aldosterone. It is possible to counteract this in two ways: by blocking the action of aldosterone at the renal tubular site and by inhibiting the biosynthesis of aldosterone. Many steroid-related compounds are known competitive antagonists of aldosterone at the mineralocorticoid receptor.

During the late 1950s, Cella, Kagawa, and associates reported (292–296) on the synthesis and structure-activity relationships of a series of steroidal spirolactones with aldosterone blocking activity. This type of biological activity was established because the compounds had no action in

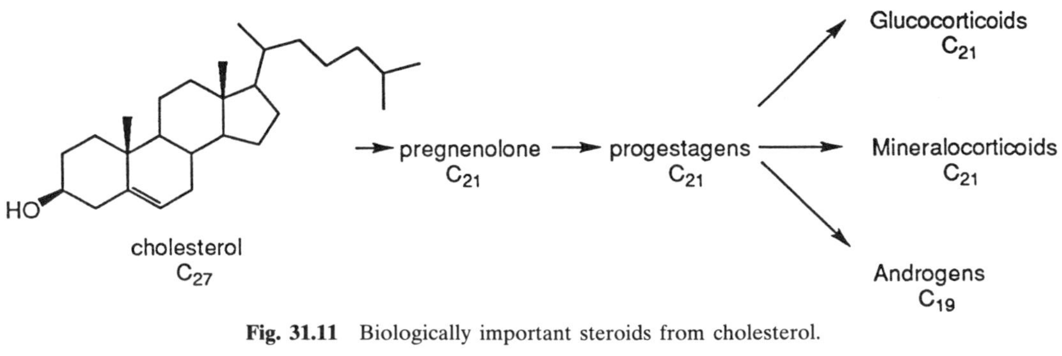

Fig. 31.11 Biologically important steroids from cholesterol.

adrenalectomized animals unless aldosterone or another mineralocorticoid was administered prior to the spirolactone (293), and because the spirolactone produced the same effect as impaired aldosterone synthesis (297).

The first compound of interest was 3-(3-oxo-17β-hydroxy-4-androsten-17-α-yl)-propanoic acid lactone (**188**), which showed aldosterone-blocking activity when administered subcutaneously to rats. Studies of related compounds established the importance, for activity, of both the five-member spirolactone and the 3-keto-Δ^4 function in ring A. An isomer of **188** with the opposite configuration at C-17 was devoid of activity (297). However, the 19-normethyl analog, **189**, was somewhat more active than **188** in rats.

Rats maintained on a low-sodium diet and treated with **189**, with consequent renal loss of sodium, compensated by increasing aldosterone secretion (299). Likewise, sodium diuresis was accompanied clinically by increased aldosterone excretion in the urine

(300). The compound has been used clinically with success in primary aldosteronism (301), in cases of nephrotic edema (302,303), and in hepatic cirrhosis (303,304). Patients with cardiac failure did not respond well (303). The effect of **189** is determined by the degree to which sodium reabsorption is controlled by aldosterone; and, thus, it is ineffective in patients with untreated Addison's disease (305) and in normal subjects on a low sodium diet (306). Because the compound blocks the effect of aldosterone on the kidney, Na^+ and Cl^- excretions are increased whereas K^+, H^+, and NH_4^+ excretions are decreased. This represents a different electrolyte excretion pattern from most other diuretics, which increase K^+ excretion.

Both **188** and **189** showed much better activity after parenteral administration than when given orally, and therefore new compounds were sought with improved absorption characteristics. Some degree of success was achieved by the introduction of additional double bonds into **188** at positions 1 and 6 to give **190**, **191**, and **192**, all of which

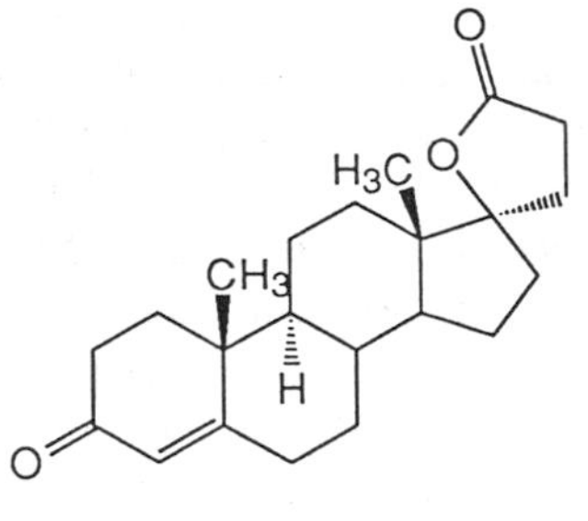

(188)

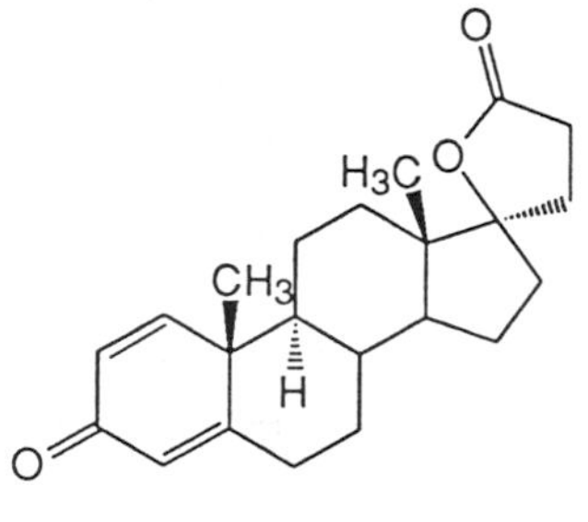

(190)

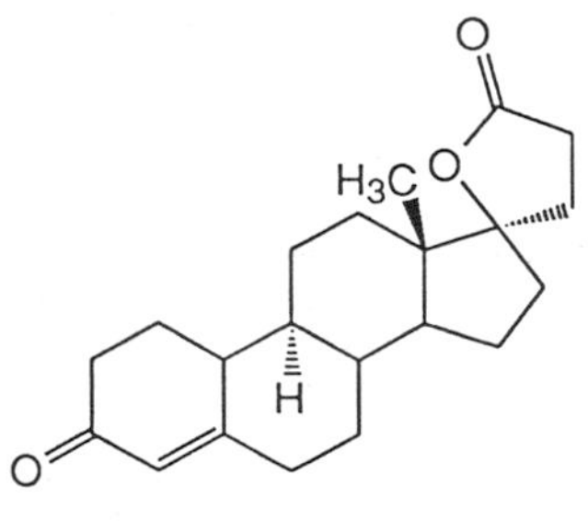

(189)

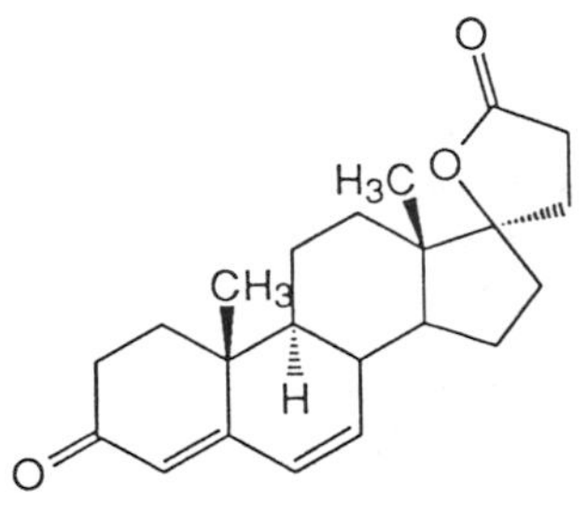

(191)

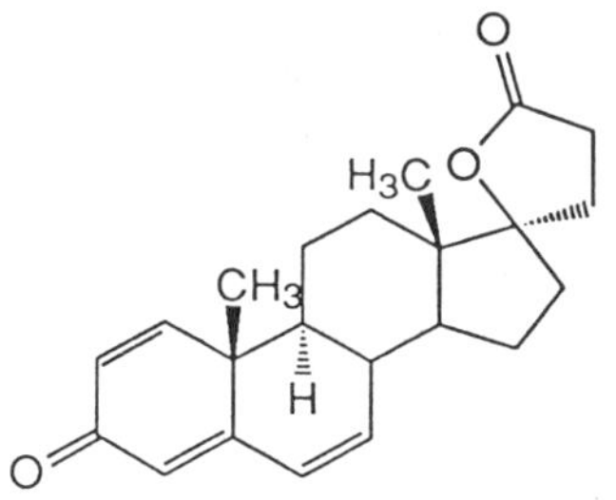

(192)

showed increased oral activity compared to the parent compound (295). Enhancement of oral activity was also noted with substitution of an acetyl thio grouping at position 1α (**193**), and even better results were obtained with this substituent at position 7 (**194**). The latter compound, 3-(3-oxo-7α-acetylthio-17β-hydroxy-4-androsten-17α-yl)propanoic acid lactone (spironolactone), has undergone extensive pharmacological and clinical studies. It is interesting that in spite of markedly increased oral activity, both **193** and **194** were less active by the parental route than the parent compound **188**. In further synthetic work on related

structures, inversion of the 7α-acetylthio group in spironolactone reduced both oral and parenteral activity by 90%, whereas 6-methylation did not change the activity significantly (307). Other structural modifications of **188**, such as introduction of methyl groups at positions 2, 4, 6, 7, or 16 or a keto or hydroxyl substituent at position 11 or a fluoro substituent at position 9, did not impart properties superior to **194** (307,308). The spirolactams **195** and **196**, corresponding to **188** and spironolactone **194**, respectively, have been synthesized (310,311), but have no significant aldosterone-blocking activity (312).

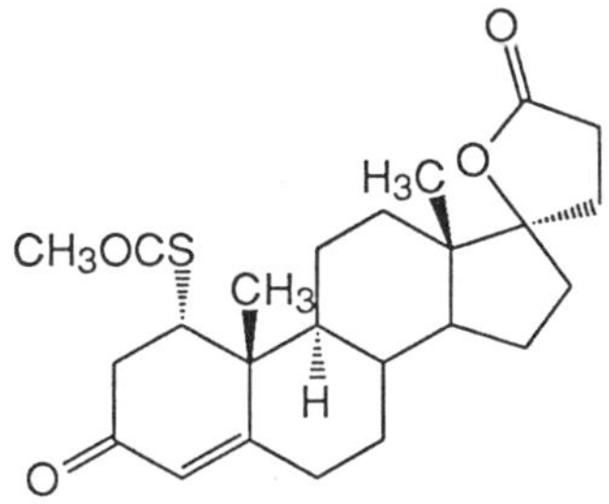

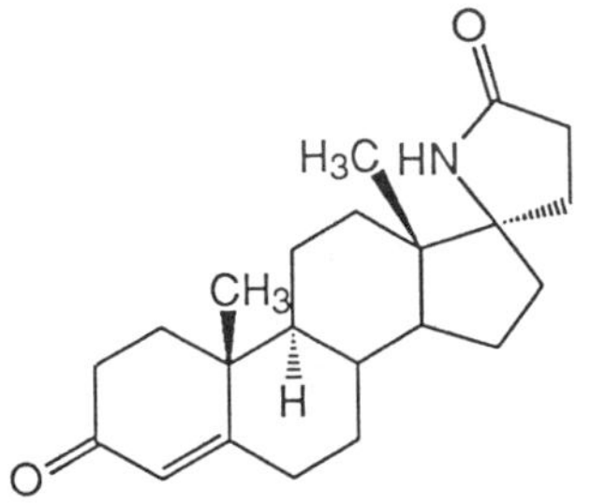

(195)

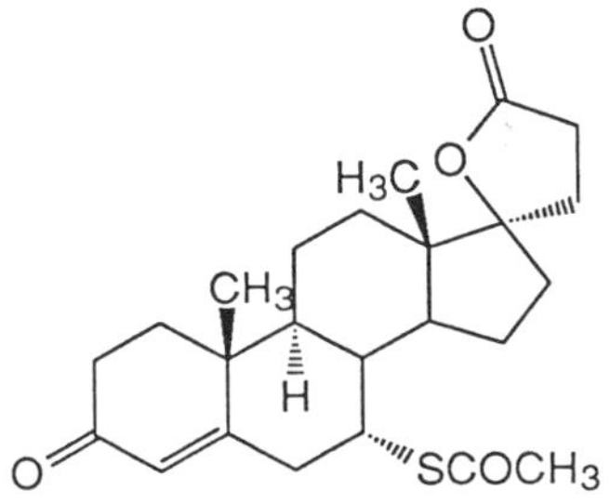

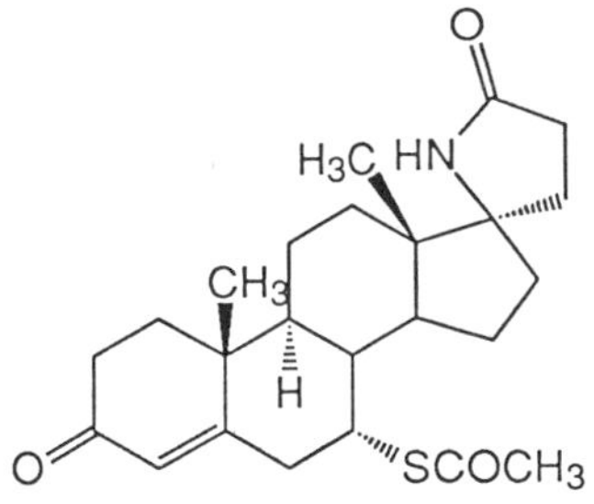

(196)

Mespirenone (**197**), a closely related analog of **192**, is three times more potent than spironolactone **194** in adrenalectomized rats treated with glucocorticoid or *d*-aldosterone (313). Studies in normal volunteers also showed that mespirenone is about six times more potent than spironolactone (314). ZK 91587, (**198**, a derivative of mespirenone) is twice as potent as spironolactone in adrenalectomized rats (315). The use of spironolactone has

(193)

(194)

been limited somewhat by its endocrine side effects, and most of the current research is this area has been directed towards finding aldosterone antagonists that

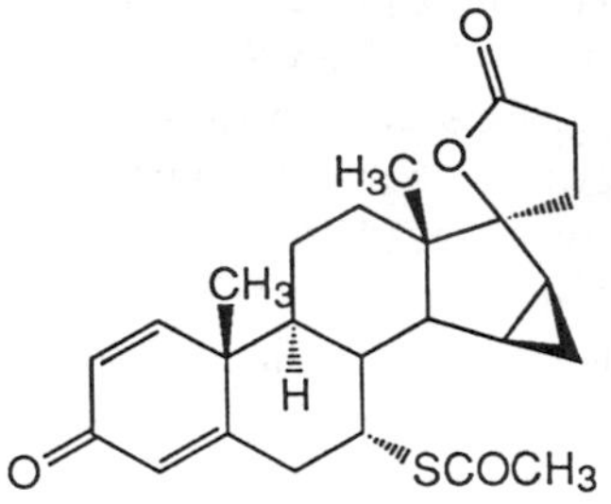

(197)

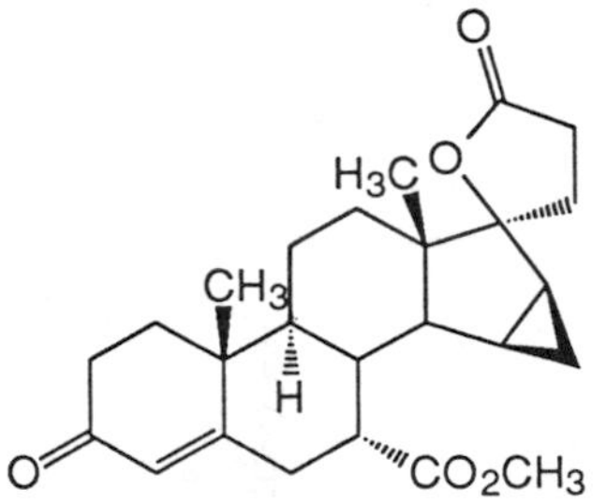

(198)

lack androgenic and progestogenic properties. Both **197** and **198** have lower relative binding affinity for the progestogen and androgen receptors when compared to spironolactone **194**.

Several metabolites of spironolactone (**199–203**) have been isolated from the urine of normal subjects (316,317).

In earlier investigations (318) potassium 3-(3-oxo-17β-hydroxyl-4,6-androstadien-17α-yl)propanoate (**204**), a water-soluble, open lactone salt, was obtained from saponification of **191**. The compound is equally effective orally and parenterally, and is approximately equipotent with spironolactone. It is relatively ineffective in the absence of mineralocorticoids. Potassium canrenoate (**204**, Table 31.17) is a specific antagonist of mineralocorticoids with pharmacodynamic properties like those of spironolactone.

Continued search for new steroidal aldosterone antagonists led to potassium prorenoate (**205**, Table 31.17). Potassium

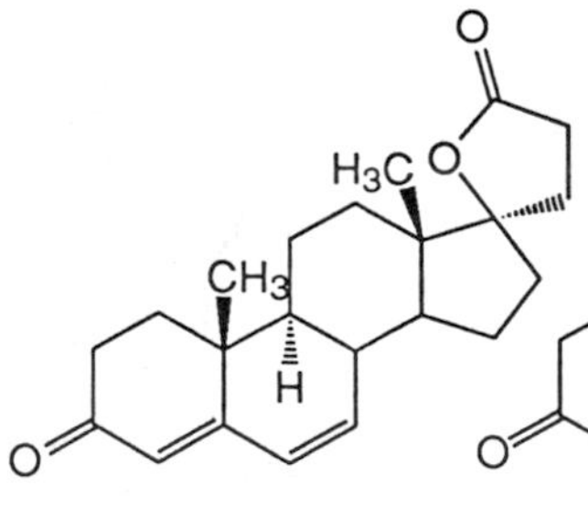

(199)

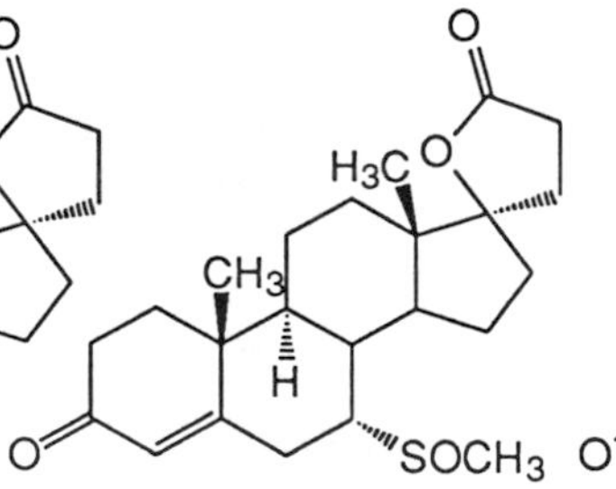

(200)

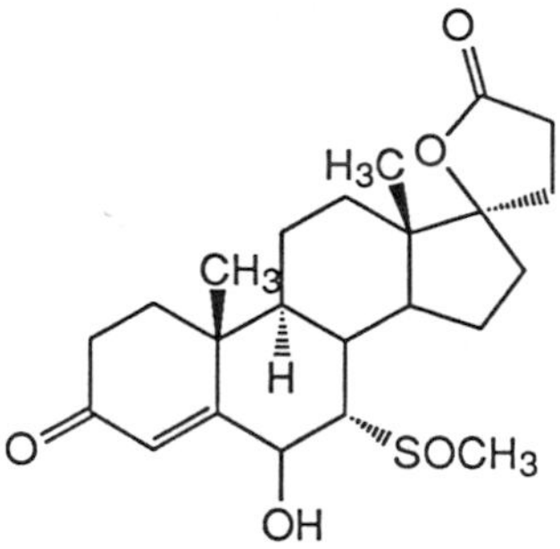

(201)

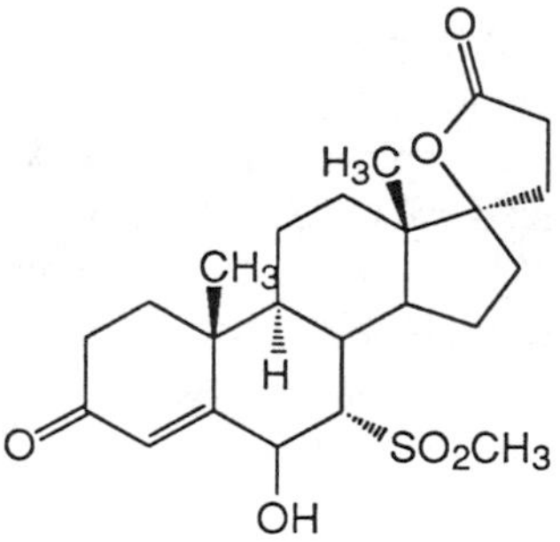

(202)

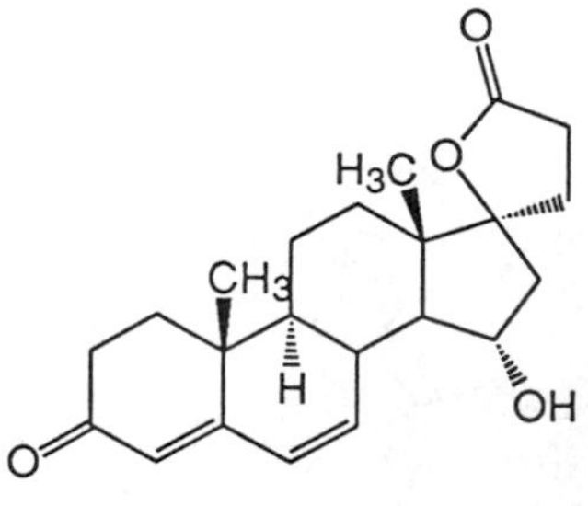

(203)

Table 31.17   Steroidal Aldosterone Antagonists

No.	Generic Name	Trade Name	Structure	Ref.
(194)	Spironolactone	Aldactone		295
(204)	Potassium canrenoate	Soldactone		318
(205)	Potassium prorenoate			319
(206)	Potassium mexrenoate			320

prorenoate is a water-soluble steroidal compound with the ability to antagonize the sodium-retaining and, when apparent, the potassium-dissipating effects of mineralocorticoids (319). In the aldosterone-treated dog, the compound had three times the potency of spironolactone. Prorenoate is relatively inactive at the renal level in adrenalectomized rats without mineralocorticoid replacement. The compound

possesses no more than 2% of the natriuretic activity of hydrochlorothiazide in
the intact animal. Clearance studies in dogs
indicate a direct renal tubular site of interaction between prorenoate and aldosterone
(319). The relative potency of prorenoate
and spironolactone was compared in a
double-blind, balanced, crossover study in
normal subjects (321). The potency of
potassium prorenoate as related to elevation of the urinary log Na^+/K^+ ratio, and
as related to potassium retention, was significantly higher than that of spironolactone. Prorenoate also produced a greater
natriuresis, but the difference was not significant. In *in vivo* experiments prorenoate
converts to the corresponding spirolactone
(321).

A new series of steroidal aldosterone
antagonists was prepared based on the
finding that introduction of a carbalkoxy
function in the 7α-position of steroidal
spirolactones enhances the activity. The
most interesting compound in this series
was potassium mexrenoate **206**, Table 31.17
(322). The oral activity of the hydroxycarboxylic acid of **206** was superior to that
of the corresponding spirolactone, **207**.
Potassium mexrenoate is a water-soluble
compound. Dose-related natriuretic responses, indexed as a reversal (increases) in
the aldosterone-depressed urinary log
Na^+/K^+ ratio, indicated that potassium
mexrenoate was between 2.1 (dog) and 4.5
(rat) times as potent as spironolactone.
Based on sodium output in intact rats,
mexrenoate was essentially inactive as a
diuretic. Diuretic potency, however, was
not indicative of antihypertensive activity.
In dogs with established hypertension
(Page model), both mexrenoate and
spironolactone exhibited equivalent antihypertensive responses (323). The isopropyl ester, potassium dicirenoate (**208**),
was also evaluated and was a potent deoxycorticosterone acetate antagonist.

Progesterone blocks aldosterone at high

(207)

(208)

doses. This effect is enhanced by the introduction of an oxygen function at C-15 and
insertion of a Δ^1 or Δ^6 double bond, e.g.,
209. Cyclopropanation of the Δ^6 double
bond to yield 15-keto-6β,7β-methylene-
progesterone, **210**, resulted in decreased
antialdosterone activity (324).

A 16β-hydroxyspironolactone (**211**) has
also been prepared, but the antimineralo-
corticoid potency is less than 14% of that of
spironolactone (**194**) (325).

Workers at Ciba-Geigy reported on a
series of 9α,11α-epoxy-derivatives of

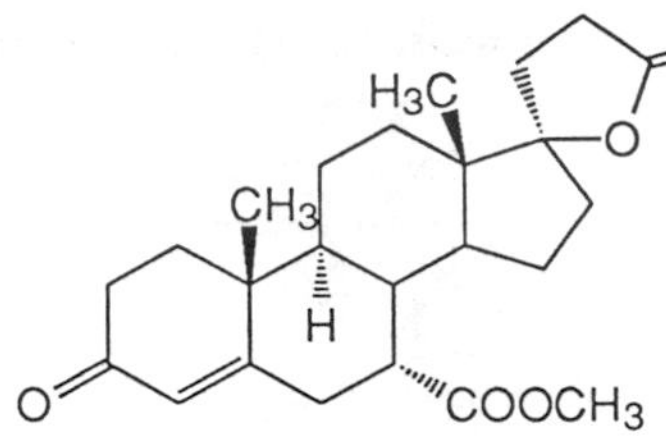

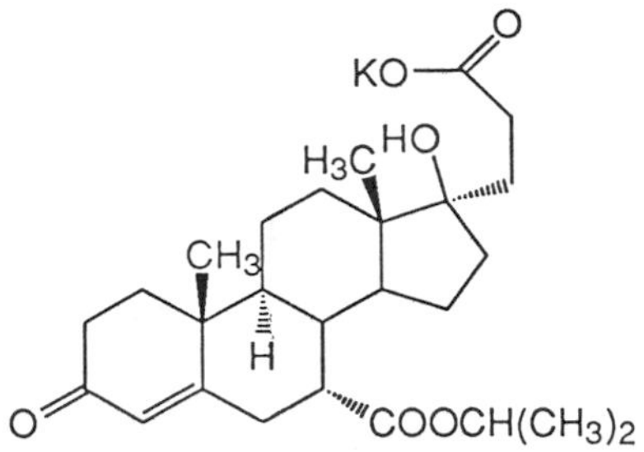

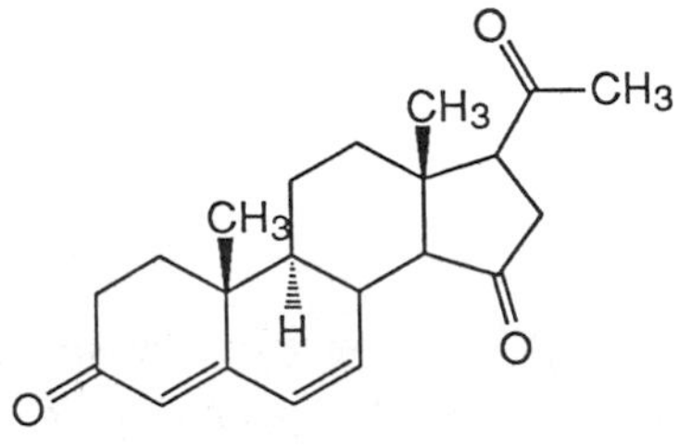

(209)

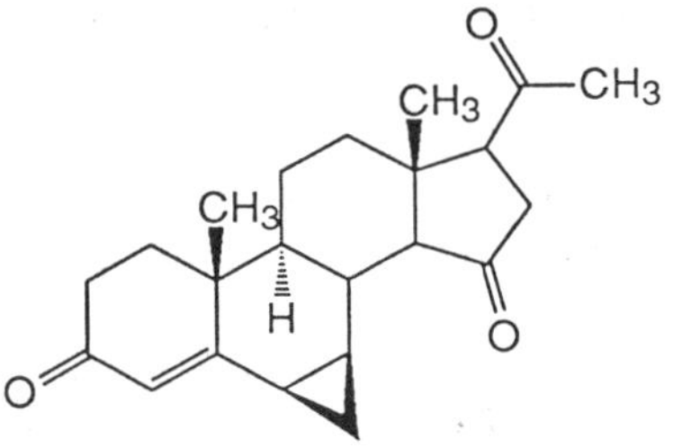

(210)

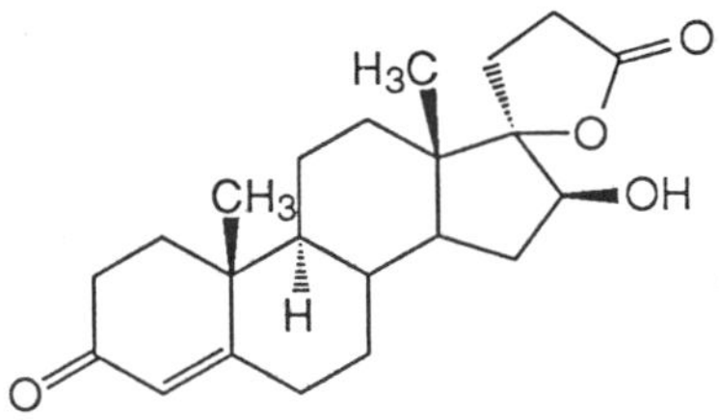

(211)

tivity. An analog of **188** lacking ring A (**215**) blocks the sodium-retaining activity of the mineralocorticoids to some extent (327). In more extensive modifications of the steroid nucleus, some 2-naphthyl-cyclopentanol ketones (**216**) were synthesized and evaluated for blocking activity (327). The hydroxylmethyl ketone, **216** (X = OH), showed the best properties in the series, being effective when administered by both the subcutaneous and oral routes. When X was F or Cl, activity is reduced.

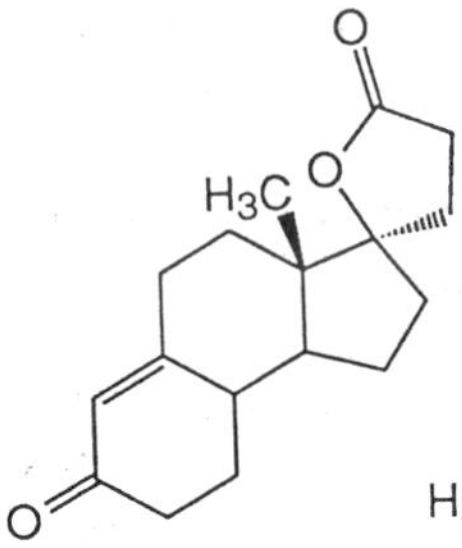

(215) 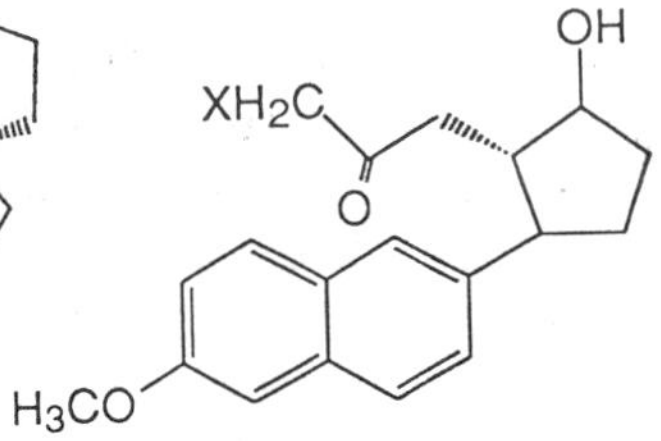   (216)

X = OH, Cl, F

spironolactone (**212**), prorenone (**213**), and mexrenone (**214**) (326). In general, introduction of the epoxide had only a small effect on the binding of these compounds to the mineralocorticoid receptor. *In vivo* at 3 mg/kg in the rat, all three were twice as potent as spironolactone (**194**). In addition, these compounds have decreased affinity for the androgen and progesterone receptors.

A number of modified steroidal type structures have been synthesized and examined for mineralocorticoid-blocking ac-

Spironolactone (**194**) is the most important compound to emerge from the many synthetic and biological investigations of aldosterone antagonists. It promotes diuresis by competing with aldosterone at receptor sites responsible for Na$^+$ reabsorption. The best results have been obtained in patients with cirrhosis and the

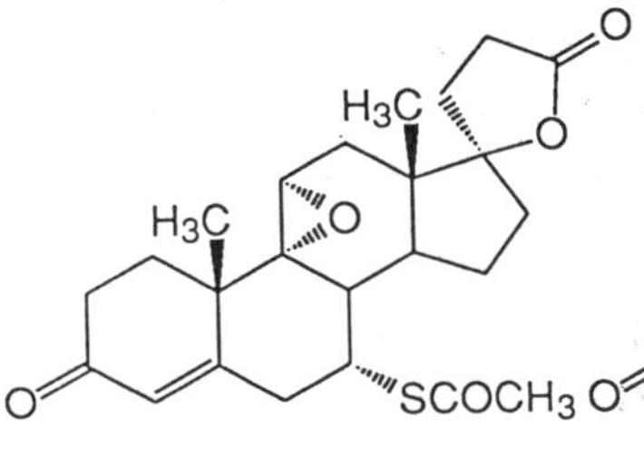

(212) 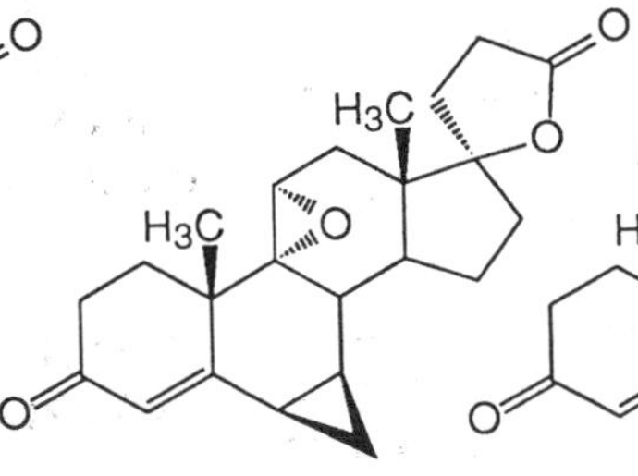   (213) 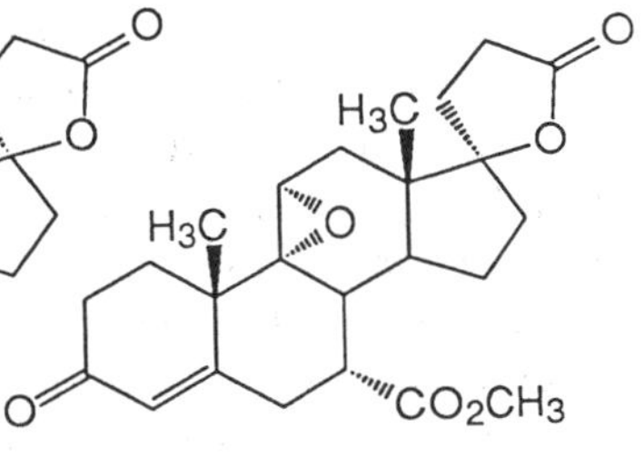   (214)

nephrotic syndrome, in whom the aldosterone secretion rate is very high. It has been much less impressive in patients with congestive heart failure, in whom aldosterone secretion is not normally raised. Spironolactone causes Na^+ diuresis without K^+ secretion, leading to particular interest in the drug in the treatment of cirrhosis with ascites, a condition in which most other diuretics would cause excretion of excessive amounts of K^+.

After oral administration, about 70% of the drug is absorbed (328). The compound undergoes extensive first-pass metabolism in the liver, and there is considerable enterohepatic circulation. The major metabolite is canrenone (**199**). Essentially no unmetabolized drug is found in the urine. Spironolactone induces hepatic cytochrome P450 and, therefore, might alter the metabolism of other drugs. The compound is also highly bound to plasma protein (95%). The usual dose of spironolactone is 25 mg four times daily, but as much as 300 mg/day may be administered safely, either alone or in combination with other diuretics. Administration of spironolactone for several weeks to hypertensive patients has a modest antihypertensive action, but no effect on blood pressure is seen in normotensive subjects. The drug also increases the antihypertensive effect of chlorothalidone and other diuretics. The effect may result from an alteration of the extracellular-intracellular sodium gradient (329).

Hyperkalemia is the most serious side effect reported with spironolactone treatment. Some additional side effects observed are loss of libido, impotence, and gynecomastia in men and menstrual irregularities in women (330). These usually disappear when treatment is discontinued.

In 1976, the FDA issued a warning because spironolactone was found to be a tumorigen in chronic toxicity studies in rats at high doses (331). It was suggested that its use should be restricted to cases only where other therapy would be inadequate or inappropriate.

11 ALDOSTERONE BIOSYNTHESIS INHIBITORS

Compounds that inhibit the synthesis of aldosterone by the adrenal cortex theoretically have potential diuretic activity, but these are not of practical usefulness. Metyrapone (**217**) exemplifies this type of compound and has undergone considerable biological and clinical study (332). In moderate doses, it blocks 11-β-hydroxylation and so inhibits the production and secretion of hydrocortisone, corticosterone, and aldosterone. However, compensatory secretion of 11-deoxycorticosterone, a potent salt-retaining hormone, negates the effect of reduced aldosterone levels.

(**217**)

Spirolactones also have an effect on aldosterone biosynthesis. Erbler found that spironolactone decreased aldosterone synthesis in the adrenal tissues of sodium-depleted rats at 10^{-5}–10^{-4} M (333). It is believed that spironolactone, carenone, and potassium caronate inhibit mitochondrial 11β and 18-hydroxylase to prevent the synthesis of aldosterone (Fig. 31.12) (334). Spironolactone may also inhibit adrenal 21-hydroxylase (335). It is believed, however, that the major site of action of these antagonists are at the receptor level, since the concentrations required to inhibit biosynthesis are much higher.

12 CYCLIC POLYNITROGEN COMPOUNDS

12.1 Xanthines

The diuretic actions of alkylxanthines, such as, caffeine (**218**) and theophyline (**219**),

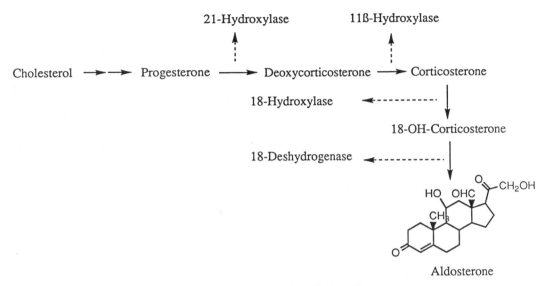

Fig. 31.12 Aldosterone biosynthetic pathway.

have been known for more than a century. They have been of limited clinical utility due to their low potency, development of tolerance after repeated administration, and side effects such as psychomotor effects and cardiac stimulation. It has been proposed that the pharmacological basis for their mechanism of action is adenosine receptor antagonism (336). Adenosine produces antidiuretic and antinatriuretic responses in several species. These effects can be competitively antagonized by theophylline (337). There are two subtypes of adenosine receptors, A_1 and A_2 (and perhaps A_3). It is believed that the renal actions of adenosine are due to its stimulation of the adenosine A_1 receptor (338). It has been shown that compounds that antagonize the A_1 receptor exhibit diuretic effects (339,340). Compounds that are selective adenosine A_2 antagonists do not

show any diuretic or natriuretic properties (341).

A series of 8-substituted 1,3-dipropylxanthines were reported by Suzuki and coworkers (342). Many of these compounds were potent adenosine A_1 receptor ligands with diuretic activity. One of the most potent analogs was 8-(dicyclopropylmethyl)-1,3-dipropylxanthine (**220**) (K_i, $A_1 = 6.4$ nM; K_i, $A_2 = 590$ nM). This compound also increased urine volume and sodium excretion in the rat after oral administration.

(**220**)

8-Cyclopentyl-1,3-dipropylxanthine (**221**) was also reported to be a potent selective adenosine A_1 receptor antagonist with diuretic action (342). At 0.1 mg/kg, i.v., **221** significantly increased urine volume and sodium excretion in the rat with no signifi-

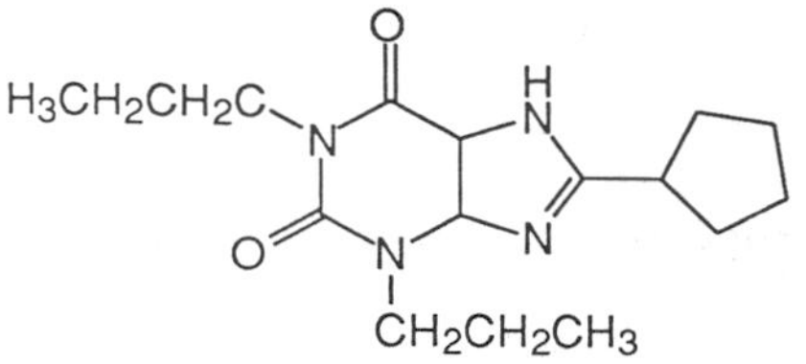

(221)

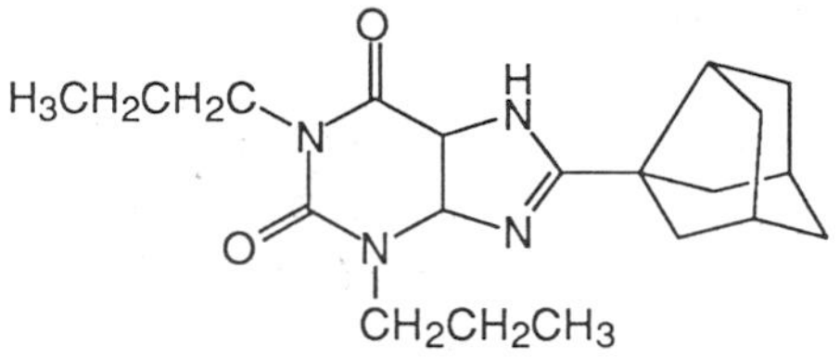

(222)

cant change in potassium excretion (343). The tubular site of action is thought to be in the proximal tubule.

8-(Noradamantan-3-yl)-1,3-dipropylxanthine (KW-3902) (222) has been recently described as being a potent adenosine A_1 receptor antagonist (K_i, $A_1 = 1.1$ nM; K_i, $A_2 = 330$ nM) (344). In saline-loaded normal rats, a 0.001–1-mg/kg oral dose of KW-3902 caused a significant increase in urine volume and sodium excretion, with little effect on potassium excretion (345). In the saline-loaded conscious dog, KW-3902 exhibited a longer lasting natriuresis than furosemide or trichlormethiazide (346). It also induced less hypokalemia and hyperuricemia in rats when compared to furosemide and trichlormethiazide (347). No attenuation in its pharmacological action was observed after repeated oral dosing (0.1 mg/kg/day) for 24 days in rats (348). KW-3902 possessed renal protective effects against glycerol-, cisplatin-, and cephaloride-induced acute renal failure (345,349,350). From stop-flow methods and lithium clearance studies, it appears that the site of action of this drug is in the proximal tubule (351). KW-3902 had little effect on renal hemodynamics.

12.2 Aminouracils

During an extensive study of compounds related to the xanthines, it was discovered that certain of the intermediate substituted 6-aminouracils were orally active diuretics in animals (352). The 1,3-disubstituted derivatives of 6-aminouracil 223 are diuretics, whereas the monosubstituted compounds are not. The 1-*n*-propyl-3-ethyl derivative, which was the most potent diuretic in the series, was unsuitable for clinical use owing to gastrointestinal side effects. Compounds

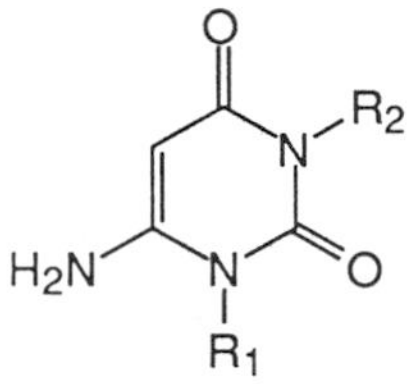

(223)

that were investigated clinically were 1-allyl-3-ethyl-6-aminouracil, aminometradine (**224**, Table 31.18), and the 1-methallyl-3-methyl analog (**225**, Table 31.18). Clinical studies in edematous subjects indicated that mercurial diuretics usually have a greater and more reliable effect (353).

At the time of their development, they represented some advance over the xanthines, but they were displaced by more effective oral diuretics within a relatively short time.

12.3 Triazines

Recognition of the triazines as a class of diuretic agents stemmed from the work of Lipschitz and Hadidian (359), who tested a group of compounds of this type in rats, (e.g., melamine (**226**) and formoguanamine (**227**)). Formoguanamine was effective orally as a diuretic in humans (361), but subsequent clinical studies revealed side effects such as crystalluria (361) and poor Na$^+$

Table 31.18 Cyclic Polynitrogen Compounds

No.	Generic Name	Trade Name	Structure	Ref.
(244)	Amiloride	Collectril		354
(224)	Aminometradine	Mictine		352
(225)	Amisometradine	Rolicton		352
(230)	Chlorazanil	Daquin, Diurazine, Orpizin		355
(219)	Theophylline (aminophylline)		(Combination with $NH_2(CH_2)_2NH_2$ Ratio 2:1)	356
(235)	Triameterene	Dyrenium		357
(250)	Clazolimine			358

(226) R$_1$ = NH$_2$, R$_2$ = R$_3$ = H
(227) R$_1$ = R$_2$ = R$_3$ = H
(228) R$_1$ = H, R$_2$ = R$_3$ = Ac

excretion (362), which precluded its further use. A structural variant, prepared in an attempt to overcome the side effects, was diacetylformoguanamine (**228**), which was active as an oral diuretic in dogs (363) but still caused crystalluria and inadequate Na$^+$ excretion. The extraordinary potency of the triazines in rats does not carry over into the dog, and the compounds are only moderately active in humans.

Among other derivatives of formoguanamine, those with other substituted amino group had a particularly favorable diuretic effect in rats (355). The most potent compounds in the series were 2-anilino-*s*-triazine (**229**) (amanozine) and 2-amino-4-(*p*-chloroanilino)-*s*-triazine (**230**) (chlorazanil). The diuretic active of the last two compounds was confirmed in dogs (364,365) and in humans (366,367). The *m*-chloro isomer of chlorazanil, 2-amino-4-(*m*-chloroanilino)-*s*-triazine, was a potent, orally effective diuretic in rats, dogs, and humans (368,369) and may be significantly more active than chlorazanil, with an en-

hanced saluretic effect. 2-Amino-4-(*p*-fluoroanilino)-*s*-triazine was twice as active as chlorazanil (366). Replacement of the halogen in chlorazanil with acetyl, carbethoxy, or sulfamoyl groups reduced activity (370), but replacement with alkylmercapto groups led to a two-fold increase in activity (371). Both the incidence and degree of crystalluria in dogs were greater with the alkylmercapto compounds than with chlorazanil, but oral toxicity in mice was reduced (371).

The only triazine to achieve any degree of clinical use is chlorazanil. It has a more pronounced effect on water excretion than on Na$^+$ and Cl$^-$ (367) and has little effect on K$^+$ excretion, which is probably linked to a lack of marked enhancement of Na$^+$ excretion. Since diuresis is not accompanied by changes in glomerular filtration rate (372), the drug probably exerts its action through inhibition of tubular reabsorption. The effects of deoxycorticosterone and chlorazanil on Na$^+$ and K$^+$ excretion are mutually antagonistic, which may mean that the natriuretic and diuretic properties of the drug are due to inhibition of Na$^+$ reabsorption in the distal segment (373).

The triazines, in particular chlorazanil, have been used clinically mainly in Europe. Interest in this type of compound declined with the advent of the more effective thiazide diuretics.

12.4 Potassium-Sparing Diuretics

12.4.1 INTRODUCTION. There are three structurally different classes of potassium-sparing diuretics: steroids, pyrazines, and pteridines. The steroidal aldosterone antagonists and inhibitors have been discussed in sections 10 and 11. These diuretic agents inhibit sodium reabsorption mainly in the principal cells of the collecting tubules. The aldosterone antagonists interact with the aldosterone cytoplasmic receptors, which cause a series of events

(229) R = H
(230) R = Cl

leading to sodium excretion and potassium reabsorption. The pyrazines and pteridines, on the other hand, inhibit the actions of the sodium channels located on the luminal side of the principal cells. These compounds are thought either to block this channel or to switch them from an open to a closed state (374). As a result, the apical cell membrane becomes hyperpolarized, and the transepithelial potential is decreased. This reduces the driving force for potassium to exit the luminal side via potassium channels, thus decreasing renal potassium loss. On average, the human body contains about 140 g of potassium. A calculation has shown that 98% of potassium is intracellular and only about 2% is in the extracellular compartment. Therefore, removal or addition of a small amount of potassium to this extracellular pool is very evident (375).

12.4.2 TRIAMTERENE. The triamterene ring system is found in many naturally occurring compounds, such as, folic acid and riboflavin. These compounds are important in the regulation of metabolism in man. The observation that xanthopterin (231) was capable of affecting renal tissue led Wiebelhaus, Weinstock, and associates to test a series of pteridines in a simple rat diuretic screening procedure (376). One compound, 2,4-diamino-6,7-dimethyl-pteridine (232), showed sufficient diuretic ac-

(231)

(232)

tivity to encourage further investigation of the diuretic potential of the pteridines. A number of related 2,4-diaminopteridines were studied, but only 232 showed good activity in both the saline-loaded and saline-deficient rat. Changes in the 2,4-diamino part of the molecule resulted in a marked decrease in diuretic activity (357).

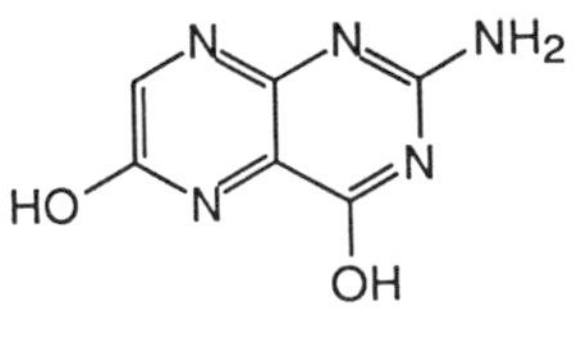

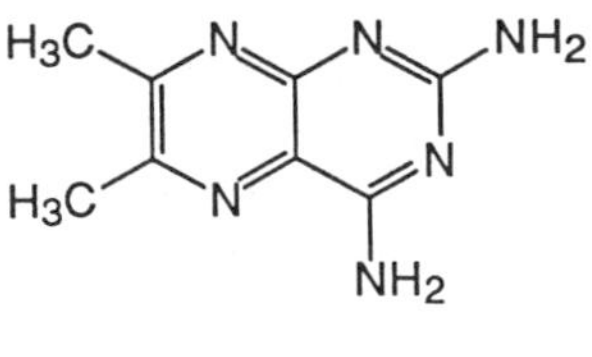

A class of related pteridines, of which 4,7-diamino-2-phenyl-6-pteridine-carboxamide (233) is the prototype, has been investigated. This derivative is active in both the saline-loaded and sodium-deficient rat, but in contrast to 232, it causes substantial potassium loss in the sodium-deficient rat. In structure–activity studies, particular attention was directed towards modification of the carboxamide function (357,377,378).

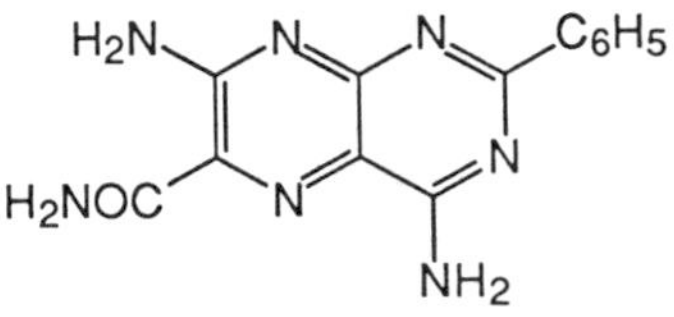

(233)

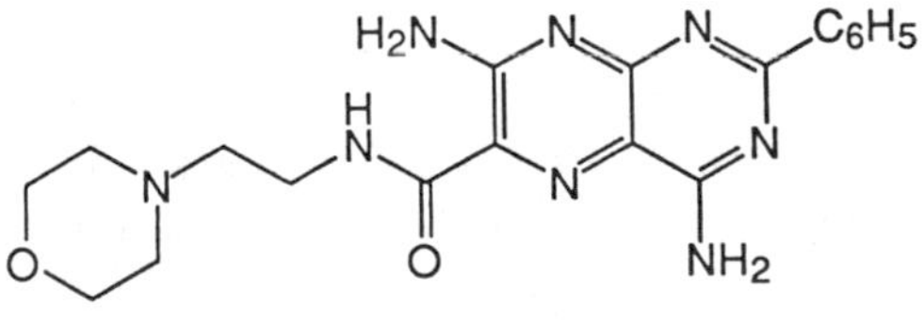

(234)

One of the more interesting compounds was 4,7-diamino-N-(2-morpholinoethyl)-2-phenyl-6-pteridinecarboxamide (234). In pharmacological investigations (379), this compound was an orally active diuretic agent, generating about the same maximum degree of response in dogs as hydrochlorothiazide. The urinary excretion of Na^+ and Cl^- was markedly enhanced, with minimal augmentation of K^+ excretion and little effect on urine pH. Onset of action was rapid, with the greatest saluretic effect occurring within 2 h of oral administration to saline-loaded dogs. The compound showed diuretic activity in both normal and adrenalectomized rats, which, with the absence of K^+ retention, indicated that aldosterone antagonism is not a major component of its saluretic activity.

A consideration of the structural features of 2,4-diamino-6,7-dimethylpteridine (232) and 4,7-diamino-2-phenyl-6-pteridinecarboxamide (233) led to the investigation of 2,4,7-triamino-6-phenylpteridine, triamterene (235), as a potential diuretic agent (357).

(235)

This compound was very potent in the saline-loaded rat, and in the sodium-deficient rat it not only caused a marked excretion of sodium but simultaneously decreased K^+ excretion. In structure–activity studies of compounds related to triamterene, replacement of one of the primary amino groups by lower alkylamino groups led to compounds that retained triamterene like diuretic activity. More extensive changes generally led to substantially less active compounds. Table 31.19 lists the activities of some 2,4,7-triamino-6-substituted pteridines. The activity of triamterene is very sensitive to substitution of the phenyl group with only small changes possible if diuretic activity is to be retained. The p-tolyl compound, for example, is only about half as active as triamterene. In general, *ortho* isomers seem to be more active than the other isomers. The p-hydroxyphenyl analog of triamterene, a metabolite of the latter, is essentially inactive. The 4-deuterophenyl analog of triamterene was prepared in an attempt to increase activity by reducing the rate of metabolism,

but its activity is very similar to that of the parent compound. When the phenyl group of triamterene is replaced by a heterocyclic nucleus, the size of the group appears to be important, and high activity is seen only in the case of small, nonbasic groups. The low activity of compounds containing basic centers in this position, such as, thiazole and pyridine, may be rationalized by assuming that the basic centers are highly solvated and are, in effect, large substituents. The 6-alkyl analogs are active diuretics; however, size is important. Although good activity is seen in the 6-n-butyl homolog, the isopropyl and cyclohexenyl derivatives have only modest activity. Isomers of triamterene were also studied; the 7-phenyl isomer was one of the most potent K^+ blockers found in the pteridines, even though it is only a weak natriuretic agent. The 2-phenyl isomer is very similar in its biological properties to triamterene. Among pyrimidopyrimidines related to triamterene, 2,4,7-triamino-5-phenylpyrimido[4,5-d]pyrimidine (236) was investigated in some detail. It resembles 233 in that it does not block K^+ excretion in the sodium-deficient rat.

(236)

The structure–activity relationships of pteridine diuretics may be rationalized by assuming that the pteridines bind to some active site at two points (357). The more important site involves a basic center of the drug, which in triamterene may be N-1, N-8, or both. Groups that decrease the base strength of the pteridine nucleus reduce activity. The other site probably involves the phenyl substituent of triamterene and may be hydrophobic in nature. There ap-

Table 31.19 2,4,7-Triamino-6-substituted Pteridines[a]

R$_6$	Diuretic Activity in Saline-Loaded Rat[b]	Diuretic Activity in Sodium-Deficient Rat[c]
C$_6$H$_5$(triamterene)	3	3
2-CH$_3$C$_6$H$_4$	2	1
3-CH$_3$C$_6$H$_4$	1	1
4-CH$_3$-C$_6$H$_4$	2	1
2-F-C$_6$H$_4$	2	
4-F-C$_6$H$_4$	1	
4-CH$_3$O-C$_6$H$_4$	1	
4-C$_6$H$_4$-C$_6$H$_4$	0	
2-Furyl	0	1
3-Furyl	1	2
2-Thienyl	2	2
3-Thienyl	1	2
4-Thiazolyl	0	
2-Pyridyl	0	
3-Pyridyl	0	
4-Pyridyl	0	
H	2	3
CH$_3$	3	2
CH(CH$_3$)$_2$	1	
(CH$_2$)$_3$CH$_3$	2	2
Δ^3-Cyclohexenyl	1	
CH$_2$C$_6$H$_5$	2	2

[a]From Ref. 357.

[b]Rating scheme for saline-loaded rat assay: maximum response at any dose, in volume percent of urine, compared to volume of 0.9% saline load, less that of untreated control. <22% = 0, 22–45% = 1, 46–69% = 2, and >69% = 3.

[c]Rating scheme for sodium-deficient rat assay: maximum response of any dose, in milligrams of sodium excreted in the urine per rat. <3 mg = 0, 3–6 mg = 2. and >9 mg = 3.

pears to be critical size limitations at the site, as shown by the change in activity in relation to methyl substitution. Because compounds, such as, 2,4,7-triaminopteridine, are active, the phenyl group is not a primary requirement for activity and apparently acts in a reinforcing capacity, such as, increasing the degree of binding and establishing the correct orientation of the molecule at the receptor site.

Triamterene (**235**) is a potent, orally effective diuretic in both the saline-loaded and sodium-deficient rat, and is accompanied by no increase in potassium excretion. Also, the effects of aldosterone on the excretion of electrolytes in the adrenalectomized rat are completely antagonized by triamterene. Similar results were obtained in dogs, and it appeared that the compound might be functioning as an aldosterone antagonist (380). Initial clinical studies (381,382) established the natriuretic properties of triamterene in humans in cases when aldosterone excretion might be at an elevated level, and evidence was obtained for inhibition of the nephrotropic effect of aldosterone. However, triamterene possessed natriuretic activity in adrenalectomized dogs and rats (383–385) and in adrenalectomized patients (386); this was inconsistent with an aldosterone antagonism mechanism. Thus, although triamterene reverses the end results of aldosterone, its activity does not depend on the displacement of aldosterone. The compound acts directly on the renal transport of sodium. Stopflow studies in dogs pointed to an effect on the distal site of Na^+/K^+ exchange, and there was no evidence for a proximal renal tubular effect (387). Triamterene acts at the apical cell membrane in the collecting tubule where it blocks sodium channels and thus leads to reduced potassium excretion (389). It is also believed to act at the peritubular side as well (389).

The overall effect of triamterene on electrolytes is to increase moderately the excretion of Na^+ and, to a lesser extent, of Cl^- and HCO_3^-, and to reduce K^+ and NH_4^+ excretion (382,390,391). Triamterene is a more active natriuretic agent than spironolactone and is well absorbed following administration of a single oral dose of 50–300 mg/day (381,382). It is about 50–60% bound to plasma proteins. Triamterene is extensively metabolized and only about 3–5% of the drug is excreted unchanged in the urine (392). The compound undergoes hepatic hydroxylation and sulfate ester formation. The sulfate ester is also biologically active and is excreted in the urine. After intravenous administration of triamterene in humans, the concentration of the sulfate ester was ten times that of the parent compound, and it has been concluded that the pharmacologically active form of the drug in humans is the ester

(393). Triamterene is also metabolized to its glucuronide adduct, which undergoes biliary elimination. The duration of action in humans is about 16 h (394).

An increased diuresis ensued when triamterene was administered to patients who were receiving the aldosterone antagonist spironolactone (395,396), thus further emphasizing the fundamental differences in the mechanism of action of these two drugs. Triamterene potentiates the natriuretic action of the thiazides while reducing the kaliuretic effect (395), and other clinical studies with this combination showed that normal serum potassium levels could be maintained without potassium supplements (396). The natriuretic potency of triamterene does not approach that of the thiazide diuretics, and the main value of the drug would appear to be its use in combination with thiazides in clinical situations where K^+ loss is a problem. Triamterene should not be prescribed to patients with renal insufficiency or hyperkalemia nor should it be administered concurrently with potassium supplements. Triamterene increases serum uric acid levels and cases of hyperuricemia have been reported. Some other side effects that are observed are skin rashes, gastointestinal disturbances, hyperkalemia, weakness, and dry mouth.

12.4.3 OTHER BICYCLIC POLYAZA DIURETICS. A group of workers at Takeda synthesized and studied the diuretic activity of a large series of polynitrogen heterocycles (397). The ring systems prepared are shown in Table 31.20, documenting the extensive effort made in this investigation. Of the 219 compounds studied in this series, two compounds, DS 210 (**237**) and DS 511 (**238**), were selected for more extensive evaluation.

The compounds were initially screened in rats, and hydrochlorothiazide was used as the reference compound. DS 210 (**237**) produces a maximal natriuretic effect similar to hydrochlorothiazide in rats without

Table 31.20   Polyaza Heterocyclic Systems Studied as Diuretics[a]

affecting potassium excretion. It shows additive activity with hydrochlorothiazide, acetazolamide, amiloride, and furosemide. Potassium excretion induced by other diuretics is not modified by DS 210. The diuretic effect is lost in adrenalectomized rats and restored by cortisol treatment (398). DS 511 (**238**) has shown diuretic activity comparable to hydrochlorothiazide in rats, dogs, and humans, and seems to have a unique mode and site of action in the nephron (399).

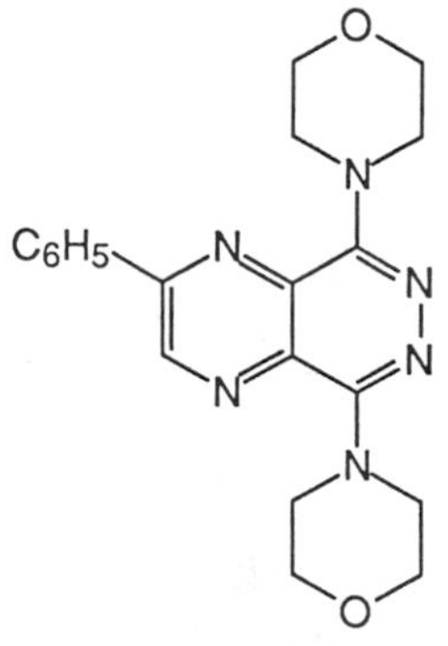

DS 210
(**237**)

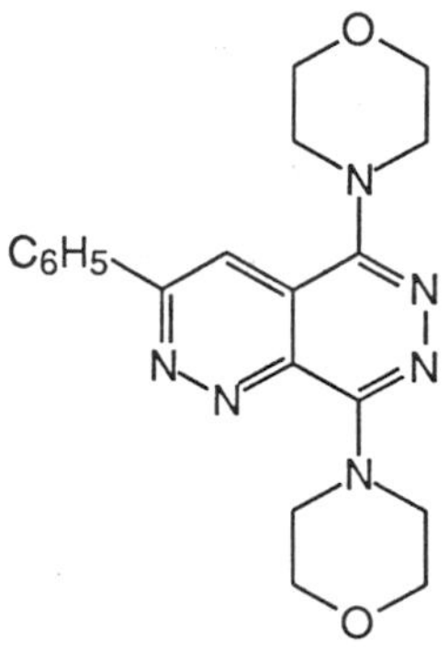

DS 511
(**238**)

Hawes and co-workers at the University of Saskatchewan prepared a series of 2,3-disubstituted 1,8-naphthyridines containing a phenyl group at the 3-, 4-, or 7-positions (400). In this study, compounds containing a phenyl group at the 7-position had no diuretic activity. Compounds **239** and **240**, which contain a phenyl group at the 4-position, have similar diuretic and natriuretic properties when compared to

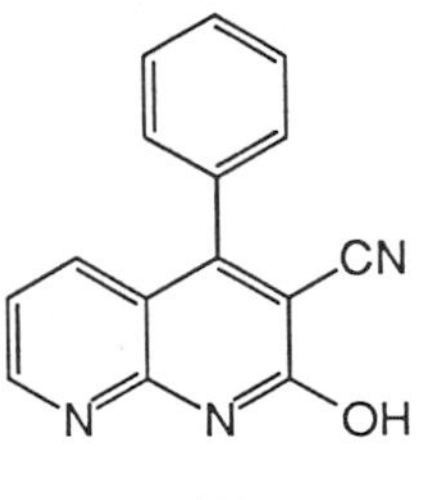

(**239**) 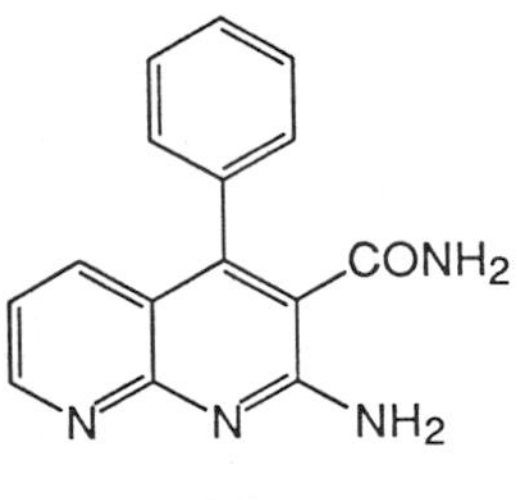   (**240**)

traimterene. These compounds also lack kaliuretic properties.

Parish and co-workers synthesized a number of 2-pyrido[2,3-d]pyrimidin-4-ones (401). 1,2-Dihydro-2-(3-pyridyl)-3H-pyrido-[2,3-d]pyrimidin-4-one (**241**) was a potent diuretic agent in the rat, however it was not as potent as hydrochlorothiazide. An oral dose of 81 mg/kg in the rat resulted in potassium excretion levels which were the same as the saline controls, while the sodium and chloride ion excretions levels had doubled. Monge and co-workers at the University of Navarra further studied the structure–activity relationships in this series (402). Placement of nitro or amino groups at the 6-position resulted in compounds that had marginal or no diuretic activity. Methylation of both the amide and amine nitrogens gave a compound with similar diuretic activity but with a poorer Na^+/K^+ ratio.

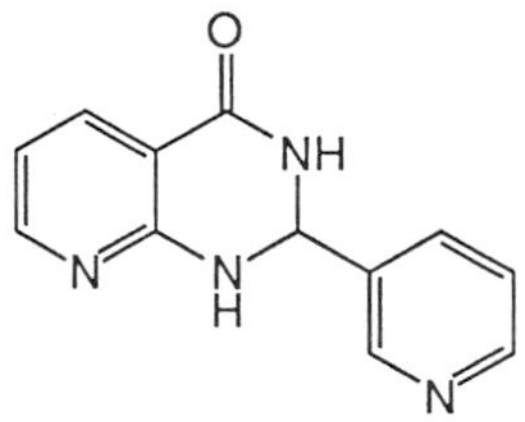

(**241**)

Hester and co-workers at Upjohn prepared a series of 1-(2-amino-1-phenylethyl)-6-phenyl-4H-[1,2,4]triazolo[4,3-a][1,4]-benzodiazepines and evaluated their diuretic activity (403). Several of these compounds possessed diuretic and natriuretic activity, after oral administration to rats, with no kaliuretic activity. The most potent benzodiazepine in this series is **242**; however it is considerably less potent than hydrochlorothiazide in the conscious rat. At 10 mg/kg, **242** begins to show significant diuretic and saliuretic activity. Hydrochlorothiazide begins to show significant activity at 0.3 mg/kg; however the efficacy of the two compounds appears to be similar.

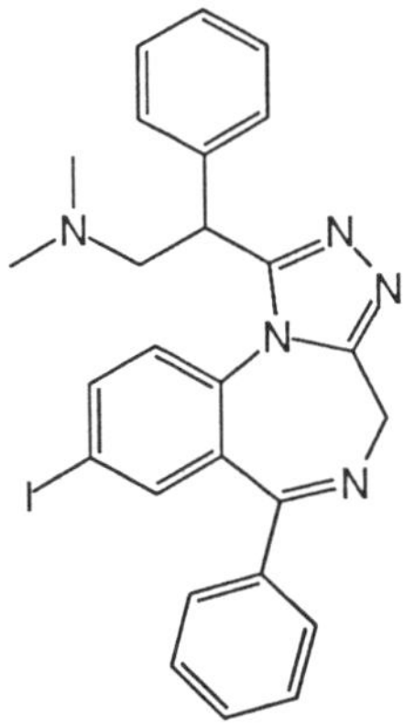

(242)

(244)

12.4.4 AMILORIDE. An empirical approach was taken by a group at Merck, Sharp, and Dohme seeking compounds with no or minimal kaluretic effects. Screening procedures indicated that *N*-amido - 3 - amino - 6 - bromopyrazinecarbox-amide (**243**), a compound available through previous work in the folic acid series, was of interest. The introduction of an amino group in the 5-position markedly increased the sodium and chloride excretion without affecting potassium excretion: *N*-amidino-3,5-diamino-6-chloropyrazine-carboxamide amiloride (**244**), was among the most promising in animals and in humans (181). The *N*-amidinopyrazinecarboximides produce a pronounced diuresis in normal rats and leaves potassium excretion unaffected or repressed. In the adrenalectomized rat they antagonize the renal actions of exogenous aldosterone, DOCA, and hydrocortisone. In dogs, the compounds are less potent, but the relative activity in the series is the same as those in rats.

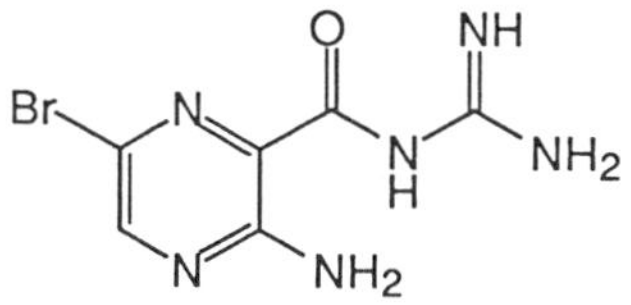

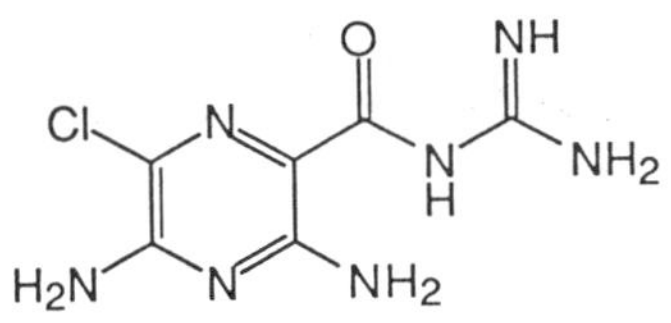

Structure–activity relationships in this series have been investigated in considerable detail and some representative compounds from these studies are listed in Table 31.21. The activities of the compounds were determined on the basis of their DOCA inhibitory activity, which closely paralleled the diuretic activity in intact rats and dogs (404–406).

Workers at Ciba investigated the replacement of the acylguanidine moiety with an 1,2,4-oxadiazol-3-amine group (407). This compound, CGS 4270 (**245**), has a similar profile to that of amiloride in rats and dogs.

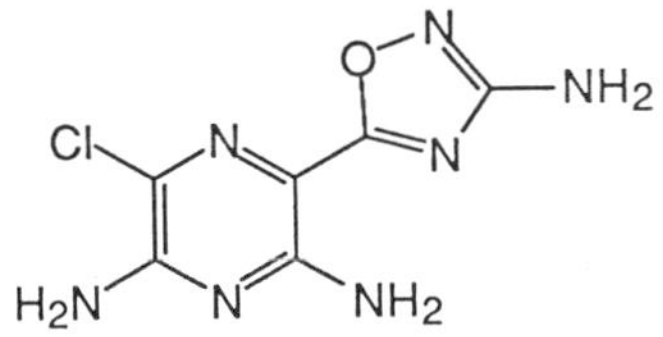

(245)

More recently, Ried and co-workers prepared a number of amiloride analogs that were modified at the 2- and 6-positions (408). In general, replacement of the chlorine at the 6-position was detrimental to the antikaliuretic, pharmocodynamic, and pharmacokinetic properties of this class. A number of the N-substituted 3,5-diamino-6-chloropyrazines-2-carboxamides were found to be as potent as triamterene. The *N*-(dimethylaminoethyl) amide analog (**246**) was more potent than triamterene as a diuretic, natriuretic, and antikaliuretic agent after oral administration to the rat. The compound was well absorbed and was excreted without significant metabolism.

(243)

Table 31.21 DOCA Inhibitory Activity in Adrenalectomized Rats of Some N-Amidino-3-aminopyrazinescarboxamides[a]

R$_1$	R$_2$	R$_3$	R$_4$	R$_5$	R$_6$	DOCA Inhibition Score[b]
H	H	H	H	H	Cl	+ + +
H	H	H	H	H	Br	+ +
H	H	H	H	H	I	+
CH$_3$	CH$_3$	H	H	H	Cl	+ + +
CH$_3$	H	H	H	H	Cl	+
H	H	H	H	H	H	0
H	H	H	H	H	CF$_3$	+ / −
H	H	H	H	H	CH$_3$	+ +
H	H	H	H	CH$_3$	H	+
H	H	H	H	H	C$_6$H$_5$	+
H	H	H	H	NH$_2$	Cl	+ + + +
H	H	H	H	CH$_3$NH	Cl	+ + +
H	H	H	H	(CH$_3$)$_2$CHCH$_2$NH	Cl	+
H	H	H	H	(CH$_3$)$_3$CNH	Cl	+ / −
H	H	H	H	(CH$_3$)$_2$N	Cl	+ + +
H	H	H	H	C$_6$H$_5$NH	Cl	+ +
CH$_3$	H	H	H	NH$_2$	Cl	+ + + +
CH$_3$	CH$_3$	H	H	NH$_2$	Cl	+ + + +
3,4-Cl$_2$C$_6$H$_5$CH$_2$	H	H	H	NH$_2$	Cl	+ + +
H	H	H	H	NH2	Br	+ + +
H	H	H	H	Cl	Cl	0
H	H	H	H	OH	Cl	+ / −
H	H	H	H	OCH$_3$	Cl	+ / −
H	H	H	H	SH	Cl	+ / −
H	H	H	H	SCH$_3$	Cl	+
H	H	H	H	SCH$_3$	Cl	+
CH$_3$CO	H	H	CH$_3$CO	H	Cl	+ / −
H	H	H	H	H	SCH$_3$	+ / −
				Aldosterone		+ / −
				Triamterene		+ +

[a]From Refs. 354, 405–407.

[b]This score is related to the dose of each compound that produces a 50% reversal of the electrolyte effect from the administration of 12 μg of DOCA to adrenalectomized rats, as follows: <10 μg (+ + + +), 10–50 μg (+ + +), 51–100 μg (+ +), 101–800 μg (+), >800 μg (+ / −).

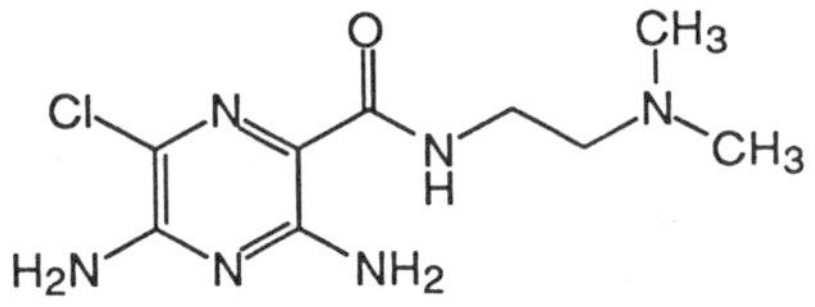

(246)

Workers at ICI have synthesized amiloride analogs that have diuretic activity combined with calcium channel blocking activity or beta-adrenoceptor blocking activity. ICI 147798 (**247**) is a single molecule derivative of amiloride which was discovered to possess both diuretic and beta-adrenoreceptor blocking properties (409). At doses of 1–20 mg/kg p.o., the natriuretic activity of ICI 147798 was similar to hydrochlorothiazide in dogs; at 1 mg/kg it produced significantly less kaliuresis than hydrochlorothiazide. ICI 147798 blocked adrenergic receptors *in vitro* and *in vivo*, and it also inhibited isoproternol-induced tachycardia in rats, guinea pigs, cats, and dogs. ICI 206970 (**248**), another analog of amiloride, possessed diuretic and calcium channel-blocking activity (410). In the dog, after oral administration, ICI 206970 was less potent than hydrochlorothiazide with respect to its diuretic and saliuretic effects. In contrast to hydrochlorothiazide, no significant changes were observed in plasma potassium levels after 14 days of dosing.

The structural similarity of amiloride and triamterene (**244** and **235**, Table 31.18) and their similar biological actions have raised the question of whether the pteridines are, in fact, closed-ring versions of the *N*-amidinopyrazinecarboxamides. The open-chain analogs of triamterene and the bicyclic analogs of amiloride that have been studied are generally less active than the drugs themselves. Triamterene is a weaker base (pK_a 6.2) than amiloride (pK_a 8.67). Amiloride as the hydrochloride is readily water-soluble, whereas triamterene is slightly soluble. After oral administration, approximately 50% of amiloride is absorbed in humans (411). It is approximately 23% bound to plasma proteins and is not metabolized in humans. It is excreted in the urine mainly unchanged.

In the usual dosage, amiloride has no important pharmacological actions except those related to the renal tubular transport of electrolytes. Clinically it is used extensively in combination with hydrochlorothiazide.

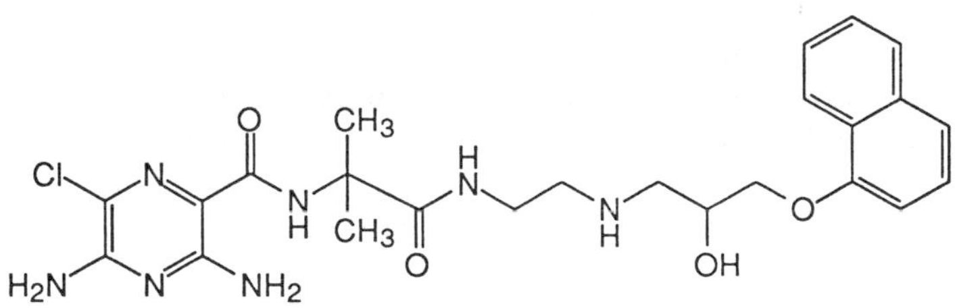

(247)

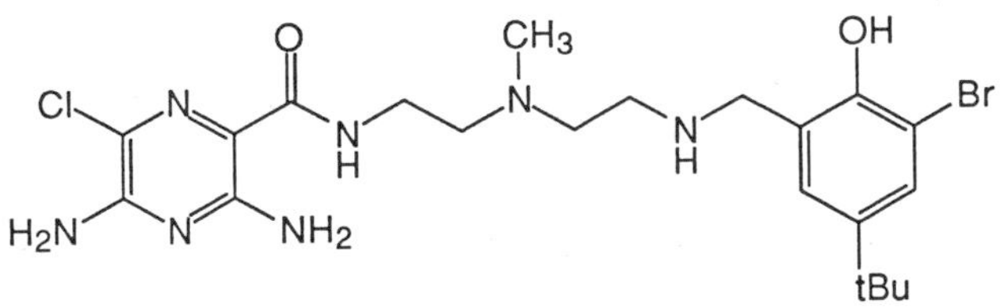

(248)

12.4.5 AZOLIMINE AND CLAZOLIMINE. A series of imidazolones was studied by a group at Lederle Laboratories in their search for a nonsteroidal antagonist of the renal effects of mineralocorticoids. Azolimine (**249**) and clazolimine (**250**) were the most interesting in this series.

(249) R = H
(250) R = Cl

Azolimine antagonized the effects of mineralocorticoids on renal electrolyte excretion in several animal models. Large doses of azolimine produced natriuresis in adrenalectromized rats in the absence of exogeneous mineralocorticoid, but its effectiveness was greater in the presence of a steroid agonist. In conscious dogs, azolimine was effective only when deoxycorticosterone was administered. Azolimine significantly improved the urinary Na^+/K^+ ratio when used in combination with thiazides and other classical diuretics in both adrenalectomized, deoxycorticosterone-treated rats and sodium-deficient rats (412). Similar effects were found for clazolimine (358). The compound may be useful in combination with the classical diuretics as an aldosterone antagonist diuretic in humans.

13 ATRIAL NATRIURETIC PEPTIDE

Atrial natriuretic peptide (ANP, **251**) is a 28-amino acid peptide that is released into circulation from the heart following atrial distension. It is synthesized and stored in specific atrial secretory granules as a 126-amino acid precursor molecule. ANP exerts natriuretic, diuretic, and vasorelaxant properties upon administration and suppresses renin and aldosterone levels (413,414). The pharmacological properties produced by this peptide suggest that it may be beneficial in the treatment of several cardiovascular disorders. The therapeutic potential of ANP, however, is limited by its poor oral absorption and extremely short biological half-life of less than 60 s in the rat and only a few minutes in humans (415,416).

The mechanism of action for the natriuretic activity of ANP has been the subject of much research over the past few years. It is believed that ANP-induced natriuresis results from its effect on renal hemodynamics. ANP is able to increase the glomerular filtration rate significantly (417,418). This effect seems to be brought about by vasodilation of the afferent arterioles with vasoconstriction of the efferent vessels. This effect increases glomerular capillary pressure and, therefore, the glomerular filtration rate. Further studies have shown that ANP may also inhibit sodium reabsorption in the collecting tubules and duct (419).

ANP has also been demonstrated to inhibit both basal and angiotensin-stimulated secretions of aldosterone *in vitro* in adrenal preparations and *in vivo* after infusion in animals and humans (420–422). Since ANP-induced natriuresis occurs very rapidly, the reduction in aldosterone secretion contributes to a longer term modulation of sodium excretion.

ANP is eliminated from the circulation

(251)

via two major pathways. Studies have shown that ANP is eliminated from the circulation by enzymatic degradation. The enzyme most responsible for its degradation is neutral endopeptidase (NEP, EC 3.4.24.11) (423,424). NEP is a zinc metallopeptidase that cleaves the alpha-amino bond of hydrophobic amino acids. Other enzymes in the renin–angiotensin and kallikrein–kinin systems have also been shown to degrade ANP.

ANP is also removed from circulation through a receptor-mediated clearance pathway (426). ANP clearance receptor (c-receptor) can be found in several tissues, including kidney cortex, vascular, and smooth muscle cells (426,427). ANP binds to this receptor, and then this receptor-ANP complex is internalized. ANP is transported to the lysosome where it undergoes extensive hydrolysis. The clearance receptor is recycled to the cell's surface, where it can repeat this process.

Since infusion of ANP was shown to produce several potentially therapeutic benefits, much of the current research in this area has been focused on methods of potentiating the activity of ANP *in vivo* by preventing its degradation (414).

13.1 ANP Clearance Receptor Blockers

Several groups have prepared ligands for the ANP clearance receptor that have prolonged the $t_{1/2}$ of ANP. SC 46542 (des-[Phe^{106}, Gly^{107}, Ala^{115}, Gln^{116}]ANP (103–126)) is a biologically inactive analog of ANP that has similar affinity for the c-receptor as ANP (428). In the normal, conscious rat and spontaneously hypertensive rat, however, SC46542 did not significantly increase immunoreactive ANP concentrations in plasma.

Two linear peptides, Ala$_7$-rat-ANP$_{8-17}$-NH$_2$ and 2-naphtoxyacetyl isonipecotyl-Arg-Ile-Asp-Arg-Ile-NH$_2$, were shown to increase plasma immunoreactive ANP con-

centrations in anesthetized rats (429). In response to the infusion of these compounds, a significant increase in glomerular filtration rate and sodium excretion was observed.

Infusion of C-ANP$_{4-23}$ was also shown to increase plasma immunoreactive ANP concentrations in anesthetized and conscious rats (430). C-ANP$_{4-23}$ increased the urinary excretion of water and sodium in the conscious DOCA/salt hypertensive rats when administered i.v. (431).

13.2 Neutral Endopeptidase Inhibitors

Originally, neutral endopeptidase (NEP) inhibitors were designed and studied for their analgesic properties, since this enzyme was known to degrade enkephalins. When it was discovered that NEP also degraded ANP, many of the known NEP inhibitor compounds, such as, thiorphan (**252**) and phosphoramidon (**253**) were evaluated for their potential diuretic and cardiovascular activities. Both thiorphan and phosphor-

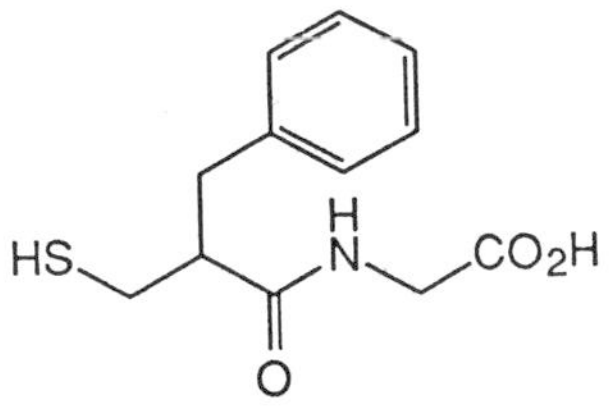

(252)

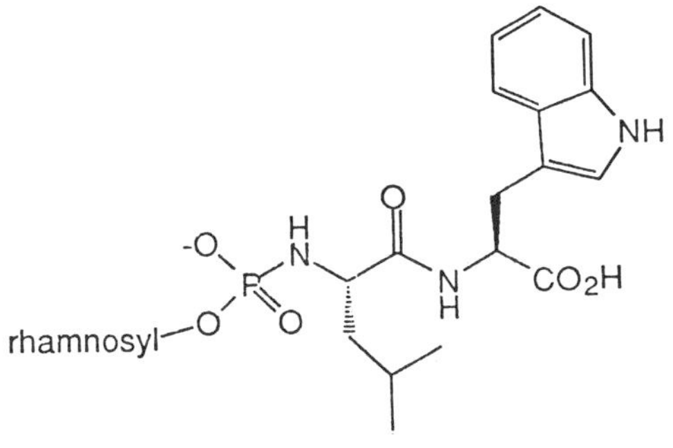

(253)

amidon increased the half-life of exogenous ANP in the rat (432). Phosphoramidon, when infused into rats with reduced renal mass significantly increased diuresis, natriuresis, and glomerular filteration rate (433). Thiorphan was also shown to increase sodium excretion in anesthetized and conscious normal rats (434).

In recent years, there have been many reports on new potent inhibitors of NEP. Many of these compounds are di- or tripeptides that contain a group that bind to the zinc atom in the active site of the enzyme. There are four different classes of NEP inhibitors: thiols, carboxylates, phosphoryl-containing, and hydroxamates.

Thiorphan was the first reported thiol inhibitor of NEP. Both the R- and S-enantiomers of thiorphan have the same enzyme inhibitory potency, $IC_{50} = 4\,nM$. Extensive structure activity relationship studies have shown that it is possible to replace the glycine residue with a O-benzyl serine and still retain potency (435). This compound, ES 37 (**254**), has an IC_{50} for NEP of $4\,nM$ and is a potent inhibitor of angiotensin-converting enzyme (ACE), $IC_{50} = 12\,nM$. Reduction of the phenyl ring also affords a potent NEP inhibitor, $IC_{50} = 32\,nM$.

Workers at Squibb replaced the glycine in thiorphan with an aminoheptanoic acid (436). This compound, SQ 29072 (**255**), is a potent NEP inhibitor with an $IC_{50} = 26\,nM$. When administered intravenously (300 μg/kg) to a conscious SHR, SQ 29072 pro-

(255)

duced a modest diuretic and natriuretic response (427). In the DOCA/salt rat, when equidepressor doses of SQ 29072 and ANF$_{99-126}$ were administered, there was a prolonged urinary excretion of sodium. Another structurally related analog, SQ 28603 (**256**), was also reported to be a potent and selective inhibitor of NEP. When infused in conscious, DOCA/salt-hypertensive rats, SQ 28603 caused an increase in plasma ANP concentration and in sodium excretion and significantly lowered mean arterial pressure (438).

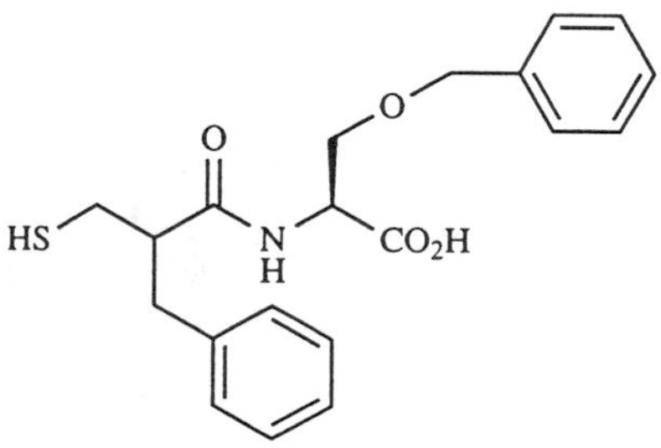

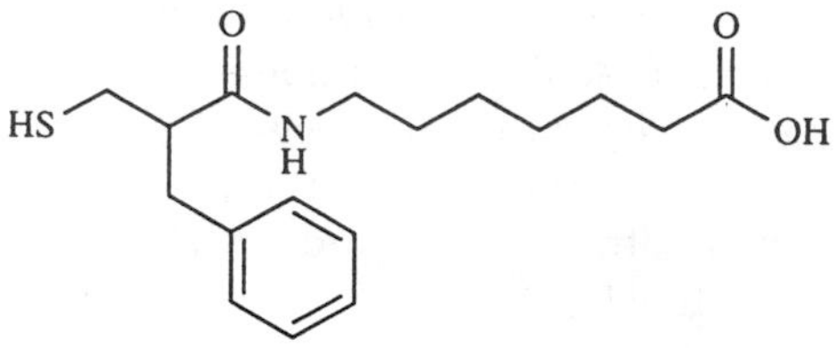

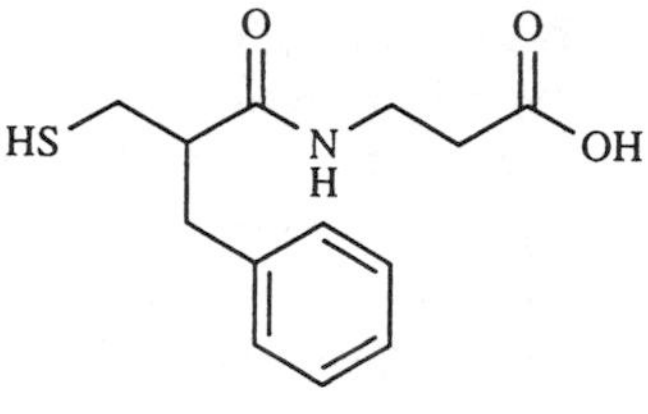

(256)

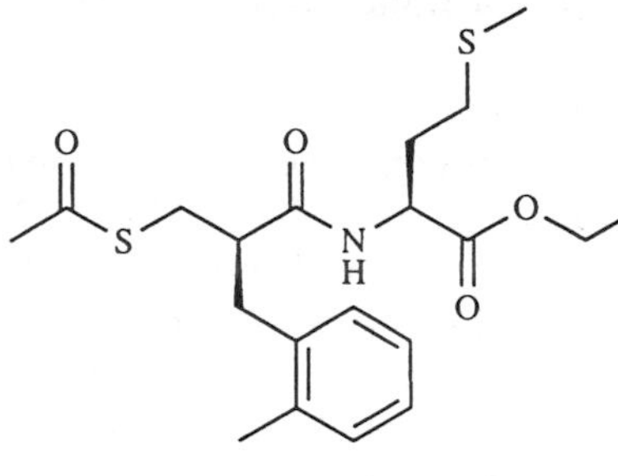

(257)

Workers at Schering-Plough also prepared thiorphan-type analogs. SCH 42495 (**257**) elevated plasma ANF concentrations in animal models (439,440). In a study with eight patients with essential hypertension, plasma ANF levels increased (+123% P <

(254)

0.01) and later remained elevated (+34, $P < 0.01$) (441). The compounds also led to a significant natriuresis in the initial 24 h of treatment. This effect attenuated over time.

A number of carboxyl-containing NEP inhibitors were prepared by workers at Schering-Plough and Pfizer. Candoxatrilat (UK 69578, **258**) was a potent NEP inhibitor ($K_i = 2.8 \times 10^{-8} M$) designed by workers at Pfizer (442). When given i.v. to mice, it increased endogenous levels of ANP and produced diuretic and natriuretic responses. When the prodrug candoxatril was administered to human subjects, doses of 10–200 mg caused a rise in basal ANP levels. Natriuresis was only observed at the highest dose (443).

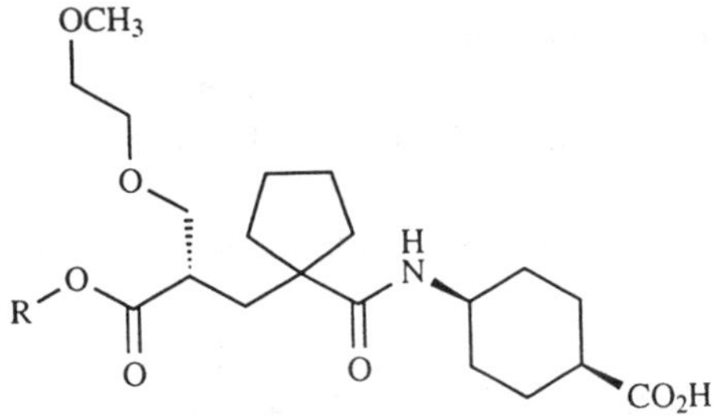

(258)

R = H, candoxatrilat

R =

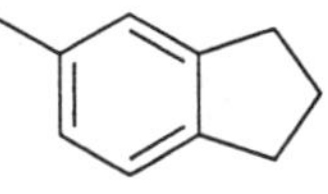

candoxatril

Workers at Schering-Plough also prepared a number of carboxyl-containing NEP inhibitors. SCH 39370 (**259**) was discovered to be a potent NEP inhibitor ($IC_{50} = 11 nM$) and, when administered to rats with congestive heart failure, it caused an increase in urinary volume and plasma ANP levels (444, 445). In an ovine heart failure model, SCH 39370, when given as a bolus injection, caused significant natriuresis and diuresis (446). A structurally related compound, SCH 34826 (**260**), produced a significant natriuretic effect in DOCA/salt-hypertensive rats (447). In normal volunteers maintained on a high

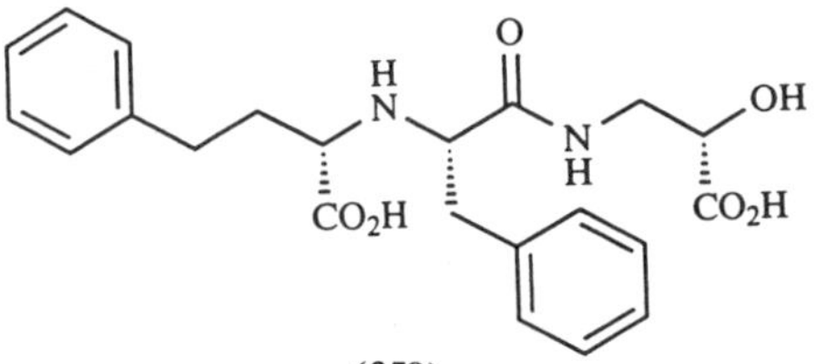

(259)

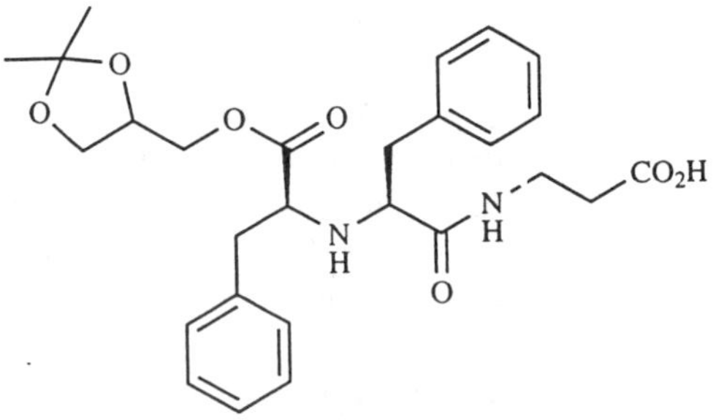

(260)

sodium diet for five days, SCH 34826 promoted a significant increase in sodium, calcium, and phosphate excretion (448).

Hydroxamates form strong bidentate ligands to zinc. Compounds containing this functional group are potent NEP inhibitors. The prototype of this class is *RS*-kelatorphan. This compound strongly inhibits NEP ($IC_{50} = 1.8 nM$) and aminopeptidase N(APN) ($IC_{50} = 380 nM$) (449). The *SS*-isomer of kelatorphan, RB 45 (**261**), is a more selective NEP inhibitor ($IC_{50} = 1.8 nM$ (NEP), $IC_{50} = 29,000$ (APN)). In rats, when given at 10 mg/kg i.v., RB 45 increased the half-life of ANP (451).

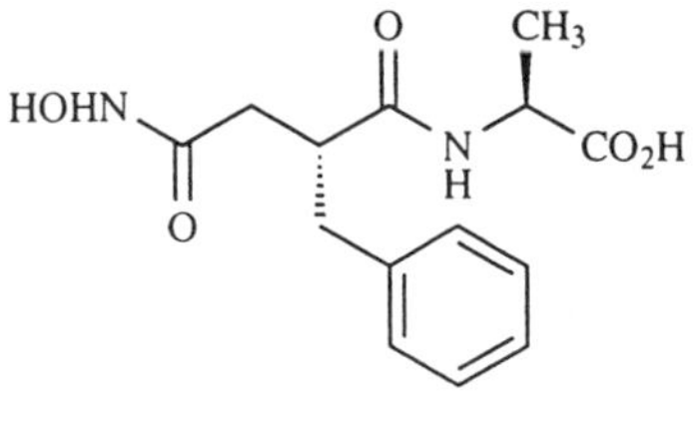

(261)

Phosphoryl-containing inhibitors also interact strongly with zinc and are potent NEP inhibitors. Phosphoramidon (**253**), which is a potent NEP inhibitor ($IC_{50} = 2 nM$), is a natural competitive inhibitor

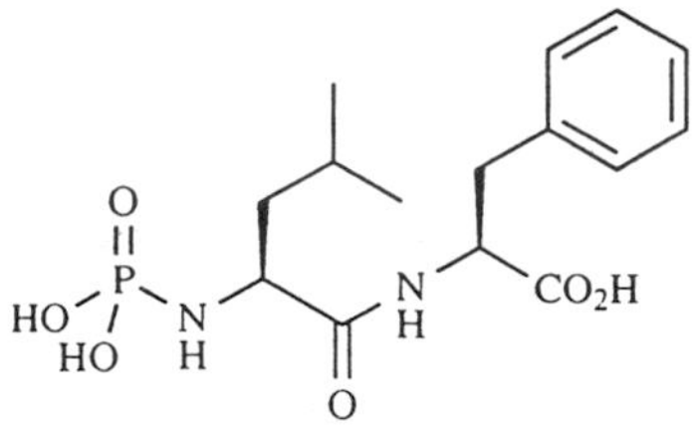

(262)

produced by *Streptomyces tanashiensis*. Vogel and co-workers reported that phosphoryl-Leu-Phe (**262**) is a potent NEP inhibitor $(IC_{50} = 0.3\,nM)$ (451). This is about an order of magnitude more potent than thiorphan.

Workers at Ciba reported on a series of potent phosphorus-containing inhibitors of NEP (452). CGS 24592 (**263**) had $IC_{50} = 1.6\,nM$. The racemic analog, CGS 24128

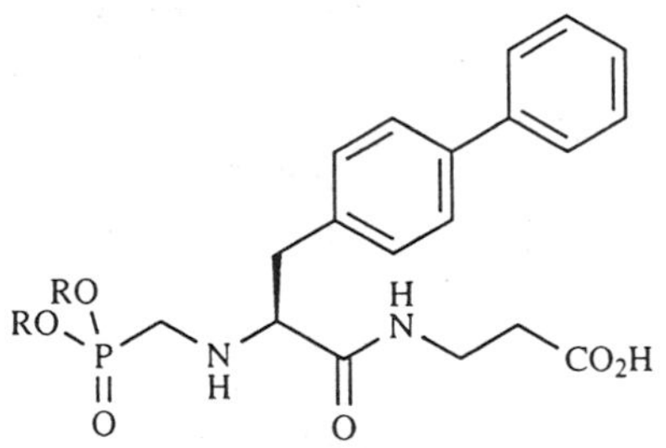

$(IC_{50} = 4.3\,nM$, NEP), increased plasma ANP immunoreactivity levels by 191% in rats administered exogenous ANP (99–126) (453). CGS 24128 also potentiated the natriuretic activity of exogenous administered ANP (99–126). Because of the poor oral bioavailability of CGS 24592, a series of prodrugs were investigated. CGS 25462, (**264**) provided significant and sustained antihypertensive effect in the DOCA/salt rat after oral administration.

14 URICOSURIC AGENTS

14.1 Introduction

In humans, one of the principal products of purine metabolism, i.e., uric acid, is implicated in several human diseases, e.g., gout. Guanine and adenine are both converted to xanthine (**265**); oxidation, catalyzed by xanthine oxidase, yields uric acid (**266**). In humans, uric acid is the excretory product and most of it is excreted by the kidney. In most mammals, uric acid is further hydrolyzed by uricase to allantoin (**267**), a more soluble excretory product. Allantoin, in turn, is further degraded to allantoic acid (**268**) by allantoinase, and then to urea and glyoxylic acid (**269**) by allantoinase. Uric acid is not the major pathway of nitrogen

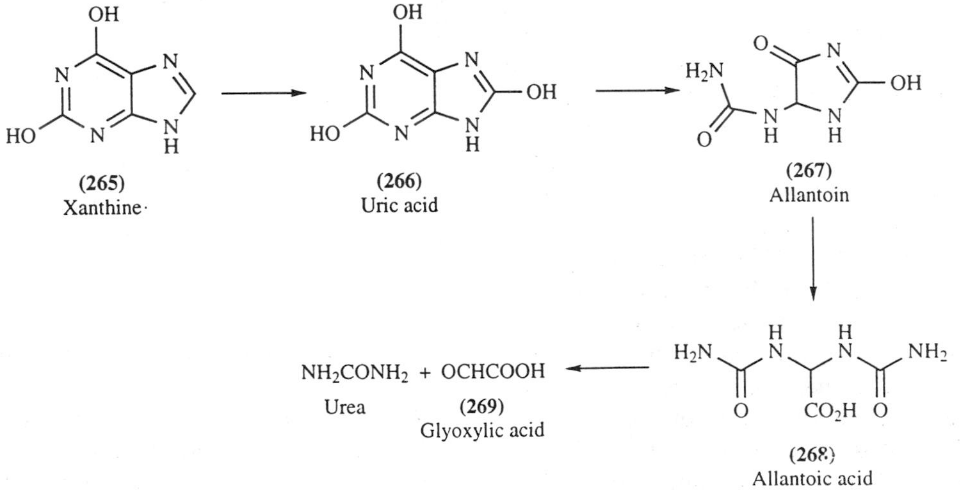

(263) R = H

(264) R = phenyl

(265)
Xanthine·

(266)
Uric acid

(267)
Allantoin

NH$_2$CONH$_2$ + OCHCOOH
　　Urea　　　　**(269)**
　　　　　　Glyoxylic acid

(268)
Allantoic acid

excretion in humans. Instead, the ammonia nitrogen of most amino acids, the major nitrogen source, is shunted into the urea cycle. Uric acid is mostly insoluble in acidic solutions, although alkalinity increases its solubility. At the pH of blood (pH 7.44), uric acid is present as the monosodium salt, which is also very slightly soluble and tends to form supersaturated solutions.

Uric acid forms from purines, which are liberated as a result of enzymatic degradation of tissue and dietary nucleoproteins and nucleotides, but it is also formed by purine synthesis (454). When the level of monosodium urate in the serum exceeds the point of maximum solubility, urate crystals may form, particularly in the joints and connective tissues. These deposits are responsible for the manifestations of gout. Serum urate levels can be lowered by decreasing the rate of production of uric acid or by increasing the rate of elimination of uric acid. The most common method of reducing uric acid levels is to administer uricosuric drugs, which increase the rate of elimination of uric acids by the kidneys.

14.2 Sodium Salicylate

The uricosuric properties of sodium salicylate (270, Table 31.22) were noted before

Table 31.22   Uricosuric Agents

No.	Generic Name	Trade Name	Structure	Ref.
(270)	Salicyclic Acid			455
(271)	Probenecid	Benemid		456
(276)	Sulfinpyrazone	Anturane		457
(279)	Allopurinol	Zyloprim		458
(281)	Benzbromarone	Desuric, Minuric Narcaricin		459

1890, and its use continued through 1950. As late as 1955, sodium salicylate was used for the long-term treatment of gout (455). For adequate uricosuric activity, however, salicylate must be administered in doses greater than 5 g/day, often resulting in serious side effects, so that its usage has gradually declined.

14.3 Probenecid

Probenecid (**271**, Table 31.22) was developed as a result of a search for a compound that would depress the renal tubular secretion of penicillin (460) at a time when the supply of penicillin was limited. Recognition of the uricosuric properties of probenecid resulted from prior experience with the uricosuric effects of the related compound carinamide (**272**) in normal subjects and in gouty subjects (456). Carinamide had been introduced as an agent for increasing penicillin blood levels by blocking its rapid excretion via the kidney. Its biological half-life was relatively short, and the search for compounds with a longer half-life that would not have to be administered so frequently led to probenecid.

(272)

In a study of a series of N-dialkylsulfamoylbenzoates (**273**), Beyer (461) found that as the length of the N-alkyl groups increased, the renal clearance of the compounds decreased. This most likely results from the enhanced lipid solubility imparted by the longer alkyl groups, which would account for their more complete back diffusion in acidic urine. Optimal activity was found in probenecid, the N-dipropyl derivative. The structure–activity relationship of probenecid congeners and that of other

(273)

R = H, CH$_3$, C$_2$H$_5$, C$_3$H$_7$

uricosuric agents has been reviewed in detail by Gutman (456).

Normally a high percentage of the uric acid filtered by the glomerulus is reabsorbed by an active transport process in the proximal tubule. It is now clear that the human proximal tubule also secretes uric acid, as does the proximal tubule of many lower animals. Small doses of probenecid depresses the excretion of uric acid by blocking tubular secretion, whereas high doses lead to greatly enhanced excretion of uric acid by depressing proximal reabsorption of uric acid (462).

Probenecid is completely absorbed after oral administration; peak plasma levels are reached in 2–4 h. The half-life of the drug in plasma for most patients is 6–12 h. The drug is 85–95% bound to plasma proteins. The small unbound portion is filtered at the glomerulus; a much larger portion is actively secreted by the proximal tubule. The high lipid solubility of the undissociated form results in virtually complete reabsorption by back-diffusion unless the urine is markedly alkaline.

Probenecid is insoluble in water, but the sodium salt is freely soluble. In the treatment of chronic gout, a single daily dose of 250 mg is given for one week, followed by 500 mg administered twice daily. A daily dose of up to 2 g may be required.

14.4 Sulfinpyrazone

Despite the therapeutic efficacy of phenylbutazone (**274**) as an antiinflammatory and uricosuric agent, its side effects were severe enough to preclude its continuous use in the treatment of chronic gout.

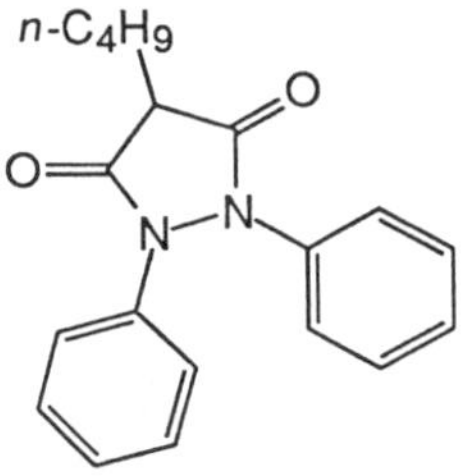

(274)

Evaluation of several chemical congeners indicated that the phenylthioethyl analog of phenylbutazone (275) had promising antiinflammatory and uricosuric activity (457). A metabolite, the sulfoxide pyrazone (276), exhibited enhanced uricosuric activity (463,464). Interestingly, the corresponding sulfone (277) does not appear to be a metabolite (457). Sulfinpyrazone lacks the clinically striking antiinflammatory and analgesic properties of phenylbutazone.

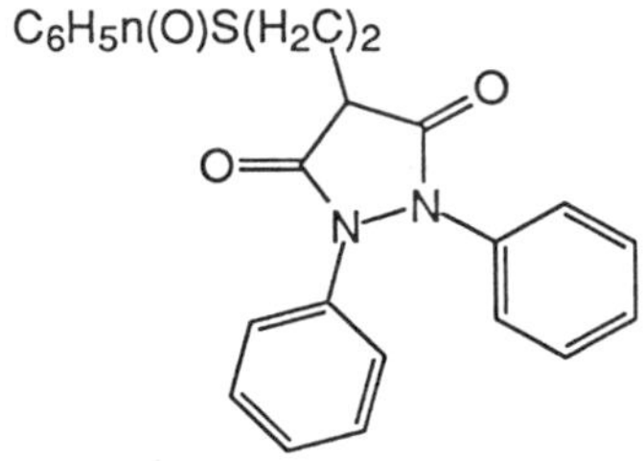

(275) n = 0
(276) n = 1
(277) n = 2

Sulfinpyrazone is a strong acid (pK_a 2.8) and readily forms soluble salts. Evaluation of a number of congeners indicated that a low pK_a and polar side chain substituents favor uricosuric activity (465) and increase the rate of renal excretion (466). The inverse relationship between uricosuric potency and pK_a has also been confirmed in a number of 2-substituted analogs of probenecid (278) All three compounds were considerably stronger acids than probenecid. Evaluation in the *Cebus albifrons* monkey indicated that these compounds were

(278)

R = OH, Cl, NO_2

about 10 times as potent as probenecid when compared on the basis of concentration of drug in plasma (467).

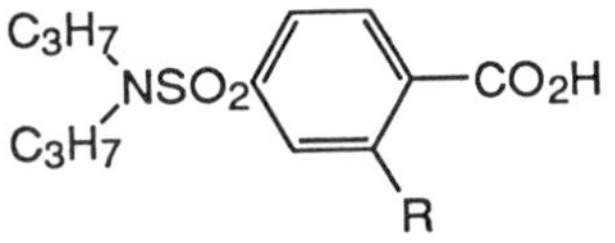

In small doses, as seen with other uricosuric agents, sulfinpyrazone may reduce the excretion of uric acid, presumably by inhibiting secretion but not tubular reabsorption. Its uricosuric action is additive to that of probenecid and phenylbutazone but antagonizes that of the salicylates. Sulfinpyrazone can displace to an unusual degree other organic anions that are bound extensively to plasma protein (e.g., sulfonamides, and salicylates), thus altering their tissue distribution and renal excretion (462,468). Depending on concomitant medication, this may be a clinical asset or liability.

For the treatment of chronic gout, the initial dosage is 100–200 mg/day. After the first week the dose may be increased up to 400 mg/day until a satisfactory lowering of plasma uric acid is achieved.

14.5 Allopurinol

Allopurinol (279) does not reduce serum uric acid levels by increasing renal uric acid excretion; instead it lowers plasma urate levels by inhibiting the final steps in uric acid biosynthesis.

Uric acid in humans is formed primarily by xanthine oxidase catalyzed oxidation of hypoxanthine and xanthine (265) to uric acid (266). Allopurinol (279) and its primary metabolite, alloxanthine (280), are inhibitors of xanthine oxidase.

Inhibition of the last two steps in uric acid biosynthesis by blocking xanthine oxidase reduces the plasma concentration and

(279)   (280)

urinary excretion of uric acid and increases the plasma levels and renal excretion of the more soluble oxypurine precursors. Normally, in humans the urinary purine content is almost solely uric acid; treatment with allopurinol results in the urinary excretion of hypoxanthine, xanthine, and uric acid each with its independent solubility. Lowering the uric acid concentration in plasma below its limit of solubility facilitates the dissolution of uric acid deposits. The effectiveness of allopurinol in the treatment of gout and hyperuricemia that results from hematological disorders and antineoplastic therapy has been demonstrated (458, 469,470).

For the control of hyperuricemia in gout, an initial daily dose of 100 mg is increased at weekly intervals by 100 mg. The usual daily maintenance dose for adults in 300 mg.

14.6 Benzbromarone

Benzbromarone (281) is a benzofuran derivative which has been reported to lower serum urate levels in animals and human studies. In normal and hyperuricaemic subjects, benzobromarone reduced serum uric acid levels by one-third to one-half

(281)

(471,472). In comparison with other urate lowering drugs, 80 mg of micronized or 100 mg of nonmicronized benzbromarone had equal urate lowering activity to 1–1.5 g of probenecid or 400–800 mg of sulfinpyrazone (471,473).

The mechanism of urate lowering activity of benzbromarone appears to be due to its uricosuric activity. In rats, benzobromarone inhibited urate reabsorption in the proximal tubules when given at 10 mg/kg i.v. (474). In isolated rat liver preparation, benzbromarone inhibits xanthine oxidase *in vitro* but not *in vivo* (475). In humans, this compound only weakly inhibits xanthine oxidase and no increase in urinary excretion of xanthine or hypoxanthine was observed (476).

After oral administration, about 50% of benzbromarone is absorbed. The drug undergoes extensive dehalogenation in the liver and is excreted mainly in the bile and feces. For control of gout the usual therapeutic dose is 100–200 mg daily. Benzbromarone has few side-effects and is usually well tolerated.

15 CONCLUSION

The development and therapeutic use of diuretic agents constitutes one of the most significant advances in medicine made during the twentieth century. Continuous progress has been made during this time on the development of safer and more effective diuretic agents. Between 1920 and 1950, a large number of organic mercurials were prepared and evaluated as diuretics. Because of lack of oral activity and toxicity of these compounds, research efforts were focused on the development of orally effective nonmercurial diuretics. The carbonic anhydrase inhibitors, developed in 1950 and later, were orally active but upset the acid-base balance and could only be given intermittently. The thiazide diuretics, developed in late 1950s, represented a true advance in the treatment of edema. They were remarkably nontoxic and effective in most cases. It very soon became apparent that not only were they effective diuretics, they were also useful in the treatment of hypertension by themselves or in combination with other antihypertensive drugs.

Four side effects were noticed following the widespread and prolonged use of the thiazide diuretics:

1. potassium depletion
2. uric acid retention
3. hyperglycemia
4. increased plasma lipids

Potassium depletion has been encountered most frequently. The kaliuretic effect of the thiazides can be compensated for by supplementary dietary potassium; nevertheless, research was directed towards the development of potassium-sparing diuretics. Amiloride (1965), spironolactone (1959), and triamterene (1965) were discovered as a result of this effort; these compounds are weak diuretics, however, and are generally used in combination with other diuretics, e.g., hydrochlorothiazide.

The next step was the discovery of the high-ceiling or loop diuretics, (e.g., ethacrynic acid (1962), furosemide (1963), and bumetanide (1971)), which are shorter acting and more potent than the thiazide diuretics. They too have the same potential side effects as the thiazides. One advantage of the loop diuretics is their efficacy in chronic renal insufficiency, particularly in cases with low glomerular filtration rates.

A large volume of highly technical information has been published over the past 15 years regarding this therapeutic area. More sensitive analytical techniques have been developed, so that data regarding bioavailability and pharmokinetics is now available for diuretics that are currently prescribed and that are in development. Advances in renal and ion-transport re-

search has led to a more precise understanding of the cellular mechanisms of actions of the various classes of diuretic agents. This has aided in the design of newer and more effective agents.

Diuretics introduced into more recent clinical studies include *(1)* newer more potent loop diuretics such as torasemide and azosemide, *(2)* development of uricosuric diuretics, *(3)* newer generation sulfamoyl diuretics, *(4)* development of neutral endopeptidase inhibitors.

Since the 1960s, diuretics have been used to treat millions of patients with hypertension. With the number of adverse effects seen with long-term diuretic treatment, such as, hypokalemia, hypercholesterolemia, and hyperglycemia, and because diuretic-based antihypertensive drug trials have failed to show a reduction in the incidence of myocardial infarction, this practice has become controversial. Today many other therapies exists for the treatment of hypertension, such as, calcium channel blockers and angiotensin-converting enzyme inhibitors. However, since diuretics are convenient, inexpensive, and generally are well tolerated, they will probably continue to play a major, though less dominant role in the treatment of hypertension.

REFERENCES

1. W. B. Schwartz, *N. Engl. J. Med.*, **240**, 173 (1949).
2. C. Rouiller, in *The Kidney, Morphology, Biochemistry, Physiology*, Academic Press, New York, 1969.
3. B. Brenner and F. Rector, in *The Kidney*, Saunders, Philadelphia, 1986, p. 9.
4. W. Foye, in *Principles of Medicinal Chemistry*, Lea and Febiger, Philadelphia, 1989, p. 447.
5. G. Giebisch, *Eur. J. Clin. Pharmacol.*, **44**(Suppl. 1), S3 (1993).
6. B. O. Rose, *Kidney International*, **39**, 337 (1991).
7. P. A. Preisig and F. C. Rector, *Am. J. Physiol.*, **255**, F461 (1988).
8. C. Sardet, A. Franchi and J. Pouyssegur, *Cell*, **56**, 271 (1989).

9. R. Green and G. Giebisch, *Am. J. Physiol.*, **257**, F669 (1989).

10. G. Giebisch, *J. Clin. Invest.*, **79**, 32 (1987).

11. C. A. Berry and F. C. Rector, *Kidney International*, **36**, 403 (1989).

12. L. Schild, G. Giebisch and L. P. Karniski, *J. Clin. Invest.*, **79**, 32 (1987).

13. W. F. Boron and E. L. Boulpaep, *Kidney International*, **36**, 392 (1989).

14. R. Greger, *Pflugers Arch.*, **390**, 38 (1981).

15. F. Bronner, *Am. J. Physiol.*, **257**, F707 (1989).

16. J. L. Borke, J. Minami, A. Verma, J. T. Penniston and R. Kumas, *Kidney International*, **34**, 262 (1988).

17. T. Shimuzu, K. Yoshitouri, M. Nakamura and M. Imai, *Am. J. Physiol.*, **259**, F408 (1990).

18. L. Palmer and I. Edelman, *Ann. NY Acad Sci.*, **372**, 1 (1981).

19. M. Martinez-Maldorado and H. Cordovc, *Kidney International*, **38**, 632 (1990).

20. K. H. Beyer, *Perspect Biol. Med.*, **19**, 500 (1976).

21. N. L. Lipschitz, Z. Hadidian and A. Kerpcsar, *J. Pharmacol. Exp. Ther.*, **79**, 97 (1943).

22. A. A. Renzi, J. J. Chart and R. Gaunt, *Toxicol. Appl. Pharmacol.*, **1**, 406 (1959).

23. D. M. Zall, D. Fisher and M. Q. Garner, *Anal. Chem.*, **28**, 1665 (1956).

24. W. E. Barrett, R. A. Rutledge, H. Sheppard and A. J. Plummer, *Toxic. Appl. Pharmacol.*, **1**, 333 (1959).

25. T. W. K. Hill and P. J. Randall, *J. Pharm. Pharmacol.*, **28**, 552 (1976).

26. G. M. Fanelli, Jr., D. I. Bohn, A. Scriabine and K. H. Beyer, Jr., *J. Pharmacol. Exp. Ther.*, **200**, 402 (1977).

27. G. M. Fanelli, D. L. Bohn and H. F. Russo, *Comp. Biochem. Physiol.*, **33**, 459 (1970).

28. B. Stavric, W. J. Johnson and H. C. Grice, *Proc. Soc. Exp. Biol. Med.*, **130**, 512 (1969); B. Stravric, E. A. Nera, W. J. Johnson and F. A. Salem, *Invest. Urol.*, **11**, 3 (1973).

29. F. W. Wolff, W. W. Parmley, K. White and R. Okun, *J. Am. Med. Assoc.*, **185**, 568 (1963).

30. I. I. A. Tabachnick, A. Gulbenkian and A. Yannell, *Life Sci.*, **4**, 1931 (1965).

31. D. Morgan and C. Davison, *Brit. Med. J.*, **280**, 295 (1980).

32. A. Amery and C. Bulpitt, *Lancet*, **1**, 681 (1978).

33. S. MacMahon and G. MacDonald, *Am. J. Med.*, **80**, 40 (1986).

34. J. F. Seely and J. H. Dirks, *J. Clin. Invest.*, **48**, 2330 (1969).

35. R. C. Blantz, *J. Clin. Invest.*, **54**, 1135 (1974).

36. E. Jendrassik, *Arch. Klin. Med.*, **38**, 499 (1886).

37. A. Vogl, *Am. Heart J.*, **39**, 881 (1950).

38. I. M. Weiner, R. I. Levy and G. H. Mudge, *J. Pharmacol. Exp. Ther.*, **138**, 96 (1962).

39. T. W. Clarkson and J. J. Vostal, in *Modern Diuretic Therapy in the Treatment of Cardiovascular and Renal Disease*, Excerpta Medica, Amsterdam, 1973, pp. 229–240.

40. E. J. Cafruny, K. C. Cho, V. Nigrovic and A. Small, in Ref. 38 pp. 124–134,

41. G. de Stevens, *Diuretics: Chemistry and Pharmacology*, Academic Press, New York, 1963, p. 38.

42. R. H. Kessler, R. Lozano and R. F. Pitts, *J. Clin. Invest.* **36**, 656 (1957).

43. R. C. Batterman, D. Unterman and A. C. De-Graff, *J. Am. Med. Assoc.*, **140**, 1268 (1949).

44. W. Modell, *Am. J. Med. Sci.*, **231**, 564 (1956).

45. L. H. Werner and C. R. Scholz, *J. Am. Chem. Soc.*, **76**, 2453 (1954).

46. R. H. Chaney and R. F. Maronde, *Am. J. Med. Sci.*, **231**, 26 (1956).

47. J. Moyer, S. Kinard and R. Herschberger, *Antibiot. Med. Clin. Ther.*, **3**, 179 (1956).

48. R. W. Berliner, J. H. Dirks and W. J. Cirksena, *Ann. N. Y. Acad. Sci.*, **139**, 424 (1966).

49. J. R. Clapp and R. R. Robinson, *Am. J. Physiol.*, **215**, 228 (1968).

50. R. L. Evanson, E. A. Lockhart and J. H. Dirks, *Am. J. Physiol.*, **222**, 282 (1972).

51. G. H. Mudge, in L. S. Goodman and A. Gilman, eds., *Pharmacological Basis of Therapeutics*, 5th ed., Macmillan, New York, Section VIII, 1975, p. 809.

52. G. M. Fanelli, Jr., D. L. Bohn, S. S. Reily and I. M. Weiner, *Am. J. Physiol.*, **224**, 985 (1973).

53. M. Laski, *Seminars in Nephrology*, **6**, 210 (1986).

54. M. B. Strauss and H. Southworth, *Bull. Johns Hopkins Hosp.*, **63**, 41 (1938).

55. T. Mann and D. Keilin, *Nature*, **146**, 164 (1940).

56. H. W. Davenport and A. E. Wilhelmi, *Proc. Soc. Exp. Biol. Med.*, **48**, 53 (1941).

57. R. F. Pitts and R. S. Alexander, *Am. J. Physiol.*, **144**, 239 (1945).

58. W. B. Schwartz, *N. Engl. J. Med.*, **240**, 173 (1949).

59. H. A. Krebs, *Biochem.*, **43**, 525 (1948).

60. R. O. Roblin, Jr. and J. W. Clapp, *J. Am. Chem. Soc.*, **72**, 4890 (1950).

61. W. H. Miller, A. M. Dessert and R. O. Roblin, Jr., *J. Am. Chem. Soc.*, **72**, 4893 (1950).

62. T. H. Maren, E. Mayer and B. C. Wadsworth, *Bull. Johns Hopkins Hosp.*, **95**, 199 (1954).

63. K. H. Beyer and J. E. Baer, *Pharmacol. Rev.*, **13**, 517 (1961).

64. A. Posner, *Am. J. Ophthalmol.*, **45**, 225 (1958).

65. D. M. Travis, *J. Pharmacol. Exp. Ther.*, **167**, 253 (1969).

66. R. W. Young, K. H. Wood, J. A. Eichler, J. R. Vaughan, Jr. and G. W. Anderson, *J. Am. Chem. Soc.*, **78**, 4649 (1956).

67. R. V. Ford, C. L. Spurr and J. H. Moyer, *Circulation*, **16**, 394 (1957).

68. J. R. Vaughan, Jr., J. A. Eichler and G. W. Anderson, *J. Org. Chem.*, **21**, 700 (1956).

69. T. H. Maren, *J. Pharmacol. Exp. Ther.*, **117**, 385 (1956).

70. R. T. Kunan, Jr., *J. Clin. Invest.*, **51**, 294 (1972).

71. T. W. K. Hill and P. J. Randall, *J. Pharm. Pharmacol.*, **28**, 552 (1976).

72. T. H. Maren, *Bull Johns Hopkins Hosp.*, **98**, 159 (1956); T. H. Maren in H. Herkin ed., *Handbook of Experimental Pharmacology*, Vol. 24, *Diuretics*, Springer, Berlin-Heidelberg-New York, 1969, p. 195.

73. T. H. Scholz, J. M. Sondey, W. C. Randall, H. Schwam, W. J. Thompson, P. J. Mallorga, M. F. Sugrue and S. L. Graham, *J. Med. Chem.*, **36**, 2134 (1993).

74. K. L. Shepard, S. L. Graham, R. J. Hudcosky, S. R. Michelson, T. H. Scholz, H. Schwam, A. M. Smith, J. M. Sondey, K. M. Strohmaier, R. L. Smith and M. F. Sugrue, *J. Med. Chem.*, **34**, 3098 (1991).

75. J. J. Baldwin, G. S. Ponticello and M. F. Sugrue, *Drugs Future*, **15**, 350 (1990).

76. E. Larson, R. Roach, R. Schoene and T. Hornbein, *J. Am. Med. Assoc.*, **248**, 328 (1982).

77. K. H. Beyer, Jr. and J. E. Baer, *Med. Clin. North Am.*, **59**, 735 (1975).

78. J. M. Sprague, *Ann. N. Y. Acad. Sci.*, **71**(4), 328 (1958).

79. F. C. Novello, S. C. Bell, L. A. Abrams, C. Ziegler and J. M. Sprague, *J. Org. Chem.*, **25**, 965 (1960).

80. F. C. Novello and J. M. Sprague, *J. Am. Chem. Soc.*, **79**, 2028 (1957).

81. F. J. Lund and W. Kobinger, *Acta Pharmacol. Toxicol.*, **16**, 297 (1960).

82. H. Horstmann, H. Wollweber and K. Meng, *Arzneim-Forsch.*, **17**, 653 (1967).

83. K. Meng and G. Kroneberg, *Arzneim-Forsch.*, **17**, 659 (1967).

84. B. Duhm, W. Maul, H. Medenwald, P. Patzchke and L. A. Wengner, *Arzneim-Forsch.*, **17**, 672 (1967).

85. C. B. Wilson and W. M. Kirkendall, *J. Pharmacol. Exp. Ther.*, **173**, 422 (1970).

86. R. J. Santos, V. Paz-Martinez, J. K. Lee and J. H. Nodine, *Int. J. Clin. Pharmacol.*, **3**, 14 (1970).

87. C. B. Wilson and W. M. Kirkendall, *J. Pharmacol. Exp. Ther.*, **171**, 288 (1970).

88. W. H. R. Auld and W. R. Murdoch, *Brit. Med. J.*, **4**, 786 (1971).

89. S. J. Jachuck, *Brit. Med. J.*, **3**, 590 (1972).

90. K. H. Beyer, Jr., *Perspect. Biol. Med.*, **19**, 500 (1976).

91. F. J. Lund and W. Kobinger, *Acta Pharmacol. Toxicol.*, **16**, 297 (1960).

92. R. L. Hauman and J. M. Weller, *Clin. Pharmacol. Ther.*, **1**, 175 (1960).

93. S. Y. P'an, A. Scriabine, D. E. McKersie and W. M. McLamore, *J. Pharmacol. Exp. Ther.*, **128**, 122 (1960).

94. G. deStevens, *Diuretics: Chemistry and Pharmacology*, Academic Press, New York, 1963, p. 100.

95. E. H. Wiseman, E. C. Schreiber and R. Pinson, Jr., *Biochem. Pharmacol.*, **11**, 881 (1962).

96. T. H. Maren and C. E. Wiley, *J. Pharmacol. Exp. Ther.*, **143**, 230 (1964).

97. E. J. Cragoe, Jr., J. A. Nicholson and J. M. Sprague, *J. Med. Pharm. Chem.*, **4**, 369 (1961).

98. J. G. Topliss, M. H. Sherlock, H. Reimann, L. M. Konzelman, E. P. Shapiro, B. W. Pettersen., H. Schneider and N. Sperber, *J. Med. Chem.*, **6**, 122 (1963).

99. A. A. Rubin, F. E. Roth, R. M. Taylor and H. Rosenkilde, *J. Pharmacol. Exp. Ther.*, **136**, 344 (1962).

100. G. deStevens, L. H. Werner, A. Halamandaris and S. Ricca, Jr., *Experientia*, **14**, 463 (1958).

101. J. E. Baer, H. F. Russo and K. H. Beyer, *Proc. Soc. Exp. Biol. Med.*, **100**, 442 (1959).

102. A. F. Esch, I. M. Wilson and E. D. Freis, *Med. Ann. Dist. Columbia*, **28**, 9 (1959).

103. C. W. H. Havard and J. C. B. Fenton, *Brit. Med. J.*, **1**, 1560 (1959).

104. H. Losse, H. Wehmeyer, W. Strobel and H. Wesselkock, *Muench Med. Wochenschr.*, **101**, 677 (1959).

105. W. Hollander, A. V. Chobanian and R. W. Wilkins in J. H. Moyer ed., *Hypertension*, Saunders, Philadelphia, 1959, p. 570.

106. L. H. Werner, A. Halamandaris, S. Ricca, Jr., L. Dorfman and G. deStevens, *J. Am. Chem. Soc.*, **82**, 1161 (1960).

107. E. J. Cragoe, Jr., O. W. Woltersdorf, Jr., J. E. Baer and J. M. Sprague, *J. Med. Chem.*, **5**, 896 (1962).

108. J. G. Topliss, M. H. Sherlock, F. H. Clarke, M. C. Daly, B. W. Pettersen, J. Lipski and N. Sperber, J. Org. Chem., **26**, 3842 (1961).

109. J. Klosa and H. Voigt, J. Prakt. Chem., **16**, 264 (1962).

110. J. Klosa, J. Prakt. Chem., **18**, 225 (1962).

111. J. Klosa, J. Prakt. Chem., **33**, 298 (1966).

112. J. Klosa, J. Prakt. Chem., **21**, 176 (1963).

113. W. J. Close, L. R. Swett, L. E. Brady, J. H. Short and M. Vernsten, J. Am. Chem. Soc., **82**, 1132 (1960).

114. J. H. Short and U. Biermacher, J. Am. Chem. Soc., **82**, 1135 (1960).

115. J. H. Short and L. R. Swett, J. Org. Chem., **26**, 3428 (1961).

116. C. T. Holdrege, R. B. Babel and L. C. Cheney, J. Am. Chem. Soc., **81**, 4807 (1959).

117. U. S. Pat. 3,111,517 (Nov. 19, 1963), W. M. McLamore and G. D. Laubach.

118. U. S. Pat. 3,009,911 (Nov. 21, 1961), J. M. McManus.

119. C. W. Whitehead, J. J. Traverso, H. R. Sullivan and F. J. Marshall, J. Org. Chem., **26**, 2814 (1961).

120. C. W. Whitehead and J. J. Traverso, J. Org. Chem., **27**, 951 (1962).

121. F. J. Lund and W. Kobinger, Acta Pharmacol. Toxicol., **16**, 297 (1960).

122. H. Stanton, R. Hanson and R. Rosenberger, Pharmacologist, **29**, 154 (1986).

123. H. J. Lang, B. Knabe, R. Muschaweck, M. Hropot and E. Linder in E. J. Cragoe, ed., Diuretic Agents, ACS Symposium Series 83, American Chemical Society, Washington D. C., p. 24.

124. L. Hofman, Arch. Int. Pharmacodyn. Ther., **169**, 189 (1967).

125. B. Johnson, Clin. Pharmacol. Ther., **11**, 77 (1970).

126. B. Beerman, M. Groschinski-Grind and B. Lindstrom, Eur. J. Clin. Pharmacol., **11**, 203 (1977).

127. D. A. Garcia and E. R. Yendt, C. M. A. J., **103**, 473 (1970).

128. J. M. Tran, M. A. Farrel and P. P. Fanestil, Am. J. Physiol., **258**, F908 (1990).

129. D. B. Hunninghake, S. King and K. LaCroix, Int. J. Clin. Pharmacol., **20**, 151 (1982).

130. M. Hohenegger, Advances Clin. Pharmacol., **9**, 1 (1975).

131. M. Goldberg in J. Orloff and R. W. Berlinger, eds., Handbook of Physiology, American Physiology Society, Washington D. C., Section 8, 1973, pp. 1003–1031.

132. E. Perez-Stable and P. V. Caralis, Am. Heart J. **106**, 245 (1983).

133. D. H. Kaump, R. L. Fransway, L. T. Blouin and D. Williams, J. New Drugs, **4**, 21 (1964).

134. W. Graf, E. Girod, E. Schmid and W. G. Stoll, Helv. Chim. Acta, **42**, 1085 (1959).

135. E. Jucker, A. Lindenmann, E. Schenker, E. Fluckiger and M. Taeschler, Arzneim-Forsch., **13**, 269 (1963).

136. E. J. Cornish, G. E. Lee and W. R. Wragg, J. Pharm. Pharmacol., **18**, 65 (1966).

137. L. T. Blouin, D. H. Kaump, R. L. Fransway and D. Williams, J. New Drugs, **3**, 302 (1963).

138. B. V. Shetty, L. A. Campanella, T. L. Thomas, M. Fedorchuk, T. A. Davidson, L. Michelson, H. Volz and S. E. Zimmerman, J. Med. Chem., **13**, 886 (1970).

139. E. Cohen, B. Klarberg and J. R. Vaughan, Jr., J. Am. Chem. Soc., **82**, 2731 (1960).

140. P. Milliez and P. Tcherdakoff, Curr. Med. Res. Opin., **3**, 9 (1975).

141. E. G. Stenger, H. Witz and R. Pulver, Schweiz Med. Wochenschr., **89**, 1130 (1959).

142. R. Veyrat, E. F. Arnold and A. Duckert, Schweiz Med. Wochenschr., **89**, 1133 (1959).

143. F. Reutter and F. Schaub, Schweiz Med. Wochenschr., **89**, 1158 (1959).

144. W. Leppla, H. Buch and G. A. Jutzler, Ger. Med. Monthly, **5**, 402 (1960).

145. M. Fuchs, B. E. Newman, S. Irie, R. Maranoff, E. Lippman and J. H. Moyer, Curr. Ther. Res., **2**, 11 (1960).

146. W. E. Bowlus and H. G. Langford, Clin. Pharmacol. Ther., **5**, 708 (1964).

147. B. Terry and J. B. Hook, J. Pharmacol. Exp. Ther., **160**, 367 (1968).

148. V. Parsons and R. Kemball Price, Practitioner, **195**, 648 (1965).

149. M. L. Hoefle, L. T. Blouin, H. A. DeWald, A. Holmes and D. Williams, J. Med. Chem., **11**, 970 (1968).

150. E. V. Mackay and S. K. Khoo, Med. J. Aust., **1**, 607 (1969).

151. V. Anania, M. S. Desole and E. Miele, Riv. It. Biol Med., **2**, 135 (1982).

152. G. Cignarella, P. Sanna, E. Miele, A. Anania and M. Desole, J. Med. Chem., **24**, 1003 (1981).

153. G. Cignarella, D. Barlocco, D. Landriania, G. Pinna, G. Andrivoli and G. Dona, 11. Farmaco, **46**, 527 (1991).

154. A. F. Lant, W. I. Baba and G. M. Wilson, Clin. Pharmacol. Ther., **7**, 196 (1966).

155. W. I. Baba, A. F. Lant and G. M. Wilson, *Clin. Pharmacol. Ther.*, **7**, 212 (1966).

156. J. L. Verbov, D. S. Tunstall-Pedoe and T. J. C. Cooke, *Brit. J. Clin. Pract.*, **20**, 351 (1966).

157. K. Corbett, S. A. Edwards, G. E. Lee and T. L. Threlfall, *Nature*, **208**, 286 (1965).

158. R. H. Sellers, M. Fuchs, G. Onesti, C. Swartz, A. N. Brest and J. H. Moyer, *Clin. Pharmacol. Ther.*, **3**, 180 (1962).

159. W. N. Suki, F. Dawoud, G. Eknoyan and M. Martinez-Maldonado, *J. Pharmacol. Exp. Ther.*, **180**, 6 (1972).

160. J. W. Smiley, G. Onesti and C. Swatz, *Clin. Pharmacol. Ther.*, **13**, 336 (1972).

161. M. F. Michelis, F. DeRubertis, N. P. Beck, R. H. McDonald, Jr. and B. B. Davis, *Clin. Pharmacol. Ther.*, **11**, 821 (1970).

162. E. J. Belair, A. I. Cohen and J. Yelnoski, *Brit. J. Pharm.*, **45**, 476 (1972).

163. B. J. Materson, J. L. Hotchkiss, J. S. Barkin, B. H. Rietberg, K. Bailey and E. C. Perez-Stable, *Curr. Ther. Res.*, **14**, 545 (1972).

164. R. M. Pilewski, E. T. Scheib, J. R. Misage, E. Kessler, E. Krifcher and A. P. Shapiro, *Clin. Pharmacol. Ther.*, **12**, 843 (1971).

165. F. Costa, R. Caldari, E. Ambrosion and B. Magnani, *Curr, Ther. Res.*, **32**, 359 (1982).

166. J. J. Mencel, J. R. Regan, J. Barton, P. R. Menard, J. G. Bruno, R. R. Calvo, B. E. Kornberg, A. Schwab, E. S. Neiss and J. T. Suh, *J. Med. Chem.*, **33**, 1606 (1990).

167. V. Cecchetti, A. Fravolini, F. Schiaffella, O. Tabarrini, G. Bruni and G. Segre, *J. Med. Chem.*, **36**, 157 (1993).

168. *Drugs of Fut.*, **10**, 298 (1985).

169. S. M. O'Grady, H. C. Palfrey and M. Field, *J. Membrane Biol.*, **96**, 11 (1987).

170. W. N. Suki, J. J. Yium, M. VonMinden, C. Saller-Hebert, G. Eknoyan and M. Martinez-Maldonado, *N. Engl. J. Med.*, **283**, 836 (1970).

171. L. P. Rybak, *J. Otolaryngol.*, **11**, 127 (1982).

172. P. W. Feit, *J. Med. Chem.*, **14**, 432 (1971).

173. D. L. Davies, A. F. Lant, N. R. Millard, A. J. Smith, J. W. Ward and G. M. Wilson, *Clin. Pharmacol. Ther.*, **15**, 141 (1974).

174. E. M. Schultz, E. J. Cragoe, Jr., J. B. Bicking, W. A. Bolhofer and J. M. Sprague, *J. Med. Chem.*, **5**, 660 (1962).

175. K. Strum, W. Siedel, R. Weyer and H. Ruschig, *Chem. Ber.*, **99**, 328 (1966).

176. W. Merkel, D. Bormann, D. Mania, R. Muschaweck and M. Hropot, *Eur. J. Med. Chem.*, **11**, 399 (1976).

177. W. Liebenow and F. Leuschner, *Arzneim-Forsch.*, **25**, 240 (1975).

178. F. Krueck, W. Bablok, E. Bensenfelder, G. Betzien and B. Kaufmann, *Eur. J. Clin. Pharm.*, **14**, 153 (1978).

179. J. Delarge and C. Lepiere, *Ann. Pharm. Francaises*, **36**, 369 (1978).

180. D. E. Duggan and R. M. Noll, *Arch. Biochem. Biophys.*, **109**, 388 (1965).

181. J. M. Sprague, *Ann Rep. Med. Chem.*, **5**, X1 (1970).

182. J. B. Bicking, W. J. Holtz, L. S. Watson and E. J. Cragoe, Jr., *J. Med. Chem.*, **19**, 530 (1976).

183. D. E. Duggan and R. M. Noll, *Arch. Biochem. Biophys.*, **109**, 388 (1965).

184. E. M. Schultz, J. B. Bicking, A. A. Deana, N. P. Gould, T. P. Strobaugh, L. S. Watson and E. J. Cragoe, Jr., *J. Med. Chem.*, **19**, 783 (1976).

185. K. H. Beyer, J. E. Baer, J. K. Michaelson and H. F. Russo, *J Pharmacol. Exp. Ther.*, **147**, 1 (1965).

186. M. Goldberg, *Ann. N. Y. Acad. Sci.*, **139**, 443 (1966).

187. C. T. Dollery, E. H. O. Parry and D. S. Young, *Lancet*, **1**, 947 (1964).

188. J. F. Maher and G. E. Schreiner, *Ann. Intern. Med.*, **62**, 15 (1965).

189. V. K. G. Pillay, F. D. Schwartz, K. Aimi and R. M. Kark, *Lancet*, **1969–1**, 77.

190. E. Blain, G. Fanelli and J. Irvin, *Clin. Exp. Hypertension*, **4**, 161 (1982).

191. R. McKenzie, T. Knight and E. J. Weinman, *Proc. Soc. Exp. Biol. Med.*, **153**, 202 (1976).

192. E. J. Weinman, T. Knight, R. M. McKenzie and G. Eknoyan, *Clin. Res.*, **24**, 416A (1976).

193. A. G. Zacchei, T. I. Wishousky, B. H. Arison and G. M. Fanelli, Jr., *Drug Metab. Dispos.*, **4**, 479 (1976).

194. A. G. Zacchei and T. I. Wishousky, *Drug Metab. Dispos.*, **4**, 490 (1976).

195. Z. E. Dziewanowska, K. F. Tempero, F. Perret, G. Hitzenberger and G. H. Besselaar, *Clin. Res.*, **24**, 253A (1976).

196. K. F. Tempero, G. Hitzenberger, Z. E. Dziewanowska and H. Halkin, *Clin. Pharmacol. Ther.*, **19**, 116 (1976).

197. K. F. Tempero, J. A. Vedin, C. E. Wilhelmsson, P. Lund-Johansen, C. Vorburger, C. Moerlin, A. Aaberg, W. Enenkel, J. Bolongnese and Z. E. Dziewanowska, *Clin. Pharmacol. Ther.*, **21**, 97 (1977).

198. O. Woltersdorf, S. deSolm, E. Schultz and E. Cragoe, *J. Med. Chem.*, **20**, 1400 (1977).

199. G. Shutske, L. Setescak, R. Allen, L. Davis, R. Effland, K. Ranborn, J. Kitzen, J. Wilken and W. Novick, *J. Med. Chem.*, **25**, 36 (1982).

200. W. Hoffman, O. Woltersdorf, F. Novello, E. Cragoe, J. Springer, L. Watson and G. Fanelli, *J. Med. Chem.*, **24**, 865 (1981).

201. J. Plattner, A. Fung, J. Parks, R. Pariza, S. Crowley, A. Pernet, P. Runnel and P. Dodge, *J. Med. Chem.*, **27**, 1016 (1984).

202. W. Siedel, K. Strum and W. Scheurich, *Chem. Ber.*, **99**, 345 (1966).

203. R. J. Timmerman, F. R. Springman and R. K. Thoms, *Curr. Ther. Res.*, **6**, 88 (1964).

204. R. Muschaweck and P. Hajdu, *Arzneim-Forsch.*, **14**, 46 (1964).

205. A. Haussler and P. Hajdu, *Arzneim-Forsch.*, **14**, 710 (1964).

206. A. Haussler and H. Wicha, *Arzneim-Forsch.*, **15**, 81 (1965).

207. J. B. Hook, A. H. Blatt, M. J. Brody and H. E. Williamson, *J. Pharmacol. Exp. Ther.*, **154**, 667 (1966).

208. J. H. Ludens, J. B. Hook, M. J. Brody and H. E. Williamson, *J. Pharmacol. Exp. Ther.*, **163**, 456 (1968).

209. W. Stokes and L. C. A. Nunn, *Brit. Med. J.*, **2**, 910 (1964).

210. W. M. Kirkendall and J. H. Stein, *Am. J. Cardiol.*, **22**, 162 (1968).

211. C. R. Bariso, I. B. Hanenson and T. E. Gaffney, *Curr. Ther. Res.*, **12**, 333 (1970).

212. R. G. Muth, *J. Am. Med. Assoc.*, **195**, 1066 (1966).

213. D. S. Silverberg, R. A. Ulan, M. A. Baltzan and R. B. Baltzan, *Can. Med. Assoc. J.*, **103**, 129 (1970).

214. O. H. Morelli, L. I. Moledo, E. Alanis, O. L. Gaston and O. Terzaghi, *Postgrad. Med. J.*, **47** (April Suppl.), 29 (1972).

215. P. W. Feit, H. Bruun and C. K. Nielsen, *J. Med. Chem.*, **13**, 1071 (1970).

216. P. W. Feit and O. B. Tvaermose Nielsen, *J. Med. Chem.*, **15**, 79 (1972).

217. P. W. Feit, O. B. Tvaermose Nielsen and N. Rastrup-Andersen, *J. Med. Chem.*, **16**, 127 (1973).

218. O. B. Tvaermose Nielsen, C. K. Nielsen and P. W. Feit, *J. Med. Chem.*, **16**, 1170 (1973).

219. P. W. Feit, O. B. Tvaermose Nielsen and H. Bruun, *J. Med. Chem.*, **17**, 572 (1974).

220. O. B. Tvaermose Nielsen, H. Bruun, C. Bretting and P. W. Feit, *J. Med. Chem.*, **18**, 41 (1975).

221. P. W. Feit and O. B. Tvaermose Nielsen, *J. Med. Chem.*, **19**, 402 (1976).

222. P. W. Feit, O. B. Tvaermose Nielsen and H. Bruun, *J. Med. Chem.*, **15**, 437 (1972).

223. O. B. Tvaermose Nielsen and P. W. Feit, in E. J. Cragoe, ed., *Diuretic Agents*, ACS Symposium Series 83, American Chemical Society, Washington, D. C. 1978, p. 12.

224. U. S. Pat. 4,082,851 (1978), P. W. Feit, O. B. Tvaermose Nielsen, C. Bretting and H. Bruun.

225. P. W. Feit and O. B. Tvaermose Nielsen, *J. Med. Chem.*, **20**, 1687 (1977).

226. E. H. Ostergaard, M. P. Magnussen, C. Kaergaard Nielsen, E. Eilertsen and H. H. Frey, *Arzneim-Forsch.*, **22**, 66 (1972).

227. P. W. Feit, K. Roholt and H. Sorensen, *J. Pharm. Sci.*, **62**, 375 (1973).

228. M. J. Asbury, P. B. B. Gatenby, S. O'Sullivan and E. Bourke, *Brit. Med. J.*, **1**, 211 (1972).

229. K. H. Olesen, B. Sigurd, E. Steiness and A. Leth, *Acta Med. Scand.*, **193**, 119 (1973).

230. K. H. Olessen, B. Sigurd, E. Steiness and A. Leth in A. F. Lant and G. M. Wilson, eds., *Excerpta Medica*, Amsterdam, 1973, p. 155.

231. E. Bourke, M. J. A. Asbury, S. O'Sullivan and P. B. B. Gatenby, *Eur. J. Pharmacol.*, **23**, 283 (1973).

232. S. Carriere and R. Dandavino, *Clin. Pharmacol. Ther.*, **20**, 428 (1976).

233. P. Friedman, Roch-Ramel, *J. Pharmacol. Exp. Ther.*, **203**, 82 (1977).

234. C. Brater and P. Chennavasi, *J. Clin. Pharmacol.*, **21**, 311 (1981).

235. S. Halladay, G. Sipes and D. Carter, *Clin. Pharmacol. Ther.*, **22**, 179 (1977).

236. W. Merkel, D. Bormann, D. Mania, R. Muschaweck and M. Hropot, *Eur. J. Med. Chem.*, **11**, 399 (1976).

237. W. McNabb, F. Nourmahamed, B. Brooks and A. Lant, *Clin. Pharm. Ther.*, **35**, 328 (1984).

238. B. Beerman and M. Grind, *Clin. Pharmacokin.*, **13**, 254 (1987).

239. S. Clissold and Brogden, *Drugs*, **29**, 489 (1985).

240. E. Klaus, H. Alpermann, G. Caspritz, W. Linz and B. Scholken, *Arzneim-Forsch.*, **33**, 1273 (1983).

241. D. Brater, *Clin. Pharmacol. Ther.*, **25**, 428 (1979).

242. D. Brater, B. Day, S. Anderson and R. Serwell, *Clin. Pharmacol. Ther.*, **34**, 454 (1983).

243. F. Leuschner, W. Neumann and H. Barhmann, *Arzneim-Forsch.*, **25**, 245 (1975).

244. F. W. Hempelmann, *Arzneim-Forsch.*, **25**, 259 (1975).

245. F. W. Hempelmann, *Arzneim-Forsch.*, **25**, 258 (1975).

246. F. W. Hempelmann, F. Leuschner and W. Liebenow, *Arzneim-Forsch.*, **25**, 252 (1975).

247. G. Voltz, *Arzneim-Forsch.*, **25**, 256 (1975).

248. M. Hohenegger and F. Holzer, *Int. J. Clin. Pharmacol.*, **13**, 298 (1975).

249. P. Federspil and H. Mausen., *Int. J. Clin. Pharmacol.*, **9**, 326 (1974).

250. R. Z. Gussin, J. R. Cummings, E. H. Stokey and M. A. Ronsberg, *J. Pharmacol. Exp. Ther.*, **167**, 194 (1969).

251. E. A. Lockhart, J. H. Dirks and S. Carriere, *Am. J. Physiol.*, **223**, 89 (1972).

252. R. Z. Gussin and M. A. Ronsberg, *Proc. Soc. Exp. Biol. Med.*, **131**, 1258 (1969).

253. Z. S. Agus and M. Goldberg, *J. Lab. Clin. Med.*, **76**, 280 (1970).

254. *FDC Rep.*, **36**(39), A6 (Sept. 30, 1974).

255. M. Wittner, A. DiStefano, E. Schlatter, J. Delarge and R. Greger, *Pfluger Arch.*, **407**, 611 (1986).

256. M. Lesne, F. Clerck-Braun, F. Duhoux and C. vanYpersele, *Arch. Int. Pharmacodyn.*, **249**, 322 (1981).

257. A. Ghys, J. Denef, J. deSuray, M. Gerin, J. Delarge and J. Willem, *Arzneim-Forsch.*, **35**, 1520 (1985).

258. E. Moller, H. Horstmann, K. Meng and D. Loew, *Experientia*, **33**, 382 (1977).

259. D. Loew and K. Meng, *Pharmatherapeutica*, **1**, 333 (1977).

260. H. J. Kramer, *Pharmatherapeutica*, **1**, 353 (1977).

261. A. Canton, D. Russo and R. Gallo, *Br. Med. J.*, **282**, 595 (1981).

262. W. Ritter, *Clin. Nephrol.*, **19**, 26 (1983).

263. K. J. Berg, S. Jorstad and A. Tromsdal, *Pharmatherapeutica*, **1**, 319 (1977).

264. D. Loew, *Curr. Med. Res. Opin.*, **4**, 455 (1977).

265. M. Mussche and N. Lamerie, *Curr. Med. Res. Opin.*, **4**, 462 (1977).

266. G. E. Stokker, A. A. Deana, S. J. deSolms, E. M. Schultz, R. L. Smith, E. J. Cragoe, Jr., J. E. Baer, C. T. Ludden, H. F. Russo, A. Scriabine, C. S. Sweet and L. S. Watson, *J. Med. Chem.*, **23**, 1414 (1980).

267. G. Stokker, A. Deana, S. deSolms, E. Schultz, R. Smith, E. Cragoe, J. Baer, H. Russo and L. Watson, *J. Med. Chem.*, **25**, 735 (1982).

268. G. Stokker, A. Deana, S. deSolms, E. Schultz, R. Smith and E. Cragoe, *J. Med. Chem.*, **24**, 1063 (1981).

269. R. L. Smith, G. E. Stokker and E. Cragoe, Jr., *J. Med. Chem.*

270. G. Stokker, R. Smith, E. Cragoe, C. Ludden, H. Russo, C. Sweet and L. Watson, *J. Med. Chem.*, **24**, 115 (1981).

271. D. Tocco, G. Stokker, R. Smith, R. Walker, B. Arison and W. Vandenheuvel, *Pharmacologist*, **20**, 214 (1978).

272. M. B. Affrime, D. T. Lowenthal, G. Onesti, P. Busby, C. Swartz and B. Lei, *Clin. Pharmacol. Ther.*, **21**, 97 (1977).

273. D. Lowenthal, G. Onesti, A. Pfrimem, J. Schrogie, K. Kim, D. Busby and R. Swartz, *J. Clin. Pharm.*, **18**, 414 (1978).

274. E. Schlatter, R. Greger and C. Weidtke, *Pflugers Arch.*, **396**, 210 (1983).

275. *Drugs of Fut.*, **2**, 317 (1977).

276. G. Satzinger, *Arzneim-Forsch.*, **27**, 466 (1977).

277. G. Satzinger, *Arzneim-Forsch.*, **27**, 1742 (1977).

278. M. Herrmann, J. Wiegleb and F. Leuschner, *Arzneim-Forsch.*, **27**, 1758 (1977).

279. M. Herrmann, H. Bahrmann, E. Berkenmayer, V. Ganser, W. Heldt and W. Steinbrecher, *Arzneim-Forsch.*, **27**, 1745 (1977).

280. J. Greven and O. Heidenreich, *Arzneim-Forsch.*, **27**, 1755 (1977).

281. V. Gladigau and K. O. Vollmer, *Arzneim-Forsch.*, **27**, 1786 (1977).

282. K. O. Vollmer, A. V. Hodenberg, A. Poission, A. Gladigau and H. Hengy, *Arzneim-Forsch.*, **27**, 1767 (1977).

283. A. V. Hodenberg, K. O. Vollmer, W. Klemisch and B. Liedtke, *Arzneim-Forsch.*, **27**, 1767 (1977).

284. E. Scheitza, *Arzneim-Forsch.*, **27**, 1804, (1977).

285. G. Biamino, *Arzneim-Forsch.*, **27**, 1786 (1977).

286. E. Scheitza, *Arzneim-Forsch.*, **27**, 1804 (1977).

287. Ger. Pat. 2,414,345 (1979), G. Satzinger, M. Herrman and K. Vollmer.

288. J. Greven, W. Pefrain, N. Glaser, K. Maywald and O. Heidenreich, *Pflugers Arch.*, **384**, 57 (1980).

289. J. Greven and O. Heidenreich, *Med. Welt*, **30**, 1014 (1979).

290. V. Gladigau and K. Vollmer, *Arzneim-Forsch.*, **27**, 1785 (1977).

291. P. Corvol, M. Claire, M. Oblin, K. Geering and B. Rossier, *Kidney International*, **20**, 1 (1981).

292. J. A. Cella and C. M. Kagawa, *J. Am. Chem. Soc.*, **79**, 4808 (1957).

293. C. M. Kagawa, J. A. Cella and C. G. VanArman, *Science*, **126**, 1015 (1957).

294. J. A. Cella, E. A. Brown and R. R. Burtner, *J. Org. Chem.*, **24**, 743 (1959).

295. J. A. Cella and R. C. Tweit, *J. Org. Chem.*, **24**, 1109 (1959).

296. E. A. Brown, R. D. Muir and J. A. Cella, *J. Org. Chem.*, **25**, 96 (1960).

297. G. W. Liddle in F. C. Barter, ed., *The Clinical Use of Aldosterone Antagonist*, Charles C. Thomas, Springfield, Ill., 1960, p. 14.

298. H. J. Hess, *J. Org. Chem.*, **27**, 1096 (1962).

299. B. Singer, *Endrocrinology*, **65**, 512 (1959).

300. E. Bolte, M. Verdy, J. Marc-Aurele, J. Brouiller, P. Beauregard and J. Genest, *Can. Med. Assoc. J.*, **79**, 881 (1958).

301. R. M. Salassa, V. R. Mattox and M. H. Power, *J. Clin. Endrocrinol. Metab.*, **18**, 787 (1958).

302. G. W. Liddle, *Arch. Intern. Med.*, **102**, 998 (1958).

303. J. D. H. Slater, A. Moxham, R. Hunter and J. D. N. Nabarro, *Lancet*, **1958–11**, 931.

304. D. N. S. Kerr, A. E. Read, R. M. Haslam and S. Sherlock, *Lancet*, **1959–11**, 1084.

305. G. W. Little, *Science*, **126**, 1016 (1957).

306. E. J. Ross and J. E. Bethune, *Lancet*, **1959–1**, 127.

307. R. C. Tweit, F. B. Colton, N. L. McNiven and W. Klyne, *J. Org. Chem.*, **27**, 3325 (1962).

308. J. A. Cella in J. H. Moyer and M. Fuchs, eds., *Edema*, Saunders, Philadelphia, 1960, p. 303.

309. N. W. Atwater, R. H. Bible, Jr., W. A. Brown, R. R. Burtner, J. S. Mihina, L. N. Nysted and P. B. Sollman, *J. Org. Chem.*, **26**, 3077 (1961).

310. L. N. Nysted and R. R. Burtner, *J. Org. Chem.*, **27**, 3175 (1962).

311. A. A. Patchett, F. Hoffman, F. F. Giarrusso, H. Schwam and G. E. Arth, *J. Org. Chem.*, **27**, 3822 (1962).

312. G. deStevens, *Diuretics: Chemistry and Pharmacology*, Academic Press, New York, 1963, p. 130.

313. M. Haberey, P. Buse, W. Losert and Y. Nishino, *Naunyn-Schmiedeberg's Arch. Pharmacol.*, **334** (Suppl.), Abst. 109 (1986).

314. M. Hildebrand and W. Seifert, *3rd World Conf. Clin. Pharmacol. Ther. (July 27–Aug 1)*, Stockholm, Abst. 138 (1986).

315. W. Losert, D. Buse, J. Casais-Stenzel, M. Haberey, H. Laurent, K. Nickish, E. Schillinger and R. Wiechert, *Arzneim-Forsch.*, **36**, 1583 (1986).

316. A. Karim and E. A. Brown, *Steroids*, **20**, 41 (1972). L. J. Chinn, E. A. Brown, S. S. Mizuba and A. Karim, *J. Med. Chem.*, **20**, 352 (1977).

317. U. Abshagen, H. Rennekamp, K. Koch, M. Senn and W. Steingross, *Steroids*, **28**, 467 (1976).

318. C. M. Kagawa, D. J. Bouska, M. L. Anderson and W. F. Krol, *Arch. Int. Pharmacodyn.*, **149**, 8 (1964).

319. L. M. Hofmann, L. J. Chinn, H. A. Pedrera, M. I. Krupnick and O. D. Suleymanov, *J. Pharmacol. Exp. Ther.*, **194**, 450 (1975).

320. J. W. Funder, J. Mercer and J. Hood, *Clin. Sci. Mol. Med.*, **51** (Suppl. 3), 333 (1976).

321. L. Ramay, I. Harrison, J. Shelton and M. Tidd, *Clin. Pharmacol. Ther.*, **18**, 391 (1975).

322. R. M. Weier and L. M. Hofmann, *J. Med. Chem.*, **18**, 817 (1975).

323. L. M. Hofmann, R. M. Weier, O. D. Suleymanov and H. A. Pedrera, *J. Pharmacol. Exp. Ther.*, **201**, 762 (1977).

324. L. J. Chinn and B. N. Desai, *J. Med. Chem.*, **18**, 268 (1975).

325. L. J. Chinn and L. M. Hofmann, *J. Med. Chem.*, **16**, 839 (1973).

326. L. J. Chinn, H. L. Dryden, Jr. and R. R. Burtner, *J. Org. Chem.*, **26**, 3910 (1961).

327. L. J. Chinn, *J. Org. Chem.*, **27**, 1741 (1962).

328. I. Weiner in L. S. Goodman and A. Gilman, *The Pharmacological Basis of Therapeutics*, 8th ed., Pergamon Press, New York 1990, pp. 713–731.

329. E. J. Ross, *Clin. Pharmacol. Ther.*, **6**, 65 (1965).

330. D. Loriaux, R. Menard, A. Taylor, J. Pita and R. Santin, *Ann. Int. Med.*, **85**, 630 (1976).

331. F. A. Kuehl, Jr., J. L. Humes, R. W. Egan, E. A. Ham, G. C. Beveridge and C. G. Van Arman, *Nature*, **265**, 170 (1977).

332. J. C. Frolich, T. W. Wilson, B. J. Sweetman, M. Smigel, A. S. Nies, K. Carr, J. T. Watson and J. A. Oates, *J. Clin. Invest.*, **55**, 763 (1975).

333. H. Erbler, *Naunyn-Schmeid. Arch. Pharmacol.*, **273**, 366 (1972).

334. B. Aupetit, J. Duchier and J. Legrand, *Ann. Endocrinol.*, **39**, 355 (1978).

335. J. Greiner, R. Kramer, J. Jarrel and H. Colby, *J. Pharmacol. Exp. Ther.*, **198**, 709 (1976).

336. C. Persson, I. Erjefalt, L. Edholm, J. Karlsson and C. Lamm, *Life Sci.*, **31**, 2673 (1982).

337. H. Osswald, *Naunyn-Schmied. Arch. Pharmacol.*, **288**, 79 (1975).

338. W. Spielman and L. Arend, *Hypertension*, **17**, 117 (1991).

339. M. Collis, G. Baxter and J. Keddie, *J. Pharm. Pharmacol.*, **38**, 850 (1986).

340. M. Collis, G. Shaw and J. Keddie, *J. Pharm. Pharmacol.*, **43**, 138 (1991).

341. J. Shimada, F. Suzuki, H. Nonaka and A. Ishii, *J. Med. Chem.*, **35**, 924 (1992).

342. J. Shimada, F. Suzuki, H. Nonaka, A. Karasawa, H. Mizumoto, T. Ohno, K. Kubo and A. Ishii, *J. Med. Chem.*, **34**, 469 (1991).

343. R. Knight, C. Bowmer and M. Yates, *Br. J. Pharmacol.*, **109**, 272 (1993).

344. F. Suzuki, J. Shimada, H. Mizumoto, A. Karasawa, K. Kubo, H. Nonaka, A. Ishii and T. Kawakita, *J. Med. Chem.*, **35**, 3066 (1992).

345. H. Mizumoto, A. Karasawa and K. Kubo, *J. Pharmacol. Exp. Ther.*, **266**, 200 (1993).

346. T. Kobayashi, H. Mizumoto and A. Karasawa, *Biol. Pharm. Bull.*, **16**, 1231 (1993).

347. T. Kobayashi, H. Mizumoto, A. Karasawa and K. Kubo, *Jpn. J. Pharmacol.*, **58** (Suppl. 1), 195 (1992).

348. H. Kusaka and A. Karasawa, *Jpn. J. Pharm.*, **63**, 513 (1993).

349. H. Mizumoto, T. Kobayashi, A. Karasawa,. H. Nonaka, A. Ishii, K. Kubo, J. Shimada and F. Suzuki, *Jpn. J. Pharm.*, **58** (Suppl. 1), 194 (1992).

350. K. Nagashima, H. Kusaka, K. Sato and A. Karasaw, *Jpn. J. Pharm.*, **64**, 9 (1994).

351. H. Mizumoto and A. Karasawa, *Jpn. J. Pharm.*, **61**, 251 (1993).

352. V. Papesch and E. F. Schroeder, *J. Org. Chem.*, **16**, 1879 (1951).

353. A. Kattus, T. M. Arrington and E. V. Newman, *Am. J. Med.*, **12**, 319 (1952).

354. E. J. Cragoe, Jr., O. B. Woltersdorf, Jr., J. B. Bicking, S. F. Kwong and J. H. Jones, *J. Med. Chem.*, **10**, 66 (1967).

355. O. Clauder and G. Bulcsu, *Magy. Kem. Foly.*, **57**, 68 (1951); *Chem Abst.*, **46**, 4023 (1952).

356. K. H. Beyer and J. F. Baer in E. Jucker, ed., *Progress in Drug Research*, Birkhauser, Basle-Stuttgard, Vol. 2, 1960, p. 21.

357. J. Weinstock, J. W. Wilson, V. D. Wiebelhaus, A. R. Maass, F. T. Brennan and G. Sosnowski, *J. Med. Chem.*, **11**, 573 (1968).

358. M. A. Ronsberg, A. Z. Gussin, E. H. Stokey and P. S. Chan, *Pharmacologist*, **18**, 150 (1976).

359. W. L. Lipschitz and Z. Hadidian, *J. Pharmacol. Exp. Ther.*, **81**, 84 (1944).

360. H. Ludwig, *Schweiz. Med. Wochenschr.*, **76**, 822 (1946).

361. V. Papesch and E. F. Schroeder in F. F. Blicke and R. H. Cox, eds., *Medicinal Chemistry*, John Wiley & Sons, Inc., New York, Vol. 111, 1956, p. 175.

362. T. Turchetti, *Riforma Med.*, **64**, 405 (1950).

363. E. V. Newman, J. Franklin and J. Genest, *Bull. Johns Hopkins Hosp.*, **82**, 409 (1948).

364. G. Szabo, O. Clauder and Z. Magyar, *Magy. Belorv. Arch.*, **6**, 156 (1953).

365. C. M. Kagawa and C. G. Van Arman, *J. Pharmacol. Exp. Ther.*, **124**, 318 (1958).

366. D. V. Miller and R. V. Ford, *Am. J. Med. Sci.*, **236**, 32 (1958).

367. R. V. Ford, J. B. Rochelle, A. C. Bullock, C. L. Spurr, C. Handley and J. H. Moyer, *Am. J. Cardiol.*, **3**, 148 (1959).

368. M. H. Sha, M. Y. Mhasalker and C. V. Deliwala, *J. Sci. Ind. Res.* (India), **19c**, 282 (1960); D. J. Mehta, U. K. Sheth and C. V. Deliwala, *Nature* (London), **187**, 1034 (1960).

369. K. N. Modi, C. V. Deliwala and U. K. Sheth, *Arch. Int. Pharmacodyn.*, **151**, 13 (1964).

370. L. Szabo, L. Szporny and O. Clauder, *Acta Pharm. Hung.*, **31**, 163 (1961); *Chem. Abstr.*, **55**, 24780i (1961).

371. W. B. McKeon, Jr., *Arch. Int. Pharmacodyn.*, **151**, 225 (1964).

372. D. A. LeSher and F. E. Shideman, *J. Pharmacol. Exp. Ther.*, **116**, 38 (1956).

373. H. E. Williamson, F. E. Shideman and D. A. LeSher, *J. Pharmacol. Exp. Ther.*, **126**, 82 (1959).

374. M. Burg and S. Sariban-Sohraby in J. B. Puschett, ed., *Diuretics, Chemistry, Pharmacology and Clinical Application*, Elsevier, New York, 1984, pp. 329–334.

375. J. E. Baer in A. F. Lant and G. M. Wilson, eds., *Modern Diuretic Therapy*, Excerpta Medica, Amsterdam, 1973, p. 148.

376. V. D. Wiebelhaus, J. Weinstock, A. R. Maass, F. T. Brennan, G. Sosnowski and T. Larsen, *J. Pharmacol. Exp. Ther.*, **149**, 397 (1965).

377. T. S. Osdene, A. A. Santilli, L. E. McCardle and M. E. Rosenthale, *J. Med. Chem.*, **9**, 697 (1966).

378. T. S. Osdene, A. A. Santilli, L. E. McCardle and M. E. Rosenthale, *J. Med. Chem.*, **10**, 165 (1967).

379. M. E. Rosenthale and C. G. Van Arman, *J. Pharmacol. Exp. Ther.*, **142**, 111 (1963).

380. V. D. Wiebelhaus, J. Weinstock, F. T. Brennan, G. Sosnowski and T. J. Larsen, *Fed. Proc.*, **20**, 409 (1961).

381. A. P. Crosley, Jr., L. Ronquillo and F. Alexander, *Fed. Proc.*, **20**, 410 (1961).

382. J. H. Laragh, E. B. Reilly, T. B. Stites and M. Angers, *Fed. Proc.*, **20**, 410 (1961).

383. V. D. Wiebelhaus, J. Weinstock, F. T. Brennan, G. Sosnowski, T. Larsen and K. Gahagan, *Pharmacologist*, **3**(2), 59 (1961).

384. W. Schaumann, *Klin. Wochenschr.*, **40**, 756 (1962).

385. W. I. Baba, G. R. Tudhope and G. M. Wilson, *Brit. Med. J.*, **2**, 756 (1962).

386. G. W. Liddle, *Metab. Clin. Exp.*, **10**, 1021 (1961).

387. G. M. Ball and J. A. Greene, Jr., *Proc. Soc. Exp. Biol. Med.*, **113**, 326 (1963).

388. J. Crabbe, *Arch. Int. Pharmacodyn. Ther.*, **173**, 474 (1968).

389. J. Gatry, *J. Pharmacol. Exp. Ther.*, **176**, 586 (1971).

390. A. P. Crosley, Jr., L. Ronquillo, W. S. Strickland and F. Alexander, *Ann. Intern. Med.*, **56**, 241 (1962).

391. D. J. Ginsberg, A. Saad and G. J. Gabuzda, *N. Engl. J. Med.*, **271**, 1229 (1964).

392. E. Mutschler, H. Gilfrich and H. Knauf, *Clin. Exp. Hyperten.*, **4**, 249 (1983).

393. H. Gilfrich, G. Kremer and W. Mohrke, *Eur. J. Clin. Pharmacol.*, **25**, 237 (1983).

394. P. Baume, F. J. Radcliffe and C. R. Corry, *Am. J. Med. Sci.*, **245**, 668 (1963).

395. W. R. Cattell and C. W. H. Havard, *Brit. Med. J.*, **2**, 1362 (1962).

396. R. A. Thompson and M. F. Crowley, *Postgrad. Med.* (Oxford), **41**, 706 (1965).

397. K. Nishikawa, H. Shimakawa, Y. Inada, Y. Shibouta, S. Kikuchi, S. Yurugi and Y. Oka, *Chem. Pharm. Bull.*, **24**, 2057 (1976).

398. K. Nishikawa and S. Kikuchi, *Jpn. J. Pharmacol.*, **22** (Suppl.), 103 (1972); Y. Nakai, Y. Shirakawa and T. Fujita, *Jpn. J. Pharmacol.*, **22** (Suppl.), 102 (1972).

399. H. Kawaki, R. Tsukuda, K. Nishikawa, S. Kikuchi and T. Hirano, *J. Takeda Res. Labs.*, **32**, 299 (1973); Y. Inada, K. Nishikawa, A. Nagaoka and S. Kikuchi, *Arzneim-Forsch.*, **27**, 1663 (1977).

400. Davis, R. Gedir, E. Hawes and G. Wibberley, *Eur. J. Med. Chem.*, **20**, 381 (1985).

401. H. Parish, R. Gilliom, W. Purcell, R. Browne, R. Spirk and H. White, *J. Med. Chem.*, **25**, 98 (1982).

402. A. Monge, V. Martinez-Merion, M. Simon and C. Sanmartin, *Arzneim-Forsch.*, **43**, 1322 (1993).

403. J. Hester, S. Luden, D. Emmert and B. West, *J. Med. Chem.*, **32**, 1157 (1989).

404. J. B. Bicking, J. W. Mason, O. W. Woltersdorf, Jr., J. H. Jones, S. F. Kwong, C. M. Robb and E. J. Cragoe, Jr., *J. Med. Chem.*, **8**, 638 (1965).

405. J. B. Bicking, C. M. Robb, S. F. Kwong and E. J. Cragoe, Jr., *J. Med. Chem.*, **10**, 598 (1967).

406. J. H. Jones, J. B. Bicking and E. J. Cragoe, Jr., *J. Med. Chem.*, **10**, 899 (1967).

407. J. Watthey, M. Desai, R. Rutledge and R. Dotson, *J. Med. Chem.*, **23**, 690 (1980).

408. T. Russ, W. Ried, F. Ullrich and E. Mutschler, *Arch. Pharm.*, **325**, 761 (1992).

409. S. Kau, B. Howe, J. Li, L. Smith, J. Keddie, J. Barlow, R. Giles and M. Goldberg, *J. Pharmacol Exp. Ther.*, **242**, 818 (1987).

410. S. Kau, P. Johnson, J. Li, J. Zuzack, K. Leszcznskak, C. Yochim, J. Schwartz and R. Giles, *Meth. Find. Exp. Clin. Pharmacol.*, **15**, 357 (1993).

411. B. Beerman and M. Groschinsky-Grind, *Clin. Pharmacokin.*, **5**, 221 (1980).

412. R. Z. Gussin, M. A. Ronsberg, E. H. Stokey and J. R. Cummings, *J. Pharmacol. Exp. Ther.*, **195**, 8 (1975).

413. V. Ackermann, *Clin. Chem.*, **32**, 241 (1986).

414. A. Raine, J. Firth and J. Ledingham, *Clin. Sci.*, **76**, 1 (1989).

415. T. Yandle, A. Richards, M. Nicholls, R. Cuneo, E. Espiner and J. Livesey, *Life Sci.*, **38**, 827 (1986).

416. F. Luft, R. Lang and G. Aronoff, *J. Pharmacol. Exp. Ther.*, **236**, 416 (1986).

417. M. Cogan, *Am. J. Physiol.*, **250**, F710 (1986).

418. M. Camargo, S. Atlas and T. Maack, *Life Sci.*, **38**, 2397 (1986).

419. A. Kenny and S. Stephenson, *Febs Lett.*, **232**, 1 (1988).

420. W. Oelkers, S. Kleiner and V. Bahr, *Hypertension*, **12**, 462 (1988).

421. R. Cuneo, E. Espiner, M. Nichols, T. Yandle and J. Livesey, *J. Clin. Endocrinol. Metab.*, **65**, 765 (1986).

422. K. Atarashi, P. Mulrow and R. Franco-Saenz, *J. Clin. Invest.*, **76**, 1807 (1985).

423. J. Almenoff and M. Orlowski, *Biochemistry*, **22**, 590 (1983).

424. S. Stephenson and A. Kenny, *J. Biochem.*, **241**, 237 (1987).

425. T. Maack, F. Almeida, M. Suzuki and D. Nussenzveig, *Contrib. Nephrol.*, **68**, 58 (1988).

426. P. Nussenzveig, J. Lewicki and T. Maack, *J. Biol. Chem.*, **265**, 20952 (1990).

427. P. Leitman, J. Resen, T. Kuno, Y. Kamisaki, J. Chang and F. Munad, *J. Biol. Chem.*, **261**, 11650 (1986).

428. J. Koepke, L. Tyler, A. Trapani, P. Bovy, K. Spear, G. Olins and E. Blaine, *J. Pharmacol. Exp. Ther.*, **249**, 172 (1989).

429. J. Okolicany, G. McEnroe, L. Gregory, J. Lewicki and T. Maack, *Can. J. Physiol.*, **69**, 1561 (1991).

430. T. Maack, M. Suzuki, F. Almeida, P. Nussenzveig, R. M. Scarborough, G. McEnroe and J. Lewicki, *Science*, **238**, 675 (1987).

431. S. Vemulapalli, P. Chiv, A. Brown, Grisctik and E. Sybertz, *Life Sci.*, **49**, 383 (1991).

432. R. Webb, G. Yasay, C. McMartin, R. McNeal and M. Zimmerman, *J. Cardiovas. Pharmacol.*, **14**, 285 (1989).

433. H. Lafferty, M. Gunning, P. Silva, M.

Zimmerman, B. Brenner and S. Anderson, *Circ. Res.*, **65**, 640 (1989).

434. A. Trapani, G. Smits, D. McGraw, K. Spear, S. Koepke, G. Olins and E. Blaine, *J. Cardiovas. Pharmacol.*, **14**, 419 (1989).

435. M. Fournie-Zaluski, E. Lucas, G. Waksman and B. Roques, *Eur. J. Biochem.*, **139**, 267 (1984).

436. A. Seymour, S. Fennell and J. Swerdel, *Hypertension*, **14**, 87 (1989).

437. A. Seymore, J. Norman, M. Asaad, S. Fennel, J. Swerdel, D. Little and C. Dorso, *J. Cardiovasc. Pharmacol.*, **16**, 163 (1990).

438. A. Seymore, J. Norman, M. Assaad, S. Fennel, D. Little, V. Kratunis and W. Rogers, *J. Cardiovasc. Pharmacol.*, **17**, 296 (1991).

439. S. Vemulapalli, P. J. S. Chiu, R. W. Watkins, C. Foster and E. J. Sybertz, *Am. J. Hypertens.*, **4**, 15A–16A (1991).

440. R. W. Watkins, P. J. S. Chiu, S. Vemulapalli, C. Foster, M. Chatterjee, E. M. Smith, B. Neustadt, M. Hastlanger and E. Sybertz, *Am. J. Hyperten.*, **4**, 32A (1991).

441. A. M. Richards, I. Crozier, T. Kosoglou, M. Rallings, E. Espiner, M. G. Nicholls, T. Vandle, H. Ikram and C. Frampton, *Hypertension*, **22**, 119 (1993).

442. J. Danilewicz, P. Barclay, I. Barnish, P. Brown, S. Campbell, K. James, G. Samuels, N. Terrett and M. Wythes, *Biochem. Biophys. Res. Comm.*, **164**, 58 (1989).

443. J. O'Connell, A. Jardine and G. Davidson, *J. Hypertension*, **10**, 271 (1992).

444. E. Sybertz, *J. Pharmacol. Exp. Ther.*, **250**, 624 (1989).

445. K. Helin, I. Tikkanen, T. Tikkanen, O. Saijonmaa, E. Sybertz, S. Vemulapall, H. Sariolatt and F. Fyhrquist, *Eur. J. Pharmacol.*, **198**, 23 (1991).

446. M. Fitzpatrick, M. Rademaker, C. Charles, T. Vandle, E. Espiner, H. Ikram and E. Sybertz, *J. Cardiovas. Pharm.*, **19**, 635 (1992).

447. S. Vemulapalli, P. Chiu, A. Brown, K. Griscti and E. Sybertz, *Life Sci.*, **49**, 383 (1991).

448. M. Burnier, M. Ganslmayer, F. Perret, M. Porchet, T. Kosoglou, A. Gould, J. Nussberger, J. Waeber and H. Brunner, *Clin. Pharmacol. Ther.*, **50**, 181 (1991).

449. M. Fournie-Zaluski, A. Coulaud, R. Bouboton, P. Chaillet, J. Devin, G. Waksman, J. Costentin and B. Roques, *J. Med. Chem.*, **28**, 1158 (1985).

450. G. Olins, P. Krieter, A. Trapani, K. Spear and P. Bovy, *Mol. Cell. Endocrinol.*, **61**, 201 (1989).

451. M. Altstein, S. Blumberg and Z. Vogel, *Eur. J. Pharm.*, **76**, 299 (1982).

452. S. DeLombaert, M. Erion, J. Tan, L. Blanchard, L. El-Chehabi, R. Ghai, C. Berry and A. Trapani, *J. Med. Chem.*, **37**, 498 (1994).

453. A. J. Trapani, M. E. Beil, D. T. Cote, S. DeLombaert, M. D. Erion, T. E. Gerlock, R. D. Ghai, M. F. Hopkins, J. V. Peppard, R. L. Webb, R. W. Lappe and M. Worcel, *J. Cardiovas. Pharm.*, **23**, 358 (1994).

454. R. Walter and P. L. Hoffman in J. R. Brobeck, ed., *Best and Taylor's Physiological Basis of Medical Practice*, 9th ed., Section 1, (6), Williams and Wilkins, Baltimore, 1973.

455. F. G. W. Marson, *Lancet*, **1955–11**, 360.

456. A. B. Gutman, *Advanc. Pharmacol.*, **4**, 91 (1966).

457. R. Pfister and F. Hafliger, *Helv. Chim. Acta.*, **44**, 232 (1961).

458. T. F. Yu and A. B. Gutman, *Am. J. Med.*, **37**, 885 (1964).

459. R. Heel, R. Brogden, T. Speight and G. Avery, *Drugs*, **14**, 349 (1977).

460. K. H. Beyer, H. F. Russo, E. K. Tillson, A. K. Miller, W. F. Verwey and S. R. Gass, *Am. J. Physiol.*, **166**, 625 (1951).

461. K. H. Beyer, *Arch. Int. Pharmacodyn.*, **98**, 97 (1954).

462. P. Brazeau in L. S. Goodman and A. Gilman, eds., *The Pharmacological Basis of Therapeutics*, 5th ed., Section VIII, Macmillan, New York, 1975, p. 860.

463. J. J. Burns, T. F. Yu, A. Ritterband, J. M. Perel, A. B. Gutman and B. B. Brodie, *J. Pharmacol. Exp. Ther.*, **119**, 418 (1957).

464. T. F. Yu, J. J. Burns and A. B. Gutman, *Arth. Rheum.*, **1**, 352 (1958).

465. J. J. Burns, T. F. Yu, P. Dayton, L. Berger, A. B. Gutman and B. B. Brodie, *Nature*, **182**, 1162 (1958).

466. A. B. Gutman, P. G. Dayton, T. F. Yu, L. Berger, W. Chen, L. E. Sicam and J. J. Burns, *Am. J. Med.*, **29**, 1017 (1960).

467. K. C. Blanchard, D. Maroske, D. G. May and I. M. Weiner, *J. Pharmacol. Exp. Ther.*, **180**, 397 (1972).

468. A. H. Anton, *J. Pharmacol. Exp. Ther.*, **134**, 291 (1961).

469. R. W. Rundles, E. N. Metz and H. R. Silberman, *Ann. Intern. Med.*, **64**, 229 (1966).

470. D. M. Woodbury and E. Fingl in L. S. Goodman and A. Gilman, eds., *The Pharmacological Basis of Therapeutics*, 5th ed., Section 11, Macmillan, New York, 1975, p. 352.

471. T. Yu, *J. Rheumatology*, **3**, 305 (1976).

472. F. Matzkies, F. Berg and R. Minzlaff, *Fortschritte de Medizin*, **95**, 1748 (1977).

473. N. Zoller, W. Dofel and W. Grobner, *Klinische Wochenschrift*, **48**, 426 (1970).

474. R. Kramp, *Eur. J. Clin. Invest.*, **3**, 245 (1973).

475. R. Kramer, M. Muller, *Experientia*, **29**, 391 (1973).

476. J. Broekhuysen, M. Pacco, R. Sion, L. Demeulenaere and M. van Hee, *Eur. J. Clin. Pharm.*, **4**, 125 (1972).

PART III
CHEMOTHERAPEUTIC AGENTS

CHAPTER THIRTY-TWO

Aminoglycoside, Macrolide, Glycopeptide, and Miscellaneous Antibacterial Antibiotics

HERBERT A. KIRST

Lilly Research Laboratories
Greenfield, Indiana, USA

CONTENTS

1 Introduction, 464
2 Aminoglycoside Antibiotics, 465
  2.1 Streptomycin, 466
  2.2 4,5-Di-O-glycosyl-2-deoxystreptamine aminoglycosides, 467
  2.3 4,6-Di-O-glycosyl-2-doexystreptamine aminoglycosides, 469
  2.4 Semisynthetic derivatives of 2-deoxystreptamine aminoglycosides, 471
  2.5 Monosubstituted 2-deoxystreptamine aminoglycosides, 472
  2.6 Other aminocyclitol aminoglycosides, 473
  2.7 Miscellaneous aminoglycosides, 475
  2.8 Summary of the aminoglycosides, 476
3 Macrolide Antibiotics, 477
  3.1 12-Membered ring macrolides, 478
  3.2 14-Membered ring macrolides, 478
  3.3 Semisynthetic derivatives of 14-membered ring macrolides, 480

Burger's Medicinal Chemistry and Drug Discovery,
Fifth Edition, Volume 2: Therapeutic Agents,
Edited by Manfred E. Wolff.
ISBN 0-471 57557-7 © 1996 John Wiley & Sons, Inc.

3.4 16-Membered ring macrolides, 484
3.5 Semisynthetic derivatives of 16-membered ring macrolides, 487
3.6 Summary of the macrolides, 489
4 Lincosaminide and Streptogramin Antibiotics, 490
4.1 Lincomycin family, 490
4.2 Streptogramins, 491
5 Glycopeptide Antibiotics, 492
5.1 Vancomycin family, 493
5.2 Teicoplanin family, 494
5.3 Ristocetin family, 495
5.4 Avoparcin family, 496
5.5 Summary of the glycopeptides, 496
6 Peptide Antibiotics, 498
6.1 Bacitracin, 498
6.2 Tyrothricin and gramicidins, 498
6.3 Polymyxin and colistin, 500
6.4 Miscellaneous lipopeptide antibiotics, 500
6.5 Glycolipopeptides: ramoplanin, 501
6.6 Thiostrepton and related thiopeptides, 501
6.7 Nisin and other lantibiotics, 502
6.8 Semisynthetic derivatives: daptomycin, 503
7 Miscellaneous Fermentation-Derived Antibiotics, 503
7.1 Fusidic acid, 504
7.2 Mupirocin, 504
7.3 Everninomicin and avilamycin, 505
7.4 Bicyclomycin, 505
7.5 Pleuromutilin and tiamulin, 507
7.6 Nucleoside antibiotics, 508
7.7 Novobiocin, 509
8 Conclusion, 510

1 INTRODUCTION

One of the more significant achievements of the 20th century has been the discovery and commercial development of the numerous therapeutic agents that now provide reliably effective treatments for many infectious diseases that had previously caused extensive mortality, morbidity, and fear. Medicinal chemists have played an important role in discovering and synthesizing these antibiotics. The remarkable successes in this endeavor are such that this century may be regarded as the "Antimicrobial Era," although concerns about the increasing prevalence of antibiotic-resistant pathogens have suggested that we may

already be entering a "Post-Antimicrobial Era" (1–3).

Modern antibiotics have been obtained from several sources, but fermentation-derived natural products have been especially important. Beginning with the discoveries of Drs. Rene Dubos and Alexander Fleming, secondary metabolites produced by soil organisms under defined aerobic growth conditions became an abundant source of new antimicrobial substances during the past half-century. The principles underlying the discovery of penicillin were subsequently recognized and applied to the search for other classes of antibiotics produced by other organisms, creating an antibiotic industry and establishing antibiotics as a critical part of modern chemotherapy in both human and veterinary medicine (4,5). An early pioneer in screening programs was Dr. Selman Waksman, whose many discoveries included streptomycin in 1944, the first widely useful antibiotic obtained from random screening of culture broths. From these beginnings, the number of distinct chemical entities from fermentation has been estimated to be many thousands (6–9).

Fermentation products have also provided a wide array of starting materials for medicinal chemists to use in preparing semisynthetic derivatives and learning structure-activity relationships (10–14). These secondary metabolites are usually difficult to obtain from any alternative source in any useful quantity or reasonable cost. Among the synthetic challenges facing chemists working with fermentation products are their highly complex structures, extensive stereochemistry, multiple functional groups, and chemical instabilities. However, advances in spectroscopy and X-ray crystallography now permit elucidation of structures whose complexity had previously defied solution. Furthermore, medicinal chemists have generally overcome the complex synthetic challenges through the knowlege and insight gained after years of experience with the molecules and through the discovery and application of milder and more selective chemical transformations.

This chapter covers some of the most important fermentation-derived antibiotics and their semisynthetic derivatives that exhibit antibacterial activity. β-Lactam antibiotics, tetracyclines, and fermentation products having activities other than antibacterial (antifungal, antiviral, antiprotozoal, antituberculosis, antitumor) are not included here, since they are covered in the chapters for those activities. Even after all of these exclusions, far too many antibacterial agents have been isolated to permit an all-inclusive compilation, so this chapter is limited to those that are used in chemotherapy or as research tools, have some historical significance, or illustrate the diversity of structures and biosynthetic pathways within antibiotic-producing organisms.

Fermentation products have often been isolated by different research groups, so different names have been given to identical compounds. Duplication may arise because identical compounds are produced by different genera or species of microorganisms. Cultures often yield multiple related factors, whose numbers increase as more minor factors are isolated and as mutant strains of the producing organism are generated and analyzed for changes in secondary metabolite production. On the other hand, different strains of the same species may produce unrelated structures, so one-to-one correlations do not necessarily exist between secondary metabolite structures and species of producing organisms. For simplicity, this chapter uses the most common names of compounds and producing organisms that have become generally accepted.

2 AMINOGLYCOSIDE ANTIBIOTICS

The most important aminoglycosides contain an aminocyclitol moiety (Fig. 32.1) to which aminosugars are glycosidically linked, so they may be more correctly called aminocyclitol antibiotics. However, aminoglycoside is the name traditionally applied to this class whose structures contain aminosugars and/or aminocyclitols as principal components. Within the wide diversity of saccharide and cyclitol constituents are common features, such as, multiple amino and hydroxyl groups that render them strongly basic molecules and confer good water solubility.

Many books and comprehensive reviews cover the chemistry, microbiology, pharmacology, and medical use of aminoglycosides (15–25). Even though many of them were published 10–15 years ago, much of their

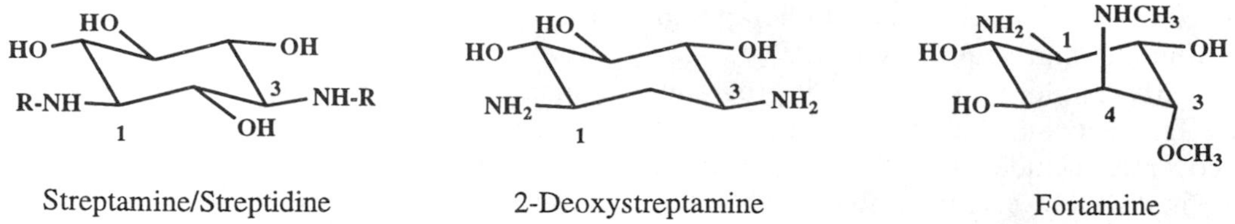

Fig. 32.1 Structures of representative aminocyclitols. Streptamine [R = H], streptidine [R = C(=NH)NH₂], 2-deoxystreptamine, and fortamine.

material describing structures, synthesis, structure–activity relationships, antimicrobial profiles, pharmacological features, and clinical experience is still generally applicable.

2.1 Streptomycin

Streptomycin, the first aminoglycoside to be discovered, was isolated from culture broths of *Streptomyces griseus* (26). It contains the aminocyclitol streptidine, which is substituted on its C-4 hydroxyl group by

the branched sugar, α-L-streptose, which itself is substituted on its C-2′ hydroxyl group by α-L-N-methylglucosamine (Fig. 32.2) (27,28). The discovery of streptomycin was a key factor in validating the concept of fermentation products as a productive source of new antibiotics (29). At the time of its discovery, its gram-negative spectrum complemented the gram-positive spectrum of penicillin and its *in vitro* activity against tubercle bacilli offered promise for treating tuberculosis. However, microbial resistance rapidly developed to the extent that its use as a single agent was

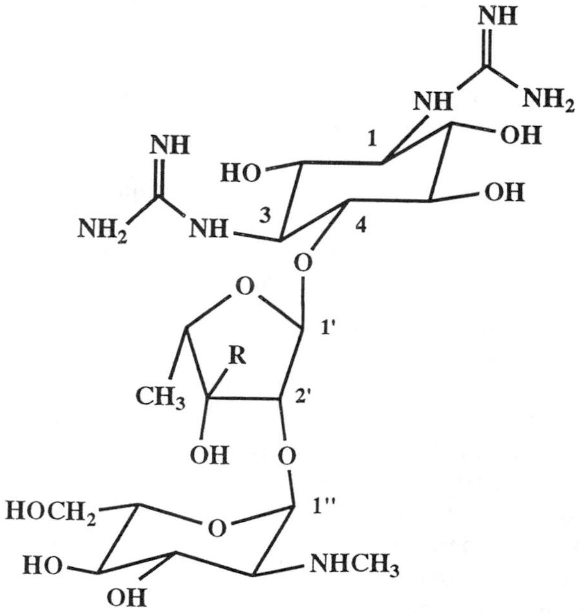

Fig. 32.2 Structures of streptomycin (R = CHO) and dihydrostreptomycin (R = CH₂OH).

obviated and it is now generally used in combination with other antibiotics.

After the isolation of streptomycin, related compounds such as bluensomycin (glebomycin) were isolated from other culture broths. In addition, many semisynthetic derivatives were prepared in efforts to improve the antimicrobial properties and overcome the liabilities of streptomycin (24,25,30–33). Of these, the most extensively evaluated derivative has been dihydrostreptomycin (Fig. 32.2), which was obtained from both chemical synthesis and fermentation sources (*S. humidus*) (34–37). However, its clinical utility declined after it was found to have a greater potential than streptomycin to cause delayed ototoxicity (deafness) (38). Very little research has been recently directed toward the synthesis and evaluation of new derivatives of streptomycin.

2.2 4,5-Di-*O*-Glycosyl-2-Deoxystreptamine Aminoglycosides

Most useful aminoglycosides contain 2-deoxystreptamine (2-DOS) as a core component that is variously substituted by saccharide entities. One large family comprises compounds in which the hydroxyl groups at C-4 and C-5 of 2-DOS are glycosylated. Among the more important members of this family are the pseudotetrasaccharides: neomycin, paromomycin, and lividomycin (Fig. 32.3); and the pseudotrisaccharides: ribostamycin, xylostasin, and butirosin (Fig. 32.4).

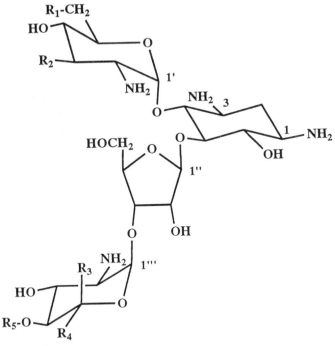

Fig. 32.3 Structures of 4,5-di-*O*-glycosyl-2-DOS pseudotetrasaccharides. Neomycin B: $R_1 = NH_2$, $R_2 = OH$, $R_3 = CH_2NH_2$, $R_4 = H$, $R_5 = H$; neomycin C: $R_1 = NH_2$, $R_2 = OH$, $R_3 = H$, $R_4 = CH_2NH_2$, $R_5 = H$; paromomycin I: $R_1 = OH$, $R_2 = OH$, $R_3 = CH_2NH_2$, $R_4 = H$, $R_5 = H$; paromomycin II: $R_1 = OH$, $R_2 = OH$, $R_3 = H$, $R_4 = CH_2NH_2$, $R_5 = H$; lividomycin A: $R_1 = OH$, $R_2 = H$, $R_3 = CH_2NH_2$, $R_4 = H$, $R_5 = \alpha$-D-mannosyl; lividomycin B: $R_1 = OH$, $R_2 = H$, $R_3 = CH_2NH_2$, $R_4 = H$, $R_5 = H$.

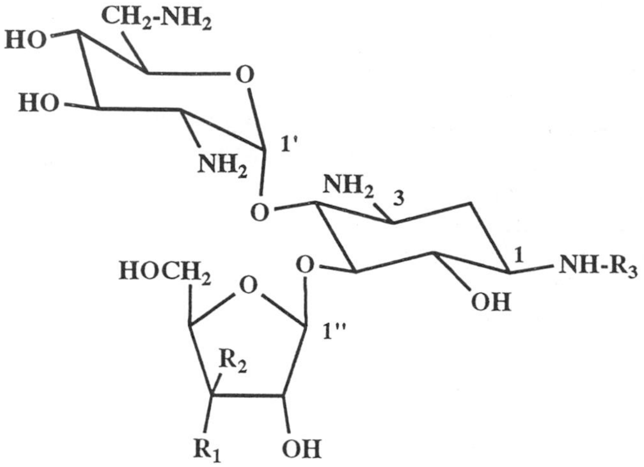

Fig. 32.4 Structures of 4,5-di-O-glycosyl-2-DOS pseudotrisaccharides. Ribostamycin: $R_1 = OH$, $R_2 = H$, $R_3 = H$; xylostasin: $R_1 = H$, $R_2 = OH$, $R_3 = H$; butirosin A: $R_1 = H$, $R_2 = OH$, $R_3 = \alpha$-hydroxy-γ-aminobutyryl; butirosin B: $R_1 = OH$, $R_2 = H$, $R_3 = \alpha$-hydroxy-γ-aminobutyryl.

The neomycin complex, isolated from culture broths of *S. fradiae*, was the first series of aminoglycosides to be discovered that contained 2-DOS (39). The major factors are neomycin B and C, which differ only in their stereochemistry at C-5‴ (Fig. 32.3) and have similar antibiotic activity (40,41). Although neomycin's broad spectrum against gram-positive and gram-negative bacteria and mycobacteria was initially promising, toxicity was encountered following parenteral adminstration. The mixture of neomycin B and C is now used for nonsystemic applications, such as, gut sterilization and as a topical agent often combined with other antibiotics and corticosteroids.

Paromomycin I, initially isolated from culture broths of *S. rimosus*, differs from neomycin B in that glucosamine rather than neosamine is linked to the C-4 hydroxyl group of 2-DOS (42). Paromomycin II was later isolated and shown to be the analog of neomycin C. Aminosidin I and II, produced by *S. chrestomyceticus*, are among several antibiotics identical to paromomycin I and II (43). Unlike neomycin, paromomycin displayed antiprotozoal activity as well as

antibacterial activity, illustrating the effect of relatively small changes in structure on antimicrobial activity (44). Paromomycin is clinically used to eradicate intestinal parasites, although toxicity prohibits its use against systemic infections. From the lividomycin complex produced by *S. lividus*, lividomycin B differs from paromomycin I in that the 3′-hydroxyl group of glucosamine is absent, while lividomycin A is a pseudopentasaccharide in which D-mannose is glycosidically linked to the 4‴-hydroxyl group (45,46). Lividomycin A has been developed as a clinical antibiotic in Japan.

Ribostamycin, produced by *S. ribosidificus*, corresponds to a pseudotrisaccharide in the neomycins (Fig. 32.4) (47). It is less toxic than neomycin and is used in Japan and some other countries to treat gram-negative infections. Its 3″-epimer, xylostasin, was later isolated from culture broths of *Bacillus subtilis* (48). Butirosin A and B were found in culture filtrates of *B. circulans* and shown to possess the unusual 4-amino-2(*S*)-hydroxybutyryl group (AHBA) as an amide on N-1 of 2-DOS (49,50). In contrast to its nonacylated parent, ribostamycin, the butirosins exhibited significant activity against *Pseudomonas aeruginosa*, a clinically important gram-negative bacterial pathogen (51). This novel discovery was especially important because it was subsequently exploited in the discovery of amikacin and its analogs (*vide infra*).

2.3 4,6-Di-*O*-Glycosyl-2-Deoxystreptamine Aminoglycosides

In contrast to the 4,5-di-*O*-glycosyl substitution pattern of the neomycin family, another large family of aminoglycosides exhibits 4,6-di-*O*-glycosyl substitution of 2-DOS. This family can be divided into the kanamycin–tobramycin and gentamicin groups. Kanamycin A (Fig. 32.5), the first member of this family to be discovered, is produced by *S. kanamyceticus* (52). X-ray analysis of kanamycin A sulfate confirmed the structure that had been previously elucidated by several research groups (53). Kanamycin B and C were initially isolated as minor factors (54–56). Kanamycin B contains 2,6-diaminoglucose attached to the 4-hydroxyl group of 2-DOS, whereas hydroxyl groups replace amino groups at the

2'- and 6'-positions in factors A and C, respectively (Fig. 32.5) (57,58). Kanamycin B is twice as active as kanamycin A, whereas kanamycin C is the least active, illustrating the antibacterial effects of both the number and position of amino groups. Both kanamycin A and B are used for parenteral treatment of gram-negative infections.

The nebramycin complex was isolated from culture broths of *S. tenebrarius* (59,60). Its most potent component, tobramycin (3'-deoxykanamycin B, Fig. 32.5), is formed during isolation by hydrolysis of the carbamate from nebramycin factor 5' (61). Kanamycin B (nebramycin factor 5) was obtained by analogous hydrolysis of factor 4 (62). Tobramycin is more potent than the kanamycins against *Enterobacteriaceae*, but more importantly, it is more efficacious against infections caused by *P. aeruginosa*. Tobramycin has achieved wide use for parenteral treatment of these gram-negative infections, although it was preceded in its development by gentamicin, the aminoglycoside which pioneered this class to its modern position of therapeutic significance.

The other group of 4,6-disubstituted-2-DOS aminoglycosides includes gentamicin, the most widely used aminoglycoside, which is produced by *Micromonospora pur-*

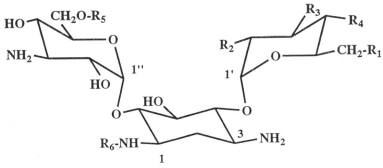

Fig. 32.5 Structures of the kanamycin–tobramycin group. Kanamycin A: $R_1 = NH_2$, $R_2 = OH$, $R_3 = OH$, $R_4 = OH$, $R_5 = H$, $R_6 = H$; kanamycin B: $R_1 = NH_2$, $R_2 = NH_2$, $R_3 = OH$, $R_4 = OH$, $R_5 = H$, $R_6 = H$; kanamycin C: $R_1 = OH$, $R_2 = NH_2$, $R_3 = OH$, $R_4 = OH$, $R_5 = H$, $R_6 = H$; tobramycin: $R_1 = NH_2$, $R_2 = NH_2$, $R_3 = H$, $R_4 = OH$, $R_5 = H$, $R_6 = H$; nebramycin 4: $R_1 = NH_2$, $R_2 = NH_2$, $R_3 = OH$, $R_4 = OH$, $R_5 = CONH_2$, $R_6 = H$; nebramycin 5': $R_1 = NH_2$, $R_2 = NH_2$, $R_3 = H$, $R_4 = OH$, $R_5 = CONH_2$, $R_6 = H$; dibekacin: $R_1 = NH_2$, $R_2 = NH_2$, $R_3 = H$, $R_4 = H$, $R_5 = H$, $R_6 = H$; amikacin: $R_1 = NH_2$, $R_2 = OH$, $R_3 = OH$, $R_4 = OH$, $R_5 = H$, $R_6 = COCH(OH)CH_2CH_2NH_2$; habekacin (arbekacin): $R_1 = NH_2$, $R_2 = NH_2$, $R_3 = H$, $R_4 = H$, $R_5 = H$, $R_6 = COCH(OH)CH_2CH_2NH_2$.

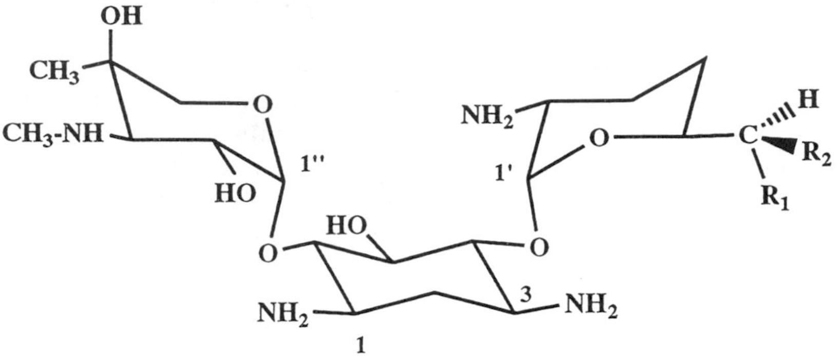

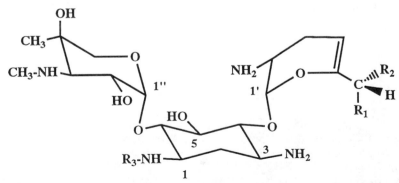

Fig. 32.6 Structures of gentamicin C components. Gentamicin C_1: $R_1 = NHCH_3$, $R_2 = CH_3$; gentamicin C_{1a}: $R_1 = NH_2$, $R_2 = H$; gentamicin C_2: $R_1 = NH_2$, $R_2 = CH_3$; gentamicin C_{2a}: $R_1 = CH_3$, $R_2 = NH_2$; gentamicin C_{2b}: $R_1 = NHCH_3$, $R_2 = H$.

purea and *M. echinospora* (63). The suffix "micin" denotes its origin from the genus *Micromonospora* whereas "mycin" denotes production by *Streptomyces*. The gentamicin C complex is characterized by the branched aminosugar, garosamine, linked to the C-6 hydroxyl group of 2-DOS, while individual gentamicin factors differ in their C- and N-substitution pattern at C-6' (Fig. 32.6) (64,65). Commercial gentamicin is a mixture of factors C_1, C_{1a}, and C_2. Gentamicin C_{2b} is identical to sagamicin (micronomicin), which was obtained from *M. sagamiensis* and developed in Japan (66–68). Analogous to other aminoglycoside complexes, many related factors have been isolated (69). Gentamicin exhibits much

greater potency than kanamycin against gram-negative bacteria, including especially *P. aeruginosa*. The commercial development of gentamicin greatly expanded the medical importance and use of aminoglycosides and their market size, such that gentamicin has been generally used as the standard against which all other aminoglycosides are compared.

Sisomicin was the first of a related group of aminoglycosides isolated from culture broths of *M. inyoensis* (70). Characteristic of this group is an unsaturated aminosugar attached to the 4-hydroxyl group of 2-DOS (Fig. 32.7) (71). Although it is more potent than gentamicin and has undergone extensive clinical evaluation, sisomicin has only

Fig. 32.7 Structures of sisomicin and related compounds. Sisomicin: $R_1 = NH_2$, $R_2 = H$, $R_3 = H$; verdamicin: $R_1 = NH_2$, $R_2 = CH_3$, $R_3 = H$; G-52: $R_1 = NHCH_3$, $R_2 = H$, $R_3 = H$; netilmicin: $R_1 = NH_2$, $R_2 = H$, $R_3 = CH_2CH_3$.

achieved a niche role and has not supplanted gentamicin or the newer aminoglycosides. Two groups of aminoglycosides that have a pentose rather than a hexose attached to the C-6 hydroxyl group of 2-DOS are the gentamicin A and seldomycin complexes, the latter produced by *S. hofunensis* (72–75). Since their most potent factors are only comparable to kanamycin, they have not been clinically developed.

2.4 Semisynthetic Derivatives of 2-Deoxystreptamine Aminoglycosides

Extensive efforts have been expended to discover semisynthetic derivatives which increase the clinical utility of aminoglycosides, especially by either circumventing microbial resistance and/or improving the therapeutic index of efficacy to toxicity. This chapter cannot cover the numerous derivatives that have been synthesized over decades of research from the many different aminoglycosides available to chemists as starting materials. Instead it focuses only on the two most successful approaches: 3′ and 4′-deoxygenation and 1-N-substitution.

During structure elucidation of the natural aminoglycosides, comparisons of their antibiotic activity illustrated the critical role of certain amino and hydroxyl groups, prompting the synthesis of derivatives in which amino and hydroxyl groups were interchanged (76,77). A more successful approach was the synthesis of deoxy derivatives in which hydroxyl groups that were substrates for aminoglycoside-inactivating enzymes were removed. This effort fostered the development of synthetic routes and transformations for selective deoxygenation in polyhydroxy molecules (78–80). The greater activity of tobramycin (3′-deoxykanamycin B) relative to kanamycin B emphasized the significance of the 3′-hydroxyl group in terms of potency and activity against resistant bacteria. Structural

modifications of the kanamycins in this aminosugar ultimately led to the synthesis of 3′,4′-dideoxykanamycin B (Fig. 32.5), a compound that exhibited potent activity against gram-negative bacteria including *P. aeruginosa* (81). Subsequently named dibekacin, it has become one of the important commercially used semisynthetic aminoglycosides.

N-Acylation and N-alkylation of amino groups has been the most extensively explored approach to modification of aminoglycosides. The discovery of butirosin with its novel 4-amino-2(*S*)-hydroxybutyryl amide (AHBA) at N-1 of 2-DOS and its enhanced antimicrobial activity prompted the synthesis of many analogous amides of other aminoglycosides. The analog of kanamycin A (Fig. 32.5) was an especially important derivative due to its activity against resistant bacteria and its diminished acute toxicity (82). These favorable features provided a therapeutic alternative to the older aminoglycosides, such that 1-N-AHBA-kanamycin A, subsequently named amikacin, has become the most widely used semisynthetic aminoglycoside (83).

The success of amikacin prompted the synthesis of many analogs of other aminoglycosides. This endeavor has expanded into studies of structure–toxicity relationships in addition to more traditional SAR studies of activity against sensitive and resistant bacteria (84–86). The most successful of many analogs have been 1-N-AHBA-dibekacin (named habekacin, Fig. 32.5) and 1-N-AHBA-gentamicin B (named isepamicin, Fig. 32.8) (87,88). The logical extension from 1-N-acyl to 1-N-alkyl derivatives led to the synthesis of 1-N-ethylsisomicin, named netilmicin (Fig. 32.7) (89). Netilmicin inhibits a broad range of resistant bacteria, although not as broad as amikacin, and has less chronic toxicity than sisomicin and gentamicin (90–92). Netilmicin, habekacin, and isepamicin are commercial semisynthetic aminoglycosides, although amikacin still remains the most

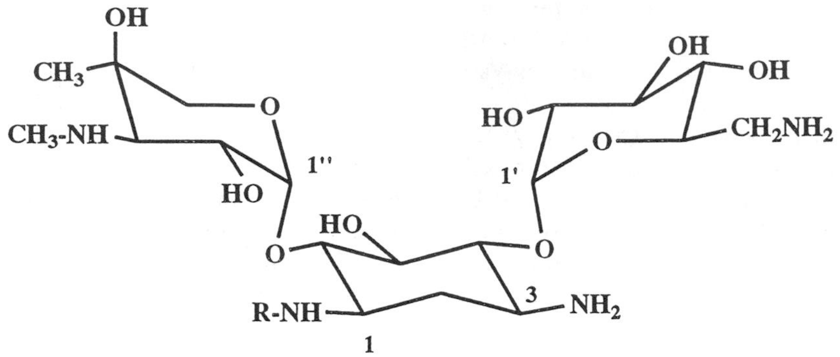

Fig. 32.8　Structures of gentamicin B (R = H) and isepamicin (R = COCH(OH)CH$_2$CH$_2$NH$_2$).

successful one. 1-*N*-Alkyl analogs have been synthesized from the kanamycins, including 1-*N*-(4-amino-2(*S*)-hydroxybutyl)-kanamycin A (butikacin) and 1-*N*-(1,3-dihydroxy-2-propyl)kanamycin B (propikacin) (93,94). These compounds have fostered the development of procedures to derivatize only one of the four or five amino groups found in most aminoglycosides selectively. Among new developments in synthetic methodology arising from these efforts were the use of transition metal cations as temporary protecting groups (95–98).

Other modifications of aminoglycosides include epimerization of substituents, such as, 5-epi-sisomicin (Sch 22591) (99,100). Hydroxyl groups have been replaced by halogens (101–103), and C-1 hydroxyalkyl derivatives have been made (84,104,105). Many programs had been directed toward aminoglycoside modification in the 1960s and 1970s, but these extensive efforts declined over the past decade with only a few exceptions (106). Factors contributing to the decline were the inability to find breakthroughs that significantly reduce the toxicity of aminoglycosides and the emergence of other antibiotics to treat gram-negative infections, most notably later generation cephalosporins and fluoroquinolones. A few laboratories have continued to prepare semisynthetic aminoglycosides to determine the molecular mechanisms by which these antibiotics exert their antimicrobial and pharmacological effects (84,107,108).

2.5 Monosubstituted 2-Deoxystreptamine Aminoglycosides

Many pseudodisaccharides comprising 2-DOS and one aminosugar have been isolated from culture broths or after hydrolysis of saccharide constituents from larger antibiotics. The lesser activity of these fragments compared to the corresponding pseudotetra- or trisaccharides indicates that all parts of the aminoglycoside structure are important for maximizing antibiotic activity. The pseudodisaccharides are also useful starting materials for synthesis. Apramycin is a unique aminoglycoside in which 2-DOS is monosubstituted on its 4-hydroxyl group by a bicyclic amino-octose, which is further substituted by 4-aminoglucose via a 1,1'-bis-glycosyl bond (Fig. 32.9). Apramycin and 3'-hydroxyapramycin were isolated from the same nebramycin complex as tobramycin (109). The lesser activity of apramycin against the critical clinical pathogen, *P. aeruginosa*, directed its development to gram-negative enterobacterial infections within veterinary medicine (110). Aminoacyl derivatives, such as, its 1-*N*-AHBA amide (analog of amikacin), and glycosyl derivatives (analogs of ribosta-

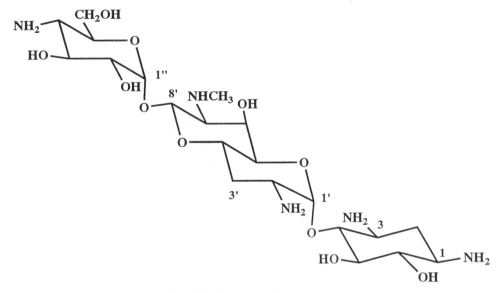

Fig. 32.9 Structure of apramycin.

mycin and tobramycin) retain antibiotic activity (111,112).

2.6 Other Aminocyclitol Aminoglycosides

Screening of fermentation broths for new antibiotics resulted in the discovery of new families of aminoglycosides during the 1970s. In contrast to 1,3-diaminocyclitols, the fortimicin family is a complex of pseu-

dodisaccharides produced by *M. olivoas-terospora* and distinguished by a 1,4-diaminocyclitol, named fortamine (Fig. 32.1) (113). In the most active factor, fortimicin A, the methylamino group at C-4 of fortamine is acylated by glycine, causing the aminocyclitol to rearrange into a different chair conformation (Fig. 32.10) (114). The structurally related sporaricins, obtained from culture broths of *Saccharopolyspora hirsuta* subsp. *kobensis*, are 1-epi-

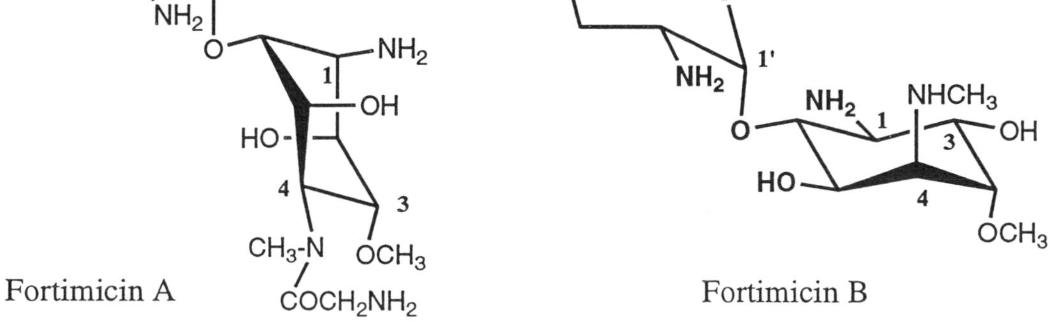

Fortimicin A

Fortimicin B

Fig. 32.10 Structures of fortimicin A and B.

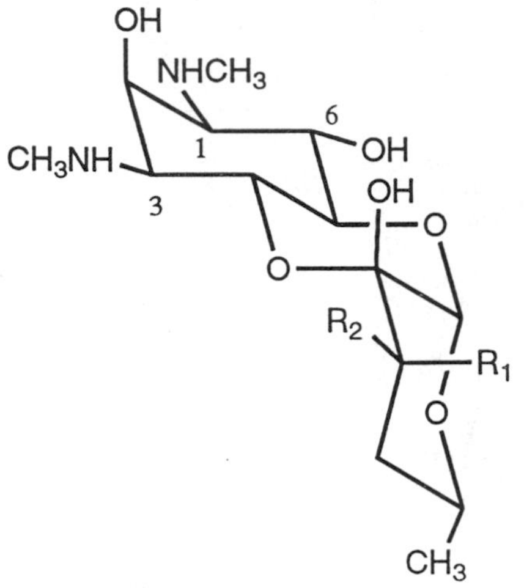

Fig. 32.11 Structures of spectinomycin hydrate ($R_1 = R_2 = OH$) and dihydrospectinomycin ($R_1 = OH$, $R_2 = H$).

2-deoxy analogs of fortimicin (115). Other related aminoglycosides include: the istamycins, produced by *S. tenjimariensis*; the sannamycins, formed by *S. sannanensis*; and dactimicin, obtained from *Dactylosporangium matsuzakiense* (116–118). Structural variations among aminosugars in this family are similar to those found in the gentamicins, including unsaturated aminosugars similar to that in sisomicin. Even though fortimicin A is only comparable to kanamycin A against susceptible bacteria and lacks strong activity against *P. aeruginosa*, it has useful activity against bacteria

resistant to other aminoglycosides and has been developed under the name, astromicin.

Structural variations within 1,3-diaminocyclitols have been found that are accompanied by changes in antimicrobial spectrum and commercial utility. Spectinomycin is an older aminocyclitol produced by *S. spectabilis* (119). Its core aminocyclitol is 1,3-di-*N*-methyl-2-epi-streptamine (actinamine) onto which a bicyclic moiety is fused through its C-4 and C-5 hydroxyl groups (Fig. 32.11). Its stereochemistry was established by X-ray crystallography (120). Dihydrospectinomycin has been isolated from fermentation as well as chemically synthesized (121,122). Spectinomycin plays an important role in the treatment of gonorrhea, including strains of *Niesseria gonorrheae* resistant to other antibiotics (123). Its activity and good therapeutic index have prompted several research groups to synthesize derivatives, but none has yet offered any clear advantages.

Hygromycin B is an older aminocyclitol produced by *S. hygroscopicus*. Its aminocyclitol is (+)-*N*-methyl-2-deoxystreptamine (hyosamine), which is substituted on its C-5 hydroxyl group by a disaccharide containing an orthoester linkage between the C-2′ and C-3′ hydroxyl groups and the C-1″ carbon atom (Fig. 32.12) (124). Its terminal sugar is a unique aminoheptose

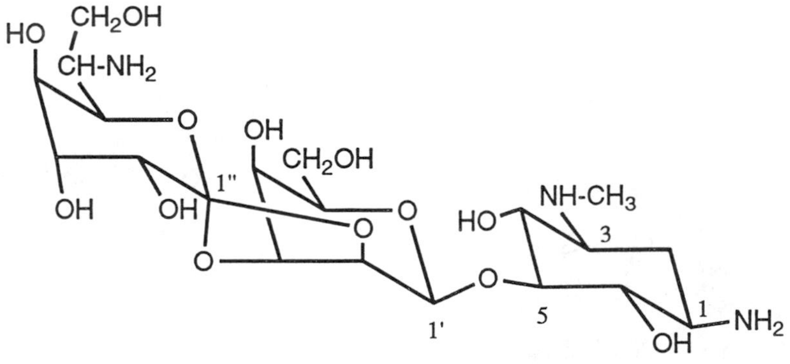

Fig. 32.12 Structure of hygromycin B.

called destomic acid. A related series named the destomycins was obtained from *S. rimofaciens*, in which the isomeric 1-*N*-methyl and the 1,3-di-*N*-methyl analogs were isolated along with a C-4',4" epimer (125,126). The non-*N*-methylated 2-DOS analog was later isolated from culture broths of *Streptoverticillium eurocidicus*, and other epimeric and hydroxylated analogs have been found (127,128). Resistance to hygromycin B in *S. hygroscopicus* occurs by enzymatic inactivation to its 4-*O*-phosphoryl derivative (129,130). Hygromycin B inhibits prokaryotes, eukaryotes, and mammalian cells, and resistance to hygromycin B is a selective marker in recombinant genetic experiments. This series has antifungal and anthelmintic activity in addition to antibacterial activity, and hygromycin B is commercially used for the control of internal parasites in poultry and pigs.

Hygromycin A is produced by the same culture as hygromycin B, but the two compounds have completely different structures (131). Hygromycin A contains an aminocyclitol in which the amine is acylated by a glycosyl derivative of 3,4-dihydroxycinnamic acid (Fig. 32.13) (132). Its efficacy against swine dysentery has led to the synthesis of several novel analogs (133–135).

The validamycins are novel aminocyclitols isolated from culture broths of *S. hygroscopicus* var. *limoneus* (136). Validamycin A is a pseudotrisaccharide in which a saturated cyclitol (validamine) and

an unsaturated one (valienamine) are joined via a secondary amino group to form a dimeric moiety (validoxylamine A), which is further glucosylated (Fig. 32.14) (137,138). Validamycin is used in some countries to control diseases caused by species of *Rhizoctonia*, such as, sheath blight on rice (139). Other monoaminocyclitols include minosaminomycin, obtained from an unspeciated *Streptomyces* (140), and LL-BM123α, obtained from a *Nocardia* species (141); both contain short peptide chains attached to their aminoglycoside moieties.

2.7 Miscellaneous Aminoglycosides

Although far fewer than the aminocyclitol antibiotics, a few aminoglycosides have a neutral cyclitol. Kasugamycin, an older antibiotic produced by *S. kasugaensis*, contains (+)-inositol and a substituted diamino sugar (Fig. 32.15) (142–144). Although it possesses some gram-negative activity, it has become an important agent for controlling rice blast (145). Other examples within this group are the myomycin and BM782 complexes, produced by strains of *Nocardia*, which contain an inositol that is further substituted by an *N*-(aminoacyl)amino sugar (146,147).

Some compounds containing acylic moieties instead of cyclitols are usually included in the aminoglycoside class. Different factors of the sorbistin complex that are

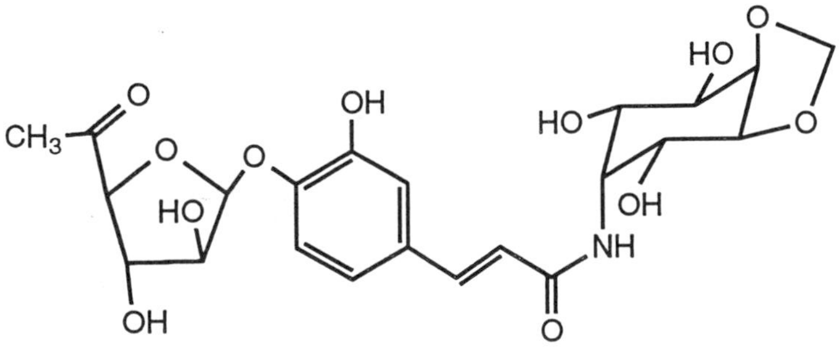

Fig. 32.13 Structure of hygromycin A.

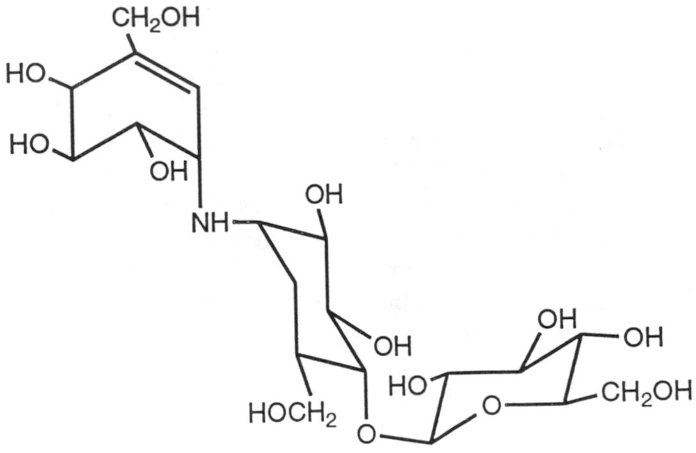

Fig. 32.14   Structure of validamycin A.

comprised of *N*-acyl-4-aminoglucose and a diaminohexitol have been isolated from culture broths of different organisms (148–150). The glycocinnamoylspermidines, such as, cinodine (LL-BM123β, γ_1 and γ_2), are a series of substituted aminotrisaccharides glycosidically linked to *p*-hydroxycinnamoylspermidine (151). Iprocinodine, a semisynthetic *N*-isopropyl derivative obtained by reductive amination, exhibited less toxicity and a better therapeutic index than cinodine; although too toxic for clinical use, both compounds had potential applications as veterinary antibiotics (152). The glysperin complex has an aminotetrasaccharide linked to *p*-hydroxybenzoylsper-

midine (153). Finally, aminoglycosides, such as, 4-trehalosamine and prumycin, contain only saccharides (154,155).

2.8 Summary of the Aminoglycosides

Key objectives guiding medicinal chemists in aminoglycoside research have been a higher therapeutic index of efficacy to toxicity and activity against resistant bacteria. The mechanisms of aminoglycoside resistance have been thoroughly studied and examples are known for all major mechanisms of resistance, including altered antibiotic target sites, antibiotic-inactivating

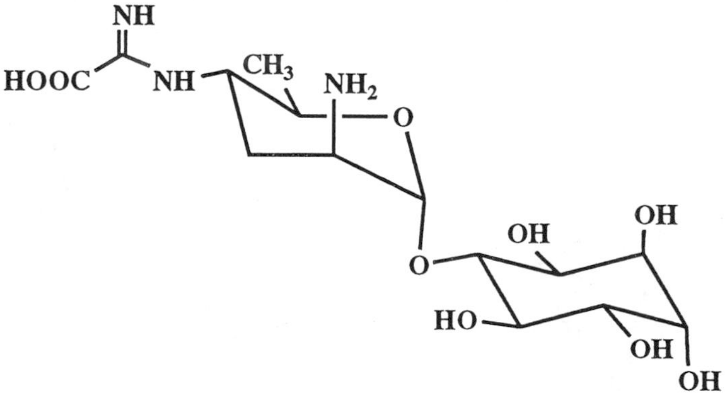

Fig. 32.15   Structure of kasugamycin.

enzymes, and reduced microbial uptake of antibiotic. Aminoglycoside-inactivating enzymes are well characterized and known to destroy antibiotic activity via N-acetylation, O-phosphorylation, and O-adenylylation of their substrates at a number of critical sites within the molecules (156,157). Both natural and semisynthetic aminoglycosides have been found that overcome bacterial resistance to older aminoglycosides, although other classes of antibiotics also fill this important medical need.

The major clinical concern about aminoglycosides is their potential toxicity, principally nephrotoxicity and ototoxicity. Many studies have been directed to finding their molecular basis and explanatory models have been developed (158–161). Clinical experience with aminoglycosides has steadily evolved whereby optimized dosing schedules coupled with appropriate monitoring of predictive clinical markers of kidney function permit effective use of these agents while managing possible nephrotoxicity, which tends to be reversible upon cessation of therapy (162,163). Clinical studies have focused on once-daily administration of aminoglycosides, based upon apparent better efficacy from high peak serum concentrations and longer post-antibiotic effects combined with less severe toxicity caused by once-daily dosing (164–166). Although the problems of resistance and toxicity are certainly not solved, the successful research efforts that addressed these issues have made the aminoglycosides one of the important classes of fermentation-derived antibiotics.

The aminoglycosides have also provided chemists with targets for total synthesis, which has fostered advances in glycosylation reactions and functional group manipulations in carbohydrates (167). However, despite advances in carbohydrate chemistry, structure–activity relationship (SAR) studies have largely relied on modifications of fermentation-derived compounds. The biosynthesis of amino-glycosides has been extensively studied, and new derivatives have been made by altering those biosynthetic processes in mutant strains of antibiotic-producing organisms. In a process called mutasynthesis (mutational biosynthesis), a block in the biosynthetic pathway leading to 2-DOS was created which was then overcome by supplying 2-DOS or an analog (168). If the latter was incorporated by the mutant strain and further processed along the usual biosynthetic pathway, analogs of the natural aminoglycosides were formed (169). This approach has not yielded a commercial product, but it illustrates another way to exploit fermentation processes to produce novel derivatives.

3 MACROLIDE ANTIBIOTICS

Macrolide was the term originally given to antibiotics produced by species of *Streptomyces* that contained a macrocyclic lactone (170). Macrolide antibiotics contain a highly substituted monocyclic lactone, named an aglycone, to which a variety of amino and neutral deoxysugars are attached. Microorganisms produce a large number of polyketide-derived secondary metabolites whose structures contain macrocyclic lactones, but this chapter covers only the non-polyene antibacterial macrolides whose aglycones range from 12- to 16-membered rings. Like the aminoglycosides, the macrolides are an old and well-established class of antibiotics. Although their antibacterial spectrum, oral efficacy, and safety had given them an important role in treating infectious diseases since the 1950s, the macrolides enjoyed a renaissance during the 1980s when several new semisynthetic derivatives were discovered and commercially developed. Books and comprehensive reviews are available which thoroughly cover those developments (171–175).

3.1 12-Membered Ring Macrolides

Methymycin was the first macrolide whose structure was fully elucidated (176,177). Its aglycone (methynolide) is the smallest ring found in conventional macrolides and is substituted by an aminosugar (desosamine) attached to the 3-hydroxyl group (Fig. 32.16). Neomethymycin is an isomer in which the hydroxyl group is moved from C-10 to C-12 (178). Only a few 12-membered macrolides have been discovered and their antibiotic activity has been too weak for practical use.

3.2 14-Membered Ring Macrolides

Erythromycin has become the most widely used macrolide antibiotic since its isolation in 1952 from culture broths of *Streptomyces erythreus* (now reclassified as *Saccharopolyspora erythraea*) (179,180). The principal factor is erythromycin A, although the common name of erythromycin is often used synonymously. Its structure consists of a substituted 14-membered aglycone (erythronolide A), whose hydroxyl groups at C-5 and C-3 are linked respectively to the aminosugar, desosamine, and a neutral sugar, cladinose (Fig. 32.17) (181). Stereochemical assignments were finally established by X-ray crystallography of erythromycin A hydroiodide dihydrate

(182). Minor factors in the fermentation include erythromycin B, C, and D, which differ in their hydroxylation at C-12 and/or O-methylation of cladinose at C-3″ (Fig. 32.17).

Erythromycin has a moderately broad spectrum of activity against gram-positive bacteria, gram-negative cocci, and *Mycoplasma* species, and is orally absorbed and relatively safe. It is used to treat infections of the respiratory tract, skin and soft tissues, and genital tract. However, its propensity to degrade in the stomach requires a means to protect it until it reaches the upper intestine where absorption occurs. To overcome this problem, erythromycin is given orally in protected forms such as enterically coated tablets or capsules that maintain their integrity in stomach acid and subsequently release the antibiotic in the more alkaline environment of the upper intestine.

The only other 14-membered macrolide in commercial use is oleandomycin, the fermentation product of *S. antibioticus* (183). It has a different 14-membered aglycone (oleandolide), which is distinguished by an epoxide at C-8 and a different neutral sugar (oleandrose) (Fig. 32.18) (184,185). Pikromycin, produced by *S. felleus*, was the first macrolide to be discovered (186). It is characterized by the lack of a neutral sugar, a ketone at C-3, and unsaturation at C-10,11 of its aglycone

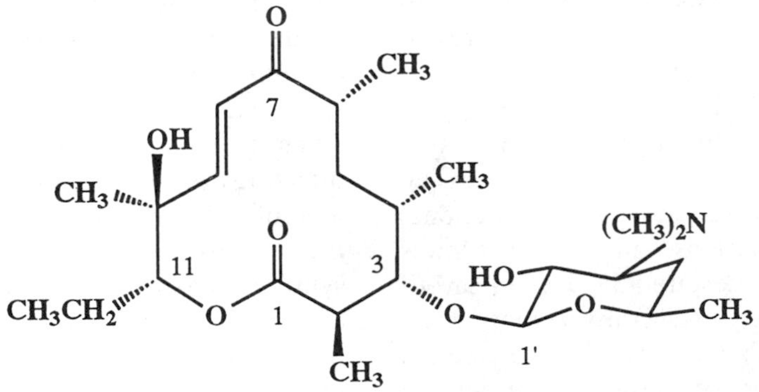

Fig. 32.16 Structure of methymycin.

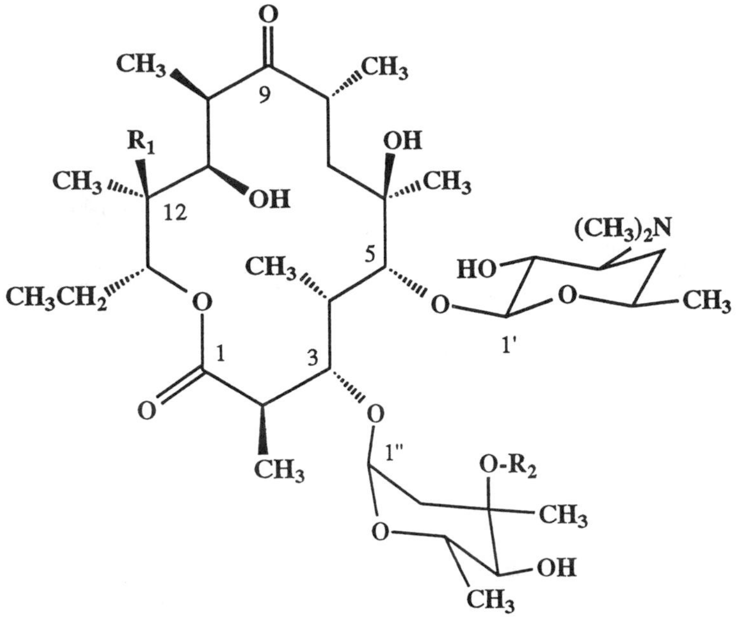

Fig. 32.17 Structures of erythromycin A ($R_1 = OH$, $R_2 = CH_3$), erythromycin B ($R_1 = H$, $R_2 = CH_3$), erythromycin C ($R_1 = OH$, $R_2 = H$), and erythromycin D ($R_1 = H$, $R_2 = H$).

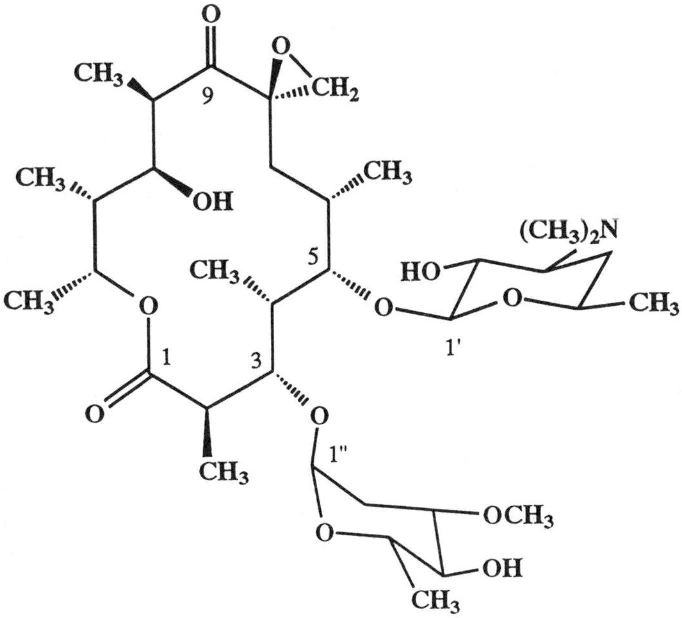

Fig. 32.18 Structure of oleandomycin.

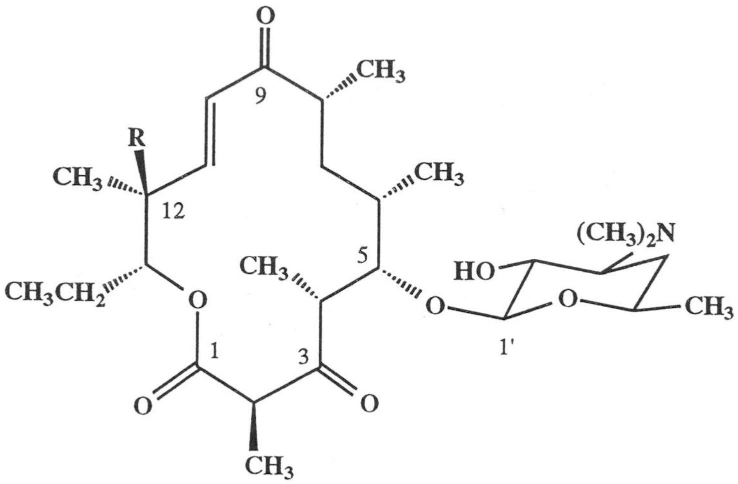

Fig. 32.19   Structures of pikromycin (R = OH) and narbomycin (R = H).

(pikronolide) (Fig. 32.19) (187). A 12-deoxy analog, narbomycin, is produced by *S. narbonensis* (Fig. 32.19) (188). The megalomicins are unique 14-membered macrolides in that they contain two aminosugars, desosamine at C-5 and the uncommon megosamine at C-6; in addition, their neutral sugar is mycarose or its acyl derivative (189). They were the first macrolides obtained from *Micromonospora* (190). 14-Membered macrolides, such as, lankamycin and kujimycin A, contain only neutral sugars (191,192). In these neutral macrolides, desosamine has been replaced by a 3-methoxysugar (chalcose) in addition to several changes in the aglycone (193). A macrolide having three neutral sugars is illustrated by 23672 RP (194).

3.3 Semisynthetic Derivatives of 14-Membered Ring Macrolides

Since its discovery, erythromycin has been an important starting material for medicinal chemists searching for semisynthetic derivatives that possess improved antimicrobial and pharmacokinetic properties (195). A wide variety of salts, esters, and salt-ester combinations have been investigated as

means to prevent degradation of erythromycin in the stomach, to increase its oral bioavailability and antibiotic concentrations in serum, and to decrease the inter- and intrasubject variability of absorption. Even though the water solubility of erythromycin base is low, its *in vivo* stability and oral bioavailability are enhanced by acid-addition salts prepared from lipophilic acids, such as, stearic acid, that yield adducts possessing even greater water-insolubility.

Another approach has been esterification of the hydroxyl group at C-2′ of desosamine by short-chain acyl groups, such as, acetyl, propionyl, ethyl succinyl, and ethyl carbonyl. These 2′-esters are prodrugs that facilitate oral absorption, but are subsequently hydrolyzed *in vivo* to liberate erythromycin as the antimicrobially active substance (196,197). The pharmacokinetic features of erythromycin have also been improved by combinations of 2′-esters and acid-addition salts of lipophilic acids, such as, erythromycin estolate (2′-propionate, laurylsulfate salt) and, more recently, erythromycin acistrate (2′-acetate, stearate salt) (198). For intravenous administration, the low water solubility of erythromycin is overcome with highly water-soluble acid-addition salts of hydrophilic acids, such as, erythromycin lactobionate and gluceptate. Each of these approaches to derivatization has yielded successful products that improve the appropriate therapeutic delivery of erythromycin. Among the other 14-membered macrolides, oleandomycin was converted to a 2′,4″, 11-tri-O-acetyl derivative (TAO) that improved oral bioavailability and taste, but its use has been limited, in part due to its greater induction of P-450 metabolic complexes (199–201).

The instability of erythromycin in acidic media arises from a facile intramolecular cyclization in which the hydroxyl group at C-6 adds to the carbonyl group, forming a 6,9-hemiketal that undergoes irreversible dehydration (Fig. 32.20) (202). Participation of the hydroxyl group at C-12 in a second intramolecular cyclization yields the 6,9;9,12-spiroketal, either through the intermediate anhydrohemiketal or directly from erythromycin (Fig. 32.20) (202–204). Because these intramolecular cyclization products lack useful antibiotic activity, several innovative strategies to derivatize erythromycin have been pursued that may be rationalized as ways to modify the functional groups involved in these transforma-

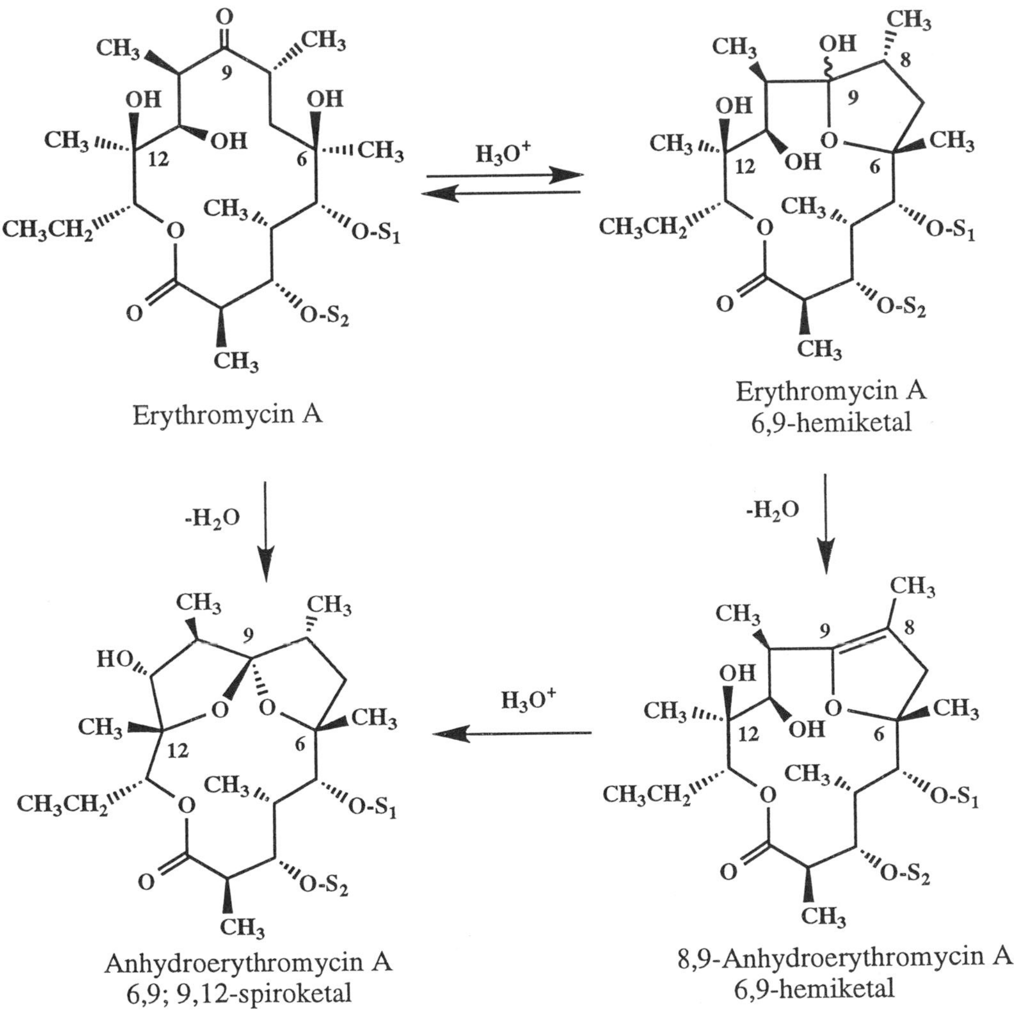

Fig. 32.20 Intramolecular cyclizations of erythromycin ($S_1 = \beta$-desosaminyl, $S_2 = \alpha$-cladinosyl).

tions and thereby diminish or prevent the intramolecular reactions.

One successful approach has been conversion of the ketone into oximes and hydrazones that are less prone to intramolecular cyclization. This strategy led to the selection of roxithromycin from a series of O-substituted oximes (Fig. 32.21) (205). Roxithromycin exhibits increased chemical stability and higher concentrations of antibiotic in serum after oral administration compared to erythromycin (206). Roxithromycin was the first example to be commercially introduced of what may be designated as a second-generation 14-membered macrolide derivative.

The ketone of erythromycin has been replaced by functional groups unable to

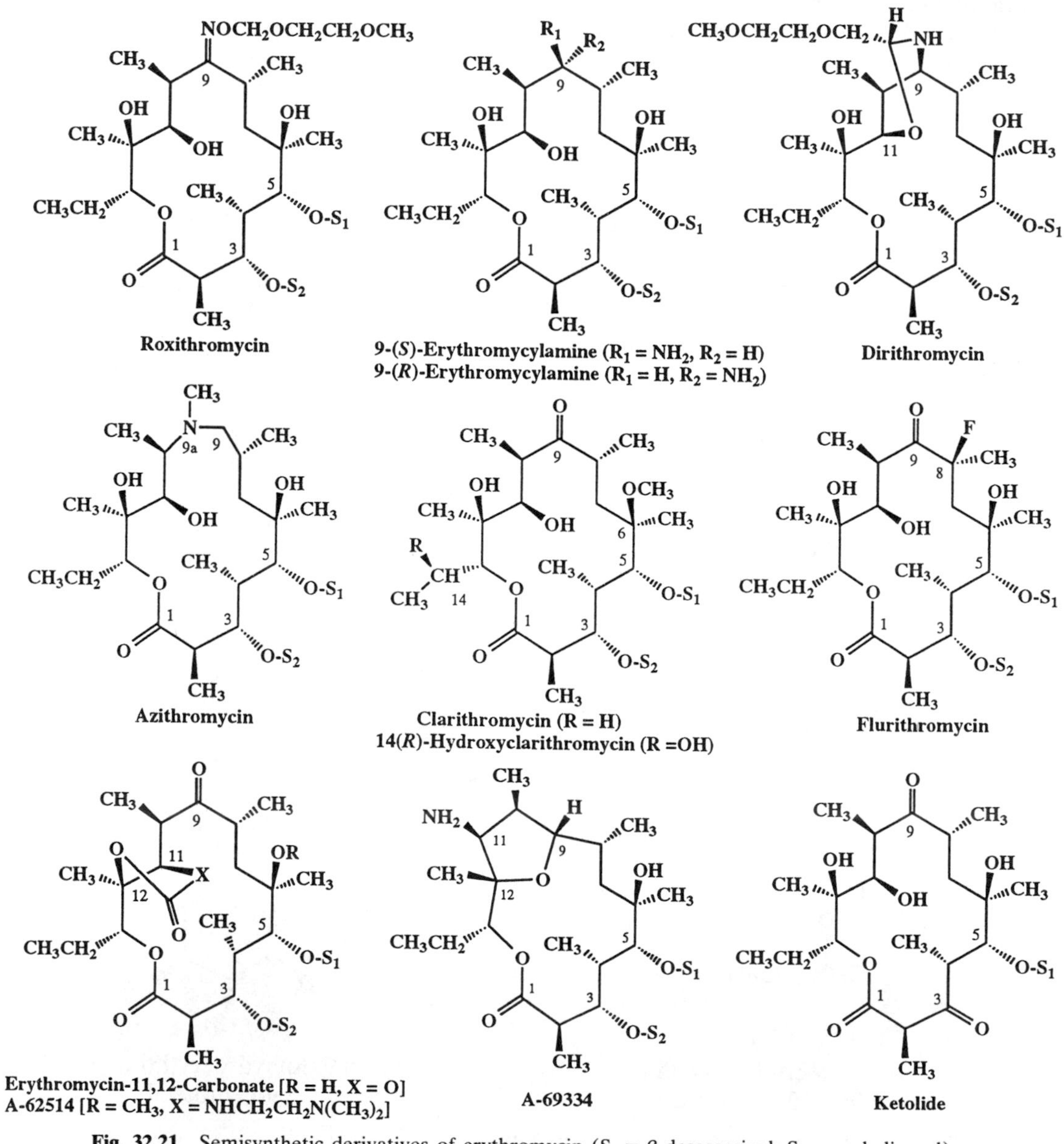

Fig. 32.21 Semisynthetic derivatives of erythromycin ($S_1 = \beta$-desosaminyl, $S_2 = \alpha$-cladinosyl).

participate in intramolecular cyclization reactions. Although 9-dihydroerythromycin is a less active derivative, reduction of the oxime or hydrazone of erythromycin gave 9(*S*)-erythromycylamine (Fig. 32.21), a potent antibiotic but with low oral bioavailability (207). Dirithromycin is its 9-*N*-11-*O*-oxazine adduct with 2-(2-methoxyethoxy) acetaldehyde (Fig. 32.21), a modification that improves oral delivery of erythromycylamine and permits once daily dosing due to high and prolonged concentrations of antibiotic in tissues and intracellular sites (208–210). Epidirithromycin has the alternative *S* configuration of the methoxyethoxymethyl substituent and is kinetically formed in the condensation (211). Derivatives of both 9(*S*)- and 9(*R*)-erythromycylamine have been recently prepared as means to increase the efficacy and oral absorption of their parent antibiotics (212–215).

Beckmann rearrangement of erythromycin 9-oxime gave ring expansion to a 15-membered intermediate, whose subsequent reduction and N-methylation produced a novel macrolide, named azithromycin, in which a basic nitrogen atom was embedded into the aglycone (Fig. 32.21) (216,217). A 6,9-imino ether isomer of the 15-membered intermediate has been recently reported (218). The term azalide was coined for derivatives of erythromycin having a nitrogen atom incorporated in the traditional macrolide ring. These derivatives are notable for their greater activity against many gram-negative bacilli and their high concentration in tissues, prolonged *in vivo* half-life, and intracellular accumulation (219–221). These advantageous pharmacokinetics permit the treatment of sexually transmitted diseases, such as, gonococcal and chlamydial infections, with a single dose, thereby eliminating the need for patients to return after initial therapy (222,223). Analogs of azithromycin have been synthesized, including 15-membered azalides, in which the 9-carbon and 9a-

nitrogen atoms are interchanged, and 14-membered azalides (224–228).

New macrolides have been obtained by modifying functional groups other than the ketone. The most important one is clarithromycin (6-*O*-methylerythromycin, Fig. 32.21), for which several syntheses have been reported (229–232). Its stability and oral bioavailability are significantly improved over erythromycin (233–236). 14(*R*)-Hydroxyclarithromycin, its principal human metabolite (Fig. 32.21), is more potent against the important respiratory pathogen, *Haemophilus influenzae* (237). Clarithromycin has significantly expanded the antimicrobial spectrum of macrolides, as exemplified by its clinical efficacy against certain intramolecular organisms, such as, the *Mycobacterium avium-intracellularie* (MAC) complex (233,238). Infections caused by opportunistic pathogens, such as, MAC, are increasing especially in immunocompromised patients.

Flurithromycin is 8-fluoroerythromycin (Fig. 32.21), a modification in which the fluorine atom at C-8 prevents the irreversible 8,9-dehydration (239). More recently, the 2′-ethylsuccinate of flurithromycin has been used in clinical trials (240). The well known 11,12-carbonate of erythromycin (Fig. 32.21) also prevents dehydration of the 6,9-hemiketal, presumably due to increased strain imposed by the cyclic carbonate (241). This strategy was recently extended to a series of 11-*N*-12-*O*-cyclic carbamates (242). Many combinations of these and other modifications involving carbon atoms 6-12 of erythromycin have been synthesized and a few of these derivatives have been in preclinical studies, but none appear to have advanced into clinical trials (242–245).

New derivatives of erythromycin have been recently discovered following strategies other than those outlined above. The 6,9;9,11-spiroketal (Fig. 32.20) was converted into a series of 9,12-bicyclic-epoxy macrolides, exemplified by the 11-amino derivative A-69334 (Fig. 32.21) (246). In addition to its chemical intermediacy in the synthesis of A69334, the same bicyclic ring system was independently discovered in the fermentation product sporeamicin A (247). The antibiotic activity of these novel compounds was unexpected since such bicyclic structures were presumed to lose antimicrobial activity, based on generalized results from the cyclized products depicted in Fig. 32.20. Another strategy hydrolyzed cladinose and oxidized the free hydroxyl group at C-3 to produce 3-keto derivatives, named ketolides, which display good antibiotic activity (Fig. 32.21) (248). These recent examples indicate that potentially useful modifications of erythromycin can still be discovered, despite the previous decades of extensive research, and open possibilities of additional strategies for modifications since some previous conclusions about structure–activity relationships will be found erroneous if they have been based on unwarranted over-generalizations.

3.4 16-Membered Ring Macrolides

Macrolides containing a 16-membered aglycone are the second large family of nonpolyene macrolides. They exhibit a wide diversity of structural variations and are usually subdivided into the leucomycin- and tylosin-related groups, based upon differences in the substitution patterns of their aglycones. Many of them are produced as multicomponent complexes, so the same compounds have often been isolated by different research groups and given different names or corporate letter-number designations. The most widely used 16-membered macrolides are josamycin and tylosin in human and veterinary medicine, respectively.

Josamycin is identical to leucomycin A$_3$, one of ten factors in the leucomycin complex (kitasamycin) produced by *S. kitasatoensis* (reclassified in the genus *Strep-*

toverticillium) (249–251). Structures for the leucomycin factors were elucidated over many years by extensive chemical studies and X-ray crystallography (252). Josamycin was isolated from culture broths of *S. narbonensis* var. *josamyceticus* (Fig. 32.22) (253). Although not registered in the United States, josamycin has become the most important 16-membered macrolide in human medicine.

A generalized figure for the leucomycin group is shown in Fig. 32.22, in which the functionality from C-9 to C-13 of the aglycone can be a dienone, dienol, epoxyenone, or epoxyenol moiety, and the hydroxyl groups at C-3 of the aglycone and C-4″ of the terminal neutral sugar (mycarose) can be substituted by short-chain acyl groups. Many of these structural permutations and combinations are known for individual factors from the many 16-membered macrolide complexes, such as, the leucomycin, midecamycin, carbomycin, platenomycin, maridomycin, deltamycin, espinomycin, and other series that are produced by different species of *Streptomyces* (171,172,175). Some of these macrolides have undergone preclinical investigation and others have achieved limited success in human or veterinary medicine. SAR studies from the large number of individual compounds in these many complexes have guided useful approaches to semisynthetic derivatives (*vide infra*).

The spiramycin complex (also discovered as foromacidine) is a unique structural variation characterized by β-O-glycosylation of the aglycone's 9-α-hydroxyl group with the aminodeoxy sugar, forosamine (Fig. 32.23) (254,255). The three principal factors in spiramycin produced by *S. ambofaciens* differ in their substituent on the hydroxyl group at C-3 (256). This bisamino macrolide containing three saccharides is used in both clinical and veterinary medicine (257,258).

The second large group of 16-membered macrolides has the more highly substituted

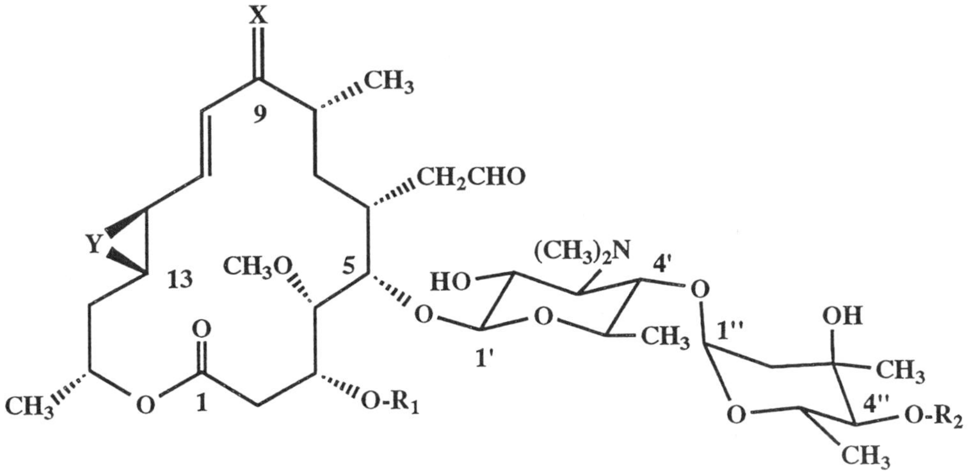

Fig. 32.22 Generalized structure of leucomycin-related macrolides. X = O or α-OH, β-H; Y = O or C-C bond; R₁ = H or acetyl; R₂ = H, acetyl, propionyl, butyryl, or isovaleryl. Josamycin: X = α-OH, Y = C-C bond, R₁ = acetyl; R₂ = isovaleryl.

aglycone exemplified by tylosin (Fig. 32.24), an important veterinary antibiotic produced by *S. fradiae* (259). Its structure was initially determined by chemical methods, with its absolute configuration later confirmed by X-ray studies of analogs (256,260–262). Many biosynthetic intermediates and shunt metabolites have been isolated that differ from tylosin in the

degree of oxidation of the aglycone and the number and structure of sugar substituents (263,264). Some related macrolides have a 12,13-epoxide, such as, angolamycin (shincomycin A) and cirramycin A₁ (Fig. 32.25).

Analogous to the aminoglycosides and megalomycins, the genus *Micromonospora* has been a valuable source of 16-membered macrolides. The first macrolide to be found

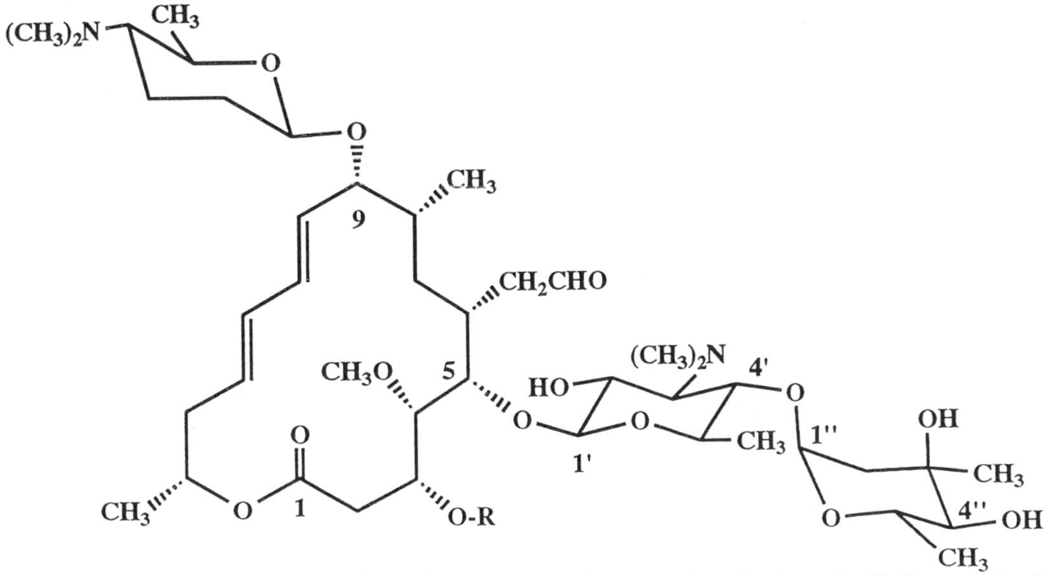

Fig. 32.23 Structures of spiramycin I (R = H), spiramycin II (R = acetyl), and spiramycin III (R = propionyl).

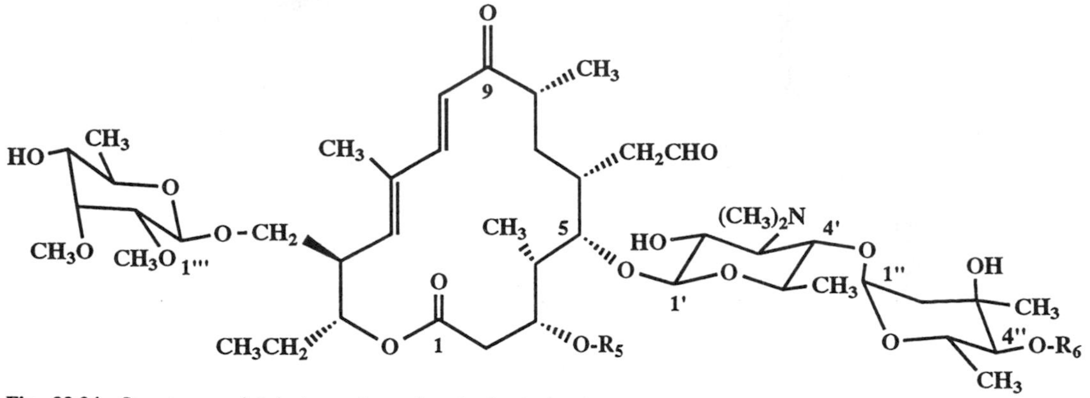

Fig. 32.24 Structures of tylosin and semisynthetic derivatives. Tylosin (R_5 = H, R_6 = H); AIV-tylosin (R_5 = acetyl, R_6 = isovaleryl); YM133 (R_5 = H, R_6 = 4-methoxyphenylacetyl).

from organisms of this genus (*M. rosaria*) was rosaramicin, whose aglycone is glycosylated by desosamine rather than mycaminose (Fig. 32.25) (265,266). Extensive clinical trials were conducted with rosaramicin, but it was ultimately not commercially developed. Structural variations within this group include the juvenimicin, izenamicin, and M-4365 complexes from other species of *Micromonospora* (Fig. 32.25) (267–269). Another important group is the mycinamicin complex, isolated from culture broths of *M. griseorubida*, whose novel structural characteristics include an unsaturated aglycone at C-2,3, a methyl group rather than a two-carbon substituent at C-6, and occasional hydroxylation at C-14 (Fig. 32.26) (270,271). Mycinamicin II (miporamicin) has been in clinical study in Japan.

Some neutral 16-membered macrolides have been isolated, such as, chalcomycin and neutramycin, that contain the novel methoxysaccharide chalcose, a methyl group or no substituent at C-6, and a hydroxyl group at C-8 (272,273). The aldgamycins are characterized by the unique bicyclic saccharide aldgarose (274). The

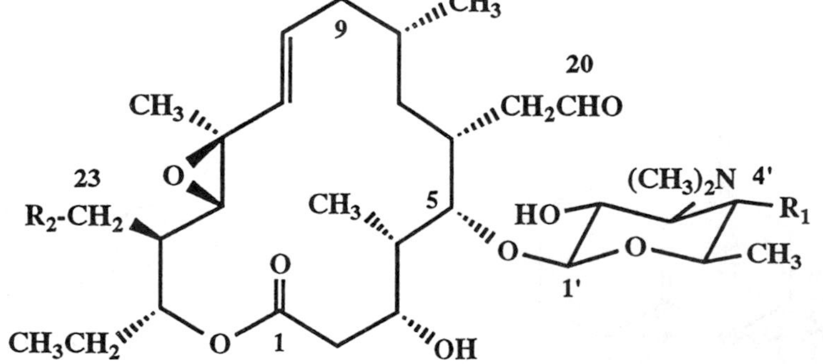

Fig. 32.25 Structures of cirramycin A_1 (R_1 = OH, R_2 = H), rosaramicin (R_1 = H, R_2 = H), and izenamicin A_3 (R_1 = H, R_2 = OH).

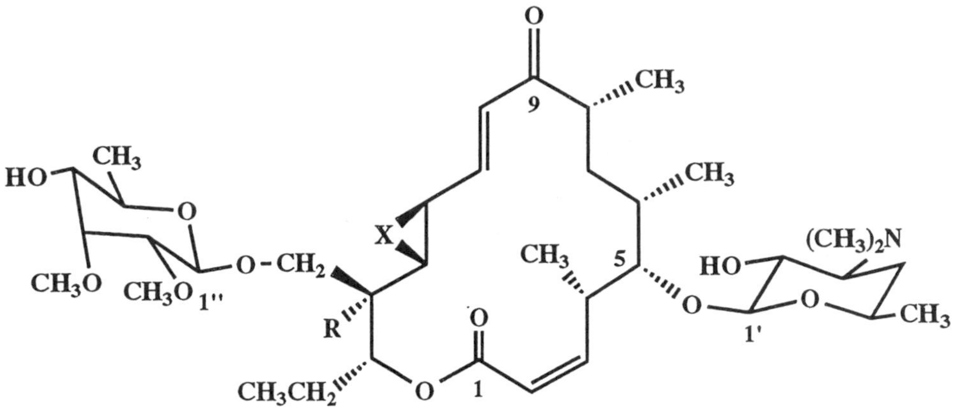

Fig. 32.26 Structures of mycinamicin I (R = H, X = O), mycinamicin II (R = OH, X = O), mycinamicin IV (R = H, X = C-C bond) and mycinamicin V (R = OH, X = C-C bond).

antibacterial activity of these neutral macrolides is lower than macrolides containing an aminosugar at the analogous position.

3.5 Semisynthetic Derivatives of 16-Membered Ring Macrolides

The 16-membered macrolides have also been the subject of extensive SAR studies and structural modifications in the quest for derivatives with superior features (275). As a result of these endeavors, some semisynthetic 16-membered macrolides were commercially introduced during the macrolide renaissance of the 1980s. During the earlier period when structures of 16-membered macrolides were being determined, their high degree of acylation, variable substitution patterns, and different acyl groups prompted the synthesis of many additional acyl derivatives. However, no substantial improvements were noted as a result of acylation of hydroxyl groups at C-3, C-9, C-2', and C-4'' (276). A key discovery was the longer *in vivo* half-life of antibiotic activity exhibited upon acylation of the tertiary hydroxyl group at C-3'' of the leucomycins, a more hindered site not acylated in natural products (277). After

synthesis of many 3''-O-acyl derivatives, 3''-O-propionyl-leucomycin A$_5$ (Fig. 32.27) was chosen for development because of its good *in vitro* activity and high concentrations of antibiotic in serum of laboratory animals (278). It has been introduced in several countries under the generic name of rokitamycin (279,280).

Prior to the synthesis of rokitamycin, other researchers had synthesized the 9,3''-di-O-acetyl derivative of midecamycin A$_1$, one factor in a complex of 16-membered macrolides produced by *S. mycarofaciens* (Fig. 32.27) (281). This diester exhibited greater efficacy in treating experimental infections in laboratory animals; subsequent investigation revealed that metabolic deacylation of the 4''-O-acyl group was followed by a facile 3''-O- to 4''-O-acyl migration thereby regenerating an active 4''-O-acyl derivative and accounting for the prolonged *in vivo* half-life of antibiotic activity and efficacy (282). 9,3''-Di-O-acetyl-midecamycin was also commercially introduced into many countries during the last decade under the generic name of miokamycin (283).

The beneficial effects of acylation in the leucomycin family ultimately prompted the synthesis of analogous esters of tylosin.

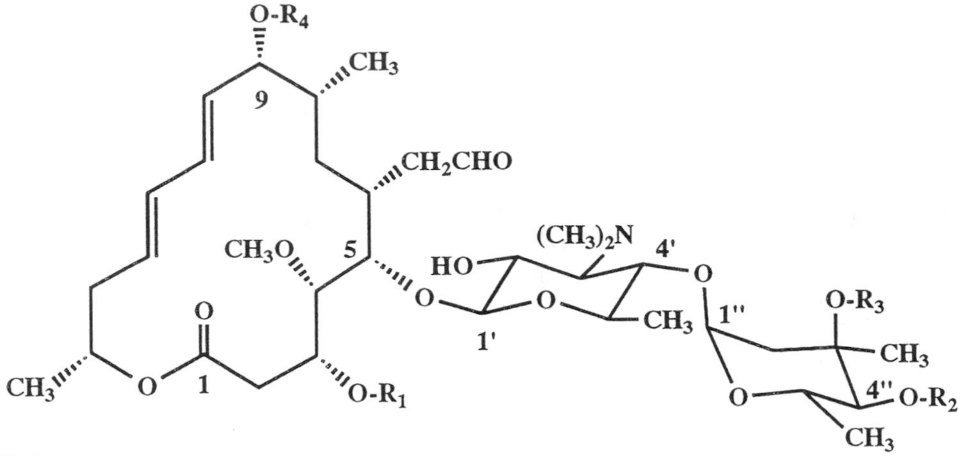

Fig. 32.27 Structures of miokamycin, midecamycin A_1, rokitamycin, and leucomycin A_5. Miokamycin (R_1 = propionyl, R_2 = propionyl, R_3 = acetyl, R_4 = acetyl); midecamycin A_1 (R_1 = propionyl, R_2 = propionyl, R_3 = H, R_4 = H); rokitamycin (R_1 = H, R_2 = butyryl, R_3 = propionyl, R_4 = H); leucomycin A_5 (R_1 = H, R_2 = butyryl, R_3 = H, R_4 = H).

Initial studies used microbial transformations by organisms, such as, *S. thermotolerans*, the producer of carbomycin, which converted tylosin into 3-*O*-acetyl-4″-*O*-acyl derivatives (284). These derivatives inhibited some macrolide-resistant staphylococci and *Mycoplasma* and produced higher concentrations of antibiotic in serum after oral administration (285–287). Based upon these advantages, 3-*O*-acetyl-4″-isovaleryltylosin (AIV-tylosin) (Fig. 32.24) was developed as a veterinary antibiotic. The biotransformations of tylosin were extended to the chemical synthesis of a wider range of derivatives. From SAR studies of *in vitro* and *in vivo* antibiotic activity and stability to mouse liver esterase, 4″-*O*-(4-methoxyphenyl-acetyl)tylosin (YM133) (Fig. 32.24) has been selected as one candidate for preclinical evaluation (288,289).

Another approach to semisynthetic derivatives arose from improved oral bioavailability after certain modifications of the aldehyde group of tylosin and desmycosin (290). From a series of reductive amination products, tilmicosin (Fig. 32.28) was selected on the basis of optimum efficacy against infections caused by the vet-

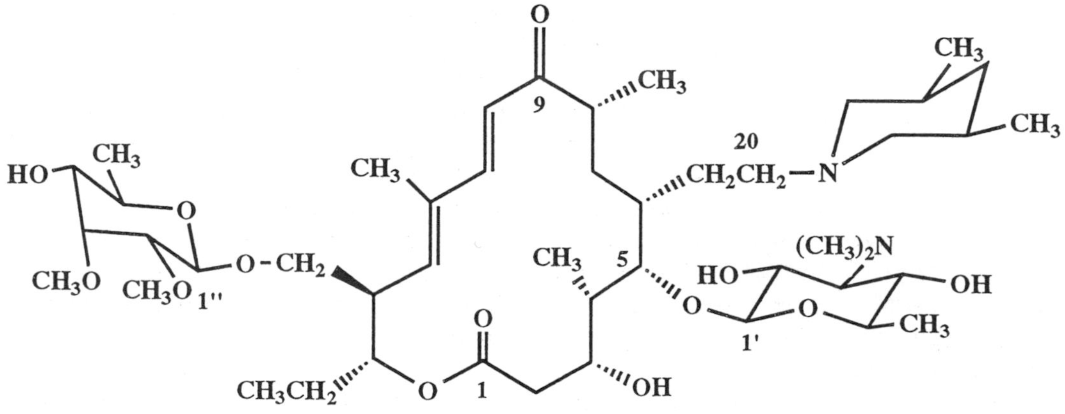

Fig. 32.28 Structure of tilmicosin.

erinary respiratory pathogens, *Pasteurella multocida* and *P. haemolytica* (291–293). Tilmicosin has a prolonged *in vivo* half-life that permits treating bovine respiratory disease by a single injection and is used to treat respiratory diseases in cattle (294,295). It is also being developed as a feed medication for the control of pneumonia in pigs (296).

Improved oral pharmacokinetics have been reported for other aldehyde-modified derivatives of tylosin as well as the natural product, mycinamicin (290,297,298). Combining this effect with improved potency generally exhibited by 4'-deoxy derivatives, desmycosin was converted to its 19-deformyl-4'-deoxy derivative (TMC-016) (298,299). Another new semisynthetic derivative being clinically evaluated is 3,4'-dideoxy-OMT (MC-352, YM17K) (300–302). Many derivatives of OMT with potent *in vitro* activity have been synthesized in which nonsaccharide substituents were introduced at C-23 by substitution or displacement reactions, but none have apparently progressed into extensive preclinical studies (303–305).

3.6 Summary of the Macrolides

Macrolides are a valuable class of antibiotics whose oral bioavailability and relative safety have given them an important clinical role. The older macrolides were well established in their niche, but a renaissance developed during the 1980s when macrolide-susceptible pathogens, such as, *Legionella*, became more prominent and semisynthetic derivatives with a broader antimicrobial spectrum and better oral pharmacokinetics emerged. Each derivative exhibits different features and the newer macrolides have been extensively reviewed, either as individual compounds or more broadly (171–174). They act as inhibitors of microbial protein synthesis, and microbial resistance is well documented (306–309).

Being lipophilic, they are readily taken up and concentrated in tissues and intracellular sites, often achieving tissue concentrations many times in excess of those in serum; to varying degrees, they achieve high and prolonged concentrations of *in vivo* efficacy and exhibit activity against many intracellular pathogens (310–314).

Macrolides are perceived as safe antibiotics whose principal side effects are gastrointestinal. A recent body of research has revealed that erythromycin acts as a nonpeptide agonist of the gut peptide, motilin, which governs gastrointestinal (GI) contractility during the interdigestive phase (315). As a result, erythromycin is being considered for management of motility-related GI disorders (316). The antibiotic activity of erythromycin has been separated from its prokinetic effects; the 8,9-anhydro-6,9-hemiketal and its ring-contracted analog (Fig. 32.29) are weaker antibiotics but more potent prokinetic agents than erythromycin (317–320). This research has been extended to SAR studies from which EM-523 and EM-574 (Fig. 32.29) have emerged as clinical candidates (321,322). Based on their mechanism of action, these macrolides have been named motilides. Other research groups have more recently entered this field with other prokinetic macrolides (323,324). Clinical trials are currently in progress to determine if a GI prokinetic derivative of erythromycin can be developed as a novel therapeutic agent for these nonantimicrobial applications.

The complex structures of macrolides have provided challenging synthetic targets that were deemed hopeless in 1956 (325). However, synthetic methodology has advanced to a degree that rapid assembly of aglycones is being realized, and understanding their three-dimensional stereochemistry has been greatly aided by computer-assisted conformational analysis (11–13,326–328).

Their biosynthesis has been studied since they were first isolated, and derivatives

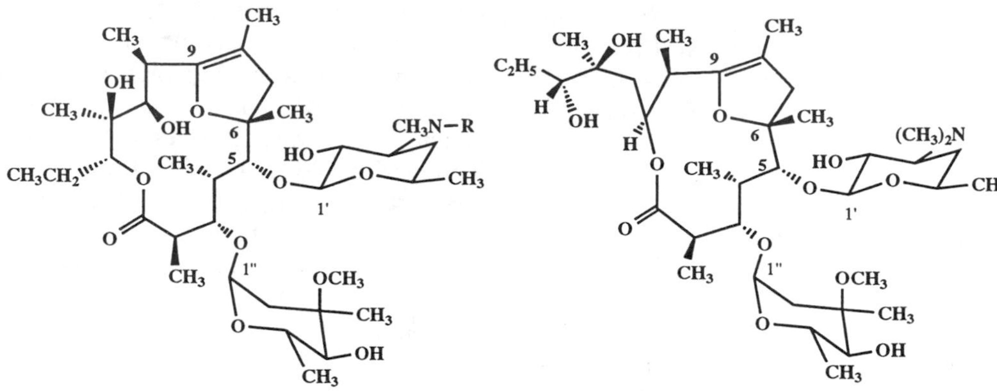

Fig. 32.29 Structures of GI prokinetic agents derived from erythromycin (motilides). 8,9-Anhydroerythromycin A 6,9-hemiketal [R = CH$_3$], EM-523 [R = CH$_2$CH$_3$], EM-574 [R = CH(CH$_3$)$_2$], and the ring-contracted analog LY267108.

have been prepared by traditional methods of biotransformations and mutation of producing organisms (329,330). Another approach to hybrid macrolides used cerulenin, an inhibitor of fatty acid biosynthesis, to block endogenous aglycone formation while precursing with an analog of the aglycone (331,332). Building on this foundation, genetic engineering of macrolide-producing organisms and the directed formation of some modified macrolides has become achievable. Specific genes governing late biosynthetic events have been identified that can direct formation of predicted derivatives by targeted gene disruption (333–335). A conceptual breakthrough in the genetic basis of macrolide biosynthesis was the modular composition for polyketide synthases, the large multienzyme complexes on which macrolides are formed (336–338). This concept not only produced a cohesive theory, but also allowed rational planning to produce novel modified macrolides by genetic reprogramming of the sequence and specificity of events in polyketide synthases, such as, the formation of 6,7-anhydroerythromycin C, a new derivative obtained by changing the enoyl reductase in the fourth cycle of erythronolide biosynthesis in *S. erythraea* (339–342). A partnership of mo-

lecular biology with microbiology, fermentation processes, and chemistry should create an innovative and synergistic environment for future macrolide research.

4 LINCOSAMINIDE AND STREPTOGRAMIN ANTIBIOTICS

The structures of the lincomycin family (lincosaminides) and streptogramin B-type antibiotics are very different from each other and from the 14- and 16-membered macrolides, but these compounds are often considered together due to their similar mechanisms of action and patterns of cross-resistance, called MLS-resistance (macrolide-lincosaminide-streptogramin B resistance) (306–309). This phenomenon occurs as a result of their overlapping binding sites on ribosomes, where they compete with each other to bind and thereby to inhibit protein synthesis.

4.1 Lincomycin Family

Lincomycin is an amide of an unusual amino acid and an amino-octose (lincosamine) isolated from culture broths of *S. lincolnensis* (Fig. 32.30) (343). It inhibits

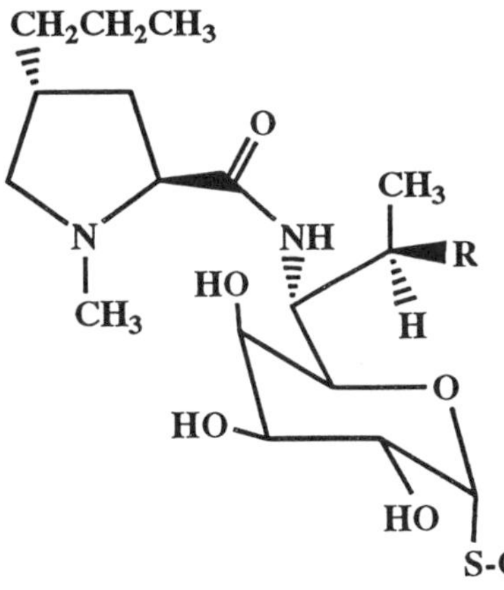

Fig. 32.30 Structures of lincomycin (R = OH) and clindamycin (R = Cl).

gram-positive cocci, *Mycoplasma* species, and some anaerobic bacteria and is orally bioavailable. It is relatively lipophilic and is distributed into tissues and intracellular sites. Lincomycin is used as both an oral and injectable antibiotic to treat gram-positive and anaerobic infections in clinical and veterinary medicine. A related antibiotic, celesticetin, was isolated from culture broths of *S. caelestis* (344).

Treatment of lincomycin with thionyl chloride converted the hydroxyl group at C-7 of lincosamine to a chlorine substituent, producing a more potent and better absorbed semisynthetic antibiotic named clindamycin (345). Both lincomycin and clindamycin are weakly basic and form orally absorbed, water-soluble, acid addition salts. Analogous to the macrolides, the 2'-*O*-palmitate and 2'-*O*-phosphate of clindamycin are pro-drugs for oral pediatric and parenteral administration, respectively. Both hydrolyze *in vivo* to release clindamycin as their active antimicrobial ingredient. Its activity is comparable to erythromycin against gram-positive cocci, but clindamycin has excellent activity against many anaerobic bacteria including the important anaerobic pathogen, *Bacteroides fragilis* (346). The association of clindamycin with pseudomembranous colitis has

been a limitation on its even wider use in antibiotic chemotherapy.

Pirlimycin is another semisynthetic lincosaminide that is being developed for veterinary medicine. It is a derivative of clindamycin, in which the natural five-membered cyclic amino acid has been replaced by a six-membered cyclic amino acid (347). Its principal application appears to be treatment of bovine mastitis caused by staphylococci and streptococci (348).

4.2 Streptogramins

The streptogramins are highly modified peptide antibiotics that are functionally but not structurally related to the macrolides and lincosaminides. The name streptogramin was originally applied to a mixture isolated from culture broths of *S. graminofaciens* (349). A related multicomponent complex of peptide antibiotics, named virginiamycin, was later discovered in broths of *S. virginiae* (350). Other research groups have found several series of related peptides, such as, the pristinamycin, mikamycin, vernamycin, ostreogrycin, staphylomycin, and patricin complexes (351). An unusual feature of this family is its mix of two structurally unrelated cyclic depsipeptides: an unusually substituted 23-membered unsaturated macrolactone (type A) and a more diverse group of cyclic hexadepsipeptides (type B) (Fig. 32.31) (352–354). Although the latter group by themselves have greater bacteriostatic activity, the combination of the two types (approximately 7:3 A:B) creates a synergistic bactericidal effect (355,356). These combinations are orally active against gram-positive and some gram-negative bacteria. Virginiamycin and mikamycin are commercially used as feed additives for improving feed efficiency and promoting growth of poultry and pigs.

The pristinamycin complex, obtained from *S. pristinaespiralis*, has been used as

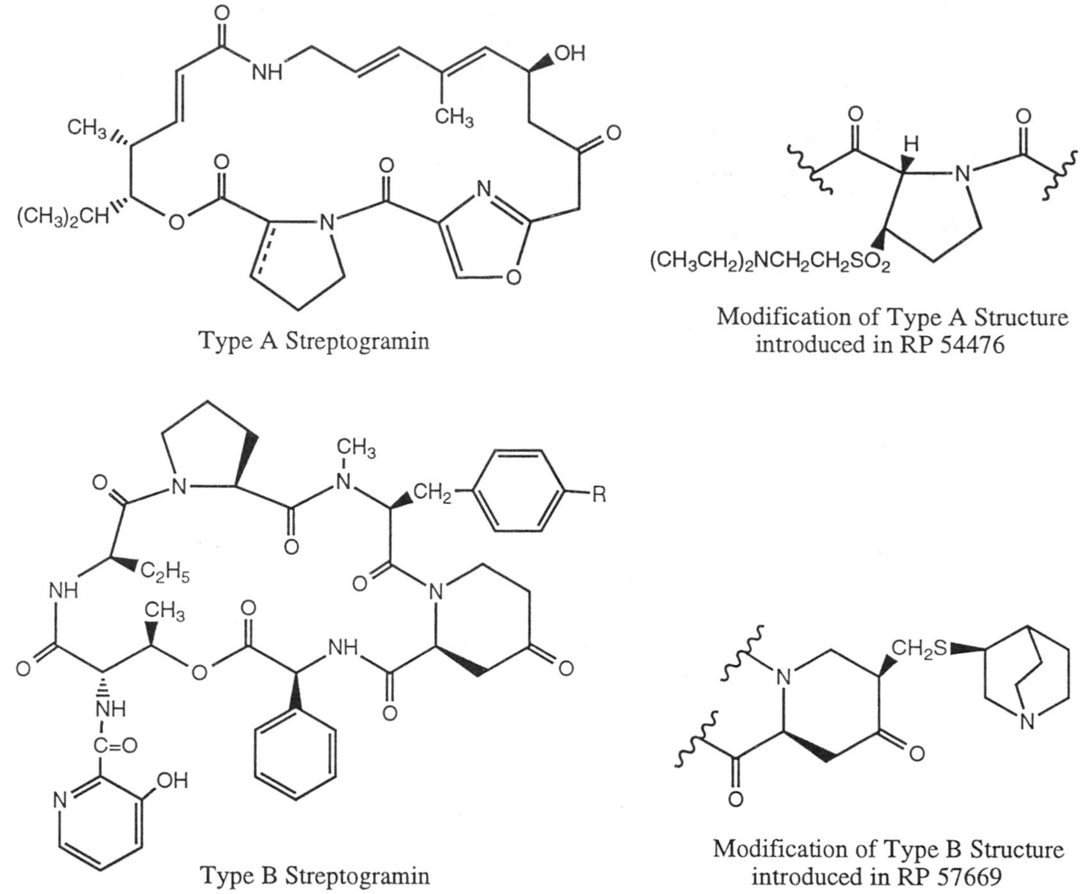

Type A Streptogramin

Modification of Type A Structure
introduced in RP 54476

Type B Streptogramin

Modification of Type B Structure
introduced in RP 57669

Fig. 32.31 Structures of natural and semisynthetic streptogramin-type cyclic depsipeptides. Type A: strepto-gramin A (virginiamycin M_1, pristinamycin II_A): dashed line represents double bond; virginiamycin M_2 (pristinamycin II_B): dashed line represents single bond; Type B: virginiamycin S_1 (R = H); Pristinamycin IA [R = $N(CH_3)_2$]; Pristinamycin IB (R = $NHCH_3$) and modifications introduced into RP 54476 and RP 57669 components of RP 59500.

an oral clinical antibiotic in France (357). However, the pristinamycins are poorly water-soluble, which has limited their clinical utility. A modification program has yielded new water-soluble semisynthetic derivatives of both streptogramin A and B that enable their combined use as an injectable antibiotic (358). Denoted as RP 59500, it is a 3:7 mixture of RP 57669 (quinupristin; pristinamycin I_A modified by an amino-thiomethyl substituent α to the carbonyl of oxopipecolic acid) and RP 54476 (dalfopristin; pristinamycin II_A modi-

fied by an amino-sulfonyl substituent added to the dehydro-proline residue) (Fig. 32.31). This combination exhibits bactericidal activity against strains resistant to other antibiotics, and RP 59500 is being clinically evaluated (359,360).

5 GLYCOPEPTIDE ANTIBIOTICS

The term "glycopeptide" is commonly used to denote a class of antibiotics whose structure is comprised of a polycyclic, cross-linked array of mostly aromatic amino acids that may be glycosylated by amino and neutral sugars. They are water-soluble, polar antibiotics that are not orally absorbed and many are zwitterionic. Their antimicrobial spectrum is almost exclusively gram-positive aerobic and anaerobic bacteria. They are valuable for treating infections caused by gram-positive bacteria, especially staphylococci that are resistant to other antibiotics. Vancomycin is also important for treating antibiotic-induced enterocolitis associated with *C. difficile*.

Vancomycin-resistant strains had not been found among susceptible species until recently. During the last decade, use of vancomycin has greatly increased, teicoplanin has become available, bacterial resistance to glycopeptides has appeared, molecular mechanisms of glycopeptide resistance have been uncovered, and research to find new derivatives that overcome this resistance has intensified. Reviews covering this expansion of glycopeptide research have been published (361–367). Fermen-

tation products research contributed many new glycopeptides, particularly through very specific mechanism-based assays to detect their presence in fermentation broths and rapid selective methods to isolate them, such as bioaffinity chromatography (368–370). Structures of new glycopeptides are now rapidly determined using modern techniques and comparisons with many available analogs.

5.1 Vancomycin Family

The prototypic glycopeptide is vancomycin, an old antibiotic isolated from culture broths of *S. orientalis* (371), an organism now reclassified as *Amycolatopsis orientalis* (372). Its structure is a tricyclic ring system of aliphatic and aromatic amino acids in which the latter are cross-linked by aryl ether and biphenyl bonds (Fig. 32.32). A disaccharide, α-vancosaminyl-β-glucosyl, is linked to the central hydroxyl group of the triaryl ether system. Efforts to elucidate the

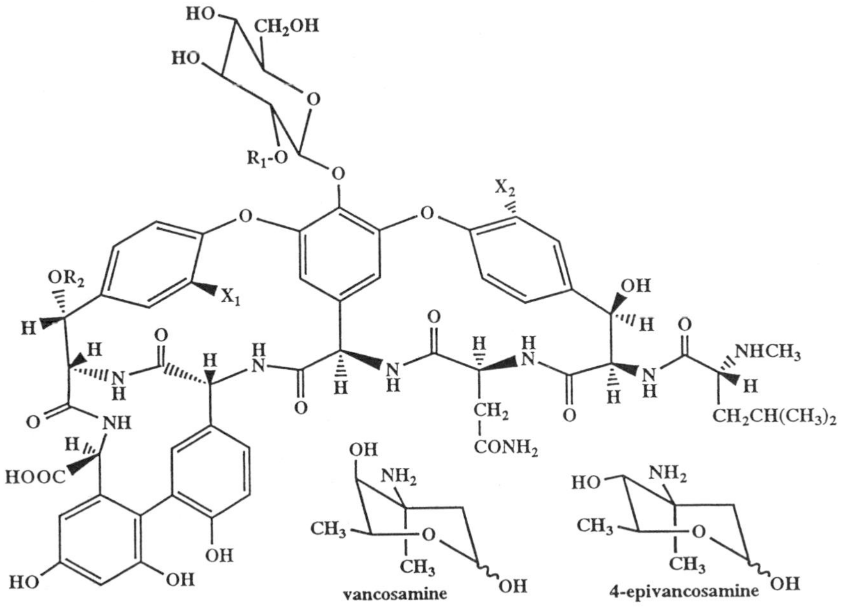

Fig. 32.32 Structures of vancomycin (X_1 = Cl, X_2 = Cl, R_1 = α-vancosaminyl, R_2 = H), A82846B (X_1 = Cl, X_2 = Cl, R_1 = α-epivancosaminyl, R_2 = α-epivancosaminyl), eremomycin (X_1 = H, X_2 = Cl, R_1 = α-epivancosaminyl, R_2 = α-epivancosaminyl), orienticin A (X_1 = Cl, X_2 = H, R_1 = α-epivancosaminyl, R_2 = α-epivancosaminyl), and decaplanin (X_1 = H, X_2 = Cl, R_1 = α-rhamnosyl, R_2 = α-epivancosaminyl).

structure of glycopeptides extended over many years and the breakthrough was accomplished in 1978 by X-ray crystallography of a degradation product of vancomycin (373). This pioneering result provided the necessary clues to assemble the structural pieces for not only vancomycin, but all glycopeptides, resulting in the rapid structure elucidation of others by modern techniques, such as, high resolution nuclear magnetic resonance (NMR) spectroscopy. An explanation of the stereochemical ambiguity about the two aryl chlorine atoms finally defined the complete structure of vancomycin in 1983 (374).

Several series of glycopeptides have been discovered that have the same cyclic peptide core as vancomycin. The A51568 and M43 complexes differ mainly in their degree of N-methylation of isoleucine (375,376). Several series have been found that contain two aminosugars, both of which are 4-epivancosamine, in which one of them replaces vancosamine and the other is attached to the benzylic hydroxyl group (Fig. 32.32). These series include the A82846, eremomycin, orienticin, and chloroorienticin complexes, which differ among themselves mostly in the number and position of aryl chlorine atoms; not unexpectedly, some factors in each complex are the same (377–381). Decaplanin (Fig. 32.32) is another member of this family that was isolated from culture broths of *Kibdelosporangium deccaensis* (382). It contains a neutral disaccharide attached to the central hydroxyl group of the triaryl ether and epivancosamine attached to the benzylic hydroxyl group. Most of these glycopeptides are potent antibiotics and some have entered preclinical studies. Many pseudoaglycones derived from partial hydrolysis of saccharides retain good antibacterial activity. Several glycopeptides have served as starting material for semisynthetic derivatives, such as, N-alkyl derivatives that have shown improved activity against vancomycin-resistant bacteria (383).

5.2 Teicoplanin Family

Teicoplanin (formerly teichomycin) is the only other glycopeptide presently used in human medicine. It was isolated from culture broths of *Actinoplanes teichomyceticus* as a complex of five factors that differ in the nine- or ten-carbon acyl group attached as an amide to glucosamine, which itself is attached to the central hydroxyl group of the triaryl ether system (Fig. 32.33) (384). The cyclic peptide core differs from vancomycin by a third cross-linked aromatic amino acid moiety in place of the aliphatic amino acid component in vancomycin. In addition, the saccharides in teicoplanin and vancomycin are completely different (Fig. 32.33) (385–388). Four additional minor factors have been isolated that differ only in their lipophilic acyl group (387).

Teicoplanin and vancomycin have a similar antimicrobial spectrum and the two glycopeptides are used in a similar manner for the parenteral treatment of infections caused by gram-positive bacteria, especially methicillin-resistant staphylococci. Teicoplanin is the more lipophilic, probably due to the lipophilic acyl group, which gives it higher tissue penetration and a longer *in vivo* half-life than vancomycin. As a result, teicoplanin can be administered as a single daily dose by either intravenous or intramuscular routes (389–391). Teicoplanin has been the subject of many structure modifications to improve its antimicrobial features as well as overcome the emerging antibiotic resistance in enterococci and other bacteria (392–394).

Series of related glycopeptides have been isolated, such as, the kibdelin, parvodicin, and ardacin complexes, whose factors differ both in their saccharides and the degree of chlorination, hydroxylation, and N-methylation of their aglycones (395–397). Although the natural products and semisynthetic derivatives have revealed structure-activity relationships, no other glycopeptide has yet been found whose

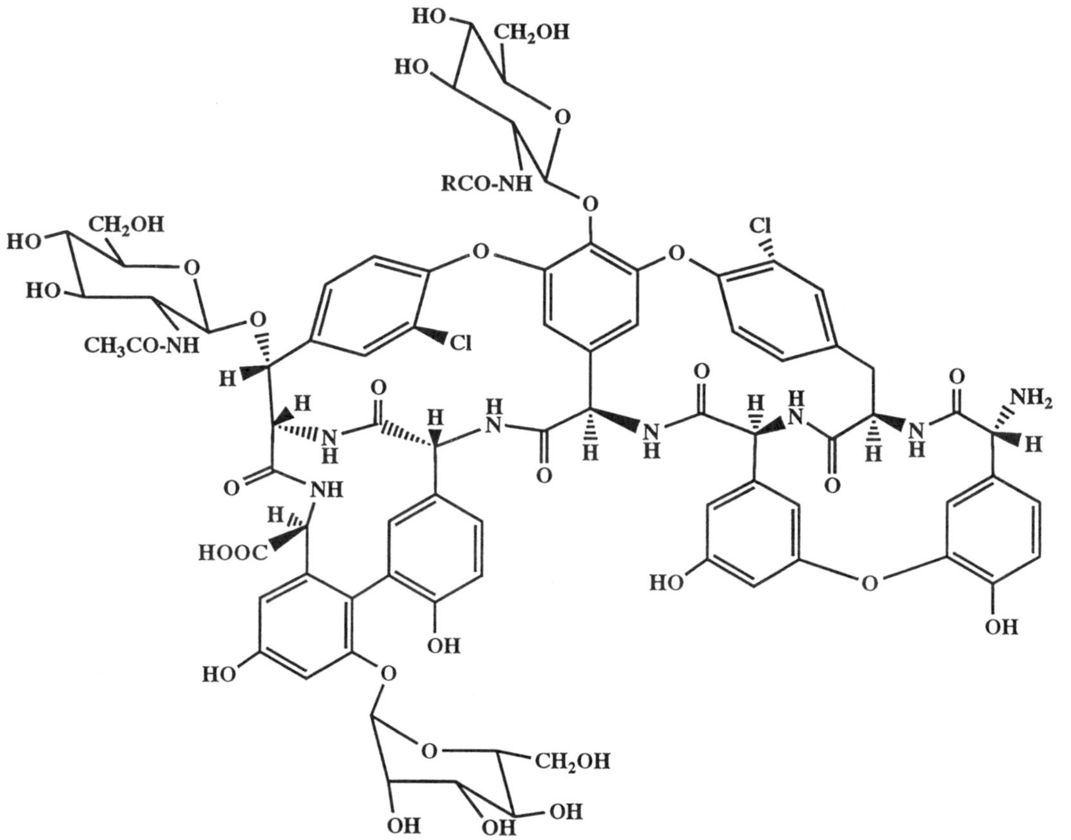

Fig. 32.33 Structures of teicoplanin factors: A_2-1 [R = $(CH_2)_2CH=CH(CH_2)_4CH_3$], A_2-2 [R = $(CH_2)_6$-CH$(CH_3)_2$], A_2-3 [R = $(CH_2)_8CH_3$], A_2-4 [R = $(CH_2)_6CH(CH_3)CH_2CH_3$], A_2-5 [R = $(CH_2)_7CH(CH_3)_2$].

antimicrobial profile is sufficiently enhanced to warrant its development.

5.3 Ristocetin Family

Ristocetin is an old glycopeptide isolated from culture broths of *Nocardia lurida*, an organism reclassified as *Amycolatopsis orientalis* subsp. *lurida* (398,399). Ristocetin A has the same polycyclic peptide core as teicoplanin, but differs in its substituents by an ester rather than a carboxylic acid, the aminosugar L-ristosamine, lack of aromatic chlorine atoms and the lipophilic acyl group, and a tetrasaccharide attached to the central hydroxyl group of the triaryl ether system (Fig. 32.34) (400).

Ristomycin, isolated from culture broths of *Proactinomyces fructiferi*, is identical to ristocetin (401). Although effective as an antibiotic, ristocetin was withdrawn from clinical use because of a high incidence of thrombocytopenia (402). Subsequent investigations of its hemostatic effects indicated that it caused platelet aggregation through its interaction with a plasma factor missing in patients suffering from one form of von Willebrand's disease. This result has been exploited by the use of ristocetin for the differential diagnosis of this disorder (403).

Several related series have been isolated, including the actaplanin, A35512, and A41030 complexes, whose factors differ in both their saccharides and aglycone substituents (404–406). Three factors of

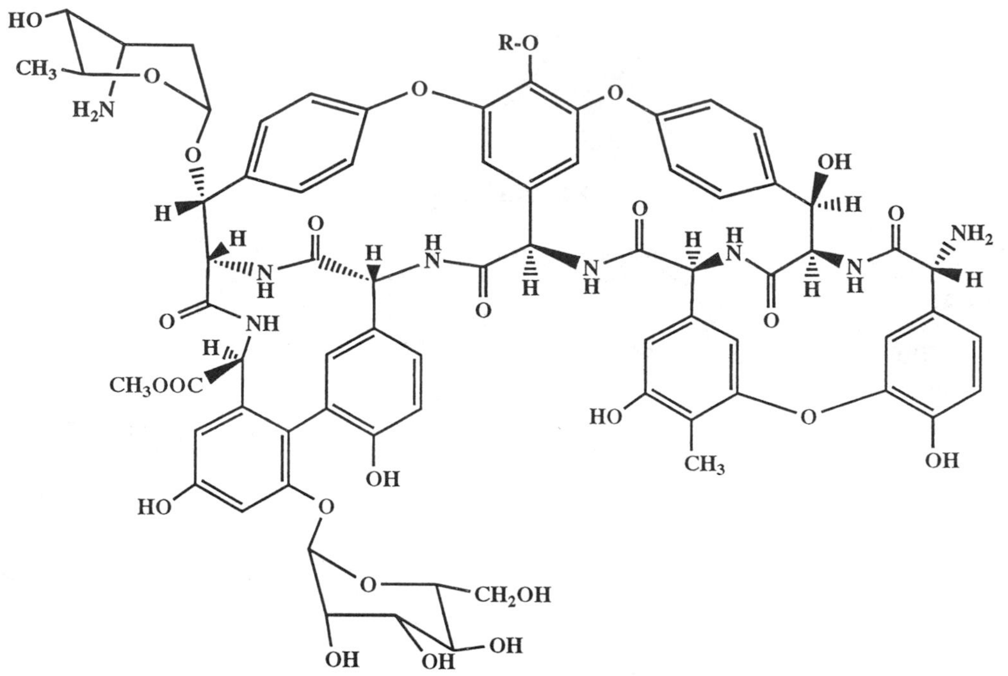

Fig. 32.34 Structure of ristocetin A [R = 2-*O*-α-(2-*O*-α-D-arabinosyl-D-mannosyl)-6-*O*-α-(L-rhamnosyl)-β-D-glu-cosyl.

A41030 are natural aglycones lacking any sugars that retain good antibacterial activity. A47934 and UK-68,597 are naturally occurring O-sulfated analogs (407–409).

5.4 Avoparcin Family

The fourth family of glycopeptides is represented by avoparcin, isolated in the 1960s from culture broths of *S. candidus*, an organism reclassified as *Amycolatopsis coloradensis* (410,411). Its aglycone is comprised of only aromatic amino acids, but two of them are not cross-linked as the bis-aryl ethers found in teicoplanin and ristocetin (Fig. 32.35) (412–414). Related series include the actinoidin, chloropolysporin, and helvecardin complexes (415–418). Synmonicin differs from the other members of this family by substitution of a thioamino acid for an aromatic amino acid (419). Avoparcin, the only

glycopeptide used in veterinary practice, has been used for many years as a feed additive to promote growth and improve feed efficiency in poultry and ruminant animals. It has recently been approved for increasing milk production in dairy cattle (420).

5.5 Summary of the Glycopeptides

Glycopeptides play a critical role in chemotherapy, especially due to their efficacy against methicillin-resistant staphylococci (365,366,421,422). Vancomycin was the sole clinical glycopeptide for many years and although once relegated to a secondary role, its use expanded due to increasing antibiotic resistance and newer indications, such as, treatment of antibiotic-induced colitis. For most of this period, there was no resistance to vancomycin, making it a drug of last resort, but this lack of resist-

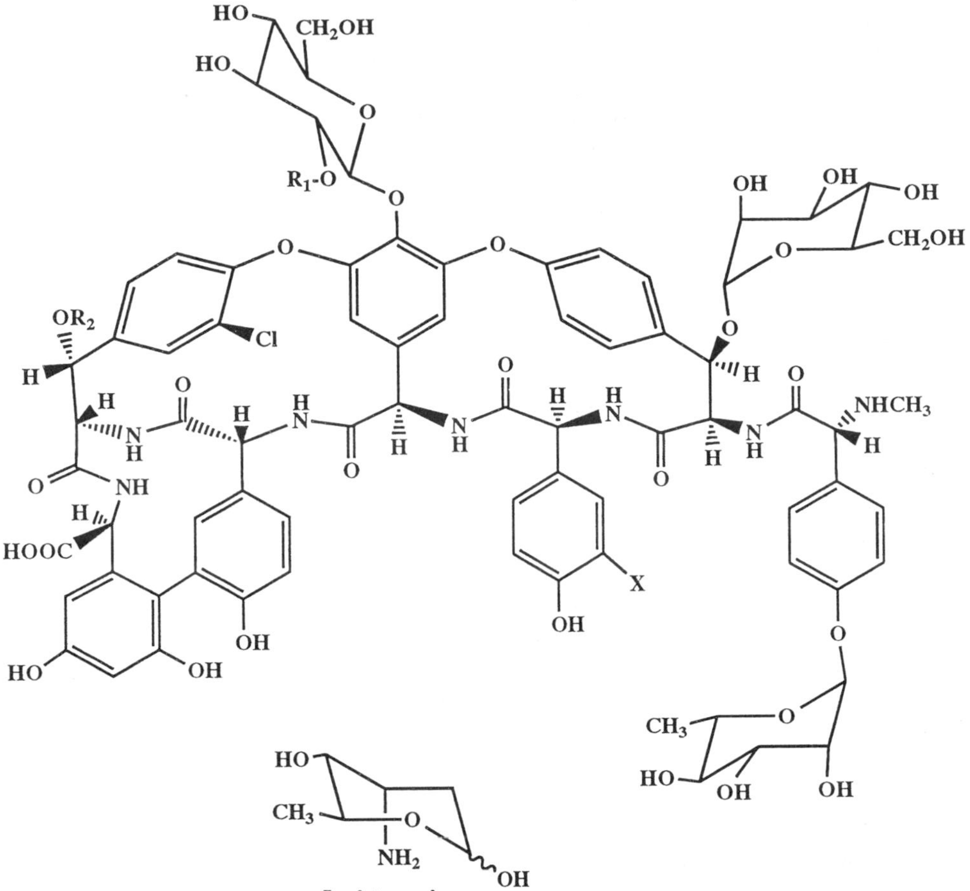

Fig. 32.35 Structures of α-avoparcin (X = H, $R_1 = R_2 = \alpha$-ristosaminyl) and β-avoparcin (X = Cl, $R_1 = R_2 = \alpha$-ristosaminyl).

ance ended with the discovery of acquired resistance in enterococci (423–426).

Vancomycin inhibits cell wall biosynthesis by binding to the terminal D-alanyl-D-alanyl sequence of peptidoglycan precursor. Based on NMR spectroscopy and molecular modeling, vancomycin fits into a pocket in which binding is stabilized by multiple hydrogen bonds (427). As a result, trans-glycosylase and transpeptidase enzymes are inhibited from constructing and cross-linking the rigid bacterial cell wall that maintains the structural integrity of the bacterium (364,373,428–430). Dimerization of glycopeptides and their interactions with membranes near the enzymes have been proposed to help explain the strong binding and selective affinity (431).

Certain gram-positive bacteria are intrinsically resistant to vancomycin, but the significance of acquired resistance in enterococci was immediately recognized (423–426). Two phenotypes of inducible resistance are known (VanA and VanB); the first is plasmid-mediated and confers high level resistance to both vancomycin and teicoplanin, whereas the second confers resistance only to vancomycin (432). Five separate genes (vanR, vanS, vanH, vanA, and vanX) are implicated in the VanA phenotype that work in concert to change the peptidoglycan sequence from D-ala-D-ala to D-ala-D-lactate. Vancomycin's inability to tightly bind the D-ala-D-lac peptidoglycan precursor is attributed to the loss of one critical hydrogen bond in the complex, a change sufficient to diminish binding and thereby reduce enzyme inhibition to ineffective levels (433–436). Clinical strains of *S. aureus* possessing the VanA mechanism of resistance have not yet been found, but the ability to transfer resistance from enterococci to *S. aureus* has been demontrated in a laboratory (437). The mechanisms of glycopeptide resistance are being vigorously studied, but the results already known emphasize the need for medicinal chemists to discover novel antibiotics that operate by new mechanisms of action and do not possess cross-resistance with other antibiotics.

6 PEPTIDE ANTIBIOTICS

The peptide antibiotics are a large diverse class of natural products, but relatively few of them have attained a significant role in chemotherapy. What to include under the designation of peptide antibiotics is somewhat arbitrary. Large proteins, such as, lysozyme, are not generally included with the smaller fermentation products. β-Lactams, glycopeptides, and streptogramins are peptide-derived antibiotics whose amino acids have undergone extensive modifications, but these compounds are often put in separate classes because of their therapeutic importance. Some peptide antibiotics contain only amino acids joined by amide bonds, whereas others contain nonamino acid constituents joined in ways other than conventional peptide linkages. The amino acids range from those commonly found in proteins to uncommon ones with highly modified structures. Finally, the peptide array may be linear or cyclic or various combinations thereof.

Comprehensive reviews of this class have been published (438–443). This chapter cannot provide an exhaustive coverage, so it will focus on a few representatives of medicinal significance. Some synthetic and SAR studies have been conducted, but nonpeptide chemists tend to avoid projects with peptides as starting materials due to their complex structures and a lack of selective chemical transformations. Many of the more complex structures have only been elucidated with the advent of modern physico-chemical and spectroscopic methods. More of these compounds may be investigated as they are recognized as an unexploited resource for new antibiotics.

6.1 Bacitracin

Bacitracin A is the main component of a multifactor complex isolated in 1943 from culture broths of *Bacillus licheniformis* (444,445). It is a dodecapeptide in which six amino acids form an intramolecular hexapeptide ring (Fig. 32.36) (446–448). Bacitracin inhibits cell wall formation in gram-positive bacteria by complexing with polyisoprenyl pyrophosphate (449,450). Because of toxicity, clinical use is generally restricted to topical applications, often combined with polymyxin B and neomycin (triple ointment) to prevent infections in skin wounds and burns. Bacitracin is used more as a veterinary antibiotic and a feed additive for improved growth and feed efficiency.

6.2 Tyrothricin and the Gramicidins

Tyrothricin is a peptide mixture produced by *Bacillus brevis* composed of two main entities, tyrocidin and gramicidin, in an approximate 4:1 ratio (451). Tyrocidin is a mixture of three cyclic decapeptides (Fig. 32.37) whereas gramicidin is a more complex mixture of linear and cyclic peptides.

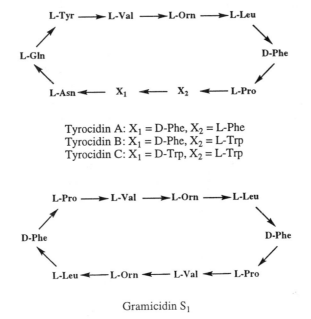

Fig. 32.36   Structure of bacitracin A.

Gramicidin S is a cyclic decapeptide similar to the tyrocidins, whereas gramicidins A–D are linear pentadecapeptides that are N-formylated and whose C-terminus forms an amide with ethanolamine (Fig. 32.37) (452). Gramicidin inhibits gram-positive bacteria, while tyrocidin is less potent but also has some activity against gram-negative bacteria. They appear to form channels or pores in bacterial membranes, causing leakage of cellular contents and loss of the cell's structural integrity (453–455). Due to toxicity, these compounds are used only by topical administration.

Tyrocidin A: X_1 = D-Phe, X_2 = L-Phe
Tyrocidin B: X_1 = D-Phe, X_2 = L-Trp
Tyrocidin C: X_1 = D-Trp, X_2 = L-Trp

Gramicidin S_1

CHO--(L-Val)--Gly--(L-Ala)--(D-Leu)--(L-Ala)--(D-Val)--(L-Val)--(D-Val)--(L-Trp)--
(D-Leu)--(L-Trp)--(D-Leu)--(L-Trp)--(D-Leu)--(L-Trp)--$NH_2CH_2CH_2OH$

Gramicidin A

Fig. 32.37   Structures of tyrocidins A-C and gramicidins S and A.

6.3 Polymyxin and Colistin

The large polymyxin family of decapeptides is produced by *B. polymyxa* and characterized by a cyclic heptapeptide, several units of α,γ-diaminobutyric acid, and a nine-carbon fatty acid (Fig. 32.38) (456–458). Its structure was determined after a long series of degradative and synthetic studies (459). Colisin A and B and circulin A are obtained from *B. colistinus* and *B. circulans*, and colistin A and B are the same as polymyxin E_1 and E_2, respectively (460). Related series include the octapeptin, stendomycin, brevistin, and polypeptin complexes (461).

The polymyxins form water-soluble salts, but only a few attempts have been made to prepare derivatives. Their activity is bactericidal but restricted to gram-negative bacteria, including strains resistant to other antibiotics. They complex with membrane phospholipids and disrupt cell membranes similar to cationic detergents (462). Resistance is acquired by a slow complex process (463). Injectable forms of polymyxin B and colistin (polymyxin E) had previously been important for treating gram-negative infections, but their principal application is now topical in combination with bacitracin and neomycin to provide greater coverage against gram-negative bacteria encountered in skin infections (464).

6.4 Miscellaneous Lipopeptide Antibiotics

Many peptide antibiotics, such as, the polymyxins, contain a lipophilic acyl group linked as an amide to a linear and/or cyclic assembly of amino acids. Many other examples of lipopeptides have been isolated, such as, amphomycin, produced by *S. canus* as a pair of undecapeptides having a 12- and 13-carbon branched-chain unsaturated acyl group and several uncommon amino acids (465,466). Amphomycin inhibits gram-positive bacteria by complexing with carrier isoprenoid phospholipids, blocking transmembrane transfer of peptidoglycan precursors and disrupting cell wall synthesis (467–469). It has been used as a topical antibiotic, but has since been withdrawn.

Enduracidin is a complex produced by *S. fungicidicus* and consists of a macrocyclic ring from 16 aromatic and aliphatic D and L amino acids via 15 amide bonds and one ester bond with L-threonine (470). The latter is *N*-acylated by L-aspartic acid, which itself is *N*-acylated by a unsaturated 12- or 13-carbon branched-chain fatty acid (471). Among its uncommon amino acids are several hydroxy- and chlorohydroxy-phenylglycines and a pair of cyclic guanidinyl amino acids. It is bactericidal against gram-positive bacteria and inhibits transglycosylation during cell wall synthesis (472). Enduracidin is used as a feed additive in veterinary practice in Japan.

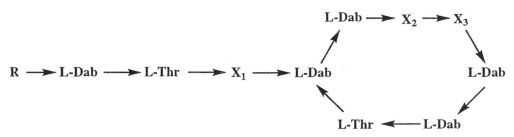

Fig. 32.38   Polymyxin A_1 (X_1 = D-Dab, X_2 = D-Leu, X_3 = L-Thr, R = (+)-6-methyloctanoyl); polymyxin B_1 (X_1 = L-Dab, X_2 = D-Phe, X_3 = L-Leu, R = (+)-6-methyloctanoyl); polymyxin D_1 (X_1 = D-Ser, X_2 = D-Leu, X_3 = L-Thr, R = (+)-6-methyloctanoyl); colistin A (X_1 = L-Dab, X_2 = D-Leu, X_3 = L-Leu, R = (+)-6-methyloctanoyl); polymyxin B_2 (X_1 = L-Dab, X_2 = D-Phe, X_3 = L-Leu, R = 7-methyloctanoyl); colistin B (X_1 = L-Dab, X_2 = D-Leu, X_3 = L-Leu, R = 7-methyloctanoyl).

Globomycin is a unique lipopeptide produced by four different actinomycetes that may be included in this family (473). It is a cyclic entity comprised of five amino acids and 3-hydroxy-2-methylnonanoic acid in which this lipophilic acid is incorporated into the cyclic core rather than located at the terminus of a peptide array (474). It inhibits gram-negative bacteria by interfering with the normal processing of lipoproteins and inhibiting bacterial cell wall synthesis (475).

6.5 Glycolipopeptides: Ramoplanin

Ramoplanin is the principal component of a complex isolated from culture broths of a species of *Actinoplanes* (476). It contains a macrocyclic ring of 16 aromatic and aliphatic amino acids joined via 15 amide bonds and one ester bond involving β-hydroxyaspartic acid. The latter is *N*-acylated by asparagine, which in turn is *N*-acylated by an unsaturated 9-carbon branched-chain fatty acid (Fig. 32.39). Other factors in the complex possess acyl groups from either 8- or 10-carbon fatty acids. Its lipodepsipeptide core bears gross resemblance to enduracidin, but ramoplanin is more complex in that one *p*-hydroxyphenylglycine is glycosylated by the disaccharide, 2-*O*-(α-mannosyl)-α-mannosyl (477,478). It may be more appropriately named a glycolipodepsipeptide.

Ramoplanin inhibits gram-positive bac-

teria, including strains resistant to other antibiotics, by blocking biosynthesis of peptidoglycan components of bacterial cell walls (479,480). Although it was evaluated as a new parenteral antibiotic, its clinical use now appears to be restricted to topical applications as the result of unacceptable side effects following parenteral administration.

6.6 Thiostrepton and Related Thiopeptides

Thiostrepton was isolated from culture broths of *S. azureus* (481). Its complex structure was elucidated by chemical degradative, X-ray crystallographic, and NMR spectroscopic studies and consists of a bicyclic array of highly aromatic moieties, many of which are thiazole and thiazolidine rings along with other nitrogen-containing heterocycles (Fig. 32.40) (482–485). Thiostrepton inhibits gram-positive bacteria in a bacteriostatic manner by inhibiting ribosomal protein synthesis (486). It is used as a veterinary antibiotic and has also become useful in genetic recombination studies where resistance to thiostrepton is employed as a marker.

Many antibiotics containing similar subunits of high sulfur content have been isolated that may be classified as thiopeptide antibiotics, such as thiopeptin, siomycin A, micrococcin P, nosiheptide, and A10255 (487–493). The first three are used

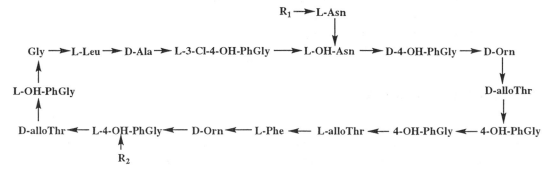

Fig. 32.39 Structure of ramoplanin A_2: R_1 = 7-methyl-2,4-octadienoyl; R_2 = α-D-mannosyl-α-D-mannosyl.

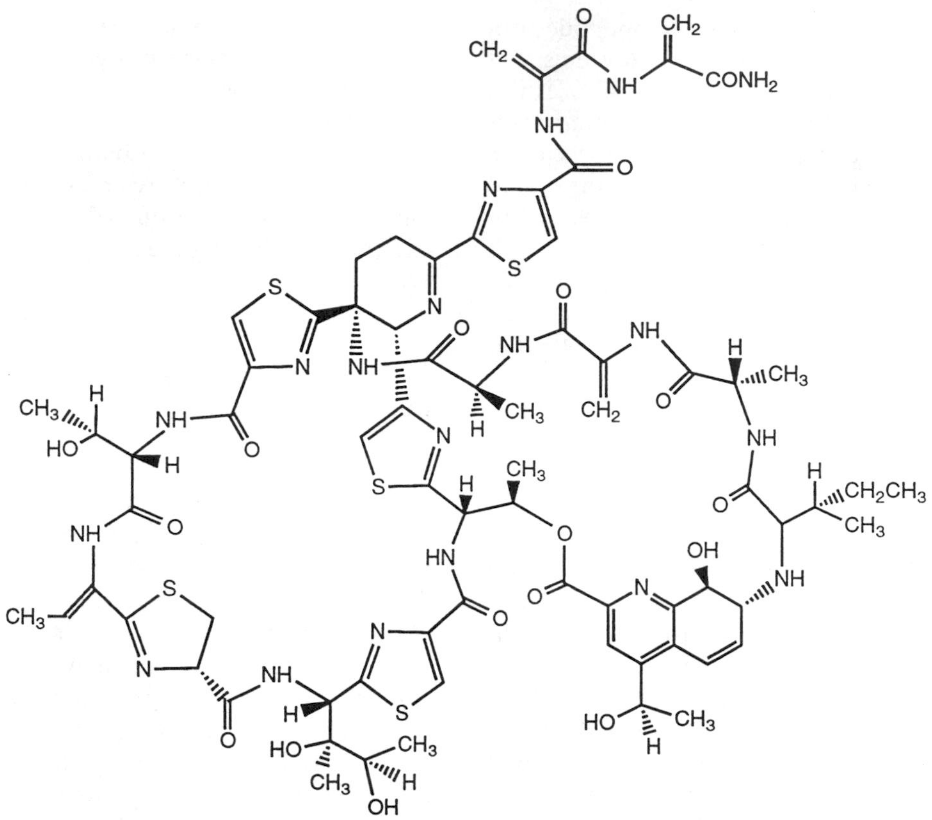

Fig. 32.40 Structure of thiostrepton A_1.

as feed additives for animals. Althiomycin is a smaller analog produced by *S. althioticus* (494,495). GE2270 A, produced by *Planobispora rosea*, is being evaluated as a new clinical antibiotic for infections caused by resistant gram-positive bacteria (496). In contrast to the other thiopeptides that inhibit bacterial protein synthesis by binding to ribosomes, GE2270 A inhibits protein synthesis by complexing with an elongation factor (EF-Tu) necessary for ribosomal binding of aminoacyl tRNA (497).

6.7 Nisin and Related Lantibiotics

Lantibiotics is the name of a family of peptides containing lanthionine, an unusual dimeric amino acid in which two amino acids are joined by a sulfur bridge (498). In

contrast to the previously discussed peptide antibiotics whose biosynthesis occurs on large multienzyme complexes, lantibiotics are assembled on ribosomes (analogous to protein biosynthesis) and undergo substantial posttranslational modifications of their amino acids (499). In particular, dehydration of serine and threonine residues to dehydroalanine and dehydrobutyrine is followed by conjugate addition of thiols from neighboring cysteine residues to form the lanthionine moieties.

The most important lantibiotic is nisin, a 34-amino acid peptide produced by *Streptococcus lactis* (500–502). It is bactericidal against gram-positive bacteria and acts by permeabilizing cell membranes (503). Subtilin is a related peptide produced by *B. subtilis* (504). Nisin and subtilin are commercially used as food preservatives

(505,506). Epidermin, duramycin, and cinnamycin are other lantibiotics, all of which have similarities to bacteriocins, the name for a group of peptide antibiotics produced by bacteria that inhibit other bacteria (498,507,508).

6.8 Semisynthetic Derivatives: Daptomycin

Daptomycin, a lipopeptide derivative, is the foremost recent example of a semisynthetic peptide antibiotic. The parent complex (A21978C) was isolated as a series of tridecadepsipeptides from culture broths of *Streptomyces roseosporus*, in which the major factors differed in the lipophilic acyl group attached to a three-amino acid chain emanating from a ring of ten amino acids that was closed by an ester bond to a threonine residue (Fig. 32.41) (509). The acyl group was most conveniently cleaved by biotransformation using *Actinoplanes utahensis*; selectivity in reacylating the resultant peptide nucleus was enhanced by first blocking the ornithine residue of A21978C as its t-BOC derivative before conducting the bio-deacylation and reacylation procedures (510). Removal of the t-BOC protecting group yielded the desired semisynthetic lipopeptides. The *n*-decanoyl analog, named daptomycin, was selected from a series of N-acyl derivatives on the basis of its antimicrobial activity and better therapeutic ratio of efficacy to acute toxicity (510).

Daptomycin inhibits gram-positive bacteria including strains resistant to other antibiotics and requires Ca^{+2} for activity (511). Two mechanisms of action have been proposed: disruption of bacterial membrane potential or inhibition of lipoteichoic acid synthesis (512–514). Daptomycin progressed into clinical trials but has been withdrawn. Nevertheless, structural modifications of such complex peptide antibiotics may yield useful semisynthetic derivatives, and combined chemical and biochemical procedures may provide more selective and efficient structural modifications.

A complex of related lipopeptides, denoted as A54145, has been isolated from culture broths of *S. fradiae* (515). The factors were tridecapeptides in which a threonine residue formed a cyclic decadepsipeptide, the N-terminus was acylated with various lipophilic acyl groups, and alternate amino acids were located at two positions in the cyclic peptide (516). Semisynthetic derivatives have been prepared by the same microbial deacylation process as employed for daptomycin (517). In addition, the ratio of factors was regulated by controlled feeding of lipid precursors to the fermentation in a biosynthetic approach to produce specific compounds (518).

7 MISCELLANEOUS FERMENTATION-DERIVED ANTIBIOTICS

In addition to the fermentation-derived antibiotics described above and those discussed in other chapters, many other fermentation products with antibacterial activity have been isolated over decades of research in the field. This section cannot provide an all-inclusive compilation of the numerous types of antibiotics that have been isolated. Consequently, it focuses on only a few of the more important repre-

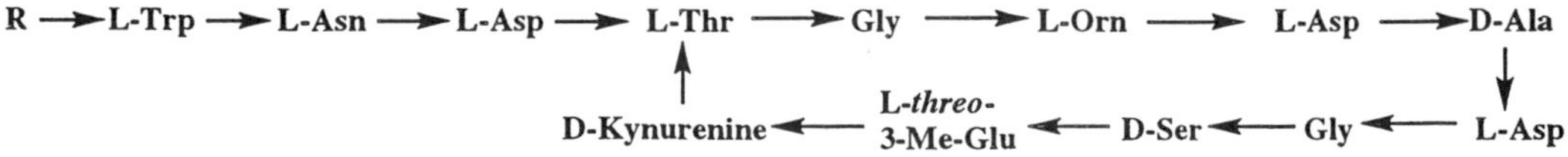

R —→ L-Trp —→ L-Asn —→ L-Asp —→ L-Thr —→ Gly —→ L-Orn —→ L-Asp —→ D-Ala

D-Kynurenine ←— L-*threo*-3-Me-Glu ←— D-Ser ←— Gly ←— L-Asp

Fig. 32.41 Structure of daptomycin.

sentatives that have either attained commercial or scientific significance or exemplify the wide structural diversity exhibited by microbial secondary metabolism.

7.1 Fusidic Acid

Fusidic acid is a fermentation product of the fungus, *Fusidium coccineum* (519). It was shown by chemical methods and X-ray crystallography to be a tetracyclic triterpene having an unsaturated carboxylic acid-containing substituent on the cyclopentyl D ring (Fig. 32.42) (520,521). Related compounds, such as, helvolic acid, helvolinic acid, and cephalosporin P$_1$, that have different oxygenation patterns in the tetracyclic skeleton have been isolated from related fungi (522,523). Many semisynthetic derivatives have been prepared during the course of SAR studies within this fusidane family of natural products (524).

Fusidic acid inhibits gram-positive bacteria, especially staphylococci, as well as some gram-negative cocci and many anaerobic bacteria (525,526). It acts by inhibi-

tion of protein synthesis, and although resistance to fusidic acid is known, cross-resistance between it and other antibiotics has not been observed. Fusidic acid has acquired a niche role primarily in the treatment of staphylococcal infections, although not as a first-choice therapy. It may be administered by intravenous, oral, or topical routes and is well absorbed by the latter two routes (527). It has not been registered for use in the United States, although it is used in many other countries. However, it may play a greater role as problems increase with staphylococci resistant to other antibiotics.

7.2 Mupirocin

Mupirocin (previously named pseudomonic acid) is obtained from fermentation of the bacterium *Pseudomonas fluorescens* (528). It is an ester of 9-hydroxynonanoic acid and a unique oxylipophilic carboxylic acid named monic acid (Fig. 32.43) (529). Mupirocin has a narrow but useful spectrum of activity against certain gram-posi-

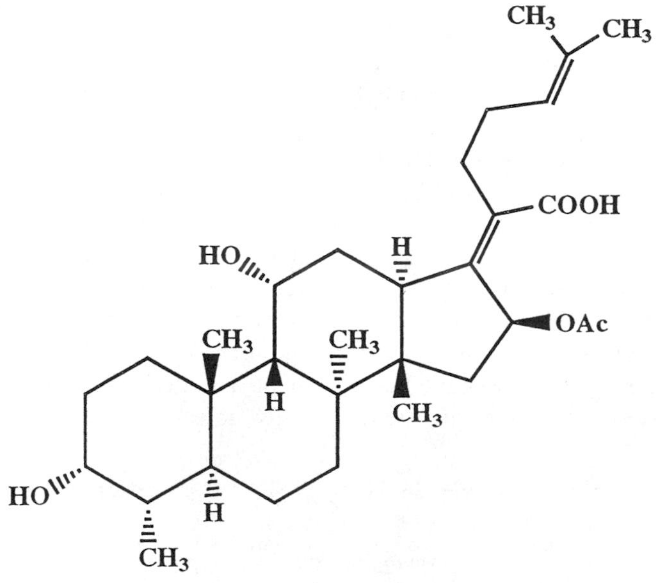

Fig. 32.42 Structure of fusidic acid.

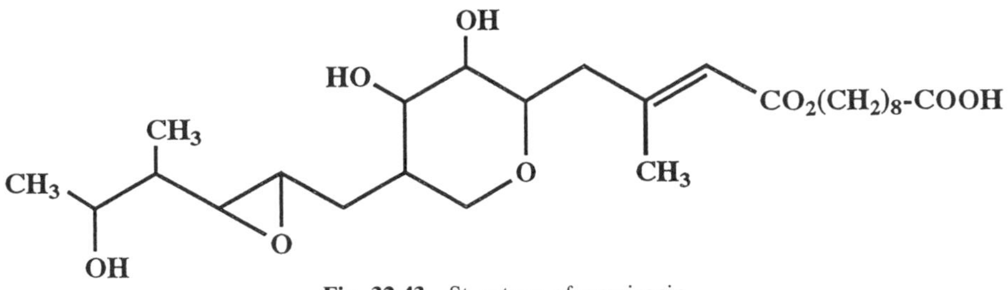

Fig. 32.43 Structure of mupirocin.

tive and gram-negative bacteria (530). Its good activity against important dermal pathogens combined with weaker activity againt normal skin flora makes it a useful topical antibiotic to treat skin infections (531–534). Mupirocin inhibits bacterial protein synthesis by binding to isoleucyl-tRNA synthetase and preventing incorporation of isoleucine into growing proteins (535). As a result of its unique structure and mechanism of action, it does not show cross-resistance to other antibiotics, and resistance to mupirocin itself was initially uncommon (536). However, bacteria have subsequently developed resistance, and high-level resistance to mupirocin is now becoming a more frequent problem (537,538).

7.3 Everninomicin and Avilamycin

The oligosaccharide orthosomycins are hepta- or octasaccharides characterized by a pair of orthocsters and a hexasubstituted benzoyl group (539). The everninomicins were isolated as a multifactor complex produced by *Micromonospora carbonacae* (540). The principal constituent is everninomicin D, whose structure was established by chemical and spectroscopic methods and later confirmed by X-ray studies of different components (Fig. 32.44) (541, 542). Structures have been assigned to several other factors and correlated with structures of related compounds (543).

The everninomicins have a moderate antibacterial spectrum and potency comparable to other antibiotics. They bind to bacterial ribosomes and inhibit protein synthesis, although full details of these processes have not been elucidated (544). Being a different structural class, cross-resistance to other antibiotics is low (545). Many semisynthetic derivatives have been prepared and although some of these as well as everninomicin B exhibited a better pharmacokinetic profile than factor D, none of these compounds were further developed due to the perceived lack of clinical need at that time. However, as a result of increasing emphasis on new antibiotics to treat resistant gram-positive bacteria, interest in the everninomicins has recently bcen renewed with the selection of SCH 27899 for reevaluation and further investigation in preclinical studies (546).

The avilamycin complex consists of related compounds produced by *S. viridochromogenes* (547). Their structures were established by chemical, spectroscopic, and X-ray crystallographic studies and correlations with the everninomicins (548–550). Avilamycin is used as a growth promotant in poultry and pigs. Several other members of the orthosomycin class have been isolated, including flambamycin, curamycin, and sporacuracin (539,543,551,552).

7.4 Bicyclomycin

Bicyclomycin is a small cyclic dipeptide isolated from culture broths of *S. sapporonensis* and *S. aizunensis* (553,554). The overall structure and stereochemistry of this

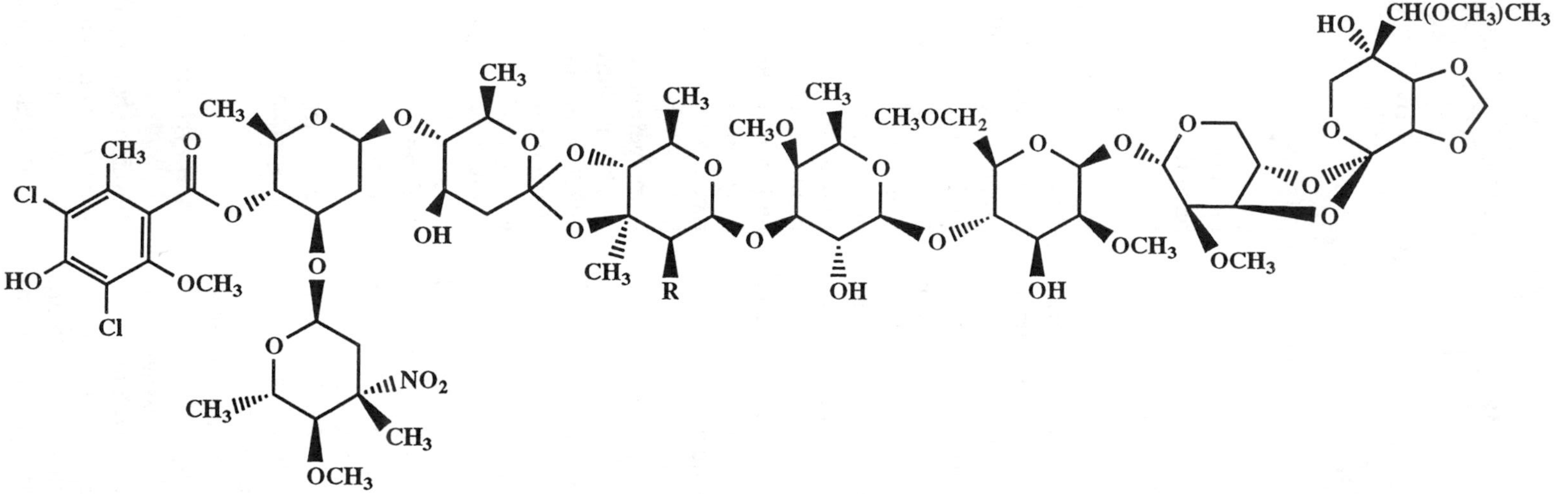

Fig. 32.44 Structures of everninomicin B (R = OH) and D (R = H).

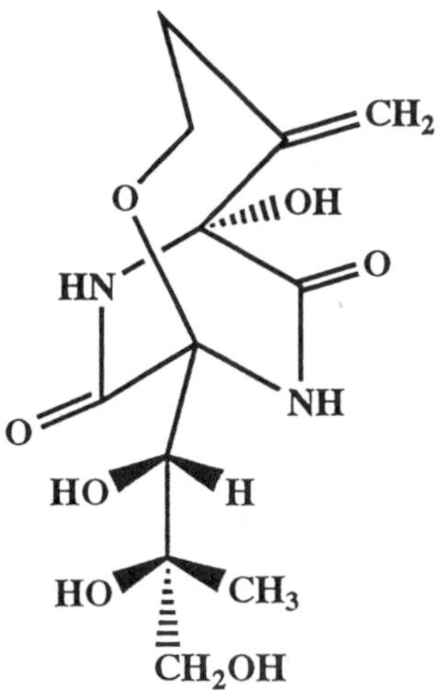

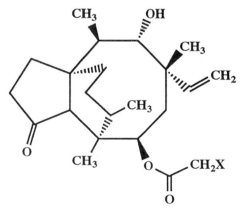

Fig. 32.45 Structure of bicyclomycin.

diketopiperazine were established by X-ray crystallography and synthesis (Fig. 32.45) (555,556). It is derived from leucine and isoleucine, although full details of its novel biosynthesis have not been elucidated (557). Also named bicozamycin, it has a narrow spectrum of activity against gram-negative bacteria (558). Its mode of action is not well understood, but it appears to be different from other known mechanisms and therefore displays no cross-resistance to other antibiotics (559,560). It is not highly metabolized and is well absorbed after injection, which may account for its good *in vivo* efficacy despite its relatively low *in vitro* activity. It is not orally absorbed and is used in Japan and other countries to treat bacterial gut infections, such as, diarrhea in humans and animals, and to treat gram-negative infections in calves and pigs by parenteral administration (561–563).

Bicyclomycin has been the subject of extensive chemical derivatization and total synthesis, much of which has been oriented toward structure–activity relationships and learning its mechanism of action (562–564). One useful derivative is the benzoate ester of the primary hydroxyl group, a modification that improves oral bioavailability of bicyclomycin (554,565). The benzoate ester is used in Japan for treating bacterial disease in yellow tail tuna (566).

7.5 Pleuromutilin and Tiamulin

Pleuromutilin is a tricyclic diterpenoid fungal metabolite isolated from culture broths of *Pleurotus mutilus* (567–569). Tiamulin (Fig. 32.46) is an important semisynthetic derivative arising from SAR studies (570). Its spectrum of activity includes gram-positive bacteria, *Mycoplasma* species, and certain spirochetes (570,571). It is extensively used in veterinary medicine for treating *Mycoplasma* infection in poultry and swine dysentery and respiratory diseases in pigs (572–574). It is administered as the free base by intramuscular injection or as its fumarate salt in the animal's feed and drinking water. Toxic interactions between tiamulin and polyether ionophores have been ascribed to selective inhibition of certain cytochrome P450 enzymes by tiamulin, causing inhibition of oxidative metabolism of polyethers and their subsequent toxic accumulation (575). A glycosyl derivative of pleuromutilin (A40104A) has been isolated from culture broths of *Clitopilus pseudo-pinsitus* (576).

Fig. 32.46 Structures of pleuromutilin [X = OH] and tiamulin [X = $SCH_2CH_2N(CH_2CH_3)_2$].

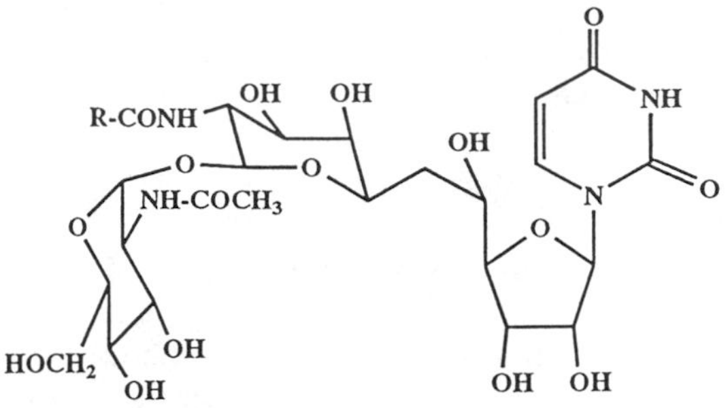

Fig. 32.47 Structures of tunicamycin V: R = $(CH_3)_2CH(CH_2)_9CH=CH-$; tunicamycin VII: R = $(CH_3)_2CH(CH_2)_{10}CH=CH-$; tunicamycin II: R = $(CH_3)_2CH(CH_2)_8CH=CH-$.

7.6 Nucleoside Antibiotics

Many nucleosides have been isolated from microbial fermentations (577–583). A few have been of interest because of their antibacterial activity, but those having antifungal, antiviral, antitumor, herbicidal, or other activities are not covered in this chapter. Nucleosides occur in all biological systems, so a diverse range of activity is not unexpected. Nucleoside antibiotics have not achieved therapeutic importance as antibacterial agents, but some of them are useful biological probes to discover not only their own mechanisms of action but also those of other antibiotics.

The tunicamycins are a multicomponent complex of lipophilic acyl nucleosides produced by *S. lysosuperificus* (584). They contain a modified uridine whose 5'-carbon atom is extended by a substituted tetrahydropyran ring system which is further substituted by a lipophilic amide and N-acetylglucosamine (Fig. 32.47) (585). Related families of nucleosides include the streptovirudins, corynetoxins, liposidomycins, septacidins, and spicamycins (586–589). As expected for such nucleoside structures, these compounds mimic several important biological molecules. In bacteria, they interfere with cell wall formation by inhibiting synthesis of peptidoglycan and teichoic acids. They also inhibit synthesis of glycoproteins by acting as analogs of UDP-N-acetylglucosamine to inhibit transglycosylation by dolichyl-P:N-acetylglucosamine-1-P transferase (585).

Puromycin is another important nucleoside isolated from culture broths of *S. alboniger* (590). Its structure was established by chemical degradation studies and total synthesis (Fig. 32.48) (591). It inhibits gram-positive bacteria, but as a potent in-

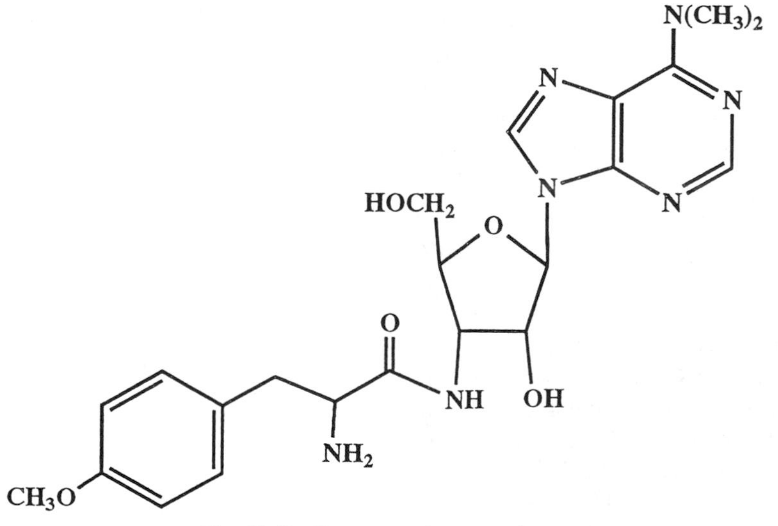

Fig. 32.48   Structure of puromycin.

hibitor of protein synthesis, it was too toxic for medicinal applications. It mimics amino-acyl-tRNA and binds in its place at the A-site of the peptidyltransferase, with subsequent transfer of a growing peptide chain to the amino group of puromycin, thereby terminating protein synthesis. This transformation has been widely used to study ribosomal protein synthesis and has been named the puromycin reaction (592). Many analogs of puromycin have been synthesized in order to find useful derivatives (593).

A related nucleoside isolated from culture broths of *S. capreolus* is A201A, an adenosine derivative in which its 3'-amino group is acylated by a cinnamoyl moiety that is *para*-substituted by a disaccharide (594). It is an inhibitor of bacterial protein synthesis whose structure bears a steric resemblance to puromycin, but it lacks the latter's α-amino group and thus cannot accept a growing peptide chain (594,595). It also bears a structural resemblance to hygromycin A, a feature that has recently prompted the synthesis of hybrid analogs (596).

The mureidomycins and pacidamycins are another family of peptidyl-nucleoside antibiotics. The mureidomycins are produced by *S. flavidovirens* and contain a complex amino acid array extending from the C-5' end of the nucleoside component (597). The pacidamycins are a related series produced by *S. coeruleorubidus* (598). Even though these compounds have a relatively narrow antibacterial spectrum, it includes *P. aeruginosa*, a pathogen that is usually among the most difficult for other antibiotics to inhibit (599). These compounds inhibit the synthesis of peptidoglycan by blocking the translocase reaction, but in a manner more specific to bacteria than mammalian cells, a desirable feature that may make potential host toxicity less likely (600). The therapeutic potential of these novel peptidyl-nucleoside antibiotics is still being investigated.

7.7 Novobiocin

Novobiocin is a relatively old antibiotic isolated by several different research groups in the 1950s (601,602). It is a glycosylated aminocoumarin whose amino group is acylated by a substituted benzoyl group (Fig. 32.49) (603). It has good antibacterial activity against *S. aureus* and certain gram-negative bacteria and acts by inhibiting the type B subunit of DNA gyrase, an enzyme responsible for maintaining the proper in-

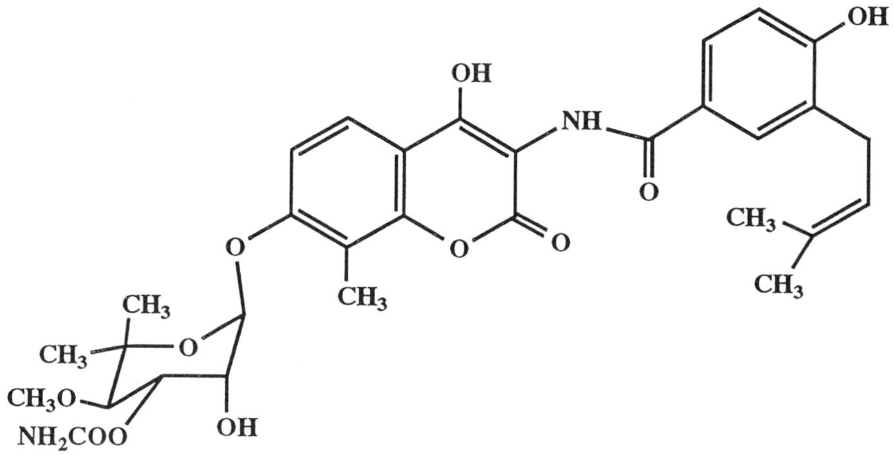

Fig. 32.49   Structure of novobiocin.

tracellular balance of supercoiled, double-stranded DNA (604). Novobiocin is used in veterinary medicine to treat staphylococcal infections, including bovine mastitis (605). Several related compounds have been isolated, including a chlorinated analog named chlorobiocin (606,607). Dimeric analogs named coumermycin have been discovered, in which the amino groups of two glycosylated aminocoumarin moieties are acylated by 3-methylpyrrole-2,4-dicarboxylic acid (608,609). Although an extensive program of structural modification was conducted that yielded potential clinical candidates, these were abandoned because of undesired side effects (610).

8 CONCLUSION

Medicinal chemists have made numerous contributions to the discovery and development of the antibiotics that presently play a vital role in clinical and veterinary medicine. Many of these antibiotics are natural products isolated from culture broths of soil microorganisms grown under aerobic fermentation conditions. Despite their often extremely complex structures, most of them can be fully elucidated by the modern tools of spectroscopy and crystallography. These antibiotics have also served as valuable biological probes for elucidating the mechanisms of antibiotic action, microbial resistance, pharmacology, and toxicology. Finally, these compounds are starting materials for SAR studies leading to semisynthetic derivatives, some of which have become commercial products.

However, pathogenic microorganisms have evolved and acquired or developed new modes of resistance. In some cases, resistance developed rapidly and the useful life of the antibiotic was short; in other cases such as vancomycin, development of resistance required decades to appear. In either event, the continual emergence of new forms of resistance will require medici-

nal chemists continually to seek new classes of antibiotics lacking cross-resistance with the older agents.

In addition to scientific and medical issues, the rising expense of research and development, the rapidly changing pharmaceutical industry, and the cost pressures on medical care create other challenges to drug discovery. Overcoming these challenges through advances in science and technology that creatively solve problems in infectious diseases will provide many opportunities for medicinal chemists to advance the antibiotic field innovatively into the 21st century.

ACKNOWLEDGMENTS

The author gratefully appreciates the invaluable assistance provided by Ms. L. W. Crandall for literature searches and critical review of the manuscript.

REFERENCES

1. M. L. Cohen, *Science*, **257**, 1050–1055 (1992).
2. J. Travis, *Science*, **264**, 359–362 (1994).
3. S. B. Levy, *The Antibiotic Paradox: How Miracle Drugs are Destroying the Miracle*, Plenum Press, New York, 1992.
4. G. L. Mandell, J. E. Bennett, and R. Dolin, Eds., *Principles and Practice of Infectious Diseases*, 4th ed., Vol. 1–2, Churchill Livingstone, New York, 1995.
5. J. F. Prescott and J. D. Baggot, Eds., *Antimicrobial Therapy in Veterinary Medicine*, 2nd ed., Iowa State University Press, Ames, Iowa, 1988.
6. B. W. Boycroft, Ed., *Dictionary of Antibiotics and Related Substances*, Chapman and Hall, London, 1988.
7. J. Berdy, CRC *Handbook of Antibiotic Compounds*, CRC Press, Boca Raton, 1987.
8. H. Umezawa, Ed., *Index of Antibiotics from Actinomycetes*, Vol. 1, University Park Press, State College, Penn., 1967.
9. H. Umezawa, Ed., *Index of Antibiotics from Actinomycetes*, Vol. 2, University Park Press, Baltimore, Md., 1978.

10. D. Perlman, Ed., *Structure-Activity Relationships among the Semisynthetic Antibiotics*, Academic Press, New York, 1977.

11. G. Lukacs and M. Ohno, Eds., *Recent Progress in the Chemical Synthesis of Antibiotics*, Springer-Verlag, Berlin, 1990.

12. G. Lukacs, Ed., *Recent Progress in the Chemical Synthesis of Antibiotics and Related Microbial Products*, Vol. 2, Springer-Verlag, Berlin, 1993.

13. K. Krohn, H. A. Kirst, and H. Maag, Eds., *Antibiotics and Antiviral Compounds: Chemical Synthesis and Modification*, VCH Verlagsgesellschaft, Wienheim, 1993.

14. P. H. Bentley and R. Ponsford, Eds., *Recent Advances in the Chemistry of Anti-Infective Agents*, Royal Society of Chemistry, Cambridge, UK, 1993.

15. H. Umezawa and I. R. Hooper, Eds., *Aminoglycoside Antibiotics*, Handbook of Experimental Pharmacology Series, Vol. 62, Springer-Verlag, Berlin, 1982.

16. A. Whelton and H. C. Neu, Eds., *The Aminoglycosides: Microbiology, Clinical Use, and Toxicology*, Marcel Dekker, Inc., New York, 1982.

17. D. N. Gilbert in ref. 4, Vol. 1, chapter 20, pp. 279–306.

18. D. McGregor in *Kirk-Othmer Encyclopedia of Chemical Technology*, 4th ed., Vol. 2, John Wiley & Sons, Inc., New York, 1992, pp. 904–926.

19. P. J. L. Daniels in *Kirk-Othmer Encyclopedia of Chemical Technology*, 3rd ed., Vol. 2, John Wiley & Sons, Inc., New York, 1978, pp. 819–852.

20. K. E. Price, J. C. Godfrey, and H. Kawaguchi in Ref. 10, pp. 239–395.

21. W. G. Barnes and G. R. Hodges, Eds., *The Aminoglycoside Antibiotics: A Guide to Therapy*, CRC Press, Boca Raton, Fla., 1984.

22. S. J. Pancoast, *Med. Clin, N. Am.*, **72**, 581–612 (1988).

23. K. L. Rinehart, Jr. and T. Suami, Eds., *Aminocyclitol Antibiotics*, ACS Symposium Series, Vol. 125, American Chemical Society, Washington, D. C., 1980.

24. D. A. Cox, K. Richardson, and B. C. Ross, *Topics Antibiot. Chem.*, **1**, 1–90 (1977).

25. J. Reden and W. Durckheimer, *Topics Current Chem.*, **83**, 105–170 (1979).

26. A. Schatz, E. Bugie, and S. A. Waksman, *Proc. Soc. Exp. Biol. Med.*, **55**, 66–69 (1944).

27. S. Neidle, D. Rogers, and M. B. Hursthouse, *Tetrahedron Lett.*, 4725–4728 (1968).

28. R. U. Lemieux and M. L. Wolfram, *Adv. Carbohyd. Chem.*, **3**, 337–384 (1948).

29. S. A. Waksman, *Science*, **118**, 259–266 (1953).

30. H. Heding and O. Lutzen, *J. Antibiot.*, **25**, 287–291 (1972).

31. T. Usui, T. Tsuchiya, and S. Umezawa, *J. Antibiot.*, **31**, 991–996 (1978).

32. T. Tsuchiya and T. Shitara, *Carbohyd. Res.*, **109**, 59–72 (1982).

33. J. P. Abad, G. Leon, and R. Amils, *J. Antibiot.*, **40**, 685–691 (1987).

34. R. L. Peck, C. E. Hoffhine, Jr., and K. Folkers, *J. Am. Chem. Soc.*, **68**, 1390–1391 (1946).

35. Q. R. Bartz, J. Controulis, H. M. Crooks, Jr., and M. C. Rebstock, *J. Am. Chem. Soc.*, **68**, 2163–2166 (1956).

36. S. Tatsuoka, T. Kusaka, A. Miyake, M. Inoue, H. Hitomi, Y. Shiraishi, H. Iwasaki, and M. Imanishi, *Chem. Pharm., Bull.*, **5**, 343–349 (1957).

37. F. Kavanagh, E. Grinnan, E. Allanson, and D. Tunin, *Appl. Microbiol.*, **8**, 160–162 (1960).

38. L. P. Garrod and F. O'Grady, *Antibiotic and Chemotherapy*, 2nd ed., Williams & Wilkins, Baltimore, 1968, pp. 100–101.

39. S. A. Waksman and H. A. Lechevalier, *Science*, **109**, 305–307 (1949).

40. K. L. Rinehart, Jr., M. Hichens, A. D. Argoudelis, W. S. Chilton, H. E. Carter, M. P. Georgiadis, C. P. Schaffner, and R. T. Schillings, *J. Am. Chem. Soc.*, **84**, 3218–3220 (1962).

41. K. L. Rinehart, *The Neomycins and Related Antibiotics*, John Wiley & Sons, Inc., New York, 1964.

42. T. H. Haskell, J. C. French, and Q. R. Bartz, *J. Am. Chem. Soc.*, **81**, 3482–3483 (1959).

43. H. Taniyama, Y. Sawada, and S. Tanaka, *Chem. Pharm. Bull.*, **21**, 609–615 (1973).

44. G. L. Coffey, L. E. Anderson, M. W. Fisher, M. M. Galbraith, A. B. Hillegas, D. L. Kohberger, P. E. Thompson, K. S. Weston, and J. Ehrlich, *Antibiot. Chemother.*, **9**, 730–738 (1959).

45. T. Oda, T. Mori, Y. Kyotani, and M. Nakayama, *J. Antibiot.*, **24**, 511–518 (1971).

46. T. Mori, Y. Kyotani, I. Watanabe, and T. Oda, *J. Antibiot.*, **25**, 149–150 (1972).

47. E. Akita, T. Tsuruoka, N. Ezaki, and T. Niida, *J. Antibiot.*, **23**, 173–183 (1970).

48. S. Horii, I. Nogami, N. Mizokami, Y. Arai, and M. Yoneda, *Antimicrob. Agents Chemother.*, **5**, 578–581 (1974).

49. P. W. K. Woo, H. W. Dion, and Q. R. Bartz, *Tetrahedron Lett.*, 2625–2628 (1971).

50. H. Tsukiura, K. Fujisawa, M. Konishi, K. Saito, K. Numata, H. Ishikawa, T. Miyaki, K. Tomita, and H. Kawaguchi, *J. Antibiot.*, **26**, 351–357 (1973).

51. C. L. Heifetz, M. W. Fisher, J. A. Chodubski, and M. O. DeCarlo, *Antimicrob. Agents Chemother.*, **2**, 89–94 (1972).

52. H. Umezawa, M. Ueda, K. Maeda, K. Yagashita, S. Konodo, Y. Okami, R. Utahara, Y. Osato, K. Nitta and T. Takeuchi, *J. Antibiot.*, **A10**, 181–188 (1957).

53. G. Koyama, Y. Iitaka, K. Maeda, and H. Umezawa, *Tetrahedron Lett.*, 1875–1879 (1968).

54. H. Schmitz, O. B. Fardig, F. A. O'Herron, M. A. Rousche, and I. R. Hooper, *J. Am. Chem. Soc.*, **80**, 2911–2912 (1958).

55. J. W. Rothrock, R. T. Geogelman, and F. J. Wolf, *Antibiot. Annual 1958–1959*, Medical Encyclopedia Inc., New York, 1959, pp. 796–803.

56. M. Murase, T. Wakazawa, M. Abe, and S. Kawaji, *J. Antibiot.*, **A14**, 156–157 (1961).

57. T. Ito, M. Nishio, and H. Ogawa, *J. Antibiot*, **A17**, 189–193 (1964).

58. M. Murase, *J. Antibiot.*, **A14**, 367–368 (1961).

59. R. Q. Thompson and E. A. Presti, *Antimicrob. Agents Chemother.-1967*, American Society for Microbiology, Washington, D. C., 1968, p. 332–340.

60. K. F. Koch, K. E. Merkel, S. C. O'Connor, J. L. Occolowitz, J. W. Paschal, and D. E. Dorman, *J. Org. Chem.*, **43**, 1430–1434 (1978).

61. K. F. Koch and J. A. Rhoades, *Antimicrob. Agents Chemother.–1970*, American Society for Microbiology, Washington, D. C., 1971, pp. 309–313.

62. K. F. Koch, F. A. Davis, and J. A. Rhoades, *J. Antibiot.*, **26**, 745–751 (1973).

63. M. J. Weinstein, G. M. Leudemann, E. M. Oden, and G. H. Wagman, *Antimicrob. Agents Chemother.-1963*, American Society for Microbiology, Washington, D. C., 1964, pp. 1–7.

64. D. J. Cooper, P. J. L. Daniels, M. D. Yudis, H. M. Marigliano, R. D. Guthrie, and S. T. K. Bukhari, *J. Chem. Soc. (C)*, 3126–3129 (1971).

65. D. J. Cooper, *Pure Appl. Chem.*, **28**, 455–467 (1971).

66. R. S. Egan, R. L. DeVault, S. L. Mueller, M. I. Levenberg, A. C. Sinclair, and R. S. Stanaszek, *J. Antibiot.*, **28**, 29–34 (1975).

67. P. J. L. Daniels, C. Luce, T. L. Nagabhushan, R. S. Jaret, D. Schumacher, H. Reimann, and J. Ilavsky, *J. Antibiot.*, **28**, 35–41 (1975).

68. M. Ohkoshi, K. Okada, and N. Kawamura, *Jpn. J. Antibiot.*, **35**, 703 (1982).

69. J. Berdy, J. K. Pauncz, Z. M. Vajna, G. Horvath, J. Gyimesi, and I Koczka, *J. Antibiot.*, **30**, 945–954 (1977).

70. G. H. Wagman, R. T. Testa, and J. A. Marquez, *J. Antibiot.*, **23**, 555–558 (1970),

71. H. Reimann, D. J. Cooper, A. K. Mallams, R. S. Jaret, A. Yehaskel, M. Kugelman, H. F. Vernay, and D. Schumacher, *J. Org. Chem.*, **39**, 1451–1457 (1974).

72. H. Maehr and C. P. Schaffner, *J. Am. Chem. Soc.*, **92**, 1697–1700 (1970).

73. T. L. Nagabhushan, W. N. Turner, P. J. L. Daniels, and J. B. Morton, *J. Org. Chem.*, **40**, 2830–2834 (1975).

74. R. S. Egan, A. C. Sinclair, R. L. De Vault, J. B. McAlpine, S. L. Mueller, P. C. Goodley, R. S. Stanaszek, M. Cirovic, R. J. Mauritz, L. A. Mitscher, K. Shirahata, S. Sato, and T. Iida, *J. Antibiot.*, **30**, 31–38 (1977).

75. J. B. McAlpine, A. C. Sinclair, R. S. Egan, R. L. De Vault, R. S. Stanaszek, M. Cirovic, S. L. Mueller, P. C. Goodley, R. J. Mauritz, N. E. Wideburg, L. A. Mitscher, K. Shirahata, H. Matsushima, S. Sato, and T. Iida, *J. Antibiot.*, **30**, 39–49 (1977).

76. R. Benveniste and J. Davies, *Antimicrob. Agents Chemother.*, **4**, 402–409 (1973).

77. H. Umezawa, H. Iwasawa, D. Ikeda, and S. Kondo, *J. Antibiot.*, **36**, 1087–1091 (1983).

78. S. W. McCombie in B. Trost and I. Fleming, Eds., *Comprehensive Organic Synthesis*, Vol. 8, Pergamon Press, Oxford, UK, 1991, chapt. 4.2, pp. 811–833.

79. D. Crich and L. Quintero, *Chem. Rev.*, **89**, 1413–1432 (1989).

80. W. Hartwig, *Tetrahedron*, **39**, 2609–2645 (1983).

81. H. Umezawa, S. Umezawa, T. Tsuchiya, and Y. Okazaki, *J. Antibiot.*, **24**, 485–487 (1971).

82. H. Kawaguchi, T. Naito, S. Nakagawa, and K. Fujisawa, *J. Antibiot.*, **25**, 695–708 (1972).

83. H. Kawaguchi and T. Naito in J. S. Bindra and D. Lednicer, Eds., *Chronicles of Drug Discovery*, Vol. 2, John Wiley & Sons, Inc., New York, 1983, pp. 207–234.

84. A. Van Schepdael, R. Busson, H. J. Vanderhaeghe, P. J. Claes, L. Verbist, M. P. Mingeot-Leclercq, R. Brasseur, and P. M. Tulkens, *J. Med. Chem.*, **34**, 1483–1492 (1991).

85. T. Yamasaki, Y. Narita, H. Hoshi, S. Aburaki, H. Kamei, T. Naito, and H. Kawaguchi, *J. Antibiot.*, **44**, 646–658 (1991).

86. H. Hoshi, S. Aburaki, S. Imura, T. Yamasaki, T. Naito, and H. Kawaguchi, *J. Antibiot.*, **43**, 858–872 (1990).

87. S. Kondo, K. Iinuma, H Yamamoto, K. Maeda, and H. Umezawa, *J. Antibiot.*, **26**, 412–415 (1973).

88. T. L. Nagabhushan, A. B. Cooper, H. Tsai, P. J. L. Daniels, and G. H. Miller, *J. Antibiot.*, **31**, 681–687 (1978).

89. J. J. Wright, *J. Chem. Soc., Chem. Commun.*, 206–208 (1976).

90. G. H. Miller, G. Arcieri, M. J. Weinstein, and J. A. Waitz, *Antimicrob. Agents Chemother.*, **10**, 827–836 (1976).

91. C. O. Solberg, D. Reeves, and I. Phillips, Eds., *J. Antimicrob. Chemother.*, **13** (Suppl. A), (1984).

92. D. M. Campoli-Richards, S. Chaplin. R. H. Sayce, and K. L. Goa, *Drugs*, **38**, 703–756 (1989).

93. K. Richardson, S. Jevons, J. W. Moore, B. C. Ross, and J. R. Wright, *J. Antibiot.*, **30**, 843–846 (1977).

94. K. Richardson, K. W. Brammer, S. Jevons, R. M. Plews, and J. R. Wright, *J. Antibiot.*, **32**, 973–977 (1979).

95. T. L. Nagabhushan, A. B. Cooper, W. N. Turner, H. Tsai, S. McCombie, A. K. Mallams, D. Rane, J. J. Wright, P. Reichert, D. L. Boxler, and J. Weinstein, *J. Am. Chem. Soc.*, **100**, 5253–5254 (1978).

96. S. Hanessian and G. Patil, *Tetrahedron Lett.*, 1035–1038 (1978).

97. T. Tsuchiya, Y. Takagi, and S. Umezawa, *Tetrahedron Lett.*, 4951–4954 (1979).

98. H. A. Kirst, B. A. Truedell, and J. E. Toth, *Tetrahedron Lett.*, 295–298 (1981).

99. J. A. Waitz, G. H. Miller, E. Moss, Jr., and P. J. S. Chu, *Antimicrob. Agents Chemother.*, **13**, 41–48 (1978).

100. D. L. Boxler, R. Brambilla, D. H. Davies, A. K. Mallams, S. W. McCombie, J. B. Morton, P. Reichert, and H. F. Vernay, *J. Chem. Soc. Perkin Trans. I*, 2168–2185 (1981).

101. A. Van Schepdael, J. Delcourt, M. Mulier, R. Busson, L. Verbist, H. J. Vanderhaeghe, M. P. Mingeot-Leclercq, P. M. Tulkens, and P. J. Claes, *J. Med. Chem.*, **34**, 1468–1474 (1991).

102. E. Umemura, T. Tsuchiya, Y. Koyama, and S. Umezawa, *Carbohyd. Res.*, **238**, 147–162 (1993).

103. R. Albert, K. Dax, and A. E. Stutz, *Tetrahedron Lett.*, **24**, 1763–1766 (1983).

104. K. Igarashi, T. Sugawara, T. Honma, Y. Tada, H. Miyazaki, H. Nagata, M. Mayama, and T. Kubota, *Carbohyd. Res.*, **109**, 73–88 (1982).

105. J. Hildebrandt, H. Loibner, E. Schutze, W. Streicher, P. Stutz, and A. Wenzel, *24th Intersci. Conf. Antimicrob. Agents Chemother.*, Washington, D. C., 1984, abstr. 310.

106. K. E. Price, *Antimicrob. Agents Chemother.*, **29**, 543–548 (1986).

107. M. P. Mingeot-Leclercq, A. Van Schepdael, R. Brasseur, R. Busson, H. J. Vanderhaeghe, P. J. Claes, and P. M. Tulkens, *J. Med. Chem.*, **34**, 1476–1482 (1991).

108. E. Wilmotte, P. Maldague, P. Tulkens, R. Baumgartner, F. Schmook, H. Walzl, and H. Obenaus, *Drugs Exptl. Clin. Res.*, **9**, 467–477 (1983).

109. S. O'Connor, L. K. T. Lam, N. D. Jones, and M. O. Chaney, *J. Org. Chem.*, **41**, 2087–2092 (1976).

110. J. E. Mortensen, J. Nanavaty, M. Veenhuizen, and T. R. Shryock, *Vet. Med.*, submitted for publication.

111. N. E. Allen, W. E. Alborn, Jr., H. A. Kirst, and J. E. Toth, *J. Med. Chem.*, **30**, 333–340 (1987).

112. Y. Abe, S. Nakagawa, T. Naito, and H. Kawaguchi, *J. Antibiot.*, **34**, 1434–1446 (1981).

113. R. S. Egan, R. S. Stanaszek, M. Cirovic, S. L. Mueller, J. Tadanier, J. R. Martin, P. Collum, A. W. Goldstein, R. L. De Vault, A. C. Sinclair, E. E. Fager, and L. A. Mitscher, *J. Antibiot.*, **30**, 552–563 (1977).

114. T. Iida, M. Sato, I. Matsubara, Y. Mori, and K. Shirahata, *J. Antibiot.*, **32**, 1273–1279 (1979).

115. T. Deutshi, M. Nakayama, I. Watanabe, T. Mori, H. Naganawa, and H. Umezawa, *J. Antibiot.*, **32**, 187–192 (1979).

116. D. Ikeda, Y. Houriuchi, M. Yoshida, T. Miyasaka, S. Kondo, and H. Umezawa, *Carbohydrate Res.*, **109**, 33–45 (1982).

117. I. Watanabe, T. Deushi, T. Yamaguchi, K. Kamiya, M. Nakayama, and T. Mori, *J. Antibiot.*, **32**, 1066–1068 (1979).

118. K. Ohba, T. Tsuruoka, K. Mizutani, N. Kato, S. Omoto, N. Ezaki, S. Inouye, T. Niida, and K. Watanabe, *J. Antibiot.*, **34**, 1090–1100 (1981).

119. P. F. Wiley, A. D. Argoudelis, and H. Hoeksema, *J. Am. Chem. Soc.*, **85**, 2652–2659 (1963).

120. T. G. Cochran, D. J. Abraham, and L. L. Martin, *J. Chem. Soc., Chem. Commun.*, 404–495 (1972).

121. H Hoeksema and J. C. Knight, *J. Antibiot.*, **28**, 240–241 (1975).

122. J. C. Knight and H. Hoeksema, *J. Antibiot.*, **28**, 136–142 (1975).

123. W. J. Holloway, *Med. Clin. North Am.*, **66**, 169–173 (1982).

124. N. Neuss, K. F. Koch, B. B. Molloy, W. Day, L. L. Huckstep, D. E. Dorman, and J. D. Roberts, *Helv. Chim. Acta*, **53**, 2314–2319 (1970).

125. S. Kondo, K. Iinuma, H. Naganawa, M. Shimura, and Y. Sekizawa, *J. Antibiot.*, **28**, 79–82 (1975).

126. M. Shimura, Y. Sekizawa, K. Iinuma, H. Naganawa, and S. Kondo, *J. Antibiot.*, **28**, 83–84 (1975).

127. J. Shoji and Y. Nakagawa, *J. Antibiot.*, **23**, 569–571 (1970).

128. S. Inouye, T. Shomura, H. Watanabe, K. Totsugawa, and T. Niida, *J. Antibiot.*, **26**, 374–385 (1973).

129. J. Leboul and J. Davies, *J. Antibiot.*, **35**, 527–528 (1982).

130. R. N. Rao, N. E. Allen, J. N. Hobbs, Jr., W. E. Alborn, Jr., H. A. Kirst, and J. W. Paschal, *Antimicrob. Agents Chemother.*, **24**, 689–695 (1983).

131. R. L. Mann, R. M. Gale, and F. R. Van Abeele, *Antibiot. Chemother.*, **3**, 1279–1282 (1953).

132. K. Kakinuma and Y. Sakagami, *Agric. Biol. Chem.*, **42**, 279–286 (1978).

133. A. Nakagawa, T. Fujimoto, S. Omura, J. C. Walsh, R. L. Stotish, and B. George, *J. Antibiot*, **40**, 1627–1635 (1987).

134. B. H. Jaynes, N. C. Elliott, and D. L. Schicho, *J. Antibiot.*, **45**, 1705–1707 (1992).

135. S. J. Hecker, C. B. Cooper, K. T. Blair, S. C. Lilley, M. L. Minich, and K. M. Werner, *Bioorg. Med. Chem. Lett.*, **3**, 289–294 (1993).

136. T. Iwasa, Y. Kameda, M. Asai, S. Horii, and K. Mizuno, *J. Antibiot.*, **24**, 119–123 (1971).

137. T. Suami, S. Ogawa, and N. Chida, *J. Antibiot.*, **33**, 98–99 (1980).

138. Y. Kameda, N. Asano, K. Matsui, S. Horii, and H. Fukase, *J. Antibiot.*, **41**, 1488–1492 (1988).

139. T. Iwasa, E. Higashide, H. Yamamoto, and M. Shibata, *J. Antibiot.*, **24**, 107–113 (1971)

140. K. Iinuma, S. Kondo, K. Maeda, and H. Umezawa, *J. Antibiot.*, **28**, 613–615 (1975).

141. G. A. Ellestad, J. H. Martin, G. O. Morton, M. L. Sassiver, and J. E. Lancaster, *J. Antibiot.*, **30**, 678–680 (1977)

142. H. Umezawa, Y. Okami, T. Hashimoto, Y. Suhara, M. Hamada, and T. Takeuchi, *J. Antibiot.*, **18A**, 101–103 (1965).

143. Y. Suhara, K. Maeda, H. Umezawa, and M. Ohno, *Tetrahedron Lett.*, 1239–1244 (1966).

144. T. Ikekawa, H. Umezawa, and Y. Iitaka, *J. Antibiot.*, 49–50 (1966).

145. T. Ishiyama, I. Hara, M. Matsuoka, K. Sato, S. Shimada, R. Izawa, T. Hashimoto, M. Hamada, Y. Okami, T. Takeuchi, and H. Umezawa, *J. Antibiot.*, **18A**, 115–119 (1965).

146. J. C. French, Q. R. Bartz, and H. W. Dion, *J. Antibiot.*, **26**, 272–283 (1973).

147. W. J. McGahren, B. A. Hardy, G. O. Morton, F. M. Lovell, N. A. Perkinson, R. T. Hargreaves, D. B. Borders, and G. A. Ellestad, *J. Org. Chem.*, **46**, 792–799 (1981).

148. M. Konishi, S. Kamata, T. Tsuno, K. Numata, H. Tsukiura, T. Naito, and H. Kawaguchi, *J. Antibiot.*, **29**, 1152–1162 (1976).

149. K. Nara, K. Katamoto, S. Suzuki, and E. Mizuta, *Chem. Pharm., Bull.*, **26**, 1091–1099 (1978).

150. J. P. Kirby, G. E. Van Lear, G. O. Morton, W. E. Gore, W. V. Curran, and D. B. Borders, *J. Antibiot*, **30**, 344–347 (1977).

151. G. A Ellestad, D. B. Cosulich, R. W. Broschard, J. H. Martin, M. P. Kunstmann, G. O. Morton, J. E. Lancaster, W. Fulmor, and F. M. Lovel, *J. Am. Chem. Soc.*, **100**, 2515–2524 (1978).

152. W. J. McGahren, F. Barbatschi, N. A. Kuck, G. O. Morton, B. Hardy, and G. A. Ellestad, *J. Antibiot.*, **35**, 794–799 (1982).

153. T. Tsuno, M. Konishi, T. Naito, and H. Kawaguchi, *J. Antibiot.*, **34**, 390–402 (1981).

154. H. Naganawa, N. Usui, T. Takita, M. Hamada, K. Maeda, and H. Umezawa, *J. Antibiot.*, **27**, 145–146 (1974).

155. S. Omura, M. Katagiri, K. Atsumi, T. Hata, A. A. Jakubowski, E. B. Springs, and M. Tishler, *J. Chem. Soc. Perkin Trans. I*, 1627–1631 (1974).

156. K. J. Shaw, P. N. Rather, R. S. Hare, and G. H. Miller, *Microbiol. Rev.*, **57**, 138–163 (1993).

157. J. E. Davies in V. Lorian, Ed., *Antibiotics in Laboratory Medicine*, 3rd ed., Williams & Wilkins, Baltimore, 1991, Chapt. 18, pp. 691–713.

158. N. D. Weiner and J. Schacht in S. A. Lerner, G. J. Matz, and J. E. Hawkins, Jr., Eds., *Aminoglycoside Ototoxicity*, Little, Brown & Co., Boston, 1981, pp. 113–124.

159. T. Hutchin and G. Cortopassi, *Antimicrob, Agents Chemother.*, **38**, 2517–2520 (1995).

160. D. N. Gilbert and W. M. Bennett, in R. K. Root and M. A. Sande, Eds., *New Dimensions in Antimicrobial Therapy*, Churchill Livingstone, New York, 1984, pp. 121–152.

161. G. Laurent, B. K. Kishore and P. M. Tulkens, *Biochem. Pharmacol.*, **40**, 2383–2392 (1990).

162. J. P. McCormack and P. J. Jewesson, *Clin. Infect. Dis.*, **14**, 320–339 (1992).

163. S. M. Watling and J. F. Dasta, *Ann. Pharmacother.*, **27**, 351–357 (1993)

164. M. L. Barclay, E. J. Begg, and K. G. Hickling, *Clin. Pharmacokin.*, **27**, 32–48, (1994)

165. R. D. Bates and M. C. Nahata, *Ann. Pharmacother.*, **28**, 757–766 (1994).

166. C. R. Kumana and K. Y. Yuen, *Drugs*, **47**, 902–913 (1994)

167. S. Umezawa, *Adv. Carbohydrate Chem. Biochem.*, **30**, 111–182 (1974).

168. K. L. Rinehart, Jr., *Pure Appl. Chem.*, **49**, 1361–1348 (1977).

169. K. Nagaoka and A. L. Demain, *J. Antibiot.*, **28**, 627-635 (1975).

170. R. B. Woodward, *Angew. Chem.*, **69**, 50–58 (1957).

171. A. J. Bryskier, J.-P. Butzler, H. C. Neu, and P. M. Tulkens, Eds., *Macrolides: Chemistry, Pharmacology and Clinical Uses*, Arnette Blackwell, Paris, 1993.

172. H. A. Kirst in *Kirk-Othmer Encyclopedia of Chemical Technology*, 4th ed., Vol. 3, John Wiley & Sons, Inc., New York, 1992, pp. 169–213.

173. H. C. Neu, L. S. Young, S. H. Zinner, and J. Acar, Eds., *The New Macrolides, Azalides, and Streptogramins in Clinical Practice*, Marcel Dekker, Inc., New York, 1995.

174. H. C. Neu, L. S. Young, and S. H. Zinner, Eds., *The New Macrolides, Azalides, and Streptogramins: Pharmacology and Clinical Applications*, Marcel Dekker, Inc., New York, 1993.

175. S. Omura, Ed., *Macrolide Antibiotics: Chemistry, Biology, and Practice*, Academic Press, Orlando, Fla., 1984.

176. C. Djerassi and J. A. Zderic, *J. Am. Chem. Soc.*, **78**, 6390–6395 (1956).

177. D. G. Manwaring, R. W. Rickards, and R. M. Smith, *Tetrahedron Lett.*, 1029–1032 (1970).

178. C. Djerassi and O. Halpern, *Tetrahedron*, **3**, 255–268 (1958).

179. J. M. McGuire, R. L. Bunch, R. C. Anderson, H. E. Boaz, E. H. Flynn, H. M. Powell, and J. W. Smith, *Antibiot. Chemother.*, **2**, 281–283 (1952).

180. D. P. Labeda, *Int. J. Syst. Bacteriol.*, **37**, 19–22, (1987).

181. P. F. Wiley, K. Gerzon, E. H. Flynn, M. V. Sigal, O. Weaver, U. C. Quarck, R. R. Chauvette, and R. Monahan, *J. Am. Chem. Soc.*, **79**, 6062–6070 (1957).

182. D. R. Harris, S. C. McGeachin, and H. H. Mills, *Tetrahedron Lett.*, 679–685 (1965).

183. B. A. Sobin, A. R. English, and W. D. Celmer, *Antibiot. Annu. 1954–1955*, Medical Encyclopedia, Inc., New York, 1955, pp. 827–830.

184. W. D. Celmer, *J. Am. Chem. Soc.*, **87**, 1797–1799 (1965).

185. H. Ogura, K. Furuhata, Y. Harada, and Y. Iitaka, *J. Am. Chem. Soc.*, **100**, 6733–6737 (1978).

186. H. Brockmann and W. Henkel, *Naturwissenschaften*, **37**, 138–139 (1950).

187. A. Furusaki, T. Matsumoto, K. Furuhata, and H. Ogura, *Bull. Chem. Soc. Jpn.*, **55**, 59–62 (1982).

188. R. Corbaz, L. Ettlinger, E. Gaumann, W. Keller, F. Kradolfer, E. Kyburz, L. Neipp, V. Prelog, R. Reusser, and H. Zahner, *Helv. Chim. Acta*, **38**, 935–942 (1955).

189. P. Bartner, D. L. Boxler, R. Brambilla, A. K. Mallams, J. B. Morton, P. Reichert, F. D. Sancilio, H. Surprenant, G. Tomalesky, G. Lukacs, A. Olesker, T. T. Thang, L. Valente, and S. Omura, *J. Chem. Soc., Perkin Trans. 1*, 1600–1624 (1979).

190. M. J. Weinstein, G. H. Wagman, J. A. Marquez, R. T. Testa, E. Oden, and J. A. Waitz, *J. Antibiot.*, **22**, 253–258 (1969).

191. E. Gaumann, R. Hutter, W. Keller-Schierlein, L. Neipp, V. Prelog, and H. Zahner, *Helv. Chim. Acta*, **43**, 601–606 (1960).

192. S. Omura, T. Muro, S. Namiki, M. Shibata, and J. Sawada, *J. Antibiot.*, **22**, 629–634 (1969).

193. R. S. Egan and J. R. Martin, *J. Am. Chem. Soc.*, **92**, 4129–4130 (1970).

194. B. Arnoux, C. Pascard, L. Raynaud, and J. Lunel, *J. Am. Chem. Soc.*, **102**, 3605–3608 (1980).

195. H. A. Kirst, *Prog. Med. Chem.*, **30**, 57–88 (1993).

196. P. L. Tardrew, J. C. H. Mao, and D. Kenney, *Appl. Microbiol.*, **18**, 159–165 (1969).

197. J. Taskinen and P. Ottoila, *J. Antimicrob. Chemother.*, **21** (Suppl. D), 1–8 (1988).

198. P. Davey and R. Williams, Eds., *J. Antimicrob. Chemother.*, **21** (Suppl. D), (1988).

199. W. D. Celmer, H. Els, and K. Murai, *Antibiot. Annu. 1957–1958*, Medical Encyclopedia Inc., New York, 1958, pp. 476–483.

200. T. M. Ludden, *Clin. Pharmacokin.*, **10**, 63–79 (1985).

201. D. Pessayre, *Int. J. Clin. Pharm. Res.*, **3**, 449–458 (1983).

202. P. Kurath, P. H. Jones, R. S. Egan, and T. J. Perun, *Experientia* **27**, 362 (1971).

203. C. Vinckier, R. Hauchecorne, T. Cachet, G. Van den Mooter, and J. Hoogmartens, *Int. J. Pharmaceutics*, **55**, 67–76 (1989).

204. R. J. Pariza and L. A. Freiberg, *Pure Appl. Chem.*, **66**, 2365–2368 (1994).

205. J.-C. Gasc, S. G. d'Ambrieres, A. Lutz, and J.-F. Chantot, *J. Antibiot.*, **44**, 313–330 (1991).

206. A. Markham and D. Faulds, *Drugs*, **48**, 297–326 (1994).

207. E. H. Massey, B. S. Kitchell, L. D. Martin, and K. Gerzon, *J. Med. Chem.*, **17**, 105–107 (1974).

208. R. N. Brogden and D. H. Peters, *Drugs*, **48**, 599–616 (1994).

209. R. G. Finch, J. M. T. Hamilton-Miller, and A. M. Lovering, Eds., *J. Antimicrob. Chemother.*, **31** (Suppl. C) (1993).

210. J. R. Prous, Ed., *Drugs of Today*, **31** (Suppl. C) (1995).

211. H. A. Kirst, J. M. Greene. J. G. Amos, R. L.

Clemens, K. A. Sullivan, L. C. Creemer, J. W. Paschal, and F. T. Counter, *32nd Intersci. Conf. Antimicrob. Agents Chemother.*, Anaheim, Calif., 1992, abstr. 1365.

212. H. A. Kirst, J. A. Wind, J. P. Leeds, K. E. Willard, M. Debono, R. Bonjouklian, J. M. Greene, K. A. Sullivan, J. W. Paschal, J. B. Deeter, N. D. Jones, J. L. Ott, A. M. Felty-Duckworth, and F. T. Counter, *J. Med. Chem.*, **33**, 3086–3094 (1990).

213. P. A. Lartey, S. L. De Ninno, R. Faghih, D. J. Hardy, J. J. Clement, J. J. Plattner, and R. L. Stephens, *J. Med. Chem.*, **34**, 3390–3395 (1991).

214. P. A. Lartey, S. L. DeNinno, R. Faghih, D. J. Hardy, J. J. Clement, and J. J. Plattner, *J. Antibiot.*, **45**, 380–385 (1992).

215. K. Shankaran and T. A. Blizzard, *Bioorg. Med. Chem. Lett.*, **2**, 1555–1558 (1992).

216. G. M. Bright, A. A. Nagel, J. Bordner, K. A. Desai, J. N. Dibrino, J. Nowakowska, L. Vincent, R. M. Watrous, F. C. Sciavolino, A. R. English, J. A. Retsema, M. R. Anderson, L. A. Brennan, R. J. Borovoy, C. R. Cimochowski, J. A. Faiella, A. E. Girard, D. Girard, C. Herbert, M. Manousos, and R. Mason, *J. Antibiot.*, **41**, 1029–1047 (1988).

217. S. Djokic, G. Kobrehel, N. Lopotar, B. Kamenar, A. Nagl, and D. Mrvos, *J. Chem. Res.* (S), 152–153 (1988).

218. B. V. Yang, M. Goldsmith, and J. P. Rizzi, *Tetrahedron Lett.*, **35**, 3025–3028 (1994).

219. R. G. Finch, G. L. Ridgway, and K. J. Towner, Eds., *J. Antimicrob. Chemother.*, **31** (Suppl. E) (1993).

220. D. H. Peters, H. A. Friedel and D. McTavish, *Drugs*, **44**, 750–799 (1992).

221. R. H. Drew and H. A. Gallis, *Pharmacotherapy*, **12**, 161–173 (1992).

222. H. Lode, *Eur. J. Clin. Microbiol, Infect. Dis.*, **10**, 807–812 (1991).

223. M. A. Waugh, *J. Antimicrob. Chemother.*, **31** (Suppl. E), 193–198 (1993).

224. R. R. Wilkening, R. W. Ratcliffe, G. A. Doss, K. F. Bartizal, A. C. Graham, and C. M. Herbert, *Bioorg. Med Chem. Lett.*, **3**, 1287–1292 (1993).

225. A. B. Jones, *J. Org. Chem.*, **57**, 4361–4367, (1992).

226. S. T. Waddell and T. A. Blizzard, *Tetrahedron Lett.*, **34**, 5385–5388 (1993).

227. A. B. Jones, J. J. Acton III, and G. A. Doss, *Tetrahedron Lett.*, **34**, 4913–4916 (1993).

228. K. Shankaran, R. R. Wilkening, T. A. Blizzard, R. W. Ratcliffe, J. V. Heck, A. C. Graham, and C. M. Herbert, *Bioorg. Med. Chem. Lett.*, **4**, 1111–1116 (1994).

229. S. Morimoto, Y. Takahashi, Y. Watanabe, and S. Omura, *J. Antibiot.*, **37**, 187–189 (1984).

230. Y. Watanabe, S. Morimoto, T. Adachi, M. Kashimura, and T. Asaka, *J. Antibiot.*, **46**, 647–660 (1993).

231. Y. Watanabe, M. Kashimura, T. Asaka, T. Adachi, and S. Morimoto, *Heterocycles*, **36**, 243–247 (1993).

232. Y. Watanabe, T. Adachi, T. Asaka, M. Kashimura, T. Matsunaga, and S. Morimoto, *J. Antibiot.*, **46**, 1163–1167 (1993).

233. L. B. Barradell, G. L. Plosker, and D. McTavisah, *Drugs*, **46**, 289–312 (1993).

234. D. J. Hardy, D. R. P. Guay, and R. N. Jones, *Diagn. Microbiol. Infect. Dis.*, **15**, 39–53 (1992).

235. M. G. Sturgill and R. P. Rapp, *Ann. Pharmacother.*, **26**, 1099–1108 (1992).

236. R. G. Finch and D. C. E. Speller, Eds., *J. Antimicrob. Chemother.*, **27** (Suppl. A), (1991).

237. D. J. Hardy, R. N. Swanson, R. A. Rode, K. Marsh, N. L. Shipkowitz, and J. J. Clement, *Antimicrob. Agents Chemother.*, **34**, 1407–1413 (1990).

238. V. St. Georgiev, *Int. J. Antimicrob. Agents*, **4**, 247–270 (1994).

239. L. Toscano, G. Fioriello, R. Spagnoli, L. Cappelletti, and G. Zanuso, *J. Antibiot.*, **36**, 1439–1450 (1983).

240. N. Colombo, A. Depaoli, M. Gobetti, and M. G. Saorin, *Arz.-Forsch./Drug Res.*, **44 (II)**, 850–855 (1994).

241. A. Neszmelyi and H. Bojarska-Dahlig, *J. Antibiot.*, **31**, 487–489 (1978)

242. W. B. Baker, J. D. Clark, R. L. Stephens, and K. H. Kim, *J. Org. Chem.*, **53**, 2340–2345 (1988).

243. J. M. Wilson, P. C. T. Hannan, C Shillingford, and D. J. C. Knowles, *J. Antibiot.*, **42**, 454–462 (1989).

244. Y. Kawashima, Y. Yamada, T. Asaka, Y. Misawa, M. Kashimura, S. Morimoto, T. Ono, T. Nagate, K. Hatayama, S. Hirono, and I. Moriguchi, *Chem. Pharm. Bull.*, **42**, 1088–1095 (1994).

245. G. Kobrehel, G. Lazarevski, S. Dokic, L. Kolacny-Babic, N. Kucisec-Tepes, and M. Cvrlje, *J. Antibiot.*, **45**, 527–534 (1992).

246. D. J. Hardy, R. N. Swanson, N. L. Shipkowitz, L. A. Freiberg, P. A. Lartey, and J. J. Clement, *Antimicrob. Agents Chemother.*, **35**, 922–928 (1991).

247. A. Morishita, K. Ishizawa, N. Mutoh, T. Yamamoto, M. Hayashi, and S. Yaginuma, *J. Antibiot.*, **45**, 607–617 (1992).

248. C. Agouridas, Y. Benedetti, A. Denis, C. Fromentin, S. Gouin d'Ambrieres, O. Le Martret, and

J. F. Chantot, *34th Intersci. Conf. Antimicrob. Agents Chemother.*, Orlando, Fla., 1994, abstr. F-164.

249. T. Hata, Y. Sano, N. Ohki, Y. Yokoyama, A. Matsumae, and S. Ito, *J. Antibiot.*, **A6**, 87–89 (1953).

250. T. Arai, S. Kuroda, and Y. Mikami in T. Arai, Ed., *Actinomycetes: The Boundary Microorganisms*, Toppan Co. Ltd., Tokyo, 1976, p. 632.

251. S. Omura, Y. Hironaka, and T. Hata, *J. Antibiot.*, **23**, 511–513 (1970).

252. A. Ducruix, C. Pascard, A. Nakagawa, and S. Omura, *J. Chem. Soc., Chem. Commun.*, 947–948 (1976).

253. T. Osono, Y. Oka, S. Watanabe, Y. Numazaki, K. Moriyama, H. Ishida, K. Suzaki, Y. Okami, and H. Umezawa, *J. Antibiot.*, **20**, 174–180 (1967).

254. S. Pinnert-Sindico, L. Ninet, J. Preud'Homme, and C. Cosar, *Antibiot. Annual 1954–1955,* Medical Encyclopedia Inc., New York, 1955, pp. 724–727.

255. R. Corbaz, L. Ettlinger, E. Gaumann, W. Keller-Schierlein, F. Kradolfer, E. Kyburz, L. Neipp, V. Prelog, A. Wettstein, and H. Zahner, *Helv. Chim. Acta*, **39**, 304–317 (1956).

256. S. Omura, A. Nakagawa, A. Neszmelyi, S. D. Gero, A.-M. Sepulchre, F. Piriou, and G. Lukacs, *J. Am. Chem. Soc.*, **97**, 4001–4009 (1975).

257. P. Davey, J.-C. Pechere, and D. Speller, Eds., *J. Antimicrob. Chemother.*, **22** (Suppl. B), (1988).

258. E. Bergogne Berezin and J. M. T. Hamilton-Miller, Eds., *Drug Investigation*, **6** (Suppl. 1), 1–54 (1993).

259. J. M. McGuire, W. S. Boniece, C. E. Higgens, M. M. Hoehn, W. M. Stark, J. Westhead, and R. N. Wolfe, *Antibiot. Chemother.*, **11**, 320–327 (1961).

260. R. B. Morin, M. Gorman, R. L. Hamill, and P. V. Demarco, *Tetrahedron Lett.*, 4737–4740 (1970).

261. S. Omura, H. Matsubara, A. Nakagawa, A. Furusaki, and T. Matsumoto, *J. Antibiot.*, **33**, 915–917 (1980).

262. N. D. Jones, M. O. Chaney, H. A. Kirst, G. M. Wild, R. H. Baltz, R. L. Hamill, and J. W. Paschal, *J. Antibiot.*, **35**, 420–425 (1982).

263. H. A. Kirst, G. M. Wild, R. H. Baltz, E. T. Seno, R. L. Hamill, J. W. Paschal, and D. E. Dorman, *J. Antibiot.*, **36**, 376–382 (1983).

264. K. Kiyoshima, K. Takada, M. Yamamoto, K. Kubo, R. Okamoto, Y. Fukagawa, T. Ishikura, H. Naganawa, T. Sawa, T. Takecuchi, and H. Umezawa, *J. Antibiot.*, **40**, 1123–1130 (1987).

265. G. H. Wagman, J. A. Waitz, J. Marquez, A. Murawski, E. M. Oden, R. T. Testa, and M. J. Weinstein, *J. Antibiot.*, **25**, 641–646 (1972).

266. A. K. Ganguly, Y.-T. Liu, O. Sarre, R. S. Jaret, A. T. McPhail, and K. K. Onan, *Tetrahedron Lett.*, 4699–4702 (1980).

267. T. Kishi, S. Harada, H. Yamana, and A. Miyake, *J. Antibiot.*, **29**, 1171–1181 (1976).

268. H. Imai, K. Suzuki, M. Morioka, T. Sasaki, K. Tanaka, S. Kadota, M. Iwanami, T. Saito, and H. Eiki, *J. Antibiot.*, **42**, 1000–1002 (1989).

269. A. Kinumaki, K. Harada, T. Suzuki, M. Suzuki, and T. Okuda, *J. Antibiot.*, **30**, 450–454 (1977).

270. K. Kinoshita, S. Satoi, M. Hayashi, K. Harada, M. Suzuki, and K. Nakatsu, *J. Antibiot.*, **38**, 522–526 (1985).

271. K. Kinoshita, S. Satoi, M. Hayashi, and K Nakatsu, *J. Antibiot.*, **42**, 1003–1005 (1989).

272. J. Krc, Jr. and R. B. Scott, *Microscope*, **23**, 15–19 (1975).

273. L. A. Mitscher and M. P. Kunstmann, *Experientia*, **25**, 12–13 (1969).

274. S. Mizobuchi, J. Mochizuki, H. Soga, H. Tanba, and H. Inoue, *J. Antibiot.*, **39**, 1776–1778 (1986).

275. H. A. Kirst, *Prog. Med. Chem.*, **31**, 265–295 (1994).

276. S. Omura, M. Katagiri, I. Umezawa, K. Komiyama, T. Maekawa, K. Sekikawa, A. Matsumae, and T. Hata, *J. Antibiot.*, **21**, 532–538 (1968).

277. H. Sakakibara, O. Okekawa, T. Fujiwara, M. Aizawa, and S. Omura, *J. Antibiot.*, **34**, 1011–1018 (1981).

278. H. Sakakibara, O. Okekawa, T. Fujiwara, M. Otani, and S. Omura, *J. Antibiot.*, **34**, 1001–1010 (1981).

279. K. Hara, *Jpn. J. Antibiot.*, **40**, 1851–1866 (1987).

280. J. R. Prous, Ed., *Drugs of the Future*, **10**, 486–489 (1985).

281. S. Omoto, K. Iwamatsu, S. Inouye, and T. Niida, *J. Antibiot.*, **29**, 536–548 (1976).

282. T. Shomura, S. Someya, S. Murata, K. Umemura, and M. Nishio, *Chem. Pharm. Bull.*, **29**, 2413–2419 (1981).

283. S. M. Holliday and D. Faulds, *Drugs*, **46**, 720–745 (1993).

284. R. Okamoto, T. Fukumoto, H. Nomura, K. Kiyoshima, K. Nakamura, A. Takamatsu, H. Naganawa, T. Takeuchi, and H. Umezawa, *J. Antibiot.*, **33**, 1300–1308 (1980).

285. R. Okamoto, M. Tsuchiya, H. Nomura, H. Iguchi, K. Kiyoshima, S. Hori, T. Inui, T. Sawa, T. Takeuchi, and H. Umezawa, *J. Antibiot.*, **33**, 1309–1315 (1980).

286. M. Tsuchiya, T. Sawa, T. Takeuchi, H. Umezawa, and R. Okamoto, *J. Antibiot.*, **35**, 673–679 (1982).

287. H. A. Kirst, M. Debono, K. E. Willard, B. A. Truedell, J. E. Toth, J. R. Turner, D. R. Berry, B. B. Briggs, D. S. Fukuda, V. M. Daupert, A. M. Felty-Duckworth, J. L. Ott, and F. T. Counter, *J. Antibiot.*, **39**, 1724–1735 (1986).

288. T. Yoshioka, K. Kiyoshima, M. Maeda, M. Sakamoto, T. Ishikura, Y. Fukagawa, T. Sawa, M. Hamada, H. Naganawa, and T. Takeuchi, *J. Antibiot.*, **41**, 1617–1628 (1988).

289. T. Terasawa, M. Watanabe, T. Okubo, and S. Mitsuhashi, *Antimicrob. Agents Chemother.*, **35**, 1370–1375 (1991).

290. H. A. Kirst, J. E. Toth, M. Debono, K. E. Willard, B. A. Truedell, J. L. Ott, F. T. Counter, A. M. Felty-Duckworth, and R. S. Pekarek, *J. Med, Chem.*, **31**, 1631–1641 (1988).

291. H. A. Kirst in M. I. Choudhary, Ed., *Studies in Medicinal Chemistry*, Vol. 1, Harwood Academic Publishers, Amsterdam, 1995, in press.

292. M. Debono, K. E. Willard, H. A. Kirst, J. A. Wind, G. D. Crouse, E. V. Tao, J. T. Vicenzi, F. T. Counter, J. L. Ott, E. E. Ose, and S. Omura, *J. Antibiot.*, **42**, 1253–1267 (1989).

293. H. A. Kirst, K. E. Willard, M. Debono, J. E. Toth, B. A. Truedell, J. P. Leeds, J. L. Ott, A. M. Felty-Duckworth, F. T. Counter, E. E. Ose, G. D. Crouse, J. M. Tustin, and S. Omura, *J. Antibiot.*, **42**, 1673–1683 (1989).

294. P. E. Gorham, L. H. Carroll, J. W. McAskill, L. E. Watkins, E. E. Ose, L. V. Tonkinson, and J. K. Merrill, *Can. Vet. J.*, **31**, 826–829 (1990).

295. D. W. Morck, J. K. Merrill, B. E. Thorlakson, M. E. Olson, L. V. Tonkinson, and J. W. Coster- ton, *J. Am. Vet. Med. Assoc.*, **202**, 273–277 (1993).

296. M. R. Langley, D. R. Brown, and M. E. Tarrant, *Proc. 13th Int. Pig Vet. Soc. Cong.*, Bangkok, 1994, p. 333.

297. A. G. Fishman, A. K. Mallams, M. S. Puar, R. R. Rossman, and R. L. Stephens, *J. Chem. Soc., Perkin Trans I*, 1189–1209 (1987).

298. T. Fujiwara, H. Watanabe, Y. Kogami, Y. Shiritani, and H. Sakakibara, *J. Antibiot.*, **42**, 903–912 (1989).

299. T. Fujiwara, A. Sakai, and H. Sakakibara, *J. Antibiot.*, **43**, 327–330 (1990).

300. S. Kageyama, T. Tsuchiya, S. Umezawa, and T. Takeuchi, *J. Antibiot.*, **45**, 144–146 (1992).

301. N. Chin and H. C. Neu, *Antimicrob. Agents Chemother.*, **36**, 1699–1702 (1992).

302. S. Kageyama, T. Tsuchiya, and S. Umezawa, *J. Antibiot.*, **46**, 1265–1278 (1993).

303. H. Nakura, H. Ohi, T. Miura, T. Fujiwara, H. Ishizone, T. Yokoi, M. Kitada, and T. Kamataki, *J. Pharmacobio-Dyn.*, **14**, 377–383 (1991).

304. S. Sakamoto, T. Tsuchiya, A. Tanaka, S. Umezawa, M. Hamada, and H. Umezawa, *J. Antibiot.*, **38**, 477–484 (1985).

305. H. A. Kirst, J. E. Toth, J. A. Wind, M. Debono, K. E. Willard, R. M. Molloy, J. W. Paschal, J. L. Ott, A. M. Felty-Duckworth, and F. T. Counter, *J. Antibiot.*, **40**, 823–842 (1987).

306. B. Weisblum, *Antimicrob. Agents Chemother.*, **39**, 797–805 (1995).

307. B. Weisblum, *Antimicrob. Agents Chemother.*, **39**, 577–585 (1995).

308. R. Leclercq and P. Courvalin, *Antimicrob. Agents Chemother.*, **35**, 1267–1272 (1991).

309. R. Leclercq and P. Courvalin, *Antimicrob. Agents Chemother.*, **35**, 1273–1276 (1991).

310. J. D. Butts, *Clin. Pharmacokinet.*, **27**, 63–84 (1994).

311. M. Barza, *Clin. Infect. Dis.*, **19**, 910–915 (1994).

312. D. E. Nix, S. D. Goodwin, C. A. Peloquin, D. L. Rotella, and J. J. Schentag, *Antimicrob. Agents Chemother.*, **35**, 1947–1959 (1991).

313. P. Periti, T. Mazzei, T. Mini, and A. Novelli, *Clin. Pharmacokinet.*, **16**, 193–214 (1989).

314. P. Periti, T. Mazzei, T. Mini, and A. Novelli, *Clin. Pharmacokinet.*, **16**, 261–282 (1989).

315. Z. Itoh, Ed., *Motilin.*, Academic Press, San Diego, 1990.

316. M. Camilleri, *Am. J. Gastroenterol.*, **88**, 169–171 (1993).

317. S. Omura, K. Tsuzuki, T. Sunazuka, S. Marui, H. Toyoda, N. Inatomi, and Z. Itoh, *J. Med. Chem.*, **30**, 1941–1943 (1987).

318. K. Tsuzuki, T. Sunazuka, S. Marui, H. Toyoda, S. Omura, N. Inatomi, and Z. Itoh, *Chem. Pharm. Bull.*, **37**, 2687–2700 (1989).

319. T. Sunazuka, K. Tsuzuki, S. Marui, H. Toyoda, S. Omura, N. Inatomi, and Z. Itoh, *Chem. Pharm. Bull.*, **37**, 2701–2709 (1989).

320. H. A. Kirst, B. Greenwood, and J. S. Gidda, *Drugs of the Future*, **17**, 18–20 (1992)

321. I. Depoortere, T. L. Peeters, and G. Vantrappen, *Peptides*, **11**, 515–519 (1990).

322. M. Satoh, T. Sakai, I. Sano, K. Fujikura, H. Koyama, K. Ohshima, Z. Itoh, and S. Omura, *J. Pharmacol. Exptl. Ther.*, **271**, 574–579 (1994).

323. P. A. Lartey, H. Nellans, R. Faghih, L. L. Klein, C. Edwards, L. Freiberg, A. C. Peterson, and J. J. Plattner, *12th Int. Symp. Med. Chem.*, Basel, Sept. 13–17, 1992, abstr. OC-01.4.

324. H. Koga, T. Sato, K. Tsuzuki, H. Onoda, H. Kuboniwa, and H. Takanashi, *Bioorg. Med. Chem. Lett.*, **4**, 1347–1352 (1994).

325. R. B. Woodward in A. Todd, Ed., *Perspectives in Organic Chemistry*, John Wiley & Sons, Inc., New York, 1956, p. 160.

326. R. K. Boeckman and S. W. Goldstein in J. ApSimon, Ed., *The Total Synthesis of Natural Products,* John Wiley & Sons, Inc., New York, Vol. 7, 1988, pp. 1–139.

327. I. Paterson, *Pure Appl. Chem.,* **64,** 1821–1830 (1992).

328. S. F. Martin, W.-C. Lee, G. J. Pacofsky, R. P. Gist, and T. A. Mulhern, *J. Am. Chem. Soc.,* **116,** 4674–4688 (1994).

329. P. F. Wiley, *Dev. Indust. Microbiol.,* **26,** 97–116 (1985).

330. R. H. Baltz and E. T. Seno, *Ann. Rev. Microbiol.,* **42,** 547–574 (1988).

331. S. Omura, *Bacteriol. Rev.,* **40,** 681–697 (1976).

332. S. Omura, N. Sadakane, Y. Tanaka, and H. Matsubara, *J. Antibiot.,* **36,** 927–930 (1983).

333. J. M. Weber, J. O. Leung, S. J. Swanson, K. B. Idler, and J. B. McAlpine, *Science,* **252,** 114–117 (1991).

334. D. Stassi, S. Donadio, M. J. Staver, and L. Katz, *J. Bacteriol.,* **175,** 182–189 (1993).

335. R. H. Lambalot, D. E. Cane, J. J. Aparicio, and L. Katz, *Biochem.,* **34,** 1858–1866 (1995).

336. S. Donadio, M. J. Staver, J. B. McAlpine, S. J. Swanson, and L. Katz, *Science,* **252,** 675–679 (1991).

337. S. Donadio and L. Katz, *Gene,* **111,** 51–60 (1992).

338. J. Cortes, S. F. Haydock, G. A. Roberts, D. J. Bevitt, and P. F. Leadlay, *Nature,* **348,** 176–178 (1990).

339. L. Katz and S. Donadio, *Ann. Rev. Microbiol.,* **47,** 875–912 (1993).

340. L. Katz and C. R. Hutchinson, *Ann. Rep. Med. Chem.,* **27,** 129–138 (1992).

341. S. Donadio, J. B. McAlpine, P. J. Sheldon, M. Jackson, and L. Katz, *Proc. Natl. Acad. Sci. USA,* **90,** 7119–7123 (1993).

342. C. M. Kao, L. Katz, and C. Khosla, *Science,* **265,** 509–512 (1994).

343. B. J. Magerlein, R. D. Birkenmeyer, R. R. Herr, and F. Kagan, *J. Am. Chem. Soc.,* **89,** 2459–2464 (1967).

344. H. Hoeksema, *J. Am. Chem. Soc.,* **90,** 755–757 (1968).

345. R. D. Birkenmeyer and F. Kagan, *J. Med. Chem.,* **13,** 616–619 (1970).

346. A. S. Klainer, *Med. Clin. North Am.,* **71,** 1169–1175 (1987).

347. R. D. Birkenmeyer, S. J. Kroll, C. Lewis, K. F. Stern, and G. E. Zurenko, *J. Med. Chem.,* **27,** 216–223 (1984).

348. C. Thornsberry, J. K. Marler, J. L. Watts, and R. J. Yancey, *Antimicrob. Agents Chemother.,* **37,** 1122–1126 (1993).

349. J. Charney, W. P. Fisher, C. Curran, R. A. Machlowitz, and A. A. Tytell, *Antibiot. Chemother.,* **3,** 1283–1286 (1953).

350. P. De Somer and P. Van Dijck, *Antibiot. Chemother.,* **5,** 632–639 (1955).

351. C. Cocito, *Microbiol. Reviews,* **43,** 145–198 (1979).

352. F. Durant, G. Evrard, J. P. Declercq, and G. Germain, *Cryst. Struct. Commun.,* **3,** 503–510 (1974).

353. J. P. Declercq, G. Germain, M. Van Meerssche, S. E. Hull, and M. J. Irwin, *Acta Crystallogr.,* **34,** 3644–3648 (1978).

354. E. Surcouf, I. Morize, D. Frechet, M. Vuilhorgne, A. Mikou, E. Guitet, and J. T. Lallemand, *Stud. Phys. Theor. Chem.,* **71,** 719–726 (1990).

355. M. Di Giambattista, G. Chinali, and C. Cocito, *J. Antimicrob. Chemother.,* **24,** 485–507 (1980).

356. D. Videau, *Pathol. Biol.,* **30,** 529–534 (1982).

357. J. Preud'homme, P. Tarridec, and A. Belloc, *Bull. Soc. Chim. France,* 585–591 (1968).

358. J. M. Paris, J. C. Barriere, C. Smith, and P. E. Bost in Ref. 11, pp. 183–248.

359. R. G. Finch, P. M. Hawkey, and K. J. Towner, Eds., *J. Antimicrob. Chemother.,* **30** (Suppl. A), (1992).

360. J. C. Barriere and J. M. Paris, *Drugs of the Future,* **18,** 833–845 (1993).

361. B. Cavalleri and F. Parenti in *Kirk-Othmer Encyclopedia of Chemical Technology,* 4th edition, John Wiley & Sons, Inc., New York, Vol. 2, 1992, pp. 995–1018.

362. G. C. Lancini, *Prog. Indust. Microbiol.,* **27,** 283–296 (1989).

363. R. Nagarajan, Ed., *Glycopeptide Antibiotics,* Marcel Dekker, Inc., New York, 1994.

364. J. C. J. Barna and D. H. Williams, *Annu. Rev. Microbiol.,* **38,** 339–357 (1984).

365. G. L. Cooper and D. B. Given, Eds., *Vancomycin, A Comprehensive Review of 30 Years of Clinical Experience,* Park Row Publishers, Inc., New York, 1986.

366. J. F. Levine, *Med. Clin. N. Am.,* **71,** 1135–1145 (1987).

367. R. Wise and D. Reeves, Eds., *J. Antimicrob. Chemother.,* **14** (Suppl. D), (1984).

368. R. C. Yao and L. W. Crandall in Ref. 363, pp. 1–27.

369. G. Cassani, *Prog. Indust. Microbiol.,* **27,** 221–235 (1989).

370. R. D. Sitrin and G. Folena-Wasserman in Ref. 363, pp. 29–61.

371. M. H. McCormick, W. M. Stark, G. E. Pittenger, R. C. Pittenger, and J. M. McGuire, *Antibiot.*

Annu. 1955–1956, Medical Encyclopedia Inc., New York, 1956 pp. 606–611.

372. M. P. Lechevalier, H. Prauser, D. P. Labeda, and J.-S. Ruan, *Int. J. Syst. Bacteriol.*, **36**, 29–37 (1986).

373. G. M. Sheldrick, P. G. Johes, O. Kennard, D. H. Williams, and G. A. Smith, *Nature*, **271**, 223–225 (1978).

374. C. M. Harris, H. Kopecka, and T. M. Harris, *J. Am. Chem. Soc.*, **105**, 6915–6922 (1983).

375. A. H. Hunt, G. G. Marconi, T. K. Elzey, and M. M. Hoehn, *J. Antibiot.*, **37**, 917–919 (1984).

376. R. Nagarajan, K. E. Merkel, K. H. Michel, H. M. Higgins, Jr., M. M. Hoehn, A. H. Hunt, N. D. Jones, J. L. Occolowitz, A. A. Schabel, and J. K. Swartzendruber, *J. Am. Chem. Soc.*, **110**, 7896–7897 (1988).

377. R. Nagarajan, D. M. Berry, A. H. Hunt, J. L. Occolowitz, and A. A. Schabel, *J. Org. Chem.*, **54**, 983–986 (1989).

378. R. Nagarajan, D. M. Berry, and A. A. Schabel, *J. Antibiot.*, **42**, 1438–1440 (1989).

379. G. F. Grause, M. G. Brazhnikova, N. N. Lomakina, T. F. Berdnikova, G. B. Fedorova, N. L. Tokareva, V. N. Borisova, and G. Y. Batta, *J. Antibiot.*, **42**, 1790–1799 (1989).

380. N. Tsuji, M. Kobayashi, T. Kamigauchi, Y. Yoshimura, and Y. Terui, *J. Antibiot.*, **41**, 819–822 (1988).

381. N. Tsuji, T. Kamigauchi, M. Kobayashi, Y. Yoshimura, and Y. Terui, *J. Antibiot.*, **41**, 1506–1510 (1988).

382. M. L. Sanchez, R. P. Wendzel, and R. N. Jones, *Antimicrob. Agents Chemother.*, **36**, 873–875 (1992).

383. R. Nagarajan, *J. Antibiot.*, **46**, 1181–1195 (1993).

384. F. Parenti, G. Beretta, M. Berti, and V. Arioli, *J. Antibiot.*, **31**, 276–283 (1978).

385. A. H. Hunt, R. M. Molloy, J. L. Occolowitz, G. G. Marconi, and M. Debono, *J. Am. Chem. Soc.*, **106**, 4891–4895 (1984).

386. J. C. J. Barna, D. H. Williams, D. J. M. Stone, T.-W. C. Leung, and D. M. Doddrell, *J. Am. Chem. Soc.*, **106**, 4895–4902 (1984).

387. A. Borghi, P. Antonini, M. Zanol, P. Ferrari, L. F. Zerilli, and G. C. Lancini, *J. Antibiot.*, **42**, 361–366 (1989).

388. S. L. Heald, L. Mueller, and P. W. Jeffs, *J. Magn. Res.*, **72**, 120–138 (1987).

389. R. N. Brogden and D. H. Peters, *Drugs*, **47**, 823–854 (1994).

390. D. M. Campoli-Richards, R. N. Brogden, and D. Faulds, *Drugs*, **40**, 449–486 (1990).

391. D. Speller and D. Greenwood, Eds., *J. Antimicrob. Chemother.*, **21** (Suppl. A), (1988).

392. A. Malabarba, R. Ciabatti, J. Kettenring, R. Scotti, G. Candiani, M. Berti, R. Pallanza, and B. P. Goldstein, *J. Antibiot.*, **46**, 668–675 (1993).

393. A. Malabarba, R. Ciabatti, R. Scotti, and B. P. Goldstein, *J. Antibiot.*, **46**, 661–667 (1993).

394. A. Malabarba, R. Ciabatti, J. Kettenring, P. Ferrari, R. Scotti, B. P. Goldstein, and M. Denaro, *J. Antibiot.*, **47**, 1493–1506 (1994).

395. G. Folena-Wasserman, B. L. Poehland, E. W.-K. Yeung, D. Staiger, L. B. Killmer, K. Snader, J. J. Dingerdissen, and P. W. Jeffs, *J. Antibiot.*, **39**, 1395–1406 (1986).

396. P. W. Jeffs, L. Mueller, C. DeBrosse, S. L. Heald, and R. Fisher, *J. Am. Chem. Soc.*, **108**, 3063–3075 (1986).

397. S. B. Christensen, H. S. Allaudeen, M. R. Burke, S. A. Carr, S. K. Chung, P. DePhillips, J. J. Dingerdissen, M. DiPaolo, A. J. Giovenella, S. L. Heald, L. B. Killmer, B. A. Mico, L. Mueller, C. H. Pan, B. L. Poehland, J. B. Rake, G. D. Roberts, M. C. Shearer, R. D. Sitrin, L. J. Nisbet, and P. W. Jeffs, *J. Antibiot.*, **40**, 970–990 (1987).

398. J. E. Philip, J. P. Schenck, and M. P. Hargie, *Antibiot. Annu.* 1956–1957, Medical Encyclopedia Inc., New York, 1957, pp. 699–705.

399. M. P. Lechevalier, H. Prauser, D. P. Labeda, and J.-S. Ruan, *Int. J. Syst. Bacteriol.*, **36**, 29–37 (1986).

400. C. M. Harris and T. M. Harris, *J. Am. Chem. Soc.*, **104**, 363–365 (1982).

401. G. F. Gauze, E. S. Kudrina, R. S. Ukholina, and G. V. Gavrilina, *Antibiotiki*, **8**, 387–392 (1963).

402. M. A. Howard and B. G. Firkin, *Thromb. Diathesis Haemorrh.*, **26**, 363–369 (1971).

403. M. E. Rick, *Clinics Lab. Med.*, **14**, 781–794 (1994).

404. A. H. Hunt, T. K. Elzey, K. E. Merkel and M. Debono, *J. Org. Chem.*, **49**, 641–645 (1984).

405. A. H. Hunt, *J. Am. Chem. Soc.*, **105**, 4463–4468 (1983).

406. A. H. Hunt, D. E. Dorman, M. Debono, and R. M. Molloy, *J. Org. Chem.*, **50**, 2031–2035 (1985).

407. L. D. Boeck and F. P. Mertz, *J. Antibiot.*, **39**, 1533–1540 (1986).

408. N. J. Skelton, D. H. Williams, R. A. Monday, and J. C. Ruddock, *J. Org., Chem.*, **55**, 3718–3723 (1990).

409. N. J. Skelton, D. H. Williams, M. J. Rance, and J. C. Ruddock, *J. Am. Chem. Soc.*, **113**, 3757–3765 (1991).

410. M. P. Kunstmann, L. A. Mitscher, J. N. Porter, A. J. Shay, and M. A. Darken, *Antimicrob. Agents Chemother.-1968*, American Society for Microbiology, Washington, D. C., 1969, pp. 242–245.

411. D. P. Labeda, *Int. J. Syst. Bacteriol.*, **45**, 124–127 (1995).

412. W. J. McGahren, J. H. Martin, G. O. Morton, R. T. Hargreaves, R. A. Leese, F. M. Lovell, G. A. Ellstad, E. O'Brien, and J. S. E. Holker, *J. Am. Chem. Soc.*, **102**, 1671–1684 (1980).

413. G. A. Ellstad, R. A. Leese, G. O. Morton, F. Barbatschi, W. E. Gore, W. J. McGahren, and I. M. Armitage, *J. Am. Chem. Soc.*, **103**, 6522–6524 (1981).

414. G. A. Ellestad, W. Swenson, and W. J. McGahren, *J. Antibiot.*, **36**, 1683–1690 (1983).

415. G. Bata, F. Sztaricskai, J. Csanadi, I. Komaromi, and R. Bognar, *J. Antibiot.*, **39**, 910–913 (1986).

416. S. L. Heald, L. Mueller, and P. W. Jeffs, *J. Antibiot.*, **40**, 630–645 (1987).

417. T. Takatsu, S. Takahashi, M. Nakajima, T. Haneishi, T. Nakamura, H. Kuwano, and T. Kinoshita, *J. Antibiot.*, **40**, 933–940 (1987)

418. M. Takeuchi, S. Takahashi, M. Inukai, T. Nakamura, and T. Kinoshita, *J. Antibiot.*, **44**, 271–277 (1991).

419. A. K. Verma, R. Prakash, S. A. Carr, G. D. Roberts, and R. D. Sitrin, *26th Intersci. Conf. Antimicrob. Agents Chemother., New Orleans* (1986) abstr. 940.

420. V. R. Carruthers, C. W. Holmes, K. A. MacDonald, D. H. Norton, A. Alexander, and W. R. Dodemaide, *N. Zealand J. Agric. Res.*, **35**, 171 (1992).

421. R. Fekety in Ref. 4, pp. 346–354.

422. J. C. Rotschafer, M. W. Garrison, and K. A. Rodvold, *Pharmacother.*, **8**, 211–219 (1988).

423. R. Leclercq, E. Derlot, J. Duval, and P. Courvalin, *N. Eng. J. Med.*, **319**, 157–161 (1988).

424. A. H. C. Uttley, R. C. George, J. Naidoo, N. Woodford, A. P. Johnson, C. H. Collins, D. Morrison, A. F. Gilfillan, L. E. Fitch, and J. Heptonstall, *Epidemiol. Infect.*, **103**, 173–181 (1989).

425. P. Courvalin, *Antimicrob. Agents Chemother.*, **34**, 2291–2296 (1990).

426. A. P. Johnson, A. H. C. Uttley, N., Woodford, and R. C. George, *Clin. Microbiol. Rev.*, **3**, 280–291 (1990).

427. D. H. Williams and J. P. Waltho, *Pure Appl. Chem.*, **61**, 585–588 (1989).

428. T. D. H. Bugg and C. T. Walsh, *Nat. Prod. Rep.*, **9**, 199–215 (1992).

429. P. E. Reynolds, *Eur. J. Clin. Microbiol, Infect. Dis.*, **8**, 943–950 (1989).

430. T. I. Nicas and N. E. Allen in Ref. No. 363, pp. 219–241.

431. D. A. Beauregard, D. H. Williams, M. N. Gwynn, and D. J. C. Knowles, *Antimicrob. Agents Chemother.*, **39**, 781–785 (1995).

432. M. Arthur and P. Courvalin, *Antimicrob. Agents Chemother.*, **37**, 1563–1571 (1993).

433. N. E. Allen, J. N. Hobbs, Jr., J. M. Richardson, and R. M. Riggin, *FEMS Microbiol. Lett.*, **98**, 109–115 (1992).

434. S. Handwerger, M. J. Pucci, K. J. Volk, J. Liu, and M. S. Lee, *J. Bacteriol.*, **174**, 5982–5984 (1992).

435. J. Messer and P. E. Reynolds, *FEMS Microbiol. Lett.*, **94**, 195–200 (1992).

436. C. T. Walsh, *Science*, **261**, 308–309 (1993).

437. W. C. Noble, Z. Virani, and R. G. A. Cree, *FEMS Microbiol. Lett.*, **93**, 195–198 (1992).

438. E. M. Wise, Jr. in *Kirk-Othmer Encyclopedia of Chemical Technology*, 4th ed., Vol. 3, John Wiley & Sons, Inc., New York, 1992, pp. 266–306.

439. H. Kleinkauf and H. Von Dohren in M. Moo-Young, Ed., *Comprehensive Biotechnology* Vol. 2, Pergamon Press, Oxford, U. K. 1985, pp. 95–135.

440. H. Kleinkauf and H. Von Dohren, *Ann. Rev. Microbiol.*, **41**, 259–289 (1987).

441. H. G. Boman, J. Marsh, and J. A. Goode, Eds., *Antimicrobial Peptides*, Ciba Foundation Symposium Series 186, John Wiley & Sons, Inc., Chichester, U. K., 1994.

442. U. Hollstein in M. E. Wolff, Ed., *Burger's Medicinal Chemistry*, 4th edition, Part II, John Wiley & Sons, Inc., New York, 1979, pp. 173–270.

443. E. Katz and A. L. Demain, *Bacteriol. Rev.*, **41**, 449–474 (1977).

444. B. A. Johnson, H. Anker, and F. L. Meleney, *Science*, **102**, 376–377 (1945).

445. H. Oka, Y. Ikai, N. Kawamura, M. Yamada, K. Harada, Y. Yamazaki, *J. Chrom.*, **462**, 315–322 (1989).

446. C. Resler and D. V. Kashelikar, *J. Am. Chem. Soc.*, **88**, 2025–2035 (1966).

447. R. E. Galardy, M. P. Printz, and L. C. Craig, *Biochem.*, **10**, 2429–2436 (1971).

448. Y. Ikai, H. Oka, J. Hayakawa, K. Harada, and M. Suzuki, *J. Antibiot.*, **45**, 1325–1334 (1992).

449. G. Siewert and J. L. Strominger, *Proc. Nat. Acad. Sci. USA*, **57**, 767–773 (1967).

450. K. J. Stone and J. L. Strominger, *Proc. Nat. Acad. Sci USA*, **68**, 3223–3227 (1971).

451. R. J. Dubos, *J. Exptl. Med.*, **70**, 1–17 (1939).

452. R. Sarges and B. Witkop, *J. Am. Chem. Soc.*, **87**, 2011–2030 (1965).

453. B. A. Wallace and K. Ravikumar, *Science*, **241**, 182–187 (1988).

454. D. A. Langs, *Science*, **241**, 188–191 (1988).

455. B. A. Wallace, *Ann. Rev. Biophysics Biophys. Chem.*, **19**, 127–157 (1990).

456. G. C. Ainsworth, A. M. Brown, and G. Brownlee, *Nature*, **160**, 263 (1947).

457. R. G. Benedict and A. F. Langlykke, *J. Bacteriol.*, **54**, 24–25 (1947).

458. P. G. Stansly, R. G. Shepherd and J. White, *Bull. Johns Hopkins Hosp.*, **81**, 43–54 (1947).

459. K. Vogler and R. O. Studer, *Experientia*, **22**, 345–354 (1966).

460. S. Wilkinson and L. A. Lowe, *J. Chem. Soc.*, 4107–4125 (1964).

461. D. R. Storm, K. S. Rosenthal, and P. E. Swanson, *Ann. Rev. Biochem.*, **46**, 723–763 (1977).

462. D. S. Feingold, C.-C. H. Chen, and I. J. Sud, *Ann. N. Y. Acad. Sci.*, **235**, 480 (1974).

463. R. A. Moore, L. Chan, and R. E. W. Hancock, *Antimicrob. Agents Chemother.*, **26**, 539–545 (1984).

464. J. Horton and G. A. Pankey, *Med. Clin. N. Amer.*, **66**, 135–142 (1982).

465. B. Heinemann, M. A. Kaplan, R. D. Muir, and I. R. Hooper, *Antibiot. Chemother.*, **3**, 1239–1242 (1953).

466. M. Bodanszky, G. F. Sigler, and A. Bodanszky, *J. Am. Chem. Soc.*, **95**, 2352–2357 (1973).

467. H. Tanaka, R. Oiwa, S. Matsukura, J. Inokoshi, and S. Omura, *J. Antibiot.*, **35**, 1216–1221 (1982).

468. D. K. Banerjee, M. G. Scher, and C. J. Waechter, *Biochem.*, **20**, 1561–1568 (1981).

469. J. H. Lakey, R. Maget-Dana, and M. Ptak, *Biochem. Biophys. Res. Commun.*, **150**, 384–390 (1988).

470. S. Goto, S. Kuwahara, N. Okubo, and H. Zenyoji, *J. Antibiot.*, **21**, 119–125 (1968).

471. H. Iwasaki, S. Horii, M. Asai, K. Mizuno, J. Ueyanaga, and A. Miyake, *Chem. Pharm. Bull.*, **21**, 1184–1191 (1973).

472. T. Tamura, H. Suzuki, Y. Nishimura, J. Mizoguchi, and Y. Hirota, *Proc. Nat. Acad. Sci. USA*, **77**, 4499–4503 (1980).

473. M. Inukai, M. Nakajima, M. Osawa, T. Haneishi, and M. Arai, *J. Antibiot.*, **31**, 421–425 (1978).

474. M. Nakajima, M. Inukai, T. Haneishi, A. Terahara, M. Arai, T. Kinoshita, and C. Tamura, *J. Antibiot.*, **31**, 426–432 (1978).

475. M. Inukai, M. Takeuchi, K. Shimizu, and M. Arai, *J. Antibiot.*, **31**, 1203–1205 (1978).

476. B. Cavalleri, H. Pagani, G. Volpe, E. Selva, and F. Parenti, *J. Antibiot.*, **37**, 309–317 (1984).

477. R. Ciabatti, J. K. Kettenring, G. Winters, G. Tuan, L. Zerilli, and B. Cavalleri, *J. Antibiot.*, **42**, 254–267 (1989)

478. R. Ciabatti and B. Cavalleri, *Prog. Indust. Microbiol.*, **27**, 205–219 (1989).

479. C. C. Johnson, S. Taylor, P. Pitsakis, P. May, and M. E. Levison, *Antimicrob. Agents Chemother.*, **36**, 2342–2345 (1992).

480. E. A. Somner and P. E. Reynolds, *Antimicrob. Agents Chemother.*, **34**, 413–419 (1990).

481. J. F. Pagano, M. J. Weinstein, H. A. Stout, and R. Donovick, *Antibiot. Annual 1955–1956*, Medical Encyclopedia Inc., New York, 1956, pp. 554–559.

482. M. Bodanszky, J. A. Scozzie, and I. Muramatsu, *J. Am. Chem. Soc.*, **91**, 4934–4936 (1969).

483. B. Anderson, D. C. Hodgkin, and M. A. Viswamitra, *Nature*, **225**, 233–235 (1970).

484. K. Tori, K. Tokura, K. Okabe, M. Ebata, H. Otsuka, and G. Lukacs, *Tetrahedron Lett.*, 185–188 (1976).

485. O. D. Hensens, G. Albers-Schonberg, and B. F. Anderson, *J. Antibiot.*, **36**, 799–813 (1983).

486. T.-P. Hausner, U. Geigenmuller, and K. H. Nierhaus, *J. Biol. Chem.*, **263**, 13103–13111 (1988).

487. O. D. Hensens and G. Albers-Schonberg, *J. Antibiot.*, **36**, 814–831 (1983).

488. K. Tokura, K. Tori, Y. Yoshimura, K. Okabe, H. Otsuka, K. Matsushita, F. Inagaki, and T. Miyazawa, *J. Antibiot.*, **33**, 1563–1567 (1980).

489. T. Endo and H. Yonehara, *J. Antibiot.*, **31**, 623–625 (1978).

490. K. Tori, K. Tokura, Y. Yoshimura, Y. Terui, K. Okabe, H. Otsuka, K. Matsushita, F. Imagaki, and T. Miyazawa, *J. Antibiot.*, **34**, 124–129 (1981).

491. F. Benazet, M. Cartier, J. Florent, C. Godard, G. Jung, J. Lunel, D. Mancy, C. Pascal, J. Renaut, P. Tarridec, J. Theilleux, R. Tissier, M. Dubost, and L. Ninet, *Experientia*, **36**, 414–416 (1980).

492. B. W. Bycroft and M. S. Gowland, *J. Chem. Soc. Chem. Commun.*, 256–258 (1978).

493. M. Debono, R. M. Molloy, J. L. Occolowitz, J. W. Paschal, A. H. Hunt, K. H. Michel, and J. W. Martin, *J. Org. Chem.*, **57**, 5200–5208 (1992).

494. H. A. Kirst, E. F. Szymanski, D. E. Dorman, J. L. Occolowitz, N. D. Jones, M. O. Chaney, R. L. Hamill, and M. M. Hoehn, *J. Antibiot.*, **28**, 286–291 (1975).

495. T. Shiba, K. Inami, K. Sawada, and Y. Hirotsu, *Heterocycles*, **13**, 175–180 (1979)

496. E. Selva, G. Beretta, N. Montanini, G. S. Saddler, L. Gastaldo, P. Ferrari, R. Lorenzetti, P. Landini,

F. Ripamonti, B. P. Goldstein, M. Berti, L. Montanaro, and M. Denaro, *J. Antibiot.*, **44**, 693–701 (1991).

497. J. Kettenring, L. Colombo, P. Ferrari, P. Tavecchia, M. Nebuloni, K. Vekey, G. G. Gallo, and E. Selva, *J. Antibiot.*, **44**, 702–715 (1991).

498. G. Jung, *Angew. Chem. Int. Eng. Ed.*, **30**, 1051–1068 (1991).

499. H. G. Sahl in Ref. 441, pp. 42–53.

500. E. Gross and J. L. Morell, *J. Am. Chem. Soc.*, **93**, 4634–4635 (1971).

501. M. Barber, G. J. Elliott, R. S. Bordoli, B. N. Green, and B. W. Bycroft, *Experientia*, **44**, 266–270 (1988).

502. K. Fukase, M. Kitazawa, A. Sano, K. Shimbo, H. Fukita, S. Horimoto, T. Wakamiya, and T. Shiba, *Tetrahedron Lett.*, **29**, 795–798 (1988).

503. E. Ruhr and H.-G. Stahl, *Antimicrob. Agents Chemother.*, **27**, 841–845 (1985).

504. E. Gross, H. H. Kiltz, and E. Nebelin, *Hoppe-Seyler's Z. Physiol. Chem.*, **354**, 810–812 (1973).

505. A. Hurst, *Adv. Appl. Microbiol.*, **27**, 85–123 (1981).

506. G. Jung and H. G. Sahl, Eds., *Nisin and Novel Lantibiotics*, Escom, Leiden, 1991.

507. J. N. Hansen, *Ann. Rev. Microbiol.*, **47**, 535–564 (1993).

508. T. R. Klaenhammer, *FEMS Microbiol. Rev.*, **12**, 39–85 (1993).

509. M. Debono, M. Barnhart, C. B. Carrell, J. A. Hoffmann, J. L. Occolowitz, B. J. Abbott, D. S. Fukuda, R. L. Hamill, K. Biemann and W. C. Herlihy, *J. Antibiot.*, **40**, 761–777 (1987).

510. M. Debono, B. J. Abbott, R. M. Molloy, D. S. Fukuda, A. H. Hunt, V. M. Daupert, F. T. Counter, J. L. Ott, C. B. Carrell, L. C. Howard, L. D. Boeck, and R. L. Hamill, *J. Antibiot.*, **41**, 1093–1105 (1988).

511. G. M. Eliopoulos, C. Thauvin, B. Gerson, and R. C. Moellering, Jr., *Antimicrob. Agents Chemother.*, **27**, 357–362 (1985).

512. W. E. Alborn, Jr., N. E. Allen, and D. A. Preston, *Antimicrob. Agents Chemother.*, **35**, 2282–2287 (1991).

513. J. H. Lakey, R. Maget-Dana, and M Ptak, *Biochim. Biophys. Acta*, **985**, 60–66 (1989).

514. M. Boaretti, P. Canepari, M. del Mar Lleo, and G. Satta, *J. Antimicrob. Chemother.*, **31**, 227–235 (1993).

515. D. S. Fukuda, R. H. DuBus, P. J. Baker, D. M. Berry, and J. S. Mynderse, *J. Antibiot.*, **43**, 595–600 (1990).

516. A. H. Hunt, R. E. Chance, M. G. Johnson, J. L. Occolowitz, J. S. Mynderse, D. S. Fukuda, and

H. A. Kirst, *28th Intersci. Conf. Antimicrob. Agents Chemother.*, Los Angeles, 1988, abstr. 969.

517. D. S. Fukuda, M. Debono, R. M. Molloy, and J. S. Mynderse, *J. Antibiot.*, **43**, 601–606 (1990).

518. L. D. Boeck and R. W. Wetzel, *J. Antibiot.*, **43**, 607–615 (1990).

519. W. O. Godtfredsen, S. Jahnsen, H. Lorck, K. Roholt, and L. Tybring, *Nature*, **193**, 987 (1962).

520. W. O. Godtfredsen, W. von Daehne, S. Vangedal, A. Marquet, D. Arigoni, and A. Melera, *Tetrahedron*, **21**, 3505–3530 (1965).

521. A. Cooper, *Tetrahedron*, **22**, 1379–1381 (1966).

522. W. von Daehne, H. Lorch, and W. O. Godtfredsen, *Tetrahedron Lett.*, 4843–4846 (1968).

523. S. Okuda, Y. Sato, T. Hattori, and M. Wakabayashi, *Tetrahedron Lett.*, 4847–4850 (1968).

524. W. von Daehne, W. O. Godtfredsen, and P. R. Rasmussen, *Adv. Appl. Microbiol.*, **25**, 95–146 (1979).

525. H. B. Drugeon, J. Caillon and M. E. Juvin, *J. Antimicrob. Chemother.*, **34**, 899–907 (1994).

526. I. Phillips and D. Speller, Eds., *J. Antimicrob. Chemother.*, **25** (Suppl. B), (1990).

527. D. S. Reeves, *J. Antimicrob. Chemother.*, **20**, 467–476 (1987).

528. A. T. Fuller, G. Mellows, M. Woodford, G. T. Banks, K. D. Barrow, and E. B. Chain, *Nature*, **234**, 416–417 (1971).

529. E. B. Chain and G. Mellows, *J. Chem. Soc. Perkin I*, 294–309 (1977).

530. R. Sutherland, R. J. Boon, K. E. Griffin, P. J. Masters, B. Slocombe, and A. R. White, *Antimicrob. Agents Chemother.*, **27**, 495–498 (1985).

531. J. J. Leyden, *Clin. Pediatr.*, **31**, 549–553 (1992).

532. M. W. Casewell and R. L. R. Hill, *J. Antimicrob. Chemother.*, **19**, 1–5 (1987).

533. M. A. Parenti, S. M. Hatfield, and J. J. Leydon, *Clin. Pharmacy*, **6**, 761–770 (1987).

534. A. Ward and D. M. Campoli-Richards, *Drugs*, **32**, 425–444 (1986).

535. J. Hughes and G. Mellows, *Biochem. J.*, **191**, 209–219 (1980).

536. B. D. Cookson, *J. Antimicrob. Chemother.*, **25**, 497–503 (1990).

537. J. Gilbart, C. R. Perry and B. Slocombe, *Antimicrob. Agents Chemother.*, **37**, 32–38 (1993).

538. D. A. Janssen, L. T. Zarins, D. R. Schaberg, S. F. Bradley, M. S. Terpenning, and C. A. Kauffman, *Antimicrob. Agents Chemother.*, **37**, 2003–2006 (1993).

539. V. M. Girijavallabhan and A. K. Ganguly in *Kirk-Othmer Encyclopedia of Chemical Technology*,

4th edition, Vol. 3, John Wiley & Sons, Inc., New York, 1992, pp. 259–266.

540. M. J. Weinstein, G. M. Luedemann, E. M. Oden, and G. H. Wagman, *Antimicrob. Agents Chemother.-1964*, American Society for Microbiol., Washington, D. C., 1965, pp. 24–32.

541. A. K. Ganguly, O. Z. Sarre, D. Greeves, and J. Morton, *J. Am. Chem. Soc.*, **97**, 1982–1985 (1975).

542. A. K. Ganguly, O. Z. Sarre, A. T. McPhail, and R. W. Miller, *J. Chem. Soc., Chem. Commun.*, 22–24 (1979).

543. A. K. Ganguly, *Topics Antibiot. Chem.*, **2**, 59–98 (1978).

544. H. Wolf, *FEBS Lett.*, **36**, 181–186 (1973).

545. R. N. Jones, M. S. Barrett, and M. E. Erwin, *34th Intersci. Conf. Antimicrob. Agents Chemother.*, Orlando, Fla., 1994, abstr. F127.

546. C. Lin, A. Nomeir, J. Lim, D. Loebenberg, R. Hare, and G. H. Miller, *34th Intersci, Conf. Antimicrob, Agents Chemother.*, Orlando, Fla., 1994, abstr. F128.

547. F. Buzzetti, F. Eisenberg, H. N. Grant, W. Keller-Schierlein, W. Voser, and H. Zahner, *Experientia*, **24**, 320–323 (1968).

548. J. L. Mertz, J. S. Peloso, B. J. Barker, G. E. Babbitt, J. L. Occolowitz, V. L. Simson, and R. M. Kline, *J. Antibiot.*, **39**, 877–887 (1986).

549. W. Keller-Schierlein, W. Heilman, W. D. Ollis, and C. Smith, *Helv. Chim. Acta*, **62**, 7–20 (1979).

550. E. Kupfer, K. Neupert-Laves, M. Dobler, and W. Keller-Schierlein, *Helv. Chim. Acta*, **65**, 3–12 (1982).

551. D. E. Wright, *Tetrahedron*, **35**, 1207–1237 (1979).

552. N. F. Capuccino, A. K. Bose, J. B. Morton, and A. K. Ganguly, *Heterocycles*, **15**, 1621–1641 (1981).

553. T. Kamiya, S. Maeno, M. Hashimoto, and Y. Mine, *J. Antibiot.*, **25**, 576–581 (1972).

554. S. Miyamura, N. Ogasawara, H. Otsuka, S. Niwayama, H. Tanaka, T. Take, T. Uchiyama, H. Ochiai, K. Abe, K. Koizumi, K. Asao, K. Matsuki, and T. Hoshino, *J. Antibiot.*, **25**, 610–612 (1972).

555. Y. Tokuma, S. Koda, T. Miyoshi, and Y. Morimoto, *Bull. Chem. Soc. Jpn.*, **47**, 18–23 (1974).

556. H. Maag, J. F. Blount, D. L. Coffen, T. V. Steppe, and F. Wong, *J. Am. Chem. Soc.*, **100**, 6786–6788 (1978).

557. M. Iseki, T. Miyoshi, T. Konomi, and H. Imanaka, *J. Antibiot.*, **33**, 488–493 (1980).

558. M. Nishida, Y. Mine, T. Matsubsara, S. Goto, and S. Kuwahara, *J. Antibiot.*, **25**, 582–593 (1972).

559. N. Tanaka, M. Iseki, T. Miyoshi, H. Aoki, and H. Imanaka, *J. Antibiot.*, **29**, 155–168 (1976).

560. A. Zweifka, H. Kohn, and W. R. Widger, *Biochem.*, **32**, 3564–3570 (1993).

561. C. D. Ericsson, H. L. DuPont, P. Sullivan, E. Galindo, D. G. Evans, and D. J. Evans, *Ann. Intern. Med.*, **98**, 20–25 (1983).

562. R. M. Williams, *Chem. Rev.*, **88**, 511–540 (1988).

563. R. M. Williams, *Studies Nat. Prod. Chem.*, **12**, 63–112 (1993).

564. Z. Zhang and H. Kohn, *J. Am. Chem. Soc.*, **116**, 9816–9826 (1994).

565. B. W. Muller, O. Zak, W. Kump, W. Tosch, and O. Wacker, *J. Antibiot.*, **32**, 689–705 (1979).

566. N. Ise, H. Shibatani, M. Oshita, N. Osaki, M. Ueki, and M. Fujisaki, *J. Liq. Chrom.*, **16**, 2399–2414 (1993).

567. F. Kavanagh, A. Hervey, and W. J. Robbins, *Proc. Nat. Acad. Sci. USA*, **37**, 570–574 (1951).

568. D. Arigoni, *Pure Appl. Chem.*, **17**, 331–348 (1968).

569. A. J. Birch, C. W. Holzapfel and R. W. Rickards, *Tetrahedron*, (Suppl. 8), 359–387 (1966).

570. H. Egger and H. Reinshagen, *J. Antibiot.*, **29**, 915–927 (1976).

571. G. Hogenauer in F. E. Hahn, Ed., *Antibiotics*, Vol. 5, Part 1, Springer Verlag, Heidelberg, 1979, pp. 344–360.

572. C. O. Baughn, W. C. Alpaugh, W. H. Linkenheimer, and D. C. Maplesden, *Avian Dis.*, **22**, 620–626 (1978).

573. R. F. W. Goodwin, *Vet. Record*, **104**, 194–195 (1979).

574. M. D. Anderson, *Vet. Med. Small Animal Clin.*, **78**, 98–101 (1983).

575. R. F. Witkamp, S. M. Nijmeijer, G. Csiko, and A. S. J. P. A. M. Van Miert, *J. Vet. Pharmacol. Ther.*, **17**, 317–322 (1994).

576. K. H. Michel, D. E. Dorman, and J. L. Occolowitz, *19th Intersci. Conf. Antimicrob. Agents Chemother.*, Boston, 1979, abstr. 1036.

577. R. J. Suhadolnik and N. L. Reichenbach in *Kirk-Othmer Encyclopedia of Chemical Technology*, 4th ed., John Wiley & Sons, Inc., New York, Vol. 3, 1992, pp. 214–259.

578. K. Isono, *J. Antibiot.*, **41**, 1711–1739 (1988).

579. K. Isono, *Pharmacol. Therapeut.*, **52**, 269–286 (1991).

580. J. Goodchild, *Topics Antibiot. Chem.*, **6**, 99–227 (1982).

581. J. G. Buchanan and R. H. Wightman, *Topics Antibiot. Chem.*, **6**, 229–339 (1982).

582. R. J. Suhadolnik, *Nucleosides as Biological Probes*, Wiley Interscience, New York, 1979.

583. R. J. Suhadolnik, *Nucleoside Antibiotics*, Wiley Interscience, New York, 1970.

584. A. Takatsuki, K. Arima, and G. Tamura, *J. Antibiot.*, **24**, 215–223 (1971).

585. G. Tamura, Ed., *Tunicamycin*, Japan Scientific Society Press, Tokyo, 1982.

586. K. Eckardt, W. Ihn, D. Tresselt, and D. Krebs, *J. Antibiot.*, **34**, 1631–1632 (1981).

587. J. A. Edgar, J. L. Frahn, P. A. Cockrum, N. Anderton, M. V. Jago, C. C. J. Culvenor, A. J. Jones, K. Murray, and K. J. Shaw, *J. Chem. Soc., Chem. Commun.*, 222–224 (1982).

588. K. Isono, M. Uramoto, H. Kusakabe, K. Kimura, K. Izaki, C. C. Nelson, and J. A. McCloskey, *J. Antibiot.*, **38**, 1617–1621 (1985).

589. Y. Hayakawa, M. Nakagawa, H. Kawai, K. Tanabe, H. Nakayama, A. Shimazu, H. Seto, and N. Otake, *Agric. Biol. Chem.*, **49**, 2685–2691 (1985).

590. J. N. Porter, R. I. Hewitt, C. W. Hesseltine, G. Krupka, J. A. Lowery, W. S. Wallace, N. Bohonos, and J. H. Williams, *Antibiot. Chemother.*, **2**, 409–410 (1952).

591. C. W. Waller, P. W. Fryth, B. L. Hutchings, and J. H. Williams, *J. Am. Chem. Soc.*, **75**, 2025 (1953).

592. R. J. Suhadolnik in Ref. 583, pp. 3–50.

593. R. Vince, S. Daluge, and J. Brownell, *J. Med. Chem.*, **29**, 2400–2403 (1986).

594. H. A. Kirst, D. E. Dorman, J. L. Occolowitz, N. D. Jones, J. W. Paschal, R. L. Hamill, and E. F. Szymanski, *J. Antibiot.*, **38**, 575–586 (1985).

595. J. K. Epp and N. E. Allen, *16th Intersci. Conf. Antimicrob. Agents Chemother.*, Chicago, 1976, abstr. 63.

596. S. J. Hecker, S. C. Lilley, M. L. Minich, and K. M. Werner, *Bioorg. Med. Chem. Lett.*, **3**, 295–298 (1993).

597. F. Isono, M. Inukai, S. Takahashi, T. Haneishi, T. Kinoshita, and H. Kuwano, *J. Antibiot.*, **42**, 667–673 (1989).

598. R. H. Chen, A. M. Buko, D. N. Whittern, and J. B. McAlpine, *J. Antibiot.*, **42**, 512–520 (1989).

599. F. Isono, K. Kodama, and M. Inukai, *Antimicrob. Agents Chemother.*, **36**, 1024–1027 (1992).

600. M. Inukai, F. Isono, and A. Takatsuki, *Antimicrob. Agents Chemother.*, **37**, 980–983 (1993).

601. J. Berger and A. D. Batcho in M. J. Weinstein and G. H. Wagman, Eds. *Antibiotics: Isolation, Separation and Purification*, Elsevier, Amsterdam, 1978, pp. 101–158.

602. H. Hoeksema and C. G. Smith, *Prog. Indust. Microbiol.*, **3**, 91–139 (1961).

603. M. O. Boles and D. J. Taylor, *Acta Crystallogr.*, **B31**, 1400–1406 (1975).

604. R. J. Reece and A. Maxwell, *CRC Crit. Rev. Biochem. Molec. Biol.*, **26**, 335–375 (1991).

605. A. H. Hamdy, N. L. Olds, and B. J. Roberts, *Am. J. Vet. Res.*, **36**, 259–262 (1975).

606. L. Ninet, F. Benazet, Y. Charpentie, M. Dubost, J. Florent, D. Mancy, J. Preud'homme, T. L. Threlfall, B. Vuillemin, D. E. Wright, A. Abraham, M. Cartier, N. De Chezelles, C. Godard, and J. Theilleux, *C. R. Acad. Sci. Paris*, **275** (series C), 455–458 (1972).

607. L. Dolak, *J. Antibiot.*, **26**, 121–125 (1973).

608. H. Kawaguchi, H. Tsukiura, M. Okanishi, T. Miyaki, T. Ohmori, K. Fujisawa, and H. Koshiyama, *J. Antibiot.*, **18A**, 1–10 (1965).

609. J. Berger, A. J. Schocher, A. D. Batcho, B. Pecherer, O. Keller, J. Maricq, A. E. Karr, B. P. Vaterlaus, A. Furlenmeier, and H. Speigelberg, *Antimicrob. Agents Chemotherapy-1965*, American Society for Microbiology, Washington, D.C., 1966, pp. 778–785.

610. J. C. Godfrey and K. E. Price in Ref. 10, pp. 653–718.

CHAPTER THIRTY-THREE

Sulfonamides and Sulfones

NITYA ANAND

Central Drug Research Institute
Chattar Manzil Palace
Lucknow, India

Burger's Medicinal Chemistry and Drug Discovery,
Fifth Edition, Volume 2: Therapeutic Agents,
Edited by Manfred E. Wolff.
ISBN 0-471-57557-7   © 1996 John Wiley & Sons, Inc.

CONTENTS

1 Introduction, 528
2 Development of Sulfonamides and
  Sulfones, 529
  2.1 Historical background, 529
  2.2 Nomenclature and classification, 531
  2.3 Earlier sulfonamides, 532
        2.3.1 N^1-Alkyl, -arylalkyl, -cycloalkyl, and
              carbo-aryl derivatives, 532
        2.3.2 N^1-Acyl derivatives, 532
        2.3.3 N^1-Heteroaromatic sulfanilamides, 533
        2.3.4 4-Substituted derivatives, 535
  2.4 Later sulfonamides, 535
  2.5 Naturally occurring sulfonamides, 537
  2.6 Sulfones, 537
  2.7 Antimicrobial spectrum, 539
  2.8 Side effects of sulfonamides as leads for new
      drugs in other areas, 541
        2.8.1 Carbonic anhydrase inhibitors, 541
        2.8.2 Saluretics, 541
        2.8.3 Insulin-releasing sulfonamides, 542
        2.8.4 Antithyroid agents, 542
        2.8.5 Tubular transport inhibitors, 542
        2.8.6 Endothelin antagonists, 543
        2.8.7 Other activities, 543
3 Action of Sulfonamides and Sulfones, 543
  3.1 Mode of action, 543
        3.1.1 Selectivity of action, 548
  3.2 Synergism with dihydrofolate reductase
      inhibitors, 548
  3.3 Drug resistance, 549
4 Structure and Biological Activity, 551

4.1 Structure–action relationships, 551
4.2 Physicochemical properties and
    chemotherapeutic activity, 552
    4.2.1 Sulfones, 559
    4.2.2 Water solubility, 560
    4.2.3 Lipid solubility, 560
4.3 Protein binding, 561
4.4 Pharmacokinetics and metabolism, 562
    4.4.1 Sulfonamides, 562
    4.4.2 Sulfones, 563
    4.4.3 Half-life, 564
5 Present Status in Therapeutics,   564
    5.1 In clinical practice, 564
    5.2 Adverse reactions, 565

1 INTRODUCTION

The development of sulfonamides is one of the most fascinating and informative chapters in medicinal chemistry, highlighting the roles of skillful planning and serendipity in drug research. Although the word chemotherapy was coined by Ehrlich to articulate the concept of the selective inhibition of microbes by chemicals in the early years of this century, the discovery of the antibacterial activity of prontosil, a sulfonamide, in the early 1930s was the beginning of the present era of chemotherapy. The subsequent recognition of a relationship between the chemical structure of these compounds and their pharmacological response highlighted the effect of substructures (pharmacophores) on biological activity, which brought into sharp focus the potential power of molecular modification in the process of drug discovery. The standardization of a suitable method for the estimation of sulfonamides in body fluids and tissues added a new dimension, that of pharmacokinetic studies, to drug research. Careful observation of side effects in pharmacological and clinical studies of the early sulfonamides revealed new and unanticipated activities; successful exploitation of these leads opened up new areas in chemotherapy such as oral antidiabetics, carbonic anhydrase inhibitors, and diuretics. This also highlighted the importance of side effects of drugs as a source of new leads

in drug design. The elucidation of the relationship between sulfonamides and p-aminobenzoic acid provided the long sought after mechanistic basis for a rational approach to chemotherapy. All these studies had a strong impact on developments in medicinal chemistry and influenced much of the later work in drug research in general and chemotherapy in particular. The rapidity with which new developments took place from 1935 to 1940, from the discovery of the antibacterial activity of prontosil to the enunciation of the theory of antimetabolites by Fildes indicates that the time was just ripe for major developments and needed only a catalyst, which was provided by this discovery.

Interest in sulfonamides has continued unabated. About 20 sulfonamides are now used in clinical practice. These vary widely in their absorption, distribution, and excretion patterns. Some remain largely unabsorbed after oral administration and are therefore considered useful for gastrointestinal tract infections. Sulfonamides of another group characterized by high solubility, quick absorption, and rapid excretion, mainly in the unaltered form, are widely used in urinary tract infections. Those belonging to yet another group are absorbed rapidly but excreted slowly, resulting in maintenance of adequate levels in the blood for long periods; these sulfonamides require less frequent administration and are particularly useful for chronic conditions and for prophylaxis. The wide differences in the absorption and excretion properties of different sulfonamides, coupled with their ease of administration, wide spectrum of antimicrobial activity, noninterference with the host-defense mechanisms, and relative freedom from problems of superinfection, are responsible for their extensive use in clinical practice even six decades after their discovery. The use of sulfonamides and sulfones presently extends from the treatment of acute and chronic gram-positive and gram-negative bacterial infections, including leprosy, to trachoma, lymphogranuloma venereum, malaria, nocardiosis, coccidiosis, and toxoplasmosis. Sulfonamides, and especially sulfones, also find use in the treatment of dermatologic inflammatory conditions; dapsone is the main stay of the treatment for dermatitis herpetiformis.

2 DEVELOPMENT OF SULFONAMIDES AND SULFONES

2.1 Historical Background

The story of sulfonamides goes back to the early years of this century when Hörlein, Dressel, and Kethe of I.G. Farbenindustrie (1) found that introduction of a sulfamyl group imparted fastness to acid wool dyes, thus indicating affinity for protein molecules. However, none of these sulfonamides was investigated for antibacterial activity. The interest in dyes as possible antimicrobials was stimulated by Ehrlich's studies on the relationship between selective staining by dyes and their antiprotozoal activity, which led to the testing of azo dyes for antibacterial activity, and some of them indeed showed such activity. In an attempt to improve the bactericidal properties of quinine derivatives, Heidelberger and Jacobs (2) prepared dyes based on dihydrocupreine, which included *p*-amino-

(1)

benzenesulfonamido–hydrocupreine **(1)**. Although the latter was reported to have bactericidal activity, it did not arouse much interest because the activity, having been tested *in vitro*, was of a low order and no further work was published on these compounds. Mietzch and Klarer at I.G. Farbenindustrie synthesized a variety of azo dyes, a continuation of Ehrlich's interest in dyes as antibacterials. With the hope of imparting to azo dyes the property of specific binding to bacterial proteins, comparable to the binding to wool proteins, Mietzch and Klarer (3) synthesized a group of such dyes containing a sulfonamide radical, which included prontosil **(2)**. Domagk, also at I.G. Farbenindustrie, carried out the testing of these dyes. The lack of correlation between *in vitro* and *in vivo* tests prompted Domagk to take up bactericidal screening *in vivo* in mice in place of *in vitro* testing, a very fortunate decision, since otherwise the fate of sulfonamides might have been different. Domagk (4) observed in 1932 that prontosil protected mice against streptococcal infections and rabbits against staphylococcal infections, though it was without action *in vitro* on bacteria. It is reported that the first patient to be success-

(2)

fully treated with prontosil was Hildegarde Domagk, the daughter of its discoverer, who had septicemia due to needle prick. Foerster (5) published the first clinical success with prontosil in a case of staphylococcal septicemia in 1933.

These studies aroused worldwide interest and further developments took place at a very fast rate. One of the earliest systematic investigation on sulfonamides was by Trefouel, Nitti, and Bovet (6) working at the Pasteur Institute in Paris. Under a program of structural modification of this class of compounds, they prepared a series of azo dyes by coupling diazotized sulfanilamide with phenols, with or without amino or alkyl group. They observed that variation in the structure of the phenolic moiety had very little effect on antibacterial activity whereas even small changes in the sulfanilamide component abolished the activity. These observations pointed to the sulfonamides group as the active structural unit and led to the conclusion that metabolic cleavage of the azo linkage generates p-aminobenzenesulfonamide, sulfanilamide, (3), which may be responsible for the antibacterial activity. They suggested that prontosil was converted to sulfanilamide in animals and showed that sulfanilamide was as effective as the parent dyestuff in protecting mice infected with streptococci. They also showed that sulfanilamide exerted a bacteriostatic effect *in vitro* on susceptible organisms. Soon after, Colebrook and Kenny (7) observed that although prontosil was inactive *in vitro*, the blood of patients treated with it had bacteriostatic activity. They also reported the dramatic cure of 64 cases of puerperal sepsis by prontosil, while Buttle, Gray, and Stephenson (8) showed that sulfanilamide could

cure streptococcal and meningococcal infections in mice. Fuller's (9) demonstration of the presence of sulfanilamide in the blood and its isolation from urine of patients (and mice) under treatment with prontosil firmly established that prontosil is reduced in the body to form sulfanilamide, a compound synthesized as early as 1909 by Gelmo (10). Fuller concluded that the therapeutic action of prontosil was very likely due to its reduction *in vivo* to sulfanilamide. Among the early patients to be treated with sulfanilamide was Franklin D. Roosevelt Jr, the son of the President of the United States. His recovery from a streptococcal throat infection helped to overcome early reservations on the medicinal value of antibacterial chemotherapy with sulfonamides. Ehrlich's concept of a relationship between the affinity of dyes for a parasite and their antimicrobial activity, which focused attention on sulfonamide azo dyes, was found to be irrelevant to the activity of the latter. Nevertheless, sulfanilamide proved to be the "magic bullet" of Ehrlich, and heralded the era of "chemotherapy," a term coined by Ehrlich to emphasize the concept of selective action of chemicals on microbes as opposed to the action on host cells. The era of modern chemotherapy had now begun. Domagk was awarded the Nobel Prize for Medicine in 1939 primarily for the discovery of the antibacterial activity of sulfonamides.

These discoveries had a tremendous impact not only on the development of sulfonamides as antimicrobials, but also on developments in chemotherapy in general. Sulfanilamide, being easy to prepare, cheap, and not covered by patents, became available for widespread use and brought a new hope for the treatment of microbial infections. Recognizing the potential of sulfonamides, almost all major research organizations the world over initiated research programs for the synthesis and study of analogs and derivatives of sulfanilamide particularly with a view to improve its antimicrobial spectrum, therapeutic ratio,

$$H_2\overset{4}{N}-\!\!\left\langle\!\!\bigcirc\!\!\right\rangle\!\!-SO_2\overset{1}{N}H_2$$

(3)

and pharmacokinetic properties. New sulfonamides were introduced in quick succession until about 1945 when interest gradually shifted to antibiotics after the introduction of penicillin.

A decade of problems encountered with antibiotics, such as emergence of resistant strains, superinfection, and allergic reactions, brought about a revival of interest in sulfonamides. The knowledge obtained during this period about the selectivity of action of sulfonamides on the parasite, the relationship between their solubility and toxicity, and their pharmacokinetics, gave a new direction to further developments in this field. New sulfonamides with modified properties began to appear. Thus there are two distinct phases in the development of sulfonamides, the pre-1945 and the post-1957.

Some of the other developments in the field of sulfonamides that had far reaching effects on future progress of chemotherapy and drug research in general may be mentioned. The standardization by Bratton and Marshall (11,12) of a simple method for the assay of sulfonamides in body fluids and tissues permitted precise determination of the absorption, distribution, and excretion of these drugs, thus providing a rational basis for calculating proper dosage requirements. Pharmacokinetic studies thus became an integral part of drug development programs. Wood's (13) observation of the competitive reversal of the action of sulfanilamide by p-aminobenzoic acid (PABA) was the first definitive demonstration of metabolite antagonism as a mechanism of drug action. This led Fildes (14) to propose his classical theory of antimetabolites, which has stimulated numerous and intensive studies in chemotherapy and pharmacology and forms the basis of rational drug design. The observation of certain clinical side effects in the early sulfonamides and the success in dissociating these effects from antimicrobial activity by structural modification (*loc.cit*), resulted in a proper appreciation of molecular modification and of the role of side effects of drugs as an important source of leads for drug design (15).

2.2 Nomenclature and Classification

The general term "sulfonamides" has been used for derivatives of p-aminobenzenesulfonamide (sulfanilamide), whereas specific compounds are described as N^1- or N^4-substituted sulfanilamides (3), depending on whether the substitution is on the amido or aromatic amino group, respectively. Most of the sulfonamides used currently are N^1-derivatives. The generic name of a sulfonamide is build up by adding the prefix "sulfa-" to an abbreviated form of the chemical name of the N^1-residue. This is done in two ways: either the amido nitrogen is taken as part of the "sulfa" residue, as in the case of N^1-heterocylic sulfonamides, e.g., sulfapyridine, or the amido nitrogen is taken as apart of the N^1-residue as in sulfaguanidine.

Similarly, the term "sulfone" is used for all derivatives of 4,4′-diaminodiphenylsulfone, dapsone, DDS, whereas specific compounds are described as substituted diaminodiphenylsulfones.

Sulfonamides have been classified in many different ways and the one based on rate of absorption and half-life (the time needed for the concentration of the drug in the blood to be reduced to one-half) appears to be most logical and clinically relevant. Sulfonamides that have a half-life of less than 10 h are termed short-acting, between 10 and 24 h are considered to be medium-acting, and longer than 24 h are long-acting. "Long-acting" denotes slow excretion and/or reabsorption of the drug into the system from the excretory routes and is to be differentiated from the depot form. Some are very highly ionized and poorly absorbed and their effect is limited to the gastrointestinal tract after oral administration; some are only topically applied.

2.3 Earlier Sulfonamides

In an effort to improve on the efficacy of sulfanilamide, more than 5000 derivatives and analogs were investigated right in the first decade of research in this field. This work was described in 1948 in a very exhaustive and well-presented monograph by Northey (16), which may be consulted for details of the work of this period. The analogs studied included position isomers, compounds having substituents on the functional groups, isosteres and ring annellates of the benzene ring. It was realized quite early on that the p-aminobenzenesulfonyl unit is inviolate for maintaining the activity; therefore most of the emphasis was laid on preparing N^1-substituted derivatives. The activity shown by some N^4-derivatives reported was found to be due either to regeneration *in vivo* of the free aromatic amino group or to mechanism other than antagonism of PABA.

By variation of the N^1-substituent a variety of N^1-substituted sulfonamides having antimicrobial activity have been obtained, all acting by PABA antagonism and include most of the sulfonamides used in clinical practice today. These developments are discussed below briefly according to the type of N^1-substitution introduced.

2.3.1 N^1-ALKYL, -ARYLAKYL, -CYCLOKYL, AND -CARBOARYL DERIVATIVES. Introduction of these substituents in general leads to lowering or loss of activity, and none of these derivatives is used clinically.

2.3.2 N^1-ACYL DERIVATIVES. N^1-Acyl substitution leads to compounds with enhanced biological activity and useful physicochemical properties required for special pharmaceutical formulations. Several clinically useful compounds belong to this group. These derivatives are strong acids and form highly soluble neutral sodium salts, giving nonirritant solutions. Such salts are useful for local treatment where high

NH$_2$

SO$_2$NHCOCH$_3$

(4)

alkalinity of the sodium salts of other sulfa drugs would be unacceptable. The most important member of this group, widely used in ophthalmic practice, is sulfacetamide (4) (17,18), which is more active than sulfanilamide and is as active as sulfadiazaine against some organisms *in vivo*. It is quickly absorbed and rapidly excreted without danger of crystallization in the kidney. N^1-Aryl-sulfonamides, though active, do not show any advantage over sulfacetamide or other commonly used sulfonamides (19).

In contrast to sulfacetamide, N^1-higher acyl derivatives are deacylated very substantially in the body and are not clinically useful. N^1-Acetyl derivatives of N^1-heterocyclic sulfanilamides are inactive *in vitro* but have the advantage of being tasteless as compared to the bitter taste of the parent compounds; these are quantitatively deacetylated in the intestine and find use in liquid medications (20,21).

N^4-Carboxyacyl derivatives of sulfacetamide such as N^4-phthalyl- and N^4-succinyl-sulfacetamides (5) and (6) (22) and of certain N^1-hererocyclic sulfonamides, such as N^4-phthalyl and N^4-succinyl-sulfathiazoles (7) and (8) (23), form highly soluble neutral sodium salts and were developed for parenteral administration. Because of rapid excretion, however, high blood levels are not built up. These compounds are inactive and remain practically unabsorbed after oral administration, but they undergo gradual intestinal deacetylation to release the active compound. Since they are poorly absorbed and rapidly excreted, large doses can be administered orally without danger of toxic effects. It has

NHR / SO$_2$NHCOCH$_3$

(5)

NHR / SO$_2$NH—thiazole

(7)

$$R = \begin{array}{c} CH_2CO_2H \\ | \\ CH_2CO^- \end{array}$$

(6)

$$R = \text{(benzene ring with } CO_2H \text{ and } CO^-)$$

(8)

NH$_2$ / SO$_2$NHCONH$_2$

(9)

NH$_2$ / SO$_2$NH$\overset{\displaystyle \|}{\underset{NH}{C}}NH_2$

(10)

been proposed that because of their high solubility under alkaline conditions, they mix well with the intestinal contents and undergo slow hydrolysis. Because much of the parent sulfonamide would be released in the lower intestine where absorption is poor, the drug would not appear in the blood in toxic concentration, giving a relatively high local concentration of the parent sulfonamide which is the effective agent. It has been suggested that local action in the gut may be the main reason for their effectiveness. Since these derivatives have been considered to produce mainly localized changes in the bacterial flora, they have been extensively used for presurgical sterilization of the gut and for intestinal infections; phthalylsulfathiazole is the most commonly used agent among those. The concept of the advantage of local action in the gut for intestinal infections is, however, questioned by some.

Among the N^1-carbamyl derivatives, sulfacarbamide (**9**) (24,25) and sulfaguanidine (**10**) (26) have been clinically used. Sulfacarbamide, though not highly active *in vivo* on account of its quick absorption and rapid excretion, finds limited use in pyelitis and urinary tract infections. Sulfaguanidine is more active than sulfanilamide *in vitro* and was earlier extensively used for intestinal infections. It is

now realized that it is not as poorly absorbed from the gastrointestinal tract as was considered earlier and can cause toxicity when used at a high dose, and has been withdrawn in many countries.

2.3.3 N^1-HETEROAROMATIC SULFANILAM-IDES. Most of the therapeutically used sulfonamides belong to this class. By varying the N^1-heterocyclic residue, it has been possible to greatly enlarge the usefulness of sulfonamides by widening their antimicrobial spectrum, increasing the inhibition index, and modifying their pharmacokinetic properties.

Sulfapyridine [M&B 693, (**11**)] (27), reported in 1938, was one of the earliest of the new sulfonamides to be used in clinical practice for the treatment of pneumonia and remained the drug of choice until it was replaced by sulfathiazole. Sulfapyridine was used on Winston Churchill to cure pneumonia in 1943 during a trip to Africa.

NH$_2$ / SO$_2$NH—pyridine

(11)

Apart from sulfapyridine, all the other clinically used N^1-heterocyclic sulfonamides possessing six-membered heterocyclic ring have more than one hetero atom. Diazines (pyrimidines, pyridazines and pyrazines) have shown the best activity; of these, the derivatives of pyrimidine have received the

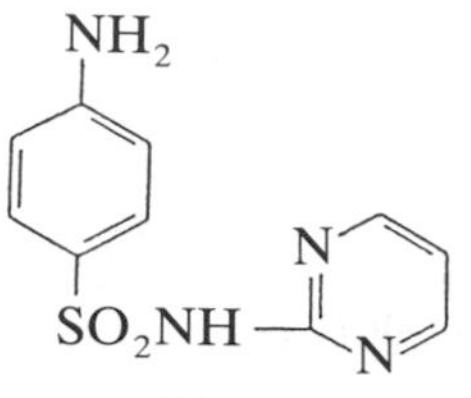

(12)

maximum attention and have provided a large number of clinically useful drugs (28). A sulfonamide linkage at either the 2- or the 4-position gives active compounds. Sulfadizaine (**12**) (29) was the first of these derivatives of pyrimidine to be introduced in clinical practice and it has retained preeminence among sulfonamides ever since, because of its broad antibacterial spectrum, high *in vivo* activity, low toxicity, high tissue penetration, and (subsequently discovered) long duration of action. In much of the later work sulfadiazine has been used as a reference standard for comparison of activity. Two methylated derivatives of sulfadiazine, sulfamerazine (**13**) (29–33), and sulfamethazine (**14**) (29–33) were soon introduced in therapeutics. Sulfamerazine has greater solubility in water at pH 7.0, and it is more quickly absorbed and more slowly excreted than sulfadiazine. Sulfamethazine is even more soluble than sulfamerazine at pH 5.5; though less active than sulfadiazine and sulfamerazine; its

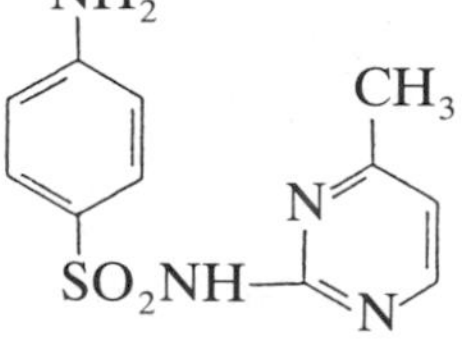

(13)

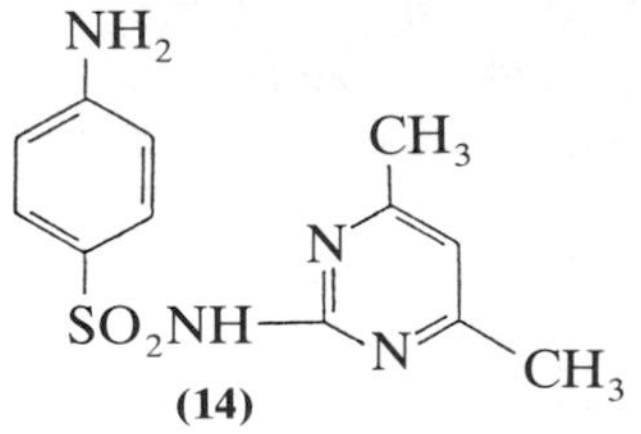

(14)

higher solubility gives it greater tolerance and clinical advantage over the other two. Since these quantitative differences in the properties of three pyrimidine derivatives seem to complement one another, these have been used in combination as a "triple sulfa." This made it possible to: (a) lower the dose of each individual component, thus reducing the toxicity of each and reduce significantly the tendency to crystalluria (a major toxicity hazard of earlier sulfonamides), thereby improving the tolerance and activity; (b) maintain higher concentrations in plasma and tissue over a long period. The use of combination sulfas was a significant development in this field (34,35).

Among the 4-sulfapyrimidine derivatives, sulfisomidine (**15**) was shown to possess useful activity (36). Its main advantage was high solubility and low tendency to crystalluria, which gives it better tolerance than other pyrimidine derivatives. A useful "disulfa" used was a combination of sulfisomidine and sulfadiazine.

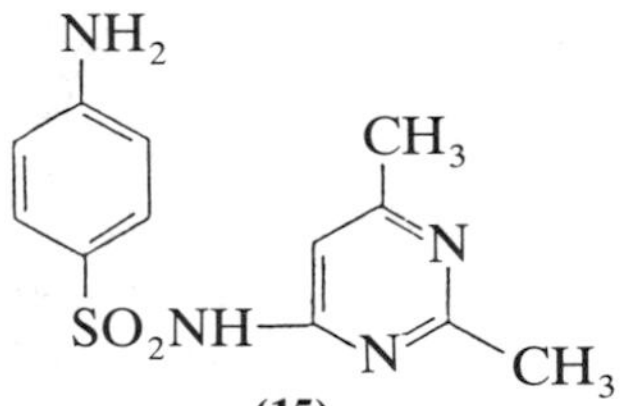

(15)

Clinically used sulfas containing five-membered heterocycles have two or more hetero atoms in the rings. The more active compounds belong to the thiazole, oxazole, isoxazole, 1,3,4-thiadiazole, and pryazole groups. Azoles having an NH in the ring have little activity, but *N*-aryl and *N*-alkyl substituted compounds are active, and some of them were of clinical interest for some time. The corresponding hetero- or benzo-annellates are less active. Most of the clinically used five-membered heterocyclic compounds (except sulfaphenazole) possess one or two CH$_3$ groups, which must increase their hydrophobicity, and impart

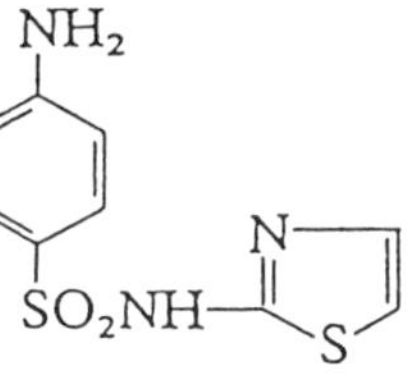

(16)

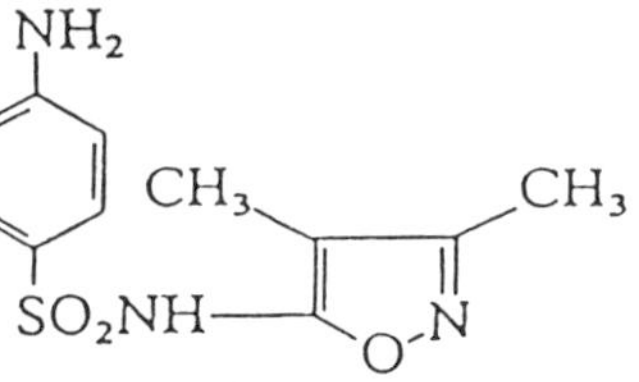

favorable properties for binding to proteins and perhaps to folate enzymes. The position of substituents in the hetero ring also has a significant effect on the activity pattern of these compounds. Sulfathiazole (16) (37), the first member of this group to be introduced in clinical practice, replaced sulfapyridine because of its wider antibacterial spectrum and higher therapeutic index. Substitution of the thiazole ring by alkyl groups did not improve the activity, while a 4-phenyl residue enhanced both the activity and the toxicity. An additional nitrogen atom in the ring at position 4, to give 1,3,4-sulfathiadiazole, did not enhance the activity; however a methyl substituent in the 5-position gave sulfamethizole (17), which exhibited an increased antibacterial activity (38,39), high solubility, quick absorption, and rapid excretion. Sulfamethizole has, in addition hypoglyccmic activity (*loc.cit*), which was increased when the methyl group was replaced by an ethyl or an isopropyl group. Another sulfonamide introduced during this period was sulfisoxazole (18); though less active than sulfadiazine, it is better tolerated because of its high solubility and rapid excretion. Its significant activity against *Proteus vulgaris* and *Escherichia coli* makes it useful in urinary tract infections (40,41).

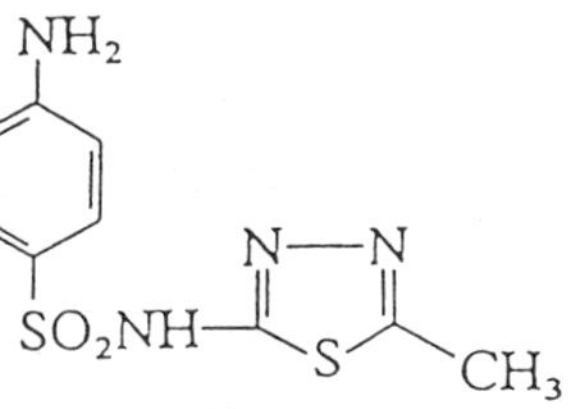

(17)

(18)

2.3.4 4-SUBSTITUTED DERIVATIVES. The only compound of any importance in this group is 4-aminomethylbenzenenesulfanilamide, homosulfanilamide, mafenide, (19) (42). Among other organisms, it is active against anaerobic bacteria, and therefore finds use for topical application in burn infected wounds and gas gangrene. It does not act through the PABA mechanism and can be used against organisms resistant to other sulfonamides (43).

(19)

2.4 Later Sulfonamides

A major advance in sulfonamide therapy came with the proper appreciation of the role of pharmacokinetic studies in determining the dosage schedule of these drugs. It was realized that some of the "earlier" sulfonamides such as sulfadizaine and sulfamerazine had a long half-life (17 and 24 h respectively) and required less frequent administration than was normally prescribed. In fact, a mixture of sulfamerazine and sulfaproxyline was one of the earliest long-acting sulfonamides to be used; it was administered three times a day. The era of newer long-acting sulfonamides started in 1956 with the introduction of sulfamethoxypyridazine (20) (44), having a half-life of around 37 h, the longest known at that time, which needed to be administered only once a day (45). Sulfachlorpyridazine (21) (46), a related sulfonamide

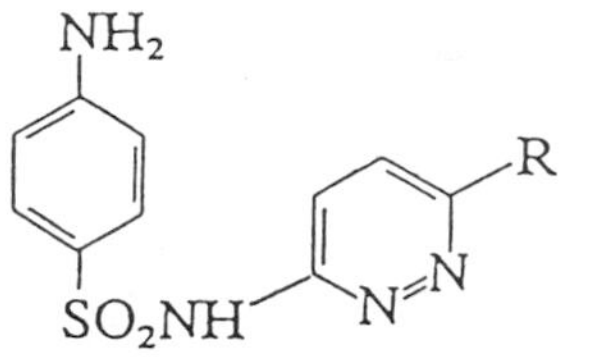

(20) R = OCH₃
(21) R = Cl

and an intermediate in the synthesis of
(**20**), was, however, found to have a half-
life of about 7 h and is useful for urinary
tract infections. In 1959 sulfadimethoxine
(**22**) was introduced with a half-life of
approximately 40 h (28,47,48). The related
4-sulfonamidopyrimidine, sulfomethoxine
(**23**) (49,50), having the two methoxyl radicals in 5,6-positions, has by far the longest
half-life, about 150 h, and needs administration once a week. Sulfamethyldiazine (**24**)
(51) and sulfamethoxydiazine (**25**) (51)
each having a half-life of 36 h, and sulfamethoxypyrazine (**26**) (52), with a half-
life of 65 h, were the other long-acting
sulfonamides introduced during this period.

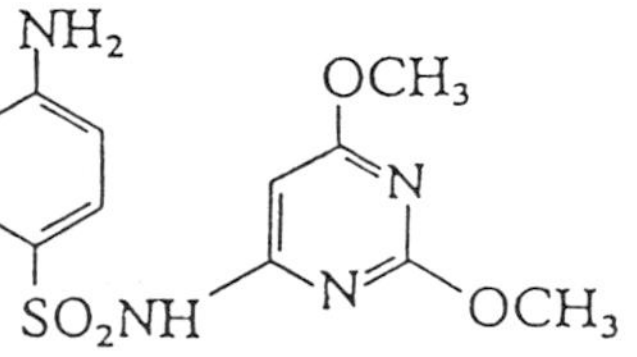

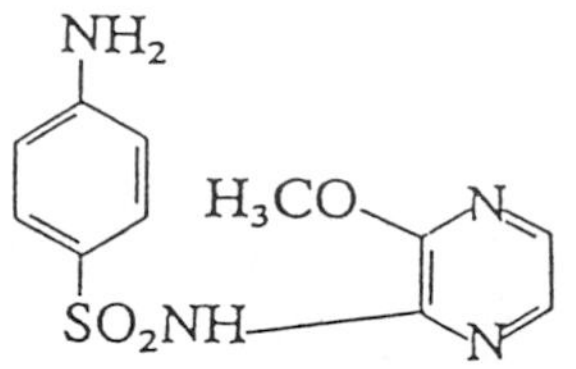

(26)

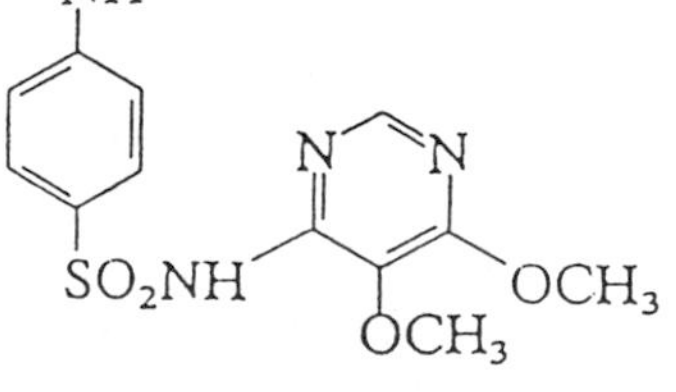

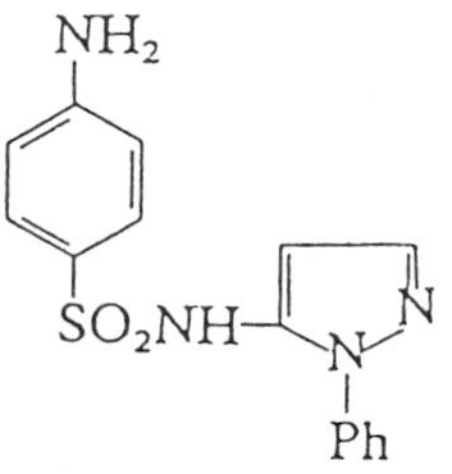

(27)

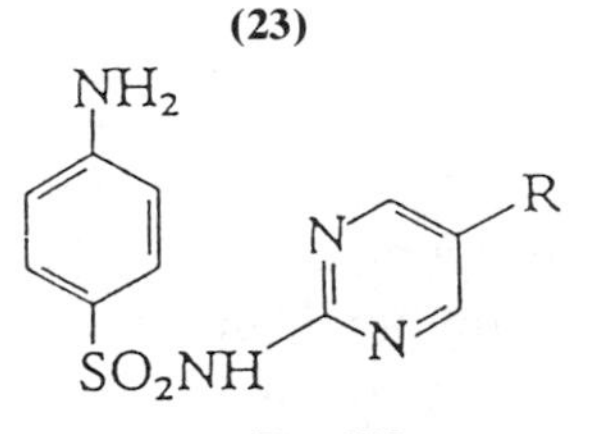

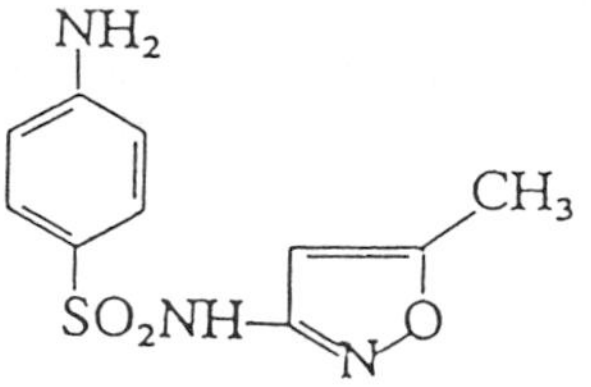

(28)

Sulfaphenazole (**27**) (53), sulfamethoxazole (**28**) (54), and sulfamoxol (**29**) (55),
the other sulfonamides put into clinical
practice during this period, are relatively
shorter acting, having a half-life of about
11 h.

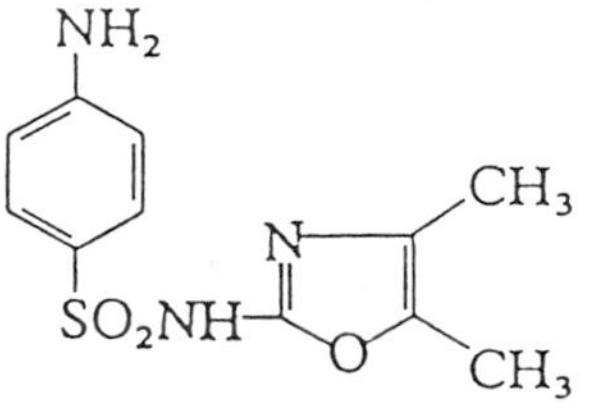

(22)

(23)

(24) R = CH₃
(25) R = OCH₃

(29)

A new broad spectrum sulfonamide,
sulfaclomide (**30**) (56), was reported in
1970. It is relatively nontoxic and gives
higher serum levels than other sulfonamides and patients do not need extra fluid
intake or alkalization. Sulfacytine, *N*-sul-
fanilyl-1-ethylcytosine, (**31**) (57) also reported in 1970 is 3–10 times more potent
in vivo than sulfisoxazole, sulfachlorpyridazine and sulfisomidine. It is highly
soluble, rapidly absorbed, and excreted
almost unchanged, and is well suited for
urinary tract infections. No major new
antimicrobial sulfonamide has been introduced into clinical practice since 1970.

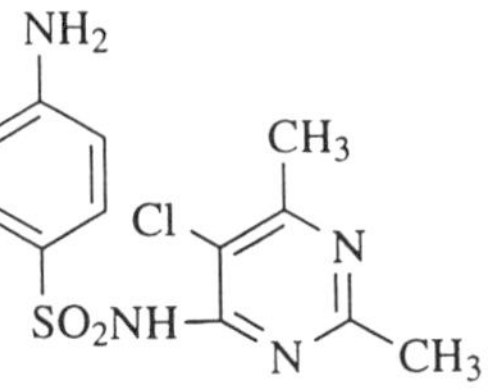

(30)

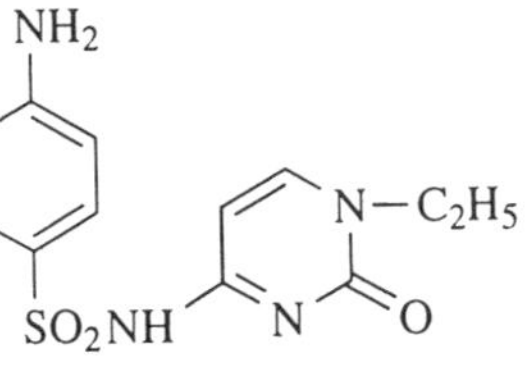

(31)

2.5 Natural Occurring Sulfonamides

An interesting development has been the isolation of some nucleoside antibiotics possessing a sulfonamide moiety. These include nucleocidin (**32**) (58) and ascamycin (**33**) (59) obtained from *Streptomyces* sp. isolated from soil samples. Nucleocidin was shown to exhibit a rather broad antibacterial spectrum and a high order of activity against trypanosomes. Its practical use however, is seriously limited by its toxicity; the LD$_{50}$ in mice is 0.2 mg/kg by intramuscular or intraperitoneal administration. Ascamycin showed a very selective action as compared to desalanylascamycin. This selectivity is very likely determined by the presence of a dealanylating enzyme present in the bacterial surface

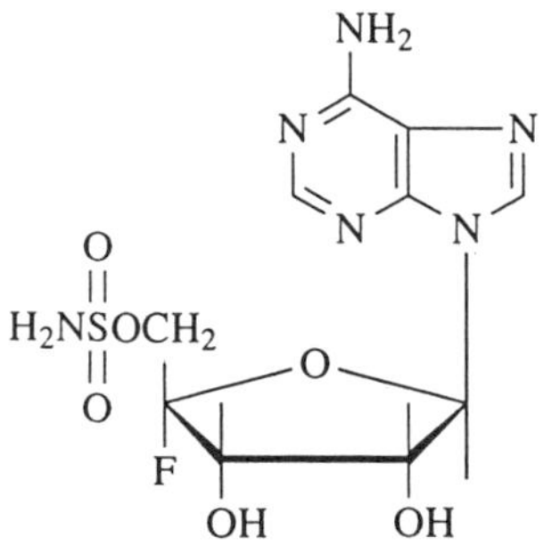

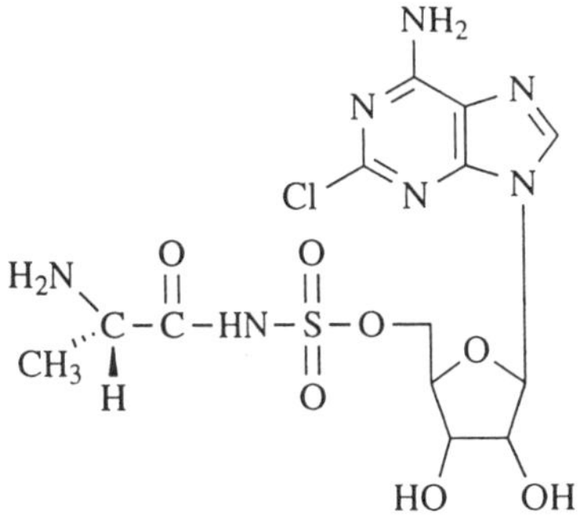

(33)

which facilitates the selective uptake of this nucleoside. Nucleocidin has been shown to be a highly potent inhibitor of protein biosynthesis. Though their precise mechanism of action has not been clearly defined, not possessing a *p*-aminophenyl structure, it is certain that these antibiotics would be acting by non-PABA mechanism. Their structures do provide interesting leads for follow up. Their structures have been confirmed by synthesis and some analogs reported.

2.6 Sulfones

The demonstration that experimental tuberculosis could be controlled by 4,4′-diaminodiphenylsulone, dapsone, (**34**) (60) and disodium 4.4′-diaminodiphenylsulfone -*N*,*N*′-didextrose sulfonate, promin (**35**) (61), was a major advance in the chemotherapy of mycobacterial infections. Although dapsone and promin proved disappointing in the therapy of human tuberculosis, the interest aroused in the possibility of treatment of mycobacterial infec-

(32)

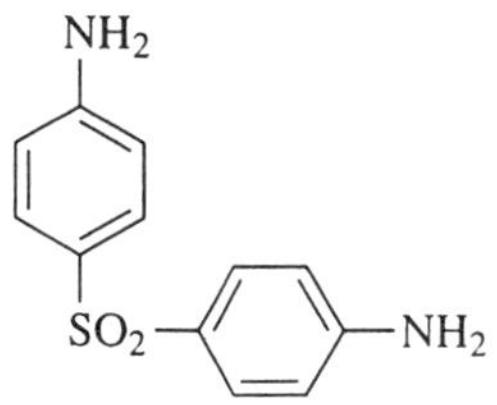

(34)

SO$_3$Na
|
NHCH(CHOH)$_4$CH$_2$OH

SO$_3$Na
|
SO$_2$—⟨ ⟩—NHCH(CHOH)$_4$CH$_2$OH

(35)

CH$_3$COHN—⟨ ⟩—SO$_2$—⟨ ⟩—N=CH

CH$_3$COHN—⟨ ⟩—SO$_2$—⟨ ⟩—N=CH

(37)

tions with sulfones led to the demonstration of the favorable effect of promin in rat leprosy (62). This was soon followed by the successful treatment of leprosy patients, first with promin and later with dapsone itself. Since then dapsone has remained the main stay for the treatment of human leprosy (63). It has now been shown that *Mycobacterium leprae* is unusually sensitive to dapsone (64) and that its growth can be inhibited by very low concentration of the latter. A successful clinical trial with a weekly dose of daspone as low as 1–5 mg has been reported (65). A number of long acting sulfonamides such as sulfamethoxine have also been found useful in the treatment of leprosy (66,67).

An important advance in the use of sulfones took place with the demonstration that *N,N'*-diacyl derivatives and certain Schiff bases of dapsone are prodrugs and have a repository effect and release dapsone slowly; *N,N'*-diacetamidodiphenylsulfone, acedapsone, **(36)** and the Schiff base DSBA **(37)** are particularly useful as repository forms (68,69). After a single intramuscular injection of 225 mg of acedapsone, a therapeutic level of DDS (20–

25 ng/mL) is maintained in the blood for as long as 60–68 days, and is useful in the prophylaxis and treatment of leprosy.

With a view to improve upon the antimycobacterial and antiprotozoal (especially antimalarial) activities of dapsone a variety of substituted sulfones have been prepared which include: (a) *N*-Alkyl- and *N,N'*-dialkyl-diaminodiphenylsulfones (70–74); (b) 4-aminodiphenylsulfones carrying one or more substituents, such as amino, alkylamino, alkyl, alkoxy, hydroxy and halogens, at the 2',3'- and/or 4'-positions (75–79); (c) compounds in which one or both benzene rings have been replaced by 2- or 4-pyridyl, 2- or 5-thiazolyl, or 8-quinolyl residue. In view of the lack of a suitable rapid screening model for antileprosy activity, most of the compounds in the earlier studies were tested against *M. tuberculosis*, either *in vitro* or in experimental infection (a correlation has been reported between MIC against Mycobacterial species 607 and *M. leprae* infection in mouse footpad, (73) and against *P. berghei* infection in rodents. In more recent studies testing has been carried out against isolated dihydropteroate synthase enzyme in cell free systems from *Escherichia coli* (79–81), *Mycobacterium lufu* (82) and *Plasmodium berghei* (82). Some mono lower *N*-alkyl derivatives of DDS such as *N*-ethyl, and propyl, though less active than DDS *in vitro* tests against *Mycobacterium* species 607 (73) were found to be somewhat more active than dapsone in experimental tuberculosis (83). These were also active against *P. berghei* infection in mice and

NHCOCH$_3$

SO$_2$—⟨ ⟩—NHCOCH$_3$

(36)

showed synthase inhibitory activity in cell-free systems from *P. berghei* and *M. lufu* (84). Some of the substituted 4-aminodiphenylsulfones carrying electron donating substituents (CH_3, OCH_3 and especially OH) were much more active than DDS in the isolated cell free enzyme dihydropteroate (81,84), but comparable in activity with DDS in the *in vitro* tests using *Mycobacterium* sp. 607 (73) or in experimental tuberculosis (78,83). With heterocyclic sulfones, compounds in which one benzene ring has been replaced by a heterocyclic residue possess significant activity, but replacement of both benzene rings by these heterocycles lead to inactive compound. As in the case of sulfonamides, the *p*-aminophenylsulfonyl unit seems to be essential for the activity of sulfones also. Overall none of the substituted diaminodiphenylsulfones was significantly more active *in vivo* than DDS and did not offer significant advantage. Even after almost 50 years of use in clinical practice, DSS alone, or in combination with other drugs (as in multidrug therapy) continues to be the main stay of chemotherapy of leprosy.

4,4′-Diaminodiphenylsulfoxide (**38**) showed significant antileprosy activity in clinical trial but was found to be more toxic than dapsone (85,86). The activity is very likely due to its oxidation *in vivo* to dapsone, as both the sulfoxide (**38**) and its *N*-methyl derivatives have been shown to undergo such oxidation (87,88). 4,4′-Diaminodiphenylsulfide also showed antituberculosis activity in experimental tuberculosis of mice, but was inactive *in vitro* (89). The activity appears to be due to the oxidation of sulfide to sulfone; sulfide is oxidized *in vivo* to the sulfoxide and sulfone (90).

(**38**)

2.7 Antimicrobial Spectrum

Following the initial dramatic results obtained with sulfonamides in the treatment of streptococcal infections, studies with these drugs were extended to other bacteria, viruses, protozoa, and fungi. It was found that many gram-positive and gram-negative bacteria, mycobacteria, some large viruses, protozoa and fungi are suspectable to the action of sulfonamides and sulfones (Table 33.1). In almost all cases their action is related to PABA antagonism.

The sulfonamides and sulfones have a relatively broad antibacterial spectrum. Individual sulfonamides do differ in their antibacterial spectrum, but these differences are more quantitative than qualitative. The bacteria most susceptible to sulfonamides include pneumococci, streptococci, meningococci, staphylococci, some coliform bacteria and shigellae, and lepra bacilli to sulfones (Table 33.1). One limitation of sulfonamides is their weak activity against bacteria responsible for typhoid fever, diphtheria, and subacute bacterial endocarditis, and have practically no activity against *P. aeruginosa*. Another limitation with sulfonamides is the rising incidence of resistant isolates in the community. Synergism of their action by dihydrofolate reductase inhibitors, and the introduction of combination therapy with them has to some extent helped to remedy this problem.

Little notice was taken of the antimalarial activity of sulfonamides and sulfones which had been reported as early as 1937 (91,92), until Archibald and Ross (93), investigating the cause of the lower prevalance of malaria in leprosy patients, showed that dapsone could clear the blood of trophozoites of both *Plasmodium falciparum* and *P. malariae*, though somewhat more slowly than chloroquine. It was soon found that dapsone potentiated the action of pyrimethamine. A combination of the two drugs markedly delayed the development of resistance (94–96), and certain

Table 33.1   Antimicrobial Spectrum of Sulfonamides and Sulfones

Gram-Positive Acid Fast	Gram-Negative	Others
	Highly Sensitive	
Bacillus anthracis (some strains)	*Calymmatobacterium granulomatis*	*Actinomyces bovis*
Corynebacterium diphtheriae (some stains)	*Hemophilus ducreyi*	*Chlamyia trachomatis*
Mycobacterium leprae (to sulfones)	*H. influenzae*	*Coccidia*
	Neisseria gonorrheae	*Lymphogranuloma venereum virus*
Staphylococcus aureus	*N. Meningitidis*	*Plasmodium falciparum*[a]
Streptococcus pneumoniae	*Pasteurella pestis*	*P. malariae*[a]
S. pyogenes (group A)	*Proteus mirabilis*	*Pneumocystis carinii*
	Shigella flexneri	*Nocardia* sp.
	S. sonnei	*Toxoplasma*
	Vibrio cholerae	*Trachoma* viruses
	Weakly Susceptible	
Clostridium welchii	*Aerobacter aerogenes*	*Plasmodium vivax*[a]
Mycobacterium tuberculosis	*Brucella abortus*	
Streptococcus viridans	*Escherichia coli*	
	Klebsiella pneumoniae	
	Proteus vulgaris	
	Pseudomonas aeruginosa	
	Salmonella	

[a]Only blood schizontocide.

lines of *P. berghei*, *P. cynomologi*, and *P. gallinaceum* made resistant to either of the drugs were still susceptible to the combination (97,98). A further advance in therapy took place with the demonstration of the repository effect of *N,N'*-diformyl and -*N,N'*-diacetyl derivatives of dapsone; in the host they are slowly hydrolyzed to release dapsone (99,100). In the rhesus monkey an intramuscular dose of 50 mg/kg of acedapsone prevented patent *P. cynomologi* infection for an average of 158 days and suppressed the parasitemia for several weeks longer (101,102). Human malaria parasites vary greatly in their sensitivity to sulfones; *P. falciparum* is very sensitive, and *P. vivax* is less so. A number of long acting sulonamides such as sulformethoxine (**23**) and sulfamethoxypyrazine (**26**) have been found to possess high an-

timalarial activity, and their effect is also potentiated by pyrimethamine (103). The action of sulfones and sulfonamides is mainly against the blood forms, with marginal activity against primary (pre-erythrocytic) tissue forms and no activity against sexual forms and latent tissue forms.

The observation of the activity of sulfonamides and sulfones against experimental toxoplasmosis (104,105) and the synergization of this action by pyrimethamine (106,107) has led to wide use of this combination in human toxoplasmosis (108–110).

Sulfonamides have been shown to be effective against *Eimeria* infection in chickens (111–113) and are now commonly used, alone or preferably in combination with pyrimethamine, for the treatment of coccidiosis (114).

McCallum and Findlay (115) showed that experimental *Lymphogranuloma venereum* virus infection in mice was cured by sulfonamides. Later, other Chlamydiae were also found to be inhibited by sulfonamides, which led to the successful clinical use of these drugs in the treatment of trachoma (116,117).

Sulfonamides were also found to have marked activity against *Nocardia asteroides* and are largely used for the treatment of systemic nocardiosis (118–122).

Sulfonamides have also high activity against *Pneumocystis carinii* (123) and combined with trimethoprim are largely used for the treatment of *P. carinii* pneumonia in AIDS patients.

2.8 Side Effects of Sulfonamides as Leads for New Drugs in Other Areas

The action of sulfonamides is not restricted to antimicrobial activity; they have other weak to strong effects (side effects), unrelated to their antimicrobial activity. Some of these side effects have provided useful leads for developing drugs in other areas (15,124). The more important of these are discussed in the following sections.

2.8.1 CARBONIC ANHYDRASE INHIBITORS. The clinical acidosis and alkaline urine observed following sulfanilamide administration (125), its carbonic anhydrase inhibiting activity (126), together with the demonstration of high concentration of carbonic anhydrase in the kidney (127), established the causal relationship between the enzymes inhibition and observed physiological changes and suggested that sulfonamides may have a potential as diuretics. Schwartz (128) showed that sulfanilamide did indeed have diuretic action. Further studies led to the synthesis of highly specific carbonic anhydrase inhibitors which included acetazolamide (**39**). These were

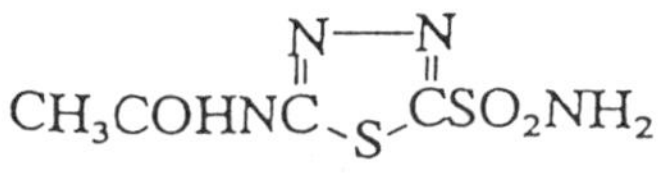

(39)

among the earliest nonmercurial diuretics. This was the beginning of the era of modern diuretics. Although carbonic anhydrase inhibitors as diuretics have been replaced by saluretics, they are used for treatment of glaucoma.

2.8.2 SALURETICS. While purusing the lead provided by sulfonamides having carbonic anhydrase inhibiting activity, it was observed that the introduction of an additional sulfamyl group in the *m*-position led to compounds that, although having diuretic action, had low carbonic anhydrase inhibiting activity and caused increased renal excretion of Na$^+$ and Cl$^-$ ions in almost equimolar proportions. Ring closure of one of these sulfonamide residues to a thiadizazine ring as in hydrochlorthiazide (**40**) (129) yielded more potent diuretics.

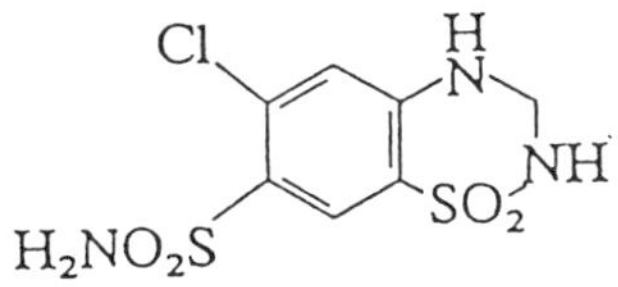

(40)

Replacement of one sulfamyl group by a carboxyl group led to the high ceiling saluretics such as furosemide (**41**) (130).

These two classes of diuretics seem to have different modes of action, since even after blocking the renal action of hydrochlorothiazide by a specific antagonist EX 4877 the effect of furosemide was retained (131).

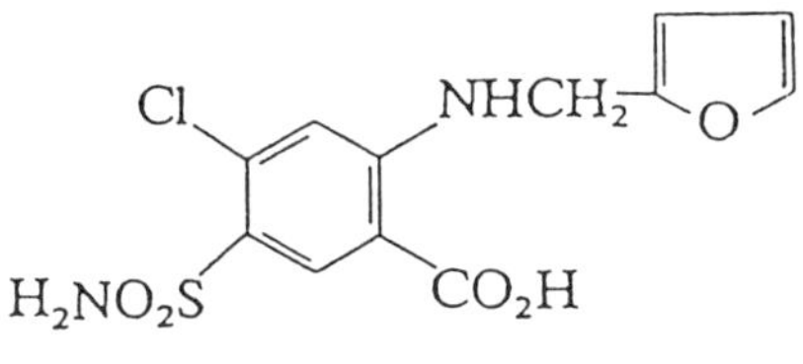

(41)

2.8.3 INSULIN-RELEASING SULFONAMIDES. That a sulfonamide (**42**) could induce hypoglycemia as a side effect was observed in 1942 by Janbon et al. (132); two patients under treatment for typhoid died of hypoglycemic shock. Loubatieres (133), as a result of extensive studies in experimental animals, showed that the compound exerted its hypoglycemic effect by stimulation of the pancreas to secrete insulin. This led to a search for oral hypoglycemic agents among sulfonamides, and it was found that a variety of aromatic sulfonylureas and thioureas could reduce blood sugar in animals and humans with potential insulin reserves. The search culminated in the introduction into clinical practice in 1955 of carbutamide (**43**), which possesses high oral hypoglycemic activity with weak antibacterial action (134). These findings opened a new chapter in the treatment of diabetes. Subsequent work has shown that

there is scope for considerable flexibility in the p-substituent; a striking change in activity is brought out by replacing the p-amino group by a p-acylaminoalkyl moiety as in glybenclamide (**44**). Sulfonylurea antidiabetics have been in clinical use for almost 50 years now (135).

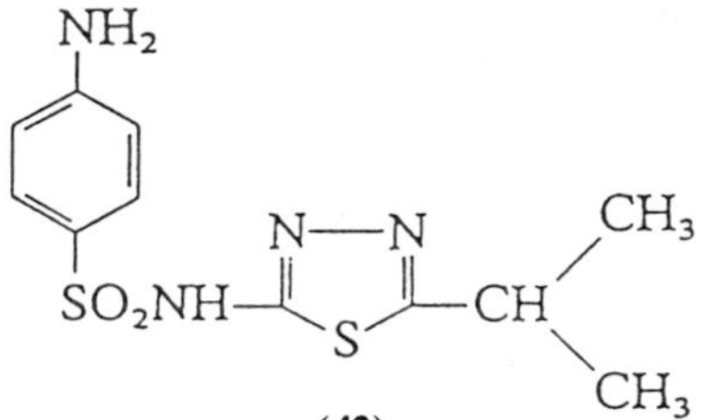

2.8.4 ANTITHYROID AGENTS. Mackenzie et al. (136) reported in 1942 the goiter-inducing action of sulfaguanidine in rats under treatment for experimental intestinal infections. These results, together with reports of incidence of goiter in factory workers engaged in sulfonamides production, and the chance observation of antithyroid activity of phenylthiourea, set off the trail that resulted in the development of the presently used antithyroid agents such as methimazole. It was, however, realized quite early that a sulfonamide group was not essential for antithyroid activity.

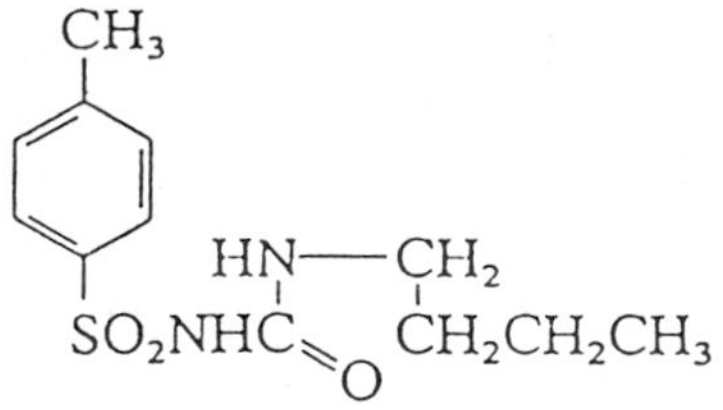

2.8.5 TUBULAR TRANSPORT INHIBITORS. The development of probenecid (**45**), the first synthetic drug for the control of uric acid excretion, was a result of the observation that some sulfonamides, particularly sulfamyl derivatives of PABA, had a penicillin-sparing effect by decreasing its renal clearance (137). Probenecid has found use both for the treatment of gout and also for prolonging the half-life of many drugs, specially penicillin in the body.

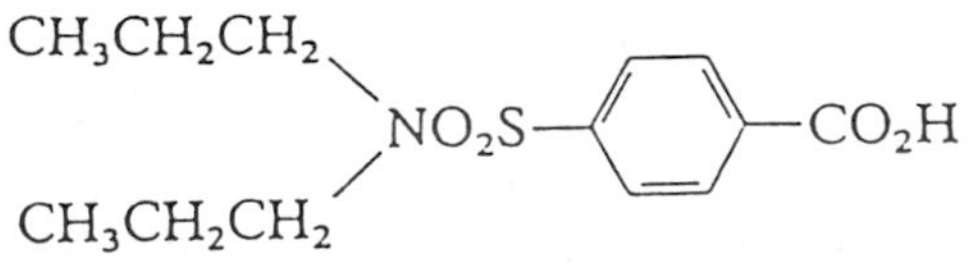

(42)

(43)

(45)

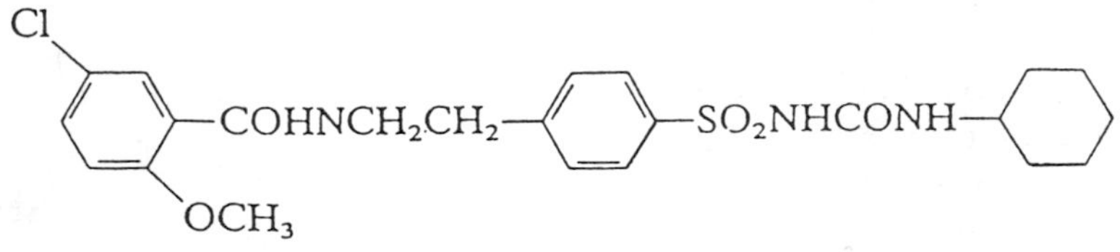

(44)

2.8.6 ENDOTHELIN ANTAGONISTS The endothelins (ETs) are a family of potent vasoconstrictor peptides originally isolated from endothelial cells (138). ETs exert their biological effects through interactions with specific receptors of which three subtypes ET-A, ET-B and ET-C have so far been characterized. ETs have been suggested to play a role in the pathophysiology of a large number of diseases especially hypertension. The ET-A subtype which is selective for ET-1 over ET-3 appears to be the predominant vascular smooth muscle receptor. Selective ET-A receptor antagonists are the subject of much current interest both to uncover their pathophysiological role and to develop new therapeutic agents. Structurally diverse ET antagonists of varying subtype selectivity have been discovered which include benzene sulfonamides. The first such sulfonamide ET receptor antagonists discovered was RO-462005 (**46**), which was ET-A/ET-B non-selective (139). Subsequently, based both on random screening and pharmacophore directed screening highly selective ETA receptor antagonist have been identified which include the naphthalene–sulfonamide (**47**) (140) and sulfisoxazole (**48**) (141). These sulfonamides represent a new class of low molecular weight, structurally simple, sub-

(48)

type selective, non-peptide endothelin antagonists, and should prove useful as lead compounds for the development of ET-A-selective endothelin antagonists.

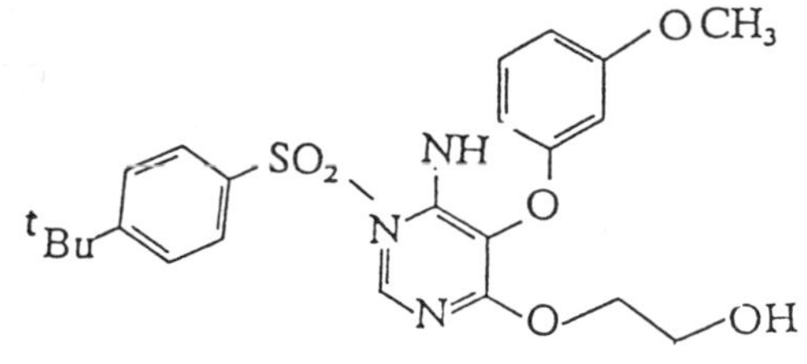

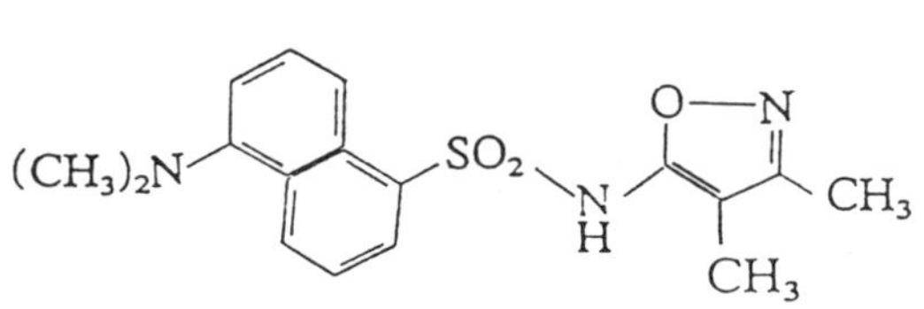

2.8.7 OTHER ACTIVITIES. A number of other activities that have been observed with sulfonamides include antiinflammatory activity especially against dermal conditions, antithrombotic, platelet aggregation inhibiting, fibrinogen receptor antagonist, angiotensin II antagonist, human renin inhibiting and K^+ channel opening activities. These activities provide useful leads in different areas of drug research.

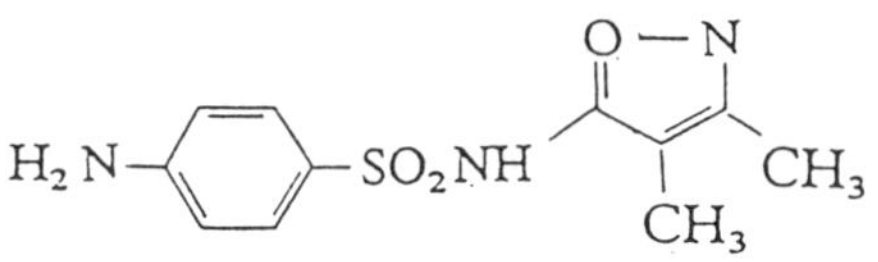

3 ACTION OF SULFONAMIDES AND SULFONES

Sulfonamides do have other actions (*vide supra*) apart from their antimicrobial activity. This present discussion, however, is restricted to their antimicrobial action.

3.1 Mode of Action

Sulfonamides are one of the few groups of drugs whose mechanism of antimicrobial action is known at the molecular level. Their action is characterized by a competitive antagonism with *p*-aminobenzoic acid (PABA), an essential growth factors vital to the metabolism of the microorganisms. Evidence for this antagonism started coming soon after the discovery of sulfonamides. It was found that substances antagonizing their action are present in peptones (142), various body tissues and fluids, especially after autolysis or acid hydrolysis (143), pus

(46)

(47)

(144), bacteria (145,146), and yeast extract (13,147). Woods (13) obtained evidence that PABA is the probable antagonistic agent in yeast extract and showed that synthetic PABA could completely reverse the bacteriostatic activity of sulfanilamide against various bacteria *in vitro*. Selbie (148) and Findlay (149) soon after found that PABA could antagonize the action of sulfonamides *in vivo* as well. Rubbo and Gillespie (150) isolated PABA as its benzoyl derivative, and Kuhn and Schwartz (151) obtained it as the methyl ester from yeast extract. Blanchard (152), McIllwain (153), and Rubbo et al. (154) finally isolated PABA from these sources. This led Woods (13) to suggest that because of its similarity of structure with PABA, sulfanilamide interfered with the utilization of PABA by the enzyme system necessary for the growth of bacteria. Based on these observations, a more general and clear enunciation of the theory of metabolite antagonism to explain the action of chemotherapeutic agents was given by Fildes in 1940 (14) in his now famous paper entitled "A rational approach in chemotherapy".

Further studies showed that the inhibition of growth by sulfonamides in simple media can be reversed not only competitively by PABA, but also noncompetitively by a number of compounds not structurally related to PABA, such as L-methionine, L-serine, glycine, adenine, guanine, and thymine (155,156). The relationship of sulfonamides to purines was uncovered by the finding that sulfonamides-inhibited cultures accumulated 4-amino-5-imidazolecarboxamide ribotide (157), a compound later shown by Shive et al. (158) and Gots (159) to be a precursor of purine biosynthesis.

With the concurrent knowledge gained in the field of bacterial physiology and metabolism, these isolated facts could be gradually fitted into a pattern. The determination of the structure of folic acid by Angier et al. (160) and Mowat et al. (161) revealed that PABA was an integral part of

the structure. Following this, Tschesche (162) made the suggestion that folic acid is formed by the condensation of PABA or *p*-aminobenzoylglutamic acid (PABG) with a pteridine and that sulfonamides compete in this condensation. Soon the structure of the active coenzyme form of folic acid, leucoverin (folinic acid, citrovorum factor), was established and its involvement in biosynthetic steps where 1-carbon units are added was elucidated (163,164). The amino acids, purines, and pyrimidines that are able to replace or spare PABA are precisely those whose formation requires 1-carbon addition catalyzed by folic acid.

Direct evidence of inhibition of folic acid synthesis by sulfonamides was soon obtained by studies on bacterial cultures. It was already known that a number of organisms could use PABA and folic acid as alternative essential growth factors (165). Lampen and Jones (166,167) found that the growth of some strains of *Lactobacillus arabinosus*, and *L. plantarum* in media containing PABA was inhibited competitively by sulfonamides, whereas folic acid caused a noncompetitive type of reversal of this inhibition, suggestive of its being the product of the inhibited reaction. Inhibition of folic acid synthesis by sulfonamides was demonstrated by Miller and Miller et al. (168,169) in growing cultures of *E. coli* and by Lascelles and Woods (170) in cultures of *Staphylococcus aureus*, a PABA-requiring mutant of *E. coli* and its parent wild strain. Nimmo-Smith et al. (171) had observed a similar inhibition of folic acid synthesis by sulfonamides and its competitive reversal by PABA in nongrowing suspensions of *L. plantarum*.

The demonstration of the enzymic synthesis of dihydropteroate and dihydrofolate (Fig. 33.1) in cell-free extracts by Shiota and Disraely (172), and Shiota et al. (173), using enzymes mainly from *L. plantarum* and *Veillonella* and with an enzyme system from *E. coli* by Brown et al. (174) and Brown and Weisman (175), set the stage for

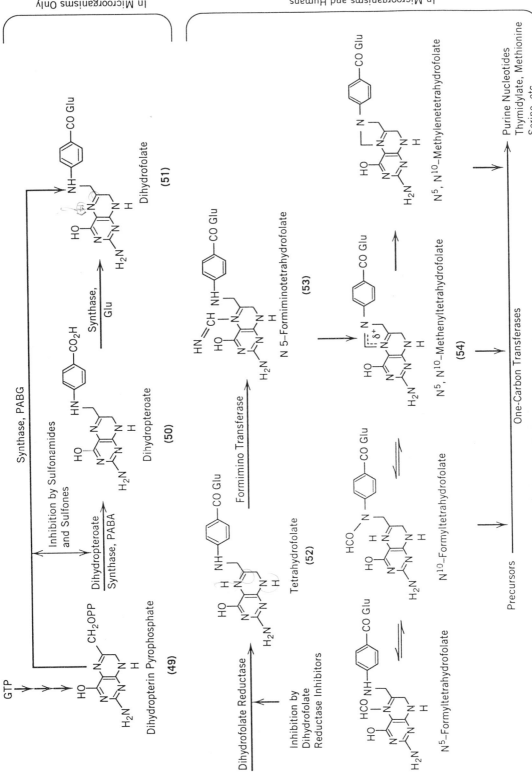

Fig. 33.1. Folate metabolism sites of action of sulfonamides/sulfones and dihydrofolate reductase inhibitors.

545

examining the action of sulfonamides at the enzyme level. Brown (176) demonstrated that the synthesis of dihydropteroate from PABA is sensitive to inhibition by a number of sulfonamides. Brown (176) in the *E. coli* system and Shiota et al. (173) in the *Veillonella* system found that the relation between a sulfonamide and PABA remained strictly competitive as long as the two compounds were added simultaneously or PABA was added first. If the enzyme and sulfonamide are preincubated with a low concentration of pteridine, the subsequent addition of PABA fails to reverse the inhibition. If, however, a high pteridine concentration is used, preincubation results in a much lesser degree of inhibition. Brown showed that the enzyme was not irreversibly inactivated. These results were explained by assuming that the sulfonamides react enzymatically with the pteridine substrate during preincubation and were suggestive of their incorporation. Thus when PABA is added after preincubation with a sulfonamide at a low concentration of pteridine, very little pteridine remains, but at a high pteridine concentration there is a very little inhibitor left. Using ^{35}S-labeled sulfanilic acid in this enzyme system, Brown (175,176) obtained evidence of incorporation. Though the nature of the resultant product was not elucidated, it was suggested that it may be a sulfanilic acid analog of dihydropteroate or the fully aromatic form. Block et al. (177) have observed similar results with sulfathiazole and sulfamethoxazole. Using ^{35}S-labeled sulfamethoxazole, they confirmed this incorporation by isolating the product formed with partially purified folate-synthesizing enzyme system and in growing cultures of *E. coli* and identified the product as N^1-3-(5-methylisoxazolyl)-N^4-(7,8-dihydro-6-pterinylmethyl) sulfanilamide (**55**). Metabolic incorporation of antimetabolites is quite common. Roland et al. (178) assessed the significance of this incorporation to the antimicrobial action of sulfonamides, and concluded that inhibition by dihydropterinsulfonamide of H_2-pteroate synthase, or of other folates enzymes tested, was not of physiological significance.

Brown (176) observed that the enzymic synthesis was much more sensitive to inhibition by sulfonamides than bacterial growth, suggestive of impeded permeability of the intact organisms to sulfonamides as compared to PABA. The more potent inhibitors of folate biosynthesis were, in general, better growth inhibitors also. Hotchkiss and Evans (179) have suggested that differences in the response of various organisms to sulfonamides may be due to quantitative differences in the ability of individual isoenzymes to produce folic acid from PABA in the presence of sulfonamides.

Richey and Brown (180) purified this enzyme from *E. coli*, virtually free from pyrophosphatase, and proposed that it may be called dyhydropteroate (H_2-pteroate) synthetase (also called synthase). This enzyme can also use PABG as the substrate to form dihydrofolate (**51**), (Fig. 33.1) directly, though PABA is used more efficiently. On the basis of the nonadditive and competitive utilization of these two substrates, it appears that H_2-pteroate synthetase is one single enzyme that can utilize either PABA or PABG (181,182), and it has been suggested (181) that the enzyme may be

(55)

allosteric. Toth-Martinez et al. (183) proposed this enzyme as a multiple proteins complex, composed of a glutamate pickup protein, reversibly attached to a PABA pickup protein, which in association with a dihydropteridine pickup protein functions as the enzyme H_2-pteroate synthetase. There is however, little evidence in the literature that PABG is the natural substrate for this enzyme, except in *M. tuberculosis*, which forms directly dihydrofolate from dihydropterin pyrophosphate (**49**) (Fig. 33.1). Brown (176), Shiota et al. (173), and Ortiz and Hotchkiss (184) have shown that the utilization of both the substrates, PABA and PABG, is competitively inhibited by sulfonamides.

The cell-free H_2-pteroate synthesizing system isolated from *E. coli* has become a very useful tool for studying structure-activity correlations among agonists and antagonists of PABA and the inhibitory effect of sulfonamides (185). Earlier the amount of H_2-pteroate formed was determined microbiologically using *Streptococcus faecalis* (ATCC 8043) as test organism. Bock et al. developed a rapid and simple thin-layer chromatographic method to follow these enzymatic reactions quantitatively (186), and used it to measure the kinetics of product formation, including sulfonamides incorporation by the folate-synthesizing enzymes.

Based on consideration of the formal electronic charges on sulfanilamide (187) and PABA (188), determined by the Huckel molecular orbital approach, Moriguchi and Wada (189) have proposed that there are two binding sites on the enzyme, located 6.7–7.0 Å apart, one being specific for the 4-NH_2 group and the other nonspecific where the acidic group of PABA or sulfamyl group of sulfonamides binds. Shefter et al. (190), on the basis of the X-ray crystallographic data and study of molecular models, noted certain structural similarities between N^1-substituted sulfonamides and *p*-aminobenzoyl glutamate and

have suggested that that N^1-substituent of sulfonamides may be competing for a site on the enzyme surface reserved for the glutamate residue, either by directly influencing the linkage of aminobenzoic acid–glutamate with the pteridine or for the coupling of glutamate to the dihydropteroic acid.

The mechanism of action of dapsone (and other sulfones) is similar to that of sulfonamides, since the action is antagonized by PABA in mycobacteria (191,192), other bacteria (193), and protozoa (98). The exceptionally highly antibacterial activity of DDS against *M. leprae* has attracted special attention (194). There is evidence that as with sulfonamides (177), DDS is also incorporated to form an analog of dihydropteroate, but the significance of this is not clear. In *M. kansasii*, Panitch and Levy found a 14–15 fold accumulation of DDS within the bacterial cells after 8 days of treatment (195). It has been suggested that there may be similar accumulation within the bacterial cells of *M. leprae*. Additional sites of action outside of the folate synthesizing enzyme system have also been proposed. DDS has unique beneficial effects on some dermal inflammatory conditions (*loc. cit.*), especially dermatologic, and it is likely that this action may contribute to its activity against *M. leprae*. Cenedella and Jarrell (196) suggested an additional mechanism for the antimalarial action of dapsone involving inhibition of glucose utilization by the intraerythrocytic parasite; this inhibition was shown to be antagonized by raising the glucose concentration of the medium.

Similar mode of action of action of sulfonamides and sulfones has been demonstrated in most of the microbes tested which are susceptible to their action. For example in the case of chlamydia it has been shown that the sulfonamide-sensitive members of this group, such as trachoma inclusion conjunctivitis viruses, have a folic acid metabolism similar to that of bacteria, and that the action of sulfonamides against them is competitively antagonized by PABA (197–199).

Thus sulfonamides and sulfones, by competing for the enzyme site for PABA, inhibit the biosynthesis of dihydrofolate and thereby of tetrahydrofolate, which is involved in 1-carbon transfer processes. This would prevent or slow down the formation of a number of building blocks of protein, DNA, and RNA biosynthesis, thereby affecting a number of synthetic processes of the organisms concurrently. Sulfonamides inhibit only growing organisms, and the bacteriostasis of the latter is preceded by a lag phase. The lag phase can now be explained as being due to stored PABA/folic acid, and its duration is dependent on the quantity stored.

3.1.1 SELECTIVITY OF ACTION. The presence of the folate-synthesizing system has been demonstrated in a number of bacteria (174,177,182,184,200), protozoa (201–203), yeasts (204), and plants (185,205,206), and this serves to explain the broad spectrum of action of sulfonamides. However since higher organisms (e.g., mammals) do not possess this biosynthesis system and require preformed folic acid, they are unaffected by sulfonamides. This selectivity of action offered for the parasite by the difference in the metabolic pathway between the microbes and humans makes sulfonamides "ideal" chemotherapeutic agents. It is not clearly understood how, in spite of the presence of folic acid in blood and tissues, sulfonamides exert their bacteriostatic action. Folic acid in animal tissues normally occurs linked to polyglutamate conjugates or to proteins and very likely cannot be used in this form by microbes. Moreover, the concentration of PABA and other products of the action of folic acid coenzymes, which are able to reverse the action of sulfonamides, may be too low in tissues to prevent the action of sulfonamides when used in adequate dosage.

3.2 Synergism with Dihydrofolate Reductase Inhibitors

In any attempt to synergize the action of sulfonamides and to avoid the development of resistance to them, the most logical approach is to combine them with agents that block the same metabolic pathway as that blocked by sulfonamides, but at different sites. The elucidation of the folic acid pathway and the demonstration of its inhibition by both sulfonamides and dihydrofolate reductase inhibitors (antifolates) Fig. 33.1) provided this possibility (207).

Greenberg (208) and Greenberg and Richesan (209,210) reported around 1950 that antifolates, aryloxyrimidines and diaminopteridines, potentiate the action of sulfonamides in experimental *P. gallinaceum* infection. Since then there has been extensive use of combinations of pyrimethamine (**56**) and sulfonamides for the treatment of plasmodial, toxoplasmal, and coccidial infections, and of trimethoprim (**57**) and sulfonamides for a number of bacterial infections. This synergism is now recognized as a general occurrence, and therapy with a combination of dihydrofolate reductase inhibitors and sulfonamides has added a new dimension to treatment with these agents (211). This synergism is clearly a consequence of the sequential occurrence of the twin loci of inhibition in

(**56**)

(**57**)

the folic acid biosynthesis (Fig. 33.1). Factors that contribute to the usefulness of such combinations include a severalfold increase in chemotherapeutic indices, better tolerance of the drugs, ability to delay development of resistance, and ability to produce cures where the curative effects of the individual drugs are minimal (207).

The classical explanation for this synergism was that the two components bind independently to two enzymes in a linear folic acid pathway, namely H_2-pteroate synthase and H_2-folate reductase. Later evidence, however, indicated that this synergism may at least in part be a result of the binding of both the inhibitors to H_2-folate reductase (212). In a recent paper Bowden et al. (213) have confirmed the existence of a moderately potent sulfonamide-binding site on wild type *E. coli* M.R.E. 600 and *L. casei* dihydrofolate reductases that appears to overlap the benzoylglutamate binding site of the substrate. It has been suggested that for clinically useful regimens the sulfonamides should be capable of binding positively to the site together with the dihydrofolate reductase inhibitors. By NMR studies it has been shown that though sulfonamides compete with PABG, the binding site for PABG and sulfonamides are not the same but overlap. The involvement of Arg-31 in this binding has recently been proposed (214).

The choice of the individual drugs in the combination used is based on the best pharmacokinetic fit (215) (e.g., trimethoprim with sulfamethoxazole, both having a half-life of about 11 h, is a commonly used combination), observed synergism in *in vitro* studies, and the inhibitory index of the antifolates for the particular organism, which in turn is related to the binding affinity to dihydrofolate reductase. Dihydrofolate reductases from various sources differ markedly in their binding ability to various inhibitors; pyrimethamine is bound much more strongly to the enzyme from plasmodia than from bacteria, and the

converse is true for trimethoprim (216,217). This explains the choice of trimethoprim for bacterial infections and of pyrimethamine for antimalarial chemotherapy.

Thus while pronounced differences in structural dependency have been observed for inhibitors of the dihydrofolate reductases of the different microbial species the structural requirements for inhibitors of the dihydropteroate synthase for the various species studied are similar (218).

3.3 Drug Resistance

Emergence of drug-resistant strains is a serious problem with sulfonamides as with many other antimicrobials. Due to long use of this group of drugs the incidence of drug resistant isolates in the community has become quite alarming. The majority of isolates of *N. meningitides* of serogroups B and C in the United States and of group A isolated from other countries are now resistant. A similar situation prevails with respect to Shigella and strains of *E. coli* isolated from patients. Since the mode of action of sulfonamides and sulfones involves the same basic mechanisms, different sulfonamides usually show cross-resistance, but not to antimicrobials of other classes acting by different mechanisms.

Resistance can arise by one or more of the following mechanisms: (a) increased production of PABA by the pathogen (219,220); (b) altered sensitivity of the target enzyme, making it selectively more sensitive to the natural substrate (221–224); (c) gene amplification of the enzyme so that more enzyme is produced, thus rendering its saturation by antagonist difficult; (d) by-pass mechanism by which the microorganism develops an ability to utilize more effectively the folic acid present in the host (225); (e) reduction in permeability of the cell to sulfonamides so that less drug is transported in (221). In the case of plasmodia a by-pass mechanism (i.e., ability to use performed folic acid) seems to be operative. Bishop has described strains resistant to both sulfonamides and pyrimethamine (225,226) that can presumably utilize the reduced forms of folic acid available in the host erythrocytes.

In bacteria the first two mechanisms, viz, over-production of PABA and/or reduced sensitivity of the dihydropteroate synthease seem to be most common. Resistant strains develop by random mutation and selection or more commonly by transfer of resistance factors (R-factors) by plasmids (224). It has been noticed that multiple drug resistance involving streptomycin, chloramphenicole, tetracycline and sulfonamides could be transferred between Shigella and E. coli in mixed cultivation in the host (227). Drug resistance acquired in this manner is usually persistent and irreversible, and can be transferred to other sensitive strains indefinitely.

In most cases drug resistance has been ascribed to the presence of R-plasmids carrying a gene for a drug-resistant dihydropteroate synthase; it was proposed that the R-factors may consist of separate transfer or resistance functions, which could be transferred separately or together (228–230). The R-plasmids conferring resistance belong to different groups. Two related R-plasmids pAr32 and pJ8102-1, which encode resistance to chloramphenicol, streptomycin, and sulfonamides (Su), have been isolated and their genetic map constructed (231). The genetic map indicated that this plasmid consists of at least three functionally different regions; one is the region with ability to transfer (tra), another for drug resistance gene (r-det) and the third encoding the replication functions (rep.) (231).

The gene for dihydropteroate (DHPS) synthase has been characterized and localized and the base sequence determined. Most of the high level sulfonamide resistance in gram-negative bacteria appears to be accounted for by only two plasmid borne genes Sul-I and Sul-II (232). In

meningococci, the sulfonamide resistance is chromosomally located, but sequence comparisons indicate that it is horizontally transferred. In a nucleotide sequence study of sulfonamide-sensitive and sulfonamide-resistant (SU-R) strains of Neisseria meningitidis of serogroups A,B and C it was shown that the sequence of four DHPS genes from sulfonamide-susceptible strains are strikingly similar (average sequence difference 4%) the sequence of three DHPS genes from resistant strains show an equally striking identity but are conspicuously different from those of the sulfonamide-susceptible strains. While the N-terminal and C-terminal parts of the genes of the two groups showed great similarity, a central 424-base pair portion was identical in genes from Su-R strains but different from Su-susceptible strains, suggesting a horizontal transfer and homologous combination of chromosomal sequences from different sources with the formation of mosaic genes (233) as described for penicillin resistance genes of S. pneuominae (234). In an earlier study by the same investigators the Su-R gene a serogroup B and serotype 15 strain of Neisseria meningitides was cloned and characterized (235). It was shown that the Su-R gene from the meningococcus strain is different from the plasmid-borne Su-R, Sul-I and Sul-II genes in members of the family Enterobacteriaceae and from the Su-R gene of pANM-1 from a serogroup C meningococcus. It was suggested that drug resistance is caused by relatively small mutational changes in the ordinary chromosomal gene for DHPS.

The chromosomal and plasmid-borne sulfonamide resistant genes from a number of other organisms and plasmids have been isolated, cloned, and characterized. The nucleotide sequence of the Su-R gene from plasmid R46 has been reported (236). In another study the base sequence of the regions coding for resistance to β-lactam antibiotics, streptomycin and sulfonamide of the IncN and IncW R46 plasmids and in

Tn 21-like transposons have been reported and found to be similar (237). From the wide dissemination of the Su-R region, it has been suggested that the DNA segment may be a mobile element and found to differ only in one base with Sul-I gene of plasmid R388 (238).

Dapsone resistance has been a problem of increasing concern. Secondary resistance is quite common in leprosy patients and presents as a clinical and bacteriologic relapse after several years of apparently successful therapy; its frequency varies from 2–30% in different countries. Mechanisms similar to those for sulfonamides would be operative for the development of resistance to sulfones also.

4 STRUCTURE AND BIOLOGICAL ACTIVITY

4.1 Structure–Action Relationship

Although the story of sulfonamides started with the discovery of their antimicrobial action, subsequent studies established their usefulness as carbonic anhydrase inhibitors, diuretics (saluretics), and antidiabetics (insulin-releasers) and more recently also as endothelin antagonists. Compounds with each type of action possess certain specific structural features in common. The present discussion, however, is restricted only to the antimicrobial sufonamides and sulfones, characterized by their ability to inhibit the biosynthesis of folic acid by competing with PABA for 7,8-dihydro-6-hydroxymethylpterin at the active site of dihydropteroate synthase. Many reviews on this area have been written and a few (315–317) of these are given at the end.

As sulfanilamide (3) is a rather small molecule and there are not too many variations that can be carried out without changing the basic nucleus, the following generalizations regarding structure–activity

relationships were arrived at quite early in the development of sulfonamides, which guided the subsequent work on molecular modification, and these generalizations still hold:

1. The amino and sulfonyl radicals on the benzene ring should be in 1,4 disposition for activity; the amino group should be unsubstituted or have a substituent that is removed readily *in vivo*.

2. Replacement of the benzene ring by other ring systems, or the introduction of additional substituents on it, decreases or abolishes the activity.

3. Exchange of the SO_2NH_2 by $SO_2C_6H_4$-p-NH_2 retains the activity, while exchange by $CONH_2$, COC_6H_4-p-NH_2 markedly reduces the activity.

4. N^1-Monosubstitution results in more active compounds with greatly modified pharmacokinetic properties. N^1-disubstitution in general leads to inactive compounds.

5. The N^1-substitution should be such whose pK_a would approximate the physiological pH.

$$NH_2$$

$$SO_2R$$

R = NHR′, OH,
Ph(NH_2-p)

(58)

The presence of a p-aminobenzensulfonyl radical thus seems essential for maintaining good activity and practically all the attention was focused on N^1-substituents. These substituents seem to affect mainly the physicochemical and the pharmacokinetic characteristics of the drugs.

The following broad generalization hold for the SAR of sulfones:

1. One *p*-aminophenylsulfonyl residue is essential for activity; the amino group in this moiety should be unsubstituted or have a substituent that is removed readily *in vivo*.

2. The second benzene ring should preferably have small substituents which will make this ring electron rich (such as CH_3, OCH_3, OH, NH_2, NHC_2H_5); *p*-substitution is most favorable for activity.

3. Replacement of the second phenyl ring by heterocycles does not improve the activity.

DDS has retained its preeminent position even after 50 years of its use.

4.2 Physicochemical Properties and Chemotherapeutic Activity

Studies to find a correlation between physicochemical properties and bacteriostatic activity of sulfonamides have been pursued almost since their discovery. The substituents that attracted the attention of investigators quite early were the amino and the sulfonamido groups in the molecule and several groups of workers almost simultaneously noted a correlation between the bacteriostatic activity and degree of ionization of sulfonamides. The primary amino group in sulfonamides apparently plays a vital part in producing bacteriostasis, since any substituent on it causes complete loss of activity. Seydel et al. (239,240), from a study of infrared (IR) spectra and activity of a number of sulfonamides, concluded that the amount of negative charge on the aromatic amino group is important for the activity. However, variation in activity within a series of compounds cannot be attributed to a change in base strength, since all the active sulfonamides (and sulfones) have a basic dissociation constant of about 2, which is

close to that of PABA. Foernzler and Martin (187), by the linear combination of atomic orbital-molecular orbital (LCAO–MO) method, computed the electronic characteristics of a series of 50 sulfonamides and found that the electronic charge on the *p*-amino group did not vary with a change in the N^1-substituent. Thus attention has been focussed mainly on the acidic dissociation constant, which varies widely from about 3 to 11. Fox and Rose (241) noted that sulfathiazole and sulfadiazine were about 600 times as active as sulfanilamide against a variety of microorganisms, and that approximately 600 times as much PABA was required to antagonize their action as to antagonize sulfanilamide; however the same amount of PABA was required to antagonize the minimum inhibitory concentration (MIC) of each drug. This suggested that the active species in both cases was similar, and that the increase in bacteriostatic activity was due to the presence of a larger proportion of the drug in an active (ionized) form. They found that the concentration of the ionized form of each drug at the minimum effective concentration was of the same order. Thus if only the ionized fraction at pH 7 was considered instead of the total concentration, the PABA/drug ratio was reduced to 1:1.6–6.4. They also observed that with a tenfold increase in ionization of sulfanilamide on altering the pH from 6.8 to 7.8, there was an eightfold increase in bacteriostatic activity. On the basis of these observations, Fox and Rose suggested that only the ionized fraction of the MIC is responsible for the antibacterial action. Schmelkes et al. (242) also noted the effect of pH of the culture medium on the MIC of sulfonamides and suggested that the active agent in a sulfonamide solution is an anionic species.

Bell and Roblin (243), in an extensive study of the relationship between the pK_a of a series of sulfonamides and their *in vitro* antibacterial activity against *E. coli*, found that the plot of log $1/\text{MIC}$ against pK_a was a parabolic curve and that the highest points of the curve lay between pK_a 6 and 7.4; the maximal activity was thus observed in compounds whose pK_a approximated the physiological pH. Since the pK_a values are related to the nature of the N^1-substituent, the investigators emphasized the value of this relationship for predicting the MIC of new sulfonamides. The pK_a of most of the active sulfonamides discovered since then, and particularly of the long-acting ones, falls in this range (Table 33.2). Bell and Roblin correlated Woods and Fildes's hypothesis (14) regarding the structural similarity of a metabolite and its antagonist with the observed facts of ionization. They emphasized the need of polarization of the sulfonyl group of active sulfonamides, so as to resemble as closely as possible the geometric and electronic characteristics of the p-aminobenzoate ion, and postulated the "the more negative the SO_2 group of N^1-substituted sulfonamides, the more bacteriostatic the compound will be". The acid dissociation constants were considered to be an indirect measure of the negative character of the SO_2 group. The hypothesis of Bell and Roblin stated that the unionized molecules had a bacteriostatic activity too, though weaker than that of the ionized form. Furthermore it was supposed that increasing the acidity of a compound decreased the negativity of the SO_2 group, thus reducing the bacteriostatic activity of the charged and uncharged molecules.

Cowles (250) and Brueckner (251), in a study of the effect of pH of the medium on the antibacterial activity of sulfonamides found that the activity increased with increase in pH of the medium only up to the point at which the ionization of the drug was about 50%, then decreased. Brueckner assumed different intra- and extra-cellular pH values to explain the observed effects. Cowles suggested that the sulfonamides penetrate the bacterial cell in the unionized form, but once inside the cell, the bacterio-

static action is due to the ionized form. Hence for maximum activity, the compounds should have a pK_a that gives the proper balance between the intrinsic activity and penetration; the half-dissociated state appeared to present the best compromise between transport and activity. This provided an alternative explanation for the parabolic relationship observed by Bell and Roblin between pK_a and MIC.

The activity of sulfaguanidine, the sulfones and the ring N-methyl sulfanilamide heterocycles appears to be inconsistent with the ionization theory. This inconsistency is resolved by considering the availability of an electron pair, regardless of the anionic charge, in response to the electrophilic center of the enzyme, as of critical importance. Sulfaguanidine has been shown by IR and nuclear magnetic resonance (NMR) spectroscopy to prefer the tautomeric structure $H_2NC_6H_4SO_2N=C(NH_2)_2$, represented as the resonance hybrid (59) (252,253). A lower energy for this structure would be expected in view of the greater stability of symmetrical resonance forms of guanidines in general. The active N^1-heterocyclic sulfanilamides are capable of amido–imido tautomerism and do, in fact, exist substantially in the imido form (253–258). It has been shown, for example, that in aqueous solution sulfapyridine mainly exists as (60) and its anionic form $H_2NC_6H_4SO_2N^-C_5H_4N$, their ratio depending on the pH of the solution. The ring N-methyl derivatives (61) increase the electron density at the sulfonamide group by resonance. The activity of sulfones may be

(59) (60)

TABLE 33.2 Characteristics of Commonly Used Sulfonamides and Sulfones[a]

General structure: H_2N—⟨benzene⟩—SO_2NHR

Generic Name	R	Common Proprietary Names	In vitro Activity[b] Against E. coli, μmol/L	Water Solubility[c] mg/100 mL at 25°C	pK_a	Lipo-solu-bility[d], %	Protein Binding at 1.0 μmol./mL, % Bound	Plasma "Half-life h" (human)	% N^4-Meta-bolite in Urine[e] (human)
				Poorly Absorbed, Locally Acting					
Phthalysulfacetamide[f]	—COCH₃	Enterosulfon, Thalisul, Thalamyd		Very sparingly soluble	Acid				
Phthalysulfathiazole[f]	⟨thiazole⟩	Thalazole, Sulfthaladine		Insoluble	Acid				
Succinylsulfathiazole[g]	⟨thiazole⟩	Sulfasuxidine, Thiacyl		20	Acid				
Sulfaguanidine	C(=NH)NH₂	Guanicil	4[h]	100	Base			5	
				Well Absorbed, Rapidly Excreted					
Sulfamethizole, Sulfamethylthiadiazole	⟨thiadiazole–CH₃⟩	Methisul, Lucosil, Ultrasul		25 (pH 6.5)	5.5		22	2.5	6
Sulfacrbamide, Sulfanilylurea	—CONH₂	Euvernil	32.2	811 (370)	5.5		95	2.5	
Sulfathiazole	⟨thiazole⟩	Cibazole, Thiazamide	1.6	60 (pH 6)	7.25	15.3	68	4	30 (40)
Sulfisoxazole, Sulfanilylurea	⟨isoxazole, CH₃, CH₃⟩	Gantrisin, Urosulfin	2.15	350 (pH 6)	5.0	4.8	76.5	6.0	16 (30)

Name	Trade name	Structure						
Sulfamethazine Sulfadimidine	Diazil	(CH$_3$ pyrimidine)	1.7	150 (29°)	7.4	82.6	66	7	60
Sulfisomidine	Elkosin Aristamid	(CH$_3$ pyrimidine)	1.5	300 (30°)	7.4	19.0	67	7.5	4
Sulfacetamide	Albucid	—COCH$_3$	2.3	670	5.4	2.0	9.5	7	5
Sulfachloropyridazine	Sonilyn	(Cl pyridazine)		90^j (pH 5.5)	6.1	14.8	80.5	8.0	
Sulfapyridine	Eubasin Dagenan	(pyridine)	4.8	30	8.4	14	70	9	30
Sulfanilamide	Prontalbin	H	128	750	10.5	71	9	9	

Readily Absorbed, Medium Rate of Excretion

Name	Trade name	Structure						
Sulfaphenazole	Orisul	(Ph pyrazole)	1.0	150	6.09	69	87.5	10	20 (80)
Sulfamethoxazole	Ganthanol	(CH$_3$ isoxazole)	0.8	Sparingly soluble	6.0	20.5	60	11	60 (14)
Sulfamoxol Sulfadimethyloxazole	Sulfuno	(CH$_3$ oxazole)	4.0	Soluble	7.4	41.4	76.5	11	

(Continued on page 556)

Table 33.2 (*Continued*)

Generic Name	R	Common Proprietary Names	*In vitro* Activity[b] Against *E. coli*, μmol/L	Water Solubility[c] mg/100 mL at 25°C	pK_a	Lipo-solu-bility,[d] %	Protein Binding at 1.0 μmol/mL, % Bound	Plasma "Half-life h" (human)	% N^4-Meta-bolite in Urine[e] (human)
Sulfadiazine	(pyrimidine, CH₃)	Debenal Pyrimal	0.9	8	6.52	26.4	37.8	17	25
Readily Absorbed, Slowly Excreted									
Sulfamerazine	(CH₃, CH₃)	Debenal M Pyrimal M	0.95	16 (37°)	6.98	62.0	56.8	24	50[j]
Sulfamethyldiazine	(CH₃)	Pallidin	1.0	40 (pH 5.5)	6.7	69.6	74.0	35	
Sulfamethoxydiazine Sulfamonomethazine	(OCH₃)	Sulfameter Durenat	2.0	Very sparingly soluble	7.0	64.0	74.2	37	20[j] (30)
Sulfamethoxypyridazine	(OCH₃, pyridazine)	Lederkyn Kynex	1.0	147 (pH 6.5) (37°C)	7.2	70.4	77	37	50 (15)
Sulfadimethoxine	(OCH₃, OCH₃)	Madribon	0.7	29.5 (pH 6.7) (37°C)	6.1	78.7	92.3	40	15 (70)

Name	Structure / R	Solubility		pK_b				
Sulfamethoxypyrazine sulfameopyrazine	H₃CO–⟨pyrazine⟩–CH₃	Very sparingly soluble	1.85	6.1		65	65	65
Sulformethoxine Sulfadoxine Fanasil	OCH₃, H₃CO–⟨pyrimidine⟩–CH₃	—	0.8	6.1	5	95	150	60[f] (10)
Diaminodiphenylsulfone Dapsone Avolosulphon (R = H)	RHN–⟨⟩–SO₂–⟨⟩–NHR	14	44		13	50[e,j]	20	
Diacetamidodiphenylsulfone Acedapsone (R = COCH₃)		0.3					43 days i.m.	

[a] Unless otherwise stated, the data are from Rieder (244).
[b] From Struller (245).
[c] From ref. 246
[d] Determined by partition between ethylene dichloride and sodium phosphate buffer (16).
[e] From Williams and Park (247).
[f] N^4-Phthallyl.
[g] N^4-Succinyl.
[h] Unpublished results from O. P. Srivastava, Central Drug Research Institute, Lucknow, India.
[i] Estimated from Niepp and Mayer (248).
[j] From ref. 249.

557

(61)

explained similarly when related to the resonance form having a high electron availability at the position para to the amino group of the aminophenyl residue.

Seydel (259) and Cammarata and Allen (260) have cited examples of active sulfonamides whose pK_a values lie outside the optimal limits given by Bell and Roblin. It has been suggested this may be partly due to the difficulties in titration of weakly soluble compounds. Seydel et al. (239,250, 261) and Cammarata and Allen (260) also showed that if a small homologous series is used, a linear relationship of the pK_a to the MIC is obtained.

The functional relationship between acid dissociation constant and the activity of sulfonamides has not been questioned since the investigations just cited were published. This, however, does not mean that the ions of different sulfonamides are equally active; other factors also have an influence and account for the observed differences in activity of different sulfonamides, such as affinity for the concerned enzyme. The pK_a is related to solubility, distribution and partition coefficients, permeability across membranes, protein binding, tubular secretion, and reabsorption in the kidneys.

In subsequent studies on correlation of physicochemical properties with activity, additional parameters have been included such as Hammett sigma values and other electronic data for net charge calculated by molecular orbital methods, spectral characteristics and hydrophobicity constant.

Bell and Roblin laid emphasis on the polarizability of the SO_2 group and related the negative charge of this group to the MIC. Seydel et al. (240), using IR spectroscopy, and Schnaare and Martin (262) and Foernzler and Martin (187), who calculated the electron density of oxygen atom of the SO_2 group using LCAO–MO method, could not find any evidence for this correlation. They have therefore, attached greater importance to the electronic charge on the 1-NH. Rastelli et al. (263), however, in a later study of the correlation of electron indices using the symmetrical stretching mode of the sulfonyl group as determined by IR and Raman spectra, and MIC, have found a direct relationship between the S—O bond polarity and MIC, thus supporting the conclusions of Bell and Roblin. These contradictory results may be due to different conditions used for spectral determination and variability of the antibacterial activity data.

Foernzler and Martin (187) found that the electron densities at the 4-NH_2, S, and O atoms of a large number of N^1-substituted sulfonamides are essentially constant, whereas those for 1-NH vary. They found a correlation between pK_a, electron density, and MIC against *E. coli* of N^1-aryl sulfonamides; this relationship was more significant when the compounds were classified into smaller groups depending on the nature of the N^1-substitutent. Seydel and his associates confined their studies to sulfanilides and N^1-(3-pyridyl)-sulfanilamides. They have extrapolated the electron density on the 1-NH group from the study of IR and NMR data and Hammett sigma values of the parent anilines and have correlated the data with the MIC against *E. coli*. Anilines were used for studying the IR spectra because they could be dissolved in nonpolar solvents, thus giving more valid data; this was not possible with sulfanilamides because of low solubility in such solvents. Seydel (264) and Garrett et al. (265) found an approximately linear relationship between bacteriostatic activity,

Hammett sigma value, and electron density of the N^1-nitrogen of a group of *m*- and *p*-substituted sulfanilides and emphasized the possibility of predicting the *in vitro* antibacterial activity of sulfanilamides by use of this relationship. Later, Seydel (266) included in this study 3-sulfapyridines, carried out regression analyses of the data, and obtained a very acceptable correlation coefficient.

Fujita and Hansch (267), in a multiparameter, linear free energy approach, correlated the pK_a, hydrophobicity constant, and Hammett sigma values of a series of sulfanilides and N^1-benzoyl and N^1-heterocyclic sulfanilamides with their MIC data against gram-positive and gram-negative organisms and their protein binding capacity. They devised suitable equations by regression analyses for the meta- and para-substituted compounds; the correlation for the para-substituted compounds was rather poor. The hydrophobicity of the compounds was found to play a definite role in the activity. It was shown that keeping the lipophilicity of the substituents unchanged, the logarithmic plot of activity against the dissociation constant gives two straight lines with opposite slopes, the point of intersection of which corresponds to the maximal activity for a series of sulfanilamides. They suggested the optimal values of the dissociation constant and hydrophobicity for maximum activity against the organisms studied.

Yamazaki et al. (268), in the study of the relationship between antibacterial activity and pK_a of 14 N^1-heterocyclic sulfanilamides, considered separately the activities of the compounds in terms of the concentrations of their ionized and unionized forms and their total concentrations in the culture medium. They found that whereas the relationship between pK_a and activity is parabolic when total concentrations is considered, it is linear for ionized and un-ionized states giving two lines having opposite slopes and intersecting each

other, the point of intersection corresponding to the pH of the culture medium. They found the pK_a for optimal activity to be between 6.61 and 7.4.

In these studies it was noticed that some of the sulfonamides had lower antibacterial activity than expected, possibly because of their poor permeation. To define the role of permeability in the antibacterial activity of sulfonamides, Seydel et al. (269,270) extended these investigations to cell-free folate synthesizing systems and correlated the inhibitory activity of these compounds on this enzyme system and on the intact organisms to their pK_a, Hammett sigma, chemical shift, and π values. The rate determining steps for sulfonamide action in the cell free system and a whole cell system was found to have similar substituent dependencies. From a comparison of the linear free energy relationships obtained in the two systems, they suggested that the observed parabolic dependence of the antibacterial activity indicates that it is not the extracellular ionic concentration that governs the potency of sulfonamides but rather the intracellular ionic concentration, which, in turn, is limited by the permeation of un-ionized compounds, thus supporting Cowles and Brueckner's postulates (*loc. cit*). They concluded that the lipophilic factors are not important in the cell-free system or for *in vitro* antibacterial activity when permeability is not limited by ionization.

Thus intensive subsequent work in this field over the last four decades has fully supported the views expressed quite early in the development of sulfonamides of the predominant role of ionization for their antibacterial activity.

4.2.1 SULFONES. Ever since the discovery of the antimycobacterial activity of sulfones in 1940s and that they share a common biological mode of action with sulfonamides (competitive antagonism of PABA), the question of their structural similarity with sulfonamides which enables them to inhibit dihydropteroate synthase has attracted much attention. It was realized quite early in these studies that as with sulfonamides, $4\text{-}NH_2\text{-}C_6H_4\text{-}SO_2$ was inviolate for optimal activity, and the substituents in the second phenyl could modulate the activity. A number of QSAR studies have been reported (79–82, 270–274) on 4-aminodiphenylsulfones with a view to analyze the contribution of these substituents to the biological activity using Linear Free Energy, Molecular Modelling and Conformational Analysis methods. It has been shown that electronic and steric effects have the decisive role both on binding to the enzyme as also the overall biological activity. The electronic effects were rationalized in terms of electronic charge perturbations which are transmitted from the multisubstituted aryl ring to the common bifunctional moiety, $4\text{-}NH_2\text{-}C_6H_4\text{-}SO_2$, through the SO_2 group mainly via hyperconjugation (270). In conformational analysis using the MINDO semiemperical molecular orbital method, it was found that 4-aminodiphenylsulfones show multiple conformational energy minima, mainly due to the torsional freedom of the sulfur–carbon bond of the substituted aryl ring with the $4\text{-}NH_2\text{-}C_6H_4\text{-}SO_2$. The highly active derivatives were in general shown to be less flexible; inhibition potency increased as entropy decreased (271,272); a 'butterfly' structure (Fig. 33.2) was considered to best represent the active conformations.

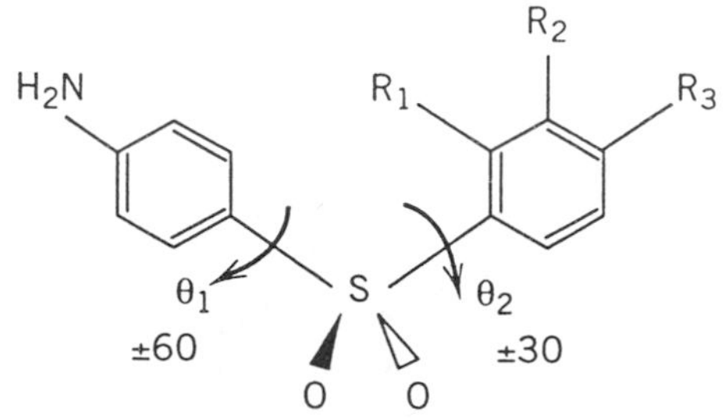

4.2.2 WATER SOLUBILITY. The clinically used sulfonamides, being weak acids, are, in general, soluble in basic aqueous solutions. As the pH is lowered, the solubility of their N^1-substituted sulfonamides decreases, usually reaching a minimum in the pH range of 3–5. This minimum corresponds to the solubility of the molecular species in water (Table 33.2). With a further decrease in pH, corresponding to that of a moderately strong acid, the sulfa drugs dissolve as cations.

Fig. 33.2. Active conformations of 4-aminodiphenyl-sulfones.

The solubility of sulfonamides is of clinical and toxicological significance because damage to kidneys is caused by crystallization of sulfonamides or their N^4-acetyl derivatives. Their solubility in the pH range of urine (i.e., pH 5.5–6.5) is, therefore, of practical interest. One of the significant advances in the first phase of sulfonamide research was the development of compounds with greater water solubility, such as sulfisoxazole, which helped to overcome the problem of crystallization in the kidney of earlier sulfonamides. However, apart from the solubility of the parent compounds, the solubility of their N^4-acetyl derivatives, which are the main metabolic products, is of great importance because these are generally less soluble than the parent compounds. For example, sulfathiazole, which itself is unlikely to be precipitated, is metabolized to its N^1-acetyl derivative, which has a poor solubility that is likely to lead to its crystallization in the kidney. The solubility of sulfonamides and their principal metabolites in aqueous media, particularly in buffered solutions and body fluids, therefore, has been the subject of many studies aimed at enhancing our understanding of their behavior in clinical situations (244,275,276).

4.2.3 LIPID SOLUBILITY. An important factor affecting the chemotherapeutic activity of sulfonamides and their *in vivo* transport is the lipid solubility of the undissociated molecule. The partition coefficients measured in solvents of different dielectric constants have been used to determine the lipid solubility and hydrophobicity constant (244,265). Chromatographic R_f values in a number of thin-layer chromatography systems have also been used as an expression of the lipophilic character of sulfonamides and found to correspond well with the Hansh values in an isobutyl alcohol–water system (277).

Table 33.2 gives the percentage of various sulfonamides passing from aqueous phase into ethylene chloride as determined by Rieder (242). Lipid solubility of different sulfonamides varies over a considerable range. These differences unquestionably influence their pharmacokinetics and antibacterial activity. It has been noted by Rieder (244) that long acting sulfonamides with a high tubular reabsorption are generally distinguished by a high degree of lipid solubility. The antibacterial activity and the half-life are also related to lipid solubility. Although a precise relationship between these factors has not been established, it has been shown in general, that as the lipid solubility increases, so does the half-life and in *vitro* activity against *E. coli*.

4.3 Protein Binding

A particularly important role in the action of sulfonamides is played by their binding to proteins. Protein binding, in general, blocks the activity of sulfonamides as of many other drugs (the bound drug is chemotherapeutically inactive), and reduces their metabolism by the liver. The binding is reversible; thus the active free form is liberated as its level in the blood is gradually lowered. The sulfonamide concentration in other body fluids too is dependent on its protein binding. Thus the unbound fraction of the drug in the plasma seems to be significant for activity, toxicity, and metabolism, whereas protein binding appears to

modulate the availability of the drug and its half-life. The manner and extent of binding of sulfonamide has been the subject of many studies (244,278–280), and the important characteristics of this binding are now reasonably clear. The binding affinity of different sulfonamides varies widely with their structure (Table 33.2) as also with the animal species and the physiological status of the animal (244,281). In plasma the drug binds predominantly to the albumin fraction. The binding is weak (4–5 kcal) and is easily reversible by dilution. It appears to be predominantly hydrophobic, with ionic binding being relatively less significant (267,280). Thus the structural features that favor binding are the same as those that increase lipophilicity, such as the presence of alkyl, alkoxy, or arly groups (242, 278,282). N^4-Acetyl derivatives are more strongly bound than the parent drugs. Introduction of hydroxyl or amino groups decreases protein binding and glucuronidation almost abolishes it. Seydel (283), in a study of the effect of the nature and position of substituents on protein binding and lipid solubility, has shown that among isomers, ortho-substituted compounds have the lowest protein binding. This would indicate that steric factors have a role in protein binding and that N^1-nitrogen atom of the sulfonamide is involved. The binding seems to take place with the basic centers of arginine, lysine, and histidine in the proteins (244). The locus of binding of several sulfonamides to serum albumin has been shown by high resolution NMR spectral studies to involve benzene ring more than the heterocycle (284).

There have been attempts to establish correlations between physicochemical properties of sulfonamides, their protein binding, and their biological activity. Martin (285) established a functional relationship between excretion and distribution and binding to albumin, and Kruger-Thiemer et al. (279) have derived a mathematical relationship. Moriguchi et al. (286) observed a parabolic relationship between protein binding and *in vitro* bacteriostatic activity in a series of sulfonamides, and suggested that too strong an affinity between sulfonamides and proteins would prevent them from reaching their site of action in bacteria; with too low an affinity, they would not be able to bind effectively with enzyme proteins to cause bacteriostasis, assuming that affinity for enzyme proteins is paralleled by affinity to bacterial proteins. In a multiparameter study of a series of N^1-heterocyclic sulfonamides, Fujita and Hansch (267) considered that in the free-state sulfonamides exist as two different species, neutral and ionized, whereas in the bound state they exist only in one form. They developed suitable equations by regression analysis and showed that for the series of sulfonamides of closely related structure, whose pK_a does not vary appreciably, the binding is governed mainly by the N^1-substituent, which supported the earlier results (278).

The implications of protein binding for chemotherapeutic activity are not fully understood. The factors favoring protein binding are also those that would favor transport across membranes, tubular reabsorption, and increased binding to enzyme protein. N^4-Acetyl derivatives are more strongly bound to proteins and yet are better excreted. No universally applicable relationship has been found between half-life of sulfonamides and protein binding. Strongly bound drugs, such as the long acting sulfonamides, do not necessarily require high dosage. However it has been established in general that protein binding modulates bioavailability and prolongs the half-life of drugs.

4.4 Pharmacokinetics and Metabolism

4.4.1 SULFONAMIDES. The sulfonamide drugs vary widely in their pharmacokinetic

properties (Table 33.2). Some of them having additional acidic or basic groups are not absorbed from the gastrointesintal tract after oral administration, leading to a high local concentration of the drug in that area; they are, therefore, used for enteric infections. A majority of the sulfonamides, however, are well absorbed, mainly from the small intestine, and insignificantly from the stomach. Absorption occurs via the unionized form, in proportion to its lipid solubility. In rate and extent of absorption, most sulfonamides behave similarly within the pK_a range 4.5–10.5. After absorption they are fairly evenly distributed in all the body tissues. High levels are achieved in pleural, peritoneal, synovial, and ocular fluids that approximate 80% of serum levels; CSF levels are effective in meningeal infections. Those that are highly soluble do not, in general, attain a high tissue concentration, show no tendency to crystallize in the kidney, are more readily excreted, and are useful in treating genitourinary infections. The relatively less soluble ones build up high levels in blood, tissues, and extravascular fluids and are useful for treating systemic infections. This wide range of solubilities and pharmacokinetic characteristics of different sulfonamides permits the access of one or the other member of the group to almost any site in the body, thus adding greatly to their usefulness as chemotherapeutic agents. The free, non-protein bound drugs and their metabolic products are ultrafiltered in the glomeruli, then partly reabsorbed. Tubular secretion also plays an important role in the excretion of sulfonamides and their metabolites. The structural features of the compounds have a marked effect on these processes and determine the rate of excretion. The renal clearance rates of the metabolites are generally higher than are those of the parent drugs.

Metabolism of sulfonamides takes place primarily in the liver and involves mainly N^4-acetylation, to a lesser extent glucuronidation and to a very small degree, C-hydroxylation of phenyl and heterocyclic rings and of alkyl substituents, and O- and ring N-dealkylation. Variation of the substituents markedly influences the metabolic fate of the fulsonamides (Table 33.2); the metabolism also differs markedly in different animal species (247,287–291). Some of the sulfonamides, such as sulfisomidine, are excreted almost unchanged; in most of them N^4-acetylation occurs to a substantial degree, but some of the newer sulfonamides, such as sulfadimethoxine and sulfaphenazole, are excreted mainly as the glucuronide. The metabolites in human urine of seven of the commonly used sulfonamides given in Figure 33.3 reveal the wide variation in the metabolic patterns of sulfonamides (292).

Fujita (293) has performed regression analyses on the rates of metabolism and renal excretion of sulfonamides in terms of their substituent constants. Equations showing the best correlation indicate that the most important factor governing the rate-determining step of the hepatic acetylation is the hydrophobicity of the drug and that pK_a does not play a significant role. The excretion phenomenon seems to be more complex and would have to take into consideration additional parameters to give an acceptable correlation.

4.4.2 SULFONES. Dapsone is well-absorbed after oral administration, and is evenly distributed in almost all the body tissues. It is excreted mainly through the kidneys. Less than 5% is excreted unchanged, very little N-acetylation takes place, and most of it is present as the mono-N-glucuronide (88,294,295). Dapsone has a half life of about 20 h. Acedapsone following intramuscular injection is very slowly absorbed and deacetylated. It has a

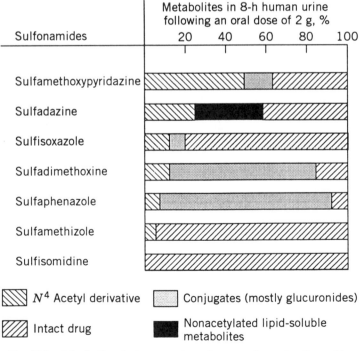

Fig. 33.3. Metabolites in human urine of commonly used sulfonamides.

half-life of about 42.6 days. It has been shown that there are marked species differences in the metabolism of dapsone; humans are relatively slow acetylators as compared to rhesus monkey (296,297). Similarly, mice deacetylate acedapsone efficiently, but rats do not (298).

4.4.3 HALF-LIFE. The half-life of sulfonamides is of great importance because the dosage regimen must be related to it. Dose schedule is a function of the molar activity and pharmacokinetic parameters. Kruger-Thiemer and his associates developed a mathematical model for these parameters and described a computer program for calculating them (299,300).

The half-life of different sulfonamides in clinical use vary widely, from 2.5 to 150 h (Table 33.2) and also show marked differences in different animal species. Reider (244) correlated the pK_a, liposolubility, surface activity, and protein binding of a group of 21 sulfonamides with their half-life in humans. He found that long-acting sulfonamides were, in general, more lipid soluble than were the short acting compounds, but no clear-cut relationship could be established; factors such as tubular secretion and tubular reabsorption are also involved. In 2-sulfapyrimidines, a 4-CH_3 group increases the half-life, 4,6-$(CH_3)_2$ reduces it to less than one-half, the corresponding methoxy derivatives have a much longer half-life, and both 5-CH_3 and 5-OCH_3 prolong half-life to the same extent. Similarly, in a 4-sulfapyrimidines, the 2,6-$(CH_3)_2$ derivative is short acting, 2,6-$(OCH_3)_2$ is long acting and the isomeric 5,6-$(OCH_3)_2$ is the most persistent sulfonamide known; sulfamethoxypyridazine has a half-life about twice as long as that of sulfapyrazine. Thus although no clear cut pattern of relationship between structure and half-life is discernible, the methoxy groups seem to prolong half-life.

5 PRESENT STATUS IN THERAPEUTICS

5.1 In Clinical Practice

Sulfonamides (and sulfones), identified as useful antibacterials, almost six decades ago, continue to be an important group of antimicrobial drugs. The number of conditions for which sulfonamides are drugs of first choice has declined on account of a gradual increase of resistance to them and the discovery of more effective antimicrobials with novel modes of action, but they still have a distinct and an important place in therapeutics (45,301–304). Although the antimicrobial spectrum of most of the sulfonamides is similar and the differences are primarily quantitative, individual sulfonamides differ markedly in their half-life and pharmacokinetic characteristics and thus provide a wide choice to meet the requirements of varied clinical situations. Other special features of sulfonamides that contribute to their clinically useful antimicrobial action are their synergistic action with dihydrofolate reductase inhibitors, their highly selective action on the microbe, relative freedom from problems of superinfection, activity by oral administration or topical application and relatively low cost. Around twenty sulfonamides and one sulfone are widely used in clinical practice.

Sulfonamides combined with trimethoprim are of value in the treatment of urinary tract infections, bacillary dysentery (particularly that caused by Shigella), salmonellosis, and chronic bronchitis. In meningococcal infections, sulfonamides are of value when the strains of N. meningitidis or H. influenzae are sensitive to them.

Sulfonamides are commonly used in preventing streptococcal infections and recurrence of rheumatic fever among susceptible subjects, especially in patients who are hypersensitive to penicillin.

Sulfacetamide sodium eye drops are employed extensively for the management of ophthalmic infections; a combination of topical and systematic application is of value in some conditions. Topical sulfonamides such as silver sulfadiazine and mafenide inhibit Enterobacteriacease, *P. aeruginosa*, staphylococci and streptococci and are extensively used to reduce the bacterial load in burn eschars.

Sulfonamides remain the drugs of choice for lymphogranuloma venereum, chancroid, and trachoma.

Sulfonamides have been found useful for the treatment of arterial infections due to *Listeria monocytogenes* and for prophylaxis of otitis media in children. A combination of sulfamethoxazole and trimethoprim has been found useful both for the prophylaxis and treatment of *Pneumocystis carinii* pneumonia, a common sequelae in patients with AIDS.

Sulfonamides, alone or combined with trimethoprim, are the drugs of choice in the treatment of infections due to Nocardia species, including cerebral nocardiosis. Sulfisoxazole, sulfamethoxazole, and sulfadiazine are the commonly used drugs. Treatment may need to be continued for some months for complete cure, and in advanced cases may need to be combined with another antibiotic.

Combined with pyrimethamine, sulformethoxine or sulfadiazine remain the drugs of choice for toxoplasmosis, including materno-fetal toxoplasmosis.

Combined with pyrimethamine, sulformethoxine, sulfamethoxy-pyridazine, and more recently, dapsone have been used both for the treatment and chemoprophylaxis of chloroquine-resistant falciparum malaria.

Dapsone remains the choice drug for all forms of leprosy and is an essential component of all multidrug therapy regimens used. Dapsone has also been reported to cure some cases of Crohn's disease, which may have a mycobacterial species origin.

5.1.1 OTHER CONDITIONS. Shortly after the introduction of sulfa drugs, sulfapyridine was found to have a unique beneficial effects on some inflammatory conditions, especially dermatologic, unrelated to their antibacterial activity (305). Later dapsone was found to share the same properties, and at a much lower dose and thus with an improved therapeutic index (306). The disorders that respond are dermatitis herpetiformis (DH), pyoderma gangrenosum, subcorneal pustular dermatosis, acrodermatitis continua, impetigo herpetiformis, ulcerative colitis, and cutaneous lesions of patients with lupus erythematosus; dapsone is the mainstay of treatment for DH (307). These disorders are characterized by edema followed by granulocytic inflammation or by vesicle or bullae formation (307). The mechanism of action is not fully understood, but it has been proposed that these drugs enter or influence the protein moiety of glycosaminoglycans and decrease tissue viscosity, resulting in prevention of edema, dilution of tissue fluid, and decrease in inflammation and vesicle and bullae formation. It is likely that this additional action of DDS may in part account for the extraordinary sensitivity of lepra bacilli to it (304). Salicylazosulfapyridine is the treatment of choice for ulcerative colitis.

5.2 Adverse Reactions

Though adverse drug reactions reported are numerous and varied, most are not serious (303,304). The minor adverse reactions reported include: GI reactions such as nausea, vomiting, and diarrhea; neurologic effects, such as peripheral neuritis, insomnia, and headache. Crystalluria, one of the earliest serious toxic reactions reported with sulfonamides, has been more or less overcome since the discovery of agents which are highly soluble at the pH of urine, or are excreted mainly as water soluble metabolites.

Blood dyscracias are rather uncommon; when they do occur, drug administration may need to be discontinued. These include hematologic reactions such as methemoglobinemia, agranulocytosis, thrombocytopenia, kernicterus in the newborn and hemolytic anemia in patients with G6PD deficiency. Kernicterus can result from administration of sulfonamides to the mother or to the newborn, because sulfonamides displace bilirubin from albumin in the newborn. Therefore, pregnant women near term or newborns should not be given sulfonamides. Hemolytic anemia is relatively more common with sulfone therapy in leprosy patients, and most often is related to the undernourished status of these patients; discontinuation of treatment is often not necessary and only supplemental therapy is required.

The most serious side effects are the hypersensitivity reactions such as urticaria, vasculitis, exfoliative dermatitis, angioedema, photosensitization, erythema nodosum and erythema multiformae exudativum (Stevens-Johnson type). The latter conditions is a particularly serious hazard of long-acting sulfonamides; when it occurs drug administration must be discontinued immediately. The morbiliform eruption is frequent in patients with AIDS being treated with pyrimethamine–sulfadoxine for *P. carinii* pneumonia and is associated with pancytopenia in some patients, and may be severe enough to require discontinuation of drugs (307). Individuals seropositive for HIV are more susceptible to developing adverse reactions to sulfonamides/sulfones; 40–80% as compared to 5% of HIV negative with other immune deficiencies (308,309).

By binding to albumin sites, sulfonamides may displace drugs such as warfarin, methotrexate, and hypoglycemic sulfonylurea drugs, and may thus potentiate the action of these drugs. Sulfonamide concentrations are increased by indomethacin, salicylates, and probenecid.

It has been suggested that adverse reactions of sulfonamides may be due to the formation of reactive hydroxylamine metabolites combined with a deficient glutathione system needed for scavenging these reactive molecules (310,311). This is supported by *in vitro* experiments in which sulfamethoxazole hydroxylamine has been found to be cytotoxic for lymphocytes, while the parent compound was not (312,313). In six healthy volunteers, the urinary recovery of sulfomethoxazole hydroxylamine accounted for 2.4% of the dose (1). With dapsone it has been shown that its hydroxylamine metabolite seems to be responsible for methaemoglobinemia; when dapsone is combined with cimetidine, an inhibitor of *N*-oxidation, the increase of methaemoglobinemia is reduced (314).

Acknowledgement

I would like to thank Dr. Ram Pratap, Scientist, Central Drug Research Institute, Lucknow and Dr. Anita Mehta, Scientist, Ranbaxy Research Centre, New Delhi for help during the preparation of the manuscript.

References

1. Hörlein, Dressel, and Kethe quoted by F. Mietzsch, *Chem. Ber.*, **71A**, 15 (1938).

2. M. Heidelberger and W. A. Jacobs, *J. Am. Chem. Soc.*, **41**, 2131 (1919).

3. F. Mietzsch and J. Klarer, Ger. Pat. 607, 537 (1935); *Chem. Abstr.*, **29**, 4135 (1935).

4. G. Domagk, *Deut. Med. Wochenschr.*, **61**, 250 (1935).

5. R. Foerster, *Zbl. Haut-u Geschlechtskr*, **45**, 549 (1933).

6. J. Trefouel, Mme J. Trfouel, F. Nitti, and D. Bovet, *C. R. Soc. Biol.*, **120**, 756 (1935).

7. L. Colebrook and M. Kenny, *Lancet* **1**, 1279 (1936).

8. G. A. H. Buttle, W. H. Grey, and D. Stephenson, *Lancet*, **1**, 1286 (1936).

9. A. T. Fuller, *Lancet*, **1**, 194 (1937).

10. P. Gelmo, *J. Prakt. Chem.*, **77**, 369 (1908); *Chem. Abstr.*, **2**, 2551 (1908).

11. E. K. Marshall, Jr., *J. Biol. Chem.*, **122**, 263 (1937).

12. A. C. Bratton and E. K. Marshall, Jr., *J. Biol. Chem.*, **128**, 537 (1939).

13. D. D. Woods, *Brit. J. Exp. Pathol.*, **21**, 74 (1940).

14. P. Fildes, *Lancet*, **1**, 955 (1940).

15. M. Tischler, "Molecular Modification in Modern Drug Research," in *Molecular Modification in Drug Design*, Advances in Chemistry Series 45, American Chemical Society, Washington, D. C. 1964, p. 1.

16. E. H. Northey, *The Sulfonamides and Allied Compounds*, American Chemical Society Monograph Series, Reinhold, New York, 1948.

17. M. L. Crossley, E. H. Northey, and M. E. Hultquist, *J. Am Chem. Soc.*, **61**, 2950 (1939).

18. U. S. Pat. 2, 311, 495 (1946) M. Dohrn and P. Diedrich; *Münch. Med. Wochenschr.*, **85**, 2017 (1938).

19. R. Pulver and H. Martin, *Arch. Exp. Pathol. Pharmacol.*, **201**, 491 (1943).

20. R. E. Flake, J. Griffin, E. Townsend, and E. M. Yow, *J. Lab. Clin. Med.*, **44**, 582 (1954).

21. U. S. Pat. 2, 833, 761 (1958) D. M. Murphy and R. G. Shepherd; *Chem. Abstr.*, **52**, 20, 216f (1958).

22. U. P. Basu, *J. Ind. Chem. Soc.*, **26**, 130 (1949).

23. M. L. Morre and C. S. Miller, *J. Am. Chem. Soc.*, **64**, 1572 (1942).

24. P. S. Winnek, G. W. Anderson, H. W. Marson, H. E. Faith, and R. O. Roblin, Jr., *J. Am. Chem. Soc.*, **64**, 1682 (1942).

25. F. Kurzer, *Chem. Rev.*, **50**, 1 (1952).

26. L. C. Leitch, B. E. Baker, and L. Brickman, *Can. J. Res.*, **23B**, 139 (1945).

27. L. E. H. Whitby, *Lancet*, **2**, 1210 (1938).

28. E. Bochni, B. Fust, J. Rieder, K. Schaerer, and L. Havas, *Chemotherapy*, **14**, 195 (1969).

29. R. O. Roblin, Jr., J. H. Williams, P. S. Winnek, and J. P. Englis, *J. Am. Chem. Soc.*, **62**, 2002 (1940).

30. W. T. Caldwell and H. B. Kime, *J. Am. Chem. Soc.*, **62**, 2365 (1940).

31. W. T. Caldwell, E. C. Kornfeld, and C. K. Donnell, *J. Am. Chem. Soc.*, **63**, 3028 (1941).

32. J. M. Sprague, L. W. Kissinger, and R. M. Lincoln., *J. Am. Chem. Soc.*, **63**, 3028 (1941).

33. R. O. Roblin, P. S. Winnek, and J. P. English, *J. Am. Chem. Soc.*, **64**, 567 (1942).

34. D. Lehr., *Proc. Soc. Exp. Biol. Med.*, **58**, 11 (1945); D. Lehr, R. Terranova, S. Blumenfeld, and M. L. Goldfarb, *Postgrad Med.*, **13**, 231 (1953).

35. E. M. Yow, *Ann. Intern. Med.*, **43**, 323 (1955).

36. T. Matsukawa, B. Ohta, and K. Shirakawa, *J. Pharm. Soc. Japan.*, **70**, 283 (1950); *Chem. Abstr.*, **45**, 2894 (1951).

37. R. J. Fosbinder and L. A. Walter, *J. Am. Chem. Soc.*, **61**, 2032 (1939).

38. J. Vonkennel, J. Kimmig, and B. Korth, *Z. Klin. Med.*, **138**, 695 (1940); *Klin Wochenschr.*, **20**, 2 (1941).

39. J. P. Bourque and J. Joyal, *Can. Med. Assoc. J.*, **68**, 337 (1953).

40. G. W. Anderson, H. E. Faith, H. W. Marson, P. S. Winnek, and R. O. Roblin, Jr., *J. Am. Chem. Soc.*, **64**, 2902 (1942).

41. E. M. Yow, *Am. Pract. Digest. Treat.*, **4**, 521 (1953).

42. E. Miller, J. M. Sprague, L.W. Kissinger, and L. F. McBurney, *J. Am. Chem. Soc.*, **62**, 2099 (1940).

43. A. Wacker, and S. Krischfeld, *Arzneim.-Forsch.*, **10**, 206 (1960).

44. R. L. Nichols, W. F. Jones, Jr., and M. Finland, *Proc. Soc. Exp. Bio. Med.*, **92**, 637 (1956).

45. L. Weinstein, M. A. Madoff, and C. M. Samet, *New Engl. J. Med.*, **263**, 793, 842, 900, 952 (1960).

46. U. S. Pat. 2,790,798 (1957) M. M. Lester and J. P. English; *Chem. Abstr.*, **51**, 15610 (1957).

47. W. Klotzer and H. Bretschneider, *Monatsh Chem.*, **87**, 136 (1956); *Chem. Abstr.*, **51**, 15610 (1957).

48. B. Fust and E. Bohni, in *Antibiotic Medicine in Clinical Therapy*, Supp. 1, Vol. 6 p. 3, 1959.

49. G. Hitzenberger, and K. H. Spitzky, *Med. Klin.* (Munich), **57**, 310 (1962); *Chem. Abstr.*, **57**, 5274 (1962).

50. M. Reber, G. Rutishauser, and H. Tholen, in H. P. Kuemmerle and R. Prezoiosi Eds., "Clearance-Untersuchungen am Menschen mit Sulfamethoxazol und Sulforthodimothoxin", 3rd International Congress on Chemotherapy, Stuttgart 1963, Vol. 1, Thieme Stuttgart, 1964, p. 648.

51. H. Horstmann, T. Knott, W. Scholtan, E. Schraufstatter, A Walter, and U. Worffel, *Arzneim-Forsch.*, **11**, 682 (1961).

52. B. Camerino and G. Palamidessi, *Gazz. Chim. Ital.*, **90**, 1815 (1960).

53. J. Tripod, L. Neipp, W. Padowtz, and W. Sackmann, *Antibiot. Chemother.*, **8**, 17 (1960).

54. B. Fust and E. Bohni, *Schweiz. Med. Wochenschr.*, **92**, 1599 (1962).

55. R. Deininger and H. Gutbrod, *Arzneim-Forsch.*, **10**, 612 (1960).

56. A. Eichhorn, *Zh, Pharm. Pharmakother, Laboratorinmsdiagn.*, **109**, 145 (1970); from *Ann. Rept. Med. Chem.*, 109 (1970).

57. L. Doub, U. Krolls, J. M. Vandenbelt, and M. W. Fisher, *J. Med. Chem.*, **13**, 242 (1970).

58. I. D. Jenkins, J. P. H. Verheyden, and J. G. Moffatt, *J. Am. Chem. Soc.*, **98**, 3346 (1976).

59. J. Castro-Pischel, M. T. Garcia-Laopez, and F. G. DeLasHeras, *Tetrahedron*, **43**, 383 (1987).

60. N. Rist, *Nature*, **146**, 838 (1940).

61. W. H. Feldman, H. C. Hinshaw, and H. E. Moses, *Am. Rev. Tuberc.*, **45**, 303 (1942).

62. E. V. Cowdry and C. Ruangsiri, *Arch. Pathol.*, **32**, 632 (1941).

63. S. G. Browne, *Advan. Pharmacol. Chemother.*, **7**, 211 (1969).

64. C. C. Shepard, D. H. McRae, and J. A. Habas, *Proc. Soc. Exp. Biol. Med.*, **122**, 893 (1966).

65. S. G. Browne, *Int. J. Lepr.*, **37**, 296 (1969).

66. J. Languillon. *Med. Trop.*, **24**, 522 (1964).

67. J. Schneider and J. Languillon, Abstracts, 8th International Congress on Leprology, Rio de Janeiro, 1963, 1964, p. 36.

68. E. F. Elslager, *Progr. Drug Res.*, **18**, 99 (1974).

69. M. I. Smith, E. L. Jackson, and H., Bauser, *Ann. N. Y. Acad. Sci.*, **52**, 704 (1949–1950).

70. B. R. Baker, M. V. Querry, and A.f. Kadish, *J. Org. Chem.*, **15**, 402 (1950).

71. N. Anand G. N. Vyas, and M. L. Dhar, *Current Science (India)*, **21**, 103 (1952); *J. Sci. Ind. Res. (India)*, **12B**, 353 (1953).

72. N. Anand, P. S. Wadia, and M. L. Dhar, *J. Sci. Ind. Res. (India)*, **12B**, 353 (1953); **13B**, 260 (1954).

73. W. J. Colwell, G. Chan, V. H. Brown, J. I. DeGraw, J. H. Peters, and N. E. Morrison, *J. Med. Chem.*, **17**, 142 (1974).

74. M. Saxena, A. K. Saxena, R. Raina, S. Chandra, A. B. Sen, N. Anand, J. K. Seydel, and M. Wiese, *Arzneim. Forsch./Drug Res.*, **39**, 1081 (1989).

75. W. H. Linnell and J. B. Stenlake, *J. Pharm. Pharmacol.*, **2**, 937 (1950).

76. E. D. Amstutz, *J. Am. Chem. Soc.*, **72**, 3420 (1950).

77. G. N. Vyas, N. Anand, and M. L. Dhar, *J. Sci. Ind. Res. (India)*, **13B**, 270 (1954); **14C**, 218 (1955).

78. J. K. Tandon, G. N. Vyas, and N. Anand, *J. Sci. Ind. Res. (India)*, **17B**, 192 (1958)

79. P. G. De Beneditti, D. Iarossi, U. Folli, C. Frassineti, M. C. Menziani, and C. Cennamo, *J. Med. Chem.*, **32**, 396 (1989).

80. P. G. De Beneditti, D. Iarossi, U. Folli, C. Frassineti, M. C. Menziani, and C. Cennamo, *J. Med. Chem.*, **30**, 459 (1987).

81. M. Wiese, J. K. Seydel, H. Pieper, G. Kruger, K. R. Noll, and J. Keck, *J. Quant. Struct-Act. Relat.*, **6**, 164 (1987).

82. M. G. Koehler, A. J. Hopfinger, and J. K. Seydel, *J. Mol. Structure (Theochem)*, **179**, 319 (1988).

83. S. K. Gupta and I. S. Mathur, *J. Sci. Ind. Res. (India)*, **16C**, 192 (1957).

84. H. Pieper, J. K. Seydel, G. Kruger, K. Noll, J. Neck, and M. Wiese, *Arzniem. Forsch./Drug Res.*, **39**, 1073 (1989).

85. Ng. Ph. Buu-Hoi, *Int. J. Lept.*, **22**, 16 (1954).

86. S. G. Browne and T. F. Davey, *Lepr. Rev.*, **32**, 194 (1961).

87. M. C. Khosla, P. S. Wadia, N. Anand, and M. L. Dhar, *J. Sci. Ind. Res. (India)*, **14C**, 152 (1955).

88. G. A. Ellard, *Brit. J. Pharmacol.*, **26**, 212 (1966).

89. S. K. Gupta, I. S. Mathur, and B. Mukherjee, *J. Sci. Ind. Res. (India)*, **18C**, 1 (1959).

90. M. C. Khosla, J. D. Kohli, and N. Anand, *J. Sci. Ind. Res. (India)*, **18C**, 51 (1959).

91. R. A. Hill and H. M. Goodwin, Jr., *5th Med. J. Nashville*, **30**, 1170 (1937); quoted by W. H. G. Richards in *Advan. Pharmacol. Chemother.*, **8**, 121 (1970).

92. L. T. Coggeshall, J. Maier, and C. A. Best, *J. Am. Med. Assoc.*, **117**, 1077 (1941).

93. H. M. Archibald and C. M. Ross, *J. Trop. Med. Hyg.*, **63**, 25 (1960).

94. S. P. Ramakrishnan, P. C. Basu, H. Singh, and N. Singh, *Bull. W. H. O.*, **27**, 213 (1962).

95. S. P. Ramakrishnan, P. C. Basu, H. Singh, and B. L. Wattal, *Indian J. Malariol.*, **17**, 141 (1963).

96. P. C. Basu, N. N. Singh, and N. Singh, *Bull. W. H. O.*, **31**, 699 (1964).

97. P. E. Thompson, A. Bayles, B. Olszewski, and J. A. Waitz, *Am. J. Trop. Med. Hyg.*, **14**, 198 (1965).

98. A. Bishop, *Parasitology*, **53**, 10p (1963).

99. E. F. Elslager, *Progr. Drug Res.*, **13**, 170 (1969).

100. A. J. Glazko, W. A. Dill, R. G. Montalbo, and E. L. Holmes, *Am. J. Trop. Med. Hyg.*, **17**, 465 (1968).

101. P. E. Thompson, B. Olszewski, and J. A. Waitz, *Am. J. Trop. Med. Hyg.*, **14**, 343 (1965)

102. E. F. Elslager, Z. B. Gavrilis, A. A. Phillips, and D. F. Worth, *J. Med. Chem.*, **12**, 357 (1969).

103. W. Peters, *Advance. Parasitol.*, **12**, 69 (1974).

104. A. B. Sabin and J. Warren, *J. Bacteriol.*, **41**, M50, 80 (1941).

105. E. Biocca, *Arg. Biol. (Sao Paulo)*, **7**, 27 (1943).

106. D. E. Eyles, *Ann. N. Y. Acad. Sci.*, **64**, 252 (1956–1957).

107. D. E. Eyles and N. Coleman, *Antibiot. Chemother.*, **5**, 529 (1955).

108. A. Werner, *Bol. Chil. Parasitol.*, **25**, 65 (1970).

109. R. E. McCabe and S. Oster, *Drugs*, **38**, (6), 937 (1989).

110. B. Fortier, F. Ajana, M. I. Pinto de Sousa, and co-worker, *Press Med.*, **20** (29), 1374 (1991).

111. P. P. Levine, *Cornell Vet.*, **29**, 309 (1939).

112. P. P. Levine, *J. Parasitol*, **26**, 233 (1940).

113. L. P. Joyner, S. F. M. Davies, and S. B. Kendall, "Chemotherapy of Coccidiosis", in R. J. Schnitzer and F. Hawking, Eds., *Experimental Chemotherapy*, Academic Press, New York, 1963, p. 445.

114. S. B. Kendall and L. P. Joyner, *Vet. Record.*, **70**, 632 (1958).

115. F. O. McCallum and G. M. Findlay, *Lancet*, **2**, 136 (1938).

116. W. G. Forster and J. R. McGibony, *Am. J. Ophithalmol.*, **27C**, 1107 (1944).

117. Centers for Disease Control, "Antibiotic-resistant strains of Neisseria gonorrhoeae: policy guidelines for detection management and control MMWR", **36**, Suppl IS-18S (1987).

118. R. E. Strauss, A. M. Kilgman, and D. M. Pillsbury, *Am. Rev. Tuberc.*, **63**, 441 (1951).

119. R. G. Connar, T. B. Ferguson, W. C. Sealy, and N. F. Conant, *J. Thorac. Surg.*, **22**, 424 (951).

120. R. A. Smego, M. B. Moeller, and H. A. Gallis, *Arch Intern. Med.*, **143**, 711 (1983).

121. P. Boiron, *Rev. Prat.*, **39** (9), 1983 (1987).

122. G. K. Herkes, J. Fryer, R. Rushworth and co-workers, *Aust. NZ J. Med.*, **19** (5), 475 (1989).

123. M. A. Fischl, G. M. Dickinson, and L. La Voie, *J. Am. Med. Assoc.*, **259**, 1185 (1988).

124. T. H. Maren, *Ann. Rev. Pharmacol. Toxicol.*, **16**, 309 (1976).

125. H. Southworth, *Proc. Soc. Exp. Biol. Med.*, **36**, 58 (1937).

126. T. Mann and D. Keilin, *Nature*, **146**, 164 (1940).

127. H. W. Davenport and A. E. Wilhelmi, *Proc. Soc. Exp. Biol. Med.*, **48**, 53 (1941).

128. W. B. Schwartz, *New Engl. Biol. Med.*, **240**, 173 (1949).

129. K. H. Beyter and J. E. Baer, *Pharmacol. Rev.*, **13**, 517 (1961).

130. P. W. Feit, O. B. T. Nielson, and H. Bruun, *J. Med. Chem.*, **15**, 437 (1972).

131. A. Small and E. J. Cafruny, *J. Pharmacol. Exp. Ther.*, **156**, 616 (1967).

132. M. Janbon, J. Chaptal, A. Vedel, and J. Schaap, *Montpellier Med.*, **21–22**, 441 (1942).

133. A Loubatieres, *Ann. N. Y. Acad. Sci.*, **71**, 4 (1957–1958).

134. H. Franke and J. Fuchs, *Deut. Med. Wochenschr.*, 1449 (1955).

135. J. C. Henquin, Fifty Years of Hypoglycemic Sulfonamides, *Rev. Fr. Endocrinol. Clin. Nutr. Metab.*, **34**, 255 (1993).

136. J. B. MacKenzie., C. G. MacKenzic, and E. V. McCollum, *Science*, **94**, 518 (1941).

137. K. H. Beyer, H. F. Russo, E. K. Tilson, A. K. Miller, W. F. Verwey, and S. R. Gass, *Am. J. Physiol.*, **166**, 625 (1951).

138. A. M. Doherty, *J. Med. Chem.*, **35**, 1493 (1992).

139. M. Clozel, V. Breu, K. Burri, J. M. Cassal, W. Fischli, G. A. Gray, G. Hirth, B. M. Loffler, M. Muller, W. Neidhart, and H. Ramuz, *Nature*, **365**, 759 (1993).

140. P. D. Stein, J. T. Hunt, D. M. Floyd and co-workers, *J. Med. Chem.*, **37**, 329 (1994).

141. M. F. Chan, I. Okun, F. L. Stavros, E. Hwang, M. E. Wolff, and V. N. Balaji, *Biochem. Biophys. Res. Comm.*, **201**, 228 (1994).

142. J. S. Lockwood, *J. Am. Med. Assoc.*, **111**, 2259 (1938).

143. C. M. MacLcod, *J. Exp. Med.*, **72**, 217 (1940).

144. D. A. Boroff, A. Cooper, and J. G. M. Bullowa, *J. Immunol.*, **43**, 341 (1942).

145. T. C. Stamp, *Lancet*, **2**, 10 (1939).

146. H. N. Green, *Brit. J. Exp. Pathol.*, **21**, 38 (1940).

147. S. Ratner, M. Blanchard, A. F. Coburn, and D. E. Green, *J. Biol. Chem.*, **155**, 689 (1944).

148. F. R. Selbie, *Brit. J. Exp. Pathol.*, **21**, 90 (1940).

149. G. M. Findlay, *Brit. J. Exp. Pathol.*, **21**, 356 (1940).

150. S. D. Rubbo and J. M. Gillepsie, *Nature*, **146**, 838 (1940).

151. R. Kuhn and K. Schwarz, *Berichte*, **74B**, 1617 (1941); *Chem. Abstr.*, **37**, 357 (1943).

152. K. C. Blanchard, *J. Biol. Chem.*, **140**, 919 (1941).

153. H. McIllwain, *Brit. J. Exp. Pathol.*, **23**, 265 (1942).

154. S. D. Rubbo, M. Maxwell, R. A. Fairbridge, and J. M. Gillespie, *Aust. J. Exp. Biol. Med. Sci.*, **19**, 185 (1941).

155. E. A. Bliss and P. H. Long, *Bull. John Hopkins Hosp.*, **69**, 14 (1941).

156. E. E. Snell and H. K. Mitchell, *Arch. Biochem.*, **1**, 93 (1943).

157. M. R. Stetten and C. L. Fox, Jr., *J. Biol. Chem.*, **161**, 333 (1945).

158. W. Shive, W. W. Ackermann, M. Gordon, M. E. Getzendaner, and R. E. Eakin, *J. Am. Chem. Soc.*, **69**, 725 (1947).

159. J. S. Gots, *Nature*, **172**, 256 (1953).

160. R. B. Angier, J. H. Boothe, B. L. Hutchings, J. H. Mowat, J. Semb, E. L. R. Stokstad, Y. Subba Row, C. W. Waller, D. B. Cosulich, M. J. Fahrenbach, M. E. Hultquist, E. Kuh, E. H. Northey, D. R. Seeger, J. P. Sickless, and J. M. Smith, Jr., *Science*, **103**, 667 (1946).

161. J. H. Mowat, J. H. Boothe, B. L. Hutchings, E. L. R. Stokstad, C. W. Waller, R. Angier, J. Semb, D. B. Consulich, and Y. Subba Row, *Ann. N. Y. Acad. Sci.*, **48**, 279 (1946–1947).

162. R. Tschesche, *Z. Naturforsch.*, **26b**, 10 (1947).

163. A. D. Welch and C. A. Nichol, *Ann. Rev. Biochem.*, **21**, 633 (1952).

164. M. Friedkin, *Ann. Rev. Biochem.*, **32**, 185 (1963).

165. D. D. Woods, "Relation of p-aminobenzoic Acid in Micro-organisms", in *Chemistry and Biology of Pteridines* (Ciba foundation Symposium), Little Brown, Boston, 1954, p. 220.

166. J. O. Lampen and M. J. Jones, *J. Biol. Chem.*, **166**, 435 (1946).

167. J. O. Lampen and M. J. Jones, *J. Biol. Chem.*, **170**, 133 (1947).

168. A. K. Miller, *Proc. Soc. Exp. Biol. Med.*, **57**, 151 (1944).

169. A. K. Miller, P. Bruno, and R. M. Berglund, *J. Bacteriol.*, **54**, G20, 9 (1947).

170. J. Lascelles and D. D. Woods, *Brit. J. Exp. Pathol.*, **33**, 288 (1952).

171. R. H. Nimmo-Smith, J. Lasceles, and D. D. Woods, *Brit. J. Exp. Pathol.*, **29**, 264 (1948).

172. T. Shiota and M. M. Disraely, *Biochim. Biophys. Acta.*, **52**, 467 (1961)

173. T. Shiota, M. N. Disraely, and M. P. McCann, *J. Biol. Chem.*, **239**, 2259 (1964).

174. G. M. Brown, R. A. Weisman, and D. A. Molnar, *J. Biol. Chem.*, **236**, 2534 (1961).

175. R. Weisman and G. M. Brown, *J. Biol. Chem.*, **239**, 326 (1964).

176. G. M. Brown, *J. Biol. Chem.*, **237**, 536 (1962).

177. L. Bock, G. H. Miller, K. J. Schaper, and J. K. Seydel, *J. Med. Chem.*, **17**, 23 (1974).

178. S. Roland, R. Ferone, R. J. Harvey, Y. L. Styles, and R. W. Morrison, *J. Biol. Chem.*, **254**, 10337 (1979).

179. R. D. Hotchkiss and A. H. Evans, *Fed. Proc.*, **19**, 912 (1960).

180. D. P. Richey and G. M. Brown, *J. Biol. Chem.*, **244**, 1582 (1969).

181. T. Shiota, C. M. Baugh, R. Jackson and R. Dillard, *Biochemistry*, **8**, 5022 (1969).

182. P. J. Ortiz, *Biochemistry*, **9**, 355 (1970).

183. B. L. Toth-Martinez, S. Papp, Z. Dinya, and F. J. Hernadi, *Biosystems*, **7**, 172 (1975); from H. H. W. Thijssen, *J. Med. Chem.*, **20**, 233 (1977).

184. P. J. Ortiz and R. D. Hotchkiss, *Biochemistry*, **5**, 67 (1966).

185. O. Okinaka and K. Iwai, *Anal. Biochem.*, **31**, 174 (1969).

186. L. Bock, W. Butte, M. Richter, and J. K. Seydel, *Anal. Biochem.*, **86**, 238 (1978).

187. E. C. Foernzler and A. N. Martin, *J. Pharm. Sci.*, **56**, 608 (1967).

188. B. Pullman and A. Pullman, *Quantum Biochemistry*, Wiley-Interscience, New York, 1963, p. 108.

189. I. Moriguchi and S. Wada, *Chem. Pharm. Bull. (Tokyo)*, **16**, 734 (1968).

190. E. Shefter, Z. F. Chmielewicz, J. F. Blount, T. F. Brennan, B. F. Sackman, and P. Sackman, *J. Pharm. Sci.*, **61**, 872 (1972).

191. R. Donovick, A. Bayan and D. Hamre, *Am. Rev. Tuberc.*, **66**, 219 (1952).

192. G. Brownlee, A. F. Green, and M. Woodbine, *Brit. J. Pharmacol.*, **3**, 15 (1948).

193. C. Levaditi, *C. R. Soc. Biol.*, **135**, 1109 (1941).

194. J. K. Seydel, M. Richter, and E. Wemple, *Int. J. Leprosy*, **48**, 18 (1977).

195. M. L. Panitch and L. Levy, *Lep. Rev.*, **49**, 131 (1978).

196. R. J. Cenedella and J. J. Jarrell, *Am. J. Trop. Med. Hyg.*, **19**, 592 (1970).

197. H. R. Morgan, *J. Exp. Med.*, **88**, 285 (1948).

198. J. W. Moulder, *The Biochemistry of Intracellular Parasitism*, University of Chicago Press, Chicago, 1962, p. 105.

199. L. M. Kurnosova and M. M. Lenkevich, *Acta Virol.*, **8**, 350 (1964).

200. L. P. Jones and F. D. Williams, *Can. J. Microbiol*, **14**, 933 (1968).

201. R. Ferone, *J. Protozool.*, **20**, 459 (1973).

202. R. D. Walter and E Konigk, *Hoppe-Seyler's Z. Physiol Chem.*, **355**, 431 (1974).

203. J. L. McCullough and T. H. Maren, *Mol. Pharmacol.*, **10**, 140 (1974).

204. L. Jaenicke and P. H. C. Chan, *Angew. Chem.*, **72**, 752 (1960).

205. H. Mitsuda and Y. Suzuki, *J. Vitaminol. (Kyoto)*, **14**, 106 (1968).

206. K. Iwai and O. Okinaka, *J. Vitaminol. (Kyoto)*, **14**, 170 (1968).

207. G. H. Hitchings, *Med. J. Aust.* (suppl), **1**, 5 (1973).

208. J. Greenberg, *J. Pharmacol, Exp. Ther.*, **97**, 484 (1949).

209. J. Greenberg and E. M. Richeson, *J. Pharmacol. Exp. Ther.*, **99**, 444 (1950).

210. J. Greenberg and E. M. Richeson, *Proc. Soc. Exp. Biol. Med.*, **77**, 174 (1951).

211. L. P. Garrod, D. G. James, and A. A. G. Lewis, *Postgrad. Med. J.* (Suppl. 45), 1 (1969).

212. M. Poe, *Science*, **194**, 533 (1976).

213. K. Bowden, A. D. Hall, B. Birdsall, J. Feeney, and G. C. K. Roberts, *Biochem. J.*, **258**, 335 (1989).

214. K. Bowden and A. D. Hall, *J. Chem. Res. Synop.*, 390 (1992).

215. J. K. Seydel and E. Wempe, *Chemotherapy*, **21**, 131 (1975).

216. S. R. M. Bushby and G. H. Hitchings, *Brit. J. Pharmacol. Chemother.*, **33**, 72 (1968).

217. J. J. Burchall and G. H. Hitchings, *Mol. Pharmacol.*, **1**, 126 (1965).

218. J. K. Seydel, M. Wiese, M. Sathish, K. Visser, M. Kansy, R. Haller, G. Kruegger, H. Pieper, K. R. Noll, and J. Keck, *Pharmacochem. Lib.*, **10**, 77 (1987).

219. M. Landy, N.W. Larkun, E. J. Oswald, and F. Strightoff, *Science*, **97**, 265 (1943).

220. P. J. White and D. D. Woods, *J. Gen. Microbiol.*, **40**, 243 (1965).

221. M. L. Pato and G. M. Brown, *Arch. Biochem. Biophys.*, **103**, 443 (1963).

222. B. Wolf and R. D. Hotchkiss, *Biochemistry*, **2**, 145 (1963).

223. R. Ho and L. Cormen, *Antimicrob. Agents Chemother.*, **5**, 388 (1974).

224. E. M. Wise. Jr. and M. M. Abou-Donia, *Proc. Nat. Acad. Sci. US*, **72**, 2621 (1975).

225. A. Bishop, *Biol. Rev.*, **34**, 445 (1959).

226. A. Bishop, in L. G. Goodwin and R. N. Nimmo-Smith, Eds., Drug, Parasites and Hosts, Little Brown, Boston, 1962, p. 98.

227. T. Watanabe, *Bacteriol. Rev.*, **27**, 87 (1963).

228. T. Watanabe and T. Fukasawa, *J. Bacterial.*, **82**, 202 (1961).

229. E. S. Anderson, in *Ecology and Epidemiology of Transferable Drug Resistance. Bacterial Episomes and Plasmids* (Ciba Foundation Symposium) G. W. W. Wolstenholme and M. O'Connor, Eds., Churchill. London, 1969, p. 102.

230. S. N. Cohen and C. A. Miller, *Proc. Nat. Acad. Sci. US*, **67**, 510 (1970); *J. Mol. Biol.*, **50**, 671 (1970).

231. T. Aoki, Y. Mitoma, and J. H. Crosa, *Plasmid*, **16**, 213 (1986).

232. P. Radstrom, G. Swedberg, and O. Skold, *Antimicrob. Agents Chemother.*, **35**, 1840 (1991).

233. P. Radstrom, C. Fermer, B. E. Kristiansen, A. Jenkins, O. Skold, and G. Swedberg, *J. Bact.*, **174**, 6386 (1992).

234. J. Maynard-Smith, C. G. Dowson, and B. G. Spratt, *Nature (London)*, **349**, 29 (1991).

235. B. E. Kristiansen, P. Radstrom, A. Jenkins, E. Ask, B. Facinelli, and O. Skold, *Antimicrob. Agents Chemother.*, **34**, 2277 (1990).

236. F. Guerineau and P. Mullineaux, *Nucleic Acid Res.*, **17**, 4370 (1989).

237. R. M. Hall and C. Vockler, *Nucleic Acid Res.*, **15**, 7491 (1987).

238. L. Sundstrom, P. Radstrom, G. Swedberg, and O. Skold, *Mol. Gen. Genet.*, **213**, 191 (1988).

239. J. K. Seydel and E. Wempe, *Arzneim-Forsch.*, **14**, 705 (1964).

240. J. K. Seydel, E. Kruger-Thiemer, and E. Wempe, *Z. Naturforsch.*, **15b**, 620 (1960).

241. C. L. Fox. Jr. and H. M. Rose, *Proc. Soc. Exp. Biol. Med.*, **50**, 142 (1942).

242. F. C. Schmelkes, O. Wyss, H. C. Marks, B. J. Ludwig, and F. B. Stranskov, *Proc. Soc. Expl. Biol, Med.*, **50**, 145 (1942).

243. P. H. Bell and R. O. Roblin, Jr., *J. Am. Chem. Soc.*, **64**, 2905 (1942).

244. J. Rieder, *Arzneim-Forsch.*, **13**, 81, 89, 95 (1963).

245. T. Struller, "Progress in Sulfonamide Research", *Prof. Drug Res.*, **12**, 389–457 (1968).

246. S. Budavari, ed., *The Merck Index*, Merck & Co., Rahway, N. J., 1989.

247. R. T. Williams and D. V. Parke, *Ann. Rev. Pharmacol*, **4**, 85 (1964).

248. L. Neipp and R. L. Meyer, *Ann. N. Y. Acad. Sci.*, **69**, 447 (1957).

249. J. E. F. Reynolds, Ed., *Martindale, The Extra Pharmacopoea*, 29th ed., Pharmaceutical Press, London, 1989.

250. P. B. Cowles, *Yale J. Biol. Med.*, **14**, 599 (1942).

251. A. H. Brueckner, *Yale J. Biol. Med.*, **15**, 813 (1943).

252. G. Schwenker, *Arch. Pharm. (Weinheim)*, **295**, 753 (1962).

253. A. Rastelli, P. G. DeBenedetti, A. Albasini, and P. G. Pecorari, *J. Chem. Soc. Perkin Trans.*, **2**, 522 (1975).

254. R. G. Shepherd, A. C. Bratton, and K. C. Blanchard, *J. Am. Chem. Soc.*, **64**, 2532 (1942).

255. R. C. Shepherd and J. P. English, *J. Org. Chem.*, **12**, 446 (1947).

256. T. A. Mastrukova, Y. N. Sheinker, I. K. Kuznetsova, and M. I. Kabachnik, *Tetrahedron*, **19**, 357 (1963).

257. T. Uno, K. Machida, K. Hanai, M. Ueda, and S. Sasaki, *Chem. Pharm. Bull. (Tokyo)*, **11**, 704 (1963).

258. T. Uno, K. Machida, and K. Hanai, *Chem. Pharm. Bull. (Tokyo)*, **14**, 756 (1966).

259. J. K. Seydel, *J. Pharm. Sci.*, **57**, 1455 (1967).

260. A. Cammarata and R. C. Allen, *J. Pharm. Sci.*, **56**, 640 (1967).

261. J. K. Seydel, *Arzneim-Forsch.*, **16**, 1447 (1966).

262. R. S. Schnaare and A. N. Martin, *J. Pharm. Sci.*, **54**, 1707 (1965).

263. A. Rastelli, P. G. DeBenedetti, G. G. Battistuzzi, and A. Albasini, *J. Med. Chem.*, **18**, 963 (1975).

264. J. K. Seydel, *Mol. Pharmacol.*, **2**, 259 (1966).

265. E. R. Garrett, J. B. Mielck, J. K. Seydel, and H. J. Kessler, *J. Med. Chem.*, **14**, 724 (1971).

266. J. K. Seydel, *J. Med. Chem.*, **14**, 724 (1971).

267. T. Fujita and C. Hansch, *J. Med. Chem.*, **10**, 991 (1967).

268. M. Yamazaki, N. Kakeya, T. Morishita, A. Kamada, and A. Aoki, *Chem. Pharm. Bull. (Tokyo)*, **18**, 702 (1970).

269. G. H. Miller, P. H. Doukas, and J. K. Seydel, *J. Med. Chem.*, **15**, 700 (1972).

270. A. Koch, J. K. Seydel, A. Gasco, C. Tirani, and R. Fruttero, *Quant. Struct-Act. Relat.*, **12**, 373 (1993).

271. R. L. Lopez de Compadre, R. A. Pearlstein, A. J. Hopfinder, and J. K. Seydel, *J. Med. Chem.*, **30**, 900 (1987); A. J. Hopfinder, R. L. Lopez de Compadre, M. G. Koshler, S. Emery, and J. K. Seydel, *Quant. Struct-Act. Relat.*, **6**, 111 (1987).

272. P. G. De Beneditti, *Prog. Drug. Res.*, **36**, 361 (1991).

273. M. Cocchi, D. Iarossi, M. C. Menziani, P. G. De Beneditti, and C. Frassineti, *Struct. Chem.*, **3**, 129 (1992).

274. A. Yu. Sokolov, M. C. Menziani, M. Cocchi, and P. G. Beneditti, *Theo. Chem.*, **79**, 293 (1991).

275. Ref. 16, p. 458.

276. D. Lehr, *Ann. N. Y. Acad. Sci.*, **69**, 417 (1957).

277. G. L. Biagi, A. M. Barbaro, M. C. Guerra, G. C. Forti, and M. E. Fracasso, *J. Med. Chem.*, **17**, 28 (1974).

278. W. Scholtan, *Arzneim-Forsch.*, **14**, 348 (1964); **18**, 505 (1968).

279. E. Kruger-Thiemer, W. Diller, and P. Bunger, *Antimicrob Agents Chemother.*, 183 (1965).

280. K. Irmscher, D. Gabe, K. Jahnke and W. Scholtan, *Arzneim-Forsch.*, **16**, 1019 (1966).

281. W. Scholtan, *Chemotherapia*, **6**, 180 (1963).

282. J. A. Shannon, *Ann. N. Y. Acad. Sci.*, **44**, 455 (1943).

283. J. K. Seydel, "Physicochemical Approaches to the Rational Development of New Drugs", in E. J. Ariens, Ed., *Drug Design*, Vol. 1, Academic Press, New York, 1971, p. 343.

284. O. Jardetzky and N. G. Wade-Jardetzky, *Mol. Pharmacol.*, **1**, 214 (1965).

285. B. K. Martin, *Nature*, **207**, 274 (1965).

286. I. Moriguchi, S. Wada, and T. Nishizawa, *Chem. Pharm. Bull (Tokyo)*, **16**, 601 (1968).

287. H. Nogami, A. Hasegawa, M. Hanano, and K. Imaoka, *Yakugoku Zazzi*, **88**, 893 (1968).

288. M. Yamazaki, M. Aoki, and A. Kamada, *Chem. Pharm. Bull (Tokyo)* **16**, 707 (1968).

289. K. Kakemi, T. Arita, and T. Koizumi, *Arch. Pract. Pharm.*, **25**, 22 (1965).

290. T. Koizumi, T. Arita, and K. Kakemi, *Chem. Pharm. Bull. (Tokyo)*, **12**, 428 (1964).

291. R. H. Adamson, J. W. Bridges, M. R. Kibby, S. R. Walker and R. T. Williams, *Biochem. J.*, **118**, 41 (1970).

292. G. Zbinden, "Molecular Modification in the Development of Newer Anti-Infective Agents: The Sulfa Drugs", in R. F. Gould, Ed., *Molecular Modification in Drug Design*. American Chemical Society, Washington, D. C., 1964, p. 25.

293. T. Fujita, "Substituent-Effect Analyses of the Rates of Metabolism and Excretion of Sulfonamide Drugs", in R. F. Gould, Ed., *Biological Correlations The Hansch Approach*, American Chemical Society, Washington, D. C. 1972, p. 80.

294. S. R. M. Bushby and A. J. Woiwood, *Am Rev. Tuberc. Pulmo Dis.*, **72**, 123, (1955).

295. S. R. M. Bushby and A. J. Woiwood, *Biochem. J.*, **63**, 406 (1956).

296. H. B. Hucker, *Ann. Rev. Pharmacol.*, **10**, 99 (1970).

297. G. R. Gordon, J. H. Peters, R. Gelber, and L. Levy, *Proc. West. Pharmacol. Soc.*, **13**, 17 (1970).

298. P. E. Thompson. *Int. J. Lepr.*, **35**, 605 (1967).

299. E. Kruger-Thiemer and P. Bunger, *Arzeim-Forsch.*, **16**, 1431 (1961).

300. E. Kruger-Thiemer and P. Bunger, *Arzneim-Forsch.*, **11**, 867 (1961).

301. L. P. Garrod, D. G. James, and A. A. G. Lewis, *Postgrad. Med. J.* (suppl.), **45**, 1 (1969).

302. FDA Drug Efficacy Reports, through *J. Am. Pharm. Assoc.*, **NS9**, 535 (1969).

303. R. Berkow, Ed., *The Merck Manual*, Merck & Co. Inc. N. J. Parkway, 1987, pp. 43–44.

304. H. C. Neu, in *Harrison's Principles of Internal Medicine*, 12th Ed., McGraw Hill, Inc. New York, 1991, pp. 491–492.

305. M. J. Costells. *Arch. Dermatol. Syph.*, **42**, 161 (1940).

306. A. L. Lorinez and R. W. Pearson, *Arch. Dermatol.*, **85**, 2 (1962).

307. O. J. Stone, *Medical Hypothesis.*, **31**, 99 (1990).

308. A. J. M. Van Der Ven, P. P. Koopmans, T. B. Vree, and J. W. M. Van Der Mer, *J. Antimicrob. Chemother.*, **34**, 1 (1994).

309. J. A. Kovacs, J. W. Hiemens, and A. M. Macher, *Ann. Int. Med.*, **100**, 663 (1984).

310. N. H. Sheer, S. P. Speilberg, D. M. Grant, B. K. Tang and W. Kalow, *Ann. Int. Med.*, **105**, 179 (1986)

311. M. J. Rieder, J. Uetrecht, N. H. Shear, M. Cannon, M. Miller, and P. Spielberg, *Ann. Int. Med.*, **110**, 286 (1989).

312. M. J. Rieder, J. Uetrecht, N. H. Shear, and S. P. Spielberg, *J. Pharmacol. Expt. Ther.*, **244**, 724 (1988).

313. M. J. Rieder, E. Sisson, I. Bird, and W. Y. Almawi, *Int. J. Immunopharmacol.*, **14**, 1175 (1992).

314. M. D. Coleman, A. K. Scott, A. M. Breckenridge, and B. K. Park, *Brit. J. Clin. Pharmaco.*, **30**, 761 (1990).

315. J. K. Seydel, "Molecular Basis for the Action of Chemotherapeutic Drugs, Structure-Activity Studies on Sulfonamides", in E. J. Ariens, Ed., *Physico Chemical Aspects of Drug Action*, Pergamon Press, New York, 1968, p. 169; "Kinetics of Antibacterial Effects", in *Handbook Exp. Pharmacol.*, **64**, 129 (1983).

316. P. G. De Beneditti, *Advances Drug Res.*, **16**, 227 (1987); *Prog. Drug Res.*, **36**, 361 (1991).

317. P. G. Sammes, "Sulfonamides and Sulfones", in C. Hansch, Ed., *Comprehensive Medicinal Chemistry*, 1st ed., Vol. II, Pergamon Press, Oxford, 1990, pp. 255–270.

CHAPTER THIRTY-FOUR

Antimycobacterial Agents

PIERO SENSI

University of Milan
Milan, Italy

GIULIANA GIALDRONI GRASSI

University of Pavia
Pavia, Italy

CONTENTS

1 Introduction, 576
2 Mycobacteria, 576
3 Pathogenesis and Epidemiology, 582
4 Laboratory Models for Screening and Evaluating
  Antimycobacterial Agents, 586
5 Search for and Discovery of Antimycobacterial
  Drugs, 588
6 Available Antimycobacterial Drugs and Related
  Products, 591
  6.1 Synthetic products. 592
      6.1.1 Sulfones, 592
      6.1.2 p-Amino-salicylic acid and analogs, 593
      6.1.3 Thioacetazone, thiocarlide, and
            thiambutosine, 594
      6.1.4 Isoniazid, ethionamide, and
            pyrazinamide, 595
      6.1.5 Clofazimine, 600
      6.1.6 Ethambutol, 601
  6.2 Antibiotics, 603
      6.2.1 Aminoglycoside antibiotics (streptomycin,
            kanamycin, and amikacin), 603
      6.2.2 Viomycin and capreomycin, 607
      6.2.3 Cycloserine, 609
      6.2.4 Rifamycins: rifampicin (rifampin),
            rifabutin, and rifapentine, 610
  6.3 Other drugs under investigation, 616
      6.3.1 Fluoroquinolones, 616
      6.3.2 Macrolides, 618
      6.3.3 β-Lactams, 621
7 Present Status of the Chemotherapy of
  Tuberculosis and Leprosy, 622

Burger's Medicinal Chemistry and Drug Discovery,
Fifth Edition, Volume 2: Therapeutic Agents,
Edited by Manfred E. Wolff.
ISBN 0-471-57557-7   © 1996 John Wiley & Sons, Inc.

1 INTRODUCTION

Some species of mycobacteria are pathogenic for several animal species and are responsible for two important human chronic diseases, tuberculosis and leprosy, as well as for other less widespread but severe infections, traditionally although improperly called atypical mycobacterioses. *Mycobacterium leprae* (identified by Hansen in 1871) and *M. tuberculosis* (identified by Koch in 1882) were among the first bacteria recognized as causative agents, respectively, of leprosy and tuberculosis. The dramatic importance of these two still present illnesses is well known: over the past 200 years, tuberculosis was responsible for the death of 1000 million people (1) and leprosy has been one of the most terrifying diseases since antiquity, and its stigma persists in virtually all cultures in some form (2).

More recently, diseases caused by other mycobacteria, like *M. avium* complex in immunodepressed patients, have become of increasing importance (3). In spite of the early discovery of the etiological agents of the infections, only in the past five decades have drugs highly effective in the treatment of mycobacterial diseases been discovered.

The introduction of chemotherapy for mycobacterial infections brought about a dramatic decrease in the mortality and morbidity of the illnesses. These successes notwithstanding, mycobacterial infections still require particular attention as a worldwide, challenging health problem. One important area of research in fighting these diseases is to determine the best way to use available drugs. Clinical evaluation of the various chemotherapeutic regimens is a complex problem.

Although some general guidelines for the therapy of these diseases are accepted everywhere, great effort must be made to adapt these guidelines to different socio-economical situations and to pathological variants. It is still necessary to search for new antimycobacterial agents for many reasons, e.g., infecting organisms became resistant to present drugs, some drugs have side effects, and the present treatment of mycobacterioses in immunodepressed patients is unsatisfactory.

Research programs for the discovery of new antimycobacterial drugs and for improving the evaluation criteria are under way in many laboratories. In addition, knowledge of specific constituents of the mycobacterial cell and their biochemical roles has advanced considerably in the recent years and may permit a more rational approach to the design of new drugs acting on specific targets. Also, recent improvements in the knowledge of the mechanism of action of the available drugs and the biochemical mechanisms of resistance to them may be used as a basis for designing new and better weapons to fight the mycobacterial diseases.

2 MYCOBACTERIA

Mycobacteria are transition forms between bacteria and fungi. The genus *Mycobacterium* belongs to the order Actinomycetales and the family Mycobacteriaceae; it is characterized by nonmotile, nonsporulating rods that resist decolorization with acidified organic solvents (4). For this reason, they are also called "acid fast" bacteria. Some mycobacterial species are pathogenic for humans. Among these, *M. tuberculosis hominis*, *M. tuberculosis bovis*, and *M. leprae* are the most important. Another variety, *M. africanum*, endowed with characteristics intermediate between those of *M. tuberculosis hominis* and *M. bovis*, is also pathogenic for humans (5).

Other mycobacterial species that resemble *M. tuberculosis* in some morphologic aspects and cultural requirements but show little or no pathogenic effect in humans (at least in absence of some underlying conditions) were improperly defined in the past as "atypical mycobacteria." They were classified by Runyon (6) into four groups according to their growth rates and pigment production. This classification, however, is inadequate to define the different species in a clear-cut way. The development of recombinant DNA technology has allowed us to define better the species belonging to these groups, but the general definition still does not seem satisfactory. At present, the most commonly adopted definitions are mycobacteria other than tuberculosis (MOTT) and nontuberculosis mycobacteria (NTM), while the term *mycobacteriosis* is proposed for the diseases caused by these organisms (7–10). A classification that takes into consideration the capacity of these species in producing disease in humans is given in Table 34.1 (11).

M. tuberculosis hominis is a nonmotile bacillus 1–2 μm long and 0.3–0.6 μm wide. It can be demonstrated in pathologic specimens by means of specific staining procedures, the most widely used being the Ziehl-Neelsen method. It is acid fast and acid alcohol fast. *In vitro* culture of tubercle

bacilli is slow and sometimes difficult. The nutritional requirements are not particularly complex, but the content of the medium greatly influences the composition of the mycobacterial cell. The most common media used for the isolation of *M. tuberculosis* from pathological specimens and for its maintenance are solid media, with egg yolk as a base (Petragnani, Lowenstein-Jensen, and IUTM media). In these media, the culture begins to appear 12–15 days after inoculation, but full growth is obtained after 30–40 days. When inoculation is made with pathological material from patients, observation must be prolonged. The most widely used maintenance liquid media are synthetic media containing albumin (Dubos, Youmans, and 7HT media). They usually allow rapid growth (8–10 days), and addition of Tween 80 makes it possible to obtain uniformly dispersed growth (12,13).

Recent laboratory diagnostic methods of mycobacterial infections and of susceptibility patterns to antimicrobial drugs include, in addition to culture in solid and

Table 34.1 Mycobacteria Other Than *M. tuberculosis*

Group	Species
Pathogenic in humans	*M. leprae*
Potentially pathogenic in humans	*M. avium-intracellulare, M. kansasii, M. fortuitum-chelonae* complex, *M. scrofulaceum, M. xenopi, M. szulgai, M. malmoense, M. simiae, M. marinum, M. ulcerans, M. haemophilum*
Saphrophytic mycobacteria rarely causing disease in humans	
Slow growth rate	*M. gordonae, M. asiaticum, M. terrae-triviale* complex, *M. gastri, M. nonchromogenicum, M. paratuberculosis*
Intermediate growth rate	*M. flavescens*
Rapid growth rate	*M. thermoresistible, M. smegmatis, M. vaccae, M. parafortuitum* complex, *M. phlei*

liquid media, radiometric methods that measure the evolution of $^{14}CO_2$ from a ^{14}C-labeled substrate, antigenic assays by enzyme-linked immunosorbent assays (ELISA), DNA probes, nucleic acid amplification methods, and restriction fragment length polymorphism (RFLP) analysis of genomic DNA (14).

In vivo pathogenic activity of tubercle bacilli is demonstrated in guinea pigs. Infection in the rabbit, which is susceptible to *M. bovis* infections but scarcely or not at all to *M. hominis*, is employed to differentiate the two infections. *M. leprae*, or the Hansen bacillus, is resistant to acid and to alcohol and requires the Ziehl-Neelsen method of staining to be recognized. It is found in lepromatous lesions, where it is arranged mostly in clumps. It is 1–8 μm long, nonmotile, and nonsporeforming. The most important drawback for the accumulation of knowledge about the biology, the susceptibility to antibiotics and the epidemiology of the disease is the impossibility of cultivating this mycobacterium *in vitro*. Tests in animals have been improved in the recent years, but they are complex and can be performed only in the specialized laboratories.

The biochemical constitution of mycobacteria is complex, and an enormous amount of work has been done in this field. Novel chemical structures have been discovered, but the relationship between these and the pathogenic and biological activities of mycobacteria have not yet been satisfactory elucidated (15–21). Information about metabolism of mycobacteria is extremely voluminous, but the overall picture of the mycobacterial metabolism is far from complete.

Metabolism of carbohydrates and lipids, electron transport, and oxidative phosphorylation have been extensively studied; research into nucleic acids and protein synthesis is at a less advanced stage. In those mycobacteria that have been studied in detail, the genome has been found to consist of a single length of DNA in the form of a closed loop. The genome is not contained by a nuclear membrane, although the tightly packed DNA is recognizable on electron microscopy as a nuclear body. Genome size of mycobacteria is larger than that of other prokaryotes, in the range of 2.8×10^9 to 4.5×10^9 bp. The DNA of most mycobacteria have between 64 and 70 mol% guanine plus cytosine (22).

To go deeply into these subjects is beyond the scope of this book. However, it is appropriate to indicate here some metabolic aspects and functional structures specific to mycobacteria that might be potential targets for antimycobacterial drugs.

Some interesting differences exist in the metabolic properties of tubercle bacilli grown *in vivo* and *in vitro*, which must be considered when the practical value of antimycobacterial agents designed and tested in laboratory is assessed in practice. Populations of *M. tuberculosis* H37 Rv grown in the lungs of mice and populations grown *in vitro* show two different phenotypes, Phe I and Phe II, with marked differences in the metabolism of certain energy sources, in the production of detectable sulfolipids, and in immunogenicity (Phe II is a better immunogen than Phe I) (23–25). Since the shift from Phe I to Phe II is readily reversible, it can be deduced that the genome of H37 Rv remains the same (26). Phe I is unable to bind neutral red (27,28) and is resistant to 4% sodium hydroxide at 37°C for 4 h (29–31), suggesting that a modification in its surface has occurred, probably because of the presence of a coating layer, rendering the surface components of Phe II that react with neutral red unavailable (28,32).

Further differences have been observed between virulent and avirulent strains of mycobacteria. The differences in oxidative metabolism are quantitative and cannot account for the capacity of virulent mycobacteria to grow in host tissue, where the oxygen tension is low. By contrast, there are some qualitative differences in amino acid metabolism. The virulent strains (H37 Rv) possess one type of asparaginase, and the avirulent one (H37 Ra) possesses two asparaginases. The avirulent strain has an aspartotransferase that transfers the aspartyl moiety of asparagine to hydroxyl-amine, whereas the avirulent strain does not (33–35).

A great deal of research effort has been focused on the constituents of the mycobacterial cell wall, since they are responsible for many of the pathogenic effects of tubercle bacilli. The cores of the mycobacterial cell wall is the mycolylarabinogalactanpeptidoglycan constituted by three covalently attached macromolecules (peptidoglycan, arabinogalactan, and mycolic acid).

A schematic structure, simplified from (36), is outlined in structure **1**. The peptidoglycan (or murein) consists of a repeating disaccharide unit, in which N-acetyl-D-glucosamine (G) is linked in a 1–4 linkage to N-glycolyl-D-muramic acid (M) attached to L-alanine-D-glutamic acid-NH_2-*meso*-diaminopimelic acid-NH_2-D-alanine. This unit is linked to a polysaccharide unit, the arabinogalactan, via a disaccharide phosphate (rhamnose–galactose phosphate). The arabinogalactan is connected to a glycolipidic region constituted of esters of mycolic acids.

Mycolic acids are α-branched, β-hydroxylated long-chain fatty acids, of which three principal groups are known: the

$$
\begin{array}{ccc}
\text{Myc} & & \text{Myc} \\
| & & | \\
-\text{Gal}-\text{Ara}-\text{Gal}-\text{Ara}-\text{Gal}- \\
& & | \\
& & \text{\textcircled{P}} \\
& & | \\
-\text{M}-\text{G}-\text{M}-\text{G}-\text{M}- \\
| & | & | \\
\text{L-Ala} & \text{L-Ala} & \text{L-Ala} \\
| & | & | \\
\text{D-Glu} & \text{D-Glu} & \text{D-Glu} \\
| & | & | \\
m\text{-DAP} & m\text{-DAP} & m\text{-DAP} \\
| & | & | \\
\text{D-Ala} & \text{D-Ala} & \text{D-Ala}
\end{array}
$$

(1)

corynomycolic acids, ranging from C_{28} to C_{36}; the nocardic acids, ranging from C_{40} to C_{60}; and the mycobacterial mycolic acids, ranging from C_{60} to C_{90} (37,38). Mycolic acids can also be detected in the skin lesions of patients suffering from lepromatous leprosy, indicating that the agent of leprosy is a mycobacterium containing it (39). The chemical structures of methoxylated mycolic acid and β-mycolic acid extracted from *M. tuberculosis* var. *hominis* are shown in structures **2** and **3** (40). The mycolic acids are linked through their carboxy groups to the end terminal 5-OH groups of the D-arabinofuranose (Ara_f) molecules, branches of the arabinogalactan

$$
CH_3-(CH_2)_{17}-\underset{\underset{CH_3}{|}}{CH}-\overset{\overset{OCH_3}{|}}{CH}-(CH_2)_{10}-\underset{\underset{CH_2}{\diagup}}{CH}\overset{}{\diagdown}CH-(CH_2)_{17}-\overset{\overset{OH}{|}}{CH}-\underset{\underset{C_{24}H_{49}}{|}}{CH}-COOH
$$

(2)

$$
CH_3-(CH_2)_{17}-\underset{\underset{CH_3}{|}}{CH}-\overset{\overset{O}{\|}}{C}-[C_{17}H_{34}]-\underset{\underset{CH_2}{\diagup}}{CH}\overset{}{\diagdown}CH-(CH_2)_{19}-\overset{\overset{OH}{|}}{CH}-\underset{\underset{C_{24}H_{49}}{|}}{CH}-COOH
$$

(3)

of the cell wall (41–43). Mycolic acids are known to be acid fast, since they bind fuchsin and the binding is acid fast. Thus it seems that the acid fastness of mycobacteria depends on two mechanisms: the capacity of the mycobacterial cell to take fuchsin into its interior and the capacity of mycolic acids to form a complex with the dye (21,44,45).

In addition to the lipid murein part of the rigid structure, there is a series of soluble lipid compounds that seem to be located in or on the outer part of the cell wall: lipoarabinomannan, waxes D, cord factor, mycosides, sulfolipids, and phospholipids (16–18). Lipoarabinomannan is an essential part of the cell envelop, which lacks covalent association with the cell wall core. In this macromolecule, arabinan chains are attached to a mannan backbone, which is, in turn, attached to a phosphatidylinositol esterified with fatty acids like palmitate and 10-methyloctadecanoate (tuberculostearate).

The so-called waxes D are ether-soluble, acetone-insoluble, chloroform-extractable peptidoglycolipid components of the mycobacterial cell, probably autolysis products of the cell wall (46). Since from a chemical point of view they are esters of

mycolic acids with arabinogalactan linked to a mucopeptide containing N-acetyl-glucosamine, N-glycolylmuramic acid, L- and D-alanine, meso-diaminopimelic acid, and D-glutamic acid, it has been suggested that they are materials synthesized in excess of those needed for insertion into the cell wall (18,21,47). The constitution of wax D differs in different varieties and strains of mycobacteria.

Cord factor is a toxic glycolipid (6,6'-dimycolate of trehalose; structure **4**), which has been deemed to be responsible for the phenomenon of cording (48,49), i.e., the capacity of *M. tuberculosis* to grow in serpentine cords, a capacity that is correlated with its capacity to kill guinea pigs (50,51). The detergent properties of cord factor and its location on the outer cell wall have led to the suggestion that it may play a role in facilitating the penetration of certain molecules necessary for growth of mycobacteria (21).

Mycosides are glycolipids and peptidoglycolipids type specific of mycobacteria (52) that often have in common terminal saccharide moietes containing rhamnoses O-methylated in various positions (53). They can be divided into two main categories (38): phenolic glycolipids with

(4)

branched-chain fatty acids, and peptido-glycolipids consisting of a sugar moiety, a short peptide, and a fatty acid. The biological activity of mycosides is still obscure. They probably have a role in cellular permeability (54). Glycolipids or peptidoglycolipids are responsible for the ropelike appearance that is evident in one of the outer layers of mycobacteria when they are visualized by the technique of negative staining (21).

The sulfolipids (which are 2,3,6,6'-tetra-esters of trehalose; structure **5**) (18–27), to which are attributed the cytochemical neutral-red fixing activity of viable, cord-forming tubercle bacilli, seem to play a role in conferring virulence to tubercle bacilli and influencing their pathogenicity (55), acting synergically with the cord factor (56).

The phospholipids (cardiolipin, phosphatidylethanolamine-glycosyl diglyceride, and phosphatidylinositol-monomannosides and -oligomannosides) were considered to be antigenic substances elaborated by *M. tuberculosis*, but the most purified preparations have been shown to behave only as haptens (18,37,57).

Even though some suggestions have been made about the biological activities of the lipids of the tubercle bacillus, a clear structure–function relationship has not yet been delineated. Nor has it been deter-

mined which structural features can produce favorable or detrimental effects. Also, the biosynthetic pathways leading to the formation and assembly of the cell wall have not been clarified, and the enzymes involved have not been purified.

Other interesting substances isolated from mycobacteria are the mycobactins, which are a group of bacterial growth factors that occur only in the genus *Mycobacterium* (15). The isolation of the mycobactins was followed by the identification of growth factors in other microorganisms, the sideramines, which differ from mycobactins but share with them some common properties, the most relevant being strong chelating capacity for ferric ions. At least nine mycobactin groups have been isolated from different mycobacteria. They have the same basic constitution, with some variations in details of structure. They consist of an octahedral iron-binding site (containing two secondary hydroxamate groups, an oxazoline ring, and a phenolic hydroxyl group) and a hydrophobic chain (containing up to 20 carbon atoms).

Mycobactin P (**6**), isolated in 1946, was the first example of a natural product with an exceptional iron-chelating activity, and its structure was the first to be determined. Mycobactin S (**7**) is the most active of these factors, showing growth stimulation at con-

$$R_1 = C_{15}H_{31} - CH - \left[CH - CH_2\right]_7 - CH -$$
with OH, CH_3, CH_3 substituents

$$R_2 = C_{15}H_{31} - CH_2 - \left[CH - CH_2\right]_6 - CH -$$
with CH_3, CH_3 substituents

$$R_3 = C_{15}H_{31} -$$

(5)

$R_1 = C_{17}H_{34}$, $R_2 = CH_3$, $R_3 = H$, $R_4 = C_2H_5$, $R_5 = CH_3$ **(6)**

$R_1 = C_{17}H_{34}$, $R_2 = H$, $R_3 = H$, $R_4 = CH_3$, $R_5 = H$ **(7)**

$R_1 = CH_3$, $R_2 = H$, $R_3 = CH_3$, $R_4 = C_{17}H_{34}$, $R_5 = CH_3$ **(8)**

centrations as low as 0.3 ng/mL. Mycobactin M (8) is a representative of the structure of M-type factors.

The most biochemically unusual product in the structure of mycobactins is N^6-hydroxylysine, which is present in the molecule in both acyclic and cyclic forms. All the known mycobactins contain either a salicyclic acid or a 6-methysalicyclic acid moiety. The oxazoline rings derive from 3-hydroxy amino acids, either serine or threonine.

The mycobactins are powerful iron chelators with an association constant of $>10^{30}$ and are essential growth factors for the mycobacteria. Therefore, it seems likely that they are involved in mycobacterial iron metabolism (58). Mycobacteria respond to iron deficiency by producing salicyclic or 6-methylsalicyclic acid, together with mycobactins that have a great affinity for ferric iron. It has been suggested that in the mycobacteria, salicyclic or methylsalicyclic acid mobilizes the iron in the environment and that mycobactins are concerned with the active transport of iron into the cell.

3 PATHOGENESIS AND EPIDEMIOLOGY

Tuberculosis is sometimes an acute but more frequently a chronic communicable disease that derives its character from several properties of the tubercle bacillus, which in contrast with many common bacterial pathogens, multiplies slowly, does not produce exotoxins, and does not stimulate an early reaction from the host. The tubercle bacillus is also an intracellular parasite, living and multiplying inside macrophages.

The structure and evolution of lesions caused by tubercle bacilli are determined by a host-aspecific defense system, by immunologic response and by genetic factors. Most commonly, mycobacteria reach the lung alveoli by inhalation. If the number of bacilli is low, the strain is of moderate virulence, and the intrinsic microbicidal activity (genetically determined) of resident alveolar macrophages is of good level, the tubercle bacilli that are ingested are rapidly killed and no further evolution of infection follows. However mycobacteria may evoke a variable immunological response, leading the host organism either to a state of resistance or to a state of disease, which depends on the number of infecting organisms and the already noted host defense capacity and genetic factors. To outline the events that follow the penetration of tubercle bacilli in the organism, four stages have been described in the pathogenesis of tuberculosis, based mainly on the research of Lurie (59) and Dannenberg (60) in rabbits. The results of animal research can be only partially applied to human disease. However, they have contributed to a better understanding of the main aspects of its course (61).

The first stage was described above, and due to the rapid destruction of the infecting organism, it can be self-limiting. However if the number of bacilli is high or if the macrophage microbicidal activity is inadequate to kill them, bacilli multiply inside the cell, causing it to burst. Through the production of chemotactic factors from the released bacilli, this event attracts monocytes from the circulation and produces to the so-called second stage.

Bloodborne monocytes are not yet activated, so they cannot destroy the bacilli, which grow within the cell in a logarithmic rate. At this time the immunological response is triggered: mycobacterial antigens processed by macrophages elicit the subset lymphocytes (CD4 + T cells) primarily involved in cell-mediated immunity, the main mechanism of protective immunity in tuberculosis. These cells secrete a number of lymphokines (particularly interferon-γ) capable of inducing macrophages to kill intracellular mycobacteria (62,63).

At this stage, the key event in the immunological response to mycobacterial antigens, i.e., the formation of granuloma (tubercle), is initiated. Stage 3 starts when the logarithmic phase of bacillary growth stops. It is characterized by the emergence of another subset of lymphocytes, CD8 + T

cells, that probably play some role also in the late phases of stage 2, and through their action, caseous necrosis is produced. CD8 + T cells are endowed with cytotoxic activity and mediate delayed-type hypersensitivity, leading to caseous necrosis and contemporaneously to interruption of logarithmic bacillary growth. CD8 + T cells in fact destroy the nonactivated macrophages, thus eliminating the intracellular environment favorable to bacillary growth. The released bacilli are then phagocytized by activated macrophages (whose activation is accelerated by previously sensitized lymphocytes). It is still debated how CD8 + T cells contribute to the production of caseous necrosis: among the different hypothesis is has been suggested that CD8 + T cells induce local sensitivity to tumor necrosis factor (TNF) (63–65).

Now the granuloma assumes its typical elementary structure: a solid caseous center is surrounded by immature macrophages (allowing intracellular bacillary growth) and activated macrophages (epithelioid cells), that kill the bacilli. The development or the arrestment of the disease depends on the balance between the two types of macrophages. Caseous material is not favorable to the bacillary growth because of acidic pH, low oxygen tension, and presence of inhibitory fatty acids; mycobacteria do not multiply but can survive for years in this environment.

The disease proceeds when stage 4 is reached, in which liquefaction of the caseous center occurs, often followed by formation of cavities where bacilli multiply extracellularly reaching high density. They can be discharged into the airways, thus diffusing in other parts of the lung and in the environment. Liquefaction and cavitation as markers of disease progression occur in human disease but not in the infection model in rabbits (61). However, the schematized patterns of infection in animals find some counterpart in the course of human infection and disease. In fact, the first contact of the human organism with tubercle bacilli, which usually takes place in infancy or adolescence, normally does not produce any clinical manifestation. The anatomic lesion induced by proliferation of mycobacteria and the reactive regional adenitis are called primary complexes. At that moment, the subject shows a positive tuberculin test (a cutaneous reaction obtained by injection or percutaneous application of culture filtrates of mycobacteria or of their purified protein content), which indicates a state of hypersensitivity to the tubercle bacillus, not necessarily a state of immunity. Usually the primary complex remains clinically silent, but it can also progress and evolve to a state of disease.

According to the definition of the American Thoracic Society, tuberculosis infection does not mean disease (66). The state of disease is defined by the appearance of clinically, radiologically, and bacteriologically documentable signs and symptoms of infection.

Chronic pulmonary tuberculosis in adults may be due to reactivation of the primary infection or to exogenous reinfection. A typical characteristic of tuberculosis is the formation in the infected tissue of nodular formations called tubercles, which can have different sizes and different modes of diffusion, giving rise to various clinical forms called miliary, infiltrate, lobar tuberculosis, and so on. The disease progresses by means of ulceration, caseation and cavitation, with bronchogenic spread of infectious material. Healing may occur at any stage of the disease by processes of resolution, fibrosis, and calcification.

Control of the disease has been achieved in part through mass vaccination with BCG (the bacillus of Calmette and Guèrin, an attenuated strain of *M. tuberculosis bovis*), but above all, control occurs through correct application of active chemotherapeutic agents. Chemotherapeutic treatment now available enables one to stop the propaga-

tion of the disease in a high percentage of cases, by killing the pathogenic bacilli, thus permitting the organism to repair or to confine the pathological alterations. Another important consequence of chemotherapeutic treatment is the prevention of dissemination of virulent bacilli to other persons.

In spite of the efficacious drugs now available for the treatment of tuberculosis, this illness is present all over the world. According to the estimates of World Health Organization in 1990, each year 8 million people develop tuberculosis and 2.9 million die from it (67). The highest incidence is in Africa (272 cases per 100,000 persons).

In developed countries the trend to a decrease in tuberculosis morbidity has been constant, probably related to better general hygienic condition, preventive measures, and in the last decades to the introduction of effective chemotherapy. Unfortunately, as indicated by recent reports of WHO, there is a resurgence of tuberculosis, even in advanced countries of Europe and North America. In United States, where the incidence of tuberculosis declined at an average rate of 6% each year until 1985, the case rate has risen from 9.3 per 100,000 in 1985 to 10.3 in 1990 (68–70). Indeed, in April 1993, WHO took the extraordinary step of declaring tuberculosis a "global emergency." More than 30 million people are expected to die of tuberculosis in the next decade, and WHO predicts that there will be 90 million more cases worldwide by 2000 (71).

The increase in morbidity occurs particularly in some ethnic minorities, in immigrants from foreign countries, and in some age classes. In fact the global percentage increase has been 25.6% among blacks and 72.3% among Hispanics. In these ethnic groups the increased incidence in subjects between the ages of 25 and 44 years has been attributed to the increased prevalence of HIV infection in this age group. In fact, recrudescence of tuberculosis coincided with the spread of acquired immunodeficiency syndrome (AIDS) (72). AIDS is the highest risk factor, that favors the progression of latent infection to active tuberculosis. In HIV-infected subjects, cell-mediated immunity is depressed, since the major targets of this virus are CD4-positive T lymphocytes, whose number reduction and functional impairment prevents them to mount the essential immunological response to the *M. tuberculosis* challenge. At present, it is estimated that less than 5% of cases of tuberculosis throughout the world are associated with HIV infection, with rates varying widely among geographic areas within the same country. The majority of cases are concentrated in sub-Saharan African countries where the AIDS epidemic caused about a 100% increase in reported tuberculosis cases in recent years (67).

HIV-infected subjects may develop tuberculosis by direct progression of exogenous infection to disease or by recrudescence of previously acquired latent infection. They are more likely to acquire tuberculosis when exposed to tubercle bacilli and to progress more rapidly than non-HIV-infected subjects from latent infection to active disease (73). Severe outbreaks of tuberculosis have been reported when groups of HIV-infected persons have been exposed to a person with infectious tuberculosis (74–76).

Tuberculosis tends to occur in the early stage of HIV infection, probably because of the virulence of tubercle bacilli. It is sometimes the first manifestation of HIV infection. The clinical course of tuberculosis is determined by the degree of immunosuppression. Very often it is a severe disease, frequently with extrapulmonary localization (73).

A further consequence of spreading HIV infection has been an increase of infections due to mycobacteria other than tuberculous (MOTT) or nontuberculous mycobacteria (NTM), previously defined as atypical

mycobacteria. They only partially resemble the tubercle bacillus and cause diseases less frequently than it. In some geographical areas, their incidence is rather high, such as, *M. kansasii*, in the central United States and *M. ulcerans* in Australia and South Africa. Unfortunately, the drugs developed for the treatment of tuberculosis have poor efficacy in these infections. Among MOTT, the *M. avium intracellulare* complex (MAC), in the past seldom recognized as cause of infection, is a frequent cause of severe disseminated infection in AIDS patients, occurring most frequently in the late stages of the disease. However, at present a growing number of patients have MAC infection as the initial manifestations of AIDS. In recent years, MAC has been isolated from 20 to 80% of patients fully followed from initial diagnosis of AIDS to death (77–79).

The resurgence of tuberculosis has been accompanied by another phenomenon of importance: the emergence of multiple drug resistant (MDR) strains of *M. tuberculosis* in the etiology of tuberculosis. MDR organisms demonstrate *in vitro* resistance to at least two major antituberculosis drugs, usually isoniazid and rifampicin.

Leprosy is a chronic disease caused by *M. leprae*, a mycobacterium that multiplies even more slowly than the tubercle bacilli. Common belief to the contrary, the organism has low pathogenicity and infectiveness. It can live dormant for years in the invaded organism and docs not show recognizable signs during the first stages of its propagation.

There are various clinical types of leprosy. In the most severe lepromatous disease, the bacilli massively infiltrate the skin, which becomes thickened, glossy, and corrugated. Then other tissues are invaded, mainly peripheral nerves and bones, and as a consequence, atrophy of skin and muscle, ulcerations, and amputation of small bones occur. The microorganism is detectable in smears from skin and mucosa.

The tuberculoid leprosy is characterized by the presence of skin macules with clear centers, which are insensitive to pain stimuli. *M. leprae* is generally detectable only during reactive phases. The final stages are similar to the lepromatous leprosy. There are also borderline forms, which present characteristics common to lepromatous and tuberculoid leprosy.

Leprosy was an epidemic disease in Europe in medieval times but now is confined to some tropical areas, especially India, the Philippines, South America, and tropical Africa. Social and economic poverty is the main reason for the prevalence of this disease, which has the greatest distribution in underdeveloped countries. Chemotherapy offers a great possibility for eradication of the disease, and multidrug therapy has given favorable results. During the past 8 years, the numbers of estimated and registered cases of leprosy have fallen from 10–12 million to 2.7 million and from 5.4 million to 1.9 million, respectively.

The success of the multidrug therapeutic regimens recommended by WHO in 1981 led the WHO assembly in 1991 to set a goal of elimination of leprosy as a public health problem by 2000 (80). Elimination was defined as a prevalence of 1 case for 10,000 people, or less; a case was defined as a patient receiving or requiring chemotherapy.

4 LABORATORY MODELS FOR SCREENING AND EVALUATING ANTIMYCOBACTERIAL AGENTS

In the search for new antimycobacterial agents, demonstration of *in vitro* activity against the virulent strain of *M. tuberculosis* $H_{37}Rv$ is one of the simplest preliminary tests. Although much more predictive than other *in vitro* models using avirulent or fast-growing mycobacteria (*M. smegmatis*, *M. phlei*), the *in vitro* test with *M. tuberculosis* gives a large number of false-positive and, unfortunately, also some false-negative results. Despite these limitations, the *in vitro* test, with various modifications of the inoculum size, the culture media, and the observation time, is still used in many laboratories for the blind primary screening of a large number of compounds. It is also used in antibiotic screening, where thousands of fermentation broths must be tested and a primary screen using *in vivo* models is a practical impossibility.

Rapid susceptibility testing of *M. tuberculosis* can be performed using the radiometric assay of $^{14}CO_2$ produced by the action of the microorganism on a ^{14}C-palmitic acid substrate. Results from this method, which is useful for screening of new chemical compounds, are available after 4–10 days incubation instead of 20–30 days (81,82).

Another method, which avoids the use of radioactive culture medium, uses firefly bioluminescence to detect ATP during mycobacterial metabolism, comparing the ATP production in drug-containing broths with that of the control broths (83). A novel test, devised to shorten the time required for sensitivity testing, uses mycobacteria infected with a specific reporter phage expressing the firefly luciferase gene. The photoreaction produced by the reaction of luciferin with ATP allows the measurement of bacterial growth or inhibition by drugs. The assays requires small amounts of test compounds and is designed to screen a large number of antimycobacterial products (84).

The *in vitro* test gives only an indication of activity; quantitative evaluation of the potential usefulness of the new products must be obtained through *in vivo* tests. These are performed, generally speaking, by inoculating virulent mycobacteria strains into laboratory animals and comparing the course of infection in treated and untreated animals. There are a variety of procedures for performing these tests, which differ with respect to animal species (mouse, guinea pig, rabbit, etc.), mycobacterial strain and size of inoculum, route of product administration, and evaluation of the results. The most current procedure for the evaluation of antituberculous drugs uses mice infected with the human virulent mycobacterial strain, evaluating the results in terms of ED_{50}, survival time, the pathology of the lung or bacterial count. The products active in the mice are then evaluated in other *in vivo* tests using more sophisticated techniques.

The best species for extrapolation of the results to humans is the rhesus monkey, *Macaca mulata* (85,86). Experimental tuberculosis in this species closely parallels the human disease, and in spite of the difficulties in terms of time, space, and cost of the test, it is advisable to perform it, especially when doubtful results have been obtained from other species. In any case, extrapolation of the results obtained in animals to the human disease requires comparison of the kinetics and metabolism of the product in different animal species and in humans. Differences in activity are sometimes clearly related to differences in metabolic behavior.

In the case of leprosy, for a long time no screening or evaluation models were available using the pathogenic agent *M. leprae*, which could not be cultivated *in vitro* or transmitted to animals. Therefore, the experimental infection of rodents with *M. lepraemurium* was used for evaluating the effect of potential drugs, although this model shows a low predictivity for activity in humans. For example, dapsone is inactive and isoniazid very active in this test, whereas the opposite is true for human leprosy. Only in 1960 was local infection in the mouse footpad with *M. leprae* set up (87). This model has been successfully used, with various procedural modifications, for screening and evaluation of drugs (88,89). Thymectomy and body irradiation

of mice inoculated with *M. leprae* provokes dissemination of bacilli, and this may be a model for a generalized infection (90).

Another model of experimental leprosy was set up using the armadillo, which develops a severe disseminated lepromatoid disease several months after inoculation with a suspension of human leprosy bacilli (91,92). Limitations in the use of this model for chemotherapeutic evaluation derive not only from its cost but also from the importance to reserve the infected animals as a source of bacillary material for the development of a vaccine.

All the available animal models for the evaluation of antileprotic agents are too time-consuming for the screening of a large number of compounds. The search for short-term models is necessary, and there are indications that it may be feasible to use some *in vitro* tests on both *M. leprae* (93) and *M. lepraemurium* (94), but more extensive confirmation is needed. It has been also proposed to test compounds against *Mycobacterium* species 607 (95), because a correlation has been observed between minimum inhibitory concentration for this species *in vitro* and the mouse footpad test with *M. leprae* (96).

Activity of products against MOTT is generally tested ad hoc, *in vitro*, or *in vivo*, and the compounds selected for this purpose are initially and temptatively those showing antitubercular action, although their activity is quite variable against the various mycobacterial species. In recent years, there has been increasing interest in identifying reliable *in vitro* susceptibility test methods against *M. avium* complex due to the severity of the infections caused by this microorganism and the increase of their frequency (97).

In addition to the standard *in vitro* assays, some *in vitro* studies have been carried out in presence of macrophages to determine the activity of drugs against *M. avium* complex and other MOTT when they are located intracellularly (98–100). A model of chronic disseminated *M. avium* complex infection, developed in beige mice, proved to be a useful tool in studying the *in vivo* activity of antimycobacterial drugs (101). The inclusion of a number of various mycobacterial species (MOTT) in the large antibacterial screening programs of new compounds could provide leads for finding new, more active chemotherapeutic agents for the diseases caused by these microorganisms. On the other hand, there is certainly need for more knowledge on the biochemistry and physiology of MOTT and on their pathogenic behavior in laboratory animals.

5 SEARCH FOR AND DISCOVERY OF ANTIMYCOBACTERIAL DRUGS

Using the laboratory models available for testing and evaluating products, the medicinal chemist has fundamentally two possible approaches in the search for antimycobacterial drugs. The first approach is the blind screening of a large number of compounds, which permits the detection of a certain number of structures endowed with antimycobacterial activity. Chemical modification of these "lead" structures, accompanied by careful studies of structure–activity relationships, can yield the optimal derivative for therapeutic use. The second approach, more challenging from the scientific point of view, is based on designing drugs to act selectively on biochemical targets specific for the particular microorganism.

The antimycobacterial drugs presently in therapeutic use (Table 34.2) have been obtained mainly by the first approach. Dapsone (4,4'-diaminodiphenylsulfone) (102, 103), a breakthrough in the chemotherapy of mycobacterial infections and still the drug of choice for the treatment of leprosy, was synthesized as an analog of the antibacterial sulfonamides. Thioacetazone

Table 34.2 Drugs Used for the Therapy of Mycobacterial Diseases

Drug	Indications(s)
Dapsone	Leprosy
Streptomycin	Tuberculosis
p-Amino-salicylic acid	Tuberculosis
Thioacetazone	Tuberculosis, leprosy
Viomycin	Tuberculosis
Pyrazinamide	Tuberculosis
Isoniazid	Tuberculosis
Thiambutosine	Leprosy
Cycloserine	Tuberculosis
Ethionamide	Tuberculosis
Kanamycin	Tuberculosis
Clofazimine	Leprosy
Capreomycin	Tuberculosis
Ethambutol	Tuberculosis
Rifampicin (rifampin)	Tuberculosis, leprosy
Rifabutin	Tuberculosis, MDR[a] tuberculosis, and prevention of MAC[b] infections
Rifapentine	Tuberculosis

[a]MDR = Multi Drug Resistant.
[b]MAC = *Mycobacterium avium* complex.

(104), thiambutosine (105), and thiocarlide (106) are chemical modifications of the thiosemicarbazone structure, which had proved to have antituberculous activity. Isoniazid (107) is the result of intensive research built around the finding of weak antituberculous activity of nicotinamide (108), and pyrazinamide (109) also originated from nicotinamide through chemical modification of the heterocyclic nucleus. Analogously, the observation of tuberculostatic activity of thioisonicotinamide (110) gave rise to ethionamide (111).

The observation that salicylic acid increases the oxygen consumption of tubercle bacilli and the hypothesis that related substances might have a reverse effect (112) yielded *ρ*-aminosalicylic (PAS) acid (113), which does not, in fact, affect the respiration of the mycobacterium. Ethambutol (114) was discovered as a result of chemical modifications of the *N,N'*-dialkylethylenediamine structures, which had shown antituberculous activity (115).

Streptomycin (116), cycloserine (117), viomycin (118), and kanamycin (119) were discovered in the course of screening for new antibiotics, and it is certainly not unimportant that they show activity also against other microorganisms commonly used in the primary tests of fermentation broths. Amikacin is a chemical derivative of kanamycin with improved properties. Rifampicin (120), rifabutin (121), and rifapentine (122) are semisynthetic antibiotics derived from the natural fermentation product rifamycin B, which, having limited antibacterial activity, was submitted to extensive studies of chemical modification to improve its properties (123,124). Clofazimine (125), now extensively used in the therapy of leprosy, was selected among several diaminophenazines known to be antituberculous agents *in vitro*, as are many other basic dyes (126).

Some antibiotics and chemotherapeutic agents developed for other indications have shown activity also against mycobacteria *in vitro* and *in vivo*. The products listed in Table 34.3 represent alternatives for the treatment of human infections resistant to the specific drugs.

The development of new antimycobacterial drugs is made difficult by a number of factors (127). First, large-scale screening systems for the detection of new antimycobacterial agents are particularly time-consuming and entail some problems related to the handling of the pathogens. Second, the development of an antimycobacterial drug takes more time and human resources than the development of other antimicrobial agents. Third, and probably most important, tuberculosis and leprosy are predominant in developing countries with low economic resources, and industrial laboratories are reluctant to in-

Table 34.3 Antimicrobial Agents Under Clinical Investigation as Antimycobacterials

Drug	Indication(s)
Amikacin	MAC infections
Ciprofloxacin	MRD tuberculosis, MAC infections, MDR tuberculosis, MAC infections, leprosy
Clarithromycin	MAC infections, leprosy
Azithromycin	MAC infections
Minocycline	Leprosy
Clavulanate/amoxicillin	MDR tuberculosis

vest in research for new products to be used in those geographic areas, where an additional drawback is the lack of patent protection. However, the actual situation makes the search for new and more effective antimycobacterial drugs necessary.

Although screening for antimycobacterial agents will continue to be a possible way to discover useful new drugs, the increasing knowledge of the biochemistry of the mycobacteria makes possible a more rational approach to the problem. In particular, the studies of the biosynthesis of the unique constituents of the mycobacterial cell will indicate the targets for inhibitors of the biosynthetic pathways present specifically in the microorganisms of the mycobacterium species.

Selective inhibitors could open new horizons in the therapy of mycobacterial infections because they will be ineffective on the eukaryotic cells of the host and, therefore, potentially will have little toxicity. Furthermore, since they will also be ineffective on the other microorganisms, they will not alter the normal microbial flora. From the available information on the biochemistry of mycobacteria, which has been previously summarized, some potential targets for specific antituberculous agents can be listed. For example, the mycolic acids seem to be acids present only in mycobacteria, suggesting that the mycolate synthetase might be a target for specific tuberculostatics.

The mycobacterial cell wall has certain characteristics that are of particular relevance to the search for potential inhibitors of its biosynthesis. Although the cell wall peptidoglycan represents a common feature in the bacterial world, in mycobacteria its structure presents some peculiarity with respect to the corresponding units in most other bacteria. For example, the muramic acid is N-acylated with glycolic acid rather than the usual acetic acid. This variation, apparently of little relevance to the general structure of the cell wall, could be of interest in the search of specific inhibitors of the enzymatic glycolylation of muramic acid. Another characteristic of the mycobacterial peptidoglycan is that the peptidic cross-links include a proportion of bonds between two residues of meso-diaminopimelic acid. It may be that products resembling the terminal meso-diaminopimelic-D-alanine are selective inhibitors of the mycobacterial cell wall, in analogy with the hypothesis that the activity of penicillins resides in their structural similarity to D-alanyl-D-alanine, the substrate of transpetidase reaction in most bacteria. Also the synthesis of arabinogalactan could be another target for potential new antimycobacterial agents. Studies on the identification of the enzymatic system involved in the biosynthesis of D-arabinose (which is much less diffuse in nature than the L-isomer) and in the polymerization of arabinogalactan are essential for the design of products acting on these targets. Also, analogs of L-rhamnose could interfere with the biosynthetic pathway for the assembly of the arabinogalactan and its attachment to peptidoglycan.

Another specific constituent of pathogenic mycobacteria is the cord factor. More extensive studies on its biosynthesis might indicate the possibility of inhibiting its formation.

The present knowledge of the chemistry and function of mycobactins could be used in various approaches to the search for antimycobacterial drugs (58). One approach would involve screening for compounds interfering at some point in the pathway of biosynthesis of mycobactins and selecting those compounds that inhibited bacterial growth under iron-limiting conditions. It is likely that the primary role of action of ρ-aminosalicylic acid is the inhibition of mycobactin synthesis.

In another approach, synthetic or semisynthetic iron ligands could be prepared and combined with macromolecules to block the outer membrane receptor proteins at the cell surface. The total synthesis of a mycobactin has been described and could be the basis for research in this direction (128). A third avenue of research could be an investigation on the synthesis of compounds that are structurally similar to mycobactins but are poor iron carriers; they could interfere with iron transport. A fourth approach could be related to the use of microbial iron ligands as carriers for antituberculosis drugs. For example, mycobactins could be synthesized with an isoniazid residue. The drug bound to mycobactin could be transported inside the cell, probably more efficiently that the drug alone, via the iron channel.

Another specific target for antimycobacterial drugs is the DNA-dependent RNA polymerase (DDRP), the enzyme that synthesizes RNA by using DNA as template. Rifamycins are potent and specific inhibitors of bacterial DDRP without having an effect on the mammalian DDRP (129,130), and the same mechanism is the basis of their action against mycobacteria (131–133).

Folate biosynthesis in mycobacterial organisms could also be a target for potential new drugs. In the search for potential antileprotic agents, several compounds of the diphenylsulfone class have been examined for their ability to suppress growth of *Mycobacterium* species 607, presumably by way of inhibition of the synthesis of the dihydrofolate, and some 2,4-diamino-6-substituted pteridines have been studied as potential inhibitors of the reduction of dihydrofolate to tetrahydrofolate in the same organism (134).

For many drugs in therapeutic use, preliminary or definitive indications of their mechanisms of action have usually been reached *a posteriori*. The active substances, besides being important as chemotherapeutic agents, often constitute tools for a better understanding of the biochemistry of mycobacteria, consequently for defining specific targets for the design of new drugs.

Although some important clues have appeared in the fields of specific biochemical pathways inside the mycobacteria and the mechanism of action of mycobacterial drugs, little progress has been made until now in overcoming the problem of drug resistance in mycobacteria. From the "fluctuation test" it appears that *M. tuberculosis* mutates spontaneously and randomly to resistance to isoniazid, streptomycin, ethambutol, and rifampicin (135), and that is in a genotype form.

The knowledge of the mechanism of mycobacterial resistance to a drug could indicate the direction in which to search for new products that specifically overcome this mechanism. In the case of aminoglycoside antibiotics, the mechanism of resistance in Enterobacteriaceae and some *Pseudomonas* strains has been extensively studied, and the inactivating enzymes have been identified. This information has allowed a rational design of chemical modifications of aminoglycoside antibiotics at the site of attack of inactivating enzymes, giving new products active against resistant strains (136).

Unfortunately, strides toward understanding the molecular basis of mycobacterial resistance to various drugs have been made only rather recently. The state of knowledge of the resistance mechanisms is not sufficient to conceive a rational design of molecules acting on mutants resistant to the specific drugs. Also, knowledge of the genetic mechanisms and genotypical systems governing resistance in mycobacteria is limited because of the difficulty of using mycobacteria in classical genetic techniques (slow growth, lipid-rich cell walls impermeable to DNA uptake). Only during the last decade has the application of recombinant DNA technologies to the mycobacteria opened new doors to obtain information on some basic mechanisms of resistance to drugs. However, at present, the delay or prevention of the evolution toward resistance of mycobacterial flora is accomplished only through combination therapy, based on the complementary action of the constituents. It is well known that the probability of a concomitant mutation toward resistant strains is much lower than the probability of mutation to resistance to a single drug.

6 AVAILABLE ANTIMYCOBACTERIAL DRUGS AND RELATED PRODUCTS

Among the several thousand compounds screened for antimycobacterial activity, only few have had therapeutic indices sufficient to warrant introducing them into clinical use. These drugs are described in some detail, together with general information about chemical analogs, mechanism of action, pharmacokinetics and metabolism, clinical use, and untoward effects. The information, in summary form, is intended to cover the aspects that are useful for understanding the role of each product in current therapy, the limitations of use, and the need for improvement or for further studies. The drugs have been subdivided arbitrarily into synthetic products and antibiotics, and within these two categories they are listed in a quasi-chronological order, with products having structural similarities grouped with the representative first introduced into therapy.

6.1 Synthetic Products

6.1.1 SULFONES. The sulfones were synthesized by analogy to sulfonamides, which had no antimycobacterial activity. The first sulfones prepared, 4.4′-diaminodiphenylsulfone (dapsone) (9) and its glucose bisulfite derivative, glucosulfone sodium (10), were found to be active in suppressing experimental tuberculous infections (102, 103).

$$R-HN-\!\!\!\langle\bigcirc\rangle\!\!\!-SO_2-\!\!\!\langle\bigcirc\rangle\!\!\!-NH-R$$

R = H

(9)

R = CH(CHOH)$_4$CH$_2$OH
|
SO$_3$Na

(10)

Their usefulness in the chemotherapy of human tuberculosis was limited, but the discovery of some effect of compound 10 in leprosy experimentally induced in rats (137) opened the way to their successful introduction into the treatment of human leprosy. Since it is thought that compound 10 is active after metabolic conversion to dapsone, intensive studies have been carried out that vary the structure of the latter to find optimal activity and to improve its low solubility. Various substitutions on the phenyl rings yielded products less active than dapsone. The only product of this type that has some clinical use is acetosulfone sodium, which contains one SO$_2$N(Na)-COCH$_3$ substituent in the ortho position to

the sulfone group. Its antibacterial effect seems to be due to the unchanged drug. Substitutions in both the amino groups gave rise to products that are in general active only if they are converted metabolically to the parent dapsone.

Substitution on the amino groups to improve solubility yielded products such as the above-mentioned glucosulfone, the methanesulfonic acid derivative, sulfoxone sodium, aldesulfone (11), and the cinnamaldehyde-sodium bisulfite addition product, sulfetrone sodium, solasulfone (12), which have limited use in leprosy treatment. They act through their metabolic conversion to dapsone in the body. Several methanesulfonic acid derivatives of 4,4′-diaminodiphenylsulfone have been tested for their ability to be metabolized to dapsone (138).

R = CH$_2$SO$_2$Na

(11)

R = CH−CH$_2$−CH−C$_6$H$_5$
      | |
      SO$_3$Na   SO$_3$Na

(12)

The 4,4′-diacetyldiaminodiphenylsulfone, acedapsone (13), has low activity *in vitro*, but is used as an injectable depot sulfone, which releases dapsone at a steady rate over several weeks. The advantage of this type of depot usage was also considered for the prophylaxis of people exposed to risk (139).

R = COCH$_3$

(13)

Although a large number of sulfones have been synthesized and tested as potential antileprotic agents, dapsone continues to be the basic therapeutic agent for *M. leprae* infections, often as a component of multidrug programs. Dapsone is weakly bactericidal against *M. leprae* at concentration estimated to be the order of 0.003 μg/mL (140); when the microorganism has been isolated from untreated patients. *M. leprae* becomes resistant to dapsone and its congeners.

It is assumed that dapsone interferes with incorporation of *p*-aminobenzoic acid into dihydrofolate, in analogy with the action of sulfonamides in other bacterial systems. This mechanism of action of dapsone has been proved for *E. coli* (141), and the finding of cross-resistance to sulfones and sulfonamides in *Mycobacterium* species 607 indirectly confirms their similarity of action (95). Unfortunately, the inability to cultivate *M. leprae in vitro* makes it difficult ultimately to verify that dapsone acts through the proposed mechanism on this organism.

Dapsone is usually administered orally at a daily dose of 100 mg (80). It is nearly completely absorbed from the gastrointestinal tract, well distributed into all tissues, and excreted in high percentage in the urine as the mono-*N*-sulfamate and other unidentified metabolites (140). It is monoacetylated in humans. The characteristics of dapsone acetylation parallel those of isoniazid and sulfametazine, thereby establishing genetic polymorphism for the acetylation of dapsone in humans (142). Two metabolic factors, greater acetylation and greater clearance of dapsone from the circulation, may contribute to emergence of dapsone-resistant *M. leprae* (143).

Acedapsone is administered intramuscolarly in oily suspension at a dose of 225 mg every 7 weeks. The product is slowly released from the site of injection, and the plasma contains mainly dapsone and its monoacetyl derivative; the ratio of these two products depends on whether the patient is a slow or rapid acetylator, as in the case of dapsone.

Several severe untoward effects are caused by dapsone and its analogs. Besides frequent gastrointestinal and central nervous system disturbances, the most com-

mon is hemolysis. There are indications that individuals with a glucose-6-phosphate dehydrogenase deficiency are more susceptible to hemolysis during sulfone therapy, although this is controversial (144).

6.1.2 *p*-AMINO-SALICYLIC ACID AND ANALOGS. The observation that benzoates and salicylates have a stimulatory effect on the respiration of mycobacteria (112) suggested that analogs of benzoic acid might interfere with the oxidative metabolism of the bacilli. When various compounds structurally related to benzoic acid were tested, it was found that some of them had limited antituberculous activity; *p*-amino-salicylic acid (**14**) was the most active (113). The discovery of PAS cannot be quoted as an example of biochemically oriented chemotherapeutic research, because in. fact, its mechanism of action is not by way of the respiration of mycobacteria. The *in vitro* and *in vivo* antimycobacterial activity of a simple molecule such as PAS stimulated the synthesis and testing of many derivatives. This extensive research failed to give rise to better drugs but did provide knowledge of the structural requirements for activity in the series.

Modification of the position of the hydroxy and amino groups with respect to the carboxy group resulted in a sharp decrease in activity. The amino group confers a distinct pharmacodynamic property to the molecule, eliminating the antipyretic and analgesic activities of salicylic acid and giving the specific tuberculostatic activity. Further nuclear substitution and replacement of the amino, hydroxy, or carboxy groups with other groups yielded inactive or poorly active products. Also, functional derivatives on the amino, hydroxy, and carboxy groups of PAS are generally inactive, unless they are converted *in vivo* into the active molecule. Among the latter, the phenylester (**15**) and benzamidosalicylic acid (**16**) must be mentioned.

PAS has bacteriostatic activity *in vitro* on

R	R'	
H	H	(14)
C_6H_5	H	(15)
H	COC_6H_5	(16)
H	$COCH_3$	(17)
H	$COCH_2NH_2$	(18)

M. tuberculosis at a concentration of the order of 1 μg/mL. It is active only against growing tubercle bacilli, being inactive against intracellular organisms. Other microorganisms are not affected by the compound. Most of the other mycobacteria are insensitive to the drug. Although active against *M. leprae* in the mouse footpad test (145), it is not used in the current treatment of leprosy.

The mechanism of action of PAS is not clear. It was demonstrated earlier that *p*-aminobenzoic acid antagonizes the antibacterial activity of PAS *in vitro* and *in vivo*, and therefore, it could act by competitively blocking the synthesis of dihydrofolic acid. On the other hand the formation of mycobactin, an ionophore for iron transport, is strongly inhibited by PAS, and the bacteriostatic activity of PAS might be due to the inhibition of the metabolic pathway for iron uptake (146–148).

PAS is readily absorbed by the gastrointestinal tract and well distributed throughout the body. It is quickly eliminated through the urine in form of inactive metabolites, the *N*-acetyl (**17**) and *N*-glycyl (**18**) derivatives. For this reason it is administered orally in a daily dose of 8–12 g. The side effects of PAS in the gastrointestinal tract and the poor acceptance by the patients are due to the high dosage. The search for derivatives with more adequate pharmacokinetic properties has given products such as **15** and **16** (the second used as calcium salt), which seem to release PAS slowly, giving more prolonged blood levels. However their use in the therapy is limited.

The initial use of PAS alone in the treatment of tuberculosis was followed by the emergence of mutants resistant to it. It is now used in combination with other and more potent antituberculous agents to increase their efficacy and prevent or delay development of bacterial resistance to them. In the case of the combination of PAS with isoniazid, the higher plasma concentration of the latter is due to competition in the acetylation reaction.

Due to its low cost, PAS is still used mainly in developing countries, although is becoming less favored because of poor compliance.

6.1.3 THIOACETAZONE, THIOCARLIDE, AND THIAMBUTOSINE. The synthesis of a series of thiosemicarbazones as intermediates in the preparation of analogs of sulfathiadiazole (104,149), which has weak antituberculous activity (150), led to the discovery of the thiosemicarbazone of *p*-acetamidobenzaldehyde (thioacetazone, amithiozone; structure **19**), the most active substance *in vitro* and *in vivo* of the series (151–153). In the search for better products, many modifications of the thioacetazone molecule have been made. The

$H_2NCSNHN = HC$ —⟨ ⟩— $NHCOCH_3$

(19)

studies clearly indicate that the activity resides in the thiosemicarbazone structure of the aromatic aldehydes. In fact, several products with modified aromatic moieties, including some heterocyclic nuclei, have been found to be active. Some of them have been tested clinically with positive results, but thioacetazone is the only thiosemicarbazone still in clinical use. Its good activity *in vitro* and *in vivo*, and the lack of cross-resistance to isoniazid and streptomycin, indicate the use of thioacetazone as a drug to combine with these drugs to delay bacterial resistance.

The mechanism of action of thioacetazone is not known. Its tuberculostatic activity is not counteracted by *p*-aminobenzoic acid (154), and there is a partial cross-resistance to antituberculous thioureas. Treatment with thioacetazone may produce thioacetazone-resistant strains, some of which are also resistant to ethionamide (155).

Despite relatively low toxicity in laboratory animals, thioacetazone has limitations in clinical use because of serious side effects (such as gastrointestinal disorders, liver damage, and anemia) when administered to humans at the initially proposed daily dose of 300 mg. A review of the side effects and efficacy of thioacetazone in relation to the dosage indicates that the drug has an activity comparable to that of PAS and an acceptable toxicity when administered at lower dosage (156).

A dose of 300 mg of isoniazid plus 150 mg of thioacetazone is a cheap and acceptable combination for long-term therapeutic treatment after the initial treatment with three drugs. This schedule is used in developing countries (157), although there are considerable differences in side effects among patients from different geographic areas (158). The reported effectiveness in leprosy (159) notwithstanding, thioacetazone is of limited usefulness in the treatment of this disease because of its side effects and the emergence of resistance of *M. leprae* to the drug.

Several thioureas that are structurally related to thiosemicarbazones have been found to be active *in vitro* and *in vivo* against *M. tuberculosis*, the most active being the 1,-diphenylthioureas with *p*-

alkoxy groups in one or both the aromatic nuclei. Of these, thiocarlide (**20**) and thiambutosine (**21**) have been introduced into clinical use.

Thiocarlide shows some efficacy against experimental tuberculosis in mice (160) and is active against tubercle bacilli resistant to PAS, streptomycin, and isoniazid (161,162); but several clinical trials have indicated little or no usefulness in the treatment of human tuberculosis, even in combination with other drugs (163). Its use is now limited. From a theoretical point of view, however, it is interesting that thiocarlide acts through inhibition of mycolic acid synthesis, as part of the more general inhibition of the free lipids of *M. tuberculosis* (164).

Thiambutosine has given disappointing results in clinical use for tuberculosis, but it was found to be active in the treatment of leprosy (165), though inferior to dapsone (166). It was recommended for patients who do not tolerate the latter (167), but the last WHO report on chemotherapy of leprosy does not mention thiambutosine among the currently available antileprosy drugs (80).

6.1.4 ISONIAZID, ETHIONAMIDE, AND PYRAZINAMIDE. After the early report that nicotinamide possesses tuberculostatic activity (107,168), several compounds related to it were examined. Attention was aimed at the isonicotinic acid derivatives, and in view of the already established antituberculous activity of thiosemicarbazones, the thiosemicarbazone of isonicotinyl aldehyde was prepared. When isonicotinyl hydrazide (iso-

iso-C$_5$H$_{11}$O —⟨ ⟩— NHCSNH —⟨ ⟩— OC$_5$H$_{11}$-iso

(20)

n-C$_4$H$_9$-O —⟨ ⟩— NHCSNH —⟨ ⟩— N(CH$_3$)$_2$

(21)

niazid, **22**), described in the chemical literature in 1912 (169), was prepared as an intermediate in the synthesis of the aldehyde and tested, it proved to be a potent antitubercular agent *in vitro* and *in vivo* (108,170,171).

The outstanding antituberculous activity of isoniazid in experimental infections, confirmed by the clinical trials, stimulated the study of chemical modifications of this simple molecule. At least 100 analogs were prepared, but any structural change caused a reduction in or loss of activity. Among the modified forms that retained appreciable activity, the N^2-alkyl derivatives should be mentioned. In particular the N^2-isopropyl derivative (iproniazide, **23**), was found to be active *in vivo*. Extensive clinical trials proved the therapeutic effectiveness of iproniazid and revealed its psychomotor stimulant effect, caused by the inhibition of monoamine oxidase (172). Use of iproniazid in the treatment of tuberculosis or of psychotic and neurotic depression was discontinued because of the hepatic toxicity of the drug. Although acetyl isoniazid is inactive, its hydrazones constitute a group of isoniazid congeners that have activity of the same order as the parent compound. The activity of these compounds generally is related to the rate of their hydrolysis to the parent compound. Some hydrazones have been introduced into therapeutic use, such as the 3,4-dimethoxybenzylidene (verazide, **24**) and the 3-methoxy-4-hydroxy-benzylidene (phthivazid, **25**) derivatives; but their utility is questionable, and products of this kind now have limited or no application. The injectable streptomycylidene isoniazid is an example of an incorrect medicinal chemical approach, because the compound in fact acts as a combination of the two components, which in addition are not in the right ratio for proper therapeutic use.

Isoniazid has bacteriostatic and bactericidal activity *in vitro* against *M. tuberculosis* and also against strains resistant to

other antimycobacterial drugs. The minimal inhibitory concentration for the human strain is of the order of 0.05 μg/mL. It acts on growing cells and not on resting cells and is effective also against intracellular bacilli. Its effect on the nontuberculous mycobacteria is marginal or nonexistent.

Several hypotheses have been put forward concerning the mechanism of action of isoniazid. The most convincing ones take into account that the activity of this drug is specific against mycobacteria at low concentrations. Investigations in this direction have indicated that the action of isoniazid is on the biosynthetic pathway to the mycolic acids (173,174). Apparently, isoniazid blocks the synthesis of fatty acids longer than C_{26} in chain length (175). Scanning electron microscopy shows impressive changes in mycobacterial cells exposed to concentrations of isoniazid that inhibit mycolic acid synthesis. There is complete loss of some areas of outer membrane as well as development of thin spots in the cell wall associated with bulging (176). A further proof of the interaction of isoniazid with mycolic acid synthesis appears to be the loss of acid fastness of cells susceptible to isoniazid (177).

Another hypothesis suggests that isonicotinic acid is responsible for the inhibitory effect of isoniazid on mycobacteria (178–180). Isoniazid is said to penetrate the cell, where it is oxidized enzymatically to isonicotinic acid, which at the intracellular pH is nearly completely ionized and cannot return across the membrane; consequently, it accumulates inside the cell. Isonicotinic acid is then quaternized and competes with nicotinic acid through the formation of an analog of the nicotinamide–adenine dinucleotide, which does not have the activity of the natural coenzyme. Alteration of the metabolic functions of the cell follows particularly with respect to lipid metabolism. A study of quantitative structure–activity correlations among a series of 2-substituted isonicotinic acid hydrazides gave evidence that the reactivity of the pyridine nitrogen atom, measured through the reaction rates for quaternization, is essential for the biological activity of compounds of this kind and seems to support the hypothesis that isonicotinic acid derivatives are incorporated into a nicotinamide–adenine dinucleotide analog (180).

These data could be reconciled with the previous hypotheses, including that of the formation of yellow pigments (181), and with the findings of other authors indicating a decrease in nicotinamide–adenine dinucleotide synthesis following treatment with isoniazid. However, it is not clear why the inhibition of the synthesis of nicotinamide–adenine dinucleotide by isoniazid is of the same order in sensitive and resistant strains (182).

Recent data give new light on the previous hypotheses. It is now known that the activity of isoniazid results from a peroxidative reaction catalyzed by catalase-peroxidase, which is encoded by the *Kat G* gene (183). In fact, clinical isolates of *M. tuberculosis* highly resistant to isoniazid lack *Kat G* (184).

The nature of the active derivative of isoniazid has not been yet defined, but isonicotinic acid is probably one of the reaction products and could act as an antimetabolite by poisoning the nicotinamide biosynthetic pool. The discovery of a novel gene, *inh A*, may complete the understanding of the mechanism of action of isoniazid. The product of *inh A* has a certain analogy with the enterobacterial *env M* enzyme, which is associated with the biosynthesis of fatty acids, phospholipids, and lipopolysaccharides (185). The *inh A* protein could be the primary target of action of isoniazid, but there is also the possibility that *inh A* requires NAD(H) as a coenzyme and that its activity can be affected of incorporation of iso-NAD produced by action of catalase-peroxidase on isoniazid, leading to a block of the synthesis of mycolic acids and loss of acid fastness. Still to be clarified is the fact that *inh A* and *Kat G* are also present in the mycobacteria naturally resistant to isoniazid. In *M. tuberculosis*, the two genes are altered in isoniazid resistant isolates (186).

Isoniazid is active in the various models of experimental tuberculosis in animals. It showed limited activity against *M. leprae* in the mouse footpad test (187), but it is essentially inactive in human leprosy (188). The drug is readily absorbed from the gastrointestinal tract and diffuses well into all organs, in various degrees. It is commonly used in adult patients at doses of the order of 5–8 mg/kg, but this dosage can be increased in children or in severe cases.

Isoniazid is one of the most effective antituberculous drugs, but when it is administered alone, a quick emergence of resistant strains follows. Therefore, isoniazid is administered with other antituberculous drugs to delay the emergence of resistant tubercle bacilli. It is recommended as a single drug in tuberculosis chemoprophylaxis, especially in household contacts and close associates of patients with isoniazid sensitive tuberculosis.

Isoniazid is usually tolerated for a long period of treatment. Hepatic side effects consist of frequent subclinical asymptomatic enzyme abnormalities (increase of SGOT and bilirubin), and occasionally severe clinical hepatitis occurs, especially in patients with previous hepatobiliary diseases and in alcoholics. Peripheral neuritis and stimulation of the central nervous system are common side effects. Isoniazid seems to compete with pyridoxal phosphate, and the concurrent administration of the latter has been suggested to prevent isoniazid toxicity (189,190).

Isoniazid is readily absorbed from the gastrointestinal tract in humans. Peak serum levels of the order of 5 μg/mL are obtained 1–2 h after administration of a 5 mg/kg dose. After absorption, isoniazid is well distributed in body fluids and tissues and penetrates into the macrophages.

Isoniazid is excreted mainly in the urine in unchanged form, together with various inactive metabolites: N-acetyl isoniazid, monoacetyl hydrazine, diacetyl hydrazine, isoniazid hydrazones with pyruvic and α-ketoglutaric acid, isonicotinic acid, and isonicotinylglycine. The primary metabolic route that determines the rate at which isoniazid is eliminated from the body is acetylation in the liver to acetyl isoniazid. There are large differences among individuals in the rate at which isoniazid is acetylated. The acetylation rate of isoniazid appears to be under genetic control (191,192), and individuals can be slow or rapid acetylators of the drug. The serum half-lives of isoniazid in a large number of subjects show a bimodal distribution; the isoniazid half-lives of rapid metabolizers range from 45 to 110 min and those of slow metabolizers from 2 to 4.5 h (193). The rate of acetylation appears to be conditioned by race. The proportion of slow acetylators varies from 10% among the Japanese and Eskimos to 60% among blacks and Caucasians. The isoniazid acetylator status of tuberculosis patients treated with isoniazid-containing regimens seems to be relevant only for once-weekly treatments with the drug (194).

As mentioned earlier, the discovery of isoniazid was the consequence of research based on the weak antitubercular activity of nicotinamide. This lead was pursued in various directions, and among the earliest modifications, thioisonicotinamide (26) (110,195,196) appeared to have the intriguing property of an in vivo efficacy superior to that expected from the in vitro activity. The hypothesis that some metabolic product of the drug was responsible for the activity in vivo stimulated the synthesis as well as the testing of a series of potential thioisonicotinamide metabolites and various other derivatives. Among the latter, increased activity was observed for the 2-alkyl derivatives (111). 2-Ethyl thioisonicotinamide (ethionamide, 27) and the 2-n-propyl analog (prothionamide, 28) were selected for clinical use. Of these two drugs, ethionamide has been more extensively studied, and prothionamide seems to possess biological properties similar to it.

CSNH$_2$

R = H   (26)
R = C$_2$H$_5$ (27)
R = n-C$_3$H$_7$ (28)

At concentrations of the order of 0.6–2.5 μg/mL, ethionamide is active in vitro against M. tuberculosis strains, either sensitive or resistant to isoniazid, streptomycin, and p-aminosalicylic acid. It shows also activity against other mycobacteria especially M. kansasii. The antibacterial action of ethionamide seems to be due to an inhibitory effect on mycolic acid synthesis, with a concomitant effect on nonmycolic acid–bound lipids (164). This pattern is like that shown by isoniazid.

However, recent studies indicate that ethionamide disturbs the synthesis of mycolic acid in both resistant and susceptible mycobacteria while isoniazid inhibits the synthesis of all kinds of mycolic acid in the same way in all susceptible strains and has no effect on mycolic acid synthesis in resistant strains (197). On the other hand, a missense mutation within the mycobacterial inh A gene was shown to confer resistance to both isoniazid and ethionamide in *M. smegmatis* and *M. bovis* (185).

Administered orally, ethionamide is effective in the treatment of experimental tuberculosis in animals. Although activity against *M. leprae* in animal infections has been reported, ethionamide is rarely used in the therapeutic treatment of leprosy. In the case of tuberculosis, the drug is given orally at doses varying from 125 mg to a maximum of 1 g daily. It is rapidly absorbed and widely distributed in the body. It provokes various side effects at the gastric level and has some hepatic toxicity. Bacterial resistance develops quickly when ethionamide is given alone; therefore, it is used in combination with other antimycobacterial drugs.

Ethionamide has a short half-life and is rapidly excreted in the urine, but only a minor percentage is in the form of unaltered product. A series of metabolites has been found in the urine: the active sulfoxide, the 2-ethyl isoniconitic acid and amide. and the corresponding dihydropyridine derivatives (198).

In studies of chemical modifications of the nicotinamide structure, other heterocyclic nuclei have been investigated, and the 2-carboxamidopyrazine (pyrazinamide, **29**) was synthesized and tested for antituberculous activity (109,199,200). The *in vitro* activity of pyrazinamide against *M. tuberculosis* is negligible at neutral pH and of the order of 5–20 μg/mL at pH 5.5. The best activity of pyrazinamide is against intracellular mycobacteria in monocytes, probably owing to the intracellular pH. This observation parallels the fact that pyrazinamide is inactive in tuberculosis in guinea pigs, predominantly extracellular, and active in murine tuberculosis, which has an important intracellular component. Apart from these observations, the mechanism of pyrazinamide remains unknown. According to some authors, the activity of pyrazinamide is due to its intracellular conversion into pyrazinoic acid (201), but

the mechanism of action of the latter is unknown. The observation that pyrazinamide blocks the increase of NAD-ase in the organisms infected with the sensitive bacilli but not with the resistant ones has suggested a mechanism of interaction between host and parasite (202).

Hepatotoxicity is the most common and serious side effect of pyrazinamide and is related to the dose and length of treatment. Another untoward effect is arthralgia, due to elevation of plasma uric acid levels. Pyrazinamide is absorbed well from the gastrointestinal tract and is excreted in the urine, mainly in the form of inactive metabolites: pyrazinoic acid, 5-hydroxypyrazinoic acid, and pyrazinuric acid (203). It was introduced into therapy in 1952 for the treatment of tuberculosis. After a period of declining use due to concern regarding its hepatotoxic effects, renewed interest has been shown in this drug for its role in short-term chemotherapy regimens. Due to its bactericidal effect on intracellular mycobacteria, pyrazinamide is recommended especially in the first 2 months of treatment of tuberculosis in combination with rifampicin and isoniazide (204). The usual daily doses are 20 to 35 mg/kg given orally in three or four equally spaced doses.

Compounds with other substitutions on the pyrazine nucleus or other carboxamido heterocycles were found to be inactive or less active than pyrazinamide. The only active analog developed because of its potential superiority over pyrazinamide was morphazinamide (*N*-morpholinomethyl-amide of pyrazinoic acid, **30**) (205). Interest in this drug ceased when it was ascertained that the activity and toxicity parallel those of pyrazinamide, to which morphazinamide is converted *in vivo* (206). Some pyrazinoic acid esters (207) and some *N*-pyrazinylthioureas (208) were found to be more active *in vitro* against *M. tuberculosis* than pyrazinamide, but no data *in vivo* have been reported.

6.1.5 CLOFAZIMINE. Clofazimine belongs to a peculiar class of phenazines, the so-called riminophenazines. Studies on these compounds derived from the original observation that treating a solution of 2-aminodiphenylamine with ferric chloride produced a red crystalline precipitate that completely inhibited the growth of tubercle bacilli H37 Rv strain *in vitro*, and was not inactivated by human serum (209–211).

The *in vivo* activity was moderate and the toxicity low. Such a compound, named B-283 (**31**), was the leading structure for a series of riminophenazines, which are alkyl or arylimino derivatives. Among the first compounds synthesized, B-663, later named clofazimine (**32**), was the most active (212,213). Its *in vitro* inhibitory activity against *M. tuberculosis* is at concentrations of 0.1–0.5 μg/mL. Strains resistant to isoniazid, and/or streptomycin, PAS, and thioacetazone, are susceptible to the drug. Some MOTT are also susceptible to this compound. Minimal inhibitory concentrations for *M. avium* complex (MAC) is 0.125–1 μg/mL (214).

R	R$'$	R$''$	R$'''$	
H	H	H	H	(**31**)
p-ClC$_6$H$_4$	CH(CH$_3$)$_2$	*p*-ClC$_6$H$_4$	H	(**32**)
C$_6$H$_5$	CH(CH$_2$–CH$_2$)$_2$CH$_2$	C$_6$H$_5$	Cl	(**33**)

This activity has rendered the drug eligible for the treatment of infections due to this organism. Clofazimine is not only bacteriostatic but also bactericidal (but the latter action is rather slow) and only on multiplying mycobacteria. The mechanism of action of riminophenazines has not been clearly elucidated because of the low solubility of these compounds in aqueous media. The activity seems to allow correlation with the *p*-quinoid system; in fact, when this is removed by reductive acylation, activity disappears. The mycobacteria, under anaerobic conditions, reduce the quinoid system. It has been shown that 20% of the respiratory hydrogen can be transferred from a respiratory enzyme to clofazimine (211). Its action has been also related to iron chelation, with resulting production of nascent oxygen radicals intracellularly (215).

In the treatment of murine tuberculosis, clofazimine was found to be more active than isoniazid on a weight-for-weight basis, but much more active on molar basis (216). This high activity of clofazimine in tuberculosis infections and other mycobacterioses was confirmed in other experiments with mice, hamsters, and rats. Much higher doses of clofazimine were necessary to achieve a therapeutic effect of the drug in guinea pigs and monkeys. Limited trials in chronic human pulmonary tuberculosis indicated that clofazimine had no significant effect on the disease at doses up to 10 mg/kg. A placebo controlled trial to evaluate clofazimine as prophylaxis against *M. avium* complex in AIDS patients showed a lack of efficacy (217). Some trials are on the way, including clofazimine in the multidrug regimen for treatment of *M. avium* complex infection (218).

In experimental infections with mycobacteria, clofazimine was found to be much more active than isoniazid against *M. kansasii* (216). Other studies have shown that clofazimine is also active in experimental infections with *M. johnei* (219), *M.*

ulcerans (220), *M. lepraemurium* (221,222), and *M. leprae* (223–230). In particular, *M. leprae* seems to be about 10 times more susceptible to clofazimine than *M. tuberculosis* (187). The marked activity against leprosy was confirmed in clinical trials (223–229). Generally speaking, the activity of clofazimine is similar to that of dapsone.

Dapsone-resistant mutants are susceptible to clofazimine. During treatment of lepromatous leprosy with clofazimine, the characteristic inflammatory reaction erythema nodosum leprosum (ENL) seldom develops; if clofazimine is combined with dapsone, the latter agent no longer causes ENL. This makes the concurrent use of corticosteroids unnecessary. It was suggested that clofazimine would have a corticosteroidlike antiinflammatory action (230). In a dye-hyaluronidase spreading test, clofazimine had a hyaluronidase-inhibitory effect, after single oral application of 100–200 mg in humans (231). In agreement with these results, it was found that clofazimine (50–100 mg/kg day) inhibited rat adjuvant arthritis and the inflammatory paw swelling following an adjuvant injection (232). It did not inhibit the primary antibody response to sheep erythrocytes or the tuberculin skin response. Thus clofazimine seems to have antiinflammatory but not immunosuppressive activity.

Clofazimine is an effective alternative drug in the therapy of leprosy. It is most active when administered twice weekly or daily, but it can be administered at monthly intervals as well, permitting monitoring of the treatment. The daily dose should not exceed 100 mg.

In laboratory animals, clofazimine has low acute and subacute toxicity (233). In clinical use, treatment with clofazimine is not accompanied by relevant toxicity. The main and sometimes unacceptable side effect is a red-purple coloration of skin, particularly within skin lesions (230). Gastrointestinal intolerance may also occur.

Clofazimine has a peculiar pharmaco-kinetic pattern, characterized by slow absorption, low blood concentration, high macrophage concentration, and extremely slow excretion. All riminophenazines are soluble in fats and in micronized form are well absorbed by the intestine. Derivatives with hydrophilic groups are either less active or inactive. Riminophenazines, once absorbed by the intestine, are carried by lipoprotein in the blood and ingested by macrophages. After continued oral administration, the macrophages appear as red-orange phagosomes. Therefore, the compounds have a diffusion that is mainly intracellular. They are stored in the body for a long time, and this confers some prophylactic action on them (216).

To obtain compounds with better pharmacodynamic properties, which would be cheaper and possibly more useful than clofazimine in the therapy of tuberculosis, a number of riminophenazines were designed. Compound B-1912 (**33**) was selected, showing a higher serum level and lower level in the tissue (except than in lipids) than clofazimine. In *M. leprae* infection experimentally induced in mice, it was shown that clofazimine and B-1912 have the same activity (234).

6.1.6 ETHAMBUTOL. Extensive studies of the chemical and biological properties of compounds related to alkylenediamine were carried out after the discovery, during the screening of randomly selected compounds, of the antimycobacterial activity of *N,N'*-diisopropylethylendiamine (**34**) (114,235). This compound was found to be active both *in vitro* and *in vivo*, with a therapeutic index of the same order as that of streptomycin.

Chemical modifications from this lead, attempted with the aim of obtaining the product with the highest therapeutic index, gave indications of the structural requirements for antimycobacterial activity. More relevant are the following: the presence of the two basic amine centers; the distance between the two carbons; and presence of a simple, small branched alkyl group on each nitrogen. A correlation between metal chelation of compounds of this structure and antimycobacterial activity suggested the utility of synthesizing products with hydroxy substitution of the *N*-alkyl groups as more effective metal chelators and possibly more active antimycobacterial agents. The most active of these derivatives was the *dextro* isomer of *N,N'*-bis-(1-hydroxy-2-butyl) ethylenediamine (ethambutol, **35**). The *meso* isomer is less active, the *levo* almost inactive. Furthermore, hydroxy substitution on other alkyl groups (isopropyl, *t*-butyl) or in other positions of the butyl group gave inactive products. These data appear in contrast with the working hypothesis of a correlation between metal chelation and antimycobacterial activity. Various other modifications of the structure of ethambutol gave inactive products, with few exceptions. The OCH_3, OC_2H_5, and $HNCH_3$ derivatives have the same activity *in vivo* as the parent compound because dealkylation occurs in the body. In addition, the monohydroxy unsymmetrical analog has activity equal to ethambutol but is more toxic.

Ethambutol inhibits *in vitro* the growth of most of the human strains of *M. tuberculosis* at concentrations of the order of 1 μg/mL. Strains resistant to other antimycobacterial agents are just as sensitive to ethambutol. Among the other mycobacteria, *M. bovis*, *M. kansasii*, and *M.*

RNHCH$_2$CH$_2$NHR

R = CH(CH$_3$)$_2$ **(34)**

R = CHC$_2$H$_5$ **(35)**
   |
   CH$_2$OH

R = CHC$_2$H$_5$ **(36)**
   |
   CHO

R = CHC$_2$H$_5$ **(37)**
   |
   COOH

marinum are usually susceptible. The MICs for *M. avium intracellulare* range between 0.95 and 15 μg/mL, 63.1% of strains being inhibited by 1.9 μg/mL or less (236,237).

The effect of ethambutol on mycobacteria is primarily bacteriostatic, with a maximum inhibitory effect at neutral pH. The primary mechanism of action of ethambutol on mycobacteria is not understood. The growth inhibition by ethambutol is largely independent of concentration, being more related to the time of exposure. It seems that most of the ethambutol taken up by mycobacteria has no direct role in growth inhibition, and there is no information on the subcellular components responsible for critical ethambutol binding (238,239).

Treatment of mycobacteria with ethambutol results in inhibition of protein and DNA synthesis, and it was proposed that ethambutol interferes with the role of polyamines and divalent cations in ribonucleic acid metabolism (240–242). Other authors have found an inhibitory effect of ethambutol on phosphorylation of specific compounds of intermediary metabolism, under conditions of endogenous respiration (243).

Other proposed primary sites of inhibition include arabinogalactan biosynthesis (244) and glucose metabolism (245). However, further studies are necessary to clarify the primary action of the drug.

Ethambutol is not active in the experimental mouse infection with *M. leprae* (187), and is not used in the treatment of human leprosy. The efficacy of ethambutol against *M. tuberculosis in vivo* was proved in various experimental models in animals and confirmed in the clinical trials in human tuberculosis. In current therapeutic use, the drug is administered at daily doses of 15–25 mg/kg, in combination with other antituberculous agents, to prevent emergence of resistant strains (246,247). The drug is well absorbed from the gastrointestinal tract. About half the ingested dose is excreted as active drug in the urine, where there are also minor quantities of two inactive metabolites: the dialdehyde (**36**) and the dicarboxylic acid (**37**) derivatives (248,249). Ethambutol is rather well tolerated. The main side effect of concern is ocular toxicity consisting of retrobulbar neuritis, with various symptoms, including reduced visual acuity, constriction of visual fields, and color blindness. Ocular toxicity appears to be dose related (250). At a daily dose of 25 mg/kg visual impairment occurs in about 3% of patients, rising to 20% at doses higher than 30 mg/kg/day.

6.2 Antibiotics

6.2.1 AMINOGLYCOSIDE ANTIBIOTICS (STREP-TOMYCIN, KANAMYCIN, AND AMIKACIN). Streptomycin was discovered in 1944 as a fermentation product of *Streptomyces griseus* (116). It belongs to the family of so-called aminoglycoside antibiotics, which includes kanamycin, gentamicin, neomycin, amikacin, nebramycin, paromomycin, kasugamycin, and spectinomycin. The chemical structure of streptomycin is *N*-methyl-L-glucosaminidostreptosidostreptidine. It is made up of three components: streptidine, streptose, and *N*-methyl-L-glucosamine (**38**). The intact molecule is necessary for antibacterial action.

Mannosidostreptomycin (**39**) is another antibiotic, produced together with streptomycin by *S. griseus*, which has not found clinical application because it is less active than streptomycin itself. Hydroxystreptomycin (**40**), produced by *S. griseocarneus*, has biological properties similar to those of streptomycin, with no advantages over it. In the attempt to improve activity and/or decrease toxicity of streptomycin, some chemical modification have been performed (e.g., on aldehyde or guanidino functions), which yielded generally less active products. One chemical derivative of streptomycin, dihydrostreptomycin (**41**), obtained by catalytic hydrogenation of the carbonyl group of streptose, has almost the

same antibacterial activity as the parent compound and investigators hoped that it would differ from the parent in having lower toxicity. Later clinical experience did not confirm this evaluation (251).

Streptomycin is both bacteriostatic and bactericidal for the tubercle bacillus *in vitro*, according to the concentration of antibiotic. Concentrations of streptomycin of the order of 1 μg/mL inhibit the growth of *M. tuberculosis* H37 Rv. MOTT are not susceptible to streptomycin. The antibacterial activity of streptomycin is not restricted to *M. tuberculosis* but includes a variety of Gram-positive and Gram-negative bacteria as well.

R^1	R^2	R^3	
—CHO	H	H	(**38**)
—CHO	H	(image: CH$_2$OH sugar)	(**39**)
—CHO	—OH	H	(**40**)
CH$_2$OH	H	H	(**41**)

The investigation of the mechanism of action of streptomycin has involved a number of elegant studies in microbiological chemistry and molecular biology that have led to a succession of hypotheses and to a continuous increase in knowledge not only of the mode of action of the antibiotic but also of the biology of the bacteria. After the first hypotheses, which attributed the activity of streptomycin to some effects on terminal respiration (252) or on the bacterial membrane (253) or to an interaction with DNA (254), it was finally ascertained that the drug is a specific inhibitor of protein biosynthesis in intact bacteria (255,256).

The ribosome, and particularly its 30S subunit, is the site of action of the antibiotic (257,258), and after careful disassemblage of 30S ribosomes, a protein designated P10 was determined to be the genetic locus responsible for the phenotypic expression of sensitivity and resistance to and dependence on streptomycin (259). The antibiotic induces a misreading of the genetic code, as demonstrated through studies of the erroneous incorporation of amino acids in cell-free ribosome systems (260). It was deduced that the misreading *in vivo* was the cause of the bactericidal effect of streptomycin, since it resulted in "flooding the cell" with erroneous, hence nonfunctional, proteins. However, it was subsequently demonstrated that this could not be the case because in the intact bacteria the antibiotic inhibits the synthesis of proteins (261).

The ultimate mode by which streptomycin exerts its bactericidal activity has not been elucidated. Two hypotheses have been put forward: one suggesting that streptomycin specifically inhibits initiation of protein synthesis (262) (this is supported by the involvement of protein P10, the site of action of streptomycin, in the initiation reaction), the other suggesting that it inhibits peptide chain elongation, that is, the synthesis of peptide bonds at any time during the growth of the peptide chain (263,264).

As noted before, sensitivity and resistance to and dependence on streptomycin all seem to be expressed in the ribosome and apparently are multiple alleles of a single genetic locus. Streptomycin-resistant mutant cells arise spontaneously in a bacterial culture, with a frequency of the order of 1.10^{-6} (265).

In the phenomenon of streptomycin dependence, bacteria require streptomycin to grow; these bacteria also arise by spontaneous mutations (266), and the mechanism of their behavior is also related to the reading of codons. This can be done correctly only in the presence of streptomycin, which overcomes an undiscriminating restriction (caused by mutation), leading to a mutant in which the ribosomal screen does not allow normal translation for growth (261–267). In addition, resistance to streptomycin can be transferred by means of R-factors or plasmids, namely, by extrachromosomal DNA carrying multiple antibiotic resistance (268).

The mode of transmission of resistance is particularly frequent among enterobacteria. Enzymatic inactivation is a frequent cause of resistance to streptomycin in eubacteria. The aminoglycoside-inactivating enzymes are phosphotranferases, acetyltransferases, and adenyltransferases. Since they act by inactivating a chemical group that is common to different aminoglycosides, bacterial strains that produce only one of them can be resistant to all aminoglycosides possessing the same chemical group (cross-resistance). Streptomycin can be inactivated by some streptomycin–adenyltransferase and streptomycin–phosphotransferase, which usually do not affect other aminoglycosides except spectinomycin (169).

In mycobacteria, mutations in the *rpsL* gene, which encodes the ribosomal protein S12 have been shown to confer resistance to streptomycin. Analysis of the primary structure of the ribosomal protein S12 in *M. tuberculosis* has revealed that mutations in the gene replacing Lys 43 or Lys 88 by arginine are frequently associated with resistance to streptomycin (270–272). A second type of mutation conferring resistance has been identified in streptomycin-resistant strains of *M. tuberculosis* that have a wild-type *rpsL* gene. These strains have point mutations in the 16S rRNA clustered in two regions around nucleotides 530 and 915 (273,274). In those isolates with a wild-type 16S rRNA and *rpsL* gene, other mechanism of drug resistance can be hypothesized, such as modifications of other components of the ribosome or alterations in cellular permeability.

From a pharmacokinetic point of view, streptomycin, like all other aminoglycoside antibiotics, is not absorbed from the gastrointestinal tract, and therefore, it must be administered parenterally. Serum peak levels are reached in 1–2 h, and the values are 9–15 μg/mL after administration of 1 g. Its half-life is 2–3 h. The serum protein binding of streptomycin is 25–35% that of dihydrostreptomycin 15% (275). Streptomycin diffuses slowly into the pleura, better into the peritoneal, pericardial, and synovial fluids. It does not penetrate into spinal fluid, unless the meninges are inflamed. Urinary elimination is rapid, and 70% of the drug is excreted in unmodified form in the first 24 h.

The most important toxic effects of streptomycin involve the peripheral and central nervous system. The eighth cranial nerve is the most frequently injured by prolonged administration of streptomycin, especially in its vestibular portion, causing equilibrium disturbances to appear. Treatment with 2–3 g a day for 2–4 months produces this type of side effect in about 75% of patients, but the incidence is much less at doses of 1 g a day. Other side effects are a hypersensitivity reaction and renal damage.

Dihydrostreptomycin was thought to be less toxic that streptomycin, but clinical use showed that it caused severe damage to the cochlear portion of the eighth cranial nerve, often inducing irreversible impairment of auditory function. For this reason dihydrostreptomycin has been discarded.

The most important clinical use of streptomycin is in the therapy of tuberculosis, and it was the first really effective drug for this disease. It is now used in combination with other antituberculous drugs, according to the schedules commonly accepted for treatment of this disease. Its importance declined after the introduction of other powerful oral antituberculous agents. Since the introduction of other broad-spectrum antibiotics, the use of streptomycin in the treatment of infections is limited to diseases in which other alternatives are lacking and the sensitivity of the infective organism indicates the choice and eventually the use of this drug in combination with other antibiotics. Thus it is still a drug of first choice for enterococcal endocarditis (in combination with penicillin or ampicillin), in brucellosis (in combination with tetracycline), in plague, and in tularemia.

Another aminoglycoside antibiotic used in the therapy of tuberculosis was isolated as a fermentation product of *Streptomyces kanamyceticus* in 1957 and named kanamycin. It consist of three components: kanamycins A, B, and C (**42–44**). Kanamycin A (**42**) is the largest part of the mixture (98%). The molecule contains deoxystreptamine (instead of the streptidine present in the streptomycin molecule) and two amino sugars: kanosamine and 6-glucosamine. It is water soluble and stable at both acid and basic pH as well as at high temperature. It has a quite broad spectrum of activity, including Gram-positive cocci and Gram-negative bacteria as well as *M. tuberculosis* and some other mycobacteria. Its activity against *M. tuberculosis* is weaker than that of streptomycin (276). The bactericidal concentrations are close to the bacteriostatic ones, but they are

	R_1	R_2	R_3
(42)	OH	NH_2	H
(43)	NH_2	NH_2	H
(44)	NH_2	OH	H
(45)	NH_2	NH_2	CO-$CHOH$-CH_2-CH_2-NH_2

hard to achieve *in vivo* (277).

The mechanism of action of kanamycin is similar to that of streptomycin, since it produces a misreading of the genetic code, interacting with 30S ribosomal subunit in more than one site (whereas streptomycin is bound only to one site), and inhibits protein synthesis (278–280). All aminoglycoside antibiotics that cause miscoding contain a 2-deoxystreptamine residue (streptomycin, kanamycin, amikacin, neomycin, paromomycin, gentamicin, and hygromycin B); kasugamycin and spectinomycin, which lack this residue, do not induce miscoding (281). Kanamycin, like streptomycin and other aminoglycosides, blocks both initiation and elongation of peptide chains (282). The mechanism of action of kanamycin was confirmed in mycobacteria. *In vitro* studies on cell-free preparations of *M. bovis* have shown that kanamycin inhibits polypeptide synthesis, followed by breakdown of polysomes and detachment of mRNA (283). A kanamycin-induced increase in ^{14}C-isoleucine incorpo-

ration by poly-U-directed ribosomes indicated a misreading of the genetic code, but this did not seem to be directly related to the bactericidal action of the antibiotic. Resistance to kanamycin can be acquired *in vitro* in a stepwise fashion by subculturing bacteria in increasing concentrations of antibiotic.

In addition to chromosomal resistance, resistance to kanamycin can be acquired by conjugation, through the transfer of extra-chromosomal DNA, the so-called R-factors or plasmids, coding aminoglycoside-inactivating enzymes (phosphotransferases, acetyltransferases, and nucleotidyltransferases). Kanamycin A can be inactivated by neomycin–kanamycin phosphotransferases I and II, kanamycin acetyltransferase, and gentamicin adenyltransferase. Cross-resistance will appear to any other aminoglycoside antibiotic that may be inactivated by the same enzyme (269). Cross-resistance is total with neomycin and paromomycin. With streptomycin, a "one-way resistance" is observed namely strains resistant to kanamycin and neomycin are usually resistant to streptomycin, whereas streptomycin-resistant strains are usually susceptible to kanamycin and gentamicin. It has been suggested that this is due to different sites of action of the antibiotics on the ribosomes (284). Thus in therapy it is advisable to administer streptomycin before kanamycin.

From a pharmacokinetic point of view, kanamycin behaves similarly to the other aminoglycosides: it is not absorbed when given by the oral route, but it is rapidly absorbed after intramuscolar administration, reaching high peak serum levels 2–3 h after administration (285). There is almost no serum protein binding (275). Diffusion into cerebrospinal fluid is poor when meninges are normal but increases when they are inflamed. Kanamycin diffuses quite well into pleural, peritoneal, synovial, and ascitic fluids (286,287), but poorly into bile, feces, amniotic fluid, and so on It is excreted by the kidney, mainly by glomerular filtration (50–80%), and for the most part in unmodified form. Kanamycin is used in therapy of infections caused by penicillin-resistant *Staphylococcus aureus* (now less frequently used because of the availability of penicillinase-resistant penicillins and other antistaphylococcal antibiotics) or to Gram-negative bacilli, such as *E. coli, Klebsiella, Enterobacter*, and *Proteus*. It has no activity against *Pseudomonas*. It has been used in the therapy of tuberculosis, in combinations with other antituberculous drugs. The common dose is 15 mg/kg/day, but a total dose of 1.5 g must not be exceeded.

Drawbacks to the use of kanamycin are its ototoxicity and nephrotoxicity (287). Cochlear and vestibular functions are damaged in about 5% of patients, but the percentage increases proportionally to the total dose administered. Thus in prolonged treatments, as in the case of tuberculosis, patients must be closely followed. Nephrotoxicity can be prevented if dosage to patients with decreased renal function is reduced in accord with the increase in creatinine clearance or the increase in creatininemia (285). Kanamycin can produce neurotoxicity, with curarelike effects due to neuromuscular blockade.

Amikacin (**45**) is a semisynthetic analog of kanamycin in which the C-1-aminogroup is amidated with a 2-hydroxy-4-aminobutyric acid moiety. This unusual amino acid was first found as a natural component of the antibiotics butirosins. When kanamycin is converted to amikacin there is a broadening of antimicrobial spectrum, because amikacin is not inactivated by many of the R-factor–mediated enzymes that attack kanamycin. In particular, useful activity against *Pseudomonas aeruginosa* results. *In vitro* and in animal trials amikacin is among the most active if not the most active aminoglycoside against *M. tuberculosis* (288).

Amikacin is not used for the initial treatment of susceptible tuberculosis, mainly due to its cost, but appears to have merit as an alternative drug for the retreatment of resistant *M. tuberculosis* infections. Amikacin seems to have a role in the treatment of MOTT infections, expecially if the infections are due to rapidly growing mycobacteria (*M. fortuitum* and *M. chelonae*) (289). It is one of the most bactericidal agents against *M. avium* complex both *in vitro* and in beige mouse model (290,291). In some clinical studies with AIDS patients, *M. avium intracellulare* complex bacteremia was cleared by combination of amikacin with other drugs (clarithromycin and ciprofloxacin) (292).

6.2.2 VIOMYCIN AND CAPREOMYCIN. Viomycin (**46**), a basic peptide antibiotic, was discovered independently by two groups of investigators (118,293) from an actinomycete named *Streptomyces puniceus* by one group and *S. floridae* by the other. Viomycin is relatively more active against the mycobacteria than against other bacteria. It inhibits protein synthesis (294) but has little or no miscoding activity (279). Viomycin reduces the amounts of dihydrostreptomycin bound to ribosomes of *M. smegmatis*, although they have different modes of action, perhaps because of a significant interaction between the binding sites of viomycin and streptomycin on ribosomes (295). Viomycin-resistant mutants isolated from *M. smegmatis* have altered ribosomes (296): one of these mutants had altered 50S subunits, and others had altered 30S subunits. The genetic locus for viomycin–capreomycin resistance (*vic* locus) in *M. smegmatis* consisted of two groups: *vic-A* and *vic-B*. It is likely that alterations in the 30S subunit conferred by *vic-B* and alterations in the 50S subunit conferred by *vic-A* interact in response to viomycin (297). There is a one-way cross-resistance with kanamycin: viomycin-resistant strains may retain their susceptibility to kanamycin, but kanamycin-resistant strains are

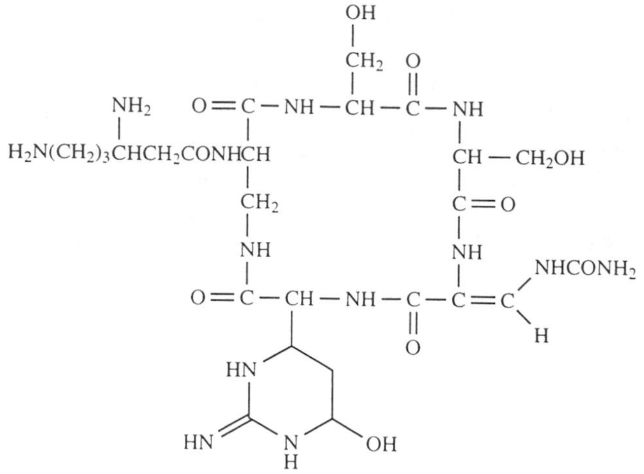

(46)

also resistant to viomycin. It is interesting to note that in *M. smegmatis*, the genetic locus for neomycin–kanamycin resistance (*nek* locus) is not linked to *str* locus (for streptomycin resistance) as in *E. coli* but is linked to *vic* locus (297).

On the whole, although viomycin is a peptide antibiotic, it behaves like the aminoglycosides from both a bacterial and a pharmacological point of view. The absorption and excretion of viomycin are similar to those of streptomycin. The usual daily dose is 1 g i.m. Side effects produced by viomycin are severe and frequent: ves-tibular and auditory impairment, renal damage, and disturbance in the electrolyte balance have a high incidence during vio-mycin therapy. For these reasons the drug is seldom used.

Capreomycin is a polypeptide complex isolated from *Streptomycin capreolus* (298). The structure of some components of the complex (**47**), indicates its similarity with viomycin. It is active only against *M. tuber-culosis* and some other mycobacteria, par-ticularly *M. kansasii*. In its antibacterial and pharmacological activities it is similar to the aminoglycosides.

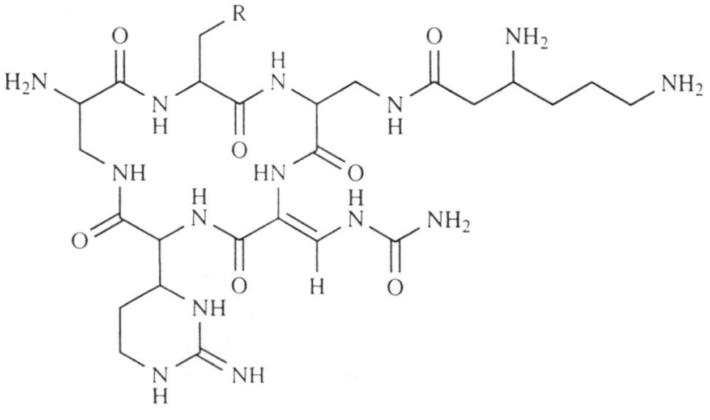

(47)

R = H, OH

Cross-resistance to kanamycin, neomycin, and viomycin has been described, and a phenomenon of partial "one-way resistance" to kanamycin has been demonstrated. Capreomycin-resistant strains are not always fully resistant to kanamycin, but kanamycin-resistant strains are always resistant to capreomycin.

The most common side effects are some damage to the auditory and renal functions and, sometimes, the appearance of anorexia with vomiting. The common daily dosage is 1 g i.m. Its use is restricted to retreatment of chronic cases with resistant mycobacterial flora.

6.2.3 CYCLOSERINE. D-Cycloserine was isolated independently by several groups of workers from cultures of *Streptomyces garyphalus*, *S. orchidaceus*, and *S. lavendulae* (117,299,300). On the basis of degradation studies and physicochemical properties, the structure of D-4-amino-3-isoxazolidone (**48**) was assigned to this antibiotic (301,302).

(**48**)

The structure was confirmed by synthesis (303), and various methods of preparation have been reported subsequently, which have also been used to prepare L-cycloserine from L-serine. Cycloserine inhibits *M. tuberculosis* at concentrations of 5–20 μg/mL. Strains resistant to other antimycobacterial drugs have the same sensitivity to cycloserine. The antibiotic is also active *in vitro* against a variety of Gram-positive and Gram-negative microorganisms, but only in culture media free of D-alanine, which antagonizes the antibacterial activity of cycloserine.

Concerning the mechanism of action of cycloserine, it has been proved in some bacterial species that this antibiotic interferes with the synthesis of the cell wall. In fact, cycloserine induces the formation of protoplasts in *E. coli*. Microorganisms treated with cycloserine accumulate a muramic-uridine-nucleotide-peptide, which differs from that produced by penicillin in the absence of the terminal D-alanine dipeptide. The inhibition of alanine racemase, which converts L-alanine into D-alanine, is probably the primary action of cycloserine (304–306). The mechanism of action in mycobacteria is likely the same. It is rather surprising that synthetic L-cycloserine has antibacterial activity also, but presumably it acts through a different mechanism of action that has not been clarified.

In vivo, cycloserine was ineffective against experimental tuberculosis in mice and was marginally effective in guinea pigs, but same activity was found against the disease induced in the monkey. The drug is more effective in humans than in animals. The explanation of the different responses resides in the different pharmacokinetic properties of cycloserine in the various animal species. When given orally to humans, cycloserine is well and quickly absorbed from the gastrointestinal tract and is well distributed in the body fluids and tissues. The usual dose for adults is 250 mg twice a day orally, always in combination with other effective tuberculostatic agents. About half the ingested dose is excreted unchanged in the urine in 24 h. A part of the antibiotic is metabolized into products not yet identified.

Cycloserine produces in the central nervous system severe side effects that can also generate psychotic states with suicidal tendencies and epileptic convulsion. Therefore, its use is limited only to cases in which other drugs cannot be used. Some chemical variations of the structure of cycloserine failed to yield products with improved antimycobacterial activity.

6.2.4 RIFAMYCINS: RIFAMPICIN (RIFAMPIN), RIFABUTIN, AND RIFAPENTINE. The rifamycin antibiotics were discovered as metabolites of a microorganism originally considered to belong to the genus *Streptomyces* and subsequently reclassified as a *Nocardia* (*N. mediterranea*) (307–309) and more recently as *Amycolatopsis mediterranea*. The crude material extracted from the fermentation broths contained several rifamycins (rifamycin complex) (124). Only rifamycin B was isolated as a pure crystalline substance, and it is essentially the only component found when sodium diethylbarbiturate is added to the fermentation media (310).

Rifamycin B (**49**) has the unusual property that, in oxygenated aqueous solutions, it tends to change spontaneously into other products with greater antibacterial activity (rifamycin O, **50**; rifamycin S, **51**). Rifamycin SV (**52**), was obtained from rifamycin S (123,124,311) by mild reduction. The structures of rifamycin B and of the related compounds involved in the "activation" process have been elucidated by chemical and x-ray crystallographic methods (312–314).

The rifamycins are the first natural substances to have been assigned an *ansa* structure consisting of an aromatic moiety spanned by an aliphatic bridge. At present, several natural substances with *ansa* structures are known, and for those that are metabolites of actinomycetales, the general name of "ansamycins" has been proposed (315). Streptovaricins (316,317), tolypomycins (318,319), and halomycins (320) are other ansamycins, structurally and biologically related to rifamycins, which also have in common activity against mycobacteria, Gram-positive, and to a lesser extent, Gram-negative bacteria.

Among the first rifamycins, the sodium salt of rifamycin SV was considered for clinical use because of its high antibacterial activity *in vitro* and good local tolerance when dissolved in polyvinylpyrrolidone solutions (321–324). Rifamycin SV is partially and irregularly absorbed from the gastrointestinal tract. When administered parenterally, the drug is rapidly eliminated through

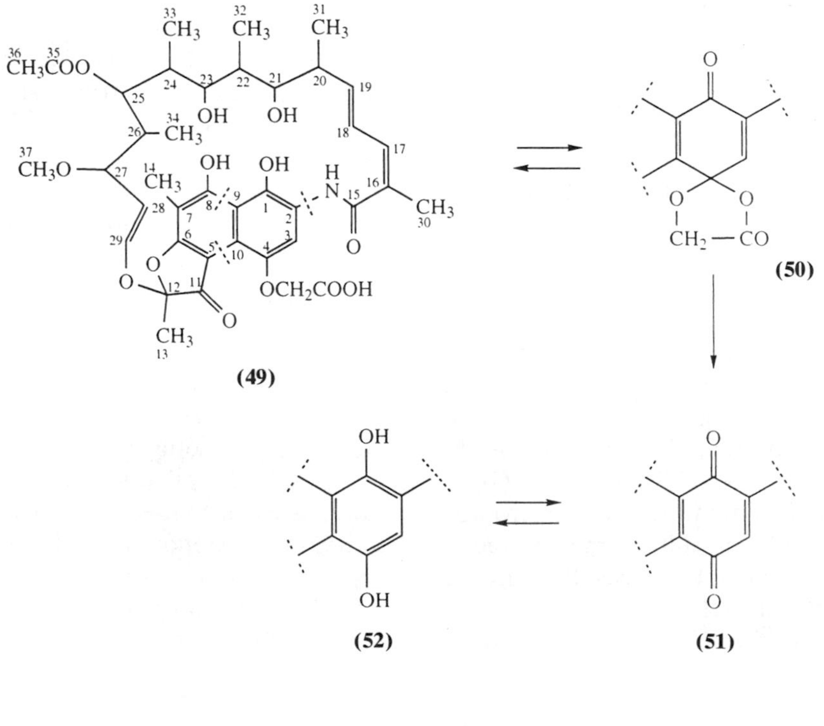

(49)

(50)

(51)

(52)

the bile (323). Thus in spite of the high *in vitro* activity against *M. tuberculosis* (of the order of 0.1 μg/mL (324), the doses of rifamycin SV required to control experimental tuberculosis in the animal are high (322). Also in human tuberculosis, rifamycin SV gave modest results (325–327). Only topical treatment of pleuropulmonary and extrapulmonary tuberculosis gave good results in the majority of cases (328).

Several authors (329–335) obtained encouraging results in short-term treatments of leprosy: the results obtained with 1–6 months of treatment were comparable to those obtained with 9 months of treatment with other antileprosy drugs, including sulfones (330). Improvements from the clinical and bacteriological standpoints were observed in patients treated for 8 months with rifamycin SV, 0.5–1 g daily (334), but some patients developed erythema nodosum leprosum. These preliminary results notwithstanding, rifamycin SV is not in current therapeutic use for leprosy, mainly because the need for frequent intramuscular injections makes the treatment unacceptable.

Several hundred derivatives of rifamycin B have been prepared with the aim of obtaining a compound that would have the following advantages over rifamycin SV: oral absorption and more prolonged therapeutic levels in the host, higher activity in the treatment of mycobacterial infections and higher activity in infections caused by Gram-negative bacteria (120).

Chemical modifications have been made on the glycolic chain of rifamycin B, on the aliphatic ansa, and on the chromophoric nucleus. A great deal of information about the structure–activity relationships has been obtained (120,336–338). All rifamycins possessing a free carboxy group are partially or totally inactive because they do not enter into bacterial cell. The minimal requirements for activity appear to be the presence of two free hydroxyls in positions

C_{21} and C_{23} on the ansa chain and of two polar groups (either free hydroxyl or carbonyl) at positions C_1 to C_8 of the naphthoquinone nucleus, together with a conformation of the ansa chain that results in certain specific geometrical relationship among these four functional groups.

This conclusion is based on the following data. First, substitution or elimination of the two free hydroxyls in position C_{21} and C_{23} gives inactive products. Inversion of the configuration at C_{23} leads to an inactive compound, and the inversion at C_{21} strongly reduces the activity (339,340). Second, modifications of the ansa chain that alter its conformation (e.g., C_{16}–C_{17} and C_{18}–C_{19} monopoxy and diepoxy derivatives) give inactive or less active products. Also, the stepwise hydrogenation of the ansa chain double bonds results in a gradual decrease in activity as a consequence of the increase in flexibility of the ansa diverging from the most active conformation.

Third, the oxygenated functions at C_1 and C_8 must be either free hydroxyl or carbonyl to maintain biological activity. Fourth, the four oxygenated functions at C_1, C_8, C_{21}, and C_{23} not only must be unhindered and underivatized but must display well-defined relationships with one another. The absolute requirements for these four functions to be in a correct geometrical relationship suggest that they are involved in the noncovalent attachment of the antibiotic to the bacterial target enzyme. This contention is supported by the observation that in the active derivatives these four functional groups lie on the same side of the molecule and almost in the same plane with identical interatomic distances between the four oxygens (341), as can be seen in the spatial model derived from x-ray studies (342). However, conformational differences have been observed at the junction of the ansa chain to the naphtoquinone chromophore, according to the nature of the substituents in position 3 (343). ^{1}HNMR studies of rifamycin S have confirmed that the conformation of the molecule in solution corresponds well to that obtained in the solid state (344). Finally, all modifications at C_3 and/or C_4 position that do not interfere with the previous requirements do not affect the general activity of the products (120,336–338).

A great number of active derivatives have been obtained by modifications on the glycolic moiety of rifamycin B and on position C_3 and/or C_4 of the aromatic nucleus of rifamycin S or SV, leaving unaltered the structure and the conformation of the ansa chain. Among the earlier derivatives of the glycolic side chain, the diethylamide of rifamycin B, rifamide (**53**), had a better therapeutic index than rifamycin SV (345). It was introduced into clinical use in some countries, but it still suffers from most of the limitations of use of rifamycin SV (346).

OH

OCH$_2$CON(C$_2$H$_5$)$_2$

(**53**)

Chemical modifications of the chromophoric nucleus of rifamycin on C_3 and/or C_4 positions gave a large number of derivatives. The nature of the substituents at positions C_3 and/or C_4 influences the physicochemical properties of the derivatives, especially lipophilicity. The various derivatives show a minor degree of variation in antibacterial activity against intact cells, because the transport through bacterial wall and membrane is the major factor affected by these substituents (347). Other biological characteristics influenced by the various modifications of the positions C_3 and/or C_4 are the absorption from the gastrointestinal tract and the kinetics of elimination. Rifamycin derivatives with substitutions at position C_4 include a series

of 4-aminoderivatives. Modifications at positions C_3 and C_4 include rifamycins with heterocyclic nuclei fused on these positions (e.g., 3,4-phenazinerifamycins, phenoxazinerifamycins, pyrrolerifamycins, thiazolerifamycins, and imidazorifamycins) with various groups on the heterocyclic nuclei. Rifamycins with substituents only in position C_3 are represented by 3-thioethers, 3-aminomethylrifamycins-, 3-carboxyrifamycins-, 3-aminorifamycins-, and 3-hydrazynorifamycins. When the 3-formyl-rifamycin SV (**54**) was prepared by oxidation of N-dialkyl-aminomethyl rifamycin SV (348), it was found that many of its N,N-disubstituted hydrazones had high activity, both *in vitro* and *in vivo*, against *M. tuberculosis* and Gram-positive bacteria and moderate activity against Gram-negative bacteria. For some of these derivatives the *in vivo* activity in animal infections was of the same order whether the products were administered orally or parenterally, indicating good absorption from the gastrointestinal tract (120). The hydrazone of 3-formylrifamycin SV with N-amino-N'-methylpiperazine (rifampicin, **55**) (120,349) was the most active *in vivo* and was selected for clinical use. Rifampicin is active *in vitro* against *M. tuberculosis* at concentrations below 1 μg/mL in semisynthetic media. It is active against other Gram-positive bacteria at lower concentrations and against Gram-negative bacteria at concentrations of 1–20 μg/mL. Rifampicin is active at the same concentrations against strains resistant to other antibiotics and antimycobacterials.

The bactericidal activity of rifampicin is demonstrated at concentrations close to the static ones. It is possible to isolate strains resistant to rifampicin from mycobacterial cultures exposed to the antibiotic, but the frequency of resistant mutants to rifampicin in sensitive populations of *M. tuberculosis* is lower than to other antimycobacterial drugs (350).

Rifampicin is active against *M. leprae*,

(54)   (55)

suppressing the multiplication and the viability of the bacilli infecting laboratory animals (90,351–354). It is also active against many other mycobacteria, although at concentrations generally higher than those effective against *M. tuberculosis.* Among the various mycobacteria tested, *M. kansasii* and *M. marinum* are the most sensitive (355). There is no evidence of cross-resistance among rifampicin and other antibiotics or antituberculous drugs (356). Transfer of resistance to rifampicin could not be obtained.

Rifampicin, like the other rifamycins, acts on the bacteria by inhibiting specifically the activity of the enzyme DNA-directed RNA-polymerase (DDRP). The mammalian DDRP is resistant even to high concentrations of rifamycins (129,130). The inhibition of the action of the bacterial RNA polymerase by rifampicin is due to the formation of a rather stable, noncovalent complex between the antibiotic and the enzyme with a binding constant of $10^{-9} M$ at 37°C (357). One molecule of rifampicin (mol. wt. 823) is bound with one molecule of the enzyme (mol. wt. 455,000) (358). The drug is not bound covalently to the protein because the complex dissociates in presence of guadinium chloride (359). The binding site of rifampicin to the RNA polymerase has been particularly well studied in *E. coli* (359–362). The inhibitory effect of rifampicin on DDRP as the cause of its bactericidal activity has been verified on several mycobacterial species, e.g., *M. smegmatis,* *M. bovis BCG,* and *M. tuberculosis* (131–133).

The DDRP comprises five subunits (α, α, β, β', and σ) and rifampicin binds to the

β subunit. When rifampicin is bound to the enzyme, the complex can still attach to the DNA template and catalyze the initiation of RNA synthesis with the formation of the first phosphodiester bond, e.g., of the dinucleotide pppApU. However, the formation of a second phosphodiester bond and, therefore, the synthesis of long-chain RNAs is inhibited. The action of rifampicin is to lead to an abortive initiation of RNA synthesis (363,364).

Resistance to rifampicin arises spontaneously in strains not exposed previously to the antibiotic at a rate of one mutation per 10^7 to 10^8 organisms. In *E. coli,* but also in *M. leprae* and in *M. tuberculosis,* resistance to rifampicin results from missense mutations in the *rpoB* gene, which encodes the β subunit of RNA polymerase. These mutations are all located in a short region of 27 codons near the center of *rpoB* and consist predominantly of point mutations, although in-frame deletions and insertion also occur. In most of *M. tuberculosis* rifampicin-resistant clinical isolates, changes have occurred in the codons for Ser 531 or His 526 (365–367).

Although the main mechanism of resistance is the modification of the target enzyme, some mutagenic treatments yield resistant mutants in which the RNA polymerase is still highly sensitive to the drug but the rate of rifampicin uptake is reduced. The mechanism of this permeability mutation is not yet clear (368). Preliminary *in vivo* studies (369) showed that the antituberculous efficacy of rifampicin in experimental infections in mice was comparable to that of isoniazid and markedly superior to that of streptomycin, kanamycin, and ethionamide. In guinea pigs it was comparable to streptomycin. Rifampicin was also remarkably active in experimental infections due to Gram-positive and Gram-negative bacteria.

The excellent antituberculous activity of rifampicin *in vivo* was confirmed by other experiments in many laboratories using various animal models (mice, guinea pigs, rabbits) and various schedules of treatment and criteria of evaluation (370–380). The overall results indicate that rifampicin has a high bactericidal effect and a therapeutic efficacy of the order of that of isoniazid and superior to all the other antituberculous drugs. The combination of rifampicin plus isoniazid has been shown to produce a more rapid, complete, and durable sterilization of infected animals that other combinations (373). Rifampicin has low toxicity according to acute, subacute, and chronic toxicity studies in several animals species (381). Results of animal and human studies showed no toxic effect on the ear (382) and eye. After oral administration, rifampicin is well absorbed in animals and humans (383). After administration of 150 and 300 mg to humans, serum levels reach maximum values around the 2nd h and persist as appreciable values beyond the 8th and 12th h, respectively. When the dose is increased, serum levels are high and long lasting.

Generally, the serum levels found at the beginning of treatment are higher than the levels that gradually set in as treatment continues. This phenomenon occurs during the first few weeks of treatment (384,385). The half-life of rifampicin is of the order of 3 h and increases in patients with biliary obstruction or liver disease (386–388). Rifampicin shows an extensive distribution in the tissues and crosses the blood–brain barrier. It reaches good antibacterial levels in cavern exudate and in pleural fluid.

Rifampicin is eliminated through both the bile and the urine. It appears rapidly in the bile, where it reaches high concentrations that last even when the antibiotic is not measurable in the serum. After reaching the threshold of hepatic eliminatory capacity, biliary levels do not increase with an increased dosage, but serum and urinary levels do. In humans, rifampicin is mainly metabolized to 25-*O*-desacetylrifampicin (389), which is only slightly less active than the parent drug against *M. tuberculosis* but considerably less active against some other bacteria. Both rifampicin and desacetylrifampicin are excreted in high concentrations in the bile. Rifampicin is reabsorbed from the gut, forming an enterohepatic cycle, but the desacetyl derivative is poorly absorbed and thus is excreted with the feces. Rifampicin has a stimulating effect on microsomal drug metabolizing enzymes (390), which leads to a decreased half-life for a number of compounds, including prednisone, norethisterone, digitoxin, quinidine, ketoconazole, and the sulfonylureas.

Many clinical trials have confirmed the efficacy of rifampicin in a variety of bacterial infections (391). In particular, rifampicin is now largely used for the treatment of human tuberculosis both in newly diagnosed patients and in patients whose primary chemotherapy has failed. Normally, rifampicin is administered orally in a dose of 600 mg daily in combination with other antituberculous drugs, mainly isoniazid. Adverse reactions to daily rifampicin are uncommon and usually trivial. The most frequent ill effects are cutaneous reactions and gastrointestinal disturbances. Rifampicin can also disturb liver function, but the risk of its causing serious or permanent liver damage is limited, particularly among patients with no previous history of liver disease.

In the case of intermittent therapy, when the drug is given three times, or once a week, a high incidence of severe untoward effects (such as the "flulike" syndrome and cytotoxic reactions) may result. These adverse reactions seem to be associated with rifampicin-dependent antibodies, suggesting an immunological basis. However with proper adjustment of the size of each single dose, of the interval between doses, and of the length of treatment, it has been possible to develop intermittent regimens with rifampicin that are highly effective and acceptably safe (392).

In human leprosy (90,351–354,393,394), treatment with rifampicin alone or in combination with other antileprosy drugs gave favorable results in almost all the patients. In particular, the variations of the morphological index and, when carried out, the mouse footpad inoculation, showed the rapid and constant bactericidal action of rifampicin. The same effect was observed in many patients who had become resistant to other antileprosy drugs.

Among the many other rifamycin derivatives prepared and tested as antimycobacterial agents, two show interesting characteristics that have been recently exploited for chemotherapeutic use: rifabutin and rifapentine.

(56)

Rifabutin (56) belongs to the group of spiroimidazorifamycins obtained by condensation of 3-amino-4-iminorifamycin SV with N-butyl-4-piperidone (121,395,396). Rifabutin has an antibacterial spectrum similar to rifampicin, but appears to possess incomplete cross-resistance with rifampicin *in vitro*. In fact some strains resistant to rifampicin are still moderately sensitive to

rifabutin. Another characteristic of rifabutin is that its activity against *Mycobacterium avium* complex is higher than that of rifampicin (397–401).

Rifabutin is rapidly absorbed by mouth, but its oral bioavailability is only 20%, and there are considerable interpatient variations. The elimination half-life is long (35–40 h) but, as a result of a large volume of distribution, average plasma concentrations remain relatively low after repeated administration of standard doses. Binding to plasma proteins is about 70%. Rifabutin is slowly but extensively metabolized, possibly to more than 20 compounds in humans, the 25-desacetyl derivative being the main metabolite (402).

Rifabutin has been proven of value in preventing or delaying mycobacterial infections in immunocompromised patients (403). It has been approved in the USA and in other countries for the prevention of MAC infections in AIDS patients.

(57)

Rifapentine (57) is an analogue of rifampicin in which a cyclopentyl group substitutes for a methyl group on the piperazine ring (122). Its activity is similar to that of rifampicin, but is slightly superior against mycobacteria, including MAC (404–406). It is more liphophilic and has a serum half-life about five times longer than rifampicin. This is due to its stronger binding to serum proteins than rifampicin (407,408).

In experimental tuberculous infections, rifapentine administered once a week has practically the same therapeutic efficacy as rifampicin administered daily (122). It is in clinical trial as a drug for the therapy of tuberculosis and leprosy at a lower dosage and frequency of administration than rifampicin.

In consideration of the high activity of practically all the semisynthetic rifamycins with substituents in positions C_3 and/or C_4 of the aromatic nucleus, a great number of derivatives have been synthesized and tested in the search for products with improved therapeutic properties in comparison to rifampicin, rifabutin, and rifapentine. A recently developed group of derivatives is the benzoxazino-rifamycins, whose first examples were reported in the late 1960s (409,411). To search for derivatives with potent *in vivo* efficacies against *M. tuberculosis* and MAC, many new 5-amino-benzoxazino-rifamycin derivatives having hydroxyl, alkyl, or other functional groups substituted at 3′, 4′, and 6′ positions in the benzoxazine ring were synthesized. Among them, the 3′-hydroxy-5′-(4-isobutyl-1-piperazinyl)benzoxazino-rifamycin (KRM-1648) was selected as the most promising (411,412). The activity of this product against *M. tuberculosis* in a murine model suggests that it is a good candidate for clinical development as a new antituberculosis agent (413).

6.3 Other Drugs under Investigation

In recent years, the need for new drugs to meet the problems connected with the emergence of multidrug-resistant tuberculosis and mycobacteriosis, particularly in AIDS patients, in the absence of any really new antituberculous compound has led to the careful evaluation of the antituberculous activity of antimicrobial agents developed for their activity against common Gram-positive and Gram-negative bacteria. Some promising drugs have been found among the fluoroquinolones, macrolides, and β-lactams (in combination with β-lactamase inhibitors).

6.3.1 FLUOROQUINOLONES.   Flouroquinolones represent an improvement of the earlier analogs (nalidixic acid, oxolinic acid, pipemidic acid, and cinoxacin) in being more potent *in vitro* and in having broader antibacterial spectrum, which includes Gram-positive and Gram-negative organisms. They have also improved pharmacokinetic properties. While the old agents distribute poorly in body tissues and fluids so that they can be employed only as urinary antiseptics, the new derivatives distribute much better in all tissues, penetrate cells, and can be efficaciously used in the treatment of systemic infections. Their oral bioavailability is excellent (414). Key points in determining these characteristics are the attachment of a fluorine atom at C_6 and of piperazinyl or *N*-methylpiperazinyl groups at C_7 and of alkyl or cycloalkyl groups at N_1 of the 1,4-dihydro-4-oxo-3-quinoline carboxylic acid (415).

Derivatives so far recognized with activity against mycobacteria are ofloxacin (**58**) and ciprofloxacin (**59**), which have been studied broadly (416–419). Sparfloxacin

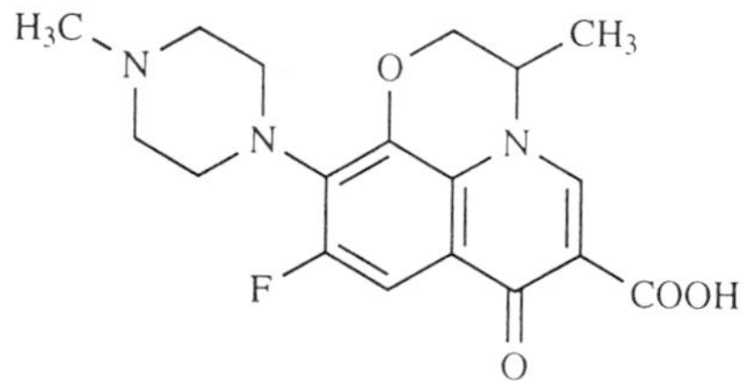

(58)

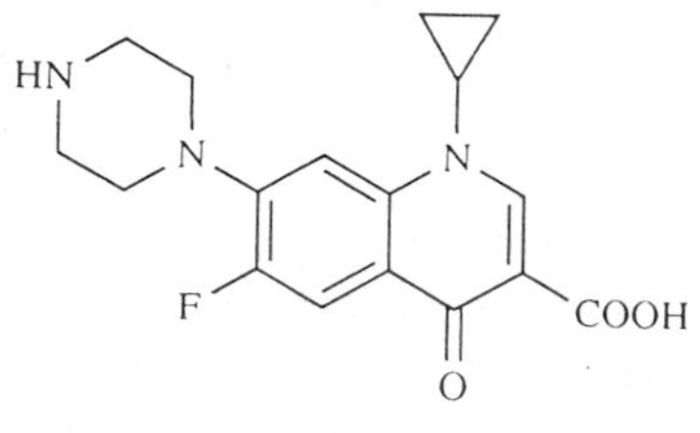

(59)

(420,421), levofloxacin (422), lomefloxacin (423,424), WIN 5723 (425), and AM-1155 (426) have been more recently investigated and are at an earlier stage of development.

MICs of ofloxacin and ciprofloxacin for *M. tuberculosis* range from 0.12 to 2 μg/mL. Both drugs are bactericidal, the MBC:MIC ratio being from 2 to 4. Levofloxacin, the levoisomer of ofloxacin, is twice as active as the parent drug, the bactericidal concentration corresponding to MIC. Ciprofloxacin and ofloxacin are active also against *M. avium* complex. MICs of ciprofloxacin for 50% of strains range from 1 to 16 μg/mL, and MIC for 90% ranges from 2 to 16 μg/mL (with data from some series exceeding 100 μg/mL). MICs of ofloxacin are a little higher, ranging from 2 to 16 μg/mL for 50% of strains and from 8 to 16 μg/mL for 90% of strains (again with values from a few studies exceeding 100 μg/mL). The MBC:MIC ratio most commonly ranges from 1 to 8. In addition, ciprofloxacin and ofloxacin show quite a good *in vitro* activity against *M. bovis*, *M. kansasii*, and *M. fortuitum* but not against *M. chelonae*.

Common favorable features of fluoroquinolones active against mycobacteria are their bactericidal action, as documented by the low MBC:MIC ratio, often ranging between 1 and 4; by a determination of growth curves in liquid medium (417,427–431); and by the independence of their activity against organisms resistant to other antimycobacterial agents (427).

The target of quinolone action in bacterial cell is topoisomerase II, a DNA gyrase that contains two A subunits (Gyr A) and two B subunits, encoded by the Gyr A gene. DNA gyrase functions within the viable bacterial cell include introduction of negative supercoils and separation of interlocked replicated daughter chromosomes. It is essential for DNA recombination and repair. The main effects of fluoroquinolones are the inhibition of DNA supercoiling and damage to DNA, whose synthesis is rapidly interrupted (432). At high quinolone concentrations, RNA and protein synthesis are also inhibited and cell filamentation occurs. It has been suggested, however, that quinolones may have other effects in the bacterial cells; for instance, the formation of an irreversible complex of drug, DNA, and enzyme functioning as a "poison." In fact, nalidixic acid reduces the burst size of bacteriophages T7 at permissive temperatures in *E. coli* strain containing a thermosensitive Gyr A subunit, but not at nonpermissive elevated temperatures, suggesting that the inhibitory action of the drug on phage T7 depends on DNA inhibition, even if the gyrase function is not required for phage growth (433). Moreover, quinolone concentrations that inhibit DNA supercoiling and decatenating activity of purified DNA gyrase are several-fold higher than those required to inhibit bacterial growth. Again, this discrepancy has suggested that other targets for quinolones exist in the bacterial cell and several interpretations of this behavior have been put forward.

The activity of fluoroquinolone on susceptible bacterial species is bactericidal, but the inhibition of DNA synthesis does not seem sufficient to explain bacterial killing. All the steps of this effect have not yet been elucidated. Cells in the logarithmic phase of growth are rapidly killed, and the rate of killing increases with an increase of drug concentration, up to a maximum above which, paradoxically, the killing is reduced (434,435). This effect was observed with penicillin and is known as the "Eagle" effect (436) after the author who first described it. It is possible that this effect is due to the inhibition of protein synthesis occurring in the presence of high concentrations of quinolones. This hypothesis is supported also by the observation that bacteria exposed to antibiotics that inhibit protein synthesis (chloramphenicol, rifampicin) or to aminoacid starvation are less efficiently killed by fluoroquinolones.

In vitro combinations of fluoroquinolones with other antituberculous drugs have given variable results. The combination of ciprofloxacin with major antituberculous drugs did not show any synergistic effect but only indifference against *M. tuberculosis*. Synergism between ciprofloxacin or ofloxacin and ethambutol was seen for some strains of *M. avium*, while the effect was less evident when rifampicin replaced ethambutol. The combination of sparfloxacin with ethambutol, but not with rifampicin, had a synergistic effect against *M. avium*. Triple combinations of sparfloxacin, rifampicin, and ethambutol resulted in synergism (416,427,437). The triple combinations comprising isoniazid; rifabutine; and either ciprofloxacin, ofloxacin, or norfloxacin, and isoniazid; rifapentine; and either pefloxacin or ciprofloxacin were the most active ones against resistant strains of *M. tuberculosis*. Rifampicin in combination with ciprofloxacin and amikacin in combination with isoniazid and ciprofloxacin gave the best results against *M. fortuitum* (438). The synergistic interaction of antituberculous drugs *in vitro*, when obtained with *in vivo* achievable concentrations, may have possible clinical efficacy.

Further information for the clinical application of fluoroquinolones can also be drawn from the activity of ofloxacin (439), ciprofloxacin (440), sparfloxacin (441), levofloxacin (442) both singly or in combination with other antituberculous drugs in some experimental models of *M. tuberculosis* infection in mice. Fluoroquinolones have been introduced in several multidrug regimens, particularly in retreatment regimens, for MDR tuberculosis both in immunocompetent and AIDS patients and in MAC disease. Ofloxacin, ciprofloxacin and more recently sparfloxacin have been administered either with ethambutol, pyrazinamide, or isoniazid and rifabutin and in other combinations. Even if some results seem promising, no definite conclusion on the clinical value of these new approaches to treatment can be drawn.

The daily dosage varies according to the different derivatives as follows: ofloxacin 400 mg twice daily, ciprofloxacin 750 mg once or twice daily, or 500 mg three times daily, and sparfloxacin 200–400 mg once daily (79,292,443,444).

Fluoroquinolones are usually well tolerated. The most common side effects are gastrointestinal reactions, central nervous disturbances, and skin and hypersensitivity reactions, particularly photosensitivity reactions (445). So far, these side effects have been observed during therapies of limited duration. However, some concern arose about the use of fluoroquinolones in mycobacterial infections that require prolonged treatment. Moreover, the necessity of multidrug therapy in mycobacterial infections increases the risk of toxicity. So far, however, no data indicate an increase of adverse events connected with the use of fluoroquinolones in the treatment of tuberculosis and mycobacteriosis. Sporadic cases of hepatotoxicity have been reported, probably due to the combination with other hepatotoxic drugs. It is suggested to monitor closely patients with concomitant liver impairment to stop the medication, if necessary (446,447).

6.3.2 MACROLIDES. Macrolides are a group of antibiotics, characterized by a large lactonic structure of 12, 14, or 16 atoms. The lactone ring has hydroxyl, alkyl, and ketone groups in various positions. In 16-membered ring macrolides an aldehydic group also can be present. Sometimes two of three hydroxyl groups are substituted by sugars, which can be neutral (6-deoxy-hexoses), bound through a α-glycosidic linkage, or basic (3-amino-sugars), bound through a β-glycosidic linkage. The first derivative introduced into therapy was erythromycin, isolated in 1952 from the fermentation broths of *Streptomyces erythreus*. It was followed by other natural derivatives, such as spiramycin and oleandomycin, produced by other *Streptomyces* species. For several decades no other derivative was added to this antibiotic family; in the 1980s josamycin and myocamycin (a semisynthetic product) entered clinical use. Since then, new semi-synthetic derivatives have become available, namely roxithromycin, dirithromycin, flurithromycin, clarithromycin, and azithromycin. Except for the last one, which has a 15-membered ring with a *N*-methyl group between C_9 and C_{10}, they are 14-membered ring derivatives. For this reason the term *azalides* has been proposed for azithromycin and other 15-membered ring macrolides (448).

Macrolides are usually administered by the oral route, but due to their physico-chemical properties, they are poorly and erratically absorbed. Passage through the stomach can result in degradation and loss of activity. To obviate this inconvenience, different salts and esters as well as adequate pharmaceutical formulations have been prepared (449). Once absorbed, macrolides diffuse well in tissues and penetrate cells, so that can be active in intracellular infections (450). They are metabolized in the liver to different extents, according to the derivative. Due to their induction of cytochrome P-450 activity, they can interfere in the activity of other drugs metabolized by the same enzyme such as theophylline, antipyrine, and methylprednisolone (451). Macrolides are eliminated mainly by the gastrointestinal route.

Among side effects, gastrointestinal disturbances prevail and some derivatives show a certain degree of hepatotoxicity. However, they are considered to be among the best tolerated antibiotics (452).

The general spectrum of activity of macrolides includes Gram-positive cocci, some Gram-negative species, and intracellular pathogens (453). The most recent macrolide derivatives show better pharmacokinetic characteristics, particularly a longer half-life, and higher tissue and intracellular concentrations than the old derivatives.

They show better activity against some Gram-negative species, such as *H. influenzae*, and in addition, clarithromycin and azithromycin have some action against mycobacteria (453). It was known that erythromycin could be active in infections due to rapidly growing mycobacteria, *M. fortuitum* and *M. chenolae* and probably also *M. smegmatis* infection. New macrolides (clarithromycin, azithromycin, roxithromycin) have a similar or higher activity (417,454).

The present interest is to determine which derivative can have clinically exploitable activity against MAC and *M. tuberculosis*.

Clarithromycin (**60**), as well as its hydroxy metabolite (14-hydroxy-clarithromycin), are active *in vitro* against MAC. However, MICs are quite variable from study to study (455–458). Several *in vitro* experiences were carried out in presence of human macrophages, where clarithromycin accumulates (459–470). MIC is 1 μg/mL and MBCs range from 16 to 64 μg/mL (460); these values are similar to those obtained in broth cultures at pH 7.4 and lower than those found in broth culture at pH 6.8 and 5. Intracellularly, mycobacteria grow at pH 6.8 and pH 5 in phagosomes and phagolysosomes, respectively. The fact that MIC and MBC against them were

(60)

lower than those found in broth cultures is an indirect indication of concentrations of clarithromycin exceeding those in the extracellular medium (460). In fact it was previously demonstrated that clarithromycin accumulates in cells at concentrations 10- to 16-fold higher than those in the extracellular fluid (461–463).

Moreover, it was shown that a single 2-h pulsed exposure of macrophages infected with *M. avium* to clarithromycin at 3 μg/mL completely inhibited the intracellular bacterial growth during the first 4 days of observation. This finding suggests that *in vivo* intracellular bacteria can be inhibited after a short period of exposure to the high concentrations of drug, which can be reached at the time of blood peak level, rather than by prolonged exposure to low concentrations (464).

In vivo experience confirms the *in vitro* results: clarithromycin showed a good activity in the beige mouse model of disseminated *M. avium* infection, inducing a dose-related reduction in spleen and liver microbial cell counts for treatment with doses of 50, 100, and 200 mg/kg. Also, in patients with infection due to *M. avium* susceptible to 2 μg/mL or less, a dramatic decline in intensity of bacteremia up to a negative blood culture was achieved after 4–10 weeks of therapy (417,457). Combination of clarithromycin with amikacin, ethambutol, or rifampicin did not result in activity superior to that of clarithromycin alone, while the combinations with clofazimine or rifabutin were more active than clarithromycin alone. The triple combinations clarithromycin–rifampicin–clofazimine, and clarithromycin–clofazimine–ethambutol were significantly more active than the macrolide alone (465).

These findings are of extreme interest, because in the therapy of tuberculosis and mycobacteriosis it is imperative to use drug combinations that, if synergistic, ensure better activity in terms of both mycobac-

terial eradication and prevention of emergence of drug resistance. Also, if macrolides are used alone, resistance rapidly appears (466,467).

Some limited clinical trials with multidrug regimens that include clarithromycin have been performed or are still under way, but more extensive trials are necessary to assess its definite place in therapy of mycobacteriosis (292,468,469). The *in vitro* activity of clarithromycin against *M. tuberculosis* is much lower than against *M. avium*, and so far it does not seem to have any clinical relevance in the treatment of tuberculosis (454).

The activity of some macrolides has been recently tested against *M. leprae in vitro* using an ATP assay and *in vivo* using a foot pad model. Erythromycin is active *in vitro* but not *in vivo* (470,471). Clarithromycin is active both *in vitro* (MIC = 0.1–2 μg/mL) and in *in vivo*, also demonstrating a bactericidal activity (472). Similar data were obtained with roxithromycin. The acid stability of these new macrolides, together with their long half-life, better penetration, and longer persistence in tissues, is probably the reason of their activity, and deserves to be further investigated (470–472).

Among the new macrolide derivatives, azithromycin (**61**) has a peculiar place due to its chemical structure, comprising a 15-membered lactone ring, and its pharmacokinetic characteristics. In fact, blood levels are low, while high concentrations are reached in tissues and cells: in polymorphonuclear neutrophyles they can be up to 200- to 300-fold higher than extracellular concentrations. The antibiotic is slowly released from the cells, a phenomenon that assumes a particular relevance in the infections were PMNs accumulate. The half-life is about 60 h (473,474). Due to these features, azithromycin can be administered for 3 days, ensuring therapeutic tissue and cell concentrations for about 7–

(61)

10 days (475–478). Azithromycin is endowed with good activity against some rapidly growing mycobacterial species and MAC. Its MICs on the latter species vary quite widely: the values range between 17 and 94 μg/mL (479,480), but the MIC_{90} (62 μg/mL) exceeds concentrations in tissues after oral dosing (481,482). It is bactericidal against *M. avium* and *M. xenopi* in macrophages. The addition of another active antimycobacterial drug such as amikacin or rifabutin enhanced the intracellular killing (483). Uptake of azithromycin by macrophages was increased by the addition of tumor necrosis factor (TNF) or γ-interferon, but it is not known if the stimulation of azithromycin uptake accounts by itself for the increased killing of the macrolide (484).

The activity of azithromycin against MAC was confirmed in infections of rats treated with cyclosporine (485) and in the beige mouse model of *M. avium* infection (480,486). The drug is also efficacious when administered intermittently. Azithromycin given in combination with rifapentine on a once weekly basis for 8 weeks showed promising activity (487). Combination with amikacin and clofazimine (488) or with clofazimine and ethambutol enhanced the efficacy of treatment; combination with rifabutin did not seem to be significantly superior to rifabutin alone (486). In the same experimental model in beige mice, azithromycin has activity comparable to clarithromycin and rifampicin against *M. kansasii, M. xenopi, M. simiae,* and *M. malmoense*. Combinations with amikacin and clofazimine were more effective than the single drug (489). Azithromycin was less active than clarithromycin and roxithromycin against *M. leprae* (490).

A few clinical trials have been carried out with azithromycin in *M. avium* infection in AIDS patients: after a few weeks of treatment (0.5–0.6 g once daily per os), a consistent reduction of bacteremia and improvement of clinical signs were observed (491). As noted, macrolides also cannot be employed alone in the therapy of mycobacteriosis but must be included in polychemotherapeutic regimens.

6.3.3 β-LACTAMS. The resistance of *M. tuberculosis* to β-lactams has been attributed to the production of β-lactamases, with the characteristics both of penicillinases and cephalosporinases (492). One of the ways shown to be useful in overcoming the problem of β-lactamase resistance in other microorganisms has been to combine an inhibitor of the enzyme with penicillin, protecting the activity of the antibiotic susceptible to the enzymatic action. Inhibitors of β-lactamase such as clavulanic acid, sulbactam, and tazobactam have been combined with amoxycillin, ampicillin, ticarcillin, and piperacillin and proved to be active not only *in vivo* but also in the clinical setting in the treatment of infections due to β-lactamase–producing strains (493,494). A number of β-lactamases have been characterized; they have different properties and different susceptibilities to β-lactamase inhibitors (495). The β-lactamase produced by *M. tuberculosis* is susceptible to clavulanic acid; therefore, interest has been focused to ascertain whether its combination with a penicillin derivative could have some activity against *M. tuber-*

culosis (496). *In vitro* the amoxycillin–clavulanic acid combination was shown to be remarkably more active than amoxycillin alone and to be endowed with bactericidal activity (497,498).

Two patients with multidrug-resistant tuberculosis responding poorly to multidrug regimens (including ethionamide, capreomycin, cycloserine in addition to streptomycin, ethambutol or pyrazinamide) were successfully treated following the addition of amoxycillin–clavulanic acid to these regimens (499). Imipenem and meropenem, two carbapenems, highly resistant to β-lactamases, were shown to have antimycobacterial activity only when a peculiar technique of *in vitro* dosing (daily addition of the drug) was adopted, compensating for the loss of activity of the antibiotics during the test incubation period. In fact, the instability of the two compounds, particularly of imipenem, in the culture medium can mask their antimycobacterial activity (500). Other β-lactams resistant to the β-lactamase action should be investigated for activity against mycobacteria and for their possible inclusion in multidrug regimens.

7 PRESENT STATUS OF THE CHEMOTHERAPY OF TUBERCULOSIS AND LEPROSY

Chemotherapy has been the most potent and useful tool for modifying the evolution and prognosis of tuberculosis and has almost reached the goal of eradicating tuberculosis from some Western countries. This positive trend was interrupted around 1985 when the incidence of tuberculosis started to increase for the reasons outlined earlier. Moreover, the rate of disease due to resistant strains of mycobacteria, which fell steadily in previous decades (being about 5% in industrialized countries), has risen to a varying extent, depending on the country

and the socioeconomic status of some population classes. Data have not been systematically monitored in each country or region; therefore, they have only an indicative value. For instance, in New York City the primary drug resistance (i.e., the resistance observed in mycobacteria isolated from previously untreated individuals) increased from 10% in 1982 to 23% in 1991 (501). The situation is similar in developing countries. Also, the data on acquired resistance, i.e., the resistance occurring when originally susceptible mycobacteria switch to resistance during therapy or after a prior therapy in the same individual, are not completely reliable due to the difference in control policy from country to country. Again some data from New York indicate a high incidence (44%) of resistant strains but in other areas of the United States as well as in Europe the situation seems not to be so dramatic (501,502). However, it is of utmost importance to know the epidemiology of resistant strains in the area where a patient is treated to apply the most adequate therapeutic regimen. This commonly must be prescribed on an empirical basis, since the results of susceptibility tests require some weeks to be available.

The success of chemotherapy is strictly related to the use of drug combinations. In the initial treatment, it is a fundamental rule to use at least three drugs, to avoid the emergence of bacterial resistance, which is the principal cause of therapeutic failure. In fact mutants resistant to one drug are easily found in the bacterial population. The use of a triple combination prevents the practitioner from involuntarily carrying out a monotherapy that would allow the emergence of resistant variants when sensitive organisms have been killed by the active drug. If multiple resistance is suspected, a four- or five-drug regimen is recommended. After the introduction of rifampicin into therapy, the duration of tuberculosis treatment, then 18–24 months, could be considerably reduced: the current length of a standard course is 6–9 months (503,504).

Drugs of first choice for the initial treatment are isoniazid and rifampicin, combined with streptomycin or ethambutol, but the combination of isoniazid, rifampicin, and pyrazinamide has also been adopted. Drugs should be administered daily for a 2-month period. In the meantime, the results of susceptibility tests should be available, which together with the clinical observation of the patient will give indications for further treatment. A 4- to 7-month period follows in which drugs can be administered intermittently, two to three times a week. If the strain is susceptible to isoniazid and rifampicin, the maintenance treatment can continue with these two drugs (503,504). However, many other options are to be considered, according to patient condition and requirements. An array of regimens have been suggested and applied with sucess for the maintenance period: one of the most commonly employed regimens includes isoniazid plus rifampicin plus pyrazinamide plus streptomycin or ethambutol administered two or three times a week. The same therapeutic regimen can be followed in HIV patients, but is should be extended to at least 9 months and a fourth drug should be added initially for severe disease. When isoniazid-resistant tuberculosis is suspected, a three-drug regimen can be adopted for the first 2 months (rifampicin plus ethambutol plus pyrazinamide), followed by 7 months of daily rifampicin and ethambutol. An alternative treatment is a four-drug regimen (rifampicin plus isoniazid plus pyrazinamide plus ethambutol or streptomycin), applied for the first 2 months, followed by rifampicin plus ethambutol twice a week for 7 months or more. For patients responding to treatment in which tubercle bacilli turn out to be susceptible to all administered drugs, ethambutol (or streptomycin) can be withdrawn and isoniazid and rifampicin are continued for 4 or 6 months.

For patients living in areas where tuberculosis sustained by multidrug-resistant mycobacteris is widespread, a five-drug regimen is recommended initially, including both ethambutol and rifampicin. Later, the treatment will be modified on the basis of results of susceptibility tests. In such patients, other antituberculous drugs, the so-called second-line drugs (ethionamide, cycloserine, capreomycin, kanamycin, para-aminosalicylic acid, and sometimes clofazimine) are variably combined with each other and with first-line drugs with the aim to overcome resistance. The activity of other drugs showing *in vitro* activity against *M. tuberculosis* is under clinical investigation: amikacin, ciprofloxacin, ofloxacin. Rifabutin has been included in some drug combinations, and it seems to have some efficacy also in the treatment of cases associated with low level rifampicin resistance. Multidrug-resistant tuberculosis often requires prolonged treatment, sometimes up to 24 months, to reach sputum conversion (502–507).

The therapy of leprosy has reached significant achievements since 1983, when WHO promoted new protocols of treatment, providing for short-term multidrug therapy (MDT) (508,509). In this period also, estimates of the total number of leprosy cases in the world decreased. This tendency is not attributed only to implementation of MDT but to a series of other factors, such as a revision in the definition of patients. Patients who have completed their course of chemotherapy are no longer counted as registered cases, even if they have residual disabilities. In addition, better living conditions and better nutrition greatly contributed to the decline in incidence of the disease in some countries. There has been a reduction of 42% in cases of leprosy in the world since 1985; the global estimates in 1991 were 5.5 million cases (of which 2 million were not receiving treatment) compared with 10–12 million in the past (508–511).

Two principal regimens have been followed according to WHO recommendations in developing countries: one directed to treat paucibacillary disease, the other for multibacillary disease. The former consists in the administration of dapsone 100 mg dialy plus rifampicin 600 mg monthly for 6 months; the other in the administration of dapsone 100 mg daily plus clofazimine 50 mg daily plus rifampicin 600 mg monthly and clofazimine 300 mg monthly for a minimum of 2 years (508,509). Some regimens used for treatment of leprosy in the United States differ from those in developing countries. Often a daily treatment with rifampicin and dapsone is privileged for 1 year in paucibacillary leprosy, for 2 years in multibacillary disease (508,509). Reactions during chemotherapy are quite frequent and are of two main types, namely erythema nodosum leprosum and reversal reactions, and are usually controlled with corticosteroids. In the most severe cases thalidomide, azothioprine, and cyclosporine have been employed, but due to serious side effects, their use has to be restricted and closely monitored (507).

New drugs showing bactericidal activity in experimental *M. leprae* infection have been submitted to clinical trials. The most promising ones seem to be ofloxacin (512), clarithromycin, minocycline (513,514), and sparfloxacin (515).

A field that attracts a particular interest is the possibility of developing a vaccine in position to prevent the disease. Three vaccines have been prepared combining BCG with *M. leprae* or other mycobacterial species and are currently under investigation, but results of these trials will not be available for 8–10 years (516). A preliminary report on a trial performed in Venezuela with a vaccine constituted by BCG and killed *M. leprae* does not demonstrate any advantage over the use of BCG alone (517). In fact, BCG alone has shown some

protective activity against both tuberculosis and leprosy (518).

REFERENCES

1. L. S. Young and G. P. Wormser, *Scand. J. Infect. Dis. (Suppl.)*, **93,** 9 (1994).
2. W. H. lopling, *Lepr. Rev.*, **62,** 1 (1991).
3. C. B Inderlied, C. A. Kemper, and L. E. M. Bermudez, *Clin. Microbiol. Rev.*, **6,** 266 (1993).
4. G. S. Wilson and A. A. Miles, *Topley and Wilson's Principles of Bacteriology and Immunity*, Arnold, London, 1975.
5. M. Castets, N. Rist, and H. Boisvert, *Bull. Soc. Med. Afr. Noire*, **16,** 221 (1969).
6. E. H. Runyon, *Adv. Tuberc. Res.*, **14,** 235 (1965).
7. G. D. Roberts, E. Koneman, and Y. K. Kim, in A. B. Balows, W. J. Hausler, K. L. Hermann, H. D. Isenberg, and H. J. Shadomy, Eds., *Manual of Clinical Microbiology*, 5th ed., ASM, Washington D. C., 1991, p. 304.
8. C. Wolinsky, *Clin. Infect. Dis.*, **15,** 1 (1992).
9. J. M. Grange, *Am. Rev. Respir. Dis.*, **140,** 161 (1969).
10. R. S. Wallace Jr., R. O'Brien, R. Glassroth, J. Raleigh, and A. Dutt, *Am. Rev. Respir. Dis.*, **142,** 940 (1990).
11. G. K. Woods and J. A. Washington II, *Rev. Inf. Dis.*, **9,** 275 (1987).
12. R. Buttiaux, H. Beerens, and A. Tacquet, *Manuel de Techniques Bactériologiques*, 4th ed., Flammarion Médecine Sciences, Paris, 1974.
13. G. Canetti and J. Grosset, *Techniques et Indications des Exames Bactériologiques en Tuberculose*, Editions de la Tourelle, St. Mandè, 1968.
14. B. Watt, A. Kayner, and G. Harris, *Rev. Med. Microbiol.*, **4,** 97 (1993).
15. G. A. Snow, *Bacteriol. Rev.*, **34,** 99 (1970).
16. E. Lederer, *Pure Appl. Chem.*, **25,** 135 (1971).
17. E. Lederer, A. Adam, R. Ciorbarn, J. F. Petit, and J. Wietzerbin, *Mol. Cell. Biochem.*, **7,** 87 (1975).
18. M. B. Goren, *Bacteriol. Rev.*, **36,** 33 (1972).
19. T. Ramakrishnan, M. Suryanarayana Murthy, and K. P. Gopinathan, *Bacteriol. Rev.*, **36,** 65 (1972).
20. L. P. Macham and C. Ratledge, *J. Gen. Microbiol.*, **89,** 379 (1975).
21. L. Barksdale and K. S. Kim, *Bacteriol. Rev.*, **41,** 217 (1977).
22. J. E. Clark-Curtiss, in J. J. McFadden, Ed.,

Molecular Biology of the Mycobacteria, Surrey University Press, London, 1990, pp. 77–96.

23. W. Segal and U. Block, *J. Bacteriol.*, **72**, 132 (1956).

24. W. Segal and U. Block, *Am. Rev. Tuberc. Pulm. Dis.*, **75**, 495 (1957).

25. W. Segal and W. T. Miller, *Proc. Soc. Exp. Biol. Med.*, **118**, 613 (1965).

26. W. Segal, *Proc. Soc. Exp. Biol. Med.*, **118**, 214 (1965).

27. G. Middlebrook, C. M. Coleman, and W. B. Schaefer, *Proc. Nat. Acad. Sci, U. S. A.*, **45**, 1801 (1959).

28. W. Segal, *Am. Rev. Respir. Dis.*, **91**, 285 (1965).

29. K. Kanai, *Jpn. J. Med. Sci. Biol.*, **20**, 401 (1967).

30. K. Kanai, *Jpn. J. Med. Sci. Biol.*, **20**, 73 (1967).

31. K. Kanai, *Jpn. J. Med. Sci. Biol.*, **20**, 91 (1967).

32. E. Kondo, K. Kanai, K. Nishimura, and T. Tsumita, *Jpn. J. Med. Sci. Biol.*, **23**, 315 (1970).

33. K. Kanai, E. Wiegeshaus, and D. W. Smith, *Jpn. J. Med. Sci. Biol.*, **23**, 327 (1970).

34. H. N. Jayaram, T. Ramakrishnan, and C. S. Vaidyanathan, *Arch. Biochem. Biophys.*, **126**, 165 (1968).

35. H. N. Jayaram, T. Ramakrishnan, and C. S. Vaidyanathan, *Indian J. Biochem.*, **6**, 106 (1969).

36. M. R. McNeil and P. J. Brennan, *Res. Microbiol.*, **142**, 451 (1991).

37. J. Asselineau, *The Bacterial Lipids*, Hermann, Paris, 1966, p. 176.

38. E. Lederer, *Chem. Phys. Lipids*, **1**, 294 (1967).

39. A. H. Etémadi and J. Convit, *Infect. Immun.*, **10**, 236 (1974).

40. A. H. Etémadi, *Exp. Ann. Biochem. Med.*, **28**, 77 (1967).

41. J. Azuma and Y. Yamamura, *J. Biochem.*, **52**, 200 (1962).

42. J. Azuma and Y. Yamamura, *J. Biochem.*, **53**, 275 (1963).

43. N. P. V. Acharya, M. Senn, and E. Lederer, *C. R. Acad. Sci. Paris*, **264**, 2173 (1967).

44. J. W. Berg, *Proc. Soc. Exp. Biol. Med.*, **84**, 196 (1953).

45. J. W. Berg, *Yale J. Biol. Med.*, **26**, 215 (1953).

46. J. Markovits, E. Vilkas, and E. Lederer, *Eur. J. Biochem.*, **18**, 287 (1971).

47. F. Kanetsuma, *Biochim. Biophys. Acta*, **98**, 476 (1965).

48. H. Bloch, *J. Exp. Med.*, **91**, 197 (1950).

49. H. Noll, H. Bloch, J. Asselineau, and E. Lederer, *Biochim. Biophys. Acta*, **20**, 299 (1956).

50. E. Dàrzins and G. Fahar, *Dis. Chest*, **30**, 642 (1956).

51. G. Middlebrook, R. J. Dubos, and C. Pierce, *J. Exp. Med.*, **88**, 521 (1948).

52. D. W. Smith, H. M. Randall, A. P. McLennan, and E. Lederer, *Nature*, **186**, 887 (1960).

53. A. P. McLennan, D. W. Smith, and H. M. Randall, *Biochem. J.*, **80**, 309 (1961).

54. G. Lanéelle and J. Asselineau, *Eur. J. Biochem.*, **5**, 487 (1968).

55. P. R. S. Gangadharam, M. L. Cohn, and G. Middlebrook, *Tubercle*, **44**, 452 (1963).

56. M. Kato and M. B. Goren, *Jpn. J. Med. Sci. Biol.*, **27**, 120 (1974).

57. P. Pigretti, E. Vilkas, E. Lederer, and H. Bloch, *Bull. Soc. Chim. Biol.*, **47**, 2039 (1965).

58. V. Braun, and K. Hantke, in L. Ninet, P. E. Bost, P. M. Bouanchaud, and J. Florent, Eds., *The Future of Antibiotherapy and Antibiotic Research*, Academic Press, London, 1981, pp. 285–296.

59. M. B. Lurie, *Resistance to Tuberculosis: Experimental Studies in Native and Acquired Defensive Mechanisms*, Harvard University Press, Cambridge, Mass., (1964).

60. A. M. Dannenberg, *Immunol. Today*, **12**, 228 (1991).

61. E. A. Nardell, in L. B. Reichman and E. S. Hershfield, Eds., *Tuberculosis: A Comprehensive International Approach*, Marcel Dekker, Inc., New York, 1993, p. 103.

62. J. L. Flinn, J. Chan, K. J. Triebold, D. K. Dalton, T. A. Stewart, and B. R. Bloom, *J. Exp. Med.*, **178**, 2249 (1993).

63. J. M. Orme, *Curr. Opin. Immunol.*, **5**, 497 (1993).

64. L. E. M. Bermudez and L. S. Young, *J. Immunol.*, **140**, 3006 (1988).

65. M. Dems, *Clin, Exp. Immunol.*, **83**, 466 (1991).

66. American Thoracic Society, *Am. Rev. Respir. Dis.*, **123**, 343 (1981).

67. A. Kochi, *Tubercle*, **72**, 1 (1991).

68. G. W. Comstock and G. M. Cauthen in ref. 61, pp. 23–48.

69. Centers for Disease Control, *MMWR*, **39**(52), 7–20 (1990).

70. Centers for Disease Control, *Tuberculosis Statistics in United States 1989*, *HH Publication No. (CDC)* 91–8322, Public Health Service, Atlanta, Ga. (1991).

71. Anonymous, *World Health Forum*, **14**, 438 (1993).

72. P. F. Barnes, and S. A. Barrows, *Ann. Intern. Med.*, **119**, 400 (1993).

73. P. C. Hopwell in ref. 61, p. 369.

74. G. Di Perri, M. C. Danzi, G. De Cecchi, G. Pizzighella, M. Solbiati. M. Cruciani, R. Luzzati, M. Malena, R. Mazzi, E. Concia, and D. Bassetti, *Lancet*, **2**, 1502 (1989).

75. C. L. Daley, P. M. Small, G. F. Schecter, G. K. Shoolnik, R. A. McAdam, W. R. Jacobs, and P. C. Hopewell, *N. Engl. J. Med.*, **326**, 231 (1992).

76. Centers for Disease Control, *MMWR*, **39**, 718 (1990).

77. C. R. Horsburgh Jr., *N. Engl. J. Med.*, **324**, 1332 (1991).

78. L. E. Bermudez, C. B. Inderlied, and L. S. Young, *Curr. Clin, Topics Infect. Dis.*, **12**, 257 (1992).

79. L. S Young, *J. Antimicrob. Chemother.*, **32**, 179 (1993).

80. WHO, *Chemotherapy of Leprosy, Technical Report Series, N. 847*, WHO, Geneva, 1994.

81. G. Vinckè, O. Yegers, H. Vanachter, P. A. Jenkins, and J. P. Butzler, *J. Antimicrob. Chemother.*, **10**, 351 (1982).

82. A. Laszlo, P. Gill, V. Handzel, M. M. Hodgkin, and D. M. Helbecque, *J. Clin. Microbiol.*, **18**, 1335 (1983).

83. L. E. Nillson, S. E. Hoffner, and S. Anséhn, *Antimicrob. Agents Chemother.*, **32**, 1208 (1988).

84. R. C. Cooksey, J. T. Crawford, W. R. Jacobs, and T. M. Shinnick, *Antimicrob. Agents Chemother.*, **37**, 1348 (1993).

85. L. H. Schmidt, *Am. Rev. Tuberc.*, **74**, 138 (1956).

86. L. H. Schmidt, *Am. N. Y. Acad. Sci.*, **135**, 747 (1966).

87. C. C. Shepard, *J. Exp. Med.*, **112**, 445 (1960).

88. C. C. Shepard and Y. T. Chang, *Proc. Soc. Exp. Biol. Med.*, **109**, 636 (1962).

89. R. J. W. Rees, *Br. J. Exp. Pathol.*, **45**, 207 (1964).

90. R. J. W. Rees, J. M. H. Pearson, and M. F. R. Waters, *Br. Med. J.*, **1**, 89 (1970).

91. W. F. Kirchheimer and E. E. Storrs, *Int. J. Lepr.*, **39**, 693 (1971).

92. E. E. Storrs, G. P. Walsh, H. P. Burchfield, and C. H. Binford, *Science*, **183**, 851 (1974).

93. T. Murohashi and K. Yoshida, *Acta Leprol.*, **54**, 31 (1974).

94. E. E. Camargo, S. M. Larson, B. S. Teper, and H. N. Wagner, *Int. J. Lepr.*, **43**, 234 (1975).

95. W. T. Colwell, G. Chan, V. H. Brown, J. I. de Graw, J. H. Peters, and N. E. Morrison, *J. Med. Chem.*, **17**, 142 (1974).

96. N. E. Morrison, *Int. J. Lepr.*, **39**, 34 (1971).

97. C. B. Inderlied, L. S. Young, and J. K. Yamada, *Antimicrob. Agents Chemother.*, **31**, 1697 (1987).

98. A. J. Crowle, A. Y. Tsang, A. E. Vatter, and M. H. May, *J. Clin. Microbiol.*, **24**, 812 (1986).

99. L. E. Bermudez and L. S. Young. *Braz. J. Med. Biol. Res.*, **20**, 191 (1987).

100. C. B. Inderlied, L. Barbara-Burnham, M. Wu, L. S. Young, and L. E. M. Bermudez, *Antimicrob. Agents Chemother.*, **38**, 1838 (1994).

101. L. E. Bermudez, P. Stevens, P. Kolonwski, M. Wu, and L. S. Young, *J. Immunol.*, **143**, 2996 (1989).

102. N. Rist, *C. R. Soc Biol.*, **130**, 972 (1939).

103. W. H. Feldman, H. C. Hinshaw, and H. E. Moses, *Proc. Staff Meet. Mayo Clin.*, **15**, 695 (1940).

104. R. Behnisch, F. Mietzsch, and H. Schmidt, *Am. Rev. Tuberc.*, **61**, 1 (1950).

105. C. F. Huebner, J. L. Marsh, R. H. Mizzoni, R. P. Mull, D. C. Schroeder, H. A. Traxell, and C. R. Scholtz, *J. Am. Chem. Soc.*, **75**, 2274 (1953).

106. N. P. Buu-Hoi and N. D. Xuong, *C. R. Acad. Sci. Paris*, **237**, 498 (1953).

107. H. H. Fox, *Science*, **116**, 129 (1952).

108. V. Chorine, *C. R. Acad. Sci. Paris*, **220**, 150 (1945).

109. S. Kushner, H. Dalalian, J. L. Sanjurio, F. L. Bach, Jr., S. R. Safir, V. K. Smith, Jr., and J. H. Williams, *J. Am. Chem. Soc.*, **74**, 3617 (1952).

110. T. S. Gardner, E. Wenis, and J. Lee, *J. Org. Chem.*, **19**, 753 (1954).

111. F. Grumbach, N. Rist, D. Libermann, M. Moyeaux, S. Cals, and S. Clavel, *C. R. Acad. Sci. Paris*, **242**, 2187 (1956).

112. F. Bernheim, *J. Bacteriol.*, **41**, 385 (1941).

113. J. Lehmann, *Lancet*, **1**, 15 (1946).

114. J. P. Thomas, C. O. Baughn, R. G. Wilkinson, and R. G. Shepherd, *Am. Rev. Respir. Dis.*, **83**, 891 (1961).

115. R. G. Wilkinson, R. G. Shepherd, J. P. Thomas, and C. Baughn, *J. Am. Chem. Soc.*, **83**, 2212 (1961).

116. A. Schatz., E. Bugie, and S. A. Waksman, *Proc. Soc. Exp. Biol. Med.*, **55**, 66 (1944).

117. R. L. Harned, P. H. Hidy, and E. K. La Baw, *Antibiot. Chemother.*, **5**, 204 (1955).

118. A. C. Finlay, G. L. Hobby, F. Hochstein, T. M. Lees, T. F. Lenert, J. A. Menas, S. Y. P'An, P. P. Regna, Y. B. Routien, B. A. Sabin, K. B. Tat, and Y. H. Kane, *Am. Rev. Respir. Dis.*, **63**, 1 (1951).

119. H. Umezawa, M. Ueda, K. Maeda, K. Yagishita, S. Korido, Y. Okami, R. Utahara, Y. Osato, K. Nitta, and T. Tacheuchi, *J. Antibiot. (Tokyo)*, **10**, 181 (1957).

120. P. Sensi, N. Maggi, S. Furesz, and G. Maffii, *Antimicrob. Agents Chemother.*, **1966**, 699 (1967).

121. L. Marsili, C. R. Pasqualucci, A. Vigevani, B. Gioia, G. Schioppacassi, and G. Oronzo, *J. Antibiot.*, **34**, 1033 (1981).

122. V. Arioli, M. Berti, G. Carniti, E. Randisi, E. Rossi, and R. Scotti, *J. Antibiot.*, **34**, 1026 (1981).

123. P. Sensi, M. T. Timbal, and G. Maffii, *Experientia*, **16**, 412 (1960).

124. P. Sensi, A. M. Greco, and R. Ballotta, *Antibiot. Ann.*, **1959–1960**, 262 (1960).

125. V. C. Barry, J. C. Belton, M. L. Conalty, J. M. Denneny, D. W. Edward, J. F. O'Sullivan, D. Twomey, and F. Winder, *Nature*, **179**, 1013 (1957).

126. P. D'Arcy Hart, *Br. Med. J.*, **2**, 849 (1946).

127. P. Sensi, *Rev. Infect. Dis.*, **11**(Suppl 2), S467 (1989).

128. P. J. Maurer and M. J. Miller, *J. Am. Chem. Soc.*, **105**, 240 (1983).

129. G. Hartmann, K.O. Honikel, F. Knussel, and J. Nuesch, *Biochim. Biophys. Acta*, **145**, 843 (1967).

130. H. Umezawa, S. Mizuno, H. Yamazaki, and K. Nitta, *J. Antibiot. (Tokyo)*, **21**, 234 (1968).

131. R. J. White, G. C. Lancini, and L. Silvestri, *J. Bacteriol.*, **108**, 737 (1971).

132. K. Konno, K. Oizumi, and S. Oka, *Am. Rev. Respir. Dis.*, **107**, 1006 (1973).

133. K. Konno, K. Oizumi, F. Arji, J. Yamaguchi, and S. Oka, *Am. Rev. Respir. Dis.*, **107**, 1002 (1973).

134. J. I. De Graw, V. H. Browa, W. T. Colwell, and N. E. Morrison, *J. Med. Chem.*, **17**, 144 (1974).

135. H. L. David, *Appl. Microbiol.*, **20**, 810 (1970).

136. K. E. Price, D. R. Chrisholm, M. Misiek, F. Leitner, and Y. H. Tsai, *J. Antibiot. (Tokyo)*, **25**, 709 (1972).

137. E. V. Cowdry, and C. Ruangsiri, *Arch.. Pathol.*, **32**, 632 (1941).

138. Y. Kurono, K. Ikeda, and K. Uekama, *Chem. Pharm. Bull. (Tokyo)*, **22**, 1261 (1974).

139. Editorial, *Lancet*, **2** (7723), 534 (1971).

140. C. C. Shepard, *Ann. Rev. Pharmacol.*, **9**, 37 (1969).

141. J. L. McCullough and T. H. Moren, *Antimicrob. Agents Chemother.*, **3**, 665 (1973).

142. R. Gelber, J. H. Peters, G. R. Gordon, A. J. Glazko, and L. Levy, *Clin. Pharmacol. Ther.*, **12** (2), 225 (1971).

143. J. H. Peters, *Am. J. Trop. Med. Hyg.*, **23**, 222 (1974).

144. J. H. S. Pettit and J. Chin, *Lancet*, **2** (7628), 1014 (1969).

145. R. J. Rees, *Trans. R. Soc. Trop. Med. Hyg.*, **61**, 581 (1967).

146. C. Ratledge and B. J. Marshall, *Biochim. Biophys. Acta*, **279**, 58 (1972).

147. C. Ratledge and K. A. Brown, *Am. Rev. Respir. Dis.*, **106**, 774 (1972).

148. K. A. Brown and C. Ratledge, *Biochim. Biophys. Acta*, **385**, 207 (1975).

149. R. Behnisch, F. Mietzsch, and H. Schmidt, *Angew. Chem.*, **60**, 113 (1948).

150. G. Domagk. R. Behnish, F. Mietzsch, and H. Schmidt, *Naturwissenschaften*, **33**, 315 (1946).

151. G. Domagk, *Am. Rev. Tuberc.*, **61**, 8 (1950).

152. R. Donovik and J. Bernstein, *Am. Rev. Tuberc.*, **60**, 539 (1949).

153. D. M. Spain, W. G. Childress, and J. S. Fisher, *Am. Rev. Tuberc*, **62**, 144 (1950).

154. G. Domagk, *Schweiz. Z. Pathol. Bakteriol.*, **12**, 575 (1949).

155. Anonymous, *Br. Med. J.*, **1** (5948), 33 (1975).

156. H. Blaha, *Med. Welt.*, **25**, 915 (1974).

157. WHO, Comité OMS d'Experts de la Tuberculose, *Technical Report*, **552**, WHO, Geneva (1974).

158. Anonymous, *Bull. WHO*, **47**, 211 (1972).

159. J. Lowe, *Lancet*, **2**, 1065 (1954).

160. E. Freerksen and M. Rosenfeld, *Beitr. Klin. Tuberk.*, **127**, 386 (1963); *Chem. Abstr.*, **60**, 4659c (1964).

161. L. Trnka, R. Urbancik, and H. Polenska, *Rozhl. Tuberk.*, **23**, 147 (1963); *Chem. Abstr.*, **59**, 8016 (1963).

162. S. Oka, K. Konno, M. Kudo, K. Oizumi, and S. Yamaguchi, *Jpn. J. Chest Dis.*, **23**, 326 (1964).

163. Editorial, *Tubercle*, **46**, 298 (1965).

164. G. Winder, P. B. Collins, and D. Whelan. *J. Gen. Microbiol.*, **66**, 379 (1971).

165. T. F. Davy, *Lepr. Rev.*, **29**, 25 (1958).

166. J. Stecca and P. Homen de Mello, *Int. J. Lepr.*, **31**, 548 (1963).

167. WHO, *Technical Report, No. 459,* Comité OMS d'Experts de la Lèpre, WHO, Geneva, 1970.

168. D. McKenzie, L. Malone, S. Kushner, J. J. Oleson, and Y. Subbarow, *J. Lab. Clin. Med.*, **33**, 1249 (1948).

169. H. Meyer and J. Mally, *Monatshefte für Chemie Monatsh. Chem.*, **33**, 392 (1912).

170. J. Bernstein, W. A. Lott, B. A. Steinberg, and H. L. Yale, *Am. Rev. Tuberc.*, **65**, 357 (1952).

171. H. A. Offe, W. Siekfen, and G. Domagk, *Z. Naturforsch*, **7b**, 446, 462 (1952).

172. E. A. Zeller, J. Barsky, J. R. Fouts, W. F. Kirchheimer, and L. S. van Orden, *Experientia*, **8**, 349 (1952).

173. F. G. Winder and P. B. Collins, *Am. Rev. Respir. Dis.*, **100**, 101 (1969).

174. K. Takayama, *Ann. N. Y. Acad. Sci.*, **235**, 426 (1974).

175. K. Takayama and H. K. Schones, *Fed. Proc.*, **33**, 1425 (1974).

176. K. Takayama, L. Wang, and R. S. Merkal, *Antimicrob. Agents Chemother.*, **4**, 62 (1973).

177. D. Kock-Weser, R. H. Ebert, W. R. Barclay, and V. S. Lee, *J. Lab. Clin. Med.*, **42**, 828 (1953).

178. J. K. Seydel, E. Wempe, and H. J. Nestler, *Arzneim. Forsch.*, **18**, 362 (1968).

179. H. J. Nestler, *Arzneim. Forsch.*, **16**, 1442 (1966).

180. J. K. Seydel, K. J. Schaper, E. Wempe, and H. P. Cordes, *J. Med. Chem.*, **19**, 483 (1976).

181. J. Youatt and S. Tham, *Am. Rev. Respir. Dis.*, **100**, 25 (1969).

182. K. S. Sriprakash and T. Ramakrishnan, *Indian J. Biochem.*, **6**, 49 (1969).

183. D. B. Young, *Curr. Biol.* **4**, 351 (1994).

184. Y. Zhang, B. Heym, B. Allen, D. Young, and S. Cole, *Nature*, **358**, 591 (1992).

185. A. Banerjee, E. Dubman, A. Quemard, V. Balasubramanian, K. Sun Um, T. Wilson, D. Collins, G. de Lisle, and W. R. Jacobs Jr., *Science*, **263**, 227 (1994).

186. Y. Zhang and D. Young, *J. Antimicrob. Chemother.*, **34**, 313 (1994).

187. C. C. Shepard and Y. T. Chang, *Int. J. Lepr.*, **32**, 260 (1964).

188. J. A. Doull, J. N. Rodriguez, A. R. Davidson, J. G. Tolentino, and J. V. Fernandez, *Int. J. Lepr.*, **25**, 173 (1957).

189. R. R. Ross, *JAMA*, **168**, 273 (1958).

190. J. M. Robson and F. M. Sullivan, *Pharmacol. Rev.*, **15**, 169 (1963).

191. W. Mandel, D. A. Heaton, W. F. Russel, and G. Middlebrook, *J. Clin. Invest.*, **38**, 1356 (1959).

192. S. Sunahara, M. Urano, and M. Ogawa, *Science*, **134**, 1530 (1961).

193. H. Tiitinen, *Scand. J. Respir. Dis.*, **50**, 110 (1969).

194. G. A. Ellard, *Clin. Pharmacol. Ther.*, **19**, 610 (1976).

195. R. I. Meltzer, A. D. Lewis, and J. A. King, *J. Am. Chem. Soc.*, **77**, 4062 (1955).

196. N. Rist, F. Grumbach, and D. Libermann, *Am. Rev. Tuberc.*, **79**, 1 (1959).

197. A. Quemard, G. Lanéelle, and C. Lacave, *Antimicrob. Agents Chemother.*, **36**, 1316 (1992).

198. A. Bieder, P. Brunel, and L. Mazeau, *Ann. Pharm. Fr.*, **24**, 493 (1966).

199. L. Malone, A. Schurr, H. Lindh, D. McKenzie, J. S. Kiser, and J. H. Williams, *Am. Rev. Tuberc.*, **65**, 511 (1952).

200. E. F. Rogers, W. J. Leanza, H. J. Becker, A. R. Matzuk, E. C. O'Neill, A. J. Basso, G. A. Stein, M. Solotorovsky, F. G. Gregory, and K. Pfister, *Science*, **116**, 253 (1952).

201. K. Konno, F. M. Feldmann, and W. McDermott, *Am. Rev. Respir. Dis.*, **95**, 461 (1967).

202. I. Toida, *Am. Rev. Respir. Dis.*, **107**, 648 (1973).

203. G. A. Ellard, *Tubercle*, **50**, 144 (1969).

204. W. Fox, *Chest*, **76**, 785 (1979).

205. E. Felder, D. Pitrè, and U. Tiepolo, *Minerva Med.*, **53**, 1699 (1962).

206. L. Trnka, J. Kuska, and A. Havel, *Chemotherapia*, **9**, 158 (1965).

207. L. Heifets and P. Lindholm-Levy, *Am. Rev. Respir. Dis.*, **145**, 1223 (1992).

208. K. Wisterowicz, M. Foks, J. Janowiec, and Z. Zwolska-Krewk, *Acta. Pol. Pharm.*, **46**, 101 (1989).

209. V. C. Barry, J. G. Belton, J. F. O'Sullivan, and D. Twomey, *J. Chem. Soc.*, **896** (1956).

210. V. C. Barry, M. L. Conalty, and E. E. Graffney, *J. Pharm. Pharmacol.*, **8**, 1089 (1956).

211. V. C. Barry, J. G. Belton, M. L. Conalty, and D. Twomey, *Nature*, **162**, 622 (1948).

212. V. C. Barry, *Sci. Proc. R. Dubl. Soc. Ser. A*, **3**, 153 (1969).

213. W. A. Vischer, *Arzneim. Forsch.*, **18**, 1529 (1968).

214. P. J. Lindholm-Levy and L. B. Heifets, *Tubercle*, **69**, 186 (1988).

215. Y. Niva, T. Sakance, Y. Miyachi, and M. Ozaki, *J. Clin. Microbiol.*, **20**, 837 (1984).

216. W. A. Vischer, *Arzneim. Forsch.*, **20**, 714 (1970).

217. D. Abrams, T. F. Mitchell, C. C. Child, S. C. Shiboski, C. L. Brosgart, M. M. Mass, and Community Consortium, *J. Infect. Dis.*, **167**, 1459 (1993).

218. R. Dautzenberg, C. Truffor, A. Mignon, W. Rozenbaum, C. Katlame, C. Perronne, R. Parrot and the GETIM (Groupe d'Etudies et de Traitement des Infections à Mycobacteries Resistantes), *Tubercle*, **72**, 168 (1991).

219. N. J. L. Gilmour, *Br. Vet. J.*, **122**, 517 (1966).

220. H. F. Lunn and R. J. W. Rees, *Lancet*, **1**, 247 (1964).

221. Y. T. Chang, *Antimicrob. Agents. Chemother.*, **1962**, 294 (1963).

222. Y. T. Chang, *Int. J. Lepr.*, **34**, 1 (1966).

223. Y. T. Chang, *Int. J. Lepr.*, **35**, 78 (1967).

224. R. Y. W. Rees, *Int. J. Lepr.*, **33**, 646 (1965).

225. J. M. Gangas, *Lepr. Rev.*, **38**, 225 (1967).

226. J. H. S. Pettit and R. J. W. Rees, *Int. J. Lepr.*, **34**, 391 (1966).

227. J. H. S. Pettit, R. J. W. Rees, and D. S. Ridley, *Int. J. Lepr.*, **35**, 25 (1967).

228. A. G. Warren, *Lepr. Rev.*, **39**, 61 (1968).

229. F. M. Imkamp, *Lepr. Rev.*, **39**, 119 (1968).

230. S. G. Browne, *Adv. Pharmacol. Chemother.*, **7**, 211 (1969).

231. H. Mathies and U. Ress, *Arzneim. Forsch.*, **20**, 1838 (1970).

232. H. L. F. Currey and P. Fowler, *Br. J. Pharmacol.*, **45**, 676 (1972).

233. E. G. Steuger, L. Aeppli, E. Peheim, and P. E. Thomann, *Arzneim. Forsch.*, **20**, 794 (1970).

234. C. C. Shepard, L. L. Walker, R. M. van Landingham, and M. A. Redus, *Proc. Soc. Exp. Biol. Med.*, **137**, 728 (1971).

235. E. G. Wilkinson, M. B. Cantrall, and R. G. Sheperd, *J. Med. Chem.*, **5**, 835 (1962).

236. A. G. Karlson, *Am. Rev. Respir. Dis.*, **84**, 905 (1961).

237. L. B. Heifets, M. D. Iseman, and P. J. Lindholm-Levy, *Antimicrob. Agents Chemother.*, **30**, 927 (1986).

238. W. H. Beggs and F. A. Andrews, *Am. Rev. Respir. Dis.*, **108**, 691 (1973).

239. W. H. Beggs and F. A. Andrews, *Antimicrob. Agents Chemother.*, **5**, 234 (1974).

240. M. Forbes, N. A. Kuck, and E. A. Peets, *J. Bacteriol.*, **84**, 1099 (1962).

241. M. Forbes, N. A. Kuck, and E. A. Peets, *J. Bacteriol.*, **89**, 1299 (1965).

242. M. Forbes, N. A. Kuck, and E. A. Peets, *Ann. N. Y. Acad. Sci.*, **135**, 726 (1966).

243. H. Reutgen and H. Iwainsky, *Z. Naturforsch*, **27b**, 1405 (1972).

244. K. Takayama and J.O. Kilburn, *Antimicrob. Agents Chemother.*, **33**, 1493 (1989).

245. G. Silve, P. Valero-Guillen, A. Quemard, M. A. Dupont, M. Daffè, and G. Lanèelle, *Antimicrob. Agents Chemother.*, **37**, 1536 (1993).

246. R. F. Corpe and F. A. Blalcock, *Dis. Chest*, **48**, 305 (1965).

247. I. D. Bobrowitz and K. S. Go Kulanathan, *Dis. Chest*, **48**, 239 (1965).

248. V. A. Place and J. P. Thomas, *Am. Rev. Respir. Dis.*, **87**, 901 (1963).

249. E. A. Peets, W. M. Sweeney, V. A. Place, and D. A. Buyske, *Am. Rev. Respir. Dis.*, **91**, 51 (1965).

250. J. E. Leihold, *Ann. N. Y. Acad. Sci.*, **135**, 904 (1966).

251. H. Shubin, *Antibiot. Ann.*, **1954–1955**, 437 (1955).

252. W. B. Geiger, *Arch. Biochem.*, **15**, 227 (1947).

253. D. Anand, B. D. Davies, and A. K. Armitage, *Nature*, **185**, 23 (1960).

254. S. S. Cohen, *J. Biol. Chem.*, **166**, 393 (1946).

255. F. E. Hahn and J. Ciak, *Bacteriol. Proc.*, **1959**, 131 (1959).

256. C. R. Krishna Murthi, *Biochem. J.*, **76**, 362 (1960).

257. C. R. Spotts and R. Y. Stamier, *Nature*, **192**, 633 (1961).

258. E. C. Cox, J. R. White, and J. G. Flakes, *Proc. Nat. Acad. Sci. U. S. A.*, **51**, 703 (1964).

259. M. Ozaki, S. Mizushima, and M. Nomura, *Nature*, **222**, 333 (1969).

260. J. Davies, D. S. Jones, and H. G. Khorana, *J. Mol. Biol.*, **18**, 48 (1966).

261. D. Elseviers and L. Gorini, in S. Mitsuhashi, Ed., *Drug Action and Drug Resistance in Bacteria, Vol. 2, Aminoglycoside Antibiotics*, University Park Press, Baltimore, Md., 1975, p. 147.

262. L. Luzzato, D. Apirion, and D. Schlessinger, *Proc. Nat, Acad, Sci., U. S. A.*, **60**, 873 (1968).

263. J. Modollel and B. D. Davies, *Proc. Nat. Acad. Sci. U. S. A.*, **61** (115), 1270 (1968).

264. J. Modollel and B. D. Davies, *Nature*, **224**, 345 (1969).

265. H. B. Newcombe and M. H. Nyholm, *Genetics*, **35**, 603 (1950).

266. H. B. Newcombe and R. Haxizko, *J. Bacteriol.*, **57**, 565 (1949).

267. J. G. Flakes, E. C. Cox, M. L. Witting, and J. R. White, *Biochem. Biophys. Res. Commun.*, **7**, 390 (1962).

268. H. Umezawa, M. Okamishi, S. Kondo, K. Hamona, R. Utahara, and K. Maeda, *Science*, **157**, 1559 (1967).

269. S. Mitsuhashi, H. Kawabe, in A. Whelton and H. C. Neu, Eds., *The Aminoglycosides. Microbiology, Clinical Use and Toxicology*, Marcel Dekker, Inc. New York, 1982, p. 97.

270. M. Finken, P. Kirschner, A. Meier, and E. C. Böttger, *Mol. Microbiol.*, **9**, 1239 (1993).

271. J. Nair, D. A. Rouse, G. H. Bai, and S. L. Morris, *Mol. Microbiol.*, **10**, 521 (1993).

272. N. Honorè and S. T. Cole, *Antimicrob. Agents Chemother.*, **38**, 238 (1994).

273. J. Douglas and L. M. Stein, *J. Infect. Dis.*, **167**, 1506, (1993).

274. A. Meier, P. Kirschner, F. C Bange, U. Vogel, and E. C. Böttger, *Antimicrob. Agents Chemother.*, **38**, 228 (1994).

275. R. C. Gordon, C. Regamey, and W. M. M. Kirby, *Antimicrob. Agents Chemother.*, **2**, 214 (1972).

276. E. M. Weyer Ed., "Kanamycin: Appraisal after Eight Years of Clinical Application", *Ann. N. Y. Acad. Sci.*, **132**, 2, 771 (1966).

277. E. M. Yow and H. Abu-Nassar, *Antibiot. Chemother.*, **11**, 148 (1963).

278. D. Apirion and D. Schlessinger, *J. Bacteriol.*, **96**, 768 (1968).

279. J. Davies, L. Gorini, and D. B. Davies, *Mol. Pharmacol.*, **1**, 93 (1965).

280. H. Masukawa, N. Tanaka, and H. Umezawa, *J. Antibiot. (Tokyo)*, **21**, 517 (1968).

281. J. Davies, P. Anderson, and B. D. Davis, *Science*, **149**, 1096 (1965).

282. N. Tanaka, Y. Yoshida, K. Sashikata, H. Yamaguchi, and H. Umezawa, *J. Antibiot. (Tokyo)*, **19**, 65 (1966).

283. K Konno, K. Oizumi, N. Kumano, and S. Oka, *Am. Rev. Respir. Dis.*, **108**, 101 (1973).

284. D. H. Starkey and E. Gregory, *Can. Med. Assoc. J.*, **105**, 587 (1971).

285. J. T. Doluisio, L. W. Dittert, and J. C. La Piana, *J. Pharmacokinet. Biopharmaceut.*, **1**, 253 (1973).

286. L. L. McDonald and J. W. St. Geme, *Antimicrob. Agents Chemother.*, **2**, 41 (1972).

287. E. Kuntz, *Klin. Wochenschr.*, **40**, 1107 (1962).

288. W. E. Sanders, R. Cacciatore, H. Valdez, N. Schneider, and C. Hartwing, *Am. Rev. Respir. Dis.*, **113** (Suppl. 4), 59 (1976).

289. W. E. Sanders Jr., C. Hartwig, N. Schneider, R. Cacciatore, and H. Valdez, *Tubercle*, **63**, 201 (1982).

290. L. B. Heifets and P. Lindholm-Levy, *Antimicrob. Agents Chemother*, **33**, 1298 (1989).

291. C. B. Inderlied, P. T. Kolonoski, M. Wu, and L. S. Young, *Antimicrob. Agents Chemother.*, **33**, 176 (1989).

292. F. De Lalla, R. Maserati, P. Scarpellini, P. Marone, R. Nicolin, F. Caccamo, and R. Rigoli, *Antimicrob. Agents Chemother.*, **36**, 1567 (1992).

293. Q. R. Bartz, J. Ehrlich, J. D. Mold, M. A. Penner, and R. M. Smith, *Am. Rev. Tuberc. Pulm. Dis.*, **63**, 4–6 (1951).

294. N. Tanaka and S. Igusa, *J. Antibiot. (Tokyo)*, **21**, 239 (1968).

295. K. Masuda and T. Yamada, *Biochim. Biophys. Acta*, **435**, 333 (1976).

296. T. Yamada, K. Masuda, K. Shoj, and M. Hari, *J. Bacteriol.*, **112**, 1 (1972).

297. T. Yamada, K. Masuda, Y. Mizuguchi, and K. Suga, *Antimicrob. Agents Chemother.*, **9**, 817 (1976).

298. E. B. Herr Jr. and M.O. Redstone, *Ann. N. Y. Acad. Sci.*, **135**, 940 (1966).

299. D. A. Harris, R. Ruger, M. A. Reagan, F. J. Wolf, R. L. Peck, H. Wallick, and H. B. Woodruff, *Antibiot. Chemother.*, **5**, 183 (1955).

300. G. Shull and J. Sardinas, *Antibiot. Chemother.*, **5**, 398 (1955).

301. F. A. Kuehl, F. J. Wolf, N. R. Trenner, R. L. Peck, R. H. Buhs, I. Putter, R. Ormond, J. E. Lyons, L. Chaiet, E. Howe, B. D. Hunnewell, G. Downing, E. Newstead, and K. Folkers, *J. Am. Chem. Soc.*, **77**, 2344 (1955).

302. P. H. Hidy, E. B. Hodge, V. V. Young, R. L. Harned, G. A. Brewer, W. F. Phillips, W. F. Runge, H. E. Stavely, A. Pohland, H. Boaz, and H. R. Sullivan, *J. Am. Chem. Soc.*, **77**, 2345 (1955).

303. C. H. Stammer, A. N. Wilson, C. F. Spencer, F. W. Bachelor, F. W. Holly, and K. Folkers, *J. Am. Chem. Soc.*, **79**, 3236 (1957).

304. J. L. Strominger, E. Ito, and R. H. Threnn, *J. Am. Chem. Soc.*, **82**, 998 (1960).

305. J. L. Strominger, R. H. Threnn, and S. S. Scott, *J. Am. Chem. Soc.*, **81**, 3083 (1959).

306. F. C. Neuhaus and J. L. Lynch, *Biochemistry*, **3**, 471 (1964).

307. P. Sensi, P. Margalith, and M. T. Timbal, *Farmaco [Sci.]*, **14**, 146 (1959).

308. P. Margalith and G. Beretta, *Mycopathol. Mycol. Appl.*, **8**, 321 (1960).

309. J. E. Thiemann, G. Zucco, and G. Pelizza, *Arch. Microbiol.*, **67**, 147 (1969).

310. P. Margalith and H. Pagani, *Appl. Microbiol.*, **9**, 325 (1961).

311. P. Sensi, R. Ballotta, A. M. Greco, and G. G. Gallo, *Farmaco [Sci.]*, **16**, 165 (1961).

312. W. Oppolzer, V. Prelog, and P. Sensi, *Experientia*, **20**, 336 (1964).

313. M. Brufani, W. Fedeli, G. Giacomello, and A. Vaciago, *Experientia*, **20**, 339 (1964).

314. J. Leitich, W. Oppolzer, and V. Prelog, *Experientia*, **20**, 343 (1964).

315. W. Oppolzer and V. Prelog, *Helv. Chim. Acta*, **56**, 2287 (1973).

316. P. Siminoff, R. M. Smith, W. T. Sokolski, and G. M. Savage, *Am. Rev. Respir. Dis.*, **75**, 579 (1957).

317. K. L. Rinehart, M. L. Maheshwari, K. Sasaki, A. J. Schacht, H. H. Mathur, and F. J. Antosz, *J. Am. Chem. Soc.*, **93**, 6273 (1971).

318. T. Kishi, M. Asai, M. Muroi, S. Harada, E. Mizuta, S. Teroo, T. Miki, and K. Mizuno, *Tetrahedron Lett.*, **1969**, 91 (1969).

319. T. Kishi, S. Harada, M. Asai, M. Muroi, and K. Mizuno, *Tetrahedron Lett.*, **1969**, 97 (1969).

320. A. K. Ganguly, S. Szmulervicz, O. Z. Sarre, D. Greeves, J. Morton, and J. Glotten, *J. Chem. Soc. Commun.*, **1974**, 395 (1974).

321. G. Maffii, P. Schiatti, G. Bianchi, and M. G. Serralunga, *Farmaco [Sci.]*, **16**, 235 (1961).

322. M. T. Timbal and A. Brega, *Farmaco [Sci.]*, **16**, 191 (1961).

323. G. Maffii, G. Bianchi, P. Schiatti, and G. G. Gallo, *Farmaco [Sci.]*, **16**, 246 (1961).

324. M. T. Timbal, R. Pallanza, and G. Carniti, *Farmaco [Sci.]*, **16**, 181 (1961).

325. B. Rescigno, *Arch. Tisiol.*, **18**, 238 (1963).

326. V. Monaldi, *Chemotherapia*, **7**, 569 (1963).

327. A. Tacquet, *Chemotherapia*, **7**, 492 (1963).

328. N. Bergamini and G. Fowst, *Arzneim. Forsch*, **796**, 953 (1965).

329. L. De Souza Lima and D. V. A. Opromolla, *Chemotherapia*, **7**, 668 (1963).

330. D. V. A. Opromolla and L. De Souza Lima, *El Dia Medico*, **37**, 700 (1965).

331. F. P. Merklen and F. Cottenot, *Bull. Soc. Fr. Dermatol.*, **70**, 528 (1963).

332. D. V. A. Opromolla, L. De Souza Lima, and G. Caprara, *Lepr. Rev.*, **36**, 123 (1965).

333. M. Silva and J. B. Risi, *Rev. Bras. Med.*, **20**, 194 (1963).

334. G. Farris and A. Baccaredda-Boy, *Int. J. Lepr.*, **31**, 560 (1963).

335. F. P. Merklen and F. Cottenot, *Presse Med.*, **72**, 48 (1964).

336. P. Sensi, *Pure Appl. Chem.*, **35**, 383 (1973).

337. G. Lancini and W. Zanichelli, in D. Perlman, Ed., *Structure-Activity Relationship among the Semisynthetic Antibiotics.*, Academic Press, Inc. New York, 1977, p. 531.

338. P. Sensi, and G. Lancini, in C. Hansch Ed., *Comprehensive Medicinal Chemistry*, Vol. 2; Pergamon Press, New York, 1990, pp. 793–811.

339. M. Brufani, G. Cecchini, L. Cellai, M. Federici, M. Guiso, and A. Segre, *J. Antibiot.*, **38**, 259 (1985).

340. M. Brufani, L. Cellai, L. Cozzella, M. Federici, M. Guiso, and A. Segre, *J. Antibiot.*, **38**, 1359 (1985).

341. S. K. Arora, *Mol. Pharmacol.*, **23**, 133 (1983).

342. M. Brufani, S. Cerrini, W. Fedeli, and A. Vaciago, *J. Mol. Biol.*, **87**, 409 (1974).

343. M. Brufani, L. Cellai, S. Cerrini, W. Fedeli, A. Segre, and A. Vaciago, *Mol. Pharmacol.*, **21**, 394 (1981).

344. G. G. Gallo, E. Martinelli, V. Pagani, and P. Sensi, *Tetrahedron*, **30**, 3093 (1974).

345. P. Sensi, N. Maggi, R. Ballotta, S. Furesz, R. Pallanza, and V. Arioli, *J. Med. Chem.*, **7**, 596 (1964).

346. G. Maffii and P. Schiatti, *Toxicol. Appl. Pharmacol.*, **8**, 138 (1966).

347. G. Pelizza, G. C. Lancini, G. C. Allievi, and G. G. Gallo, *Farmaco [Sci.]*, **28**, 298 (1973).

348. N. Maggi, R. Pallanza, and P. Sensi, *Antimicrob. Agents Chemother.*, **1965**, 765.

349. N. Maggi, C. R. Pasqualucci, R. Ballotta, and P. Sensi, *Chemotherapia*, **11**, 285 (1966).

350. L. Verbist and A. Gyselen, *Am. Rev. Respir. Dis.*, **98**, 923 (1968).

351. G. R. F. Hilson, D. K. Banerjee, and J. B. Holmes, *Int. J. Lepr.*, **39**, 3499 (1971).

352. C. C. Shepard, L. L. Walker, R. M. Van Landingham, and M. A. Redus, *Am. J. Trop. Med. Hyg.*, **20**, 616 (1971).

353. S. R. Pattyn, *Int. J. Lepr.*, **41**, 489 (1973).

354. S. R. Pattyn and E. J. Saerens, *Ann. Soc. Belg. Med. Trop.*, **54**, 35 (1974).

355. J. K. McClatchy, R. F. Wagonner, and W. Lester, *Am. Rev. Respir. Dis.*, **100**, 234 (1969).

356. P. W. Steinbruck, *Acta Tuberc. Pneumol. Belg.*, **60**, 413 (1969).

357. W. Wehrli, *Eur. J. Biochem.*, **80**, 325 (1977).

358. R. J. White and G. C. Lancini, *Biochim. Biophys. Acta*, **240**, 429 (1971).

359. U. J. Lill and G. R. Hartmann, *Eur. J. Biochem.*, **38**, 336 (1973).

360. S. Riva and L. G. Silvestri, *Ann. Rev. Microbiol.*, **26**, 199 (1972).

361. W. Wehrli and M. Staehelin, in J. W. Corcoran and H. Hahn, Eds., *Antibiotics*, Vol. 3, *Mechanism of Action of Antimicrobial and Antitumor Agents*, Springer-Verlag, Berlin, **1975**, pp. 252–268.

362. W. Stender, A. A. Stutz, and K. H. Scheit, *Eur. J. Biochem.*, **56**, 129 (1975).

363. W. Schulz and W. Zillig, *Nucleic Acids Res.*, **9**, 6889 (1981).

364. C. Kessler, M. Huaifeng, and G. R. Hartmann, *Eur. J. Biochem.*, **122**, 515 (1982).

365. N. Honore and S. T. Cole, *Antimicrob. Agents Chemother.*, **37**, 414 (1993).

366. A. Telenti, P. Imboden, F. Marchesi, D. Lowrie, S. Cole, M. J. Colston, L. Matter, K. Schopper, and T. Bodmer, *Lancet*, **341**, 647 (1993).

367. L. Miller, J. T. Crawford, and T. M. Shinnick, *Antimicrob. Agents Chemother.*, **38**, 805 (1994).

368. G. R. Hartmann, P. Heinrich, M. C. Kollenda, B. Skrobranck, M. Tropschug, and W. Weiss, *Angew. Chem.*, **24**, 1009 (1985).

369. R. Pallanza, V. Arioli, S. Furesz, and G. Bolzoni, *Arzneim Forsch.*, **17**, 529 (1967).

370. F. Grumbach, and N. Rist, *Rev. Tuberc.* (Paris), **31**, 749 (1967).

371. L. Verbist, *Acta Tuberc. Pneumol. Belg.*, **60**, 390 (1969).

372. F. Grumbach, G. Canetti, and M. Le Lirzin, *Tubercle*, **50**, 280 (1969).

373. F. Grumbach, G. Canetti, and M. Le Lirzin, *Rev. Tubercle Pneumol. (Paris)*, **34**, 312 (1970).

374. F. Kradolfer and R. Schnell, *Chemotherapy*, **15**, 242 (1970).

375. F. Kradolfer, *Am. Rev. Respir. Dis.*, **98**, 104 (1968).

376. F. Kradolfer, *Antibiot. Chemother.*, **16**, 352 (1970).

377. V. Nitti, E. Catena, A. Ninni, and A. Di Filippo, *Arch. Tisiol.*, **21**, 867 (1966).

378. V. Nitti, E. Catena, A. Ninni, and A. Di Filippo, *Chemotherapia*, **12**, 369 (1967).

379. V. Nitti, *Antibiot. Chemother.*, **16**, 444 (1970).

380. M. Lucchesi and P. Mancini, *Antibiot. Chemother.*, **16**, 431 (1970).

381. S. Furesz, *Antibiot. Chemother.*, **16**, 316 (1970).

382. P. Kluyskens, *Acta Tuberc. Pneumol. Belg.*, **60**, 323 (1969).

383. S. Furesz, R. Scotti, R. Pallanza, and E. Mapelli, *Arzneim Forsch.*, **17**, 726 (1967).

384. G. Curci, A. Ninni, and F. Iodice, *Acta Turberc. Pneumol. Belg.*, **60**, 276 (1969).

385. G. Acocella, V. Pagani, M. Marchetti, G. B. Baroni, and F. B. Nicolis, *Chemotherapy*, **16**, 356 (1971).

386. G. Acocella, *Clin. Pharmacokinet.*, **3**, 108 (1978).

387. G. Acocella, L. Bonollo, M. Garimoldi, M. Mainardi, L. T. Tenconi, and F. B. Nicolis, *Gut*, **13**, 47 (1972).

388. L. Dettli and F. Spina, *Farmaco [Sci.]*, **23**, 795 (1968).

389. N. Maggi, S. Furesz, R. Pallanza, and G. Pelizza, *Arzneim. Forsch.*, **19**, 651 (1969).

390. A. M. Baciewicz, T. H. Self, and W. B. Bekemeyer, *Arch. Intern. Med.*, **147**, 565 (1987).

391. G. Binda, E. Domenichini, A. Gottardi, B. Orlandi, E. Ortelli, B. Pacini, and G. Fowst, *Arzneim Forsch.*, **21**, 796 (1971).

392. D. J. Girling, *J. Antimicrob. Chemother.*, **3**, 115 (1977).

393. D. L. Leiker, *Int. J. Lepr.*, **39**, 462 (1971).

394. R. J. W. Rees, M. F. R. Waters, H. S. Helmy, and J. M. H. Pearson, *Int. J. Lepr.*, **41**, 682 (1973).

395. C. Della Bruna, G. Schioppacassi, D. Ungheri, D. Iabes, E. Morvillo, and A. Sanfilippo, *J. Antibiot.*, **36**, 1502 (1983).

396. D. Ungheri, C. Della Bruna, and A. Sanfilippo, *G. Ital. Chemioter.*, **31**, 211 (1984).

397. C. L. Woodley and J.O. Kilbum, *Am. Rev. Respir. Dis.*, **126**, 586 (1982).

398. H. Saito, K. Sato, and H. Tomioka, *Tubercle*, **69**, 187 (1988).

399. L. B. Heifets and M. D. Iseman, *Am. Rev. Respir. Dis.*, **132**, 710 (1985).

400. M. H. Cynamon, *Antimicrob. Agents Chemother.*, **28**, 440 (1985).

401. S. P. Klemens, M. A. Grossi, and M. H. Cynamon, *Antimicrob. Agents Chemother.*, **38**, 234 (1994).

402. M. H. Skinner and T. F. Blaschke, *Clin. Pharmacokinet.*, **28**(2), 115 (1995).

403. S. D. Nightingale, D. W. Cameron, F. M. Gordin, P. M. Sullam, D. L. Cohn, R. E. Chaisson, L. J. Eron, P. D. Sparti, B. Bihari, D. L. Kaufman, J. J. Stern, D. D. Pearce, W. G. Weinberg, A. La Marca, and F. P. Siegal, *N. Engl. J. Med.*, **329**, 828 (1993).

404. J. M. Dickinson and D. A. Mitchison, *Tubercle*, **68**, 113 (1987).

405. C. S. F. Easmon and J. P. Crane, *J. Antimicrob. Chemother.*, **13**, 585 (1984).

406. S. P. Klemens and M. H. Cynamon, *J. Antimicrob. Chemother.*, **29**, 555 (1992).

407. A. Assandri, T. Cristina, and L. Moro, *J. Antibiot.*, **31**, 894 (1978).

408. A. Assandri, A. Perazzi, and M. Berti, *J. Antibiot.*, **30**, 409 (1977).

409. H. Bickel, F. Knusel, W. Kamp, and L. Neipp, *Antimicrob. Agents Chemother.*, **1966**, 352 (1967).

410. F. Kradolfer, L. Neipp, and W. Sackmann, *Antimicrob. Agents Chemother.*, **1966**, 359 (1967).

411. T. Yamane, T. Hashizume, K. Yamashita, K. Hosoe, T. Hidaka, K. Watanabe, H. Kawaharada, and S. Kudoh, *Chem. Pharm. Bull.*, **40**, 2707 (1992).

412. T. Yamane, T. Hashizuma, K. Yamashita, E. Konishi, K. Hosoe, T. Hidaka, K. Watanabe, H. Kawaharada, T. Yamamoto, and F. Kuze, *Chem. Pharm. Bull.*, **41**, 148 (1993).

413. S. P. Klemens, M. A. Grossi, and M. H. Cynamon, *Antimicrob. Agents Chemother.*, **38**, 2245 (1994).

414. N. Karabalut and G. Drusano, in D. C. Hooper and J. S. Wolfson, Eds., *Quinolone Antimicrobial Agents*, 2nd ed., *American Society for Microbiology*, Washington, D. C., 1993, p. 195.

415. L. A. Mitscher, P. Devasthole, and R. Zadov in ref. 414, p. 3.

416. G. M. Eliopoulos and C. T. Eliopoulos in ref. 414, p. 161.

417. L. B. Heifets, *Semin. Respir. Infect.* **9**, 84 (1994).

418. N. Khardory, K. Rolston, B. Rosenbaum, S. Hayat, and G. T. Bodey, *J. Antimicrob. Chemother.*, **24**, 667 (1989).

419. J. A. Garcia-Rodriguez and A. C. Gomez-Garcia, *J. Antimicrob. Chemother.*, **32**, 797 (1993).

420. N. Rastogi and K. S. Goh, *Amtimicrob. Agents Chemother.*, **35**, 1933 (1991).

421. B. Ji, C. Truffot-Pernot and J. Grosset, *Tubercle*, **72**, 181 (1991).

422. N. Mor, J. Vanderkolk, and L. Heifets, *Antimicrob. Agents Chemother.*, **38**, 1161 (1993).

423. C. Piersimoni, V. Morbiducci, S. Bornigia, G. De Sio, and G. Scalise, *Am. Rev. Respir. Dis.*, **146**, 1445 (1992).

424. D. K. Benerjee, J. Ford, and S. Markanday, *J. Antimicrob. Chemother.*, **30**, 236 (1992).

425. L. B. Heifets and P. J. Lindholm-Levy, *Antimicrob. Agents Chemother.*, **34**, 770 (1990).

426. H. Tomioka, H. Saito, and K. Sato, *Antimicrob. Agents Chemother.*, **37**, 1259 (1993).

427. D. C. Leysen, A. Haemers, and S. R. Pattyn, *Antimicrob. Agents Chemother.*, **33**, 1 (1989).

428. L. B. Heifets and P. J. Lindholm-Levy, *Tubercle*, **68**, 267 (1987).

429. D. M. Yajko, P. S. Nassos, and W. K. Hadley, *Antimicrob. Agents Chemother.*, **31**, 117 (1987).

430. D. M. Yajko, C. A. Sanders, P. S. Nassos, and K. Hadley, *Antimicrob. Agents Chemother.*, **34**, 2442 (1990).

431. N. Rastogi, La Brousse, K. S. Goh, and J. P. Carvalho de Sousa, *Antimicrob. Agents Chemother.*, **35**, 2473 (1991).

432. D. C. Hooper and J. S. Wolfson in ref. 414, p. 53.

433. K. N. Kreuzer and N. R. Cozzarelli, *J. Bacteriol.*, **140**, 424 (1979).

434. G. C. Crumplin and J. T. Smith, *Nature*, **260**, 643 (1976).

435. J. T. Smith, *Pharm., J.*, **233**, 299 (1984).

436. H. Eagle and A. D. Musselmann, *J. Exp. Med.*, **88**, 99 (1948).

437. M. Casal, J. Gutierrez, J. Gonzales, and P. Ruiz, *Chemioterapia*, **6**, 437 (1987).

438. M. Casal, F. Rodriguez, J. Gutierrez, and P. Ruiz, *Rev. Infect. Dis.*, **11**(Suppl. 5), S1042 (1989).

439. M. Tsukamura, *Am. Rev. Respir. Dis.*, **132**, 915 (1985).

440. M. Chadwich, G. Nicholson, and H. Gaya, *Am. J. Med.*, **87**(Suppl. 5A), 35s (1989).

441. V. Lalande, C. Truffot-Pernot, A. Paccaly-Moulin, J. Grosset, and B. Ji, *Antimicrob. Agents Chemother.*, **37**, 407 (1993).

442. S. P. Klemens, C. A. Sharpe, M. C. Roggie, and M. H. Cynamon, *Antimicrob. Agents Chemother.*, **38**, 1476 (1994).

443. Hong Kong Chest Service and British Medical Research Council, *Tuberc. Lung Dis.*, **73**, 59 (1992).

444. R. J. O'Brien in ref. 414, p. 207.

445. D. C. Hooper and J. S. Wolfson in ref. 414, p. 489.

446. P. Ball, *Rev. Infect. Dis.*, **11**(Suppl. 5), S1365 (1989).

447. N. Kennedy, R. Fox, L. Viso, F. I. Ngowi, and S. H. Gillespie, *J. Antimicrob. Chemother.*, **32**, 897 (1993).

448. A. Bryskier, J. Gasc, and E. Agoridas in A. Bryskier, J. P. Butzler, H. C. Neu, and P. M. Tulkens, Eds., *Macrolides: Chemistry, Structure, Activity*, Arnette Blackwell, Oxford, UK, 1993, p. 5.

449. H. Lode, M. Boeckh, and T. Schaberg in ref. 448, p. 409.

450. M. T. Labro in ref. 448, p. 379.

451. D. Mansuy and M. Delaforge in ref. 448, p. 635.

452. R. E. Polk and D. Israel in ref. 448, p. 647.

453. H. C. Neu in ref. 448, p. 167.

454. C. B. Inderlied, L. M. Bermudez, and L. S. Young in ref. 448, p. 285.

455. L. B. Heifets in L. B. Heifets, Ed., *Drug Susceptibility in the Chemotherapy of Mycobacterial Infections*, CRC Press, Inc. Boca Raton, Fla., 1991.

456. C. Truffot-Pernot and J. B. Grosset, *Antimicrob. Agents Chemother.*, **35**, 1677 (1991).

457. S. Naik and R. Ruck, *Antimicrob. Agents Chemother.*, **33**, 1614 (1989).

458. P. B. Fernandes, D. J. Hardy, D. McDaniel, C. W. Hanson, and R. N. Swanson, *Antimicrob. Agents Chemother.*, **33**, 1531 (1989).

459. Y. Cohen, C. Perronne, C. Truffot-Pernot, J. Grosset, J. I. Vilde, and J. J. Pocidalo, *Antimicrob. Agents Chemother.*, **36**, 2104 (1992).

460. N. Mor and L. Heifets, *Antimicrob. Agents Chemother.*, **37**, 111 (1993).

461. N. Mor, J. Vanderkolk, and L. Heifets, *Pharmacotherapy*, **14**, 100 (1994).

462. R. Anderson, G. Joone, and C. E. J. Van Rensburg, *J. Antimicrob. Chemother.*, **22**, 923 (1988).

463. M. Ishiguro, H. Koga, S. Kohno, T. Hayashy, K. Yamaguchi, and M. Hirota, *J. Antimicrob. Chemother.*, **24**, 719 (1989).

464. N. Mor and L. Heifets, *Antimicrob. Agents Chemother.*, **37**, 1380 (1993).

465. S. P. Klemens, M. S. De Stefano, and M. H. Cynamon, *Antimicrob. Agents Chemother.*, **36**, 2413 (1992).

466. R. E. Chaisson, C. Benson, M. Dube, J. Korvick, A. Wu, S. Lichter, M. Dellerson, T. Smith, and F. Sattler, *Program Abstracts 32nd Intersci. Conf. Antimicrob. Agents Chemother.* (ICAAC), Am. Soc. Microb., Washington, D.C., 1992, abstract 891.

467. L. B. Heifets, N. Mor, and J. Vanderkolk, *Antimicrob. Agents Chemother.*, **37**, 2364 (1993).

468. P. Prokocima, M. Dellerson, C. Crafs, A. Pernet, B. Ruff, and J. Grosset, *Program Abstracts 30th Intersci. Conf. Antimicrob. Agents Chemother.* (ICAAC), Am. Soc. Microb., Washington, D.C., 1990, abstract 634.

469. B. Dautzenberg, C. Truffot-Pernot, S. Legris, M. C. Meyohas, H. C. Berlie, A. Mercat, S. Chevret, and J. Grosset, *Am. Rev. Respir. Dis.*, **144**, 564 (1991).

470. S. G. Franzblau and R. C. Hasting, *Antimicrob. Agents Chemother.*, **32**, 1758 (1988).

471. R. Gelber, *Prog. Drug Res.*, **34**, 421 (1990).

472. N. Ramasel, J. Krahenbuhl, and R. C. Hasting, *Antimicrob. Agents Chemother.*, **33**, 657 (1989).

473. H. Lode, *Eur. J. Clin. Microbiol. Infect. Dis.*, **10**, 807 (1991).

474. P. J. McDonald and H. Pruul, *Eur. J. Clin. Microbiol. Infect. Dis.*, **10**, 828 (1991).

475. S. Schonwald, V. Skerk, I. Petricevic, V. Car, L. Majerus-Misic, and M. Gunjaca, *Eur. J. Clin. Microbiol. Infect. Dis.*, **10**, 877 (1991).

476. A. I. M. Hoepelman, A. P. Sips, J. L. M. Van Helmond, P. W. C. Van Barneveld, A. J. Neve, M. Zwinkels, M. Rozenberg-Arska, and J. Verhoef, *J. Antimicrob. Chemother.*, **31**(Suppl. E), 147 (1993).

477. F. Bradbury, *J. Antimicrob. Chemother.*, **31**(Suppl. E), 153 (1993).

478. J. Myburgh, G. J. Nagel, and E. Petschel, *J. Antimicrob. Chemother.*, **31**(Suppl. E), 163 (1993).

479. O. G. Berlin, L. S. Young, S. A. Floyd-Reising, and D. A. Bruckner, *Eur. J. Clin. Microbiol. Infect. Dis.*, **6**, 486 (1987).

480. C. B. Inderlied, P. T. Kolonoski, M. Wu, and L. S. Young, *J. Infect. Dis.*, **159**, 994 (1989).

481. A. E. Girard, D. Girard, and J. A. Retsema, *J. Antimicrob. Chemother.*, **25**(Suppl. A), 61 (1980).

482. J. A. Retsema, A. E. Girard, D. Girard, and W. B. Milisen, *J. Antimicrob. Chemother.*, **25**(Suppl. A), 83 (1990).

483. M. J. Gevaudan, C. Bollet, M. N. Mallet, G. Giulian, H. Tissot-Dupont, and P. De Micco, *Pathol. Biol.*, **35**, 413 (1990).

484. L. E. Bermudez, C. Inderlied, and L. S. Young, *Antimicrob. Agents Chemother.*, **35**, 2625 (1991).

485. S. T. Brown, F. F. Edwards, E. M. Bernard, W. Tong, and D. Armstrong, *Antimicrob. Agents Chemother.*, **37**, 398 (1993).

486. M. H. Cynamon and S. P. Klemens, *Antimicrob. Agents Chemother.*, **36**, 1611 (1992).

487. S. P. Klemens and M. H. Cynamon, *Antimicrob. Agents Chemother.*, **38**, 1721 (1994).

488. P. T. Kolonoski, J. Martinelli, M. L. Petrosky, M. Wu, and C. B. Inderlied, paper presented at *the 89th Annual Meeting of the American Society for Microbiology*, New Orleans, 1989.

489. S. P. Klemens and M. H. Cynamon, *Antimicrob. Agents Chemother.*, **38**, 1455 (1994).

490. R. H. Gelber, P. Siu, M. Tsang, and L. P. Murray, *Antimicrob. Agents Chemother.*, **35**, 760 (1991).

491. L. S. Young, L. Wiviott, M. Wu, P. Kolonoski, R. Bolan, and C. B. Inderlied, *Lancet*, **338**, 1107 (1991).

492. J. E. Kasic, *Am. Rev. Respir. Dis.*, **91**, 117 (1965).

493. H. C. Neu, A. P. R. Wilson, and R. N. Grunberg, *J. Chemother.*, **5**, 67 (1993).

494. K. Bush, *Antimicrob. Agents. Chemother.*, **33**, 259 (1989).

495. Y. Zhang, V. A. Steingube, and R. J. Wallace Jr., *Am. Rev. Respir. Dis.*, **145**, 657 (1992).

496. G. N. Rolinson, *J. Chemother.*, **6**, 283 (1994).

497. M. H. Cynamon and G. S. Palmer, *Antimicrob. Agents. Chemother.*, **24**, 429 (1983).

498. M. Casal, F. Rodriguez, M. Benavente, and M. Luna, *Eur. J. Clin. Microbiol.*, **5**, 453 (1986).

499. J. P. Nadler, J. Berger, J. A. Nord, R. Cofsky, and M. Saxena, *Chest*, **99**, 1025 (1991).

500. B. Watt, J. R. Edwards, A. Rayner, A. J. Grindey, and G. Harvis, *Tuberc. Lung Dis.*, **73**, 134 (1992).

501. T. R. Frieden, T. Sterling, A. Pablos-Mendez, J.O. Kilburn, G. M. Cauthen, and S. W. Dooley, *New. Engl. J. Med.*, **328**, 521 (1993).

502. M. D. Iseman and J. A. Sbarbaro, *Curr. Clin. Top. Infect. Dis.*, **12**, 188 (1992).

503. R. J. O'Brian in ref. 61, p. 207.

504. C. Grassi, *Medit. J. Infect. Parasit. Dis.*, **7**, 17 (1992).

505. American Thoracic Society, *Am. J. Respir. Crit. Care Med.*, **149**, 1359 (1994).

506. R. F. Jacobs, *Clin. Infect. Dis.*, **19**, 1 (1994).

507. M. D. Iseman, *New Engl. J. Med.*, **329**, 784 (1993).

508. L. J. Yoder, *Curr. Opin. Infect. Dis.*, **4**, 302 (1991).

509. H. J. Yoder, *Curr. Opin. Infect. Dis.*, **6**, 349 (1993).

510. S. K. Nordeen, *Lepr. Rev.*, **62**, 72 (1991).

511. S. K. Nordeen, *Lepr. Rev.*, **63**, 282 (1992).

512. Anonymous, *Trop. Dis. Res. News*, **38**, 2 (1992).

513. B. Jt, E. G. Perani, and J. H. Grosset, *Antimicrob. Agents Chemother.*, **33**, 579 (1991).

514. R. H. Gelber, S. Byrd, L. P. Murray, P. Siu, and M. Tsang, *Br. Med. J.*, **304**, 91 (1992).

515. M. Gidoh and S. Tsutsumi, *Lepr. Rev.*, **63**, 108 (1992).

516. M. D. Gupte, *Indian J. Lepr.*, **63**, 342 (1991).

517. J. Convit, C. Sampson, M. Zuniga, P. G. Smith, J. Plata, J. Silva, J. Molinie, M. E. Pinardi, B. R. Bhoom, and A. Salgado, *Lancet*, **339**, 446 (1992).

518. J. M. Ponninhans, P. E. M. Fine, J. A. C. Sterne, R. J. Wilson, E. Msosa, P. J. K. Gruer, P. A. Jenkins, S. B. Lucas, N. G. Liomba, and L. Buss, *Lancet*, **339**, 6363 (1992).

CHAPTER THIRTY-FIVE

Antifungal Agents

EUGENE D. WEINBERG

Department of Biology and
    Program in Medical Sciences
Indiana University
Bloomington, Indiana, USA

CONTENTS

1 Introduction, 637

2 Fungal Diseases, 638

3 Antifungal Substances, 640
  3.1 Treatment of dermatophytoses, 641
      3.1.1 Griseofulvin, 642
      3.1.2 Tolnaftate, 644
      3.1.3 Allylamines, 644
      3.1.4 Azoles, 644
  3.2 Treatment of systemic mycoses, 647
      3.2.1 Azoles, 647
      3.2.2 Amphotericin B, 648
      3.2.3 5-Fluorocytosine, 649
      3.2.4 Antifolates, 650
      3.2.5 Diamidines, 650
      3.2.6 Echinocandins, 650
      3.2.7 Iron chelators, 650

4 Conclusions, 652

1 INTRODUCTION

Fungi are heterotrophic, eukaryotic microorganisms that are distinguished from algae by lack of photosynthetic ability. Fungi differ from bacteria by greater size and possession of such intracellular structures as nuclear membranes and mitochondria. Cells of fungi pathogenic for animals are nonmotile and generally have a rigid cell wall containing chitin and polysaccharide. Fungal cell membranes, other than those of *Pneumocystis*, contain ergosterol and zymosterol whereas animal cell membranes have cholesterol. Bacterial cell membranes, other than those of *Mycoplasma*, lack sterols.

Burger's Medicinal Chemistry and Drug Discovery,
Fifth Edition, Volume 2: Therapeutic Agents,
Edited by Manfred E. Wolff.
ISBN 0-471-57557-7   © 1996 John Wiley & Sons, Inc.

As with bacteria, fungi occupy an amazing variety and number of ecological niches. They fulfill a critical function in nature by converting such polymers as lignin, chitin, and cellulose into humus. But this "destructive" function is quite undesirable when the substrates under attack are contained in fabrics, electrical insulation, leather, wooden posts, or food-stuffs. Fortunately, relatively few fungi cause infectious disease in animals or humans.

The appellation *fungi* is a general term that includes both yeasts and molds. Yeast cells are spherical to ellipsoidal, 3–15 μm in diameter. Yeasts usually reproduce by budding and form mucoid colonies on agar media. Molds consist of tubular branching cells, 2–10 μm in diameter, called *hyphae*. Tangled masses of hyphae constitute *mycelia*, and these appear as dry colonies on agar surfaces or in natural habitats. Some yeastlike fungi can be induced, by environmental or nutritional factors, to grow in a filamentous way; conversely, some molds may be stimulated to grow in a yeastlike manner. Such dimorphism can be important in pathogenesis. For example, in histoplasmosis, infection is initiated by inhalation of hyphal conidia. The latter are transformed in the host into yeast cells which then parasitize the macrophages (1).

The true fungi are grouped into four classes. Phycomyceters (algalike), Ascomycetes (sac-fungi), Basidiomyctes (mushrooms), and Deuteromycetes (imperfect fungi). The fourth class contains many of the 50,000 fungal species. When a sexual reproductive stage is demonstrated for a species in this group, the organism is reassigned to its proper class. Approximately 90% of human fungal infections are caused by less than a score of genera (Table 35.1).

Pneumocystosis now is considered to be caused by a fungus rather than a protozoan (see Table 35.1). The agent, *Pneumocystis carinii*, is a non-budding organism related by sequence data of rRNA and mitochondrial DNA to ascomycetous yeasts (2).

Moreover, its thymidylate synthase and dihydrofolate reductase gene sequences have a higher homology with those of fungi than of protozoa. Nevertheless, the lack of ergosterol in its cell membranes as well as its greater response to antiprotozoal rather than antifungal drugs indicates that *Pneumocystis carinii* might lie on the taxonomic border between fungi and protozoa.

In addition to true fungi, an order of bacteria termed Actinomycetales is sometimes included in the science of mycology. Cells of species in this order can grow into branching filaments and produce dry colonies on agar surfaces that superficially resemble mold colonies. However, the diameter of the cells, 0.5–1.0 μm, is similar to that of bacteria rather than fungi. Moreover, the cells contain neither nuclear membranes nor mitocohondria, and the "hyphae" tend to fragment into bacillary elements. Strains of Actinomycetales are susceptible to antibacterial rather than to antifungal chemotherapeutic agents. Members of the order are important in humus formation, and a number of clinically useful antibiotics are produced by specific strains of these bacteria.

Although most fungi of medical interest can be cultured, diagnoses in the absence of culture often can be made by evaluation of the clinical condition, the inflammatory response, and the appearance of the pathogen in tissue biopsy. Slides of biopsies are more satisfactorily mailed to consultants than are cultures; the latter may become contaminated or lose viability in transit. Furthermore, a positive biopsy indicates that the fungus actually has invaded the tissue whereas a positive culture may have been derived from a contaminant of the lesion.

2 FUNGAL DISEASES

Many infections of plants are caused by fungi; the lesions produced include various

Table 35.1   Mycotic Infectious Diseases of Humans

Disease	Etiologic Agents	Main Tissues Affected
Contagious, Superficial Disease		
Dermatophytoses (ringworm/tinea)	*Epidermophyton, Microsporum, Trichophyton* spp.	Skin, hair, nails
Noncontagious, Systemic Diseases		
Aspergillosis	*Aspergillus* spp.	External ear, lungs, eye, brain
Blastomycosis	*Blastomyces dermatitidis*	Lungs, skin, bone, testes
Candidiasis	*Candida* spp.	Respiratory, gastrointestinal and urogenital tracts; skin
Chromomycosis	*Cladosporium, Fonsecaea* and *Phialophora* spp.	Skin
Coccidioidomycosis	*Coccidioides immitis*	Lungs, skin, joints, meninges
Cryptococcosis	*Cryptococcus neoformans*	Lungs, meninges
Histoplasmosis	*Histoplasma capsulatum*	Lungs, spleen, liver, adrenals, lymph nodes
Mucormycosis	*Absidia, Mucor, Rhizopus* spp.	Nasal mucosa, lungs, blood vessels, brain
Paracoccidioidomycosis	*Paracoccidioides brasiliensis*	Skin, nasal mucosa, lungs, liver, adrenals, lymph nodes
Pneumocystosis	*Pneumocystis carinii*	Lungs
Pseudallesheriasis iasis	*Pseudallescheria boydii*	External ear, lungs, eye
Sporotrichosis	*Sporothrix schenkii*	Skin, joints, lungs

types of blights, mildews, leafspots, and galls, and such systemic diseases as rusts and smuts. In contrast to plants, far fewer infectious diseases of animals and humans are caused by fungi than are initiated by bacteria or viruses. Mycotic infections of humans are conveniently divided into two groups (Table 35.1). (*1*) The dermatophytoses consist of contagious superficial skin infections that are limited to the epidermal region. These lesions of skin, hair, and nails are caused by specialized saprophytic molds that digest keratin in soil as well as in hosts. (*2*) The systemic mycoses are not contagious except in the case of neonates who have acquired the etiologic agent of candidiasis from their mother or other attendants. The agent of candidiasis is a normal commensal of humans that usually is prevented from overgrowth by bacterial commensals and by well-functioning defense mechanisms. The agents of the other systemic mycoses are free-living saprophytes that invade the skin, lungs, lymphatic tissue and various organs of susceptible hosts who have accidentally been in contact with the fungal environment.

Severe toxic and hemorrhagic diseases can occur in animals and humans after ingestion of harvested grains and nuts in which *saprophytic* fungi have grown and produced toxins. Likewise, hemolytic, neurotoxic, and gastrotoxic diseases are associated with ingestion of certain types of *saprophytic* mushrooms. Antifungal drugs are powerless in the treatment of mycotoxic diseases. Fortunately, the pathogenic fungi listed in Table 35.1 produce no known toxins in infected hosts. Often, however, a marked hypersensitivity to chemical components of the invading organism is induced. The lesions caused by dermatophytes that utilize dead cornified hair, skin, and nails are essentially allergic reactions to the fungi and their metabolites. The typical tissue reaction in the systemic mycotic and actinomycotic diseases is a chronic granuloma with necrosis and abscess formation.

The discovery that specific fungi are the etiologic agents of specific infectious diseases preceded by three decades the work of Pasteur and Koch on pathogenic bacterial. Nevertheless, medical mycology has not received as much attention as have other areas of medical microbiology, and at present there exist neither clinically available vaccines nor useful antisera for mycotic diseases. Moreover, comparatively few safe and effective drugs are available for treatment of systemic mycoses, especially in patients whose cell-mediated immune system and/or iron-withholding defense system is/are depressed. The number of patients with compromised defense systems continues to increase. In such cases, the presently available drugs for long-term

Table 35.2 Predisposing Factors for Systemic Mycotic Infections

Suppression of cell-mediated immunity
  Immunosuppressive therapy
  Deficient T lymphocyte quantity or quality
  Last trimester of pregnancy

Nosocomial contamination
  Parenteral alimentation
  Indwelling intravascular catheter

Hematological deficits
  Deficient quantity or activity of neutrophiles
  Deficient activity of macrophages
  Hematologic malignancy

Trauma
  Extensive burns
  Extensive surgery

Suppression of antifungal action of bacterial
    commensals
  Administration of antibacterial drugs

Suppression of iron withholding defense system
  Diabetic induced ferriuria
  Kwashiorkor (insufficient synthesis of transferrin)
  Ketoacidosis (insufficient activity of transferrin)
  Oral contraceptive use (insufficient synthesis of cervical lactoferrin)
  Deferoxamine therapy for iron or aluminum overload (drug can function as a fungal siderophore)

therapy often are unable to eradicate the fungal infection. Predisposing factors that favor establishment and progression of systemic fungal infections are listed in Table 35.2.

3 ANTIFUNGAL SUBSTANCES

The screening tests for *in vitro* and *in vivo* antifungal action are quite similar to those employed for antibacterial potency. It is not difficult in the *in vitro* procedures to discover a reasonable number and variety of synthetic and natural compounds that are active in small quantities. But many of the substances detected by such screening must be eliminated from practical consideration after examination in *in vivo* systems. In the case of prospective plant fungistats, such factors as particle size, combination with suitable wetting agents, and resistance to weathering and to microbial deterioration are of great importance. In the case of both plant and animal fungistats, toxicity as well as inability to penetrate to the site of fungal invasion can cause new candidates to be eliminated.

With animal or human infections, the test substance may appear to be inactive because the hosts are unable to provide sufficient natural defense factors to prevent relapses after the substance has been metabolized. Animal models are instrumental in distinguishing between fungicidal and fungistatic compounds and between efficacy in normal and immunocompromised hosts.

The route of introduction of the test substance also can be critical. For example, griseofulvin is active against dermatophytes via oral, but not topical, administration.

3.1 Treatment of Dermatophytoses

During the earlier decades of the 20th century, scores of synthetic compounds were developed for topical administration in dermatophytic lesions. Examples include fatty acids such as zinc 10-undecylenate (**1**), salicylates such as *N*-phenylsalicyl amide (**2**), aniline dyes such as methylrosaniline chloride (**3**), and quinolines such as 5-chloro-7-iodo-8-quinolinol (**4**). The compounds

$$(CH_2{=}CH(CH_2)_8CO_2)_2Zn$$

(**1**)

(**2**)

Table 35.3 Potentially Effective Antifungal Compounds

Disease	Compounds
Dermatophytoses	Azoles (butoconazole, clotrimazole, econazole, itraconazole, miconazole, oxiconazole, sulconazole); griseofulvin; naftifine; terbinafine; tolnaftate
Aspergillosis	Amphotericin B $\pm$ 5-fluorocytosine, itraconazole
Blastomycosis	Amphotericin B, itraconazole, ketoconazole
Candidiasis	Amphotericin B $\pm$ 5-fluorocytosine; nystatin; azoles (butoconazole, clotrimazole, econazole, fluconazole, itraconazole, ketoconazole, miconazole, terconazole, tioconazole)
Chromomycosis	5-Fluorocytosine, itraconazole, ketoconazole
Coccidioidomycosis	Amphotericin B, fluconazole, itraconazole, ketoconazole
Cryptococcosis	Amphotericin B $\pm$ 5-fluorocytosine, fluconazole
Histoplasmosis	Amphotericin B, itraconazole, ketoconazole
Mucormycosis	Amphotericin B
Paracoccidioidomycosis	Itraconazole, ketoconazole
Pneumocystosis	Trimethoprim/sulfamethoxasole, pentamidine isethionate, LY-303, 366, deferoxamine
Pseudallescheriasis	Amphotericin B, miconazole
Sporotrichosis	Amphotericin B, itraconazole, potassium iodide

Table 35.4   Examples of Trade Names of Antifungal Drugs

Chemical Nature	Generic Name	Trade Names
Allylamine	Naftifine	Naftin
	Terbinifine	Lamisil
Antifolate	Trimethoprim/ Sulfamethoxasole	Septra
Azole	Butoconazole	Femstat
	Clotrimazole	Canesten, Lotrimin, Mycelex
	Econazole	Spectazole
	Fluconazole	Diflucan
	Itraconazole	Sporanox
	Ketoconazole	Nizoral
	Miconazole	Micatin, Monistat, Nibustat
	Oxiconazole	Oxistat
	Sulconazole	Exelderm
	Terconazole	Terazole
	Tioconazole	Vagistat
Benzofuran cyclohexane	Griseofulvin	Fulcin, Fulvicin, Grifulvin, Grisactin, Grisovin, Gris-Peg, Lamoryl, Likuden, Poncyl, Spirofulvin, Sporostatin
Diamidine	Pentamidine isethionate	Pentam
Polyene	Amphotericin B	Fungizone
	Nystatin	Fungicidin, Moronal, Mycostatin, Nilstat, Nystan, O-V Statin
	Pimaricin	Myprozine, Natamycin, Pimifucin, Tennectin
Pyrimidine	5-Fluorocytosine	Ancobon, Flucytosine
Thiocarbamate	Tolnaftate	Focusan, Hi-Alazin, Sporilene, Tinactin, Tinaderm, Tonoftal

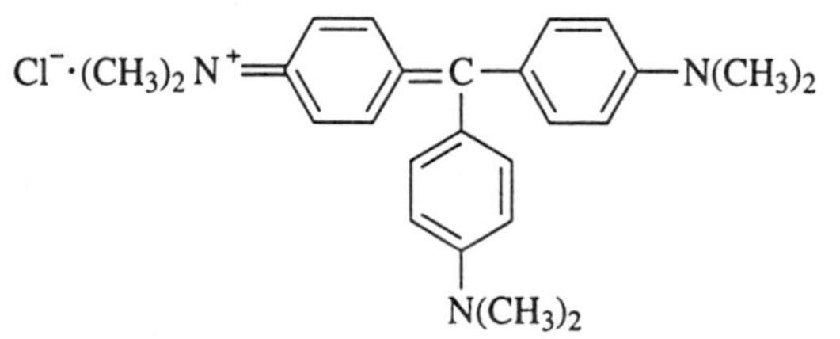

(3)

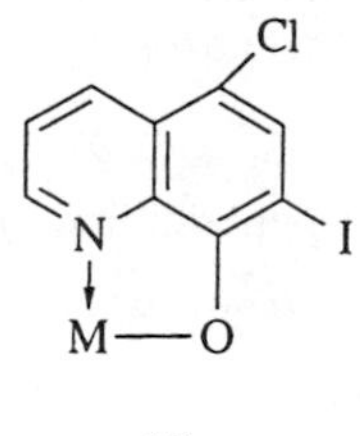

(4)

employed in earlier decades largely have been replaced by the more recently developed substances listed in Tables 35.3 and 35.4. Mechanisms of action of presently used compounds are summarized in Table 35.5.

3.1.1 GRISEOFULVIN (3). Although this antibiotic, a product of *Penicillium griseofulvin*, was discovered in 1939, its antifungal activity was not utilized until 1951. At that time, the compound was shown to be active against plant pathogenic fungi on systemic as well as on topical administration. Of the four stereoisomers, only griseofulvin (5) itself is active. The fluoro analog retains the

Table 35.5 Mechanisms of Action of Antifungal Compounds

Compounds	Mechanisms of Action
I. Interference with Ergosterol	
Allylamines and thiocarbamates	Suppression of fungal squalene epoxidase results in accumulation of squalene and decreased synthesis of ergosterol
Azoles	Inhibition of cytochrome P-450 that catalyzes 14 α-demethylation of lanosterol to ergosterol; accumulation of 14-methylated sterols causes permeability disruption
Polyenes	Binding to ergosterol in fungal cell membranes results in membrane disorganization
II. Interference With Other Metabolic Processes	
Antifolates	Inhibition of dihydropteroate synthetase and dihydrofolate reductase interferes with purine and pyrimidine synthesis
Benzofuran cyclohexenes	Binding to fungal proteins involved in tubulin assembly causes malformtion of spindle and cytoplasmic microtubules
Diamidines	Binding to DNA interferes with replication
Echinocandins	Inhibition of β-(1,3)-glucan synthetase interferes with cell-wall formation
Pyrimidines	Deamination of 5-fluorocytosine by fungal cells to 5-fluoro-uracil which is incorporated into RNA in place of uracil or is converted to 5-F-2′-deoxyuridylic acid which inhibits thymidine synthetase

(5)

potency of the original drug but the bromo and dechloro analogs are ineffective. Replacement of the methoxy substitutent on the cyclohexane ring either with propoxy or butoxy increases in vitro activity whereas substitution of an amino acid group at this site eliminates activity.

Griseofulvin is neither active against bacterial nor systemic fungal diseases in humans. Moreover, when used topically, the drug is inactive in dermatophytic infections. However, in 1958, the antibiotic was found to be effective against dermatophyes when administered orally. Apparently, a small amount of the drug can diffuse from the blood to the site of fungal multiplication in the skin and hair. This low concentration sufficiently slows the rate of hyphal penetration so that the outward thrust of keratinized host cells is able to deprive the fungus of access to nutrients. Because of low toxicity, the drug can be given to most patients for a period of many weeks. However, griseofulvin induces hepatic microsomal enzymes that decrease the activity of warfarin-type oral anticoagulants and of oral contraceptive agents (4) (Table 35.6). Moreover, the drug has demonstrated teratogenic and carcinogenic activities in animal models. Thus it should be reserved for treatment of dermatophytes that are known to be resistant to the topical

Table 35.6 Possible Adverse Interactions of Antifungal Drugs with Other Medications

Drug	Possible Adverse Effect
Amphotericin B	May increase digitalis toxicity by inducing hypokalemia
	May increase toxicity of 5-fluorocytosine by impairing renal function
	May be increased in renal toxicity in presence of aminoglycosides, cyclosporine, cisplatin
Azoles	May enhance activity of cyclosporine, warfarin, phenytoin, sulfonylureas
	May enhance toxicity of terfenadine/astemizole
	May have decreased activity in presence of rifampin
	May have decreased absorption (ketoconazole) in presence of antacids/cimetidine
Griseofulvin	May lower activity of warfarin, oral contraceptive agents
	May potentiate effects of ethanol
	May be more rapidly metabolized in presence of phenobarbital

antidermatophytic compounds listed in Table 35.3.

In susceptible fungi, griseofulvin binds to RNA and to microtubule-associated proteins to result in inhibition of nucleic acid synthesis and of microtubule function. In addition to causing the formation of abnormal fungal cells, the drug can inhibit mitosis of animal and plant cells to cause multipolar mitosis and production of abnormal nuclei.

3.1.2 TOLNAFTATE (5). The O-2-naphthyl ester of N-dimethylthiocarbanilic acid was found in 1960 to have good activity against dermatophytes when applied topically. Because of poor permeability, tolnaftate (6) is inactive against most other fungi. Antidermatophytic potency is maintained if the aromatic methyl group is removed or replaced by hydroxy or methoxy, or if the entire tolyl group is replaced by an α- or β-naphthyl substituent. Activity is lost if the aromatic methyl group is replaced by

halogen, or by a carboxyl or nitro group. The carbamate methyl group is essential for activity. The other naphthalene isomers are inert. In susceptible fungi, tolnaftate suppresses squalene epoxidase to result in accumulation of squalene with consequent decrease in ergosterol synthesis.

3.1.3 ALLYLAMINES (6). In the past decade, specific allylamines also have been found to be useful against dermatophytes by inhibiting squalene epoxidase. Naftifine (7) and terbinafine (8) are each active on topical administration. Additionally, terbinafine is effective when given orally. The affinity of terbinafine for mammalian squalene epoxidase is low and the drug is well-tolerated. It accumulates in sebum and hair and diffuses into the nail plate. In these various sites of dermatophytic lesions, the efficacy of terbinafine is comparable to that of griseofulvin.

3.1.4 AZOLES. A variety of azoles are useful when applied directly to dermatophytic lesions (Table 35.3). Topically applied azoles are relatively free of side effects other than occasional irritation. The mode of action of the azoles and their utility in systemic mycoses is described below.

(6)

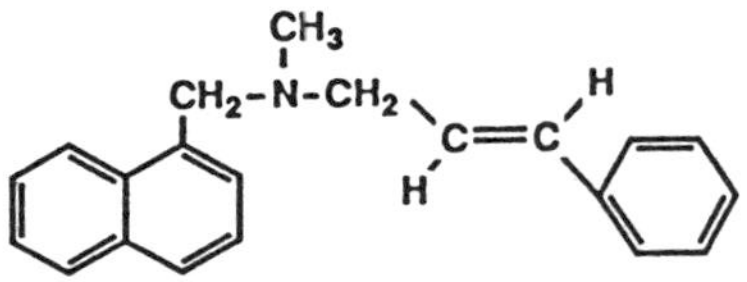

(7)

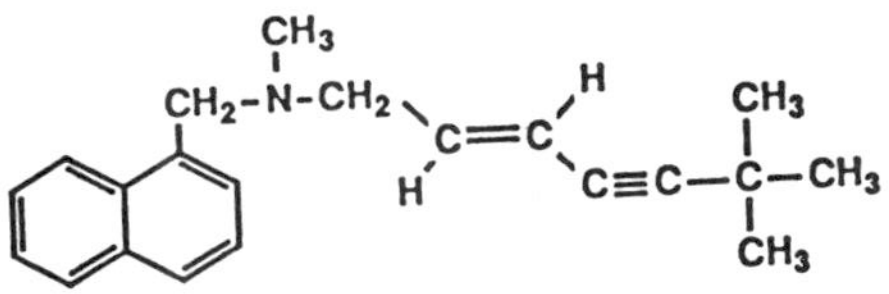

(8)

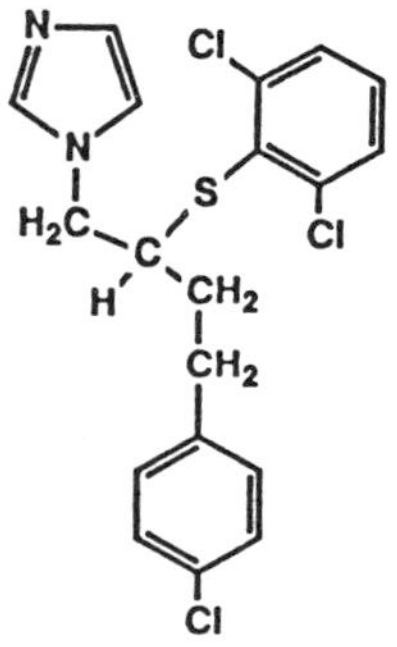

(9)

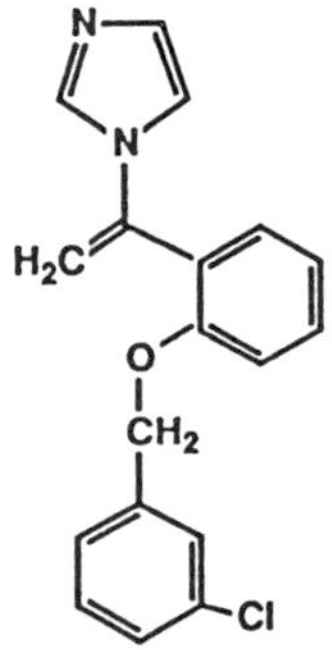

(10)

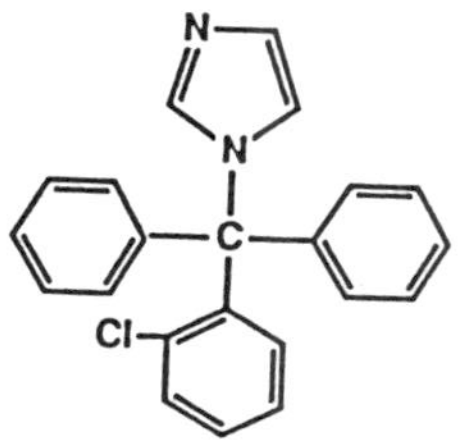

(11)

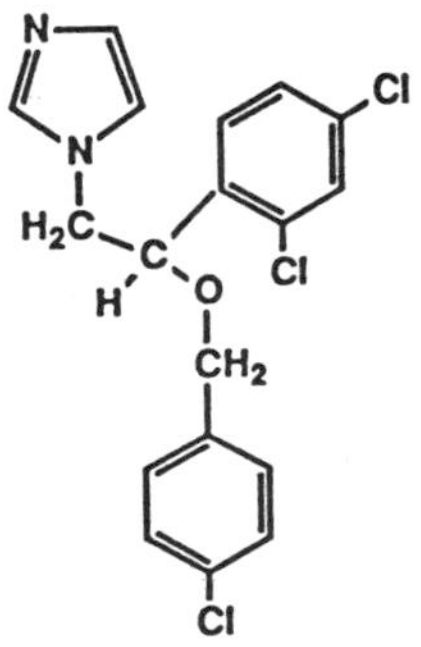

(12)

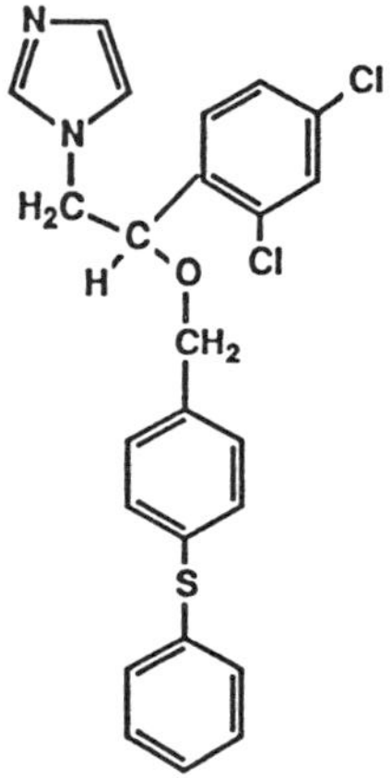

(13)

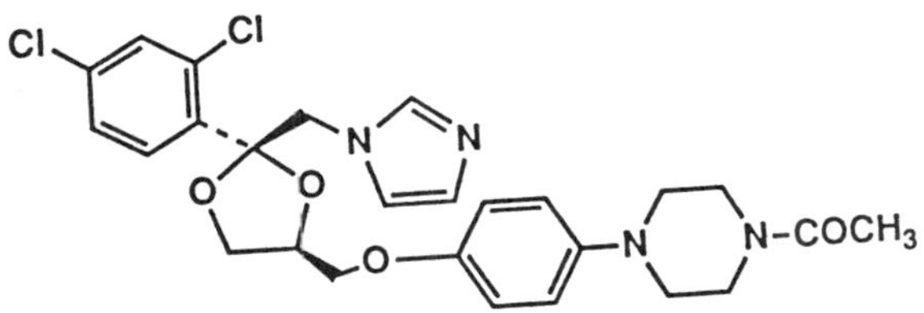

(14)

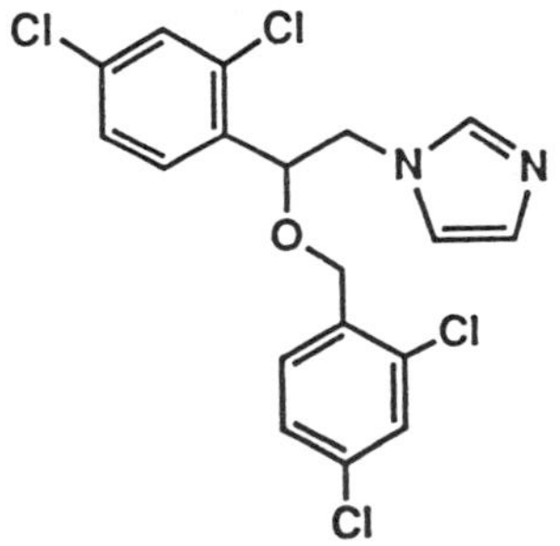

(15)

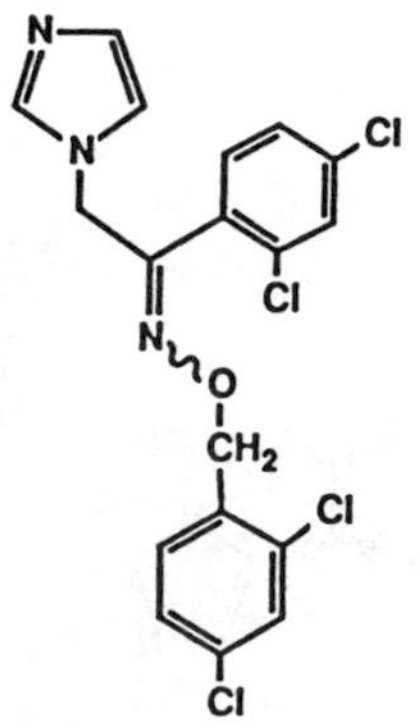

(19)

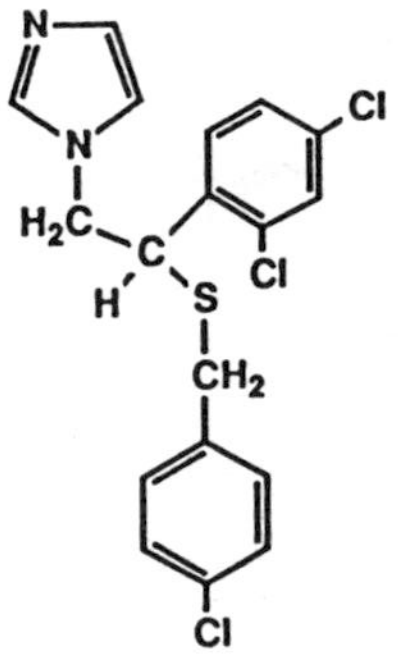

(18)

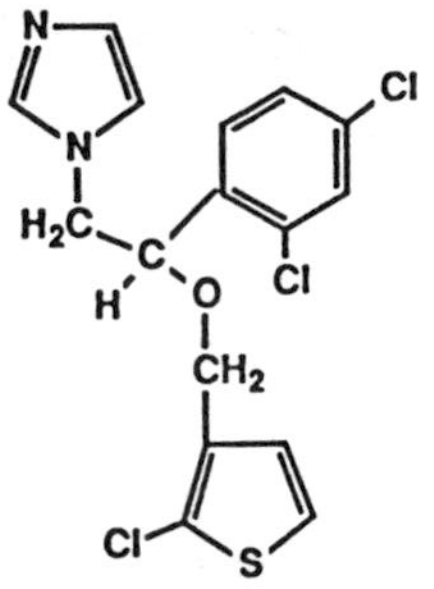

(17)

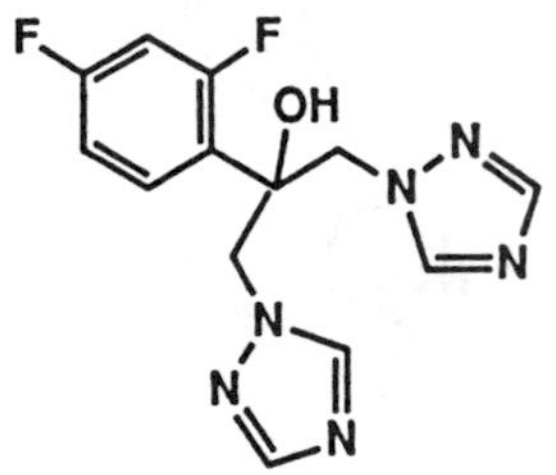

(16)

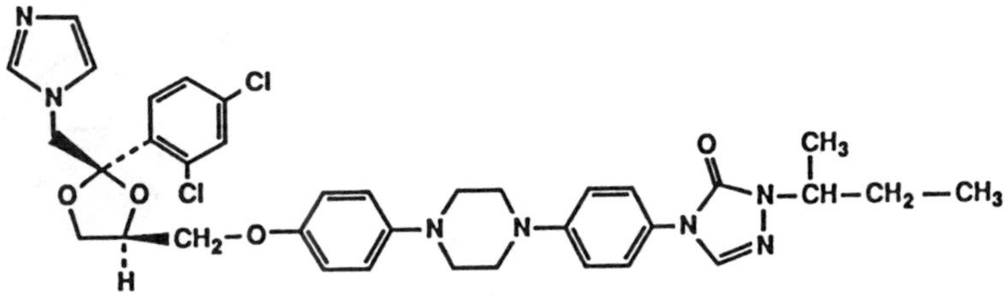

(21)

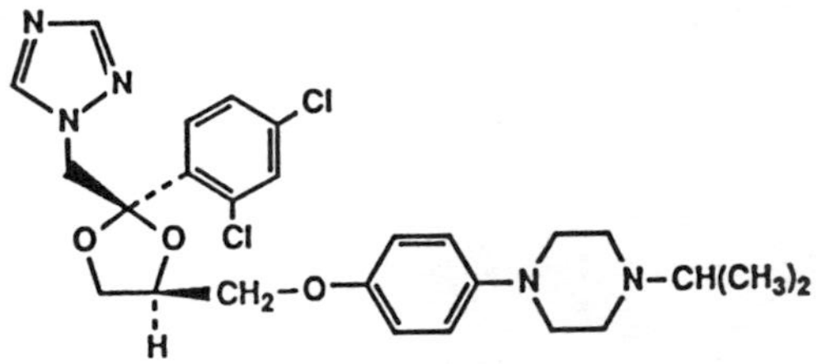

(20)

3.2 Treatment of Systemic Mycoses

3.2.1 AZOLES (7–9) (9–21). During the past quarter century, a series of imidazoles and triazoles have been synthesized and examined for antifungal activity. Various members of the series inhibit a broad spectrum of fungal pathogens. Some of the drugs are employed topically for cutaneous yeast as well as dermatophytic lesions. Several are available for oral administration to patients with systemic fungal infections. One of the azoles, fluconazole, also can be administered intravenously.

The antimycotic azoles have a five-membered azole ring with the N-1 atom linked via an aliphatic carbon atom to other aromatic rings. The imidazoles (two nitrogen atoms in azole ring) include butoconazole (9), cloconazole (10), clotrimazole (11), econazole (12), fenticonazole (13), ketoconazole (14), miconazole (15), oxiconazole (16), sulconazole (17), and tioconazole (18). Triazoles (three nitrogen atoms in the ring) include fluconazole (19), itraconazole (20), and terconazole (21).

Azoles exert antifungal activity by inhibiting a cytochrome P-450 that catalyzes the 14 α-demethylation of lanosterol to ergosterol. The N-3 atom of the azole ring binds to the ferric ion atom in the heme prosthetic group to prevent the activation of oxygen for insertion into lanosterol. The efficacy of specific azoles is defined not only by the strength of binding to the heme iron but also by the affinity of the N-1 substituent for the apoprotein of the cytochrome. Fortunately, fungal cytochrome P-450 enzymes are ≤ 1000 times more sensitive to the azole drugs than are the analogous enzymes in mammalian cells. Inhibition of the fungal 14 α-demethylase causes accumulation of 14 α-methylated sterols that induce membrane leakiness, unbalanced chitin synthesis, cell disorganization, and death of the pathogens.

Although the azoles are considered to have a wide antifungal spectrum, individual differences in activity occur. For instance, itraconazole is much more active than ketoconazole against the agents of aspergillosis and sporotrichosis. Similarly, fluconazole is active against *Candida tropicalis* whereas ketoconazole is not. Conversely, ketoconazole is active against *Candida krusei*, an agent resistant to fluconazole.

The antifungal azoles also show marked differences in pharmacokinetics. Clotrimazole and miconazole, for example, are relatively insoluble in aqueous solution and are poorly absorbed from the intestine. The former is excessively rapidly metabolized in the liver. The latter, although more slowly metabolized, was dissolved in polyethoxylated castor oil to obtain satisfactory intestinal absorption. Unfortunately, the vehicle caused toxic effects sufficient to result in discontinuance of enteral administration. Ketoconazole has improved water solubility as well as a satisfactory half life in the patient. However, it requires gastric acidity for absorption and it penetrates poorly into cerebrospinal fluid (CSF).

Intestinal absorption of itraconazole likewise requires gastric acidity but, unlike that of ketoconazole, is not suppressed by H_2-blockers. As with ketoconazole, the drug fails to accumulate in CSF. However, itraconazole can persist in other tissues for long periods. Gastrointestinal absorption of fluconazole is not dependent on gastric acidity; moreover, the drug can also be administered intravenously. It readily crosses the blood brain barrier and thus is useful in cases of fungal meningitis.

Despite the low affinity of azole drugs for mammalian cytochrome P-450 enzymes, inhibition of gonadal and adrenal steroidogenesis has been observed with ketoconazole. Some loss of male libido and sexual potency, as well as development of gynecomastia, has been reported. Neither itraconazole nor fluconazole appear to cause endocrinopathies. Ketoconazole has been used clinically to reduce steroidogenesis in such conditions as female acne and hirsutism, precocious puberty, prostatic and male breast carcinoma, and Cushing's syndrome.

The cytochrome P-450 enzymes involved in the metabolism of the azoles also are responsible for metabolizing various other drugs. Accordingly, employment of azoles necessitates consideration of possible drug–drug interactions (10) (Table 35.6). The ability of azoles to slow catabolism of other drugs increases the activity of the latter. Conversely, induction of hepatic P-450 oxidative enzymes by rifampin depresses activity of the azoles (11).

3.2.2 AMPHOTERICIN B (7,12). This antibiotic, produced by *Streptomyces nodosus*, was discovered in 1955. It became the first systemic antifungal drug and remains useful for treatment of many of the diseases listed in Table 35.3. Amphotericin B (**22**) is a polyene macrolide with seven conjugated double bonds (a heptaene), an internal ester, a free carboxyl group, and a glycoside side chain with a primary amino group. The drug is amphoteric, insoluble in water, and is not absorbed via oral or intramuscular routes. Amphotericin B is commercially available for intravenous infusion as a deoxycholate micellar suspension.

Approximately 60 polyene antifungal compounds have been described. Each contains a macrocyclic lactone ring with a rigid lipophilic portion. The rings have a chromophore of four to seven conjugated double bonds, often linked to an amino sugar; in amphotericin B, nystatin (**23**), and pimaricin (**24**), this is D-mycosamine. Nystatin and pimaricin are tetraenes; the former has been employed for therapy of cutaneous, oral, and vaginal candidiasis, and the latter for mycotic keratoses.

Polyenes essentially are insoluble in water and, because of extensive unsaturation, are rather unstable. They act against fungal, algal, and animal cells by combining with membrane sterols. The combination results in altered permeability with consequent leakage of sodium, potassium and hydrogen ions, eventually leading to cell death.

Amphotericin B binds approximately tenfold more strongly to fungal cell mem-

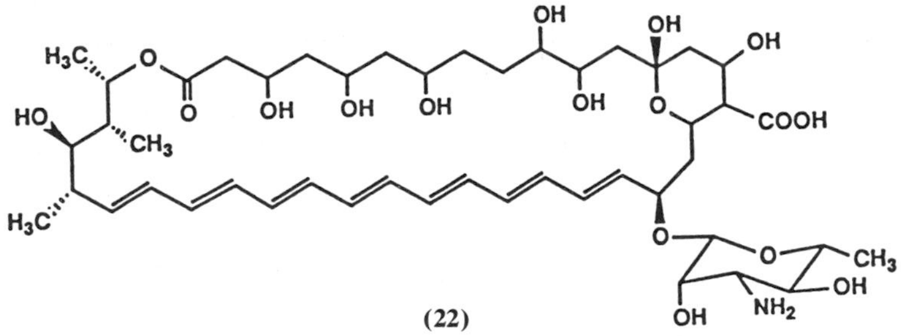

(22)

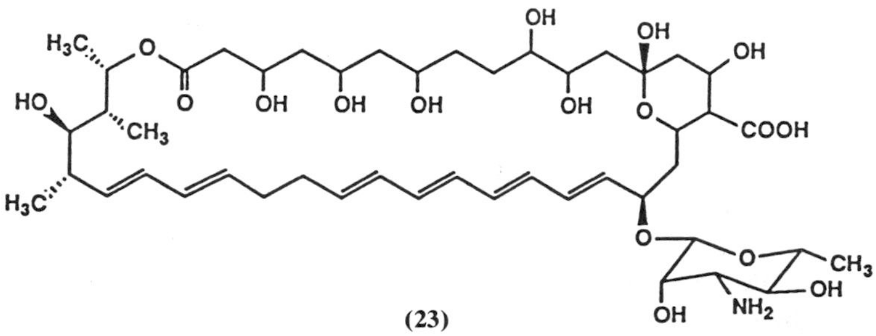

(23)

(24)

brane ergosterol than to animal cell membrane cholesterol. Nevertheless, the interaction with and disruption of mammalian cells by the drug can result in adverse side effects, especially in nephrotoxicity. Complications include hypokalemia; hypomagnesemia; renal tubular acidosis; nephrocalcinosis; and, because of reduced erythropoetin synthesis, anemia. Moreover, the renal toxicity caused by aminoglycosides and cyclosporin can be enhanced by amphotericin B.

To lessen renal toxicity, the tissue distribution of amphotericin B can be altered by changing the vehicle from deoxycholate to various lipid formulations including packaging in liposomes. Such formulations aid the drug to accumulate in the monocyte–macrophage system rather than to be bound to proteins. Amphotericin B generally is not used in therapy of dermatophytoses because of the effective and comparatively safer drugs listed in Table 35.3. Although the polyene antibiotic has a moderately broad antifungal spectrum and resistance rarely develops, not all fungal pathogens are susceptible. Furthermore, the drug may not penetrate readily into all infected tissues. Even when administered intrathecally, amphotericin might not reach isolated areas in patients who have obstructive hydrocephalus.

3.2.3 5-FLUOROCYTOSINE (7). This fluorinated pyrimidine was first synthesized in

the 1950s as a potential antileukemic drug. 5-Fluorocytosine (**25**) is not effective against neoplastic cells but does inhibit sensitive strains of *Aspergillus*, *Candida*, *Cladosporium*, and *Cryptococcus*. The drug is well-absorbed through the gastrointestinal tract, has fine penetration into CSF, and is low in toxicity. Unfortunately, except for chromomycosis caused by *Cladosporium*, 5-fluorocytosine cannot be used alone because it readily selects for resistant fungal strains. Accordingly, in *Aspergillus*, *Candida*, and *Cryptococcus* infections, the drug is combined with amphotericin B. The polyene suppresses the emergence of clones resistant to 5-fluorocytosine. In combination with the latter drug, lower doses of amphotericin B can be employed.

(25)

Susceptible fungal cells deaminate 5-fluorocytosine to 5-fluorouracil. The latter is metabolized to 5-fluoro-2'-deoxyuredylic acid which suppresses DNA synthesis by inhibition of thymidylate synthetase. The drug can be incorporated also into RNA to cause inhibition of protein synthesis. Human cells have little or no deaminase activity and 90% of the ingested drug is excreted unchanged in urine. However, the concurrent administration of amphotericin B might suppress renal excretion; thus plasma levels of 5-fluorocytosine should be monitored. Excessive amounts in plasma can be hemolytic. For *in vitro* assays of the drug in plasma, the culture medium must be devoid of purines and pyrimidines. The nucleic acid bases neutralize the drug and their presence would result in falsely low readings.

3.2.4 ANTIFOLATES (13). The combination of two antifolates, pyrimethamine and sulfadiazine was shown in 1966 in rat models to be effective against *Pneumocystis carinii* pneumonia. At a ratio of 1:20 (Fansidar), the combination has been used prophylactically in infants and in AIDS patients. More recently, the 1:6 combination of trimethoprim (26) and sulfamethoxasole (27) has become the standard therapy. As with protozoa, sequential blockade of dihydropteroate synthetase and dihydrofolate reductase by sulfonamides and by pyrimethamine or trimethoprim, respectively, shows superior activity to that of members of either set of drugs alone.

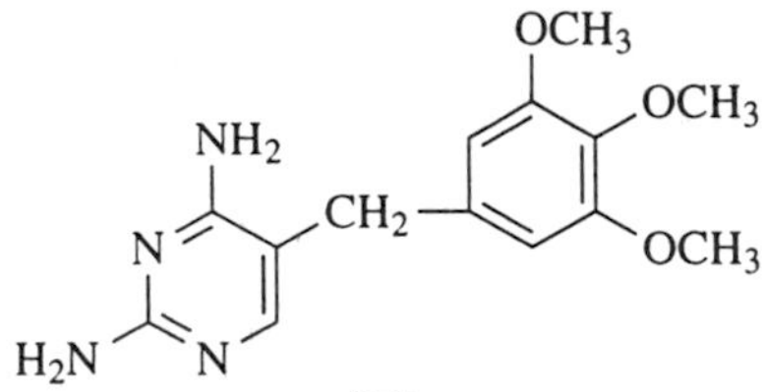

3.2.5 DIAMIDINES (13). In earlier decades, the isethionate of 2-hydroxystilbamidine (28) was employed in therapy of North American blastomycosis. Since 1970, pentamidine isethionate (29) has been a useful drug for pneumocystosis. However, systemic administration of the drug can be associated with severe hypotension, hypoglycemia, cardiac arrythmia, and renal and hepatic dysfunction. Aerosolized pentamidine isethionate allows the drug to be inhaled and deposited at infected sites without the risk of full systemic toxicity. Still, some areas of the lung, as well as extra-pulmonary sites, are not completely protected by this procedure.

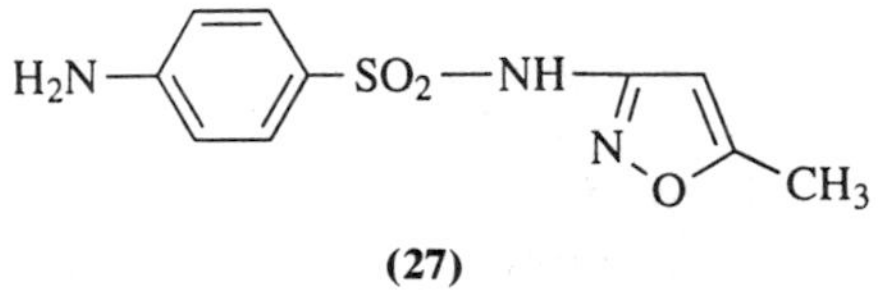

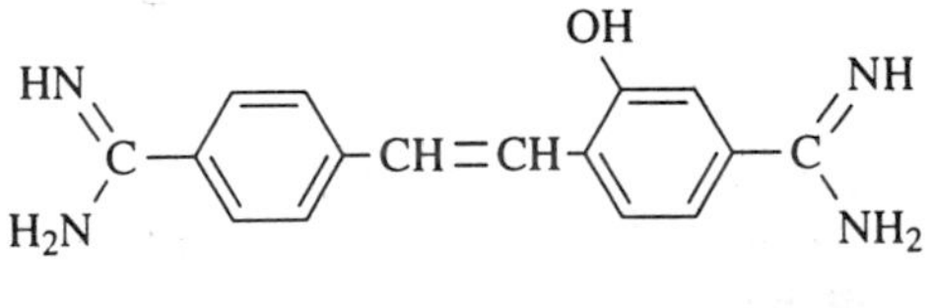

3.2.6 ECHINOCANDINS (13,14). Pathogenic fungi, including *Pneumocystis carinii*, have units of β-(1,3) glucans in their cell walls. Biosynthesis of the glucans is inhibited by echinocandins. In echinocandin B (30), a linoleoyl group acylates the N terminus of the cyclic hexapeptide. In cilofungin (31), an *n*-octyloxybenzoyl group has been substituted for the linoleoyl side chain; this compound is less hemolytic than echinocandin B. Compounds L-693,989 (32) and LY-303,366 (33) have improved water solubility. The latter compound has shown good prophylactic and therapeutic activity against *Pneumocystics carinii* in rodent models. Inasmuch as mammalian cells lack walls, the echinocandins appear to be attractive compounds for development as clinical antifungal drugs.

3.2.7 IRON CHELATORS (15). The ability of fungi, protozoa, and bacteria to obtain growth essential iron from the host is an important facet of their virulence. Accord-

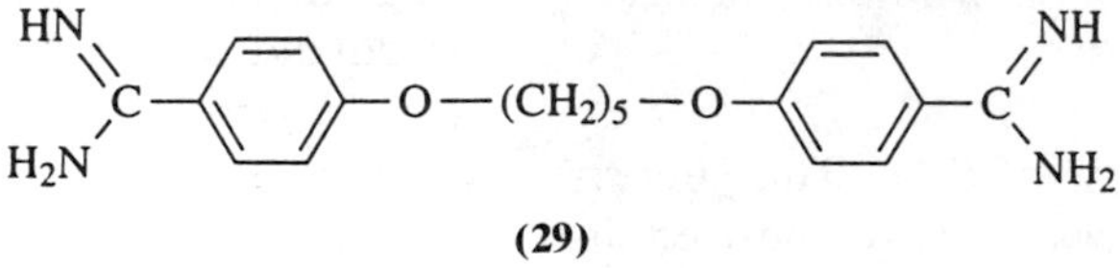

(26)

(27)

(28)

(29)

(30) Echinocandin B R_1 = linoleoyl, $R_2 = R_3 = CH_3$, $R_4 = R_5 = H$

(31) Cilofungin (LY121019) R_1 = *p*-(*n*-octyloxy)benzoyl, $R_2 = R_3 = CH_3$, $R_4 = R_5 = H$

(32) L-693,989 R_1 = 10, 12 dimethylmyristoyl, R_2 = H, $R_3 = CH_2CONH_2$, $R_4 = PO_3Na$, $R_5 = H$

(33) LY303,366 R_1 = $R_2 = R_3 = CH_3$, $R_4 = R_5 = H$

ingly, iron chelating drugs increasingly are being tested for anti-infective potency. Deferoxamine (DFO) (**34**), an actinomycete hydroxamate siderophore, has been used for several decades to reduce iron overload. This drug is an effective adjunct in prophy- laxis and therapy of human malaria; it also markedly reduces the intensity of *Pneumo- cystis carinii* infections in rodent models. The iron-dependent transformation of conidia of *Paracoccidioides brasiliensis* to the pathogenic yeast form within macro-

(**34**)

(35)

phages likewise is inhibited by DFO (16).

A series of hydroxypyridinones contains members such as CP20 (**35**) that can lower iron overload via oral administration and which are inexpensive to manufacture. These compounds also are antimalarial and have *in vitro* activity against *Pneumocystis carinii*. The anti-infective activity of both DFO and the hydroxypyrinidones are suppressed by iron but not by other metals (15).

When using iron chelators as anti-infectives, care must be taken to ensure that the drugs do not act as iron-transporting siderophores for other opportunistic pathogens such as *Rhizopus*. Accumulation of Fe-DFO in dialysis patients has resulted in some cases of mucormycosis. With hydroxypyrinidones, microbial iron acquisition has not yet been reported.

4 CONCLUSIONS

Recent decades have witnessed a marked improvement in the variety and selective toxicity of commercially available antifungal drugs. Enhanced knowledge of fungal metabolism and architecture is resulting in the rational design and development of diverse, potentially effective compounds. Unfortunately, the numbers and kinds of patients who are severely stressed by deficits in cell-mediated immunity and/or other defense factors continues to soar. Moreover, immunological aspects of prevention and therapy of fungal diseases have failed to materialize. Thus the search for novel, effective, and less toxic antifungal drugs must continue.

ACKNOWLEDGMENT

Supported in part by a Grant-in-Aid from the Office of Research and the University Graduate School, Indiana University, Bloomington, Indiana.

REFERENCES

1. S. E. Vartivarian, *Clin. Infec. Dis.*, **14**(Suppl 1), S30–36 (1992).
2. K. J. Kwon-Chung, *Clin. Infec. Dis.*, **19**(Suppl 1), S1-7 (1994).
3. M. R. McGinnis and M. G. Rinaldi in V. Lorian, Ed., *Antibiotics in Laboratory Medicine*, Williams & Wilkins, Baltimore, 3rd ed., 1991, pp. 198–257.
4. P. S. Wells, A. M. Holbrook, N. R. Crowther, and J. Hirsch, *Ann. Intern. Med.*, **121**, 676–683 (1994).
5. H. Vanden Bossche and P. Marichal, *Amer. J. Obstet. Gynecol.*, **165**, 1193–1199 (1991).
6. J. A. Balfour and D. Faulds, *Drugs*, **43**, 259–284 (1992).
7. C. A. Lyman and T. J. Walsh, *Drugs*, **44**, 9–35 (1992).
8. G. P. Bodey, *Clin. Infec. Dis.*, **14**(Suppl 1), S161–169 (1992).
9. C. A. Kauffman, *Clin. Infec. Dis.*, **19**(Suppl 1), S28–32 (1994).
10. R. M. Tucker, D. W. Denning, L. H. Hanson, M. G. Rinaldi, J. R. Graybill, P. K. Sharkey, D. Pappagianis, and D. A. Stevens, *Clin. Infec. Dis.*, **14**, 165–172 (1992).
11. S. M. Borcherding, A. M. Baciewicz, and T. H. Self, *Arch. Intern. Med.*, **152**, 711–715 (1992).
12. D. R. Bennett, *Drug Evaluations Annual*, American Medical Association, Milwaukee, 1991, pp. 1479–1492.
13. S. F. Queener, *J. Medic. Chem.*, **38**, 4739–4759 (1995).
14. M. Debono and R. S. Gordee, *Annu. Rev. Microbiol.*, **48**, 471–497 (1994).
15. G. A. Weinberg, *Antimicr. Agents Chemother.*, **38**, 997–1003 (1994).
16. L. E. Cano, B. Gomez, E. Brummer, A. Restrepo, and D. A. Stevens, *Infec. Immun.*, **62**, 1494–1496 (1994).

Index

Note: Index terms followed by an "*i*" indicate illustrations. Terms followed by "(table)" indicate material in tables. Whenever possible generic names of compounds were used.

A10255, 501
A201A, 509
A21978C, 503
A35512, 495
A40104A, 507
A41030, 495, 496
A47934, 496
A51568, 494
A54145, 503
A56234, 354–355
A62514, 482*i*, 483
A69334, 482*i*, 484
A82846, 493*i*, 494
AA-193, 356–360
A band, 158, 159*i*
Abbott 49816, 326, 327*i*, 328
  structure-activity relationship, 354
Abbott 53385, 405
Acedapsone, 538, 540, 592, 593
  metabolism, 563–564
ACE inhibitors:
  and blood pressure control, 268, 269, 276, 289–292
  chemistry, 290–292
  for congestive heart failure, 223, 225, 226–228
  discovery, 289
  and NO, 210
  pharmacology, 289–290
Acetazolamide, 281, 324, 325*i*, 380*i*, 382, 541
Acetbutolol, 216*i*, 219, 220
Acetosulfone sodium, 592
4-Acetoxybutyltrimethylammonium, 10, 40
Acetoxymethyltrimethylammonium, 10
3-Acetoxyquinuclidine, 22–23
Acetoxytrimethylammonium, 10
Acetylarsenocholine, 7
Acetylcholine, 4. *See also*
  Anticholinergic drugs
  analogs, derivatives, and

congeners, 6–7
  acyl group variations, 9–10
  ester group substitutions, 11–13
  ethylene bridge variations, 10–11
  quaternary ammonium group variations, 7–9
  conformations, 37–39, 42*i*, 98–99, 99*i*
  "five atom rule," 6
  lack of asymmetric carbon, 102
  muscarinic action, 82
  as poor therapeutic agent, 6
  prodrug, 13
  release on vagal stimulation, 171
  role as neurotransmitter, 60–61
  role in blood pressure control, 268
  role in gastric acid secretion, 62, 121, 122–123
  structure compared to atropine, 79
Acetylcholine-derived anticholinergics, 104
Acetylcholine receptors, 5–6
Acetylcholine-release modulators, 52–53
  therapeutic indications, 4
Acetylcholinesterase, 43*i*. *See also*
  Anticholinesterases
Acetyl coenzyme A, 4
Erythro-Acetyl-α,β-dimethylcholine, 11
Acetyl-α-methylcholine, 10, 11 (table), 38
Acetyl-β-methylcholine, 10–11, 11 (table), 15, 38
Acetyl γ-homocholine, 10, 40
Acetylphosphonocholine, 7
Acetylselenocholine, 10
Acetylsulfonocholine, 7
Acetylthiocholine, 10
Acetylthionocholine, 10

Acquired immune deficiency syndrome (AIDS), *see* AIDS
Acrylylcholine, 9
Actaplanin, 495
Actin, 157, 158–159, 175, 177*i*
  F-actin, 159, 160
  G-actin, 159, 160
Actinamine, 474
Actinoidins, 496
Actinomycetales, 638
Action potentials, 164–165, 165*i*
Activity-structure relationships, *see*
  Structure-activity relationships
trans-ACTM, 38, 40–41
Acute myocardial infarction, *see*
  Myocardial infarction
5-Acyl-2,3-dihydrobenzofuran-2-carboxylic acids, 335
Adenosine, 306–307
Adenosine receptor antagonists, for hypertension, 306–307
Adenosine receptors, 427
Adenylyl cyclase, and cardiac function, 173, 186
Adibendan, 230
Adiphenine, 108
β-Adrenergic blocking drugs, *see*
  β-Blockers
Adrenergic drugs, heart rate increase mechanism, 165
α₁-Adrenergic receptor antagonists, for hypertension, 295–297
α₂-Adrenergic receptor antagonists, for hypertension, 297–298
Adrenergic receptors, and autonomic cardiac regulation, 186–187
AF-DX-116, 83
Afferent artioles, 365, 367*i*
Afterload, 207
Aglycones, 232, 477
AIDS, 615
  drug resistance in, 616

653

AIDS (*Continued*)
  prophylactic antifungals, 650
  sulfonamides for, 541, 566
  and tuberculosis, 584–585, 623
AIV-tylosin, 486*i*, 488
Alacepril, 291*i*, 292
Aldesulfone, 592
Aldgamycins, 486–487
Aldgarose, 486
Aldosterone:
  and ANP, 440
  role in renal function, 372, 419
  role in renin-angiotensin-
    aldosterone, *see* Renin-
    angiotensin-aldosterone system
Aldosterone antagonists, steroidal,
  419–426, 423 (table)
Aldosterone biosynthesis
  inhibitors, 426, 427*i*
Alipamide, 393*i*, 395
Allantoic acid, 444
Allantoin, 444
Allantoinase, 444
Allopurinol, 445*i*, 447–448
Alloxanthine, 447
Allylamine antifungals, 644
Almokalant, 196*i*
Althiomycin, 502
Alverine citrate, 112
Alzheimer's disease:
  acetylcholine-release modulators
    for, 53
  anticholinergics for, 86
  anticholinesterases for, 42, 45,
    50
  cholinergics for, 4, 15, 22, 26, 29
  and M1 receptors, 5, 6
AM-1155, 617
Amanozine, 430
N-Amido-3-amino-6-bromo-
  pyrazinecarboxamide, 437,
  438 (table)
Amikacin, 469, 471, 588, 623
  antimycobacterial activity, 603,
    607
  combination with azithromycin,
    621
Amiloride, 325, 429*i*, 437, 438, 439
  discovery, 282, 283, 284
Amino acids:
  renal reabsorption, 366, 368
  and urea cycle, 445
p-Aminobenzenesulfonamido-
  hydrocupreine, 529
p-Aminobenzoic acid (PABA):
  antagonism by sulfonamides,
    528, 539, 543–544, 549, 550
  antagonism by sulfanilamide,
    531

p-Aminobenzoylglutamic acid
  (PABG), 544, 547, 549
2[(2-Aminobenzyl)sulfinyl]-1*H*-
  benzimidazoles, gastric proton
  pump inhibitors, 127, 129*i*
4-Amino-6-chloro-1,3-benzene-
  disulfonamide, 383
Aminocyclitol antibiotics, 465,
  473–475
Aminoglycoside antibiotics,
  465–477. *See also specific*
  amines, antibiotics, and
  sugars
  with antimycobacterial activity,
    603–607
  biosynthesis, 477
  interaction with amphotericin B,
    649
  resistance to, 476–477, 591,
    604–605
  structure-toxicity relationship,
    471
  toxicity, 477
Aminoglycoside-inactivating
  enzymes, 477, 604–605
14β-Amino-20-hydroxy-5β*H*-
  pregnane-3β-rhamnoside, 234*i*,
  244
4-Aminomethylbenzene-
  sulfanilamide, 535
2-Aminomethyl-3,4,6-
  trichlorophenol, 417
Aminometradine, 428, 429*i*
1-(2-Amino-1-phenylethyl)-6-
  phenyl-4*H*-[1,2,4]triazolo[4,3-
  a][1,4]benzodiazepine
  diuretics, 436
Aminopteridines, 548–549
4-Aminopyridine, 171
p-Amino-salicylic acid, 588–589,
  593–594, 623
3-Amino-5-sulfamoylbenzoic acid
  diuretics, 406
Aminouracil diuretics, 428
Amiodarone, 196*i*, 199, 201–202,
  203–204
Amisometradine, 428, 429*i*
Amlodipine, 215*i*, 216–217, 294
Ammonium chloride, 375*i*
Ammonium groups, quaternary,
  see Quaternary ammonium
  groups
Amoxicillin, 589, 621, 622
cAMP:
  gastric acid secretion inhibition,
    63–64, 122
  and phosphodiesterase
    inhibitors, 229–230

Amphotericins, 641 (table),
  648–649
Ampicillin, 621
Amprotropine, 108
Amrinone, 230
Amyl nitrite, 211*i*
Analgesics, anticholinergic side
  effects, 108
Anatoxin-a, 50
Anesthesia premedication,
  anticholinergics, 66–67, 107
Angina pectoris, 156, 204, 207–208
Angiotensin, 301*i*, 419
Angiotensin-converting enzyme,
  419
Angiotensin-converting enzyme
  inhibitors, *see* ACE inhibitors
Angiotensin II, 227–228, 278
  and blood pressure control,
    268–269
  synthesis, 301*i*, 419
Angiotensin II receptor
  antagonists, 276, 300–304
Angolamycin, 485
6,7-Anhydroerythromycin C, 490
Aniline dyes, as antifungals, 641
Animal fungistats, 641. *See also*
  Antifungal substances
Ansamycins, 610
ansa structure, rifamycins, 610
Antacids, toxicology, 140
9-Anthracene carboxylate, 145
Antiarrhythmic drugs:
  discovery, 155
  mechanism, 197–202
  receptors for, 199
  Singh-Vaughan-Williams
    classification, 193, 194 (table)
    criticism, 193–199
Antiarrhythmic drugs, class I, 193,
  194 (table), 195*i*
  increased mortality with, 199,
    200, 202
  mechanism, 197–200
Antiarrhythmic drugs, class Ia,
  194 (table), 195*i*
  mechanism, 198
Antiarrhythmic drugs, class Ib,
  194 (table), 195*i*
  mechanism, 198
Antiarrhythmic drugs, class Ic,
  193, 194 (table), 195*i*
  for arrhythmia prophylaxis,
    195–196
  mechanism, 198
Antiarrhythmic drugs, class II, *see*
  β-Blockers
Antiarrhythmic drugs, class III,
  171, 193, 194 (table), 196*i*

mechanism, 200–202
reverse use dependence, 201
for secondary MI prevention,
206
and *Torsade de pointes*, 191, 200,
201
Antiarrhythmic drugs, class IV,
see Calcium channel blockers
Antibiotics, 464–465, 510. *See also*
Aminoglycoside antibiotics;
Glycopeptide antibiotics;
Lincosaminide antibiotics;
Macrolide antibiotics; Peptide
antibiotics; Streptogramin
antibiotics
with antimycobacterial activity,
603–616. *See also*
Antimycobacterial drugs
miscellaneous, 503–510
Anticholinergic drugs, *see also*
Solanaceous alkaloids
acetylcholine-derived: dual
action, 104
action of, 61
assays:
antispasmodic activity, 68–70
antiulcer activity, 70–71
mydriatic/cycloplegic activity,
71–72
classification, 61, 67
conformations, 98–99, 99*i*
dissociation constants, 102–103,
104 (table)
muscarinic receptor interaction,
104–106
receptor selectivity/specificity,
67, 68 (table)
receptor subtype selective, 82–87
relative activities, 109–110 (table)
side effects, 107–108
structure-activity relationship:
basicity, 93
cationic head, 87, 93–96
cyclic moieties, 87, 96–98
epoxy group, 101
esteratic linkage, 99–100
hydroxyl group, 100–101
main chain length, 87, 98–99
stereoisomerism, 101–103
synthetic, 79–87, 80 (table),
88–92 (table)
therapeutic uses, 106–107
Anticholinesterases, 4, 42–44
carbamate-derived inhibitors,
45–47
organophosphorus-derived
inhibitors, 47–50, 48–49 (table)
other inhibitors, 51–52
reversible inhibitors, 44–45

1,2,3,4-tetrahydro-9-
aminoacridine-related
reversible, noncovalent
inhibitors, 50–51
therapeutic indications, 4
Anticoagulants, for secondary MI
prevention, 206
Antidepressants, gastric acid
secretion inhibition, 144
Antifolates:
as antifungals, 650, 651
combination with sulfonamides,
548–549
Antifungal substances, 641 (table),
642 (table)
for dermatophytoses, 641–646
drug interactions, 644 (table)
mechanisms, 643 (table)
screening tests for, 640–641
for systemic mycoses, 647–652
trade names, 642 (table)
Antihistamines, anticholinergic
side effects, 108
Antihypertensive drugs:
antisense molecules, 311–312
coronary heart disease, failure
to reduce mortality from, 274
future, 313–314
history of discovery, 277–278,
279 (table)
"ideal," 276–277, 277 (table)
major classes, 278, 280 (table),
280–298. *See also* ACE
inhibitors; β-Blockers;
Calcium channel blockers;
Diuretics
miscellaneous drugs, 298–300,
299 (table)
new drugs, 300–309
novel mechanistic approaches,
309–311
overview, 266–267
treatment guidelines, 275–276
Antileprotic drugs, 588 (table),
589 (table), 623–624. *See also*
Leprosy
Antimalaria drugs, 539–540
Antimicrobial drugs:
with antimycobacterial activity,
589 (table)
targets of action: cell wall, etc.,
589–591
Antimuscarinic drugs, 61
for gastric acid secretion
inhibition, 121
Antimycobacterial drugs, 576,
588 (table)
antimicrobials used as,
589 (table), 603–607

combinations, 618, 620
discovery, 588–591
"Eagle" effect, 617
screening/evaluation laboratory
models, 586–588
Antioxidants, 221
Antipyrine, 619
Antisense molecules, 311–312
Antispasmodics, anticholinergics,
61–62, 106
Antithyroid agents, 542
Antiulcer drugs:
anticholinergics, 61, 62–66
"ideal" agent, 64
nonanticolinergics, 111–112
Apical membrane
cardiac, 123, 144–145
renal, 368*i*, 370
Apramycin, 472–473
Aprikalim, 171
AQ-RA 741, 83
Arabinogalactan, 579–580
Ardacin, 494
Arecaidine, 21
Arecoline, 5, 72
for Alzheimer's, 22
derivatives, 20–28
Arin, 49
Aromatic sulfonamide diuretics,
383–384, 385 table
Arotinolol, 216*i*, 222
1-Aroyl-3-[(1-benzyl-4-
piperidinyl)ethyl]thiourea
derivatives, 52
Arrhythmia, *see* Cardiac
arrhythmia
Arteriodilation, 208
Aryloxyrimidines, 548–549
Ascamycin, 537
Ascomycetes, 638
ASL-8123, 288
Aspergillosis, 639, 641
Aspirin:
after myocardial infarction,
205
for secondary MI prevention,
206
Astromicin, 474
Atenolol, 216*i*, 219, 274, 285, 286,
288*i*
Atheroma, 204
Atherosclerosis, 204
pathogenesis, 204–205
Ca^{2+},Mg^{2+}-ATPase, 161, 176, 177,
178
H^+/K^+-ATPase, 123. *See also*
Gastric proton pump
inhibitors half-life, 137
Na^+,K^+-ATPase, 161–163

Na⁺,K⁺-ATPase (*Continued*)
and autonomic cardiovascular
regulation, 186
enzymology, 177–180
inhibition by cardiotonic
steroids, 179, 232, 235–245
[major discussion], 238*i*,
248–253 [in passing]
isoforms, 185–186
and renal function, 366, 368–372
sequencing, 180–184
Atrial natriuretic factor (ANF),
174, 306, 440–441
and blood pressure control, 268
Atrinositol, 308–309
Atrioventricular (AV) node, 156,
165*i*
Atropine, 187
absorption, 75
activity, 100, 110
analogs, 79–82
for anesthesia premedication, 66
as antidote for mushroom
poisoning, 73
antispasmodic activity, 61
cholinergic properties at low
doses, 104
excretion, 76
gastric acid secretion inhibition,
62, 123
metabolism, 75–76
as parasympatholytic, 61
selectivity, 105
semisynthetic derivatives, 77–78
stereoisomerism: asymmetric
carbon, 101, 102
structure, 73, 74
compared to acetylcholine, 79
synthesis, 74–75
therapeutic uses, 106–107
toxic doses, 101
Atropine oxide, 77
Atropinesterase, 76–77
Atropinic agents, 61
Atroscine, 73
Australian Trial, 274, 276
Automaticity, in myocytes, 163–164,
165, 190
Autonomic regulation, of cardiac
function, 186–188
Avilamycin, 505
AV node, 156
action potentials, 165*i*
Avoparcins, 496, 497*i*
AX-RA 513, 83
Azalides, 483
Aziridinium cation, 51
Azithromycin, 482*i*, 483, 589

antimycobacterial activity,
620–621
Azo dyes, 529
Azole antifungals:
for dermatophytoses, 641 (table),
644
for systemic mycoses, 647–648
Azolimine, 440
Azosemide, 401*i*, 412

B-283, 600
B-1912, 601
Bacitracin, 498, 499*i*
Banthine, 82
B-283, 600
B-1912, 601
Bacitracin, 498, 499*i*
Banthine, 82
Baroreceptors, 267–268
Basidiomycetes, 638
Bay K8644, 294, 295
BCG vaccine, 584, 624
Belladonna, 72. *See also*
Solanaceous alkaloids
Bemitradine, 399
Benazepril, 291*i*, 292
Bendroflumethiazide, 281, 390*i*
Benzbromarone, 445*i*, 448
Benzimidazoles, 230
gastric proton pump inhibitors,
127
Benzodiazipine anticholinergics,
82–83
2-Benzoforanylquinuclin-2-diene,
84
Benzolamide, 380*i*, 382
Benzothiadiazine diuretics, animal
studies, 373
Benzothiazepine anticholinergics,
83–84
Benzothiazepine calcium channel
blockers, 214, 215*i*, 218
Benzthiazide NF, 390*i*
β-Blockers, 193, 194 (table), 196*i*,
216*i*
adverse effects, of first
generation drugs, 221
antihypertensive effects, 220
for arrythmia, 187
and blood pressure control, 268,
269, 276, 278, 285–288
chemistry, 287–288
for congestive heart failure,
222–223, 226
discovery, 285–287
mechanism, 200, 218–220
newest compounds: vasodilators,
221 (table), 221–222

pharmacology, 287
as preferable to class I
antiarrhythmics, 199, 200
for secondary MI prevention,
206
Beta lactams, *see* β-Lactams
(under Lactam)
Betaxololi, 287, 288
Bethanechol, 11
BIBR277, 302, 303*i*
BIBS22, 303, 304
BIBS39, 303, 304
Bicarbonate, renal reabsorption,
368, 379
Bicozamycin, 505, 507
Bicyclomycin, 505, 507
Bioaffinity chromatography, 493
Bisguanylhydrazones, 234*i*,
242–243, 245, 253
BL-6341, 140
Blastomycosis, 639, 641
Blood pressure control, 267–269.
See also Antihypertensive
drugs; Hypertension
cellular sites, 269 (table)
variability, 271
Bluensomycin, 467
BM782, 475
BMY-15037-1, 398–399
Bowman's capsule, 365, 367*i*
BQ123, 308
Bradykinin, and NO, 210
Bretylium, 196*i*, 201
Brevistin, 500
Bromoacetylcholine, 9
Brush border membrane vesicles,
357
Bufadienolides, 232, 233*i*, 235
Bufalin, 233*i*, 235
Bumetanide, 284, 401*i*,
406–411
bioavailability, 400
pharmacology, 411
Bundle of His, 156
Burimamide, 63
Butenolides, 232, 233*i*, 234–235
Buthiazide, 390*i*
Butikacin, 472
Butirosin, 467, 468–469
Butoconazole, 647
Butrylcholine, 104
3-Butylamino-4-chloro-5-
sulfamoylbenzoic acid, 406,
408
BY 308, 127

Caffeine, 174, 426–427
Calcitrol, 370

Calcium:
  renal reabsorption, 370
  transport in cardiac function,
    174–177
Calcium channel blockers, 193,
  194 (table), 196i
  chemistry, 294–295
  discovery, 292
  for hypertension, 276, 278,
    292–295
  mechanism, 202
  pharmacology, 214–217, 292–294
  receptor sites, 217–218
  types, 214, 215i
Calcium channels, 157i, 158–164,
  166i, 166–167
  gating, 167–168
  types, 169–170
Calcium$^+$-influx factor, 175
Calcium^{2+}-phosphoinoside, 172,
  173
Calcium^{2+} pump, 175–177, 178.
  See also Ca^{2+}.Mg^{2+}2-ATPase
  (under ATP)
Calcium regulating hormones, and
  hypertension, 310
Calmodulin, and cardiac function,
  173, 177, 209
Calomel, 387
Calpain, 176
cAMP, and heart function, 172, 173
Candidiasis, 639–640, 641
Candoxatrilat, 443
Canrenone, 426
Capillary fluid shift, 268
Capreomycin, 588, 623
  antimycobacterial activity,
    608–609
N_7-Caproylchlorothiazide, 386
Captopril, 289, 290
Carbachol, tertiary amine
  congener, 7
Carbamate-derived
  anticholinesterases, 45–47
Carbamoyl choline, 9
  conformation, 38
Carbapenems, 622
Carbenoxolone, 111
Carbomycins, 484, 488
Carbonic anhydrase, 368, 372,
  378–379
Carbonic anhydrase inhibitors,
  123, 281, 324, 378–379,
  380 (table)
  clinical application, 382–383
  history, 379–380
  structure-activity relationship,
    380–382

sulfonamide side effects as leads
  for new, 541
p-Carboxybenzenesulfonamide,
  383
Carbutamide, 542
Cardiac arrhythmia, 187. *See also*
  Antiarrhythmic drugs
  mechanisms, 189–193
  reentry, 191–193, 192i
  Working Group on Arrhythmias,
    193–199
Cardiac glycosides, 230, 232
  stereochemistry, 235i, 240
Cardiac muscle, 156, 158
  contraction, 175, 177i
Cardiac physiology:
  autonomic regulation, 186–188
  endothelial-derived factors and,
    174, 188–189
  excitation and contraction
    coupling, 160–165
  intracellular Ca$^+$ signaling,
    174–177
  ion channels, *see* Ion channels
  ligand-mediated membrane
    transduction, 171–174
  myocardial contractility,
    158–160, 159i
  Na$^+$,K$^+$-ATPase, *see* Na$^+$,K$^+$-
    ATPase (under ATP)
  preload and afterload, 207–208
  summary, 156–158, 157i
Cardiogenic shock, 206
Cardiotonic steroids, 233–234i. *See
  also* Digitalis
  Na$^+$,K$^+$-ATPase inhibition, 179,
    232, 235–245 [major discussion],
    238i, 248–253 [in passing]
  biological activity, 242,
    243 (table)
  chemistry, 232–235
  for congestive heart failure, 226
  history, 155
  pharmacology, 235–239
  receptor selectivity, 246–247
  sources, 231
  structure-activity relationship:
    binding and physical
      properties, 247–248
    binding thermodynamics,
      246–247
    electrostatic interactions,
      245–246
    hydrophobic bonding, 245
    17β side chain, 241i, 248, 249
    steroid ring system, 241i,
      250–251
    sugar moieties, 241i, 251–253

summary, 240–242, 241i
  uses, 231–232
Cardiovascular disease, 155, 266.
  See also Hypertension
Cardiovascular risk factors, 324
Carinamide, 446
Carrageenin, 111
Carteolol, 286
Carvedilol, 216i, 221, 222, 286
  for congestive heart failure, 223
Cassaine, 233i, 235, 245
Catecholamines, 228–229
CD4 + T cells, 583
CD8 + T cells, 583
Celesticetin, 491
Celiprolol, 221, 287
Central anticholinergic syndrome,
  67
Central ischemic receptors, 268
Cephalosporinases, 621
Cephalosporins, 472, 504
Cerulenin, 490
cGMP, and cardiac function, 172,
  173–174
CGP42112A, 301–302
CGS 4270, 437
CGS 21680, 307
CGS 24128, 444
CGS 24592, 444
CGS 25462, 444
Chalcomycin, 486
Chalcose, 486
Channel proteins, 166–167
Channels, *see* Ion channels
Charybdotoxin, 171
Chemistry, *see* Medicinal
  chemistry
Chloramphenicol, 550
Chlorazanil, 429i, 430
Chloride:
  movement across apical
    membrane, 123, 144–145
  renal reabsorption, 368–369, 370
Chloride channels, 171
Chloride/formate exchanger, 369
Chlormadinone acetate, 234i, 244
6-Chlorobenzene-1,3-
  disulfonamide, 383, 384
Chlorobiocin, 510
5-Chloro-7-iodo-8-quinolinol, 641
Chloromerodrin, 377i
[5-Chloro-3-(2-methylphenyl)-4-
  oxo-4H-1-benzopyran-7-
  yl]oxyactic acid, 343
Chloroorienticins, 494
α-Chloro-β-phenethylamine, 51
Chloropolysporins, 496
Chloroquine, 539

Chlorothiazide, 281, 282, 324, 325*i*, 383, 390*i*, 391–392
Chlorthalidone, 274, 282, 393*i*, 394
  with spironolactones, 426
Cholecystokinin, 63
Cholesterol:
  and fish oils, 207
  and ischemic heart disease, 204, 205, 206
Choline, *see also* Acetylcholine, analogs, derivatives, and congeners for choline modifications
  activity *vs.* acetylcholine, 9
  esters, 8 (table)
Choline acetyltransferase, 4
Choline ethers, 12
Cholinergic receptors, 5–6
  and autonomic cardiac regulation, 187–188
Cholinergics:
  acetylcholine and analogs, *see* Acetylcholine
  anticholinesterases, *see* Anticholinesterases
  arecoline and related compounds, 20–28
  conformations, 98–99, 99*i*
  defined, 4
  dissociation constants, 102–103, 104 (table)
  heart rate decrease mechanism, 165
  muscarine, muscarone, and related compounds, 15–19
  nicotine and related compounds, 13–15
  oxotremorine and related compounds, 28–36
  pilocarpine and related compounds, 20
  structure (conformation)-activity relationship, 37–42
  therapeutic indications, 4
  unique muscarinic agonists, 36
Cholinomimetic therapy, 4
Chromomycosis, 639, 641
Cilazapril, 291*i*, 292
Cilofungin, 650, 651*i*
Cimetidine, 111, 120–121, 123–124
Cinnamycin, 503
Cinodine, 476
Ciprofloxacin, 589, 616–617, 618, 623
Circulins, 500
Cirramycin, 485, 486*i*
Cladinose, 478
Clarithromycin, 482*i*, 483, 589, 624

antimycobacterial activity, 619–620
Clavulanate, 589
Clavulanic acid, 621, 622
Clazolimine, 429*i*, 440
Clindamycin, 491
Cloconazole, 647
Clofazimine, 588, 600–601, 623, 624
  combination with azithromycin, 621
Clofilium, 201
Clonidine, 297, 298
Clopamide, 393*i*, 394–395, 408
Clorexolone, 393*i*, 396
Clotrimazole, 647
Coccidioidomycosis, 639, 641
Conduction block, 191
Congestive heart failure, 155
  β-blockers for, 222–223
  diuretics for, 364, 374
  as maladaptive hypertrophy, 223, 225
  pathophysiology, 223–225
  symptoms/signs, 225–226
  treatment, 226–228
Convoluted tubules, 366, 367*i*, 369*i*, 370
Cord factor, 580, 590
Coronary arteries, 156
Coronary artery bypass grafts, 207
Coronary artery disease, 156, 204
  lack of success of diuretic therapy, 324
Cortex (kidney), 365, 366*i*, 367*i*
Cortical nephrons, 282*i*, 365, 366, 367*i*
Corynetoxins, 508
Corynomycolic acids, 579
CP20, 652
Cromakalim, 171, 304–305
Cryptococcosis, 639, 641
Curamycin, 505
CV-11974, 303
Cyclic-AMP, *see* cAMP (under AMP)
Cyclic-GMP, *see* cGMP (under GMP)
Cyclic polynitrogen diuretics, 426–440, 429 (table), 434–436. *See also* Potassium-sparing diuretics
Cyclopenthiazide, 390*i*
Cyclopentolate, 107, 108
5-Cyclopentyl-1,3-dipropylxanthine, 427–428
Cycloplegics, anticholinergics, 61, 66, 106–107
Cycloserine, 588, 589, 623

antimycobacterial activity, 609
Cyclosporin, interaction with amphotericin B, 649
Cyclothiazide NF, 390*i*
Cytochrome P-450:
  and antifungal action, 647
  and NO production, 212
Cytoprotection, 125, 137–138

Dactmicin, 474
Dalfopristin, 492
Dapsone, 588, 592–593, 624. *See also* Sulfones
  combination with pyrimethamine, 539–540
  mechanism of action, 547–548
  metabolism, 563
  resistance to, 551
  side effects, 593
Daptomycins, 503
Datura, 72
D-cells, 121
Deanol, 13
Decahydroquinoline, 39–40
Decahydroquinoliniums, 40
Decalin, 39–40
Decaplanin, 493*i*, 494
Deferoxamine, 651–652
dl-Dehydromuscarine, 16
4,5-Dehydromuscarone, 16
Delayed after-depolarization (DADs), 191
Deltamycins, 484
Dendrotoxin, 171
2-Deoxystreptamine, 466*i*, 467
  derivative aminoglycoside antibiotics, 466–473
Depolarization, myocytes, 156, 157*i*, 158, 160–165
  EADs and DADs, 191
Dermatophytoses, 639
  antifungal substances for, 641–646
Desethermuscarine, 19
Desethermuscarone, 19
Desethylamiodarone, 203
Desosamine, 480, 486
Destomic acid, 475
Destomycins, 475
Deuteroiomycetes, 638
Diacetylformoguanamine, 430
Diacylglycerol, 173
Dialkylaminoalkanols, 79, 80*i*
N-Dialkylsulfamoylbenzoate uricosurics, 446
Diamidine antifungals, 650
3,5-Diamino-6-chloropyrazine-2-

carboxamide diuretics,
437–438
2,4-Diamino-6,7-dimethyl
pteridine, 431, 432
4,4'-Diaminodiphenylsulfone, *see*
Dapsone
4,4'-Diaminodiphenylsulfoxide,
539
4,7-Diamino-*N*-(2-morpholinoethyl)-
2-phenyl-6-pteridine-
carboxamide, 431
4,7-Diamino-2-phenyl-6-pteridine-
carboxamide, 431, 432
2,4-Diaminopteridine diuretics, 431
Diapamide, 393*i*, 395
Diastolic pressure, 271
Diazoxide, 284, 388
Dibekacin, 471
6,7-Dichloro-2,3-dihydro-
benzofuran-2-carboxylic acid
diuretics, 338*i*, 342 (table)
Dichloroisoproterenol, 285
Dichlorophenamide, 380*i*, 382, 383
8-(Dicyclopropylmethyl)-1,3-
dipropylxanthine, 427
2-Diethylaminoethyl
phenylacetate, 100
Digitalis, *see also* Cardiotonic
steroids; Digoxin
endogenous, 224, 253–255
as generic terms, 231
neurohormonal action, 231–232,
248
as positive inotropic agent, 228
Digitalis receptors, 239–240, 241*i*,
243, 246–248, 254–255
Digitoxigen, C17*β*-modified
derivatives, 233*i*, 249
Digitoxigenin, 233*i*
glycosidation, 250, 252
structure *vs.* chlormadinone
acetate, 244
Digitoxigenin monodigitoxoside,
253
Digitoxigenin rhamnoside, 252,
253
Digitoxin, 233*i*
Diglucomethoxane, 377–378
4,5-Di-*O*-glycosyl-2-
deoxystreptamine
aminoglycosides, 467–469
4,6-Di-*O*-glycosyl-2-
deoxystreptamine
aminoglycosides, 469–471
Digoxigenin, 233*i*
glycosidation, 252
Digoxin, 230, 231, 233*i*, 234, 239
reflex action and, 232

renin-angiotensin system
suppression, 232
Dihydrocupreine, 529
Dihydroethacrynic acid, 326, 327*i*
Dihydrofolate reductase, 544, 545*i*,
546–547, 559, 560
Dihydrofolate reductase inhibitors,
539, 548–549
5,6-Dihydrofuro[3,2*f*]-1,2-
benzisoxazole-6-carboxylic
acid diuretics, 331, 332 (table),
405
7,8-Dihydrofuro[2,3*g*]-1,2-
benzisoxazole-7-carboxylic
acid diuretics, 331–332
7,8-Dihydrofuro[2,3*g*]-benzoxazole-
7-carboxylic acid diuretics,
332–333, 334 (table)
Dihydrofuroxanthone, structure-
activity relationship, 347–354
Dihydrofuroxanthone-2-carboxylic
acid diuretics, 330 (table), 331
Dihydropteroate synthase, 544,
545*i*, 546–547, 549
Dihydropyridine calcium channel
blockers, 214, 215*i*, 217 (table),
218
Dihydropyridine receptors, 160,
161, 174
1,2-Dihydro-2-(3-pyridyl)-3*H*-
pyrido[2,3-*d*]pyrimidin-4-one,
436
Dihydrospectinomycin, 474
Dihydrostreptomycin, 466*i*, 467,
603, 605
N,*N*'-Diisopropylethylendiamine,
601
Dilevalol, 221
Diltiazem, 202, 215*i*, 217, 292*i*, 295
N,*N*-Dimethylaminoethanol, 13
N,*N*-Dimethylethanolamine,
acetate ester, 7
Dimethylpyrindene, 85, 86
1,2,3-Dioxaphosphorinane, 49–50
1,1-Dioxide diuretics, 387 (table)
Dioxolanes, 17, 18, 38
1,3-Dioxolo[4,5*f*]-1,2-
benzisoxazole-6-carboxylic
acid diuretics, 333, 336 (table)
1,3-Dioxolo[4,5*g*]-1,2-
benzisoxazole-7-carboxylic
acid diuretics, 333, 335,
337 (table)
Diphenylacetic acid, 100–101
Diphenylacetyloxy
anticholinergics, 85–86
4-[(Diphenylacetyl)oxy]-1,1-
dimethylpiperidium, 86

Diphenylamine-2-carboxylate, 145
1,1-Diphenyl-5-diethylamino-
pentane, 100
1,1-Diphenyl-3-piperidino-1-
butanol, 101
1,3-Dipropylxanthine diuretics,
427–428
Dirithromycin, 482*i*, 483, 619
Disopyramide, 195*i*
Distal convoluted tubule, 366,
367*i*, 370
Disulfa drugs, 534
Disuprazole, 127, 128*i*
Diuretics, *see also numerous specific
compounds and classes of
diuretics*; terms beginning
"Renal"
chemistry, 283
classification, 324–326, 372
clinical aspects, 374
combinations, 392
for congestive heart failure, 226
discovery, 280–282
evaluation, 372–374
history, 324–325, 326, 365,
448–449
"ideal," 324
pharmacology, 282–283, 372–374
side effects, 324, 374, 449
sulfonamide side effects as leads
for new, 541
DNA-directed RNA-polymerase
(DDRP), 613
Dobutamine, 228–229
L-Dopa, 228
Dopamine, 228–229
Dopaminergic agents, 229
Dopexamine, 229
Doxazosin, 297
Doxepin, 144
DR-3438, 343–344, 345
Drug design, 588. *See also*
Medicinal chemistry;
Structure-activity relationships
molecular modification and,
528, 531
new drug evaluation, 372
purpose of, 107
side effects as lead for, 531
Drug resistance, 591, 616, 622–623.
See also specific drugs for
information on resistance
MLS-resistance, 490
and revival of interest in
sulfonamides, 531
R-plasmids and, 550
DS 210, 434, 436
DS 511, 434, 436

DSBA, 538
Dubos medium, 577
Duodenal ulcers, *see* Peptic ulcers
DuP-996, 52–53
Duramycin, 503
Dyes:
  as antifungals, 641
  and sulfonamide development,
    529

E3174, 300, 301
"Eagle" effect, 617
Early after-depolarization (EADs),
  191
Echinocandin antifungals, 650,
  651*i*
E. coli, sulfonamide resistance,
  549, 550
Econazole, 647
Ectopic beats, 164, 190, 206
Ectothiophate, 49
Edema, diuretics for, 324, 364,
  374
Edrophonium, 44
Efferent artioles, 365, 367*i*
Eicosanoid biosynthesis, and
  hypertension, 311
Eicosapentaenoic acid, 206
Electrocardiagram (ECG), 164,
  165*i*
Electrolytes, 365, 366, 368–372. *See
  also specific electrolytes*
EM-523, 489, 490*i*
EM-574, 489, 490*i*
Enalapnilat, 398
Enalapril, 289, 290
Enalkiren, 305
Encainide, 195*i*
Endocardial endothelium factors,
  189
Endocardin, 189
Endogenous digitalis-like factors
  (EDLF), 224, 253–255
Endothelial cells, 188
Endothelial-derived factors, and
  cardiovascular function, 174,
  188–189
Endothelial-derived
  hyperpolarizing factor
  (EDHF), 209
Endothelial-derived relaxing factor
  (EDRF), 174, 188, 224, 310
  nitric oxide as, 174, 189,
    208–210, 209*i*
Endothelial-leukocyte adhesion
  molecule-1, 188
Endothelin antagonists, 189, 543
  for hypertension, 307–308

Endothelins, 188–189, 543
  and blood pressure control, 268
Enduracidin, 500
Enoximone, 230
Enprostil, 141
Enterochromaffin-like cell
  proliferation, 140–141
Enterogastrones, 63
Epidermin, 503
Epidirithromycin, 483
Epinine, 229
5-Epi-sisomicin, 472
4-Epivancosamine, 493*i*, 494
Eremomycins, 493*i*, 494
Ergosterol, 637, 643
Erythema nodosum leprosem, 601,
  624
Erythromycin, 478, 479*i*, 480–481
  antimycobacterial activity, 618,
    619, 620
  intramolecular cyclization, 481*i*
  as motilide, 489
  prodrugs for, 480–481
Erythromycin acistrate, 480
Erythromycin-11,12-carbonate,
  482*i*, 483
Erythromycin estolate, 480
Erythromycin gluceptate, 481
Erythromycin lactobionate, 481
9-Erythromycylamines, 482*i*, 483
Erythronolide A, 478
Erythrophleum alkaloids, 232,
  233*i*, 240, 245
ES 37, 442
Esmolol, 288
Espinomycins, 484
Essential hypertension, *see*
  Hypertension
Ethacrynic acid, 281, 282*i*, 324–325,
  326, 400, 401*i*, 403
  analogs, 402 (table)
  animal studies, 373
  structure-activity relationship,
    345–347
Ethambutol, 588, 589, 601–603, 623
  resistance to, 591
  side effects, 603
  synergism with ciprofloxacin/
    ofloxacin, 618
Ethanolamine, acetate ester, 7
Ethionamide, 588, 598–599, 623
Ethoxzolamide USP, 380*i*, 382
Etozolin, 417–418
Eucatropine, 107
Everninomicin, 505, 506*i*

Famotidine, 120
Fatty acids, as antifungals, 641

Felodipine, 215*i*, 217 (table), 294
Fenquizone, 398
Fenticonazole, 647
Fermentation products, 464–465
Fibrillation, 156
  after AMI, 202
  and reentry, 192*i*, 193
Fish oils, 206–207
Flambamycin, 505
Flecainide, 195*i*
Fluconazole, 647
5-Fluorocytosine, 649
Fluoroquinolones, 472, 616–618
Flurithromycin, 482*i*, 483, 619
Folate biosynthesis:
  inhibition by sulfonamides, 544,
    545*i*, 546–548
  as target for antimycobacterial
    drugs, 590–591
Folic acid, 544
Formoguanamine, 428, 430
Formylcholine, 9, 104
Foromacidine, 484
Forosamine, 484
Forskolin, 124
Fortamine, 466*i*, 473
Fortimicin, 473–474
Fosenopril, 291*i*, 292
Foxglove, 231. *See also* Digitalis
Free-radical scavengers, 221
Fungal diseases/pathogens,
  638–640, 641 (table)
  predisposing factors, 640 (table)
Fung*i*, 637–638. *See also*
  Antifungal agents
Furosemide, 281, 282*i*, 324–325,
  336, 405–406, 541
  animal studies, 373
  bioavailability, 400
Fusidic acid, 504

Gallopamil, 295
  gastric acid secretion inhibition,
    144
Ganglionic blocking agents, for
  hypertension, problems with,
  278
Gap junctions, 158, 174
Gastric acid, 120
Gastric acid inhibitors, 63–66,
  123–124, 124 (table)
  need for, 120–121
  nonanticholinergics, 111–112
  in vitro test assays, 124–126
  in vivo test assays, 124, 125
Gastric acid secretion, 62–63, 65*i*,
  120, 121–123, 122*i*
  inducing, 124

inhibition, 123–124. *See also*
    Gastric acid inhibitors; Gastric
    proton pump inhibitors
Gastric inhibitory polypeptides
    (GIP), 63
Gastric juice, 62
Gastric parietal cells, 120, 122*i*
Gastric proton pump, 123
Gastric proton pump inhibitors,
    121, 123
  irreversible, 126–142
    mechanism, 127–134
    metabolism, 139–140
    pharmacokinetics, 139
    pharmacology, 135–138
    structure-activity relationship,
        130, 134–135
    toxicology, 140–142
  miscellaneous inhibitors,
      144–145
  reversible, 142–144
Gastric ulcers, *see* Peptic ulcers
Gastrin, 63, 121–122, 140
  binding, 122–123
Gastrin receptor antagonists, 121
Gastroesophageal reflux disease,
    121
Gating, ion channels, 166, 167–168
G-cells, 121
GE2270 A, 502
Gemfibrozil, 276
Genetic linkage experiments, 277
Genins, 232
Genoscopolamine, 77, 78
Gentamicins, 469–470, 471, 603
16β-Gitoxin, 251
Glaucoma, 383
  β-blockers for, 220
  cholinergics for, 4
Globomycin, 501
Glomerular filtrate, 366, 367*i*
Glomerulus, 365, 367*i*
Glucocorticoids, 310
Glucose, 375*i*
  in cardiac glycosides, 232, 236*i*
  renal reabsorption, 366, 368
Glucose-6-phosphate
    dehydrogenase, 379
Glucosulfone sodium, 592
Glutathione-*S*-transferases, and
    NO production, 212
Glybenclamide, 542
Glycerin, 375*i*
Glyceryl dinitrate, 213
Glyceryl trinitrate (nitroglycerin),
    207, 211*i*
  bioactivation, 213
  thiol exposure, 211–212

Glycocinnamoylspermidine
    antibiotics, 476
Glycolipopeptide antibiotics, 501
Glycopeptide antibiotics, 492–498
  resistance to, 496–498
  structure-activity relationship,
      497
Glyoxylic acid, 444
Glysperins, 476
Gomphoside, 250, 253
Gout, 365, 374, 444
G-proteins, 67, 172, 187
G-protein signaling systems,
    172–174
GR117289, 303
Gramicidins, 498–499
Griseofulvin, 642–644
Growth factors, and hypertension,
    310–311
Guanabenz, 297, 298
Guanadrel, 299*i*
Guanethidine, 299*i*
Guanfacine, 297, 298
Guanylhydrazones, 234*i*, 242–243,
    245, 253
Guanylyl cyclase, and cardiac
    function, 173–174
Guy's Hospital pill, 376

Habekacin, 471–472
Halomycins, 610
Hansen bacillus, *see Mycobacterium
    leprae*
Heart, *see* Cardiac physiology;
    Coronary entries; Myocardial
    entries
Heart attack, *see* Myocardial
    infarction
Heart block, 191
Heart failure, *see* Congestive heart
    failure
Heidenhain pouch, 71
Heliobacter pylori infection, 71,
    111, 120, 138
Hellebrin, 233*i*, 235
Helsinki Heart Study, 276
Helvecardins, 496
Helvolic acid, 504
Helvolinic acid, 504
Heparin, 174, 205
Heptylphysostigmine, 46
Hexahydrosiladifenidol, 86
H$_2$-folate reductase, 544, 545*i*,
    546–547, 559, 560
High-ceiling diuretics, 281, 282*i*,
    325, 399–419, 401–402
    (table), 409–410 (table),
    414–415 (table)

pharmacology, 399–400, 411
side effects, 400
High-density lipoproteins (HDLs),
    204
High resolution nuclear magnetic
    resonance spectroscopy, 494
Himbacine alkaloids, 87
His bundle, 156
Histamine-H$_2$-receptor antagonists,
    63
  as antiulcer agents, 120–121, 137
  enterochromaffin-like cell
      proliferation, 140
Histamine-H$_2$-receptors, role in
    gastric acid secretion, 63, 121,
    122
Histoplasmosis, 638, 639, 641
HIV infection, *see* AIDS
HMG Co-A reductase inhibitors,
    for secondary MI prevention,
    206, 207
Homatropine, 100
Homatropine bromide, 107
Homatropine methylbromide, 77
Homeostasis, 365
Homosulfanilamide, 535
HP 522, 326, 327*i*, 328, 405
  structure-activity relationship,
      345–347
H$_2$-pteroate synthase, 544, 545*i*,
    546–547, 549
7HT medium, 577
Hydralazine, 299*i*
Hydrochlorothiazide, 281, 390*i*,
    391–392, 408, 541
Hydroflume thiazide, 390*i*
Hydrogen ion secretion, 379
Hydrothiazide diuretics, 388–389,
    391
  structure-activity relationship,
      388, 390
Hydroxyalkylpyrrolidines, 79, 80*i*
3'-Hydroxyapramycin, 472
14-Hydroxyclarithromycin, 482*i*,
    483
N^6-Hydroxylysine, 582
4-Hydroxy piperidines, 79, 80*i*
14β-Hydroxy-5β*H*-pregnane-3β-
    rhamnoside, 244
Hydroxypyridinone antifungals,
    652
16β-Hydroxyspironolactone, 424
2-Hydroxystilbamidine isethionate,
    650
Hydroxystreptomycin, 603
3-Hydroxytropine, 73
Hygromycins, 474–475
Hyoscine, *see* Scopolamine

(−)-Hyoscyamine, 73, 75
    cholinergic properties at low
        doses, 104
    synthesis, 75
(±)-Hyoscyamine, *see* Atropine
Hyoscyamus, 72
Hypergastrinaemia, 140, 141
Hyperpolarization, 209–210
Hypertension, 155, 225. *See also*
        Antihypertensive drugs
    classification, 270–272
    defined, 267–268
    diuretics for, 364, 374
    drug therapy rationale, 272–275
    and endogenous digitalis-like
        factors, 254
    epidemiology, 270–272
    experimental models of genetic,
        312–313
    genetic aspects, 310
    risk of uncontrolled, 270 (table)
    and secondary MI prevention,
        207
Hypertension Detection and Follow
    Up Program, 283
Hyperuricemia, 325, 326, 365
Hyphae, 638
Hypolipidemic drugs, for
    secondary MI prevention, 206

Ibopamine, 229
Ibutilide, 196*i*
ICI 206970, 439
ICI 147798, 439
Idazoxan, 295
Imidazole antifungals, 647
Imidazoline receptors, 298
Imipenem, 622
Indacrinone, 326, 327*i*, 336,
        403–404
    structure-activity relationship,
        345–347, 358
Indapamide, 283, 284, 392, 393*i*,
        395
Indecainide, 195*i*
Indene anticholinergics, 85–86
Inflammatory cytokins, 210
Inositol, 176
Inositol-1,4,5-triphosphate, 173
Inotropic agents, 228–230
    for congestive heart failure, 226
Insulin-releasing sulfonamide, 542
Intercalated cells, kidneys, 371,
        372*i*
Intercalcated discs, 158, 174
Intercellular adhesion molecules,
        188
Interferon-γ, 583

Ion channels, 165–167. *See also*
        Calcium channels; Chloride
        channels; Potassium channels;
        Sodium channels
    gating (voltage-dependent
        channels), 166, 167–168
    inward voltage-dependent
        channels, 168–170
    ligand-gated, 172–174
    outward voltage-dependent
        channels, 170–171
    rectifying, 170
    subunits, 166–167
    vascular smooth muscle, and
        hypertension, 311
IPPPSH Trial, 274, 276
Ipratropium bromide, 78
Iprocinodine, 476
Iron chelating antifungals, 651–652
Iron metabolism, mycobacteria,
        582, 590
Irreversible proton pump
        inhibitors, *see* Gastric proton
        pump inhibitors, irreversible
Ischemic heart disease, 155. *See
        also* Angina pectoris
        Myocardial infarction;
        Myocardial ischemia
    pathophysiology, 204–205
Isepamicin, 471–472
Isobutyl methylxanthine, 124
Isoidide dinitrate, 211*i*
Isometheptene hydrochloride, 112
Isoniazid, 588, 595–598, 623
    combination with rifampcin, 614
    drug resistance to, 585, 591
    resistance to, 597
    side effects, 597–598
Isosorbide, 375*i*
Isosorbide dinitrate, 211*i*
Isradipine, 215*i*, 217 (table), 218
Istamycins, 474
Itraconazole, 647
Izenamicins, 486

Josamycin, 484, 485*i*
JT 246, 234*i*
JT 253, 234*i*
Juvenimicins, 486
Juxtamedullary nephrons, 365,
        367*i*

Kaliuretic diuretics, 325
Kanamycins, 469, 588, 589, 623
    antimycobacterial activity, 603,
        605–607
    resistance to, 606
Kasugamycin, 475, 603

RS-Kelatorphan, 443
Ketanserin, 306
Ketoconazole, 647
Ketolide, 482*i*, 484
15-Keto-6β,7β-methylen-
        progesterone, 424
Kibdelin, 494
Kidneys, 268, 365–373, 366*i*, 375. *See
        also* terms beginning "Renal"
KRM-1648, 616
Kujimycin A, 480
KW-3902, 428

L-159,066, 298
L-163,017, 304
L-693,989, 650, 651*i*
Labetalol, 286
Labetolol, 216*i*, 220, 222
β-Lactamases, 621
β-Lactams, 498, 621–622
    resistance to, 550–551
Lactoylcholine, 38
Lankamycin, 480
Lansoprazole, 121, 127, 128*i*, 137
    active form of, 130, 134, 135
    cytoprotection, 137–138
    drug interactions, 139–140
    pharmacokinetics, 139
    toxicology, 140, 141
Lantibiotics, 502–503
Left-sided heart failure, 224, 225
Legionella, 489
Leminoprozole, 127, 129*i*
    pharmacology, 136
Lepromatous leprosy, 586
Leprosy, 576. *See also* Dapsone
    drugs for, 588 (table), 589 (table),
        623–624
    pathogenesis, 582–586
    screening/evaluation of
        antimycobacterial agents, 587
    vaccine, 624
Leucomycins, 484, 485*i*
Leucoverin, 544
Leukocyte adhesion molecules, 188
Levodopa, 229
Levofloxacin, 617, 618
Licorice extracts, 120
Lidocaine, 195*i*
    after AMI, 202–203
    mechanism, 197–198
    replacement by sotalol as drug
        of choice for arrhythmia, 200,
        203
Ligand-gated channels, 172–174
Lincomycins, 490–491
Lincosaminide antibiotics,
        490–491

Lipids, and ischemic heart disease, 204
Lipoarabinomannan, 580
Lipopeptide antibiotics, 500–501
Lipopolysaccharides, 210
Lipoproteins, and ischemic heart disease, 204
Liposidomycins, 508
Lisinopril, 290, 292
Lividomycin, 467, 468
LL-BM123α, 475
LND 623, 234i, 244
Lomofloxacin, 617
Loop diuretics, see High-ceiling diuretics
Loop of Henle, 365–366, 367i, 369, 370i
Losartan, 278, 300, 301
Low-ceiling diuretics, 281
Low-density lipoproteins (LDLs), 204, 206
Loxtidine, 140
L-type calcium channels, 169, 170i, 293
  and autonomic cardiovascular regulation, 186
LY-303,366, 650, 651i
Lymphokines, 583
Lysozyme, 498

M-43, 494
M-4365, 486
MAC complex, see Mycobacterium avium intracellulare (MAC) complex
Macrolide antibiotics, 477–490
  antimycobacterial activity, 618–621
  biosynthesis, 489
  drug interactions with, 619
  12-membered ring, 478
  14-membered ring, 478–480
    semisynthetic derivatives of, 480–484
  16-membered ring, 484–487
    semisynthetic derivatives of, 487–489
  as motilides, 489, 490i
  side effects, 489
Mafenide, 535
Malaria drugs, 539–540
Mandeloylchalines, 103
Mandelyl tropeine, 77
Mannitol, 375–376
Mannosidostreptomycin, 603
Maridomycins, 484
Mast-cell deganulating peptide, 171

MC-352, 489
McN-A-343, 34–35
Medicinal chemistry, see also Drug design; Structure-activity relationships
  antimicrobial vs. post-antimicrobial eras, 464
  as "science of serendipity," 155
Medroxolol, 216i, 222
Medulla (kidney), 365, 366i, 367i
Mefruside, 384–385, 393i
Megalomicins, 480
Megosamine, 480
Melamine, 428, 430
Membrane-stabilizing anti-arrhythmic drugs, see Antiarrhythmic drugs, class I
Meralluride, 281, 377i
Merbaphen, 324, 325i, 376
Mercaptoacetic acid, 377
Mercurial diuretics, 281, 324, 376–378, 377 (table)
  animal studies, 373
  pharmacology, 378
  side effects, 378
  structure-activity relationship, 376–378
Mercurophylline NF XII, 377i
Meropenem, 622
Mersalyl, 378
Mespirenone, 421
Metabolite antagonism, 531
Metalazone, 393i, 397–398
Methantheline, 108
Methazolamide USP, 380i, 382
Methimazole, 542
Methoctramine, 84–85
Methscopolamine, 110
Methscopolamine bromide, 77
Methscopolamine nitrate, 77
Methyclothiazide, 390i
N-Methylarecoline, 21
Methylatropine nitrate, 77
Methyldopa, 297, 298
N-Methyldopamine, 229
N-Methylethanolamine, acetate ester, 7
15-(R)-Methyl-PGE$_{2+}$, 112
Methylprednisolone, 619
Methylrosaniline chloride, 641
N-Methyl scopolamine (NMS), 67
5-Methylsulfonylbenzoic acid diuretics, 408
2-Methyl-4-trimethylammonium-methyl-1,3-dioxolane, 17
Metiamide, 63
Metolazone, 283, 284
Metoprolol, 216i, 219, 220, 285,

286, 288i
  for congestive heart failure, 222
Mexiletine, 195i
Mexrenone, 425
Micins, 470. See also numerous specific Micins
Miconazole, 647
Micrococcin P, 501–502
Micropuncture, 366, 374
Midecamycins, 484, 487
Migraine, β-blockers for, 220
Mikamycins, 491
Milrinone, 230
Minimal topological difference (MTD), 248
Minocycline, 589, 624
Minoxidil, 299i
Miokamycin, 487
Mixanpril, 306
MK 447, 417
MK 473, 404
M line, 158, 159i
MLS-resistance, 490
Molds, 638
Molecular modification, 528, 531
Monic acid, 505, 506i
Monoguanylhydrazones, 243
Moricizine, 195i
Morphazinamide, 599
Morphine, after myocardial infarction, 205
Morpholiniums, 40
Motilides, macrolides as, 489, 490i
Motilin, 489
Motility-related GI disorders, 489
MOTT, see Mycobacteria other than tuberculosis (MOTT)
Moxonidine, 298
MRC Trial, 274
Mucormycosis, 639, 641
Mucosal resistance, 64
Multiple Risk Factor Intervention Trial, 283
Mupirocin, 505–506
Mureidomycins, 509
Muscarine, 5, 15–16
  acetylcholine mimicry, 61
  related compounds, 15–19
Muscarine iodide, 38
Muscarinic agonists, 36
Muscarinic receptors, 5–6
  and anticholinergic drugs, 104–106
  and cardiovascular function, 187
  role in gastric acid secretion, 121
m muscarinic receptors, 5–6
M$_1$ muscarinic receptors, 5–6

M$_2$ muscarinic receptors, 5–6
and cardiovascular function, 187
M$_3$ muscarinic receptors, 5–6
Muscarone and related
compounds, 15–16, 19, 38
Musculotropic spasmolytics, 62
Mushrooms, 640
atropine as antidote for, 73
Mutasynthesis, 476
Muzolimine, 416–417
Myasthenia gravis, cholinergics
for, 4
Mycarose, 480
Mycelia, 638
Mycinamicins, 486, 487i, 489
Mycins, 470. *See also* numerous
specific Mycins
Mycobacteria, 576–582. *See also*
Antimycobacterial drugs;
Leprosy; Tubercule bacilli
drug resistance, 585, 591
iron metabolism, 582
media for, 577
metabolism, 578–579
structure, 578, 579–582
Mycobacterial cell wall, 579–581
as target for antimycobacterial
drugs, 590
Mycobacterial mycolic acids, 579
Mycobacteria other than
tuberculosis (MOTT), 577
(table)
increase in infection, and AIDS,
585
screening/evaluation of
antimycobacterial agents,
587–588
Mycobacteriosis, 577
Mycobacterium avium intracellulare
(MAC) complex, 483, 576, 615
and AIDS, 585
screening/evaluation of
antimycobacterial agents,
587–588
Mycobacterium kansasii, 585
Mycobacterium leprae, 576, 585. *See
also* Leprosy
testing models, 578
Mycobacterium tuberculosis, see
Tubercle bacilli
Mycobacterium ulcerans, 585
Mycobactins, 581–582, 590
Mycolic acids, 579–580
Mycology, 640
D-Mycosamine, 648
Mycoses, systemic, 639–640
antifungal substances for,
647–652

Mycosides, 580–581
Mycotic infections, *see* Fungal
diseases
Mycotoxic diseases, 640
Mydriatics, anticholinergics, 61,
66, 106–107, 108
Myocamycin, 619
Myocardial hypertrophy, 223, 225
Myocardial infarction, 156, 202,
205
secondary prevention, 206–207
therapy, 205–206
Myocardial ischemia, 156, 171. *See
also* Ischemic heart
disease
Myocardial pacemaker cells, 160
Myocardium, 156, 158
Myocytes, 157i, 160i, 162
automaticity, 163–164, 165, 190
depolarization, 156, 157i, 158,
160–165, 191
Myofibrils, 157i, 158, 161
Myofilaments, 157i, 158
Myomycins, 475
Myosin, 157, 158–159, 160, 175, 177i

Nadolol, 216i, 220
Naftifine, 644, 645i
Nalidixic acid, 617
1,8-Naphthpyridine diuretics, 436
2-Naphthylcyclopentanol ketone
diuretics, 425
Narbomycin, 480
NC-1300, 127, 129i
pharmacology, 136, 137
NC-1300-B, 127, 129i
pharmacology, 136–137
Nebramycins, 469, 603
Negative inotropes, 228
Neomycin, 467, 468, 603
in triple ointment, 498
Nephrons, 282i, 365, 366, 367i
Nerve gases, 47, 49
Netilmicin, 470i, 471–472
Neuroleptics, gastric acid secretion
inhibition, 144
Neuropeptide Y receptor
antagonists, for hypertension,
308–309
Neutral endopeptidase, 441
Neutral endopeptidase inhibitors,
441–444
for hypertension, 306
Neutramycin, 486
Nicardipine, 217 (table), 294
Nicorandil, 171
Nicotinamide, 588
Nicotine and related compounds, 5

as cholinergics, 13–15
Nicotinic receptors, 5, 6
Nifedipine, 214, 215i, 217, 218,
292, 294–295
Nimodipine, 215i, 217 (table)
Nisin, 502–503
Nisoldipine, 217 (table)
Nitrates:
for angina, 207–208
bioactivation, 212–213
chemistry of, 210–212
history, 155
pharmacokinetics, 214
thiol exposure, 211–212
tolerance, 213–214
Nitrendipine, 217 (table), 218,
294–295
Nitric oxide:
and blood pressure control, 268,
310
as negative inotrope, 210
as probable endothelial relaxing
factor, 174, 189, 208–210, 209i
redox forms, 213
release by nitrates, 208
Nitric oxide synthases (NOS), 209
Nitriso compounds, as vasodilators,
see Nitrates
Nitrites:
musculotropic spasmolytic, 62
vasodilators, *see* Nitrates
Nitroglycerin, *see* Glyceryltrinitrate
Nitroprusside, 211, 212
Nitrovasodilators, *see* Nitrates
Nizatidine, 120
N. meningitides, sulfonamide
resistance, 549, 550
Nocardic acids, 579
Nonsteroidal antiinflammatory
drugs, 71
Nontuberculosis mycobacteria
(NTM), 577
increase in infection, and AIDS,
585
8-(Noradamantan-3-yl)-1,3-
dipropylxanthine, 428
Norarecoline, 21, 23–25
Norepinephrine, 285
and blood pressure control, 268
Nornicotine, 14
Nosiheptide, 501
Novobiocin, 509–510
N-type calcium channels, 169, 293
Nuclear magnetic resonance
spectroscopy, high resolution,
494
Nucleocidin, 537
Nucleoside antibiotics, 508–509

Nystatin, 648

Octapeptins, 500
Ofloxacin, 616–617, 618, 623, 624
Oleandolide, 478
Oleandomycin, 478, 479i, 481, 618
Oleandrose, 478
Omega-3 polyunsaturated acids, 206–207
Omeprazole, 121, 123–124
    activation, 128, 130, 131i
    chemistry, 125–126
    cytoprotection, 137–138
    drug interactions, 139–140
    metabolism, 139
    pharmacokinetics, 139
    pharmacology, 135–137
    toxicology, 140, 141
OMPA, 49
Onchidal, 51
OPC-21268, 307
Opportunistic infections, see
    Mycobacterium avium
    intracellulare complex
Organic nitrates, as vasodilators,
    see Nitrates
Organic nitrites, as vasodilators,
    see Nitrates
Organomercurial diuretics, see
    Mercurial diuretics
Organophosphorus-derived
    anticholinesterases, 47–50,
    48–49 (table)
Orienticins, 493i, 494
Orthophosphoric acid,
    anticholinesterase derivatives,
    47
Orthosomycins, 505
Osmotic diuretics, 375 (table),
    375–376
Ostreogrycins, 491
Ouabain, 231, 233i, 239, 242, 244
    as endogenous digitalis-like
    factor, 254
Oxiconazole, 647
(4-Oxo-4H-1-benzopyran-7-
    yl)oxyacetic acid diuretics,
    338–339, 343–345, 344 (table)
3-(3-Oxo-17β-hydroxy-4-
    androsten-17-α-yl)propanoic
    acid lactone, 420
1-Oxoisoindoline diuretics,
    395–397
Oxotremorine and related
    compounds, 28–36
    muscarinic actions/activities,
    30–31 (table)
Oxotremorine-M, 34

Oxprenolol, 219, 220
Oxyntic gland, 121
Ozolinone, 418–419

PABA, see p-Aminobenzoic acid
    (under Amino entries)
Pacidamycins, 509
Pantoprazole, 121, 127, 128i, 135
    activation, 130, 132
    cytoprotection, 137
    drug interactions, 140
    pharmacokinetics, 138, 139
Papaverine, 62
Paracoccidioidomycosis, 639, 641
Parasympathic nervous system, 60
    and blood pressure control, 268
    and cardiac function, 171
Parasympatholytic drugs, 61
Parasympathomimetics, see
    Cholinergics
Parathion, 49
Parathyroid hormone, 370
Parkinson's disease, 107, 108
Paromomycin, 467, 468, 603
Parvodicin, 494
Patricins, 491
Pavlov gastric pouch, 70–71
PD123319, 301–302
Pelvis (kidney), 365, 366i
Penicillin, 621
    "Eagle" effect, 617
Penicillinases, 621
Pentadienolides, 232, 233i, 235
Pentaerythritol tetranitrate, 211i
Pentagastrin, 63
Pentamidine isethionate, 650
Pentaquine, 278
Pentolamine, 295
Pepsin inhibitors, 111
Pepsinogens, 62, 120
Pepsins, 62
Peptic ulcers, 62, 64, 138
    antispasmodics for, 106
    healing, 120
    risk factors, 71
Peptide antibiotics, 498–503
Peptidoglycan, 579, 590
Peptidoglycolipids, 581
Perirenal fat, 365, 366i
Peroxynitrate, 210, 213
Phenolic glycolipids, 580–581
Phenotyping, 277
Phenoxyacetic acid diuretics,
    325–328, 327i. See also related
    derivatives
    structure-activity relationship,
    345–354
Phenoxybenzamine, 296

Phentolamine, 296–297
Phenylalkylamine calcium
    channel blockers, 214, 215i,
    217–218
Phenylbutazone, 446–447
N-Phenylsalicyl amide, 641
Phenytoin, 195i
Phosphate, renal reabsorption, 368
Phosphatidylinositol-4,5-
    biphosphate, 173
Phosphodiesterase inhibitors,
    229–230, 230 (table)
Phosphoinositides, 82
Phospholipase C, and cardiac
    function, 173, 187
Phospholipids, in mycobacterial
    cell wall, 580, 581
Phosphonic acid anticholinesterase
    derivatives, 47
Phosphoramidon, 441–442,
    443–444
Phosphorus-derived anti-
    cholinesterases, 47–50, 48–49
    (table)
Phosphoryl-Leu-Phe, 444
Phthaylsulfathiazole, 533
Phthivazid, 596
Phycomycetes, 638
Physostigmine, 45–47, 67
Picoprazole, 125, 126
Pikromycin, 478, 480
Pikronolide, 480
Pilocarpine, 5, 20, 72
Pimaricin, 648
Pimobendan, 230
Pinacidil, 171, 304, 305
Pindolol, 216i, 219, 220, 287
Piperarcillin, 621
Piperidinium, 40
Piprozoline, 417
Pirenzepine, 82, 141
Piretanide, 401i, 412
Pirlimycin, 491
Plant fungistats, 641. See also
    Antifungal substances
Platenomycins, 484
Pleuromutilin, 507
Pneumocystis carinii, 541, 566, 638,
    650, 651–652
Pneumocystosis, 638, 639, 641
Poldine, 111
Polyene antifungals, 648
Polyketide synthases, 490
Polymethylene tetramine
    anticholinergics, 84–85
Polymyxins, 500
    in triple ointment (polymyxin
    B), 498

Polynitrogen heterocyclic diuretics.
426–440, 429 (table), 434–436
Polypeptins, 500
Polythiazide, 391*i*
Positive inotropes, 228
Potassium:
movement across apical
membrane, 123, 144–145
renal reabsorption, 370, 371–372
Potassium canrenoate, 422, 423*i*
Potassium channel blockers, *see*
Antiarrhythmic drugs, class
III
Potassium channel opening drugs.
171
for hypertension, 278, 304–305
Potassium channels, 166*i*, 166–167.
170–171, 187
gating, 167–168
Potassium mexrenoate, 423*i*, 424
Potassium 3-(3-oxo-17β-hydroxy-
4,6-androstadien-17-α-
yl)propanoate, 422
Potassium oxonate, 335–336
Potassium prorenoate, 422–424
Potassium-sparing diuretics, 325,
430–440
structure-activity relationship,
432–433
Potassium supplements:
for congestive heart failure, 226
in diuretic therapy, 374
Practolol, 287, 288*i*
Prazosin, 295–296, 297
Prednisolone 3,20-bisguanyl-
hydrazone, 234*i*, 242
Pregnane and derivatives, 234*i*,
243–245
Preload, 207
Prenalterol, 228
Prenylamine, 292
Primary complexes, 584
Pristinamycins, 491–492
ProBanthine, 82
Probenecid, 335, 336, 338*i*, 391,
446, 542
Procainamide, 195*i*, 203
Procyclidine, 86
Proglumide, 121, 123
Prokinetic macrolides, 489
Promethazine, 111
Promin, 537–538
Pronethalol, 285, 287
Prontosil, 528, 529–530
Propafenone, 195*i*
Propikacin, 472
Propivane, 110
Propranolol, 196*i*, 200, 220, 285,
287

as nonselective β-blocker, 219
Prorenone, 425
Prostacyclin, 188, 189
Prostaglandins, 123
as antiulcer agents, 112
gastric acid secretion inhibition,
63–64
Protein kinase C, and cardiac
function, 173
Protein kinase C inhibitors, for
hypertension, 309
Prothionamide, 598
Proximal convoluted tubule, 366,
369*i*
Prumycin, 476
Pseudallescheriasis, 639, 641
Pseudomonas aeruginosa, 469, 470,
471, 472
Pseudomonic acid, 505–506
Pseudotropine, 73–74, 102
Psychosedatives, anticholinergic
side effects, 108
Psychotomimetics, anticholinergic
side effects, 108
Pteridine diuretics, 431, 433 (table)
structure-activity relationship,
432–433
P-type calcium channels, 169
Purkinje fibers, 156, 163
Puromycins, 508–509
Pyrazinamide, 588, 623
Pyrazine diuretics, 431
Pyrazinoic acid esters, 599–600
N-Pyrazinylthioureas, 600
Pyrazone, 447
2-Pyrido[2,3-d]pyrimidin-4-one
diuretics, 436
Pyrido-1,2,4-thiadiazines diuretics,
387 (table)
2[(Pyridylmethyl)sulfinyl]-1*H*-
benzimidazoles, gastric proton
pump inhibitors, 127
Pyrimethamine, 539, 548, 549, 650
Pyrzainamide, 599–600

Quaternary ammonium
compounds, 111
as anticholinesterases, 44–45
side effects, 108
variations in cholinergics, 7–9
Quaternary tropyl acetates, 39
Quinazolinone sulfonamides,
397–398
Quinethazone, 393*i*, 397, 408
Quinidine, 195*i*
mechanism, 197–198
Quinine, 171
Quinolines, as antifungals, 641
Quinuclidine anticholinergics, 84

Quinuclidinyl benzilate, 67
3-Quinuclidinyl diphenylacetate,
100
Quinupristin, 492

Rabeprazole sodium, 127, 128*i*, 135
active form of, 132
pharmacokinetics, 139
pharmacology, 136, 138
toxicology, 141, 142
Ramipril, 291*i*, 292
Ranitidine, 120
toxicology, 140
RB 45, 443
Rectifying ion channels, 170
Reentry phenomena, 191–193,
192*i*, 202
Renal collecting tubule, 371
Renal cortical nephrons, 282*i*, 365,
366, 367*i*
Renal juxtamedullary nephrons,
365, 367*i*
Renal output, and blood pressure
control, 267
Renal papillae, 366–367
Renal physiology, 365–373, 366*i*
Renal reabsorption, 366–372
Renin-angiotensin-aldosterone
system, 419
and blood pressure control,
268–269
and congestive heart failure, 224,
227*i*
suppression by digoxin, 232
Renin inhibitors, 305–306
Reserpine, 278, 299*i*
Resistance, *see* Drug resistance
Reversible proton pump inhibitors,
142–144
Rhamnose, 251
Ribostamycin, 467, 468
Rifabutin, 588, 615, 623
Rifamide, 612
Rifampicin, 588, 623, 624
antimycobacterial activity,
612–615
combination with isoniazid, 614
resistance to, 585, 591, 613
side effects, 614–615
Rifamycins:
ansa structure, 610
antimycobacterial activity,
610–612
structure-activity relationship,
611
Rifapentine, 588, 615–616
Right-sided heart failure, 224, 225
Rilmenidine, 298
Riminophenazines, 600

Ristocetins, 495–496
Ristomycin, 495
L-Ristosamine, 495
Ro 18-5364, 127, 128*i*, 134
Ro 31-7549, 309
Ro 42-5892, 305–306
Ro 46-2005, 308
Ro 47-0203, 308
Ro-462005, 543
Rokitamycin, 487
Romoplanin, 501
Rosaramicin, 486
Roxatidine, 121
Roxithromycin, 482, 619, 620
RP 23672, 480
RP 54476, 492
RP 57669, 492
RP 59500, 492
R-plasmids, and drug resistance, 550
Rubreserine, 45–46
Ryanodiine receptors, 161, 174

S-1924, 127, 129*i*
S-3337, 127, 129*i*, 136
S-8666, 336, 355–356
Sagamicin, 470
Salbutamol, 228
Salicylates, as antifungals, 641
Salicylic acid, 445–446
Salt intake, 254
Saluretics, 541
Saluretic diuretics, 386
Sannamycins, 474
SA node, 156, 158, 160
    action potentials, 165*i*
    arrhythmias of, 190
    pacemaker role, 164
Saprophytes, 640
Saralasin, 300
Sarcolemma, 157*i*, 158
Sarcomeres, 158, 159*i*, 160, 161
Sarcoplasmic reticulum, 158
Saviprazole, 127, 129
    activation, 132–134, 135
    pharmacology, 135–136, 137
    toxicology, 140, 141
SC 46542, 441
SCH 22591, 472
SCH 27899, 505
SCH 28080, 142–144
SCH 32651, 142–144
SCH 34826, 306, 443
SCH 39370, 443
SCH 42495, 442–443
Scopine, 73
Scopolamine, 72
    as CNS depressant, 101
    stereoisomerism, 101

structure, 73, 74
    synthesis, 75
Secretin, 63
Seldomycins, 471
Selenocholine, 9
Sematilide, 196*i*, 201
Semicarbazone, 243
Septacidins, 508
Serotonin receptor antagonists, for hypertension, 306
SHEP Trial, 274–275, 276
Shigella, sulfonamide resistance, 549, 550
Sila-difenidol anticholinergics, 86
Sinoatrial (SA) node, *see* SA node
Siomycin A, 501–502
Sisomicin, 470–471
SK&F 65601, 141–142
SK&F 93479, 140
SK&F 95601, 127, 128*i*, 132, 135
    pharmacology, 136, 138
SK&F 96067, 142, 143*i*, 144
SK&F 97574, 142, 143*i*
SKF108566, 302–303
Sodium, renal reabsorption, 368–372, 374
Sodium channel blockers, *see* Antiarrhythmic drugs, class I
Sodium channels, 166*i*, 166–167, 168–169
    gating, 167–168
Sodium/hydrogen exchanger, 368
Sodium mercaptomerin, 377*i*
Sodium nitrite, 211*i*
Sodium nitroprusside, 299*i*
Sodium-potassium pump, *see* Na$^+$,K$^+$-ATPase (under ATP)
Sodium salicylate, 445–446
Sodium thiocyanate, 278
Solanaceous alkaloids, 72. *See also* Anticholinergic drugs
    fate, 75–77
    history, 72–73
    preparation, 74–75
    semisynthetic derivatives, 77–78
    structure, 73–74
Solasulfone, 592
Soman, 49
Somatostatin, 123
Sorbistins, 475–476
Sorbitol, 375*i*
Sotalol, 196*i*, 201–202, 203
    as nonselective β-blocker, 219
    replacement of lidocaine as drug of choice for arrhythmia, 200, 203
Sparfloxacin, 616, 618
Spasm, 61–62. *See also* Antispasmodics

Spectinomycin, 474, 603
Spicamycins, 508
Spiramycins, 484, 618
Spirodioxolane, 18
Spirolactam diuretics, 421
Spironolactones, 283
    and blood pressure control, 268, 282
    for congestive heart failure, 226
    discovery, 284
    side effects, 426
    steroidal, 419–426
Spiropiperidines, 27–28
Spontaneously-hypertensive rat, 278, 312
Sporacins, 473–474
Sporacuracin, 505
Sporeamicin A, 484
Sporotrichosis, 639, 641
SQ 28603, 442
SQ 29072, 442
SR47436, 302, 303*i*
Staphylomycins, 491
Staurosporine, 309
Stendomycins, 500
Steroidal aldosterone antagonists, 419–426, 423 (table)
STOP Trial, 274
Stramonium, 72
Streptamine, 466*i*
Streptogramin antibiotics, 490, 491–492, 498
Stroptokinase, 206
Streptomycin, 465, 466–467, 588, 589, 623
    antimycobacterial activity, 603–605
    dependence, 604
    resistance to, 550–551, 591, 604–605
Streptovaricins, 610
Streptovirudins, 508
Strokes, 274
Structure-activity relationships:
    aminoglycoside antibiotics (structure-toxicity), 471
    anticholinergic drugs, *see* Anticholinergic drugs, structure-activity relationship
    carbonic anhydrase inhibitors, 380–382
    cardiotonic steroids, *see* Cardiotonic steroids, structure-activity relationship
    cholinergics, 37–42
    glycopeptide antibiotics, 497
    hydrothiazide diuretics, 388, 390
    irreversible gastric proton pump inhibitors, 130, 134–135

Structure-activity relationships
(*Continued*)
mercurial diuretics, 376–378
minimal topological difference
for studying, 248
phenoxyacetic acid diuretics,
345–354
potassium-sparing diuretics,
432–433
rifamycins, 611
sulfanilamide, 551
sulfonamides, 551
properties-activity relationship,
552–553, 554–558
table, 559
sulfones, 551–552
properties-activity relationship,
559–560
thiazide diuretics, 386
tubercle bacilli, 581
Subtilin, 502–503
Subunits, ion channels, 166–167
Sucrose, 375*i*
Sugars, cardiac glycosides, 232,
234–235, 236*i*, 241*i*, 251–253
Sulbactam, 621
Sulconazole, 647
Sulfacarbamide, 533
Sulfacetamide, 532
Sulfachlorpyridazine, 535–536
Sulfaclomide, 536
Sulfacytine, 536
Sulfadiazine, 563, 650
Sulfadimethoxine, 536, 563
Sulfadizaine, 534
Sulfaguanidine, 533, 542
Sulfamerazine, 534, 535
Sulfamethazine, 534
Sulfamethizole, 535, 563
Sulfamethoxasole, 650
Sulfamethoxazole, 536, 549
Sulfamethoxine, 538, 540
Sulfamethoxydiazine, 536
Sulfamethoxypyrazine, 536, 540
Sulfamethoxypyridazine, 535, 563
Sulfamethyldiazine, 536
Sulfamoxol, 536
5-Sulfamoylanthranilic acid
diuretics, 405
Sulfamoyl-2,3-dihydrobenzofuran-
2-carboxylic acid diuretics,
335–338, 340–341 (table)
Sulfanilamide, 379–380
discovery, 365
diuretic effect, 541
as "magic bullet," 530
PABA antagonism, 531
structure-activity relationship,
551

Sulfaphenazole, 536, 563
Sulfaproxyline, 535
Sulfapyridine, 533
Sulfated amylopectin, 111
Sulfathiazole, 535
Sulfinpyrazone, 445*i*, 446–447
Sulfisomidine, 534, 563
Sulfisoxazole, 535, 543, 563
Sulfolipids, 580, 581
Sulfomethoxine, 536
Sulfonamides:
antimicrobial spectrum, 539–541,
540 (table)
assay for, 531
classification, 531
clinical practice, 528–529,
564–565
combination with dihydrofolate
reductase inhibitors, 548–549
early drugs, 532–535
half-life, 564
history, 529–531
importance of *p*-aminobenzen-
sulfonyl unit, 532
later drugs, 535–537
lipid solubility, 560–561
mechanism of action, 530,
543–547
metabolism, 562–563
natural-occurring, 537
PABA antagonism, 528, 539,
543–544, 549, 550
pharmacokinetics, 562
properties-activity relationship,
552–553, 554–558 (table), 559
protein binding, 561–562
resistance to, 549–551
selectivity, 548
side effects, 565–566
as leads for new drugs, 541–543
structure-activity relationship,
551
water solubility, 560
Sulfones, 537–539
antimicrobial spectrum, 539–541,
540 (table)
antimycobacterials, 592–593
classification, 531
clinical practice, 529, 564–565
half-life, 563–564
importance of *p*-aminobenzen-
sulfonyl unit, 539
mechanism of action, 547–548
metabolism, 563–564
properties-activity relationship,
559–560
side effects, 565–566
structure-activity relationship,
551–552

Sulfydryl-donating compounds, as
organic nitrate cofactor,
213
Sulmazole, 230
Superinfection, 531
Sympathic nervous system, and
blood pressure control, 268
Sympatholytic effects, 296
Sympatholytics, and blood
pressure control, 268
Systemic mycoses, 639–640
antifungal substances for,
647–652
Systolic pressure, 271

Tabun, 49
Tacrine, 50–51
Tazobactam, 621
TCV-116, 302, 303*i*
Technicon Autoanalyzer, 373
Teichomycins, 494–495
Teicoplanins, 494–495
Telenzepine, 83
Temocapril, 291*i*, 292
Terazosin, 297
Terbinafine, 644, 645*i*
Terconazole, 647
Tetracycline, 550
Tetraethylammonium, 171
1,2,3,4-Tetrahydro-9-aminoacridine
and related inhibitors, 44,
50–51
Tetrahydrofolate, 548
Tetramethylammonium, as
anticholinesterase, 44
Tetrodotoxin, 172, 199
THA, 50–51
Theophylline, 377, 426–427, 429*i*,
619
Thevetose, 232, 236*i*, 251
1,3,4-Thiadiazole-2-sulfonamides,
380
Thiambutosine, 588, 594, 595
Thiazesim, 292
Thiazide diuretics, 276, 324,
386–389, 390–391 (table)
clinical applications, 392
pharmacology, 391–392
structure-activity relationship,
386
Thienoimidazoles, gastric proton
pump inhibitors, 129*i*, 134, 135
Thioacetazone, 588, 594–595
Thiocarlide, 588, 594, 595
Thiocholine, 9
Thioisonicotinamide, 588, 598
Thiopeptide antibiotics, 501–502
Thiopeptin, 501–502
Thiophan, 441–442

Thiosorbitol, 377
Thiostrepton, 501, 502i
Thrombolytic enzymes, 206
Tiamulin, 507
Ticarcillin, 621
Tienilic acid, 325, 326, 327i, 336
    structure-activity relationship,
        345–347, 348
Tilmicosin, 488–489
Timolol, 216i, 220, 285, 286
Timoprazole, 134, 135
    chemistry, 125, 126
Tioconazole, 647
Tissue perfusion, blood pressure
    control, 267
recombinantTissue plasminogen
    activator (rTPA), 206
Tizolemide, 399
TMB-8, 144
TMC-016, 489
Tobramycin, 469
Tocainide, 195i, 199
Tolnaftate antifungals, 644
Tolypomycins, 610
TOMHS Trial, 274, 275
Torasemide, 284, 401i, 415–416
Torsade de pointes, 191, 200, 201
Toxic shock, and nitric oxide, 210
Toxoplasmosis, 540
recombinantTPA, 206
Transensor, 69
4-Trehalosamine, 476
Tremorine, 28
Trest, 82
2,4,7-Triamino-6-phenylpteridine,
    see Triamterene
2,4,7-Triamino-5-phenyl-
    pyrimido[4,5-d]pyrimidine,
    432
Triamterene, 429i, 432, 433–434, 439
    discovery, 283, 284
Triazine diuretics, 428, 430
Triazole antifungals, 647
Trichloromethiazide, 281, 391i
Tricyclic benzodiazipine
    anticholinergics, 82–83
Triflocin, 413, 415
Trifluoperazine, 144
Triggered (cardiac) disorders, 190
Trimethoprim, 549, 650
Trimipramine, 144
Tripitramine, 85
Triple ointment, 498
Trisdigitoxose, 234, 236i
Tropeines, 77
Tropic acid, 75
Tropicamide, 108
Tropines, 40, 73–74, 102
Tropinoylcholines, 103

Tropomyosin, 158, 159i, 160, 175
Troponin, 159–160
Troponin C, 161
T tubules, 158, 160, 165
T-type calcium channels, 169, 293
Tubercle bacilli, 576–577
    cell wall, 579–581
    drug resistance, 585, 591
    media for, 577
    metabolism, 578
    screening/evaluation of
        antimycobacterial agents,
        586–587
    structure-function relationship,
        581
    testing models, 578
    unusual characteristics of, 582
Tuberculin test, 584
Tuberculoid leprosy, 586
Tuberculosis, 576
    and AIDS, 584–585, 623
    drugs for, 588 (table), 589
        (table), 622–623
    pathogenesis, 582–585
    resurgence in, 584–585
Tubular transport inhibitors, 542
Tumor necrosis factor, 583
Tunicamycins, 508
TY-11345, 127, 128i
Tylosin, 485, 486i
Tyrocidin, 498–499
Tyrothricins, 498–499

UDP-N-acetylglucosamine, 508
UK-68,597, 496
UK-69,578, 306
Urapidil, 286, 287i
Urate uptake, 357–360, 365, 374
Urea, 375i, 444
Uric acid, 365, 374, 444
Uricase, 444
Uricosuric agents, 325–326, 365,
    444–449. See also Diuretics
    animal studies, 373
Use-dependent block, 197
Uzarigenin, 250

Vagotomy, 121
Validamycins, 475
Validoxylamine A, 475
Valienamine, 475
Vancomycins, 493–494
    resistance to, 496–498
    structure-activity relationship,
        497
Vancosamine, 493i
α-Vancosaminyl-β-glucosyl, 493
Vascular cell adhesion molecule-1,
    188

Vascular smooth muscle ion
    channels, and hypertension,
    311
Vasoactive intestinal polypeptides
    (VIP), 63
Vasoconstriction-volume
    hypothesis, of blood pressure
    control, 268
Vasoconstrictors, release by
    endothelial cells, 188
Vasodilating β-blockers, 221 (table),
    221–222
Vasodilators:
    after myocardial infarction, 205
    and angina, 207–208. See also
        Nitrates
    and blood pressure control, 268
    for congestive heart failure,
        226–227
    release by endothelial cells,
        188
Vasopressin, and blood pressure
    control, 268
Vasopressin receptor antagonists,
    for hypertension, 307
Venodilation, 208, 213
Ventricular fibrillation, see
    Fibrillation
Verapamil, 196i, 202, 214, 215i,
    217, 294, 295
    discovery, 292
    gastric acid secretion inhibition,
        144
Verazide, 596
Vernamycins, 491
Very low-density lipoproteins
    (VLDLs), 204
Vesarinone, 230
Veterans Cooperative Trial, 274
VH-AH-37, 83
Viomycin, 588, 589
    antimycobacterial activity,
        607–608
Virginiamycin, 491, 492i
Vitamin K antagonists, for
    secondary MI prevention, 206
Voltage-dependent ion channels,
    see Ion channels
von Willebrand's disease, 495

Water, renal reabsorption, 366
Waxes D, 580
WIN 5723, 617
Working Group on Arrhythmias,
    193–199

Xanthine diuretics, 426–427
Xanthonyloxyacetic acid diuretics,
    328, 329 (table), 331

Xanthonyloxyacetic acid diuretics
  (*Continued*)
  structure-activity relationship,
    347–354
Xanthopterin, 431
Xipamide, 401*i*, 412–413, 414–415
  (table)
XM953, 303–304

Xylostatin, 467, 468

Yeasts, 638
YM133, 486*i*, 488
YM17K, 489
Youmans medium, 577

Zankiren, 305, 306

Ziehl-Neelsen method, 577,
  578
Zinc 10-undecylenate, 641
ZK 91587, 421–422
Z lines, 158, 159*i*
Zollinger-Ellison syndrome, 121,
  138
Zymosterol, 637